The Twenty Amino Acids with Abbreviations and Messenger RNA Code Designations

Amino acid	One letter symbol	Three letter symbol	mRNA code designation
alanine	A	ala	GCU, GCC, GCA, GCG
arginine	R	arg	CGU, CGC, CGA, CGG, AGA, AGG
asparagine	P	asn	AAU, AAC
aspartic acid	D	asp	GAU, GAC
cysteine	C	cys	UGU, UGC
glutamic acid	E	glu	GAA, GAG
glutamine	O	gln	
glycine			
histidine			
isoleucine			
leucine			UA, CUG
lysine			
methionine			
phenylalanine			
proline			
serine			GU, AGC
threonine			
tryptophan			
tyrosine			
valine			

Cytosine (C)

Guanine (G)

Encyclopedia of Molecular Biology and Molecular Medicine

Editorial Board

Encyclopedia of Molecular Biology and Molecular Medicine

Volume 2

Denaturation of DNA to Growth Factors

Edited by
Robert A. Meyers

Weinheim • New York • Basel • Cambridge • Tokyo

Robert A. Meyers
3715 Gleneagles Drive
Tarzana, CA 91356, USA

Management Supervised by:
Chernow Editorial Services, Inc.,
1133 Broadway, Suite 721, New York, NY, USA

Cover art courtesy of Dr. Zuzana Hostomska and Dr. Zdenek
Hostomsky from Figure 6 of their article, "AIDS HIV Enzymes,
Three-Dimensional Structures of."
Art prepared by David A. Matthews.

Library of Congress Cataloging-in-Publication Data

Encyclopedia of molecular biology and molecular medicine / edited by
  Robert A. Meyers.
      p.   cm.
    Contents: v. 1. Achilles' cleavage to cytoskeleton-plasma membrane
interactions
    Includes bibliographical references and index.
    ISBN 3-527-28478-8 (set). — ISBN 3-527-28471-0 (v. 1 : alk.
paper)
    1. Molecular biology—Encyclopedias.   2. Molecular genetics—
Encyclopedias.   3. Biotechnology—Encyclopedias.   I. Meyers,
Robert A. (Robert Allen), 1936–   .
  QH506.E534   1996
  611'.01816—dc20   95-44980
                                                     CIP

    Updates to Library of Congress CIP Data for Volume 2:
      Contents: v. 2. Denaturation of DNA to growth factors
      ISBN 3-527-28472-9 (Volume 2)

    Die Deutsche Bibliothek – CIP – Einheitsaufnahme

Encyclopedia of molecular biology and molecular medicine /
[editorial board Werner Arber . . .]. - Weinheim ; New York ;
Basel ; Cambridge ; Tokyo ; VCH.
NE: Arber, Werner [Hrsg.]

Vol. 2. Denaturation of DNA to growth factors / ed. by
  Robert A. Meyers. - 1996
    ISBN 3-527-28478-8
NE: Meyers, Robert A. [Hrsg.]

Printed in the United States of America

ISBN 3-527-28472-9 (Volume 2)
ISBN 3-527-28478-8 (Set)

Printing History:
10 9 8 7 6 5 4 3 2 1

Distribution:
VCH, P.O. Box 10 11 61, D-69451 Weinheim, Federal Republic of Germany
Switzerland: VCH, P.O. Box, CH-4020 Basel, Switzerland
United Kingdom and Ireland: VCH, 8 Wellington Court, Cambridge CB1 1HZ, United Kingdom
USA and Canada: VCH, 220 East 23rd Street, New York, NY 10010-4606, USA
Japan: VCH, Eikow Building, 10-9 Hongo 1-chome, Bunkyo-ku, Tokyo 113, Japan

Contents of Volume 2

Preface

This is a six-volume encyclopedia of molecular genetics and the molecular basis of life, with a focus on molecular medicine, including genetic screening, gene therapy, molecular medicine, structural biology, and the technology and findings of the Human Genome Project. More than 90 percent of the encyclopedia articles discuss topics in human genetics or molecular medicine. Molecular biology is treated in the balance of the encyclopedia by articles covering each of the major animal and plant genomes, and by articles covering key biotechnology advances and basics. The molecular structure, interactions, and analyses of proteins, carbohydrates, lipids, and other biological components complete the coverage of the molecular basis of life. This is the most comprehensive, detailed treatment of molecular biology and molecular medicine to date.

The most active research areas are presented. Some examples of topics include PCR technology, evolution of genetic intelligence, genomic imprinting, oligonucleotides, genetic mutations in aging, transposons in the human genome, human repetitive elements, molecular chaperones, protein aggregation, protein folding, protein designs for the specific recognition of DNA, signal transduction by heterotrimeric G proteins, transgenic livestock as bioreactors, motor proteins, molecular biology of *Caenorhabditis elegans*, colon cancer, lung cancer, oncogenes, tumor suppressor genes, cystic fibrosis, molecular basis of AIDS and AIDS therapy (four articles), Huntington's disease, infectious disease testing by ligase chain reaction, immuno-PCR, biosensors, cytochrome P450, and gene transfer techniques for use in genetic vaccination.

A comprehensive treatment of the Human Genome Project is included. It consists of core articles that present overviews and perspectives (for example, nucleic acid sequencing techniques, gene distribution in the human genome, human linkage maps, physical maps of human chromosomes, cloned DNA markers, human disease gene mapping, and human genetic predisposition to disease). These core articles are supported by detailed treatments of the various mapping methods (for example, chromosome microdissection and microcloning, high speed DNA sequencing in ultrathin gels, gene mapping by fluorescence *in situ* hybridization, and robotics and automation in genome reseach). Articles on the genetic and physical mapping status of several of the human chromosomes are presented to give the reader an introduction to the results of the Human Genome Project, which will be a complete genetic and physical map of all 23 chromosomes. This is, of course, an early view, as only a small percentage of the human genome has been mapped at this time.

The overall objective of this publication is to provide a single source for the molecular basis of life, disease diagnosis, and therapy. In addition, the theory and techniques for understanding, modifying, manipulating, expressing, and synthesizing biological molecules are presented. The result is an encyclopedic reference work with coverage of the following subject areas:

- Nucleic acids,
- Structure determination techniques (DNA, RNA, and proteins),
- Purification and processing of DNA, RNA, and proteins,
- The Human Genome Project,
- Proteins, polypeptides, and amino acids,
- Lipids,
- Immunology and biomolecular interactions,
- Molecular biology of specific organisms and organ systems,
- Molecular biology of specific diseases,
- Biochemistry and pharmacology, and
- Biotechnology.

Our purpose is to provide a multivolume reference for the expanding number of scientists who will contribute to this field, many of whom, as in the past, will enter molecular biology research from majors or careers in animal, plant, and cell biology and medicine, as well as physics, chemistry, mathematics, and engineering. It is a useful source of information for preclinical and clinical research in industry and in hospitals and for physicians who will apply the advances in genetic medicine in their practices. Teachers and professors in schools and universities will utilize this six-volume set for course preparation, and students will find a comprehensive introduction to molecular biology and genetic medicine.

The nearly 300 articles that comprise the set are designed as *self-contained treatments*. Each article begins with a *key word section*, including definitions, to assist the scientist or student who is unfamiliar with the specific subject area. The *Encyclopedia* includes more than 1,900 key words, each of which is defined within the context of the particular scientific field covered by the article. Because the same word may have different meanings in different fields, some key words have more than one definition.

In addition to the key word definitions, the *glossary of basic terms* found at the back of each volume defines the most commonly used terms in molecular biology. These definitions, along with the reference materials printed on the inside front and back covers of each volume (the genetic code, the common amino acids, and the structures of the deoxyribonucleotides), should allow most readers to understand articles in the encyclopedia without referring to a dictionary, textbook, or other reference work.

Each article begins with a concise definition of the subject and its importance, followed by the body of the article and references for further reading. Each subject is presented on a first-principles basis, including detailed figures, tables, and drawings to elucidate atomic or structural features. Because of the self-contained nature of each article, some overlap among articles on related topics exists. Cross-references to related articles are provided to help the reader expand his or her range of inquiry.

The detailed *cumulative subject index* as well as the *complete list of articles* and a *compilation of key words* at the end of volume 6 provide the reader with additional tools for locating topics of particular interest.

The articles in the encyclopedia may be categorized as follows: (1) *core articles* that give perspective on the major disciplines (such as, DNA structure, DNA replication and transcription, translation of RNA to protein, genetics, human genetic predisposition to disease, molecular genetic medicine, peptides, energetics and regulation in biological catalysis of enzymes, bioprocess engineering, immunology, medicinal chemistry, and cancer); (2) *satellite core articles* that give perspectives on particularly active areas of research and importance (for example, the core article on cancer is supported by satellite core articles on oncogenes and tumor suppressor genes); and (3) *specific subject articles* on such topics as liver cancer, lung cancer, breast cancer, colon cancer, and retinoblastoma. Cross-references are given to related articles, so that articles on cancer would reference DNA in neoplastic-disease diagnosis; steroids, antitumor; and cancer chemotherapy, theoretical foundations of. Because the subjects are advancing so rapidly, sometimes a specific field is covered within core or satellite core articles, rather than in an individual subject article.

It must be admitted that in some cases readers at the beginning of the educational hierarchy, that is, university students with a year of college chemistry, physics, and biology, may need to refer to a dictionary of scientific terms to access a few of the articles. These users will find the extra effort well worthwhile. Also, readers at the top of the hierarchy—molecular biologists and bioengineers—who are experts in one or more discipline within molecular biology, might find some of the articles in their field of research to be less detailed than desired. However, these scientists will find the articles in related disciplines to be useful in planning experimentation and interpreting results within their lines of research. The hope is that the reader will find as complete and authoritative a coverage of the molecular basis of life as can be provided in a reference work.

Robert A. Meyers

Acknowledgments

PUBLISHER'S ACKNOWLEDGMENTS

The publisher gratefully acknowledges the contributions of the following people to the development of this reference work.

Abul K. Abbas
Harvard Medical School

David Abramson
Cornell Medical Center

Thomas O. Baldwin
Texas A&M University

M. J. Bishop
*University of Cambridge School
of Medicine*

Pamela Bjorkman
California Institute of Technology

Kristina Bostrom
*University of California School
of Medicine*

Lisa Brannon-Peppas
Eli Lilly and Company

Arnold L. Demain
Massachusetts Institute of Technology

Thomas Devlin
*Hahnemann University School
of Medicine*

H. Dintzis
Johns Hopkins School of Medicine

Tim Donlon
Kapiolani Medical Center

William A. Eaton
National Institutes of Health

Gadi Fibich
University of California

Ian Garner
Pharmaceutical Proteins, Ltd.

James I. Garrels
Cold Springs Harbor Laboratory

Alison Goate
*University of Washington School
of Medicine*

Paul Goodfellow
Washington University

Jay A. Gottfried
New York University

Barney Graham
Vanderbilt University Hospital

Michael D. Griswold
Washington State University

George Guilbault
University of New Orleans

Sarah Gurr
University of Oxford

Barry Hall
University of Rochester

Patrick Hess
National Health Labs

Veronica van Heyningen
Western General Hospital

John Hildebrandt
Medical University of South Carolina

George Alan Jeffrey
University of Pittsburgh

Dick Junghans
Deaconess Hospital

David Kaplan
*Natick Research, Development,
and Engineering Center*

Brian Kay
University of North Carolina

Roger Keynes
University of Cambridge

Richard A. King
University of Minnesota

Claude B. Klee
National Cancer Institute

Ted Klein
E.I. du Pont de Nemours & Co.

Joseph R. Lakowitz
*University of Maryland School
of Medicine*

Charles Langley
University of California

Jenifer Levini
Cornell University

Tim Lohman
*Washington University School
of Medicine*

Mike McKeown
Salk Institute

L. McLaughlin
Boston College

Peter B. Moore
Yale University

Krishna Murthy
*Temple University Medical
School*

R. H. Pain
Jožf Stefan Institute

Carl A. Pinkert
University of Alabama at Birmingham

Richard Porter
*State University of New York, Stony
Brook*

Clive Pullinger
University of California

Thomas Quinn
Johns Hopkins University

Francesco Ramirez
Mount Sinai Hospital

Andrew Rice
Baylor College of Medicine

David Schlessinger
Washington University School of Medicine

Dennis Shields
Albert Einstein College of Medicine

Barton Slatko
New England Biolabs

Jonathan Smith
Rockefeller University

Gary Spedding
Butler University

Robert F. Steiner
University of Maryland, Baltimore County

Lorne B. Taichman
State University of New York, Stony Brook

M. Thompson
University of Toronto

Cary Thrall
Synchrom, Inc.

Terrence Tiersch
Louisiana State University

Alan Tomkinson
University of Texas Health Center at San Antonio

Jan Vilcek
New York University Medical Center

Stephen S. Wachtel
University of Tennessee College of Medicine

Steve Warren
Emory University School of Medicine

Jean Weissenbach
Institute Pasteur

Michael J. Welsh
University of Iowa College of Medicine

Christine A. White
Scripps Memorial Hospital

Timothy Wilcosky
Center for Epidemiologic and Medical Studies

Steven A. Williams
Smith College and University of Massachusetss

Emily S. Winn-Dean
Applied Biosystems Inc.

June Hsieh Wu
Chang-Gung Medical College

EDITOR'S ACKNOWLEDGMENTS

I would like to acknowledge Robert A. Meyers, Jr., Tamara H. Cochrane, and Kenneth Cochrane, who maintained the encyclopedia article and topical outline database that I used to monitor author manuscript scheduling and to correlate the contents to assure completeness of coverage. I would also like to acknowledge my wife, Ilene Meyers, who assisted in this effort in the areas of project planning and scheduling. And finally, I thank Madeleine Lewis, who so ably assisted me with author recruitment during the concluding year of manuscript preparation.

VCH Verlagsgesellschaft mbH, Weinheim, Germany
Hans-Joachim Kraus, Ph.D., *Executive Editor, Life Sciences*

VCH Publishers, Inc., New York, NY
Camille Pecoul, *Director, Production and Manufacturing*

Michael P. Beaubien, *Managing Editor, Chernow Editorial Services, New York, NY*

Brenda Griffing, *Copy Editor*

Contributors

Michael W. W. Adams
Department of Biochemistry
University of Georgia
Athens, GA 30602, USA
Enzymes, High Temperature

Göran Akusjärvi
Department of Medical Immunology
and Microbiology
Uppsala University
S-751 23 Uppsala, Sweden
Gene Expression, Regulation of

Nicholas D. Allen
Laboratory of Developmental Genetics
and Imprinting
The Babraham Institute
Cambridge CB2 4AT, United Kingdom
Genomic Imprinting

Werner Arber
Department of Microbiology
Biozentrum
University of Basel
CH-4056 Basel, Switzerland
*Genetic Diversity in
Microorganisms*

Sanford Asher
Department of Chemistry
University of Pittsburgh
Pittsburgh, PA 15261, USA
Enzymology, Nonaqueous

Elizabeth Baker
Department of Cytogenetics
and Molecular Genetics
Centre for Medical Genetics
Women's and Children's Hospital
Adelaide 5006, Australia
Fragile Sites

Thomas O. Baldwin
Department of Biochemistry
and Biophysics
Texas A&M University
College Station, TX 77843, USA
*Enzymes, Energetics and Regulation
of Biological Catalysis by*

Shabbir Bambot
Department of Chemical Engineering
Center for Biotechnology and
Bioengineering
University of Pittsburgh
Pittsburgh, PA 15261, USA
Enzymology, Nonaqueous

Anthony H. Barnett
Department of Medicine
University of Birmingham and
Birmingham Heartlands Hospital
Birmingham B9 5SS, United Kingdom
Diabetes Mellitus

George N. Bennett
Department of Biochemistry
and Cell Biology
Rice University
Houston, TX 77251, USA
*Expression Systems for DNA
Processes*

Giorgio Bernardi
Laboratoire de Génétique Moléculaire
Institut Jacques Monod
75005 Paris, France
*Gene Distribution in the Human
Genome*

Carol Bernstein
Department of Microbiology
and Immunology
University of Arizona College of Medicine
Tucson, AZ 85724, USA
DNA Repair in Aging and Sex

Harris Bernstein
Department of Microbiology
and Immunology
University of Arizona College of Medicine
Tucson, AZ 85724, USA
DNA Repair in Aging and Sex

Ernest Beutler
Department of Molecular and
Experimental Medicine
The Scripps Research Institute
La Jolla, CA 92037, USA
Gaucher Disease

Daniel G. Bichet
Department of Medicine
Centre de Recherche
Hôpital du Sacré-Coeur de Montréal
Montréal, Quebec H4J 1C5, Canada
Diabetes Insipidus

Mariel Birnbaumer
Department of Anesthesiology
University of California School
of Medicine
Los Angeles, CA 90095, USA
Diabetes Insipidus

Richard D. Blake
Department of Biochemistry,
Microbiology, and Molecular Biology
University of Maine
Orono, ME 04469, USA
Denaturation of DNA

Miroslav Blumenberg
Ronald O. Perelman Department
of Dermatology and
Department of Biochemistry
New York University Medical Center
New York, NY 10016, USA
*Epidermal Keratinocytes,
Molecular Biology of*

Claus-Thomas Bock
Abteilung Gastroenterologie
und Hepatologie
Medizinische Hochschule Hannover
D-30625 Hannover, Germany
*Electron Microscopy of
Biomolecules*

Franklyn F. Bolander, Jr.
Department of Biological Sciences
University of South Carolina
Columbia, SC 29208, USA
Endocrinology, Molecular

Richard Bormett
Department of Chemistry
University of Pittsburgh
Pittsburgh, PA 15261, USA
Enzymology, Nonaqueous

Robert L. Brumley, Jr.
Department of Chemistry
University of Wisconsin
Madison, WI 53706, USA
*DNA Sequencing in Ultrathin
Gels, High Speed*

Bruce Budowle
Laboratory Division
Forensic Science Research
and Training Center
Federal Bureau of Investigation Academy
Quantico, VA 22135, USA
DNA Fingerprint Analysis

Antony W. Burgess
Ludwig Institute for Cancer Research
Melbourne
Victoria 3050, Australia
Growth Factors

C. R. Calladine
Department of Engineering
University of Cambridge
Cambridge CB2 1PZ, United Kingdom
DNA Structure

Donald C. Chang
Department of Biology
Hong Kong University of Science
and Technology
Clear Water Bay, Kowloon, Hong Kong
Electroporation and Electrofusion

Sudipta Chatterjee
Department of Chemical Engineering
Center for Biotechnology
and Bioengineering
University of Pittsburgh
Pittsburgh, PA 15261, USA
Enzymology, Nonaqueous

Jon A. Christopher
Department of Chemistry
Texas A&M University
College Station, TX 77843, USA
*Enzymes, Energetics and Regulation
of Biological Catalysis by*

John C. Crabbe
Veterans Affairs Medical Center and
Department of Medical Psychology
Oregon Health Sciences University
Portland, OR 97201, USA
*Drug Addiction and Alcoholism,
Genetic Basis of*

Horace R. Drew
Laboratory of Molecular Biology
CSIRO Division of Biomolecular
Engineering
North Ryde, New South Wales 2113,
Australia
DNA Structure

P. Leslie Dutton
Department of Biochemistry
and Biophysics
Johnson Research Foundation
University of Pennsylvania
Philadelphia, PA 19104, USA
Electron Transfer, Biological

Charles J. Epstein
Department of Pediatrics
University of California
San Francisco, CA 94143, USA
*Down Syndrome, Molecular
Genetics of*

Ramy S. Farid
Department of Chemistry
Rutgers University
Newark, NJ 07102, USA
Electron Transfer, Biological

Robert Feil
Laboratory of Developmental Genetics
and Imprinting
The Babraham Institute
Cambridge CB2 4AT, United Kingdom
Genomic Imprinting

Malcolm A. Ferguson-Smith
Department of Pathology
Cambridge University
Cambridge CB2 1QP, United Kingdom
Gene Order by FISH and FACS

J. K. Findlay
Prince Henry's Institute of Medical
Research
Clayton, Victoria 3168, Australia
*Female Reproductive System,
Molecular Biology of*

Ron M. Fourney
Biology Research
and Development Section
Central Forensic Laboratory
Royal Canadian Mounted Police
Ottawa, Ontario K1G 2M3, Canada
DNA Fingerprint Analysis

Frank K. Fujimura
Myriad Genetic Laboratories, Inc.
Salt Lake City, UT 84108, USA
Genetic Testing

Judith E. Grisel
Veterans Affairs Medical Center and
Department of Medical Psychology
Oregon Health Sciences University
Portland, OR 97201, USA
*Drug Addiction and Alcoholism,
Genetic Basis of*

Jeffrey C. Hall
Department of Biology
Brandeis University
Waltham, MA 02254, USA
Drosophila *Mating, Genetic Basis of*

Barry Halliwell
Neurodegenerative Disease
Research Centre
Pharmacology Group
King's College
University of London
London SW3 6LX, United Kingdom
*Free Radicals in Biochemistry and
Medicine*

Amanda C. Heppell-Parton
Clinical Oncology
and Radiotherapeutics Unit
Medical Research Council Centre
Cambridge CB2 2QH, United Kingdom
*Gene Mapping by Fluorescence In
Situ Hybridization*

A. Rus Hoelzel
Laboratory of Viral Carcinogenesis
National Cancer Institute
Frederick, MD 21702, USA
Genetic Analysis of Populations

David Jenkins
The Diabetic Centre
Worcester Royal Infirmary
Worcester WR5 1HN, United Kingdom
Diabetes Mellitus

George J. Kantor
Department of Biological Sciences
Wright State University
Dayton, OH 45435, USA
DNA Damage and Repair

Stephen G. Kayes
Department of Structural
and Cellular Biology
University of South Alabama
Mobile, AL 36688, USA
*Genetic Vaccination, Gene
Transfer Techniques for Use in*

Lorne T. Kirby (retired)
Department of Pathology
University of British Columbia
Vancouver, British Columbia V6H 3V4,
Canada
DNA Fingerprint Analysis

Akira Kobata
Tokyo Metropolitan Institute
of Gerontology
35–2 Sakaecho, Itabashi-Ku
Tokyo 173, Japan
Glycobiology

Joseph R. Lakowicz
Department of Biological Chemistry
University of Maryland School
of Medicine
Baltimore, MD 21201, USA
*Fluorescence Spectroscopy
of Biomolecules*

Daniel Lednicer (Retired)
826 Bowie Road
Rockville, MD 20852, USA
Drug Synthesis

Hisaji Maki
Department of Prokaryotic
Molecular Genetics
Graduate School of Biosciences
Nara Institute of Science and Technology
Ikoma, Nara 630–01, Japan
*DNA Replication and
Transcription*

Brian McNeil
Department of Bioscience
and Biotechnology
University of Strathclyde
Glasgow G1 1XW, United Kingdom
Fungal Biotechnology

J. E. Mercer
Department of Obstetrics
and Gynaecology
Monash Medical Centre
Monash University
Clayton, Victoria 3168, Australia
*Female Reproductive System,
Molecular Biology of*

Bradley T. Messmer
Laboratory of Molecular Genetics
and Informatics
Rockefeller University
New York, NY 10021, USA
Genetic Intelligence, Evolution of

Catherine H. Mijovic
Department of Medicine
University of Birmingham
Clinical Research Block
Queen Elizabeth Hospital
Birmingham B15 2TH, United Kingdom
Diabetes Mellitus

Hirotada Mori
Nara Institute of Science and Technology
Ikoma, Nara 630–01, Japan
E. coli Genome

Christopher C. Moser
Department of Biochemistry
and Biophysics
Johnson Research Foundation
University of Pennsylvania
Philadelphia, PA 19104, USA
Electron Transfer, Biological

Yusaku Nakabeppu
Department of Biochemistry
Medical Institute of Bioregulation
Kyushu University
Fukuoka 812, Japan
*DNA Replication
and Transcription*

Eberhard Neumann
Faculty of Chemistry
University of Bielefeld
D-33501 Bielefeld, Germany
*Electric and Magnetic
Field Reception*

Charles Onwulata
U.S. Department of Agriculture
Agricultural Research Service
Eastern Regional Research Center
Philadelphia, PA 19118, USA
Food Proteins, Interactions of

B. A. Oostra
Department of Clinical Genetics
Erasmus University
3000 DR Rotterdam, The Netherlands
*Fragile X Linked Mental
Retardation*

Nicholas Parris
U.S. Department of Agriculture
Agricultural Research Service
Eastern Regional Research Center
Philadelphia, PA 19118, USA
Food Proteins, Interactions of

Frederica P. Perera
Division of Environmental
Health Sciences
School of Public Health
Columbia University
New York, NY 10032, USA
Epidemiology, Molecular

Stephen W. Raso
Department of Chemistry
Texas A&M University
College Station, TX 77843, USA
*Enzymes, Energetics and
Regulation of Biological
Catalysis by*

Wolf Reik
Laboratory of Developmental Genetics
and Imprinting
The Babraham Institute
Cambridge CB2 4AT, United Kingdom
Genomic Imprinting

Alistair G. C. Renwick
Institute of Health
and Community Medicine
University of Malaysia Sarawak
94300 Kota Samarahan
Sarawak, Malaysia
Glycoproteins, Secretory

Robert I. Richards
Department of Cytogenetics
and Molecular Genetics
Centre for Medical Genetics
Women's and Children's Hospital
Adelaide 5006, Australia
Fragile Sites

Walter Rosenthal
Rudolf-Buchheim-Instut für
Pharmakologie der Justus-Liebig-
Universität Gießen
D-35392 Gießen, Germany
Diabetes Insipidus

Nathaniel Rothman
School of Hygiene and Public Health
Johns Hopkins University
Baltimore, MD 21202, USA
   Epidemiology, Molecular

Alan J. Russell
Department of Chemical Engineering
Center for Biotechnology
and Bioengineering
University of Pittsburgh
Pittsburgh, PA 15261, USA
   Enzymology, Nonaqueous

Ka-Yiu San
Department of Chemical Engineering
Rice University
Houston, TX 77251, USA
   *Expression Systems
   for DNA Processes*

Linda J. Sandell
Veterans Affairs Medical Center
and Departments of Orthopaedics
and Biochemistry
University of Washington
Seattle, WA 98108, USA
   Extracellular Matrix

Robert D. C. Saunders
Department of Anatomy and Physiology
University of Dundee
Dundee DD1 4HN, United Kingdom
   Drosophila *Genome*

John G. Scandalios
Department of Genetics, Box 7614
North Carolina State University
Raleigh, NC 27695, USA
   *Environmental Stress, Genomic
   Responses to*

Paul A. Schulte
National Institute for Occupational Safety
and Health
Centers for Disease Control
and Prevention
Cincinnati, OH 45226, USA
   Epidemiology, Molecular

Anita Seibold
Graduate School of Biomedical Sciences
University of Texas Health Science Center
Houston, TX 77030, USA
   Diabetes Insipidus

Mutsuo Sekiguchi
Department of Biochemistry
Medical Institute of Bioregulation
Kyushu University
Fukuoka 812, Japan
   *DNA Replication
   and Transcription*

David Sidransky
Department of Otolaryngology—Head
and Neck Surgery
Head and Neck Cancer Research Division
Johns Hopkins University School
of Medicine
Baltimore, MD 21205, USA
   *DNA in Neoplastic Disease
   Diagnosis*

Lloyd M. Smith
Department of Chemistry
University of Wisconsin
Madison, WI 53706, USA
   *DNA Sequencing in Ultrathin
   Gels, High Speed*

D. Peter Snustad
Department of Genetics and Cell Biology
University of Minnesota
St. Paul, MN 55108, USA
   Genetics

John Sommerville
School of Biological and Medical Sciences
University of St. Andrews
St. Andrews, Fife KY16 9TS,
United Kingdom
   *Electron Microscopy
   of Biomolecules*

Alfred Stracher
Department of Biochemistry
Health Science Center at Brooklyn
State University of New York
Brooklyn, NY 11203, USA
   *Drug Targeting and Delivery,
   Molecular Principles of*

Grant R. Sutherland
Department of Cytogenetics
and Molecular Genetics
Centre for Medical Genetics
Women's and Children's Hospital
Adelaide 5006, Australia
   Fragile Sites

David S. Thaler
Laboratory of Molecular Genetics
and Informatics
Rockefeller University
New York, NY 10021, USA
   Genetic Intelligence, Evolution of

A. J. M. H. Verkerk
Department of Clinical Genetics
Erasmus University
3000 DR Rotterdam, The Netherlands
   *Fragile X Linked Mental
   Retardation*

Peter J. Wirth
Laboratory of Experimental
Carcinogenesis
National Cancer Institute
National Institutes of Health
Bethesda, MD 20892, USA
   *Gel Eletrophoresis of Proteins,
   Two-Dimensional Polyacrylamide*

Jon A. Wolff
Department of Pediatrics
University of Wisconsin
Madison, WI 53705, USA
   *Genetic Vaccination, Gene
   Transfer Techniques for Use in*

Takashi Yura
HSP Research Institute
Kyoto Research Park
Kyoto 600, Japan
   E. coli *Genome*

Eugene R. Zabarovsky
Microbiology and Tumor Biology Center
Karolinska Institute
Stockholm S-171 77, Sweden
   DNA Markers, Cloned

Hanswalter Zentgraf
Applied Tumor Virology
German Cancer Research Center
D-69120 Heidelberg, Germany
   *Electron Microscopy
   of Biomolecules*

Miriam M. Ziegler
Department of Biochemistry
and Biophysics
Texas A&M University
College Station, TX 77843, USA
   *Enzymes, Energetics and
   Regulation of Biological
   Catalysis by*

Encyclopedia of
Molecular Biology and
Molecular Medicine

Denaturation of DNA

Richard D. Blake

Key Words

Conformation (DNA)   The three-dimensional arrangement of the polynucleotide chains that form the double helix; resulting from local configurational tendencies of the chains and various noncovalent interactions.

Degradation (DNA)   Breakdown of covalent bonds supporting the polynucleotide chains, caused by exposure to extremes in environmental conditions. The process is generally irreversible.

Melting Curve (DNA)   The variation with temperature of any property (e.g., absorption at 260 nm) that is sensitive to the native structure of DNA.

Stacking Forces (DNA)   The forces between stacked bases in both strands of genetic material.

T_m **(DNA)**   The melting temperature of DNA, determined from the midpoint in the change with temperature in the dependent variable of the melting curve (e.g., absorption at 260 nm).

The denaturation of double-helical DNA is brought about by any of the many artificial means of disturbing the noncovalent bonds that support the native conformation. Such disturbance causes either alterations at the level of local structure of the total collapse of the helix. The process is generally reversible.

1   DENATURATION AND DEGRADATION

Unless otherwise stated, references to DNA in this article signify the duplex form. Most physical and chemical properties of single-stranded DNA are like those of RNA.

Native duplex DNA is held together by noncovalent forces substantially weaker than the covalent forces holding polynucleotide chains together. For the most part this noncovalent support consists of short-range, temperature-sensitive dispersion forces between stacked base pairs. Watson–Crick hydrogen bonds between bases maintain the alignment of the two strands, but the vertical "stacking" forces between nearest-neighbor base pairs are the major sources of support to the helix. Stacking forces vary with the sequence of stacked pairs in the duplex, as seen in Table 1, and are responsible for local variations in the stability of DNA as well as for the strong dependence of the structural state of each pair on the state of its neighbors. Indeed, the dependence extends well beyond next neighbors. The large energy needed to free a pair from the middle of a stacked array leads to the cooperative collapse of entire stretches that may involve 250–500 contiguous pairs. Factors used to break these bonds usually operate over a narrow range of conditions and affect large stretches or "domains" of DNA.

In addition to stacking and hydrogen bonds, a number of weaker bonds, primarily from solvation effects, the occasional water bridge, and hydrogen bonds associated with certain sequence arrangements, contribute to the support of the duplex. Also, the phosphodiester groups ($\cdots\!-\!O\!-\!P[O_2]\!-\!O\!-\!\cdots$) that join nucleotides in the backbone have a negative charge; therefore the polynucleotide chain is a polyanion with strong repulsive forces between opposing chains, counteracting the stabilizing forces from stacking and hydrogen bonds. The high charge density of the helix serves to attract counterions (e.g., Na^+) to it in high concentration, as seen schematically in Figure 1 (structure **I**), reducing the effective charge on the phosphates at pH7 from -1 e^- to about -0.24 e^-. If the counterion is an alkaline earth metal (Li^+, Na^+, K^+, . . .), the counterion concentration close to the helix exceeds 1 M, regardless of the bulk concentration, and represents an integral component of the native structure.

If the weak forces stabilizing DNA are artificially challenged, the double helix collapses and the DNA is *denatured*. Denaturation leads to nonhelical structures and, if conditions are sufficiently severe, to total separation of the polynucleotide chains (Figure 1). If *all* noncovalent bonds are broken, the chains assume the conformation of random coils (structure **II**). However, denaturation is rarely if ever that complete; rather the DNA assumes a complex disordered

Table 1   Nearest-Neighbor Thermodynamics for Polymeric and Oligomeric Domains[a]

5'-ij-3'	Polymeric Domains 0.075M–Na⁺ observed[b]	Polymeric Domains .... [tentative]	Oligomeric Fragments 1.0M–Na⁺ [observed[c]]		Polymeric Domains 0.075M–Na⁺ [tentative]	
	Tij	$Tij = A \cdot log_{10}[Na^+] + B$	ΔHij^0	ΔSij^0	ΔHij^*	ΔSij^*
1 GC (GC)	135.83	$9.00 \, log_{10}[Na^+] + 146.(3)$	46.4	111.7	47.1	115.2
2 AC (GT)	108.80	$13.26 \, log_{10}[Na^+] + 123.(5)$	27.2	72.4	28.2	73.8
3 GG (CC)	99.31	$18.64 \, log_{10}[Na^+] + 94.(7)$	46.0	111.3	45.7	121.9
4 CG (CG)	88.84	$19.79 \, log_{10}[Na^+] + 88.(5)$	49.8	116.3	48.7	123.8
5 AG (CT)	85.12	$17.51 \, log_{10}[Na^+] + 100.(7)$	32.6	87.0	33.0	90.2
6 GA (TC)	80.43	$17.04 \, log_{10}[Na^+] + 87.(8)$	23.4	56.5	23.3	61.1
7 AT (AT)	72.29	$14.79 \, log_{10}[Na^+] + 115.(3)$	36.0	100.0	36.4	103.3
8 AA (TT)	66.51	$22.39 \, log_{10}[Na^+] + 74.(5)$	38.1	100.4	37.9	105.6
9 CA (TG)	64.92	$16.25 \, log_{10}[Na^+] + 107.(4)$	24.3	54.0	23.5	60.7
10 TA (TA)	50.11	$19.92 \, log_{10}[Na^+] + 87.(8)$	25.1	70.7	25.3	74.7

[a]Tij are expressed in degrees Celsius, ΔHij in kJ(mol-ij)$^{-1}$, and ΔSij in kJ(mol-ij-deg)$^{-1}$.
[b]S. G. Delcourt and R. D. Blake, Stacking energies in DNA. *J. Biol. Chem.* 266:15160–15169 (1991); R. D. Blake and S. G. Delcourt (unpublished).
[c]K. J. Breslauer, R. Frank, H. Blöcker, and L. A. Marky, Predicting DNA Duplex stability from the base sequence. *Proc. Natl. Acad. Sci. U.S.A.* 83:3746–3750 (1986).

state with many regions of single-strand stacking and double-stranded hairpins, and a few segments in which the structure may be random and dynamic. The metastable, quasi-stochastic structure of denatured DNA that has been returned to conditions of solvent or temperature that ordinarily support the native duplex (structure **III**) is also the prototypic structure for single-stranded DNAs and RNAs.

If DNA is unnaturally forced into an altered structure, one that is not everywhere helical or everywhere recognizable by the various enzymes and proteins that maintain the native structure, it is still said to be *denatured*. Denaturation is distinguished from the more severe process of *degradation,* during which covalent bonds are broken. When denaturation is induced by temperature, the process should be conducted as rapidly as is physically possible, preferably by microwave heating, to minimize the exposure of the single strands to the degradative conditions of high temperatures. The rate of depurination increases substantially above 50°C. Many procedures for denaturing DNA call for heating in boiling water for 2 to more than 10 minutes. Since denaturation is almost instantaneous, the incubation time need be long enough only to ensure that thermal equilibrium has been achieved. Long exposures of DNA to 100°C may lead to significant levels of depurination, a change that is often accompanied by spontaneous cleavage of the strand.

2   ANALYTICAL METHODS FOR FOLLOWING THE DENATURATION PROCESS

2.1   CHANGES IN ABSORPTION WITH TEMPERATURE

A popular method of monitoring the denaturation process entails the use of *electronic absorption spectroscopy* in the ultraviolet region of 200–290 nm. The mean absorption coefficient (absorptivity) for $\pi \rightarrow \pi^*$ electronic transitions of the heterocyclic bases is unusually large over this region. In the helical state these transitions are neighbor-perturbed in the close-ordered stacking of the bases. The molar absorption coefficient for the helix, $\varepsilon_{260nm}(P)$, is only $7 (\pm 0.2) \times 10^3$ L/mol·cm) under physiological solvent conditions, and it is significantly hypochromic to the constituent nucleotides. During denaturation, the bases unstack and the molar absorption co-

efficient increases to about 10^4 L/mol · cm). This value is close to but still less than the value for the constituent nucleotides, since residual stacking of the unpaired bases persists beyond the denaturation temperature of the helix. The increase in absorption at 260 nm is approximately 38%; consequently, denaturation can be monitored with considerable precision on nucleotide residue concentrations in the micromolar range.

A typical thermal denaturation or melting curve of DNA is shown in Figure 2. A melting temperature T_m, determined from the midpoint of the transition, is generally taken as a measure of the thermal stability of the DNA, while the transition breadth is sometimes interpreted as a measure of the variation in GC content. In single-stranded DNAs, the conformational states of stacked bases are essentially independent of those of their neighbors, so that "denaturation" occurs over a wide (80–100°C) range of temperatures. By contrast, duplex DNA melts within 12–16°C, and if the sequence is sufficiently biased, within 0.1°C.

If quantitative significance is required of the T_m or transition breadth, the melting curves must be reproducible and must represent true equilibria between helix (native) and coil (denatured) states at all temperatures. Usually this means that the curves should be carried out slowly, with suitable delays at each temperature to ensure that the absorption (or other dependent variable) is stable. *Measured T_m* of curves obtained when there has been insufficient time to achieve equilibrium are lower than *true T_m*; this artifact, which is more apparent at low ionic strengths, is attributable to slow re-formation of the helix in the maintenance of equilibria. Helix reformation involves the condensation of a high counterion concentration on the DNA (Figure 1) and is limited by diffusion of counterion, particularly at lower ionic strengths. This is reflected in the fact that the second-order rate constant for renaturation, k_2 [L/(mol · s)] exhibits a strong power dependence on [Na⁺]:

$$\frac{\partial \log k_2}{\partial \log [Na^+]} \bigg|_T \quad 3.6 \qquad (1)$$

Meaningful projection of this relationship to heating rates is difficult. As a rule of thumb, when absorbance values are monitored in

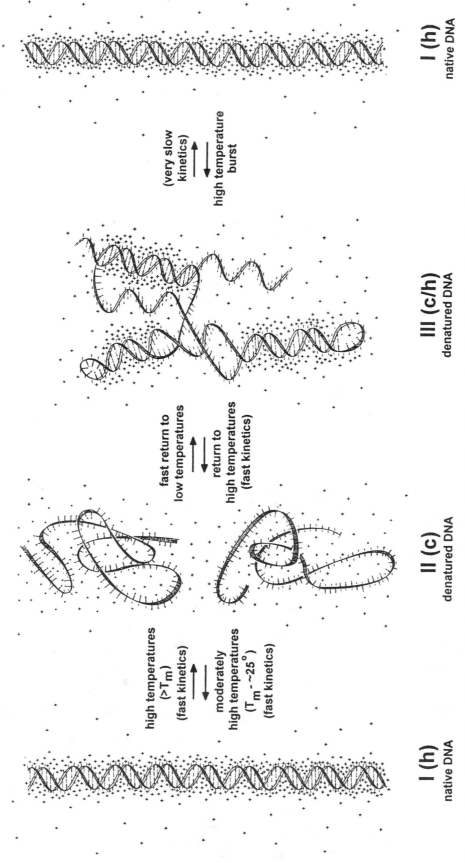

Figure 1. Schematic illustration of DNA as a polyanion. **I** represents the native (helical) state, with large numbers of condensed counterions. Thermal denaturation, represented by **I**(h) → **II**(c), is associated with loss of helical secondary structure and most of the condensed cation. Quick cooling, represented by **II**(c) → **III**(c/h), results in a metastable single-stranded denatured DNA with local regions of imperfect secondary structure.

3

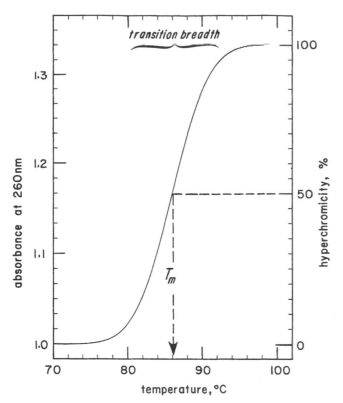

Figure 2. A typical thermal denaturation (melting) curve of DNA, obtained by the variation in absorption at 260 nm with increasing temperature. The T_m is usually defined as the temperature corresponding to half the hyperchromic effect or overall increase in absorption. The T_m has semiquantitative significance only for large heterogeneous DNAs.

continuous fashion as temperature is steadily increased, the heating rate should not exceed 10°C per hour when the sodium counterion concentration is in the neighborhood of 0.075M. Failure to reach equilibria, detectable by some irreversibility, occurs when rates exceed 15°C per hour. The rates can be faster at higher concentrations of sodium, but must be slower at lower concentrations. Below 0.02 M Na$^+$, the rate must be so slow that it is difficult to obtain equilibrium curves. A new hazard, thermally induced depurination in the denatured strands, then becomes of significant concern.

A further advantage of absorption spectroscopy is the ability to determine characteristics of the base composition of the denaturing DNA by spectral decomposition. The dissociation of A·T base pairs contributes more than four times more to the absorption change than G·C pairs in curves obtained at 260 nm, while almost the reverse is true at 282 nm. Both pairs contribute equally at 270 nm, and therefore this is the preferred wavelength for expressing the ordinate of melting curves as the fraction of base pairs that have denatured.

2.2   GRADIENT GEL ELECTROPHORESIS

Gradient gel electrophoresis, a powerful method developed in Lerman's laboratory, has the capability of detecting small differences in electrophoretic mobilities of nanogram quantities of related sequences. Two configurations are in current use: one with a gradient of denaturant (DGGE), and the other with a temperature gradient (TGGE). The denaturing gradient in DGGE consists of urea and formamide, and the gel is operated at high temperature. During the

run, the entire apparatus is immersed in a bath maintained at a constant high temperature (usually ca. 60°C). Excellent results have been obtained on standard acrylamide gels with both configurations. There are two basic orientations in both configurations:

> The gradient of denaturing conditions in *perpendicular* gels is established perpendicular to the electric field in a modified vertical slab gel apparatus.
> In *parallel* gels the gradient is formed parallel to the field.

A comparative analysis of a number of specimens is best conducted on *parallel* gels, by coelectrophoresis in vertical lanes produced with a comb in the standard fashion. The example of a *perpendicular* DGGE pattern of a 323 bp restriction fragment of a plasmid containing a 272 bp segment of the human β-globin gene is shown in Figure 3. The horizontal gradient in this experiment was 40–55% (v/v) formamide plus 7 M urea, and the DNA was stained by ethidium bromide in the usual manner. The combined effects of high temperature and denaturants lower the stability of the entire DNA, but it is the weakest melting domain that alters the mobility. An "effective denaturation temperature" of the domain, T_e, can be defined for the combined effects of denaturants and temperature. The optimum gradient for following melting is then estimated from the relationship between the bath temperature T_b and a constant X, representing the combined influence of temperature and chemical denaturants on stability:

$$T_e = T_b + X \qquad (2)$$

For the formamide–urea gradients, X has a value of approximately $-0.3°/1\%$ denaturant. DGGE is therefore a high resolution method.

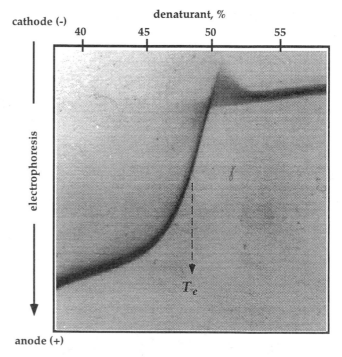

Figure 3. Perpendicular DGGE profile of the 323 bp *Eco*RI-*Hin*dIII fragment of the plasmid pCH6G1, containing a 272 bp segment from the human β-globin sequence [cf. Abrams *et al., Genomics,* 7: 463–475 (1990)]. The denaturant was a mixture of formamide and 7 M urea (v/v). (Gel picture courtesy of E. Abrams and L. Lerman.)

Under ideal circumstances, the T_e for similar domains can be distinguished within just 0.1% of the denaturant; corresponding to a δT_m of ±0.03°C. The difference in free energy for δT_m of such small magnitude is estimated from the relationship:

$$\delta \Delta G_m = \delta T_m \frac{\Delta H_m}{T_m} \quad (3)$$

Assuming a T_m of 76°C (349 K) and ΔH_m of 38 kJ/(mol · bp) (cf. Table 1), we see that the sensitivity for distinguishing related sequences is just a few joules, or smaller than the difference in energy between two 300 bp domains differing by the exchange of a single A·T bp for a G·C.

During electrophoresis the mobility of the specimen, generally a restriction fragment 200–800 bp long, remains constant until the weakest domain has denatured, causing a radical reduction in mobility. The melted domain represents a bulbous obstruction to reptational transport of the DNA through the pores of the gel, much like that which hinders "a phython that has swallowed a goat." In the perpendicular DGGE gel of Figure 3, this retardation happens immediately on the right-hand side of the gel, where the denaturant is most concentrated. On the left-hand, low denaturant side, the helix is unaltered and mobility is "normal." In between, the mobility decreases within a narrow range of denaturant concentration, signaling a subtransition. The pattern corresponds in appearance and physical significance to the absorption change in Figure 2, except that the DGGE pattern is attributable to the melting of just a single domain. The midpoint of the curve corresponds to T_e for the domain. Assignments of subtransitions to specific regions of sequence in the fragment is by reference to calculated stabilities of prototypic sequences through statistical mechanical analysis (see Section 4.1).

2.3   OTHER METHODS

Methods for monitoring the denaturation of DNA can be classified as short-range physical, long-range physical, chemical, and biological. Table 2 gives a partial list of physical methods that have been used. Chemical methods include those that probe differences in the reactivity of the bases in the helical and nonhelical states, while biological methods include such techniques as S1 nuclease sensitivity, reaction with antibodies, and transformation methods.

3   PHYSICAL AND CHEMICAL FACTORS AFFECTING THE STABILITY OF DNA

A number of factors from within and without the DNA affect its sensitivity to denaturation: pH, temperature, counterion concentration and type, GC content, distribution of nearest neighbors in the DNA, local sequence and conformational states of neighboring domains, presence of wobble or other mispairs, DNA length, presence of various salts and nonaqueous solvents, and supercoiling.

3.1   EFFECTS OF pH

The first denaturation experiments, conducted by Gulland, were followed by electrometric titration of DNA. Representative results are shown in Figure 4. Protonation and deprotonation of the bases at hydrogen ion concentrations below pH of 5 and above pH of 10, the regions considered unsafe to the native structure, alters the hydro-

gen bond donor–acceptor relationships between base pairs and places charges on the bases. DNA denatures at pH values below about 4 and above approximately 11: somewhat lower and higher, respectively, than the pK values of the free bases. The forward titrations of native DNA with acid and alkali, starting at pH 6.9 in the middle of the safe region, are represented by the open circles in Figure 4. Back titrations starting at the pH extremes take different routes (solid circles). The difference between forward and backward curves is sometimes referred to as *hysteresis,* a borrowed phrase.

The back titrations represent the properties of denatured DNA and correspond almost precisely to titration curves for the free bases. Indeed, the line through the solid circles represents the sum of calculated curves from the pK values of the free bases. The deviation of the curve for native DNA from the denatured curve is a reflection of the free energy of physically isolating the ionizable groups by Watson–Crick hydrogen bonding. Calculation of an apparent shift in pK for native DNA leads to the dotted curves in Figure 4. The experimental titration curve for native DNA deviates from the theoretical (dotted) curves midway through titration, toward that for the free bases as it becomes denatured.

There are practical reasons for avoiding pHs below roughly 5. Protonation leads to precipitation of polymeric DNA, while still lower pHs greatly increase the rates of depurination and subsequent chain cleavage of denatured DNA. Denaturation at high pH, on the other hand, appears to be less problematic, and, as first shown by Marmur, represents a good method for separating the two strands from each other. Provided the strands differ from unity in their purine/pyrimidine ratio, they can be separated by CsCl buoyant density gradient centrifugation at about pH 11, where thymine and guanine residues are titrated.

3.2   EFFECTS OF TEMPERATURE, COUNTERION CONCENTRATION, AND BASE CONTENT

It is more convenient to study the denaturing effects of temperature, described by example in Figure 2, and illustrated schematically in Figure 1. The thermodynamic effects of factors from within and without the DNA molecule have been studied with the T_m as a measure of stability. Results obtained in Doty's laboratory and reproduced in Figure 5 show the variation in T_m with the mole fraction of guanine plus cytosine residues in the DNA,

$$F_{GC} = \frac{G + C}{G + C + A + T} \quad (4)$$

where G, C, . . . , denote the numbers of each base in DNAs isolated from a wide variety of bacteria. Figure 5 also indicates that T_m depends on the counterion concentration—in this instance [Na+]. These two variables can be wed into a single empirical relationship, which, refined with the addition of more recent results, is given by the Marmur–Schildkraut–Doty (MSD) equation:

$$T_m (°C) = 193.67 - (3.09 - F_{GC})(34.64 - 2.83 \ln[Na^+]) \quad (5)$$

The reduction in quasi-randomness of nearest-neighbor frequencies at the extremes of base composition limits the applicability of this expression to large DNA specimens of $0.3 < F_{GC} < 0.7$. This limitation is apparent in Figure 5, where large deviations can be seen in the T_m for poly[d(A)·d(T)], [d(A-T)·d(A-T)], and [d(G-C)·d(G-C)], synthetic DNAs with only one or two neighbor types at F_{GC} = 0 and 100%, respectively. Also, because of limi-

Table 2   Selected Physical Methods for Following the Denaturation Process

Methods That Depend on Short-Range Physical Properties

Absorption Spectroscopy

Thomas, R. (1954) *Biochim. Biophys, Acta.* 14:231.
Blake, R. D. and Hydorn, T. G. (1985) *J Biochem. Biophys. Methods,* 11:307.
Some of the first DNA denaturation curves were carried out by taking advantage of the large hyperchromic effect associated with the destacking of base
    pairs.

Circular Dichroism

Gray, D. M., Ratliff, R. L., and Vaughan, M. R. (1992) *Methods Enzymol.* 211a:389.
CD measures the difference in molar extinction of the left and right circularly polarized components of monochromatic light in the region of 200–290 nm. As
    such, it is used to detect the spatial asymmetry and helical handedness as the chromophoric base pairs twist one way or the other around the major axis.

Infrared Spectroscopy

Taillanier, E., and Liquier, J. (1992) *Methods Enzymol.* 211a:307.

Raman Spectroscopy

Peticolas, W. L., and Evertsz, E. (1992) *Methods Enzymol.* 211a:335.
In the 500–1800 cm^{-1} region IR and Raman spectra reflect changes in *vibrational* and *rotational* modes affecterd by changes in hydrogen bonding and stack-
    ing interactions. Changes involving polar groups give the most intense absorptions and are useful for following the dissociation of hydrogen bonding. This
    method suffers from spectral interference by the solvent water.

Nuclear Magnetic Resonance

Feigon, J., Skenář, V., Wang, E., Gilbert, D. E., Macaya, R. F., and Schultze, P. (1992) *Methods Enzymol.* 211a:235.
One-dimensional NMR spectra (in the radio frequency range = ca. 10^{10}HZ) from 9.5 to 0.5 ppm are used mainly for studies of the exchangeable imino and
    amino resonances occuring during denaturation.

Differential Scanning Calorimetry

Breslauer, K. J., Freire, E., and Straume, M. (1992) *Methods Enzymol.* 211a:533.

Methods That Depend on Long-Range Physical Properties

Viscometry

Zamenhof, S., Alexander, H. E., and Leidy, G. (1953) *J. Exp. Med.* 98:373; Doty, P., and Rice S. A. (1955) *Biochim Biophys. Acta,* 16:446; Rice, S. A., and
    Doty, P. (1957) *J. Am. Chem.* 79:3937.
The first thermal denaturation profiles of DNA and bacterial "transforming activity" were obtained from changes in viscosity.

Sedimentation Velocity

Changes in these hydrodynamic properties reflect the reduction in frictional coefficient during denaturation, when DNA changes from a rigid rod into
    flexible chains. These methods are generally more difficult and cumbersome than spectral methods, requiring greater amounts of monodisperse material.

Dynamic Light Scattering

Electron Microscopy

Borovik, A. S., Kalambet, Yu. A., Lyubchenko, Yu. L., Shitov, V. T., and Golovanov, Eu. I. (1980) *Nucleic Acids Res.* 8:4165.
Partially denatured regions are fixed at intermediate temperatures within the melting region by reaction of the imino and amino groups with the
    bifunctional reagent glyoxal, thereby preventing reformation of the helix.

Gradient Gel Electrophoresis

Fischer, S. G., and Lerman, L. S. (1979) *Methods Enzymol.* 68:183; Abrams, E. S., and Stanton, V. P., Jr. (1992) *Methods Enzymol.* 212b:71; Wartell, R. M.,
    Hosseini, S. H., and Moran, C. P. (1990) *Nucleic Acids Res.* 18:2699; Ke, S.-H., and Wartell, R. M. (1993) *Nucleic Acids Res.* 21:5137.

tations in the kinetics of achieving equilibria between native and denatured states at low [Na$^+$], and the denaturing effects of Cl$^-$ at high [NaCl], this equation is applicable only between 0.02–0.4 M–Na$^+$. The accuracy of the MSD equation is ±0.15°C over the region 0.06–0.15 M Na$^+$, but beyond this region it is ±0.25°C.

The derivative of expression (5) with respect to [Na$^+$] yields

$$\frac{dT_m}{d \ln [\text{Na}^+]} = 8.75 - 2.83F_{\text{GC}} \qquad (6)$$

indicating a strong dependence of stability on counterion concentration. This expression states that the melting temperature increas-

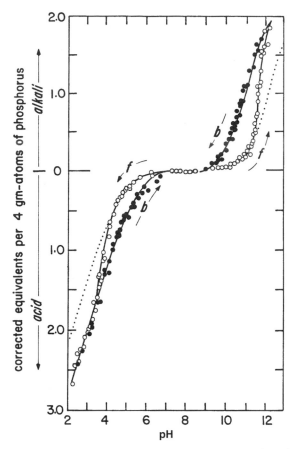

Figure 4. Electrometric titration curves of DNA, showing the dependence on hydrogen ion concentration of the number of equivalents of H⁺ bound per 4 g-atom (= mol) of phosphorus. The curve with open circles (*f*) represents forward titration, while closed circles (*b*) denote back titration. [After J. M. Gulland, D. O. Jordan, and H. F. W. Taylor, *J. Chem. Soc.* 1132 (1947).]

This expression leads to the theoretical Manning–Record equation, in which the electrostatic factors affecting denaturation are expressed explicitly:

$$\frac{\partial T_m}{\partial \ln a_{Na^+}} = \frac{RT_m^2}{\Delta H_m^\circ} \; \alpha \Delta \psi_{Na^+} \tag{10}$$

where a_{Na^+} is the activity of Na⁺ and α is the activity coefficient for NaCl; T_m is specified by the MSD equation (5), and ΔH_m is given by $\Sigma f_{ij} \Delta H_{ij}^\circ$, where f_{ij} is the fraction of the *ij*th pair for ΔH_{ij}° in Table 1. $\Delta \psi_{Na^+}$ represents the difference between the fraction of counterion bound to helix and coil states:

$$\Delta \psi_{Na^+} = \psi_{Na^+,h} - \psi_{Na^+,c} \tag{11}$$

Denaturation is rarely a two-state, all-or-none *helix → coil* transition, particularly when lengths exceed about 500 bp. If $\bar{N}_p$ represents the average number of phosphate groups in the cooperatively melting domain, then the change in fractional counterion concentration *per phosphate* is given by:

$$\Delta \psi_{Na^+} = \frac{\Delta \bar{r}_{Na^+}}{2\bar{N}_P} \tag{12}$$

According to equation (10), $\Delta \psi_{Na^+}$ is obtained directly from measured slopes of the variation in T_m with [Na⁺]. A value of 0.88 has been obtained for $c_{Na^+,h}$ in this way. The fraction of counterion on the helix involves contributions from both condensation and screening,

$$\psi_{Na^+,h} = \bar{\theta}_{Na^+,h} + (2\xi_h)^{-1} \tag{13}$$

The fraction that is condensed, $\bar{\theta}_{Na^+,h'}$

es between 13.70 and 20.15°C for each tenfold increase in [Na⁺], depending on the F_{GC} of the DNA. The physicochemical basis for this dependence is the release from the helix of large numbers of condensed counterions. The thermal denaturation of DNA, represented in Figure 1 by **I → II**, can be expressed by the following reaction scheme:

$$DNA_h - r_{Na^+,h} \cdot Na^+ \leftrightarrow DNA_c - r_{Na^+,c} \cdot Na^+ + \Delta \bar{r}_{Na^+} \cdot Na^+ \tag{7}$$

where subscripts h and c denote helix (native) and coil (denatured) states; and $\Delta \bar{r}_{Na^+} = (\bar{r}_{Na^+,h} - \bar{r}_{Na^+,c})$, the number of Na⁺ counterions released from association with the average-sized cooperatively melting domain. The variation with melting temperature of the equilibrium constant K_m for equation (7) is given by the familiar van't Hoff relationship

$$\frac{\partial(\ln K_m)}{\partial T_m} = \frac{\partial(-\Delta G_m^\circ/RT_m)}{\partial T_m} = \frac{\Delta H_m^\circ}{RT_m^2} \tag{8}$$

where R is the gas constant; the T_m is in kelvins, and ΔH_m° is the transition enthalpy per residue of phosphate at T_m. Rephrasing terms with the chain rule yields

$$\frac{\partial T_m}{\partial \ln [Na+]} = -\frac{RT_m^2}{\Delta H_m^\circ} \frac{\partial(\ln K_m)}{\partial \ln [Na+]}\bigg|_{T_m} \tag{9}$$

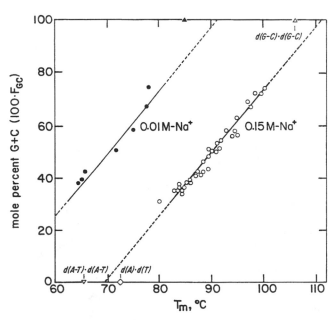

Figure 5. Dependence of T_m on GC content at two concentrations of monovalent counterion. Data points indicate T_m for genomic DNAs from 40 different bacterial sources of varying F_{GC}. [After J. Marmur and P. Doty, *J. Mol. Biol.* 5: 109–118 (1962).]

$$\bar{\theta}_{Na^+,h} = \frac{\bar{r}_{Na^+,h}}{\bar{N}_P} = 1 - \xi_h^{-1} \qquad (14)$$

has a value of 0.76. A value of 0.75 ± 0.1 was found by totally independent means, in excellent agreement with that predicted by the Manning–Record equation (10). In equations (13) and (14), ξ_h represents a dimensionless charge density parameter,

$$\xi_h = \frac{e^2}{\epsilon k T b_h} = \frac{7.14 \text{ Å}}{b_h} \qquad (15)$$

with a value of 4.2 for the helix. The quantity $e^2/\epsilon k T$ in equation (15), where ϵ is the dielectric constant, represents the characteristic length, with a value of 7.14 Å for aqueous solutions, and b_h is the mean spacing (Å) of charges along the contour axis of the helix, that is, 3.34 Å/2. Thus, b_h reflects the molecular relationships of charged phosphates to one another, a measure of considerable structural detail. If $\xi > 1$ (i.e., if $b < 7.14$ Å), counterion condensation occurs. The saturated charge fraction, which for the helix is 0.24 (0.12 for divalent counterions), is represented by $(\xi_h)^{-1}$.

From the corresponding Na^+ dependence of T_m for single-stranded DNA, b_c has a value of 4.1 ± 0.2 Å, so that $\psi_{Na^+,c} = 0.70$, $\bar{\theta}_{Na^+,c} = 0.39$, and $\xi_c = 1.7$. Thus, some condensation of counterion persists in the denatured strands. The fraction condensed to the helix is seen to decrease from 0.76 per phosphate to 0.39 on the denatured coil. This change is relatively constant over most of the salt range of applicability of the MSD equation (5). If equations (6) and (10) are set equal, it is found that the dependence of the slope, $dT_m/d \ln[Na^+]$, on F_{GC} is almost quantitatively compensated for by increases in T_m^2 and ΔH_m° with F_{GC} (Table 1). In other words, $\Delta \psi_{Na}$ is more or less constant for DNA over a wide range of sodium ion concentrations and temperatures.

The local concentration of counterion can be determined from V_p, defined as the volume element of the local region of the DNA in which the condensed counterions are constrained, and given by

$$V_P = 41.1(\xi - 1)b_h^3 \qquad (16)$$

The local concentration of Na^+ given by

$$c_{Na^+,h}^{loc} = 10^3 \bar{\theta}_{Na^+,h} V_P^{-1} = 24.3(\xi b_h^3)^{-1} \qquad (17)$$

is 1.2 M Na^+. Condensed Na^+ ions are therefore distributed as a dense cloud about the helix, as illustrated in Figure 1. To a large extent, this cloud dissipates during denaturation, where the local concentration of Na^+ about denatured coils, $c_{Na^+,c}^{loc}$, falls to 0.21 M Na^+. The large value for $\Delta c_{Na^+}^{loc}$ is responsible for a number of characteristic properties of DNA. The strong dependencies of T_m and renaturation rates on sodium ion concentration are just two examples of such properties.

3.3 EFFECTS OF NEAREST NEIGHBOR FREQUENCIES AND NEIGHBORING DOMAINS

Equation (5), the MSD equation, summarizes the relation that is found between T_m, [Na$^+$], and F_{GC} for a large mixture of DNAs, or a single large DNA such as the 4.2 Mb *E. coli* DNA. It would seem reasonable to expect that equation (5) would also apply to DNAs of more modest size, say, a 5 kb plasmid DNA, but it does not. It fails in this respect because biases in neighbor frequencies are more apparent in short DNAs or at the local level of domains, with the re-

sult that the effects of different stacking energies on the T_m of subtransitions for short domains within the DNA are similarly more apparent. Deviations from equation (5) are significant, but they are best seen in melting curves obtained at high resolution. Furthermore, it is helpful if the dependent variable, say, the absorption at wavelength λ, $A_{\lambda,nm}$, is represented in the first derivative, $dA_{\lambda,nm}/dT$, to reveal subtle changes on a more convenient scale. Figure 6a shows two derivative melting curves of pN/MCS-22, a 4619 bp plasmid with a prominent low-melting subtransition for a 200 bp AT-rich insert consisting of a tandemly repeating hexadecanucleotide sequence: [AAGTTGAACAAAAAAT]$_{12}$AAGTTGAA, sandwiched between two long GC-rich segments in the plasmid. Both curves in Figure 6a were obtained on the same plasmid, but linearized at different sites. The plasmid was constructed with the insert in the center of a 55 bp multiple-cloning site (MCS) consisting of 10 contiguous and overlapping tetra- and hexanucleotide recognition sites for various restriction endonucleases. At this resolution, it can be seen that denaturation is a complex process, taking place in piecemeal fashion by distinct domains of increasing stability with increasing temperature. A domain in this instance is a particular stretch of DNA that dissociates or melts in a more or less cooperative, all-or-none fashion, seen in profiles such as these as sharp peaks representing subtransitions. [The resolution of domains in such curves is about the same as can be achieved by gradient gel electrophoresis (GGE), $\pm 0.03°$.] The denatured domains of partially melted DNA can be readily seen in transmission electron micrographs (EM) by fixing the denatured regions with the bifunctional reagent glyoxal (cf. Table 2). Also, partially denatured regions are targets for digestion by single-strand specific endonucleases, such as S1 from *Aspergillus oryzae*. Denatured regions can also be detected by GGE.

Analysis of curves such as those in Figure 6a shows that the T_m of either the overall curve or a subtransition depends not as directly on F_{GC} as it does on the mole fraction of the 10 nearest-neighbors, $f_{ij}/\Sigma f_{i,j}$, where ij denotes the stack of pair i on its neighbor j, and i,j represents the generic pair for the different combinations of i and j. There are $4^2 = 16$ nearest stacked neighbors, but 10 are unique; each with a different stacking energy that depends intimately on the local structure of the helix. Table 1 shows that purine · pyrimidine pairs (abbreviated R·Y) stacked upon pyrimidine · purine pairs, $_{3'}^{5'}{<}{(R \cdot Y) \atop (Y \cdot R)}{>}_{5'}^{3'}$, are most stable, while $_{3'}^{5'}{<}{(R \cdot Y) \atop (R \cdot Y)}{>}_{5'}^{3'}$ are intermediate, and the 5'-pyrimidine-upon-3'-purine arrangements, $_{3'}^{5'}{<}{(Y \cdot R) \atop (R \cdot Y)}{>}_{5'}^{3'}$ are the weakest. Theoretical analyses indicate that dispersive Lennard–Jones forces dominate the interactions between stacked base pairs.

Values for T_{ij} in Table 1 accurately predict stabilities for subtransitions in the high resolution melting curves. By way of example, we note that the observed T_m of the featured insert in Figure 6a is 73.07 $\pm 0.03°$C when the plasmid is cut and linearized by *Kpn* I immediately adjacent to one end of the insert. When values for T_{ij} from Table 1 are used, the T_m calculated from the sequence of the plasmid is 73.08°C (Section 4.1). By contrast, the T_m predicted by the MSD equation for the F_{GC} of the insert is only 71.85°C, more than a degree too low. Similarly, the large deviations in T_m for poly[d(A) · d(T)], [d(A-T) · d(A-T)], and [d(G-C) · d(g-C)] from those predicted by the MSD equation and clearly visible in Figure 5, are consequences of the extreme neighbor biases in these synthetic DNAs.

Despite excellent agreement between observed and calculated T_{ij} in Table 1, doubt persists regarding whether consideration of

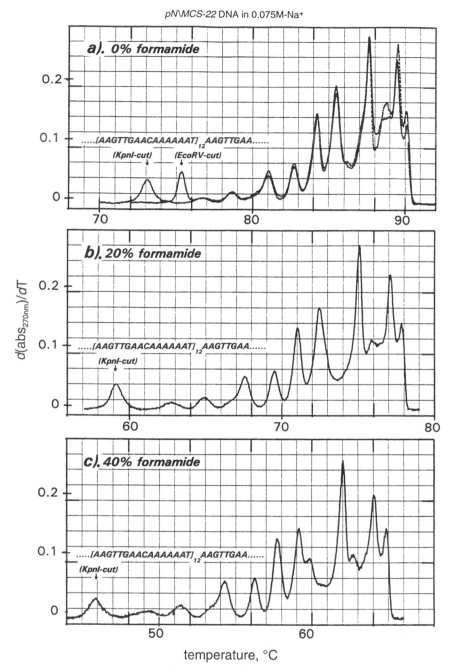

Figure 6. The thermal denaturation (melting) curve of pN/MCS-22 plasmid DNA (4619 bp), as the derivative of the change in absorption at 270 nm with temperature, $dA_{270\ nm}/dT$. At 270 nm the change in absorption for dissociation of A·T base pairs is equal to the change for G·C pairs. pN/MCS-22 is a pBR322 derivative with a multiple cloning sequence (MCS) of 55 bp at the unique *Nru* I cleavage site and contains a synthetic insert with the sequence [AAGTTGAACAAAAAAT] $_{12}$AAGTTGAA (19% GC) at the *Sma* I locus of the MCS. (This plasmid was constructed and prepared for melting by S. G. Delcourt.) The *Nru* I locus divides two very GC-rich regions that serve as strong helical boundaries for the insert. (a) The melting subtransition for the insert is seen as a peak at 75.34 ±0.03°C when the plasmid is linearized prior to melting by cutting at the unique *Eco*RV locus. This locus is 800 bp away from the insert, and as a result the insert melts as a loop. When the plasmid is linearized by cutting at the unique *Kpn* I immediately adjacent to the insert, the insert melts at 73.07 ±0.03°C. The [Na⁺] was 0.075 M, and the rate of heating was 6.00°C/h. (b) Melting curve of *Kpn* I cut pN/MCS-22 DNA in 20% formamide, 0.075 M Na⁺. (c) pN/MCS-22 DNA in 40% formamide. (R. D. Blake and S. G. Delcourt, unpublished.)

nearest-neighbor interactions is sufficient, or whether next-neighbor pairs and even beyond should be included in some instances. The magnitude of the question is immense, since the number of unique sequence elements of length N rises by 0.625×4^N. Nevertheless, there is some evidence that tracts or runs of particular pairs suspected of adopting variant structures of B-DNA, not seen at the level of the neighbor pair, have altered stabilities (Section 3.4).

Thermodynamic quantities, ΔH°_{ij} and ΔS°_{ij}, have also been determined for the 10 stacked pairs. The work was done in Breslauer's laboratory in an extensive series of studies on oligomeric duplexes in 1.0 M NaCl by differential scanning calorimetry. Differences in these quantities for polymeric specimens is probably not very large; however, inaccuracies are quite evident when the values derived from oligomeric specimens are used in a statistical mechanical model to calculate melting curves for polymeric specimens (Section 4.1). At present, it is not clear to what extent these inaccuracies are due to differences associated with oligomeric and polymeric specimens and high and low salt conditions, respectively, and/or to experimental error. In the first instance, strand separation occurs during the melting of oligomeric duplexes but not during dissociation of polymeric domains; thus the equilibrium constant for oligomeric duplexes must be amended for bimolecular strand association. Defining s_{ij} as the stability constant for stacking the ith pair onto the jth pair, the amended constant for formation of the first stacked base pair is given by:

$$s^*_{ij} = s_{ij} \exp \frac{\delta \Delta S^{\circ}_{ij}}{RT} (T - T_{ij}) = \beta s_{ij} \qquad (18)$$

where b is an association factor for formation of the first pair joining the two strands. Since the nucleation event reduces translational and rotational freedom of the oligomers, β is primarily entropic. Various thermodynamic and kinetic studies on the denaturation and nucleation of oligomers of defined length indicate β to have a value of approximately 0.0015 L/mol, or 1 Å3 per helical residue pair. Values are proposed in Table 1 for adjustments to ΔS_{ij} to reconcile the oligomer values to polymer behavior. Adjustments were made that minimize changes in measured ΔS_{ij} where

$$\Delta S^{\circ}_{ij} = \frac{\Delta H^{\circ}_{ij}}{T_{ij}} \qquad (19)$$

The average difference between ΔS_{ij} and ΔS^*_{ij} is only 15.2 63.9%, but now ΔH^*_{ij} and ΔS^*_{ij} lead to better agreement between calculated and observed melting curves. This is only a preliminary adjustment; thermodynamic relationships for all ij cannot be said to be unique because of the degree of experimental uncertainty in the calorimetric quantities for polymeric specimens. A knowledge of quantitative values for ΔG_{ij}, ΔH_{ij}, ΔS_{ij}, and ΔCP_{ij} for polymeric specimens is needed and will be immensely helpful in rationalizing the contributions of various factors to the stability of DNA, including some elusive forces associated with the aqueous solvent.

The curves illustrated in Figure 6a also show a dependence of domain stability on the states of neighboring domains. When the plasmid is linearized by cutting with *Eco*RV at a distance of 800 bp from the insert, so that the insert domain must now melt as an internal loop, the observed T_m jumps to 75.34 ±0.03°C, which is 2.27°C higher than when the insert melts from the end of the DNA. Loop formation generates two new boundaries with neighboring domains that remain in the helical state. The higher T_m for loops is due to the

greater probability that loops will renature, since the two ends of the domain are held together by the two helical domains that border it, reducing motional freedom of the intervening domain. Theory, based on a lower entropy for loops (Section 4.1), predicts a T_m of 75.04°C. This value is close to what is observed, but not as close as it is in most cases (±0.10°C). The unexpectedly great stability of this particular loop domain appears to be due to the $(A \cdot T)_6$-tracts in the insert.

These considerations show the dependence of T_m on F_{GC} in the MSD equation to be an oversimplification of more complex dependencies of DNA stability on stacking energies and the physical states of neighboring domains. As such, equation (5) is quantitatively applicable only to DNAs with suitably large and random distributions of neighbor pairs.

3.4 EFFECTS OF SOLVENTS

A large number of different salts and miscible organic solvents have been investigated for their effects on the denaturation of DNA. Using a specific immunochemical technique for monitoring denatured DNA, Levine et al. conducted an extensive survey of the concentrations of various organic solvents needed to bring about a 50% level of denaturation at 73°C. Their results are summarized in Table 3. Some immediate relationships emerge. Thus, denaturing effectiveness seems to be related to the number or size of the hydrophobic alkyl substituents on alcohols, carbonates, amides, and ureas. For example, the ranking of alcohols is as follows: n-butyl > n-propyl > ethyl > methyl. Conversely, the addition of hydrophilic groups reduces the denaturing effectiveness of a solvent; for example, ethyl alcohol > ethylene glycol; n-propyl alcohol > glycerol; cyclohexyl alcohol > inositol. These agents can affect the stability of DNA in two ways: by destabilizing the helix and/or by interacting favorably with the coil. In either case, equilibria are shifted toward denaturation. No correlation seems to exist between the denaturing ability of these solvents and other properties, such as the relative solvent dielectric constant. But in their study Levine et al. also noted that features of these solvents that increased their denaturing effectiveness also increased the solubility of adenine. The relationship between the activity coefficient of adenine and the concentration of solvent giving 50% denaturation is linear. This may indicate that the organic solvents are interacting directly with the bases, favoring the denatured state in the helix → coil equilibrium. It does not necessarily follow that these solvents are affecting hydrophobic bonds involved in the support of the helix, since the solubility of the bases may have less to do with the solvent than with interactions with other bases or the denaturant.

Formamide is a very popular denaturant, added to aqueous solutions of DNA in numerous studies for the purpose of destabilizing the duplex. This solvent is reported to lower the melting temperature of DNA by −0.65°C per 1% formamide, independent of GC content. The high resolution melting curves of pN/MCS-22 DNA in Figure 6 are, however, at some variance with these conclusions. Figure 6 shows melting curves in 0, 20, and 40% formamide. The sharpness of subtransitions at three very different formamide concentrations is unaffected by formamide, but the pattern of T_m for the many subtransitions is not conserved. The slope for $\partial T_m / \partial C_{formamide}$ decreases linearly with the fractional GC content, as shown in Figure 7.

The integrity of the double helix is also sensitive to the nature of the anion of certain concentrated salts, and to a lesser extent, the

Table 3   Organic Reagents and Their Concentrations (M) That Give a 50% Level of Denaturation of T_4 DNA ($F_{GC} = 0.34$) at 73°C and an Ionic Strength of 0.043 M

Aliphatic Alcohols					
Methyl	3.5	Allyl	0.50	n-Butyl	0.33
Ethyl	1.2	sec-Butyl	0.62	tert-Amyl	0.39
Isopropyl	0.90	tert-Butyl	0.60	Ethylene glycol	2.2
n-Propyl	0.54	Isobutyl	0.45	Glycerol	1.8
Thio Alcohols					
Dithioglycol	2.2				
Cyclic Alcohols					
Cyclohexyl	0.22	Inositol	1.5	p-Methoxyphenol	0.09
Benzyl	0.09	Phenol	0.08		
Other Cyclic Compounds					
Aniline	0.08	Purine	0.13	γ-Butyrolactone	0.55
Pyridine	0.09	1,4-Dioxane	0.64	3-aminotriazole	0.42
Amides					
Formamide	1.9	N-Ethylacetamide	0.88	Hexanamide	0.17
N-Ethylformamide	1.0	N,N-Dimethylacetamide	0.60	Glycolamide	1.1
N,N-dimethylformamide	0.60	Propionamide	0.62	Thiocetamide	0.32
Acetamide	1.1	Butyramide	0.46	δ-Valerolactam	0.34
Ureas					
Urea	1.0	Ethylurea	0.60	Thiourea	0.41
Carbohydrazide	1.0	tert-Butylurea	0.22	Allylthiourea	0.28
1,3-Dimethylurea	1.0	Ethyleneurea	0.53	Ethylenethiourea	0.32
Carbamates					
Urethane	0.50	N-Methylurethane	0.38	N-Propylurethane	0.24
Other Compounds					
Cyanoguanidine	0.21	Glycine	2.2	Tween 40	>20%
Sulfamide	1.1	Acetonitrile	1.2	Triton X-100	>10%

Source: Levine et al., *Biochemistry*, 2:168 (1963).

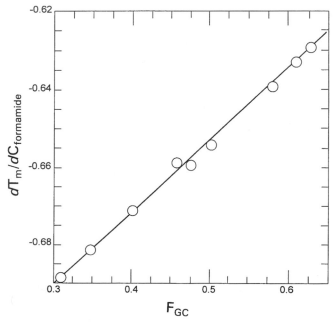

Figure 7.   Dependence of the slope $\partial T_m / \partial C_{formamide}$ on F_{GC}; [Na$^+$], 0.075 M.

cation of the salt. Salts that are most effective in shifting the helix → coil equilibrium are also most effective in solubilizing the bases. The following selected anions are listed in order of decreasing effectiveness in destabilizing the double helix (or increasing effectiveness in supporting the helix):

$$Cl_3CCOO^- > SCN^- > I^- > CH_3COO^- > Br^- > Cl^-$$

while the order for cations is:

$$Li^+ > Na^+ > K^+ \gg Mg^{2+}$$

3.5 DNA LENGTH

Studies of the effect of length on duplex stability represent a sensitive means of evaluating useful thermodynamic characteristics of DNA. Knowledge of the length dependence is also useful for estimating stringency when conducting Southern blots or for evaluating the relative stabilities of oligomeric probes for hybridization studies, or primers for the polymerase chain reaction (PCR). The decrease in stability with decreasing duplex length is the consequence of strand separation and a greater mixing entropy for short DNA fragments than for the polymeric strands. This dependence is readily seen in the relationship between T_m, oligomer length, and oligomer strand concentration obtained by integrating the van't Hoff equation (8) between $T_{m,N}$ and $T_{m,}$:

$$(T_{m,N})^{-1} = (T_{m,\infty})^{-1} + \frac{R}{(N-1)\Delta H} \ln(c_N \beta) \qquad (20)$$

where $T_{m,N}$ is the melting temperature, in kelvins, and c_N is the concentration of oligomers of length N. The temperature $T_{m,}$ represents the stability of a "corresponding" DNA of infinite length—certainly one of greater length than the cooperative length, so that any dependence of T_m on concentration is nil, and one with an identical nearest-neighbor composition. It is found that $T_{m,N}$ approximates $T_{m,}$ only when N is greater than about 1000 bp. The β in equation (20) represents the association parameter for the bimolecular reaction between complementary oligomeric fragments as described earlier (equation 18) and corresponds to the thermodynamic molal volume available to fragments said to be in the helical state. For an oligomeric specimen at two different strand concentrations, $c_{N,1}$ and $c_{N,2}$, equation (20) leads to

$$T_{m,N}(c_{N,1})^{-1} = T_{m,N}(c_{N,2})^{-1} + \frac{R}{(N-1)\Delta H} \ln \frac{c_{N,1}}{c_{N,2}} \qquad (21)$$

Equations (20) and (21) are useful for establishing the most favorable conditions for binding of oligomeric probes and primers. Qualitative estimates of $T_{m,N}$ can be made by assuming values for $T_{m,}$ from equation (5), the MSD equation, and ΔH from Table 1. (More quantitative estimates require that the fractions of neighbor pairs ij, f_{ij}, be equal in both oligomer and polymer so that from Table 1, $T_{m,} = \Sigma f_{ij} \cdot T_{ij}$, and similarly for ΔH.) Figure 8a illustrates the dependence of $T_{m,N}$ on N for two different oligomer concentrations, 0.001 M and 1.0 M. These curves were calculated from equation (20) for a DNA of $F_{GC} = 0.5$, where all $f_{ij} = 0.10$, so that in 0.075 M Na$^+$, $T_{m,} = 85.22°C$ and $\Delta H = 34.91$ kJ/mol. Figure 8b then shows the variation of the *difference* in $T_{m,N}$ with oligomer concentrations of two different oligomer sizes, $N = 10$ and $N = 20$. The T_m for $N = 10$ decreases more rapidly with decreasing oligomer concentration than it does for $N = 20$, so that when $c_N = 0.001$, the T_m for

the eicosamer is almost 21°C greater than for the decamer. Thus stringent conditions for efficient discrimination among a collection of probes of different sizes would seem to be easier to achieve at lower oligomer concentrations. On the other hand, curves in Figure 8a show that if binding involves competition between oligomeric probes and a polymeric complement strand, high concentrations are called for. Of course, the optimization of temperature, oligomer concentrations, and sizes of probes and primers for hybridization and PCR also requires consideration of kinetic factors. In many instances, the most favorable conditions can be established only by trial and error.

3.6 OTHER FACTORS

The denaturability of DNA is sensitive to a number of additional factors, including the presence of mispairs and the level of supercoiling. The effect of mispairs is discussed in Section 4.2 in the context of empirical analyses for the level of sequence divergence from the altered stability of heteroduplexes, determined by absorption spectroscopy or by gradient gel electrophoresis. Supercoiling is a significant factor in the denaturability of DNA. Closed-circular supercoiled plasmid DNAs denature only at very high temperatures, and they renature immediately if low temperature conditions are restored. However, the torsional stresses induced by supercoiling can be transmitted to certain sequences, prone to unusual structural states. Since these altered structures are devoid of any meaningful biological role, they could be said to be denatured. Inverted repeat sequences with twofold symmetry have the capacity to form cruciform structures. Each strand can dissociate from its complement and fold back on itself to form two opposing hairpin helices and, in doing so, alleviate some of the stresses of supercoiling.

The H structure is another "denatured" state of DNA that occurs with particular sequences. This is a three-stranded structure that is induced by H$^+$ together with modulation of supercoiling stresses over stretches of d(G)·d(C) sequences. The G-4 tetrameric structure is yet another altered state of DNA proposed for the guanine-rich telomeric sequences. But, by our definition, this is not a denatured state, since the hypothetical G-4 structure is considered to serve a vital biological role in the replication and maintenance of chromosomal termini.

4 UTILITY OF DENATURATION STUDIES

4.1 THERMODYNAMIC ANALYSES

Knowledge of the forces that govern the conformation and stability of DNA is needed to understand and model the behavior of this molecule in cells. The equilibrium concentrations of individual base pairs in all possible conformational substates must be known if plausible models are to be developed for the mechanics and dynamic behavior of DNA during replication, repair, and transcription. The evolution of DNA sequences, and indeed of species, will be fully understood only when it can be shown how subtle alterations in dynamic flexibility, stability, and conformation can lead to mutations. In this respect, studies of the thermal denaturation of DNA are important adjuncts to structural studies, providing the means for identifying various molecular sources of stability in the helix and for quantitating the thermodynamic characteristics of those sources.

As pointed out, the stability of DNA depends on its base sequence. The 10–12% difference in energy between G·C and A·T base pairs indicated by equation (5), the MSD equation, originates from interactions between nearest stacked neighbors. The equilibri-

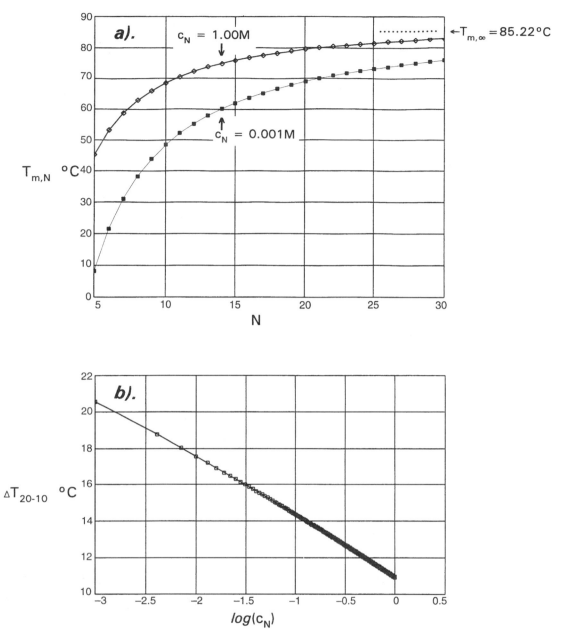

Figure 8. (a) Dependence of $T_{m,N}$ on oligomer length N calculated by equation (22). Results are shown for two oligomer concentrations, 0.001 M and 1.0 M strands/L. Values used: $T_{m,\infty}$ = 85.22°C, ΔH = 34.91 kJ/mol, β = 0.0015L/mol. (b) Dependence of the difference ($T_{m,N=20} - T_{m,N=10}$) on concentration c_N, mol-strands/L.

um thermal denaturation of highly purified DNAs of moderate lengths takes place in piecemeal fashion by dissociation of short domains. The example of the melting of a well-characterized plasmid DNA of 4619 bp, seen as discrete subtransitions in high resolution absorbance–temperature curves, was shown in Figure 6. Domains vary in size from a few tens of base pairs to stretches greater than 500 bp. Five intermediate nonhelical states can be identified, indicated as cases I–V in Table 4 and distinguishable by the physical states of neighboring domains. The simplest case is represented by melting from the ends of the helix. Here s_{ij} represents the stability constant for stacking pair i onto j at the end of a helical domain, and, in the absence of any extraordinary long-range energetic factors, by

s_{ij}^N for a sequence domain of N pairs. In equation (22), s_{ij} are related to temperature by:

$$s_{ij} = \exp \frac{-\Delta G_{ij}}{RT} = \exp \frac{-\Delta H_{ij} + T\,\Delta S_{ij}}{RT} \qquad (22)$$

where values for ΔH_{ij}^* and ΔS_{ij}^* from Table 1 are used.

In long DNAs denaturation is primarily internal with formation of closed loops, represented by case I in Table 4. At least three factors affect the thermodynamics and temperatures at which internal domains dissociate: (1) the previously mentioned costs of unstacking each base pair in the domain, (2) the energetic costs of interrupting the helix, and (3) the energetic level of the loop:

Table 4   Summary of Melting Process

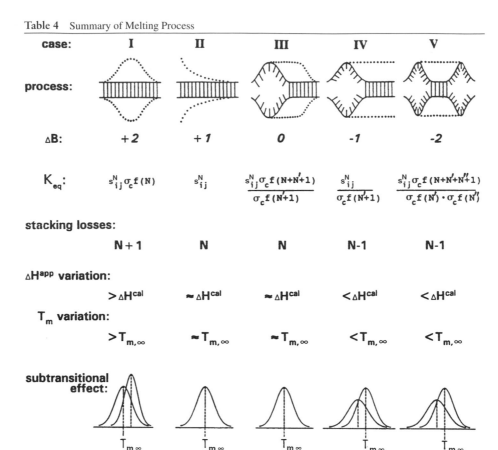

case:	I	II	III	IV	V
ΔB:	+2	+1	0	-1	-2
K_{eq}:	$s_{ij}^N \sigma_c f(N)$	s_{ij}^N	$\dfrac{s_{ij}^N \sigma_c f(N+N'+1)}{\sigma_c f(N'+1)}$	$\dfrac{s_{ij}^N}{\sigma_c f(N'+1)}$	$\dfrac{s_{ij}^N \sigma_c f(N+N'+N''+1)}{\sigma_c f(N') \cdot \sigma_c f(N'')}$
stacking losses:	N+1	N	N	N-1	N-1
ΔH^{app} variation:	$> \Delta H^{cal}$	$\approx \Delta H^{cal}$	$\approx \Delta H^{cal}$	$< \Delta H^{cal}$	$< \Delta H^{cal}$
T_m variation:	$> T_{m,\infty}$	$\approx T_{m,\infty}$	$\approx T_{m,\infty}$	$< T_{m,\infty}$	$< T_{m,\infty}$

subtransitional effect:

$$K_{eq} = s_{ij}^N \sigma_c f(N) \tag{23}$$

The cooperativity parameter σ_c represents the weighting factor for helix interruption:

$$\sigma_c = \exp \frac{-\Delta G_c}{RT_m} \tag{24}$$

which depends on GC content:

$$\sigma_c = -7.7 \times 10^{-6} F_{GC} + 8.1 \times 10^{-6} \tag{25}$$

so that $43 \geqslant \Delta G_c \geqslant 34$ kJ/(mol-interruptions). The statistical weight for rotational and translational constraints on N denatured residues of a closed loop is a significant item in the equilibrium constant (23),

$$f(N) = \zeta_b (N + D)^{-1.7 \pm 0.2} \tag{26}$$

where ζ_b represents the weight for long-range electrostatic effects at the boundaries between loop and helix. From the observed behavior of loops of $(dA \cdot dT)_N$, ζ_b has the following approximate value:

$$\zeta_b \approx \exp(2.3)[Na^+]^{2.0} \tag{27}$$

but appears to approach unity for loops of higher GC content; D represents an empirical stiffness parameter, to which a value of unity is appropriate.

With these values for the various parameters, it is possible to account for three of the five intermediates in Table 4. However, case

IV and V domains frequently show indications of nonequilibrium behavior due to the abrupt increase in translational freedom, similar to that from strand dissociation. The parameter governing strand association is given by:

$$\beta = KN_\alpha \tag{28}$$

where N_α is the length of the terminal domain with α translational degrees of freedom, and K includes all factors independent of N_α, and

$$\beta = \theta_{int} - 3 \tag{29}$$

where θ_{int} represents the average fraction of intact base pairs for DNAs with at least one intact pair and β has a value of ~0.0015 L/mol. An analogous parameter, λ_m, can be defined for the association of residues of a domain situated between two denatured loop regions of m total residue pairs (cases IV and V). The melting behavior of these isolated domains is abrupt, and generally nonequilibrium. The physical meaning of λ_m is similar to that expressed by Scheffler and Baldwin for formation of intramolecular hairpin loops in poly[d(A-T)·d(A-T)], and by Uhlenbeck and Tinoco for associations in single-stranded RNA chains.

The model for denaturation is the canonical one-dimensional lattice with loop entropy, in which linked nucleotide residues exist in either the paired or the unpaired state for each configuration of states. The free energy for the kth configuration is given in the usual way by

$$\Delta G_k = -RT \ln Z_k \qquad (30)$$

where the probability that the chain will assume the kth configuration is Z_k/Z, and Z is the partition function over all configurations. The fraction of residues in the paired state is proportional to the statistical weights,

$$\theta_{bp} = \sum \frac{[N_k Z_k/Z]}{N_{len}} \qquad (31)$$

where N_k is the number of paired residues in the kth configuration and Z_k is the product of the several weighting factors in equation (23).

A considerable effort has been made to evaluate the numerous parameters in the statistical mechanical model, and quantitative agreement with experiment can be achieved for the melting of most domains. Calculated melting curves are obtained with a program that reads in the DNA sequence and, using Poland's recursion formulas, calculates conditional probabilities at each temperature that the $(m - 1)$th residue is in the paired state, given that the mth residue is paired. Following this, the unconditional probabilities that the mth residue is in the paired state are calculated. The unconditional prob-

abilities yield $\theta_{bp}(T)$, and, thereby, $d\theta_{bp}(T)/dT$. The latter is proportional to $dA_{270\ nm}/dT$ of experimental curves.

Computation time is proportional to N_{len}^2. For computational efficiency, Fixman and Freire proposed that $f(N)$ (equation 26) be represented by an exponential series,

$$f(N) = \sum_{i=1,j} a_i \exp(-b_i N) \qquad (32)$$

since computation time becomes proportional to N_{len}. Values for a_i and b_i can be determined in subroutine by quasi-Newton least-squares fit, such that the difference between the summation of exponentials and the exact function is minimized over the range of N_{len} N 1.

The example of both experimental and calculated subtransitions for the insert $[AAGTTGAACAAAAAAT]_{12}AAGTTGAA$ in pN/MCS-22 seen in Figure 6 is magnified in Figure 9. The plasmid was linearized by Kpn I adjacent to the insert, so that the insert melts from the end of the helix (case II, Table 4). Vertical lines in Figure 9 correspond to positions of dissociated base pairs in the plas-

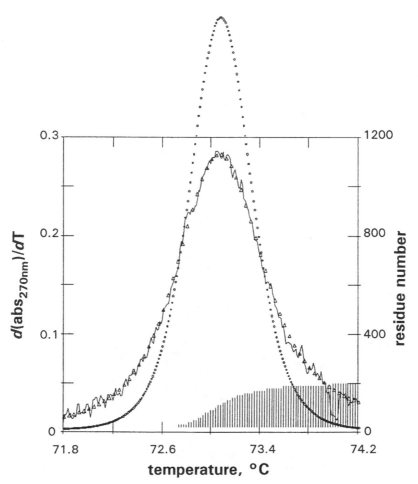

Figure 9. Subtransition for the insert [AAGTTGAACAAAAAAT] $_{12}$AAGTTGAA in pN/MCS-22. Solid line, experimental curve; triangles, statistical–mechanical curve; open circles, a two-state curve, calculated with values for ΔH_{ij}^* and ΔS_{ij}^* from Table 1. The right-hand-side vertical axis represents the denaturation map. The vertical lines denote residue locations in the DNA with probabilities <0.5 of being in the helical state. The map shows the state of the first 1200 bp at the 5′ end of the plasmid.

mid DNA according to the map on the right-hand-side vertical axis. The denaturation map indicates that the insert melts gradually from the end inward, taking 1.5°C for 184 bp of the insert to dissociate. Agreement between the experimental and calculated curves is excellent. This figure includes a curve calculated on the assumption of two-state melting, represented by the small open circles. The obvious agreement between experimental and statistical mechanical curves on the one hand, and the obvious disagreement of these and the two-state curve on the other, simply confirm that melting of the insert is multistate.

4.2   POINT MUTATIONS AND MISPAIR THERMODYNAMICS

Denaturing gradient gel electrophoresis and temperature gradient gel electrophoresis are sensitive methods for detecting small variations in local stability of the duplex. These techniques have the capacity to detect differences in methylation of C residues in otherwise identical DNA specimens. A *perpendicular* DGGE gel obtained by Abrams and Lerman is shown in Figure 10, representing the electrophoretic pattern of four almost identical fragments: the homoduplexes and heteroduplexes of the defined wild type and a point mutant of the 272 bp *Hae* III fragment of the human β-globin gene (cf. Figure 3). The mutation is a single $(A \cdot T) \rightarrow (G \cdot C)$ transition at position 378 of exon 2. The difference between specimens in Figure 10 and those in Figure 3 is that a 30 bp GC-rich tail or

"clamp" was covalently attached to the 5′ end of the β-globin fragment, changing its melting behavior and increasing the sensitivity to small differences in stability over the affected region. This increased sensitivity is explained by examination of the calculated melting curves for clamped and unclamped β-globin fragments in Figure 11. Four melting curves are shown: two for the wild-type and two for the mutant sequence [with the single $(A \cdot T) \rightarrow (G \cdot C)$ transition at position 378], each with and without a 30 bp "clamp" sequence of 65% F_{GC}. Without the clamp, the melting curves break up into two small subtransitions with T_m of 74.67 and 77.66°C. Since the domain with the mutation is the *second* of these two subtransitions, the mutant sequence cannot be distinguished by DGGE (Figure 3). However, the clamp causes both sequences to melt together as one large subtransition, so that mutant and wild type are distinguishable by their T_m and T_e. Thus, the first two closely spaced transitions in the DGGE pattern of Figure 10, at 55.7 and 55.8% denaturant, represent the wild-type and mutant heteroduplexes, respectively. The second two transitions, at 56.8 and 56.9%, represent the homoduplexes.

The occurrence of point mutations implies formation of mispairs as intermediates during replication and repair of genomic DNAs. Donohue argued early on that more than two Watson–Crick base pairs could be accommodated in the double helix. In some cases the structures of these non-Watson–Crick mispairs require little distortion of the duplex, so their occurrence may go undetected, leading

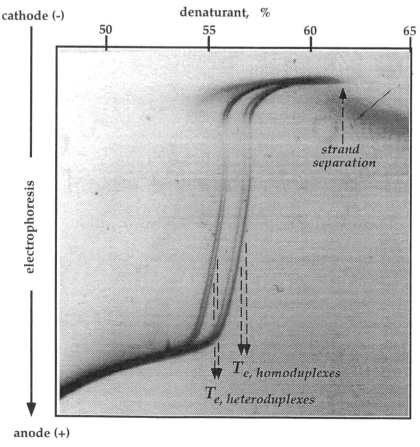

Figure 10.   A perpendicular DGGE profile of the two 323 bp β-globin-containing fragments described in Figure 3, except that in this case a 30 bp GC-rich clamp has been attached to the 5′ end of each fragment. The clamp facilitates the separation of the homo- and heterduplexes, as well as the wild type and the mutant. (Gel picture provided by L. Lerman and E. Abrams.)

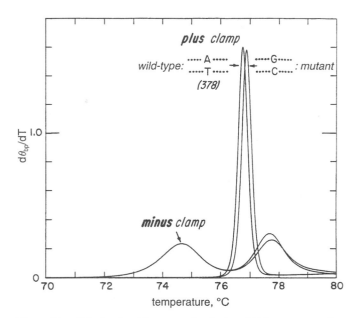

Figure 11. Calculated melting curves for the two 323 bp sequences described in Sections 2.2 and 4.2, with and without the 30 bp clamp sequence. The sodium ion concentration was adjusted to 0.0384 M to correspond to the conditions of the DGGE experiments. [From Abrams et al.., *Genomics*, 7: 463–475 (1990).]

to spontaneous substitutions. An immediate question entails the effect of local DNA structure on mispair stacking. For this reason, a number of recent studies have focused on the stability and structure of non-Watson–Crick mismatches. Other studies indicate that the fidelity of mispair detection is sensitive to the identities of the 3' and 5' neighbor pairs; therefore, the question calls for a systematic investigation of the effects of 16 different nearest neighbors on the stabilities of 10 mismatches. Wartell has approached this question

through the use of temperature gradient gel electrophoresis of many short DNA specimens differing only by the mismatch. The example of a perpendicular TGGE gel pattern of three pairs and two mispairs obtained by Ke and Wartell is shown in Figure 12. This result was obtained on five almost identical 373 bp fragments, differing only by the type of matched or mismatched pair at the fourteenth position from the 5' end. The horizontal temperature gradient was from 17 to 35.5°C. This result indicates stabilities of pairs and mispairs with 5'C·G and 3'A·T neighbors occur in the following order: G·C > C·G > T·A > G·G > C·C. The order for a large number of pairs and mispairs is given in Table 5. There is excellent agreement between relative stabilities of pairs in Table 5, and T_{ij} in Table 1.

4.3 HYBRIDIZATION

Denaturation is necessary for producing hybrids of complementary polynucleotide chains from different source materials, for forming heteroduplexes for Southern blots or detecting evolutionary or mutational differences, and for joining complementary single-stranded ends for cloning DNA fragments into linearized vectors.

4.4 STRINGENCY

Studies of denaturation are required to define the level of stringency for a particular hybridization scheme. "Stringency" refers to a set of conditions that favor kinetic discrimination between two or more competing DNA strands during hybridization. As the temperature or solvent conditions for renaturation (hybridization) approach those for denaturation, hybrids of less perfect alignment are disfavored because of their lower stability. Thus, stringency increases as $(T_m - T) \to 0$, so that chains of lesser sequence similarity are prevented from kinetically competing with more perfect regions of sequence and slowing up the hybridization process. As has been shown, the T_m is not a simple characterization of the stability of (large) DNAs. Therefore, stringency is a qualitative attribute.

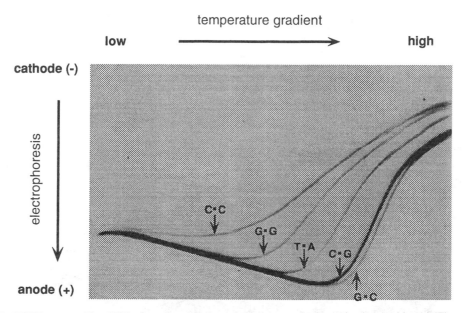

Figure 12. Perpendicular TGGE pattern of five 373 bp fragments differing only by the type of pair or mispair at position 14. [Figure 7 of S.-H. Ke and R. M. Wartell (1993) *Nucleic Acids Res.* 21: 5137–5143 (1993). Original gel photograph kindly provided by R. Wartell.]

Table 5   Relative Stabilities of Watson–Crick Pairs and Mispairs

Nearest-Neighbor Pairs		Relative Stabilities of These Intervening Pairs and Mispairs
5′	3′	
G·C	T·A	C·G > G·C > T·A > A·T >
		G·G > A·G > G·A > G·T > T·G > A·A >
		C·A > T·T ≥ A·C > T·C ≥ C·T > C·C
G·C	G·C	C·G > G·C > T·A > A·T >
		G·A > G·G > T·G > A·G > G·T > A·A =
		T·T > C·A > C·T > A·C > T·C > C·C
C·G	A·T	G·C > C·G > A·T > T·A >
		G·T > G·A > G·G > A·G ≥ T·G > A·A =
		T·T > A·C > C·A, T·C, C·T > C·C
T·A	T·A	G·C > C·G > T·A > A·T >
		G·T > G·G, A·G > G·A, T·G > A·A >
		T·T > A·C > T·C > C·C, C·A > C·T

Source: S.-H. Ke and R. M. Wartell, *Nucleic Acids Res.* 21:5137–5143 (1993).

4.5   Polymerase Chain Reaction

Repeated cycles of denaturation and synthesis with the thermal-stable *Taq* polymerase from *Thermus aquaticus* is the basis for the amplification of minute quantities of DNA by PCR, a simple alternative or complement to cloning.

4.6   Genomic Base Distributions

Denaturation is a sensitive and accurate way to determine the F_{GC} and distribution of base content in genomic DNAs. The distribution of base composition in metazoan DNAs is said to be a complex mosaic of "Bernardi isochores" and satellite families. Distribution in the warm-blooded vertebrates varies among closely related species mainly because of the different amounts and types of satellite repeating elements. A description of the distribution is useful therefore for what it reveals about the overall content and pattern of different sequence elements in the genome, as well as about differences between species. As such, the distribution reflects not only variations in content and characteristics of the active elements, but also something of the fate of those elements, or residual elements that have been discarded or otherwise found redundant during the long evolutionary history of a particular lineage. Distributions can be measured effectively by two methods, density gradient centrifugation and high resolution derivative melting. The latter offers greater intrinsic resolution, since distributions obtained this way reflect the mean F_{GC} of segments only 350 ± 150 bp long, whereas buoyant densities reflect the mean F_{GC} of entire duplex strands of intact DNA.

Figure 13 plots the distribution of F_{GC} in genomic DNAs from

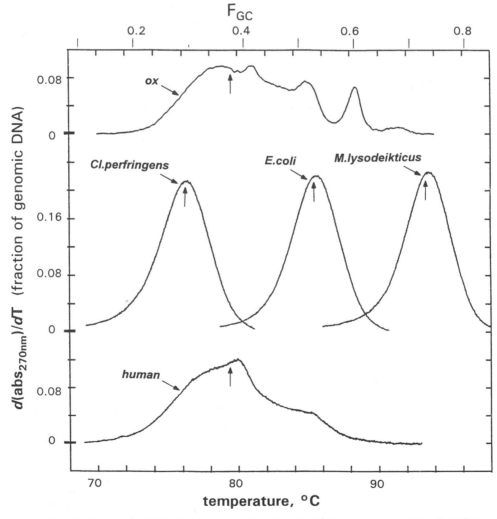

Figure 13.   Denaturation profiles of total genomic DNAs from human, ox (*Bos taurus*), and three bacterial species: *Clostridium perfringens, E. coli,* and *Micrococcus lysodiekticus.*

several species. By measuring these curves at selected wavelengths and normalizing the integrated areas, it is possible to determine the fraction of satellites of particular base contents on the basis of equation (5), the MSD equation. The bacterial curves are almost Gaussian in shape and narrowly distributed, reflecting the highly streamlined genomes expected of rapidly dividing single-celled organisms. Bacterial DNAs are almost 1000 times longer than the plasmid DNA that produced the complex melting curves in Figure 6. Melting curves of bacterial DNAs therefore reflect the superposition of some tens of thousands of subtransitions for domains of F_{GC} that fall into narrow Gaussian distributions. Profiles of metazoan DNAs are far broader and generally exhibit numerous satellitelike bands or shoulders, plainly visible in the ox profile, indicating the presence of repetitive families. The upper x-axis in Figure 9 for the F_{GC} was calculated from the MSD equation. The upward-pointing arrows in Figure 13, which represent temperatures at which integrated areas are 0.5, therefore correspond to overall base compositions. Base compositions obtained by this method are in excellent agreement with values obtained by chemical analysis.

4.7 MEASUREMENTS OF SEQUENCE DIVERGENCE IN REPETITIVE FAMILIES

Measurements of the level of sequence variation in repetitive sequence families can be readily determined from the δT_m for the difference in stability between homo- (unmelted) and hetero- (renatured) duplexes by absorption melting. Mispairs in the heteroduplex population contribute to a lower overall stability. The method is that of Bonner, from the reduction in T_m. The magnitude of δT_m is directly proportional to the frequency of mispairs in heteroduplexes. The currently accepted relationship is a 1.18% divergence in sequence for each 1.0°C lowering of the T_m. This is a semiquantitative relationship. The gradient gel electro-phoresis results described in Section 4.2 indicate that this relationship must depend on the number, type, and nearest neighbors of mispairs in heteroduplexes.

4.8 OTHERS

Denaturation of total genomic DNAs followed by careful measurement of the kinetics of renaturation is used to measure different levels of sequence complexity: the distributions of single copy, middle, and highly repetitive sequence families in genomic DNAs. Denaturation is also required for primer bonding, an important first step in PCR and DNA sequencing.

See also HYDROGEN BONDING IN BIOLOGICAL STRUCTURES; PARTIAL DENATURATION MAPPING.

Bibliography

Thermodynamics of Denaturation

Bloomfield, V. A., Crothers, D. M., and Tinoco, I., Jr. (1974) *Physical Chemistry of Nucleic Acids*. Harper & Row, New York.

Cantor, C. R., and Schimmel, P. R. (1980) *Biophysical Chemistry*, Part III: *The Behavior of Biological Macromolecules*. Freeman, San Francisco.

Fixman, M., and Freier, J. (1977) *Biopolymers,* 16:2693–2704.

Poland, D. (1974) *Biopolymers,* 13:1859–1871.

Wartell, R. M., and Benight, A. S. (1985) Phys. Rep. 126:67.

Electrostatic Effects and Denaturation

Anderson, C. F., and Record, M. T., Jr. (1990) *Annu. Rev. Biophys. Biophys. Chem.* 19:423.

Manning, G. S. (1978) *Q. Rev. Biophys.* 11:179.

Record, M. T., Jr., Anderson, C. F., and Lohman, T. (1978) *Q. Rev. Biophys.* 11:103.

Hybridization Techniques

Erlich, H. A. (1989) *PCR Technology.* Stockton Press, New York.

Hames, B. D., and Higgins, S. J (1985) *Nucleic Acid Hybridization,* IRL Press, Washington, DC.

Howe, C. J., and Ward, E. S. (1989) *Nucleic Acid Sequencing.* IRL Press, New York.

Wetmur, J. G. (1991) *Crit. Rev. Biochem. Mol. Biol.* 26:227–259.

Altered Structures of DNA and Their Detection

Lilley, D. M. J., and Dahlberg, J. E., Eds. (1992) *Methods Enzymology,* Vol. 212b. Academic Press, New York.

DIABETES INSIPIDUS

Walter Rosenthal, Anita Seibold, Daniel G. Bichet, and Mariel Birnbaumer

Key Words

Central Diabetes Insipidus (CDI)   A form of diabetes insipidus caused by inadequate or no release of vasopressin from the posterior pituitary.

Diabetes Insipidus   An acquired or inherited disease characterized by two key symptoms: polyuria and polydipsia.

Nephrogenic Diabetes Insipidus (NDI)   A form of diabetes caused by a resistance of the kidneys to vasopressin.

Vasopressin Receptors   Cell surface receptors mediating cellular responses to vasopressin. Whereas the antidiuretic response is mediated by V_2 receptors, other responses are mediated by V_1 receptors.

Vasopressins   Nonapeptides synthesized in hypothalamic nuclei and released from the posterior pituitary. Their main function is the conservation of water. Vasopressins also contract vascular smooth muscle, stimulate glycogenolysis in liver cells, and enhance the release of corticotropin from the anterior pituitary. *Synonym:* antidiuretic hormone (ADH).

The nonapeptide vasopressin acts via two different cell surface receptors. Whereas the antidiuretic response is mediated by V_2 receptors, other responses are mediated by V_1 receptors. Diabetes insipidus is a disease characterized by two key symptoms: polyuria and polydipsia. Central diabetes insipidus (CDI) is caused by insufficient or no release of vasopressin from the posterior pituitary. Nephrogenic diabetes insipidus (NDI) is caused by resistance of the kidney (lack of responsiveness) toward vasopressin. Acquired and inherited forms of CDI and NDI have been described. Recent work shows that autosomal dominant CDI is caused by defects in the vasopressin precursor gene, that the major cause of X-chromosomal recessive NDI is a defect in the V_2 receptor gene, and that mutations in the water channel aquaporin 2 are responsible for autosomal-dominant NDI. Identification of the genes responsible for these diseases facilitates carrier identification and prenatal diagnosis; it is also a step toward an improved treatment of CDI and NDI patients.

1   VASOPRESSIN AND VASOPRESSIN RECEPTORS

1.1   MAJOR ROLES OF VASOPRESSINS

Vasopressins, also referred to as antidiuretic hormones (ADHs), exert a dual action. With the development of terrestrial life, they have became the main mediator for the conservation of water: in amphibia, they increase the water permeability of skin, nephron, and bladder epithelial cells, whereas in mammals, including man, their primary target is the epithelium lining renal collecting ducts. The second action of vasopressins is to contract smooth muscle—in particular, smooth muscle of the vasculature.

1.2   VASOPRESSIN STRUCTURE, SYNTHESIS, AND RELEASE

Vasopressins are nonapeptides characterized by an amidated C-terminal glycine residue and a disulfide bridge between cysteine residues at positions 1 and 6, which is essential for their biological activity (Figure 1). In all mammals except swine, the nonapeptide 8-arginine vasopressin (AVP) is found; the porcine variant is 8-lysine vasopressin (LVP). The two vasopressins are closely related to oxytocin, from which they differ by only two amino acids and with which they share the ability to contract smooth muscle. Vasotocin, the corresponding hormone in lower vertebrates such as amphibia, reptiles, and birds, is a hybrid composed of the N-terminus of oxytocin and the C-terminus of AVP. Based on comparison of precursor amino acid sequences, vasotocin has been proposed as the ancestral molecule of AVP.

In mammals, relatively large and inactive precursors of vasopressin (and oxytocin) are synthesized in neurons of the hypothalamic supraoptic and paraventricular nuclei. The gene encoding the vasopressin precursor (vasopressin gene) contains three exons (1–3) (Figure 2). In man, it is located on chromosome 20, separated by only 8 kb from the oxytocin precursor gene, which shows a very similar organization. The three exons of the vasopressin gene encode the functional domains of the vasopressin preprohormone consisting of a signal peptide (encoded by exon 1), the hormone (vasopressin; encoded by exon 1), a peptide essential for the axonal transport of the hormone (neurophysin II; encoded by exons 1–3), and a C-terminal glycopeptide of unknown function (encoded by exon 3). The signal peptide is removed in the endoplasmic reticulum, yielding the prohormone. Further cleavage occurs in neurosecretory granules during axonal transport to the posterior pituitary gland, from which vasopressin and neurophysin II are released by an exocytotic pathway. In man, the release is triggered by minute (2%) increases in plasma osmolality (hypernatremia) that are sensed by peripheral and central osmoreceptors (osmotic regulation). The release of vasopressin is also stimulated by a more pronounced (10%) decrease in extracellular fluid volume (hypovolemia), sensed by cardiovascular baroreceptors (nonosmotic regulation). Plasma

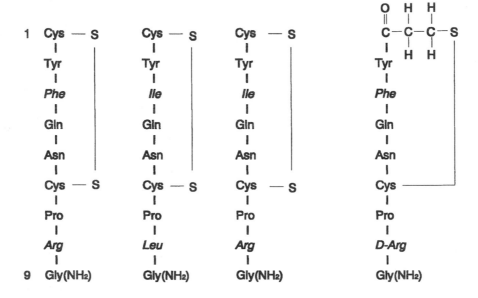

Figure 1.   Structures of arginine–vasopressin (AVP), oxytocin, vasotocin, and the synthetic vasopressin analogue 1-deamino-8-D-arginine vasopressin (dDAVP, desmopressin).

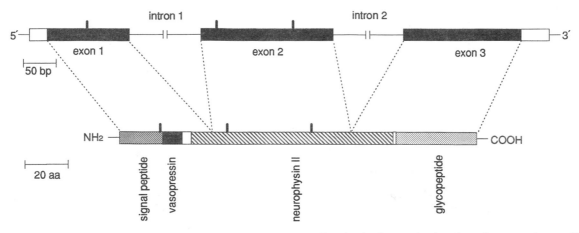

Figure 2.   Structure of the vasopressin precursor gene and the vasopressin precursor. Translated and untranslated portions of exons are shown as black or open bars, respectively. Introns are depicted as interrupted horizontal lines. Intron 1 is 1374 bp long; intron 2 165 bp. The various filled portions of the precursor indicate the different peptides; open portions are spacer peptides. Corresponding regions of the gene and the peptide are assigned to each other by dotted lines. The sites of three mutations found in patients with autosomal dominant CDI are indicated in the gene and the peptide (bold vertical lines).

levels of AVP range from below 1 pg/mL to 50 pg/mL. The biological half-life is 30–40 minutes.

1.3   VASOPRESSIN RECEPTORS AND TRANSMEMBRANE SIGNALING

In correspondence to its dual action, vasopressin acts via V_1 (subtypes V_{1a} and V_{1b}) and V_2 cell surface receptors, which bind the hormone with high affinity ($K_D \sim 1$ and 5 nM, respectively). The amino acid sequences of these receptors were derived from cloned cDNA. Hydropathy profiles based on these sequences are consistent with seven transmembrane domains (Figure 3), a characteristic feature of the large group of G-protein-coupled receptors. Within this category, vasopressin receptors together with the oxytocin receptor form a subfamily; well-conserved regions include the transmembrane domains and the extracellular loops (E2, E3, E4).

Sequencing of the human gene for the V_2 receptor revealed the existence of three expressed exons (Figure 4). The first exon contains the start codon; it is separated from the second exon by a 360 bp intron that interrupts codon 9 (positions of introns are indicated by arrowheads in Figure 3). The second exons encodes approximately 80% of receptor sequence. The second intron interrupts codon 304, located at the junction of the third extracellular loop and the seventh transmembrane domain (see Figure 3). Exon 3 encodes the seventh transmembrane domain and the C-terminus; it also contains 406 bp of untranslated region terminating at the polyadenylation splice signal AATAAA. A putative TATA box is found 100 bp upstream from the predicted translation start.

V_1 and V_2 receptors activate different signal transduction pathways (Figure 5). Like the receptor for oxytocin, the V_1 receptor stimulates phospholipase C via a G protein. In hepatocytes and possibly in other cell types, the involved G protein belongs to the G_q family; members of this group are not sensitive to ADP-ribosylating bacterial toxins. Phospholipase C-induced formation of inositol 1,4,5-trisphosphate and diacylglycerol causes a release of Ca^{2+} from intracellular stores (increase in cytosolic Ca^{2+}) and

activation of protein kinase C, respectively. V_{1a} receptors are found on hepatocytes and smooth muscle cells, mediating vasopressin-induced stimulation of glycogenolysis and contraction, respectively. In fact, at concentrations higher than those required for antidiuresis, vasopressin is a very potent vasoconstrictor and causes an increase in peripheral resistance and diastolic blood pressure. V_{1b} receptors are found on corticotropic cells of the anterior pituitary. Upon stimulation, they promote the release of corticotropin, which is also controlled by other factors, in particular by its own releasing factor. Vasopressin reaching the anterior pituitary is released from fibers that originate from neurons in the paraventricular nucleus and terminate in the median eminence. From there the hormone is transported to the anterior pituitary by the hypophyseal portal system.

V_2 receptors couple to adenylyl cyclase via the stimulatory G protein G_S, which is covalently modified (ADP-ribosylated) by cholera toxin. Receptor activation leads to an increase in cellular cAMP and subsequent activation of cAMP-dependent protein kinase. In mammals, the V_2 receptor is found in the basolateral membrane of epithelial cells lining the renal collecting duct and also—to varying degrees from species to species—in epithelial cells lining the medullary thick ascending loop of Henle. The final event following the activation of the V_2 receptor in the collecting duct is the fusion of preformed water channel-(aquaporin Z-) containing vesicles with the apical membrane. The insertion of these water channels into the apical plasma membrane dramatically increases its water permeability and allows the transfer of fluid from the tubular lumen to the interstitium. This is clearly the major mechanism by which vasopressin exerts its antidiuretic effect. By a similar mechanism, vasopressin increases the water permeability of skin, nephron, and bladder epithelial cells in amphibia. Stimulation of V_2 receptors located in the ascending loop of Henle causes an increase in the activity of the $Na^+:K^+:2Cl^-$ cotransporter and thereby an increase in salt transport across the epithelium. As a consequence, the tubular fluid is diluted, and the osmolality of the medullary interstitium is increased.

Although a direct demonstration of extrarenal V_2 receptors is lacking, evidence for their existence is provided by two sets of ob-

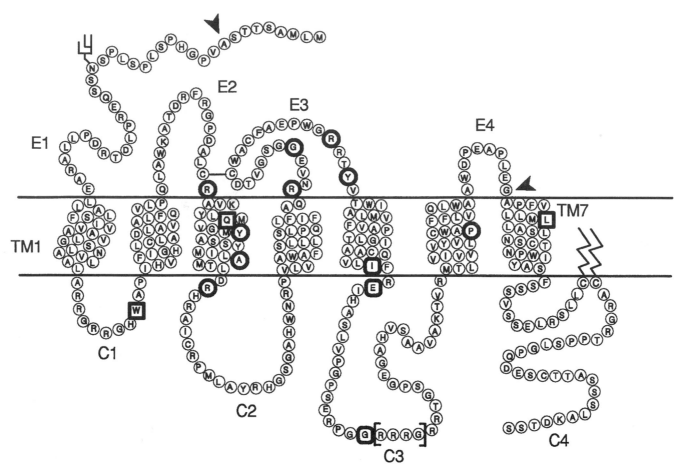

Figure 3. A secondary structure model of the human V$_2$ vasopressin receptor. The depicted membrane topology is consistent with the hydropathy profile of the polypeptide chain of the V$_2$ receptor and assumed to be analogous to that of bacteriorhodopsin, a bacterial light-driven proton pump, and to that of the vertebrate photoreceptor (rhodopsin). Similar models have been proposed for V$_1$ receptors and the oxytocin receptor. Predicted amino acids are given in the one-letter code; E1–E4, extracellular domains; C1–C4, intracellular domains; transmembrane regions (TM1–TM7) are counted from left to right. Arrowheads indicate exon–intron junctions. Based on the presense of consensus regions for modifications in the amino acid sequence, the model depicts a sugar residue at asparagine 22 of the extracellular C-terminus, a disulfide bridge between cysteines 142 and 192 of the first (E2) and second extracellular loop (E3), respectively, and fatty acid residues at cysteines 341 and 342, attaching the intracellular C-terminus to the plasma membrane. The first 16 mutations found in patients with X-linked NDI are indicated by bold circles (missense mutations), bold squares (nonsense mutations), rounded squares (deletion or insertion causing a frameshift and premature stop), and brackets (inframe deletion). For more details on these mutations, see Section 4.3 and Table 2.

servations made with a synthetic analogue of vasopressin, desmopressin (1-deamino[8-D-arginine]vasopressin), a selective V$_2$ receptor agonist:

1. In various species including man, monkey, dog, and rat, desmopressin (or the combination of vasopressin plus a se-

lective V$_1$ receptor antagonist) causes a decrease in blood pressure and peripheral resistance (vasodilatory response).

2. In humans, monkeys, and dogs, desmopressin releases coagulation factor VIIIc (antihemophilic factor) and von Willebrand factor, which facilitates platelet adhesion and serves as the plasma carrier for factor VIIIc (coagulation response).

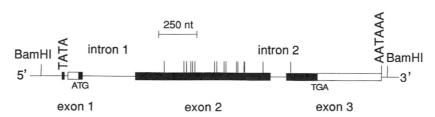

Figure 4. Structure of the human V$_2$ receptor gene: this 2.2 kb fragment obtained with the restriction enzyme BamHI contains the transcriptional unit for the human V$_2$ receptor. Translated and untranslated portions of exons are shown as black or open bars, respectively. The three exons frame two introns, which are depicted as lines. The start codon (ATG), the stop codon (TGA), the putative TATA box and the polyadenylation splice signal (AATAAA) are shown. Vertical lines indicate 16 mutations found in patients with X-chromosomal recessive NDI (see Figure 3). For more details on mutations see Section 4.3 and Table 2.

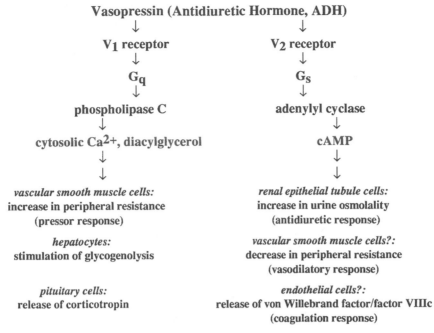

Figure 5. Transmembrane signaling by vasopressin and cellular responses.

These findings are consistent with the existence of V_2 receptors on vascular smooth muscle cells, endothelial cells (which synthesize von Willebrand factor), and—assuming that the increase in factor VIIIc is not secondary to the increase in von Willebrand factor—with their existence on hepatocytes (which synthesize factor VIIIc). Alternatively, the coagulation response may be accomplished by stimulation of V_2 receptors on blood cells (lymphocytes) and subsequent intercellular signaling. Owing to its capacity to increase plasma levels of von Willebrand factor and factor VIIIc, desmopressin is in extensive use for the diagnosis, prevention, and treatment of various bleeding disorders.

2   CLINICAL MANIFESTATIONS OF DIABETES INSIPIDUS

The main symptoms of diabetes insipidus are polyuria, excessive thirst, and polydipsia. Insufficient fluid intake causes dehydration and a high serum osmolality (hypernatremia). Affected human adults pass large volumes (2.5–30 L/day) of hypoosmotic urine (osmolality < 290 mmol/kg). If the urine output is very high (>15 L/day), the patient's life is ruled by frequent drinking of large amounts of fluid and by micturition every 30–60 minutes throughout day and night. Mild forms of the disease may not cause major inconveniences for the patient and may therefore remain undiagnosed. The disease has been described not only in man but also in dogs, rats, and mice.

3   CENTRAL DIABETES INSIPIDUS (CDI)

3.1   Pathophysiology and Etiology

CDI, also referred to as neurogenic or neurohypophyseal diabetes insipidus, is caused by inadequate or no release of vasopressin. As a consequence, vasopressin is low or not detectable in peripheral blood even after water deprivation. However patients respond normally to administered vasopressin or vasopressin analogues. Thus

CDI can be treated by hormone replacement therapy. The drug of choice is desmopressin, which is routinely used intranasally. It has a long-lasting antidiuretic activity and almost completely lacks pressor effects. For other clinical uses of desmopressin, see Section 1.3.

Frequent causes of the disease are brain tumors, pituitary or hypothalamic surgery, and severe head injuries. CDI of unknown cause (idiopathic CDI) is also common; it usually becomes apparent in childhood or early adult life. Compared to the acquired or idiopathic CDI, familial CDI is rare. Similar to idiopathic CDI, the familial form does not manifest itself immediately after birth. Patients may remain free of symptoms for months or years; they typically present with symptoms in their childhood or teens.

3.2   Molecular Genetic Defects in Autosomal Dominant CDI

Research for the molecular defect causing human familial CDI was inspired by an animal model, the Brattleboro rat, which was shown to have a single base deletion in the region of the vasopressin gene coding for neurophysin II (see Figure 2). The mutation causes a frameshift resulting in a mutated C-terminus of the precursor. Although the N-terminal two-thirds of the predicted mutant protein (including the hormone moiety) is identical to this portion of the wild type, the mutant protein is retained in the endoplasmic reticulum, and vasopressin is not detectable in the neurohypophysis of homozygous animals.

In contrast to the autosomal recessive genetic mode of inheritance of CDI in the Brattleboro rat, human familial CDI is an autosomal dominant trait. Based on genetic linkage analysis of restriction fragment length polymorphism (RFLP) haplotypes, a defect in the vasopressin gene was also suggested as the cause for the human disease. This assumption was proven by the identification of different missense mutations in the vasopressin gene of CDI patients (Table 1; see Figure 2). To obtain sufficient amounts of DNA for se-

Table 1   Mutations in the Human Vasopressin Gene Associated with Autosomal Dominant Central Diabetes Insipidus

Change in Nucleotides[a]	Type of Mutation	Exon / Moiety in Precursor	Predicted Change in Amino Acid[b]	Reference
G → A at 278	Missense	1 / signal peptide	Thr → Ala at 19	Kovacs et al., 1992; Rittig et al., 1993
G → T at 1740	Missense	2 / neurophysin II	Gly → Val at 17	Bahnsen et al., 1992
C → T at 1761	Missense	2 / neurophysin II	Pro → Leu at 24	Repaske and Browning, 1994
G → A at 1859	Missense	2 / neurophysin II	Gly → Ser at 57	Ito et al., 1991

[a]Nucleotide numbering is according to the reported sequence of the human vasopressin gene: E. Sauville, D. Carney, and J. Battey: The human vasopressin gene is linked to the oxytocin gene and is selectively expressed in cultured lung cancer cells. *J. Biol. Chem.* 260:10236–10241 (1985).
[b]Numbers refer to the codon of the individual peptides encoded on the vasopressin gene.

References

Bahnsen, U., Oosting, P., Swaab, D. F., Nahke, P., Richter, D., and Schmale, H. (1992) A missense mutation in the vasopressin–neurophysin precursor gene cosegregates with human autosomal dominant neurohypophyseal diabetes insipidus. *EMBO J.* 11:19–23.
Ito, M., Mori, Y., Oiso, Y., and Saito, H. (1991) A single base substitution in the coding region for neurophysin II associated with familial central diabetes insipidus. *J. Clin. Invest.* 87:725–728.
Kovacs, L., McLeod, J. F., Rittig, S., Gaskill, M. B., Bradley, G., Cox, N., and Robertson, G. L. (1992) A single base substitution in exon 1 of the vasopressin–neurophysin II gene is linked to familial neurogenic diabetes insipidus in a Caucasian kindred. *Am. J. Hum. Genet.* 51:A30.
Repaske, D. R., Browning, J. E. (1994) A *de novo* mutation in the coding sequence for neurophysin-II (Pro(24) - Leu) is associated with onset and transmission of autosomal dominant neurohypophyseal diabetes insipidus. *J. Clin. Endocrinol. Metab.* 79:421–427.
Rittig, S., Kovacs, L., Gregersen, N., Robertson, G. L., and Pederson, E. B. (1993) Familial neurogenic diabetes insipidus in 5 Danish kindreds. Abstract 31, Fourth International Vasopressin Conference, Berlin, May 23–27.

quencing experiments, genomic DNA obtained from white blood cells of patients was used to clone the mutant gene or to amplify portions of the mutant gene by applying polymerase chain reaction (PCR) technology. In two affected members of a Japanese family, an G → A transition was found, which results in a substitution of Gly 57 of neurophysin II for Ser. In a Dutch family, a G → T transversion, converting Gly 17 of neurophysin II to Val, cosegregates with the disease. Moreover, a G → A transition was identified in patients but not in healthy members of a Caucasian (presumably North American) family; the mutation replaces the C-terminal amino acid of the signal peptide, Ala 19, with Thr. The same mutation was found in two other kindreds, one Danish and one Japanese. More recently, a C → T transition, substituting Leu for Pro 24 of neurophysin II, was shown to be associated with onset and transmission of CDI within a family (not shown in Figure 2).

At present it is not clear how the genetic defect of one allele causes CDI. As the patients are heterozygous for the mutation, one expects that the wild-type precursor is also formed in their hypothalamic nuclei. In fact, normal amounts of vasopressin have been detected in the plasma of patients with autosomal dominant CDI before the onset of polyuria and polydipsia. In analogy to the findings in the Brattleboro rat, the mutant precursors found in patients may accumulate in the hormone-producing neurons, as a consequence of improper processing. The intracellular protein deposits may impair cell function and eventually cause cell death. Especially the G → A transition at nucleotide 278, which causes a change in the amino acid sequence at the cleavage site between the signal peptide and the vasopressin moiety, very likely interferes with processing of the precursor. The proposed mechanism is also consistent with the delayed onset of the disease.

4   NEPHROGENIC DIABETES INSIPIDUS (NDI)

4.1 Pathophysiology and Etiology

NDI (or peripheral diabetes insipidus) is a typical example of a disease caused by an end-organ resistance. In contrast to CDI patients, NDI patients show normal or even elevated plasma levels of vasopressin and an adequate increase of vasopressin levels upon dehydration. However the kidney fails to respond to the endogenous vasopressin, to administered hormone, or to hormone analogues.

NDI falls into two categories: acquired or hereditary. A common cause for the acquired form is Li$^+$ used for the treatment of bipolar manic–depressive illness. Other drugs that antagonize the antidiuretic actions of vasopressin include certain antibiotics (demethyltetracyclines, e.g., demeclocycline) and volatile fluorocarbon anesthetics (e.g., methoxyflurane). Acquired NDI may also occur in association with systemic diseases like amyloidosis, Sjögren syndrome or sarcoidosis. Similar to the familial form of CDI, hereditary NDI is generally a rare disease.

There is no causal therapy for NDI. It is most important to encourage patients to drink plenty of fluid and to advice parents not to restrict fluid intake of their affected children. Paradoxically, commonly used diuretic drugs (thiazides) reduce polyuria up to 50%; a frequently encountered side effect is hypokalemia. NDI induced by Li$^+$ may be successfully treated with amiloride, a potassium-sparing diuretic. Anti-inflammatory drugs like indomethacin may also reduce polyuria, probably by inhibiting prostaglandin synthesis in the kidney. As one would expect, NDI patients receiving a kidney transplant for renal end-stage failure are "cured."

4.2 Prevalence and Specific Features of X-Chromosomal Recessive NDI

X-Chromosomal NDI appears to occur worldwide. Generally, however, it is a rare disease. In Quebec, the prevalence of the disease is estimated to be 3.7 per million males. Yet in other defined regions in North America, the prevalence is considerably higher. An extreme example is an area on Nova Scotia, which includes two villages. Among the 2500 inhabitants, 30 male patients have been diagnosed. It is assumed that the affected families are descendants of the so-called Ulster Scot immigrants, some of whom arrived aboard the ship *Hopewell* at Halifax, Nova Scotia, in 1761. According to the "*Hopewell* hypothesis," most North American patients suffering from the disorder are progeny of an Ulster Scot ancestor.

In contrast to the familial form of CDI, congenital NDI shows an X-chromosomal recessive mode inheritance about 90% of affected families, and the concentration defect of the kidney can be demonstrated shortly after birth. However polyuria and polydipsia often remain undiagnosed in babies and, later in infancy, may be regarded as a nuisance or misbehavior. Undiagnosed babies and young children are commonly supplied with insufficient and often—for educational purposes—restricted amounts of fluid. As a consequence, repeated episodes of severe dehydration, especially if they occur during the first years of life, frequently lead to mental retardation, hypocaloric dwarfism, or even death. If patients are provided with unrestricted amounts of water from birth on, both development and life expectancy are normal. Usually heterozygous females do not present with symptoms; only a few are severely affected. Thus upon biochemical testing (administration of desmopressin), a wide range of renal and extrarenal responses—ranging from normal to absent—may be observed. Obligatory carriers are identified by definition as mothers and daughters of affected males. Rather than by biochemical testing, nonobligatory carriers (sisters of affected male patients or of obligatory carriers) may be identified by genetic linkage analysis or by sequence analysis of the gene responsible for X-chromosomal recessive NDI (see Section 4.3).

Some patient observations have been helpful in the search for the gene responsible for X-linked NDI.

1. V_1 receptor-mediated responses (pressor responses) to vasopressin or vasopressin analogues are normal.
2. With few exceptions, patients lack not only renal but also extrarenal responses to desmopressin, indicating that the disorder is not caused by a defect of a protein specifically expressed in the kidney (e.g., the water channels in renal epithelial cells).
3. Patients respond to epinephrine and parathyroid hormone (but not to vasopressin and its analogues) with an increase in plasma cAMP concentration and urinary cAMP excretion, respectively. Thus genetic defects in ubiquitous signal transduction components (the G_S–adenylyl cyclase system; see Figure 5) as they occur in pseudohypoparathyroidism type Ia are not likely.

All the foregoing observations point to a defect in the V_2 receptor gene.

4.3 Molecular Genetic Defect in X-Chromosomal Recessive NDI

Analysis of RFLPs in North American and European NDI families revealed a linkage between haplotypes for marker alleles of Xq28, the subtelomeric portion of the long arm of the human X chromosome, and the disease. In particular, highly significant (logarithm of odds) lod scores were obtained with marker allele DXS52 and probe St14-1. Based on these findings, the gene responsible for NDI was assigned to Xq28; its function, however, remained unknown. More recently, the human V_2 receptor gene was mapped to a region of 1 million bp within Xq28, between the locus for the glucose 6-phosphate dehydrogenase gene (distal of the V_2 receptor gene) and DXS52 (proximal of the V_2 receptor gene). This region also contains the genes for red and green color vision. Although the data show that DXS52 and V_2 receptor gene do not reside on the same locus, they are consistent with the assumption that the V_2 receptor gene is identical with the gene responsible for X-linked NDI.

Only few months passed between the molecular cloning of the human V_2 receptor and the identification of mutations in the V_2 receptor gene, segregating with the clinical phenotype in NDI families (Table 2). This rapid development was possible by applying PCR technology: genomic DNA was isolated from blood samples (leukocytes) of patients, and the entire V_2 receptor gene or a portion of the gene was amplified by PCR. The PCR products were sequenced directly or following subcloning into a plasmid. In some instances, the difference in the electrophoretic mobility of relatively short stretches of single-stranded DNA (PCR products) that differ in as little as a single base was exploited to screen for mutations (single-stranded conformational polymorphism: SSCP); subsequently, the mutation was identified by sequencing. Within a year, 16 mutations associated with congenital NDI had been reported in the literature (see Table 2). In the meantime, approximately 80 other mutations have been discovered. They were found in patients from various continents (America, Africa, Asia, Europe) and diverse ethnic groups. The mutations were present in affected males but not in healthy male siblings. Mothers of patients were shown to be heterozygous and thereby identified as carriers of the mutant gene.

Mutations are scattered throughout the coding region of the V_2 receptor gene. Taking into account new mutations (not shown in Table 2 or Figure 4), there is no convincing evidence for clusters. The only evidence for sites in the receptor gene prone to mutations stems from the observation that some mutants have been found in two or more unrelated families. Examples are mutations in the codons 113 (Arg), 167 (Ser) and 181 (Arg) which were found in 3, 7, and 3, respectively, unrelated families.

The most common mutations are single base substitutions (transitions or transversions) which give rise to missense mutations. All functionally analyzed missense mutations have been shown to abolish or severely impair receptor function, providing biochemical proof for a causal relationship between the mutation and the clinical phenotype. Several mutations leading to an exchange of amino acids in the extracellular loops (which form the binding pocket for the hormone) decrease the affinity for vasopressin (mutant R113W) or cause a complete loss of binding capacity (mutant R181C). Mutations causing an exchange of single amino acids in the cytoplasmic domains of the receptor protein may affect the interaction of the agonist-occupied receptor with the G_S/adenylyl cyclase system. For example, the R137H mutant shows wild type properties in terms of vasopressin binding, but is unable to stimulate the G_S/adenylyl cyclase system in response to even high doses of vasopressin. The exchange of amino acids in transmembrane domains may affect various parameters such as hormone binding (e.g., mutants Y128S and P286R) or delivery of the receptor to the cell surface (e.g., mutants S167L and L44P). Recent evidence suggests that mutants essentially not expressed on the cell surface are not properly processed and accumulate in a pre-Golgi-compartment, where they are degraded. Many mutations seem to affect more than one parameter. In particular, a decreased cell surface expression is frequently observed besides true "functional" defects. For example, the R113W mutant shows a reduced affinity for vasopressin, a diminished activity to stimulate the G_S/adenylyl cyclase system and a dimished ability to be expressed at the cell surface. Thus the clinical phenotype may be determined by a combination of defects of the receptor protein.

Nonsense mutations lead to the expression of a truncated protein that is very unlikely to be active. An extreme example is the W71X mutation. The predicted mutant protein consists of the extracellular N-terminus, transmembrane domain 1 (TM1) and the first cytosolic loop (C1), comprising together just 20% of the amino acids of the wild-type receptor. With the exception of one inframe deletion (810del12), the identified deletions or insertions cause a frameshift and introduce a premature stop codon. The predicted mutant pro-

Table 2   The First 16 Mutations Described in the Human V_2 Receptor Gene Associated with X-Chromosonal Recessive Nephrogenic Diabetes Insipidus

V_2 Receptor Mutant[a,b,c,d]	Type of Mutation	Change in Nucleotides[b]	Predicted Change in Amino Acid	Predicted Protein Domain	References
W71X*	Nonsense	G → A at 284	Trp → stop at 71	C1	Bichet et al., 1993
R113W*	Missense	C → T at 408	Arg → Trp at 113	E2	Bichet et al., 1993; Holtzman et al., 1993
Q119X*	Nonsense	C → T at 426	Gln → stop at 119	TM3	Pan et al., 1992
Y128S*	Missense	A → C at 454	Tyr → Ser at 128	TM3	Pan et al., 1992
A132D[†]	Missense	C → A at 466	Ala → Asp at 132	TM3	Rosenthal et al., 1992
R137H*	Missense	C → A at 481	Arg → His at 137	C2	Bichet et al., 1993; Rosenthal et al., 1993
(1) R181C*	(1) Missense	(1) C → T at 612	(1) Arg → Cys at 181	(1) E3	Pan et al., 1992
(2) 810del12*	(2) Inframe deletion	(2) deletion of 12 bp 3′ to 810	(2) Deletion of Arg 247-Arg 248– Arg 249-Gly 250	(2) C3	
G185C[#]	Missense	G → T at 624	Gly → Cys at 185	E3	van den Ouweland et al., 1992
R202C[#]	Missense	C → T at 675	Arg → Cys at 202	E3	van den Ouweland et al., 1992
Y205C[#]	Missense	A → G at 685	Tyr → Cys at 205	E3	van den Ouweland et al., 1992
755insC[$]	Frameshift	Insertion of C in codon 228 (nucleotides 753–755)	Frameshift 3′ to codon 228; codon 258 → stop	TM5 and C3	Merendino et al., 1993
763delA*	Frameshift	Deletion of A at 763	Frameshift 3′ to codon 231, codon 270 → stop	C3	Pan et al., 1992
804dX*	Frameshift	Deletion of G between 804 and 809	Frameshift 3′ to codon 247; codon 270 → stop	C3	Rosenthal et al., 1992
P286R[§]	Missense	C → G at 928	Pro → Arg at 286	TM6	Pan et al., 1992
L312X*	Nonsense	T → A at 1006	Leu → stop at 312	TM6	Bichet et al., 1993

[a]In case of single-base substitution, mutants are named according to the amino acid change (one-letter code) at the indicated codon: X, termination at the indicated codon. In case of a deletion (del), the nucleotide number is given first. Following "del" is the total number of deleted nucleotides or—in case of a single-base deletion—the base. For the single-base insertion (755insC), nucleotide 755 (C) may precede or follow the inserted base (C).

[b]Nucleotide numbers are given according to the sequence numbering of GenBank entry Z11687 in which Met 1 corresponds to nucleotides 72–74.

[c]Origin of families: *, North American family; [†], Iranian family; [#], Dutch family; [$], Lithuanian family; [§], El Salvadoran family.

[d]Mutations preceded by (1) and (2) were found in the same patient; only the missense mutation (1) seems to be responsible for NDI.

References

Bichet, D. G., Arthus, M.-F., Lonergan, M., Hendy, G. N., Paradis, A. J., Fujiwara, T. M., Morgan, K., Gregory, M. C., Rosenthal, W., Antaramian, A., Didwania, A., and Birnbaumer, M. (1993) X-linked nephrogenic diabetes insipidus in North America and the Hopewell hypothesis. *J. Clin. Invest.* 92:1262–1268.

Holtzman, E. J., Harris, H. W., Kolakowski, L. F., Guay-Woodford, L. M., Botelho, B., and Ausiello, D. A. (1993) Brief report: A molecular defect in the vasopressin V2–receptor gene causing nephrogenic diabetes insipidus. *New Engl. J. Med.* 328:1534–1537.

Merendino, J. J., Spiegel, A. M., Crawford, J. D., O'Carroll, A. M., Brownstein, M. J., and Lolait, S. L. (1993) Brief report: A mutation in the vasopressin V2–receptor gene in a kindred with X-linked nephrogenic diabetes insipidus. *New Engl. J. Med.* 328:1538–1541.

Pan, Y., Metzenberg, A., Das, S., Jing, B., and Gitschier, J. (1992) Mutations in the V2 vasopressin receptor gene are associated with X-linked nephrogenic diabetes insipidus. *Nature Genet.* 2:103–106.

Rosenthal, W., Antaramian, A., Gilbert, S., and Birnbaumer, M. (1993) Nephrogenic diabetes insipidus: A V2 vasopressin receptor unable to stimulate adenylyl cyclase. *J. Biol. Chem.* 268:13030–13033.

Rosenthal, W., Siebold, A., Antaramian, A., Lonergan, M., Arthus, M.-F., Hendy, G. N., Birnbaumer, M., and Bichet, D. G. (1992) Molecular identification of the gene responsible for congenital nephrogenic diabetes insipidus. *Nature* 359:233–235.

van den Ouweland, A. M. W., Dreesen, J. C. F. M., Verdijk, M., Knoers, N. V. A. M., Monnens, L. A. H., Rocchi, M., and van Oost, B. A. (1992) Mutations in the vasopressin type 2 receptor gene (*AVPR2*) associated with nephrogenic diabetes insipidus. *Nature Genet.* 2:99–102.

teins contain nonreceptor sequences 3′ of the site of mutation and are truncated to varying degrees. Similar to nonsense mutations, frameshift mutations normally abolish the function of a protein.

Of particular interest is the diversity of mutations found in North American NDI families (see Table 2). The W71X mutation was only found in the *Hopewell* kindred and in four satellites families that share the RFLP haplotype with the *Hopewell* progeny. Although this mutant causes NDI in a very large North American kindred, it does not explain the origin of the disease in many other North American families. Thus the *Hopewell* hypothesis proposing a common ancestor for most North American NDI patients ("founder effect") needs to be revised. Identification of a mutation responsible for NDI in a large pedigree [e.g., the *Hopewell* pedigree (W71X) or the Utah

pedigree (L312X)] will also be useful to elucidate whether the clinical phenotype—in particular, the severity of the disease—is determined mainly by the mutation in the V_2 receptor gene or whether other factors contribute to it.

4.4   ATYPICAL FORMS OF FAMILIAL NDI AND ANIMAL MODELS

The present data indicate that familial NDI will be most often ascribable to a defect in the V_2 receptor gene. As mentioned earlier (Section 4.2), patients with X-linked NDI typically lack extrarenal responses to desmopressin. In less than 10% of patients with congenital NDI, however, coagulation or vasodilatory responses to desmopressin are preserved. This group comprises both male and female patients, and the disease shows an autosomal recessive mode of inheritance. It was shown recently that this atypical form of congenital NDI is caused by mutations in the gene encoding the vasopressin-regulated water channel, aquaporin-2, which is located on chromosome 12. Three patients who are homozygous for a missense mutation and a patient shown to be compound heterozygous for two mutations have been described in the literature. About five other mutations have been presented at meetings. The few existing reports on an autosomal dominant form of NDI fail to provide convincing evidence for a male-to-male transmission, which would exclude an X-linked mode of inheritance.

Whereas the molecular defect causing human autosomal dominant CDI has been shown to correspond to that causing autosomal recessive CDI in the Brattleboro rat (see Section 3.2), a human counterpart of an animal model for hereditary NDI, the mouse strain DI +/+ severe, has not been found. In these mice, a constitutively activated cAMP–phosphodiesterase type III prevents a hormonal increase in cAMP content in medullary collecting ducts. Both the biochemical defect and the clinical symptoms (polyuria and polydipsia) can be corrected by administration of rolipram and cilostamide, which are cAMP–phosphodiesterase type III specific inhibitors. Although rolipram was not effective in two patients with X-linked NDI, the possibility cannot be excluded that genetic or somatic mutations in a gene encoding a protein involved in cAMP metabolism in the kidney are an additional cause of congenital NDI.

Familial (presumably X-linked) NDI has also been described in dogs. The affinity of V_2 receptors for vasopressin in kidney membranes from NDI dogs (huskies) was 10-fold lower than the affinity in kidney membranes from normal dogs. The NDI phenotype could be reversed by treatment of animals with V_2-specific agonists given at a very high dose. The data are not only consistent with a defect in the V_2 receptor gene in dogs but may also have implications for human patients with a defect in the same gene. Expression in mammalian cells and subsequent functional characterization of V_2 receptor mutants causing NDI in humans should identify those that show a moderate decrease in the affinity for vasopressin without an impaired intrinsic ability to activate the G_S–adenylyl cyclase system. This may apply to a small number of patients, since NDI patients generally do not show a renal response to desmopressin, despite the fact that the routinely administered test dose results in blood levels much higher than those of endogenous vasopressin. It remains possible that patients expressing mutant receptors with a reduced or no affinity for both vasopressin and desmopressin can be successfully treated with (non-peptide) agonists binding to domains of the receptor other than these two peptides.

5   OUTLOOK

No disease or symptom has been associated with a defect in the V_1 receptor. Patients with central diabetes insipidus are adequately treated with desmopressin, a selective V_2 receptor agonist. This shows that in man, stimulation of V_1 receptors is not essential for the control of blood pressure, glucose metabolism, and release of corticotropin but may rather contribute to the fine-tuning of these parameters. It seems therefore possible that a defect in the V_1 receptor does not result in a clinical phenotype. Likewise, putative extrarenal V_2 receptors (see Sections 1.3 and 4.2) are apparently not crucial for the physiological control of blood pressure or a normal coagulation response. This assumption is based on the finding that patients with X-linked NDI do not show an apparent defect in blood pressure control or hemostasis, although they lack vasodilatory and coagulation responses to selec-

tive V_2 receptor agonists. Therefore the only known essential physiological role for vasopressin in man and possibly other mammals is the conservation of fluid. However, the V_1 receptor-mediated strong constriction of blood vessels in the gut (splanchnic vasoconstriction) has secured vasopressin a place in the treatment of life-threatening variceal bleeding (e.g., from eso-phageal varices), a sequel of liver cirrhosis and portal hypertension. In addition, extrarenal V_2 receptors appear to be the target for desmopressin and other V_2 receptor agonists used for the diagnosis, prevention, and treatment of various hemostatic disorders (e.g., von Willebrand disease and moderately severe hemophilia). Thus both V_1 receptors and extrarenal V_2 receptors are of great pharmacological significance.

The mutations in the V_2 receptor associated with X-linked NDI were the first naturally occurring mutations found in the very large group of G-protein-coupled hormone receptors. Since then, a number of diseases were shown to be caused by mutations in genes encoding G protein-coupled hormone receptors. Most mutations decrease the ability of a receptor to be activated by the extracellular signal, to transduce the signal to G proteins or to be transported to the correct cellular site. Some mutations result in a hyperactive phenotype with a concomitant loss of receptor regulation by extracellular signals. Excepting somatic mutations in the receptor for thyroid-stimulating hormone, mutations in G protein-coupled receptors are germ line mutations. The diseases/symptoms caused by mutations in G protein-coupled hormone receptors include—besides X-linked NDI—primary adrenocortical deficiency, hypocalciuric hypercalcemia/hyperparathyroidism, hypercalcemia/metaphysical chondrodysplasia, hypocalcemia, male precocious puberty, male pseudohermaphroditism, hyperfunctioning thyroid adenoma and Hirschsprung's disease.

Prior to the detection of naturally occurring mutations in the V2 receptor, genetic defects had been identified only in one other G-protein-coupled receptor, i.e., the light receptor rhodopsin. Mutations in the rhodopsin gene—besides defects in other genes—cause autosomal dominant retinitis pigmentosa, a disease characterized by a progressive loss of vision over years or decades. About 100 rhodopsin mutations associated with autosomal dominant retinitis pigmentosa are known. Similar to the mutations in the V_2 receptor associated with X-linked NDI, mutations causing autosomal retinitis pigmentosa are scattered throughout the photoreceptor. Compared to wild-type rhodopsin, mutants expressed in fibro-blasts are not efficiently inserted in the plasma membrane but rather accumulate intracellularly. The accumulation of protein within the cell may eventually lead to cell death. Thus a pathogenic mechanism similar to that suggested for autosomal dominant CDI may apply for autosomal dominant retinitis pigmentosa.

The identification of genes responsible for familial forms of CDI and NDI will facilitate carrier identification and early diagnosis; it is also the first step toward an improved treatment of CDI and NDI patients.

See also ENDOCRINOLOGY, MOLECULAR; HUMAN GENETIC PREDISPOSITION TO DISEASE; RECEPTOR BIOCHEMISTRY.

Bibliography

Diabetes Insipidus

Fujiwara, T. M., Morgan, K., Bichet, D. G. (1995) Molecular biology of diabetes insipidus. *Annu. Rev. Med.* 46:331–343.

Knoers, N. V. A. M. (1994) Molecular characterization of nephrogenic diabetes insipidus. *Trends Endocrincol. Metab.* 5:422–428.

Schmale, H., Bahnsen, U., Richter, D. (1993) Structure and expression of the vasopressin precursor gene in central diabetes insipidus. *Ann. N.Y. Acad. Sci.* 689:74–82.

G-Protein-Coupled Receptors, G-Proteins

Birnbaumer, L. (1992) Receptor-to-effector signaling through G-proteins: roles for βγ dimers as well as α subunits. *Cell* 71:1069–1072.

Coughlin, S. R. (1994) Expanding horizons for receptors coupled to G proteins: diversity and disease. *Current Opinions in Cell Biology* 6:191–197.

DIABETES MELLITUS

Catherine H. Mijovic, David Jenkins, and Anthony H. Barnett

Key Words

Diabetes   A heterogeneous disorder characterized by persistent hyperglycemia.

Human Leukocyte Antigen (HLA) Class II   A heterodimeric cell surface protein found on antigen-presenting cells that presents antigen to CD4 T-cell receptors.

Insulin   A hormone with several metabolic effects. Deficiency of insulin results in diabetes.

Major Histocompatibility Complex (MHC)   Cluster of genes that includes several loci, many of which are involved in regulation of the immune response, particularly allograft rejection.

Maturity Onset Diabetes of the Young (MODY)   A subset of non-insulin-dependent diabetes with autosomal dominant inheritance.

Diabetes mellitus is a heterogeneous disorder. Insulin-dependent diabetes mellitus (IDDM) is determined by both genetic and environmental factors. Several genes are involved, the most important of which occur within the major histocompatibility complex (MHC). Linkage between IDDM and at least 11 other loci on nine chromosomes, including the insulin locus, has also been shown. None of the genes have been precisely identified. Non-insulin-dependent diabetes mellitus (NIDDM) also has a complex etiology. It is more difficult to investigate because of disease heterogeneity and lack of sufficient families. Abnormalities of the glucokinase gene have been identified as causes of some cases of maturity onset diabetes of the young (MODY), a subset of NIDDM. It seems likely that recent advances in molecular genetics will further our knowledge of these complex disorders.

1   INTRODUCTION

Diabetes mellitus, a heterogeneous condition of complex etiology characterized by persistent hyperglycemia, is strongly associated with a tendency to both micro- and macrovascular disease. The tendency increases with increased duration of diabetes. Diabetes mellitus consists of two major subsets. Type 1 or insulin-dependent diabetes (IDDM) is an autoimmune, pancreatic β-cell-specific disease whose clinical features are due to insulin deficiency. At diagnosis, the pancreatic islets of patients with IDDM contain a mononuclear cell infiltrate (insulitis). Type 2 or non-insulin-dependent diabetes (NIDDM) appears to be due to a combination of partial insulin deficiency and resistance to insulin action. β-Cell deficiency in NIDDM does not result from autoimmune destruction but may be related to amylin deposits within the pancreatic islets. NIDDM also differs from IDDM in several other respects (Table 1). Predisposition to both diseases appears to be determined by multiple genes.

Analysis of genetic predisposition to diabetes has confirmed that IDDM and NIDDM are different. IDDM has a lower concordance rate among identical twins than NIDDM. Subjects with IDDM are also less likely to have a diabetic relative, suggesting that the genetic predisposition to NIDDM is greater than that for IDDM.

2   METHODS OF IDENTIFYING DISEASE SUSCEPTIBILITY GENES

Certain genes are polymorphic; that is, different forms (alleles) occur in different individuals. Any allele increased in frequency among diabetic patients compared with healthy control subjects may directly predispose to diabetes, or (as is more likely) be in linkage disequilibrium with a disease susceptibility allele. Linkage disequilibrium refers to the coinheritance of combinations of alleles at different loci on the same haplotype more frequently than would be expected by chance alone. This complicates the distinction of true susceptibility alleles from those secondarily associated with disease as a result of linkage disequilibrium.

This population association approach to gene mapping was limited by the few known polymorphic genes. The polymerase chain reaction (PCR) has revolutionized genetics and has led to the discovery of a huge variety of polymorphic loci throughout the human genome. Most are noncoding repeated nucleotide sequences (minisatellites and microsatellites) of unknown function, also known as variable numbers of tandem repeats (VNTRs). Associations between such marker loci and diabetes suggest linkage disequilibrium, indicating possible locations of disease susceptibility genes.

Family studies provide complementary methods of gene map-

Table 1 Comparison of Characteristics of IDDM and NIDDM

Factor	IDDM	NIDDM
Symptoms	Thirst, polyuria, fatigue Acute onset	Varies between none and those of IDDM; gradual onset
Age of onset	Usually young but may occur at any age	Usually > 35 years but may occur at any age
Biochemical features	Severe insulin deficiency	Reduced insulin; insulin resistance implicated in etiology
	Ketosis-prone	Ketosis-resistant
Treatment	Absolute insulin requirement	Diet and/or oral hypoglycemics usually sufficient initially
Autoimmune features	ICA[a] present at diagnosis; insulitis present	ICA[a] absent; insulitis absent
Microvascular complications	Usually occur after 10 years	Often present at diagnosis
Genetic susceptibility	Polygenic	Polygenic
	Major genes in MHC and insulin gene region	Glucokinase gene implicated in some MODY patients
	35–55% concordance in identical twins	Near 100% concordance in identical twins
Environmental factors	?Viruses, ?toxins, ?nutrition	Obesity, exercise, nutrition, age fetal/neonatal environment

[a]Islet cell antibody.

ping. Conventional linkage analysis is useful for studying monogenic disorders. This method requires large, multigeneration pedigrees in which a family trait of known mode of inheritance is transmitted. Analysis of the segregation of a disease and a genetic marker allows calculation of the probability of whether the disease susceptibility gene and the marker are linked. NIDDM and IDDM are determined by several genes of uncertain mode of inheritance. Such linkage analysis has, therefore, been of little value in these disorders, with the notable exception of maturity onset diabetes of the young (MODY), a subset of NIDDM. Analysis of allele sharing by affected sib pairs can be more easily applied to polygenic disorders. This approach has been useful in IDDM, but large numbers of affected sib pairs are required. In NIDDM the late onset of the disease together with its excess mortality seriously hinder the identification of sufficient informative families for sib-pair analysis. Heterogeneity of NIDDM also confounds comparison of findings between different studies, particularly when different ethnic groups have been used.

Animal models for IDDM and NIDDM may clarify the genetics of both diseases. Models of IDDM include the nonobese diabetic (NOD) mouse and biobreeding (BB) rat (Table 2). Models of NIDDM include the ob/ob mouse. Identification of susceptibility genes in animal models produces candidate genes for the human disease.

3   IDDM AND THE MHC

The first genes to be considered as candidate susceptibility genes for IDDM, the human leukocyte antigen (HLA) genes, were implicated in the early 1970s. These occur within the human major histocompatibility complex (MHC), on the short arm of chromosome 6 (Figure 1). The class I HLA-B alleles B8 and B15 were found to be increased in frequency among Caucasian IDDM patients, while HLA-B7 was decreased. Subsequent serological typing of the HLA class II DRB1 locus showed positive associations between IDDM and DR3 and DR4 and a negative association with DR2. DR3 is in linkage disequilibrium with B8, DR4 with B15, and DR2 with B7. The DR associations with IDDM are stronger than the class I asso-

Table 2 Comparison of Characteristics of IDDM in Man, the NOD Mouse, and the BB Rat

	Man	NOD Mouse	BB Rat
Autoimmune features	Moderate insulitis; ICA[a] present	Severe insulitis; ICA[a] present	Severe insulitis; ICA[a] present
Other organs affected	Associated autoimmune endocrinopathy in small subset	Lymphocytic infiltrate in thyroid, adrenal, and salivary glands	Prone to thyroiditis and gastric parietal antibodies
Sex ratio, F:M	1:1	4:1	1:1
T-Cell lymphopenia	No	No	Severe (absence of RT6[+] cells)
T-Cell dependence	Yes	Yes	Yes
MHC-associated genetic susceptibility	Yes	Yes	Yes
Non-MHC-associated genetic susceptibility	Yes	Yes	Yes
Penetrance of diabetogenic genotype	30%	70% females 15% males	60%

[a]Islet cell antibody.

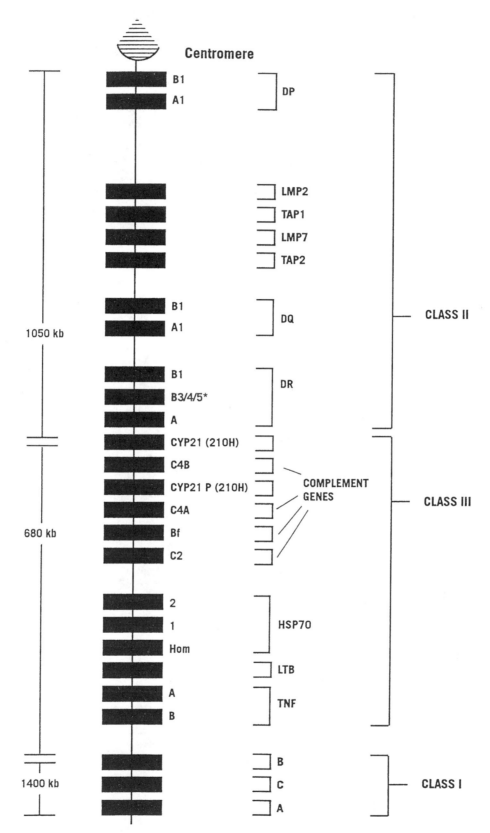

Figure 1.   Diagrammatic representation of the human major histocompatibility complex on chromosome 6p. Approximate distances between certain genes shown in kilobases (kb) on the left; diagram is not to scale. The asterisk(*) indicates that the presence or absence of *DRB* genes *DRB3, 4,* and *5* depends on the particular MHC haplotype.

ciations, and consequently the search for the diabetogenic MHC-encoded gene or genes has been concentrated within the class II region. A role for class I alleles in determining susceptibility cannot, however, be discounted.

Affected sib-pair studies have confirmed linkage of susceptibility to IDDM to the MHC. Estimates of the contribution of MHC genes to the total genetic susceptibility to IDDM vary widely between 30 and 80%.

3.1   The Structure of the MHC

The MHC comprises 4 Mb of DNA, and more than 80 genes have been mapped to the region (Figure 1). These genes fall into three groups: (1) those involved with antigen presentation (i.e., the class I and II HLA genes), (2) those involved with antigen processing [e.g., the transporter associated with antigen processing (TAP) and large multifunctional protease (LMP) genes] and other functions related to the immune response (e.g., the complement genes), and (3) those with no defined role in the immune system [e.g., the C-21 hydroxylase genes (CYP21 and CYP21P)].

There are about 20 class I loci at the telomeric end of the complex and at least 15 class IIA and IIB loci at the centromeric end. These are separated by the class III genes, which include the genes for the complement components C2, C4, and factor B, heat shock protein 70 (HSP70), and the cytokines tumor necrosis factor (TNF) and lymphotoxin (LT) (encoded by TNF-A, TNF-B, and LTB).

3.2   The Function of HLA Class I and II Molecules

The class I and II HLA gene products have a role in the presentation of processed antigen to T cells. Class I HLA molecules comprise an α-peptide chain encoded by the class I loci (e.g., HLA-A, B, C) combined with β_2-microglobulin that is encoded on chromosome 15. They show a wide pattern of tissue distribution but are absent from some cells (e.g., corneal endothelium). The level of expression shows variation according to cell type. In general, class I molecules present processed endogenous (intracellular) proteins. These may be native peptides or, in the case of a virus-infected cell, viral proteins. At the cell surface the loaded class I molecules usually interact with CD8$^+$ cytotoxic T cells via the T-cell receptor. This results in killing of the presenting cell (Figure 2). The HLA class IIA and IIB genes (e.g., HLA-DRA and HLA-DRB1) encode α- and β-chains respectively. These combine to form cell surface heterodimers and are primarily expressed by specialized antigen-presenting cells (APCs) such as macrophages, dendritic cells, and B lymphocytes. Class II molecules present extracellular proteins that have been taken up by endocytosis into endosome-related vesicles. At the surface of the APC they interact with T-cell receptors on CD4$^+$ T cells. This results in help for B lymphocytes to produce immunoglobulins specific for the exogenous protein (Figure 2).

Both HLA class I and class II molecules have now been crystallized, and their three-dimensional structures were shown to be similar. Peptides are bound in a groove composed of eight strands of antiparallel β-sheet as a floor and two antiparallel helical regions as the sides. Polymorphism of this groove allows a variety of peptides to be bound.

3.3   HLA Class II Genetic Associations with IDDM

The HLA class II molecules DR3 and/or DR4 occur in more than 90% of Caucasian IDDM patients but also occur in 60% of the healthy control population. Assuming the DR3 and DR4 antigens predispose di-

rectly to IDDM, they are nevertheless too frequent in healthy subjects to explain the observed prevalence of the disease, regardless of their penetrance. (Penetrance can be complete—i.e., every individual carrying the susceptibility gene is affected—or incomplete—i.e., some individuals carry the gene but do not develop the disease.) It thus appears that the DR associations with IDDM are secondary to linkage disequilibrium with a diabetes susceptibility gene(s), although a contributory role for DRB1 alleles cannot be excluded.

Mode of inheritance studies indicate that the susceptibility alleles associated with DR3 and DR4 may be distinct. DR3 and DR4 predispose to IDDM synergistically, DR3/4 heterozygotes have a greater risk of disease than those possessing either DR3 or DR4 alone. The susceptibility associated with DR4 appears dominant in the absence of DR3, whereas the susceptibility associated with DR3 appears recessive in the absence of DR4.

There is emerging evidence for age-dependent HLA-genetic heterogeneity among IDDM patients. The clinical heterogeneity of diabetes mellitus requires that studies of genetic heterogeneity incorporate strict classification criteria to ensure that all the patients present unambiguously with the clinical characteristics typical of IDDM. A lower frequency of DR3/4 heterozygotes and a higher frequency of non-DR3/non-DR4 subjects in adult onset IDDM (onset after age 15) compared with childhood onset IDDM (onset before age 15) has been described. These findings indicate that the DR3- and DR4-associated genetic susceptibility is greater in childhood onset than adult onset IDDM and suggest an involvement of other unidentified factors, which become more important with age. More studies are required to confirm these findings.

Some investigations have suggested that DR3-associated IDDM is distinct from DR4-associated IDDM based on differences in clinical presentation, associated autoimmune disease, and age of onset. These findings are not, however, consistent in all studies.

The DRB1 alleles are in linkage disequilibrium with alleles at the DQA1 and DQB1 loci. In the early 1980s it was shown that white Caucasoid DR4-positive haplotypes could be split into those that were and were not associated with IDDM on the basis of the DQ specificity that was present. This led to the consideration of the DQ loci as susceptibility determinants for IDDM. Recently the use of PCR amplification, DNA sequencing, and sequence-specific oligonucleotide probe hybridization has led to the identification of many more alleles at the DRB1, DQA1, and DQB1 loci than could be detected by serological typing or restriction fragment length polymorphism (RFLP) analysis. The serologically defined DR4 specificity, for example, can currently be subdivided into 12 subtypes, and only some of these are positively associated with IDDM. These DR4 subtypes are a result of polymorphism of the DRβ chains encoded by different alleles at the DRB1 locus. The coding region of the DRA gene is nonpolymorphic. In contrast, both the DQα and DQβ chains are polymorphic, with 14 alleles so far identified at the DQA1 locus and 19 alleles at DQB1.

In Caucasian populations IDDM is strongly positively asociated with the DQ alleles DQA1*0301, DQB1*0302, and DQB1*0201, and negatively associated with DQB1*0602.

3.4   Transracial Studies in Mapping Susceptibility to IDDM

Improved mapping of disease susceptibility has come from studying disease associations in different races. Although linkage disequilibrium is strong between MHC class II genes, recombination during evolution has generated combinations of MHC alleles (ex-

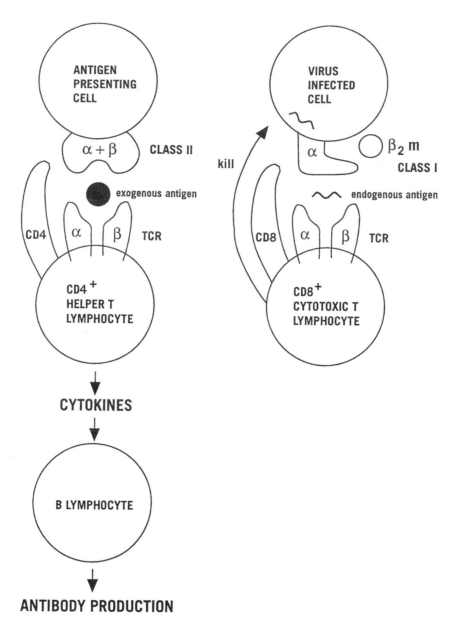

Figure 2.   Schematic representation of the functions of the MHC class I and class II molecules: TCR, T-cell receptor; β_2m, β_2-microglobulin.

tended haplotypes) that vary between populations. This has caused disease-predisposing alleles to be in linkage disequilibrium with different marker alleles in different races. Any allele that is consistently associated with a disease in all races, irrespective of the extended haplotype on which it occurs, is a candidate disease susceptibility determinant. This criterion assumes that IDDM has the same genetic basis in all races. Transracial studies therefore require rigorous application of diagnostic criteria for disease, as well as careful matching of disease and control populations for ethnic origin.

Sections 3.4.1 through 3.4.8 describe the current understanding, based on data from the different racial groups, of the role of *DRB1* and *DQ* alleles in determining susceptibility to IDDM.

3.4.1 DR3-Associated Susceptibility

IDDM is positively associated with *DR3* (encoded by the *DRB1*0301* allele) and with *DQB1*0201* in all races except the Japanese, in whom

these alleles are very rare. (The rarity of these disease-associated alleles may contribute to the low prevalence of IDDM in Japanese subjects.) *DRB1*0301* is in linkage disequilibrium with *DQA1*0501* and *DQB1*0201*. *DQA1*0501* does not show consistent disease associations across the races. These data are thus consistent with a role for *DRB1*0301* and/or *DQB1*0201* in determining *DR3*-associated susceptibility to IDDM. Study of *DR7* haplotypes also suggests a role for *DQB1*0201* in disease susceptibility. The *DR7-DQA1*0201-DQB1*0303* haplotype is protective, whereas the *DR7-DQA1*0201-DQB1*0201* haplotype is neutral only with respect to IDDM in Caucasians. In Negroes *DR7-DQA1*0301-DQB1*0201* is predisposing. These findings are consistent with the idea of *DQB1*0201* being a susceptibility allele in combination with *DQA1*0301* and *DQA1*0501*, although indicating that its effect can be altered by other alleles such as *DQA1*0201*. This is not surprising when the DQβ chain is considered in the context of the whole DQ molecule, where both the α and β chains combine to form the peptide-binding cleft.

3.4.2 *DR4*-Associated Susceptibility

DR4 is associated with IDDM in all races except the Chinese. *DR4* is in linkage disequilibrium with *DQA1*0301* and with either *DQB1*0301* or *DQB1*0302* in Caucasians. In Orientals *DR4* also occurs in linkage disequilibrium with *DQB1*0401*. *DQA1*0301* predisposes to IDDM in all races except the Chinese.

DR4 and *DQA1*0301* are not rare in the Chinese population. Prior to the finding of a lack of association with *DQA1*0301* among the Chinese it was suggested that *DQA1*0301* directly determines *DR4*-associated susceptibility. The Chinese data, however, suggest that *DQA1*0301* may not predispose to disease directly but that it may be in linkage disequilibrium with a predisposing allele at another locus. Alternatively, any effect of *DQA1*0301* may be modified by another susceptibility gene in the Chinese. (There are no data to suggest that IDDM in Chinese subjects is distinct from the disease in other races.)

The original data concerning *DQB1* alleles in *DR4*-associated susceptibility came from Caucasian populations and implicated *DQB1*0302* as the determinant of susceptibility. *DQB1*0302* combines with *DQA1*0301* to form the DQ molecule DQ8. Approximately 90% of Caucasian *DR4*-positive diabetics are DQ8 positive, compared with 50% of racially matched *DR4*-positive controls. The discovery of the existence of *DR4* subtypes, however, showed that not all *DR4* haplotypes encoding DQ8 are associated with disease. In Caucasians only *DRB1*0401* and *DRB1*0402* haplotypes encoding DQ8 are positively associated with disease; *DRB1*0403,0404* and *0405*-DQ8 positive haplotypes are not. In the Japanese the majority of *DR4*-positive diabetics are DQ4 (encoded by *DQA1*0301/DQB1*0401*) positive. DQ8 does occur in the Japanese population but in combination with *DRB1*0403* and *DRB1*0406*. Both *DRB1*0403*-DQ8 and *DRB1*0406*-DQ8 haplotypes are reduced among Japanese diabetics compared to controls; but interestingly, *DR8*-DQ8 haplotypes are increased. A modifying effect of the various *DRB1* alleles on any susceptibility encoded by the DQ8 molecule has been proposed to explain these findings in Japanese and Caucasian populations.

These data suggest that DQ8 may be a predisposing factor when it is present in association with *DRB1*0401* and *DRB1*0402*. Alternatively, the data could be interpreted as suggesting that DQ8 is in linkage disequilibrium with a true diabetes susceptibility allele that occurs only on certain haplotypes (e.g., those marked by *DRB1*0401* and *DRB1*0402*).

The role of DQ8 in determining susceptibility to IDDM is also complicated by the finding that the association of *DR4*-DQ8 with IDDM in Caucasians is confined to the *DR3/4* subgroup of patients. *DR1/4* (associated with an interactive effect on risk, as is *DR3/4* heterozygosity) and *DR4/X* patients (where *X* is not 1, 3, or 5) show a distribution of *DR4*-associated *DQ* alleles similar to that seen in *DR4* control subjects.

The situation is more complex in the Chinese, in whom *DR4* is not associated with the disease, and *DRB1*0401* and *DRB1*0402* are absent from the Chinese as well as from the Japanese. In both the Japanese and Chinese *DR4* haplotypes may carry *DQA1*0301* with *DQB1*0401*, encoding the DQ4 molecule. This combination occurs on the *DRB1*0405* subset of *DR4* haplotypes in both races. In the Japanese these *DRB1*0405*-DQA1*0301*-DQB1*0401* haplotypes are associated with susceptibility to the disease, whereas in the Chinese they are not. This finding of a contrasting disease association with the same DQ molecule in two different races has several possible explanations:

1. The DQ4 molecule is not a susceptibility determinant for IDDM.
2. The DQ4 molecule is a susceptibility determinant, but its effect is modified by another product of the Chinese genotype. This product might be MHC- or non-MHC-encoded.
3. A factor necessary for disease penetrance in the presence of the DQ4 molecule is absent from the Chinese environment.
4. There is heterogeneity between IDDM in the Japanese and the Chinese.

Currently, none of these possibilities can be excluded.

The available data are compatible with the idea of DQ8 contributing to a portion of *DR4*-associated susceptibility but indicate that it is not the sole determinant.

3.4.3 *DR9*-Associated Susceptibility

DR9 has also been positively associated with the disease in negroids, Japanese, and Chinese. In some groups of Chinese *DR3/9* heterozygosity has a synergistic effect on the risk of developing IDDM. Analysis of mode of inheritance suggests that the susceptibility associated with *DR9* is distinct from that associated with *DR3* and *DR4*. The diabetes-associated *DR9* haplotypes in Negroes carry the *DQA1*0301* and *DQB1*0201* alleles. This cis-encoded *DQ* combination is also found on the diabetes-associated negroid *DR7* haplotypes. This combination could also be encoded in trans in *DR3/4* heterozygotes and has thus been considered as a potential predisposing DQ heterodimer (see Section 3.4.7). In both the Japanese and Chinese, *DR9* haplotypes carry the *DQA1*0301* and *DQB1*0303* alleles encoding DQ9, although they differ at the *HLA-B* locus. These data may indicate that *DR9*-associated susceptibility is determined by the same factor in the Japanese and Chinese but by a distinct mechanism in Negroes.

DR9 is relatively rare in Caucasians but appears to be neutral with respect to diabetes. Caucasian and Oriental *DR9* haplotypes both carry the *DQA1*0301* and *DQB1*0303* alleles. *DQ* alleles cannot, therefore, be the sole contributors to the *DR9*-associated susceptibility to IDDM observed in the Oriental races.

3.4.4 MHC-Encoded Protection

The *DR2*-associated *DQB1*0602* allele and the very similar *DR6*-associated *DQB1*0603* allele have both been negatively associated with IDDM in races in which they are common alleles and may directly protect against the disease. In Caucasians these alleles occur on the *DRB1*1501*-DQA1*0102*-DQB1*0602* and the *DRB1*1301*-DQA1*0103*-DQB1*0603* haplotypes. The finding of a rare putative recombinant *DR2* haplotype, *DRB1*1501*-DQB1*0402*, which segregated with IDDM in a Caucasian family and was present in two French diabetic patients, suggests that the *DRB1*1501* allele alone is not protective. This indicates that the *DQB1*0602* allele alone or in combination with *DRB1*1501* may be protective. In blacks *DQB1*0602* commonly occurs with *DRB1*1503* or *DRB1*1101*. Most studies have indicated that *DQB1*0602* is a protective allele in blacks, supporting a direct role for *DQB1*0602* independent of *DRB1*.

*DQB1*0604* (found on *DRB1*1302*-DQA1*0102*-DQB1*0604* haplotypes) has also been associated with protection in Caucasians. In the Japanese, protection has additionally been associated with the *DQB1*0601* allele (found on the *DRB1*1502*-DQA1*0103*-DQB1*0601* and *DRB1*0803*-DQA1*0103*-DQB1*0601* haplotypes). There is some evidence that protective alleles may be dominant or partly dominant over predisposing alleles. This has particularly been suggested by breeding studies of the NOD mouse and BB rat models (see Section 4). In man, studies of *DR* and *DQ* genotypes heterozygous for a protective and a predisposing allele indicate that the effect of the protective allele may be partly dominant.

3.4.5 Single-Residue Hypotheses: DQβ Chain Position 57 and DQα Chain Position 52

Examination of the amino acid sequences of the first domain (peptide-binding region) of the DQβ chain in white Caucasians led to the hypothesis that susceptibility to IDDM is strongly influenced by the identity of the residue at position 57. It was proposed that the presence of aspartate at this position (Asp 57) confers protection against IDDM and that the absence of this residue (non–Asp 57, and in particular the presence of Ala, Ser, or Val) is associated with increased risk. This hypothesis was supported by population and family studies of Caucasians and by the finding that the incidence of IDDM in various countries reflects the population frequency of Asp-57-positive *DQB1* alleles. Subsequently an arginine residue at position 52 (Arg 52) of the DQα chain was also implicated in conferring susceptibility.

There are, however, many data inconsistent with the DQβ position 57 hypothesis. Various MHC haplotypes show differing degrees of association with IDDM which cannot be explained simply by the presence or absence of a single amino acid residue. Similarly, the increased risk associated with *DR3-DQ2/DR4-DQ8* heterozygosity compared to *DR3-DQ2* and *DR4-DQ8* homozygosity cannot be explained. Some alleles do not fit the correlation. The *DQB1*0604* allele, for example, is non–Asp 57 but was protective in a Caucasian population. *DQB1*0201* is non–Asp 57 but occurs on neutral Caucasian *DR7* haplotypes as well as predisposing *DR3* haplotypes. A modifying role of the DQα chain was invoked to explain this (see Section 3.4.1).

Studies of different ethnic groups also do not support a role for Asp 57. There is a positive association between the Asp-57-encoding alleles *DQB1*0401* and *DQB1*0303* and IDDM in Japanese subjects. Population studies showed 72% of Japanese diabetics and 27% of Chinese diabetics to be homozygous for Asp 57 compared to 0–7% of Caucasians. There were no findings of non–Asp 57 homozygosity in any Japanese diabetic, and it was present in only 23% of Chinese diabetics compared to more than 90% of Caucasian patients.

A non–Asp 57 DQβ chain may thus simply be a useful marker of susceptibility in Caucasians. The presence of Asp 57 on the DQβ chain does not directly mediate protection.

There are also contraindications to the Arg 52 hypothesis. The only common Arg-52-positive *DQA1* alleles in Caucasians are those that occur on *DR3* and *DR4* haplotypes, and it is possible that the disease correlation with Arg 52 is a result of the known IDDM-predisposing effects of these haplotypes.

The evidence from DQβ chains led some investigators to propose a similar hypothesis for DRβ1 chains. A primary role for position 57 in determining DRβ*1*-associated susceptibility is not, however, supported by the available data. As discussed in Section 3.4.2, susceptibility associated with DQ8 may be modified by the specificity of the *DRB1* allele present. There appears to be a hierarchy of degree of susceptibility, with *DRB1*0402* being the most diabetogenic and *DRB1*0403* the least. This hierarchy has been correlated with the identity of four amino acids between positions 67 and 74 of the

DRβ1 chain. It has been suggested that this is evidence for a structural role of the DRβ1 chain in susceptibility to IDDM.

3.4.6 DQ Heterodimer Hypotheses

The DQ heterodimer is formed by a combination of an α chain encoded by an allele of the *DQA1* gene and a β chain encoded by an allele of the *DQB1* gene. Such DQ heterodimers may consist of a combination of α and β chains encoded by *DQA1* and *DQB1* alleles on the same haplotype (these are cis-encoded DQ molecules). Alternatively, they consist of an α chain encoded by one haplotype and a β chain encoded by the other haplotype (trans-encoded DQ molecules). An individual who is heterozygous at the *DQA1* locus (possessing a different *DQA1* allele on each haplotype) and heterozygous at the *DQB1* locus can thus theoretically produce four different DQ heterodimers, two encoded in cis and two encoded in trans. An individual who is homozygous at each locus (possessing the same *DQA1* and *DQB1* alleles on both haplotypes) can only produce one species of heterodimer.

These considerations led to the suggestion that individuals able to encode DQ heterodimers (in cis or in trans) comprising both Arg-52-positive DQα and non-Asp-57-positive DQβ chains were strongly predisposed to IDDM. It is possible to encode four, two, one, or none of such Arg 52/non–Asp 57 heterodimers. Studies of European Caucasian populations have suggested that absolute risk of disease varies with the number of these heterodimers encoded. The possibility of encoding four Arg 52/non–Asp 57 heterodimers was associated with the highest absolute risk for the disease, and one heterodimer presented the lowest risk.

There are a number of contraindications to such heterodimer analysis. There are large variations in the relative risk associated with DQ molecules identical with respect to Arg and Asp at positions 52 and 57 but from different allelic origins. Similarly, a number of diabetic patients have both non–Arg 52 α and Asp 57 β chains (up to 10% of Caucasian and 22% of Japanese patients). In addition, according to the theory, *DR3/3*, *DR4/4*, and *DR3/4* can all encode four Arg 52/non–Asp 57 heterodimers, and yet the relative risk associated with the *DR3/4* genotype is much higher than for *DR3/3* or *DR4/4* individuals. A clear dose effect has not been detected in Japanese subjects. These findings suggest that factors other than the Arg 52/non–Asp 57 heterodimer have a role in determining susceptibility to the disease.

Consideration of DQ heterodimers has been expanded to consider the specificity of the whole DQ molecule classified by the combination of particular *DQA1* and *DQB1* alleles in cis or in trans. As part of the Eleventh International Histocompatibility Workshop (1991), a large collaborative analysis of population studies of Caucasians, blacks, and Japanese was made. In this work, certain DQ molecules (encoded in cis or in trans) consistently associated with predisposition and others with protection were identified. Some molecules occurred in all three ethnic groups, while others were either common only in, or rare in, the Japanese. The predisposing molecules identified were *DQA1*/DQB1*0301/0302* (cis), *0501/0201* (cis), *0301,0201* (cis or trans), *0501/0302* (trans), *0301,0303* (cis, common in Japanese, relatively rare in Caucasians), *0301,0401* (cis, Japanese only), *0301,0402* (trans), and the protective molecules, *DQA1*/DQB1*0102/0602* (cis), *0103/0603* (cis, rare in Japanese), *0103/0601* (cis, Japanese), *0501/0301*. Between 80 and 90% of the diabetic patients studied carried one or more of the predisposing combinations, and a dose effect was observed. The results were interpreted as evidence that particular DQ molecules play a dominant role in determining susceptibility or resistance.

There are a number of problems, however, associated with such analysis. As discussed earlier, it is not possible to concisely identify predisposing *DQA1* or *B1* alleles in isolation from the MHC haplotype on which they occur. This is highlighted by the following examples. There may be modification of the susceptibility associated with DQ8-encoding haplotypes by the DRβ1 chain (see Section 3.4.2). The DQ molecule DQ4 encoded by *DQA1*0301/DQB1*0401* that was identified as predisposing in the Japanese is not associated with susceptibility in the Chinese. The *DQA1*0301/DQB1*0303* combination encoding DQ9 is relatively rare in Caucasians compared to the Japanese, but it appears to have a neutral effect in Caucasians compared to the predisposing effect in the Japanese (see Section 3.4.3).

The consideration of models invoking susceptibility associated with particular *DQ* molecules is also complicated by various other factors. Counting of heterodimers is not reliable in heterozygous individuals because certain pairings of DQα and β chains do not occur, and preferential formation of certain DQ heterodimers cannot be excluded. DQ molecules encoded by genes in the trans position may be less efficiently expressed than those encoded in cis. It is thus currently not possible to accurately predict the DQ heterodimer status of heterozygous individuals. This imprecision may cause a large margin of error in counting DQ heterodimers in patients and controls, hence in the calculation of associated relative risks.

Such models should, therefore, be viewed with caution until they can be tested by direct examination of expressed DQ molecules.

3.4.7 *DR3/4* Heterozygosity and *DQ* Heterodimers

DR3/4 heterozygosity has been associated with increased risk of IDDM in all races except the Japanese, where *DR3* is rare. The risk of developing the disease is significantly greater for *DR3/4* heterozygotes than for *DR3/3* or *DR4/4* homozygotes.

The mechanism for this association is unclear, although many investigators have postulated that the possible existence of the trans-encoded DQ heterodimers *DQA1*0301/DQB1*0201* and *DQA1*0501/DQB1*0302* is important. The *A1*0501/DQB1*0302* heterodimer is, however, encoded by *DR4/11* individuals, who have a low risk of IDDM. If *DQA1*0501/DQB1*0302* is a high risk heterodimer, then it is necessary to invoke a theory whereby a factor on *DR11*-positive haplotypes confers a degree of protection from it. In addition, there is no evidence to confirm that all four possible heterodimers are expressed in *DR3/4* heterozygotes. The single structural study of the DQ heterodimers encoded by *DR3/4* heterozygotes detected the trans-encoded *DQA1*0501/DQB1*0302* molecule. The occurrence of the other possible trans-encoded heterodimer *DQA1*0301/DQB1*0201* in *DR3/4* subjects has yet to be confirmed.

Consideration of the interactive risk associated with *DR4/8* heterozygotes in the Japanese also casts doubt on a simple role for trans-encoded DQ molecules. *DRB1*0405-DQ4/DR8-DQ8* Japanese heterozygotes cannot form any additional trans-encoded species of DQ heterodimer because only the *DQA1*0301* allele is present.

3.4.8 Extended MHC Haplotypes: Evidence for Non-*HLA-DR/DQ*-Encoded Susceptibility

It remains a possibility that the associations between IDDM and *DQ* alleles simply occur as a result of linkage disequilibrium between *DQ* and other genes that encode susceptibility. Alternatively DQ molecules in combination with other HLA gene products may confer susceptibility. Evidence that polymorphism at *DQ* alone cannot account for variations in susceptibility has been provided by the study of *DR3* and *DR4* positive haplotypes from Finnish families with affected children (the DIME study). These *DR3* and *DR4* haplotypes were all *DQ2* and *DQ8* positive, respectively. They could, however, be subdivided on the basis of typing the HLA class I *HLA-A, B*, and *C* alleles (e.g., *A2, Cw4, Bw35, DR4, DQ8; A2, Cw1, Bw56, DR4, DQ8*). There were large variations in the absolute risk for IDDM associated with each *DR3-DQ2* and each *DR4-DQ8* haplotype categorized in this way. This broad range of results suggests that other genes on these haplotypes have a role in modifying susceptibility to IDDM (see Section 5.1).

3.5 SUMMARY OF STUDIES OF HLA ASSOCIATIONS WITH IDDM

None of the preceding models of the involvement of HLA molecules as susceptibility determinants of IDDM fit all the available data. It is clear that factors other than HLA-DR or DQ which may be genetically encoded or of environmental origin are involved. Class II molecules, however, currently remain the best candidates for contributory factors to disease development. Section 3.6 describes some of the theories that have been proposed for the mechanism of their involvement.

3.6 PROPOSED MECHANISMS FOR THE INVOLVEMENT OF CLASS II MOLECULES IN THE PATHOGENESIS OF IDDM

The mechanisms by which particular class II molecules might determine susceptibility or resistance to disease are speculative. The overall outcome with respect to disease will depend on the relative dominance of all the class II molecules encoded. It is unclear whether class II molecules could play a role within or outside the thymus or both. Class II molecules may have a role in the selection of T-cell repertoire. In the thymus, diabetogenic T-cell clones could be positively selected (by predisposing DQ molecules) or actively deleted (by protective DQ molecules: dominant resistance). Alternatively, protective DQ molecules could actively select T-cell clones, providing peripheral tolerance to β-cell antigens whereas predisposing DQ molecules may actively delete them (dominant susceptibility). In the periphery, dominant susceptibility would be provided by the preferential binding and presentation of diabetogenic peptides to effector cells, and dominant resistance by the presentation of such antigens to suppressor cells. Failure to bind the antigen would result in a neutral outcome or recessive resistance.

Two different models for the involvement of class II molecules in the development of IDDM have recently been proposed. One suggests an active promotion of autoimmunity and IDDM by the immune system, whereas the other suggests that IDDM arises because of a failure of the immune system to maintain tolerance to the pancreatic β-cell.

Current evidence does not allow the rejection of either model. The role of *DQ* alleles in determining susceptibility to IDDM remains unclear. The identification of a diabetogenic peptide, which may be an environmental antigen or an autoantigen (self-peptide) would be a key advance in the understanding of the mechanisms by which IDDM develops.

3.7   Candidate Diabetogenic Peptides

A number of pancreatic islet cell autoantigens have been implicated in the etiology of IDDM. These have been identified by their reactivity with islet cell autoantibodies isolated from diabetics and their prediabetic relatives. Recent studies have focused on glutamic acid decarboxylase (GAD) (which catalyzes the conversion of L-glutamate to γ-aminobutyric acid and is found in high concentrations in neurones and β-cells) as a target autoantigen. A 38 kDa polypeptide of the insulin secretory granule membrane has also been identified as a major antigen for T cells from patients with recent onset disease. Most recently, a human islet cDNA expression library has been screened with sera positive for islet cell antibody (ICA) from relatives of diabetics who themselves progressed to IDDM. This identified a novel 69 kDa peptide autoantigen (ICA69). Other candidates include insulin, proinsulin, glucagon, glycolipids, mycobacterial heat shock protein 65, and carboxypeptidase H.

The role of autoantibodies in the pathogenesis of the disease remains uncertain. It is unclear whether they initiate or contribute to destruction of the islet cell or are simply the result of an immune response to the damage.

IDDM may be triggered by an environmental stimulus. Most recently interest has focused on the role of dietary bovine serum albumin (BSA) in cow's milk. Studies of rodents and humans have suggested that early exposure to cow's milk can trigger the disease. In one study, antibodies to BSA (specifically to the ABBOS peptide, pre-BSA position 152–168) were detected in all diabetic patients but in only 2.5% of controls. These antibodies were shown to cross-react with p69, a γ-interferon-inducible pancreatic β-cell protein. (It is currently unclear whether this protein is related to the ICA69 autoantigen mentioned earlier.) It has been proposed that in genetically susceptible individuals who are able to present ABBOS to the immune system, molecular mimicry results in pancreatic β-cell damage and IDDM.

The identification of diabetogenic antigen(s) is a prerequisite for further development of therapeutic approaches to the disease. Induction of tolerance to such antigens would be an important preventive strategy.

3.8   Conclusions from Studies of HLA Associations with IDDM

A role for HLA-DQ in protection against insulin-dependent diabetes mellitus is indicated by the consistent reduction in frequency of the *DQB1*0602* and/or the similar *0603* alleles in type 1 diabetic patients in different races. Similarly, *DR3* (encoded by *DRB1*0301*) and its associated DQ2 molecule (encoded by *DQA1*0501/DQB1*0201*) are increased in frequency in diabetic patients of all races except the Japanese, in whom it is a rare allele. The available data are thus compatible with a role for DQ both in protection from IDDM and in *DR3*-associated disease susceptibility.

At present none of the proposed models of a mechanism for the involvement of DQ molecules in susceptibility to IDDM can explain all the available data. Simple models based solely on class II heterodimers are unlikely to be correct. It is possible that a component of the MHC-associated susceptibility to IDDM is neither *HLA-DR* nor *DQ*-encoded (see Section 5.1).

4   ANIMAL MODELS OF IDDM

There are two animal models of spontaneously occurring IDDM: the nonobese diabetic (NOD) mouse and the biobreeding (BB) rat. The characteristics of these animal models are compared to the human disease in Table 2. While these models closely resemble human IDDM, each shows a major discordant feature—namely, the strong female preponderance of the disease in the NOD mouse and the severe lymphopenia in the BB rat. The direct relevance of findings in these animal models to the human disease is thus questionable.

4.1   The NOD Mouse

The MHC-encoded susceptibility for diabetes in the NOD mouse is located on chromosome 17 and is termed *IDD-1*. *IDD-1* is dominant with a low degree of penetrance and is essential for the development of diabetes. The NOD mouse model has been used to extensively investigate the contribution of MHC class II molecules to development of the disease. The NOD mouse expresses a single distinctive class II molecule I-A^{nod} (*I-A* is the murine homologue of *DQ*). The second type of class II molecule, *I-E* (the murine homologue of *DR*), is not expressed owing to a nonfunctional *I-E* α-chain gene. Insulitis in NOD mice can be prevented by the introduction of *I-E* α chain or non-NOD *I-A* transgenes. These experiments indicate that class II molecules do have a role in determining susceptibility to IDDM.

The *I-A^{nod}* β chain is non–Asp 57. This finding was used as evidence supporting the proposed protective role of Asp-57-positive *DQ* β chains. Insertion of either Asp-57-positive or certain non–Asp 57 *I-A*β genes into NOD mice, however, can *both* protect against diabetes. These data from transgenic NOD mice thus further question the Asp 57 hypothesis but implicate a role for *DQ* in IDDM susceptibility. The protective mechanism by which the susceptibility to diabetes conferred by *I-A^{nod}* can be overcome by the introduction of I-E or other *I-A* molecules has yet to be identified.

The CTS strain of mouse also possesses the *I-A^{nod}* β chain sequence together with a lack of *I-E* expression and yet does not develop diabetes. Although this natural strain has the same class II genes as the NOD mouse it differs at the class I and III loci. A NOD MHC congenic mouse expressing the CTS MHC developed diabetes but at a significantly lower frequency than in the NOD strain. This indicates that the susceptibility associated with *Idd1* may be determined by multiple MHC-encoded genes.

Other MHC congenic strains of mice have shown that whilst the NOD MHC is essential for the development of diabetes it is not sufficient, thus non-MHC linked *Idd* loci are implicated. The first identified, *Idd2*, was mapped to the Thy-1 region on mouse chromosome 9. Recently a linkage map of the mouse genome has been generated using microsatellite DNA markers. Breeding experiments allowing the analysis of the cosegregation of these markers and diabetes have identified further non-MHC candidate susceptibility loci. These are *IDD-3* and *IDD-10* on chromosome 3; *IDD-4, IDD-5, IDD-6, IDD-7, IDD-8* and *IDD-9* on chromosome 11, 1, 6, 7, 14 and 4 respectively. At least two other loci are also implicated. The identities of the susceptibility genes at these loci are unknown.

Comparative mapping of the mouse and human genomes has indicated extensive regions of homology. Human DNA regions syntenic to *IDD-4* and *IDD-5* in mice are found on chromosomes 17p and 2q, respectively. These human regions have shown evidence of linkage to IDDM (Section 5.2).

The NOD mouse model also demonstrates the importance of environmental factors in the development of diabetes. Animals reared in a pathogen-free environment show a higher incidence of disease and a more even female-to-male sex ratio. Different colonies of animals show variation in incidence of disease, as well. There is also a dietary influence: diets deficient in essential fatty acids or simplified proteins have been shown to reduce the incidence of diabetes.

4.2   THE BB RAT

Susceptibility to diabetes in the BB rat has also been mapped to the MHC class II region, on chromosome 20. As in the NOD mouse, the BB rat MHC class II β chains are non–Asp 57, but this is also true of the Lewis strain, which does not develop diabetes.

The development of diabetes in the BB rat also involves at least two other genes: *Lyp,* which is tightly linked to the neuropeptide *Y(Npy)* gene on chromosome 4, and a third unmapped gene for which the Fischer rat strain carries an allele conferring resistance.

The *Lyp* gene is in sole control of the T-cell lymphopenia, which appears to be inherited as a recessive trait. Diabetes can develop in rats without T-cell lymphopenia, but it seems to be an important predisposing factor. The lymphopenia is a result of a lack of circulating RT6$^+$ T cells. The coisogenic diabetes-resistant BB rat does have such T cells and does not develop diabetes. It has been shown that the paucity of RT6$^+$ T cells is unlikely to be due to a defect in *RT6* gene expression.

Environmental factors affect the incidence of disease, as do dietary factors. Cow's milk proteins and wheat gluten have been re ported to be important for full expression of diabetes, whereas semi-purified diets (e.g., amino acids rather than proteins) prevent insulitis and diabetes and improve the lymphopenia.

5   OTHER POSSIBLE DETERMINANTS OF SUSCEPTIBILITY

5.1   OTHER MHC DETERMINANTS OF IDDM

The MHC contains several other genes that might affect susceptibility to IDDM. Some studies have suggested that alleles of *HLA-B,* tumor necrosis factor, heat shock protein 70, complement, and transporter-associated peptide genes may also alter disease predisposition. These reports should be treated with caution. Associations between a variety of MHC alleles occur simply because of linkage disequilibrium between the marker allele and the true susceptibility allele(s). MHC class II alleles appear to be the most closely associated with IDDM. Demonstration of associations between IDDM and other MHC genes that are independent of class II associations will require studies of large diabetic and control groups matched for class II alleles. The continuing discovery of more genes within the MHC may identify other candidate determinants of susceptibility.

5.2   NON-MHC DETERMINANTS OF IDDM

The recent work of John Todd and colleagues using semi-automated fluorescence-based technology and large collections of affected sib-pair families has allowed linkage analysis of the whole human genome. This has shown that although the MHC encodes the major susceptibility locus *IDDM-1,* at least 11 other loci on nine chromosomes are involved. These are termed *IDDM-2– IDDM-10* residing on chromosomes 11p15, 15q, 11q13, 6q25, 18q, 2q31, 6q27, 3q21–q25, 10p11.2–q11.2 respectively and the *DXS1068* and *GCK* loci on chromosomes Xq and 7Q respectively. Only the identity of *IDDM-2* in the insulin gene region has been resolved..

5.2.1   Insulin *(INS)* Gene

Various studies have shown IDDM to be positively associated with *INS* polymorphism on chromosome 11p15 in Caucasian populations. Linkage between *INS* and IDDM could not be demonstrated until recently because relatively little polymorphism is present at the marker sites in the vicinity of the *INS* locus. Consequently, many families with IDDM were not informative for linkage studies. Examination of *INS* allele sharing in large numbers of sibling pairs affected by IDDM using various *INS*-related polymorphisms (Figure 3) in informative families has confirmed linkage. Linkage to INS was also demonstrated by the whole genome screen of Todd and colleagues, the locus is designated as *IDDM-2.*

Current evidence indicates that *IDDM-2* is the variable number of tandem repeat (VNTR) sequence in the 5′ regulatory region of the insulin gene (Figure 3). Short (class 1) alleles confer increased susceptibility, long (class 3) alleles are protective. The mechanism by which VNTR polymorphism affects susceptibility to IDDM is uncertain. It is likely to influence level of insulin gene expression but whether this is directly relevant to its role in determining susceptibility to IDDM remains to be proven.

Interestingly the IDDM-associated class 1 VNTR polymorphism is found in virtually all Orientals, regardless of whether they have IDDM or not. If this INS polymorphism predisposes to IDDM other factors in Oriental populations must protect them from the disease, causing the low observed incidence of IDDM.

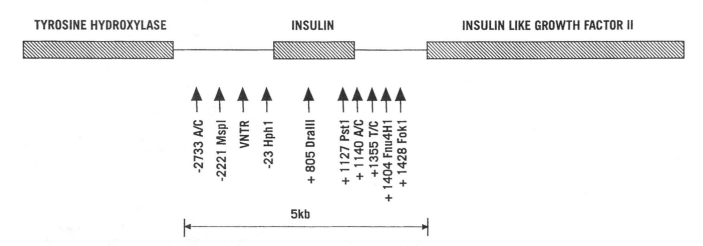

Figure 3.   Schematic representation of the insulin gene region: VNTR, variable number of tandem repeats. Names and positions of known polymorphic sites are indicated by arrows.

5.2.2 Immune Response Genes

The autoimmune nature of IDDM suggests the genes encoding products involved in the immune response as possible candidates encoding susceptibility. The immunoglobulin heavy chain region on chromosome 14q has been considered. Gm alleles have been studied as genetic markers in this region, they encode serologically detected antigens which occur on the constant portions of the Y heavy chains of IgG immunoglobulins. They have not been directly associated with IDDM and linkage with chromosome 14q has not been detected. One report indicates positive interactions between Gm alleles, HLA alleles and IDDM. These require confirmation. Weak population associations between IDDM and TCR polymorphism have been found but are not supported by linkage analysis.

6   INHERITED SUSCEPTIBILITY TO NIDDM

6.1   Evidence for Genetic Factors

Concordance rates for NIDDM among monozygotic twins were initially estimated to be approximately 90%. This was considerably higher than the concordance rate among dizygotic twins, suggesting that NIDDM is largely determined by genetic factors. Although interpretation of such studies is complicated by ascertainment bias, the genetic basis of NIDDM is supported by clustering of the disease within families. Comparison of the risks of NIDDM developing in various relatives of affected subjects suggests that two or three genes may be responsible. It has, however, long been recognized that environmental factors such as age, obesity, and lack of exercise contribute to the expression of the diabetic phenotype.

Data recently collected from a Swedish twin registry (which should be free of ascertainment bias) suggest that the concordance rate for NIDDM among monozygotic twins is actually much lower than 90%, and not much greater than the rate among dizygotic twins. It should be noted, however, that the nondiabetic co-twin was not tested for glucose intolerance in this study, an omission that could have led to underestimation of the concordance rate. Evidence is also emerging that disease susceptibility is partly determined by factors that influence growth during fetal development and the first year of life. Increased birth weight and increased weight at 1 year of age have been associated with a reduced risk of NIDDM later in life. These observations have led to speculation that high concordance rates for NIDDM among twins may simply be due to sharing of similar NIDDM-predisposing environmental factors in early life. Although these observations question the magnitude of genetic susceptibility relative to environmental susceptibility to NIDDM, they do not explain the increased risk of NIDDM to relatives in subsequent generations. It is also possible that some determinants of fetal and infant growth are themselves genetic, as suggested by ethnic differences in birth weight. There remains, therefore, considerable evidence that NIDDM is partly determined by genetic factors.

6.2   Identification of Candidate Genes

Attempts to map susceptibility to NIDDM have relied on identifying polymorphic candidate genes. NIDDM is characterized by both insulin resistance and abnormal insulin secretion. Many studies have examined populations of patients with NIDDM and healthy control subjects. Such investigations have looked for associations between NIDDM and genes that might alter insulin secretion and

action, including the genes for insulin, the insulin receptor, the various glucose transporters (GLUT), amylin, and glycogen synthetase.

A major problem with such disease association studies is that NIDDM is not a homogeneous disorder. Most of these studies have been disappointingly negative. Candidate genes that initially yielded significant associations with NIDDM have not done so in other populations. This failure to produce consistent results may reflect an initial association due to chance or to ethnic stratification; or perhaps the association was real in a subset of NIDDM present in the initial study population only.

6.3   Animal Models of NIDDM

There are number of mouse models of NIDDM encompassing different features associated with the disease. This diversity reflects the heterogeneous nature of the disease.

Single gene mutations producing obesity have been widely studied. Examples include a dominant mutation at the agouti locus (*A,* chromosome 2) and recessive mutations at the diabetes (*db,* chromosome 4), obese (*ob,* chromosome 6), and tubby and adipose (*tub, Ad,* chromosome 7) loci. These mutations produce severe obesity and various degrees of insulin resistance.

This distribution of a number of dominant and recessive genes over different chromosomes, each associated with obesity and insulin resistance in the mouse, is evidence for a similar heterogeneous genetic background to human NIDDM.

The effect of these obesity genes depends on the genetic background of the inbred strain. The *db/db* genotype, for example, results in obesity, islet hyperplasia, and marked hyperinsulinemia in all inbred strains. Some strains, however, also develop the features of IDDM (juvenile onset associated with the selective destruction of pancreatic β-cells, immunity against β-cells and severe insulinopenia). Other strains develop the milder features of NIDDM (obesity, hyperinsulinemia, severe insulin resistance). Members of a third group of strains remain diabetes resistant. A gender effect is also apparent, with only the males of some strains developing diabetes.

The importance of genetic background indicates that the obesity-induced diabetes phenotype is determined by multiple genes. This parallels the human situation, where obesity is associated with a greater risk of NIDDM if there is a family history of the disease. Identification of the susceptibility loci in mice may provide candidate genes in the human by using a comparative mapping strategy as with the NOD mouse model (see Section 4.1).

6.4   Maturity Onset Diabetes of the Young (MODY)

MODY is a well-defined subset of NIDDM with an early age of onset (usually diagnosed before 25 years of age), and an autosomal dominant mode of inheritance. The hyperglycemia usually responds well to dietary management alone, and microvascular complications may be less common. MODY contrasts, therefore, with the majority of forms of NIDDM. The existence of large MODY pedigrees has made this disorder amenable to conventional linkage analysis.

In MODY, hyperglycemia occurs as a consequence of reduced insulin secretion rather than reduced insulin action. The observation that reduced insulin secretion by pancreatic β-cells in NIDDM is partly due to chronic hyperglycemia suggested that β-cell sensing

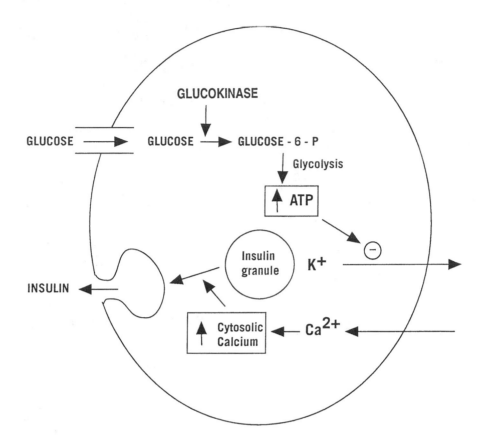

Pancreatic Beta Cell

Figure 4. Simplified diagram of the link between glycolysis and insulin release showing the central role of glucokinase. [Adapted from Figure 1 in A. Hattersley and R. Turner, Mutations of the glucokinase gene and Type 2 diabetes, in *Quarterly Journal of Medicine* 86:227–232, (1993) by permission of Oxford University Press. We are grateful to Dr. A. Hattersley for useful discussion.]

of glucose was at fault. It was postulated that glucokinase, which acts as a glucose sensor on human β-cells, might be a candidate gene for MODY.

Glucokinase is one of the family of hexokinases that catalyze the conversion of glucose to glucose 6-phosphate, the first step of glycolysis (Figure 4). Glucokinase is specific to the liver and pancreatic β-cell, differing from other hexokinases in that it has a low affinity for glucose and is not inhibited by glucose 6-phosphate. These properties enable glycolysis in β-cells to increase in response to the extracellular concentration of glucose. Increased flux through the glycolytic pathway in turn increases the intracellular ATP/ADP ratio, which closes ATP-dependent potassium channels in the β-cell membrane, and this action stimulates insulin release.

The glucokinase gene is situated on the short arm of chromosome 7 and is flanked by microsatellite sequences. These polymorphisms were used to demonstrate linkage between MODY and glucokinase in a number of families. Abnormalities of the coding region of the glucokinase gene subsequently were identified in the affected individuals. The scanning technique used is termed single-stranded conformational polymorphism (SSCP) analysis, which identifies some single base mutations in PCR-amplified DNA. It should be noted, however, that different glucokinase mutations have been identified in different MODY families, and linkage with glucokinase was not found in all MODY families. Thus MODY itself must be heterogeneous. This is further evidenced by the demonstration of linkage

between MODY and the adenosine deaminase gene in one large pedigree.

6.5   GLUCOKINASE POLYMORPHISM IN NON-MODY NIDDM

Positive associations between glucokinase polymorphism and NIDDM have been found in a black American population and in a Mauritian creole population, but not in Caucasian populations. Linkage between NIDDM and glucokinase has not been found in Caucasian families with NIDDM. The significance of the associations between NIDDM and glucokinase in non-Caucasian populations is unclear. They may have occurred because the diabetic subjects actually included large numbers of MODY patients. Alternatively, the associations may have been due simply to ethnic stratification.

One major benefit of identifying glucokinase as the determinant of some cases of MODY is the clear demonstration of NIDDM heterogeneity. Distinction of other subsets of NIDDM may allow identification of other susceptibility genes.

7   FUTURE STUDIES

The genetics of both IDDM and NIDDM remain incompletely understood. Major determinants of susceptibility to IDDM have been mapped to the *DQ* subregion of the MHC. The *INS* region is also implicated. The exact identities of the predisposing genes have yet to be established, along with the mechanisms by which these genes predispose to pancreatic β-cell destruction. Other genes are likely to be involved. The genetic defect responsible for a small subset of NIDDM has now been established.

The candidate gene approach that proved disappointing for so long has started to yield results. An alternative method is to screen systematically gene markers distributed throughout the genome for linkage with diabetes. The construction of detailed maps of the entire genome is crucial to such an approach. This method has been useful in IDDM for which large numbers of multiplex families are now available in the British Diabetic Association—Warren repository. The next challenge is to precisely characterize the newly-identified susceptibility loci. No such family resource is available for NIDDM. At present, the best approach for NIDDM seems to be to characterize the metabolic defects that identify subsets of NIDDM and then to study seemingly appropriate candidate genes. Elucidation of the molecular genetics of diabetic diseases should clarify the pathogenesis of these conditions and may improve their treatment, as well as suggesting strategies for prevention.

See also DIABETES INSIPIDUS; IMMUNOLOGY; MAJOR HISTOCOMPATIBILITY COMPLEX.

Bibliography

Field, L. L. (1991) Non-HLA region genes in insulin-dependent diabetes mellitus. In *Baillière's Clinical Endocrinology and Metabolism,* Vol. II: *Genetics of Diabetes,* L. C. Harrison and B. D. Tait, Eds. Baillière Tindall, London, Philadelphia, Sydney, Tokyo, Toronto.

Hales, C. N., and Barker, D. J. P. (1992) Type 2 (non-insulin-dependent) diabetes mellitus: The thrifty phenotype hypothesis. *Diabetologia,* 35:595–601.

Hattersley, A. T., and Turner, R. C. (1993) Mutations of the glucokinase gene and type 2 diabetes. *Q. J. Med.* 86:227–232.

Jenkins, D., Mijovic, C., Fletcher, J., Jacobs, K. H., Bradwell, A. R., and Barnett, A. H. (1990) Identification of susceptibility loci for type 1 (insulin-dependent) diabetes by trans-racial gene mapping. *Diabetologia,* 33:387–395.

Parham, P. (1990) A diversity of diabetes. *Nature,* 345:662–664.

O'Rahilly, S., Wainscoat, J. S., and Turner, R. C. (1988) Type 2 (non-insulin-dependent) diabetes mellitus: New genetics for old nightmares. *Diabetologia,* 31:407–414.

Rich, S. S. (1990) Mapping genes in diabetes. *Diabetes,* 39:1315–1319.

Thomson, G., Robinson, W. P., Kuhner, M. K., Joe, S., MacDonald, M. J., Gottschall, J. L., et al. (1988) Genetic heterogeneity, modes of inheritance, and risk estimates for a joint study of Caucasians with insulin-dependent diabetes mellitus. *Am. J. Hum. Genet.* 43:799–816.

Todd, J. A. (1995) Genetic analysis of type 1 diabetes using whole genome approaches. *Proc. Natl. Acad. Sci. USA,* 92:8560–8565.

Wicker, L. S., Todd, J. A., and Peterson, L. B. (1995) Genetic control of autoimmune diabetes in the NOD mouse. *Annu. Rev. Immunol.,* 13:179–200.

Disease, Human Genetic Predisposition to: *see* Human Genetic Predisposition to Disease.

Disease Diagnosis: *see* method, disease, substrate used; Genetic Testing; or human disease entries.

Disease Mapping, Human: *see* Human Disease Gene Mapping.

DNA DAMAGE AND REPAIR
George J. Kantor

Key Words

DNA Damage   Any modification of the structure of the DNA polymer.

DNA Repair   A restoration of damaged DNA to its original structure, usually accomplished through a multistep enzymatic process.

Endogenous Agents   Physical and chemical agents of intracellular origin that are part of the normal intracellular environment.

Exogenous Agents   Physical and chemical agents of extracellular origin that can enter the intracellular environment.

Mutation   Any change in the sequence of nucleotides in genomic DNA.

The genetic molecule DNA is continually exposed to a wide variety of chemical and physical agents that can change its structure. These changes interfere with replication and transcription of DNA and are thus referred to as DNA damage. Biological consequences include cell death and mutations—events that may cause cancers, mental retardation, and reduced growth and development. Cells defend against the effects of DNA damage by using enzymatic repair that restores the DNA to its original structure. If repair is timely, the consequences of DNA damage are eliminated. An understanding of the nature of DNA damage, its biological consequences, and its elimination by repair is an important step toward resolving the many health problems associated with exposure to DNA-damaging agents.

1   DEFINITION OF DNA DAMAGE

Modifications of the DNA molecule are caused by the action of either extra- or intracellular agents. Chemicals, electromagnetic radiation, heat, and pressure can modify the normal parts of DNA. Each agent introduces a set of modifications which are usually unique for that agent. For example, ultraviolet light (UV) causes several photochemical changes in DNA that are not caused by any other agent. DNA modifications also occur when incorrect bases inserted in one DNA strand cannot pair correctly with the corresponding base in the other strand. Modifications of these kinds can occur from a DNA replication error or by the spontaneous loss of an amino group (deamination) that converts cytosine to uracil or adenine to hypoxanthine. Several forms of DNA damage induced by specific agents have been identified, but biological effectiveness has been estimated for only a few forms. Examples of specific DNA modifications are illustrated in Figure 1.

1.1 EXOGENOUS DNA DAMAGE

1.1.1 Initial Discoveries with UV

Action spectra studies to measure the biological effects of radiation of different energies demonstrated that nucleic acids were target molecules for the induction of mutations and cell lethality caused by UV. The efficient wavelengths for these biological effects (250–270 nm) represent a range of energies that are absorbed by DNA. These observations, along with the identification of DNA as the genetic molecule, prompted scientists to examine DNA for the responsible chemical modifications. Several photochemical changes in DNA have been identified. The most common of these are cyclobutadipyrimidine (the pyrimidine dimer, Figure 1b) and the pyrimidine(6-4)pyrimidone adduct (the 6-4 photoproduct, Figure 1c). Both have significant effects in biological systems and thus represent from a biological point of view relevant DNA damage. The pyrimidine dimer is the result of a covalent joining of two adjacent pyrimidines in one of the DNA strands. This arrangement is promoted by the absorption of the energy of UV photons by the pyrimidines.

Pyrimidine dimers and other types of UV-induced DNA damage are detected in cells exposed to sunlight. Even though the portion of the solar spectrum that reaches the earth's surface is limited to wavelengths greater than about 290 nm, a part of the remaining UV component (290–320 nm) can be absorbed by DNA and induce these modifications. Direct exposure to noontime sunlight in June, at about the time of the summer solstice, induces about $1–2 \times 10^4$ pyrimidine dimers per minute per diploid human cell held in culture in Dayton, Ohio (39°17′ north latitude). The yield would be lower in the presence of any kind of shading, such as is experienced by viable cells beneath the horny layer in human skin. However, it is clear that sunlight can create changes in DNA that are proven to be lethal to cells and mutagenic to survivors. This is the reason for concern about ozone depletion in the earth's stratosphere. Ozone is the major constituent of the stratosphere that absorbs sunlight UV, thus shielding objects on the earth's surface.

1.1.2 Examples of Other DNA Damages

The success in detecting and identifying UV-induced DNA changes encouraged the investigation of DNA for modifications induced by other agents. Other exogenously induced DNA damages include breaks in the backbone structure of the DNA polymer due to ionizing radiation from X-rays and some chemicals. If a break in each of the two DNA strands occurs at the same relative position in the double-stranded structure, a double-strand break results (Figure 1c). Heat (thermal energy) promotes the removal of purines and pyrimidines, resulting in missing bases along the DNA chain. These are referred to as apurinic or apyrimidinic sites (AP sites). Chemicals known as alkylating agents add alkyl groups to the purine and pyrimidine rings. For example, dimethylnitrosamine is an alkylating agent to which humans are exposed daily through ingested foods including meats and vegetables, and in vivo formation in the stomach. It is responsible for methylating guanine and forming the high-

ly mutagenic and carcinogenic O^6-methylguanine in DNA (Figure 1b). Highly reactive chemical radicals formed through the direct action of ionizing radiation on water, such as hydroxyl radicals, interact with the purines and pyrimidines, promoting several radiolysis products. Examples of major radiolysis products of thymine are thymine glycol and 5-hydroperoxy-6-hydroxy-5,6-dihydrothymine. Because of the multiplicity of radiolysis products, the biological activity of individual products has not been well characterized. DNA strand cross-links, covalent bonds that hold the two complementary DNA strands in an inseparable state, can be created directly by radiation and by the action of some chemicals. These examples serve to illustrate the kinds of changes referred to as DNA damage. Usually, the specific DNA damage alters the chemical structure enough to prevent the appropriate hydrogen bonding that is responsible for the normal base pairing found in DNA (Figure 1a). This normal base pairing is the basis for the hereditary role of DNA.

1.2 ENDOGENOUS DNA DAMAGE

Spontaneously induced DNA damage produced by endogenous agents probably accounts for the greatest fraction of damage expe-

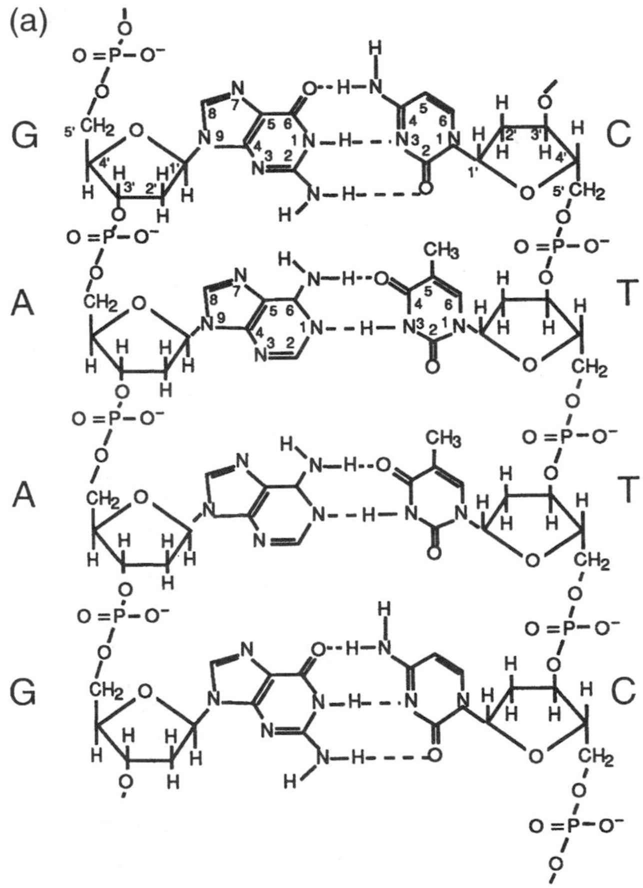

Figure 1. Examples of modifications that have been identified as biologically relevant DNA damage. (a) Representation of a portion of a double-stranded DNA molecule, using structural formulas to show parts of two antiparallel chains with complementary base pairing achieved through hydrogen bonding (———) of adenine (A) to thymine (T) and guanine (G) to cytosine (C). The numbers in the ring structures identify specific atoms in the nucleotides. (continued next page)

rienced by most cells of any biological system. The damage is caused by oxidants produced in cells as by-products of normal metabolism, by thermal energy at normal temperatures, by DNA biosynthetic errors, and by other normal events such as deaminations of the purine and pyrimidine rings.

Oxidants in cells include the superoxide anion ($O_2^{\cdot-}$), hydrogen peroxide (H_2O_2), hydroxyl radicals ($\cdot OH$), and singlet oxygen ($^1O_2$). Singlet oxygen is generated through incomplete reduction of oxygen to water during respiration, by exposure of cells to certain metals, some drugs, and radiation (including light), and by release from stimulated macrophages. Many chemical modifications of DNA by oxidants occur. A few are illustrated in Figure 1. These include thymine glycol and 5-hydroxymethyluracil. 8-Hydroxydeoxyguanosine and cyanuric acid are major products induced by singlet oxygen.

Other naturally occurring DNA modifications include AP sites located along the DNA strands. Human body temperatures (37°C) are sufficient to break the glycosidic bond between a sugar and a purine or a pyrimidine, creating such sites. Spontaneous deaminations and replication errors create sites of incorrect base pairings in the double-stranded structure.

Oxidants and oxidant-induced DNA damage and other kinds of spontaneously induced DNA damage are ubiquitous in biological systems. It has been estimated that these natural reactions cause be-

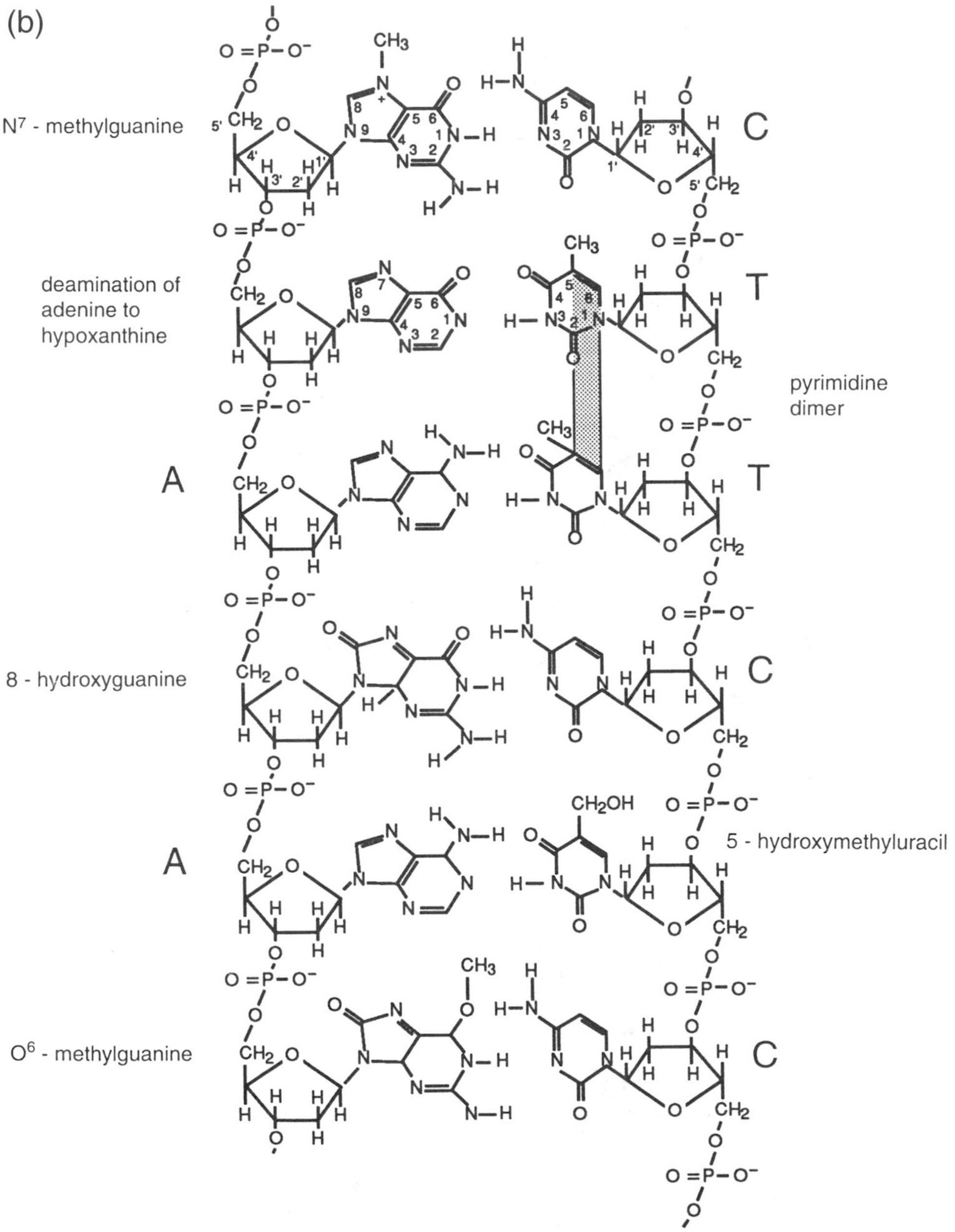

Figure 1 (continued) (b) Examples of DNA damage include the UV-induced pyrimidine dimer, the alkylation products O^6-methylguanine and N^7-methylguanine, the oxidant-induced 5-hydroxymethyluracil, the singlet-oxygen-induced 8-hydroxyguanine, and the adenine deamination product hypoxanthine. (continued next page)

tween 10^4 and 10^5 oxidant-induced damaged sites and an equal number of abasic sites per human cell per day. While these numbers represent a very small fraction (about 0.0004%) of the total number of DNA bases per diploid cell (about 6×10^9 bp), the absolute number is significant. Each damaged site has the potential of interfering with normal cellular events and resulting in a mutation or cell death. The recognition that these reactions involving natural products cause a significant level of DNA damage constitutes a major advance in our understanding of the role of DNA damage in the health of individual cells.

2   DEALING WITH DNA DAMAGE

2.1   BIOLOGICAL INDICATIONS OF REPAIR

DNA-damaging agents can kill normal cells and induce mutations in surviving cells. Scientists have discovered mutant cells that are much more sensitive than normal cells to the lethal action of DNA-damaging agents. The mutant cells that are still viable after exposure to a DNA-damaging agent have a higher mutation frequency than normal cells treated in an identical manner. Viruses or plasmids with damaged DNA usually do not multiply as well in host cells

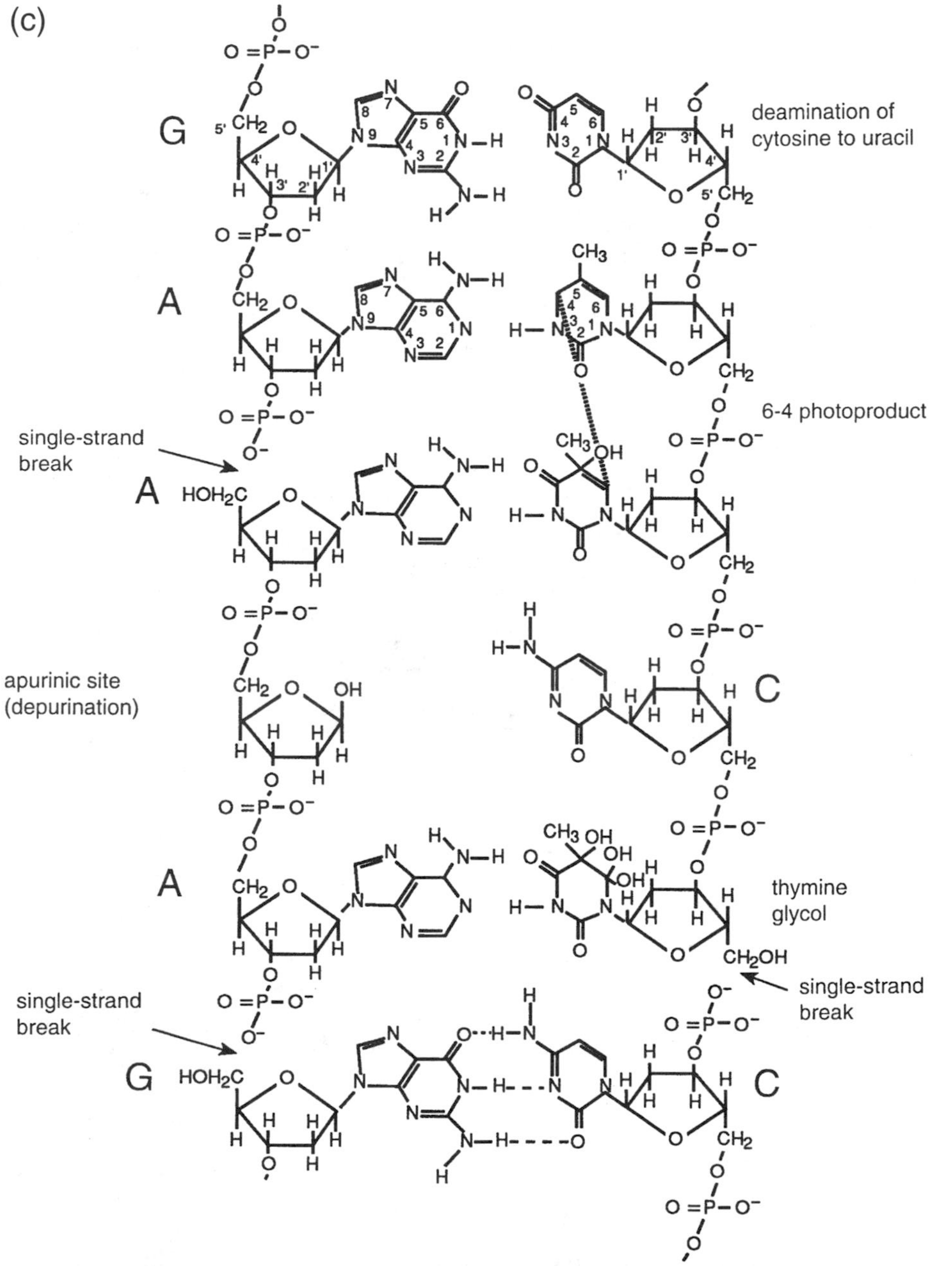

Figure 1 (continued) (c) Examples of DNA damage include the UV-induced 6-4 photoproduct, the cytosine deamination product uracil, the spontaneously induced apurinic site (AP site) caused by disruption of the glycosidic bond to guanine, the oxidant-induced thymine glycol, and the radiation-induced single- and double-strand breaks.

known to be sensitive to DNA-damaging agents as in normal cells. These traits of the sensitive mutant cells are readily observable indicators of a cellular defect in dealing with DNA damage. Normal cells have an ability to restore damaged DNA to its original structure; sensitive cells usually lack this ability. The damage in DNA blocks its replication and blocks its transcription into an RNA template, which provides the necessary information for synthesis of proteins. These essential biochemical processes cannot occur unless the DNA damage is removed.

2.2   Mechanisms of DNA Repair

Many different enzymatic processes for restoring damaged DNA to its original structure have been discovered. These processes, referred to as DNA repair, include direct reversal of the damage and damage elimination by cutting it out and restoring the section that was cut away.

2.2.1 Direct Reversal

Photoreactivation is the direct photoenzymatic reversal of pyrimidine dimers to the original pyrimidine monomeric structures. The enzyme DNA photolyase binds to DNA at the site of a pyrimidine dimer. A photon of blue light (300–500 nm) activates the dimer–photolyase complex, providing energy to disrupt the cyclobutane ring of the dimer. This restores the original pyrimidines. Photolyase is released in an unaltered form, capable of binding to another pyrimidine dimer. This repair process is outlined in Figure 2.

Another example of direct reversal of DNA damage is the removal of alkyl groups by alkyl-accepting enzymes known as transferases. The O^6-methylguanine product is restored to its original guanine structure by transfer of the methyl group to a cysteine residue of the specific enzyme O^6-methylguanine methyltransferase. In the reaction, the enzyme is altered to an inactive state and is unable to act again as an alkyl transferase.

2.2.2 Base Excision Repair

Several forms of DNA base damage are repaired by removing the damaged base and replacing it with a new and correct one. The correct one is selected by the base in the opposite strand using normal Watson–Crick base pairing rules. The repair process is initiated by the hydrolysis of the N-glycosidic bond joining the damaged base to the sugar, creating an AP site. In mammalian cells, for example, the hypoxanthine illustrated in the damaged DNA molecule in Figure 1 is removed through the action of the specific hypoxanthine–DNA glycosylase, and the uracil in Figure 1 is removed by uracil–DNA glycosylase. The AP site is further processed by an AP endonuclease that cleaves one of the DNA backbone phosphodiester bonds adjacent to the AP site. The AP sugar–phosphate group is then removed by a 5′-deoxyribophosphodiesterase. The resulting single nucleotide gap is filled in by the action of a DNA polymerase. A DNA polynucleotide ligase seals the new nucleotide in the DNA structure, completing the repair process. Several different DNA–glycosylases with unique substrates representing modified bases have been identified in both prokaryotic and eukaryotic cells.

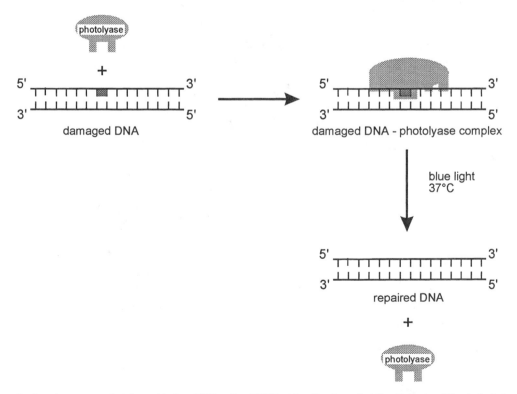

Figure 2.   Photoreactivation. An enzyme, photolyase, binds to UV-irradiated DNA at the site of a pyrimidine dimer (small box). A photon of blue light is absorbed by the photolyase-DNA complex, and the absorbed energy is used to collapse the pyrimidine dimer into pyrimidine monomers, liberating the photolyase. (DNA is represented as two parallel lines to depict the antiparallel sugarphosphate backbone shown in Figure 1. Short vertical lines depict complementary base pairs along the length of the backbone.)

2.2.3 Nucleotide Excision Repair

One of the most thoroughly studied and best understood repair mechanisms is nucleotide excision repair, a mechanism responsible for repair of many different forms of bulky DNA damage. This process employs enzymes to excise a short section of the DNA strand that contains damage. The opposite complementary strand is left intact. Subsequent DNA synthesis and rejoining steps fill in the resulting single-strand gap, thereby restoring the molecule to its original structure. Nucleotide excision repair is a general repair mechanism detected in every organism examined thus far. It is found in both prokaryotes and eukaryotes. It is a common mechanism for repair of sunlight-induced and other DNA damage in human cells.

The molecular details of nucleotide excision repair have not been characterized in human cells, although investigators have detected the general steps requiring DNA incision (i.e., excision of short damage-containing DNA regions, resynthesis of the excised regions, ligation). The best-characterized system, the one detected in the bacterium *Escherichia coli,* serves as a model for studying excision repair in other organisms including humans.

The *E. coli* excision repair mechanism was deciphered as a direct result of the isolation of three different *E. coli* mutants, all very sensitive to UV radiation, as well as to several other DNA-damaging agents. These mutants are referred to as *uvrA, uvrB,* and *uvrC.* The mechanism required for repair of the UV-induced DNA damage and the DNA damage induced by the several other agents to which the mutants are sensitive requires the concerted action of each of the three proteins that are the products of the *uvrA, uvrB,* and *uvrC* genes. These three proteins constitute a complex referred to as the ABC excinuclease, which is capable of hydrolyzing specific phosphodiester bonds on both sides of a damaged site, such as a pyrimidine dimer, in a DNA strand. This complex serves to remove a 12–13 nucleotide oligomer containing the DNA damage. The gap that is generated is filled in by DNA polymerase I and sealed by ligase. This excision repair process is illustrated in Figure 3.

An important feature of the ABC excinuclease repair process is its wide range of substrate. The excinuclease excises UV- and chemical-induced DNA damage, including pyrimidine dimers, 6-4 photoproducts, and many base damages caused by chemicals. This is in contrast to the substrate range of enzymes like photolyase (which is specific for pyrimidine dimers alone), the O^6-methyltransferase (which is specific for the O^6-methylguanine adduct), and several DNA–glycosylases with high substrate specificity.

2.2.4 Other Features of Repair

Several features of DNA repair have not yet been dealt with, and although they are beyond the scope of this review, deserve to be mentioned. To accommodate either chronic exposure or a sudden acute exposure to DNA-damaging agents, some organisms have evolved inducible repair systems that provide for an increased capability to repair DNA damage at times of unusually high damage levels. In addition, several other more subtle mechanisms for repair exist to deal with some of the DNA modifications that escape immediate excision repair. One of these, referred to as recombinational repair, uses preexisting DNA strands to fill in single-stranded DNA gaps created as a result of incomplete DNA synthesis around a site of DNA damage. Mechanisms involved in recombination, including the involvement of homologous DNA sequences and, in *E. coli,* the RecA protein, are employed in this process.

Recent advances in the understanding of DNA excision repair emphasize the heterogeneity of repair at different locations throughout the chromosomes of both prokaryotes and eukaryotes. In fact, DNA is not repaired equally well in all parts of chromosomes. Some genomic regions are repaired early and rapidly, and others slowly, over long periods of time. Regions repaired rapidly are the transcriptionally active ones, while the slowly repaired regions are inactive ones. The repair heterogeneity extends further, to specific DNA strands of the double-stranded structure. Transcribed strands of specific genes are repaired more rapidly than the nontranscribed strands. A likely explanation for this phenomenon can be proposed now that a protein that couples transcription and repair has been detected in *E. coli.* Both the protein and the gene for this protein have been identified and isolated. Apparently the transcription–repair coupling factor (TRCF) recognizes the RNA polymerase stalled at a DNA-damaged site in the transcribed DNA strand and binds to it. This interaction promotes the release of the polymerase, but the TRCF remains at the damaged site. The DNA-bound TRCF recruits the ABC excinuclease complex through affinity of TRCF for the A protein. The ABC excinuclease then performs the dual incision (i.e., on each side of the damaged site) to initiate the excision repair process.

3   DNA REPAIR IN HUMAN CELLS

3.1   Repair in Normal Human Cells

Efficient DNA repair is an important factor in preventing DNA damage from causing cell death and mutations, events that are responsible for cancers in humans and probably other maladies including some aspects of aging. Our understanding of repair mechanisms and the consequences of repair is limited because of the inherent difficulties in using humans as experimental subjects and the lack of sensitive and reproducible methods that can be applied successfully to the study of molecular events in whole animals. Most of the information comes from epidemiological studies, from studies with experimental animals other than man, and from extrapolation of data obtained using human cells maintained in cell culture systems. Transgenic mice with human repair genes represent a new model system that will provide important evidence for the significance and mechanisms of repair. Our best understanding is that for most individuals, most DNA damage is removed rapidly by a repair process that proceeds in an error-free manner in all cells.

3.1.1 Repair Rate

DNA damage in human cells is repaired by a battery of systems that include all those discussed earlier. For example, cultured human cells remove about 80% of UV-induced pyrimidine dimers in 24 hours by an excision repair process. Human skin cells in vivo may repair the same amount of damage even faster, in 2–3 hours, again by an excision repair process. Direct reversal of some pyrimidine dimers induced in living skin cells by the UV component of sunlight can occur by way of the photoreactivation process, using the blue component of sunlight. Other forms of damage such as alkylation damage are repaired rapidly through the direct reversal mechanism involving alkyl transferases.

While repair of a variety of different kinds of DNA damage in human cells is rapid, repair may not remove all damaged sites. For example, cultured human cells repair the remaining 20% of pyrimidine dimers over several days. Alkylation damage is repaired in

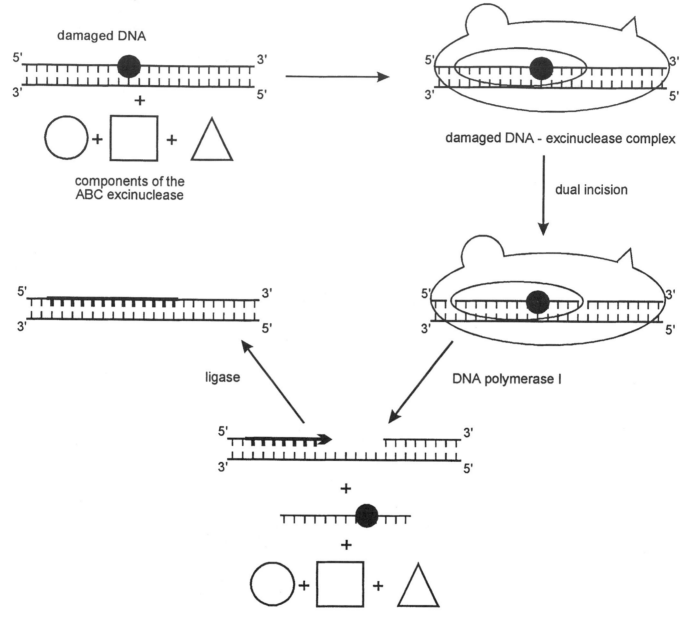

Figure 3.   A model for nucleotide excision repair in *E. coli*. The protein products of the *uvrA, uvrB,* and *uvrC* genes combine to form the ABC excinuclease complex on damaged DNA. The excinuclease breaks the phosphodiester backbone in the single strand of DNA twice, once on each side of the damaged site. DNA polymerase I initiates synthesis at one of the incision locations and displaces the excinuclease complex and the DNA fragment containing the damaged site. The gap-filling DNA synthesis step is terminated at the other incision site. Ligase joins the new DNA to the original DNA, completing the repair process.

a similar manner: most is removed at a rapid rate, but the remaining fraction is eliminated slowly over long periods. A small fraction of the DNA damage may never be repaired. Indeed, a low level of DNA damage may accumulate in some cells during the lifetime of an individual.

3.1.2 Repair Capacity

Some evidence indicates that a wide variation in DNA repair capacity occurs throughout the normal human population. In addition, a decline in the repair capacity with increased age has been detected. This decline is estimated to be about 0.6% per year after ado-

lescence, indicating that elderly individuals (60–80 years of age) may have a 20–40% decrease in the capacity to repair DNA. This wide variation and the decline with age may be in part responsible for an increased risk for cancers in some youth and in all older people.

3.2 Hereditary Diseases with Defects in Repair of DNA

Normal excision repair is defective in several human hereditary diseases including xeroderma pigmentosum (XP), ataxia telangiectasia, Cockayne's syndrome, Fanconi anemia, Bloom's syndrome, and trichothiodystrophy. Defective repair may cause the symptoms associated with these rare disorders, all of which are autosomal and recessive. A higher than normal likelihood of cancer is associated with all these except Cockayne's syndrome and trichothiodystrophy. The XP disease has been studied more extensively than the other disorders. XP patients have a high incidence of skin cancer on the parts of their body that are exposed to sunlight. Their individual cells do not repair most of the DNA damage induced by the UV radiation from sunlight and consequently are very sensitive to killing by UV and have a high mutation frequency. It is this increased mutation frequency in surviving cells that probably is responsible for the high incidence of skin cancer in XP patients. The only known effective treatment for afflicted individuals is a preventive one, the avoidance of sunlight and other UV sources. This association in the XP disorder of cancer with DNA damage and defective repair represents some of the best evidence that damage to human DNA, if not repaired in a timely fashion, can induce malignant lesions and other developmental problems.

The study of these human hereditary diseases is helpful in analyzing the normal repair process. Cells from such individuals are mutants blocked at specific steps in excision repair. The definition of a specific molecular defect in repair in a mutant cell defines a necessary step in normal repair.

3.3   HUMAN DNA REPAIR GENES

The XP disorder has been subdivided into seven different categories or complementation groups through a genetic analysis known as the complementation test. This test, in its broadest interpretation, defines the minimum number of genes involved in a specific process. Cells from all the XP patients studied have a defect in the ability to initiate nucleotide excision repair. They are unable to make the incision in DNA in the vicinity of the damage. The results of the complementation tests suggest that the protein products from at least seven genes are involved in the initial steps. Obviously the nucleotide excision repair process in human cells is much more complex than the ABC excinuclease system operable in *E. coli*. An area of active research consists of the mapping of some of the human repair genes in the genome, specifically those defined by the mutations that result in the XP disorder, and the cloning of these genes in the process of isolating and defining them. Several human repair genes have now been isolated and their DNA sequences determined.

The approach to understanding excision repair in normal human cells through the study of repair-defective mutant cells is exemplified here by an XP mutation and the Cockayne's syndrome (CS) mutation. Specifically, the XP mutation defined by the complementation group C(XP-C) renders these cells incapable of repairing most of their DNA but capable of repairing the 5–10% of the genome that represents the actively transcribed genes. In contrast, CS cells repair most of the genome efficiently but do not repair the actively transcribed genes at the preferential rate detected in normal cells. Apparently, CS cells lack the strand-specific repair mechanism. The genes responsible for the XP-C defect (the *XPCC* gene) and the CS defect (the *ERCC-6* gene) have been isolated and the DNA sequences determined. The cloning of the genes allows for the synthesis of the proteins and their isolation in quantities suitable for study. These isolations, along with the DNA sequence information, will allow the determination of the amino acid sequences of the individual proteins, which will greatly aid in determining their respective functions. Analysis of the *ERCC-6* gene has led to the sug-

gestion that its protein product is the human analogue of the *E. coli* transcription–repair coupling factor discussed in Section 2.2.4. The ERCC-6 protein may be one that promotes excision repair by recruiting repair enzymes to the damaged sites in transcriptionally active regions. While the XP-C cells probably have the normal gene for the TRCF, the XP-C mutation represented by the *XPCC* gene may define a coupling factor required for repair of genomic regions not involved in transcription.

4   PERSPECTIVES

DNA repair in human cells is necessary for the continuance of cell functions and is ultimately essential for the health of the individual. The prospect for isolating and defining the genes involved in repair processes is now very real. The molecular events involved in maintaining the stability of DNA in the somatic cells can be resolved. This resolution should lead to a better understanding of carcinogenesis and to means for preventing diseases associated with defects in the ability to process DNA damage. Known human hereditary diseases with DNA repair defects are rare diseases that in total do not account for a high fraction of health problems induced by DNA damage. The understanding of these diseases and the development of genetic and other means for therapy are very important to affected individuals and their families. However, most cancers occur in normal people with no known genetic disorders. A thorough understanding of DNA damage, its causes, and the molecular events involved in its repair, including knowledge of the genes, will promote better health in the normal human population. Better health will come about through improved risk assessments, awareness of risks, means of controlling the risk and, finally, through intervention methods that can ameliorate negative health effects due to unrepaired DNA damage.

See also ENVIRONMENTAL STRESS, GENOMIC RESPONSES TO; IONIZING RADIATION DAMAGE TO DNA; ULTRAVIOLET RADIATION DAMAGE TO DNA.

Bibliography

Ames, B. N., and Gold, L. S. (1991) Endogenous mutagens and the causes of aging and cancer. *Mutat. Res.* 250:3–16.

Bohr, V. A., Phillips, D. H., and Hanawalt, P. C. (1987) Heterogeneous DNA damage and repair in the mammalian genome. *Cancer Res.* 47:6426–6436.

Cleaver, J. E., and Kraemer, K. H. (1989) Xeroderma pigmentosum. In *The Metabolic Basis of Inherited Disease*, C. R. Scriver, A. L. Beaudet, W. S. Sly, and D. Valle, Eds., pp. 2949–2971. McGraw-Hill Information Sciences, New York.

Friedberg, E. C., Walker, G. C., and Siede, W. (1995) *DNA Repair and Mutagenesis.* ASM Press, Washington, D.C.

Lindahl, T. (1990) Repair of intrinsic DNA lesions. *Mutat. Res.* 238:305–311.

Mullaart, E., Lohman, P. H. M., Berends, F., and Vijg, J. (1990) DNA damage metabolism and aging. *Mutat. Res.* 237:189–210.

Piette, J. (1991) Biological consequences associated with DNA oxidation mediated by singlet oxygen. *J. Photochem. Photobiol. B: Biol.* 11:241–260.

Sancar, A., and Sancar, G. B. (1988) DNA repair enzymes. *Annu. Rev. Biochem.* 57:29–67.

Setlow, R. B. (1982) DNA repair, aging and cancer. *Natl. Cancer Inst. Monogr.* 60:249–255.

Sutherland, B. M., and Woodhead, A. D., Eds. (1990) *DNA Damage and Repair in Human Tissue.* Plenum Press, New York, London.

DNA:DNA, DNA:RNA and RNA:RNA Hybrids: *see* Nucleic Acid Hybrids, Formation and Structure of.

DNA FINGERPRINT ANALYSIS

Lorne T. Kirby, Ronald M. Fourney, and Bruce Budowle

Key Words

Amplified Fragment Length Polymorphism (AMP-FLP)   PCR-amplified fragment lengths consisting of VNTRs.

AP-PCR   See RAPD.

Chain of Custody (Continuity)   A record of the custody and handling of evidence material from the time it is first obtained until it is entered in a court of law.

DAF   See RAPD.

Exclusion   The elimination of the possibility that a crime suspect is the source of a specimen or that a putative relative is in fact a biological relative.

Hypervariable Region   A segment of DNA characterized by considerable variation in the number of tandem repeats, or a high degree of polymorphism due to point mutations.

Inclusion   Describing the impossibility of excluding a crime suspect as the source of a specimen or putative relative as a biological relative.

Minisatellites   Regions of tandem repeats in the genome.

Minisatellite Variant Repeat PCR (MVR-PCR)   A PCR procedure based not only on tandem repeat copy number but also on the different interspersion pattern that exists among the alleles. Sequence variation is measured along minisatellite alleles consisting of two types (a and t) of repeat unit that differ at one polymorphic site. The DNA profiles of the alleles at the same locus in a diploid organism are superimposed to generate a ternary code in which each repeat unit site is aa, tt, or at(ta).

Polymerase Chain Reaction (PCR)   In vitro replication of specific target DNA sequences.

Random Amplified Polymorphic DNA (RAPD)   RAPD, AP-PCR, and DAF are PCR procedures (differing in the length of primers used, the amplification conditions, and the resolution and visualization of the products) that usually use only one arbitrary sequence primer per reaction for the generation of DNA fingerprints. In an RAPD system, a PCR primer with an arbitrary sequence is used under reduced stringency reactions. Patterns of PCR products are observed after separation by electrophoresis.

Short Tandem Repeat (STR)   Also referred to as simple sequence length polymorphism (SSLP). STRs are tandem repeat regions two to seven base pairs long, scattered throughout genomes. PCR procedures are used to amplify these regions for the generation of DNA profiles. (See also Tandem Repeat.)

Tandem Repeat   The end-to-end duplication of a series of identical or almost identical motifs (usually 2–80 base pairs each) of DNA scattered throughout genomes.

Variable Number of Tandem Repeats (VNTR)   The variable number of tandem repeats forming alleles at a locus.

The analysis of DNA is revolutionizing the field of identification. The stability of DNA, the high degree of assay accuracy and precision, and the need for only minute quantities of tissue have contributed to the application of DNA typing. The use of this procedure is rapidly altering the manner of carrying out genetic analyses in human parentage, rape, and homicide cases; in animal poaching, parentage, breeding, and population studies; in medical analysis; and in patent disputes, clone identity assays, parentage testing, and gene bank management programs involving plants.

DNA analysis is the process of preparing and interpreting bar-code-like profiles or dots of DNA segments for individual identification. Picogram to microgram quantities of genomic DNA are isolated and polymorphic segments are directly analyzed or amplified by the polymerase chain reaction and analyzed. With the exception of identical twins and clones, profiles of the segments from humans, other animals, microorganisms, fungi, and plants are unique to each individual.

1   PRINCIPLES

1.1   SPECIMEN PROCESSING

DNA is the molecule of heredity. It is an integral component of all living matter except RNA viruses. Any living or nonliving organic matter containing relatively intact DNA fragments can be used for analysis. Common sources include human or other animal blood, semen, and solid tissues, and plant foliage and seeds.

A chain of custody (continuity) document is required for a legal case specimen. The drawing and labeling (name ID, date, location) of blood samples in paternity and immigration cases must be authenticated, as must details of specimen collection and custody in rape and homicide cases.

Specimens are stored under clean, dry, cool conditions to reduce contamination by microorganisms and DNA degradation due to cell lysis and DNAase activity. Successful analyses have been carried out on DNA extracted from dried blood stains, dried museum specimens, and frozen tissue preserved for many decades. Indeed, these studies have formed the basis of recent new theories in molecular evolution and anthropology.

A good yield of high molecular weight, duplex DNA devoid of organic and inorganic contaminants can be extracted from tissues (Figure 1). The enzyme proteinase K is generally used to assist with cell lysis and to digest proteins. The detergent sodium dodecyl sulfate (SDS) facilitates the separation of residual protein from DNA. Proteinaceous materials are removed by phenol extraction (or a nonorganic NaCl or LiCl precipitation), and chloroform is added to remove traces of phenol. If the phenol is appropriately equilibrated, most contaminating RNA can be eliminated. High molecular weight DNA is recovered in a relatively pure form by ethanol precipitation in the presence of salt. The extraction and purification steps, under certain circumstances, can be bypassed after cellular lysis. Fragments of DNA can be directly amplified by means of the polymerase chain reaction (PCR) and the amplified fraction used to construct a DNA profile. If this approach is followed, possible inhibitors of the

PCR can be removed by means of a simple nonorganic technique using an ion exchange resin, such as Chelex 100, a Centricon or Micron 100 filter, or both.

If the analyst is presented with mixed specimens, it is possible to separate components such as sperm and female cells (actually non-sperm cells) often present in vaginal swabs from rape victims (Figure 2). The female cells are lysed in an extraction buffer devoid of dithiothreitol (DTT), and the DNA is recovered from the supernatant. The sperm pellet is then lysed by adding extraction buffer containing DTT and increased detergent.

DNA quality can be evaluated by ethidium bromide staining or by Southern blotting and hybridizing with a species-specific probe. The degree of degradation is estimated by observing, under UV light, ethidium bromide stained fragment lengths that were subjected to agarose gel electrophoresis. High molecular weight DNA appears as a single large band, whereas partially degraded DNA forms a long smear of large to small fragments (>20,000 base pairs to a few hundred base pairs for human DNA). Ethidium bromide staining cannot differentiate DNA from different organisms—species-specific probes are used for this purpose.

Quantitation is the final step in preparing DNA for analysis. At least three different procedures are available. In the first technique, the ab-

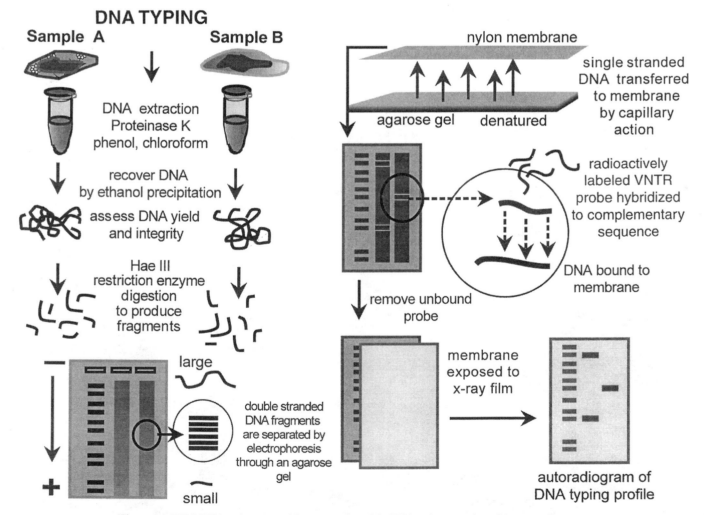

Figure 1.   RFLP DNA typing protocol from extraction of the DNA to the generation of the autoradiogram.

DIFFERENTIAL EXTRACTION

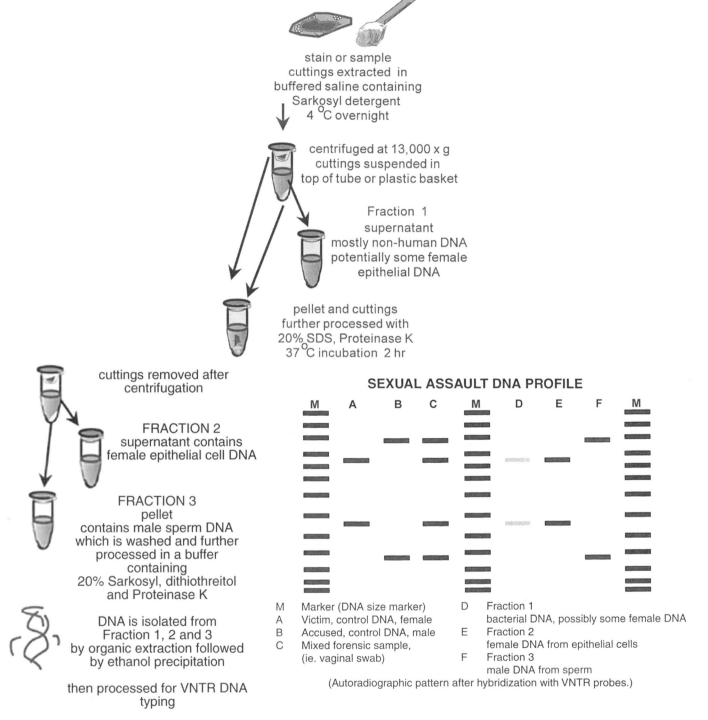

stain or sample
cuttings extracted in
buffered saline containing
Sarkosyl detergent
4 °C overnight

centrifuged at 13,000 x g
cuttings suspended in
top of tube or plastic basket

Fraction 1
supernatant
mostly non-human DNA
potentially some female
epithelial DNA

pellet and cuttings
further processed with
20% SDS, Proteinase K
37 °C incubation 2 hr

cuttings removed after
centrifugation

FRACTION 2
supernatant contains
female epithelial cell DNA

FRACTION 3
pellet
contains male sperm DNA
which is washed and further
processed in a buffer
containing
20% Sarkosyl, dithiothreitol
and Proteinase K

DNA is isolated from
Fraction 1, 2 and 3
by organic extraction followed
by ethanol precipitation

then processed for VNTR DNA
typing

SEXUAL ASSAULT DNA PROFILE

| M | A | B | C | M | D | E | F | M |

M   Marker (DNA size marker)
A   Victim, control DNA, female
B   Accused, control DNA, male
C   Mixed forensic sample,
      (ie. vaginal swab)

D   Fraction 1
      bacterial DNA, possibly some female DNA
E   Fraction 2
      female DNA from epithelial cells
F   Fraction 3
      male DNA from sperm

(Autoradiographic pattern after hybridization with VNTR probes.)

Figure 2.   Protocol for the differential extraction of sperm and nonsperm cells.

sorbance of a DNA solution is measured in a spectrophotometer at 260 nm wavelength. Second, the DNA and a series of standards of known concentration are subjected to agarose gel electrophoresis and stained with ethidium bromide. In the third procedure, an aliquot of the DNA is fixed on a nylon membrane and is hybridized with a human specific probe such as a labeled highly repetitive, primate-specific α-satellite probe p17H8 that detects locus D17Z1.

1.2   SPECIMEN ANALYSIS

Many polymorphic genetic markers (segments) consist of tandem repetitive sequences. Depending on the locus, each repeat can consist of a minimum of two to perhaps 80 base pairs; single base pair polymorphic sites are possible, as well. The number of tandem repeats varies from none to hundreds.

A locus may, therefore, be polymorphic for the number of tandem repeats and for the base pair composition at specific sites within each repeat. There are variable numbers of tandem repeats (VNTRs) and short tandem repeats (STRs), as well as minisatellite variant repeats (MVRs). DNA profile construction is based on the analysis of variations in sequence (e.g., MVRs) at one or more loci by such techniques as restriction fragment length polymorphism (RFLP), amplified fragment length polymorphism (AMP-FLP), and MVR–polymerase chain reaction (MVR-PCR), as indicated in Table 1.

Polymorphic sites that may or may not be located in tandem repeat units provide another form of variation useful in identity analysis. Hence the random amplified polymorphic DNA (RAPD), arbitrary primer PCR (AP-PCR), and DNA amplification fingerprinting (DAF) systems, which are based on the annealing of template DNA to arbitrary sequence PCR primers, such that the primer sequences are sufficiently close together on complementary DNA strands to facilitate amplification. The specifics of the systems differ with respect to length of the primers used, amplification conditions, and resolution and visualization of products.

The final DNA pattern or profile can be visualized with color dyes, fluorescent dyes, radioisotopes, or other stains such as silver. There are three general analytical approaches used for DNA typing: VNTR analysis, dot-blot analysis of sequence polymorphisms, and direct sequence analysis. For VNTR analysis, the restriction-digested or amplified DNA fragments are separated, generally based on their size, by electrophoresis in a sieving medium. After electrophoretic separation, the relative positions of the bands are determined with reference to size standards. In dot-blot analysis, the target DNA sequence is amplified by PCR and the amplified products are fixed to membranes. The DNA is then incubated with allele-specific probes. If hybridization occurs, the dot will be marked; if the DNA does not contain the particular allele, the spot will remain blank. Alternatively, dot-blots can be configured in which the allele-specific probes are fixed to the membrane and the amplified DNA is allowed to hybridize to the fixed probes. This reverse dot-blot approach is used for *HLA-DQα* typing. Finally, the sequence of DNA bases can be read by direct sequencing.

A number of analytical approaches can be used to isolate and to identify profile patterns. DNA from each of a number of loci may be analyzed separately and the results combined to form a multiple-locus profile. The DNA may be analyzed directly, or it may be amplified and then analyzed.

1.3   DATA PROCESSING

Genetic profiles are presented as (1) the presence or absence of dots, (2) the position or size of DNA fragments separated by an elec-

trophoretic approach, or (3) direct sequence information (e.g., AATCGTACCTGATCC). The results produced by any of these methods can be evaluated visually. As data become more abundant and complex, however, computer-assisted analyses are needed to aid in pattern interpretation, and in data storage and manipulation.

Interpretation of dot-blot profiles is performed by "eye," and the data are generally transcribed manually into a computer storage bank. However, since most dots are colored or visible on X-ray film, densitometric scanning systems can be used to automatically score and store the data for future analysis.

Most VNTR analyses are based on size or position. The first-generation data acquisition systems used to determine the relative size of unknown DNA fragments consisted of measurement with a ruler and comparison with known size standards or relative mobility of the DNA fragments (on autoradiograms). Later, semiautomated computer-assisted image capture systems (which are essentially electronic rulers) replaced the manual method of data analysis. With the use of fluorescent or UV detection systems, labeled DNA fragments can be detected in real-time analysis and automatically transcribed to the computer. Both VNTR typing and sequencing can be automated using this approach.

The main advantages of nonmanual data acquisition are reduction of human transcriptional errors and, potentially, reduction in labor. Once the data have been stored in the computer, various statistical analyses, both simple and sophisticated, can be performed for determinations, such as the likelihood of occurrence of DNA profiles for identity typing and linkage analysis.

1.4   QUALITY ASSURANCE

Laboratory quality assurance is the documented verification that proper procedures have been carried out by skilled and highly trained personnel to yield valid and reliable results. The validity of DNA typing results centers on correctly identifying true nonmatches and matches from DNA profile comparisons. "Test reliability" implies reproducibility under defined conditions of use of the same methods, regardless of the laboratories and practitioners doing the work. Quality assurance must encompass all significant aspects of the DNA typing process, including personnel education and training, documentation of records, data analysis, quality control of reagents and equipment, technical controls, data analysis, proficiency testing, reporting of results, and auditing of the laboratory procedures.

Quality assurance and appropriate standards for DNA typing evolved initially from the experience of clinical laboratories but more recently careful definitions have been developed through consensus from forensic laboratories represented in the North Ameri-

Table 1   Common Characteristics of DNA Typing Systems

System	Number of Alleles[a]	Minimal DNA Required (ng)	PCR-Based System
RFLP VNTR (autoradiographic)	>50 (theoretically 250–1000)	40–100.0	No
AMp-FLP	>10 (typically 10–24)	0.5–5.0	Yes
STR	>5 (typically 5–22)	0.5–5.0	Yes
MVR	> Potential 1000	10–50.0	Yes
Mitochondrial DNA sequencing	Unknown	<1.0	Yes
DQ α Amplitype[b]	6	0.2–5.0	Yes
Amplitype Polymarker[b]	12	1.0–5.0	Yes

[a]Based on representative North American databases.

[b]From Perkin-Elmer (Roche Molecular Systems Division).

can nonregulatory initiative, the Technical Working Group on DNA Analysis Methods (TWGDAM). It is important to note that quality assurance guidelines in forensic science must retain enough flexibility to accommodate the nature of forensic samples as well as future advancements in recombinant DNA technology and molecular biology. Good quality management is paramount for obtaining quality results.

2   TECHNIQUES

The procedural approaches outlined in the following text, except those for plant materials, are carried out in the forensic laboratories of the U.S. Federal Bureau of Investigation and/or the Royal Canadian Mounted Police. The methods have undergone considerable scrutiny in terms of quality assurance, and they are designed to facilitate analysis of limited quantities of tissue. (See the bibliography for references for detailed selected protocols.)

2.1   Specimen Processing

The success of DNA fingerprint analysis depends on the quality of the extracted DNA. Quality is a function of the degree of degradation (size of the DNA molecules) and the purity (as measured by the removal of inhibitors of restriction endonucleases and the PCR), the effects on fragment migration through electrophoresis gels, and the differential separation of mixtures of DNA. The forensic protocols are characterized by their relative simplicity, moderate cost, and capability of analyzing small quantities of sample.

2.1.1   Organic Extraction of DNA from Whole Blood

Liquid blood samples usually are collected in EDTA vacutainers. NaF, the anticoagulant used in collection tubes for blood alcohol analysis, also is suitable. Heparin tubes may be acceptable; however, this anticoagulant may interfere with the activity of certain restriction enzymes. Although not the samples of choice, clotted specimens can provide an adequate source of DNA.

Liquid blood can be stored for at least 5 days at room temperature, but 4°C is preferable. For longer storage, the specimen should be aliquotted (0.7 mL) into 1.5 mL screw-cap polypropylene tubes and frozen at −20 or −70°C. Liquid blood may be freeze-thawed as required, and this process is, in fact, advantageous to ensure complete lysis of the red cell membranes. Drying blood onto filter paper is an effective form of storage for later assay; thus, prior to short- or long-term storage, aliquots of the blood may be placed on filter papers such as Schleicher and Schuell 903 Blood Specimen Collection Paper and stored dry and/or frozen.

The extraction procedure involves lysis of the red cells with salt–sodium–citrate (SSC: NaCl, Na_3 citrate), followed by centrifugation, then lysis of the white cell pellet, and digestion of proteinaceous materials by incubating in a sodium acetate, SDS, and Proteinase K solution. The resultant solution is extracted with phenol, and the DNA is precipitated from the aqueous layer by the addition of cold ethanol and salt followed by centrifugation. (If the DNA is to be amplified by PCR as opposed to direct restriction analysis, a procedure involving butanol extraction and Centricon or Micron 100 dialysis of the aqueous layer can be followed.) Considerable care must be exercised not to disturb the DNA pellet when removing the ethanol. Excess ethanol usually is removed under vacuum in a specially designed centrifuge. The resultant pellet should be slightly moist to ensure resuspension in TE (Tris-EDTA) buffer. The DNA solution can be quantitated (0.7 mL of blood typically yields 20–40 μg of genomic DNA) and used immediately for DNA typing, or it can be tightly capped to prevent desiccation, and stored at 4°C.

2.1.2   Organic Extraction of DNA from Body Fluid Stains

Organic extraction is applicable to blood, saliva, semen, or other stains dried on fabrics or similar materials. Two approaches are commonly used. In the first, the stained material (approximately 1 cm²) is cut into pieces and placed into a 1.5 mL polypropylene centrifuge. Stain extraction buffer (consisting of Tris, Na_2EDTA, SDS, NaCl, and DTT, together with proteinase K) is added to saturate the stain. The tube is tightly capped and then incubated. The extraction buffer, containing the cellular DNA, is separated from the substrate material by transferring the stain cuttings to the Costar basket and centrifuging. The extracted cuttings should be sealed in a package and stored at −20°C for possible future reference and the stain extract transferred to a 1.5 mL microcentrifuge tube for phenol extraction and DNA precipitation.

In a second approach, a 1.5 mL Sarstedt tube that has a depression in the cap is used to separate the stain from its substrate. The pieces of stain material are placed in the tube, and extraction buffer with Proteinase K is added. After incubation, a hole is punched in the tube cap, the cuttings are transferred to the cap, and the unit is centrifuged. The cap with the spent cutting is removed, a new cap is added, the solution is extracted with phenol, and the DNA is precipitated. Alternatively, butanol extraction and Centricon 100 dialysis can be carried out.

A 1 cm² fabric stain contains approximately 50 μL of fluid with a yield of about 50 ng of DNA per milliliter of blood and 80 ng per milliliter of semen.

2.1.3   Nonorganic Extraction of DNA from Whole Blood and from Body Fluid Stains

Nonorganic extraction methods that use NaCl or LiCl as a substitute for phenol and chloroform to salt out protein also are available. Caution must be exercised if these methods are being considered for forensic analysis, since problems with restriction endonuclease digestion and band shifting have been reported. However, with proper care these methods can produce satisfactory results.

2.1.4   Differential Extraction of DNA from Sperm-Containing Stains

Two procedures, one with Costar Spin-X tubes and the other with Sarstedt tubes, are currently in use to isolate DNA from sperm cells on anal, vaginal, or oral swabs, and from seminal stains (Figure 2).

Three fractions are produced with the first method. Fraction 1 contains primarily nonhuman DNA with some epithelial cells; fraction 2 contains primarily epithelial cell DNA; and fraction 3 contains primarily sperm DNA. The swab or stain material is cut into pieces and the pieces are placed into a 1.5 mL tube. Phosphate-buffered saline (PBS) and Sarkosyl are added to saturate the cuttings and the resultant is incubated. Fraction 1 is collected by transferring the cuttings to a small plastic basket, centrifuging, and removing the supernatant (fraction 1). The cuttings are transferred back to the 1.5 mL tube containing the pellet; next TNE buffer (Tris, NaCl, EDTA), SDS, plus Proteinase K are added, and the product is incubated. Fraction 2 is collected by transferring the cuttings to the plastic basket, centrifuging, and removing the supernatant. The cuttings should be stored frozen. TNE, Sarkosyl, DTT, and Proteinase K are added to the pellet (fraction 3), and the product is incubated. DNA is purified from each fraction by organic extraction and ethanol precipitation.

In the second method, the cuttings together with TNE buffer, Sarkosyl, plus Proteinase K, are placed in a 1.5 mL Sarstedt tube and the resultant is incubated. A hole is punched in the tube depression cap, the unit is centrifuged, and the supernatant (lysed female vaginal cellular fraction) is removed and saved. The cuttings should be stored frozen. A new cap is placed on the tube, the sperm pellet is resuspended in TNE, Sarkosyl, DTT, plus Proteinase K, and the resultant is incubated. DNA from the female fraction and from the sperm fraction is purified by organic extraction and ethanol precipitation.

2.1.5 Chelex Extraction of DNA from Whole Blood and from Blood Stains for PCR Amplification

A 3 μL aliquot of whole blood or a 3 mm^2 blood stain is placed in a 1.5 mL microcentrifuge tube, PBS buffer is added, and the resultant is vortexed, incubated, and microcentrifuged. The supernatant is discarded and 5% Chelex 100 is added to the pellet. The pellet is incubated, boiled, and microcentrifuged. The DNA-containing supernatant is used in the PCR amplification process.

2.1.6 Chelex Extraction of DNA from Sperm-Containing Stains for PCR Amplification

Pieces of swab or stain fabric are placed in a 1.5 mL microcentrifuge tube, PBS is added, and the material is incubated. The cuttings are removed and stored frozen. The tube is microcentrifuged, the supernatant is discarded, and Proteinase K solution is added to the pellet. After incubation, the tube is microcentrifuged—the supernatant contains the lysed (generally female in origin) cell fraction, and the pellet contains the sperm. The pellet is thoroughly washed and then reacted with 5% Chelex, Proteinase K, and DTT. The supernatant (female cell fraction) is treated with 20% Chelex. After incubation, the samples are boiled and microcentrifuged, and the DNA is purified by organic extraction and ethanol precipitation.

2.1.7 Chelex Extraction of DNA from Saliva from Oral Swabs, Filter Paper, or Gauze

A 3 mm^2 portion of the sample is placed in a 1.5 mL microcentrifuge tube and 5% Chelex solution is added. After incubation, the tube is boiled, then microcentrifuged. The resulting supernatant is used in the PCR amplification process.

2.1.8 Extraction of DNA from Bone

Soft tissue is first removed, the bone then is crushed into small fragments and ground into a fine powder in the presence of liquid nitrogen. Aliquots of the powdered bone are transferred to polypropylene tubes, and EDTA is added for decalcification. The decalcification process is repeated over a number of days until the supernatant remains clear after the addition of a calcium detection solution consisting of saturated ammonium oxalate solution. The pellet is repeatedly washed with distilled water, then extracted with Tris, EDTA, NaCl, and DTT buffer and later treated with Proteinase

K. The resulting DNA is purified by organic extraction followed by n-butanol concentration and recovery, using a Centricon 30 microconcentrator tube (membrane filtration).

2.1.9 Recovery of DNA from Hair Roots

The proximal 5–10 mm of each hair root is placed in a 1.5 mL microcentrifuge tube, and stain extraction buffer is added together with Proteinase K. The tube contents are observed after the incubation period. If portions of the hair root remain, additional Proteinase K and DTT are added for another period of incubation. The DNA is purified by organic extraction and ethanol precipitation.

2.1.10   Extraction of DNA from Plant Tissues

The fortified polysaccharide cell walls of plant tissues pose an additional problem in the isolation of DNA. Disruption generally is accomplished by both mechanical and chemical techniques including freeze-thawing, grinding or homogenizing, enzymatic digestion, and treatment with hexadecyltrimethylammonium bromide (CTAB) (Figure 3).

A number of protocols are available to isolate DNA from plant tissues including broadleaf and conifer foliage, seeds, and callus. Minipreparations yielding approximately 800 μg of DNA can be obtained easily from 4.0 g of leaf tissue that is first ground with a pestle in a mortar containing sand and liquid nitrogen.

DNA suitable for PCR amplification has been isolated from foliage and callus tissue using a Mini-Beadbeater (Biospec Products, Bartlesville, OK) process. A KCl solution (including 2-mercaptoethanol for older tissues), glass beads, and 10–50 mg of tissue are added to a 2 mL microcentrifuge tube. The tube contents are homogenized in the Mini-Beadbeater, then boiled and microcentrifuged. The resulting supernatant is used in the PCR amplification reaction.

Megagametophyte DNA can be isolated from individual seeds as follows. The seeds are imbibed in water prior to megagametophyte removal. The tissue is ground in Tris, EDTA, sorbitol, and B-mercaptoethanol buffer in a microcentrifuge tube. Sarkosyl is added, the resultant is incubated, then NaCl and CTAB are added, and incubation is continued. RNA is digested by the addition of RNAase. DNA is extracted from the homogenate with phenol–chloroform–isoamyl alcohol and precipitated with ethanol. DNA yields per seed (Douglas fir) range from 300 to 500 ng.

2.1.11   Quantification of Genomic DNA

Quantification of DNA is important for enhancing successful analyses, especially when the amount of material for PCR amplification is limited. The most common practice is evaluation by comparison with DNA standards of known concentration. An aliquot of the sample and a series of standards are subjected to agarose gel electrophoresis in the presence of ethidium bromide. The result is visualized by ultraviolet fluorescence. Although useful, this approach fails to distinguish human from nonhuman genomic DNA and has limited sensitivity (the detection sensitivity limit with agarose gel–ethidium bromide staining is approximately 1–5 ng). Forensic samples can be contaminated with nonhuman DNA, and/or a portion of the DNA may be single stranded and thus have a reduced affinity for ethidium bromide, with a concomitant reduced fluorescence. The DNA in a sample may not be detected even though sufficient sample is available for PCR amplification.

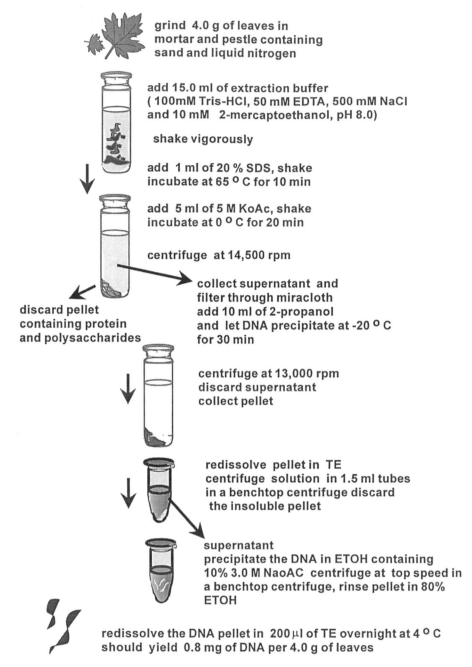

grind 4.0 g of leaves in
mortar and pestle containing
sand and liquid nitrogen

add 15.0 ml of extraction buffer
(100mM Tris-HCl, 50 mM EDTA, 500 mM NaCl
and 10 mM 2-mercaptoethanol, pH 8.0)

shake vigorously

add 1 ml of 20 % SDS, shake
incubate at 65 °C for 10 min

add 5 ml of 5 M KoAc, shake
incubate at 0 °C for 20 min

centrifuge at 14,500 rpm

collect supernatant and
filter through miracloth
add 10 ml of 2-propanol
and let DNA precipitate at -20 °C
for 30 min

discard pellet
containing protein
and polysaccharides

centrifuge at 13,000 rpm
discard supernatant
collect pellet

redissolve pellet in TE
centrifuge solution in 1.5 ml tubes
in a benchtop centrifuge discard
 the insoluble pellet

supernatant
precipitate the DNA in ETOH containing
10% 3.0 M NaoAC centrifuge at top speed in
a benchtop centrifuge, rinse pellet in 80%
ETOH

redissolve the DNA pellet in 200 μl of TE overnight at 4 °C
should yield 0.8 mg of DNA per 4.0 g of leaves

Figure 3.   Protocol for extraction of DNA from plant material.

A very sensitive (detection < 500 pg), simple procedure using a slot-blot approach with the primate-specific α-satellite probe p17H8 is available for the quantification of human genomic DNA. An aliguot of the extracted DNA and a series of standards are denatured to the single-stranded form with NaOH and applied to a nylon membrane via vacuum filtration. Denatured p17H8 probe is hybridized to the membrane-bound DNA, and the membrane is exposed to film. Isotopic or chemiluminescent labeling can be used and the sensitivity of detection enhanced by extending the film exposure time. This method is capable of quantifying both single- and double-stranded human genomic DNA. The relative intensities of the autoradiographic or chemiluminescent-based image of the DNA standards are used to access the approximate concentration of human DNA present in the sample.

2.2   SPECIMEN ANALYSIS

DNA fingerprints consist of polymorphic genetic markers that can be resolved and visualized in a number of ways. VNTR markers require restriction digestion (Figure 1) or amplification by PCR (Figure 4) and gel separation. Sequence polymorphism markers require amplification by PCR and gel separation (for the RAPD, MV-PCR, and direct sequencing systems) or simply direct fixing to membranes to determine whether a specific allele is present (Figure 5).

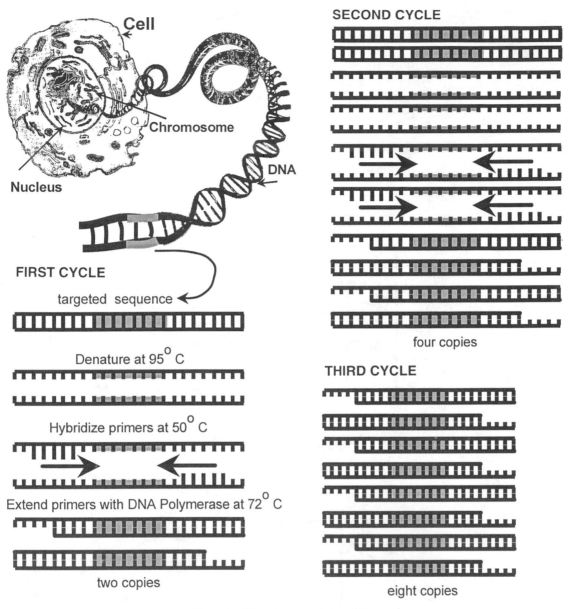

Figure 4. The polymerase chain reaction.

2.2.1 Agarose Gel Preparation and Running

Agarose gels are used for three purposes: 0.8% agarose gels are cast for yield and integrity (molecular weight) testing and restriction digest testing; and analytical 1.0% gels are cast for sample typing analysis. A number of precautions must be followed when preparing agarose gels. Water should be added after boiling if liquid was lost to evaporation, and the agarose solution should be equilibrated to 55°C prior to casting to prevent distortion of both the gel and the plastic electrophoresis tray. Samples are pipetted into the wells and topped up with buffer where required—empty wells are filled with buffer. Minigels are run for 5 minutes and large gels for 10 minutes at 40 V to facilitate movement of the samples into the gel. Buffer is then added to submerge the gel 2–3 mm. Alternatively, the buffer can be placed over the gel and then the samples can be applied to the gel. Yield and integrity, and restriction test gels can be run un-

der a variety of conditions (e.g., at 40 V for 3 h or 60 V for 2 h); analytical gels are run at 30–35 V until the dye front is 1 cm from the bottom (usually 14–18 h lapsed time).

Gels are stained with ethidium bromide, then destained and photographed under UV light (302–316 nm) using a transilluminator and a Polaroid camera system.

Yield and integrity gels include calibration size standards (e.g., λ*Hin*dIII cut DNA). The approximate sample DNA concentration is based on the relative intensities of the UV fluorescence compared with known standards. The DNA integrity is judged by the relative position of the sample band with respect to the size marker and standards. Intact DNA is represented by a band that migrates close to the origin. A smear extending from the origin to the dye front indicates that the DNA is fragmented and may not be suitable for typing.

Restriction-digested DNA generally appears as a similar smear.

DNA Typing Allele Detection: Variable Number of Tandem Repeats, MVR-PCR Profiles and ASO Reverse Dot Blot Format

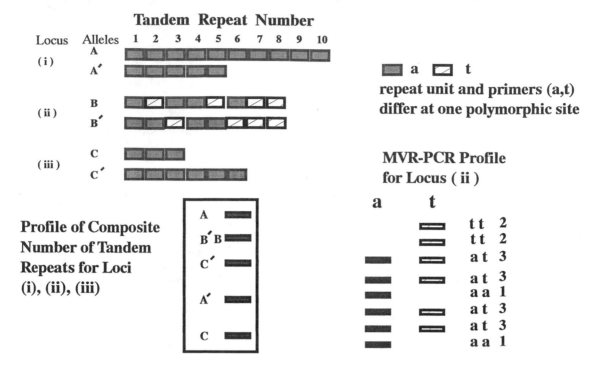

DQ α Dot Blot Format

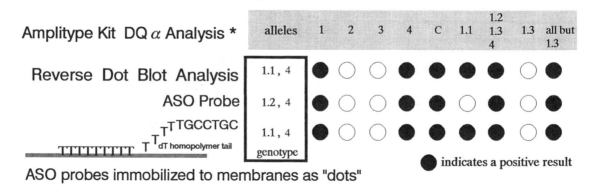

*Perkin Elmer (Division Roche Molecular Systems)

Figure 5.   PCR-based DNA typing methods.

However, depending on the DNA source and the endonuclease used, variations may be observed. For example, with *Hin*dIII-digested human DNA, a smear generally below the 3 kb size range, a band at 3.5 kb (representing the male Y chromosome), and a high molecular weight band at approximately 23 kb are commonly present. When digestion is incomplete, most of the DNA remains near the electrophoresis gel origin; the smear is not distributed as usual in the gel lane. In this case, the DNA should be reprecipitated and redigested. Careful consideration should be given to the organization of the samples on analytical gels. Generally, evidentiary and known

samples are separated by size markers (e.g., BRL 30 band size markers). No more than five sample lanes are present between a set of size markers.

2.2.2 Alkaline Southern Transfer

DNA is denatured (i.e., made single stranded) by the action of NaOH. The single-stranded DNA is directly transferred from agarose gels to nylon membranes, such as positively charged Pall Biodyne, by Southern blotting—a capillary action process (Figure 1).

2.2.3 Radioisotope and Nonradioisotope Probe Hybridization

Radioactive probes are labeled with ^{32}P to a specific activity exceeding 10^5 cpm/μL using commercially available labeling kits. Typically 50–100 ng aliquots of probe are labeled. Prior to hybridization, the probe is denatured by boiling for 5 minutes followed by quenching on ice. Unincorporated nucleotides can be removed by exclusion chromatography through a Nuc Trap push column (Stratagene).

The hybridization solution for probing VNTR sequences immobilized to nylon membranes has been considerably simplified relative to traditional formulations containing formamide, Denhardt's solution, and dextran sulfate. Only SDS, polyethylene glycol, and phosphate buffer are used, and prehybridization is eliminated because of the addition of large quantities of SDS.

Posthybridization washes are carried out to remove loosely bound probe that could lead to nonspecific membrane background staining. Wash stringency increases as the solution temperature is increased and the buffer salt concentration is decreased. As the wash stringency increases, greater amounts of mismatched probe are removed. At the completion of the wash cycle, the moist membrane is heat-baked; then the membranes are covered with Saran wrap and autoradiography is undertaken (Figure 1).

The DNA on the membrane can be analyzed for additional genetic marker loci if the radioactive probe is stripped from the moist membranes using an EDTA-SDS wash at 90°C. Because of a possible reduction in the efficiency of probe removal, complete drying of the membrane should be avoided. Although a small amount of target DNA is lost with each stripping, the procedure has been repeated up to 14 times with a minimal loss of hybridization signal.

Recently, chemiluminescent labeling methods have been greatly improved. They may well become the method of choice, which would eliminate potentially hazardous radioactive materials from the DNA typing process. Nonradioactive protocols such as chemiluminescence systems provide labeled probes with a long shelf life, rapid detection (e.g., two hours), and ease of use.

2.2.4 The Polymerase Chain Reaction

The advent of the polymerase chain reaction has greatly simplified the manner in which genetic analyses are carried out. Although there are many variations for PCR, it is basically an in vitro process that enables production of millions of copies of target DNA sequence through a cyclical enzymatic reaction that repetitively duplicates DNA strands. The products from earlier cycles of PCR are used as substrates for ensuing cycles, and an exponential increase in the target sequence is thus obtained.

A typical PCR is based on the annealing and extension of two oligonucleotide primers that flank a specific target DNA segment. Generally for PCR to work, information about the target DNA sequence (at least the flanking region) is necessary for primer synthesis. Primers are single-stranded DNA oligonucleotides 20–30 base pairs long that can be obtained commercially or synthesized in-house.

First the template DNA to be amplified by PCR is denatured by heating the sample to approximately 95°C using a thermal cycler. When the double-stranded DNA molecules have been denatured, each primer hybridizes to one of the separated strands.

Primer annealing is accomplished by lowering the thermal cycler temperature to between 37 and 72°C; the choice of annealing temperature generally is dictated by the sequence of the primers. Usu-

ally the specific annealing temperature is empirically determined for each primer set–target region combination, although computer programs are available to assist in this determination.

The next phase in the PCR cycle, primer extension, is generally carried out at 72°C, the temperature at which *Thermus aquaticus (Taq)* DNA polymerase, a commonly used thermostable DNA polymerase, can most effectively extend the primers.

These three steps (denaturation, primer annealing, and primer extension) represent a single PCR cycle. When the newly synthesized strand extends through the region that is complementary to the other primer, it can serve as a primer binding site and template for a subsequent PCR cycle. Upon repeated cycles of denaturation, primer annealing, and primer extension, an exponential accumulation of a discrete DNA fragment is generated. Repeating the cycle typically 25–30 times will yield millions of copies of target sequence.

PCR, in principle, is easy to accomplish. One needs only a DNA template sample, primers, a mixture of four 2′-deoxynucleoside triphosphates (dNTPs), buffer, and a thermostable DNA polymerase. All ingredients are placed in a reaction tube and inserted into a thermal cycler, which enables a programmable change in temperature in the reaction tubes. Routinely, PCR can be carried out in this manner in 1–2 hours (Figure 4).

2.2.5 Typing of Sequence Polymorphisms in PCR-Amplified Products

There are three basic approaches for analyzing PCR-amplified products. These are dot-blot assays using allele-specific oligonucleotide (ASO) probes to detect sequence specific alleles, sequencing assays to determine the array of nucleotides contained within a particular DNA fragment, and separation by agarose or polyacrylamide gel electrophoresis to detect differences in the size of the PCR products.

Short ASO probes can be used to detect single nucleotide differences in PCR products using a dot-blot method. Traditionally, double-stranded DNA that has been denatured with sodium hydroxide is spotted onto a nylon membrane in the shape of a spot or a small slot. Under appropriate stringency conditions, labeled ASO probes will hybridize with immobilized DNA samples that contain exact complementary sequences. If hybridization occurs, a colored dot (or, if a radioactive probe is used, a black dot) will appear, demonstrating the presence of an allele. If the sample does not carry the allele, no reaction will occur.

With this approach a different ASO probe is required for typing each of the alleles at a particular locus. A different nylon membrane is needed for each ASO probe, which can be cumbersome, especially when there are a large number of alleles at a locus. To overcome this limitation a reverse dot-blot system has been developed: When the various ASO probes are immobilized on one nylon membrane and the amplified alleles allowed to hybridize to the membrane, one sample can be typed with only one membrane strip.

Direct sequencing reveals the maximum information contained within a DNA fragment. PCR makes it possible to obtain sufficient quantities of sequencing template without the need for cloning and large-scale DNA purification. In sequencing, the DNA generally is made single stranded by any of a number of approaches, including the use of asymmetric PCR, attachment of a biotin molecule to the 5′ end of one of the PCR primers and subsequent capture of the biotinylated DNA strand using streptavidin, the use of magnetic beads to capture one of the DNA strands, and a combination of PCR and sequencing called "cycle sequencing."

The Sanger sequencing method, which utilizes dideoxynucleotide analogues, generally is the approach of choice. It is based on the elongation of a DNA primer hybridized to a single-stranded DNA template. The constituents of a typical DNA sequencing reaction are the DNA template, a DNA primer, a DNA polymerase, the four deoxynucleotides, and one of the chain terminator dideoxynucleotides. Four different reactions are allowed to proceed in different tubes; each reaction contains a different dideoxy terminator for either adenine, guanine, thymine, or cytosine (these four reactions can be carried out in one tube if different fluorescent labels are used for each nucleotide base). At various times during the sequencing reaction the dideoxynucleotides are incorporated into the growing DNA strand. However, because of the structure of these terminators, the DNA strand cannot be extended further. Because the number of chain terminators in a given reaction is limited, the dideoxynucleotide is only occasionally incorporated into the growing DNA strand. After the sequencing reaction, each tube contains a family of DNA molecules, each with the particular chain terminator at the 3′ end of the strand. The DNA fragments obtained from the four sequencing reactions are loaded into adjacent lanes in a sequencing gel. After separation by electrophoresis and detection by autoradiography or fluorescence, the DNA sequence can be read by identifying the position of each successively larger band.

2.2.6 Electrophoretic Separation and Visualization of PCR-Amplified VNTR Products

Currently, polymorphic loci whose alleles are the result of VNTRs are the most informative for identity testing. The size of the PCR product is dictated by the number of repeat sequences it contains. The general approach for separating these fragments is by agarose or polyacrylamide gel electrophoresis. Visualization of the separated PCR products is by ethidium bromide or silver staining, respectively. The size of the separated product, or its allelic designation, can be made by comparison with an allelic ladder standard in an adjacent electrophoresis lane. PCR provides a high, specific yield of target DNA, enabling fairly simple and rapid DNA typing procedures.

An alternate approach to the manual methods for VNTR typing is the automated detection of the DNA with fluorescently labeled PCR products. The primers for PCR can be labeled with fluorescein or rhodamine derivatives, resulting in a fluorescent PCR product. Electrophoresis of the PCR products is carried out in a polyacryl-amide gel housed in an automated detection machine. The labeled fragments are detected by laser excitation as they migrate past a designated window, and the signal is digitized and analyzed by computer so that a size (or allele designation) can be assigned to each DNA band.

2.2.7 RAPD, AP-PCR, and DAF Amplification, Amplicon Separation, and Visualization

Length, composition, and concentration of arbitrary primers, the type of polymerase used, the PCR stringency, agarose versus polyacrylamide gel electrophoresis, and detection using radioisotopes, ethidium bromide, or silver staining represent a wide variety of technical variations that can impact on modern amplification techniques. However, all the procedures are based on the use of arbitrary primers, with the considerable benefit of not requiring prior genetic (DNA sequence) or biochemical knowledge.

The primers are usually 10-mers containing at least four G·C nucleotides. The eight nucleotides at the 3′ end of the primer form the critical domain. One base pair change in this region results in a significantly altered profile. Only one primer is normally used per reaction; on a stochastic basis, four times as many bands are expected from an amplification with two arbitrary primers relative to two reactions with separate primers. In fact, fewer bands are observed than expected. The greatest number of different bands is obtained if two primers are used separately, then as a pair.

DNA profile reproducibility is a major concern. To ensure the greatest possible assay precision, the following should be considered:

1. *Template DNA concentration:* greater concentrations, (e.g., 1 ng/μL) are required for simpler genomes, such as bacteria.
2. *Primer concentration:* shorter primers require higher concentrations. Also, at least a 10 times greater concentration is required with the DAF protocol than the RAPD protocol.
3. *MgCl$_2$ concentration:* simpler genomes require higher concentrations of MgCl$_2$.
4. *Preparation of reagent master mixes:* the PCR reagents should be premixed on ice and dispensed, the template DNA added, and the mixture gently vortexed, touch-spun, then transferred to a thermal cycler at 94°C. This approach minimizes preincubation false priming.

RAPD protocols are characterized by agarose gel fragment separation and visualization by staining with ethidium bromide. Usually fewer than 10 bands are detected per gel run. AP-PCR and DAF protocols use polyacrylamide gels. Bands are usually visualized in AP-PCR by radioisotope labeling and in DAF by silver staining. Up to 50 bands can be detected per gel run with these procedures. The phylogenetically conserved invariant bands are useful for identification at the species level, whereas polymorphic markers are useful for individual identification. Template DNA or amplicon digestion with restriction enzymes, lower PCR annealing temperatures, and use of the Stoffel fragment in place of Ampli*Taq* DNA polymerase also enhance the detection of polymorphic markers.

2.2.8 Probes and Primers

Probes are labeled fragments of single-stranded DNA or RNA that under appropriate conditions will hybridize to their complementary DNA strands. Therefore, probes enable the detection of a unique DNA sequence among a myriad of DNA sequences. The development of probes specific for VNTR loci has revolutionized the field of human identity testing (Tables 2 and 3).

Primers are short oligonucleotide sequences that serve as initiation points for DNA synthesis. It is through the design and use of synthetic primers that flank a DNA region of interest that the PCR amplification process is enabled to occur. A large battery of loci amenable to PCR are available for human identity testing (Tables 4 and 5).

2.3   Data Processing

Data processing can be divided into two categories: capturing and recording the results, and statistical analysis of these data. Genetic profiles can be presented as the presence or absence of dots, as the position or size of DNA fragments separated by electrophoresis, or as direct sequence information. The simplest method for recording DNA typing results is by visual evaluation followed by manual transcription of the data. The visual interpretation approach can be used easily for dot blots, simple VNTR profiles, multilocus profiles, and (to a degree) reading of sequencing gels; however, certain applica-

Table 2   Characteristics of Forensic DNA Typiong Probes: VNTR Probes Used with the Restriction Enzyme *Hae* III

Common Name	Chromosome Designation	Insert Size (kbp)	Core Repeat (bp)	Heterozygosity	*Hae* III fragment Size Range (kbp)	Source of Probe[a]
MS1	D1S7	4.6	9	0.92	0.5–>12.00	Cellmark
yNH24	D2S44	2.0	31	0.95	0.7–8.5	Promega
pH30	D4S139	4.5	31	0.97	2.0–>12.0	BRL
LH1	D5S110	1.7	41	>0.95	0.6–10.0	BRL
pTBQ7	D10S28	1.9	33	0.96	0.4–10.0	Promega
pCM101	D14S13	2.2	15	0.96	0.7–12.0	Promega
3′HVR	D16S85	4.0	17	0.69	0.2–5.0	Collaborative or Promega
pEFD52	D17S26	6.4	18	0.95	0.7–11.0	Promega
V1	D17S79	4.0	38	0.68	0.5–3.0	Lifecodes

[a]Some commercial suppliers in North America.

Table 3   Characteristics of Forensic DNA Typing Probes

Common Name	Chromosome Designation	Core Repeat (bp)	*Hae* III fragment Size (bp)	Source of Probe[a]
		Monomorphic Probes		
pMGB7	D7Z2	2731 (ca. 100 copies)	2731 constant band plus others	Oncor or private laboratory
pY3.4	DYZ1	3564 (ca. 4000 copies)	3564 constant band plus others	Oncor or private laboratory
		DNA Quantitation Probe		
p17H8	D17Z1	2712 (ca. 1000 copies)	Not applicable using slot blot	Oncor or private laboratory

[a]Private laboratory: DNA Diagnostic Laboratory, Department of Pathology, McMaster University Medical Centre, Hamilton Ontario, Canada.

Table 4   Characteristics of Common Amplified Fragment Length Polymorphic Loci

Locus	Clone Designation	Repeat Size (bp)	Number of Alleles[a]	Allele Size (bp)
D1S80	pMCT118	16	27	400–750
D17S5	pYNZ22	70	13	170–1200
Col2A1	Not applicable	31/34	23	500–700
ApoB	Not applicable	15/30	25	500–100

[a]Based on representative North American databases.

Table 5   Characteristics of Common Short Tandem Repeats

Locus	Chromosome Location	Primer/ Length	GC Content (%)	Optimal Temperature (°C)	Product Size (bp)	Repeat Unit	Number of Alleles[a]
CD4	12	32/34	53/54	68	140–170	AAAG	7
vWF	12	24/28	38/36	64	102–154	TCTA	7
THO1	11	24/24	50/50	68	179–207	AATG	5
D21S11	21	18/22	50/36	64	172–264	TCTA/T CTG	12
FABP	4	25/24	44/50	64	199–223	AAT	6
FES	15	20/20	50/55	60	257–289	ATTT	10
F13A1	6	20/13	55/37	60	179–239	AAAG	11
FGA	4	20/21	50/43	60	256–284	TCTT	10
ACTBP	6	20/20	50/50	64	231–339	AAAG	44
HPRT	X	24/24	46/42	60	257–297	AGAT	8
ARA	X	24/24	54/50	68	255–315	AGC	15
Amelogenin	X,Y	24/24	50/46	60	X = 106 Y = 112	Sex typing	2

[a]Based on a preliminary study of Canadian Caucasians.

tions can make this method of data collection problematic. While pattern comparisons can be made effectively, meaningful estimates of the size of RFLP fragments cannot be done by eye. Furthermore, the amount of data for evaluation from a sequencing gel can be overwhelming. Automation of data entry can minimize transcriptional errors during data transfer and can provide easy access at a later date.

Although they depend on the DNA typing method and the number of evaluations required, automated methods for analyzing DNA profiles are desirable in some cases. Since such tests are so powerful, for example, it becomes feasible for many laboratories to carry high caseloads, benefiting from the automation of data recording to facilitate data access. Also, when there is a large quantity of data to analyze or measurements of closely spaced bands are required, more accurate location of the bands is achieved with automated systems. In addition, more objective measurements with regard to position can be made with automatic detection approaches that complement the scientist's visual evaluation of the data. Finally, since the data are transcribed directly to the computer, there is a reduction in human error during data transfer.

For VNTR profiles generated by RFLP analysis, data can be recorded using a semiautomated image analysis system. These systems are composed of common hardware that includes a personal computer, a video camera, and a light box. Image analysis software can be obtained from several commercial, government, or academic sources. The cost of such systems generally is within the budget of a DNA application laboratory. Since the complexity of sequencing interpretations favors the tendency toward automated detection of fluorescently labeled PCR products, recording of most sequencing data in the future will be facilitated by automation. Some DNA markers may not require automation for recording data; these include dot blots and VNTRs amplified by PCR.

Data analysis for identity testing can encompass many facets of the test: for example, whether the two profiles being compared are sufficiently similar to potentially have originated from the same source, and the likelihood of occurrence of a particular profile in a given situation. Standard procedures have been developed to define the various aspects involved in statistical analyses for identity testing purposes. A database is created, and then the frequency of occurrence of a DNA band (or allele) is estimated based on that database, generally by applying some multiplication rule.

3   APPLICATIONS

DNA typing is applicable to any identification problem for which a minimal quantity of relatively intact DNA is available. The requirement for intact DNA depends on the method of analysis. RFLP typing generally requires high molecular weight DNA, whereas most genetic marker typing based on PCR can make use of much smaller, even somewhat degraded, fragments.

3.1   HUMAN

Suspect or victim tissue found at the scene of a rape or a homicide provides a source of forensic evidence for determining whether the suspect (or victim) is a potential source of the crime scene specimens (Figure 2). Accident or homicide victims unidentifiable from their physical features can be identified, provided typable DNA can be isolated from the remains and matched with DNA, perhaps from hair roots from the victim's hairbrush. If the victim's putative parents, children, or other relatives are available, parentage (or other biological relationship) testing also may be carried out.

Parentage (or familial relationships) can be determined for child custody, immigration, and counseling for genetic diseases. The bands in the offspring's DNA profile must have been inherited from the biological parents (Figure 6). Barring mutations, the presence of bands not found in either parental profile is indicative of nonparentage.

Applications in medicine include genetic counseling, tracing the percentage of donor versus recipient cells in bone marrow transplants, determination of possible "contamination" of fetal chorionic villi sample tissue with maternal tissue, tissue culture cell line identification, and the confirmation of twin zygosity.

Identification monitors are feasible for security purposes in the

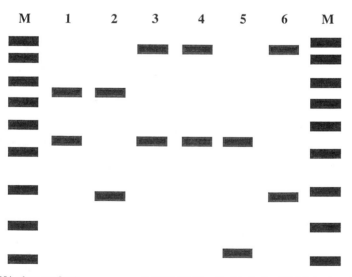

M Marker (DNA size marker)
1 Mother
2 Child 1 (paternity inclusion)
3 Child 2 identical twin to child 3 (paternity inclusion)
4 Child 3 identical twin to child 2 (paternity inclusion)
5 Child 4 (paternity exclusion)
6 Father

Figure 6.   VNTR autoradiogram displaying profiles from a paternity analysis.

armed forces and for identification in the wake of mass disasters or war. For these applications, DNA profiles could be recorded for future match comparisons.

3.2 ANIMAL

Parentage, poaching, identification for theft or loss, population studies, the detection of trait markers, and breeding programs are areas of application of DNA typing in animal husbandry. Proof of parentage may be important for pedigree registration and for documenting sale value. A DNA profile from animal remains can be compared with a profile from a frozen steak or from a trophy specimen in a poaching case. The evolution of populations can be traced, for example, by analyzing DNA from dried museum specimens or from insects housed in amber for millions of years. Base pair sequences and profile bands can be compared with those from modern relatives. Profiling is used in breeding programs for populations of endangered species with small gene pools. Animals exhibiting the greatest number of DNA profile differences and, therefore, probably the greatest genetic diversity, are chosen for breeding. Also, biological relationships can be confirmed in artificial insemination and embryo transfer programs for domesticated animals.

Forensically informative nucleotide sequencing (FINS) is a procedure in which a specific segment of the cytochrome *b* gene in the mitochondrial DNA genome is amplified by PCR. The sequence is determined and analyzed by comparing it with a phylogenetic database to identify the species. The technique has been successfully applied to wildlife management and to the identification of the species origin of animal parts and products. It also is useful for measuring the amount of genetic variation within and among species in evolutionary studies.

3.3 MICROORGANISMS

DNA typing provides a powerful tool for the identification of microorganisms. Different bacterial strains, for example, have unique profiles. Profiling also provides a basis for issuing patents for microorganisms such as yeast used in the brewing industry.

3.4 PLANTS

The field of plant identification for patent, parentage, theft, and trait marker purposes has recently received considerable impetus because of DNA analysis, especially using RAPD-type systems. Evidence from DNA typing can be used to prove or disprove the infringement of breeders' rights relating to the origin of cloned material such as cuttings, as well as parentage of plants with desirable traits; the theft of expensive trees, gene bank management for identification, and the measurement of variation are other typical botanical applications. DNA typing of plants also has been used for determining the geographic location of samples associated with crime scenes.

4 PERSPECTIVES

4.1 DNA FINGERPRINTING VERSUS OTHER IDENTIFICATION TECHNIQUES

Many of the conventional biochemical identification techniques such as isozyme electrophoresis are being replaced by DNA typing, especially where increased information is desired. Although conventional methods such as blood typing are rapid, simple, and definitive for exclusions, they are much less powerful than DNA analysis for the probability of inclusion. If a mother and a putative

father both have blood type O, their offspring cannot be type A, B, or AB. The putative father can be readily excluded as the sire of progeny having non-O blood type. If the offspring is type O, it is considerably more difficult to exclude the putative father.

There are a number of additional key factors to consider when selecting DNA as the material for analysis. These are the highly polymorphic nature of DNA, the genetic continuity of DNA in different tissues of the same organism, the stability of DNA relative to that of protein genetic markers, and the ability to amplify DNA segments in vitro when a minimal quantity of tissue is available or the DNA is moderately degraded.

4.2 FUTURE DIRECTIONS

Automation is a primary objective for future DNA fingerprinting applications. Current methodology is labor intensive because of the number of isolated steps (extraction, amplification, endonuclease digestion, electrophoresis, labeling, hybrid-ization, blot scanning, profile comparison matching, data reduction, and/or probability calculation). Automation is especially desirable for direct sequencing of highly polymorphic regions of the genome. When achieved, it will reduce tedious portions of the current techniques and become a major incentive behind the development of new procedures for the human genome sequencing initiative.

Although typing approaches using RAPD (DAF, AP-PCR) systems circumvent the need for probes, they are limited with respect to the calculation of probabilities. It is extremely difficult to determine population frequencies of these amplified fragments. Probes that identify a single locus (together with population allele frequencies) for each species are, therefore, needed. Alternatively, statistical methods for evaluating RAPD-type profiles could be developed.

DNA typing began in 1985 with the multilocus analysis of sperm and blood in a double rape–homicide case in Britain. Within a decade, applications have reached into every taxonomic kingdom— Monera, Protista, Animalia, and Planta. It appears that the potential uses of molecular biology techniques for identity testing have only begun to be realized.

See also HUMAN REPETITIVE ELEMENTS; PCR TECHNOLOGY; RESTRICTION ENDONUCLEASES AND DNA MODIFICATION METHYLTRANSFERASES FOR THE MANIPULATION OF DNA.

Bibliography

Allen, R. C., Graves, G., and Budowle, B. (1989) Polymerase chain reaction amplification products separated on rehydratable polyacrylamide gels and stained with silver. *BioTechniques,* 7:736–744.

Bartlett, S. E. and Davidson, W. S. (1992) FINS (Forensically Informative Nucleotide Sequencing): A procedure for identifying the animal origin of biological specimens. *Bio Tech,* 12:408–411.

Budowle, B., and Baechtel, F. S. (1990) Modifications to improve the effectiveness of restriction fragment length polymorphism typing. *Appl. Theor. Electro.* 1:181–187.

Dellaporta, S. L., Chomet, P. S., Mottinger, J. P., Wood, J. A., Yu, S. M., and Hicks, J. B. (1984) *Cold Spring Harbor Symp. Quant. Biol.* 49:321–328.

Erlich, H. A., Ed. (1989) *PCR Technology: Principles and Applications for DNA Amplification.* Stockton Press, New York.

Gresshoff, P. M. Ed. (1992) *Plant Biotechnology and Development.* CRC Press, Ann Arbor, MI.

Grimberg, J., Nawoschik, S., Belluscio, L., McKee, R., Turck, A., and Eisenberg, A. (1989) A simple and efficient nonorganic procedure for the isolation of genomic DNA from blood. *Nucleic Acids Res.* 17: 8390.

Hochmeister, M. N., Budowle, B., Borer, U. V., Eggmann, U., Comey, C. T., and Dirnhofer, R. (1991) Typing of DNA extracted from compact bone from human remains. *J. Forens. Sci.* 36:1649–1661.

Jeffreys, A. J., Wilson, V., and Thein, S. L. (1985) Hypervariable "minisatellite" regions in human DNA. *Nature*, 314:67–72.

Kearney, J. J., et al. (1991) Technical Working Group on DNA Analysis Methods (TWGDAM) and California Association of Criminalists Ad Hoc Committee on DNA Quality Assurance: Guidelines for Quality Assurance Program for DNA Analysis. *Crime Laboratory Digest,* Vol. 18, No. 2. U.S. Department of Justice, Federal Bureau of Investigation, Quantico, VA.

Kirby, L. T. (1992) *DNA Fingerprinting: An Introduction.* Freeman, New York.

Reynolds, R., Sensabaugh, G., and Blake, E. (1991) Analysis of genetic markers in forensic DNA samples using the polymerase chain reaction. *Anal. Chem.* 63:1–15.

Saiki, R. K., Scharf, S., Faloona, F., Mullis, K., Horn, G. T., Erlich, H. A., and Arnheim, N. (1985) Enzymatic amplification of beta-globin genomic sequences and restriction site analysis for diagnosis of sickle cell anemia. *Science,* 230:1350–1354.

Sambrook, J., Fritsch, E. F., and Maniatis, T. (1989) *Molecular Cloning: A Laboratory Manual,* Vols. 1, 2, and 3, 2nd ed. Cold Spring Harbor, Cold Spring Harbor, NY.

Singer-Sam, J., Tanguay, R. L., and Riggs, A. (1989) Use of Chelex to improve the PCR signal from a small number of cells. *Amplifications,* 3:11.

Waye, J. S., Presley, L. A., Budowle, B., Shutler, G. S., and Fourney, R. M. (1989) A simple and sensitive method for quantifying human genomic DNA in forensic specimen extracts. *BioTechniques,* 7:852–855.

Weir, B. S. (1992) Population genetics in forensic DNA debate. *Proc. Natl. Acad. Sci. U.S.A.* 89:11654–11659.

DNA Fragment Sizing, Mass Spectrometry in High Speed: *see* Mass Spectrometry High Speed DNA Fragment Sizing.

DNA in Neoplastic Disease Diagnosis

David Sidransky

1   **Mutations in DNA**

2   **Genetics of Cancer**

3   **Tumor Progression Models**

4   **Diagnostics**
4.1   Inherited Susceptibility
4.2   Tumor Diagnosis: Hematologic Malignancies
4.3   Tumor Diagnosis: Solid Tumors

5   **Future Impact**

Key Words

Allelic Loss   Loss of a paternal or maternal chromosomal allele from a tumor cell; such an event is detected by loss of a band on a DNA blot.

Cytology   Microscopic examination of cells obtained from bodily fluids or tissue for pathologic analysis.

Linkage   The likelihood that a disease allele and a chromosomal marker (usually nearby) will be inherited together.

Oncogene   Cellular gene altered in cancer progression.

Polymerase Chain Reaction (PCR)   Enzymatic amplification of DNA molecules through the use of specific primers in vitro.

Proto-Oncogene   Cellular gene capable of dominant transforming function in cancer when one copy is altered or mutated (i.e., activated).

Translocation   Physical movement of genetic material from one chromosome to another.

Tumor Suppressor Gene   Cellular gene whose normal wild-type suppressor function is lost when both copies are altered or mutated (i.e., inactivated).

DNA is the genetic code upon which all life is sustained. Much of human disease is due to errors within DNA that are either inherited (germ line) or acquired after birth (somatic). Neoplasms are now known to arise through a series of changes in specific oncogenes. Because these changes are intimately involved in tumor progression, they provide novel markers for cancer detection. The ability to detect these genetic changes within DNA allows identification of individuals at risk for the development of certain diseases such as cancer. Furthermore, the polymerase chain reaction has allowed amplification of small quantities of DNA, leading to a revolution in molecular diagnosis. Detection of DNA alterations in blood and cytologic samples can allow rapid and sensitive diagnosis of many neoplasms and has revolutionized the approach to cancer diagnosis.

1   MUTATIONS IN DNA

DNA, the double-stranded helix composed of polynucleotides, is the basis of all life processes. To ensure healthy, viable offspring, all creatures—from the fly to man—must successfully replicate DNA that is free of errors. Despite the existence of a variety of mechanisms to prevent the formation of nucleotide substitutions, however, excessive mutagens or defective repair enzymes may lead to mutation fixation. Although many DNA mutations are silent, a particular change within a regulatory or coding region of a gene may lead to human disease.

For a variety of human diseases, a specific genetic mutation has already been described. The pace of new discoveries is rapidly accelerating as new genes are found and new mutations within these genes are characterized. In particular, cancer is at the forefront of these discoveries, as genes involved in both cancer susceptibility and spontaneous tumor formation are rapidly being identified. Recent discoveries include the genes responsible for a variety of cancer predisposition syndromes (Table 1). All these diseases share a high susceptibility to particular forms of cancer, with the majority of affected patients developing tumors at an early age. Furthermore, many spontaneous (noninherited) tumors have been shown to arise from somatic mutation and inactivation of these genes.

Table 1   Inherited Cancer Susceptibility Syndromes in Which the Candidate Gene has been Identified

Disease	Gene	Clinical Manifestations	Type of Cancer[a]
Familial adenomatous polyposis	APC	Colonioc polyps	Colon
Neurofibromatosis-2	NF2	Neurofibromas	Neurofibrosarcoma
Neurofibromatosis-1	NF1	Loss of hearing	Acoustic neuroma, meningioma
Retinoblastoma	RB1	Loss of vision	Retinoblastoma
Von Hippel-Lindau disease	VHL	Renal defects	Kidney
Li-Fraumeni syndrome	p53	Multiple tumors	Sarcoma, breast
Wilms' tumor	WT1	Renal mass	Nephroblastoma
Hereditory nonpolyposis coli	MSH2, MLH1	Multiple tumors	Colon, uterus
Multiple endocrine neoplasia	RET[b]	Hypocalcemia	Medullary thyroid cancer
Hereditary breast cancer	BRCA-1	Breast cancer	Breast
Xeroderma pigmentosum	ERCC	Photosensitivity	Skin cancer
Familial Melanoma	p16 (CDKN2)	Multiple nevi	Melanoma

[a]Most common tumors in this syndrome.
[b]The only proto-oncogene implicated in these syndromes.

2   GENETICS OF CANCER

Inactivation of tumor suppressor genes was predicted by a model based on a hypothesis of Alfred G. Knudson, Jr. Knudson reasoned, after studying the kinetics of childhood tumors, that genetic susceptibility to cancer could be explained by inactivation of one allele in the germ line of affected patients. The other allele was then mutated or deleted in a somatic cell. The concept of tumor suppressor genes has spread to spontaneous tumors, where cancers are often found to have inactivated both copies of a particular gene during tumor progression. Often, one allele (gene copy) is mutated and the other allele is lost by deletion or recombination. Thus, p53 mutations are usually accompanied by loss of heterozygosity on chromosome 17p, inactivating the second copy of this tumor suppressor gene. Because these inactivation events take time, affected patients usually present with single spontaneous tumors much later in life.

The other major class of oncogenes, termed proto-oncogenes, were originally discovered in tumor viruses. Subsequently, these genes were found to be derived from normal cellular genes picked up by the viruses that then were reintroduced in activated (mutated) form during transformation of normal cells to cancer. In tumors, these genes are commonly activated by translocation or point mutation during cancer progression. Only one altered proto-oncogene has been found to be an inherited cause of cancer susceptibility (Table 1). However, both classes of oncogenes contain genetic alterations (Tables 1 and 2) and their detection facilitates DNA diagnosis. Thus, al-

Table 2   Proto-oncogenes and Tumor Suppressor Genes Found to Be Altered in the Progression of Common Cancers[a]

Cancer	Tumor Suppressor Genes	Proto-oncogenes	Chromosomal Loss
Lung	p53, Rb	k-ras, erb B	3p, 5q, 9p, 13q, 17p
Head and neck	p53, p16	cyclin D_1	3p, 9p, 13q, 17p
Breast	p53	cyclin D_1, Her2/neu	1q, 3p, 11p, 13q, 17
Colon	p53, APC	k-ras	5q, 8p, 17p, 18q
Prostate	p53, Rb		8p, 11p, 13q, 16q, 17p
Pancreas	p53, p16	k-ras	17p, 18q

[a]Areas of allelic loss on chromosomal arms represent regions thought to contain inactivated tumor suppressor genes; many have not been identified.

terations in oncogenes are useful targets for assessing patients at risk for developing cancer and for detecting the presence of cancer cells.

3   TUMOR PROGRESSION MODELS

Tumors are now known to arise through a series of genetic steps during progression involving specific activation of proto-oncogenes or inactivation of tumor suppressor genes. To understand genetic progression, a histopathologic progression model for particular tumor type serves to outline the progressive development of a normal cell into a cancer cell. Pathologic analysis has detailed various histopathologic for colorectal carcinoma, from adenoma to carcinoma. These steps have been correlated with specific genetic events that drive the progression pathway (Figure 1). Careful review of this model reveals that activation of proto-oncogenes such as k-ras and inactivation of a tumor suppressor gene such as APC on chromosome 5q occur early. Late events in progression include loss of chromosome 17p (and p53 mutations), as well as additional loss of other chromosomal arms. It is immediately obvious from this progression model that early steps involved in progression such as k-ras and APC are useful targets for cancer diagnosis. Analysis of individual tumors revealed that the accumulation of these genetic changes, not necessarily their order of progression, leads to tumor outgrowth. However, the general order of progression can provide insights into the best targets for initial screening. Patients who might harbor inherited mutations of APC could be screened for cancer susceptibility. Furthermore, identification of specific mutations of k-ras or APC in stool could lead to detection of colorectal cancer.

4   DIAGNOSTICS

4.1   INHERITED SUSCEPTIBILITY

Although a variety of tumor suppressor genes have been shown to control susceptibility to the formation of certain tumors, only a small fraction of cancers have been linked to a particular genetic susceptibility. While cancers may "run" in a certain family, often no specific inheritance pattern can be designated. This absence of pattern may be partly attributable to the tendency of many cancers to arise spontaneously instead of being inherited. Moreover, cancer is a complex disease that entails some background susceptibility, and exposure to environmental factors, as evidenced by a mouse model demonstrating variable susceptibility to lung cancer linked to a gene

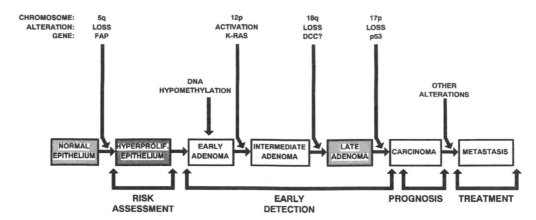

Figure 1. Genetic progression of colorectal cancer: the accumulation of genetic changes, not necessarily the order, allows progression. (Reprinted with kind permission from Fearon, E. R., Vogelstein, B., *Cell,* 61:759–767, 1990. Copyright *Cell Press,* 1990.)

on mouse chromosome 4. In families where particular cancers are common and a candidate tumor suppressor gene is known or localized to a specific chromosomal region, however, the ability to test these family members is of great importance.

In cases where the gene is not known, linkage has been the hallmark of screening strategies. Linkage involves the use of polymorphic (informative) markers from chromosomal regions that are closely associated with the diseased gene. Even if the exact gene has not been isolated, these markers allow rapid identification of anonymous alleles. If the patient is known to have inherited a nearby allele from an affected parent, then it is quite likely that the patient carries the abnormal gene and will develop the disease. If, however, the patient does not appear to carry this nearby allele, he or she is quite likely to be disease free. Depending on the number of markers used and the proximity of the markers to the gene, such linkage studies can be nearly definitive for either the absence or presence of the abnormal gene.

Several new DNA diagnostic tests have been developed to test for cancer susceptibility when the responsible gene has been identified. In retinoblastoma, the polymerase chain reaction (PCR) allows amplification of target DNA from the retinoblastoma gene in blood, followed by specific hybridization with probes recognizing the most common mutations. The ability to detect these specific mutations may be useful in preventive screening (e.g., to help test individuals prior to the development of a tumor). Alternatively, if these tumors are found to be sporadic and not due to an inherited gene mutation, other family members can be assured that their risk of developing this eye tumor is negligibly small.

Additionally, novel screening techniques have been developed for patients with familial adenomatous polyposis (FAP): symptoms include hundreds of polyps, and the likelihood of colon cancer mutations are responsible for the disease and often lead to termination codons, which in turn produce truncated protein. Patients who carry mutations will have short protein products, which can be identified by PCR of the *APC* genes, followed by in vitro translation, and separation of protein products on a gel.

Other attempts have been made to facilitate detection of mutations of p53 in patients with the Li–Fraumeni syndrome. These PCR-based techniques involve functional tests of PCR products obtained from the target p53 gene. The PCR products are cloned into vectors and introduced into bacteria or yeast, then identified as normal or mutant by a color assay. However, for most cancer suscepti-

bility syndromes, the mutations either are not known or are so diverse that direct tests cannot be done. In these cases, only a rigorous and laborious sequencing of the entire gene will identify the causative mutation, making the diagnosis expensive and not readily available outside the research setting.

4.2 Tumor Diagnosis: Hematologic Malignancies

Specific genetic alterations that activate proto-oncogenes or inactivate tumor suppressor genes have become the basis of novel diagnostic assays for cancer. In the great majority of cancer patients, no specific inherited gene mutations have been identified. Therefore, diagnosing affected patients early and accurately can lead to definitive and often life-saving therapy. Additionally, the ability to identify patients with minimal residual disease (e.g., residual cancer) as determined by novel PCR-based techniques has led to an important application for patients with hematologic disorders.

Perhaps the best-known activating mutation in hematologic malignancies involves the translocation between chromosomes 9 and 22 resulting in the Philadelphia (Ph) chromosome (Table 3). This translocation produces a *BCR/Abl* rearrangement resulting in a chimeric protein with transforming activity. Rearrangement occurs in 95% of adults with chronic myelogenous leukemia (CML) and can also be identified in 20–35% of adult patients with acute lymphoid leukemia (ALL). The presence of this translocation is a poor prognostic indicator for patients with ALL. Initially performed pre-

Table 3 Some Common Translocations in Leukemia and Lymphomas and the Putative Proto-oncogenes Identified: Many Translocations and Chimeric Gene Products Are Variable

Leukemia[a]	Cytogenetics	Proto-oncogene
CML, ALL	t(9, 22)	*BCR/Abl*[b]
APML	t(15, 17)	*PML/RARα*[b]
ALL/AML	t(4, 11)	*ALL1/AF4*[b]
AML	t(8, 21)	*AML1/ETO*[b]
Burkitt's lymphoma	t(8, 14)	c-*Myc*
Centrocytic lymphoma	t(11, 14)	*Bcl*-1
Nodular lymphoma	t(14, 18)	*Bcl*-2
Lymphocytic lymphoma	t(3, 14)	*Bcl*-6

[a]CML, chronic myelogenous leukemia; ALL, acute lymphoid leukemia; AML, acute myeloid leukemia; APML, acute promyelocytic leukemia.
[b]Chimeric (fusion) gene products.

dominantly by Southern blot techniques, fluorescence in situ hybridization (FISH) and novel PCR analysis can allow rapid and easy diagnosis of this critical rearrangement. FISH allows hybridization of genomic clones derived from the chromosomal regions near these breakpoints, followed by fluorescent detection of their immediate proximity due to the translocation. Most PCR techniques are based on identification of chimeric RNA transcripts from critical translocations (Table 3). Following isolation of RNA, cDNA is formed by reverse transcriptase followed by PCR amplification (RT-PCR), utilizing primers derived from each of the genes normally located on different chromosomes. The presence of a chimeric product immediately identifies the abnormal juxtaposition of the two genes caused by the translocation, providing a very sensitive and specific diagnostic method. Additionally, detection of the *BCR/Abl* gene rearrangement can be used to monitor CML and ALL patients following bone marrow transplants. Although there is some disagreement over the value of finding residual disease by PCR, relapses identified by PCR may precede cytogenetic or actual clinical relapses by 6–8 months.

Another major rearrangement is the reciprocal translocation between chromosomes 14 and 18 which brings together the immunoglobulin heavy chain region and the critical *Bcl*-2 gene in a large percentage of nodular and follicular lymphomas and in a smaller percentage of diffuse large cell lymphomas. The *Bcl*-2 gene is involved in cellular apoptosis (or programmed cell death), and altered expression allows cells to remain resistant to this phenomenon. Detection of rearrangements of *Bcl*-2 by Southern blot assessment is useful for determining clonality as well for differential diagnosis (of follicular lymphoma vs. a reactive lymph node). There may be additional prognostic information in that *Bcl*-2-positive patients may be less likely to enter complete remission. Recent PCR-based techniques are significantly superior to conventional diagnostic methods. As in the other PCR-based techniques, assessment of minimal residual disease can be done following bone marrow transplantation.

Additional translocations involve rearrangement of chromosomes 8 and 14 leading to overexpression of c-*myc*, a critical nuclear transcription factor, in Burkitt's lymphoma and some diffuse lymphomas. The novel *RAR*-α/*myl* rearrangement between chromosomes 15 and 17 is found in acute promyelocytic leukemia. This translocation may have prognostic importance because of the induction of complete remissions in 80% of patients with this rearrangement following treatment with retinoic acid. Additionally, there are new breakpoints constantly being cloned, present in a smaller percentage of leukemias, which can be detected by Southern blot analysis, FISH, or PCR-based techniques. These may help in establishing definitive diagnosis in some of these leukemias, and all may play a role in assessment of minimal residual disease as described. Many of these novel translocations are listed in Table 3, with the corresponding gene rearrangements.

Another molecular application for hematological malignancies lies in the assessment of clonality. Clonal rearrangement of the immunoglobulin (IG) heavy chain region indicates the presence a clonal population of B cells. Almost all mature B-cell cancers such as CLL, follicular lymphomas, and multiple myeloma contain rearrangements of both the heavy chain and adjoining regions. Detection of these rearrangements establishes the presence of clonality (and therefore malignancy) in difficult-to-diagnose proliferative lymphocytic disorders. Rearrangement is usually done by Southern blot analysis, although progress is being made in PCR-based techniques. Residual presence of a clonal population of cells can be used in a fashion similar to that described for the *BCR/Abl* rearrangement. Furthermore, rearrangements of the β chain in the T-cell receptor can be used in similar strategies for T-cell hematologic malignancies.

4.3   TUMOR DIAGNOSIS: SOLID TUMORS

Although proto-oncogenes are occasionally involved in the development of solid tumors, tumor suppressor genes are commonly inactivated in the progression of most epithelial cancers. Again, because many patients do not have inherited gene mutations, it is difficult to identify high risk patients by family history alone. Therefore, diagnosing patients with early tumors while they can still be successfully treated is of paramount importance. Recently, novel PCR-based assays were developed to identify rare cancer cells among an excess background of normal cells. These assays depend on molecular genetic techniques of increased precision, able to detect point mutations or genetic alterations within cancer-associated genes or loci that are specific for different tumor types. Cytologic samples can be analyzed by PCR both to obtain the diagnosis and, potentially, to provide prognostic information.

Initial demonstration of a PCR-based assay was used in the detection of p53 mutations in the urine from patients with bladder cancer. Investigators used the PCR to amplify a portion of the p53 gene followed by cloning to separate PCR molecules and then probing with specific oligomers able to recognize one mutant copy among 10,000 normal cells. Even though these samples were cytologically negative, 1–7% of the cells were found to be positive by this assay. This work was quickly followed by similar detection of *ras* gene mutations in the stool of patients' colorectal cancers. Moreover, *ras* and p53 genetic changes were detected in sputum months before some patients developed lung cancer. Importantly, these patients were amenable to surgical resection and could have undergone complete removal of the tumors. Other investigators have used techniques such as enriched PCR or allele-specific amplification, to further enrich for the mutant molecules from the cancer cells for accurate diagnosis. Additional techniques such as the ligase chain reaction can also provide information from the samples with a minimal number of cells. As promising as these techniques appear, they will have to be validated in prospective trials to test the sensitivity and specificity of the various assays. Identification of the high risk populations (e.g., patients who smoke for identification of lung cancer) will help validate results and eventually lead to the application of these tests for the population at large.

In addition to diagnostic information, knowing the specific genetic changes involved in various tumors can provide important prognostic information. For example, inactivation of p53 by immunohistochemical analysis or direct sequencing can provide physicians with the knowledge that many patients exhibiting these mutations will die as a result. This outcome was best demonstrated with respect to bladder cancer and appears to be probable for many other tumor types such as lung cancer. It is well known that inactivation of the retinoblastoma gene in bladder cancer also may provide a poor prognosis. As other genetic changes are identified in various tumors, they should provide important prognostic information for clinical outcome. Clinicians can then use the identification of these genetic changes to appropriately put patients into high risk groups for intensive therapy. Investigators have also used specific genetic changes to identify infiltrating tumors cells in apparently normal histologic margins and lymph nodes. PCR-based assays similar to those described detect these rare cells, not visible by light microscopy, in apparently normal margins or tissue. The identification of these infiltrating tumor cells may have important implications for prognosis and therapeutic intervention in these affected patients.

5   FUTURE IMPACT

Molecular biology has provided an entirely new approach to cancer diagnosis. Since DNA is intimately involved in every aspect of normal cell life, the ability to detect abnormal or mutated DNA has become the basis of a new rational molecular approach. The ability to detect patients susceptible to different types of cancer has profound implications for affected patients and their families. These tests will continue to aid in the diagnose of patients with hematologic disorders and in the precise tracking of their disease. Furthermore, the ability to identify tumors early, when they are still surgically resectable, can have a significant effect on the general population.

It has been mentioned that tumor suppressor genes are often found to exist in areas of chromosomal loss. Many such genes may yet be identified on these chromosomes (Table 2), and their mode of mutation is expected to be similar to that of genes previously identified. These new genes could then serve as additional markers for cancer diagnosis. New genes may be found to be the cause of other cancer susceptibility syndromes, and/or they may be implicated in the progression of many other tumors. Detection of a combination of gene alterations can then lead to improved diagnostic tests. Thus, a new era of diagnostics is at hand, based on the molecular biology of cancer and the ability to detect the fundamental DNA alterations of this deadly disease.

See also CANCER; DNA MARKERS, CLONED; GENETIC TESTING; ONCOGENES; TUMOR SUPPRESSOR GENES.

Bibliography

Ambinder, R. F., and Griffin, C. A. (1991) Biology of the lymphomas: Cytogenetics, molecular biology, and virology. *Curr. Opin. Oncol.* 3(5):806–812.

Bishop, J. M. (1991) Molecular themes in oncogenesis. *Cell,* 64:235–248.

Fearon, E. R., and Vogelstein, B. (1990) A genetic model for colorectal tumorigenesis. *Cell,*61:759–767.

Knudson, A. G., Jr. (1985) Hereditary cancer, oncogenes, and anti-oncogenes. *Cancer Res.* 45:1437–1443.

Landegren, U., Kaiser, R., Caskey, C. T., and Hood L. (1988) DNA diagnostics—Molecular techniques and automation. *Science,* 242:229–237.

Sidransky, D., Tokino, T., Hamilton, S. R., et al. (1992) Identification of *ras* oncogene mutations in the stool of patients with curable colorectal tumor. *Science,* 256:102–105.

DNA MARKERS, CLONED

Eugene R. Zabarovsky

1   Principles

2   Techniques

    2.1   General Characteristics of λ-Based Vectors Used for Construction of Genomic Libraries

    2.2   Construction of General Genomic Libraries

3   Applications and Perspectives

    3.1   Cloning of DNA Markers Specific for a Particular Chromosome

    3.2   Alu-PCR as a Tool to Clone Markers from Specific Regions of Chromosomes

    3.3   Use of Microdissection to Construct Region-Specific Libraries

    3.4   CpG Islands as Powerful Markers for Genome Mapping; CpG Islands and Functional Genes

    3.5   Linking and Jumping Libraries

    3.6   Alu-PCR and Subtractive Procedures to Clone CpG Islands from Defined Regions of Chromosomes

    3.7   Use of Linking and Jumping Clones to Construct a Physical Chromosome Map

4   Summary

Key Words

Blue-White Selection Not really selection but color identification. Vectors carrying the β-galactosidase (*lacZ*) gene (or part of it) produce blue plaques in the presence of 5-bromo-4-chloro-3-indolyl-β-D-galactopyranoside (X-gal). If this gene is located in a stuffer fragment, then all recombinants will form white plaques and parental vectors—produce blue plaques in the presence of X-gal.

Genetic Selection Usually, in cloning selection against parental, nonrecombinant molecules in favor of recombinant. For λ-based vectors used for construction of genomic libraries, the two most commonly used types of selection are *Spi-* and *supF*. Spi- phages carrying *red* and *gam* genes cannot grow in *E. coli* lysogens carrying prophage P2; since, however, the majority of the vectors contain these genes in a stuffer fragment, only recombinant phages can grow in *E. coli* strains based on *spi*. Selection for *supF* exploits λ vectors carrying amber mutations. These vectors cannot replicate without the *supF* gene, which must be present either in the host or in the cloned insert. If the insert carries the *supF* gene, only recombinant phage will be able to replicate in an *E. coli* host without the suppressor gene.

Polylinker A short DNA fragment (in the vector) that contains recognition sites for many restriction enzymes, which can be used for cloning DNA fragments into this vector.

Restriction Enzyme An enzyme that recognizes a specific sequence in DNA and can cut at or near this sequence. In cloning procedures, the most commonly used enzymes produce specific protruding (sticky) ends at the ends of DNA molecule. Each enzyme produces unique sticky ends. The DNA molecules possessing the same sticky ends can be efficiently joined with the aid of DNA ligase.

(STS) Sequence Tagged Site A short (200–500 bp) sequenced fragment of genomic DNA that can be specifically amplified using PCR. STS represents or is linked to some kind of marker (i.e., is mapped to a specific locus on a chromosome).

By virtue of the powerful technology developed in molecular biology, it is possible to isolate any DNA fragment in the genome of an organism and, after reverse transcription, any transcribed gene in the form of a complementary DNA. The isolation (cloning) procedure involves the insertion of the DNA fragment into a vector, capable of replication in a microorganism, which allows production of large quantities of the DNA fragment for physical or biological analysis. Upon determination of the location in the genome from which the particular DNA fragment was derived, that fragment acquires the property of a DNA marker. Such DNA markers are a prerequisite for physical and genetic mapping of the genome of the organism. DNA markers are also of importance for the diagnosis of genetic diseases. DNA markers can be divided into several different classes depending on the way in which the markers were selected among the fragments of genomic DNA. Examples of such classes are anonymous, micro- and minisatellites, restriction fragment length polymorphism markers (RFLP), and *Not* I linking clones.

Vectors and clone libraries of different types can be used to clone markers. Lambda-based vectors and genomic libraries of different kinds are commonly used for this purpose. Many different variants of λ-based vectors that combine features of different cloning vehicles (plasmids; M13 and P1 phages) have been created for this purpose. The use of each vector is usually limited to a specific task: the construction of general genomic libraries (which contain all genomic DNA fragments) or special genomic libraries (which contain only a particular subset of genomic DNA fragments). Among these special libraries, *Not* I linking and jumping libraries have particular value for physical and genetic mapping of the human genome.

1   PRINCIPLES

In molecular biology, "cloning" is the insertion of DNA with interesting information into a specific vector that allows replication and transfer of the cloned DNA from one host to another. The vector containing the inserted DNA is called a "recombinant vector" to distinguish it from its parental vector, which does not contain any foreign DNA. Usually "interesting information" is a piece of DNA obtained from any target organism; it can be a gene (or part of a gene) or simply anonymous DNA sequences for which no function is yet known. It can originate either directly from DNA, or it can be obtained from reverse transcription of RNA molecules. The main idea of cloning is to obtain the interesting piece of DNA in a quantity large enough for analysis and further experiments. Now the vectors and strategies used for cloning come in many different types. I will concentrate on the widely used λ-based vectors and the construction genomic libraries, which play an important role in the Human Genome Project.

A genomic library is a collection of recombinant vectors; it contains DNA fragments representing the genome of a particular organism. Genomic libraries can be either general, containing DNA fragments covering the whole genome, or special, containing only specific genomic fragments that differ in certain parameters. Some are CG rich whereas others contain only particular size fragments of DNA obtained after digestion with a particular restriction enzyme, contain specific repeats, and so on. Important special genomic libraries are the jumping and linking types (see Section 3.5). Cloned DNA fragments can be located to a specific site of a chromosome, after which they can serve as a markers for physical and genetic mapping. Different types of markers are used. The so-called anonymous markers represent randomly cloned DNA fragments whose functions or specific features are not known. Other DNA markers can possess specific features. They can contain a known

gene or expressed sequences with unknown function, CpG islands (also associated with genes, see Section 3.4) or recognition sites for rare cutting restriction enzymes convenient for long-range mapping. Such markers can be polymorphic—that is, they have different structures in different individuals (they are usually distinguished on the basis of different mobility in gel electrophoresis). Such markers are extremely important in mapping and cloning human disease genes and for construction of genetic maps.

Three types of polymorphic markers are commonly used (Figure 1). Restriction fragment length polymorphism (RFLP) markers recognize genomic fragments that contain polymorphic recognition sites for a particular restriction endonuclease (e.g., *Tag* I, *Msp* I). The same chromosomal regions in different individuals contain or lack this recognition site. Such different variants of the chromosomal locus are called alleles. Usually RFLP markers have two alleles, and they are randomly distributed throughout the human genome.

A second form of DNA polymorphism results from variation in the number of tandemly repeated (VNTR) DNA sequences in a particular locus. Usually they are divided in two classes: mini- and microsatellites. Minisatellites are DNA fragments 0.2–2 kbp long that contain many copies (from 3 to more than 40) of 15–60 bp repeats.

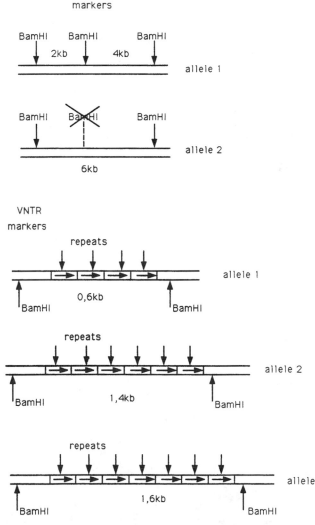

Figure 1.   The difference between RFLP and VNTR polymorphism.

All these repeats share a 10–15 bp core sequence similar to the generalized recombination signal *(chi)* of *E. coli.* When DNA from different individuals is digested with a restriction enzyme that does not cut inside these repeats, the length of the fragments produced will depend on the number of repeats at the locus. Since the minisatellites and repeats constitute a relatively large fragment, it is possible to discriminate between different alleles using ordinary nondenaturing gel electrophoresis and Southern blot analysis. Many different alleles (usually more than five) can be distinguished at a locus containing minisatellites. Minisatellites cluster around the distal ends of human chromosomes and sometimes are located near the genes.

Microsatellites are relatively short DNA fragments (usually <100 bp) that contain runs of tandemly repeated DNA with a repeat unit of 1–5 bp (A, AC, AG, AAC, AAAG, etc). It is believed that they are very numerous (5×10^5) and uniformly distributed throughout the human genome, with an estimated average spacing of about 6 kbp.

Microsatellites can be very polymorphic (more than 10 alleles of the same locus), and polymorphism usually increases with the numbers of the repeats. Microsatellites with fewer than 10 copies are usually not polymorphic. Since microsatellites are short, they can be analyzed quickly by using the polymerase chain reaction (PCR) with primers flanking each locus. Different alleles can be resolved using denaturing gel electrophoresis. Thus microsatellites may be the most useful polymorphic markers, since they are the most plentiful and have many alleles.

Polymorphic markers are less useful for physical mapping, however, and a priori are not connected with genes. Very frequently they are associated with Alu repeats (see Section 3.2), and this creates problems for their use with PCR, since Alu repeats are conserved in the human genome. Thus flanking primers for the microsatellite located in the Alu repeat are unlikely to be useful because they prime from many Alu repeats and instead of discrete bands, give a smear after electrophoretic separation of the PCR products.

An optimal marker should have features of all the different types of marker just discussed. The ideal marker should contain (1) a gene, (2) a CpG island that has been shown to be very conserved in the genome and can be used for comparative gene mapping in different species, (3) a rare cutting restriction site useful for physical mapping, and (4) polymorphic sequences (e.g., microsatellites). One of the best candidates for having all these features together are *Not* I linking clones, that is, recombinant clones containing *Not* I restriction site.

2   TECHNIQUES

2.1   GENERAL CHARACTERISTICS OF -BASED VECTORS USED FOR CONSTRUCTION OF GENOMIC LIBRARIES

Among other vectors used for construction of genomic libraries (yeast artificial chromosomes, P1), phage λ-based vectors are still the most popular. The reason is that the genetics and features of both λ phage and *E. coli* (the host for λ phage) are well known. Two basic features that make cloning work convenient with λ-based vectors:

1. Lambda phage has genes that are not critical for its replication and can be replaced by any foreign DNA fragment of a given size. The size of the phage DNA that can be packaged into viable phage particles is limited to 37.7–52.9 kbp. This means

that it is possible to biologically regulate the size range of the cloned DNA fragment.
2. The easiest way to obtain a maximal proportion of recombinant molecules is to perform ligation at high concentrations of vector and inserted (genomic) DNA, since an elevated concentration of DNA facilitates intermolecular ligation instead of self-ligation of a vector's molecules. In this case the main product is long DNA chains (>200 kbp) containing many copies of vector and genomic DNA fragments.

There exist extremely efficient in vitro systems for packaging such DNA into λ phage particles (10^9 plaque-forming units per microgram μg of DNA) to produce viable phages. To combine the features of different vector systems, extensive modifications of λ phage vectors were developed (Figure 2).

Cosmids are essentially plasmids that contain the *cos* region of phage λ responsible for packaging of DNA into the phage particle. Any DNA molecule containing this region and having a size between 37.7 and 52.9 kb can be packaged into phage particles and introduced into *E. coli.* The advantages of cosmids are easy handling (as with plasmids) and big cloning capacity. Since the plasmid body is usually small (3–6 kb), big DNA molecules (46–49 kb) can be cloned in these vectors. A problem arises if cosmids are used for the construction of genomic libraries from DNA containing numerous repeats. In this case cosmid molecules become unstable and subject to rearrangements in the *E. coli* cell.

Phasmids are λ phages that have an inserted plasmid. They have the same basic features as λ phage vectors, but inserted foreign DNA fragments can be separated from the body of phage DNA and converted into plasmid form. After the conversion, the cloned DNA fragment will exist as a recombinant plasmid. There are two main ways for such a conversion to take place: biological and enzymatic. In the first case, the phasmid vector will contain signal sequences bordering the cloned DNA fragment and plasmid body in the phage vehicle. These signal sequences (e.g., from P1, λ, M13 phage) can be recognized by specific proteins (e.g., M13), and the cloned DNA fragment together with plasmid body will be cut out.

Enzymatic conversion can occur if the inserted DNA fragment is placed between two recognition sequences of some rare cutting restriction endonuclease (e.g., I-*Sce* I enzyme, which recognizes 18 bp and probably does not have any recognition site in the human genomic DNA). This enzyme can be used to separate the cloned DNA fragment together with the plasmid body. Self-ligation of the molecules will result in the production of a recombinant plasmid that can be introduced into the *E. coli* host.

Hyphages represent another type of λ-based vector. They were constructed from M13 vectors with a built-in *cos* site of λ. Since these vectors have the main features of M13 vectors, they can be obtained in single-stranded form. Their distinctive feature is the capability to be packaged with high efficiency into λ-phage-like particles. This decreases the chance of recovering nonrecombinant vectors and opens the possibility of constructing a representative genomic library in single-strand form.

Diphasmids are even newer vectors, which offer the opportunity to combine the advantages of phages (λ and M13) and plasmids. Diphasmids can be divided into two classes: those that can replicate as phage λ (an improvement over phasmids) and those that are incapable of replicating as phage λ (i.e., cosmids capable of being packaged into phage M13 particles). In some cases it is more convenient to work with a genomic library in plasmid than in λ phage

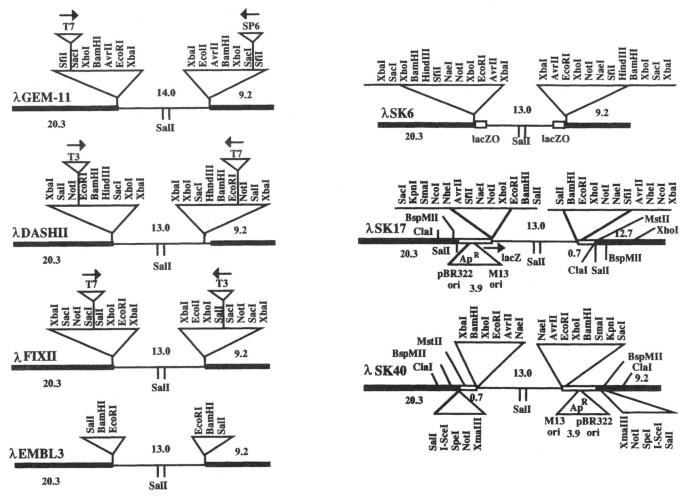

Figure 2. Examples of λ-based vectors: standard λ vectors (λGEM11, λFIXII, λDASHII, λEMBL3, λSK6) and diphasmid vectors (λSK17 and λSK40). Sizes are in kilobases. Not all restriction sites are shown. Heavy lines represent vector arms; thin lines denote the stuffer fragment; open boxes mark plasmid and M13 sequences; *lacZO* is the *lac* operator sequence; T3, T7, and SP6 are promoters for T3, T7, and SP6 RNA polymerases.

form. The construction of a representative genomic library directly in a plasmid vector has several drawbacks and difficulties. But all these problems can be easily solved with the help of phasmid and diphasmid vectors. Often a genomic library is constructed in λ phage and then converted in its entirety to plasmid form.

2.2 CONSTRUCTION OF GENERAL GENOMIC LIBRARIES

Representativity is one of the most important features of a genomic library. In a "representative" genomic library, every genomic DNA fragment will be present in one at least of the recombinant phages of the library. In practice, however, this is difficult to achieve. Some genomic fragments are not clonable because of the strategy used for construction of the genomic library. For example, if the maximal cloning capacity of the vector is 18 kb and *Eco*RI digestion is used to construct the library, no genomic *Eco*RI fragments bigger than 18 kb will be present. In some cases, genomic DNA fragments can suppress the growth of the vector or the host cell, with the result that its cloning can be restricted to specific vector systems.

The important reason for decreased representativity is the different replication potential of different recombinant phages. Most researchers work with amplified libraries. The ligated DNA molecules are packaged into the λ phage particles and plated on a lawn of

E. coli cells (usually many petri dishes are used for such plating). Then all λ phage particles are eluted from the plaques and the liquid eluted from all the petri dishes is mixed. Glycerol or dimethyl sulfoxide is added to the eluate, and aliquots are kept at −76°C. This procedure is called amplification of the library.

With amplification, a library can be kept for years and can be used for many experiments by many researchers. On the other hand, each recombinant phage present in the library gives a single plaque at the first plating, and since different recombinant phages differ in their growth potential, there may be differences of 100 times in the abundance of clones after amplification. This means that in the amplified library, some of the phages are present 100 times more often than others. In this case, to recover all recombinant phages obtained after packaging into λ phage particles, one needs to plate 100 times more phages than were obtained after the original plating. Since such quantities are difficult to achieve in practice, some recombinant phages are virtually lost from the library after amplification.

How is the representativity of a library estimated? A library is considered to be representative if after the first plating (before amplification) it contains a number of recombinant clones together containing genomic DNA fragments equal to 7–10 genome equivalents. For example, human genomic DNA contains approximately 3×10^9 bp. If the vector contains on the average 15 kb inserts, the

Table 1   Comparison of Three Basic Methods Used in Construction of Representative Genomic Libraries

Method	DNA (μg) Needed to Construct a Representative Genomic Library	DNA fractionation	Self-ligation of Vector DNA	Self-ligation of Genomic DNA	Effectiveness of Packaging Per Microgram of Vector DNA	Effectiveness of Packaging Per Microgram of Genomic DNA	Number of Packaging Reactions to Get Representative Library at Maximal Efficiency per Microgram of Genomic DNA	Genetic Selection Necessary to Remove Nonrecombinants
Classical	100–1000	Yes	Yes	Yes	$10^5–10^7$	$10^5–10^7$	1	Yes
Dephosphorylation	5–10	No	Yes	No	$10^4–10^5$	$10^5–10^6$	3–5	Yes
Partial filling in	5–10	No	No	No	$10^5–10^7$	$10^5–10^7$	1	No

representative library should contain $1.4–2.0 \times 10^6$ recombinant clones.

The way in which the genomic DNA fragments are produced for cloning is also important. The more randomly the genomic DNA is broken, the more representative the library that can be obtained. Clearly, the *Eco*RI enzyme (6 bp recognition site) will cut genomic DNA less randomly than *Sau*3AI (4 bp recognition). Probably, the shearing of DNA molecules using physical methods (e.g., syringe, sonication) is the most reliable way to obtain randomly broken DNA molecules.

An important characteristic feature of a library is the percentage of recombinants. If a library contains 100% of recombinants, 1.5×10^6 phages must be obtained to get representation of the human genome. In the case of a library with 50% of recombinants, one

needs to work with 3×10^6 phages to get the same result. On the other hand, genetic selection (to increase the proportion of recombinants) usually decreases the representativity of the library. For most purposes, if a library contains more than 80% of recombinant phages, it is better to omit the genetic selection procedures. To calculate the percentage of recombinants, one can use genetic *(Spi)* selection, as in the case of λ-based vectors from the European Molecular Biology Laboratory (EMBL), in Heidelberg. Another selection relies on blue-white color identification (e.g., λ-Charon series). A third class of vectors has both genetic selection and blue-white color identification (λSK4, λSK6).

There are three commonly used ways to construct genomic libraries (Table 1, Figure 3). The original, and still quite widespread ("classical") method, includes generation of sheared genomic DNA

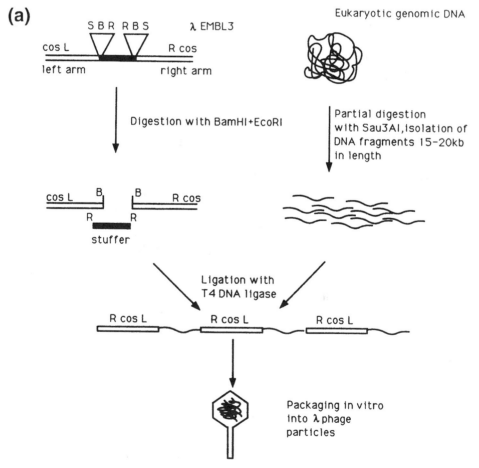

Figure 3.   Two approaches for constructing genomic libraries: (a) classical method and (b) partial filling in method; L cos and R cos, left and right parts of the *cos* site, correspondingly; B, *Bam*HI; R, *Eco*RI; S, *Sal* I; X, *Xma* III.

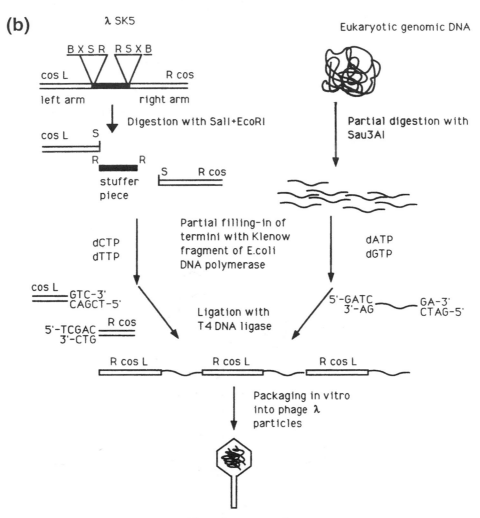

(b)

Figure 3.   (*continued*)

fragments using physical or enzymatic manipulations followed by the physical separation of fragments of a particular size using, for example, ultracentrifugation or gel electrophoresis. The vector DNA is digested with two (or even three) restriction enzymes whose recognition sites are located in the polylinker and separated by a few base pairs. The arms and the stuffer piece are purified from the small oligonucleotide molecules released after the digestion by, for example, precipitation with polyethylene glycol 6000. The stuffer piece and both arms will now have different sticky ends, preventing re-creation of the original vector molecules during subsequent ligation with genomic DNA fragments.

In the second, "dephosphorylation" approach, the phage arms are prepared by simultaneous digestion with two restriction enzymes as shown earlier. Genomic DNA is partially digested to the extent that DNA fragments with sizes in the range of 15–20 kb will represent the majority of the products. These DNA fragments are dephosphorylated to prevent their ligation to each other, and then ligated to the vector arms. If too big or too small genomic DNA fragments are ligated to the phage arms, size limitations will make it impossible for these recombinant molecules to yield viable phages. Compared to the preceding method, this procedure is quick, and representative libraries can be obtained from a small quantity of genomic DNA.

The third, "partial filling in" method, also avoids fractionation steps (Figure 3). Phage arms are prepared by double digestion as described before (in this particular case, *Sal* I and *Eco*RI are shown, but many other combinations can be used), and the sticky ends produced are partially filled in with the Klenow fragment of DNA polymerase I (or other DNA polymerase) in the presence of dTTP and dCTP. Genomic DNA partially digested with *Sau*3AI is also partially filled in, but in the presence dATP and dGTP. Under such conditions, self-ligation of vector arms or genomic DNA is impossible.

Genomic libraries can be constructed in cosmids using the approaches just described. The absence of selection against nonrecombinant vectors and the possibility of packaging into phage particles concatemers composed solely of cosmid fragments make it even more important to prevent self-ligation of vector fragments. Many similar approaches have been suggested, and one of them is shown in Figure 4, where the formation of vector-concatemers is prevented by partial filling in. A similar effect can be achieved by dephosphorylation or by digestion at the first step with *Acc* I and *Sma* I instead of *Eco*RI and *Hin*dIII. *Sma* I produces blunt ends and *Acc* I gives sticky ends with only two protruding base pairs. The ligation of these ends will be far less effective than for *Bam*HI and *Sau*3AI sticky ends (four protruding base pairs).

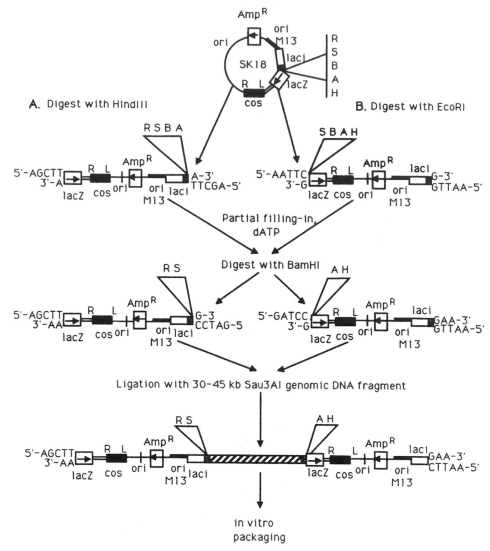

Figure 4. One approach to the construction of genomic libraries in cosmid vectors (not all restriction sites are shown): A, *Acc* I; B, *Bam*HI; H, *Hin*dIII; R, *Eco*RI; S, *Sma* I.

3   APPLICATIONS AND PERSPECTIVES

3.1   CLONING OF DNA MARKERS FOR A PARTICULAR CHROMOSOME

It is important, for many purposes, to clone individual DNA markers for a specific chromosome. The most straightforward approach is to prepare special libraries that contain recombinant clones from a particular chromosome. One approach is based on the use of the fluorescence-activated cell sorter (FACS) system. FACS operates on the principle of rapid analysis of suspended particles, single cells, or even chromosomes. For chromosome sorting, metaphase chromosomes are isolated from cells blocked in metaphase with Colcemid. A suspension of chromosomes stained with fluorescent dyes (usually Hoechst 33258 and chromomycin A3) passes through the focus of a laser beam that excites the DNA-bound fluorescent dyes.

Photomultipliers are used to monitor the emissions from each chromosomal particle, and a computer-based system evaluates the fluorescence signal of each individual chromosome. Equal-sized droplets are formed by ultrasonic dispersion, and the droplets containing the desired chromosome as indicated by the fluorescence measurement are deflected from the main stream by an electric field and collected in a tube. DNA isolated from sorted chromosomes can be used for construction of genomic libraries.

Chromosome-sorted libraries are currently available for many chromosomes. The most important characteristic property of the library is the purity of the sorted chromosomes (the fewer the recombinant clones that contain inserts from other chromosomes, the greater the purity).

Another approach is based on the use of hybrid cell lines. To obtain such somatic cell hybrids, human cells are fused with rodent cells (e.g., mouse or hamster). When the resulting hybrid cells are grown in culture, there is a progressive loss of human chromosomes until only one or a few of them are left. At this step some of the segregant hybrid cells can be quite stable. In a modification of this technique, human cell line is transfected with a plasmid vector or infected with a retroviral vector that contains selectable marker (e.g., *gpt* or *neo*). Transfected clones are screened for those that contain only a single integrated plasmid per cell (and thus contain a single marked chromosome). These transformants are micronucleated by prolonged Colcemid treatment, and the microcells, containing only one chromosome, are produced using special techniques. The microcells are fused to mouse cells, and the resulting human–mouse microcell hybrids, containing the single marked chromosome, are isolated by growth in selective medium. Hybrid cells containing only fragments of human chromosomes can be produced using human chromosomes with translocations and deletions, or the fragments can be produced experimentally by X-irradiation (radiation hybrids). Such hybrid cell lines are very useful not only for constructing genomic libraries but also for physical mapping of already isolated DNA markers and genes. Another advantage of somatic hybrid cells is that DNA is available in large amounts and can be used for different purposes. Human-specific clones can be isolated from the library by hybridization to total human DNA.

3.2 ALU-PCR AS A TOOL TO CLONE MARKERS FROM SPECIFIC REGIONS OF CHROMOSOMES

At least 50% of the human genome consists of repetitive sequences of different kinds. Among them, about 50% are repeats randomly distributed in the human genome—different kinds of short (SINE) and long (LINE) interspersed (repeating) element. The most abundant among them is the Alu repeat family. Alu repeats are present at $0.5–1.0 \times 10^6$ copies per genome, with an estimated average spacing about 4 kb. Alu repeats have a length of about 300 bp and consist of two homologous units. Related repeats exist also in other mammals. Different members of this family from the same species usually have homology of 80–90% but are only about 50% identical in different species. These repeats have conserved and variable regions. It is possible to find conserved sequences that are species-specific. These conserved sequences can be used as primers for PCR, to specifically amplify human sequences in the presence of nonhuman DNA. These features are the basis for using Alu repeats for isolation of human chromosome specific sequences from hybrid cell lines containing human and nonhuman DNA sequences. Moreover, if a hybrid cell line contains only a short piece of human chromosome, the Alu-PCR approach can be used for isolation of markers specific for a defined region of the chromosome.

The principles of the approach are shown in Figure 5. In the case of two hybrid cell lines, one containing complete human chromosome (HCL1) and the other carrying the same chromosome but with deletion (HCL2), this method offers the possibility of obtaining markers specific for the deletion. This variant can be called the differential Alu-PCR approach for obtaining DNA markers. The approach is mainly used in two variants.

In one variant, Alu-PCR is done using DNA from both cell lines, and the products of the reactions are separated by agarose gel electrophoresis. Some bands present in products from HCL1 will be absent among the products from HCL2. These bands can be excised and cloned, giving markers localized in the deletion. The disadvantage of this approach is that usually such Alu-PCR work results in a large number of products that have a very complex pattern and look like a smear on the gel. Among the solutions to this problem that have been suggested is the use of more specific primers (only for a subset of the Alu repeats), or genomic DNA digested with restriction enzyme. Another way is to use hybrid cell lines that contain only small pieces of human chromosomes.

A second variant of the Alu-PCR approach is mainly used in connection with sources that contain only a limited amount of human material: yeast artificial chromosome (YAC) clones and radiation hybrid cell lines containing small pieces of the human chromosomes. The YACs can, for example, be used for Alu-PCR and total products of the PCR reaction can serve as a probe to screen genomic libraries (e.g., in cosmids). The hybridization pattern reveals which cosmids are present in one YAC, which in other YACs, and which are present in one but absent in another. Such an approach is also useful for mapping.

3.3 USE OF MICRODISSECTION TO CONSTRUCT REGION-SPECIFIC LIBRARIES

Another approach to obtaining region-specific libraries is to use chromosome microdissection to physically remove the chromosomal region of interest; the minute quantities of microdissected DNA can be subjected to a microcloning procedure. Spreads of human chromosomes are made and stained using standard cytogenetic techniques. DNA from an individual band is then cut from the chromosome using ultrafine glass needles or isolated with the help of laser equipment. In the latter case, all other chromosomes are de-

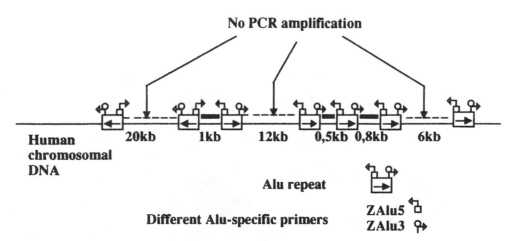

Figure 5. General scheme for Alu-PCR.

stroyed by the laser and intact DNA is present only in the chromosome of interest. DNA obtained from only a few (2–20) chromosomes is enough for constructing a region-specific library. This DNA is amplified using PCR and cloned in plasmid or λ vectors.

It is possible to divide all methods suggested for amplification and cloning of microdissected DNA into two groups. In the first group, chromosomal DNA is usually digested with frequently cutting enzymes like *Mbo* I and ligated to the linker adapter. These adapter molecules serve as a binding site for primers in a PCR amplification and may contain recognition sites for restriction enzymes. Amplified DNA is then digested (e.g., with *Mbo* I) and cloned into the vector. In another scheme, chromosomal DNA serves as a direct substrate for PCR amplification. Usually such amplification is performed in several steps: one or two are nonspecific and are performed under low stringency annealing conditions to permit random annealing of primer to chromosomal DNA. The last step of amplification is performed under high stringency conditions, in which the primers anneal only to the PCR products generated in the first step. Then amplified DNA is digested with some restriction enzyme and cloned into the vector. The biggest drawback to this techniques is connected with PCR. Any impurity will lead to unpredictable results. The average size of the inserts is usually in the range 200–600 bp, which is not convenient for most purposes. Also these inserts are random, without any linkage to genes or polymorphic sequences. About 50% of the clones are usually useless because they contain repetitive sequences like Alu repeats.

Nevertheless, this method offers the possibility of obtaining from a specific region of a chromosome DNA sequences that can serve as probes to isolate cosmid, YAC, or other clones from the same region.

3.4   CpG Islands as Powerful Markers for Genome Mapping; CpG Islands and Functional Genes

Although human DNA is highly methylated, stably unmethylated sequences (about 1% of the genome) have been observed in human chromosomal DNA. Such sequences occur as discrete "islands," usually 1–2 kb long, that are dispersed in the genome. They are usually called CpG (rich) islands because they contain more than 50% of C·G (human genome contains, on average, about 40% C·G). Their distinctive feature is the presence of CpG pairs at a predicted frequency, whereas elsewhere in the genome these pairs are present at a frequency less than 25%. Altogether, there are about 30,000 islands in the haploid genome (the average spacing is about one per 100 kb). It is now clear that the majority (if not all) CpG islands are associated with genes.

Since one of the main goals of the Human Genome Project is the isolation of all genes and the construction of a transcriptional gene map, it is clear that markers located in the CpG islands have an additional value for physical mapping. It has been shown that recognition sites for many of the rare cutting enzymes are closely associated with CpG islands. For example, at least 82% of all *Not* I and 76% of all *Xma* III sites are located in the CpG islands. More than 20% of CpG-island-containing genes have at least one *Not* I site in their sequence, while about 65% of those genes have *Xma* III site(s). Summarizing the data for all genes (with or without CpG islands), we can conclude that approximately 12% of all well-characterized human genes contain *Not* I sites, and 43% of them have *Xma* III sites. For human genome mapping, this means that by sequencing DNA fragments containing *Not* I sites it is possible to tag up to one-fifth of all expressed genes. The recombinant clone containing the *Not* I (or other rare cutting enzyme) recognition site is called linking clone.

3.5   Linking and Jumping Libraries

For long-range mapping and cloning of large stretches of genomic DNA, the two most widely used methods are construction of overlapping DNA sequences (contigs) using chromosome walking (e.g., YAC cloning) and chromosome jumping. The technique of YAC cloning is now used by many laboratories. Still, this approach is not devoid of problems and drawbacks. These problems could be diminished, however, by using jumping/linking libraries. Moreover, jumping and linking libraries can be used independently for construction of a long-range restriction map using pulsed field gel electrophoresis (PFGE) (Section 3.7). Jumping clones contain DNA sequences adjacent to neighboring *Not* I sites, and linking clones contain DNA sequences surrounding the same restriction site.

The two best-known kinds of jumping library are the *Not* I jumping library and the "general" jumping (hopping) library. The basic principle of both methods is to clone only the ends of large DNA fragments rather than continuous DNA segments, as in YAC clones. Internal DNA is deleted by controlled biochemical techniques. The main difference is that in the first type of library, complete digestion with a rare cutting enzyme (*Not* I is the most popular) is used, and the second is based on a partial digestion with a frequently cutting enzyme, followed by isolation of DNA fragments of desired size. Using the first type of library it is possible to jump over long distances (>200 kb), but only from certain starting points (i.e., those containing the recognition site for the rare cutting enzyme). Using the hopping library it is possible to start jumping from practically any point and to cover a defined but shorter distance (<200 kb). Only the first type of jumping library can be used in conjunction with linking libraries to create genomic maps, as described next.

There are two main approaches for the construction of *Not* I jumping and linking libraries. According to first method (Figure 6a), jumping libraries are constructed as follows: DNA of high molecular mass, isolated in low-melting agarose, is completely digested with *Not* I. The DNA is ligated at very low concentration, in the presence of dephosphorylated plasmid containing a marker (*supF* gene), and then circularized, trapping the *supF* gene, which acts like a marker to select clones that contain the ends of a long fragment. Another enzyme, one that has no recognition site in the plasmid, is used to digest the large circular molecules into small fragments, each of which is cloned in a vector phage carrying amber mutations. Recombinant phages containing the plasmid with the two terminal fragments are selected in an *E. coli* strain lacking the suppressor gene.

The linking library can be constructed in different ways. In the original (and still the most popular) protocol, the genomic DNA is partially digested with *Sau*3A and size selected to obtain 10–20 kb fragments. The DNA is then diluted and circularized in the presence of *supF* marker plasmid. The circular products are digested with *Not* I, ligated into a *Not* I digested suppressor-dependent vector (NotEMBL3A), and plated on a suppressor negative host.

A modification based on the use of a genomic library in circular form (e.g., plasmid) begins with the isolation of DNA from the total library. This material is digested with *Not* I and separated using specific conditions for PFGE when linear and circular DNAs segregate in two discrete zones independently of their sizes. The

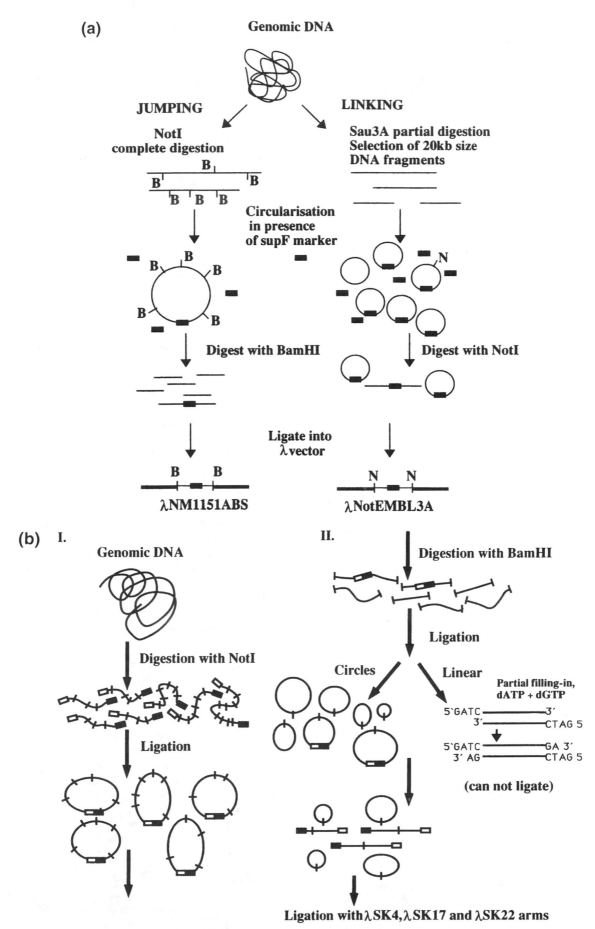

(a)

Genomic DNA

JUMPING

NotI complete digestion

B
B B
B B B

LINKING

Sau3A partial digestion Selection of 20kb size DNA fragments

Circularisation in presence of supF marker

B B
B B
B B

N

Digest with BamHI

Digest with NotI

Ligate into λ vector

B   B

λNM1151ABS

N   N

λNotEMBL3A

(b)   **I.**

II.

Genomic DNA

Digestion with BamHI

Digestion with NotI

Ligation

Circles   **Linear**

Partial filling-in, dATP + dGTP

5´GATC ———————3´
3´ ——————— CTAG 5

5´GATC ——————— GA 3´
3´ AG ——————— CTAG 5

(can not ligate)

Ligation

Ligation with λSK4, λSK17 and λSK22 arms

Figure 6. Two approaches for the construction of jumping and linking libraries: (a) using *supF* marker and (b) using partial filling in. (a) Black boxes, *supF* marker; B, *Bam*HI. (b) Black and white bars denote *Not* I sites and vertical slashes represent *Bam*HI sites; BI, BII, construction of the jumping library; BI, construction of the linking library. In this case digestion of the genomic DNA with *Bam*HI is the first step.

zone containing linear DNA fragments is cut out, and the DNA is eluted, self-ligated, and introduced into *E. coli.*

In another approach, DNA from a total genomic library is digested with *Not* I, and a selectable marker (e.g., resistance to the antibiotic) is inserted into recombinants containing this site. Then such recombinants are selected by their resistance to the antibiotic.

The most important drawback is that all these methods for the construction of linking libraries exploit strategies and vectors different from those used in constructing jumping libraries. Thus some fragments present in one library (e.g., jumping) will be absent in another (e.g., linking), which creates serious problems for the use of these libraries in mapping.

An integrated approach for construction of jumping and linking libraries is outlined in Figure 6b. The most important feature here is that the same vectors and protocol are used for construction of both libraries.

For the linking library, genomic DNA is completely digested with *Bam*HI. Subsequently, the DNA is self-ligated at a very low concentration (without a *supF* marker), to yield circular molecules as the main product. To eliminate any linear molecules, the sticky ends are partly filled in with the Klenow fragment in the presence of dATP and dGTP. Since the Klenow fragment also has exonuclease activity, all the *Bam*HI sticky ends are neutralized and nearly all ends generated upon random DNA breakage become unavailable for ligation. Subsequently, the DNA is cut with *Not* I and cloned in λSK4, λSK17, and λSK22 vectors.

The same strategy is applied in the construction of a *Not* I jumping library. One initial step is added: genomic DNA is fully digested with *Not* I and self-ligated at a very low concentration. The subsequent steps are the same as for the linking library. An obvious difference between this approach and the preceding ones is that the procedure combines a biochemical selection for *Not* I jumping fragments with improved ligation kinetics during the preparation of the libraries (see Table 2 for comparison of two methods).

Table 2 Comparison of Two Main Methods for Construction of *Not* I Jumping Libraries

Method I (with *supF* marker)	Method II (with partial filling in)
Materials Required for Construction of Jumping Library	
2 μg genomic DNA (10 mL ligation volume)	1 μg genomic DNA (5 mL ligation volume)
40 μg vector arms	1 μg vector arms
500 μL sonicated extract (SE) for in vitro packaging	15 μL SE
2000 μL freeze–thaw extract for in vitro packaging (FTL)	10 μL FTL
Cloning Capacity	
0–12 kb	0.2–23 kb
Expected Yield	
$1–5 \times 10^4$/μg genomic DNA	$1–5 \times 10^5$/μg genomic DNA
Percentage of Recombinants Before Genetic Selection	
<1%	45–60%
Maximum Sizes of Jumps	
450 kb	>1000 kb

3.6 ALU-PCR AND SUBTRACTIVE PROCEDURES TO CLONE CpG ISLANDS FROM DEFINED REGIONS OF CHROMOSOMES

The Alu-PCR approach is used successfully for cloning DNA markers, but it does result in cloning small DNA fragments (500 bp) between Alu sequences. Alu sequences are distributed in a random fashion and are not linked with genes or other markers. An obvious suggestion for making Alu-PCR more useful for mapping is to use not simply genomic DNA from different sources but linking libraries constructed from these sources. This modification has certain advantages: using isolated probes, it is easy to clone a parental linking clone (e.g., *Not* I), which is a natural marker on the chromosome, convenient for linkage with other markers. Furthermore, linking clones are located in CpG-rich islands that are associated with genes. According to this scheme, linking libraries are constructed from different hybrid cell lines containing either whole or deleted human chromosomes. Then total DNA isolated from these libraries can be used for Alu-PCR in the manner described in Section 3.2. But in this case every PCR product (either discrete bands or total product) is used as a probe to isolate linking clones from the defined region of the chromosomes.

Genomic subtractive methods represent potentially powerful tools for the identification of deleted sequences and cloning region specific markers. This approach has given rewarding results in less complex systems like yeast or cDNA libraries, but the great complexity of the human genome has generated serious problems. These problems can be overcome by reducing the complexity of the human genomic sequences. Two approaches have been suggested to achieve this aim. In one (representational difference analysis), only a subset of genomic sequences (e.g., *Bam*HI fragments <1 kb) is used for subtractive procedures; this approach will result in cloning of random sequences. In the other, *Not* I or *Xho* I (*Sal* I) linking libraries are used instead of whole genomic DNA (Figure 7). The *Not* I-linking library is at least 100 times lower in complexity than the whole human genome. It is approximately equal in complexity to the yeast genome. Since this approach is not linked with Alu repeats, it offers the possibility of isolating *Not* I linking clones that are unavailable for cloning using Alu-PCR.

3.7 USE OF LINKING AND JUMPING CLONES TO CONSTRUCT A PHYSICAL CHROMOSOME MAP

In theory, a linking library is sufficient to construct a physical chromosome map. When linking clones are used to probe a PFGE genomic blot (*Not* I-digested genomic DNA), each clone should reveal two DNA fragments, which are adjacent in the genome. Thus, in principle, one should be able to order the rare cutting sites with just a single library and one digest, although it will not generally be possible to distinguish between two fragments of the same size. To resolve such ambiguities, it is important to use several different libraries, each for a particular enzyme, and to overlap the resulting patterns just as in ordinary restriction fragment analysis. To accomplish this for the whole human genome will be very laborious. But for small stretches of the genome containing 5–10 *Not* I sites, this approach can be efficient. The use of jumping and linking libraries in a complementary fashion simplifies this approach (Figure 8). Moreover, by cross-screening the two libraries, it should be possible in principle to jump from clone to clone (—jumping—linking—jumping—), to generate an ordered map without using PFGE techniques at all. One of the main problems with this approach is that

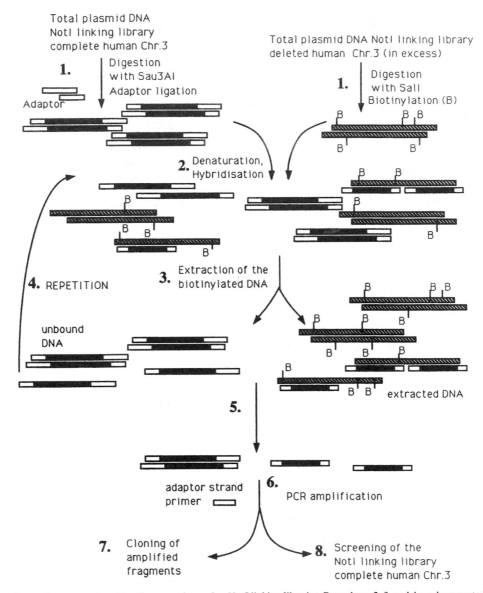

Figure 7.   One of the schemes for a genomic subtraction procedure using *Not* I linking libraries. Procedures 2, 3, and 4 can be repeated many times to remove all common sequences present in the two libraries. The sequences that are unable to hybridize to the biotinylated DNA fragments from the second library will specifically represent a region deleted in this library (i.e., in the chromosome that was used for construction of the second library).

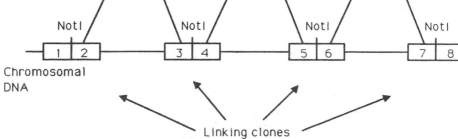

Figure 8.   Long-range mapping using jumping and linking libraries.

the presence of very CG-rich regions and repeats in the human DNA may result in cross-hybridization between different clones.

In the shotgun sequencing approach for long-range genome mapping, the hybridization technique is replaced by sequencing. In the case of individual chromosomes, the main points of the shotgun sequencing strategy are as follows:

1. Jumping and linking libraries are constructed using the same scheme and the same vectors.
2. The *Not* I sites in both libraries are adjacent to direct and reverse sequencing primers available for sequencing reactions without any recloning procedure.
3. An average human chromosome contains about 150 *Not* I sites. Random sequencing of 1000 clones from each (linking and jumping) library will give sequence information (300–500 bp) about a considerable fraction of all these *Not* I sites.

Subsequently, the linear order of the *Not* I clones on a chromosome can be established using a computer program. Even a 20 bp sequence is likely to uniquely identify a sequence in the human genome.

The sequence data provide a means to discriminate even between different instances of the same class of repeats. This approach (to use CpG islands as landmarks for genome mapping) should lead to the identification of new genes and should play a role in joining genetic and physical mapping. The approach can be used in complementary fashion for creating YACs and cosmid contigs, and it results in the generation of new sequence-tagged sites. Since the average inserts in SK vectors are bigger than 6 kb, practically every linking clone will contain some microsatellites.

4   SUMMARY

While several different strategies are available to obtain and use DNA markers for identifying and mapping DNA sequences in complex organisms, no single system is likely to suffice for obtaining a complete and accurate map of the human genome. Rather, a combination of different approaches and vector systems is needed to corroborate data from different sources.

See also GENETIC TESTING; HUMAN CHROMOSOMES, PHYSICAL MAPS OF; HUMAN LINKAGE MAPS; NUCLEIC ACID SEQUENCING TECHNIQUES; PCR TECHNOLOGY; RNA METHODOLOGIES.

Bibliography

Ausubel, F. M., Brent, R., Kingston, R. E., Moore, D. D., Sedman, J. G., Smith, J. A., and Struhl, K. (1987–1990) *Current Topics in Molecular Biology.* Wiley, New York, Chichester, Brisbane, Toronto, Singapore.

Berger S., and Kimmel, A. R., Eds. (1987) *Methods in Enzymology,* Vol. 152: *Guide to Molecular Cloning Techniques.* Academic Press, New York.

Collins, F. S. (1988) Chromosome jumping. In *Genome Analysis: A practical approach,* K. E. Davis, Ed., pp. 73–94. IRL Press, Oxford.

Gottesman, M. M., Ed. (1987) *Methods in Enzymology,* Vol. 151: *Molecular Genetics of Mammalian Cells.* Academic Press, New York.

Poustka, A., and Lehrach, H. (1988) Chromosome jumping: A long-range cloning technique. In *Genetic Engineering Principles and Methods,* Vol. 10, J. K. Setlow, Ed., pp. 169–193. Brookhaven National Laboratory, Upton, NY, and Plenum Press, New York, London.

Sambrook, J., Fritsch, E. F., and Maniatis, T. (1989) *Molecular Cloning: A Laboratory Manual,* 2nd ed. Cold Spring Habor Laboratory Press, Cold Spring Habor, NY.

Watson, J. D., Gilman, M., Witkovski, J., and Zoller, M. (1992) *Recombinant DNA,* 2nd ed. Scientific American Books, New York.

DNA REPAIR IN AGING AND SEX

Carol Bernstein and Harris Bernstein

Key Words

Aging   The progressive impairments of function experienced by many organisms throughout their life span.

Complementation   The masking of the expression of mutant genes by corresponding wild-type genes when two homologous chromosomes share a common cytoplasm.

DNA Damage   A DNA alteration having an abnormal structure that cannot itself be replicated when the DNA is replicated but may be repaired.

DNA Repair   The process of removing damage from DNA and restoring the DNA structure.

Mutation   A change in the sequence of DNA base pairs that may be replicated and thus inherited.

Sex   The process by which genetic material (usually DNA) from two separate parents is brought together in a common cytoplasm where recombination of the genetic material ordinarily occurs, followed by the passage of the recombined genome(s) to progeny.

A number of theories have been proposed to account for the biological phenomena of aging and sexual reproduction (sex). An emerging unified theory that accounts for a considerable amount of the data relating to both aging and sex is presented here. Although this theory is still somewhat controversial, it is consistent with and encompasses most other theories of aging and at least one other theory of sex.

Aging appears to be a consequence of DNA damage, while sexual reproduction (sex) appears to be an adaptation for coping with both DNA damage and mutation. DNA, the genetic material of most organisms, is composed of molecular subunits that are not endowed with any peculiar chemical stability. Thus DNA is subject to a wide variety of chemical reactions that might be expected of any such molecule in a warm aqueous medium. DNA damages are known to occur very frequently, and organisms have evolved enzyme-mediated repair processes to cope with them. In any cell, however, some DNA damage may remain unrepaired despite the existence of repair processes. Aging seems to be due to the accumulation of unrepaired DNA damage in somatic cells, especially in nondividing cells such as those in mammalian brain and muscle.

On the other hand, the primary function of sex appears to be the repair of damages in germ cell DNA, through efficient recombinational repair when chromosomes pair during the sexual process. This allows an undamaged genome to initiate the next generation. In addition, in diploid organisms, sex allows chromosomes from genetically unrelated individuals (parents) to come together in a common cytoplasm (that of progeny). Since genetically unrelated parents ordinarily would not have common mutations, the chromosomes present in the progeny should complement each other, masking expression of any deleterious mutations that might be present.

Thus, aging and sex appear to be two sides of the same coin. Aging reflects the accumulation of DNA damage and sex reflects not only the removal of DNA damage but, in diploid organisms, the masking of mutations by complementation.

1   THE DNA DAMAGE THEORY OF AGING

1.1   OCCURRENCE AND IMMEDIATE CONSEQUENCES OF UNREPAIRED DNA DAMAGE

Considerable evidence indicates that DNA damage occurs frequently in somatic cells of multicellular organisms. Table 1 lists types of spontaneous DNA damage that have been found so far. These data suggest, for instance, that in the rat about 95,000 DNA damages of various types occur per cell per day. The majority of these damages alter the structure of only a single DNA strand, so the redundant information in the complementary strand can be used to repair the damage. But it is known that repair processes are less than 100% efficient; unrepaired damages left each day gradually build up in nondividing or slowly dividing cells. These damages may then interfere with DNA transcription and replication.

A number of different types of DNA damage have been tested for their effects on transcription and DNA replication. It was found that

Table 1   Endogenous DNA Damages in Mammalian Cells

Type of Damage	Approximate Incidence (DNA damages/cell/day)
Oxidative	86,000 (rats)
	10,000 (humans)
Single-strand break	7,200 (in vitro)
O^6-Methylguanine	2,000 (in vitro)
Double-strand break	>40 (rats)
	>3 (humans)
DNA cross-link	>37 (rats)
	>3 (humans)
Glucose 6-phosphate adduct	3 (humans)

transcription is blocked by UV-induced damages (mainly pyrimidine dimers), by adducts produced by derivatives of benzo[a]-pyrene, N-acetoxy-2-fluorenylacetamide or aflatoxin B_1, and also by thymine glycol, an oxidized base. UV-induced DNA damages and thymine glycol have also been shown to block DNA replication. These findings suggest that many types of DNA damage inhibit transcription and replication.

1.2   ACCUMULATION OF DNA DAMAGE AND ITS LONG-TERM EFFECTS

About 86,000 oxidative DNA damages occur in each rat cell, on average, per day (Table 1). There are about 20 different known kinds of oxidative DNA damage. One particular kind of naturally occurring oxidative DNA damage, 8-OH-2'-deoxyguanosine (oh8Gua), has been found to accumulate in rat kidney cells at about 80 residues per genome per day. At the age of 1000 days (which is old, for a rat) there would be about 80,000 of these damaged bases in each kidney cell. We can estimate how this level of accumulation would affect the function of a kidney cell. There are about 3 billion base pairs in the genome of a rat cell. Much of this DNA is noncoding. About 1–2% of the DNA in a cell (about 10,000 genes) is transcribed in most mammalian somatic cells. If the oh8Gua damages are distributed more or less randomly among all the DNA, then 1–2% of the 80,000 damages would be in genes transcribed by the kidney cell. Thus there would be 800–1600 oh8Gua damages in transcribing genes of the kidney. Consequently, these damages should inhibit expression of roughly 8–16% of transcribing genes.

In addition to oh8Gua, another type of damage, the single-strand break, tends to accumulate in DNA of mammalian brain and muscle. These damages, found in numerous studies, probably interfere with mRNA synthesis. A reduction in ability to transcribe mRNA should lead to a decline in function of the cells. In fact, in mammalian brain, it has been shown that as single-strand DNA damages accumulate with age, mRNA synthesis and protein synthesis decline, neuron loss occurs, tissue function is reduced, and functional impairments directly related to the central processes of aging (e.g., cognitive dysfunction and decline in homeostatic regulation) occur. Similarly, it has been shown that as single-strand DNA damages accumulate in muscle cells, mRNA and protein synthesis decline, cellular structures deteriorate, and cells die; these conditions are accompanied by a reduction in muscle strength and speed of contraction. Thus for brain and muscle, accumulation of DNA damage is paralleled by declines in function, suggesting a direct cause-and-effect relationship between the accumulation of DNA damage and major features of aging. In other tissues, including liver and lymphocytes, evidence for an increase in DNA damage paralleled by a decline in gene expression and cellular function has also been observed. In general, it appears that tissues composed of nondividing or slowly dividing cells accumulate DNA damage and experience functional declines with age.

1.3   POSITIVE CORRELATION OF DNA REPAIR AND NEGATIVE CORRELATION OF DNA DAMAGE WITH LIFE SPAN

If DNA damage is the cause of aging, then the following relationships can be reasonably expected: (1) a positive correlation between ability to repair DNA and life span among different species, (2) a positive correlation between ingestion of dietary agents that reduce DNA damage and life span, (3) a negative correlation between level of endogenous DNA damaging agents and life span, (4) a negative correlation between exposure to exogenous DNA damaging agents and life span, and (5) premature aging associated with genetic conditions that allow more DNA damages to occur. The evidence bearing on these expectations is discussed point by point.

1. At least 16 studies have been performed in mammals on the correlation between the ability of cells to repair DNA and the life span of the species from which the cells were taken. The life spans of the species varied from 1.5 years for the shrew to 95 years for man. All but one of the studies showed a positive correlation between DNA repair capacity and life span.

2. Many experiments have been performed on the effect on an organism's life span of adding antioxidants to the diets. Although the results of such experiments are not entirely consistent, certain antioxidants have been found to generally increase life span. Vitamin E, for example, has been found to increase the life span of the rat, insects, rotifers, nematodes, and paramecia.

3. As indicated in Table 1, endogenous oxidative DNA damage occurs frequently in mammalian cells. Organisms that produce higher levels of reactive oxygen species that cause oxidative damage, such as the superoxide or hydroxyl free radicals, singlet oxygen, and hydrogen peroxide, should experience higher levels of oxidative DNA damage and, thus, have shorter life spans. Such a negative correlation of production of reactive oxygen species and life span was shown to hold for the six mammalian species and the two insect species that were tested.

4. More than 50 studies have examined the possible experimental acceleration of aging by externally applied DNA damaging agents. Overall, it has been found that sublethal doses of ionizing radiation or DNA-damaging chemicals in the diet shorten life span, but many specific aspects of normal aging are not accelerated. Several authors have noted that the distribution (over time and in different tissues) of DNA damages induced by external agents does not closely mimic that of natural damages. This difference could explain why the life-shortening effects induced by external agents do not closely mimic natural aging. In particular, natural damages probably accumulate gradually, so they would tend to build up in nondividing cells while being diluted out in dividing cells. Exposure to an external agent over a brief period, on the other hand, could cause equally large numbers of damages to nondividing and rapidly dividing cells. If the effects on rapidly dividing cells included interference with DNA replication, the damage could be widespread, indeed. In addition, if oxidative damages are important in normal aging, then brain cells, which have a high level of oxidative metabolism, should have more damages than most other cell types. Externally applied damages would not be expected to produce this particular type of bias. Thus the general finding that sublethal exposure to DNA-damaging agents shortens lifespan while not accelerating the natural aging process is consistent with the DNA damage theory of aging.

5. Human genetic syndromes have been evaluated with respect to whether they cause premature aging. Twenty-one criteria were used for determining whether a syndrome exhibits premature aging, including premature graying or loss of hair, dementia or relevant degenerative neuropathy, increased susceptibility to age-related neoplasms, and degenerative vascular disease. When 162 human genetic syndromes were examined, 10 were found to have 6 or more features of premature aging. Of these 10, the syndromes of Down, Cockayne and Werner, plus ataxia telangiectasia had evidence of increased DNA damage or reduced DNA repair. Down's syndrome, with 15, had the highest number of features of premature aging. Individuals with Down's syndrome usually have an extra copy of the Cu/Zn dismutase gene, which produces abnormally high levels of the DNA-damaging agents hydrogen peroxide, the hydroxyl radical, and singlet oxygen. Individuals with Werner's syndrome have a high frequency of chromosomal aberrations (usually a reflection of DNA damages). Cells of individuals with Cockayne's syndrome or ataxia telangiectasia show reduced levels of ability to repair DNA damages.

Thus, in general, the five expectations of the DNA damage theory of aging enumerated here are supported by, or are consistent with, available evidence.

1.4   COMPENSATION FOR DNA DAMAGE

Like repair, cellular redundancy may be another strategy for coping with DNA damage. As pointed out in Section 1.3, the brain has an unusually high level of oxidative metabolism compared to other organs. Brain neurons are nondividing, and there is evidence for DNA damage accumulation in the brain (Section 1.2). There is a clear loss of neurons with age. Compared to young rats, old rats have about a 50% loss of neurons in many regions of their brains. Numerous studies in humans have also shown a loss of neurons with age. The brain appears to use a strategy of cellular redundancy to compensate for the loss of neuron function. It has been estimated that the brain is twofold larger than necessary for short-term survival. Comparisons of different mammalian species indicate that the maximum life span in mammals is directly proportional to brain size. Thus, the brain appears to be protected from loss of neuronal function by cellular redundancy, and this type of redundancy may be significant in determining life span.

Cell replacement may also be a strategy for coping with DNA damage. Bone marrow and hemopoietic cells of man, guinea pig, and mouse seem to maintain their population numbers by this means. For instance, mouse bone marrow cells have a turnover time of about 1–2 days, and there appear to be no significant differences in erythrocyte production from marrow stem cell lines in old and young adult mice, suggesting that DNA damages do not accumulate in this cell population.

Other cells that seem to avoid aging are those of the germ line. While multicellular organisms ordinarily age and die, the germ line is potentially immortal. It has been proposed that meiosis is the stage of the sexual cycle during which DNA damages are removed from the germ line. This idea has been tested using single-celled paramecia, which can undergo either asexual or sexual reproduction. When they grow asexually, clones of *Paramecium tetraurelia* age (show reduced vigor) and then die. These paramecia have a macronucleus containing about 800 copies of the genome which express cellular functions and a micronucleus containing the germ line DNA.

If the macronuclei of clonally young paramecia are injected into old paramecia, the old paramecia have their life span prolonged. In contrast, cytoplasmic transfer from young to old paramecia does not prolong the life span of the old paramecia. This suggests that the macronucleus rather than the cytoplasm determines clonal aging. Asexually growing clones of paramecia have been found to accumulate DNA damage in the macronucleus over successive generations of clonal growth. Upon sexual reproduction (conjugation) or self-fertilization (automixis), a new macronucleus develops from the micronucleus, and the old macronucleus disintegrates. Both these processes (conjugation and automixis) include meiosis, which involves pairing of homologous chromosomes and the opportunity for recombinational repair of DNA (see Section 2.4). It was found that the level of DNA damage in macronuclear DNA is low at a few generations after meiosis, but as the cells undergo clonal aging, there is an increase. Thus accumulation of DNA damage in the macronucleus may account for clonal aging, and DNA repair during meiosis (principally recombinational repair) may account, in large part, for the potential immortality of the germ line.

2   REPAIR OF DNA DAMAGE

2.1   REPAIR DEPENDS ON REDUNDANCY

Most of the DNA damages listed in Table 1 are single-strand damages, and the redundant information in the opposite, undamaged strand of DNA can allow recovery of the information lost in the damaged strand. The likely major mechanism for repairing these damages is referred to as excision repair and is described in Section 2.2. Since single-strand damages, as distinct from double-strand damages, constitute the great majority of damages that occur in DNA, their repair is probably central to combating aging.

Double-strand damages are those in which both strands of the DNA are damaged in the same or nearby regions. Double-strand damages need more redundancy than is available in double-stranded DNA for their repair. For these damages, the information in a second homologous chromosome is needed to repair the damaged information. This type of repair involves the exchange of DNA segments between the two homologous chromosomes. This process of recombinational repair is described further in Sections 2.4 and 2.5.

2.2   EXCISION REPAIR

Excision repair, which removes damages in one of the two strands of DNA, is carried out by a sequence of enzymatically catalyzed steps.

Thymine glycol is one type of single-strand DNA damage that forms after exposure of DNA to oxygen-containing free radicals. Figure 1 illustrates an excision repair pathway for thymine glycol. (This pathway is derived from evidence from a number of sources but is nevertheless somewhat speculative.) Thymine glycol is represented in Figure 1, by a hat on a square with a T in it; the adenine it is paired with is represented by a rectangle with an A in it. A short (12 bp) length of DNA is shown in schematic form. The vertical lines represent deoxyribose in the backbones of the paired DNA strands and the slanted lines represent the phosphodiester bonds that connect the deoxyriboses. The larger rectangles represent purines (adenine and guanine), and the smaller squares represent the pyrimidines (thymine and cytosine) with which they are paired. The first enzyme that acts in the pathway has two activities. It has a glycosylase activity, which uncouples the thymine glycol from the deoxyribose to which it is linked. It also has an apyrimidinic endonuclease, which breaks the deoxyribose–phosphate backbone of the DNA (on the 3′ side) after removal of the thymine glycol. The actions of this enzyme are indicated near the arrow between the top DNA structure in Figure 1 and the second DNA structure down. This latter structure now has a break in the backbone of one DNA strand, and the thymine glycol formerly at that position is shown as free from the backbone. In the second step (second arrow down) an en-

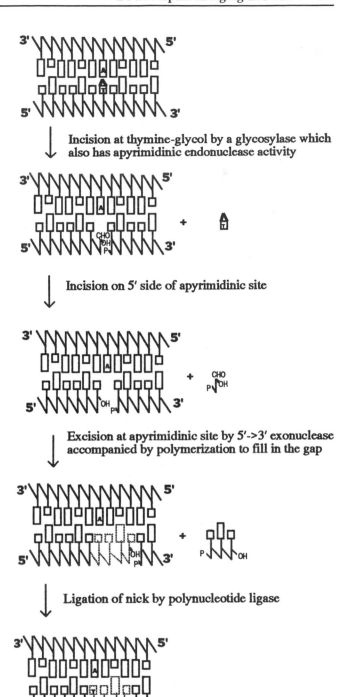

Figure 1.   Excision repair (for details, see text).

zymatic cut is made on the 5′ side of the apyrimidinic site (where the thymine glycol is no longer attached). This releases a deoxyribose–phosphate unit of the DNA. In the third step (third arrow down), a DNA polymerase employs its two enzymatic activities, a 5′ → 3′ exonucleolytic activity and a 5′ → 3′ polymerizing activity. In this step the exonuclease excises a section of the DNA single chain (shown released to the right of the rest of the DNA) and then catalyzes the extension of the remaining, cut DNA chain in a 5′ → 3′ direction until the missing nucleotides are filled in (stippled lines and boxes in Figure 1). The final step in the pathway is carried out by the polynucleotide ligase, which forms the last phosphodiester bond needed to complete the repair.

While our example deals with thymine glycol, other types of single-strand DNA damage are also common, and their repair often occurs by a process similar to that shown in Figure 1. Some of the unusual bases or fragments of bases that are left in DNA as a result of DNA damage processes are uracil, hypoxanthine, 3-methyladenine, an imidazole-ring-opened form of 7-methylguanine, urea, and hydroxymethyluracil. Specific glycosylases have been found in mammalian cells that catalyze the breakage of the bond linking each of these unusual bases or base fragments with the deoxyribose–phosphate backbone of the DNA. When this occurs, the DNA is left with a single base gap along one of the deoxyribose–phosphate backbones. Such a gap is called an apyrimidinic or apurinic site because either a pyrimidine or a purine had originally been located in it. Generically, these anomalies are referred to as AP sites. Nucleases that specifically recognize sites of base loss in DNA, called AP endonucleases, act to make a nick in the deoxyribose–phosphate backbone of the DNA. An exonuclease then degrades the DNA in a $5' \rightarrow 3'$ direction or a $3' \rightarrow 5'$ direction at the free end created by the nick. This is followed by polymerization by a DNA polymerase and ligation by DNA ligase to restore the original DNA structure.

2.3   EFFICIENCY OF REPAIR OF SINGLE-STRAND DNA DAMAGE

Even if a repair mechanism for a given type of single-strand DNA damage is available, the mechanism may not be utilized for each occurrence of the damage within the DNA of an organism. A number of kinds of preferential repair have been observed.

There appears to be a strong preferential removal of UV-induced pyrimidine dimers from transcriptionally active genes compared to removal of these dimers from total cellular DNA. In addition, transcription-blocking pyrimidine dimers occurring in the transcribed strand of a gene are much more likely to be removed than similar damages in the nontranscribed strand of the same gene.

There are also differences in efficiency with which oxidative damages are removed from different types of somatic cells. As noted in Section 1.2, the base oh8Gua, a product of oxidative damage to DNA, accumulates in rat kidney cells at the rate of 80 residues per day. A similar rate of accumulation occurs in liver cells. However, no such accumulation of oh8Gua is seen in cells of rat testes or rat brain. The lack of accumulation in the latter tissues has been attributed to higher levels of DNA repair (for this particular damage) in these tissues. The results discussed in this section indicate that repair of single-strand damage is often less than 100% efficient, and thus accumulation of such damage is expected.

2.4   RECOMBINATIONAL REPAIR

Recombinational repair appears to be effective against both single-strand and double-strand types of damage. It is probably of central importance during the meiotic stage of the sexual cycle. Meiosis appears to be an adaptation for promoting recombinational repair between the homologous chromosomes of differing parental origin that reside in a diploid cell.

Recombinational repair was first reported in 1947 and was actually the first type of DNA repair detected. This initial observation was made with a bacterial virus, referred to as bacteriophage T4. The original study and further investigations showed that recombinational repair occurs in bacteriophage T4 in response to damages induced by UV, nitrous acid, mitomycin C, ^{32}P decay, psoralen plus near UV light, X-rays, and hydrogen peroxide. These DNA-damaging agents cause a wide variety of single-strand and double-strand damages. Seven different gene products of bacteriophage T4 were shown to be required both for recombinational repair of these damages and for recombination of genetic markers. This finding suggests that recombination of genetic markers and recombinational repair share the same enzymatic steps and are largely the same process. A similar type of recombinational repair was shown to occur in several other bacterial viruses and in several animal viruses. The recombinational repair in these viruses usually accompanies a sexual process, whereby the genomes from two parental viruses enter the same cytoplasm (a cell of a bacterial or animal host), the genomes recombine with each other, and the resulting recombinant genomes are passed on to progeny viruses.

Recombinational repair has been shown to take place in bacteria, as well as in viruses. Most studies of recombinational repair in bacteria have involved daughter chromosomes of a single parental chromosome rather than homologous chromosomes from two different parents. About 135 genotoxic chemicals were shown to cause damages that could be repaired, in part, by recombinational repair in the bacterium *Escherichia coli*. Recombinational repair of two types of double-strand damage, interstrand cross-links introduced by psoralen plus near-UV light, and double-strand breaks introduced by X-rays, have been well studied in *E. coli*. These repair processes were shown to require the RecA protein, which is needed for pairing of homologous chromosomes and strand exchange.

Recombinational repair has also been shown to occur in eukaryotes—simple fungi as well as multicellular eukaryotes. In the single-celled yeast *Saccharomyces cerevisiae*, repair of either interstrand cross-links or double-strand breaks was shown to depend on the function of genes required for genetic recombination. When the yeast cells were recombination proficient and diploid (containing two homologous sets of chromosomes), it took 35 double-strand breaks or 120 cross-links to induce an average of one lethal hit per cell. When recombination genes were defective, an average of about two double-strand breaks or just one cross-link per cell was sufficient to introduce a lethal hit. Clearly, recombinational repair is highly efficient at removing double-strand damages. When about 100 heterogeneous chemicals, which caused many kinds of DNA damage, were tested in yeast, it was found that 44 of these substances stimulated mitotic recombination. Since increased recombination can reflect stepped-up recombinational repair, this result indicates that recombinational repair in yeast is used in response to many kinds of DNA damage.

In the fruit fly *Drosophila melanogaster*, defects in two genes, *mei-41* and *mus-101*, cause reduced levels of genetic recombination during meiosis as well as increased sensitivity to X-rays, UV, methyl methanesulfonate, and other DNA-damaging agents. Recombination in wild-type *D. melanogaster* is stimulated by DNA-damaging agents including UV, X-ray, and mitomycin C. These observations suggest that recombinational repair occurs in this organism as well.

Evidence consistent with the occurrence of recombinational repair has also been found in mammalian cells and in plants. At least 18 studies have shown that DNA-damaging agents applied to mammalian cells in vitro stimulate a form of recombination between homologous chromosomes referred to as sister-chromatid exchange, as would be expected if recombinational repair were carried out in response to the DNA damages induced. DNA-damaging treatments, given at meiosis, increased recombination frequencies in plants (in *Lilium, Tradescantia,* and *Chlamydomonas*), a result consistent with the use of recombinational repair in response to the occurrence of these damages.

Overall, evidence for recombinational repair has been obtained in a wide range of organisms: in viruses, bacteria, fungi, insects, mammals, and plants. Recombinational repair appears to be effective against a broad spectrum of types of DNA damage; in particular, it is highly efficient in removing double-strand damages.

2.5 LIKELY MECHANISM OF RECOMBINATIONAL REPAIR

Extensive data have been obtained on the segregation of genetic markers during meiotic recombination. These data have led to a detailed molecular model of the process of meiotic recombination, and this model (Figure 2) also provides the steps expected and required for recombinational repair of DNA damage. Figure 2 shows two homologous chromosomes, each composed of double-stranded DNA: one in gray, the other in white. The two antiparallel strands are drawn without their helical twists, and the 5′ end of each strand is represented by a circular knob, while the 3′ end is pointed. The DNA strands shown represent a region one or a few genes long.

In step A, a double-strand gap is formed at the site of a double-strand damage, removing the damage (black bar) from one chromosome. During step B an exonuclease, acting at the gap, strips back two of the single strands in a 5′ → 3′ direction. The original gap now ends in single-stranded DNA regions with 3′ ends.

In step C, one of the 3′ ends invades the homologous paired chromosome, pairing with the complementary strand of the homologous chromosome. The strand that had been at that position becomes displaced into a "displacement loop" (D-loop). In step D the D-loop is enlarged because of DNA synthesis primed by the invading strand. The newly synthesized DNA is indicated by a black region. In step E, the gap in the upper gray strand is filled by DNA synthesis primed by the 3′ end of the strand, using the D-loop region as a template. During step F, the first invading stand continues to elongate, enlarging the D-loop. The D-loop pairs with the complementary strand of the upper gray chromosome strand and displaces the lower gray strand. The displaced lower gray strand pairs with the complementary strand of the homologous white chromosome. In going from step F to step G, H, I, or J, extra lengths of single strands (shown overlapping at step F) are degraded prior to ligation of the gray displacement strand to the newly synthesized black strand.

At this point (step F), single strands of DNA cross (a white strand crosses a gray strand) at two regions, which are called Holliday junctions. Each Holliday junction can exist in two topologically equivalent forms and can be resolved in two alternative ways. This set of possibilities gives rise to four possible configurations shown in alternative steps G, H, I, and J. If both Holliday junctions are resolved in the same sense, cutting the inner strands at both junctions (G) or the outer strands at both junctions (H), the two resulting chromosomes are not recombinant for outside markers. If the junctions are resolved in opposite senses (I and J), with one inner pair of strands cut and one outer pair of strands cut, the resulting chromosomes are recombinant for outside markers (a gray chromosome section joined to a white chromosome section).

The end result of the recombinational repair process illustrated in Figure 2 is that the double-strand gap formed with the removal of damage from the "gray" chromosome in step A has been restored with

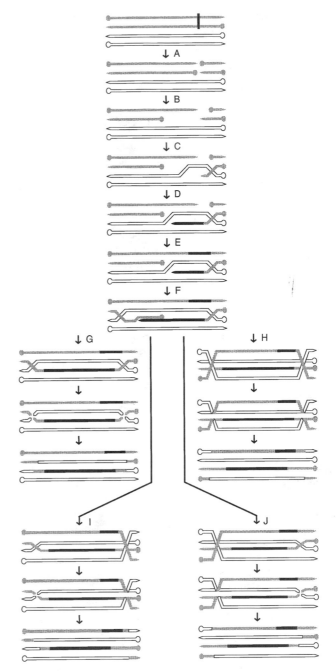

Figure 2.   Recombinational repair (for details, see text).

information copied from the "white" chromosome by complementary DNA synthesis of copies from the two "white" strands. While still somewhat speculative, this model of recombinational repair accounts for the known data on genetically observed recombination and on recombinational repair obtained with numerous organisms.

3   THE DNA REPAIR (AND COMPLEMENTATION) THEORY OF SEX

Sexual reproduction is characterized by two basic features, outcrossing and recombination. Outcrossing is the process by which two haploid genomes (each consisting of one or more chromosomes) from separate parents come together in a shared cytoplasm. Recombination is the process by which the homologous chromosomes from the two parents pair with each other and exchange genetic information, giving rise to new chromosomes, which are then passed on to progeny. In eukaryotes, recombination occurs during meiosis, a key stage of the sexual cycle.

3.1   Frequency of Sexual Reproduction

Sexual reproduction is a widespread strategy for reproduction. Of approximately a million animal species, about 99.9% are sexual. Among higher plants, the majority of species are sexual. As shown in Table 2, only 8% of higher plants are known to be apomicts (i.e., able to reproduce only by vegetative means, such as by runners or bulbs, or by asexual formation of seeds). Sex is also common among the simple eukaryotes, including fungi, algae, and protozoa. Sex has been found among bacterial species, as well, and is common in bacterial viruses and animal viruses.

3.2   Costs of Sex

Sex, while widespread, is very costly to the organism using it. For example, a sexual female lizard passes only 50% of her genes to a particular egg, while a comparable nonsexual (parthenogenetic) female lizard passes 100% of her genes to each egg. Thus, a sexual female is only half as efficient in propagating her genes as a nonsexual female (all other factors being equal). In addition, when two individuals must find each other to mate, there is a cost of searching out the other party. In animals this may be done through mating calls, while in flowering plants an insect vector may be attracted by colorful flowers or perfume scents. Not only must one find another individual of the same species, but it may be necessary to determine the appropriateness of the other individual, giving rise to costs of courting. In birds these costs include feather displays and dances.

Another type of cost of sex arises from the randomization of genetic information during meiosis. A parent organism that has met the test of survival has, by definition, a well-adapted combination of genes. The process of meiosis, which includes recombination, generates untested new combinations of genes to be passed on to progeny. These new combinations, on average, should be less successful than the parental combinations of genes because random changes in successful genetic information are more likely to be deleterious than beneficial. Clearly sex has large costs.

Table 2   Plant Reproduction

Reproduction Strategies for Higher Plant Species	Percentage
Primarily cross-fertilized	55
Partly self-, partly cross-fertilized	7
Primarily self-fertilized	15
Facultative apomicts[a]	
Can be cross-fertilized	1
Can be self- or cross-fertilized	0.2
Can be self-fertilized	0.1
Apomicts, not known if can be self- or cross-fertilized	13
Obligate apomicts	8

[a]Apomixis: reproduction by vegetative mechanisms, including propagation by runners or bulbs, or with embryos or seeds formed by asexual means.

3.3   Benefits of Sex

Sex must have a large benefit to make up for its large costs. A major function for sex is to counteract two types of "noise" in the transmission of genetic information from parent to progeny: DNA damage and mutation.

As pointed out in Section 1.1, about 95,000 DNA damages occur, on average, per day per cell in the rat. Most of these are single-strand damages that can be removed by excision repair or other repair processes that need only the redundant information present on the opposite strand of DNA. However, excision repair of single-strand damages is not 100% efficient, and double-strand damages also occur at significant frequencies (Table 1). Such DNA damages remaining in germ cells (e.g., egg and sperm, in mammals) would cause the death of zygotes and loss of potential progeny. During the meiotic stage of the sexual process, however, recombinational repair is promoted. This repair is one major benefit that can compensate for the costs of sex, since clearing the germ line of damages increases viability of progeny.

The other major type of noise in the transmission of genetic information from parent to progeny is mutation. Genes carrying mutations often code for nonfunctional proteins. If one of a pair of chromosomes in a diploid cell carries a mutation in a given gene, and the second chromosome carries the homologous gene in a functional form, this second gene can usually provide an adequate level of gene expression for the organism to function normally. This masking of the expression of mutant genes by wild-type genes is called complementation. Complementation is available in the diploid phase of the life cycle of organisms. However, complementation is most beneficial when an organism undergoes outcrossing.

To see why mutation makes it beneficial to have outcrossing, consider a hypothetical population of diploid organisms that is strictly inbreeding (i.e., all the organisms are self-fertilizing), and assume that the population has been long established. In such a population the rate at which new mutations arise will be balanced by the rate at which they are lost from the population by natural selection. In this self-fertilizing population, each mutation present in an individual will have a one in four chance of being paired with the same mutation (becoming homozygous recessive) in each progeny. If each individual had an average of one to several deleterious mutations, the cost of inbreeding in terms of defective progeny would be high. If a hypothetical outcrossing individual should arise in such an otherwise inbreeding population, any mutations in this outcrosser would very likely be complemented by the wild-type alleles from its mating partner, and thus defective progeny would be avoided. Complementation would occur because the homologous chromosomes from the two parents are not likely to carry the same mutations. Thus, loss of progeny due to expression of mutations would be greatly reduced. This gives a strong immediate selective advantage to switching to outcrossing from inbreeding.

This advantage would not last indefinitely in our hypothetical example. Because of the ability of outcrossers to mask deleterious mutations by complementation, mutations would not be weeded out by natural selection as efficiently as in inbreeding individuals. Eventually, the mutations that build up in the population would cause as much lethality to progeny of the outcrosser as occurred in the original inbreeding population. However, if the outcrosser now tried to switch back to inbreeding, there would be a great loss of progeny due to the larger number of mutations now present. In summary then, there is a large immediate benefit to switching from inbreeding to outcrossing and a large immediate disadvantage to switching back. Therefore mutation provides a selective pressure to maintain the outcrossing feature of sexual reproduction among diploids.

Overall, meiosis, with its promotion of recombinational repair, may be the only efficient way to correct endogenous double-strand damages and left over single-strand damages in diploid cells that produce germ cells. On this view, the recombination aspect of sex is an adaptation for dealing with DNA damage, a major type of genetic noise. Furthermore, outcrossing allows the masking of mutations, through complementation. The outcrossing aspect of sex deals with mutation, a second major type of genetic noise.

3.4   OTHER EXPECTATIONS OF THE THEORY

If DNA damage and its repair are important in maintaining sexual reproduction, certain expectations follow. If recombination during meiosis and sexual reproduction reflects recombinational repair of germ line DNA, there should be other evidence of avoidance of DNA damage in the germ line, as well.

As indicated in Table 1, the largest known source of DNA damage is oxidative damage due to endogenous cellular metabolism. Presumably, to avoid DNA damage, germ line cells should have evolved ways to avoid high levels of metabolism.

Eggs would seem, at first sight, to be poor candidates for avoiding metabolism. They are, in general, much larger than somatic cells of the organism (e.g., egg cells have about a thousand fold greater mass than somatic cells in humans). It requires considerable metabolism to produce large egg cells. However, much of the cytoplasmic material within an egg cell is generated by the activity of other cells. Some insects have nurse cells around each egg cell. The nurse cells, which are connected to the egg cell by cytoplasmic bridges, provide most of the ribosomes, mRNA, and proteins of the egg cell. The nurse cells themselves contain hundreds to thousands of copies of their genomes, presumably to protect the nurse cells themselves from losing function as a result of any oxidative damage incurred while providing large amounts of metabolic products for the egg cell. Vertebrate egg cells are surrounded by follicle cells rather than nurse cells. Instead of cytoplasmic bridges to the egg cell, however, the follicle cells have small gap junctions connecting them to the egg. While these gap junctions are not large enough to transmit bulky macromolecules, they do transmit precursor molecules to the egg. In addition, for chickens, amphibians, and insects, the yolk proteins accumulated by the egg are made in liver or liver-type cells. These mechanisms allow eggs to be protected from oxidative damage while they store up material to sustain the zygote in its initial growth.

Sperm or pollen cells, in contrast to egg cells, are usually the smallest cells of an animal or plant. This allows a different strategy for effective protection against oxidative damage to their DNA. Because sperm and pollen cells are so small, we can assume that minimal metabolism was used in their formation. Thus the processes of producing both egg and sperm appear to have been adapted to circumvent the production of DNA damage in their especially important germ line DNA.

If complementation of mutations is important in maintaining the outcrossing aspect of sex, this significance should be confirmed by general biological observations as well. Indeed, both in animals and plants, it is usually seen that hybrids formed from the crossing of two genetically distinct inbred lines, are more vigorous than either of the parental lines. This hybrid vigor is responsible for much of the crop improvement that has been achieved in modern agriculture. The opposite side of this observation is that consanguineous matings in humans are known to frequently result in impaired offspring. Observations in other animals and in plants suggest that close inbreeding results in the production of less vigorous progeny.

4   OVERVIEW

The occurrence of high levels of endogenous DNA damage in mammals is now well established. DNA of mammalian somatic cells is the master informational molecule, and accumulated damages in this molecule likely cause the progressive irreversible deterioration of cell, tissue, and organ function that defines aging. When an organism forms progeny via germ cells, it is important that these cells be free of DNA damages, since such damages cause inviability. To facilitate efficient DNA repair, the redundant information available in the diploid cell can be used to replace damaged information through recombinational repair.

Sexual reproduction appears to be an adaptation to promote pairing and exchange between homologous chromosomes for the purpose of efficient repair of the DNA that is passed on to germ cells. In eukaryotes this occurs during meiosis, while in bacteria and viruses it occurs during less complex, but equivalent, processes.

Mutations are another type of error in DNA. Unlike damages, mutations cannot be recognized by repair enzymes. Mutations, however, can be masked when information from two unrelated individuals (parents) is brought together through fertilization to form the progeny zygote. Thus the outcrossing aspect of the sexual cycle in diploid organisms appears to be maintained by the advantage of masking mutations. Overall, aging appears to be a consequence of the accumulation of DNA damage, and sex appears to be an adaptation for the removal of damage through recombinational repair and the masking of mutations through outcrossing.

See also AGING, GENETIC MUTATIONS IN; CELL DEATH AND AGING, MOLECULAR MECHANISMS OF; DNA DAMAGE AND REPAIR.

Bibliography

Alexander, P. (1967) The role of DNA lesions in processes leading to aging in mice. *Symp. Soc. Exp. Biol.* 21:29–50.
Ames, B. N., and Gold, L. S. (1991) Endogenous mutagens and the causes of aging and cancer. *Mutat. Res.* 250:3–16.
Arking, R. (1991) *Biology of Aging; Observations and Principles.* Prentice Hall, Englewood Cliffs, NJ.
Bernstein, C., and Bernstein, H. (1991) *Aging, Sex and DNA Repair.* Academic press, San Diego, CA.
Bernstein, H., Hopf, F. A., and Michod, R. E. (1987) The molecular basis of the evolution of sex. *Adv. Genet.* 24:323–370.
Cox, M. M. (1991) The RecA protein as a recombinational repair system. *Mol. Microbiol.* 5:1295–1299.
Darwin, C. (1889) *The Effects of Cross and Self Fertilization in the Vegetable Kingdom.* Appleton, New York.
Dougherty, E. C. (1955) Comparative evolution and the origin of sexuality. *Syst. Zool.* 4:145–190.
Holmes, G. E., Bernstein, C., and Bernstein, H. (1992) Oxidative and other DNA damages as the basis of aging: A review. *Mutat. Res.* 275:305–315.
Michod, R. E., and Levin, B. R., Eds. (1988) *The Evolution of Sex; An Examination of Current Ideas.* Sinauer, Sunderland, MA.
Mullaart, E., Lohman, P. H. M., Berends, F., and Vijg, J. (1990) DNA damage metabolism and aging. *Mutat. Res.* 237:189–210.

DNA REPLICATION AND TRANSCRIPTION

Hisaji Maki, Yusaku Nakabeppu, and Mutsuo Sekiguchi

1   DNA Replication
  1.1   Structural Aspects of DNA Replication
  1.2   Biochemistry of DNA Replication
  1.3   Regulation of DNA Replication

2.   Transcription
  2.1   RNA Polymerase and Transcriptional Apparatus
  2.2   Transcription Units
  2.3   Regulation of Transcription

Key Words

DNA Polymerase   An enzyme that catalyzes the formation of DNA chains by adding deoxyribonucleotides to the 3′-hydroxyl end of a preexisting DNA strand.

Origin (ori)   A unique site on the chromosome from which DNA replication starts in one or both directions.

Promoter   A region on a DNA molecule in which an RNA polymerase binds and initiates transcription.

Replication Fork   A moving front of DNA replication at which a double-stranded DNA helix is separated into two newly replicated helices.

Transcription Factor   A class of regulatory proteins that bind to a promoter or to a nearby sequence of DNA to facilitate or prevent initiation of transcription.

The genetic information of organisms is kept in DNA in the form of an array of nucleotide sequences, and the cell must precisely replicate its chromosomal DNA. The self-complementary nature of the double-stranded DNA molecule is the basis for accurate replication, and organisms are equipped with the enzyme DNA polymerase, which adds deoxyribonucleotides to preexisting DNA chains in the 5′ → 3′ direction, along the template strand of DNA. One strand (the leading strand) is synthesized continuously, while the other (the lagging stand) is synthesized discontinuously and is then sealed.

To extract information from DNA, a limited region of DNA, which constitutes the gene or the gene cluster (operon), is transcribed into RNA. This process is catalyzed by RNA polymerase. As is the case with DNA polymerase, this enzyme is composed of multiple subunits and catalyzes the addition of ribonucleotides to the growing RNA chain in the 5′ → 3′ direction, along the template DNA. Both the starting point and the frequency of transcription are determined by the promoter, a specific sequence of DNA to which RNA polymerase and transcription regulatory proteins can bind.

1   DNA REPLICATION

Precise transmission of chromosomal DNA from generation to generation is crucial for cell propagation, which can be achieved only when chromosomal DNA is accurately replicated, providing two copies of the entire genome, for consistent distribution to each daughter cell. To this end, the cell comes equipped with special mechanisms to maintain the high fidelity of DNA replication, to segregate the replicated DNA, and to tightly coordinate DNA replication and cell division within the cell cycle.

Interest and research in DNA replication have been spurred by increasing concerns with mutagenesis and carcinogenesis, as well as interest in evolution and cancer therapy. Spontaneous mutations, apparently a major force in evolution, arise largely from errors during DNA replication. Cancer cells, having multiple mutations in genes controlling the cell cycle, escape tight regulation of the cell cycle, and their chromosomal DNA replication gets an abnormal start. This uncoupling of cell division and DNA synthesis may frequently induce chromosome disorders and additional mutations, thus accelerating the malignancy. Most anticancer treatments involve inhibition of DNA replication in the cancer cells.

Many basic concepts and techniques in modern biotechnology are derived from research on DNA replication. The development of cloning vectors has depended on knowledge of replication mechanisms of bacterial plasmids and phages. Genetic engineering, DNA sequencing, and the polymerase chain reaction (PCR) utilize replicative enzymes, such as DNA polymerase and DNA ligase.

1.1   STRUCTURAL ASPECTS OF DNA REPLICATION

The double-stranded nature of DNA provides the basis for its semiconservative replication. The two strands of the parental duplex contain complementary sequence information. DNA synthesis by a polymerase also makes use of this complementarity in generating the daughter duplex. An exact copy can be made by complementary base pairing between the free nucleotide substrate and the parental DNA strand template. Thus, the daughter duplex consists of one parental strand and one newly replicated strand. Complementary base pairing also allows for the correction of replication errors and DNA damage by several repair mechanisms.

1.1.1 Replicon: Unit of Replication

DNA replication starts at particular sites, called origins of DNA replication. A DNA sequence whose replication starts from an origin and proceeds bidirectionally or unidirectionally to a terminus site or sites is called a replicon, a unit of DNA replication. In bacterial cells, the circular chromosome contains a unique origin, and DNA replication proceeds bidirectionally from the origin to the terminus (Figure 1A). Therefore, the entire bacterial genome (4700 kb for *Escherichia coli*) is a single replicon. This genome structure is common to most bacterial plasmids and some phages, although some are replicated unidirectionally.

On the other hand, eukaryotic cells contain multiple replication origins on each chromosome (Figure 1B). For the eukaryotic cell to replicate its huge genome (2900 Mb per human haploid genome) in a relatively short period—about 7 hours for S phase in cultured animal cells, many replicons must function simultaneously. Each replicon is 20–200 kb long, and the number of replicons utilized depends on the growth state of the cell: more rapidly growing cells use more replicons.

1.1.2 Replication Fork

In each replicon, replication is continuous from the origin to the terminus and is accompanied by movement of the replicating point,

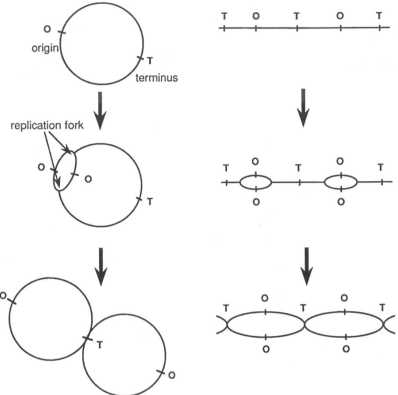

A. Prokaryote chromosome   B. Eukaryote chromosome

Figure 1.   Unit of DNA replication in (A) prokaryotic and (B) eukaryotic cells: O, replication origin; T, replication terminus.

called the replication fork. The parental DNA strands are replicated simultaneously at the fork. Since DNA polymerase can extend a DNA chain only in the $5' \rightarrow 3'$ direction, replication at a fork is semidiscontinuous (Figure 2): DNA synthesis is continuous on one strand (leading strand) and discontinuous on the other (lagging strand). Short pieces of DNA, called Okazaki fragments, are repeatedly synthesized on the lagging-strand template. These Okazaki fragments consist of a few thousand nucleotides in bacterial cells and a few hundred nucleotides in eukaryotic cells.

The velocity of fork movement has been estimated from the size of a replicon and the length of its replication period. In *E. coli,* the replication fork proceeds at 1000 bp per second; one Okazaki frag-

ment is synthesized every 1–2 seconds. In eukaryotic cells, the rate of fork movement is slow (10–100 bp/s). The distinction between prokaryotic and eukaryotic replication may be due to a difference in the replication machinery or in the structure of the chromosome.

1.2   BIOCHEMISTRY OF DNA REPLICATION

DNA replication is a complex process involving numerous enzymes at the replication fork. In *E. coli,* more than 20 different proteins participate in DNA replication. These were identified by screening mutants defective in DNA replication and by purifying enzymes required for *in vitro* DNA synthesis. From their biochemical roles at different stages of chromosomal DNA replication, these proteins are classified into some half-dozen categories:

1. Proteins involved in initiation of replication: DnaA, HU, DNA gyrase, single-stranded DNA-binding protein (SSB), DnaB, DnaC, RNaseH1, and RNA polymerase.
2. Primosome proteins, including DNA helicases and primase: PriA, PriB, PriC, DnaT, DnaB, DnaC, and DnaG (primase).
3. Proteins required for chain elongation and connecting Okazaki fragments: DNA polymerase III (Pol III) holoenzyme, SSB, DNA polymerase I (Pol I), RNaseH1, and DNA ligase.
4. A swivel to relieve torsional stress in advance of the replication fork: DNA gyrase.
5. Proteins for the termination of replication and segregation of daughter molecules: terminus binding protein (Tus), DNA gyrase, and DNA topoisomerase IV.

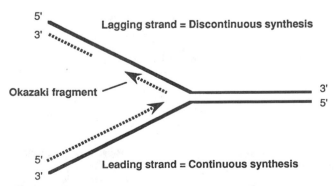

Figure 2.   Semidiscontinuous DNA synthesis at the replication fork.

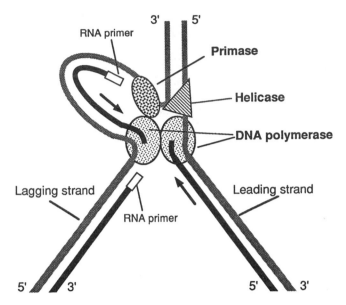

Figure 3. Hypothetical structure of replication machinery. Primase, DNA helicase, and twin DNA polymerases form a "replisome."

Among these proteins, the DnaB helicase, primase (DnaG), and Pol III holoenzyme are the basic components acting at the replication fork, probably forming a multiprotein complex called a replisome (Figure 3).

Replication of many bacterial plasmids and small phages requires most of the host-encoded replication proteins, although the initiation step of these extrachromosomal replicons differs from that at the host origin. Large phages such as T4 and T7 encode their own replication proteins, the activities of which resemble those of the *E. coli* replication proteins.

In eukaryotic cells, chain elongation, termination, and segregation appear to be carried out by activities very similar to those of the *E. coli* replication proteins. Proteins involved in the initiation of replication of the eukaryotic viruses have also been identified. Although candidates for an origin-binding protein have been found in yeast cells, mechanisms involved in initiation of eukaryotic chromosome replication are largely unknown.

1.2.1 DNA Polymerase

DNA polymerase plays a primary role in replication, catalyzing synthesis of a DNA chain complementary to the template DNA strand as follows:

$$(dNMP)_n + dNTP \rightarrow (dNMP)_{n+1} + PP_i$$

The reaction absolutely requires three components: deoxynucleosidetriphosphate (dNTP) as a substrate, single-stranded DNA as a template, and a short RNA or DNA as a primer. The primer must be annealed to the template by base pairing, and its 3' terminus must possess a free 3'-OH group. Chain growth is exclusively 5' → 3', since the polymerization mechanism is a nucleophilic attack by the 3'-OH group of the primer on the α-phosphate of the incoming dNTP. The products are a new primer, longer by one nucleotide, and an inorganic pyrophosphate. 2',3'-DideoxyNTP can be incorporated by some DNA polymerases but blocks chain elongation because it lacks a free 3'-OH group. Some procedures for DNA sequencing take advantage of this chain termination property.

Both prokaryotic and eukaryotic cells possess three or more distinct DNA polymerases (Table 1), and these differ in structures as well as biochemical and biological functions. All DNA polymerases except for the eukaryotic Pol α and Pol β are accompanied by 3' → 5' exonuclease activity. The fidelity of DNA synthesis is increased by the 3' → 5' exonuclease activity, which allows for removal of nucleotides incorrectly inserted by the polymerase.

In *E. coli*, Pol III holoenzyme is the major replicative polymerase for both leading- and lagging-strand synthesis. Pol I participates in lagging-strand synthesis by eliminating primer RNAs and also has a role in repair synthesis. The biological function of Pol II is unclear, although this enzyme is induced when the chromosome DNA is damaged. The Pol III holoenzyme is a huge multiprotein complex that consists of 10 distinct polypeptides. This enzyme extends the DNA chain with high processivity (>500 kb of DNA can be continuously synthesized without dissociation of polymerase from the template) and high catalytic efficiency (the velocity of chain elongation is 1000 nucleotides per second at 37°C). The catalytic core, composed of three subunits, contains the polymerase activity and a 3' → 5' exonuclease for proofreading. The remaining seven auxiliary subunits enhance the processivity of the core by clamping it onto the template. They also promote repeated association of the polymerase necessary for discontinuous synthesis of the lagging

Table 1   DNA Polymerases Found in Prokaryotes and Eukaryotes

Organism	DNA polymerase	Processivity	Editing Exonuclease	Function	Unique Features[a]
E. coli	Pol I	Low	Yes	Repair	5'→3'-Exonuclease associated
	Pol II	Medium	Yes	?	Structural homology to eukaryotic Pol α
	Pol III	High	Yes	Replication	Asymmetric dimer with twin active sites
Yeast	Pol I	Low	No	Replication	Primase associated
	Pol II	High	Yes	Replication	PCNA-independent processivity
	Pol III	High	Yes	Replication	PCNA-dependent processivity
	Mitochondrial Pol	High	Yes	Mitochondrial replication	Preference to ribohomopolymer template
Mammal	Pol α	Low	No	Replication	Primase associated
	Pol β	Low	No	Repair	Small single-polypeptide
	Pol γ	High	Yes	Mitochondrial replication	Preference to ribohomomopolymer template
	Pol δ	High	Yes	Replication (?)	PCNA-dependent processivity
	Pol ε	High	Yes	Replication (?)	PCNA-independent processivity

[a]PCNA, proliferative cell nuclear antigen.

strand. Structural analysis of the Pol III holoenzyme and studies on a reconstituted replication fork suggest that the holoenzyme is an asymmetric dimer with twin polymerase active sites: half of the dimer has high processivity and might be a polymerase for continuous synthesis of the leading strand; the other half has a recycling capacity needed for synthesis of the lagging strand. Thus, it seems likely that a single molecule of Pol III holoenzyme acts at the replication fork, catalyzing concurrently synthesis of both leading and lagging strands.

In eurkaryotic replication, the division of labor among the polymerases remains ambiguous. Pol β, a small single polypeptide, has been suggested to participate in repair. Pol α (Pol I in yeast) is apparently involved in DNA replication, since mutant cells defective in this polymerase activity are inviable. However, DNA synthesis by Pol α, lacking the $3' \rightarrow 5'$ exonuclease activity, is inaccurate and shows a low processivity. These enzymatic characteristics make Pol α a poor candidate for the major replicative polymerase. In addition, Pol α is unique in possessing a primase activity, the only such activity so far identified in eukaryotic cells. Thus, Pol α may play a role in the priming of DNA synthesis. On the other hand, both Pol δ and Pol ε possess a $3' \rightarrow 5'$ exonuclease activity and are highly processive polymerases. Their yeast counterparts, Pol III and Pol II, respectively, are essential for cell growth. Therefore, these polymerases are better suited for chromosome replication.

1.2.2 DNA Helicase and Primase

DNA polymerase cannot replicate a duplex DNA without the assistance of other proteins. This enzyme requires single-stranded DNA as a template and cannot synthesize a chain without an RNA or DNA primer annealed to the template. Two other enzymes enable polymerase to work on a duplex DNA. One is a DNA helicase, which opens up the duplex at the replication fork to provide a single-stranded template. The other is a primase that synthesizes a short RNA to prime DNA chain elongation.

Several DNA helicases have been identified from E. coli and its phages, including DnaB, T7 gp4, and T4 gp41 (Table 2). Biochemical characterization of these activities in in vitro DNA replication systems suggests that the primary replicative helicase binds to and moves on the lagging-strand template in the $5' \rightarrow 3'$ direction, unwinding the duplex as it goes. This helicase action requires ATP hydrolysis. In addition, genetic studies indicate that these helicases are essential for DNA replication and growth of a cell or its phages.

Another common property of the replicative helicases is a close association with a primase. The T7 gp4 has both primase and helicase activity within the same polypeptide. The helicase T4 gp41 greatly enhances primase activity of the T4 gp61. A similar interaction has been observed between DnaB protein and primase (DnaG protein) of E. coli.

Although several DNA helicase activities have been found in eukaryotic cells, the replicative helicase has not been identified. Large T antigen encoded by the SV40 virus is required for the replication of SV40 DNA and has a $3' \rightarrow 5'$ helicase activity. Moreover, human DNA Pol α, which contains a primase subunit, has a specific affinity to the large T antigen helicase.

Primases can start a new chain when copying a duplex DNA but, like DNA polymerases, require single-stranded DNA as the template. Primases can utilize dNTPs in place of rNTPs, although they absolutely require an rNTP (ATP in most cases) to initiate the primer. Under physiological concentrations, rNTPs are preferentially used during primer synthesis. In prokaryotes, primer synthesis is initiated at preferred sequences on the template: 3'-GTC for E. coli primase (DnaG), 3'-CTG(G/T) for T7 phage primase (gp4), and 3'-TTG for T4 phage primase (gp61). The sequence preference of the initiation of primer synthesis by eukaryotic primases is uncertain. All primases found in various organisms extend the primer chain in the $5' \rightarrow 3'$ direction to a particular length: 10–12 nucleotides by DnaG protein, 4 nucleotides by T7 gp4, 5 nucleotides by T4 gp61, and 8–12 nucleotides by eukaryotic primases.

A functional interaction between primase and replicative helicase has been extensively studied using E. coli replication proteins. On most templates, DnaG exhibits a feeble priming activity that can be greatly enhanced if DnaB first binds that DNA. This stimulation of primase activity is further increased when the DnaB helicase is activated to its processive form at the replication fork. The activation of DnaB protein is a key step in the initiation of DNA replication and is catalyzed in several ways (see Section 1.3.2).

E. coli has two other known helicases (PriA and Rep) involved in DNA replication. These differ from the DnaB protein in their direction of translocation and with regard to their requirement for cell growth. However, PriA and Rep helicases are required for ss $\rightarrow$ RF and RF $\rightarrow$ ss DNA replication of φX174 phage genome, respectively. The PriA protein can bind to a specific sequence, the primosome assembly site (PAS), and initiate formation of a prepriming complex with five other proteins, including the DnaB helicase.

1.2.3 Other Replication Proteins

In addition to components of the replisome, several auxiliary proteins are needed for DNA replication. Single-stranded DNA-binding protein protects the exposed template and stabilizes structure of the replication fork. E. coli mutants defective in SSB are in-

Table 2 Replicative Helicases and Primases of E. coli, and T4 and T7 Phages

Enzyme	E. coli	T4 phage	T7 phage
DNA helicase	DnaB	gp41	gp4
Mass, kDa	50	59	63, 56
Form	Hexamer	Multimer	Multimer
Polarity	$5' \rightarrow 3'$	$5' \rightarrow 3'$	$5' \rightarrow 3'$
NTPase	ATP>GTP>CTP	GTP>ATP>dATP	dTTP>dATP>ATP
Primase	DnaG	gp61	gp4
Mass, kDa	60	40	63
Form	Monomer	Monomer	Multimer
Primer length	10 ± 1	5	4 or 5
Preference site	3'-GTC	3'-TTG	3'-CTG

viable, and under nonpermissive conditions, DNA synthesis in these conditional mutants quickly ceases. *E. coli* contains about 300 molecules of SSB per cell, a level sufficient to cover about 1400 nucleotides of the lagging-strand template at each replication fork.

Although no eukaryotic counterpart to the *E. coli* SSB has been firmly assigned, a candidate has been identified in the in vitro replication system. This protein, RF-A, consists of three distinct subunits, binds tightly to single-stranded DNA, and is required for the helicase action of the virus-encoded T antigen at the SV40 origin.

Enzymes to remove primer RNA and fill the gap, such as RNaseH1 and DNA polymerase I of *E. coli,* are essential for the sealing of Okazaki fragments by DNA ligase. Under restrictive conditions, mutants defective in either DNA polymerase I or DNA ligase show a massive accumulation of short Okazaki fragments.

DNA topoisomerases function as swivels to relieve torsional stress produced ahead of the replication fork. From analyses of mutants, it appears that DNA gyrase (a type II topoisomerase from *E. coli*) acts as the swivel in prokaryotic cells. In eukaryotes, either type I or type II topoisomerases can provide the swivel action needed. Another important role of topoisomerases in replication is the decatenation of daughter molecules. This function, provided by type II topoisomerases in both prokaryotic and eukaryotic cells, allows for the separation of daughter molecules prior to segregation.

1.3   REGULATION OF DNA REPLICATION

Initiation of DNA replication is a major determinant of the cell cycle and thus is tightly regulated in all the organisms. Although signals that turn on DNA replication in eukaryotic and prokaryotic cells are known, how the signals are transduced to the initiation machinery is largely unknown. Even so, the mechanisms of initiation mechanisms have been detailed using systems that initiate and replicate a minichromosome in vitro. These various minichromosomes contain replication origins from chromosomes *(E. coli oriC),* bacteriophages (λ *ori,* P1 phage *ori*), plasmids (ColE1 *ori,* F plasmid *oriS*), and animal viruses (SV40 *ori*).

1.3.1 Isolation and Structure of Replication Origin

A DNA segment carrying a replication origin can be isolated as a minireplicon. DNA fragments can be linked to a selectable marker and introduced into the cell. Fragments that allow for maintenance of the marker are likely to contain an origin for autonomous replication of the DNA. Analysis to determine the replication capacity of deletion and base substitution mutants of the minireplicon will define a minimal region of the replication origin and essential motifs in that region. Replication origins that have been examined in this way can be classified into two groups, based on structural similarities. One class, to which most prokaryotic replicons belong, consists of four to seven repeats of an initiator protein binding site and an AT-rich region (Figure 4A). The other group (ColE1 plasmid *ori,* T7 primary *ori,* T4 primary *oriA*) carries a transcriptional promoter instead of the initiator protein binding sites (Figure 4B).

In contrast to the prokaryotic origins, the structure of eukaryotic replication origins is not so well understood. Chromosomal origins have been isolated from yeast and from several eukaryotic viruses. Attempts to identify such autonomously replicating segments from animal or plant chromosomes have been unsuccessful. More direct approaches, using electron microscopy and two-dimensional gel electrophoresis, have led to the identification of replication origins

A. Origin with initiator-protein binding site

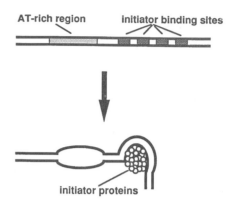

B. Origin with transcriptional promoter

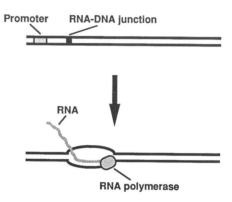

Figure 4.   Two types of replication origin and related mechanisms of initiation.

in higher eukaryotes. There are some indications that eukaryotic origins (1) carry sequences to which particular proteins bind and (2) contain transcriptional promoters and their regulatory elements.

1.3.2 Initiation Mechanisms of DNA Replication

The experimental reconstitution of initiation has elucidated several events during this process, as follows: recognition of the origin, opening of a particular region in the origin, loading of a replicative helicase onto the single-stranded DNA region, and initial priming for leading- and lagging-strand synthesis.

Some bacterial plasmids and phages are replicated unidirectionally. A major determinant for uni- or bidirectionality is the number of replicative helicases loaded at the origin. One helicase can form one replication fork. Thus, bidirectional replication involves two replicative helicases.

The mechanism of initiation at the origin carrying the initiator protein binding site has been elucidated from studies on in vitro replication of *oriC* plasmid DNA (Figure 4A). At least eight proteins are required for the in vitro reconstitution of the *oriC* minichromosome. A key event is formation of the initial complex in which the initiator protein (DnaA protein) tightly binds to its 9 bp recognition sequence in *oriC*. Following this stage, an AT-rich region is opened in an ATP-dependent manner, and the DnaB helicase and SSB are loaded onto the single-stranded region to form the prepriming complex. Primase and DNA polymerase III holoenzyme then begin their functions in DNA synthesis. A similar mechanism has been found in the in vitro initiation of SV40 DNA replication. The virus-encoded T antigen acts as the initiator protein, which specifically binds to the SV40 origin as well as the replicative helicase. Several plasmids (F-factor and R-factors) and bacteriophages (λ, P1) in this replicon class encode their own initiator proteins that specifically recognize and bind the origin sequence. Most of these origin-binding proteins can open the duplex DNA by themselves but fail to load the DnaB helicase. They contain one or two DnaA recognition sequences in their origins and utilize the DnaA protein as a landmark for entrance and activation of the replicative helicase.

Replicons without an initiator protein use transcription for origin recognition and duplex opening (Figure 4B). Initiation of ColE1 plasmid replication requires RNA synthesis, which begins 555 bp upstream from the transition point between RNA and DNA. The transcript (RNA II) forms an RNA–DNA hybrid around the transition point, is processed by RNaseH1, and serves as the initial primer for the leading-strand DNA synthesis. DNA pol I extends the leading-strand DNA about 400 nucleotides. This initial chain elongation, resulting in duplex opening, involves no helicase action but does require that the template be negatively supercoiled.

2 TRANSCRIPTION

The transfer of information from DNA to protein begins with the synthesis of RNA molecules in a process called transcription. In transcription, genetic information carried in DNA is transferred to several kinds of RNA molecule (mRNA, tRNA, and rRNA), so that an mRNA encoding a polypeptide may in its turn translate the information from the 4-letter language of the nucleic acids to the 20-letter language of the amino acids with the coordinated actions of tRNA and rRNA as part of the machinery of protein synthesis.

2.1 RNA POLYMERASE AND TRANSCRIPTIONAL APPARATUS

RNA polymerase binds to a specific DNA sequence, called the promoter, and unwinds the duplex DNA for about one turn of the helix to expose a short stretch of single-stranded DNA so that complementary base pairing with incoming ribonucleotides can proceed. The enzyme joins two of the ribonucleoside triphophate monomers and then moves along the DNA strand, extending the growing RNA

chain in the $5' \rightarrow 3'$ direction until it encounters a second special sequence, called the terminator, which signals where RNA synthesis should stop. After transcription has been completed, each RNA chain is released from the DNA template as a free, single-stranded RNA molecule. In addition to RNA polymerase, other protein factors are required for efficient transcription, and they are responsible for determining transcription efficiency of specific transcription units.

2.1.1 Prokaryotic RNA Polymerase

In prokaryotic cells, a single type of RNA polymerase synthesizes all classes of RNA molecules except primer RNA, which is synthesized by primase. The RNA polymerase of E. coli is composed of multiple subunits, with a total mass of about 450 kDa. The holoenzyme, the catalytically active form with promoter selectivity, consists of five subunits: two α subunits, and one each of β, β′, and σ (Table 3). The α subunit stably connects the β and β′ subunits, then constituting a four-subunit structure, $(\alpha)_2\beta\beta'$, namely, the core enzyme. The core enzyme has the potential to catalyze RNA synthesis but cannot initiate transcription from the proper start site in the promoter of duplexed DNA template. The β subunit has nucleotide substrate-binding activity and is thought to be part of the active site of the enzyme, while the β′ subunit contributes to template DNA binding and to association of the core enzyme with the σ subunit. The function of the σ subunit is to ensure polymerase binding to the proper site of the promoter. There are multiple forms of sigma, each responsible for recognizing a particular class of promoters. The predominant form, serving for general promoters, is σ^{70}.

As shown in Table 3, three additional proteins associate with the RNA polymerase. The 10-kDa ω subunit is consistently present in the preparation of E. coli RNA polymerase and is thought to be involved in the regulation of transcription. NusA and ρ, which influence the efficiency of elongation and termination of transcription, are also found as components of some forms of RNA polymerase. During the cycling of transcription, these components associate with or dissociate from the core enzyme (Figure 5).

2.1.2 Eukaryotic RNA Polymerases and Transcriptional Apparatus

In eukaryotic cells, there are three known distinct types of RNA polymerase. RNA polymerase I (pol I) makes ribosomal RNA and pol II mostly makes messenger RNA, while pol III makes small RNA molecules, such as transfer RNA, 5S ribosomal RNA, and a

Table 3   Subunit Composition of E. coli RNA Polymerase and Transcription Factors

Subunit	Gene	Number of Amino Acids	Molecular Weight (Da)	Function
α	rpoA	329	36,512	Connecting ββ′ subunits
β	rpoB	1342	150,618	Catalyzing RNA synthesis; substrate nucleotide binding
β′	rpoC	1407	155,163	Template binding and association with σ subunit
σ^{70}	rpoD	613	70,263	Recognition of general promotors
ω	rpoZ	91	10,230	Regulation of transcription
ρ	rho	419	46,974	Termination
NusA	nusA	494	54,536	Elongation, termination

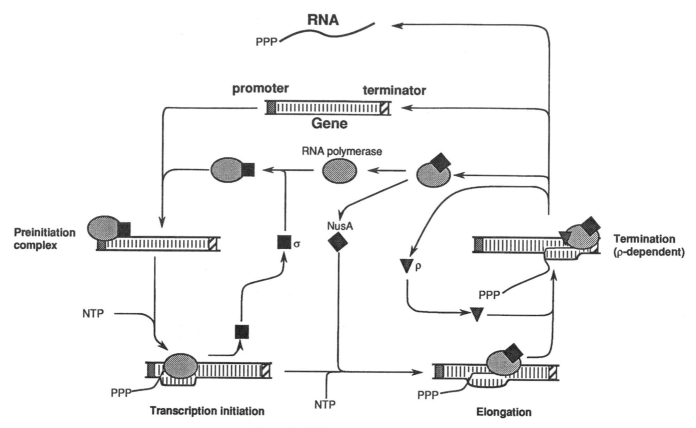

Figure 5. RNA synthesis cycle in *E. coli.*

few other small RNA molecules (Table 4). All these enzymes are large and complex, with molecular weights of about a half million.

Among eukaryotic RNA polymerases, RNA polymerases of the yeast *Saccharomyces cerevisiae* have been characterized most extensively. Each RNA polymerase molecule consists of two large subunits (ca. 160–200 and 130–150 kDa) and a collection of smaller polypeptides. The largest subunit shares amino acid sequence homology with the β′ subunit of the *E. coli* enzyme, while the second largest is related to the prokaryotic β subunit. The third largest subunit of pol I is related to the σ subunit of *E. coli* RNA polymerase. Among the three types of eukaryotic RNA polymerases, five subunits (RPB5, 6, 8, 10, and RCP10) are shared; pol I and pol III share two additional subunits (RPC19 and RPC40). Most of yeast genes for the pol II and pol III subunits have been cloned.

The largest subunit of pol II (RPB1) contains an unusual repetitive domain, which is not found in the homologous large subunits of pol I and III or in the prokaryotic β′ subunit. This carboxyl ter-

minal domain (CTD), called the "tail" of pol II, consists of multiple heptapeptide repeats of the consensus sequence TSPTSPS. The heptapeptide sequence is found in the largest subunits of most eukaryotic RNA polymerase II; 17 repeats in *Plasmodium*, 26–27 repeats in yeast, 32 repeats in *Caenorhabditis*, 40 repeats in *Arabidopsis*, 45 repeats in *Drosophila*, and 52 repeats in mouse and hamster RNA polymerase II. Deletion of most or all of the CTD is lethal for yeast, *Drosophila*, and mouse cells, thereby indicating that the CTD has an essential role in transcription. The tail of RNA polymerase II in an elongation complex is highly phosphorylated.

None of the eukaryotic RNA polymerases can initiate transcription by themselves; rather, they require a group of general transcription initiation factors for initiation at proper sites. For example, pol II requires seven or eight general initiation factors [TFIIA, IIB, IID, IIE, IIF, IIG(J), IIH, and/or II-I], and formation of a functional preinitiation complex on eukaryotic promoters involves ordered interactions of an array of general initiation factors on the pro-

Table 4   Eukaryotic RNA Polymerases[a]

Polymerase	Location	Products	Polymerase Activity of Cells (%)
pol I	Nucleolus	35*S* and 47*S* pre-rRNA	50–70
pol II	Nucleoplasm	hnRNA (pre-mRNA) snRNA (U1, U2, U4, U5)	20–40
pol III	Nucleoplasm	tRNA, 5*S* RNA, 7*S* RNA, snRNA(U6), other small RNA molecules	10

[a]Mitochondria and chloroplast contain distinctive RNA polymerases.

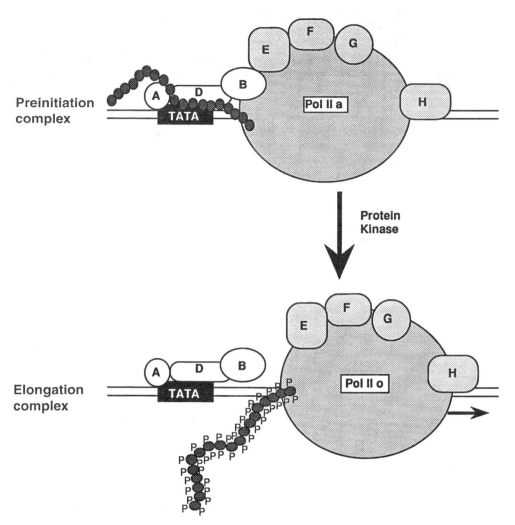

Figure 6. Formation of a functional preinitiation complex on a TATA-containing promoter by general initiation factors. After completion of the final preinitiation complex, the elongation complex may be released from the promoter to start elongation by phosphorylation of the carboxyl terminal domain by transcription factor H of polymerase II (TFIIH).

moter (Figure 6). Transcription factor D for polymerase II (TFIID) binds to the TATA box of the promoter, then TFIIA and B enter to form the DAB complex. TFII-I is likely to be involved in formation of the DIB complex, instead of the DAB complex, on a promoter with an initiator (see Section 2.2: Figure 7). The nonphosphorylated form of RNA polymerase II (pol IIa), together with TFIIF, binds to the DAB complex on the promoter, followed by association of TFIIE, G(J), and H to form a final initiation complex. In the preinitiation complex, pol II interacts with TFIID via its nonphosphorylated CTD, which is then phosphorylated by TFIIH after completion of the preinitiation complex. The RNA polymerase with phosphorylated CTD (pol IIo) loses its affinity for TFIID and prepares to start elongation (an elongation complex).

RNA polymerase III requires different combinations of transcription initiation factors to form a preinitiation complex on the promoters of 5S RNA, tRNA, and other small RNA genes. Three pol III specific initiation factors, TFIIIA, B, and C, are required for transcription of the 5S RNA genes, TFIIB and C for the tRNA genes, and TFIIIB and TFIID for the U6 gene. TFIIIA and C, as well as TFIID, recognize and bind to specific DNA elements, while TFIIIB has by itself no sequence-specific interaction with DNA. RNA poly-

merase I also requires several transcription factors to form an active initiation complex on the promoter of rDNA.

2.2   Transcription Units

Transcription takes place in limited regions of the genomic DNA, and only one of the two DNA strands is used as a template. The promoter, an oriented DNA sequence along the template DNA, determines the starting point of the region to be transcribed and also which of the two strands will be copied. A transcription unit extends to the terminator, at which the RNA polymerase stops adding nucleotides to the growing RNA chain. The critical feature of the transcription unit is that it constitutes a stretch of DNA expressed via the production of a single RNA molecule. A transcription unit may include only one gene, or several. Figure 7 shows a typical transcription unit of *E. coli* and various transcription units for three eukaryotic RNA polymerases.

2.2.1 Prokaryotic Transcription Unit

Bacterial promoters are identified by two short conserved sequences centered at −35 and −10 relative to the starting point. In *E. coli,*

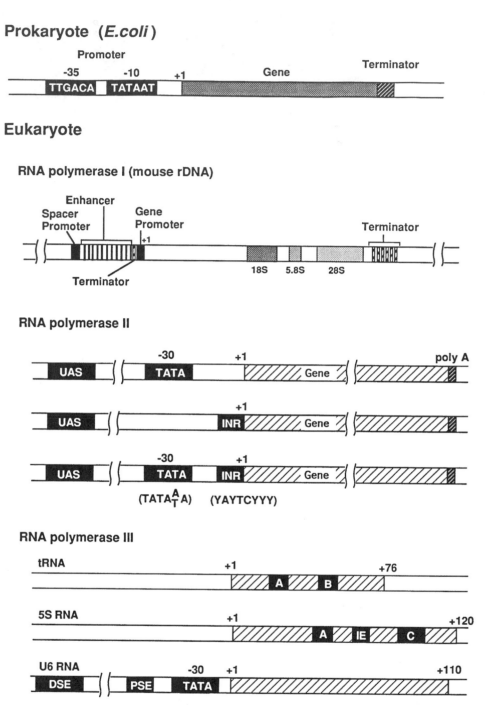

Figure 7. Various transcriptional units in prokaryotic and eukaryotic cells.

TTGACA and TATAAT are consensus sequences for the -35 and the -10 boxes of the promoters, which are recognized by the RNA polymerase holoenzyme carrying σ^{70}. Promoters whose -10 and -35 sequences closely approximate the consensus sequences are strong promoters, and the distance separating the two sequences, usually 16–18 bp, is also important. Bacterial RNA polymerase terminates transcription at sites of two types: factor-independent sites contain a GC-rich hairpin followed by a run of U residues, while ρ-factor-dependent sites require ρ-specific sequences.

2.2.2 Eukaryotic Transcription Unit

In eukaryotes, the promoters recognized by each type of RNA polymerase are distinct. The promoters for RNA polymerase I and II are located upstream of the transcription starting point, whereas the promoters of tRNA and 5S RNA genes for RNA polymerase III are downstream of the starting point.

Promoter elements that are necessary and sufficient for specific initiation by RNA polymerase II with its general factors are referred to as minimal or core promoter elements. The most common core elements are the TATA box, present about 30 bases upstream of the transcription start site, and the less well-characterized initiator (INR) elements located at the start site. Specific genes may contain either or both elements, which may function in conjunction with other cis elements.

For RNA polymerase III, there are two types of promoter: the promoters for U6 are upstream of the starting point, while those of the 5S RNA and tRNA genes lie downstream of the starting point. The external promoter for the U6 gene consists of the TATA box and the distal and proximal sequence elements (DSE and PSE), which enhance promoter activity. Intragenic A and B boxes in the tRNA gene or the A box in the 5S RNA gene can be recognized by, and bound by TFIIIC and IE; C boxes in 5S RNA are involved in interaction with TFIIIA.

In the case of RNA polymerase I, the sole product of transcription is a large precursor that contains the sequences of the major rRNA. Termination occurs at a discrete site and apparently requires few factors. The termination capacity of RNA polymerase II has not been characterized, and the 3′ ends of its transcripts are generated by cleavage of the transcript. The sequence AAUAAA, located a few bases upstream of the cleavage site, may provide the signal. Transcription by RNA polymerase III terminates within a run of four U residues embedded in a GC-rich region, as is the case with the prokaryotic factor-independent terminators.

2.3   Regulation of Transcription

Transcriptional regulation is the primary means of controlling gene expression and is essential for the developmental program of multicellular organisms, as well as the switching on and off of gene expression in response to intra- or extracellular signals. In any organism, RNA synthesis can be regulated mainly by controlling the efficiency of initiation.

2.3.1 Transcriptional Regulation in Prokaryotes

In bacterial cells, there are several different mechanisms by which modulation of transcription initiation is achieved. Various σ factors modify the ability of RNA polymerase to bind to specific promoter sequences, thus enabling it to initiate at a different set of promoters.

In a set of genes whose expression is coinduced in response to intra- or extracellular signals, a common repressor protein called an operator binds to specific DNA sequences in a group of promoters and represses their transcription by inhibiting RNA polymerase binding to the promoters. Upon responding to a specific signal, such repressor molecules are inactivated, and RNA polymerase can initiate transcription from the promoters. For example, the LexA protein, which represses the SOS-regulon, is proteolytically inactivated by an activated form of the RecA protein in response to DNA damage. Modulation of genes (an operon) involved in the biosynthesis of a specific metabolite is mediated by the repressors that directly sense the intracellular status of the product. For example, the LacI repressor protein binds to the operator region of the *lac* operon promoter in the absence of lactose or its derivative, thereby prohibiting transcription initiation on the promoter, and dissociates from the promoter upon binding of a derivative of lactose, allowing for transcription of the *lac* operon.

Promoters requiring positive activator proteins for efficient expression are characterized by a poor fit to the consensus sequence in the −35 region. The protein activators bind to the region near the −35 box, and stimulate initiation either by improving the affinity of RNA polymerase for the sequence or by increasing the rate of isomerization of initiation complex. The complex of cAMP with catabolite activator protein (CAP) enhances the affinity of RNA polymerase to the *lac* promoter about 20-fold.

Transcription efficiency can be modulated by the degree of DNA methylation and superhelicity.

In addition to the regulation of initiation, the level of transcription can be regulated by premature termination of transcription, or attenuation. For example, in most bacterial amino acid biosynthesis operons, early termination of transcription occurs at the attenuator site, which lies 100–200 bp downstream of the start site, when an abundance of the charged cognate tRNA signals a sufficiency of the amino acid. Attenuation in these operons is regulated in a complex mechanism involving charged cognate tRNA, ribosome, and ρ protein.

2.3.2 Transcriptional Regulation in Eukaryotes

In eukaryotic cells, transcription by RNA polymerase II is regulated more dynamically than transcriptions by pol I and III. An assortment of regulatory elements for RNA polymerase II transcription can be scattered both upstream and downstream of the transcription start site for a gene. Each gene has a particular combination of positive and negative regulatory cis elements that are uniquely arranged as to number, type, and spatial array. These elements are binding sites for sequence-specific transcription factors that activate or repress transcription of the gene. Usually, cis elements are arrayed within several hundred base pairs of the initiation site, but some elements can exert their control over much greater distances (1–30 kb). The region near the starting point, in particular the TATA box (if there is one), is responsible for selection of the exact starting point and is called the promoter, as described earlier. The elements further upstream determine the efficiency with which the promoter is used and are known as the upstream activating sequence (UAS) or enhancers (Figure 7).

Binding of transcription factors to specific sequences in a promoter or a UAS may be followed by protein–protein interactions with other components of the general transcription apparatus including RNA polymerase, perhaps requiring the DNA between the promoter and the UAS to be "looped out." Interaction between the activator and the general transcription factors (very likely TFIID and TFIIB) may require other factors, so-called positive cofactors, which cannot by themselves bind DNA. These protein–protein interactions enhance the formation of preinitiation complex.

In eukaryotic cells, efficiency of transcription is also regulated by attenuation of transcription elongation. Modification of DNA such as methylation of cytosine in the CpG sequence or a higher order of chromatin structure involving supercoiling and nucleosome assembly may determine the accessibility of chromatin to the transcription apparatus.

See also Chromatin Formation and Structure; DNA Structure.

Bibliography

Alberts, B., Bray, D., Lewis, J., Raff, M., Roberts, K., and Watson, J. D. (1994) *Molecular Biology of the Cell,* 3rd ed. Garland, New York.

Horiuchi, T., Maki, H., and Sekiguchi, M. (1989) Mutators and fidelity of DNA replication. *Bull. Inst. Pasteur,* 87: 309–336.

Kornberg, A., and Baker, T. A. (1992) *DNA Replication,* 2nd ed. Freeman, New York.

Lewin, B. (1994) *Gene V.* Oxford, New York, NY.

Lindahl, T., Sedgwick, B., Sekiguchi, M., and Nakabeppu, Y. (1988) Regulation and expression of the adaptive response to alkylating agents. *Annu. Rev. Biochem.* 57:133–157.

Marians, K. J. (1992) Prokaryotic DNA replication. *Annu. Rev. Biochem.* 61:673–719.

Moses, R. E., and Summers, W. C., Eds. (1988). *DNA Replication and Mutagenesis.* American Society for Microbiology, Washington, DC.

Peterson, M. G., and Tjian, R. (1992) The tell-tail trigger. *Nature,* 358:620–621.

Reznikoff, W. S., Burgess, R. R., Dahlberg, J. E., Cross, C. A., Record, M. T. Jr., and Wickens, M. P. (1987) *RNA Polymerase and the Regulation of Transcription: A Steenbock Symposium.* Elsevier, New York.

Roeder, R. G. (1991) The complexities of eukaryotic transcription initiation: Regulation of preinitiation complex assembly. *Trends Biochem. Sci.* 16:402–408.

Young, R. A. (1991) RNA polymerase II. *Annu. Rev. Biochem.* 60:689–715.

DNA Sequencing in Ultrathin Gels, High Speed

Robert L. Brumley, Jr. and Lloyd M. Smith

1   **Principles**

2   **Apparatus Design**

3   **Techniques**
    3.1   Gel Preparation and Assembly
    3.1   Sample Loading and Electrophoresis

4   **Automated DNA Sequencing**

Key Words

Anode   In an electrophoresis cell, the terminal toward which negatively charges species migrate.

Cathode   In an electrophoresis cell, the terminal toward which positively charges species migrate.

Electrophoresis   The movement of small ions and/or macromolecules (such as DNA) through a matrix or solution under the influence of an electric field.

Polyacrylamide   A polymer matrix formed by the copolymerization of acrylamide and bisacrylamide.

Horizontal ultrathin gel electrophoresis (HUGE), as applied to DNA sequencing, is used to separate DNA fragments in polyacrylamide gels that are less than 0.1 mm thick. HUGE is compatible with conventional radioisotope-based (manual) and fluorophore-based (automated) DNA sequencing. When HUGE is used with manual sequencing protocols, as many as 350–400 bases of sequence information can be resolved in less than 30 minutes of electrophoresis. By comparison, traditional vertical electrophoresis can require up to several hours to obtain the same sequence. Automat-ed DNA sequencing methods combined with HUGE increase the throughput of analysis by close to an order of magnitude compared to automated systems currently in use. Although HUGE is a relatively new technique for separating DNA, it is gaining acceptance as an effective way to significantly decrease the time required for the electrophoresis step of DNA sequencing.

1   PRINCIPLES

Large-scale projects such as the Human Genome Initiative generate substantial demands for improved DNA sequencing methodology. The response has been a worldwide effort to develop new techniques that will significantly decrease the time and cost of sequencing DNA. At present, all practical methods of sequencing DNA are based on the electrophoresis of DNA fragments through a polyacrylamide matrix. This step consumes much of the time required to sequence DNA and has remained relatively unchanged since sequencing was first developed. One of the main limitations on the speed of electrophoretic separations of DNA is the relatively thick (0.4 mm) polyacrylamide gels that are employed. It has long been known that the thinner the gel, the larger the applied electric field can be without deleterious heat effects on the gel matrix. Although methods for the preparation of ultrathin gels have been appearing in the literature since the early 1980s, the use of gels less than 0.2 mm thick has not found widespread utility, presumably as a result of the practical problems associated with their use. Recently, it has been shown that ultrathin gels (typically 50 μm) can be prepared and used for DNA sequencing using the technique of capillary gel electrophoresis (CGE), in which the efficiency of heat transfer permits the use of much larger electric fields without deleterious thermal effects. These high electric fields result in separation speeds increased as much as 26-fold over conventional electrophoresis.

Although the separation of DNA fragments is much faster using CGE, the practicality of electrophoresis by this method is limited because one gel-filled capillary must be prepared for each DNA sample that is analyzed in parallel. An alternative approach to multiple capillaries would be to design an electrophoresis system that combines the advantages of the slab gel format (ease of gel pouring and multiple samples) and CGE (increase in electrophoretic separation speed). To this end, an electrophoresis apparatus has been designed that achieves rapid separations and allows the introduction of samples onto horizontal slab gels as thin as 50 μm. We describe here the instrument and methodology used for DNA sequencing by horizontal ultrathin gel electrophoresis (HUGE).

2   APPARATUS DESIGN

Schematic diagrams of the electrophoresis apparatus are shown in Figure 1. The polycarbonate apparatus base (6 in. × 15 in. × 1 in.) is machined to provide space for a series of clamps, guides, inlet and outlet manifolds, and a water jacket. The end alignment bar and two guide blocks aid in the correct positioning of the glass components on the water jacket O-ring. In addition, pressure adjustment screws on the end alignment bars apply pressure at the ends of the cathode and anode assemblies, to ensure that a proper liquid-tight seal is created between the polyurethane pads and the top glass. The pressure screws at the anode assembly, which ensure that the comb fits properly between the glass components, are essential for the consistent formation of sample wells. Electrodes are constructed by stringing

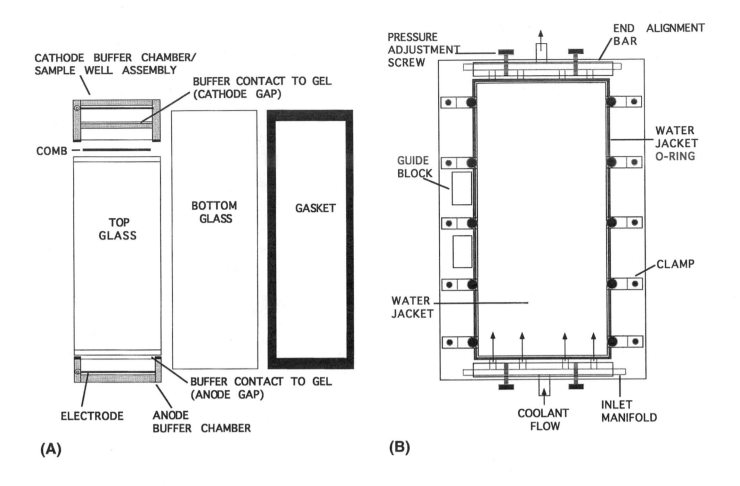

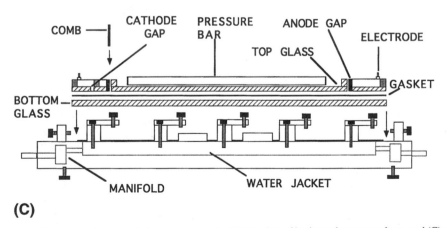

Figure 1. The HUGE apparatus: (A) top view of glass and plastic components, (B) top view of horizontal apparatus base, and (C) side view of horizontal apparatus base and glass components.

a platinum wire across the back of each buffer chamber. The wire is secured with silicon glue in small holes drilled in the sides of the buffer chambers. The wires are connected to "banana" type plugs (see Figure 1A, 1C).

The anode and cathode buffer chambers (Figure 1A) were designed to hold up to 20 mL of liquid. Figure 1A shows the slot formed by the juxtaposition of the top glass with the foam spacers on the anode buffer chamber (anode gap) as well as the slot in the cathode buffer chamber assembly (cathode gap), each of which provides electrical contact between the gel and the buffer. The comb fits in the gap created by the foam spacers between the cathode buffer chamber assembly and the top glass plate.

The comb is made of 0.75 mm poly(methylmethacrylate) (PMMA) approximately 3.8 cm high and 7.2 cm wide. The "teeth" of the comb, which form the sample wells, are 2.0 mm wide, and a 0.25 mm gap is cut between each pair of teeth. The bottom and top glass plates currently in use in our lab are $10.0 \times 30.5 \times 0.5$ and $10.0 \times 25.0 \times 0.5$ cm^3, respectively. Two types of glass are used for the two methods of DNA sequencing practiced in our lab. For radioactivity-based DNA sequencing, soda lime float glass is adequate for the top and bottom glass plates. For automated fluorescence-based DNA sequencing, optical quality (fused silica or BK-7) glass polished to 4-wave length (2μm) flatness over any 2-inch area is used. The glass faces of the cathode assembly and the top glass plate that come into contact with the comb are polished to within 5 μm. The gasket shown in Figure 1A, which determines the thickness of the gel, is cut from a polyester sheet (usually 50 or 75 μm thick).

3   TECHNIQUES

The preparation of the surfaces of the glass components used in HUGE is much the same as conventional gel electrophoresis. However, such thin gels are intolerant of foreign matter (dust particles, lint, petroleum jelly) on the glass. Therefore, pouring bubble-free ultrathin gels requires great care in maintaining a clean working area and practicing the strict cleaning protocol described in Section 3.1; the use of lint-free paper towels is essential, as well.

3.1   GEL PREPARATION AND ASSEMBLY

We currently employ two different protocols for HUGE, one for radioactive sequencing and one for automated DNA sequencing. Many of the techniques are the same for both procedures, with only minor modifications for the performance of automated DNA sequencing. The manual procedure is outlined in Sections 3.1 and 3.2. The procedure for automated DNA sequencing is discussed in detail in Section 4.

Prior to gel preparation, all glass components are cleaned and the bottom glass is treated so that the gel will bond to the glass when the "components" are disassembled after electrophoresis. Likewise the end of the top plate and the glass surface of the cathode assembly where the comb is inserted are treated to facilitate formation of divisions between the wells. All glass plates are cleaned using a common laboratory glassware detergent (Alconox), rinsed thoroughly with distilled water, and then dried with lint-free paper towels. The top and bottom glass plates are cleaned further with three ethanol wipes and one final rinse with distilled water. The components of the apparatus are assembled as shown in Figure 1C and

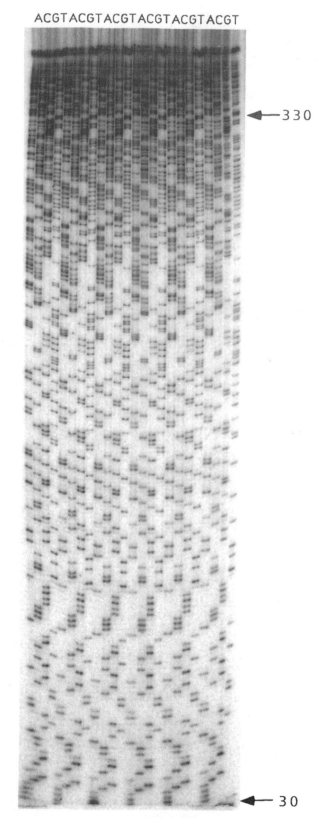

Figure 2.   Autoradiograph obtained by HUGE (six identical sets of sequencing reactions). The gel was electrophoresed for 17 minutes at 40 W constant power (128–213 V/cm). [Reprinted with permission from Brumley and Smith (1991).]

clamped into place. Then the gel is cast directly on the apparatus, as follows: about 8 mL of 6% polyacrylamide solution (7.5 M urea, 1X TBE) is poured into one of the buffer chambers, and the liquid flows down the glass within the inside border of the gel gasket; then the three vertical "gaps" (comb, anode, and cathode) are filled with gel solution, and the comb is placed in the slot between the cathode chamber and top glass plate. The gel is allowed to polymerize for about 60 minutes prior to use.

3.2 SAMPLE LOADING AND ELECTROPHORESIS

About 15 minutes before samples are to be loaded, the water in the circulating bath is preheated to the desired temperature (usually about 35–40°C for sequencing gels). After preheating, the water is allowed to circulate through the water jacket via the inlet and outlet manifolds, to prewarm the glass plates. The comb is then removed and the sample wells are rinsed thoroughly with distilled water. The electrodes are connected from the power supply to the apparatus, and electrophoresis buffer is added to the anode and cathode buffer chambers. The wells are rinsed once again with distilled water, and sequencing reaction products are dispensed into the wells with a 1.0 μL syringe. Care must be taken to dispense the samples slowly into the bottom of the wells to ensure that the sample does not splash up the sides of the wells.

Gels are usually electrophoresed at 40–50 W constant power, which results in a field strength of 150–250 V/cm. Electrophoresis time varies, depending on the amount and location of the sequence of interest. If the first 250–300 bases of sequence from the primer

are desired, electrophoresis is stopped after about 15–20 minutes, when the bromophenol blue tracking dye reaches the anode buffer chamber. (This dye is routinely included in the "stop" solution supplied with many commercial radioactive sequencing kits.) When electrophoresis is complete, the buffer is removed from the chambers and properly discarded (the buffer in the anode chamber is radioactive). The water circulator is turned off and the water jacket is drained of fluid. The "components" are removed from the apparatus base and disassembled such that the gel adheres to the bottom glass plate. The gel can then be "fixed" with an alcohol/acid solution and dried or wrapped in plastic wrap and exposed to X-ray film to visualize the separated fragments. Figure 2 is an example of a typical sequencing "ladder" produced by the Sanger dideoxy radioactive sequencing protocol.

4   AUTOMATED DNA SEQUENCING

The automated instrument shown in Figure 3 comprises, in addition to the electrophoresis apparatus, the input and output optics assemblies. The input assembly provides a line of excitation light of the desired wavelength across the width of the gel. The excitation source for the instrument described here is a multiline argon ion laser. The major wavelengths of 488 and 514 nm excite the fluorescein and rhodamine dyes commonly used in automated DNA sequencing. The input assembly is composed of the laser source, focusing optics, and light coupler. The system described here directs the laser beam through the side of the gel to provide a nearly uniform line of laser light across the gel. Because of the nature of

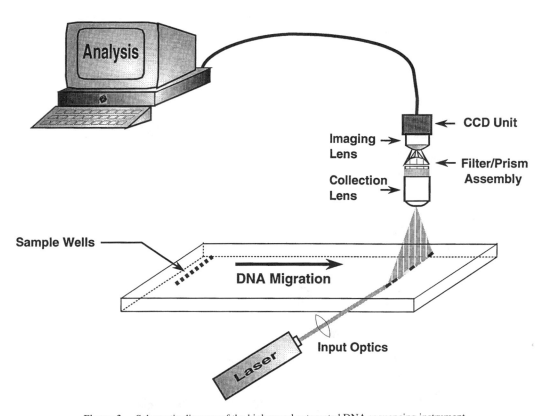

Figure 3.   Schematic diagram of the high speed automated DNA sequencing instrument.

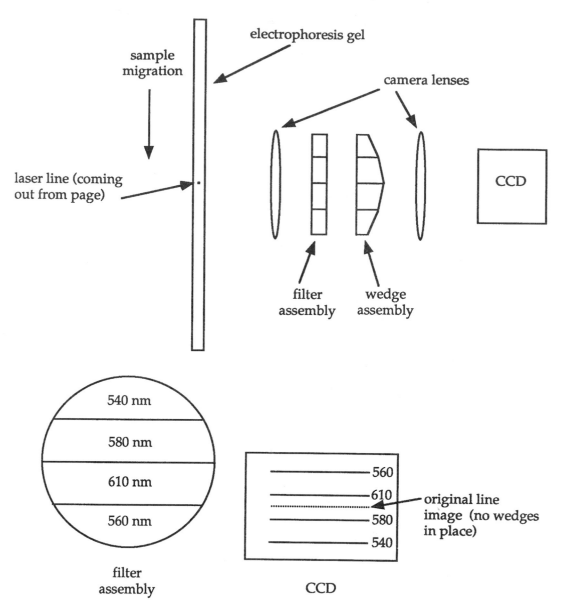

Figure 4. Diagram of the fluorescence detection system employed for imaging of the line of emitted fluorescence onto a CCD detector at four wavelengths: front views of the filter assembly and of the fourline images on the CCD detector.

Gaussian laser beam optics, there is an optimum way of using the focusing optics to obtain this uniform laser line. The fundamental limitation of Gaussian beam optics is that the more tightly the beam is focused, the shorter the distance over which that focus can be maintained. Another technical issue concerns the introduction of the laser beam into such a thin gel. An optical coupler is used to direct the beam between the closely spaced glass plates in a way that does not distort or scatter the beam. This optical interface is prepared by clamping the coupler in place prior to pouring the gel. Once the gel has been poured, the liquid-tight seal formed by the optical coupler allows the addition of polyacrylamide to the cavity of the coupler. The gel that polymerizes in the coupler and the gap in the spacer material provides a reasonable optical interface for the laser beam.

The output optics shown in Figure 4 will image the DNA fragments as they pass through the laser line (detector region). The light emitted from the fluorescent labels must be separated in the spatial and wavelength dimensions. This is accomplished by collecting the light with a wide-format 150 mm lens. The lens is used "backward" such that the detector region is analogous to the surface of the film in a camera. When the lens is used in this orientation, the rays of light exiting the lens are roughly parallel. The collimated light is then filtered with four 10 nm bandpass filters and split into four lines, each representing the light from one of the four fluorophores that is used to label the DNA. These four lines are focused with a 50 mm lens onto the surface of a two-dimensional charged-coupled device (CCD) as shown.

Data from the image is taken in frames (snapshots) every 0.2–1.0 second (usually 0.5s) on a CCD camera system that includes control electronics, a thermoelectric cooler, and a computer interface. The CCD chip is a rectangular array of square pixels divided into

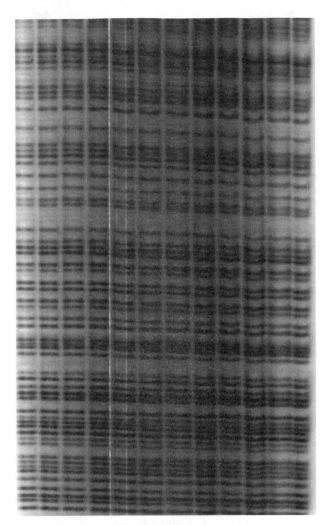

Figure 5. Portion of an "image file" obtained from a HUGE sequencing gel of 12 identical samples. The different colored bands correspond to the four different bases in the DNA sequence. (See color plate 1.)

two equal regions: an unmasked region, where the image is exposed, and a masked region where readout takes place. Each frame of data collected on the CCD system requires a series of steps: exposure, frame transfer, and readout. This type of CCD operation, referred to as "frame transfer mode," allows the rapid collection and processing of large amounts of image data. In this mode, an exposure made on the CCD is transferred to the masked region for the slower process of readout, and the CCD is able to take another image exposure while the preceding one is being processed. The result of this procedure is the collection at half-second intervals of 512 spatial pixels at each of the wavelengths indicated in Figure 4; a final image of about 17 million data points is produced and stored in 70 minutes. A portion of such an image is shown in Figure 5 (see color plate 1).

Further data processing is required to extract the DNA sequence information from this image file. The process of using raw fluorescence data from multiple runs to produce the final sequence can be broken down into several distinct tasks: lane finding, multicomponent analysis, mobility shifting, digital filtering, and base calling.

Lane finding consists of "compression" from the two-dimensional CCD data into one-dimensional data representing the signal response of each lane. This step is currently done by a human who interacts with the computer to determine the correct lane alignment for each line.

Multicomponent analysis is the determination of the concentrations of the four fluorophores present at each sample point in time from the spectral data collected at four different wavelengths. The spectral data are multiplied by a 4×4 matrix, which specifies the transform from a four-channel wavelength spectrum to a four-channel signal, where each channel represents the response to a single dye. The determination of this matrix is a fairly straightforward calibration step, and when it has been accomplished, dye concentration information can be obtained from the fluorescence intensity data.

The four dye primers used in the enzymatic sequencing reactions have slightly different electrophoretic mobilities owing to differences in their size and charge. When attached to DNA fragments that are being electrophoresed, they cause changes in the mobilities of the fragments by a constant offset that does not depend on fragment length (to a first approximation, anyhow). These *mobility shifts* mean that each of the four channels of data will be offset from the others by a small amount. To a first approximation, this shift can be compensated for by simply offsetting the channels from one another by a fixed amount. This is quite adequate in most cases. When nonlinearity in the offsets is encountered (which occurs at higher electric fields as a result of "biased reptation" effects on mobility), the offset applied can be a given fraction of the spacing at that position in the sequence.

There are two primary types of noise present in the raw signal data. The first is high frequency shot noise. The second is low frequency signal drift. Both can have a significant effect on the accuracy of base calling, and thus they need to be eliminated. Currently, a zero-area Gaussian *digital filter* is being used. The device acts as a bandpass filter and works well at eliminating both types of noise.

Base calling consists of base sequence determination from the four-channel, one-dimensional signal data provided by the first four steps. A human can determine the base sequence by visually inspecting the signal data for uniformly spaced peaks, determining which channel a given peak is in, and labeling that peak with the corresponding base. The task, however, is very time-consuming and error-prone. To address this issue, computer software has been developed which automates the base sequence determination. This software uses various parameters of the data, such as the peak intensities, spacing between peaks, and width of peaks, to correctly identify the base sequence. Currently the program uses an iterative, modular approach, determining a rough sequence based on spacing and size, then refining it by ascertaining confidences for each base and eliminating those that are low on each iteration. This approach allows compensation for poorly resolved or noisy regions by using the information from better resolved regions to guide base calling in these bad areas. Since the algorithm is adaptive, there is no need to calibrate the software to a specific instrument or data collection device. Finally, to improve performance, new modules that examine peak features other than size, spacing, or width can easily be plugged into the current framework. This flexibility is primarily due to the modular, object-oriented approach used. Figure 6 (see color plate 2) shows an example of an analyzed data set along with the inferred base sequence.

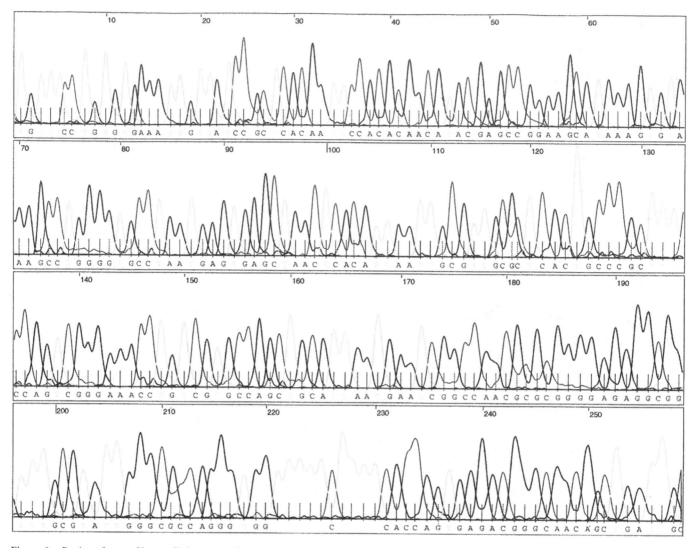

Figure 6. Portion of a set of base-called sequence data generated by fluorescence-based sequencing on the HUGE system. The region shown was obtained in about 40 minutes. (See color plate 2.)

See also NUCLEIC ACID SEQUENCING TECHNIQUES; ROBOTICS AND AUTOMATION IN GENOME RESEARCH.

Bibliography

Brumley, R. L., and Smith, L. M. (1991) Rapid DNA sequencing by horizontal ultrathin gel electrophoresis. *Nucleic Acids Res.* 19(15):4121–4126.

Giddings, M. C., Brumley, R. L., Haker, M., and Smith, L. M. (1993) An adaptive, object-oriented strategy for base calling in DNA sequence analysis. *Nucleic Acids Res.* 21(19):4530–4540.

Higher speed imaging: Frame transfer CCD's (1995) *Photometrics,* CCD newsbrief (Winter).

Hunkapiller, T., Kaiser, R. J., Koop, B. F., and Hood, L. (1991) *Science,* 254:59–67.

Kostichka, A. J., Marchbanks, M., Brumley, R. L., Drossman, H. and Smith, L. M. (1992) High speed automated DNA sequencing in ultrathin slab gels. *Bio/Technology,* 10(1):78–81.

Luckey, J. A., Drossman, H., Kostichka, A. J., and Smith, L. M. (1993) High speed DNA sequencing by capillary gel electrophoresis. *Methods Enzymol.* 218:154–172.

Smith, L. M. (1989) DNA sequence analysis: Past, present, and future. *Am. Biotechnol. Lab.* 7(5):10–25.

DNA STRUCTURE

C. R. Calladine and Horace R. Drew

1 **Review of Past Work**

2 **Helical Structure**

3 **Base Step Geometry**

4 **Bending and Curvature of DNA**

5 **Superhelical DNA**

6 **Supercoiling of Long DNA**

Key Words

Curvature and Twist   Two forms of distortion, required of the double-helical molecule in a range of biological activities. Also can be understood in terms of the stacking arrangements at base pair steps.

Double Helix   The structure adopted by two sugar–phosphate chains coiling around each other, with the bases forming pairs to connect them, like the treads of a spiral staircase.

Roll, Slide, Twist   The three main geometrical parameters required to describe the conformation of any helix step formed by the stacking of one base pair onto the next.

Supercoiling   A central feature of the behavior in biology of the long, thin molecule; can be understood in terms of the molecule's curvature and twist.

DNA occupies a central position in molecular biology: it contains the coded message that is passed on whenever a cell divides, which prescribes the detailed operation of any living organism.

To understand how DNA works in the cell, it is necessary first to understand something of its structure. The famous Watson–Crick double-helical structure of DNA (1953), with its stack of complementary base pairs A-T and G-C forming the treads of a spiral staircase held together by two sugar–phosphate chains, is now very well known. Representations of it are frequently used as motifs in advertisements for drugs and scientific magazines. But that level of knowledge of the structure of DNA imparts only a superficial understanding of the many ways by which DNA works in a cell.

Here, for reasons of limited space, we deal only with the essential aspects of what is now known about DNA structure. First we introduce the main components of DNA. Next we explain how the insolubility of the organic bases in water, together with the known sizes of these bases and of the "links" of the two sugar–phosphate chains, produces a spiral or helical structure with about 11 base pairs for each helical turn. The double-helical structure is flexible, not rigid, since the ways by which the base pairs can stack onto each other, under some constraint from the sugar–phosphate chains, allow for many different relative motions between successive base pairs. These motions are deployed whenever the DNA is required to bend or twist as in the formation of chromosomes, for example; and each kind of base pair step such as AA, TA, or CA has its own conformational preferences.

In discussing the geometrical changes that take place at the level of base stacking when straight DNA is made into a curve, we demonstrate how the base-stacking preferences of different sequences may influence how DNA will curve around proteins, such as the histone proteins in chromosomes. We also discuss the geometry that is needed to understand how DNA can form three-dimensional supercoils, such as those that produce anomalies of expected velocity in gel electrophoresis. Finally, since supercoiling appears to be very relevant to understanding how genes work in living cells, we explain the theory of DNA supercoiling in closed domains such as circles or loops, and we introduce the equation $Lk = Tw + Wr$ as an aid to understanding the behaviour of a long, twisted string.

1   REVIEW OF PAST WORK

The word *structure* has a multiplicity of meanings in many fields such as engineering or management or music, as well as in the present area of biology. The structure of deoxyribonucleic acid—DNA—can be considered at several different levels. Structure describes, in general, the way in which any object is constructed from its component parts.

It has been known for most of this century that DNA is a long chain molecule, containing the bases adenine, thymine, guanine, and cytosine (A, T, G, and C) covalently attached to a chain of repeating sugar–phosphate units. This one-dimensional chain might be called the *first-level* structure of DNA. By 1950 the chemist E. Chargaff had found that there are always approximately equal numbers of bases A and T, and also equal numbers of G and C within the DNA molecule. J. D. Watson and F. H. C. Crick clarified the matter in 1953, when they found by model building that two such chains could wind around each other in a double-helical fashion, with one chain going "up" and the other "down," and with the bases joining in pairs across the helical axis. Furthermore, on account of specific hydrogen bonding arrangements and in accord with Chargaff's rules, the base pairs were postulated to be of two types only: A-T and G-C. This three-dimensional *second-level* structure constituted a major breakthrough in understanding the function of DNA because specific base pairing meant that the two DNA chains were *self-complementary,* hence could make copies of each other as templates. Thus, if one chain carries a long, coded message for the construction of a protein in its sequence of bases A, T, G, and C, it can act as a template onto which a new chain may be constructed, by the addition of complementary bases: A with T or G with C. In that way, the self-complementary double-helical structure for DNA explains how it is possible for genes to be transmitted from generation to generation of a living cell.

In fact, Watson and Crick arrived at a plausible model for the double-helical structure of DNA mainly by interpreting the X-ray diffraction patterns taken from elongated *fibers* of DNA that had been obtained by Rosalind Franklin and M. F. Wilkins some time earlier. These workers had found two distinct kinds of X-ray diffraction pattern, depending on whether the fibrous material was dry ("A" form) or wet ("B" form). Franklin herself had decided to work on the structural interpretation of the A-form pattern, because it clearly contained more precise data from which a structure could be deduced by calculation. But Watson and Crick more quickly interpreted her simpler B-form pattern as a uniform double-helical structure, with planar base pairs perpendicular to the helix axis, and 10 base pairs per turn.

In the 40 years or so that have elapsed since the Watson–Crick structure was formulated, DNA has been subjected to intensive research. Much of this effort has had a structural motivation, on the premise that only a more thorough knowledge of the three-dimensional structure of that molecule can provide a better understanding of the many detailed ways by which DNA functions in biology.

The central focus of structural research into DNA has been provided most recently by single-crystal X-ray methods. Until almost 1980, the only X-ray data on DNA came from the diffraction of fibers. Thus for many years after the B and A forms had been worked out, detailed fiber studies were made of DNA having various sequence characteristics, such as AT-rich or GC-rich, or all A on one strand versus all T on the other. Many different detailed geometries of the double helix were deduced in that way; and they were given identifying letters (C, D, E, etc.). The culmination of that line of work came with the well-known paper of A. G. W. Leslie, S. Arnott, and colleagues of 1980, summarized in Table 1. Their work established beyond doubt that DNA is very flexible and polymorphic, depending on the solution conditions and its base sequence.

However, the amount of data available from X-ray diffraction patterns of DNA fibers is insufficient to build models at a precise, atomic level of detail. To interpret the fiber data, it must be assumed for the most part that the base pairs are roughly planar and that together the base pairs and the sugar–phosphate chains form a regular helical structure. But in 1979 A. Klug and colleagues suggested an irregular structure for DNA having the sequence TATATA . . . , with an alternation of twist angles between TA and AT steps, thereby pointing to a limitation of the fiber models.

By the early 1980s it had become possible to produce oligomers (i.e., short segments) of DNA with specific base sequences, in such quantity and of such purity that *single crystals* could be grown from them. X-ray diffraction of those crystals then produced far more data than were obtainable from fibers; and the data were sufficient for many structures to be solved to resolutions of about 0.15–0.25 nm. The early oligomer structures obtained in that way produced a number of surprises. For example, the base pairs were *not planar* as had mostly been assumed hitherto (and as shown still in many models today), but instead were "propeller-twisted." And the detailed shape of the double helix was far from being uniform: for example, the angle of helical twist of one base pair onto the next could vary over a range of 20–50° from one step to another, rather than remaining uniform at 33 or 36° as in the A or B form, respectively. Likewise, by adopting different groove widths or chain-to-chain spacings along the length of the molecule, the sugar–phosphate chains could deviate from the perfect helical symmetry that had been assumed for them. This detailed picture of an *irregular* double-helical molecule might be called part of the *third-level* structure of DNA. This third level of structure is important because it is involved in the folding or wrapping of DNA into even larger structures in chromosomes.

It soon became clear that certain ideas about DNA structure had to be abandoned. No longer did the sugar–phosphate chains form rigid "backbones" into which perfectly equivalent base pairs slotted with complete regularity. Now it was clear that the local structure of DNA was irregular, as influenced by the stacking preferences of specific base pairs onto each other. Another idea that had to be abandoned at this stage was that the helical repeat of DNA—that is, the number of base pairs in a complete turn of the double helix—was

an integer. It turns out that the packing arrangements within elongated *fibers* of DNA tend to push the molecule into conformations with integral numbers of base pairs per helical turn (or some simple fractional ratios: see Table 1); but it is now known from the independent work of J. C. Wang and D. Rhodes that in solution, the helical repeat normally averages about 10.6 base pairs per turn, although it varies between 9.9 and 11.1 depending on the base sequence.

X-ray crystallography has not been the only tool available to molecular biologists for the study of the detailed three-dimensional structure of DNA. The technique of *gel electrophoresis,* for example, which is useful for separating electrically charged biological molecules by size—since larger ones migrate more slowly through a gel than smaller ones—is another tool for studying DNA structure. It was discovered by J. Marini, D. M. Crothers, and colleagues in 1982 that DNA having a repeating sequence, and a sequence repeat of about 10 base pairs, might migrate through a gel more slowly than random sequence DNA of the same length. The general explanation for this phenomenon is that DNA may be intrinsically curved on account of its base sequence, and when the sequence repeat is close to (but not exactly equal to) the helical repeat, the curvature imparts a curved or, more generally, a superhelical form like that of a telephone cord. This might be called a *fourth level* of DNA structure. As far as the gel is concerned, such structures are generally more bulky than a straight piece of DNA of the same molecular weight; and so they migrate more slowly through the gel than normal DNA. Many other useful methods for the study of structure in DNA, such as nuclear magnetic resonance (NMR), or electric or circular dichroism, or Raman spectroscopy, are not treated here for lack of space.

So far, we have been discussing the structure of relatively short pieces of DNA. But cellular DNA is very long. Its enormous length raises some special topological problems that have been studied since the late 1960s and are familiar to anyone who has handled long ropes, or threads: the path will tend to "writhe" or coil if the rope or filament is pretwisted. In this article we regard this sort of phenomenon as a *fifth level* of structure in DNA, on account of the highly elongated shape of the molecule.

We do not discuss here many complex structural forms that may be adopted by DNA in special circumstances. These include triple and quadruple helices and four-stranded Holliday junctions (as used in recombination), as well as left-handed Z-DNA molecules or cruciforms. These various strange structures, which are of interest to many and may be important in certain aspects of biology, are omitted here mainly for reasons for space.

Table 1   Leading Geometrical Parameters[a] of Different Types of DNA Double Helix from Fibers

Conformational Type	Pitch of Helical Chains (nm)	Average Axial Rise per Step (nm)	Average Helical Twist per Step (deg)
A	2.82	0.256	32.7 (= 360/11)
B	3.38	0.338	36.0 (= 360/10)
C	3.10	0.332	38.6 (= 360 × 3/28)
D	2.43	0.304	45.0 (= 360/8)
E	2.44	0.325	48.0 (= 360 × 2/15)
Z	4.35	0.363	−30.0 (= −360/12)

[a]Determined by X-ray diffraction of DNA fibers; Z-DNA has left-handed helical twist.
Source: A. G. W. Leslie *et al., J. Mol. Biol.* 143:49–72 (1980).

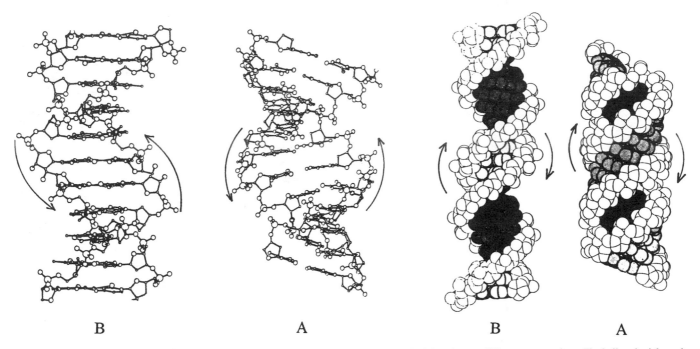

B   **A**   **B**   **A**

Figure 1.   Atomic-level models for the classical B and A forms of DNA, using both ball-and-stick and space-filling representations. The ball-and-stick models show one complete double-helical turn, while the space-filling models show two: the scales are different for the two representations. The arrows show the conventional directions along the sugar–phosphate chains for reading the sequence of bases. [Ball-and-stick models reproduced, with permission, from R. E. Dickerson, *Sci. Am.* 249:94–111 (1983). Space-filling models reproduced, with permission, from C. J. Alden and S.-H. Kim. *J. Mol. Biol.* 132:411–434 (1979).]

2   HELICAL STRUCTURE

Figure 1 contains several diagrams of a short piece of DNA in atomic detail. The molecule is shown in its well-known A and B forms, each of which appears in both ball-and-stick and space-filling representations. These pictures may appear fairly complicated to the newcomer. To begin to understand the structure of the molecule, we must first become familiar with the different parts.

Let us start by looking at the ball-and-stick model of the B form: clearly, the base pairs make a neat, parallel stack up the middle of the helix. The two ends of each base pair are connected to two helically wrapped sugar–phosphate chains. Each connection is made via the five-sided sugar ring; and these sugar or ribose rings are linked together along the chain by phosphate groups. The phosphorus atoms can be picked out readily in the ball-and-stick pictures because they are larger than the other atoms and are connected to four oxygen atoms—two forming part of the chain and two projecting outward from it.

Each of the chains is "polar"; that is, it has a definite running direction. The arrows shown in all depictions represent the conventional positive direction (known as $5' \rightarrow 3'$) for reading the sequence of bases (A, G, T, C) along each strand.

Figure 2 shows a more detailed view of the two distinct base pairs A-T and G-C, together with a schematic representation of the sugar rings that join them onto the helical chains. Each of the base pairs in Figure 1 is one of these two types; and each can go into the structure either way up, say as A-T or T-A, or else G-C or C-G. Hence an arbitrary sequence of bases can be strung along one chain, with its complementary sequence along the other chain.

The space-filling version of the B form in Figure 1 shows how the base pairs prefer to stack on each other in very close contact, like

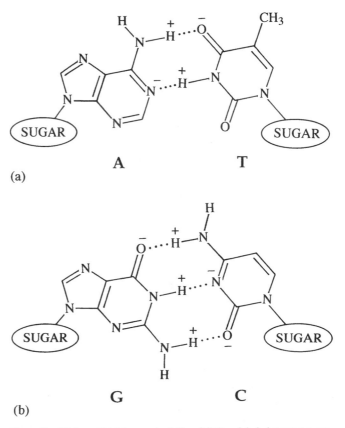

Figure 2.   Watson–Crick base pairs A-T and G-C; unlabeled atoms are carbon. [Adapted, with permission, from Calladine and Drew (1992).]

books in a pile. It is also clear from this picture that two spiral *grooves* run up either side of the molecule, between the sugar–phosphate chains. One of these is generally narrower than the other, and it is known as the *minor groove.* When you look into the minor groove at the edges of base pairs, you see the sugar–phosphate chain on the right running *down,* while the one on the left runs *up.* In contrast, if you look into the other groove—the major groove—the chains appear to run in an opposite sense. It is important to be able to recognize easily the minor- and major-groove edges of a base pair. From Figure 2 you can see that the minor-groove edge lies more or less along a straight line joining the sugar rings. The minor groove is smaller than the major because the sugar rings are attached to one side of the base pairs.

Books on the chemistry of DNA are concerned mainly with the details of chemical bonding between the atoms. The bonds shown as "sticks" in Figures 1 and 2 are all covalent bonds, which are strong. Where such bonds form a closed ring, as in Figure 2, the resulting structure is often planar and rigid; but where they form a chain there is a good deal of overall flexibility, since although atom-to-atom lengths of the bonds are fixed, as are the angles between adjacent bonds (to within a degree or so), the "torsion" angle at the midbond of any three in a chain is generally free to swivel or rotate, with only weak constraint. The five-sided sugar ring, one should note, is *not* rigid like the base rings: it has some minor flexibility, which has been studied extensively (too much, perhaps).

The pairing of bases comes about through *hydrogen bonds,* which are represented in Figure 2 by dotted lines. These are much weaker than covalent bonds, and so the pairs can be separated or "melted" relatively easily, as when the two chains of DNA come apart during the process of replication with an enzyme, or at high temperatures in water. The hydrogen-bonded connections between bases A and T, and between bases G and C, are specific for each other: thus A cannot form a satisfactory pair with any base other than T, such as with C or G, in the usual helical structure. Note also that there are two hydrogen bonds within the A-T pair, but three within the G-C. The hydrogen bonds of Figure 2 are often referred to as "Watson–Crick" hydrogen bonds, as opposed to an alternative type called "Hoogsteen," which are not much used.

When we look at a space-filling version of either the B or the A form, we can see that there are far more contacts among atoms than the ones described so far. What forces determine these additional contacts? Why does the whole arrangement form a twisted, *spiral* structure rather than an untwisted one?

To answer these questions, we must introduce another kind of chemical force. In a living cell, the DNA is always surrounded by *water* molecules, except where it is locally in contact with a protein. Such water molecules are omitted from Figure 1 for the sake of clarity. It turns out that the DNA bases A, G, T, and C are *hydrophobic*—they "hate" water and avoid contact with it, like grease or oil, while the sugar–phosphate chains are hydrophilic—they like to be surrounded by water and are soluble in it. These nonspecific water-hating and water-liking effects ensure that the bases go to the *inside* of the molecule, where they can mostly escape from contact with the water, whereas the sugar–phosphate chains prefer to stay on the outside, exposed to the water.

It is well known that the "links" of the sugar–phosphate chain are about 0.6 nm long per base, while the thickness of the bases (the van der Waals thickness, corresponding to the thickness of a "book") is about 0.33 nm. Therefore, if the chains were laid out straight, so that the structure was like that of a straight, untwisted ladder, there

would be gaps of about 0.27 nm between successive base pairs. And if there were such gaps, water could enter them. Since, however, the bases are hydrophobic, they pack tightly against one another to prevent water from getting in. For that reason primarily, the structure ends up as a twisted helix.

Figure 3 illustrates the basic geometrical idea by means of a model in which the base pairs are represented by rectangular blocks, 1.8 nm wide and 0.33 nm thick, and the links of the sugar–phosphate chain by rods 0.6 nm long. Some elementary three-dimensional geometry then gives the helical twist angle between any two blocks as about 32° for these particular dimensions; and so an extended stack of such blocks, one on top of another, would give a complete turn of double helix after $360°/32° \approx 11$ steps. Figure 3 also shows another possible scheme for stacking the blocks that excludes water, while maintaining the length of sugar–phosphate links at 0.6 nm. That model, when repeated, would produce a skewed but untwisted ladderlike structure.

Our present description is too crude to explain why nature prefers a helical structure to a skewed ladder; but we can imagine that the helical structure is more compact overall, and therefore more stable. Neither does a simple block model explain why DNA assembles, in general, into a *right*-handed spiral (like a corkscrew) rather than a left-handed one. As Watson and Crick discovered when they built their first stereochemical model of B-form DNA, the assembly of atoms as a whole seems to fit better in a right-handed than a left-handed version, for most base sequences.

So far, we have discussed the double-helical structure of DNA mainly in terms of an idealized B form. That is because the parallel stacking of base pairs in a B form is easier to understand than the less straightforward stacking arrangements of bases in an A form. But the two idealized forms nevertheless have many features in

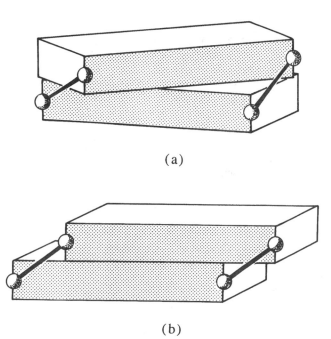

(a)

(b)

Figure 3. Two base pairs (represented as rigid blocks with the minor-groove side shaded) stack onto each other, subject to constraint from the sugar–phosphate chains (shown as rigid links), either (a) with twist, so as to build up a helix, or (b) with skew, so as to build an untwisted, skewed ladder. [Adapted, with permission, from Calladine and Drew (1992).]

common: antiparallel sugar–phosphate chains, the base pairing, the connection of bases to the chains by sugar rings, and the close packing of the base pairs onto each other. However, there are also many differences, most of which may be seen in Figure 1: the stack is shorter (see also Table 1) and has a bigger overall diameter in the A form; the bases in the A form are not perpendicular to the overall helix axis; the relative shapes and sizes of the two grooves in the "A" form are different; and the twist angle of one base pair relative to its neighbor is about 10% smaller, with the result that the number of base pairs needed for a complete turn of helix is about 10% larger in the A than in the B form.

As mentioned in Section 1, Franklin discovered that the X-ray diffraction pattern of DNA fibers would switch from an A to a B form when the moisture content was increased; and this switching effect was reversible. Later workers found that A-T-rich DNA was more reluctant to switch from B to A when the moisture content was reduced, whereas some G-C-rich DNA was more reluctant to adopt the B form when the moisture content was increased. As we shall see in Section 3, the switch between these two forms involves mainly a change in the stacking arrangements of one base pair onto another; and to understand how that comes about, we shall need to introduce some more three-dimensional geometry. That geometry will lead, in turn, to a deeper understanding of what happens when the need to wrap around histone proteins causes a DNA molecule to be bent into a curve, when packaging the molecule into chromatin, or during the recognition by a protein of a piece of DNA having a particular sequence of bases.

For many years there has been a controversy in the literature about the biological significance of A-form DNA: some workers have argued that the A form is merely an artifact of experimental preparations and that the dry atmosphere required to produce it is not to be found inside a cell. However, that view is contradicted by much evidence for A-form DNA in wet crystals, and for some sequences in solution, especially from the recent work of W. Hillen and colleagues, which shows an 11.1 repeat in solution for DNA with all G on one strand versus all C on the other. Whichever way the controversy is eventually sorted out, it will certainly be helpful to think about structural aspects of the transition from B to A because they provide the key to an understanding of the geometry of *curvature* of DNA. The same line of thought also leads to a realization that in discussions about the biological relevance of some idealized A form, we may be guilty of using an oversimplified scheme of classification of a molecule that in fact displays a very rich range of structural variations.

3   BASE STEP GEOMETRY

To understand more clearly the distortion that occurs inside a DNA molecule when it bends around a protein, when it changes from B to A form, or when it untwists so that the strands can separate during transcription, it is necessary to understand changes that take place in general within a typical dinucleotide step, consisting of two consecutive base pairs and their associated sugar–phosphate links. For those purposes the fundamental geometric unit of DNA is not a single base pair, but the imaginary *step* that relates two consecutive base pairs.

Let us begin by replacing each base pair of DNA by a rigid, domino-shaped block. Later on, we shall sometimes use two separate blocks for the two bases; but for present purposes a single block is better. Furthermore, let us remove temporarily the two connecting

sugar–phosphate chains, to permit a direct focus on the geometrical relationships between two consecutive base pair blocks. Now since the two units are rigid bodies, we can say immediately that their relative positions are described by *six degrees of freedom*, of which three are *translations* (i.e., displacements without rotation) and three are *rotations* (without translation).

Figure 4a shows two blocks that were originally in direct contact but have been moved apart by translation along the axis through their centers, and normal to their surfaces. Two other axes, each perpendicular to this first one, and aligned with the respective edges of the blocks, are also shown. Euler showed, in the eighteenth century, that the most general motions of one block relative to the other consist of a translation and a rotation along each of these three mutually perpendicular axes: three translations and three rotations in all.

The easiest rotation to envisage is *twist* (Figure 4b): the blocks are still parallel, having rotated relative to each other about the axis along which they have already been translated in Figure 4a. The relationship of the two blocks is now almost exactly like that of two consecutive base pairs in the ball-and-stick representation of B-form DNA in Figure 1. The *roll* motion (a rotation: Figure 4c) and the *slide* (a translation: Figure 4d) have the same reference axis, which runs along the long direction of base pairs.

These motions roll (R), slide (S), and twist (T) are the three degrees of freedom that vary most commonly in DNA structures, as revealed by single-crystal X-ray diffraction methods. Thus in general, the "rise" variable that is shown in Figure 4a hardly varies at all, since the base pairs stack onto each other like books of a definite thickness. Neither is there usually any significant rotation about the "front–back" axis shown in Figure 4a, for essentially the same reason. This close, water-excluding stacking of base pairs does not inhibit relative *translation* in the front–back direction; but for most sequences, there seems to be little tendency for such a motion to occur in practice. In summary, the conformation of any given dinucleotide step may be described, to a first order, by just *three* parameters R, S, and T out of a complete theoretical range of six.

Before we give any specific examples of the consequences of variation of these three key parameters, we should explain a few points. First, we have adopted the convention of shading the minor-groove edge of each base pair, as in Figure 3. This is necessary to preserve a sense of direction in drawings lacking all details of the base pairs (cf. Figure 2), and of the sugar–phosphate chains and their connections. Note that this is a different convention from that used in the space-filling models of Figure 1. Second, we have not given precise details needed to define the various translations and rotations when the base pairs are represented by rigid blocks. For example, twist T and roll R (shown separately in Figure 4b and c) in practice usually occur simultaneously; and we need to be careful whether twist is applied first and then roll, or *vice versa*. Points of that sort are important for the precise evaluation of data obtained by X-ray crystallography; but they are not critical for the present purposes of understanding the main geometrical ideas. Third, it should be noted that each of the motions R, S, and T keeps the same value if the dinucleotide step is turned over, by rotation through 180°, about its front–back axis. This is a consequence of the double-helical nature of the molecule; and in a completely regular form the DNA would have an axis of twofold rotational symmetry about the middle of any step, parallel to its front–back axis.

A more direct way of appreciating the influence of parameters R, S, and T on the shape of a DNA molecule is to examine a series of

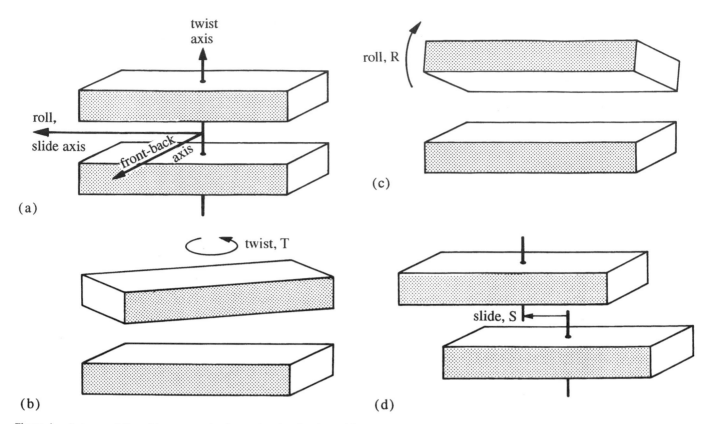

Figure 4. A step consisting of two consecutive base pairs, showing the positive sense of various relative motions: (a) coordinate system and "rise" motion, (b) pure twist, *T,* (c) pure roll, *R,* (d) pure slide, *S.* The symbols *R,S,T* used here are a simplified subset of an internationally agreed set of symbols known as the *Cambridge Accord* [R. E. Dickerson *et al., EMBO J.* 8:1–4 (1989)]. As used in this article, *R, S,* and *T* are equivalent to ρ, D_y, and ω, respectively, of the *Accord.* [Adapted—the positive sense of slide used here is opposite from that used originally—with permission, from Calladine and Drew (1992).]

helices. In Figure 5a we see the simplest possible arrangement, consisting of 10 uniform steps having $R = 0°$, $S = 0$ nm, and $T = 36°$. Since $R = 0°$, all the base pair blocks are parallel to each other, with a common normal that has been shown here as a thin line passing through the centers of the blocks. And since $T = 36°$, the 10 steps make a complete turn, so that the last block is exactly parallel to the first. In this picture the blocks are shown as thinner than those of Figure 4, and the "minor-groove edge" is black; but they are, of course, essentially blocks of the same sort.

In Figure 5b the structure is the same as in Figure 5a, as far as *R* and *T* are concerned; but now a slide $S = -0.2$ nm has been introduced at each step. To show this, the centers of individual blocks have been marked; and the common normal line has been taken out, since it no longer connects those centers (see Figure 4d). The inclusion of slide at each step in Figure 5b moves the blocks out from a central imaginary axis, and makes the overall diameter of the stack wider; but it has no effect upon the overall length or height of the stack.

Consider next the stack shown in Figure 5c. There is obviously a major change here because the blocks are now highly tilted with respect to the vertical and, except in a few cases, they cannot be viewed edge-on. These changes are a consequence of the nonzero roll angle, $R = 12°$ at each step, without any slide. Since there is no slide, we have retained the "normal" line of Figure 5a; but now this line—which we might think of as a wire that goes through the blocks normally—is slightly curved as a result of the roll angle be-

tween the blocks. And the wire is not bent into a simple circular path, because the plane of bending in each step rotates through a 36° twist angle with respect to its neighbor. Instead, the wire curves into a gentle helix that goes through (almost) one complete turn for the 10-step stack. Since the blocks are attached orthogonally to this wire, they are all tilted to the vertical axis, as shown. Thus the 12° roll angle provides, together with a twist of 36°, a 20° *tilt* for the blocks.

The final picture of the series, Figure 5d, shows what happens if a uniform slide of -0.2 nm is provided at each step, in addition to the 12° roll angle. One effect is that the stack of blocks moves out to a larger diameter, in much the same way as in Figure 5b. But there is another feature that was not present in the earlier example. Here, because the blocks are already tilted from the horizontal, the provision of (negative) slide gives an overall, "downhill," *shortening* of the stack of base pairs.

The values of *R* and *S* used in Figure 5 were chosen to achieve correspondence between helices (a) and (d) and the idealized B and A forms, as illustrated in Figure 1. Thus, when we introduce simultaneously a roll of 12° and a slide of -0.2 nm to each of the steps in idealized B-form DNA, we alter the base pair stacking arrangements to those of idealized A-form DNA. Many of the well-known attributes of the A form with respect to the B form are reproduced accurately by this model: a significant overall shortening, a tilting of the base pairs, and a movement of base pairs away from the helical axis. The only well-known feature of the transition not yet ac-

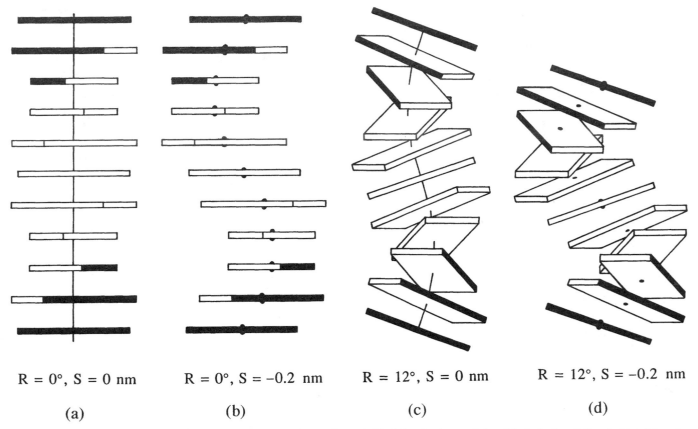

$R = 0°, S = 0$ nm

(a)

$R = 0°, S = -0.2$ nm

(b)

$R = 12°, S = 0$ nm

(c)

$R = 12°, S = -0.2$ nm

(d)

Figure 5. Diagrams showing the way in which alterations of (uniform) roll R and slide S alter the characteristics of the helix from B form in (a) to A form in (d): cf. Figure 1. [Adapted, with permission, from Calladine and Drew (1992).]

counted for is the change in helical repeat from about 10 base pairs per turn in B to about 11 in A. But that relatively minor change can be achieved by a small adjustment of twist at each step, in combination with altered roll and slide, without affecting the overall structure much.

Figure 5 makes an important point about what happens *within* the DNA helix during a B-to-A transition. All the observed outward changes are achieved by changing the values of roll R and slide S at each step, and by adjusting twist T by a few degrees: these outward changes are all a consequence of inward motions at the level of stacking of base pairs in a dinucleotide step. And the reverse is also true: the outward features of the B-to-A transition can occur only if the corresponding inward changes in base stacking take place.

Now all these geometrical relationships (between the internal base stacking on the one hand, and the external form of the DNA molecule on the other) have been established on the basis of a model from which the sugar–phosphate chains have been removed. But the chains may be put back without any difficulty; and it is well known that the sugar-to-sugar distance along the chain alters by relatively little during a B-to-A transition, say from 0.65 nm to 0.55 nm at most. While it is true that there are some local changes in sugar–phosphate conformation during the transition—for example, the phosphate group moves about a half-step in the $5' \rightarrow 3'$ direction, as may be seen clearly in Figure 1—nevertheless the most substantial changes are at base-stacking level.

Figures 1 and 2 show *uniform* pieces of DNA, that is, DNA with

equal values of R, S, and T at all the dinucleotide steps. It was pointed out at the beginning of this article that crystallized oligomers of DNA show significant diversity in base step conformation (i.e., considerable perturbations in all of R, S, and T from step to step). Nevertheless, the *mean* values of these quantities over a complete helical turn still determine the outward form of the molecule.

Some obvious questions arise at this stage. If, indeed, the structure of a double helix is fixed by its base-stacking arrangements, as quantified by the values of parameters R, S, and T, what actually *determines* those values? Or, since DNA can easily switch from one form to another, what determines the allowable *range* of values?

The answer to these questions lies in the different stacking preferences of the two types of base pair, A-T and G-C, in all the various combinations that are possible for a dinucleotide step. Now there are altogether 16 possible two-letter "words" in a four-letter "alphabet." But when we take account of the way in which a given dinucleotide step may have different readings along the two strands, it is clear that the number of physically distinct cases will be less than 16, or precisely 10, as half of a 4×4 matrix. We need to examine them all.

Let us begin with steps that are composed only of A-T base pairs. There are only three possibilities: AA (= TT), AT, and TA. Here the two letters show the order of bases, as one moves along either sugar–phosphate chain in a standard $5' \rightarrow 3'$ direction. So if a step reads AA along one strand, it will read TT along the other, since A and T are always linked across the strands by an A-T base pair. By the same token, a step that reads AT on one strand will also read AT on the other; so that, unlike the AA (= TT) step, there is no need to specify the different readings on the two strands.

Likewise, there are three distinct types of step composed only of G-C base pairs: GG (= CC), GC, and CG. Last, there are four possible dinucleotide steps consisting of one GC and one AT base pair: AG (= CT), GA (= TC), AC (= GT), and CA (= TG). Each of these reads differently along the two strands, and the alternative is given in each case. Thus we find that altogether there are 10 different dinucleotide steps as classified by their base sequences.

The electrostatic forces between base pairs that are stacked together are subtle, and not straightforward to calculate accurately. Each atom within a base has its own electrostatic charge. Also, these electrical charges produce a kind of "sandwich" structure of partial charge for each base pair, positive at the protons of atoms but negative at the electrons above and below; and consequently the computations must be done in three-dimensional space. Another complication is that the base pairs in double-helical DNA are not perfectly planar, as we have assumed so far, but are arranged in the stack with "propeller-twist" as shown schematically in Figure 6, and as may be seen in some base pairs from the ball-and-stick models of Figure 1. Propeller-twist tends to be larger in A-T base pairs than in G-C pairs, perhaps because of the smaller number of hydrogen bonds connecting A to T than G to C (see Figure 2). The larger size of the G base versus A may also be a reason for this discrepancy.

The energy of stacking of base pairs onto one another for all 10 distinct dinucleotide steps has been investigated carefully by C. A. Hunter. For each case he has computed the total energy of the stacked base pairs for a range of values of R, S, T, and propeller-twist. Although the results of the computation involve much detail, it is possible to make a relatively simple synopsis.

In summary, A-T and G-C base pairs have distinct regions over their respective surfaces that are either positively or negatively charged, in a partial sense. The G-C pair has a large region of negative charge on the major-groove edge of G, near the N and O atoms there, while there is some positive charge over the major-groove edge of the hexagonal ring of C: see Figure 2. In contrast, the A-T pair has no such major groupings of partial electric charge: instead, small packets of positive or negative charge are scattered all over its surface. One consequence of these distributions of partial charge is that for a dinucleotide step GG (= CC) or GC or CG, consisting of two highly charged G-C base pairs, there are often *two* different values of slide S which the step prefers to adopt, while various inter-

mediate values of S are hindered by repulsion between regions of strong similar charge. These two ranges may explain in part the tendency of DNA to adopt either the A or the B form. For steps AA (= TT) or AT or TA, consisting of two more-neutral A-T base pairs, there are fewer such repulsive interactions. But for those steps propeller-twist may play an important part. For example, in AA (= TT) steps the high propeller-twist of 15–25° provides a kind of mechanical interlocking action, which effectively fixes the values of R, S, and T to near (0°, 0 nm, 36°) in most cases.

For dinucleotide steps that contain both an A-T and a G-C base pair, the situation is more subtle; and there may be considerable conformational freedom, especially in CA (= TG), as seen in many crystal studies.

Hunter's calculations were concerned only with interactions between the base pairs themselves, without regard for any constraints that might be imposed by the sugar–phosphate chains. M. A. El Hassan, who recently examined the conformations adopted by different kinds of step, in single crystals of DNA oligomers, showed that each of the 10 kinds occupies a particular region of R,S,T conformational space. Figure 7 is a synopsis of El Hassan's results in the S,T projection of this space. The bistable nature of the homogeneous G,C steps [GG (= CC), GC, CG] is a prominent feature of the diagram. Another strong feature is a cigar-shaped "channel" over which the step CA (= TG) can move, involving a range of about 30° in twist coupled with a range of about 0.5 nm in slide. Various other steps, such as AA (= TT), AT, and GA (= TC), occupy small portions of the same channel in conformational space, while TA and AC (= GT) occupy "T-shaped" regions, which may signify a measure of bistability.

The "channel" in conformational space that is occupied by so many different steps indicates a firm linkage between twist T and slide S, as plotted in Figure 7. Indeed, roll R is also linked to these two variables, so that the channel constitutes, more or less, a single-degree-of-freedom path in the three-dimensional R,S,T space. This suggests that the "links" of the two sugar–phosphate chains exert some overall constraint for these steps, within which the electrostatic preferences must fit. By contrast, the homogenous G,C steps are not confined to the special channel in Figure 7: thus it appears that the electrostatic effects that produce bistable behavior are strong enough to overcome the weak constraints imposed by the sugar–phosphate chains for other kinds of step.

CA (= TG) is the most extreme example of a step that covers a wide range of the single-degree-of-freedom channel in conformational space. Over its entire range the sugar–phosphate chains adopt the well-known B pattern. With the homogenous G,C steps, the observed bistability also involves a switch in the conformation of the sugar–phosphate chains between A and B patterns. Finally, TA and AC (= GT) are the only other steps, apart from the homogenous G,C steps, that adopt in their wide range of conformations the two distinct sugar–phosphate types of chain.

Before moving on to the next section, we summarize the many kinds of chemical force that may play a part in determining the conformation of DNA. First we have the two sugar–phosphate chains and their attached nucleic acid bases. The connections there are all *covalent*—the strongest type of chemical bonding. But strength alone does not confer rigidity, unless there is a closed ring with double-bond character, as in the bases themselves (and as in benzene). Hence the sugar–phosphate chains have much freedom of motion, apart from an almost constant length. Next, the *hydrogen bonds* that connect the bases together in pairs provide a specific means of

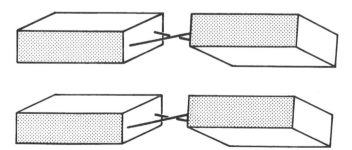

Figure 6. Two base pairs with propeller-twist: each base is represented by a separate block. The agreed sign convention for propeller-twist (cf. legend for Figure 4) is that the sense of propeller-twist shown—and almost invariably found in crystal structures—is *negative*. [Adapted, with permission, from Calladine and Drew (1992).]

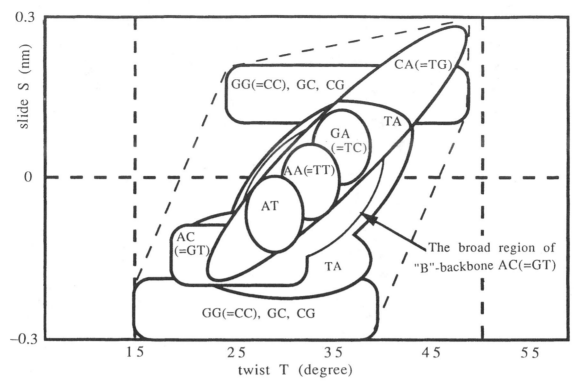

Figure 7.   Schematic representation of the different regions of a two-dimensional twist *T,* slide *S* conformational space that are occupied by the different dinucleotide steps in many crystallised oligomers. (Reproduced, with permission, from M. A. El Hassan, Ph.D. dissertation, University of Cambridge, 1995.)

recognition between A and T, or between G and C; but they are flexible enough to allow some propeller-twisting of base pairs. Finally, the overall helical structure is attributable mainly to the *hydrophobic* nature of the bases, which encourages them to stack onto each other closely in the center of the molecule, to escape from the surrounding water. By contrast, the hydrophilic sugar–phosphate chains lie on the outside of the molecule, in contact with water.

This simple picture of the structure-forming forces is satisfactory to a first-order approximation, but it does not include any features that would allow us to understand how the sequence of bases could have a significant effect on the overall structure. For example, some sequences might confer special rigidity to the double helix, while others might impart particular flexibility; some sequences might facilitate the B-to-A transition, while others might resist it; and some sequences might predispose the DNA to untwist easily so that the strands could come apart, ready for the beginning of transcription or replication, or recombination. All effects of that sort must be attributable ultimately to stacking preferences of the base pairs in the 10 possible kinds of dinucleotide step. Such preferences are influenced by the distribution of partial electric charges over the surfaces of the bases; but they are also constrained by the sugar–phosphate chains, which allow values of *R, S,* and *T* only in certain regions of "conformational space." Indeed, it seems likely that the conformation of the sugar–phosphate chains in one step tends to influence the stacking conformations in neighboring dinucleotide steps, by inducing a similar slide *S* in the neighboring steps.

4   BENDING AND CURVATURE OF DNA

In this section we shall explain the geometrical aspects of what happens when the DNA double helix becomes *curved.* Clearly, if a

straight piece of DNA—like any of those shown in Figures 1 and 5—is obliged to bend around a corner, changes in the stacking geometry are inevitable. In a cell, the DNA may be intrinsically curved on account of its base sequence; or it may be forcibly curved by being bent around a protein. But as far as the *geometry* of bending is concerned, the two situations are closely related.

If you were to ask an engineer what happens inside a straight elastic bar when it is bent into a circular arc, you would hear about the Bernoulli–Euler theory of bending for elastic bars. There the key idea, which is underpinned both by experiment and by considerations of symmetry, is that planar cross sections of the bar in its original, straight configuration remain not only planar but also perpendicular to the axis of the bar when it is in its deformed, curved configuration. Such a simple geometrical idea provides a useful approach to the problem because it describes internal strain within the bar, from which the stresses and the "bending moments" may be computed.

Unfortunately, this classical theory of bending is of no use whatsoever when we come to consider the curvature of DNA! One might suppose, of course, that planar blocks representing the base pairs would bear some relation to planar sections of the classical theory; but upon realizing that the A form of DNA—which is undoubtedly straight—has all its base pairs nonorthogonal to the straight axis, one must abandon the classical bending theory as irrelevant for thinking about the curvature of DNA.

In fact, there is *no unique way* of getting DNA to bend around a curve. For example, Figure 8 shows four different ways in which a 20-step piece of DNA can be bent through a total angle of 90°. The rigid blocks that represent the base pairs are of exactly the same kind as those shown in Figure 5. As before, it has been convenient to have *T* = 36° to ensure that there are 10 steps in any complete helical turn.

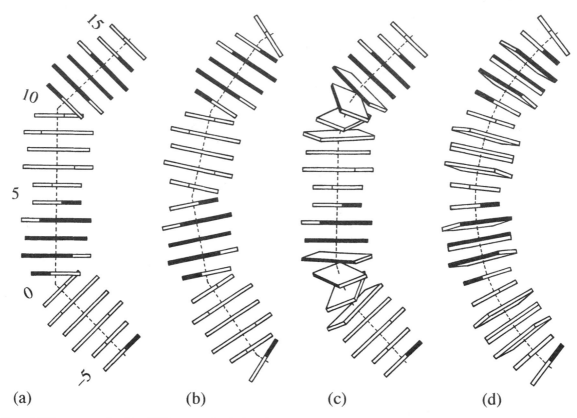

Figure 8. Several different ways of making DNA to curve by the introduction of roll angles at the steps according to various patterns. [Adapted with permission from Calladine and Drew (1992).]

The steps are numbered sequentially from bottom to top, the bottom step being designated −5. The minor grooves at steps −5, 5, and 15 are facing to the right (i.e., toward the center of curvature).

In Figure 8a the total bend of 90° has been achieved in a rather crude way, by the insertion of $R = +45°$ at two particular steps, 0 and 10. Recall from Figure 4c that the sense of roll is positive when the minor-groove edges are more widely separated than the major-groove edges. One might argue, of course, against this crude arrangement: first, the DNA in Figure 8a can hardly be described as *curved,* since it contains two discrete *kinks*; and second, a roll angle of 45° at an isolated step represents a serious unstacking of the bases concerned, which is not commonly observed. Such objections are largely overcome in Figure 8b, where the turning is now achieved at four steps instead of two, as $R = +22.5°$, $−22.5°$, $+22.5°$, and $−22.5°$ at steps 0, 5, 10, and 15, respectively. Note that the roll axes of all these steps are parallel to one another, and perpendicular to the plane of the diagram; hence all four roll angles contribute equally and directly to the total bend angle of 90°. But also note that the minor groove changes its orientation by 180° after every five steps; for this particular method of curving, therefore, the required roll angles must alternate in sign.

The next example is different again. Comparison with Figure 8a indicates that about half the blocks in Figure 8c are in exactly the same orientation as before, but that the other half are oriented very differently. Somehow the isolated "kink" at step 0 in Figure 8a has been replaced by positive roll over the five steps from −2 to +2, inclusive. In fact, these five steps have been given roll $R = +14°$, and the effect is to introduce a turn of 45° overall. The best way of understanding what is happening here is to examine the upper five

steps shown schematically in Figure 5c and compare them with the five steps from −2 to 2 in Figure 8c. The key point is that the bend angle in Figure 8c is made up by the change in the sense of *tilt* that is seen in the model of Figure 5c over five steps.

These three different patterns of roll, which equally achieve the same amount of curvature in Figure 8(a–c), are displayed in the corresponding parts (a)–(c) of Figure 9. In these graphical diagrams, roll R is plotted against step position along the length of the molecule.

Yet another pattern (Figure 9d) also produces the same overall curvature; and the corresponding stack of base pairs is shown in Figure 8d. The sinusoidal pattern of roll R in Figure 9d is the sine wave that gives the "best overall fit" to each of the other three patterns. Or, in technical language, the pattern in Figure 9d is the *Fourier component* of each of the other three patterns at a period of 10 (i.e., at a period equal to the helical repeat of the molecule).

Still with reference to Figure 9, it might seem strange that the pattern (d) is claimed to be the Fourier component of the pattern (c), since the *mean* value of R in (c) is not zero. However, a change in the mean value of roll R by addition to or subtraction from the roll angle at every step of a constant number makes *no change* to the overall curvature. Note that we have already seen that a transition from the B to the A form of DNA gives no change in curvature— since both are straight; and the same is true in general, whether the DNA is straight or curved.

In summary, if the pattern of roll angles repeats itself along the molecule, with a period equal or close to the helical repeat, then the overall curvature imparted to the DNA is proportional to the main Fourier component of the roll angles. There are many possible pat-

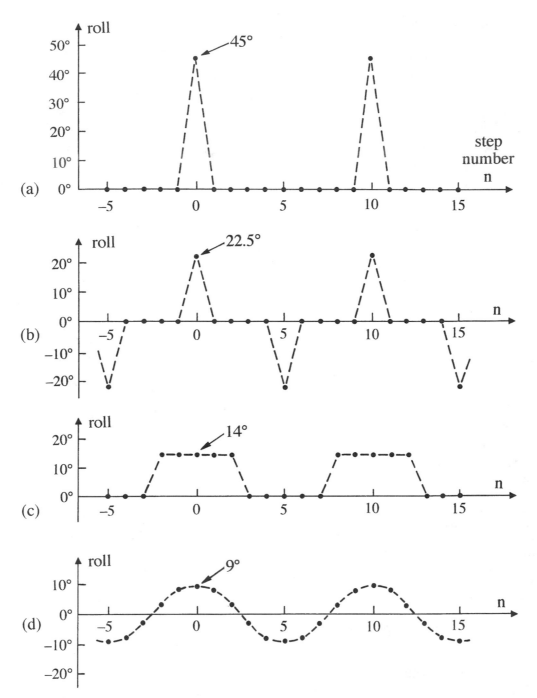

Figure 9. Cartesian plots of roll angle against step number *n*, corresponding to the four arrangements shown in Figure 8. All these patterns have equal Fourier components (with amplitude 9°) at a period of 10 steps. [Adapted, with permission, from Calladine and Drew (1992).]

terns of roll that have the same Fourier component; and so there are many different ways of achieving the same overall curvature. In particular, the addition of a constant roll angle at every step has no effect on the overall curvature.

The level of curvature that is produced by any of the *R* distributions shown in Figure 9 is the same as that required for the wrapping of DNA around the histone "spools" that play a major role in the packing of DNA into chromosomes. Since there are about 80 base pairs of DNA in any single turn around a spool, the curvature is 360° per 80 base pairs; or 45° per 10 base pairs as here; or 4.5°

per base pair on average. Figure 9 shows that to bend the DNA in that way, the roll angles at individual dinucleotide steps must have a *range* of values of at least 14° (i.e., three times the mean curvature per base pair).

It seems clear that it should be easier to bend a piece of DNA around such a spool if the base pairs in positions where the minor groove faces *inward*, toward the center of curvature, prefer to stack at a low or negative roll angle, while those lying at positions where the minor groove points *outward* are content to adopt a high or positive roll angle. There is now much evidence from DNA sequences that have been wrapped around such protein spools, through the independent studies of S. Satchwell, S. Nedospasov, and colleagues, that certain sequences such as AA (= TT) prefer to lie near low roll positions (−5, 5, 15, etc. of Figure 8), while other steps such as GC prefer high roll locations (0, 10, etc.). The step CA (= TG), which has an especially wide range of allowable roll angles, has little preference for being located at any of the positions 0, 5, 10, 15, . . . : it is content to lie at either low roll or high roll positions.

All these positional preferences are consistent with the scheme shown in Figure 7, which has been constructed on the basis of dinucleotide step conformations seen in crystallized oligomers. Thus the range of conformational states at the level of a dinucleotide step that are needed for curving DNA around proteins is similar to that found in crystals. The forces required to pack DNA oligomers into crystals thus seem to be of the same order as (or less than) those needed to curve DNA around proteins.

5   SUPERHELICAL DNA

In Section 1 we mentioned a strange phenomenon discovered by Marini, Crothers, and co-workers in 1982: namely, DNA having certain repeating sequences, and with a repeating unit of 10 or 11 base pairs, ran more slowly through an electrophoretic gel than normal, random sequence DNA. Clearly, the repeating base sequence must impart to the DNA some sort of higher level structure that hinders DNA's passage through a gel, as first noted by P. J. Hagerman in 1985. Many experiments with DNA of different sequence repeat characteristics have shown that for an overall length of about 50 base pairs, the sequence-dependent retarding effect is small. But for longer lengths, in the region of 50–200 base pairs, the special retarding effect increases with length; and as the DNA length increases still further, the retarding effect "saturates," typically at around 200–250 base pairs; and then it remains mostly unchanged for DNA of longer length.

The likely and obvious explanation of such phenomena is that the sequence repeat imparts to the DNA a *superhelical* structure, something like that of a telephone cord. The "plateau" length of around 200–250 base pairs corresponds to one complete turn of superhelix, and the retarding effect comes about because the overall diameter of the superhelix is larger than that of an ordinary, double-helical DNA "rod." While the overall diameter of a DNA superhelix does not increase for contour lengths longer than one superhelical turn, it becomes somewhat smaller for any fraction of a single superhelical turn; and indeed for only a *small* fraction of a superhelical turn, the diameter is only a little larger than that of the straight DNA rod itself.

In Section 4 we saw what is required, in the way of roll angles, for a piece of DNA to bend into a plane curve. Although it was not stated explicitly there, DNA can form a perfectly *plane* curve only if the repeating pattern of roll angles is of exactly the same length as the helical repeat of the DNA. Now such a helical repeat will not generally be an integer, since it is equal to $360°/T$, where T is the average base-pair step-twist angle, which we know can vary continuously. Thus the helical repeat varies continuously from 9.9 to 11.1 base pairs per helical turn. Hence, if some repeat sequence DNA of 10 or 11 has roll angles that depend on sequence, then the DNA will usually form a shape that is different from a plane curve. What kind of curve will it adopt? If we assume that each sequence repeat has the same geometrical form, then we may conclude that the overall shape will be a regular superhelix, since that is the shape of a general uniform "space curve."

The most straightforward way of understanding the three-dimensional geometry of a superhelix is to imagine an end-to-end assembly of a set of identical bricks, of the kind shown in Figure 10. In the simplest case (Figure 10a), the brick is an ordinary one, but it has a square cross section; and when several bricks are put together end to end, they build a straight and untwisted rod. (The rule for assembly is that square ends are abutted, with the hatched faces adjoining.) In case (b) the block is curved, with the hatched surface lying on the outside of the curve. A set of blocks like that will build into a perfect plane circle.

Now each of these bricks represents 10 dinucleotide steps of DNA: the straight brick (Figure 10a), for example, represents a straight piece of DNA like that shown in Figure 5a. In Figure 10 all internal detail has been omitted: in effect, the single helical turn of DNA has been disguised by means of an enclosing "box." Nevertheless, the ends of the box correspond exactly to the parallel ends of the stack in Figure 5a.

The brick labeled "$k = 1°$" (Figure 10b) has an overall curvature

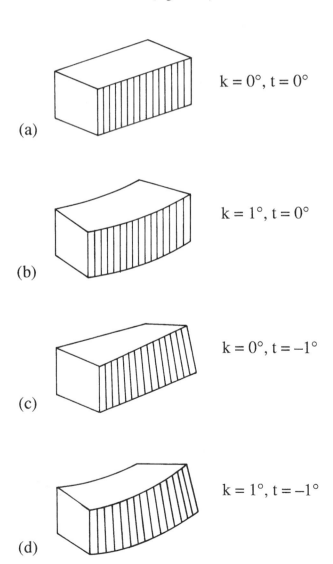

Figure 10. "Bricks" containing one double-helical turn of DNA, for use in an explanation of supercoiled DNA. Curvature k and twist t are specified in units of degrees per base pair. [Adapted, with permission, from Calladine and Drew (1992).]

k of 1° per step. Thus the brick represents a curve like the lower half of any of the four variants in Figure 8, except that the overall curvature is 1° per step rather than the 4.5° per step as shown there.

The brick in Figure 10c is straight (i.e., the two ends are parallel planes), but it is *twisted* in a left-handed sense: $t = -1$° per step. This does not correspond directly to any of the detailed pictures of DNA presented earlier, but it is still like the stack shown in Figure 5a except that here $T = 35$°, so that the two ends have a left-handed relative twist of 10°. An end-to-end assembly of such bricks will build a straight, but twisted, rod.

Finally, Figure 10d is the brick we are really interested in, because it combines the attributes of the two preceding blocks (Figures 10b and 10c): it is simultaneously curved and twisted. It is like the curved DNA of Figure 8, but without the artificial restriction imposed there, whereby the helical repeat matches *exactly* the sequence repeat.

What sort of shape is constructed if many identical bricks of type (d) are assembled end to end? The answer is a left-handed helix—or superhelix—which climbs around an imaginary cylinder at an angle of 45° to the superhelical axis.

Simple geometry enables us to compute the physical dimensions of the superhelix for any given pair of values of curvature k and twist t:

$$
\left.
\begin{aligned}
\tan \alpha &= \frac{t}{k} \\[6pt]
\left(\frac{360°}{N}\right)^2 &= k^2 + t^2 \\[6pt]
p &= \frac{360° \, t}{k^2 + t^2} \\[6pt]
2\pi r &= \frac{360° \, k}{k^2 + t^2}
\end{aligned}
\right\} \tag{1}
$$

Here, α is the angle made by the superhelix with any cross section of the imaginary cylinder around which it wraps; N is the total number of base pairs in one superhelical turn; p is the pitch (= wavelength) of the superhelix; and r is the radius of the superhelix: N, p, and r are all expressed in terms of a base pair as a unit of length. It can be seen from equation (1) that in the case $t = 0$°, the superhelix collapses to a plane circle—which would have a circumference of 360 base pairs for $k = 1$°.

Many available data on the retardation in gel electrophoresis of DNA having a sequence repeat of 10 to 11 base pairs, relative to normal DNA of the same length, are consistent with a "telephone cord" superhelix model, in which values of roll and twist are assigned to specific dinucleotide steps in approximate accordance with Table 2. In that model the various kinds of step have particular roll angles that impart an *intrinsic* curvature to the DNA. The roll angles concerned have a small range, by comparison with those that are required for the wrapping of DNA around histone proteins in the nucleosome. In the analysis that was conducted to obtain Table 2, it was essential to consider variations in twist as well as those in roll, to ensure that the observed retardation of gel velocity was correlated with the overall diameter of the superhelix. The correlation thus supports a telephone coil model of DNA, as distinct from a model involving a flat, circular coil—which would migrate in a different way through the gel. In fact, repeating-sequence

Table 2 Roll and Twist Angles for Dinucleotide Steps to First Order, without Magnesium Ions

Dinucleotide Step	Roll R (deg)	Twist T (deg)
AA (= TT)	0	35
TA	6.6	34
Others	3.3	34

Note: These values of R and T [from C. R. Calladine, H. R. Drew, and M. J. McCall, *J. Mol. Biol.* 201:127–137 (1988)] for relaxed, stress-free DNA, provide a simple scheme for explaining data on gel-running anomalies in repeating-sequence DNA from many studies. The repeat sequences concerned have step TA only in the context ..TTAA.. . Other workers have produced comparable schemes; but the one shown here fits the data reasonably well with a small number of disposable parameters and is consistent with a much wider range of phenomena, such as X-ray diffraction of DNA in crystals, wrapping of DNA in nucleosomes, and specific cutting of DNA by nucleases such as DNAase I.

DNA with a repeat of 10.5 base pairs per turn—an actual repeat of 21 base pairs but with two halves of almost-identical sequence—is retarded in gels in a way that does *not* involve a plateau or saturation after 200–250 base pairs; rather, the plateau occurs at much longer values of 300–350 base pairs, where measured, consistent with a shape close to a plane circle. Both the supercoiled and the large plane–circular shapes, as deduced indirectly from gel running, have been confirmed directly by electron microscopy by various workers.

In all these experiments, the strongest retardation effects are found when the repeating sequence is of a kind (AAAAAANNNN)$_n$, where A = adenine and N = any of G, C, or T. That comes about because the AA (= TT) step has an unusually *low* value of roll, near zero—as shown convincingly by X-ray crystallography—in contrast to a positive roll of a few degrees in other steps. The situation is much like that shown in Figures 8c and 9c, where the curvature comes from a phase contrast between tracts with zero roll and intermediate tracts having roll of 14°—except that here the *intrinsic* positive roll of these non-AA steps is much less, perhaps +3° to +4°. In more recent experiments, I. Brukner and colleagues have shown that GGCC sequences may actually impart as much curvature as AAAA, but in the opposite direction, because of their high preferred roll angles. The curvature produced by GGCC is most noticeable in buffers that contain cations such as magnesium, and it correlates well with strong curvature of the GGCC sequence around histone proteins of chromatin.

Most of our treatment of DNA structure in this article has implicitly assumed that the DNA may be considered as if it were *static*, although it is well known that it is subject to constant thermal agitation or Brownian motion. The longer the DNA, the more likely are dynamic effects to be important. Many experimental studies have shown that the "persistence length" of DNA is around 150–200 base pairs in typical buffers. So DNA shorter than that length may behave more or less as a rigid rod, whereas DNA longer than that may behave more like a flexible piece of string. It seems obvious, then, that a long "telephone cord" of DNA will be extremely floppy under thermal agitation, say at size 1000–2000 base pairs overall. However, any single superhelical turn contains just one persistence length of DNA; hence the *local radius* of the telephone cord superhelix may be well conserved, even though the cord as a whole in a global sense is floppy.

6   SUPERCOILING OF LONG DNA

In preceding sections we have considered many different aspects of the structure of DNA and have illustrated some of the key ideas by means of examples involving short pieces of DNA. But the real DNA in any cell is very long—on the order of 10,000 times the diameter of a cell—and some special structural features are needed to achieve the required level of compaction.

In general, most long DNA in cells is found to be "underwound" by about 6%. In other words, if the DNA were stretched out straight, the average angle of helical twist per step would be 6% less than the value for which the same DNA is relaxed and stress-free. We know, of course, that if a short elastic rod is twisted (i.e., its ends are given a relative rotation about the axis along the rod), it remains straight, and some torsional stress in proportion to the angle of twist is generated. But when we apply twist to a long, slender, wirelike rod, a quite different response may be found: the rod tends to writhe or curl around itself, as if it were trying to escape from the stress of twisting.

For purposes of illustration it is useful to consider a *closed circle* of DNA, of the kind that is found commonly in bacterial cells carrying "plasmids." Such an arrangement models well the important features of very long DNA in animal or plant cells. Figure 11 shows a closed circle that has been made from a long, flexible strip of white rubber, having a square cross section and one face painted black. Any long piece of DNA contains a great many double-helical turns, but those are not shown here: instead, the double-helical turns are disguised, just as in Figure 10, for the sake of simplicity. And so our model rubber rod appears *untwisted* when it is relaxed. Since at present we are not concerned with local detail, however, the DNA may be represented by a featureless rod.

In Figure 11a this strip has been curved around and joined to itself, then placed onto a flat table. The black face of the strip is vis-ible everywhere, which means that no twisting has occurred. The corresponding piece of DNA is neither underwound nor overwound.

In Figure 11b, three turns of left-handed twist have been introduced to the circle. This operation involves cutting the rod, giving the ends three turns of relative rotation, and then rejoining the ends. The rod now rests on a table in the same circular shape as before, but it is now underwound by three turns. These three turns of left-handed twist can be seen clearly: the black face is visible at just three places around the circumference.

If the ring is next lifted up and shaken, it will tend to adopt a form like that shown in Figure 11d, but it may still be placed onto a table in any of the conformations shown in (c), (d), or (e). All these forms have an overall shape that is no longer circular; and they contain one, two, or three "crossovers," respectively. Note that in the last sketch (Figure 11e), the black face of the strip is visible everywhere, so there is no twist. Evidently there is some sort of tradeoff between the twist and the number of crossovers: as we go from left to right through (c), (d), and (e), the magnitude of twist decreases, while the number of crossovers increases.

To describe more clearly what we observe in these pictures, we must now introduce some technical terms. First, linkage Lk describes *how many times* the black face of the rod entwines around the opposite white face, those two faces being regarded as separate for this purpose. In Figure 11a, the two faces do not intertwine at all—they would come apart completely if the rod were split. But in Figure 11b, after the rod has been cut and the three turns wound in, the two faces are linked three times. Linkage is a *topological* quantity: it does not change, however the ring is distorted. In four of the present cases (Figures 11b–11e); $Lk = -3$. Lk changes only if there is cutting and relative rotation of the ends, as in the change from (a) to (b).

The next quantity that needs to be defined is twist Tw. The essential feature of twist is shown in Figure 10. If we move along the

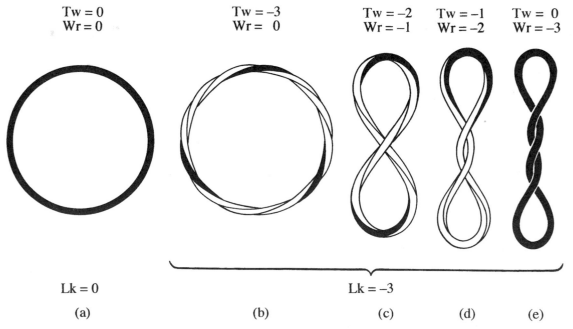

$$Tw = 0 \qquad Tw = -3 \qquad Tw = -2 \qquad Tw = -1 \qquad Tw = 0$$
$$Wr = 0 \qquad Wr = 0 \qquad Wr = -1 \qquad Wr = -2 \qquad Wr = -3$$

$$Lk = 0 \qquad\qquad\qquad\qquad\qquad Lk = -3$$

$$\text{(a)} \qquad\qquad \text{(b)} \qquad\qquad \text{(c)} \qquad\quad \text{(d)} \qquad\quad \text{(e)}$$

Figure 11.   A closed circular molecule of DNA, represented as a rubber rod of square cross section, which undergoes changes in twist *Tw,* linkage *Lk,* and writhe *Wr.* (Adapted with permission from Calladine and Drew, 1992.)

block of Figure 10c we undergo a small rotation about the longitudinal axis of the rod—which in this case has a negative value, since it is a left-handed twisting motion. The twist of a rod is measured by a rotation per unit length; and in an elastic rod a change of twist is always accompanied by torque or twisting force. In the present example of Figure 11, Tw is a *total* of the many small twist angles within small portions of rod, integrated over the whole length of the rod, and expressed as a number of turns. Thus it is clear that $Tw = -3$ in Figure 11b has a negative value because the twist is left-handed, and that $Tw = 0$ in both Figures 11a and 11e. And $Tw = -2$ and -1 in Figures 11c and 11d, respectively.

A third important quantity is known as writhe Wr. In a crude sense Wr is equal to the number of crossovers as marked on Figure 11, but it may be defined in general by means of the equation

$$Tw + Wr = Lk \qquad (2)$$

which relates all these three quantities to each other. For a given value of Lk, there is a trade off between Tw and Wr, as we have already seen; and to satisfy the equation, we must assign a negative sign to Wr in the examples shown, say $Wr = -3$ for $Tw = 0$, $Lk = -3$. Writhe is negative in the present example for right-handed supercoiling of rods running in opposite senses throughout the supercoil; and it turns out that to define the sense of Wr correctly, it is necessary to assign a sense of direction or polarity to the rod so that the crossovers may be assigned positive or negative values.

Note that Lk is by definition an integer. But neither Tw or Wr is confined to integer values. They do indeed have integer values in Figure 11; but we can easily construct other examples in which that is not so. For instance, if one took the lower loop of any of the examples in Figures 11c–11e and rotated it about the overall longitudinal axis by 90°, values for both Tw and Wr would be found midway between integers.

Equation (2), which was put forward in about 1970 independently by the mathematicians J. H. White and F. B. Fuller, is precisely the tool we need to understand the three-dimensional coiling of long DNA within a cell. What, then, is the explanation for the 6% underwinding normally found in cellular DNA?

The packaging of DNA inside a cell into compact chromosomes involves several different levels of structure, within a complex hierarchy of foldings that is not yet fully understood. But some aspects of the early stages of the compaction process are clear. As we have already mentioned, DNA in any higher cell is wrapped around protein spools or histone protein octamers to form "nucleosomes." The spools are built from four sorts of histone protein called H2A, H2B, H3, and H4; the DNA wraps around each spool for almost two turns in a left-handed sense. There are about 80 base pairs in each of these turns, and the spools are positioned along the DNA at a spacing that may differ from species to species or within different parts of a chromosome, but is typically 200 base pairs plus or minus 50. Thus, when a piece of DNA is fully loaded with its protein spools, the nucleosomes are like a set of beads on a string. The next level of compaction involves the wrapping of this string of beads into some higher order spiral of diameter about 300 Å. And in turn, those 300 Å diameter fibers form still other, higher level structures that are less well understood.

Figure 12a shows a highly schematic drawing of DNA wrapped twice around a single nucleosome spool as a left-handed supercoil (the true wrapping is 1¾ turns). Here, the DNA is represented by a ribbon with one side shaded, and the histone spool by a wine-bottle cork. (Physical models are often useful for clarifying situations of this kind.) The ends of the ribbon are attached to two blocks that cannot rotate relative to each other and represent connections to neighboring DNA. If the cork is now removed (Figure 12b), and the ribbon pulled out straight (Figure 12c), the ribbon gains two complete turns of left-handed twist. The diagrams are labeled with val-

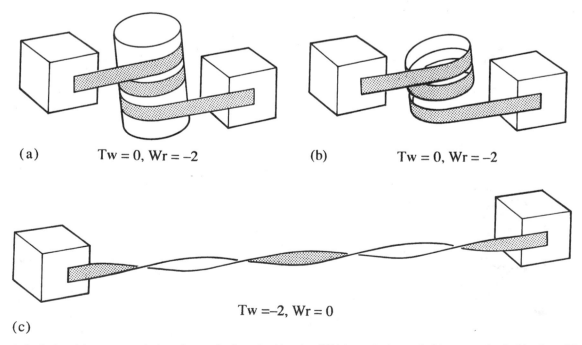

(a) Tw = 0, Wr = –2   (b) Tw = 0, Wr = –2

(c) Tw =–2, Wr = 0

Figure 12. A physical model to represent the interchange of twist and writhe when DNA is wrapped around a histone spool as in (a), released from the constraint as in (b), and then stretched out as in (c). (Adapted with permission from Calladine and Drew, 1992.)

ues of both twist and writhe—clearly there is no change of linkage during these manoeuvers—and we can see that the left-handed twist in Figure 12c facilitates writhing, hence favors wrapping into a left-handed supercoil around a spool in Figure 12a. Replacing the ribbon now by real DNA, we can see that underwinding by two turns for each 200 base pairs might help in the process of forming nucleosome cores. Any 200 base pairs might then correspond to about 17 double-helical turns instead of the expected $200/10.6 = 19$ turns; and so we have established here a possible underwinding of 2/19 or near 10%. This is close to, but somewhat larger than, the observed underwinding of 6% in living cells. The difference is due mainly to the tendency of DNA, when it curves around the histone proteins, to *overtwist* by about 4%, from 10.6 to 10.2 base pairs per turn. So the net unwinding is about 6%.

Figure 12 also helps to explain how DNA can be untwisted in preparation for the separation of the two strands of bases, as the transcription apparatus of the cell tries to read the coded message on just one strand of the molecule. If you grasp and hold straight between your two hands a piece of rope that has one strand twisted around the other, and then rotate your hands relative to each other about the axis of the rope, the two strands of the rope begin to come apart. If nature wants to do something similar, and to untwist a short piece of DNA so that its strands will come apart, there is a problem: How can the DNA be "gripped" firmly enough? And how can one or two turns of untwisting—the level required for a typical polymerase protein—be imparted to the molecule?

One answer lies in Figure 12. If the DNA is bound to a series of protein spools, and then one individual spool is taken away and the DNA is stretched straight, the required left-handed twisting will have been achieved if $Tw = -2$ locally. So it is no accident that the *right*-handed double-helical DNA is wrapped in *left*-handed supercoils around nucleosome cores. There is now good evidence from the recent work of D. M. J. Lilley and colleagues that negative supercoiling of DNA activates genes in bacteria, where there are rather loosely binding histones, although the situation remains unclear in higher organisms such as yeast or humans, where there are histones that bind DNA tightly.

See also ELECTRON MICROSCOPY OF BIOMOLECULES; NUCLEAR MAGNETIC RESONANCE OF BIOMOLECULES IN SOLUTION; NUCLEIC ACID HYBRIDS, FORMATION AND STRUCTURE OF; X-RAY DIFFRACTION OF BIOMOLECULES.

Bibliography

Blackburn, G. M., and Gait, M. J., Eds. (1990) *Nucleic Acids in Chemistry and Biology.* IRL Press, Oxford.
Calladine, C. R., and Drew, H. R. (1992) *Understanding DNA: The Molecule and How It Works.* Academic Press, London.
Holde, van K. (1988) *Chromatin.* Springer-Verlag, New York.
Hunter, C. A. (1993) Sequence dependent DNA structure: The role of base stacking interactions. *J. Mol. Biol.* 230:1025–1054.
Judson, H. F. (1979) *The Eighth Day of Creation.* Simon & Schuster, New York.
Saenger, W. (1984) *Principles of Nucleic Acid Structure.* Springer-Verlag, New York.
Sinden, R. R. (1994) *DNA Structure and Function.* Academic Press, San Diego, CA.
Travers, A. A. (1993) *DNA–Protein Interactions.* Chapman & Hall, London.
Wolffe, A. (1992) *Chromatin.* Academic Press, London.

DOWN SYNDROME, MOLECULAR GENETICS OF

Charles J. Epstein

1 **Phenotype of Down Syndrome**
2 **Cytogenetics**
3 **Structure of Chromosome 21**
4 **Maps**
5 **Pathogenesis**
6 **Animal Models**

Key Words

Acrocentric   Describing a chromosome in which the centromere is very close to one end.

Autosomes   All the chromosomes in the genome except for the sex chromosomes (X and Y).

Centromere   The structure within each chromosome at which the fibers required to move the chromosomes during meiosis or cell division (mitosis) attach.

Meiosis   The process within the germ cells during which genetic recombination occurs and the number of chromosomes within the egg or sperm is reduced from two homologous sets found in somatic cells to one.

Phenotype   The characteristics or traits that result from a genetic mutation or other genetic alteration.

Trisomy   The presence in the genome of three rather than two copies of a specific chromosome.

Down syndrome is the commonest of the genetically caused forms of mental retardation. It occurs with a frequency of approximately one per 800–1000 live births, and its incidence increases with increasing maternal age. Down syndrome is caused by the presence of an extra chromosome 21 within the genome, which, in turn, results in a 50% increase in the expression of the genes contained on the chromosome. By mechanisms currently undefined, the increased expression of several genes on human chromosome 21 results in a syndrome characterized by mental and growth retardation, a distinctive set of major and minor congenital malformations, a variety of cellular abnormalities, and, later in life, by the development of Alzheimer's disease.

1   PHENOTYPE OF DOWN SYNDROME

The most immediately apparent, if not the most serious, manifestations of Down syndrome (DS) are the many minor dysmorphic features that collectively constitute its distinctive physical phenotype. Salient among these are upslanting palpebral fissures, epicanthic folds, flat nasal bridge, brachycephaly, short broad hands, incurved fifth fingers, loose skin of the nape of the neck, open mouth with protruding tongue, and transverse palmar creases. Although any individual with DS will have many of the characteristic features and can be easily recognized as having the disorder, none of these features is present in all persons with DS.

Down syndrome is manifested in the nervous system in three principal ways: hypotonia, which occurs in virtually all affected newborns and infants; delayed psychomotor development in infancy and mental retardation throughout life; and neuronal degeneration during the adult years. The latter process, which is pathologically identical to Alzheimer's disease (presenile and/or senile dementia), results in significant pathologic changes in the brain and may further compromise the already impaired mental functioning.

Down syndrome is associated with two types of major congenital malformation. Most frequent (about 40%) is congenital heart disease, usually of the endocardial cushion type or one of its variants. Gastrointestinal tract abnormalities occur in about 4.5% of individuals, more than half of whom have duodenal stenosis or atresia. Structural abnormalities of the thymus and functional defects in T-cell function leading to an increased susceptibility to infection are present, and there is a 10–18 times normal incidence of childhood leukemia, with the frequent occurrence of acute megakaryoblastic leukemia.

2   CYTOGENETICS

Down syndrome is the phenotypic manifestation of trisomy 21. As such, it occurs when a third copy of chromosome 21 is present in the genome, either as a free chromosome or as part of a Robertsonian fusion chromosome (in which the long arms of two acrocentric chromosomes are joined at the centromeres). Except in the 2–4% of cases in which there is mosaicism—with two populations of cells, one diploid and one trisomic—all cells of the body are trisomic. Although these cytogenetic abnormalities involve most or all of chromosome 21, in rare cases translocations occur in which only part of the long arm of the chromosome is triplicated. Depending on the region of the chromosome that is involved, such cases may or may not express the classical DS phenotype.

Analyses using DNA markers have shown that maternal nondisjunction, the failure of paired chromosomes to separate properly, accounts for 95% of all cases of trisomy 21, with 69% percent of the informative cases occurring at meiosis I and 28% at meiosis II, and 3% at mitosis. Furthermore, there appears to be a reduced rate of recombination in the maternal meioses that gave rise to trisomic offspring, findings that are consistent with the hypothesis that absence of pairing and/or reduced chiasma frequency and recombination predispose to nondisjunction.

3   STRUCTURE OF CHROMOSOME 21

Chromosome 21 is the smallest of the human autosomes, constituting approximately 1.7% of the length of the haploid genome. It consists of approximately 51×10^6 base pairs, with a sex-equal genetic length of about 56 centimorgans (cM). The correspondence between DNA and genetic lengths is reasonable, since it is assumed that 1 cM is roughly equivalent to 10^6 base pairs. In physical terms, chromosome 21 is an acrocentric chromosome; the centromere is very close to one end, and the short arm is very small. The short arm (21p) terminates in a satellite region, which varies in size from molecule to molecule. Proximal to the satellite is the stalk (secondary constriction) which, as the nucleolar organizer region (NOR), contains multiple copies of the ribosomal RNA genes (*RNR4*) and stains characteristically with silver. The degree of silver staining appears

to be a representation of the molecular activity of the ribosomal RNA genes that the chromosome contains. The region of 21p adjacent to the centromere contains highly repeated DNA sequences, which consist of the satellite (including alphoid) and the "724" families of sequences. None of these gene families is unique to chromosome 21. It is believed that these families of repeated gene sequences may be involved in the juxtaposition or association of the satellite regions (satellite association) of the acrocentric chromosomes during mitotic metaphase and with formation of the nucleolus during interphase.

The major part of chromosome 21 is the long arm (21q), which has a characteristic banding pattern consisting of three or four bands at low resolution and as many as 11 dark and light bands resolvable by prometaphase banding. All genes of known function (other than for ribosomal RNA) are located on this arm of chromosome 21, and only this arm is essential for normal development and function. The presence of a Robertsonian fusion chromosome in which the short arms of two acrocentric chromosomes (sometimes both chromosomes 21) are deleted does not cause detectable abnormalities if the genome is otherwise balanced.

4   MAPS

A list of the loci of known function or with known products mapped to human chromosome 21 is presented in the legend to Figure 1. In addition to these genes, the database of the Human Genome Project contains more than 400 anonymous DNA marker segments derived from chromosome 21 and more have been described since the generation of the maps in Figure 1. While the functions, if any, of the regions from which these segments are derived are unknown, their sequences provide valuable probes for mapping the chromosome and for following the segregation of loci at meiosis.

Two types of chromosome 21 map are now under construction. The first is a *physical* map on which the physical locations of individual loci are placed, using features of the banding pattern of the chromosome as specific landmarks. Construction of a physical map is based on the use of a variety of tools, including somatic cell hybrids, DNA dosage studies of individuals with chromosome 21 duplications or deletions, in situ hybridization of labeled DNA probes to metaphase chromosomes, sequence tagged sites (STS) that are detectable by means of the polymerase chain reaction, and yeast artificial chromosomes. It has been possible with these techniques to map regionally many of the defined and anonymous loci on chromosome 21; the results of these efforts are contained in Figure 1.

The second type of chromosome 21 map is a *genetic* map, which shows the linkage relationships between loci and the inferred genetic distances of loci from one another. This type of map is derived from studies of the segregation of polymorphic loci within families. Because of intensive efforts devoted to the mapping of chromosome 21, the genetic map has rapidly evolved, and a recent version of the map is presented in Figure 1: the majority of the genetic distance, in physical terms, clearly appears to be located in the distal tenth of the chromosome.

Cases in which only part of chromosome 21 is triplicated have been intensively studied to arrive at a phenotypic map that will permit the correlation of particular phenotypic features of DS with specific regions or loci of the chromosome. On the basis of both molecular and cytogenetic analyses, it appears that many of the phenotypic features of DS are caused by triplication of the region around *D21S55* (which has been localized to band 21q22.2). The

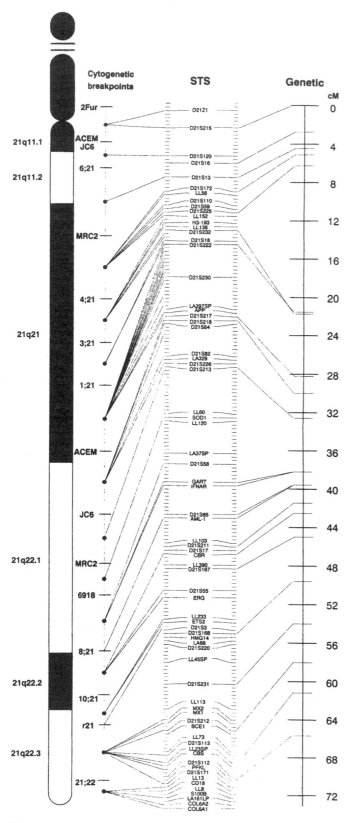

Figure 1.   Physical and genetic maps of human chromosome 21. *Left:* a low resolution G-banding pattern next to which is a physical map of a large number of sequence tagged sites (STS) located with regard to a series of chromosomal breakpoints defined by translocations. *Right:* genetic map showing the locations of several of the STSs; the loci of known function in

full extent of the region, however, remains undefined. More proximal triplication may result in some degree of mental retardation without the visible phenotype of DS. The region associated with the development of Alzheimer's disease has not been defined.

5   PATHOGENESIS

In searching for mechanisms to relate the components of the DS phenotype to specific genes on chromosome 21, it is important to realize that the immediate consequence of having a third copy of all or part of a chromosome is a gene dosage effect for each of the loci on the triplicated chromosome or chromosome segment. Such gene dosage effects, in which the concentration of gene products is 1.5 times normal, as would be expected from the presence of three rather than two copies of each gene, occur for all but one of the eight tested chromosome 21 loci. The only significant exception is in the levels of the APP (amyloid precursor protein) mRNA in brains from fetuses with DS, in which increases to about four times normal levels have been found. It thus appears that some type of dysregulation of the expression of the APP locus is superimposed on the gene dosage effect.

On the basis of studies in animal models (see Section 6), decreased platelet serotonin uptake and prostaglandin synthesis, both of which occur in DS, have been ascribed to increased activity of CuZn-superoxide dismutase (SODI). With these exceptions, it has not been possible to relate any component of the DS phenotype to overexpression of specific loci. However, it is possible that the development of Alzheimer's disease in DS is related to the overproduction of APP, perhaps superimposed on an intrinsically defective nervous system.

6   ANIMAL MODELS

Many of the consequences of aneuploidy in humans arise during the period of morphogenesis, placing a special stumbling block in the way of their investigation. Research on events occurring during gestation, especially early gestation, is both technically impractical and, at the present time and for the foreseeable future, ethically and legally impossible. For these reasons, interest has turned to the development of animal models that will duplicate the human condition—in developmental and functional terms—as closely as possible. The

---

the STS list are APP, amyloid-β (A4) precursor protein; SOD1, superoxide dismutase 1, soluble (CuZnSOD), the same locus as ALS (amyotrophic lateral sclerosis); GART, a trifunctional enzyme with phosphoribosylglycinamide synthetase, phosphoribosylaminoimidazole synthetase, and phosphoribosylglycinamide formyltransferase activities; IFNAR, interferon alfa (and beta) receptor; AML1, acute myeloid leukemia 1 oncogene; CBR, carbonyl reductase (NADPH); ERG, avian erythroblastosis virus E26 (v-*ets*) oncogene-related sequence; ETS2, avian erythroblastosis virus E26 (v-*ets*) oncogene homologue 2; HMG 14, high mobility group protein 14; MX2, MX1, homology to murine myxovirus (influenza) resistance 2 and 1; BCEI, estrogen-inducible sequence expressed in breast cancer; CBS, cystathionine-β-synthetase; PFKL, phosphofructokinase, liver type; CD18, lymphocyte function-associated antigen, β subunit; S100B, S100 protein, β subunit; COL6A2, COL6A1, collagen, type VI, α-2 and α-1. Not shown are GLUR5, glutamate receptor subunit 5 (near SOD1); IFNGT1, interferon gamma transducer 1 (near FNAR); EPM1, progressive, myoclonic epilepsy (Unverricht–Lundborg type); and CRYA1, crystallin, α-polypeptide 1 (between CBS and PFKL). (Modified and reprinted with permission from *Nature, 359:*380–287. Copyright 1992 Macmillan Magazines Limited.)

mouse has been the model animal of choice for several reasons: ease of manipulation, genetic control, and similarities to the human in the processes of morphogenesis and probably of central nervous system function, in neurobiological if not psychological terms.

Three types of animal models for DS now exist. In one type, the effects of increased concentrations of specific gene products are analyzed in transgenic mice carrying one or more copies of individual human chromosome 21 genes. Strains of transgenic animals carrying the genes for human CuZnSOD, APP, and S100β have been made, and some features of DS were observed in the CuZnSOD-transgenic animals. At the other end of the extreme, mice with trisomy 16 have been bred. Mouse chromosome 16 carries many of the genes present on human chromosome 21q (as far distal as the loci for the Mx proteins), in addition to a large number of unrelated genes. Trisomy 16 mice, which survive only until the end of gestation, have many features of interest, including congenital heart disease (endocardial cushion defects), abnormalities of the thymus and brain, dysregulation of APP expression, and atrophy of transplanted neurons. Postnatally viable mice with an extra copy of just the region of mouse chromosome 16 that is homologous to 21q have also been generated. Because they reproduce the genetic imbalance found in the cases of partial trisomy 21 that exhibit the DS phenotype, these animals should be particularly valuable for studies of the pathogenesis of DS.

See also ALZHEIMER'S DISEASE; HUMAN GENETIC PREDISPOSITION TO DISEASE.

Bibliography

Epstein, C. J. (1993) Down syndrome, in *The Metabolic Basis of Inherited Disease,* 7th ed., C. R. Scriver, A. L. Beaudet, W. S. Sly, and D. Valle, Eds. McGraw-Hill, New York.

———, Ed. (1993) *The Phenotypic Mapping of Down Syndrome and Other Aneuploid Conditions.* Wiley-Liss, New York.

Epstein, C. J., Hassold, T., LoH, I. T., Nadel, L. and Patterson, D., Eds. (1995) *Etiology and Pathogenesis of Down Syndrome.* Wiley-Liss, New York.

DROSOPHILA Genome

Robert D. C. Saunders

Key Words

Balanced Lethal Stocks   Recessive lethal stocks are maintained as balanced stocks, in which the population consists of flies heterozygous for the mutant chromosome and for a corresponding balancer chromosome. Balancer chromosomes possess multiple inversions, which suppress recombination, and several genetic markers, including a dominant visible, and are themselves recessive lethal or sterile, so that balancer homozygotes are either inviable or sterile.

Euchromatin   Corresponds to the single copy and middle repetitive DNA of the genome, in which most of the genes are located. Unlike heterochromatin, euchromatin does polytenize.

Heterochromatin   Describing regions of the chromosome that are highly condensed in metaphase chromosomes, and undergo late replication in S phase. Heterochromatin generally corresponds to regions of the genome containing simple sequence repetitive, or satellite, DNA. There are few genes in heterochromatic regions.

In Situ Hybridization   Cloned sections of DNA can be accurately mapped to polytene chromosomes by in situ hybridization, in which chemically or radioactively labeled nucleic acid is hybridized to denatured chromosomes on a microscope slide. Signals are detected immunochemically or by application of a radioactive emulsion.

Polytene Chromosome   Highly replicated chromosome in an interphase-like chromatin conformation, with each chromatid precisely synapsed with its sisters. This results in a banded structure, permitting high resolution mapping by light microscopy.

Drosophila melanogaster is arguably the genetically best characterized metazoan organism. Since its adoption for genetic analysis, its many advantages as a laboratory animal have led to its use as a model system for a wide variety of biomedical researches. The *Drosophila* genome is relatively small (170×10^6 base pairs) and is distributed between four pairs of chromosomes. The polytene chromosomes of *Drosophila,* which have provided a physical map for genetic studies since they were mapped in the 1930s, continue to be a powerful tool in molecular genetic analysis.

This article describes the features of the *Drosophila* genome relevant to genetic and molecular studies in *Drosophila*.

1   INTRODUCTION

In the early years of this century, *Drosophila melanogaster* (then known as *Drosophila ampelophila*) was adopted by Thomas Hunt Morgan as an experimental organism in the then new science of genetics. *Drosophila* proved to be an ideal organism for genetic research: it has a short life cycle (10–14 days at 25°C), and it has very

Table 1   Haploid Genome Sizes

Organism	Genome Size ($\times 10^6$ bp)
Escherichia coli	4
Saccharomyces cerevisiae	16
Schizosaccharomyces pombe	15
Caenorhabditis elegans	100
Drosophila melanogaster	170
Homo sapiens, Mus musculus	3000

Table 2   The Composition of the *Drosophila melanogaster* Genome

Component	Proportion (%)	Number ($\times 10^6$)
Single-copy sequences	61	103.7
Satellite DNA	21	35.7
rDNA, histones, etc.	3	5.1
Transposable elements	9	15.3
Foldback DNA	6	10.2
Total genome size		170

modest requirements in terms of husbandry. In fact a modern *Drosophila* laboratory may maintain several thousand genetically distinct stocks. Many of the basic concepts of genetics were established using *Drosophila* species, including important advances in population genetics and evolutionary theory, particularly with *Drosophila pseudoobscura* and related species. However it is *D. melanogaster* which has made the biggest impact on modern molecular genetic research. With the advent of molecular cloning technology, *Drosophila* found itself ideally placed to benefit from integrated genetics and molecular biology.

As a model organism, *Drosophila* has been an essential tool for the elucidation of basic genetic processes, common to many eukaryotes:

> For many areas of biomedical research. Examples include neurobiology, embryology, pattern formation, signal transduction, and cell cycle regulation. In particular, research in these areas is facilitated by the sophisticated genetic analysis possible with *Drosophila*.
>
> For genome mapping and analysis. *D. melanogaster* has a genome of moderate size: 170×10^6 base pairs, in comparison with those of yeasts, nematode worms, mice, humans, and other species in which large-scale molecular mapping is carried out. Table 1 shows the haploid genome sizes of some much-studied species. Thus, the *Drosophila* genome is ideal for use as a model for genome mapping in higher eukaryotes, and additionally has features such as polytene chromosomes, which simplify mapping.
>
> For studying the many medically and economically important pests and disease vectors in the order Diptera. Examples are anopheline mosquitoes (malaria vectors), aedine mosquitoes (vectors of yellow and dengue fevers), and tsetse flies (vector of trypanosomiasis, or sleeping sickness) to name but a few. As a result of several features of their biology and life cycle, these insects are not good research organisms, from the point of view of laboratory culture and genetics. *Drosophila* is a model system for the molecular genetic analysis of these insects.

This article reviews the genetics and molecular biology of *D. melanogaster,* with respect to its use as a model system.

2   THE *DROSOPHILA* GENOME

The haploid genome of *Drosophila melanogaster* is relatively small, 170×10^6 base pairs, compared with the typical mammalian genome of approximately 3×10^9 base pairs. *Drosophila* genomic DNA is unmethylated, in contrast to most other higher eukaryotes. The composition of the *Drosophila* genome in terms of single copy and repetitive DNA fractions is presented in Table 2.

About 21% of the genome comprises highly repetitive simple sequence DNA known as satellite DNA. Satellite DNA is present in large blocks of repeated short (5–10 bp) units principally located close to centromeric regions. There are several classes of satellite, distinguishable by the repeat unit sequence, and these classes often have chromosome-specific distribution. One satellite, the 1.688 or 359 bp satellite, is atypical, comprising arrays of repeated 359 bp units confined to the X chromosome. In general, satellite DNA is concentrated in heterochromatin, which is discussed in Section 2.1.

A substantial proportion, 18%, of the genome is moderately repetitive. Some of this repetition is due to the ribosomal RNA and histone gene families, but much is accounted for by mobile DNA, or transposable elements. Transposable elements are found in variable, dispersed locations within the genome. Several transposable elements have proven to be very useful in the experimental manipulation and analysis of *Drosophila,* particularly the *P* element, which is described in detail in Section 2.3.

2.1   THE CHROMOSOMES OF *DROSOPHILA*

The mitotic chromosome complement of *Drosophila melanogaster* is illustrated in Figure 1. There are two metacentric autosomes, chromosomes 2 and 3; one tiny dotlike autosome, chromosome 4; and the heteromorphic sex chromosomes, the acrocentric X chromosome and the submetacentric Y chromosome. Sex determination in *D. melanogaster* is based on the ratio of X chromosomes to autosomes. XX individuals have a ratio of 1 and develop as females, while XY individuals have a ratio of 0.5 and develop as males. Thus, male *Drosophila* are hemizygous for X-linked loci. XO individuals

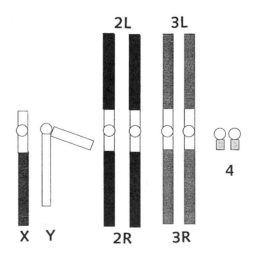

Figure 1.   Diagrammatic representation of the diploid complement of mitotic chromosomes of *Drosophila melanogaster.* The karyotype shown is XY (male). Circles and open boxes represent centromeres and heterochromatin, respectively.

are male, but sterile, since the Y chromosome carries factors essential for male fertility.

Chromatin can be subdivided on microscopic criteria into heterochromatin, which is characterized by having a highly condensed state and late replication in S phase, and euchromatin. Modern molecular genetic research has shown that in general the genes lie in the euchromatin, while the heterochromatin is composed principally of simple sequence repetitive DNA. Indeed, the term "heterochromatin" is often used as a synonym for "satellite DNA." There are exceptions to this rule, including the male fertility factors of the Y chromosome just mentioned, and a few autosomal loci.

In common with many dipterans, *Drosophila* possesses polytene chromosomes in several tissues, most notably the larval salivary glands. The salivary glands of third instar larvae are easy to identify and dissect, and their chromosomes are simple to prepare for analysis. Polytene chromosomes are very large, highly replicated chromosomes that are invaluable in genetic research using *Drosophila* (and many other dipteran species) as a model. Polytene chromosomes and their experimental use are described in Section 4.

2.2 GENES

It is estimated that there are approximately 15,000 genes in the *Drosophila* genome. Since genes are principally located in the euchromatic fraction of the genome, about 100 kb, this represents a gene density of one gene per 6700 base pairs of euchromatin. Pseudogenes appear to be quite rare in the *Drosophila* genome.

The *Drosophila* genetic database FlyBase lists, at the time of writing, 8676 genes represented by a total of 23,878 alleles.

Because *Drosophila* males have one copy of the X-linked gene while females have two, some form of compensation is required. Unlike mammals, in which one of the X chromosomes in the female is inactivated, X-linked genes in *Drosophila* are twice as active in males than in females.

2.3 TRANSPOSABLE ELEMENTS

Transposable elements, or transposons, are segments of DNA that are able to excise from the genome, or replicate, and insert at another location. These elements fall into the middle repetitive fraction of the genome, since they are generally present in copy numbers varying from a few tens to several hundred. *Drosophila* possesses a great many transposable elements, classified into several families on the basis of structure and transposition mechanism. Thus different strains of *D. melanogaster* have different chromosomal locations for each type of transposable element. The transposition activity of transposable elements can result in a genetic syndrome known as hybrid dysgenesis: high mutation rates, high frequency of chromosome rearrangements, sterility, and male recombination (male recombination does not normally occur in *Drosophila*). In general, each type of transposable element displays particular preferred sites of insertion, determined by DNA sequence and other, less well understood, factors.

The *P* element is the best-characterized transposable element. Genetically, the activity of this element is seen as hybrid dysgenesis resulting from the cross of *P* element containing (P strain) males with females lacking *P* elements (M strain). Crossing M males with P females does not cause hybrid dysgenesis. Molecular analysis revealed the *P* element to be a 2.9 kb segment of DNA that encodes two proteins: a transposase and a transposase repressor, depending on alternative splicing of four exons. Transposase is normally expressed in the germ line only. Additionally, the 31 bp terminal repeats of the element and other factors (not encoded by the element) are also required for transposition.

Cloned *P* elements may be injected into *Drosophila* embryos and will subsequently integrate into the genome. By in vitro genetic manipulation, a rich variety of artificial *P* elements have been constructed for use in inserting exogenous DNA into the *Drosophila* germ line. These elements do not themselves encode transposase (hence are stable once integrated into the genome) and therefore require a source of transposase, which can be supplied by a defective *P* element encoding transposase, but lacking one or both of the terminal repeats. A more elegant method is to inject embryos of a *Drosophila* strain carrying a stably integrated element expressing transposase. Examples of *P*-element derivatives are described in Sections 2.3.1 to 2.3.6.

2.3.1 Transformation Vectors

One use of germ line transformation is to provide direct experimental evidence of the identity of a cloned gene. This can be done by determining whether a given segment of DNA identified in a cloning project can rescue a mutant phenotype when introduced into the genome by *P*-element-mediated germ line transformation. Transformation vectors are defective elements that lack transposase but possess the terminal repeats. In addition, they have a selectable marker gene that permits the selection or recognition of genetically transformed flies. Typical marker genes include the *Drosophila* gene *white*$^+$ and the bacterial antibiotic resistance gene *neo*^{R}. *white*$^+$ restores wild-type eye color in individuals of *white*$^-$ background, and *neo*^{R} confers resistance to the antibiotic G418, to which *Drosophila* are normally sensitive.

2.3.2 Mutagenesis

Insertion of a *P* element or another transposable element often affects that gene's pattern of expression. It is therefore possible to make use of transposon mobilization as a mutagen. Mutants induced in this way present a considerable advantage when the mutated gene is to be cloned (see Section 2.4).

2.3.3 Enhancer Trapping

P elements that contain the *Escherichia coli lacZ* gene lacking a functional promoter sequence have been constructed. Upon insertion, these elements express *lacZ* only if there is a neighboring gene with an enhancer able to direct expression of the *lacZ* transgene. *lacZ* expression is conveniently detected with a histochemical or an immunological assay. Partly because of the high gene density of the *Drosophila* genome, a large proportion of enhancer trap elements do express *lacZ*, often in an intricate temporal and spatial pattern that is frequently informative about the nearby gene's identity or function. Enhancer trapping is therefore an efficient means of detecting genes.

2.3.4 The *GAL4*–UAS System

Two distinct element insertions comprise the *GAL4*–UAS system. The first carries the *Saccharomyces cerevisiae GAL4* gene, which encodes a transcription factor that activates transcription from an upstream activating sequence (UAS$_{GAL4}$). The second element carries the transgene of interest, under the control of the UAS$_{GAL4}$, and which is therefore expressed only in cells expressing *GAL4*. The *GAL4* expression may be controlled by a specific promoter or by neighboring genomic enhancers (in which case the *GAL4* element functions as an enhancer trap). This bipartite system is extremely flexible, since a large collection of lines expressing *GAL4* in distinct spatial and temporal patterns is available. These can be combined with any UAS element carrying a transgene, which may be a reporter gene such as *lacZ*, or a mutant gene predicted to have a dominant phenotype. One exciting possibility is to express toxin genes (such as ricin A chain) in specific cell types, to investigate the developmental consequences of ablation of particular cells.

2.3.5 FLP-FRT

Another two-component system consists of one element that carries the gene encoding the *Saccharomyces cerevisiae* 2 μm plasmid FLP recombinase and another the target sequences for this enzyme (FRT). The expression of FLP can be controlled in several ways like the *GAL4* just described. Cells in which FLP is expressed will rearrange sequences flanked by the FRT sites. Depending on the orientation of the two FRT sequences, this can be an inversion or an excision, enabling a gene between the sequences to turned on or off.

2.3.6 Visible Marker Genes

Elements bearing visible marker genes can be used as dominant markers for the purpose of generating chromosomal deficiencies. For example a w^- stock (w mutants have white eyes) bearing a *P* [w^+] element inserted at, for example, 96A will have normal reddish-brown eyes. Following X irradiation, flies with this chromosome region deleted will have lost the *P* [w^+] element and will have white eyes. These w^- stocks can be analyzed by polytene cytogenetics to reveal which, if any, carry a chromosomal deletion.

2.4 Cloning Genes in Drosophila

Drosophila genes may be cloned by virtue of their homology to genes isolated from other organisms. The approaches used in this strategy [(e.g., the polymerase chain reaction (PCR), low stringency hybridization] are similar to those in studies of other organisms, and since they are not specific to *Drosophila* research, I shall not describe them here. Instead, I shall consider cloning techniques unique to, or particularly advantageous in, *Drosophila*. Some of these approaches will be made redundant by the development of the molecular physical maps of *D. melanogaster* but will still be useful in other *Drosophila* and dipteran species.

Mapping, by recombination and cytogenetics, reveals a gene's physical location on the polytene chromosome map to, optimally, within a few tens of kilobases. The corresponding clones can be identified within one of the molecular physical maps. If that is not possible, the region can be entered using conventional means: chromosome walking, in which sequential screening of a genomic library is used to isolate a series of overlapping cloned DNA segments. Together these represent a larger DNA sequence than can be accommodated in a single clone. Progress and orientation of the walk can be determined by in situ hybridization to polytene chromosomes. Initiation of a chromosome walk requires a starting clone, or entry point, in the target gene's locality.

If no entry point in the appropriate region is available, other

strategies can be used. First, direct microdissection of the locus followed by PCR amplification can yield a suitable probe for initiating a chromosome walk. As applied to polytene chromosomes, this technique is less demanding and offers significantly higher resolution than is obtainable with mitotic chromosomes. Fragments corresponding to a few tens of kilobases may be microdissected, compared with a resolution of $1–10 \times 10^6$ base pairs possible with metaphase chromosome preparations. Second, advantage can be taken of chromosome rearrangements. A clone mapping close to one breakpoint of a chromosome deficiency whose other breakpoint lies near to the gene of interest can enable one to make a "chromosome jump" to the vicinity of that gene. This is achieved by switching to and from standard and inversion genomic libraries as one enters the inversion during the chromosome walk. Other chromosome rearrangements useful in this way include deficiencies and translocations.

Mutations induced by transposable element mobilization are generally due to insertion of the transposon into a gene. A library constructed using DNA extracted from, for example, a *P*-element-induced mutant can be probed with *P* element DNA to isolate clones with the *P* insertion. Elements other than *P* can be used for transposon tagging, such as *hobo* and *I*, though these systems are less well developed. In practice, using *hobo* or *I* may be a practical alternative to *P* when attempts to recover *P* insertions in a particular gene repeatedly fail because of insertion site preferences. In cases of a *P* element that has been engineered to contain plasmid sequences, the insertion site can be directly cloned from genomic DNA by plasmid rescue.

A considerable problem in gene cloning projects lies in the recognition of the target gene within the cloned DNA. In some cases, a distinct mutagenic event will be recognized, such as gross or minor chromosome rearrangement (particularly in the case of radiation-induced mutations). Often, however, no such recognizable event exists. It is always advantageous to have several mutant alleles, induced by different mutagens, available. Ultimately, direct proof, such as rescue of mutant phenotype by *P*-element-mediated germ line transformation with a defined segment of DNA, as described in Section 2.3, must be demonstrated.

3 GENETIC ANALYSIS IN DROSOPHILA

3.1 Mutagenesis

A typical genetic screen for mutations, in this case autosomal recessive lethals, is shown in Figure 2. A large variety of mutagens are available, which can differ widely in the nature of the lesions produced. For example, ethylmethanesulfonate (EMS), administered in the food, causes predominantly single base pair changes, whereas X-rays induce chromosomal deletions and other rearrangements. The use of transposable elements as mutagens was described in Section 2.3.

Male flies are treated with mutagen and mated en masse to virgin females heterozygous for two balancer chromosomes. All subsequent matings are single pair matings, again with virgin females. The target chromosome is marked with recessive markers, to facilitate later genetic manipulations (the mutagenized chromosome is labeled with an asterisk in subsequent crosses in Figure 2). Note that the F1 and F2 crosses are identical; this is necessary when the mutagen, such as EMS, causes lesions requiring one or more cell divisions to become fixed. The F3 cross yields a balanced stock in which

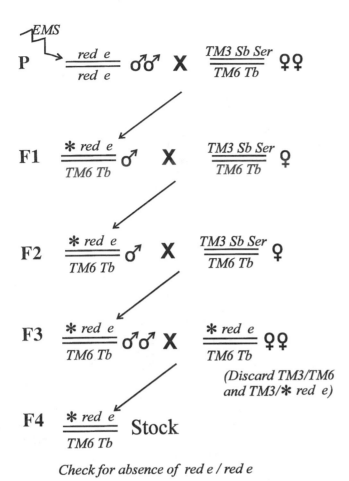

Figure 2. A mutagenesis scheme designed to recover autosomal recessive lethal mutations. With minor modification, however, other types of phenotype may be screened for. Full descriptions of marker mutations and balancer chromosomes can be found in *The Genome of* Drosophila melanogaster (Lindsley and Zimm). All females in crosses are virgin. The markers in this figure are: *red*: red Malpighian tubules (wild-type tubules are whitish; this recessive mutation also results in brown eyes in the adult); *ebony* (*e*): adult has black body, recessive; *Stubble* (*Sb*): dominant mutation affecting adult bristle morphology; *Serrate* (*Ser*): dominant mutation causing serrated wing margins; *Tubby* (*Tb*): dominant causing tubby larvae and adults, but more easily scored in larvae. The mutagenized chromosome is indicated by an asterisk.

flies homozygous for the mutagenized chromosome (easily recognized because of the recessive markers) can be examined for the phenotype of interest. In the case of recessive lethals, this class would be absent. Balancer chromosomes have multiple inversions, which prevent recombination from occurring, and carry a recessive lethal mutation to render balancer chromosome homozygotes inviable. Balancer chromosomes can be followed in genetic crosses because, in general, they carry a dominant marker. Because balancer chromosomes suppress recombination, these features will not be separated by recombination. Some X chromosome balancer chromosomes carry a female sterile mutation rather than a lethal, with the result that the balancer chromosome will be male viable.

Many classes of mutant phenotype can be discovered, using appropriate genetic systems, from dominant temperature-sensitive lethals to simple visible mutations, from female sterile mutants to behavioral mutations.

3.1.1 Reverse Genetics

Often the *Drosophila* researcher is in the position of having cloned a gene but not having a corresponding mutant. For some model organisms, there is an efficient means of targeted mutagenesis, but not in *Drosophila*. A useful technique in this regard is to use *P* element mutagenesis as described in Section 2.3 and detect insertions in the gene of interest using PCR with one gene-specific primer and one *P*-element-specific primer.

As an alternative to the PCR screening of *P*-element insertions, investigators can undertake saturation mutagenesis of a chromosomal region known to include the gene of interest, and spanned by a small chromosome deficiency. Mutants displaying an appropriate phenotype then may be examined molecularly, to determine whether the gene in question is mutated.

3.1.2 Classes of Mutant Allele

Several classes of mutant allele may be recovered following mutagenesis. The simplest is the loss-of-function mutation, either complete (null mutation or amorph) or partial (hypomorph). Amorphs and hypomorphs are generally recessive and can be formally distinguished by comparing their phenotypes in combination with a deficiency (see Section 4.2.1). Other allelic states are hypermorphs, in which there is a gain of gene function, and neomorphs, in which a novel function is acquired. Neomorphs are generally dominant alleles. Mutants of any or all classes can be recovered for a given gene, using appropriate mutagenesis protocols.

As an additional complication, an allele's phenotypic effect may be temperature sensitive. At its extreme, this can be seen as a true conditional mutant, in which environmental conditions can entirely suppress the mutant phenotype.

3.1.3 Position Effect Variegation

As described earlier, most genes are located within euchromatin. In comparison with euchromatin, heterochromatin has unusual properties, which are connected with differences in chromatin organization and composition. These properties influence gene expression. When a gene (e.g., *white*) that has been relocated by chromosome rearrangement is now found close to heterochromatin, its expression is affected by the heterochromatin, which appears to "heterochromatinize" its neighboring euchromatin. This phenomenon may occur to different extents in different clonal cell lineages within the developing organism. Thus a *white* allele, mutated by an inversion that brings it into close proximity of the centric heterochromatin, may be active in some ommatidia of the compound eye and inactive in others, resulting in a mottled eye with red and white facets (*white* mutants have white eyes rather than the wild-type reddish color). The *white* allele w^{m4} is an example. There are loci that display the opposite effect; an example is *light*, which maps to the pericentric heterochromatin of chromosome arm 2L: rearrangements that relocate *light* to euchromatic regions result in reduced levels of expression. Mutants that enhance or suppress PEV are known, and characterization of these genes is illuminating with respect to chromatin structure.

3.2 GENETIC MAPPING

Genetic, or recombination, mapping in *Drosophila melanogaster* has a long history. The theoretical basis of recombination mapping was devised and implemented using *Drosophila* in 1913. The map unit, or centimorgan (cM), corresponds to one percent recombination. These values are additive over short map distances, but over longer distances, complications arise from multiple crossovers. Correlating the cytogenetic and recombination map positions of a number of loci has demonstrated that the recombination rate varies along the chromosome: Recombination is effectively suppressed close to telomeres and centromeres. Thus one centimorgan represents a larger stretch of DNA near the telomere or centromere than in the middle of a chromosome arm.

Under normal circumstances, there is no recombination in *D. melanogaster* males. The special circumstances in which male recombination is observed include hybrid dysgenesis. Additionally, recombination in the tiny fourth chromosome does not occur, except in triploid females.

Recombination mapping is now generally undertaken as start point to cytological mapping, the localization of a gene on the polytene chromosome map, and its subsequent molecular cloning. Any correlation between the polytene chromosome map and the recombination map permits the approximate cytogenetic map position to be determined from recombination map data. The recombination map order of loci can be important in cloning projects.

A typical scheme for mapping a recessive lethal mutation is shown in Figure 3. The first two crosses in this scheme are carried

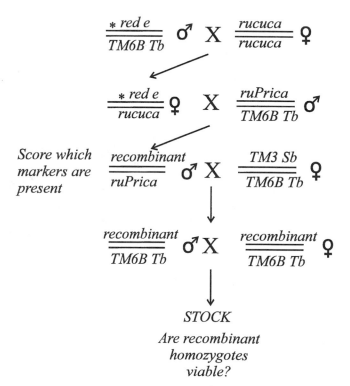

Figure 3. Recombination mapping scheme. Markers are as described in the legend to Figure 2, and as follows. The chromosome *rucuca* is multiply marked with the following recessives: *ru* (*roughoid*: eyes are rough in appearance); *h* (*hairy*: extra hairs on wing veins); *th* (*thread*: aristae are thread-like, instead of branched); *st* (*scarlet*: adult eye is bright red); *cu* (*curled*: wings curled upward); *sr* (*stripe*: stripe on dorsal surface of thorax); *e*; *ca* (*claret*: adult eyes are ruby red). Flies homozygous for both *st* and *ca* have orange eyes. The chromosome *ruPrica* is as *rucuca*, but in addition contains the dominant marker *Prickly* (*Pr*), which causes the adult bristles to be shortened, with twisted tips. As in Figure 2, all females are virgin.

out en masse, while subsequent crosses are single pair matings. Each recombinant stock generated in this scheme can be classified by which markers are present on the recombinant chromosome, and whether the lethal is present. This system makes it easy to assign the lethal to one interval between markers. Further recombination mapping may be undertaken, concentrating on that interval, to yield a more precise map position.

Genetic mapping in *Drosophila* has rarely used microsatellites and restriction fragment length polymorphism (RFLP) markers, since the availability of many visible markers and chromosome rearrangements permits the easy localization of genes to a molecularly relevant precision. However, microsatellites and RFLPs are of central importance for recombination mapping in genetically less tractable insects, such as mosquitoes.

4   POLYTENE CHROMOSOMES AND CYTOGENETIC MAPPING

4.1   POLYTENE CHROMOSOMES

Polytene chromosomes are found in many species of Diptera including *Drosophila*, where the easiest to work with are found in the salivary glands of the third instar larva. The polytene chromosomes of *Drosophila* represent the euchromatic portion of the genome, in a highly amplified condition, in which each chromatid is replicated approximately a thousandfold and each copy precisely aligned, resulting in a highly reproducible transverse banding pattern along the chromosomes (Figure 4). The pericentromeric heterochromatin and the heterochromatic Y chromosome remain unpolytenized and are located together with the centromeres in the chromocenter, to which each of the chromosome arms is attached. Thus the polytene chromosome complement is roughly star shaped; the major chromosome arms are of approximately equal length and radiate from the chromocenter. The arm corresponding to chromosome 4 is about 5% of the length of the others. This arrangement is shown in Figure 4, with a diagrammatic interpretation in the inset.

The importance of polytene chromosomes for genetic research was realized in the 1930s, and highly detailed cytogenetic maps were created, based on the banding pattern of the chromosomes. The map used today is basically the one devised by Calvin Bridges in the mid-1930s; it provides a reference system of map locations understandable by all researchers possessing a copy of the map. Each major arm is divided into 20 divisions numbered 1–20 (X chromosome), 21–60 (second chromosome), and 61–100 (third chromosome). The tiny fourth chromosome is allocated two divisions, 101–102. The telomeric divisions are 1, 21, 60, 61, 100, and 102. Each numbered division is further broken down into six subdivisions, lettered A–F. This basic map was further refined by Bridges and his son to include a number for each band on the map. Thus each band of the polytene chromosome complement can be unambiguously identified (at least in principle). Reproductions of these maps can be found in Lindsley and Zimm's *The Genome of* Drosophila.

Genes identified by mutation can be accurately mapped to the polytene chromosomes using a huge collection of chromosome rearrangements, such as deficiencies, inversions, duplications, transpositions, and translocations. At the simplest level, this involves the determination of whether a mutant allele is located within a given deficiency (see Section 4.2.1).

The resolution afforded by the use of polytene chromosomes is very high; features of 5–100 kb can be recognized on the map by

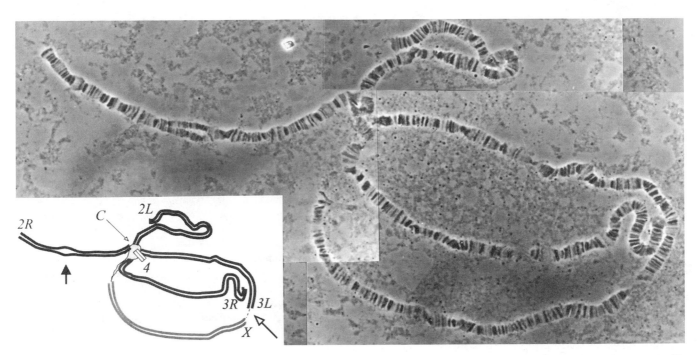

Figure 4. The polytene chromosomes of *Drosophila melanogaster*: phase contrast micrograph of a preparation of *D. melanogaster* polytene chromosomes. The unstained specimen has been fixed in 17% lactic acid, 50% acetic acid. Each chromosome arm extends from the chromocenter. *Inset*: Diagrammatic representation of the photograph; each chromosome arm is labeled. The X chromosome is shaded in light gray, chromosome 2 in dark gray, and chromosome 3 in black, with chromosome 4 in white. *C*, chromocenter; *X*, X chromosome; *2L, 2R*, left and right arms of chromosome 2; *3L, 3R*, left and right arms of chromosome 3; *4*, chromosome 4. The solid arrow points to a region of asynapsis in arm 2L. The open arrow points to ectopic pairing between the telomeres of 3L and X.

light microscopic analysis, which can rapidly provide positional information at a precision useful by molecular cloning standards. In situ hybridization can map cloned segments of DNA to single-band resolution, several orders of magnitude better than is easily possible with metaphase chromosomes, and this technique is consequently invaluable for many gene cloning strategies and physical genome mapping projects.

4.2   CHROMOSOME ABERRATIONS

A great many chromosome aberrations, affecting all chromosomes, have been recovered and analyzed. The gene and chromosomal aberration data sets of the *Drosophila* database FlyBase contain details of 10,863 chromosomal aberrations. These include the following:

1. *Deficiencies* (Figure 5a). "Deficiency" is the term used in *Drosophila* research for a chromosomal deletion. In a deficiency chromosome, a segment of the chromosome is deleted. In practice, the largest deficiency of practical utility is about one polytene map division. Flies carrying deficiencies larger than this have severely reduced fitness and viability. Furthermore, there are a number of haploinsufficient loci in the genome, which are inviable when hemizygous (e.g., when uncovered by a deficiency). Unless the specialized telomeric sequences have been replaced, telomeric deficiencies can be unstable.

2. *Inversions* (Figure 5b). Inversion chromosomes are those in which a section has been inverted with respect to the wild-type gene order. Inversion and normal chromosomes synapse to form inversion loops, in both meiotic and polytene chromosomes. Chromosomes bearing inversions suppress recombi-

nation within the inverted region when heterozygous with wild-type sequence chromosomes. Balancer chromosomes are chromosomes carrying several inversions, together with suitable dominant and recessive markers, and are used to effectively suppress recombination along the whole chromosome. The use of balancer chromosomes is described in Section 3. Chromosome inversions that contain the centromere

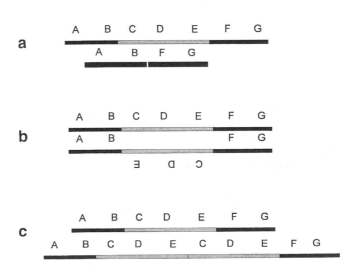

Figure 5. Chromosome rearrangements. In each panel, a standard order chromosome bearing loci A–G appears above the rearranged chromosome. The region involved in the rearrangement is shaded. (a) Deficiency: loci C, D, and E have been deleted. (b) Inversion: loci C, D, and E have been inverted. (c) Duplication: loci D, D, and E have been duplicated in tandem.

are referred to as pericentric inversions; those that do not are paracentric inversions.

3. *Duplications* (Figure 5c). These chromosomes carry an additional segment derived from a homologous or other chromosome.

4. *Translocations.* At least two chromosome breaks, and the reciprocal interchange of chromosome sections between two or more nonhomologous chromosomes, are involved in a translocation. More than two chromosomes can be rearranged.

5. *Transpositions.* Essentially, a transposition is the insertion of a chromosome section into another chromosome, involving three breaks.

6. *Compound chromosomes.* These chromosomes are derived from the normal chromosome complement by breakage and fusion of chromosome arms. Compound X chromosomes (two X chromosomes sharing a single centromere) affect the pattern of inheritance of normal X and Y chromosomes: in flies of normal karyotype, Y chromosomes are patroclinous (inherited from the father), while in a typical compound X stock, they are matroclinous (inherited from the mother). Compound chromosomes involving the attachment of X and Y chromosomes, whole autosome fusions, and whole autosome arm fusions can be constructed.

The genetic technologies for constructing novel aberrations and compound chromosomes are well established for *D. melanogaster.* Translocations can be combined to created segmental aneuploids, synthetic deficiencies, or duplications, when necessitated by the lack of conventional deficiencies for a particular chromosome region. Recombination between inversion chromosomes can also be used to effect synthetic deficiencies and duplications. Complete discussions of the structure, construction, and experimental use chromosome rearrangements can be found in the literature.

The principle of deficiency mapping a recessive mutant allele is to test whether a deficiency chromosome "uncovers" the mutation by examining the phenotype of a stock carrying both the deficiency chromosome and the chromosome bearing the recessive allele. If the mutant phenotype is expressed, the locus must lie within the region deleted in the deficiency chromosome (Figure 6). In addition to simple deficiency chromosomes, the *Drosophila* researcher can create "synthetic deficiencies" by combining other, often more complex, rearrangements. If a recessive allele's homozygous phenotype is identical to its hemizygous phenotype, it is likely to be an amorphic (null) allele, while if the hemizygous phenotype is stronger, it is likely to be a hypomorphic allele (see Section 3.1.2).

In some cases, a chromosome rearrangement inactivates or otherwise affects the expression of a gene, hence can be used to pinpoint that gene's location both cytologically and molecularly.

The polytene chromosome map location of a gene is often a vital

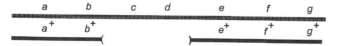

Figure 6. The principle of deficiency mapping. A chromosome bearing recessive mutations at several loci (*a–g*) is heterozygous with a deficiency chromosome. The region deleted is indicated by parentheses. In this case mutants at loci *c* and *d* will be uncovered by the deficiency and will display a mutant phenotype. Conversely, loci *a*, *b*, *e*, *f*, and *g* are heterozygous with wild-type alleles on the inversion chromosome and do not display mutant phenotype.

prelude to both further genetic analysis and its molecular cloning, as discussed further in Section 2.4.

5 PHYSICAL MAPS

Many chromosome walks have been undertaken in *Drosophila* ranging from 50 kb to more than 1000 kb, using cosmid and phage lambda cloning techniques. These are dispersed throughout the genome. However, from the mid-1980s, the technologies for large-scale physical genome mapping were developed, such as clone fingerprinting, and novel cloning methods, such as yeast artificial chromosome (YAC) and P1 vectors. These technical advances have been used for map construction in many organisms, including *D. melanogaster.*

At the time of writing, several physical maps are under assembly, using vectors featuring different characteristics and cloning capacities. Of these, the map assembled using YAC clones has been essentially completed, while those using cosmid and P1 vectors are still under construction. These physical maps comprise overlapping contiguous segments of cloned DNA, termed contigs. An important feature of the physical maps is that they are tied to the polytene chromosome map, and, in the cases of the P1 and cosmid maps, to the recombination map.

5.1 YACs

A YAC library has been analyzed by mapping each clone by in situ hybridization to polytene chromosomes. This approach has yielded a map that covers 90% of the euchromatic genome. Since this approach relies solely on hybridization to polytene chromosomes, contigs are inferred rather than directly demonstrated as molecularly overlapping clones. Additionally, the map covers only the euchromatic parts of the genome, since the in situ hybridization analyses were carried out using polytene chromosomes.

The insert sizes of the YAC library used are in the range 120–500 kb. This is moderate for YAC clones (which can be as large as 1 Mb), but it means that the number of chimeric clones in the library is small. Nevertheless, even these clones hybridize to an average of 10 bands on the polytene chromosomes. The advantage of YAC cloning is that fewer clones need to be analyzed for the map, but these clones are comparatively hard to propagate and analyze.

5.2 COSMIDS

The cosmid-based genome map is being assembled by a consortium of laboratories in Europe. The polytene chromosomes are the starting point in the strategy, with probes specific to each polytene map division prepared by chromosome microdissection. These probes are used to identify clones derived from each division, which are then fingerprinted and assembled into overlapping arrays (i.e., contigs) by computer analysis. Further stages of analysis include in situ hybridization to polytene chromosomes and the determination of sequence-tagged sites (STSs) at regular intervals of approximately 40 kb along the physical map. STSs are short single-copy sequences, of known map location. They can be used to recover corresponding DNA sequences from genomic DNA or genomic libraries, by PCR, and thus make the map independent of the clones used to generate it.

Cosmid clones are easy to prepare and work with, not only because of the moderate insert size (≈ 40 kb) but because they are propagated in *E. coli.* Their main disadvantage is their small cloning capacity, which necessitates the analysis of a larger number to cover the genome.

5.3   P1 Clones

Libraries constructed using P1 vectors, also known as pacmids (by analogy with cosmids) or P1 artificial chromosomes (PACs), have advantages over both cosmid and YAC libraries. The American *Drosophila* genome project is engaged in mapping the *Drosophila* genome with P1 clones, using a strategy of STS content mapping. In this approach, P1 clones are screened by PCR to identify those containing each STS. STSs may be derived from a number of sources, including sequence database entries, and sequenced termini of P1 and cosmid clones. This project, which is a model for genome map assembly using STSs and PCR, is highly amenable to automation and general application. Because many of the STSs correspond to cloned and sequenced genes, it is easy to integrate genetic information with the physical map.

The P1 cloning system is in some ways a compromise between the YAC and cosmid systems: insert sizes are large (80–100 kb), while the clones themselves are propagated as plasmids in *E. coli* and are thus relatively simple to work with, both during mapping and in subsequent use.

5.4   Overview

Within most genomic libraries, certain genomic sequences may be overrepresented, underrepresented, or even absent. In some cases such imbalances are due to the nature of the genomic DNA, either in terms of its genetic content, distribution of restriction sites, or presence of repetitive DNA, which can permit rearrangement by recombination. In many cases these effects differ between cloning systems. In practice, mapping with three different cloning vector types and three different mapping strategies will result in a more representative combined map resource. Ultimately, all three maps can be related by their correlation with the polytene chromosome map.

6   *DROSOPHILA* AS A MODEL FOR DISEASE VECTORS

Many diseases of humans and animals are transmitted by hematophagous insects, frequently Diptera. Additionally, some insects are pests in their own right. These insects often share many genetic and biological features with *Drosophila,* but they can be difficult to maintain in the laboratory because of preferred habitat or source of blood meals. Even species that are colonized relatively easily may be difficult to use for genetic experiments. *Drosophila* has been used as a model for many of these dipteran disease vectors. Of these, some may benefit more than others from the crossover of molecular biological technology developed for *Drosophila,* such as *Simulium* species and anopheline mosquitoes, which have excellent polytene chromosomes.

6.1   Cloning Genes

The DNA sequence of genes identified by mutation in *Drosophila* can be used to devise strategies by which homologous genes may be cloned in other dipteran insects. For example, the *Anopheles gambiae* homologue of the *white* gene has been cloned by PCR amplification on the basis of sequence conservation, with the practical objective of acquiring a useful marker for genetic transformation, using *white* mutants of *A. gambiae.*

6.2   Transfer of Methodology

6.2.1 Germ Line Transformation

P-element-mediated germ line transformation has become the paradigm for the manipulation of insect genomes. Attempts have been made to use *Drosophila P*-element vector systems in other dipterans, such as mosquitoes. Observed integrations were at low frequency and due to random integration rather than *P*-element activity, presumably because of the absence of the *Drosophila* nuclear encoded components of the *P*-element transposition machinery. Much current research activity is aimed at discovery of transposons that may be utilized for germ line transformation of medically important insects. Beyond the obvious laboratory uses for such systems, it is possible that transposons might be used to drive certain genetic characteristics through wild populations. One character of interest in this respect is the inability to support development of malaria parasites.

6.2.2 Genome Mapping

Genetic analysis of disease vectors is often limited by the organisms' intractability for these experiments. Great advances have recently been made in the genetic mapping of insects, using microsatellites and RFLPs. A good example of this is the malaria vector *Anopheles gambiae,* for which a high resolution map of microsatellite markers is under construction. In this case a direct transfer of technology—that of polytene chromosome microdissection—has also been made, to achieve a low resolution physical map of the target species.

7   PROSPECTS

To some extent, *Drosophila* is a late starter with respect to the development of molecular physical maps. This state of affairs is undoubtedly due to the availability, since the 1930s, of an excellent physical map, the polytene chromosome map. It is an enduring legacy to the mapping talents of Calvin Bridges that his polytene chromosome map has continued to be relevant to researchers undertaking detailed molecular genetic studies undreamt of by the early geneticists. As a result, however, of the recent work concerning molecular mapping, a complete, multilevel molecular physical map will soon be constructed, and attention is now turning toward large-scale sequencing of the *Drosophila melanogaster* genome. We can look forward to the day when at least the euchromatic regions of the *Drosophila* genome have been cloned and sequenced, bringing a wealth of genetic, cytogenetic, and molecular data into one unified map.

See also Chromatin Formation and Structure; Drosophila Mating, Genetic Basis of; Genetics; Homeodomain Proteins; Hybridization for Sequencing of DNA; Recombination, Molecular Biology of; Sequence Analysis.

Bibliography

Ashburner, M. (1989) Drosophila, *A Laboratory Handbook.* Cold Spring Harbor Laboratory Press, Cold Spring Harbor, NY.

FlyBase (1994). The *Drosophila* Genetic Database. *Nucleic Acids Res.* 22:3456–3458.

Hartl, D. L., Ajioka, J. W., Cai, H., Lohe, A., Lozovskaya, E. R., Smoller, D. A., and Duncan, I. W. (1992) Towards a *Drosophila* genome map. *Trends Genet.* 8:70–75.

Kafatos, F. C., Louis, C., Savakis, D. M., Glover, M., Ashburner, A., Link, J., Siden-Kiamos, I., and Saunders, R. D. C. (1991) Integrated maps of the *Drosophila* genome: Progress and prospects. *Trends Genet.* 7:155–161.

Lindsley, D. L., and Zimm, G. G. (1992) *The Genome of* Drosophila melanogaster. Academic Press, San Diego, CA.

DROSOPHILA MATING, GENETIC BASIS OF

Jeffrey C. Hall

1 **Introduction**

2 **Mutants and Genes**
   2.1 Sensory Transduction Mutants
   2.2 Neural Membrane (and Related) Gene Products
   2.3 Neurochemical Mutants and Gene Manipulations
   2.4 Transcription Factors
   2.5 Known and Putative RNA-Processing Factors
   2.6 Small-Peptide-Related Genes and Manipulations Thereof
   2.7 Pheromones

3 **Perspectives**

Key Words

Biological Rhythms Oscillating "parameters" involving physiology, biochemistry, and behavior (the latter being an element of this review). The periodicities of such rhythms have a wide range, the most well known of which is "circadian" ($\approx$ 24 h periods, even in constant conditions). Some rhythms are short-term, such as that involving the *Drosophila* male's courtship song.

Ionic Channels Several of these key proteins (operating in membranes of excitable cells) were identified molecularly in *Drosophila* via physiological—including in certain instances courtship–behavioral—genetics. In particular, potassium channel genes and polypeptides were discovered, starting with genetics, permitting them to be widely manipulated in molecular terms.

Neuropeptides Small molecules functioning as "modulatory factors," within and without the nervous systems of many organisms. Genetic studies of genes encoding such molecules are in a primitive state, although manipulation of certain peptide-encoding clones, particularly in the context of reproductive studies, are on the upswing.

Neurotransmitters Against a background of near-ubiquity of the form and the function of these small intercellular signaling molecules, genetic and molecular experiments involving genes encoding neurotransmitter-metabolizing enzymes have led to interesting disruptions of *Drosophila* courtship, among the

various areas of physiology and behavior in which these manipulations have been performed.

Pheromones Descriptions and experimental applications of such interanimal communication molecules in *Drosophila* have particularly concerned cuticular hydrocarbons, which appear to trigger and modulate interactions between males and females. Most of these substances are rather nonvolatile; but there are hints that odorants having longer range activity play a role in communication between flies of the two sexes. Pheromonal studies are being augmented by the application of mutants involved in the production and processing of these molecules or in the flies' responses to them. In addition, certain molecular–genetic forays are an emerging a part of the courtship pheromone story.

Reverse Genetics Regarding courtship neurogenetics and molecular neurobiology, most of the genes were originally identified by mutations. This permitted the "forward genetic" isolation of such genetic loci as molecular entities (this being one of *Drosophila*'s fortes). Increasing numbers of genes whose expression and informational content impact on the nervous system are originally known from the "pure" molecular angle. One aim would be to manipulate or damage the expression or function of the "wild-type" form of such a "neuro gene," and then assess the phenotypic consequences of such engineered (or perhaps luckily encountered) mutants. It is not uncommon—in such reverse genetics of (at least) *Drosophila*—for even the "gene knockout" in question to cause little in the way of overt biological defects. Thus, behavioral, including courtship-related, analyses of such variants have increasing importance.

RNA Splicing Factors Such proteins influence "everything." In *Drosophila* behavior, certain behaviorally relevant mutations are in genes that encode RNA-processing factors; a subset of such variants was discovered in courtship contexts. Among the most conspicuously pertinent genes and products are those positioned at quite "central control points" with regard to the mechanisms by which a *Drosophila* embryo develops into a male, a female, or (anomalously) something else. In turn, a subset of the biologically significant mutant applications and gene manipulations has involved assessing the degree to which a variant fly behaves in a male- and/or femalelike manner in various courtship contexts.

Sensory Transduction Processes operating "proximally" to the initial perception of an incoming stimulus have been particularly analyzable in *Drosophila,* especially with regard to photo-transduction mutations and cloned factors defined by them. Several other gene products expressed in the eye are now being manipulated by the aforementioned reverse genetic strategies. An "internally" disrupted fly, expressing some kind of sensory-impairing mutation, is an especially valuable tool to apply in courtship contexts (as opposed to sledgehammer, or meticulously difficult, approaches that can accompany outside interventions or environmental manipulations). However, an intriguing problem associated with sensory mutant applications is that several of the relevant "sensory genes" have turned out to be expressed (in their normal form) much more widely than in the "input organ" alone; which could mean that the courtship defects uncovered have in part been misinterpreted with respect to the precise "internal etiology" of the male's or female's reproductive abnormalities.

Transposons   Mobile genetic elements that have revolutionized *Drosophila* genetics, beginning with population biology and genetic characterizations of transposable elements; followed by the cloning and subsequent manipulations of these transposons. One of the singular accomplishments in this area involved manipulation of the relevant "vectors" such that germ line transformants could be generated in routine manner—followed by a host of further manipulations of the transposon-derived vectors, aimed at asking a wide array of biological questions. Certain of the "powerful" ways in which the animal has been interrogated by virtue of these gene manipulations, and by bioassaying their effects in transgenic strains, are nicely exemplified by courtship-related experiments.

Visual System Ganglia   Among the most extensive characterizations of regions within dipteran nervous systems are those that involve "visual processing." *Drosophila* has made substantial contributions here, in regard to the several mutations that disrupt optic "lobe" development in informative ways—both from the standpoint of how the assembly of such ganglia are genetically controlled and in terms of impinging on the neural substrates underlying complex responses to moving visual stimuli. Certain of the experimental questions asked attempt to elucidate the courtship significance of female movements, including relatively long-term modulations thereof, with regard to the male's "desire" to tract her.

Glossary

Ace   Codes for acetylcholinesterase, which breaks down the neurotransmitter acetylcholine. Loss of function in the haplo-X portions of sexually dimorphic (mosaic) brains still allows male behavior in spite of neurochemical deficit and anatomical abnormalities of the male tissue; causes courtship song rhythm abnormality when expressed in thoracic nervous system.

cac   Putatively encodes calcium channel; abnormal courtship song (defective "pulses").

CaM kinase   Dominant-negative transgenic males or females impaired in experience-dependent features of courtship.

Cha   Acetylcholine synthesis; in conditional mutants, courtship "turned off" well before general locomotion ceases.

cpo   RNA-processing; sluggish, including in courtship context.

Ddc   Serotonin and dopamine synthesis; learning-defective? (problematical in "shock–odor" paradigms; conditioned-courtship impairment still stands).

dnc   Encodes cAMP phosphodiesterase; learning and memory defective, including in courtship contexts.

dsx   A gene of the sex determination cascade, which encodes male- and female-specific transcription factors as a result of alternative RNA splicing, under the control of the functions coded for by the *tra* and *tra-2* genes. XX and XY *dsx* mutants are each transformed into intersexes; courtship of the XY version of this mutant is mildly impaired, although certain sex-specific neural functions remain intact. Loss of *dsx* function leads to sex-pheromone anomalies.

e   Anomalous levels of amine-containing substances; visual, locomotor, and courtship defects.

eag   Potassium channel subunit (which polypeptide is a substrate for CaM kinase); impaired in conditioned courtship.

Est-6   A pheromone-metabolizing esterase gene (in part); influences postmating behavior of female.

fru   Putatively involved in sex determination; bisexual courtship and behavioral sterility.

iav   Octopamine-depleted; sluggish; impaired in general courtship vigor and in conditioned courtship.

kété   Sex–pheromonal anomalies; female sterility.

nap   No-action-potential-specific allele of *mle* gene (involved in dosage compensation). Controls, in part, RNA processing of *para* gene's primary transcripts; in males expressing this "specific" allele (which is conditional), turnoff of nerve conduction stops song clock

nbA   Nightblindness and aberrant electroretinogram; seemingly mutated at *cac* locus.

nerd   Mediocre male mating success; sex–pheromonal anomalies.

Ngbo   Sex–pheromonal anomalies.

nonA   Putative RNA-processing factor; visual defects and courtship song ones for some variants involving the locus.

norpA   Phospholipase C-encoding phototransduction factor, which, however, is expressed well beyond the visual system; blind; courtship-abnormal (as mutant male or female); mild circadian rhythm defect.

omb   Encodes gene regulatory factor; optomotor blind (including in courtship context); or abnormal in external appearance; or developmental "neural-lethal" (depending on the mutant allele).

para   Sodium channel-encoding; rapid heat-induced paralysis and/or developmental lethality (depending on the mutant allele). "Olfaction-specific" defects caused by two special alleles; in males expressing a conditional allele, turn off of nerve conduction stops song clock.

per   Encodes putative negatively acting transcription factor, which possesses at least the circadian clock function associated with the gene. Mutations also disrupt or eliminate short-term defects in courtship song oscillations.

rut   Adenylate cyclase encoding; learning and memory defects, including in courtship contexts.

Sh   Encodes potassium channel subunits; learning-impaired, including in courtship context.

s-p   Encodes "Sex Peptide," transferred from male to female during mating and influences her reproductive behavior thereafter.

Sxl   RNA processing at early "step" in sex determination "hierarchy"; in the face of mutations permitting adult viability, transformation of sexual characters (including behavioral, pheromonal).

tra   RNA processing (influences *dsx*); target of *Sxl* action; transforms XX fly into one that looks and behaves as a male.

tra-2   RNA processing (influences *dsx*); transformations similar to those caused by *tra;* also, a conditionally mutant *tra-2* allele permits male behaviors to be turned on in XX adults.

w   Tryptophan/guanine transmembrane transporter protein; poor visual acuity, including in courtship contexts (owing to lack of "screening pigment" from mutant eyes). Ectopic expression induces males to behave in a homosexual-like manner.

Reproductive behavior in *Drosophila* is among the more important of these fruit flies' actions, both from the organism's perspective and in terms of how its nervous system can be investigated. That courtship and mating are "adaptive" need not be argued. These behaviors are also rich with complexity; yet observing and recording the actions of, and interactions among, courting flies is a relatively straightforward procedure (albeit sometimes deceptively so).

One feature of courtship's complex richness is the involvement of several sensory modalities and many motor actions. This serves to introduce reproductive behavioral genetics, a part of which has involved the application of sensory and motor mutants, to "dissect" the significance of a given observable courtship component, or an inter-fly communication cue that could be hypothesized to trigger one or more steps of the courtship sequence. Other types of mutants, involving defects that can be deemed more sophisticated behavioral problems than mere faulty "reflex" responses, have become courtship mutants in addition to whatever they may have been initially. Some notable examples of such intersections between "higher" behavior and courtship are in the area of *Drosophila* mnemogenetics. Here, certain of the most behaviorally interesting ways that learning and memory have been studied genetically involve the sharply defined courtship abnormalities exhibited by conditioning mutants. Such findings have also helped us scrap the naive view that *Drosophila*'s male–female interactions are merely "hard wired."

Thus, the combination of reproductive and general behavior genetics has not only deepened our understanding of how courtship actions are initiated and sustained, but also has broadened the perspective from which one views the relevant gene actions. These actions mean, for this review, more than "effects of mutants," regardless of whether such phenotypic assessments have delved beneath the fly's "surface," hence into its nervous system. In addition to studies of such mutational effects, there have been increasingly intense analyses of how the mutationally defined genes are expressed and what they encode. These molecularly based studies have also consisted of more than "characterizing" clones and gene products. In addition, the molecular neurogenetics of reproduction nicely exemplifies certain state-of-the art manipulations of "neuro genes" and of the cells and tissues in which they are expressed.

A further molecular bonus is provided by gene descriptions and manipulations performed with not only courtship and mating but also evolution in mind. *Drosophila*'s reproductive biology connects with a large number of fascinating and often exquisitely well-defined differences among species, which is to be expected, if, for example, a male fly should be able to recognize that a potential courtship object is of his species, and if it is similarly useful for the female to determine whether her suitor is a suitably intraspecific one before she allows mating to occur. It follows, perhaps, that some of the more beguiling courtship-related genes would exhibit interspecific differences in informational content and expression. So the combination of an evolutionary perspective, genetic dissection of courtship, and isolation of the genes defined by the pertinent mutations (in one species) stimulates the investigator to tap into naturally occurring, molecularly definable interspecific variations. The un-

covering of such information can then be turned back in the direction of reproduction, by bioassaying the courtship–behavioral significance of such sequence divergences.

"Courtship-specific" mutants comprise a key category of behavioral and neural mutants, which one would expect to play a prominent role in the genetic approach to *Drosophila* reproduction. Yet, even in cases of reproductive behavioral variants that were isolated by encountering a mating defect, or by systematically searching for mutants of this sort, the genetic changes in question turned out to have more widespread immediate effects and overall biological impact. They thus revealed genes possessing broader than expected significance in terms of how the fly's nervous system is constructed and operated. One such example involves a set of courtship–behavioral mutants that has led to an expanded view of a certain developmental–genetic area of inquiry, which established the "sex determination hierarchy" in *Drosophila*. This sequence of gene actions is now believed to contain more factors, paths, and branchpoints than had been appreciated. With the relevant horizon-expanding discoveries having come from analysis of the animal's inner workings and of the fly's behavior, the study of the biological significance of sex-determining factors is increasingly being turned toward the analysis of courtship and mating.

1   INTRODUCTION

Reproductive behavior in *Drosophila* has long been a heavily genetic subject. In fact, some of the first behavioral genetic experiments (ever) involved *Drosophila* courtship and mating, and these genetic studies in turn were the first behavioral ones involving this organism. Since that time (80 years ago), a large number of further genetic variants have been applied to analyze the fruit fly's reproduction, including to some extent the neural substrates of the relevant behaviors. Thus, reproductive "behavior genetics" became in part "neurogenetics."

Over the past 10 years or so, a biochemical and molecular genetic component has arisen within this area of inquiry. Thus, neurochemical correlates of behavioral defects exhibited by "courtship mutants" were uncovered, or mutants initially identified neurochemically were applied in courtship experiments. Moving closer to the level of gene action, reproductive molecular genetic investigations have come to the fore, with regard to (1) the isolation, as cloned genes, of several genetic loci defined by mutations that cause courtship abnormalities and (2) molecular manipulations of factors that were then bioassayed in part by studying the effects of such DNA alterations on reproductive behavior.

Findings at the intersections of behavioral, genetic, and molecular areas have recently been reviewed (more than once: see Bibliography); these summaries have usually been organized around the various behavioral and neurobiological features of *Drosophila*'s reproduction. Instead of following that kind of outline, this article discusses courtship and mating, via application of mutants, biochemical assays, and manipulation of cloned nucleic acids, from the standpoint of various kinds of courtship-related gene products and biochemical phenomena.

2   MUTANTS AND GENES

2.1   Sensory Transduction Mutants

Most of the mutations in the area of sensory transduction have to do with *Drosophila*'s adult vision. In turn, many of the genes defined by blinding or visually impairing mutations have been cloned and sequenced (some such genes were originally defined molecularly, followed in a few instances by assessing their significance via reverse genetics). The only molecularly defined phototransduction mutant studied for courtship defects is *no-receptor-potential-A (norpA)*. Mutant males have difficulty tracking a moving female within a given moment or minute of courtship, and their mating-initiation kinetics are considerably slower than normal. This male behavioral defect is worse than in cases of olfactorily defective mutants (see Section 2.7), implying that visually mediated sensory courtship cues could be more important to the male than are female-produced pheromones.

When groups of females and males each express *norpA,* the mating-initiation latency is actually improved—compared to a situation in which only the males are genetically blind in this manner; it is as if females might use visual cues to "fend off" male advances. This idea may be a reflection of the normal (and phylogenetically universal?) courtship phenomenon of female coyness, whereby she may wish to avoid immediately mating. In so doing, she gives herself time to assess whether the male is the "right one"—a member of her species, and also a robust behaver.

norpA encodes a phospholipase C (PLC) in one of at least two forms of this intracellular signaling related enzyme in *Drosophila.* Whereas some have imagined that the *norpA*-encoded PLC is "eye specific," this has been shown not to be the case: the gene's product is found in a variety of regions in the central and peripheral nervous systems. It has also been shown that certain olfactory responses are defective in *norpA* mutants, which would not be surprising: the gene encodes not "visionase" but, instead, a factor that can easily be thought of as involved in a variety of different excitable-cell types. Thus, the courtship defects just noted may not have only a visual, or even a solely sensory, etiology. In this regard (i.e., that *norpA* mutants could be "CNS-impaired"), it is notable that the centrally controlled biological clock in the adult fly's brain runs at a faster (< 24 h) pace than normal.

2.2 Neural Membrane (and Related) Gene Products

One of the great success stories in *Drosophila* neurogenetics was its identification, via hyperexcitable mutants and forward-genetic cloning of the relevant genetic loci, of potassium channel encoding genes. The mutants that defined these genes in the first place have been shown to be more than mere pathophysiological variants, via courtship studies: Thus, *Shaker* and *ether-à-go-go* mutants are, as males, impaired in "conditioned courtship." This designation refers to suppressive effects exerted on males in two courtship situations: with mated females or very young-adult males (the former tend to inhibit courtship, in part by emanating an antiaphrodisiac; the latter elicit remarkably high levels of courtship, as performed by older males, which chemically stimulated behavior wanes by the time a male is 1–2 days old). A genetically normal male's courtship declines when he is in the presence of either kind of unreceptive fly (this is defective in *eag,* in mated-female experiments); and there are depressive aftereffects, meaning low courtship vigor when the "experienced" male is placed with a subsequent female or immature male (this is defective in *Sh,* again in tests involving mated females). That modulation of potassium channel function is a part of the cellular "learning machinery" is well appreciated in general; this, then, would include *Drosophila* learning, on the one hand, and the flies' courtship behaviors on the other.

The hypoexcitable mutants in *Drosophila* include those for which paralysis can be induced by high temperature treatments. Among the most salient of the genes in this area are *paralytic,* which encodes the major sodium channel in *Drosophila;* and *no-action-potential,* which encodes an RNA helicase (see Section 2.5.4 for the latter). In courtship studies, *para^TS1*, which passes out immediately when the temperature is raised and recovers very soon after it is lowered, was used to "turn off" nerve conduction in courting males. When this treatment was effected, the biological clock that underlies a certain high frequency rhythm accompanying the male's courtship song was stopped; the clock began to "run" again as soon as the male was returned to a cooler temperature. [The "song" is produced by unilateral wing vibrations generated by the male as he follows a female or orients toward her (Figure 1); the rhythm concerns systematic oscillations of rate at which the male generates "tone pulses," which are the most salient feature of his acoustical output when he courts.]

Another mutant, whose effects ultimately impinge on a membrane-related function, is *shibire^TS*. The *shi* gene encodes a dynamin protein involved in synaptic vesicle recycling (specifically, in the endocytotic end of this process). Compared to *para^TS1*, *shi^TS* flies pass out and recover with somewhat slower kinetics (when the temperature is raised and lowered, respectively). Nevertheless, the latter gene was useful as a tool to assess the effects on male courtship behavior on turning off, or on, the female's movements. In the short term, such movements seem to stimulate the male's ability (or willingness?) to follow her. This seems to jibe with the courtship decrements exhibited by *norpA* mutants (and other visual ones: see Section 2.4.1), though perhaps her movements promote the male's detection of chemosensory as well as visually stimulating cues. A curious feature of the *shi^TS* experiments was that the relatively feeble (but well above zero) courtship performance observed when a male was in the presence of a heated, immobile *shi^TS* improved approximately threefold after the temperature was lowered and she regained mobility; whereas in the reciprocal kind of temperature manipulation, turning off female movements led to a quick 50-fold plummeting of male courtship vigor.

A further and not necessarily expected feature of the female movement story was the finding that longer term slowdowns in such movements are correlated with improved mating propensities. The wild-type female becomes relatively immobile several minutes after the initiation of courtship, as if she is a better target for male mating attempts. One of the cues stimulating a female to "slow down" seems to be an olfactory one emanating from the courting male. This was inferred because of the effects of some intriguing mutations at the *para* locus, that is, olfactorily defective mutants. These were originally isolated in a screen for odor response defects per se, or in learning experiments involving *Drosophila's* ability to associate an (artificial) odorant with a noxious (unconditioned) stimulus and to modify its behavior accordingly (see Sections 2.3.2 and 2.3.3). The "*olfactory-D*" and "*smellblind*" mutations so isolated turned out to be special *para* alleles, which do not under ordinary (genetic) circumstances lead to heat-induced paralysis. At mild temperatures, either mutant seems unable to smell any kind of odorant; indeed, as males, they do not respond to the female-produced aphrodisiac pheromone. In any case, it was found that a *para^olfD* or *para^sbl* female tends to keep moving, inappropriately, during later stages of courtship; hence she mates relatively poorly in short-term (< 1 h) experiments (though either mutant female type mates fairly well when left with males for several days).

Figure 1. Elements of courtship behavior in *Drosophila melanogaster,* indicating part of the insect's reproductive complexity. *Left:* A male (left) orienting toward a female and extending one wing, preparatory to vibrating it in order to produce his courtship song (see Figure 2). *Right:* A copulating pair. Higher resolution views of these flies would reveal their external sexual dimorphisms, which include obvious ones with regard to the external genitalia and the male's sex combs (on his anteriormost legs) plus darkened abdominal pigmentation. Less obvious are sex-specific elements of sensory organ morphologies, certain of which may be involved in the reception of courtship-related cues. Internally, there is also an increasing experimental appreciation of sexual dimorphisms within both afferent pathways aimed at the central nervous system, and in certain of the central ganglia of males and females. These features of sex-specific anatomy, and other as-yet undiscovered ones, are likely to be the substrates for components of the behavioral dimorphisms that are such salient elements of the courtship sequence. The "singing" (and wing display) stage depicted on the left is preceded by the male orienting toward the female, tapping her abdomen with his forelegs, and following her when she moves. Much of the time spent orienting or following includes singing. After at least a few seconds, up to 2–5 minutes, of these preliminaries, the male extends his proboscis and contacts the female's genitalia (known as "licking"); this usually proceeds directly to the first copulation attempt. If it succeeds, the duration of mating is about 20 minutes. If it does not, the male cycles back to an earlier courtship stage (usually following/singing) and continues once more through the remainder of the sequence.

When *para^{olfD}* or *para^{sbl}* is expressed in males only, in a courtship context, the moment-to-moment behavior of the flies is minimally impaired; and mating-initiation latencies are not stretched out very much (e.g., they are less anomalously long in comparison to tests of males expressing a given visual mutation, as noted earlier).

Among the few mutants isolated in *Drosophila* as courtship song defective—meaning blatantly abnormal in this character (as opposed to the more subtle abnormalities that can occur with respect to the song rhythm)—is the *cacophony* variant (Figure 2). This *cac* mutation defines a locus that was also identified independently: by embryonic lethal mutations and also by a certain kind of visual physiology mutant called *night-blind-A*. [Note: *cac* males or females exhibit only a subtle visual defect; and *nbA* males are barely if at all abnormal in the quantitatively determined features of their song pulses.] This collection of genetically related mutations is mentioned because of the strong working hypothesis that all three types of variant are mutated in a calcium channel encoding gene (one of at least two such loci in *Drosophila,* with the other one having been mutated, so far, only in regard to lethal genetic variants).

One of the all-out classical behavioral mutants in *Drosophila* is *white,* referring to this mutant's lack of eye pigment. Other *w* mutations cause a graded series of reductions in eye pigment levels. The *w*-mutant males have long been known to court subnormally in the short term, at least in part because of their mediocre "visual acuity," caused by the lack of ommatidial "screening pigment" in the compound eye. These abnormalities are ameliorated, in the relative

sense, by testing pigmentless or pigment-depleted *w* mutants in the dark, along with parallel mating-ability assessment of wild-type males in the same conditions: mutant and normal males were rendered equivalent in their mating successes.

The *w* gene encodes an ATP-binding transmembrane transporter, whose absence (in *w⁻*) blocks the passage of pigment precursors across membranes in eye cells (and others in the testis). What amounts to the opposite kind of genetic abnormality of *w* expression, in comparison to the straightforward mutant applications just discussed, has been mediated molecularly; this involved construction of a *heat shock promoter,* fused to *w*-coding sequences. In the ensuing transgenic flies, heat shocks lead not only to production of the *w⁺*-encoded transporter, but also to its spatially ubiquitous expression (owing to where the *hsp* factor is normally active under such high temperature conditions). Astonishingly, the ectopic expression of the *w* product—probably because this membrane protein is anomalously present in the brain of the relevant transgenic males—causes such flies to court each other in an extremely vigorous manner. This can include large groups of "intercourting" males, who, for instance, snake around their container in long chains or other more geometrically complicated arrangements. The "induced-white" males also do not repel "homosexual" advances (unlike the behavior of wild-type males). The adjective just quoted is overstated somewhat, because these engineered males will mate with females (unlike the case of an analogous "bisexual" mutant: see Section 2.4.3). How the *w*-encoded membrane protein, including what it may be "transporting" in terms of brain cells, leads to such a

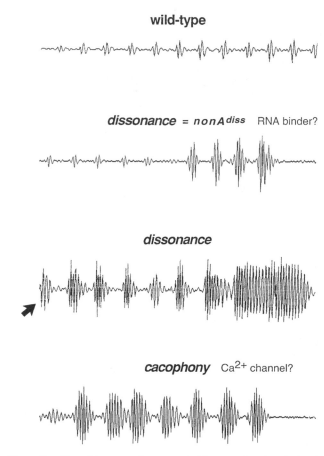

wild-type

dissonance = *nonA*^{diss}   RNA binder?

dissonance

cacophony   Ca²⁺ channel?

Figure 2. Courtship song pulses generated by normal and mutant males. These so-called pulses of tone, resulting from male courtship wing vibrations, are based on recordings of such sounds as produced by genetically normal (wild-type) *D. melanogaster* and by the two mutants in this species isolated on the basis of singing abnormalities. The top trace shows normal pulses, each one of which contains only a few cycles (nominally ≈ 2/pulse). The pulses are generated at the rate of about 30/second in this species. Song trains generated by males expressing the *dissonance* mutation at the *nonA* locus usually commence with normal-appearing (and normal-sounding) signals; but these very often proceed into pulses that are anomalously polycyclic—as exemplified by the last 4 pulses in the second trace and by all the pulses depicted in the third one, the plot for which commences at pulse no. 11 (arrow) in this song train (most of pulses 1–10 were normal). The songs generated by *cacophony* mutant males include mostly polycyclic pulses, among which are interspersed occasional normal ones (but there is no tendency for song trains to begin with wild-type-like signals). Macromolecules of the types hypothesized to be encoded within the *nonA* and *cac* loci are indicated in the second and fourth traces.

breakdown in the usual (for this species) absence of intermale courtship is unknown.

2.3 NEUROCHEMICAL MUTANTS AND GENE MANIPULATIONS

2.3.1 Acetylcholine

The genes encoding the acetylcholine (ACh) synthesis and breakdown enzymes, cholineacetyltransferase (*Cha*) and acetylcholinesterase (*Ace:* see Figure 3; see color plate 3), provided among the first examples of reverse neurogenetics in *Drosophila:* find the nor-

mal gene initially, then mutate it, and assess the biological effects of the mutation. With regard to behavior, the problem with "basic" mutations at each of these two loci is that they are embryonic lethals. Thus, conditional (usually heat-, but in one case also a cold-sensitive) *Cha* and *Ace* mutations were induced. These, incidentally, have been found to be caused by intragenic amino acid substitutions, which has turned out by no means always to be the etiology of a TS mutant in this organism. The former were studied in courtship experiments; intriguingly, the lengthy heating times necessary for the induction of paralysis were appreciably longer than the times at which male courtship actions were completely turned off; in other words, heated *Cha*^{TS} males were still normally mobile at a time when their ability to sense a female, direct courtship actions at her, or both, had been inactivated. It was as if a cessation of ACh synthesis in a particular brain region, which could "go" well in advance of a neurally more global inactivation of the enzyme, somehow ruins the function of a "courtship-controlling center," while semi-redundant and more widespread motor control centers were left running.

A subsequent *Cha* experiment merged genetic mosaic technology with a *Cha*^{TS} mutant. Here, the mutant tissues were designed to be in a more limited portion of the brain, owing to the mosaicism—in this case, involving flies of mixed sex–chromosomal genotype: part XX and female, part X"O" and male. Indeed, malelike courtship of these gynandromorphs was turned off at high temperature, while the XX//XO mosaics' general ability to move was left intact.

In these mosaics, the extent of brain tissue that was XO and *Cha*^{TS} (set up to be this way by the design of the mosaic-producing crosses) was unknown. However, in mosaics produced with regard to unconditionally *Ace*⁻ lethal mutations, those mutant versus *Ace*⁺ brain regions were readily assessable in each gynandromorph, after it had been tested for the behavior or piece of physiology in question (see Figure 3 for the expression pattern in the adult CNS of the normal *Ace* allele). By the way, this mosaic strategy permits a developmentally early-lethal mutation to be "carried to adulthood," ironically allowing such a genetic factor to be investigated for effects on phenotypes that necessarily are performed by the flies later in their life cycle. For the courtship components of these mosaic experiments, it was found that XO, *Ace*⁻ tissues in dorsal, posterior brain regions were correlated with a gynandromorph's ability or willingness to court a female—provided the patch of *Ace*⁻ brain tissue was not too large, in which case the mosaic was generally debilitated and uninformative with respect to the demands of courtship performance. These results have two implications:

1. The brain region that had to be mono-X for such a mosaic to "think it was a male" was the same as that determined in gynandromorph experiments using a benign enzyme marker for XX versus XO tissues (see Section 2.4.3 for further implications of this result in the context of sex determination genes).
2. Even though the XO, *Ace*⁻ tissues are badly damaged morphologically, as well as of course neurochemically, by the effects of the mutation just noted, genetic maleness still had behavioral meaning if that genotype within the CNS was in the relevant brain region. It is as if the key requirement is the following: this dorsoposterior region of the brain had to be "non-female (XX)" as opposed to XO and male per se, since the latter genotype did not have to be associated with normality for neural structure and function.

$Ace^-//Ace^+$ mosaics were studied in one further circumstance; here a *transformer* mutation was introduced into these flies, so that they were thoroughgoing males and not gynandromorphs (see below). Instead of "basic behavioral maleness," these mosaics were studied with regard to detailed courtship features such as singing behavior. Intriguingly, such mosaics were able to generate normal-sounding courtship song sessions from either wing, at different moments of their courtship session; but the rhythmic component of pulse rate production was knocked out with regard to one of the wings. It turned out that that side of the ventral, thoracic nervous

system was thoroughly Ace^- in genotype (and in terms of the neuroanatomical damage just noted), suggesting that song rhythmicity might be under thoracic control. This supposition falls loosely within the general context of song control by those ganglia (or perhaps one of the three such pairs), inasmuch as a basic XX//XO gynandromorph with (undamaged) male brain tissue will court and perform wing display at a female but will not generate normal species-specific song sounds unless at least a portion of the thoracic central nervous system is XO.

Most of the recent work on the *Ace* and *Cha* mutants and genes has involved cloning, sequencing, and manipulation of regulatory sequences at these loci, followed by bioassaying such DNA alterations in transgenic flies. Such assays have tended to determine merely "alive or dead?"; but it would be possible to extend phenotypic characterizations into the courtship realm, against a background of mutant and mosaic results of the kinds described here.

2.3.2 Amine-Containing Putative Transmitters and Related Small Molecules

An enzyme-gene involved in serotonin and dopamine synthesis, *Dopa-decarboxylase,* was found and mutationally defined, again by reverse genetics. The *Ddc* mutants isolated included temperature-sensitive ones; since Ddc^- mutations are developmental lethals, the conditional alleles provide one way to assess the behavioral significance of the two neurotransmitters. These were tested not only for "general learning" abilities but also for the type of conditioned courtship involving aftereffects of prior male experiences with mated females: both kinds of learning were reported to be subnormal, after heat treatments of the Ddc^{TS} adults. However, the odor-mediated, shock-avoidance conditioning side of the story has been called into question in conjunction with attempts to replicate these aspects of the Ddc^{TS}-related behavioral findings. By analogy to the results described in Section 2.3.1, molecular manipulations of *Ddc* have

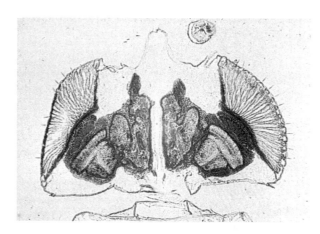

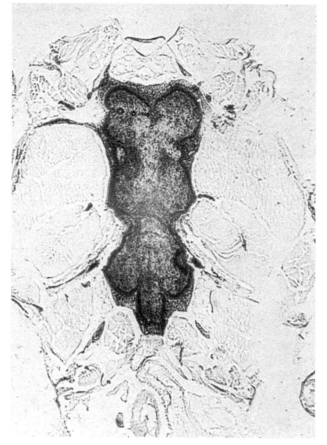

Figure 3. Acetylcholinesterase (AChe) in the central nervous system of the adult. *D. melanogaster;* the major features are shown by staining for the activity of this neurotransmitter-metabolizing enzyme. The histochemical (non-antibody-related) procedure was performed on horizontal sections through the head (top panel) and the thorax (bottom). Within each panel, anterior is to the top. This particular brain section was taken at the level of the esophagus (the empty-appearing channel running through the head's CNS); flanking this part of the alimentary system are ganglia of the CNS per se; in more lateral locations are four pair of optic ganglia, situated between the brain ganglia and the compound eyes. The latter tissue appears red because of its inherent pigment (i.e., the photoreceptor and other cells within the eye are not AChE-positive). The scale of this organism's anterior CNS is such that the distance between the outermost optic ganglia (those just underneath the eyes proper) is about 450 μm. The bottom panel shows (at approximately the same scale) the ventral, thoracic CNS (and a few AChE-positive neuronal processes, which were in part caught within this plane of section). This portion of the fly's nervous system consists of four pair of fused ganglia, the three thoracic ones per se, and (at the bottom of the image), the "abdominal" ganglion (located in the posterior thorax, but projecting axons into the abdomen). (See color plate 3.)

been performed; one such transgenic type permitted the generation of adult flies lacking the enzyme on the nervous system, but maintaining it (as they did during development) in epidermal locations (otherwise, a DDC-null condition in those tissues is lethal). It would be valuable to reexamine the possible involvement of serotonin and/or dopamine in courtship-learning abilities by testing these CNS DDC-less males.

Among the many behaviorally sluggish mutants known in *Drosophila* is *inactive;* this mutant also exhibits a severe ($\approx$ 80–90%) depletion in levels of octopamine. Not only are the general courtship abilities of *iav* males impaired (i.e., in straightforward observations involving poor mating success with virgin females), but courtship suppression in the "mated-female" system of courtship-related learning is also subnormal. The latter behavioral defect means, ironically, that *iav* males court in an inappropriately vigorous manner, after a prior experience with a fertilized female. Young *iav* males show an attractiveness to older courting males that is extended beyond the usual period mentioned earlier ($\approx$ 1 day), as if this gene plays some sort of role in posteclosion (adult emergence) maturation.

One of the *period* clock mutants was once noted to be markedly subnormal in its rate of octopamine synthesis; whether this old and unconfirmed observation will prove to have any connection with the particular courtship abnormality exhibited by *per* mutants remains to be determined.

Dark body-colored *ebony* mutants have long been studied behaviorally; both visually mediated responses and courtship performance have been assessed. As in the case of several body color mutants in *Drosophila,* catecholamines are aberrant in *e*: the level of *N*-β-alanyldopamine is dramatically lowered; β-alanine and dopamine accumulate to abnormally high levels (in pupae and pharate adults); the gene has been postulated to encode β-alanyldopamine synthetase. In courtship experiments carried out from different perspectives, *e* males have been (variously) shown to (1) behave in a manner that is subnormal (including reduced mating success) in a way that is similar to the case of blind *norpA* males; though the latter lack all light-elicited electrical responses of photoreceptors, while *e* is "missing" only the light-on and light-off transient spike in the electroretinogram (ERG; see Section 2.5.3 for further ERG/courtship mutants); (2) have their relative mating success increased in the dark and in fact outperform wild-type males in that condition; (3) be deficient in wing vibrations, as exemplified by production of infrequent bouts of the courtship "hums" as well as singing with longer-than-normal intervals between tone pulses; and yet (4) initiate matings faster than normal as *e/+* heterozygous males.

2.3.3 Kinase-Related Genes

All these factors to be discussed in the context of modulation by phosphorylation involve genetically and/or molecularly defined entities that have been biologically studied primarily in regard to learning and memory. Thus, the *dunce* gene, known from biochemical experiments and clone-and-sequence studies to encode a cAMP-phosphodiesterase, leads to higher than normal cAMP levels when mutated and, presumably, to overactivation of cAMP-dependent kinase activity. When mutated, *rutabaga* involves the opposite kind of defect owing to this gene's product, an adenylate cyclase (again, revealed both biochemically and molecularly). In-

cluded in the learning-related problems exhibited by *dnc* and *rut* mutants are courtship defects: mutant males are subnormal in their degree of courtship suppression, in tests involving prior experiences with either mated females or immature males. It is now noted that the former kind of testing regime involves associative learning, whereas as the latter seems to be a case of a nonassociative habituation-like behavioral aftereffect (related, in turn, to the immature male pheromone). Another type of nonassociative "learning" is affected by *dnc* and *rut,* in a courtship context that focuses mainly on female responsiveness to the male's actions. The phenomenon involves an "acoustical priming" that occurs when a group of females is exposed to song components and subsequently mixed with a group of (naive) males. This enhancing aftereffect (i.e., decreased mating-initiation latency) normally lasts 3–5 minutes. But homozygous *dnc* or *rut* mutant females exhibit no priming effect, and the effect for *dnc/+* or *rut/+* heterozygotes last only about a minute: that is, these are "dominant memory" mutations in the sensitization-like phenomenon related to this courtship-related aftereffect. [Note: *dnc* and *rut* also have this genetic property in tests of associative learning involving the shock–odor paradigm mentioned in Sections 2.2 and 2.3.2.]

Calcium/calmodulin-kinase is a factor that has been sensed to be involved in learning and memory, in terms of its "stable switch" properties, related to the autophosphorylation that is a salient feature of the enzyme's action. This quasi-theoretical underpinning seems to have been buttressed recently by some reverse genetic studies, involving one form of CaM-kinase and experiments on the learning abilities of mice. In parallel, analogous gene manipulations and learning experiments have occurred in *Drosophila;* but here, instead of a "gene knockout," a "dominant-negative" transgenic type was created, involving fusion of *hsp* to an "autoinhibitory" peptide. In a CaM-kinase-normal genetic background, constitutive expression (at mild temperatures) of the dominant inhibitor (depleter, really), and especially heat-induction of it, was found to impinge on the fly's learning abilities in two ways: in the mated female–related associated-learning tests, and in the acoustic priming experiments. Thus, the effects of this fusion gene construct as expressed in males and females, respectively, were assessed in these two different kinds of courtship experiment. In an extension of this work, at the behavioral level, CaM-kinase-depleted males did not even show a decrement in courtship when in the presence of mated females, let alone a suppressive aftereffect. Moreover, biochemically it was shown that one substrate of *Drosophila*'s CaM-kinase is a particular domain of the *eag*-encoded potassium channel subunit, which would seem to fit with the fact that *eag,* in males, is a courtship-learning mutant (see Section 2.2).

2.4 Transcription Factors

2.4.1 A Visual-System and Embryonic-Developmental Gene

Visual mutants have been searched for not only by relatively crude phototaxis tests (which led to recovery of mutants such as *norpA*) but also by monitoring such optomotor responses of adult flies as the "turning" adult-body movements that are elicited by moving visual stimuli. A classical mutant of this sort is *optomotor-blind.* The behavioral defect just stated (by the mutant's name) has a neurodevelopmental etiology whereby the adult ends up being devoid of certain motion-detecting neurons deep in the optic ganglia. Subsequent experiments involving *omb*—which have been very extensive from the standpoint of genetics, molecular cloning, and the pleiotropic morphological phenotypes caused by the by-now wide array of alleles that are mutated at this locus—have included the determination that this is a developmentally vital gene, which encodes a putative (trans-acting) gene regulatory factor. In courtship experiments on males expressing the original, viable *omb* mutation, it was found that these flies are just as severely subnormal in mating-initiation latencies as are thoroughly blind mutants. One could conclude that it is less important that a male sense the gestalt of a female by merely seeing her than that he respond to her movement by "optomotoring" toward her. However—given the pleiotropy now known to be associated with the *omb* gene's action—it could have been that this reproductive–behavioral impairment was occurring for "other reasons" (e.g., more extensive or deep-seated ones than in regard to the optic–ganglion defect). That *omb* males court females in such a mediocre manner for nonvisual reasons was, however, controlled for by showing that mutant males placed in the dark with genetically normal females initiate matings with the same kinetics measured for wild-type male–female pairs in that condition. Curiously, this sameness meant faster "dark kinetics" for *omb* males + normal females in comparison to parallel tests in lighted conditions (i.e., the mutant situation was boosted up to the wild-type level). This prompted the experiment, described in Section 2.1 that disclosed (via application of *norpA*) that "sighted" females exhibit greater coyness than do those that are in the dark or genetically blinded.

2.4.2 A "Clock" Gene

The canonical biological clock mutations at the *period* locus of *D. melanogaster* (Figure 4) have led to a still-burgeoning story about how this gene's action seems to in effect be a timekeeper, at least in terms of how it controls the fly's circadian rhythms. It does so—goes the strongest and most current hypothesis about *per* gene expression and function of the encoded protein (PER)—by being a negatively acting, non-DNA-binding transcription factor. For example, PER represses its own mRNA synthesis, which fluctuates in accordance with a circadian rhythm, and can influence the daily oscillations of mRNA production emanating from certain other loci—"clock-controlled" genes (hypothetical "output factors" of the central, *per*-controlled pacemaker).

The *per* gene, however, has entered some distinctly nondaily time domains. In the first set of relevant observations, it was demonstrated that *per* mutations that speed up, slow down, or apparently abolish circadian clock function (per^S, per^L, and per^O) lead to parallel timing abnormalities with regard to the aforementioned courtship song rhythm, whose normal periodicity is about 1 minute (in *D. melanogaster*: Figure 4). How *per*'s inferred function—and the 24-hour rhythmicities of mRNA and protein oscillations that are such a salient feature of the gene's expression—may or (more likely) may not connect with regulation of a clock function whose pace is a thousand times faster than the circadian one is a matter of befuddling incomprehension.

Undaunted, however, investigators have recently asked further courtship-related questions about the effects of *per* mutations. Knowing that the normal song rhythmicity is somehow detected and processed by (wild-type) females, whose acceptance of copulation attempts is optimal when the pulse rate production cycles

have approximately 60-second periods, investigators wondered whether these rhythm variants would provide support to the old "sender–receiver" coupling hypothesis. Specifically, the idea was that per^S males (circadian period, 19 h; song period, 40 s) might do relatively poorly in the presence of wild-type females. But perhaps two wrongs would make a right—if *per*'s action also controlled song perception as carried out by the female—in a case where per^S males were matched with similarly mutated females. The same might occur in the mating behavior of the per^L (29 h/80 s) flies. However, the hypotheses failed, at least insofar as these rhythm factors are concerned: the two mutant female types just noted responded best to pulse songs varying with normal, 60-second periodicities.

The *period* gene has been imagined to affect every area of *Drosophila* biology having anything to do with time. Learning and memory are in this ballpark, and in fact it was claimed several years ago that long-period rhythm mutants, including per^L, impair conditioned courtship in terms of both the mated-female and immature male paradigms. These results proved to be largely irreproducible; in tests of "classical conditioning" involving electric shocks and artificial odorants, moreover, none of the three types of *per* mutation caused any decrement in initial learning or memory time course parameters.

Another facet to the *period* gene story, which is a gem, is in the area of molecular evolutionary studies. For example, *per* seems to be among the more rapidly diverging genes known, with regard to sequence comparisons among *Drosophila* species and, in more recent experiments, among a variety of insect species (well beyond diptera). One of the relatively conserved regions of the protein, however, is that which provided one of the "transcription factor hints" about PER's function. In another region of the polypeptide are runs of threonine–glycine pairs; these Thr-Gly repeats, and surrounding amino acid sequences, are relatively nonconserved. Could this region of the protein have something to do with rhythm differences among species? Those interspecific phenotypic variations could be expected to participate "mainly" in the control of courtship song rhythmicity. Indeed, period values range from less than 40 seconds to more than 70 seconds among closely related species (including *D. melanogaster*). In fact, an in-vitro-effected intragenic deletion of *D. melanogaster*'s Thr-Gly repeat caused the resulting transgenic males' song rhythm to be non-*melanogaster*-like, with periodicities distinctly shorter than normal.

Other *per* gene studies have involved interspecific behavioral differences in a more direct experimental sense. Song rhythm testings of reciprocally hybrid males, resulting from crosses between *D. melanogaster* and *D. simulans,* showed that the genetic etiology of the rhythm difference mapped to the X chromosome. This is where *per* is located. These findings led to interspecific transgenic experiments: the *D. simulans per* gene (Figure 4) was put into a per^O *D. melanogaster* "host;" the latter not only had rhythmic singing restored, but this courtship character had *simulans*-like (35–40 s) periodicity (Figure 4). However, one's attention is again focused on the puzzling matter of whether PER-as-transcription factor has anything to do with the basic control of—and intra- plus interspecifically determined variations in—this short-term rhythmicity.

2.4.3 Sex Determination Factors

A signal success story in developmental molecular genetics of *Drosophila* involves studies of genes controlling sex determination.

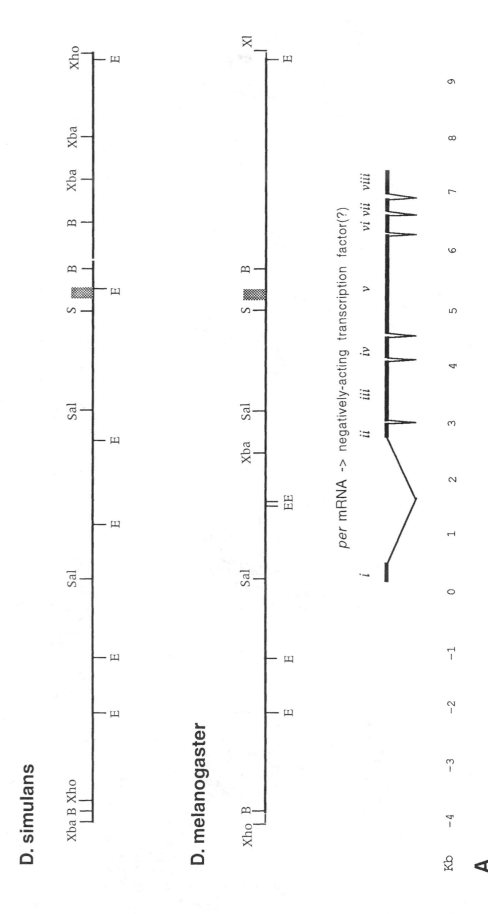

Figure 4. Molecular biology of the *period* clock gene and its control of courtship song rhythmicities in two *Drosophila* species. (A) DNA cloned from the *per* locus of two closely related species (including "restriction sites," indicated by abbreviated names above and below the horizontal lines); roughly in the middle, the *per*-encoded mRNA is lined up below its actual "genomic sources." The 8 *per* exons are designated by lowercase Roman numerals. (Exon *i*, the left-hand portion of *ii*, and the right-hand portion of *viii* are noncoding.) The polypeptide translated from this mRNA contains approximately 1200 amino acids and seems to function as a "negative transcription factor" in terms of its control of circadian rhythms (see text). (B) The behavioral effects of transducing various forms of the *per* gene—including interspecific "chimeras"—into a *D. melanogaster* host whose endogenous allele at this locus is a "null." When the entire 13.2 kb DNA "construct" was from *D. melanogaster* (13.2m-TGm), arrhythmicity for courtship singing behavior was "rescued" (in that cyclical song variations were produced by the transgenic males) and defined by a 60-second (approximately) rhythm.

139

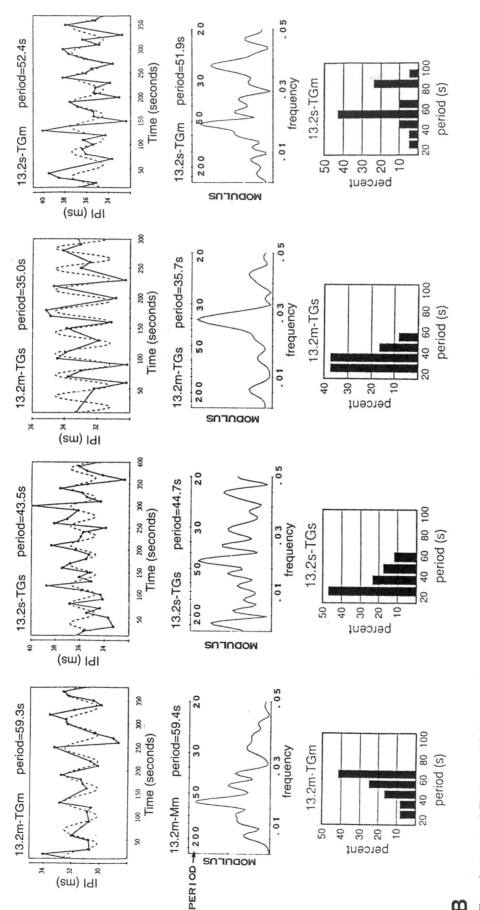

B

Figure 4. (*continued*) This is exemplified by an actual song plot, showing a systematic fluctuation of the pertinent "interpulse interval" (IPI) (see ordinates; cf. Figure 2); below the plot is the result of spectrally analyzing this particular time series; and at the bottom is a summary of such analyses (for all males of this transgenic genotype). The second trio of (vertically arranged) results is from a transgenic type called 13.2s-TGs, which means that the entire *per* gene came from *D. simulans*—and note that the rhythm example, as well as those in the summary histogram, are *simulans*-like, having periods of roughly 40 seconds. "TG" refers to a central (700 base pair) region within the *per* gene of these two species, including nucleotide sequences that encode threonine-glycine (TG) repeats—that is, approximately 17–24 pairs of alternating T-then-G residues. Thus, the third set of results involves transgenic males most of whose *per* gene was from *D. melanogaster* (13.2m), but whose central 0.7 kb region was from *D. simulans* (TGs). The right-hand trio of panels involves the reciprocal chimera–transgenic type. These two chimeric genes mediated, respectively, *simulans*- and *melanogaster*-type song rhythms (i.e., IPI oscillations defined by periodicities of ≈ 40 and 60 s).

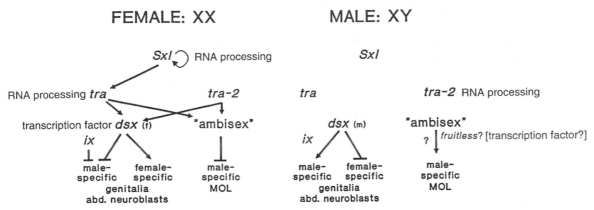

Figure 5.   Genetics and molecular biology of the sex determination hierarchy: a brief outline of the series of gene actions occurring in conjunction with developmental determination of female versus male *Drosophila*. *Sex-lethal*—whose expression "responds" to the ratio of X chromosomes to autosomes—thus produces functional RNA-processing proteins (SXL) in females but not in males. SXL controls not only its own expression (circle-arrow), but also sex-specific RNA splicing of primary transcripts of the (third-chromosomal) *transformer (tra)* gene; thus, a functional RNA-processing protein (TRA) is generated from this gene, again in females only. That factor, along with the product of another *transformer* gene (located on a separate chromosome from *tra* itself, i.e., no. 2), mediates sex-specific splicing of RNA transcribed from the *doublesex* locus. This leads to two differentially functional products, DSX(f) in females and DSX(m) in males. The role of the *intersex (ix)*, as it is surmised to act "downstream" of and in conjunction with *dsx* (in females), has been established only in formal terms. More generally, the other downstream elements of the hierarchy are depicted in terms some of the sex-specific features of developmental differentiation known in this species ("abd." is shorthand for abdominal, referring to presumptive neurons that exhibit a sexual dimorphism in terms of numbers of these cells and form within the posteriormost portion of the developing CNS). The hypothetical "ambisex" gene is presumed to define a "new branch" within the sex determination hierarchy. This gene's action would, like that of *dsx*, be controlled in part by the action of the *tra* and *tra-2* gene products; yet, downstream of ambisex are, in the main, internal and neurally related aspects of sexual differentiation, such as the developmental appearance of a male-specific structure called the muscle of Lawrence. The formation (and maintenance? function?) of this MOL is dependent on a mono-X genotype within neurons innervating the relevant abdominal region; such an XY (or X "O") chromosomal constitution would eventually influence expression of ambisex. It can be suggested that the *fruitless* gene could be ambisex, owing to this gene's influences on neurally related phenotypes; these involve (at least) MOL formation and sex-specific courtship behaviors.

The result has been the establishment of a "hierarchy" of gene actions and interactions (Figure 5). The genes were identified initially by mutations that "transform" on chromosomal sex type (e.g., XX) at least part way into the opposite type (mono-X; i.e., XY or XO). Most analysis of these altered phenotypes has concerned sexually dimorphic characters visible on the external surface of the adult. Thus, for example, recessive *doublesex* mutations cause an XX or XY fly to appear to be intersexual. The *dsx* gene is located at a rather "downstream" position in the hierarchy. It encodes a DNA-binding transcription factor; the protein has sex-specific qualities, and this is controlled by differential splicing, mediated by RNA-binding proteins acting further upstream (see Section 2.5.). *dsx* mutations do affect certain tangible "internal phenotypes" (yolk protein and pheromone production; the elaboration of particular neurons in the posterior-most ganglion of the thoracic CNS: cf. Figure 3). Further such effects were inferred from the superficially "intersexual-like" quality of sexual behavior, given that an XY;*dsx* mutant fly courted females very feebly and elicited, as a mature (chromosomal) male, anomalously high levels of courtship (performed, as usual, by wild-type males). The firmness of these conclusions has been called into question: reexamination of *dsx* effects on courtship showed, by the application of several independently isolated mutant alleles at the locus, that XY;*dsx⁻* flies can court rather vigorously, albeit still less so than wild-type males. Moreover, there is no "transformation" of courtship song quality, regarding the wing vibrations that such mutant XY flies direct at females. [Analysis of the aforementioned gynandromorphs indicated that a thoroughly XX thorax deranges song quality (see earlier discussion,

including Figure 3 for a depiction of the thoracic CNS's basic anatomy). A partial, *dsx*-induced transformation of XY tissue there, in the direction of femaleness, would likely result in at least some sort of pulse or hum quality anomalies.] Finally (for the moment) consider the dsx^D mutation, which (in the absence of any other *dsx* allele) causes only the male form of DSX to be produced, thus making an XX animal end up looking like a normal male; yet this fly exhibited no malelike courtship actions whatsoever.

Another internal sexual dimorphism in *Drosophila* involves a pair of large abdominal muscles, called the muscles of Lawrence (MOL). Their development in that male manner is dependent on the mono-X genotype of nerves innervating the pertinent region of the abdomen. By analogy to the courtship studies just noted, *dsx* was found not to affect development of the MOL. Mutations at the *fruitless* locus, originally identified as a male-sterile mutant, do affect MOL development, however. *Fruitless's* sterility has a behavioral etiology: mutant males court females in a reasonably avid manner but do not even attempt to mate with them, let alone succeed. The courtship block before the attempted copulation stage means that *fru* males essentially never curl their abdomen in the direction of the female's genitalia. An equally dramatic defect exhibited by this mutant is very vigorous courtship by *fru* males of other males (mutant or wild-type ones). *fruitless* is now known also to be an anatomical mutant, hypothetically of the "neuromuscular" type: *fru* males (and especially mutant types carrying chromosome aberrations with *fru* locus lesions) are largely devoid of the MOL structures, as if those muscles might be involved in the abdominal bending that describes attempted copulation. However, a particular *fru*-mutant allele (of the five extant ones), which does cause intermale courtships and eliminates the MOL, permits fertility. Thus, the MOL's biological role is unknown.

Nevertheless, the foregoing findings—pitted against the relatively minor role played by *dsx* in "determining" *Drosophila*'s nervous system such that sex-specific behaviors occur and that neurally controlled sexually dimorphic structures are formed—led to the hypothesis that the sex determination hierarchy branches "below" the transcript splicing controllers. The *fru* gene may be positioned at this newly defined branch point, at roughly the same downstream position in the hierarchy as *dsx* (Figure 5), which should now be regarded as controlling only some features of male-versus-female development. Furthermore, it can be hypothesized that *fru* will be found to encode a transcription factor of its own, which would regulate genes farther downstream, primarily concerned with sexual dimorphisms in the nervous system and sex differences in adult behavior.

2.5 KNOWN AND PUTATIVE RNA-PROCESSING FACTORS

2.5.1 Sex Determination Factors

The upstream factors in the hierarchy of gene action discussed in the foregoing section include the RNA-binding proteins alluded to there. Again, all three (of the relevant) genes were originally known from mutations that altered sex determination. One of these, *Sex-lethal*, is located near the top of the hierarchy (Figure 5), where it—for example—"responds" to chromosomal sex. Farther downstream, the SXL protein modulates sex-specific splicing of the RNAs encoded by one of the *transformer* genes (which in turn regulate *dsx*-transcript splicing: again see (Figure 5 and the associated discussion). "Everything" about the fly's sex would appear to funnel through *Sxl, tra,* and *tra-2,* because mutations at these loci transform the sexual behavior of the relevant fly, in addition to its external appearance. (Regarding *Sxl,* such experiments had to be performed with "partial loss-of-function" mutations at this locus, if viable adults were to be obtained for behavioral testing.) Thus, one hypothesizes that not only *dsx,* but also *fru,* might be "right downstream" of *tra/tra-2* actions.

One of the more intriguing behavioral studies related to these the RNA-binding proteins encoded by these genes exploited a TS allele of *tra-2.* Here, it was found that low temperature development, which causes an XX;*tra-2^TS* mutant to develop as an externally appearing intersex, is not the end of the story insofar as sexual behavior is concerned: heat-treating such a fly, posteclosion, was found to "turn on" the ability of this quasi-female to court other females in a largely malelike manner. It is as if there are certain very late components of neural development in this species, which are still genetically and otherwise malleable after the animal emerges as, one would have thought, a by-now "hard-wired" male or female.

The (original) *transformer* gene (located on chromosome 3 and symbolized simply by *tra*) has had its expression manipulated not temporally, as discussed for the second-chromosomal *tra-2* mutation, but spatially. These *tra* experiments drew on "enhancer trap" principles and methodologies, which are so much a part of *Drosophila* genetics and molecular genetics. Here, the so-called second-generation enhancer-trapped strains were utilized. These involve genetic crosses in which a certain transposon is mobilized and lands in a host of different chromosomal locations; the transposon was engineered to carry a gene cloned from yeast, which encodes a

trans-acting transcription factor, GAL4. If the vector's landing site has "trapped an enhancer," this event can be revealed by crossing in a transgene (already established in separate strains) carrying the GAL4 cis-acting target (known as UAS), fused to the *lacZ* reporter gene from *E. coli.* β-Galactosidase activity then reports the nature of what one hopes will be a (*Drosophila* chromosomal) regulatory factor that mediates expression in certain tissues only. This system has been well known to work in general terms for a few years. It is now being exploited biologically; and one of the first such usages involved the *tra* gene and studies of courtship behavior. For this, a series of strains was found (from the pertinent β-gal-staining screens) in which GAL4 production was robust in a variety of brain regions within the adult head. Then, each of these particular enhancer-trapped strains was crossed to flies carrying UAS fused to a cDNA construct encoding the female "splice form" of the *tra*-produced protein (i.e., TRA$_F$). Intriguingly, a subset of the resulting XY flies, which in principle would possess "partially feminized brain," behaved aberrantly: they courted other males in a much more vigorous than normal manner (though, by now, not to be regarded as a behavioral miracle). The flies that behaved in this odd way expressed the reporter, and by inference TRA$_F$, in brain regions known to be involved, at least in part, in the processing of olfactory inputs. It was as if these particular transgenic types were responding in an aberrant manner to sex pheromones—interpreting male odors as femalelike aphrodisiacs, or failing to internally sense the "courtship-inhibitory" substance believed to emanate from wild-type males (see Sections 2.2 and 2.7).

2.5.2 A Sluggish Mutant

Certain putative RNA-binding proteins seem to be rather global in their biological significance. One of these is encoded by the *couch-potato* gene, defined initially by mutants stemming from the first-generation sort of enhancer trapping (i.e., mobilization of a straightforward *lacZ*-containing transposon). In this particular instance, the "insert" reported expression in many tissues, including the nervous system; and homozygosity for a transposon at this (newly identified) genetic locus led to the sluggish behavior implied by the gene's name. More severely mutated forms of the gene lead to developmental lethality. But the original, viable mutants were tested in terms of male courtship actions as well as their general mobility; as expected, *cpo* males courted females in a desultory manner, more or less as do mutant *iav* males (see Section 2.3.2). This is not particularly informative, but the *cpo* short story is mentionable here not only because there is a certain degree of molecular knowledge about the gene and its product; but also to serve as a warning: in a sense, any subnormally behaving mutant in this organism is likely to be a "courtship mutant" as well, in that the motor demands of reproductive behavior might cause this aspect of the fly's repertoire to be the "first to go." [Recall the effects of heat-treating the *Cha^TS* mutants, whereby courtship was affected before any general locomotor defects were noticeable.]

2.5.3 A Song and Visual Behavior Gene

Another gene that encodes what seems (from nucleic acid sequencing) to be another RNA-binding protein is *no-on-transient-A.* Here the product is, in a manner loosely similar to the case of *cpo,* widely expressed in tissues and life cycle stages (including adulthood). Yet, *nonA* is not a vital gene: *nonA$^-$* flies exhibit reasonable (albeit subnormal) viability and primarily can be termed behavioral mutants: certain visual responses are absent, and—as a male—this mutant type generates an extremely abnormal courtship song. This gets ahead of the story somewhat: *nonA* mutants were originally isolated as visual variants; later examination of their ERGs disclosed the absence of the light-on and light-off transient spikes (hence the gene designation is something of a misnomer). The light-coincident photoreceptor potential is normal, and indeed these flies are not thoroughly blind but only "optomotor-blind." A unique type of *nonA* allele was isolated many years later in a courtship song mutant screen; the resulting *dissonance* mutation (Figure 2) turned out to cause visual defects as well (of the types just described) and to be at the *nonA* locus. Visual mutants located there, in retrospective testings, were found to sing normally.

Cloning of *nonA* led to the inference that NONA is an RNA binder; in fact, one such set of studies started with screens for antibodies against *Drosophila* chromatin—leading to the isolation of certain polypeptides and then to partial sequence data. These results led, in turn, to cloned nucleic acid material, which turned out to correspond to *nonA*. The NONA protein is now regarded as being in a "family" of RNA-binding proteins, whose other members are encoded mainly by mammalian genes: not only are the putative motifs that per se interact with RNA present among the four family members, but also "conserved" is a rather lengthy (and relatively C-terminal) charged region. It turned out that the "visual-only" *nonA* mutations in *Drosophila* are in or near the RNA-binding regions, whereas the visual + song mutation, called *nonA-dissonance,* has an amino acid substitution within the charged region. In the narrow sense, the significance of this gene and its mutations is that the latter, *nonA^{diss}*, brings into focus a potential biological role for this not-very-feature-laden charged region; in fact, one infers this to be a near loss-of-function mutant, owing to the *dissonance*-like (visual + song) defects observed in *nonA$^-$* flies. In the broader sense, it seems interesting that a globally expressed and possibly generic ("housekeeping"?) sort of cellular function can lead to rather specific and limited—that is, behavioral—defects when it is mutated or even eliminated from the genome.

2.5.4 A Factor Interacting with a Sodium Channel Gene

It was noted in Section 2.2 that *para* encodes the "main" sodium channel protein in *Drosophila* and that a certain phenotypic study of a *para^{TS}* mutant involved courtship singing behavior. In reality, *para* encodes a set of similar but distinct Na$^+$ channel polypeptides (by alternative splicing, naturally). The functional significance of these different *para* products is only beginning to be uncovered. (Recall in this respect that certain *para* mutations are "olfactory specific" in the phenotypic defects caused by them.) One feature of the emerging findings about *para* involves regulation of the pertinent "RNA processing" (considered in the broadest sense). In this regard, a prominent *para* regulator is specified by the *male-lethal* gene. Historically, this locus had been studied from two very different perspectives: (1) in regard to the phenotype implied by the gene's name, and the further inference that *mle* is involved in "dosage compensation" (whereby genes on the male's one X are expressed at roughly twice the level as a given X-linked gene on one of the female's two Xs); but (2) also because *mle* had been independently mutated to cause temperature-sensitive paralysis (of either males or females, actually), by the *no-action-potential-TS* allele. It is now inferred that *mle* encodes an RNA helicase and that at least one (no

doubt far from the only one) of the X-gene-encoded transcript sets influenced by this MLE protein consists of those coming from *para*. One consequence, then, of the *mle^{napTS}* would be to lower the overall amount of *para*-encoded Na$^+$ channel molecules available for insertion into neuronal membranes (this makes a long, and emergingly complicated, story short). Owing to the general temperature sensitivity of sodium channel function, a mere lowering of channel density in these membranes can nevertheless lead to "sharp" heat-sensitive paralysis. Indeed, *mle^{napTS}* adults pass out very quickly when heated—as do *para^{TS1}* flies—with both mutant types also exhibiting rapid recovery when cooled. This property permitted *nap^{TS}* males to "become a behavioral mutant" (in addition to a pathophysiological one): causing courting males of this type to pass out, followed by their reintroduction to females at the time of cooling-induced recovery showed that the aforementioned courtship song clock had stopped for the duration of the paralysis and nerve conduction turnoff. This is the same kind of clock stoppage induced by similar manipulations of *para^{TS1}* males.

2.6 SMALL-PEPTIDE-RELATED GENES AND MANIPULATIONS THEREOF

2.6.1 The Sex Peptide

Several male-produced substances are transferred to the female during mating. One of the best known is "the" sex peptide. It and other materials in the male's seminal fluid influence postmating female behavior, including such elementary activities as her egg-laying and also the courtship (qua courtship) activities that she exhibits when fertilized. To test the behavioral significance of the S-P alone, cloned nucleic acid sequences encoding this peptide were ectopically expressed in unmated females, under *hsp* control. Heat activation of S-P production turned on female rejection behaviors and nonreceptivity to further copulation attempts. These attributes are naturally a part of a fertilized *Drosophila* female's behavior (remating soon after a mating has occurred can be regarded as a waste of time or even a danger). Such behaviors are usually sustained for several days, but ectopic *hsp* induction of S-P causes their appearance for only one day. This could be because other male-transferred entities were of course lacking from the *hsp-s-p* transgenic females—such as the male's sperm, which (from a variety of older experiments) have been implicated as influencing the female's postmating behavior, or other seminal fluid substances, which could enhance S-P effects or stabilize its presence. Support for the latter notion was provided by another S-P-related transgenic type; for this, an enhancer from a *yolk-protein* gene was included in the construct, and the supposed boost in or increased longevity of S-P production in virgin females made them behave for several days as if they had mated.

2.6.2 A Specific Esterase

Another substance passed from male to female during mating is "esterase-6" (EST-6). This enzyme is found mainly in the male and principally in his anterior ejaculatory duct; but this protein (in its natural form: see Section 2.7) is also detectable in the hemolymph (blood) of both sexes. After a female has copulated, male-derived EST-6 is rapidly transferred to her hemolymph. The enzyme is believed to play a role in sperm utilization and remating behavior (see Section 2.7). The significance of that role was downplayed as a result of earlier studies indicating that EST-6 does not persist in a fertilized female for a "long enough" postmating time (during which her behavior continues to be non-virgin-like); but more sensitive methods subsequently revealed that the protein remains in females for a period that can account for this enzyme's behavioral effects.

Cloning of the *Est-6* gene in principle would permit an even higher sensitivity detection of the enzyme's tissue distribution. This molecular accomplishment indeed seemed to prompt spatially more widespread searching for the gene's product's presence. Surprisingly, perhaps, *Est-6* transcripts were found not only in the bodies but also the heads (including the eyes) of both male and female adults. A subsequent series of expression-related experiments fused various subsets of "5′-flanking DNA" of the *Est-6* gene with a reporter factor (*lacZ*, as usual). In this way, various independently acting gene regulatory regions were identified; two of them directed *Est-6* expression in the male's ejaculatory duct and bulb. Other cis-acting (5′-flanking) sequences controlled production of the enzyme-encoding mRNA in a variety of tissues, such as salivary glands, the respiratory and digestive systems, and appendages of the adult head. These results involve tissues in which the *Est-6* gene is transcriptionally active (as opposed to sites where the naturally secreted enzyme is transported), and thereby draw attention to intrafly locations (e.g., salivary glands, head appendages, and even the ejaculatory bulb) where EST-6's presence was previously unknown. The possibility can be entertained that certain of these expression locations—particularly those involving sensory appendages—could influence perception of signals passing between a male–female pair, hence the degree to which they actively court in a given set of circumstances (e.g., whether the female has mated). Thus, in the narrow sense, the *Est-6* gene may affect reproduction for reasons beyond its presence in, and transfer from, the male reproductive system. More generally, this combination of descriptive and gene-manipulative studies strongly suggests that EST-6 functions well beyond the fly's "mating system."

2.6.3 Molecular Ablation of Peptide-Producing Cells

One of the many new molecular tools for studying *Drosophila* neurobiology and behavior involves "toxin genes" derived from other organisms. The idea is that fusing such factors to cell- or tissue-specific promoters can aid in interpreting the biological significance of the piece of anatomy that would be ablated by this kind of "internal fine surgery." One of the first exploitations of this approach involves, once again, mating-transferred substances. Many of these materials, including S-P, are generated in what are called the main cells of the male's accessory gland (part of his somatic reproductive system). Using a main-cell-specific promoter, a diphtheria toxin subunit was expressed—or at least designed to be—only in those locations. The transgenic males so ablated mated normally with females (by outward appearance), but left them in a virginlike state behaviorally. A sour note that sounded in conjunction with this positive result was that the "toxin construct" inexplicably made these engineered males spermless as well as main cell–less. It was as if the cell-killing substance somehow leaked into other parts of the male's reproductive system (which this particular diphtheria toxin subunit is not supposed to be able to do). Thus, the effects on females of mating with these males were not cleanly interpretable. The investigators had to "subtract" sperm effects by performing their own controls involving a spermless mutant that was, however, normal with respect to the accessory gland. They eventually concluded that accessory gland substances are involved in only a short-

term inhibition of remating. This could imply that the effects of S-P production enhanced by *yolk-protein* in the *hsp*-related ectopic expression experiment were overinterpreted (at least as stated earlier by this reviewer).

2.7 PHEROMONES

Pheromones must be considered to be a part of the "molecular" story surrounding *Drosophila* courtship, if only because certain of the courtship-stimulating and -inhibiting chemical cues are precisely known. Moreover, the actions of certain genes are implicated as influencing the production of these reproductively important, chemosensory-related stimuli.

Adults of *D. melanogaster* and its close relatives ($n = 7$ additional species) produce at least a dozen cuticular hydrocarbons, certain of which vary according to sex and species. The long chain substances are quite nonvolatile, a property superficially consistent with consistent reports that such fly pheromones (when bioassayed in seminatural contexts) act only over short ranges. Thus, these materials may be received in the main as contact chemosensory cues. But certain experiments have shown chemical stimuli to be effective over distances up to, and in a couple of instances much greater than 1 cm. Those materials would seem almost certainly to be sensed as "odors," which could mean that as-yet unidentified pheromones play a role in male–female attractiveness or the opposite kinds of behaviors.

The latter interactions include strong inferences that mated females have an antiaphrodisiac quality. That the relevant substances would be "smelled" by the male is inferred from *para^{olfD}* and *para^{sbl}* males' inability to sense that a female has mated (though perhaps those two Na$^+$ channel mutations not only produce olfactory impairment but also disrupt contact chemoperception). The antiaphrodisiac story is somewhat controversial: it has been argued, variously, that "the" substance in question is *cis*-vaccenyl acetate (CVA), a rather long chain molecule that is yet another of the seminal materials transferred during copulation. Moving from male to female, as well, is the aforementioned EST-6; and it was suggested that this enzyme acts on CVA to generate *cis*-vaccenyl alcohol, "the" actual courtship-inhibitory substance. This hypothesis was knocked down, but CVA could "still" play a key antiaphrodisiac role; though it also (or instead) can act as a non-sex-specific aggregation pheromone. Another courtship inhibitor may be 7-tricosene (7-T), the main hydrocarbon associated with the male cuticle. Apparently 7-T is passively rubbed onto the female's body surface during courtship, where it may exert an inhibitory influence on a subsequent male's advances, mask the (main) female aphrodisiac—called 7,11-heptacosadiene (HCD)—or both.

Studies of pheromonal genetic control have included application of sex determination defective mutants. The "upstream" factors transform the nature of an XX or mono-X fly into a "surface chemical female," as would be expected from results presented earlier. Once the action of *tra* and the like has reached a gene regulator such as *dsx* (or *fru?*), one imagines that the transcription factors in question could control expression of enzyme-encoding genes that would be different in the two sexes and would mediate hydrocarbon biosynthesis. Nothing is yet known about what these "sexually chemical" effector genes could be. Moreover, the case of the "downstream" *dsx* gene is something of a puzzle in its pheromonal context: Could it simply be that a *dsx*$^-$ mutation would cause both XX and XY flies to be pheromonal intersexes? Or instead (as in the case of the MOL), would a *dsx*$^-$ mutation have no effect at all on the pheromones that usually accompany the two chromosomal types just noted? The apparently contrary set of results were that XY;*dsx⁻* (chromosomal) males generate male-specific cuticular hydrocarbons but no female ones; XX;*dsx⁻* flies are sex chemically defective, however, in that they do not generate the female HCD aphrodisiac and are "transformed" in that they produce appreciable quantities of malelike 7-T.

Other mutations causing pheromonal anomalies include the following:

1. *nerd,* with its diminished production of 7-T; but the mutant also (and by definition) initiates mating with poor kinetics. It is not clear how this connects with the depleted levels of the hydrocarbon—unless 7-T somehow induces a certain female rejection behavior (which such flies do perform, when courted by *nerd* males) as well as (hypothetically) affecting male behavior when 7-T is on the female's body.
2. The original *fruitless* mutant; though not the genetic lesion in it that causes behavioral sterility, intermale courtship, and so on. *fru¹* is in fact a short inversion for which the "other" chromosomal breakpoint causes mutant males passively to elicit courtship as performed by any other type of male. This genetic lesion also leads to a decrement in overall hydrocarbon production (e.g., 7-T levels are at most half-normal), but a slight increase in 7-pentacosene (7-P), which now becomes the main hydrocarbon on the *fru¹* male's cuticle. This relative switch in 7-T versus 7-P content could explain the courtship elicitation in that *D. melanogaster* males will court other males that are rich in 7-P; but the notion is undermined by the fact that *nerd* males have a similar 7-T/7-P ratio, yet are not "attractive" to other males.
3. The *Ngbo* locus in *D. simulans,* which influences natural geographic variation in relative 7-T/7-P proportions. Most strains have a high such ratio, with this mutation causing an increase in 7-P and a decrease in 7-T in males that are either heterozygous or homozygous for the mutation (the pheromonal effect being accentuated in the latter genetic situation).
4. *kété,* which was chemically induced on the X chromosome of *D. simulans* and causes subnormal production of 7-P in males and a quite substantial decrement of 7-T levels in both sexes—against a background of the latter hydrocarbon seeming to be an important sex pheromone in females of this species (as opposed to HCD playing that role in *D. melanogaster*). *kété* mutant males appear to court normally, but wild-type *D. simulans* females produce diminished amounts of their hydrocarbons after mating with this mutant type (compared to their production levels ≈ 1 week after mating with wild-type males). The *kété* mutant is distinctly pleiotropic, albeit still in a reproductive context: homozygous mutant females lay eggs that do not develop.

Given this kind of pleiotropy, which also seems to be a part of (for example) *nerd*'s action, it is a stretch to imagine that all (or any?) of these four loci will permit investigators, perhaps via cloning of the factors defined by the lesions or apparent "point mutations" in question, to delve into biochemical and molecular genetics of pheromone production. Yet, a pleiotropically acting factor may still play a material role in any given one of the phenomena it influences. This means, for instance, that certain of these pheromone-related genes might eventually be viewed as being at least marginally a part of the sex determination machinery—located at well downstream positions within the relevant hierarchy of gene actions. In that manner, *nerd, kété,* and the like might play mul-tiple roles with respect to reproductively related features of the animal's development and function, and they would do so by responding in their own fashions to the more "centrally" located (upstream) control elements.

3   PERSPECTIVES

These summary remarks and musings stem from the particular-cum-general points just made about pheromonal mutants, constituting in addition a compressed recapitulation of the opening paragraphs.

The genetics and molecular biology of courtship involve the discovery and application of a very broad range of *Drosophila* mutants and genes. This is to be expected, because so many aspects of the fly's neurobiology feed into the multifaceted excitable-cell functions and whole-animal actions that sum up to mate recognition, courtship actions, mating, and its immediate (especially behavioral) consequences. Also revealed by the genetic—and, increasingly, molecular—investigations in this area is the matter of how pleiotropic are the relevant mutants and molecules: few behavioral and neural mutants, no matter how they may have been initially isolated, are without interestingly informative effects on courtship behavior. Few or none of the so-called courtship mutants are limited to defects in that area alone. Here is a corollary: as the study of a vast horde of neuro genes in *Drosophila* and other organisms becomes increasingly molecular, precious few such factors are seen to be expressed either in a limited portion of the nervous system or at restricted stages of the life cycle. Instead, one infers that all these genes are more versatile than might have been imagined when one naively started out by searching for genes that would function only during development or in the adult and, at either of these times, in only one kind of sensory structure or subset of the CNS. The tacit but incorrect assumption seemed to be that mutations of these genes (effected by reverse genetics, or in some instances by the application of preexisting variants) would lead only to certain specific, tractable—and experimentally comforting—defects.

[A further irony of sorts is that several reverse genetically engineered mutants, including those involving neurally expressed genes, are said to have "no phenotype"; what this usually means, however, is simply that the mutant developed in a superficially normal manner and yet failed to be observed for any but its most elementary behaviors, as opposed to complex ones that might include assessing the "gene knockout's" reproductive abilities.]

This versatility and the concomitant mutational pleiotropy notwithstanding, it is hoped that the molecularly related examples discussed do not imply that "all genes are expressed 'everywhere,' and their mutations affect 'everything,' so we learn nothing from studying and manipulating the clones and the mutants." In contrast, delving into the reproductively related subset of a given gene's significance, in terms of both its normal and mutated forms, informs the courtship neurogeneticist about the meaning and control of these fascinating features of the fly's behavioral repertoire.

See also DROSOPHILA GENOME; SEX DETERMINATION.

Bibliography

Baker, B. S., Gorman, M., and Marín, I. (1994) Dosage compensation in *Drosophila. Annu. Rev. Genet.* 28:491–521.

Burtis, K. C., and Wolfner, M. F. (1992) The view from the bottom: Sex-specific traits and their control in *Drosophila. Semin. Dev. Biol.* 3:331–340.

Carlson, J. (1991) Olfaction in *Drosophila:* Genetic and molecular analysis. *Trends Neurosci.* 14:520–524.

Cobb, M., and Ferveur, J.-F. (1995) Evolution and genetic control of mate recognition and stimulation in *Drosophila. Behav. Process.* in press.

———, Venard, R., and Jallon, J.-M. (1994) From genes to smell: Studies of olfaction in *Drosophila melanogaster.* In *Sensory Systems in Arthropods,* K. Wiese, E. G. Gribakin, A. V. Popov, and G. Reminger, eds., pp. 462–468. Birkhäuser Verlag, Basel.

DeZazzo, J., and Tully, T. (1995) Dissection of memory formation: From behavioral pharmacology to molecular genetics. *Trends Neurosci.* 18:212–218.

Ferveur, J.-F., Cobb, M., Oguma, Y., and Jallon, J.-M. (1994). Pheromones: The fruit fly perfumed garden. In *The Difference Between the Sexes,* R. V. Short and E. Balaban, Eds., pp. 363–378. Cambridge University Press, Cambridge.

Greenspan, R. J. (1995) Understanding the genetic construction of behavior. *Sci. Am.* 272(4): 88–93.

Hall, J. C. (1994) The mating of a fly. *Science,* 264:1702–1714.

———. (1996). Genes and Behavior in *Drosophila* In *Encyclopedia of Neuroscience,* G. Adelman and B. Smith, Eds. Elsevier, Amsterdam, in press.

———. (1996). Genetics of biological rhythms in *Drosophila.* In *Handbook of Behavioral of Biology: Circadian Clocks,* J. S. Takahashi, F. W. Turek, and R. Y. Moore, Eds. Plenum Press, New York, in press.

———. (1995) Tripping along the trail to the molecular mechanisms of biological clocks. *Trends Neurosci.* 18:218–230.

Hirsch, H. V., and Tompkins, L. (1994) The flexible fly: Experience-dependent development of complex behaviors in *Drosophila melanogaster. J. Exp. Biol.* 195:1–18.

Kaiser, K. (1993) Second generation enhancer traps. *Curr. Biol.* 3:560–562.

Kubli, E. (1992) My favorite molecule: The sex peptide. *BioEssays,* 14:779–784.

Ranganathan, R., Malicki, D., and Zuker, C. S. (1995) Signal transduction in *Drosophila* photoreceptors. *Annu. Rev. Neurosci.* 18:283–317.

Restifo, L. L., and White, K. (1990) Molecular and genetic approaches to neurotransmitter and neuromodulator systems in *Drosophila. Adv. Insect Physiol.* 22:115–219.

Sentry, J. W., Yang, M. M., and Kaiser, K. (1993) Conditional cell ablation in *Drosophila. BioEssays,* 15:491–493.

Taylor, B. J., Villella, A., Ryner, L. C., Baker, B. S., and Hall, J. C. (1994) Behavioral and neurobiological implications of sex-determining factors in *Drosophila. Dev. Genet.* 15:275–296.

Tully, T. (1994) Gene disruption of learning and memory: A structure–function conundrum. *Sem. Neurosci.* 6:59–66.

Wu, C.-F. and Ganetzky, B. (1992) Neurogenetic studies of ion channels in Drosophila. In *Ion Channels,* Vol. 3, T. Narahashi, Ed. pp. 237–244. Plenum Press, New York.

DRUG ADDICTION AND ALCOHOLISM, GENETIC BASIS OF

Judith E. Grisel and John C. Crabbe

Key Words

Congenic Strain   Strain created through the transfer of a gene of interest from a donor or mutant strain onto the genetic background of an inbred strain.

Dependence   The hallmark of addiction; characterized by frequent, regular, and excessive drug use and craving. Accompanied by tolerance and withdrawal.

Genotype   The genetic constitution of an organism.

Inbred Strain   A group of animals that are genetically identical (and homozygous at all genes) except for gender because mating of close relatives (usually siblings) has occurred for at least 20 generations.

Phenotype   The observable properties of an organism, resulting from the interaction of genotype and environment.

Psychoactive Drug   A substance that changes cognition or affect.

Quantitative Trait   A continual graded phenotype (i.e., one that is not all or none).

Quantitative Trait Locus   A gene influencing a quantitative trait.

Selected Line   Animals bred to display a marked degree of a particular phenotype.

Tolerance   Decrease in drug effect with repeated use; conversely, condition manifesting in the need to increase drug dose to obtain the same effect.

Withdrawal   Abstinence syndrome, characterized by responses opposite to those initially induced by the drug.

Although a genetic influence on the etiology of alcoholism has been suspected for centuries and was empirically determined several decades ago, its precise neurobiological underpinnings have only recently begun to be elucidated. Moreover, similar corroboration now accumulating suggests genetic predispositions to abuse other drugs. This article provides a brief overview of some of the evidence supporting the existence of a molecular genetic predisposition to abuse alcohol and drugs and explains the methods employed in studies using genetic animal models.

1   INTRODUCTION

1.1   ETIOLOGY OF GENETIC PREDISPOSITIONS TO SUBSTANCE ABUSE IN HUMANS

People have used distilled spirits, opiates, and other natural or synthetic drugs since the beginning of recorded history. But despite the use of psychoactive substances to varying degrees in many societies, addiction occurs among only a small percentage of those who expose themselves to these substances. In the absence of a definitive biological marker for addiction, the condition is defined behaviorally. Broadly speaking, behavior is labeled addictive when it is excessive, compulsive, and destructive psychologically or physically. The United States has the highest level of psychoactive drug use of any industrialized society, and according to one U.S. Surgeon General, 30% of all deaths in the country are premature because of alcohol and tobacco use.

Determinants of substance abuse include an interaction of biological and environmental factors. For instance, drug availability, a family history of drug abuse, and low economic status have all been associated with the prevalence of addiction. Furthermore, alcoholism runs in families. One-third of alcoholics have at least one alcoholic parent, and children of alcoholics are several times more likely to become alcoholic than children of nonalcoholics, even when they are adopted by nonalcoholic parents. That alcoholism and substance abuse tend to run in families does not itself imply a genetic basis for increased risk (speaking French also runs in families). However, studies with twins and other close genetic relatives have clearly shown that risk increases with genetic similarity to an alcoholic or drug-abusing relative, even if individuals do not share family environments. The tendency to abuse alcohol appears to be codetermined with the tendency to abuse other drugs.

Several lines of research indicate that people who develop problems of substance abuse differ from others even before their drug use begins. Such findings are critical because they indicate that the abnormalities found in alcoholics and addicts are not all due to the effects of chronic drug use, but may predate drug exposure. Characteristics in this subset could then be used as markers to signal a risk for future development of abuse, and possibly to help target individuals for preventive intervention. Psychological studies show that prealcoholics (those tested before drinking begins, who subsequently develop alcoholism) are more independent, impulsive, undercontrolled, and nonconformist than control subjects. This tendency toward what in extreme forms is called "antisocial personality" has also been shown to be heritable, and in addition to a positive family history is a major predictor of subsequent alcoholism.

Some biomarkers of risk are also suspected. Brain electrophysiological evidence suggests that a specific, genetically determined anomaly exists in both alcoholics and their nondrinking offspring. When exposed to novel stimuli, about 35% of 7- to 13-year-old sons of alcoholic fathers, who presumably have not had alcohol themselves, have evoked potentials that are reduced in magnitude, like those of chronic alcoholics. It is not currently known whether this neurological marker (a particular pattern of brain waves) is specifically linked to chemical abnormalities that generate alcohol craving, is associated with emotional or behavioral problems that lead to compulsive drinking, or relates to the behavioral changes associated with alcoholism in some other way. In addition, however, there is evidence that the cellular response to ethanol may differ between alcoholic and nonalcoholic subjects. Lymphocytes from alcoholic subjects, cultured in the absence of alcohol, show enhanced cyclic adenosine monophosphate (cAMP) signal transduction compared to lymphocytes from nonalcoholic subjects. As a cellular second messenger, cAMP plays a critical role in the transmission of information in neurons by activating protein kinases, thereby leading to changes in the neuronal membrane potential. These results indicate that the regulation of cAMP signal transduction may be altered in subjects at risk for alcoholism. Although the possible relationships of these in vitro findings to a predisposition for drug abuse are not

currently clear, the differences may constitute useful markers for alcoholism liability.

1.2 CHARACTERIZING RELEVANT RESPONSES TO DRUGS OF ABUSE

Aspects of drug responses that may be relevant to addiction can be divided into four general classes. The first evaluates sensitivity to the acute effects of a drug. A second category is the response to chronic drug exposure: for example, the neuroadaptation underlying drug tolerance or sensitization, as well as phenomena such as dependence and withdrawal. The third group of responses are those related to drug reward (or aversion), reflecting the reinforcing properties of the drug. Finally, varied patterns of metabolism may lead to differential susceptibility to drugs among individuals. The rather specialized topic of drug metabolism is beyond the scope of this review, but studies in the other domains will be addressed. Regardless of which domain is studied, responses can also be classified as qualitative (i.e., all-or-none), or quantitative (i.e. continually varying). Most drug response traits are quantitative, presumably because they reflect the aggregate influence of multiple genes and multiple environmental influences.

2 GENETIC ANIMAL MODELS FOR STUDYING DRUGS OF ABUSE

The use of animal models has promoted studies of the genetic basis of drug and alcohol abuse. Genetic animal models offer several advantages over studies using human subjects. In particular, the experimenter is in control of every subject's genotype. Whereas aside from monozygotic twins, no two humans have identical genotypes, numerous stable genotypes of rat and mouse are available, which facilitates the sharing of information among laboratories. These animal models are currently being used to assess the role of specific genes in a particular phenotype.

The first step is to study these special genetic strains (usually rats and mice) to see how they differ in response to drugs. Once strain differences in behavior have been determined, attempts can be made to find out which gene products mediate behavior and the mechanisms whereby they exert these effects. Animal models are based on specific phenotypes that may model human responses—genetic studies in animals thus can yield identification of genetic markers that may be useful in predicting drug abuse susceptibility in humans.

Within the general areas of drug effects described in Section 1.2, different animal models have been employed. Some involve selective breeding to produce lines of animals that differ markedly with respect to a particular drug effect. Others take advantage of the ready availability of inbred mouse strains and the frequent variation of drug responses among different strains. Since genetic animal models were first employed in the 1950s, hundreds of studies have used them to study the processes underlying addiction, and reviews of this literature are available (see Bibliography). This article discusses representative examples from the animal literature to give a general overview of this field of research with respect to methodology, progress, and applications.

2.1 SELECTIVELY BRED LINES

One way to study a specific drug effect in order to isolate the genes critical for that response is to breed animals selectively for a trait of interest. Selection is often bidirectional in that animals are chosen to breed based on their extreme phenotype in either of two poles on some behavioral or pharmacological dimension. With increasing generations of selection, the two lines diverge phenotypically as a result of the selection pressure on the genes underlying the behavior of interest. Gene frequencies at genes affecting the selected trait increase (or decrease) until they eventually reach homozygosity. Table 1 lists some selectively bred rodent lines and the drug responses for which they were developed. Behavioral and pharmacological traits that covary with the selected trait are assumed to do so because of the influence of genes in common, which implies a common mechanism of action. Because they differ dramatically, oppositely selected lines not only are useful for understanding the trait of interest, but provide a valuable tool for further hypothesis generation and testing. Selective breeding in the past had been based solely on a phenotypic response and required from 20 to 40 generations of breeding, preferably in duplicate lines, before populations of animals homozygous at all genes affecting the trait of interest were obtained. More recently, as sophistication in statistical and molecular tools has advanced, it has been possible to selectively breed animals on the basis of genotype rather than phenotype (see Section 3.5).

One of the most extensively studied sets of selected lines in behavioral pharmacogenetics consists of the long-sleep (LS) and short-sleep (SS) mice. In these lines, selection was based on the duration of loss of the righting reflex in response to an acute high dose injection of ethanol. Following 14 generations of selection, the ratio of the ED_{50}s (50% effective dose) was over two times greater in the LS mice, and this response differential has remained stable for more than 50 generations. These differences do not appear to be due to differences in ethanol metabolism between the lines; it may be, rather, that they are attributable to differences in the effect a given concentration of ethanol produces in the brain. Research with these lines suggests that the differences between LS and SS mice in response to ethanol involve alterations at the γ-aminobutyric acid (GABA) receptor complex. GABA is the principal inhibitory neural transmitter. Acti-

vating or blocking the activity at this complex (with GABA agonists or antagonists) affects ethanol sensitivity. LS mice are more susceptible to these manipulations than SS mice, suggesting that the effects of ethanol may be different in these lines because of differences in the GABA system. Neural cell preparations from SS mice demonstrate an attenuated flux of chloride ions relative to those from LS mice in response to pharmacological manipulations involving ethanol. Thus our knowledge that the GABA receptor complex operates by gating entry of chloride ions to cells suggests a mechanism to explain the lesser sensitivity of SS mice (vs. LS mice) to the depressant effects of ethanol. The GABA receptor is composed of several protein subunits, and studies involving oocytes expressing various combinations of cloned GABA receptor subunits showed that ethanol sensitivity in this preparation depends on the presence of the γ_{2L} subunit, which differs from the alternative γ_{2S} subunit by only eight amino acids. This region contains a consensus phosphorylation site for the intracellular second messenger, protein kinase C(PKC), and may contribute to the differences seen in these selected lines, although this hypothesis has not been confirmed.

Of the effects of chronic drug treatment, seizure susceptibility following withdrawal from a number of depressant drugs has been intensively studied using genetic animal models. Selected lines of withdrawal seizure-prone (WSP) and withdrawal seizure-resistant (WSR) mice have been bred with respect to their reaction following ethanol dependence and withdrawal (Figure 1). Each generation, these mice were exposed to ethanol vapor for 72 hours and scored for withdrawal severity for 25 hours after removal from the vapor chambers. After 11 generations of selection, the WSP mice demonstrated withdrawal scores about 10 times more severe than WSR mice; this difference has been maintained for nearly 50 generations. WSP mice also exhibit significant dependence following just a single injection of ethanol, while the WSR mice are highly resistant to these effects. Interestingly, these lines show similar differences when withdrawn from diazepam, phenobarbital, pentobarbital, and nitrous oxide, indicating the likelihood of a common neural substrate for

Table 1 Lines Selected for Drug Abuse Related Traits

Substance and Traits	Selection Phenotype
Mouse	
Ethanol	
LS/SS (long/short sleep)	Duration of loss of righting reflex after ethanol
Hot/cold	Body temperature response to acute ethanol
Fast/slow	Ethanol-stimulated open field activity
WSP/WSR (withdrawal seizure-prone and -resistant)	Handling-induced convulsions after ethanol vapor inhalation
Diazepam	
DS/DR (diazepam sensitive/resistant)	Sensitivity to diazepam-induced ataxia
DHP/DLP (diazepam high/low performers)	Sensitivity to diazepam-induced ataxia
Opiates	
HAR/LAR (high/low analgesic response)	Opiate analgesia in the hot-plate test
HA/LA (high/low analgesia)	Analgesia following cold water swim
Rat	
Ethanol	
AT/ANT (alcohol tolerant/nontolerant)	Ethanol impairment of tilting-plant performance
P/NP (preferring/nonpreferring)	Preference for drinking 10% ethanol
HAD/LAD (high/low alcohol drinking)	Preference for drinking 10% ethanol

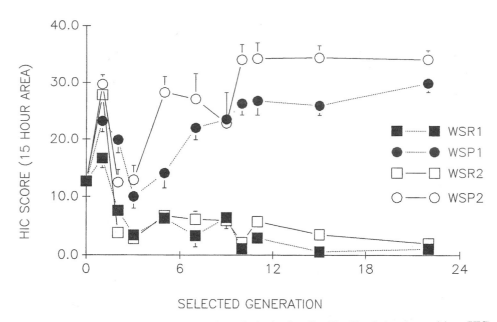

Figure 1. Mean ± SE area under the withdrawal curve over generations of selective breeding. Handling-induced convulsions (HIC) scores indicate seizure susceptibility. In each generation, mice were exposed to identical concentrations of ethanol vapor. Selection was relaxed for two generations ($S_{22}G_{23}$ and $S_{22}G_{24}$), so the $S_{16}G_{16}$ and $S_{22}G_{24}$ data points are before and after relaxation, respectively. (Adopted from A. E. Kosobud and J. C. Crabbe, Genetic influences in the development of physical dependence and withdrawal in animals. In *Genetic Factors and Alcoholism,* H. Begleiter and B. Kissin, Eds., Oxford University Press, Oxford, 1995.)

withdrawal seizures from depressant drugs. This finding may be related to the propensity for multiple addictions in human subjects.

The neural changes associated with chronic exposure to ethanol have also been studied in WSR and WSP mice. There are differences in the number of binding sites for dihydropyridine-sensitive calcium channels after ethanol: WSP mice show a very large increase, while WSR mice show little increase. Because calcium is an important intermediary in neuronal excitability, this variation could help to explain the difference in seizure susceptibility between the lines. In addition, the amount of zinc found in hippocampal mossy fibers is much lower in WSP than in WSR mice. Lower zinc levels are a factor in many animal models of seizure disorders. In addition, there is evidence that chronic ethanol alters GABA function differentially in WSP and WSR mice (by modulating mRNA levels for specific subunits of the GABA receptor). These changes may contribute to the difference in neural excitability seen following drug withdrawal, through an alteration of GABA's inhibitory activity. Interestingly, these lines do not differ with respect to many other ethanol effects such as psychomotor stimulation or tolerance development, suggesting that the mechanisms underlying sensitivity, tolerance, and dependence are at least partially independent.

In some cases selected lines have been used to identify a specific gene. Korpi, Kleingoor, Kettenmann, and Seeburg, for instance, recently identified a point mutation responsible for the difference in benzodiazepine sensitivity in rats selectively bred to be alcohol nontolerant and alcohol tolerant (ANT and AT). The ANT line is highly susceptible not only to alcohol-induced impairment of coordination but to impairment of postural reflexes by both benzodiazepine and barbiturate agonists. In this case, divergence of the two lines was very rapid, reaching statistical significance after just a single generation of selection. This rapid divergence may reflect a phenotype that is influenced by only one or a few genes. In fact, these authors showed that a single mutation in the α6 subunit of the GABA

receptor complex was responsible for this difference. Although the α6 gene of ANT rats is expressed at wild-type levels, it carries a point mutation generating an arginine-to-glutamine substitution at position 100. The arginine at this particular location had been shown to be necessary for the diazepam insensitivity of the wild-type GABA receptor, and in general it is thought to play a critical role in modulating chloride flux at the GABA receptor. The vast majority of drug-related responses are unlikely to be mediated by a single residue difference, and it does not appear that this mutation is responsible for the differences between these two lines in alcohol sensitivity, but it is interesting nonetheless. For many of the more complex genetic influences on these drug-related phenotypes, more sophisticated possibilities are being investigated.

A common method for studying the "abuse liability" for a particular substance in humans utilizes self-administration experiments in animals. Self-administration models for the evaluation of genetic differences measure either the amount of drug-adulterated water an animal chooses to consume compared to plain water or the likelihood that an animal will emit a response (e.g., bar pressing) that results in drug delivery. The conceptual background for these studies derives from research involving electrical self-stimulation of particular brain regions. From initial experiments in which rats were shown to work in order to stimulate their own brains electrically, a model of the "neurobiology of craving" has been developed. Drugs that have addictive potential activate the neurotransmitter dopamine in a specific area of the brain called the mesolimbic pathway. Opioids, stimulants, nicotine, barbiturates, some benzodiazepines, alcohol, and phencyclidine (PCP) are all self-administered by animals, release dopamine in this midbrain area, and have high abuse liability in humans. Of particular research interest are the neural substrates that determine differences in the likelihood that an individual will self-administer a particular drug.

A number of selected rodent lines have been developed to prefer or reject alcohol solutions: preferring (P) and nonpreferring (NP), high (HAD) and low (LAD) alcohol drinking rats, among others. The P and HAD lines typically drink about 5–9 g of ethanol per kilogram (kg) body weight per day, while the NP and LAD lines consistently self-administer less than 1 g/kg/day in the same free-choice paradigm. Research from the laboratories of Li and his colleagues indicates that differences in the reward circuit involving mesolimbic dopamine may contribute to these differences in alcohol preference. The alcohol-preferring HAD and P rats have lower levels of dopamine and dopamine metabolites in the nucleus accumbens. Furthermore, administration of dopamine agonists has been shown to decrease ethanol consumption in P rats. There are fewer dopamine D_2 receptors in P than in NP rats in several brain regions, including the nucleus accumbens and the related ventral tegmental area. Differences in endogenous opioids (i.e., endorphins and enkephalins) as well as in other monoamine levels (including serotonin) have also been related to these phenotypic differences. For instance, shortly following ethanol administration there is a transient (gone by 24 h) fourfold increase in mRNA for pro-opiomelanocortin, the precursor protein that gives rise to β-endorphin in P but not NP rats. P rats also have higher levels of proenkephalin, the precursor peptide to leu- and met-enkephalin.

2.2 INBRED STRAINS

Inbreeding is a process involving several generations of sister–brother matings, resulting in a reduction in genetic variance within a population to virtually zero. All animals within a strain become virtually identical twins. Inbred strains are useful because differences among strains are primarily of genetic origin, while differences among individuals within a strain are nongenetic. Comparisons of strain means for multiple traits can therefore be used to estimate genetic correlation in a manner analogous to the comparison of oppositely selected lines. Over 30 years ago it was found that certain inbred mouse strains (e.g., DBA/2 and BALB/c) tend to avoid alcohol while others (C57BL/6 in particular) readily self-administer an ethanol solution in preference to water. Although the precise causes of this variation are not known, this behavior appears to be polygenically determined, and it involves differences in the salience of sensory stimuli (such as taste and odor), monoamine levels in particular brain regions, and endogenous opioid levels. For example, Gianoulakis and her colleagues have found that endogenous levels of β-endorphin increase in a dose-dependent manner after ethanol administration and do so to a greater degree in C57BL/6 mice than in DBA/2 mice. These differences were also present in human subjects who were either at a high (three-generation history of alcoholism) or low risk for future development of alcoholism: individuals in the high risk group had lower basal levels of β-endorphin-like immunoreactivity than members of the low risk group. Furthermore, in response to a low dose of ethanol, only subjects in the high risk group showed an increase in β-endorphin.

Amphetamine and cocaine greatly increase the activity of rodents (and people), and this effect is known to depend in part on genetic determinants. One reason researchers are interested in the stimulating effects of these drugs is that they may be related to the rewarding effects (and thus the abuse potential) of the drug. Inbred strains vary widely in their response to an injection of amphetamine, with some strains (e.g., C57BL/6) exhibiting hyperactivity and others, such as BALB/c mice, showing amphetamine-induced inhibition of locomotion. It has been determined that the locomotor response to amphetamine is polygenic. F_1 hybrids derived from these progenitor strains have also been evaluated for their locomotor response to an acute dose of amphetamine. In this case, offspring behaved similarly to their C57BL/6 parent, so the allele(s) responsible for an excitatory response to amphetamine appear to be generally dominant to those leading to an inhibitory effect. Similar differences exist between these two strains with regard to the acute effects of cocaine on locomotor activity. However, in this instance, F_1 hybrid strains exhibit an intermediate response to cocaine; thus the cocaine response is incompletely dominant. This phenotype is also polygenic. However, different strain distribution patterns regarding the psychomotor effects of the two drugs indicate that the loci underlying the responses to amphetamine and cocaine are not the same, a conclusion consistent with the different patterns of dominance.

3   NEUROBIOLOGICAL SUBSTRATES/ CANDIDATE GENES

In 1993 Nestler, Hope, and Widnell reviewed much of the molecular evidence for the neuroadaptation that underlies addiction to stimulants and opiates. They focused their paper on two brain regions, the mesolimbic dopamine system (see Section 2.1) and the locus ceruleus. The locus ceruleus, the major noradrenergic nucleus in the brain, has been implicated in opiate dependence and withdrawal since the 1970s. It is now known that opiates inhibit locus ceruleus neurons through regulation of G proteins, leading to an inhibition of adenylate cyclase, reduced levels of cAMP, and reduced cAMP-dependent protein kinase activity. Since a primary function of cAMP-dependent kinase is to phosphorylate cellular proteins, acute opiate administration results in a general decrease in phosphorylation, thereby affecting many neuronal processes. More specifically, in addition to inhibiting the activity of these neurons, there is decreased catecholamine synthesis (caused by a reduction in phosphorylation of tyrosine hydroxylase, the rate-limiting factor in catecholamine synthesis), alterations in gene expression via decreased phosphorylation of transcription factors such as the cAMP response element-binding protein (CREB), and reduced expression of immediate early genes such as c-fos. Following chronic morphine, there is adaptation to all these effects, including an increase in adenylate cyclase, cAMP, and cAMP-dependent kinase, and subsequent recovery of neuronal firing rates toward control levels. CREB phosphorylation returns to normal levels, and binding activity in this brain region increases for CRE (part of the DNA sequence that binds CREB). Withdrawal from opiates in the dependent organism results in a marked elevation in locus ceruleus firing rates relative to control levels, and this increase in cell excitability is both necessary and sufficient for producing many of the behavioral signs of opiate withdrawal.

There is evidence for genetic differences in the activity of locus ceruleus cells that may relate to differences in drug preference. Basal levels of adenylate cyclase and cAMP-dependent protein kinase activity are higher in the locus ceruleus of Lewis rats (an inbred strain that shows a high preference for drugs of abuse) as compared to nonpreferring Fischer 344 rats, for example.

The mesolimbic dopamine system, important for the reinforcing effects of drugs of abuse (see Section 2.1), is also altered by chronic drug administration, through similar mechanisms. Chronic opiates result in an upregulation of adenylate cyclase and cAMP-dependent protein kinase in the nucleus accumbens, but not in the ventral tegmentum. However, chronic morphine (or cocaine or ethanol) increases levels of tyrosine hydroxylase, an important enzyme for catecholamine biosynthesis, and decreases levels of the three major neurofilament proteins (NF-200, NF-160, and NF-68) in this brain region. These findings point to the changes in dopaminergic transmission and neuronal structure seen in the addicted state, which may contribute to the functional change (supersensitivity) seen in dopamine (D_1) receptors with chronic use.

Research by this group of investigators has supported the contention that differences in reactivity of this brain region to drugs of abuse may underlie some of the genetic differences that contribute to a vulnerability toward drug addiction. Lewis rats have much lower levels (only in the ventral tegmentum) of the three neurofilament proteins just mentioned than do Fischer rats. Furthermore, while chronic morphine decreases these protein levels in the ventral tegmentum of Fischer rats, it does not affect the neurofilaments of Lewis rats. These results were found to be generalizable to the P (preferring) and NP (nonpreferring) selected lines as well, in that the P rats have lower levels of these neurofilaments in the ventral tegmental area than NP rats. Also in this brain region, there appears to be a strain difference with respect to tyrosine hydroxylase: morphine increased the activity of this enzyme in Fischer rats but not in Lewis rats. Because these differences have been established in two rat lines only, it will be necessary to replicate them in other strains that prefer or avoid drugs of abuse.

Not only do we now know that a liability to drug addiction is partially genetically transmitted, but we have some idea of ways in which drug responses may differ between abuse-prone and non-prone subjects. Molecular techniques have recently been brought to bear on these findings, and the precise genetic influences are beginning to be uncovered. Genetic strategies such as quantitative trait locus (QTL) mapping and chemical induction of mutations are particularly useful in discovering the role of unknown genes. Other techniques, such as the use of transgenic and null mutant populations, or antisense procedures, are valuable for testing the role of particular candidate genes by permitting direct manipulation of the genetic material in a cell by specific deletions or additions, which are studied in light of their influence on addictive processes.

3.1   TRANSGENIC AND NULL MUTANT LINES

Transgenic mice are currently being developed to enable the testing of specific hypotheses related to alcohol and drug abuse. These mice are produced to either overexpress or underexpress particular gene sequences; thus they vary in their levels of protein products compared to control animals. Techniques commonly employed in the development of these strains include retroviral infection of embryos, manipulation of embryonic stem cells, and direct alteration of fertilized mouse eggs (through ribonuclease injection).

Transgenic mice overexpressing human transforming growth factor α (TGFα) have been studied for alcohol sensitivity by Hilakivi-Clarke and her colleagues. These mice are known to differ from controls in a number of ways. Most notably, they show high levels of aggression and appear to be more sensitive to the physiological effects of alcohol as measured by the loss of righting reflex induced by an injection of ethanol. However, there are no differences in ethanol metabolism or in ethanol self-administration between these transgenics and controls. These mice may serve as a useful model in the search for the mechanism underlying sensitivity to ethanol.

Jeanne Wehner is one of the researchers studying the γ-PKC null mutant mice developed by Abeliovich and his colleagues. This enzyme is involved in a wide array of second messenger activities, but homozygous PKC knockouts are viable and exhibit normal fertility. These mice show decreased behavioral sensitivity to acute ethanol injections as well as an attenuated effect of ethanol on GABA-stimulated chloride flux, although they do not appear to differ from wild-type controls in terms of benzodiazepine or barbiturate responsiveness. Wehner and her co-workers also have been studying sensitivity and tolerance to ethanol in c-fos null mutants, which were found to be much more sensitive to the acute effects of ethanol as well as lacking evidence of tolerance in response to repeated daily injections. Heterozygotes with respect to the c-fos gene demonstrated intermediate levels of initial sensitivity and tolerance (compared to homozygotes or wild-type mice), indicating that the c-fos gene may be important for the expression of some of ethanol's actions.

Another transgenic mouse line makes use of the reporter gene bacterial chloramphenicol acetyltransferase (CAT) to monitor tyrosine hydroxylase transcription in the rat. Chronic morphine was found to increase CAT expression in these mice, providing further support for the suggestion that chronic opiates regulate dopaminergic transmission. René Hen has developed transgenics deficient in serotonin 1B receptors. This is another mutation likely to have implications for substance abuse, since these receptors may play a role in drug responses such as the locomotor effects of cocaine and ethanol self-administration. Behavioral testing of these mice is just getting under way.

3.2   CONGENIC STRAINS AND SUBTRACTIVE HYBRIDIZATION METHODS

Congenic strains are produced through successive generations of backcrossing and selection so that a small region of chromosome containing a gene of interest is transferred onto the genetic background of an inbred strain. Both the drug responses and the genotype of the inbred strain are likely to be well characterized for the phenotype of interest, and thus the differences between the congenics and known inbreds can be attributed to the influence of the backcrossed mutation, acting either directly or through its interaction with the rest of the genome.

When the genes of interest are unknown, the technique of polymerase chain reaction (PCR) differential display is proving useful. This approach compares cDNA fragments from control and experimentally treated animals (e.g., those given a drug). Products differentially represented in the two groups signify genes whose expression has been increased or decreased by the experimental manipulation. The cDNA sequences can then be used to develop genetic probes, which in turn can be used to clone the genes affected by the drug.

The portion of the genome engineered by congenics or identified by differential display may be small enough to permit chromosome walking to a target gene or large enough to contain restriction fragment length polymorphisms (RFLPs). Subtractive techniques such as genetically directed representational difference analysis (GDRDA) can be used to identify previously uncharacterized differences between two DNA samples that are genetically identical except for the region of interest. In this way new polymorphic markers linked to a phenotype can be identified in the absence of any a priori knowledge of their influence. GDRDA resolution is promising because it should be limited only by the density of polymorphisms, estimated to be 1 to 2 per megabase.

3.3 ANTISENSE STRATEGIES

The use of antisense oligonucleotides to block the activity of specific genes is another technique that has been applied to the addiction field. It can be used in the developing zygote to create a transgenic lines (as discussed in Section 3.2) or, more readily, it can be applied to transiently arrest synthesis of a particular protein. In the latter application, protein translation is blocked by treatment with short deoxyoligonucleotides that are complementary to the ribonucleotide sequences of specific mRNAs. As this is a relatively straightforward technique, it is more easily utilized by pharmacologists and behavioral researchers who do not have an extensive background in molecular strategies. For example, injection of D_2 antisense into the nucleus accumbens greatly reduces the reinforcing properties of cocaine as measured by operant self-administration in rats. Another recent study using antisense oligonucleotides suggests a role for c-*fos* in amphetamine withdrawal. However, a number of technical issues are involved, both in successful application of the method and in the interpretation of antisense results, and antisense experiments may be successful in only a minority of attempted applications.

3.4 RECOMBINANT INBRED STRAINS AND QUANTITATIVE TRAIT LOCUS GENE MAPPING

Quantitative traits such as the ones just discussed can be effectively studied in recombinant inbred (RI) strains, and genetic loci contributing to such phenotypes are currently being identified using these animals. Originally developed to detect major gene effects (i.e., qualitative traits), RI strains are currently being exploited to identify the genes underlying quantitative, polygenic traits, in which the combined influence of many genes interacting with the environment produces a particular phenotype, as is the case in the vast majority of drug responses. Some of the major characteristics distinguishing qualitative and quantitative traits are listed in Table 2.

RI strains typically are inbred strains derived from the F_2 cross originating from two inbred progenitor strains. In the F_2 generation, recombination (crossing over) leads to the production of many genetically distinct individuals according to the Mendelian principles of segregation and independent assortment. For each gene for which the two progenitor strains possess different alleles, an F_2 mouse can possess one of three genotypes. A pair of F_2 mice is then mated to

Table 2 Major Distinguishing Characteristics of Qualitative and Quantitative Traits

Qualitative Traits	Quantitative Traits
Discrete distribution according to Mendelian ratios	Continual distribution (i.e., normal distribution)
Monogenic or single-locus determination	Polygenic or multilocus determination
Genotype often determinable from phenotype	Genotype only probabilistically related to phenotype
Relatively rare in nature	Very common in nature
Origin in macromutations	Origin in micromutations

Source: Adapted from S. D. Tanksley, Mapping polygenes. *Annu. Rev. Gene.* 27:205–233 (1993).

"fix" its unique pattern of recombinations. Following many generations of inbreeding, the resulting animals in each such RI strain become genetically homozygous at all loci. Of necessity, each such animal possesses two copies of the allele from one or the other progenitor parent. Drug responses can then be tested in both parent strains and a battery of their RIs. Under controlled conditions, these mice will differ according to their unique genotypes, and strain means can be compared to previously typed genetic maps. When a relationship between a strain distribution pattern (SPD) and a pattern for a mapped gene is determined, a locus associated with the trait is assumed to be linked to the previously identified marker. Loci such as these are called quantitative trait loci. The QTL method of gene mapping allows identification of possible chromosomal positions of genes that influence polygenic traits. Because of the large degree of linkage conservation between the mouse and human genomes, most QTLs mapped in the mouse will indicate likely map locations in the human genome suitable for testing.

The principle of linkage (the tendency of genes close to each other to be inherited together) is a critical component of QTL mapping. The unit of measurement of chromosome length, the centimorgan (cM), is based on linkage as well and is defined as the chromosome length showing a 1% frequency of recombination, which represents about a million DNA base pairs. Meiotic chromosome recombinations (crossing over), which occur during the development of RI strains, interrupt linkage of loci along the chromosome. The frequency of recombination of the progenitor genes found in the homozygous RI strains is therefore directly related to their map distance apart. Correlation of the RI strain means for a given behavioral or pharmacological trait with the allelic distribution of previously mapped marker loci permits provisional mapping of QTLs that may exert an influence on the trait. It should be obvious that one of the keys for QTL mapping is having a large marker array. Marker identification has grown considerably with the recent advances in the characterization of RFLPs and simple sequence length polymorphisms (microsatellites). The essence of QTL mapping is simply to correlate allelic variation at a marker with phenotypic variability in a population. A significant correlation means that a possible QTL influencing the trait is near the marker locus—in other words, that they are linked. No correlation means that there is no QTL linked to the marker. A consequence of the large number of correlations evaluated in these analyses is that false positives (type I errors) are likely to be generated. Therefore, these QTLs are considered to be candidate sites until they can be confirmed in an F_2 population that has been screened for the provisional QTL. Individual F_2 animals are phenotyped and genotyped for markers at the locations indicated by the provisional QTL analysis in RI strains. Only associations seen in both analyses are considered to represent QTL affecting the trait.

The BXD RI series, derived from the cross of C57BL/6J and DBA/2J parent strains, is presently one of the most extensively studied. There are 26 RI strains resulting from this cross, each of which is fixed genetically. What has so greatly empowered this mouse model is the vast library of known markers on these strains. Indeed, over 1300 loci have been mapped in these RI strains, making QTL analysis particularly productive.

Most drug responses appear to be coordinately determined by several QTLs. Ethanol acceptance, one measure of how readily an ethanol solution will be consumed, was found to be associated with multiple distinct QTLs. These QTLs were located on several regions of several chromosomes, but together they accounted for 95%

of the variability associated with ethanol acceptance. Other phenotypes have been related to these particular QTLs, and so some areas of a chromosome appear to be associated with multiple traits. In some cases, multiple QTLs even appear to codetermine responses to different drugs. For example, two QTL markers, *Car-2* and *Ly-9* (found on chromosomes 3 and 1, respectively) have been found provisionally to be correlated with both ethanol withdrawal severity and amphetamine hyperthermia. Occasionally the correlation between a QTL and a phenotype is so high that the technique may be identifying the precise locus of influence. When evaluated across 20 inbred strains, one of the QTLs for ethanol acceptance is apparently identical or very closely linked to the gene *Ltw-4*, on chromosome 1. The next step here would be to determine the mechanisms by which the products of this locus influence the behavioral phenotype. Table 3 lists some drug-related traits for which QTL analyses have suggested the presence of QTLs in two regions of mouse chromosome 9.

Combined with other advances in molecular biology, these techniques show promise for even greater application. For example, several QTL candidates that may influence the severity of ethanol withdrawal have been identified. To confirm the influence of one of these sites, near *Pmv-7* on mouse chromosome 2, F_2 offspring derived from C57 and DBA parents were individually tested for acute ethanol withdrawal, and genetic markers in this region were tested using PCR amplification. A significant association of this phenotype with a very closely linked (4 cM distance) PCR marker locus, *D2Mit9*, was found. This confirmation in an independent, genetically segregating population is particularly convincing and establishes that there is a QTL in the *Pmv-7/D2Mit9* region of chromosome 2 which accounts for about 40% of the variance in acute ethanol withdrawal. Furthermore, chronic ethanol withdrawal severity and nitrous oxide withdrawal are linked to markers in the same region, indicating that genes near this locus confer susceptibility to withdrawal from multiple drugs of abuse. Candidate genes

Table 3 Drug-Related Phenotypes Significantly Correlated with Genetic Materials on Chromosome 9 in the BXD RI Strains[a]

Phenotypes Correlated with the Dopamine D2 Receptor Gene Locus Drd_2 at 28 cM

Quinpirole (D_2 agonist) hypothermia*
Haloperidol (D_2 antagonist) catalepsy*
Voluntary consumption of methamphetamine/saccharin
Methamphetamine hypothermia, 4 mg/kg
Methamphetamine place preference
Ethanol hypothermia, 3 g/kg
Voluntary consumption of 10% ethanol*
Voluntary consumption of morphine/saccharin
Locomotor activity after saline

Phenotypes Correlated with Markers at 37–39 cM

Morphine-induced activity
Morphine-induced analgesia*
Morphine-induced Straub tail (behavioral index of morphine effect)
Conditioned place preference to ethanol*
Ethanol hypothermia
Ethanol anesthesia (loss of righting reflex)

[a]Phenotypes marked with an asterisk have been confirmed by F_2 analysis; all others are tentative at present.

near *D2Mit9* include *Gad-1,* which is related to neural excitability in that it codes for glutamic acid decarboxylase, the rate-limiting enzyme catalyzing synthesis of the inhibitory neurotransmitter GABA. Other candidate genes near this locus are those that encode the α subunit of brain voltage-dependent sodium channels, also related to neural excitability and possibly to withdrawal severity.

Similar to the effects of alcohol, the response to opiates such as morphine is highly variable within a species. For instance, some people are insensitive to the analgesic effect of morphine, while others (about 30%) experience substantial pain relief from a placebo. One early study tested inmates of a Boston jail for their subjective response to morphine. While many found the experience pleasurable, as would be expected (euphoria is typically thought of as one of the primary effects of this drug, contributing to its high abuse liability), others reported that the drug was aversive and that they would not wish to repeat the experience. There is also evidence for high variability from animal models. C57BL/6J mice will consume 200–300 mg/kg/day of morphine in 0.2% saccharin. This dosage is clearly psychoactive and results in withdrawal upon removal of drug. In sharp contrast, DBA/2J mice will not consume more than 10 mg/kg/day under the same free-choice conditions. Oral morphine preference values have been established in two-bottle choice studies in the BXD recombinant inbred strains. These morphine preference scores show continuous distributions that are consistent with polygenic traits.

Berrettini and his colleagues recently employed QTL analysis in attempts to understand the underpinnings of such differences. In their experiment, 606 F_2 generation mice were assessed for morphine consumption. Three QTLs influencing oral morphine preference were identified on mouse chromosomes 10, 6, and 1. These three QTLs together accounted for about 85–90% of the genetic variance, and about 50% of the total variance in morphine consumption between the two parental strains. The markers identified on chromosomes 10 and 6 have been independently correlated with morphine consumption, confirming their influence on this phenotype. Delineation of these loci may reveal important clues to human genetic vulnerability to drug addiction. At present only one of the genes mapped is near a candidate locus that might explain the differences related to morphine consumption. The μ-opioid receptor gene (*Oprm*) has recently been mapped to this same region of chromosome 10, raising the possibility that the QTL and *Oprm* are one and the same. The authors suggest that these QTLs could mediate a genetic predisposition in C57BL/6J mice to experience drug-mediated euphoria that is more pronounced than that experienced by DBA/2J mice.

A recent review attempted to synthesize the available QTL data for responses to multiple drugs of abuse. Several regions in which provisional QTLs for multiple drug effects were seen were identified. Figure 2 depicts some of the marker loci for chromosome 9 that were significantly associated with provisional drug response QTLs. Chromosome 9 is relatively rich with respect to genes influencing drug-related phenotypes compared to other chromosomes. As is evident from Figure 2, the QTLs are not randomly distributed across the chromosomes but seem to be clustered at particular chromosomal regions. It is important to note that most of these QTLs have not been subject to confirmation testing in F_2 mice, hence are provisional at this time.

The principles of QTL analysis can also be applied to other crosses, such as those derived from selectively bred lines, although at pre-

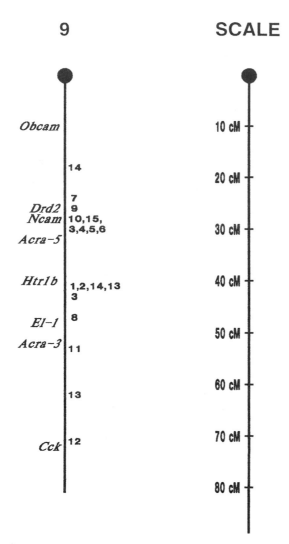

Figure 2. Schematic representation of mouse chromosome 9 depicting locations of provisional quantitative trait loci that affect drug responses. Scale readings are in centimorgans from the centromere (large circle at top). Candidate genes are indicated to the left of the chromosome: *Obcam,* opioid-binding protein; *Drd2,* dopamine D$_2$ receptor; *Ncam,* neural cell adhesion molecule; *Acra-5,* α5 subunit of acetylcholine nicotinic receptor; *Htr1b,* serotonin 1B receptor; *El-1,* mouse epilepsy locus; *Acra-3,* α3 subunit of acetylcholine nicotinic receptor; *Cck,* cholecystokinin. Numbers represent drug-related traits associated with nearby markers (not shown) in BXD RI strains ($p < .01$): 1, loss of righting reflex sensitivity; 2, chronic ethanol withdrawal severity; 3, sensitivity to ethanol ataxia; 4, tolerance to ethanol ataxia; 5, tolerance to ethanol hypothermia; 6, ethanol-conditioned place preference; 7, consumption of 10% ethanol; 8, ethanol-conditioned taste aversion; 9, methamphetamine locomotor stimulation; 10, methamphetamine consumption in saccharin; 11, methamphetamine anorexia; 12, methamphetamine temperature sensitivity; 13, cocaine-induced locomotor activity; 14, morphine-induced Straub tail; 15, haloperidol-induced catalepsy. [Adapted from Crabbe, et al., (1994), which gives the source of data for the drug response traits.]

sent most of these lines are not as well characterized with regard to marker loci: thus this technique has not been widely employed in these lines. One obvious advantage of using RI strains from selected lines is that the parental stocks differ specifically on the trait of interest. A project is under way to analyze RIs developed from the

LS and SS mice that differ on ethanol-induced loss of righting reflex as a measure of sensitivity to ethanol.

3.5   FUTURE RESEARCH AND CONCLUSIONS

Genotypic (or marker-assisted) selection can also be carried out when there is strong evidence supporting the influence of particular genes on a behavior. In these cases it is possible to determine breeding based on PCR genotyping. This approach takes only a single generation to produce a population genetically homozygous for the putative genes of influence (flanked by the selected markers), and it obviates the need for extensive behavioral testing in each generation. In addition, the method precludes the problem of genetic drift, which often arises after many generations of selection. With phenotypic selection, the gene pool is altered in ways that were not intended (i.e., inbreeding inadvertently occurs for genes other than those affecting the trait of interest because of the fiscal necessity to limit breeding populations). Of course, the main limitation of genotypic selection is that the genes influencing a phenotype are often unknown. Furthermore, as is evident in the phenotypic selections, in most cases the high and low responding lines demonstrate polygenic determination, further evincing the complex nature of these phenotypes. When one or even a few genes are thought to influence a trait, genotypic selection will enable relatively rapid and definitive hypothesis testing, but in more obscure cases less direct methods will be necessary.

Genotypic selection is practiced to develop congenic lines by backcrossing a particular allele of interest onto an inbred strain. This strategy would be particularly useful with respect to subtractive-based techniques like subtractive hybridization, or GDRDA. It would also be of interest to produce lines with multiple QTLs (e.g., one could load a group of QTLs known to influence alcohol preference onto DBA mice). Mice with the desired alleles at microsatellites flanking a QTL of interest can be selectively bred, allowing regional transfer of DNA containing risk-associated or protective QTLs onto background strains, thus further enabling characterization of their mechanisms of influence. This, like strategies utilizing multiple knockout mice, would forward understanding of the ways in which many gene products interact to result in different phenotypes.

Recent progress in mapping strategies suggests that a bridge is being built, effectively linking molecular techniques with the use of genetic animal models. This merger, as applied to the study of drug abuse, is likely to result in significant advances in our understanding of the specific relationship between genetics and behavior. Addiction researchers working at the level of behavior and those whose focus has been on the genome are beginning to find common ground in this interdisciplinary neurobiology, enhancing our understanding of the addictive process.

See also MOUSE BEHAVIOR, GENETIC APPROACHES TO; NEUROPSYCHIATRIC DISEASE.

Bibliography

Alcohol and Health. (1993) Eighth Special Report to the U.S. Congress. U.S. Department of Health and Human Services, Washington, DC.

Crabbe, J. C., and Harris, R. A., Eds. (1991) *The Genetic Basis of Alcohol and Drug Actions.* Plenum Press, New York.

———, Belknap, J. K., and Buck, K. J. (1994) Genetic animal models of alcohol and drug abuse. *Science,* 264:1715–1723.

Kalivas, P. W., and Samson, H. H., Eds. (1992) The neurobiology of drug and alcohol addiction. *Annals of the New York Academy of Sciences*, Vol. 264.

Koob, G. F., and Bloom, F. E. (1988) Cellular and molecular mechanisms of drug dependence. *Science*, 242:715–723.

Lee, T. N. H., Ed. (1995) *Molecular Approaches to Drug Abuse Research*, Vol. III. NIDA Research Monograph, U.S. Department of Health and Human Services, Rockville, MD.

Nestler, E. T., Hope, B. T., and Widnell, K. L. (1993) Drug addiction: A model for the molecular basis of neural plasticity. *Neuron*, 11:995–1006.

Trends in Pharmacological Sciences. (1992) Whole issue, 13(5).

DRUG SYNTHESIS

Daniel Lednicer

Key Words

Agonist   A substance that elicits the same response as the hormone secreted by the body (i.e., the endogenous hormone).

Antagonist (sometimes, **Blocker**)   A substance that prevents the response of the endogenous hormone.

Antibiotic   A compound that kills bacteria without affecting the host. Usually a compound not present in the host, which blocks a mechanism involved in the bacterial life cycle.

Antidiabetic   A compound that lowers blood sugar in non-insulin-dependent (adult onset) diabetes.

β-Adrenergic   The subdivision of the adrenergic system that influences such responses as heart rate, blood pressure, and bronchial constriction.

Chiral   Describing a molecule consisting of nonsuperimposable isomers that are mirror images of each other.

Diuretic   A drug that reduces the volume of body fluid and associated minerals (Na^+, K^+, etc.) by increasing urine flow; used to treat edema and hypertension.

Drug Metabolism   The chemical modification of drugs by the host after administration; may increase or decrease activity and sometimes toxicity.

Enantiomer   One of two isomers of a chiral compound.

Gram Positive (or Negative)   Reaction of bacteria to a dye known as Gram stain. Ability to accept the dye is roughly correlated with sensitivity to antibiotics. Early penicillins were effective only on Gram-positive organisms.

Hormones   Chemical substances, synthesized by various structures in the body, that serve to elicit biological responses at ei-

ther a proximate or a remote site; can be (simple molecules (e.g., adrenaline) or complex peptides (e.g., growth hormone).

Lead   Usually the first compound to display activity on a bioassay; The basis for the design of all related compounds.

Medicinal Chemistry   The area of chemistry that deals with drugs; includes the design and preparation of drugs as well as related areas such as the study of their disposition, or metabolism.

Receptors   Specific structures on cells that bind hormones and/or drugs; the binding event results in biological response. Binding of antagonist blocks response to hormone.

Screen   A bioassay used to test compounds to determine initial biological activity; designed to serve as a model for the desired therapeutic effect. May be carried out either in vitro (e.g., antibacterial) or in small animals (e.g., analgetic). Many initial screens are designed for high throughput.

Structure–Activity Relations (SAR)   Studies of the effect on biological activity of modification of chemical structure within a series of active compounds. Used extensively in optimizing the structures of leads.

Drug synthesis encompasses the organic chemical manipulations involved in the preparation of pure compounds intended to have useful biological activity for the treatment of disease states. The design of target molecules involves the closely allied area often called drug design. The term "drug synthesis" is closely associated with and often regarded as a segment of medicinal chemistry; it also encompasses the organic synthetic methodology used for synthesizing drugs on a commercial scale.

1   INTRODUCTION

The use of chemical compounds for treating disease predates chemistry by at least a millennium. Empirical ingestion of plants and plant extracts as potential cures and palliations for mankind's ills led quite early on to a sizable pharmacopeia. With the rise of technology came attempts to improve on the specificity of those botanicals by use of various purification techniques. The isolation of the pure organic compound morphine from opium in 1803 thus actually predates Wohler's synthesis of urea (1828), an event often held to mark the beginning of organic chemistry.

With the rise of organic chemistry it became recognized that most if not all items in the pharmacopeia owed their activity to constituents that were usually organic compounds. The shortcomings of the early drugs combined with the development of techniques for the manipulation of the structure of organic compounds led to the nascence of drug synthesis as a technique for arriving at effective therapeutic agents.

2   SYNTHESIS IN DRUG DISCOVERY

2.1   LEADS FOR SYNTHETIC TARGETS

2.1.1   Natural Products

Opioids   Several overlapping strategies have been used over the years to develop useful synthetic drugs. One of the earliest and still prominent approaches starts with a natural product that has demonstrated biological activity. Observation of such activity may range all the way from folkloric usage, as in the case of opium, to complex in vitro or in vivo screens assays, as in the case of many antibacterial compounds. The first identified natural product more often than not suffers some serious shortcoming that prevents its use as a therapeutic agent: lack of oral activity, perhaps, or undesirable side effects, too short a duration of action due to rapid metabolism, or even serious supply problems. In a typical example, the structure of the active natural product is used as a starting point in designing simplified structures to attempt to find the smallest portion of the molecule that retains activity.

The natural product morphine (Figure 1) found early and widespread application in medicine because of its potent analgetic (i.e., pain-killing) activity. The serious shortcomings of this drug—most prominently, addiction potential and respiratory depression—provided some of the impetus for synthesis programs aimed at better-tolerated drugs (Figure 1). Analgetic activity, it was found, was retained in the face of deletion of major parts of the molecule as in the totally synthetic molecule levallorphan; activity is retained even when yet another ring (benzomorphan A) is deleted. Meperidine, a simple phenylpiperidine, actually prepared a good many years prior to the two bridged polycyclic compounds, represents the almost ultimate simplification from morphine. The simplified molecules all show the same shortcomings as the parent; replacement of the N-methyl group by N-dimethylallyl in a benzomorphan leads to pentazocine, an analgetic with reduced addiction potential.

β-Lactam Antibiotics   Three prime examples are the penicillins, the cephalosporins, and the monobactams.

The serendipitous discovery that the mold metabolite *penicillin* is a selective toxin to many classes of pathogenic bacteria but has virtually no effect on healthy mammalian cells led to its large-scale use as soon as technology had been developed for its production in quantity. This first antibiotic, later renamed penicillin G, had some very serious shortcomings, which included activity against a limited selection of bacterial strains (Gram positive), very quick elimination from the body, poor stability, and lack of activity when taken orally. The search for more stable congeners, which at first

involved careful examination of related substances produced by the mold, led to the isolation of penicillin V. This compound, in which the phenylacetyl side chain ($C_6H_5CH_2CO$) is replaced by phenoxyacetyl ($C_6H_5OCH_2CO$), proved to be more stable and had a longer biological half life. Fermentations were steered to produce this compound in preference to penicillin G by adding the precursor phenoxyacetic acid, required for the side, chain to the medium. This process has been so highly optimized that the compound is now classed as a tonnage chemical. Its basic structure is shown in Figure 2.

The subsequent availability of the parent molecule, 6-aminopenicillanic acid (6-APA), also shown in Figure 2, either by modified fermentation conditions or by a highly efficient process for removing the side chain of penicillin V, opened the way to the preparation of a host of analogues that included modified side chains. These are all obtained by reaction of 6-APA from fermentation or deacylation of penicillin V with synthetic acyl precursors for the new side chain in activated form. Many of these analogues show improved properties over the parent molecule, such as better stability, much longer half-life, oral activity, and efficacy against a broader selection of pathogens. It is important to note that all these compounds retain the same antibacterial mechanism as the parent molecule. All have as their target the bacterial cell wall protein synthesizing process. The compound is not toxic to mammals because higher organisms lack this system for the synthesis of cell wall proteins. Figure 3 shows the acyl groups of some of the more important modified penicillins licensed for clinical use.

The availability of penicillin and its congeners revolutionized the treatment of infectious disease and led to a worldwide search for other microorganisms capable of producing metabolites that display antibiotic properties. Many of today's familiar antibacterial drugs are the result of the worldwide search. For example, examination of a penicillinlike complex from organisms found in water near a sewage outlet in Sardinia led to the isolation of cephalosporin C, whose nucleus is roughly isomeric with that of the penicillins. Though this antibiotic is active against a broader selection of bacteria than the then-available penicillins, its activity was not quite

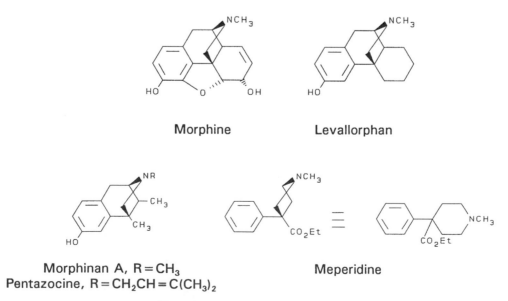

Figure 1.   Opioid analgetic compounds.

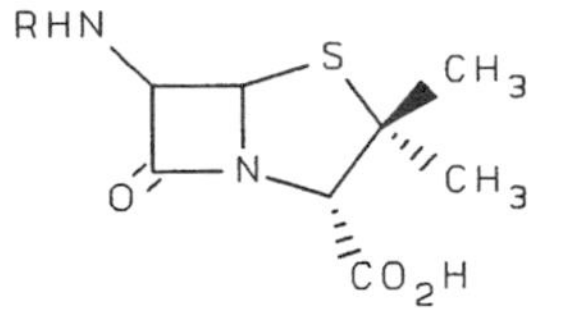

Penicillin G; R=C$_6$H$_5$CH$_2$CO
Penicillin V; R=C$_6$H$_5$OCH$_2$CO
6-Aminopenicillanic acid (6-APA); R=H

Figure 2.   Penicillins produced by fermentation.

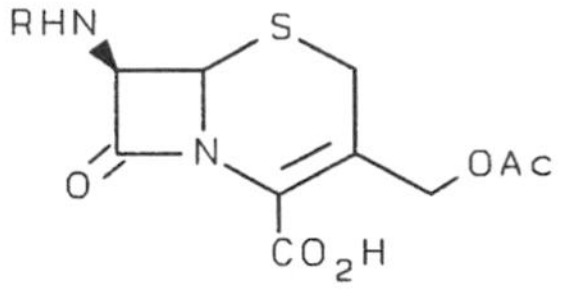

R= HO$_2$CH(NH$_2$)(CH$_2$)$_3$CO; Cephalosporin C
R= H; 7-Aminocephalosporanic Acid

Figure 4.   Cephalosporins produced by fermentation.

good enough to warrant its use as a drug. The development of a very selective reaction for removing the aminopimelic acid side chain made available the nucleus, 7-aminocephalosporanic acid. This substance provided the active nucleus on which new analogues could be based. Work on such analogues was rendered more practical by the development of a chemical sequence for converting the very readily available penicillin V to a cephalosporin derivative. (see later-Section 3.2.1). Figure 4 shows the cephalosporins just discussed.

Whereas 6-APA has only one position at which modifications can be easily prepared, 7-ACA has two such centers; in addition to the amino group, available for amide formation, the allylic acetoxy group can be displaced by a variety of other functions. The mechanism of action of compounds in this class is the same as that of the penicillins: inhibition of bacterial cell wall synthesizing enzymes. Cephalosporins as a class tend to be active against a broader selection of bacteria than the penicillins; this is especially true of Gram-negative organisms, which are largely resistant to the latter. Activity of penicillins is often limited by the presence or development of a series of bacterial enzymes, β-lactamases, which deactivate the drug by opening the four-membered lactam ring. Cephalosporins tend to be more resistant to this enzyme.

Cephalothin (Figure 5), the amide of 7-ACA with thiophene-acetic acid, is a very active representative of the group of deriva-

tives in which only the amide is modified. Displacement of the allylic acetate group by pyridine affords the antibiotic cephaloridine, also shown in Figure 5. The CH$_2$OAc group can be replaced in its entirety by a chloro group by a set of complex synthetic manipulations. The amide of the resulting nucleus with phenylglycine leads to the antibiotic cefaclor (Figure 5). Extensive investigation of the structure–activity relationships in this series of compounds has led to some derivatives, which include quite complex substituents at both the amide and allylic methylene groups. The antibiotic ceftiofur (Figure 5) is representative of some of the more recent modifications.

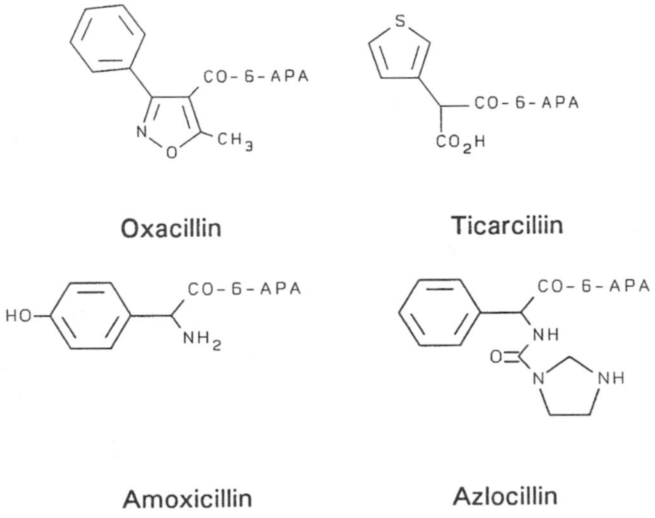

The lead natural products for the great majority of antibiotics including the penicillins and cephalosporins are metabolites of molds. Investigators turned their attention to metabolites of other microorganisms, including bacteria, as this ground came to be ever more intensively covered. The discovery of a weak antibiotic with an unusual spectrum of activity (Gram-negative bacteria) soon followed. The structure of one of these agents, SQ 26,180 (Figure 6) surprisingly consists of the prototype for β-lactams, a freestanding azetidone. Subsequent structural modification led to a small series of clinically effective antibiotics, the *monobactams*; aztreonam (Figure 6) is a typical example of this class. The scarcity of the natural material mandated production of aztreonam and its analogues by total or partial synthesis from β-lactam precursors available by degradation of penicillin V. The methoxyl group on nitrogen of the lead compound, SQ 26,180, was omitted in aztreonam to facilitate synthetic manipulations.

2.1.2 Endogenous Hormones

Adrenergic Agonists and Antagonists   A closely related approach starts with the structure of endogenous hormones or messenger substances. The targets of synthetic manipulations in such cases include compounds that will actually block the action of the endogenous substance and/or show improved absorption on oral administration. The familiar hormones epinephrine (adrenaline) and norepinephrine (noradrenaline) (Figure 7) are intimately involved in regulation of smooth muscles involved in such basic functions as heart rate, blood pressure, and bronchial function. These endogenous hormones are not suitable for use as drugs per se because of the multiplicity of end points each affects as well as their short duration of action. Synthetic programs based on these α-amino alcohols succeeded in providing structurally related subseries that showed quite discrete activities. It was later recognized that the structural modifications result in changed affinities to subsets of adrenergic receptors, hence the wide variability in drug activity. The availability of

Oxacillin

Ticarcilin

Amoxicillin

Azlocillin

Figure 3.   Semi-synthetic penicillin antibiotics.

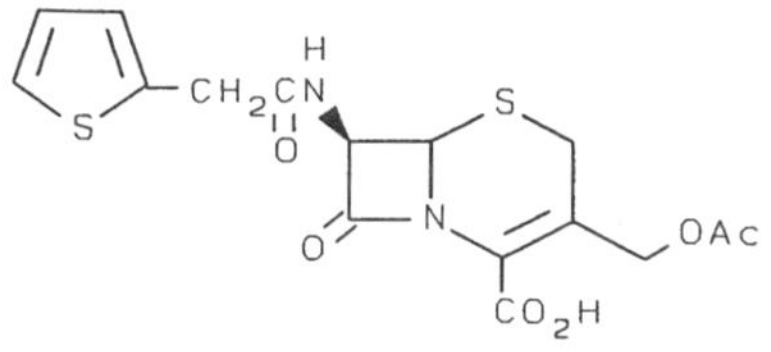

Cephalothin

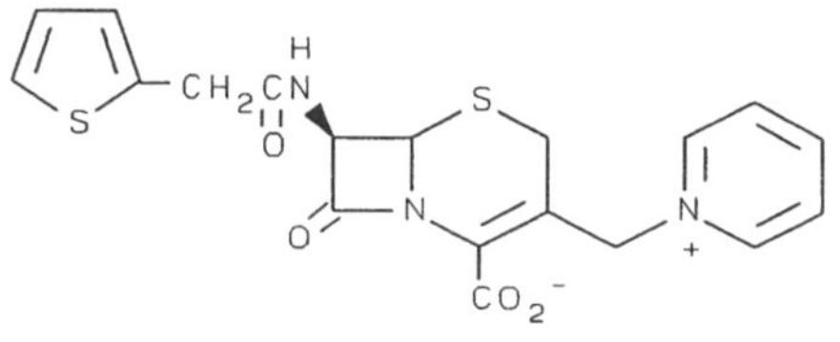

Cephaloridine

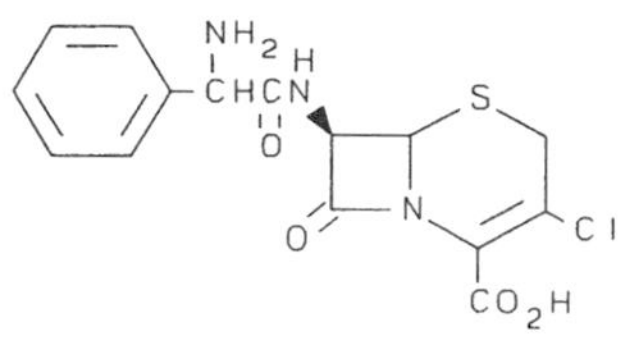

Cefaclor

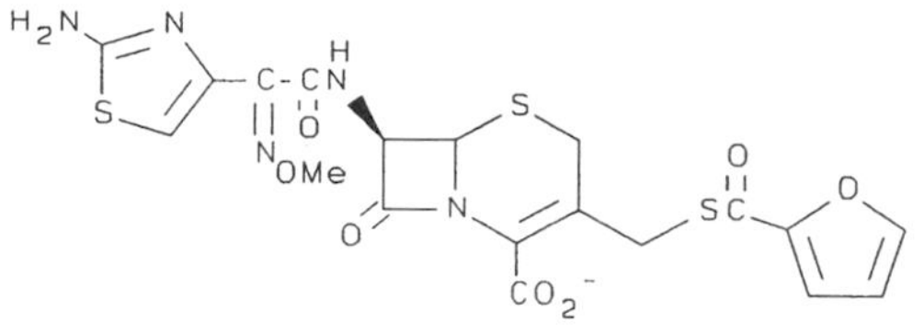

Ceftiofur

Figure 5.   Semisynthetic cephalosporin antibiotics.

the compounds was in fact circumstantial to the recognition of those subsets. Replacement of the two phenolic hydroxyls in epinephrine by a *p*-acetamido group, insertion of an oxymethylene in the side chain, and replacement of *N*-methyl by *N*-isopropyl gives the β-adrenergic blocking agent atenolol (Figure 7), which is very widely used as an antihypertensive agent. A β-adrenergic agonist is produced on the other hand by interposing a methylene group between oxygen and carbon in the *m*-phenol and replacement of *N*-methyl by a *tert*-butyl group. This compound, albuterol (Figure 7), has been used as a bronchodilator for treatment of asthma.

Emerging structure–activity relations (SAR) that reveal the dependency of activity and/or potency on presence or absence of various structural features; thus relative solubility in oil versus water (lipophilicity ratio) and comparisons of electronic distribution or shape help guide further synthesis. Systematic application of a study of those factors (quantitative SAR: QSAR) may provide added insight. The recent wide availability of microcomputers has had a profound effect on drug design by facilitating SAR and QSAR studies. The development of software for modeling organic compounds has

allowed the expansion of SAR studies to three dimensions by providing a tool for readily studying the overlay of active compounds.

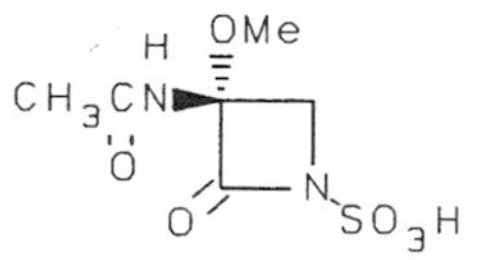

Steroid Hormones   The tetracyclic steroid nucleus provides the framework for a series of endocrine hormones intimately involved in mammalian reproductive function and maintenance of secondary sex characteristics. Appropriate substitution leads variously to the estrogens and progestins associated with female function, or the androgens associated with the male. Somewhat more structurally complex steroids, the so-called corticosteroids, are involved in homeostatic regulation including mineral and energy balance. Because the compounds are so difficult to obtain from natural sources, the initial clinical uses of these complex molecules necessarily involved replacement therapy. As steroids became more readily available by partial synthesis from plant sources, they were found to have unanticipated activities when used in supraphysiological concentrations. Corticosteroids are, for example, used extensively as anti-inflammatory drugs.

The steroid hormones estrone and estradiol, its corresponding alcohol, were the first in the *estrogen* series to be assigned chemical structures. Though it has been firmly established that estradiol is the endogenous molecule that interacts with the tissue receptor, this compound is not particularly active when administered orally because it is poorly absorbed and readily undergoes metabolic inactivation. The addition of a substituent at the 17 position, such as the ethynyl group in ethynylestradiol, leads to a drug that has potent oral activity; mestranol, the corresponding methyl ether, retains the good activity of its parent (Figure 8).

The difficulty in obtaining estrogens either from natural sources or from total synthesis led to extensive investigations of molecules that somewhat resembled those compounds but were not based on the steroid nucleus. Early on it was discovered that stilbenes possess appreciable estrogenic activity; one of these, diethylstilbestrol (DES) (Figure 9), was used quite widely for a number of years before its bizarre second-generation tumorigenic effect was recognized, leading to removal of the drug from the market. Further synthesis in this series showed that activity was retained when of one of the ethyl groups in DES was replaced by phenyl. Conversion of one phenolic group to a basic dialkylaminoethoxy group led to es-

SQ 26,180

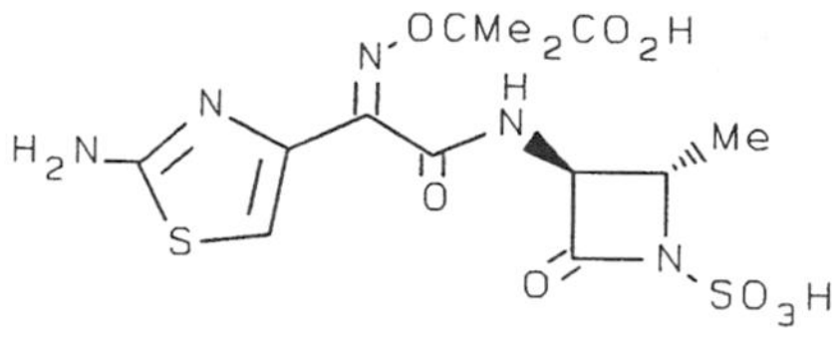

Aztreonam

Figure 6.   Monobactam from fermentation, SQ 26,180 and semi-synthetic antibiotic aztreonam.

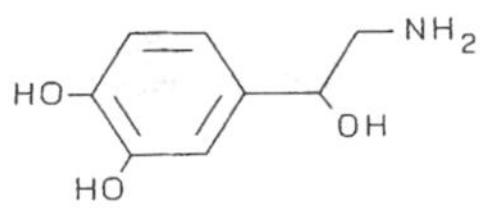

Norepinephrine

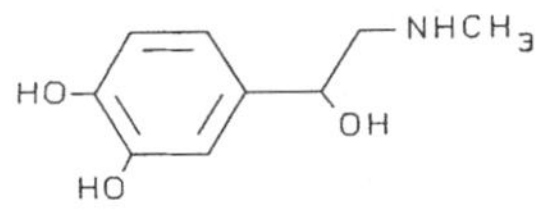

Epinephrine

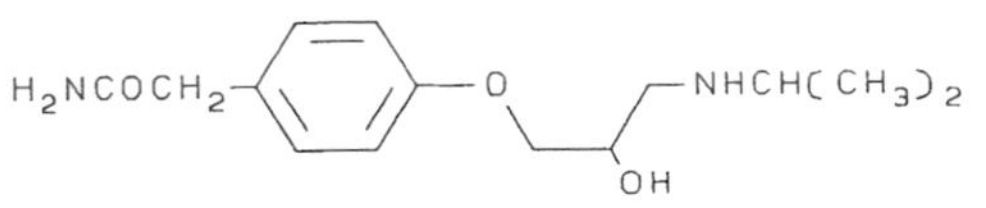

Atenolol

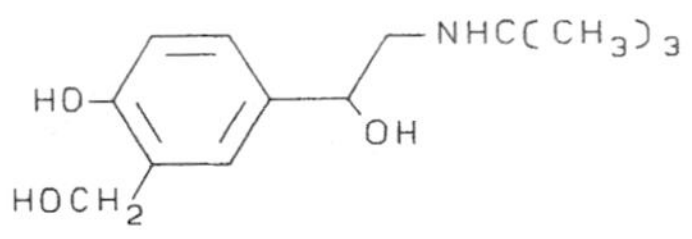

Albuterol

Figure 7. Adrenergic agonists and antagonists.

trogen antagonists. One of these antiestrogens, tamoxifen (Figure 9), is finding increasing use in the treatment of estrogen receptor positive breast cancer.

Progesterone (Figure 10) is a steroid that includes the partially saturated A ring and angular methyl group more typical of the steroid hormones. This *progestin* is closely coupled with estradiol in regulation of the female reproductive cycle; the compound is also essential for early endocrine support of the developing fetus. The high blood levels present early in pregnancy are associated with suspension of ovulation. Initial applications of progesterone involved largely replacement therapy for cases of progesterone deficiency. As in the case of estradiol, this compound was found to be poorly ab-

sorbed up on oral administration. The preparation of an orally effective progestin offered an attractive target, since it would facilitate replacement therapy and, more importantly, might provide a method for inhibiting ovulation by mimicking early pregnancy. Synthetic investigations, made possible by the availability of steroid starting materials from plant sources (see Section 3.2.2), revealed that compounds devoid of the methyl group (numbered 19) at the 10 position but including an ethynyl group at 17 showed good oral activity. Two of these, so-called 19-nor steroids, norethynodrel and norethindrone, were eventually approved for use as oral contraceptives (Figure 10). It has been found subsequently that the very small amount of mestranol present in each of these compounds as

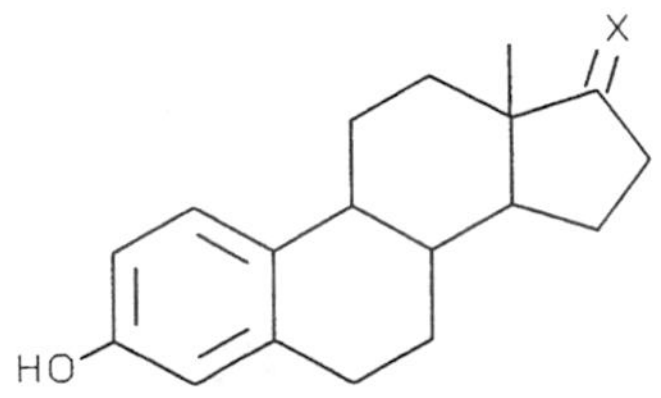

X = O; Estrone
X = -OH, ⋯H; Estradiol

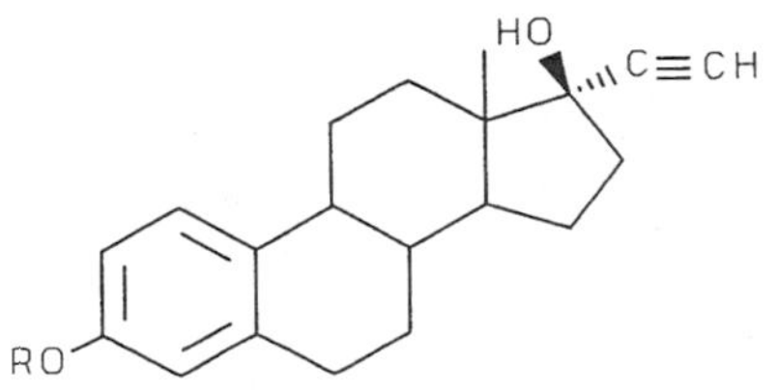

R = H; Ethynylestradiol
R = Me; Mestranol

Figure 8. Estrogenic steroids.

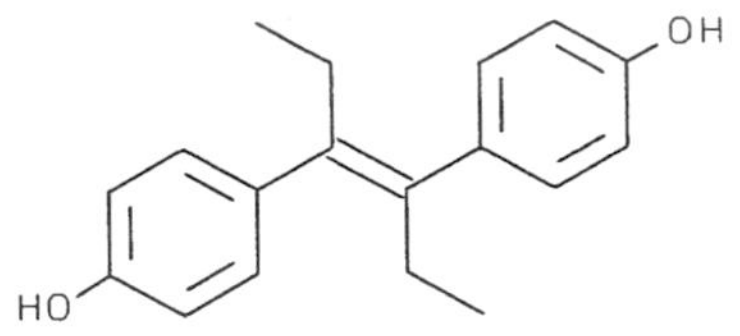

Diethylstilbestrol

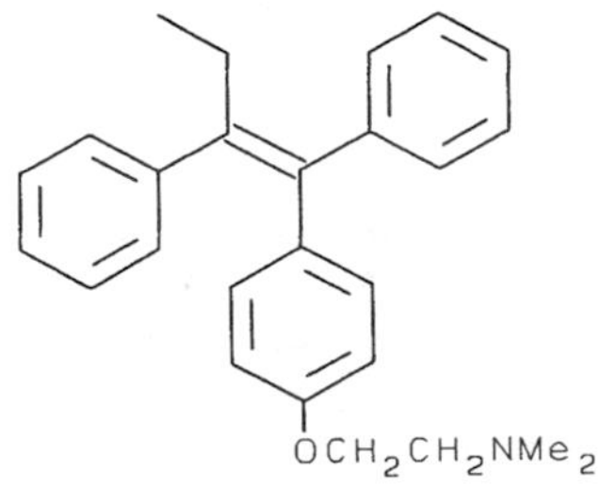

Tamoxifen

Figure 9. Synthetic estrogen diethylstilbestrol and antitumor estrogen antagonist tamoxifen.

Progesterone

4,5-olefin; Norethindrone
5,10-olefin; Norethynodrel

Figure 10. Progesterone and semi-synthetic orally active contraceptive progestins.

a synthetic impurity played a role in their contraceptive activity. Most oral contraceptive today consist of such estrogen–progestin combinations.

2.1.3 Random Screening

One very common and undeniably successful approach relies initially on random screening. This strategy starts with the development of an assay, historically performed in vivo but increasingly done in vitro, which is intended to mimic a disease state. Since typically there are no compounds of known activity against the selected disease, the screen is used to test as wide a structural array of compounds as possible. Test candidates usually are obtained from a number of sources, including most prominently internal collections. Once an active compound has been found, related structures are synthesized to systematically modify the structure of the initial active lead. The SAR developed by this procedure is used to optimize the compound's activity, much as described earlier.

One of the first compounds to show activity in a then-new bioassay intended to identify anxiolytic agents, known at the time as minor tranquilizers, was the benzodiazepine chlordiazepoxide (Figure 11). The finding that this compound, under the trade name of Librium, was quite active as an antianxiety agent in humans occasioned an enormous amount of work on analogues, both in the laboratory that made the discovery and among its competitors. The number of benzodiazepines prepared ranks in the tens of thousands, some dozens of which found clinical use. The spectrum of activity of a typical benzodiazepine is quite complex and includes hypnotic and muscle relaxant properties. A significant number of drugs have been developed historically by modifying the structures of active leads to emphasize one particular aspect of a lead compound's side effects. Sulfonamide antibiotics were the among the first compounds to exhibit documented therapeutic activity for treating bacterial infections; these agents pointed the way to drugs that were toxic to bacteria but relatively innocuous to a mammalian host. Some of these were in found to exhibit diuretic effects in the clinic, while other seemed to lower blood sugar in patients. Systematic modification of those compounds led to the structurally related thiazide diuretic drugs and oral sulfonylurea antidiabetic agents. Corresponding structural modifications of benzodiazepines provided drugs that emphasize specific facets of the lead's spectrum of activities. Triazolam, for example, is used mainly for its hypnotic activity, while the related analogue midazolam finds use mainly as an injectable anesthetic (Figure 11). Deletion of the pendant ring typical of the classical benzodiazepine led to flumezanil (Figure 11). This com-

pound, actually a benzodiazepine antagonist, has potential utility in treating drug overdoses.

2.2   Drug Target Based Design

2.2.1 Structures of Drug Receptors

A large body of evidence indicates that many drugs owe their effect to the binding of drug molecules to specific structures on cells called receptors or, alternatively, with enzymes. Detailed knowledge of the three-dimensional structure of a receptor should in theory allow the design of better drugs by optimizing the fit of the drug molecule to its receptor by appropriate structural modification. Increased understanding of the molecular biology and biochemistry associated with cell function today often allows the identification and isolation of enzymes or cell receptors that are presumed drug targets. Detailed three-dimensional chemical structures for many such receptors and enzymes are now available from data provided by X-ray crystallographic studies combined with computerized molecular modeling programs. Models derived from those studies have in several cases provided detailed structures of the drug–receptor interaction site. The use of this potentially powerful tool for designing de novo molecules to fit those sites, and thus ideally, coming up with new drugs, is currently being pursued quite intensively.

2.2.2 Computer Modeling

Many biopolymers that form receptors cannot be obtained in a form suitable for detailed structure determination. Information on those receptor sites can be gleaned by using an alternative strategy, however. Computer modeling of overlays of drugs that are known to in-

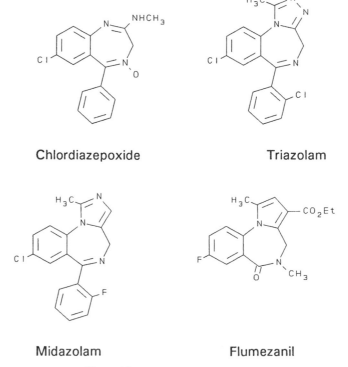

Chlordiazepoxide   Triazolam

Midazolam   Flumezanil

Figure 11. Typical benzodiazepines.

teract with a common receptor will often reveal structural feature shared by the various compounds. It is assumed that those shared groups define interaction sites with the putative receptor. New compounds can then be designed which will position the critical functions at the interaction sites. This approach to drug design is also the focus of much current attention.

3   SYNTHESIS FOR BULK DRUG SUBSTANCE PRODUCTION

The actual synthetic methods used to obtain specific compounds in the discovery phase of drug synthesis are of no special concern with respect to practicality. The chemist has at his or her disposal at this point the full repertoire of manipulations—that is, reactions—that comprise organic synthesis. The objective of that phase of the work consists in obtaining the wide array of structures required to find the one that has the best set of biological properties.

The emphasis reverses once the ideal compound has been identified. Practical considerations such as availability of starting materials, ease of scaling up the synthesis, safety of the individual steps on a larger scale, and, not least, economics play an increasingly important role. The larger samples required for detailed pharmacology and toxicology may be prepared by simply repeating the original synthesis on a larger scale. The amounts of drug required when a compound gets to the market will almost certainly be prepared by a quite different set of reactions. The sponsor will however have to use the most sensitive analytical techniques that are applicable to demonstrate that the final products, from the two processes, and sometimes even significant by-products, are identical. It should be noted too that development work goes on well beyond the point of commercialization in the cases of drugs that gain widespread use, leading to ever-changing proprietary methodology. There is consequently no assurance that the syntheses outlined in Sections 3.1 and 3.2 truly represent current processes.

3.1   SYNTHESIS FROM STARTING MATERIALS DERIVED FROM PETROCHEMICALS

The so-called coal tar chemicals, which usually were obtained from coal coking distillates, provide the basis for all drug synthesis except for the structurally complex drugs that are derived from natural products. Petroleum products have in fact largely replaced coal tar as a source for the same starting materials.

3.1.1 Carbocyclic Compounds

Two drugs, the structures of which have a benzene ring as the central element, represent differing levels of synthetic complexity. Aspirin is probably today's most frequently ingested drug, annual con-

sumption being reckoned in tons. Starting material for this drug is phenol (itself probably prepared from benzene by any one of several highly specialized commercial oxidation reactions). Under high pressure, the phenol is reacted with carbon dioxide and a base such as sodium hydroxide, in a process difficult to carry out on the laboratory scale but quite practical in large-scale process equipment, to give salicylic acid. Acetylation of the hydroxyl group on the product with acetic anhydride leads to aspirin (Figure 12).

The more recently developed anti-inflammatory agent ibuprofen is coming to rival aspirin in popularity. A large number of methods for its preparation have consequently been devised, all of which are necessarily more complex than that required for aspirin. One of the more straightforward starts with the acetylation of the petrochemical product isobutylbenzene (1) to the methyl ketone 2 (Figure 13). A mechanistically complex transformation, accomplished by treating 2 with sulfur and morpholine, transposes the oxygen to the end of the chain to give the phenylacetic acid 3 (Figure 13). The methyl group on the carbon adjacent to the acid cannot be added directly, since reactions to accomplish this lead to the formation of a by-product that has added two methyl groups to the carbon adjacent to the acid. The acid is consequently first transformed to the *bis*carboxylic ester (4), which perforce gives a monomethyl derivative 5 (Figure 13). (Conversion to the *di* ester also increases reactivity of the remaining hydrogen.) The ester is then hydrolyzed to the *di* acid; upon warming, the heat-labile acid loses carbon dioxide to give ibuprofen.

3.1.2 Heterocyclic Compounds

Close to half of all drugs in clinical use today are based on ring systems that include at least one atom other than carbon. The structurally simpler heterocyclic compounds are often prepared from heterocycles obtained as petroleum or coal tar products. In other cases, the synthesis of the drug involves building up the heterocyclic ring. The routes used to prepare the benzodiazepine tranquilizers diazepam and triazolam illustrate this last approach. Acylation of *p*-chloroaniline (6) with benzoyl chloride leads to the aminobenzophenone 7 (Figure 14); reaction of that intermediate with chloroacetyl chloride converts the primary aromatic amino group to the corresponding β-chloroamide. Exposure of that product to ammonia may be envisioned to result initially in displacement of chlorine by ammonia; the basic nitrogen on that intermediate then condenses with the benzophenone ketone to form the seven-membered ring of compound 8 (Figure 14). Methylation on the amide nitrogen atom completes the synthesis of diazepam (Figure 14).

A quite similar sequence starting with *p*-chloroaniline (6) and *p*-benzoyl chloride leads to the analogue of 8, which contains an additional chlorine atom. The amide group is then converted to a thioamide (9) to increase its reactivity. Condensation of 9 with the

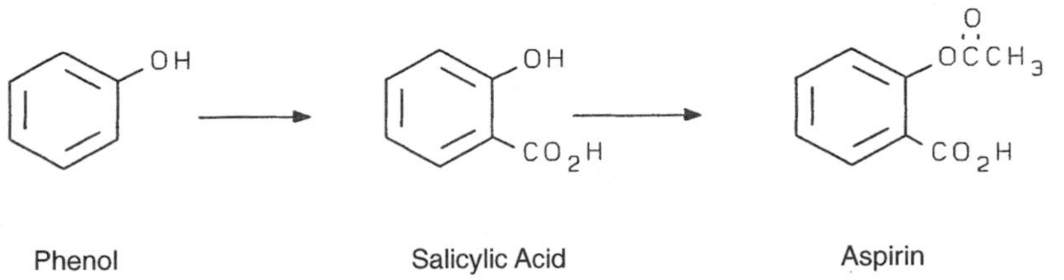

Figure 12.   A synthetic route to aspirin.

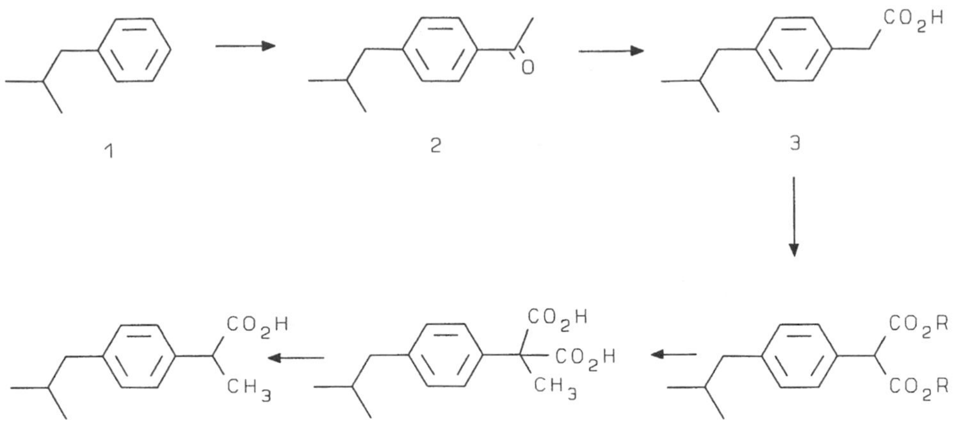

Figure 13. A synthetic scheme for ibuprofen.

acetic acid amide of hydrazine can be imagined as starting with attack of the basic terminal atom of the hydrazide on the thioamide; the hydrazide amide carbonyl group then reacts with the nitrogen on the seven-membered ring to close the fused five-membered ring; the product is triazolam (Figure 14).

3.1.3 Synthesis of Chiral Drugs

The various structures such as receptors and enzymes, which are the end targets of drug action, are composed of biopolymers whose constituent units comprise chiral compounds. Eukaryotic proteins, for example, consist virtually entirely of L amino acids. Significant differences will consequently exist in binding affinity, and thus potency, between enantiomers of drugs that are chiral; it has been demonstrated in many cases that drug action is due entirely to a single enantiomer. Because of the lack of chiral discrimination in organic reactions, the majority of purely synthetic drugs are, however, available only as racemic mixtures. The inactive isomer is at best an inactive diluent. Until recently, the only way to isolate the active enantiomer involved relatively inefficient resolution processes, further discouraging the use of enantiomerically pure drugs.

Increasing recognition of the importance of chirality to activity,

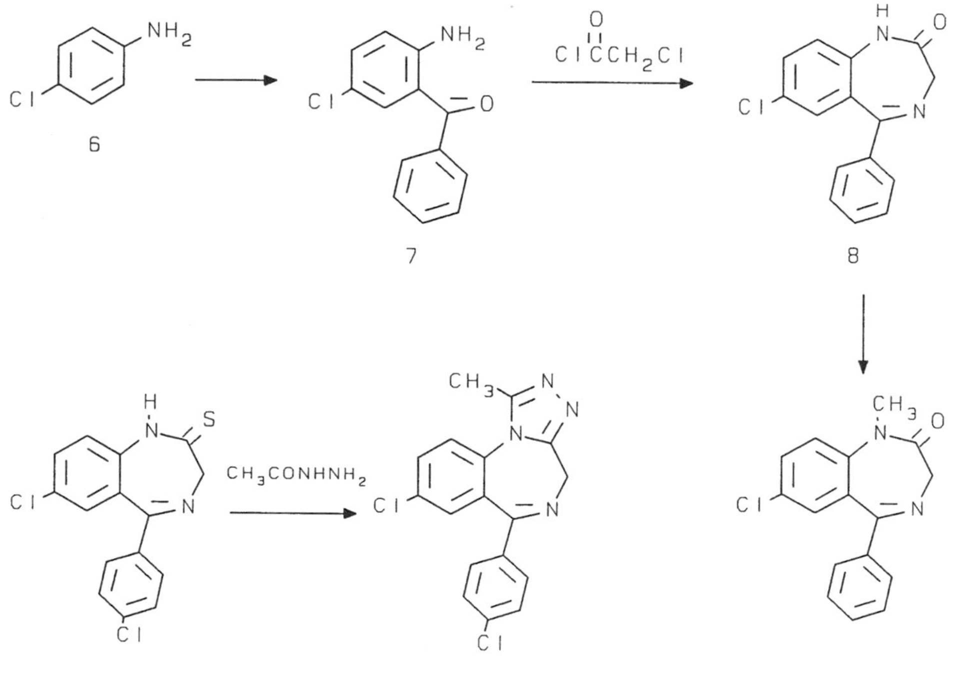

Figure 14. Typical synthetic routes for producing benzodiazepines.

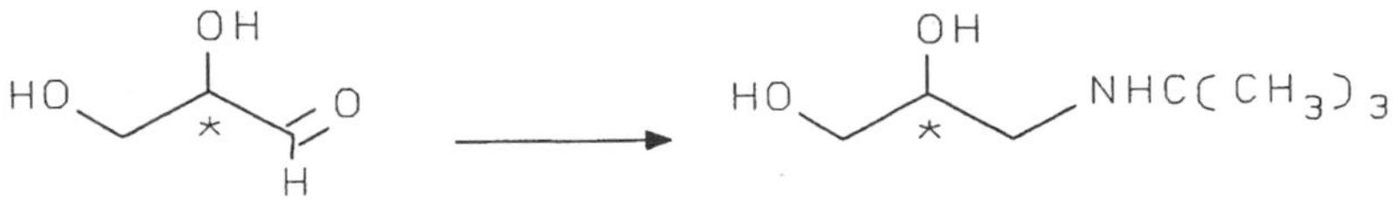

Figure 15. Scheme for preparing chiral β-blocker side chain from chiron, D-glyceraldehyde.

coupled with new developments in synthetic methodology, led to the introduction of a number of drugs as single enantiomers. Nature is, as noted earlier, essentially chiral. Natural products thus provide a rich source of chiral intermediates for synthesis; these have been termed "chirons." For example, D-glyceraldehyde, obtainable in a very few steps from the tonnage carbohydrate mannitol, has the same absolute configuration, S, at the chiral central carbon (*) as does the side chain carbon in β-adrenergic agonists and antagonists. This carbohydrate-derived aldehyde can be readily transformed into an intermediate for those side chains (**10**) by reductive alkylation with *tert*-butylamine (Figure 15).

The intermediate **10** is used in one of the published procedures for the construction of the β-adrenergic blocker timolol (Figure 16). This compound differs further from most β-adrenergic blockers in that the nucleus consists of a heterocycle instead of a benzene ring. As shown in Figure 16, that thiadiazole, **11**, is built in a single step by reaction of carbocyanidic amide with sulfuryl chloride. The hydroxyl group in the product is then converted to its corresponding *p*-toluenesulfonate, **12**. Displacement of that last group with oxygen of the side chain **10** as its alcoholate salt gives the intermediate **13**. Displacement of the ring chlorine with the nitrogen atom on morpholine affords chiral S timolol.

3.2 PARTIAL SYNTHESIS FROM NATURAL PRODUCTS

A significant number of drugs have their antecedents in natural product chemistry, being based on leads provided by compounds first isolated from plants, microorganisms, or animals. These agents as a class tend to be structurally more complex than those derived

from purely synthetic leads. It is often more expeditious to prepare such bulk drug substance by structural modification of related, easily obtained natural products.

3.2.1 β-Lactams

β-Lactam antibiotics, which include penicillins, cephalosporins, and monobactams, are probably the dominant class of drugs for treating bacterial infections. It is thus noteworthy that virtually all the currently marketed drugs represent structural modifications of the substances first isolated from nature. The presence of at least two chiral centers in the four-membered ring β-lactam moiety proper adds a level of complexity to the preparation of these compounds by total synthesis from petrochemical intermediates. Though a number of these drugs have been prepared by total synthesis, the availability of penicillin V as a bulk chemical has probably given partial synthesis an economic advantage.

The detailed structure–activity studies mentioned earlier in connection with the β-lactams revealed that the most fruitful area for structural modification for improving antibacterial activity in the penicillin area involves side chain replacement. Chemical reactivity rules against direct removal of the phenoxyacetamide side chain from the very abundant penicillin V (**14**, free acid): it will be noted that the molecule in fact contains two amide functions, one in the β-lactam ring and that which connects the side chain (Figure 17). The strained nature of the ring amide will lead that portion of the molecule to cleave in preference to the open-chain function. Conversion of the side chain amide to an imidoyl chloride (**15**) provides a way to selectively activate the desired function. (The ring amide

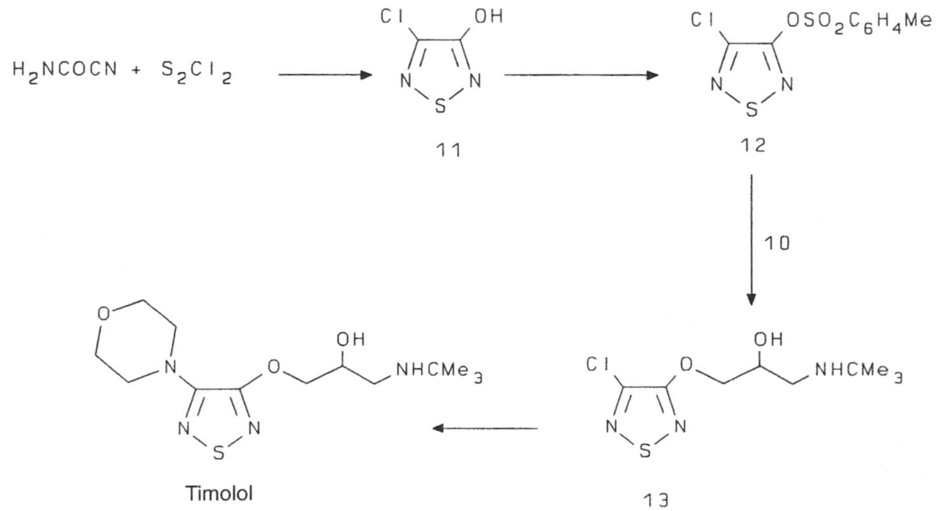

Figure 16. A synthetic scheme for timolol.

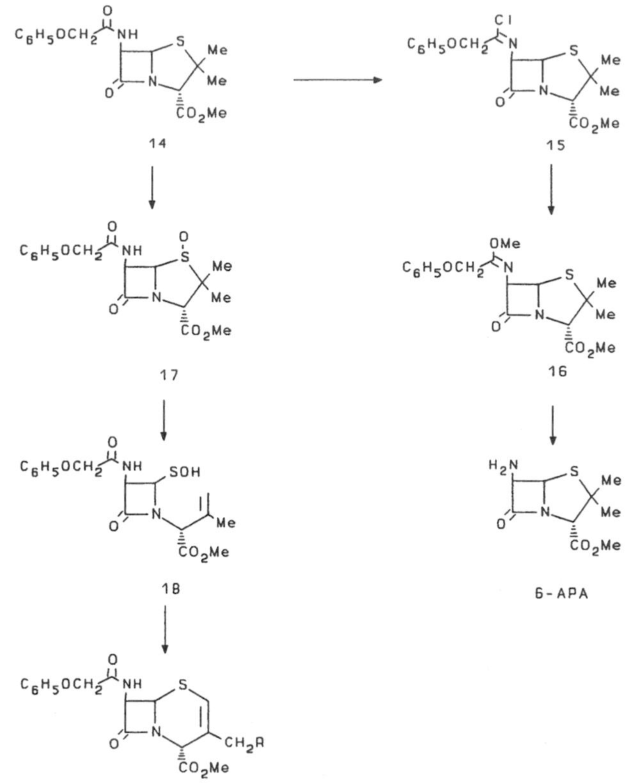

R=H, 7-ADCA,(phenoxyacetamide)

R=OAc; 7-ACA,(phenoxyacetamide)

Figure 17. Syntheses of key intermediates for cephalosporin (7-ADCA, 7-ACA) and pencillin (6-APA) antibiotics from penicillin V (**14**).

cannot form on imidoyl chloride since it is tertiary.) Reaction of **15** with methanol gives the imidate **16**. This last readily hydrolyzes under neutral conditions to give the desired **6**-aminopenicillanic acid as its methyl ester (6-APA); this has served as starting material for a host of modified penicillins.

Fermentation yields of cephalosporins tended to be much lower than those of penicillins; this to some extent hindered the development of this class of antibiotics. The discovery of the penicillin-to-cephalosporin rearrangement made penicillin V available as a starting material for those compounds as well. Careful oxidation of **14** gives the corresponding sulfoxide **17** (Figure 17). This compound undergoes ring opening upon heating to give the sulfenic acid intermediate **18,** which can be cyclized to a cephalosporin under acid conditions. The direct transformation from **17** to the phenoxyacetamide of 7-ADCA (7-aminodeacetyl cephalosproranic acid) can be achieved under the right conditions without isolation of **18.**

The SAR of the cephalosporins indicates that activity can be improved by modification both at the amide side chain and at the 7 position on the six-membered ring (R in the 7-ADCA structure). As in the case of the penicillins, removal of the side chain gives starting materials for the former. The methyl group on the six-membered ring is activated toward a number of reactions by its position on a double bond. As an allylic group, it can for example be functionalized to give the phenoxyacetamide of 7-ACA (7-amino cephalosporanic acid) (Figure 17). Ring-opened intermediates such as **18** have utility beyond cephalosporin production. One or both former parts of the six-membered ring can be modified or even removed to give

entrée to monobactams or β-lactams with highly modified fused five- and six-membered rings.

3.2.2 Steroids

The relatively complex carbon skeleton of the steroids coupled with the multiplicity of chiral centers make these molecules difficult targets for total syntheses. The elegant published total syntheses for the steroid hormones are too lengthy and complicated to be suitable for preparation of bulk drug substances. Instead, virtually all steroid drugs are prepared by partial synthesis from one of two relatively abundant naturally occurring steroids. It should be noted, however, that norgestrel (Figure 18), a synthetic progestin used in some oral contraceptives, is in fact produced by total synthesis; the presence of the "unnatural" *ethyl* group on the 5–6 ring junction makes the compound inaccessible from natural starting materials.

Diosgenin (Figure 18) isolated from Mexican yam roots, was the first important source for a starting material for the synthesis of steroids. This was eventually surpassed by stigmasterol. The relative ease of obtaining stigmasterol from the soybean sterol fraction worked in its favor as a basic starting material; the availability of soybean sterols may be virtually without limit. The side chains at the 17 position (see norgestrel for numbering) of each molecule must be considerably simplified to provide a suitable starting material for steroid-based drugs. In the case of stigmasterol, this leads to progesterone; diosgenin affords the corresponding compound, which has an additional double bond at the 17 position.

To remove the excess side chain carbons in stigmasterol, the reactivity of the two double bonds must first be differentiated. This is accomplished by oxidizing the alcohol at the 3 position. The double bond at C-5 shifts into the conjugated 4,5 position, lowering its electron density. Reaction with ozone, on an industrial scale, proceeds preferentially at the more reactive side chain double bond; the bulk of the side chain is cleaved upon workup to give **20** (Figure 19). The aldehyde carbon is then removed by an oxidative step to

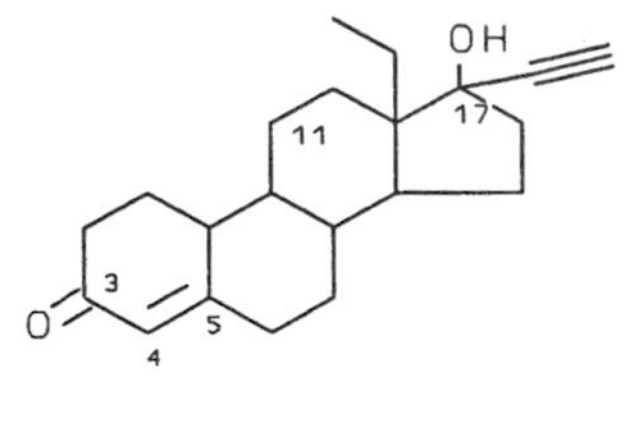

Norgestrel

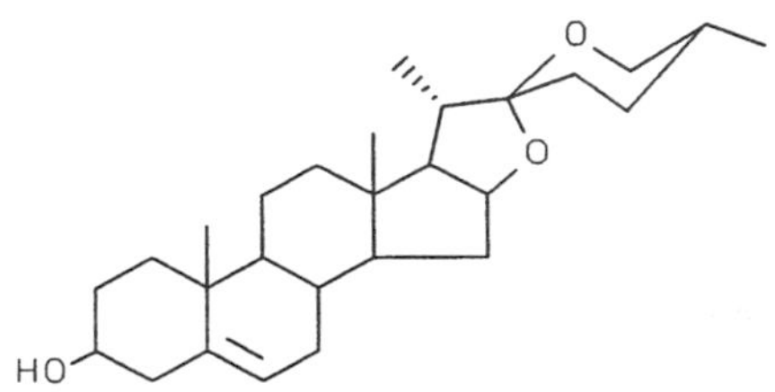

Diosgenin

Figure 18. Totally synthetic progestin, norgestrel and disogenin, steroid starting material from Mexican yams.

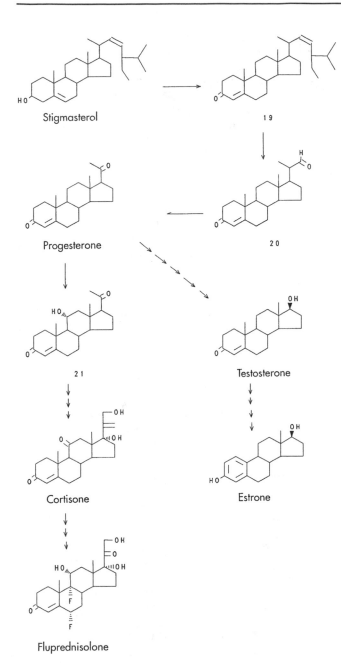

Figure 19. Outline of schemes used to prepare steroid drugs from abundant soybean sterol, stigmasterol.

lead to the key intermediate progesterone. This can be used as starting material for the preparation of testosterone (number of arrows in Figure 19 does *not* indicate the number of steps). Technology has been developed for obtaining estrone from testosterone or its precursors by degradation chemistry.

The key step for preparing corticosteroids involves introduction of oxygen at the 11 position of progesterone. The remoteness of that position from other functionality almost precludes the introduction of that function by purely chemical means. Instead, that oxygen is introduced by fermentation of progesterone with soil molds such as *Rhizopus nigricans*. It should be noted that the new hydroxyl group is introduced on the α face; this is the opposite configuration required for corticosteroid activity, and adjustment by chemical manipulation is necessary. The acetyl side chain at the 17 position is

then converted to a dihydroxyacetone function to give finally cortisone (Figure 19). This sequence, which is quite complex and requires addition and removal of protecting groups, is beyond the scope of this article. Further elaboration of cortisone or one of its precursors leads to a host of other anti-inflammatory corticosteroids, such as fluprednisolone.

See also Biotransformations of Drugs and Chemicals; Medicinal Chemistry; Steroids, Antitumor.

Bibliography

Albert, A. (1985) *Selective Toxicity.* Chapman & Hall, New York.

Hansch, C., Sammes, P. G., and Taylor, J. B. (1990) *Comprehensive Medicinal Chemistry.* Pergamon Press, Elmsford, NY.

Lednicer, D., and Mitscher, L. A. *Organic Chemistry of Drug Synthesis*, Vol. 1 (1977), Vol. 2 (1980), Vol. 3 (1984), Vol. 4 (1990). Wiley, New York.

Wilson, C. O., Gisvold, O., Delgado, J. N., and Remers, W. A. (1991) *Wilson and Gisvold's Textbook of Organic and Pharmaceutical Chemistry.* Lippincott, Philadelphia.

Wolf, M., Ed. (1980) *Burger's Medicinal Chemistry,* 4th ed., Wiley, New York.

Drug Targeting and Delivery, Molecular Principles of

Alfred Stracher

Key Words

Bioavailability   That portion of an administered drug which becomes available to a cell for biological action.

Drug Delivery   The means, either chemical or mechanical, by which drugs are sent to their intended destination.

Drug Targeting   The use of carrier systems to deliver a drug to its intended area of action.

Epitope   An antigenic determinant or a group recognized by an antibody.

Liposomes   Microparticulate colloidal systems that are capable of encapsulating a variety of drugs.

Monoclonal Antibodies   Antibodies synthesized by a population of identical antibody-producing cells (clones).

Parenteral   Describing the administration of a drug by means other than by way of the intestines, such as intramuscular, intravenous, or subcutaneous.

Permeation Enhancers   Small molecules, usually lipophilic, that enhance the penetrability of proteins or polypeptides across mucosal membranes.

Polymers   Large molecules composed of individual monomers linked in long chains and held together by covalent or noncovalent forces. Can be either naturally occurring (e.g., starch) or synthetic (e.g., nylon).

Toxins   A colloidal proteinaceous poisonous substance that is produced by a lower organism; usually very unstable and toxic to tissue.

Conventional drug delivery technology, which in the past has concentrated on improvements in mechanical devices such as implants or pumps to achieve more sustained release of drugs, is now advancing on a molecular level. Recombinant technology has produced a variety of new potential therapeutics in the form of peptides and proteins, and these successes have spurred the search for newer delivery and targeting methods. The development of new polymeric materials, the discovery of monoclonal antibodies, and the achievement of greater knowledge of cellular receptors have resulted in new attempts to deliver and target drugs via conventional as well as unconventional routes of administration. Delivery and/or targeting across the blood-brain barrier remains an area of intense interest. Microencapsulation of drugs within biodegradable polymers and liposomes has already achieved nominal successes in improving the pharmacodynamics of a variety of drugs such as antibiotics and chemotherapeutic agents. Challenges in the delivery of large molecules such as proteins and genes lie ahead, but already some progress has been made in these areas as well. Although drug delivery technology is still in its infancy, and drug targeting even more so, the burgeoning interest in this field provides ample evidence that further progress is forthcoming.

1   INTRODUCTION

1.1   General Principles of Drug Targeting/Delivery

When a pharmaceutical agent is administered to a patient, either orally or by injection, the drug distributes itself in most of the whole-body water and tissues, while only a small portion of the administered dose goes to the diseased area where it is expected to have its curative effect. Not only is this wasteful of expensive drugs, since larger doses have to be given to ensure an effect in only a part of the body, but the remainder of the medication, now in general circulation, can produce severe undesirable effects in organs for which it was not intended. Thus, the means by which a drug reaches its target site takes on increasing significance, as well as its delivery at the right moment and frequency.

Recent developments have fueled an increased intensity in research aimed at creating new drug delivery systems. Much of this interest has stemmed from the new advances in biotechnology and immunology, which have resulted in the creation of a new class of peptide and protein drugs. Concurrent attempts to overcome the barriers that limit the availability of these macromolecules has led to an exploration of nonparenteral routes, with the goals of achieving systemic delivery as well as finding means to overcome the enzymatic and absorption barriers for the purpose of increasing their bioavailability. Although for conventional drugs the oral route has been the most convenient and popular, most peptide and protein drugs have low oral uptake as a result of proteolytic degradation in the gastrointestinal tract and poor permeability of the intestinal mu-

cosa to high molecular weight substances. Several approaches to overcome these obstacles are now under intense investigation: (1) inhibiting proteolytic degradation, (2) increasing the permeability across the relevant membrane, (3) achieving structural modification to improve resistance to breakdown or to enhance permeability, and (4) creating specific pharmaceutic formulation to prolong the retention time of drugs at the site of administration using so-called controlled-delivery systems.

1.2   Controlled-Release Systems

A number of combinations and variations on the foregoing themes have also been investigated. These include linkage of drugs to monoclonal antibodies (Mabs), encapsulation of drugs within liposomes, modification of the liposome surface to alter the pharmacokinetics, coating of proteins and/or liposomes with polymers or polysaccharides, and fusion of toxins to antibodies via recombinant technology. All these modifications are designed to accelerate and control the transport of pharmacologically active agents from sites of administration to organs and tissues by increasing residence time, bioavailability, and penetrability.

1.3   Site-Specific Delivery (Targeting)

The alterations in drug structure just listed are not limited entirely to enhancing the stability of the drug but are also designed to improve the targeting of the drug to a specific organ or tissue. By taking advantage of a feature on the cell membrane that becomes a focal point for incorporating a specific carrier into the design of the drug to carry it to its designated goal, targeting or site-specific delivery can be improved. The carriers generally utilized to the present have been monoclonal antibodies to specific cell membrane etiopes or receptors. A greater understanding of membrane-specific features, however, might enable one to design small molecular carriers attached to drugs for enhanced uptake.

Thus, new drugs in the form of peptides, proteins, oligonucleotides, and genes are now on the horizon. Our limitations at this juncture may be summarized by a question: How do we deliver the drugs, intact, to preferred sites within the cell, to achieve maximal physiologic effectiveness?

2   GENERAL METHODOLOGY

2.1   Routes of Administration

Classically, drugs have been administered orally or parenterally, with the oral route being the most convenient and popular. To achieve maximal effectiveness, most of the new proteinaceous therapeutics have been delivered parenterally: intramuscularly, intravenously, or subcutaneously. Low patient compliance for long-term treatments, however, has led to investigation of nonparenteral routes for the systemic delivery of peptide and protein drugs. Thus, there have been studies of administration by other routes (nasal, buccal, rectal, ocular, pulmonary, vaginal), using a variety of so-called permeation enhancers, to increase bioavailability. The major barriers to be overcome are significant proteolytic activities and the presence of various epithelia at different locations. Since the skin is relatively impermeable, the nasal route appears to be the most attractive alternative to parenteral administration for the delivery of polypeptides, such as insulin.

The epithelial membranes, which cover the organ tissues that are used in nonparenteral drug administration, constitute a highly efficient barrier to drug absorption. Drugs must penetrate the barrier by active transport, vesicle transport, or concentration-dependent diffusion. Alternatively, the drug must navigate through the tight junctions between cells. Studies have indicated that the pore radius cannot accommodate molecules of a size greater than a tripeptide. Absorption enhancers have been used to increase pore size, however, enabling molecules the size of insulin to be transported across the mucosal membrane. Some commonly used enhancers are sodium deoxycholate, sodium glycocholate, dimethyl-β-cyclodextrin, lauroyl-l-lysophosphatidylcholine, to mention but a few. In clinical applications, however, chronic use of these additives has been accompanied by changes in the mucosal surface, and new enhancers are being evaluated to overcome such difficulties. The mechanism by which enhancers exert their effect is not well known, although it is thought that they function by disrupting the ordered membrane phospholipid domain, thereby lowering barrier function and increasing permeability. Most effective enhancers have structural similarities to the phospholipid domains of the mucosal membrane.

2.2 DELIVERY AND TARGETING OF DRUGS

At the same time that new routes of administration are under investigation, other studies to deliver drugs by conventional routes, in a more selective manner, continue. Still dominating current delivery methodology are mechanical methods (pumps, patches, osmotic devices, etc.) to improve release over longer periods, as well as the older "sustained release" technology, which includes formulating drugs in suspensions, emulsions, slowly dissolving coatings, and compressed pills. As a result of the development of new biomaterials, as well as new advances in molecular biology, immunology, and membrane biology, however, novel approaches to drug delivery and targeting are already having an impact on the pharmaceutical industry and will continue to do so in the years ahead. The most advanced of these methods incorporate the use of liposomes, polymers, and monoclonal antibodies.

Liposomes, which are capable of encapsulating a variety of drugs and delivering them via the circulatory system, are microparticulate colloidal systems composed of lipids. They are the most widely studied vesicles to date, and they can be formulated with a variety of lipid types and compositions, which can alter their stability, pharmacokinetics and biodistribution. A major disadvantage of liposomes as delivery systems is their size, which prevents them from crossing most normal membrane barriers and limits their administration to the intravenous route. In addition, their tissue selectivity is limited to the reticuloendothelial cells, which recognize them as foreign microparticulates and are concentrated in tissues such as the liver and spleen. Nevertheless, their biodistribution can be altered to some degree by attaching various ligands to the surface, such as monoclonal antibodies. Other modifications, such as the simultaneous attachment of polyethylene glycol and monoclonal antibodies, the so-called stealth liposomes, have led to improved circulation time as well as improved site-directed delivery. More recently, so-called cationic liposomes, constructed from positively charged lipids or by the attachment of cationic ligands to the surface, have been used for constructing DNA/liposome ternary complexes for

the purpose of delivering genes, leading to increased transfection of cells for purposes of gene therapy.

Polymers have also been used as drug delivery systems. They generally release drugs by (1) polymeric degradation or chemical cleavage of the drug from the polymer, (2) swelling of the polymer to release drug trapped within the polymeric chains, (3) osmotic pressure effects, which create pores that release a drug, which is dispersed within a polymeric network, and (4) simple diffusion of the drug from within the polymeric matrix to the surrounding medium.

Polyesters are the most widely studied biodegradable synthetic polymers for drug delivery. They degrade by hydrolysis of the ester bonds to form nontoxic organic alcohols and acids. Because of their low toxicity and biocompatibility, polyesters of lactic acid and glycolic acid are the most commonly used materials. They are now being used for the slow release of large molecules such as proteins, polysaccharides, and oligonucleotides. Release usually occurs by one of the mechanisms just enumerated. Release rates are affected by polymer size and composition as well as by macro-molecule size, solubility, and stability. Polymer systems are now being used to deliver many large molecules, including insulin, growth factors, and oligonucleotides, as well as smaller drugs such as anticancer agents, and steroids.

Monoclonal antibodies have offered the greatest potential for selective targeting to specific cells. Thus, bioactive agents such as drugs, radioisotopes, or toxins have been chemically linked to the desired antibody with the intent of targeting these agents to selective cells for treatment. Toxins such as ricin have been used in this manner in cancer chemotherapy in the hope that the toxin will accumulate at the tumor site and kill the cancer cells selectively without damaging surrounding tissue. However, this technology is limited by the low availability of antibodies selective for a given tumor cell and by the changing surface configuration of the tumor cell.

Nevertheless this attractive approach is still under intense investigation. A number of variations on this theme have been used as well, for example, monoclonal antibodies have been attached to liposomes in an attempt to target these vesicles to a desired site. Additional attachment of polyethylene glycol to MAb-modified liposomes has given rise to the term "stealth" liposomes, which are more stable and have longer circulation times than unmodified forms. It is likely that these types of modification will achieve increasing clinical applicability in the near term.

See also ANTIBODY MOLECULES, GENETIC ENGINEERING OF; BIOTRANSFORMATIONS OF DRUGS AND CHEMICALS; DRUG SYNTHESIS; ELECTROPORATION AND ELECTROFUSION; GENETIC VACCINATION, GENE TRANSFER TECHNIQUES FOR USE IN; MEDICINAL CHEMISTRY; MOLECULAR GENETIC MEDICINE.

Bibliography

Gregoriadis, G. (1989) *Targeting of Drugs: Implications in Medicine.* Wiley, New York.
Langer, R. (1990) New method of drug delivery. *Science,* 249:1527–1533.
Pardridge, W. M. (1991) *Peptide Drug Delivery to the Brain.* Raven Press, New York.

E. coli GENOME

Hirotada Mori and Takashi Yura

Key Words

DNA Databank   Databank for experimentally determined DNA sequences.

Genetic Map   Linear or circular diagram showing the relative position of genes in a chromosome.

Genome   The total complement of genetic material in a cell or individual.

Protein Databank   Databank for amino acid sequences of proteins determined experimentally or deduced from DNA sequences.

Escherichia coli is one of the best studied organisms and has contributed extensively to understanding the fundamentals of molecular biology. *E. coli* strain K12 is most widely used as a host–vector (EK) system for recombinant DNA and other experiments. Thus knowledge of the complete nucleotide sequence of its genome is important not only to gain insights into the structural and functional organization and evolution of the genome, but to make the best use of *E. coli* as an experimental system.

1   THE *E. coli* MOLECULE

The chromosome of *E. coli* is a single circular DNA molecule that consists of about 4700 kilobases (kb), in which some 1500 genes have been identified. The number of entries in databanks in the United States, Europe, and Japan of the *E. coli* DNA sequence is about 2300, amounting to 4000 kb (January 1993); this includes data from diverse strains and mutants. When all repeats and overlaps are eliminated from the total entries specifically for *E. coli* K12, the actual segments of chromosomal DNA that have been sequenced account for more than 40% of the genome (Figure 1).

2   MAPPING THE GENOME

The construction of a physical map of an entire genome depends on the availability of ordered clones of chromosomal segments in the form of recombinant plasmids or phages. The current physical map of the *E. coli* genome was established particularly by such clones of lambda phage ("Kohara clones") that carry the overlapping DNA segments covering virtually the entire genome. This makes the *E. coli* genome an ideal subject for systematic sequencing. Such a "genome project" consists of three phases of operation: (1) compilation of DNA sequence data currently available at the databank, (2) systematic DNA sequencing, and (3) theoretical and experimental analyses of DNA sequence data.

2.1   COMPILATION OF DNA SEQUENCE DATA

The genetic linkage map of *E. coli* K12 with all known genes, together with information on gene products and comprehensive literature, has been compiled and updated by continuing efforts of the *E. coli* Genetic Stock Center. Attempts to incorporate the genetic data into the DNA sequence data from databanks have been made, and this result along with the *E. coli* protein index are open to the public through the European Molecular Biology Library as "ECD" (*E. coli* database). Novel algorithms were developed to correlate and integrate the *E. coli* genetic map with the physical map, including DNA sequence data, thus integrating adjacent DNA sequences into contiguous structures and eliminating redundancies. Furthermore, a relational database, GeneScape, became available on personal computers (Macintosh) to permit easy inspection and manipulation of genomic DNA sequences.

2.2   SYSTEMATIC DNA SEQUENCING

With recent advances in sequencing strategies and technical improvements including the use of automatic sequencers and the polymerase chain reaction (PCR), goals of determining complete genome sequences have been set for bacteria and lower eukaryotes. In *E. coli,* a set of 476 lambda recombinant phage clones that covers more than 99% of the genome provides most useful material for such endeavors. Each clone carries a DNA fragment of 15 to 20 kb, which was prepared by digesting DNA with *Sau*3AI or *Eco*RI, and alignment was achieved by analyzing restriction fragments obtained by each of eight restriction enzymes (Figure 2). These and other recombinant clones have been used for systematic sequencing of the *E. coli* genome. Starting from 0 minutes of the genetic map, sequences for 111.4 kb (0–2.4 min) have been determined (see Figure 2). Other regions that have been sequenced include the 84.5–86.5 minute region (near the chromosomal replication origin, *ori C*) of 91.4 kb and a 150 kb sequence around the replication terminus *(ter C)*. The map of about 90 genes and putative genes predicted from the contiguous sequence data for the 0–2.4 minute region is shown in Figure 3.

2.3   ANALYSIS OF SEQUENCE DATA

In general, novel nucleotide sequences are first subjected to analysis for open reading frames (ORFs) with searches in three phases

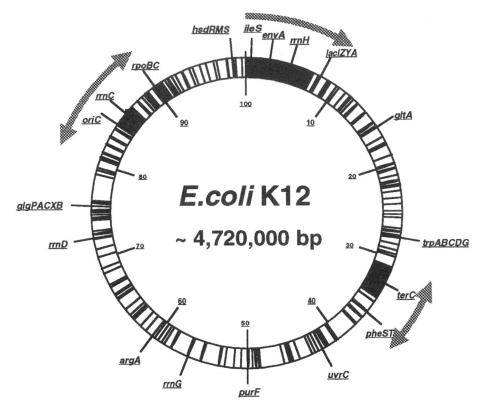

Figure 1. Sequenced areas of the *E. coli* K12 chromosome. Arrows indicate positions and direction of systematic sequencing projects currently in progress. Black bars on the genetic map indicate regions that have been sequenced.

and for both orientations. Possible ORFs thus obtained are analyzed for sequence similarity against a large collection of protein sequence data (e.g., the database of the Protein Identification Resource in the United States). Through such analyses, the possible nature or functions of putative ORFs can be predicted and the results can then be utilized for further investigation, either theoretical or experimental. For example, analysis of the 0–2.4 minute region

led to the prediction of the occurrence of several novel clusters of genes similar to genes of other organisms that are involved in fatty acid metabolism, iron transport, and symbiotic nitrogen fixation (Figure 3). In view of the tight arrangements of genes on the chromosome with little spacing between, it seems that about 4000 genes may constitute the genome of *E. coli*.

Extensive analyses of *E. coli* chromosomal DNA, combined with

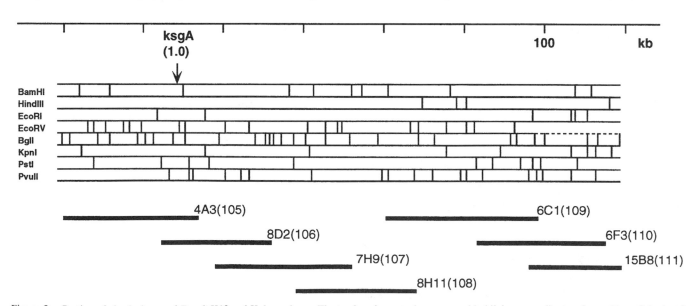

Figure 2. Portion of physical map of *E. coli* K12 and Kohara clones. The top bar shows scales expressed in kilobase coordinates; the position of *thr A* at 0 minutes on the genetic map is taken as 0 kb. Horizontal bars under the restriction map represent individual clones.

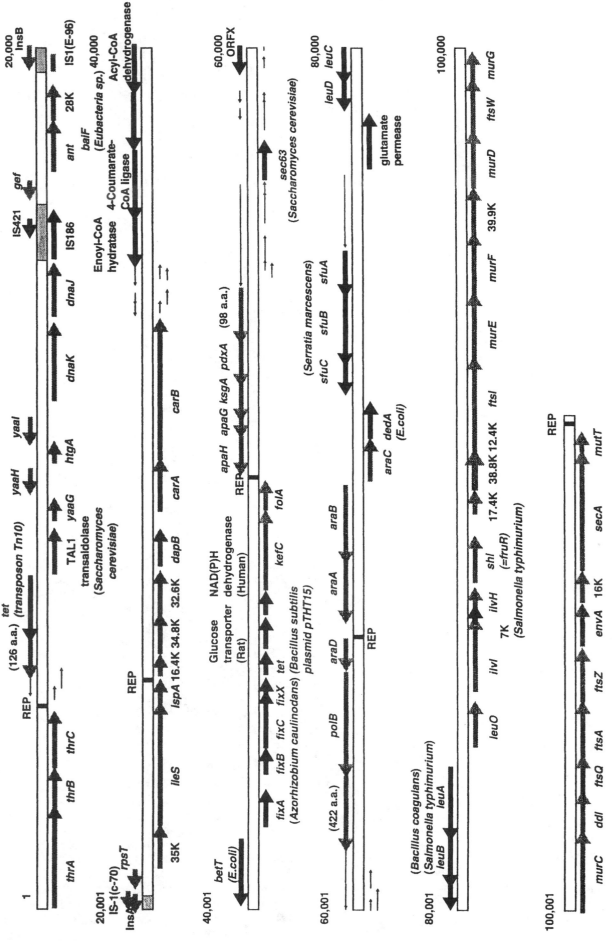

Figure 3. Physical map of ORFs on the 0–2.4 minute region of the *E. coli* chromosome: thick, black arrows, putative genes predicted by homology analysis; thin black arrows, uncharacterized ORFs; lighter arrows, previously sequenced genes.

currently available sequence data, revealed some interesting features, such as repeated sequences, gene families, functional organization of genes, and clues to the evolution of the genome. It is anticipated that an increasing number of important novel features that are essential for deeper understanding of the genome will be unraveled as contiguous sequence data accumulate.

See also BACTERIAL GROWTH AND DIVISION; EXPRESSION SYSTEMS FOR DNA PROCESSES; NUCLEIC ACID SEQUENCING TECHNIQUES.

Bibliography

Bouffard, G., Ostell, J., and Rudd, K. E. (1992) *Comput. Appl. Biosci.* 8:563–567.

Daniels, D. L., Plunkett III, G., Burland, V., and Blattner, F. R. (1992) *Science,* 257:771–778.

Drlica, K., and Riley, M., Eds. (1990) *The Bacterial Chromosome.* American Society for Microbiology, Washington, DC.

Kröger, M., Wahl, R., Schachtel, G., and Rice, P. (1992) *Nucleic Acids Res.* 20:2119–2144 (Suppl.).

Neidhardt, F. C., Ingram, J. L., Low, B. K., Magasanik, B., Schaechter, M., and Umbarger, H. E., Eds. (1987) Escherichia coli *and* Salmonella typhimurium: *Cellular and Molecular Biology.* American Society for Microbiology, Washington, DC.

Yura, T., Mori, H., Nagai, H., Nagata, T., Ishihama, A., Fujita, N., Isono, K., Mizobuchi, K., and Nakata, A. (1992) *Nucleic Acids Res.* 20:3305–3308.

ELECTRIC AND MAGNETIC FIELD RECEPTION

Eberhard Neumann

Key Words

Chemical Electric and Magnetic Field Effects   The consequences of the primary field effects on ions and dipoles, which cause the rate and equilibrium constants of reactive chemical processes and phase transitions to depend on the electric field strength and, to a weaker extent, on the magnetic flux density.

Cooperativity   A mechanism for the simultaneous state transition of a cluster or domain of strongly coupled molecules in a narrow field strength range.

Electric and Magnetic Field Strength   Measures of the force on electric charges and electric and magnetic dipoles, respectively. The targets are either freely mobile or fixed ionic and magnetic groups of larger macromolecules.

Interfacial Electric Polarization   Field-induced ion accumulation at interfaces of media with different dielectric constants, leading to ionic charging up of membranes and to large induced transmembrane voltages by small external fields.

Magnetic Induction of Electric Fields   The main effect of time-varying magnetic fields, causing ionic (ac) currents and dipolar displacement currents, both of which interfere with chemical reactivity.

Magnetosomes   Small elongated magnetite (Fe_3O_4) organelles, which in magnetobacteria serve for geomagnetic field sensing and direction finding and are also found in cells of higher organisms.

Electric field reception *(E)* and magnetic field reception *(B)* are defined as the mutual interaction of external fields with the relatively high internal fields of the ionic and the (electric and magnetic, respectively) dipolar components, including the cellular magnetosomes, involved in sensory processes. The natural *E* fields are particularly apparent in the electrophysiological function of nerves. Physicochemically, the *E* and *B* fields primarily act on charges, changing their positions and motional states. As a consequence, the extent and the rate of chemical reactions, conformational changes, and phase transitions of the field-receiving molecules are dependent on the strength of the *E* and *B* fields. Although there are *always* finite chemical shifts in rate and equilibrium constants, weak external fields require amplification processes (e.g., interfacial polarization of larger particles, cooperativity in domain structures).

1   CONCEPTS

In biology, electric and magnetic field reception is commonly associated with sensory processes. The electric fish have developed special electric organs to sense the environment for orientation, navigation, and food search. Electric eels and torpedo fish utilize large electric discharge organs for prey capture. Magnetobacteria contain small magnetite (Fe_3O_4) organelles, which serve for direction finding in the earth's magnetic field. Such magnetosomes are also found in higher organisms. It is important to note that biological weak-field reception and behavioral responses are intimately coupled to the nervous system. The receptive and sensing fields of the biosphere are either static (dc) fields or short field pulses of 0.1–1 ms duration with low pulse frequencies (≤ 3 kHz), producing direct currents (dc) or current pulses.

Electrotechnologically, man-made electric and magnetic field devices have become indispensable tools of life. In particular in medicine, we observe growing application of electric and magnetic field apparatus for both diagnostic and therapeutic purposes. On the other hand, it is not yet clear whether the technical fields of power transmission lines and electrical household devices bear potential health risks, even under normal safety conditions. To estimate possible beneficial or negative effects, it is necessary to understand the physical and chemical interaction mechanisms of electric and magnetic field reception.

Generally, electric and magnetic fields act as forces on charges: electrolyte ions and fixed ionic groups, as well as free electric and magnetic dipolar molecules and fixed dipolar groups of macromolecules, including membrane lipids and cellular magnetosomes. As a consequence of these primary field effects, reactive chemical interactions are dependent on the electric field strength $\mathbf{E}$ (V/m) and, to a much weaker extent, on the magnetic field strength $\mathbf{H}$ (A/m).

1.1 NATURAL FIELDS

All life processes have evolved in the natural electric and magnetic force fields of the earth. These geofields are quasi-static or only slowly time varying. Compared with possible technical fields, they are classified as weak. The average geoelectric field strength is $|\mathbf{E}_0| \approx 200$ V/m, the average geomagnetic flux density $|\mathbf{B}_0| = \mu_0\,\mu|\mathbf{H}_0| \approx 50\ \mu\text{T}$, where $\mu_0 = 1.256 \times 10^{-6}$ Vs/(Am) is the vacuum permeability and the magnetic susceptibility χ_{ma} is $\mu - 1$.

Even in its lowest forms, life makes use of electric and magnetic field effects, most obviously, on the microscopic level, in the rapid signal transmission of neurons, involving in particular the gated transport of Na^+, K^+, Ca^{2+}, and Cl^- ions.

1.2 DEFINITION OF FIELD RECEPTION

In molecular biology the term "field reception" appears meaningful only if it comprises more than just the direct interactions of the force fields (and their spatial gradients) with molecules, changing their positional and motional states. An adequate definition of this term must specify at least the biochemical or biophysical process following the preceding direct force effects. Thus field reception indispensably includes field effects on the reactive interactions, the molecular orientations, and the transport processes of the "field-receiving" ions and electric and magnetic dipoles. Moreover, amplification mechanisms for both small external fields and locally small chemical shifts are important features of the molecular characterization of field reception.

Briefly, the effects of external E and B fields may be summarized as a cascade of mutual interaction steps:

$$(E, B)_{\text{ext}} \leftrightarrow (E, B)_{\text{int}} \leftrightarrow \left\{ \begin{array}{l} \Delta c, \Delta\varphi_m \\ \Delta K/K_0, \Delta\beta/\beta_0 \\ \Delta k/k_0 \end{array} \right\}$$

The primary step in field reception (and emission) is the mutual interference of the external E and B fields, $(E,B)_{\text{ext}}$, with the local (body and cell) internal fields $(E,B)_{\text{int}}$, provided the external static and low frequency E fields can penetrate the systems—for instance, via direct (metal) electrode contact. The subsequent steps may occur in parallel (Figure 1): (1) mutual changes in electrolyte flows (resulting in local ion concentration changes Δc) and in the internal fields, including the relatively very strong E fields of the cell membranes ($E_m \approx 10^3$–10^4 kV/m) corresponding to electric membrane potential differences in the order of $|\Delta\varphi_m| \approx 0.01$–0.2 V, and (2) field-induced changes of ligand-binding processes, of conformational transitions, and of cooperative domain phase transitions (e.g., in membranes).

Theoretically the field-induced changes may be quantified as rel-

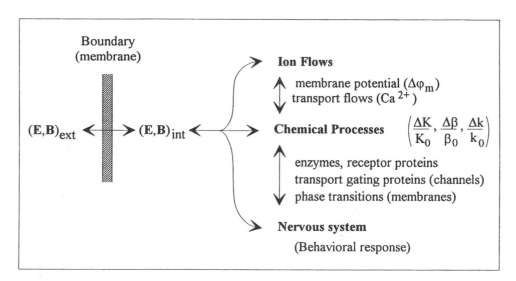

Figure 1.   Interaction scheme for electric *(E)* and magnetic *(B)* field reception. Direct current, static, and low frequency external E fields ($\mathbf{E}_{\text{ext}}$) induce noticeable changes in the body's internal fields only if there is direct contact (e.g., via metal electrodes). The surfaces of conducting matter represent no barriers for high frequency E fields and all magnetic fields. The $(\mathbf{E, B})_{\text{ext}}$-induced changes in $(\mathbf{E, B})_{\text{int}}$ modulate ion flows and chemical processes, in particular in and on membranes. Weak-field reception and emission by special skin organs always involve the nervous system.

ative changes ($\Delta K/K_0$, $\Delta\beta/\beta_0$, $\Delta k/k_0$) in the equilibrium constants (*K*), in the extents of ligand binding and of transitions (β), and in the rate coefficients (*k*).

2   FIELD AND MOLECULAR MOTION

2.1   INTERNAL AND EXTERNAL FIELDS

To understand the effects of external fields on freely mobile ions, dipolar charge pairs, and "fixed" ionic or dipolar groups of larger molecules, it is important to realize that these ions and dipoles are themselves the sources of local fields, which near the charges have an impressively high field strength (in the order of 3000 kV/m at body temperature). The local electric fields are largely screened by mobile counterions (ionic atmosphere) or by neutralizing ion pairing. The motion of charges in moving fluids (like the electrolyte ions in the blood) always produces magnetic fields, which in turn give rise to electric potential gradients.

External electric and magnetic fields have to compete with the internal biofields. The **E** (dc) fields and the low frequency (LF) alternating (ac) **E** fields of the air are effectively reduced by conducting material like the skin of the human body. At low frequencies (≤ 3 kHz), the internally induced electric fields in conductive bodies are therefore extremely small. On the other hand, external magnetic fields are not screened; they permeate bodies. The main effect of time-varying magnetic fields is the induction of time-varying electric fields, which cause dissipative ionic (ac) currents as well as electric dipolar displacement (capacitive) currents.

On the level of cells and organelles, the dielectric material of the plasma membrane has condenser properties for the storage of electric energy. External fields cause changes in the ion accumulations at the membrane–solution interfaces. Externally induced interfacial polarizations can lead to enormous amplification of the transmembrane electric field, which in turn can appreciably affect the reactions of membrane proteins and the various transmembrane transport phenomena.

2.2   THERMAL MOTION

Another important aspect is that the electric and magnetic interaction energy ($\Delta\hat{G}$) must compete with the randomizing thermal motion of the species. The thermal noise energy is given by $k_B T$, or on a molar scale, by RT, where k_B is the Boltzmann constant, $R (= N_A \cdot k_B)$ is the gas constant, N_A the Avogadro constant, and *T* is the thermodynamic (Kelvin) temperature. To be relevant, the electric and magnetic field effects must emerge from thermal noise and the normal concentration fluctuations.

2.3   FREELY MOBILE CHARGES

The **E** and **B** fields are a measure of the respective force line densities. **E** and **B** are defined by the electric and magnetic forces $\mathbf{F}_{el}$ and $\mathbf{F}_{ma}$ exerted on the charge *q*, respectively, changing the velocity **v** of free charges and the position of fixed and dipolar charge pairs.

The direction of change is globally given by the charge number z_i (with sign) of the ion or ionic group *i*: $q = z_i e_0$, where $e_0 = 1.6 \times 10^{-19}$, As is the positive elementary charge, and by the drift speed

$$\mathbf{v}_i = \frac{z_i}{|z_i|} u_i \mathbf{E} \qquad (1)$$

where u_i is the electric mobility. The Coulomb law $\mathbf{F}_{el} = q\mathbf{E}$ and the Lorentz law $\mathbf{F}_{ma} = q(\mathbf{v} \times \mathbf{B})$ yield the total force on *q*:

$$\mathbf{F} = q(\mathbf{E} + \mathbf{v} \times \mathbf{B}) \qquad (2)$$

The vector cross product $\mathbf{v} \times \mathbf{B} = |\mathbf{v}|\,|\mathbf{B}| \sin\theta$, where θ is the angle between the vectors **v** and **B**, describes the circular motion of a charge around the **B** lines. Note that $\mathbf{F}_{ma}$ is maximal if **v** and **B** are at right angles and $\mathbf{F}_{ma} = 0$ if **v** and **B** are parallel or if the charge is fixed ($\mathbf{v} = 0$).

2.4   DIPOLE ORIENTATION AND DIPOLOPHORESIS

Electric and magnetic dipoles are oriented in homogeneous **E** and **B** fields, respectively. The fields exert a torque against the thermal energy $k_B T$. In spatially inhomogeneous fields, dipoles not only rotate into the field direction but also move translationally in the direction of higher or lower field strengths (dipolophoresis, sometimes inappropriately called dielectrophoresis). The total force $\mathbf{F}(r)$ exerted by, say, a radially inhomogeneous field is given by the dot products:

$$\mathbf{F}(\mathbf{r}) = m_{el} \cdot \frac{\partial \mathbf{E}}{\partial \mathbf{r}} + \mathbf{m}_{ma} \cdot \frac{\partial \mathbf{B}}{\partial \mathbf{r}} \qquad (3)$$

where the partial space derivatives are a measure of the inhomogeneity of the respective fields.

2.5   ELECTRIC CURRENTS

The motions of charges and dipoles may be measured as ohmic and capacitive currents. The current density **J** (A/m²) is related to the field strength by

$$\mathbf{J} = \kappa \cdot \mathbf{E} + \frac{\partial \mathbf{D}}{\partial t} \qquad (4)$$

where κ is the ionic conductivity of the medium and $\mathbf{D} = \varepsilon_0\, \varepsilon\mathbf{E}$ is the dielectric displacement vector (covering dipole reorientations), where $\varepsilon_0 = 8.854 \times 10^{-12}$ As/(Vm) is the vacuum permittivity and ε is the (relative) dielectric permittivity of the homogeneous medium.

In polyelectrolytes like DNA and on polyionic membrane surfaces, an external field can cause displacements of the counterion atmospheres, thereby inducing stationary "ionic flow dipoles" having appreciably high dipole moments.

The circular current induced by a homogeneous magnetic field acting on moving charges q_i ($\mathbf{v}_i$) is given by the sum of all ion contributions:

$$\mathbf{J} = \kappa \mathbf{E}_{ind} = \kappa \, \Sigma_i (-\mathbf{v}_i \times \mathbf{B}) \qquad (5)$$

Time-varying magnetic fields induce ion flows by magnetic induction. The voltage V_{ind} induced around a closed path of length **l** $= \int d\mathbf{s}$, where $d\mathbf{s}$ is a differential path element, is given by

$$V_{ind} = \oint \mathbf{E} \cdot d\mathbf{l} = -\iint \frac{\partial \mathbf{B}}{\partial t} \cdot d\mathbf{s} \qquad (6)$$

where $\partial\mathbf{B}/\partial t$ is the time derivative of **B.** Since the dot product $(\partial\mathbf{B}/\partial t) \cdot d\mathbf{s}$ is $(\partial|\mathbf{B}|/\partial t)d|\mathbf{s}| \cos\theta$, V_{ind} is maximal if **B** is perpendicular to the path (surface), that is, parallel to **s.** The magnitude of the $E(t)$ field of frequency f at a perpendicular distance **r** from the center of a path around the **B** direction is $|\mathbf{E}_{ind}| = \pi f |\mathbf{B}||\mathbf{r}|$ and the current density $|J| = \kappa \pi f |B| |r|$ increases linearly with frequency. Magnetically induced currents are generally about 10^3 times larger than those induced directly by external electric fields $\mathbf{E}_0$ (f).

3   CHEMICAL REACTION EQUILIBRIA

In biochemistry we frequently encounter ligand (L)-induced changes in the conformation of macromolecular binding sites (B). This induced fit is described by the scheme

$$\begin{array}{ccccc} (K_1) & & (K_2) & \\ L + B & \rightleftharpoons & LB & \rightleftharpoons & LB' \end{array} \qquad (7)$$

where the conformational transition $B \rightleftharpoons B'$ is inherent. The physically more realistic reaction scheme comprises two binding steps and two structural transitions:

$$\begin{array}{ccc} & (K_1) & \\ L + B & \rightleftharpoons & LB \\ (K_4) \ \Updownarrow & & \Updownarrow \ (K_2) \\ L + B' & \rightleftharpoons & LB' \\ & (K_3) & \end{array} \qquad (8)$$

The individual equilibrium constants $K_1 = [L][B]/[LB]$, $K_3 = [L][B']/[LB']$ and $K_2 = [LB']/[LB]$, $K_4 = [B']/[B]$, where the brackets denote molar concentrations, are related by $K_1/K_3 = K_2/K_4$.

The apparent equilibrium constants K, defined as equilibrium concentration ratios, may vary with concentration and with the ionic strength because of the screening effects of ions and other nonreactive interactions. When a reaction is written in the suggestive form $0 = \Sigma_i v_i J$, where v_j is the stoichiometric coefficient of the reaction partner J (positive for a product, negative for an educt), the "really constant" thermodynamic equilibrium constant K^θ is defined as the equilibrium activity ratio:

$$K^\theta = \Pi_j a_j^{vj} = |K| \cdot Y \qquad (9)$$

The thermodynamic activity a_j is a dimensionless quantity expressed as $a_j = |c_j|y_j$, where $|c_j|$ is the numerical value of the concentration c_j and y_j is the thermodynamic activity coefficient. Hence, from $K^\theta = |K|Y$, we see that $K = \Pi_j c_j^{y_j}$ and $Y = \Pi_j y_j^{y_j}$.

3.1 DIPOLE REACTION

For the general analysis of field-induced changes, the scheme $0 = \Sigma_j v_j J$ is rewritten such that the reaction partners are specified as reactive educts J_r and as products J_p according to

$$\Sigma_r |v_r| J_r \underset{\overleftarrow{k}}{\overset{\overrightarrow{k}}{\rightleftharpoons}} \Sigma_p v_p J_p \qquad (10)$$

The molar reaction dipole moment ΔM relevant for field effects is now defined as the difference

$$\Delta M = M_p - M_r = N_A \langle \Delta m \rangle \qquad (11)$$

where M_p and M_r are the average molar moments of the products and reactants, respectively. The molecular average $\langle \Delta m \rangle$ is

given by the individual contributions of the reactant and product molecules:

$$\langle \Delta m \rangle = \Sigma_r |v_r| |\mathbf{m}_r| \langle \cos \theta \rangle - \Sigma_p v_p |\mathbf{m}_p| \langle \cos \theta \rangle \qquad (12)$$

The angle brackets $\langle \ \rangle$ denote orientational averages, and θ is the angle between the dipolar axis of a molecule J and the field direction.

Introducing now $F_{(E,B)} = F_E = |\mathbf{E}|$ or, in the other case, $F_{(E,B)} = F_B = |\mathbf{B}|$ for the amount of the electric field strength vector **E** and that of the magnetic flux density vector **B,** respectively, the general expression of the (isothermal, isobaric) field-induced change is given by

$$\left(\frac{\partial \ln|K|}{\partial F_{(E,B)}} \right)_{p,T} = \frac{\Delta M}{RT} \qquad (13)$$

For $F = |E|$, ΔM is the electric reaction moment; for $F = |B|$, ΔM refers to the molar difference in the magnetic moments.

Note that $K = \overrightarrow{k} / \overleftarrow{k}$, where $\overrightarrow{k}$ and $\overleftarrow{k}$ are the rate constants for the forward and reverse reaction steps, respectively. Integration of equation (13) yields:

$$K = K_0 \cdot e^X \qquad (14)$$

where K_0 is the K value at zero field. The "field effect exponent" X is given by the ratio of the electric or magnetic free energy $\Delta \hat{G}_{(E,B)}$, respectively, to the thermal energy

$$X = \frac{-\Delta \hat{G}_{(E,B)}}{RT} = \frac{\int \Delta M \, dF_{(E,B)}}{RT} \qquad (15)$$

The integration boundaries are $F_{(E,B)}$ and $F_{(E,B)} = 0$, respectively. If the apparent equilibrium constant K and the reaction moment ΔM are not independent of the concentrations of the interaction partners or of the ionic strength, the relations

$$K^\theta = K_0^\theta \cdot e^{X^\theta} \qquad (16)$$

and

$$X^\theta = \frac{-\Delta \hat{G}^\theta}{RT} = \frac{\int \Delta M^\theta \, dF_{(E,B)}}{RT} \qquad (17)$$

are used, where $\Delta \hat{G}\theta$ and ΔM^θ are the standard values.

From $K^\theta = |K|Y$ we obtain:

$$X = X^\theta - \ln \left(\frac{Y_E}{Y} \right) \qquad (18)$$

and

$$\Delta M = \Delta M^\theta - RT \cdot \frac{\partial \ln (Y_E/Y)}{\partial F_{(E,B)}} \qquad (19)$$

where Y is the normal activity factor at $F_{(E,B)} = 0$ and, purely formally, Y_E is that at $F_{(E,B)}$.

3.2 THE WEAK-FIELD APPROXIMATION

When $\Delta \hat{G}_{(E,B)} \ll RT$, then $X \ll 1$ and the factor e^X in $K = K_0 e^X$ can be expanded into a power series $e^X = 1 + X + \cdots$, where higher order terms are negligibly small. Therefore, for small field effects equation (14) takes the form:

$$K = K_0(1 + X)$$

For practical purposes a relative change $\Delta K/K_0$ is introduced which readily permits us to express field induced changes in percent. With $K = K_0 e^X$ we obtain

$$\frac{\Delta K}{K_0} = \frac{K - K_0}{K_0} = e^X - 1 \qquad (20)$$

The small-field approximation ($X \ll 1$) now reads:

$$\frac{\Delta K}{K_0} = X \qquad (21)$$

For small field effects in a random ensemble, $\langle \Delta m \rangle$ is proportional to $F_{(E,B)}$. Hence, X is proportional to $F^2_{(E,B)}/2\,RT$ and thereby $\Delta K/K_0$ is proportional to $F^2_{(E,B)}/2\,RT$; see Figure 2.

In summary, if $\Delta M\,F_{(E,}\,nB_) \ll RT$, the field-induced changes in the equilibrium concentrations are proportional to the *square* of the field strength.

We also see in $K = K_0\,(1 + X)$ that no matter how small the difference $\langle \Delta m \rangle$ in the electric or magnetic dipole moments of the interaction partners, there is *always a finite change* caused by a field. Of course, the inevitable thermal motion leads to local activity fluctuations: $\delta a_j = y_j \delta c_j + c_j \delta y_j$. We must determine the relevance of the small, field-induced changes Δc_j and Δy_j in concentration and activity coefficient, respectively, compared with the random thermal fluctuations $\pm \delta c_j$ and $\pm \delta y_j$.

3.3   Conformational State Transition

In a field, an intramolecular state transition or conformational change of the type

$$B \rightleftharpoons B' \qquad (22)$$

proceeds in the direction of B' if the dipole moment $m\,(B') > m\,(B)$; hence the following equation holds:

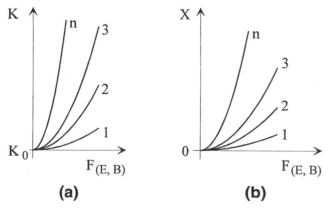

(a)   **(b)**

Figure 2. Field dependences of (a) the equilibrium constant $K = K_0\,e^X$ and (b) the field effect exponent $X = \int \Delta M\,dF_{(E,B)}/(RT) \propto F^2_{(E,B)}$. The steepness $(\partial K/\partial F)_{p,T} = K \cdot \Delta M/RT)$ and $(\partial X/\partial F)_{p,T} = \Delta M/RT$ increase with ΔM and with the domain size n. Since the relative change $\Delta \beta/\beta_0$ in the advancement of a chemical process is proportional to $X \propto F^2_{(E,B)}$ we obtain $\Delta \beta/\beta_0 \propto F^2_{(E,B)}$. The rate constants k exhibit qualitatively the same field dependences as K; therefore, generally $\Delta k/k_0$ is proportional to $F^2_{(E,B)}$.

$$\langle \Delta m \rangle = \mathbf{m}(B')\,\langle \cos \theta \rangle - \mathbf{m}(B)\,\langle \cos \theta \rangle > 0$$

When the degree of transition is defined as $\beta = [B']/[B_T]$, where $[B_T]$ is the concentration of all binding sites (or all states $[B_T] = [B] + [B']$), we obtain:

$$K = \frac{[B']}{[B]} = \frac{\beta}{1 - \beta} = K_0 \cdot e^X \qquad (23)$$

$$\beta = \frac{K_0 e^x}{1 + K_0 e^x} \qquad (24)$$

The weak-field approximation ($X \ll 1$, $e^X = 1 + X$) yields $\Delta K/K_0 = X$ and

$$\frac{\Delta \beta}{\beta_0} = \frac{\beta - \beta_0}{\beta_0} = X \frac{1}{1 + K_0} \qquad (25)$$

Thus the relative shifts in both K and β are linearly dependent on X. If $K_0 \ll 1$, we obtain $\Delta \beta/\beta_0 = X$.

3.4   Ligand Binding

For the analysis, the ligand-binding equilibrium is written in the direction of dissociation:

$$LB \rightleftharpoons L + B \qquad (26)$$

The degree of ligand association is defined by

$$\beta = \frac{[LB]}{[B_T]} = \frac{[L]}{[L] + K} \qquad (27)$$

where K is given by the total concentrations $[L_T]$ and $[B_T]$ of ligand and binding site, respectively:

$$K = \frac{[L][B]}{[LB]} = ([L_T] - \beta[B_T])\,\frac{1 - \beta}{\beta} \qquad (28)$$

Using $(1 - a)^{1/2} \approx 1 - 1/2\,a$ if $a < 1$, we obtain

$$\beta \approx \frac{[L_T]}{[L_T] + [B_T] + K} \qquad (29)$$

The field-induced relative change $\Delta \beta/\beta_0$ then is

$$\frac{\Delta \beta}{\beta_0} \approx \frac{K_0(1 - e^X)}{[L_T] + [B_T] + K_0\,e^X} \qquad (30)$$

The weak-field approximation ($X \ll 1$) yields

$$\frac{\Delta \beta}{\beta_0} \approx -X \left[\frac{1 + ([L_T] + [B_T])}{K_0} \right] \qquad (31)$$

Here, too, the relative change in β is proportional to X. Note that for equation (26), we have $\langle \Delta m \rangle = (\langle m_L \rangle + \langle m_B \rangle) - \langle m_{LB} \rangle$.

3.5   Dipole Pairing

Frequently the pair association of dipolar species (B) is found to involve major dipole moment changes $\langle \Delta m \rangle$ if the pairing reduces the dipole moment to zero by antiparallel alignment. Therefore $\langle \Delta m \rangle$ for a dimerization reaction of the type

$$BB \rightleftharpoons 2B \qquad (32)$$

is given by $\langle \Delta m \rangle = 2\langle m(B) \rangle - \langle m(BB) \rangle = 2\langle m(B) \rangle$. For this case we have

$$\beta = \frac{2[B \cdot B]}{[B_T]} = \frac{[B]}{[B] + K/2} \qquad (33)$$

$$K = \frac{[B]^2}{[BB]} = \frac{(1 - \beta)^2}{\beta}[B_T] \qquad (34)$$

$$\beta \approx \frac{[B_T]}{[B]_T + K/2} \qquad (35)$$

The weak-field approximation reads:

$$\frac{\Delta\beta}{\beta_0} \approx -X \left(\frac{1 + 2[B_T]}{K_0} \right)^{-1} \qquad (36)$$

describing the decrease in the pair complex with increasing field strength.

4   RATE CONSTANTS

When time courses of field-induced processes can be resolved, the time constants or experimental rate coefficients can be analyzed in terms of individual rate constants. Generally when equilibrium constants are field dependent, the two rate constants must depend differently on the field strength.

4.1   DIPOLE TRANSITIONS

The equilibrium constant for the general equation (10) is given by

$$K = \frac{\Pi_p c_p^{\nu_p}}{\Pi_r c_r^{|\nu_r|}} = \frac{\overrightarrow{k}}{\overleftarrow{k}} \qquad (37)$$

The reaction moment ΔM is related to the kinetic activation moments M_a by

$$\Delta M = \overrightarrow{M}_a - \overleftarrow{M}_a \qquad (38)$$

Similarly, the field effect exponents are connected by

$$X = \overrightarrow{x}_a - \overleftarrow{x}_a \qquad (39)$$

Analogous to $K = K_0 e^X$, we obtain

$$k = k_0 \cdot e^{x_a} \qquad (40)$$

where the field effect factor is defined by

$$x_a = \frac{\hat{G}_a}{RT} = \frac{\int M_a dF_{(E,B)}}{RT} \qquad (41)$$

The weak-field approximation ($x_a \ll 1$) yields $k = k_0(1 + x_a)$. The field-induced relative change of k is generally given by

$$\frac{\Delta k}{k_0} = e^{x_a} - 1 \qquad (42)$$

Hence, the weak-field approximation reads

$$\frac{\Delta k}{k_0} = x_a \qquad (43)$$

Like the field-induced shifts in K and β, the relative change in the rate coefficients of dipolar equilibria are linearly dependent on x_a; frequently x_a is proportional to $F^2_{(E,B)}$.

If the experimental rate coefficient k is analyzed in terms of the Eyring transition state theory, we have $k = k\ddagger \cdot K\ddagger$, where $k\ddagger = k_B T/h$ is a universal decay constant and $K\ddagger$ and $\Delta M\ddagger$ are the equilibrium constant and the reaction moment, respectively, of the assumed rapid preequilibrium. For this case M_a associated with k is given by $M_a = \Delta M\ddagger$.

On the other hand, we may equally well assume that the activation moments M_a are determined by the actual dipole moments (**m**) of the educts and products. Hence

$$\overrightarrow{M}_a = N^A \cdot \Sigma_p \nu_p |\mathbf{m}_p| \langle \cos \theta \rangle$$
$$\overleftarrow{M}_a = N^A \cdot \Sigma_r \nu_r \|\mathbf{m}_r| \langle \cos \theta \rangle \qquad (44)$$

Nonidealities are covered by the activity coefficients. With $K^\theta = |\overrightarrow{k}^\theta|/|\overleftarrow{k}^\theta|$ and $Y = \Pi_p y_p^{\nu_p}/\Pi_r y_r^{|\nu_r|}$ we formally have

$$\overrightarrow{k} = \overrightarrow{k}^\theta \cdot \Pi_p y_p^{\nu_p} \quad \text{and} \quad \overleftarrow{k} = \overleftarrow{k}^\theta \cdot \Pi_r y_r^{|\nu_r|} \qquad (45)$$

The field effect factors are given by

$$x_a = x_a^\theta + \ln \Pi y^\nu \qquad (46)$$

where Πy^ν is the respective activity coefficient term.

If, by chemical relaxation spectrometry, the rate coefficients can be determined separately, the *dipole moments* of the interacting reaction partners can be estimated from the field dependence of the k values.

4.2   ION EQUILIBRIUM

According to Onsager, it is only the dissociation rate constant k_{dis} of the separation of an ion pair that is primarily affected by an electric field **E**:

$$L^{z+} \cdot B^{z-} \underset{k_{ass}}{\overset{k_{dis}}{\rightleftharpoons}} L^{z+} + B^{z-} \qquad (47)$$

In the linear range, Onsager's theory of very diluted weak electrolytes (z_+/z_-) yields

$$\left(\frac{\partial \ln |k_{dis}^\theta|}{\partial |\mathbf{E}_0|} \right)_{p,T} = \frac{(z_L u_L - z_B u_B)|z_L z_B|e_0^3}{(u_L + u_B)\, 8\pi\varepsilon_0\varepsilon(k_B T)^2} = \gamma \qquad (48)$$

where u_L and u_B are the electric mobilities of the separated ionic species. For symmetric electrolytes where $z_L = -z_B = |z|$, equation (48) is reduced to $\gamma = |z^3|e_0^3/[8\pi\varepsilon_0\varepsilon(k_B T)^2]$.

If the actual physical dipole moment of the ion pair $L \cdot B$ is not zero (such as the electric dipole moments of the $MgSO_4$ ion pairs), and if the reaction occurs at higher salt concentrations, we use the expressions

$$k_{dis} = k_{dis}^\theta y_{LB} \quad \text{and} \quad k_{ass} = k_{ass}^\theta y_L y_B \qquad (49)$$

Specifically, with $x_a^\theta = \gamma |\mathbf{E}_0|$, we have:

$$\overrightarrow{x}_a = \gamma |\mathbf{E}_0| + \ln \left(\frac{y_{LB,E}}{y_{LB}} \right) \qquad (50)$$

$$\overleftarrow{x}_a = \overleftarrow{x}_a^\theta + \ln \left(\frac{y_E}{y} \right)^2 \qquad (51)$$

When the dipole moment of the ion pair $L \cdot B$ is dependent on the ionic strength because of screening by ionic "half-clouds," we may use

$$\overline{M}_a = \overline{M}_a^\theta + RT \; \frac{\partial \ln(y_E/y)^2}{\partial |\mathbf{E}_0|}$$

where $y = y_L = y_B$.

Often it is useful to define a degree of dissociation by $\alpha = [B]/[B_T]$. If the reaction is symmetric ($[L_T] = [B_T]$), we have:

$$K = \frac{\alpha^2}{1 - \alpha} [B_T] \qquad (52)$$

The small-field approximation of the relative change in the degree of dissociation is derived in analogy to equation (31), resulting in:

$$\frac{\Delta\alpha}{\alpha_0} \approx x \left(3 + \frac{K_0}{[B_T]} + \frac{2[B_T]}{K_0} \right)^{-1} \qquad (53)$$

Formally the field dependence of K can be calculated from k_{dis} and k_{ass} and described as $\Delta K/K_0 = \overline{x}_a - \overline{x}_a$.

4.3 ACTIVITY COEFFICIENTS

Thermodynamic activity coefficients and the theoretical approaches are defined for equilibrium conditions only. The $\mathbf{E}$ and $\mathbf{B}$ field effects on electrolytes, however, represent nonequilibrium situations. Under stationary conditions we may consider the field-induced distortion of screening ionic atmospheres and other local changes in the ion concentration to be actual steady state changes Δc_i. Purely formally, we introduce the relative change $\Delta y_i/y_i$ in the (theoretically individual) activity coefficient y_i of the ion (or ionic group) of type i.

The Debye–Hückel approximation for diluted solution is given by

$$\ln y_i = -z_i^2 A' I_c^{1/2} \qquad (54)$$

where A' is a (temperature-dependent) constant and

$$I_c = \frac{1}{2} \sum_i z_i^2 c_i \qquad (55)$$

is the ionic strength (including all ions i).

For elementary reactions between ionic partners like those in equation (47) we have in general:

$$\ln Y = 2 z_+ z_- A' I_c^{1/2} \qquad (56)$$

The small-field approximation is specified as

$$\ln \left(\frac{Y_E}{Y} \right) = \ln \left(1 + \frac{\Delta Y}{Y} \right) = \frac{\Delta Y}{Y} \qquad (57)$$

Because $\ln Y$ is proportional to $I_c^{1/2}$, we obtain for $\Delta I_c \ll I_c$,

$$\ln \left(\frac{Y_E}{Y} \right) = \left(1 + \frac{\Delta I_c}{I_c} \right)^{1/2} \approx 1 + \frac{\Delta I_c}{2I_c} \qquad (58)$$

where $\Delta I_c = \frac{1}{2} \sum_i z_i^2 \Delta c_i$ covers the field-induced local changes in the screening ions.

Similarly, field effects on rate coefficients may also involve changes in the y coefficients. Using the association of L^{z+} and B^{z-} as an example and applying $y_L = y_B = y$, we have $\ln(y_E/y)^2 \approx 2 \Delta y/y$ and the suggestive form

$$\frac{\Delta y}{y} \approx 2 + \frac{\Delta I_c}{I_c} \qquad (59)$$

The concentration changes Δc_i entering into ΔI_c and thus into the relative changes of K, α and k, may also contain contributions from the ionic currents induced by the fields.

4.4 ION FLOWS

The electrodiffusive flows of the cellular ions Na^+, K^+, Ca^{2+}, and Cl^- through electrochemically gated channels of membranes are dependent on the local ion concentrations c_i and on induced local field changes.

The flow density vector $\mathbf{i}_i$ (mol/sm^2) of the ion type i is given by the Nernst–Planck equation for the flow direction $\mathbf{r}$:

$$\mathbf{i}_i = -D_i \left(\frac{\partial c_i}{\partial \mathbf{r}} + \frac{z_i c_i F}{RT} \frac{\partial \varphi}{\partial \mathbf{r}} \right) \qquad (60)$$

where D_i is the diffusion coefficient, $F = e_0 N_A$ is the Faraday constant, and $\partial \varphi/\partial \mathbf{r} = -\mathbf{E}$ is the electric potential gradient at site $\mathbf{r}$.

The current density is then

$$\mathbf{J} = F \sum_i z_i \mathbf{i}_i = \kappa \cdot \mathbf{E} \qquad (61)$$

The relative changes $\Delta J/J_0$ are caused by a change in E or in the ion concentrations, or in both. Finally, the change in the transmembrane voltage (or membrane field E_m) may directly change the conformation of a membrane transport-gating protein or of a membrane-associated enzyme.

5   AMPLIFICATION MECHANISMS

Natural electric and magnetic field reception operates at small external field strengths. The molecular reaction moments of the interacting, directly "field-receiving" charges and electric and magnetic dipoles are relatively small ($|\Delta \hat{G}|/RT \ll 1$). Obviously, there must be amplification mechanisms for both enhancement of the field (resulting in larger internal fields and ion flows) on the one hand, and enhancement of local chemical reactivities, on the other hand.

If a chemical amplification cascade enhances a substrate signal from the level $[S_0]$ to the product level $[P_0]$, a relevant external field effect must already have changed the substrate concentration by a significant amount $\Delta[S]$. The cascade, which may contain a network of nonlinear couplings of reaction and diffusion flows, will process the relative change $\Delta[S]/[S_0]$ to the output $\Delta[P]/[P_0] = \Delta[S]/[S_0]$. Hence the relative amplification will not change unless there are significant field effects on the local elements of the cascade itself.

The distributions of electric organs in rows of larger skin areas of electric fish or the strings of magnetosomes in magnetobacteria and other species suggest that geometrical factors are essential in adding up signals from many inputs. The feedback system of many inner-vated sensor and emittor organs even enables the organisms to differentiate between inanimate ohmic-dissipative matter and capacitively storing biomatter of cells (organelles and lipid/protein membranes). Besides the special structure, form, and shapes of the sensory receptors, there are some general principles for molecular enhancement and amplification processes.

5.1   FIELD ENHANCEMENT

At interfaces of matter of differing relative permittivity (dielectric constant, ε) such as plasma membranes, applied fields cause ion accumulations that charge up cells, organelles, and larger macromolecules (like condensers).

If a macromolecule is considered to be a spherical microdroplet continuum (ε_2) in aqueous solution ($\varepsilon_1 = 80$ at 20°C), the induced field across the center is given by:

$$\mathbf{E}_{ind} = \frac{3\varepsilon_1}{2\varepsilon_1 + \varepsilon_2} \mathbf{E}_0 \qquad (62)$$

Since usually $\varepsilon_2 < \varepsilon_1$, we obtain $\mathbf{E}_{ind} \leq 1.5 \, \mathbf{E}_0$. Note that equation (62) is applicable if the conductivity term in the complex dielectric permittivity is negligible.

A much larger field amplification occurs when single cells or organelles or even clusters of cells and tissue pieces are exposed to external fields ($\mathbf{E}_0$). The ionic interfacial polarization causes in-

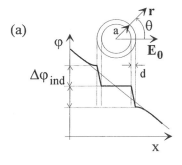

(a)

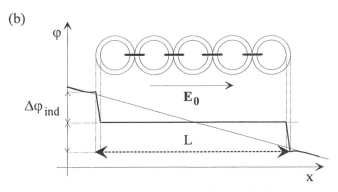

(b)

Figure 3.   Field amplification by interfacial polarization through dc and low frequency **E** fields. Electric potential profile $\varphi(x)$ in the direction x of the external field vector $\mathbf{E}_0$ through the pole caps. (a) Single cell sphere of radius a, membrane thickness d, and radius vector **r**. The steep potential drop across the membranes in the direction of $\mathbf{E}_0$ is given by $\Delta\varphi_{ind}(\theta) = 1.5 \, E_0 a$ |cos θ|, if the conductivity of the membrane is very small compared to the cell inside and outside. The field amplification factor at the pole caps is $f(a) = (E_{ind}/E_0) = 1.5 \, a/d$. (b) Array of cells connected by conducting gap junctions. The field amplification factor is $f(L) = (E_{ind}/E_0) = 0.5 \, L/d$.

duced electric potential differences $\Delta\varphi_{ind}$ which, as a result of the conducting cell interior, occur only across the plasma membranes (see Figure 3a). For spherical cells, the induced transmembrane potential difference in the direction of the external field is given by

$$\Delta\varphi_{ind}(\theta) = -\frac{3}{2} af(\lambda)|\mathbf{E}_0||\cos\theta| \qquad (63)$$

if the radius a ($\ll d$) is large compared with the membrane thickness d (≈ 10 nm). In equation (63), θ is the angle between $\mathbf{E}_0$ and the radius vector, and the conductivity factor $f(\lambda) \leq 1$ is $f = 1$ for nonconducting membranes. At the pole caps [i.e., at $\theta = 0°$ (right side) and $\theta = 180°$ (left side in $\mathbf{E}_0$ direction, Figure 3) hence cos $\theta = \pm 1$], the induced transmembrane voltage is maximum. The induced part $\mathbf{E}_{ind}$ of the mean transmembrane field is

$$|\mathbf{E}_{ind}| = -\frac{\Delta\varphi_{ind}}{d} = \frac{3}{2}\frac{a}{d}|\mathbf{E}_0| \qquad (64)$$

The amplification factor is defined by

$$f(a) = \frac{|\mathbf{E}_{ind}|}{|\mathbf{E}_0|} = \frac{3a}{2d} \qquad (65)$$

For a cell radius of $a = 10 \, \mu$m and $d = 10$ nm, $f(a) = 1.5 \times 10^3$. Hence, interfacial polarization can very appreciably amplify weak external fields; this type of amplification mechanism is the basis for all the electroporation phenomena.

Interfacial polarization not only amplifies external fields. In addition, the induced potential difference is superposed to the natural potential difference of cell membranes, $\Delta\varphi_m \approx -100$ mV, where the electric potential of the outside is arbitrarily taken to be zero. The actual transmembrane potential drop in the direction of $\mathbf{E}_0$ is given by

$$\Delta\varphi = -\left[\frac{3}{2}af(\lambda)|\mathbf{E}_0| + \frac{\Delta\varphi_m}{\cos\theta}\right]|\cos\theta| \qquad (66)$$

It is seen that $\Delta\varphi_{ind}$ adds up to $\Delta\varphi_m$ at the left pole cap (cos $\theta = -1$) and at the right pole cap (cos $\theta = 1$) $\Delta\varphi_{ind}$ reduces $\Delta\varphi_m$. Since the left cap is hyperpolarized and the right cap is depolarized, the membrane proteins and transport processes in the two caps are differently affected by $\mathbf{E}_0$. Therefore a longer lasting dc field may produce appreciably large asymmetric field effects.

At the pole caps of oriented bacteria of length L, the induced transmembrane potential difference is

$$\Delta\varphi_{ind} = -\frac{1}{2}L|\mathbf{E}_0| \qquad (67)$$

and the amplification factor is $f(L) = L/2d$. Here, too, the field effect is asymmetric at the two pole caps.

Another important example of field amplification by geometrical factors consists of a series of cells connected by gap junctions. Such a string of cells in series resembles a single elongated cell. In an applied external field the outer pole cap of the head cell and that of the tail cell experience the induced potential drop $\Delta\varphi_{ind} = -\frac{1}{2}L|\mathbf{E}_0|$, where L is the length of the string (Figure 3b). The amplification factor $f(L)$ increases with L and may become very large. Here, too, the field effect is asymmetric with respect to the membrane voltage in the head and tail cell.

5.2   Cooperativity and Phase Transition

In systems with domain structures, the molecules of a domain are cooperatively coupled. A domain changing state or, more specifically, molecular orientation as a whole, is described by

$$B_n \rightleftharpoons B'_n \qquad (68)$$

where n is the number of cooperatively coupled molecules (domain size) and B' represents the (orientational) state favored in the presence of an applied field because of higher moment.

In the simplest case, the collective transition of a domain is described by the field effect exponent

$$X_n = \frac{-\Delta \hat{G}_n}{RT} = \frac{\int \Delta M_n \, dF_{(E,B)}}{RT} \qquad (69)$$

where $X_n = n X$ and $\Delta M_n = n \Delta M$.

For an ensemble of independent domains of different sizes, we may formalize a distribution constant

$$K(\bar{n}) = K_0 \exp(X_{\bar{n}}) \qquad (70)$$

where $\bar{n}$ is the mean domain size, representing a kind of cooperative amplification factor for the reaction moment ΔM.

If $\bar{n} \gg 1$, very steep phase transitions can occur in a very narrow field strength range. Even a weak external field thus may have a large effect (see Figure 2).

The orientational rearrangement within a domain may be subjected to a continuum approach, where the polarization P has different values in the two states B_n and B'_n. If $\Delta P = P(B'_n) - P(B_n)$ and V_n is the domain volume,

$$\Delta M_n = V_n \cdot \Delta P \qquad (71)$$

Specifically, the electric and magnetic polarization differences are given by

$$\begin{aligned} \Delta P_{el} &= \varepsilon_0 \, \Delta \chi_{el} |E| \\ \Delta P_{ma} &= \mu_0^{-1} \Delta \chi_{ma} |B| \end{aligned} \qquad (72)$$

where the $\Delta \chi_{el}$ and $\Delta \chi_{ma}$ are the electric and magnetic susceptibility differences, respectively. The field effect exponents at constant domain volume are derived from equation (69):

$$x_{el} = \frac{\varepsilon_0 \, V_n \cdot \Delta \chi_{el}}{2k_B T} \, g|E_0|^2 \qquad (73)$$

where g is a shape factor and

$$x_{ma} = \frac{V_n \cdot \Delta \chi_{ma}}{2 \mu_0 \, k_B T} \, |B|^2 \qquad (74)$$

In an ensemble of domains, V_n represents the size average $V_{\bar{n}}$.

More specifically, a domain orientation function ϕ^{OR} may be used, which is different for permanent (m_p) and induced (m_{ind}) dipole mechanisms. For uniaxial permanent dipoles we have

$$\phi_{perm}^{OR} = 1 - \frac{3}{x_p} \left(\coth x_p - \frac{1}{x_p} \right) \qquad (75)$$

where $x_p = |m_p| F_{(E,B)dir} / (k_B T)$. For $F = E$, E_{dir} is the Onsager directing field. If $x \ll 1$, the weak-field approximation yields $\phi_p^{OR} = x_p^2/15$ as the first term of a power series.

The orientation function for induced electric dipoles $m_{ind} = \alpha \cdot E_{int}$, where α is the polarizability tensor and E_{int} is the internal field, is given by

$$\phi_{ind}^{OR} = \frac{3}{4} \left(\frac{e^{x_{ind}}}{x_{ind}^{1/2} \int dy \, e^{y^2}} \right) - \frac{1}{2} \qquad (76)$$

where $x_{ind} = \alpha \cdot g_i |E_0|^2 / (2k_B T)$ and $g_i = |E_{int}| / |E_0|$ is the shape factor.

If $x_{ind} \ll 1$, the weak-field expression yields

$$\phi_{ind}^{OR} = 2 x_{ind} / 15 \propto |E_0|^2$$

In the case of domain systems, we have

$$x_p(n) = \frac{n |m_p| F_{(E,B)dir}}{k_B T} \qquad (77)$$

and

$$x_{ind}(n) = \frac{n \cdot \alpha \cdot g_i |E_0|^2}{2 \, k_B T} \qquad (78)$$

If $n \gg 1$, the amplification of small molecular field effects may be appreciably large (see Figure 2).

5.3   Amplification at High Basis Fields

A particular amplification mode resides directly in the field dependence of equilibrium constants K. This enhancement mode appears to be especially important for all membrane processes occurring in the strong natural transmembrane fields.

From equations (14) and (15) it is seen that the steepness of the $K(X)$ relationship, given by

$$\left(\frac{\partial K}{\partial F_{(E,B)}} \right)_{p,T} = \frac{K(E) \Delta M}{RT} \qquad (79)$$

depends not only on ΔM but also on $K(E)$.

If an external field causes an induced field $|E_{ind}| = \Delta E_0$ at the level E_m of the resting membrane potential difference ($\Delta \varphi_m = E_m d$), the relative change $\Delta K / K(E_m)$ is larger than $\Delta K' / K_0$ at $E_m = 0$ for the same ΔE_0 value (see Figure 4). The inequality $\Delta K / K(E_m) > \Delta K' / K_0$ is due to the field dependence of the reaction moment ΔM ($\propto F_{(E,B)}$). Therefore, in terms of the field effect exponents, we have for the different integration boundaries:

$$X(E_m \to E_m + \Delta E_0) > X(O \to \Delta E_0)$$

If the external force is an oscillatory (ac) field, say, $E_0(f) = \hat{E}_0 \sin(2\pi ft)$ where $\hat{E}_0$ is the peak amplitude, the enforced oscillatory change $\Delta K \pm (f) / K(E_m)$ in the equilibrium constant of a reaction is asymmetric: $\Delta K_+ / K(E_m) \neq \Delta K_- / K(E_m)$; see Figure 4. For instance, the induced concentration changes Δc_+ are larger for one ac field direction ($+\Delta E_0$) compared with the Δc_- values of the other direction ($-\Delta E_0$). In a similar way, the forward rate change is different from the reverse rate change.

5.4   Comparison of Ac/Dc Effects

In dc fields, freely mobile charges and dipoles are translationally or rotationally net-displaced. In ac fields, however, the motions are os-

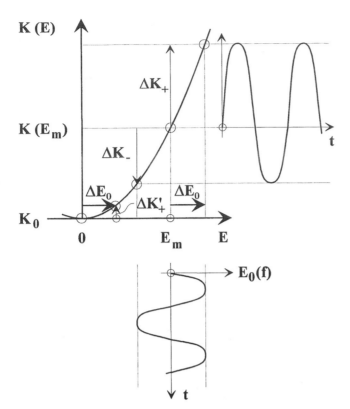

Figure 4. Amplification at high field and asymmetric field effect for oscillatory external fields (ac). Application of an external field, inducing $E_{ind} = \Delta|E_0|$, at $E = 0$ causes only a small change $\Delta K'_+$. If applied at $E = E_m$ (e.g., at the resting potential $\Delta\varphi \approx -100$ mV), the same external field causes a much larger change ΔK_+, because $\Delta K/\Delta E = K(E)\Delta M/(RT)$ depends on $K(E)$. The asymmetry of the $K(E)$ dependence leads to an asymmetric amplification of K: $\Delta K_+ > \Delta K_-$, if the symmetric external field $\mathbf{E}_{0(f)} = \hat{\mathbf{E}}_0 \sin(2\pi ft)$ operates at E_m. At $E = 0$, $\Delta K_+ = \Delta K_-$. In a similar way, the rate constant k is asymmetrically dependent on $\mathbf{E}_0(f)$.

cillatory around a mean position. If the frequency of the ac field is too high, inertia and viscosity may prevent the species from following the variation of the external fields. On the other hand, fields acting on an ensemble of charges and dipoles may enforce and synchronize the individually random motions into a concerted collective motion of all the interaction partners. Such an externally enforced cooperation may bias in one direction an otherwise random fluctuation and produce the small net concentration changes involved in weak-field reception.

The interpretation of experimental data on weak-field effects is still very controversial. The formalism outlined here may help in two ways: (1) in the design of experiments that use both species and sample size as variables and (2) in the analysis of data in terms of thermodynamic and kinetic parameters consistent with the first principles of chemical reactivity in electric and magnetic fields.

There is continued public concern about possible health risks of weak electromagnetic fields (i.e., field intensities that are small compared with those of the natural geofields).

As indicated here, the problem is not that there are always finite effects which, in principle, may either be beneficial or bear a risk. The still unresolved question is *whether and how,* if at all, the small and insignificant local changes in the concentrations of the biochemical reaction partners may be amplified such that they emerge

from the natural thermal fluctuations. Despite the various experimental and theoretical activities aimed at accumulating knowledge to identify and to understand these mechanisms, no conclusive and convincing answers can yet be given.

See also FREE RADICALS IN BIOCHEMISTRY AND MEDICINE; SYNAPSES.

Bibliography

Böttcher, C. J. F. (1973) *Theory of Electric Polarization,* Vol. 1, *Dielectrics in Static Fields.* Elsevier, Amsterdam.

Eigen, M., and DeMaeyer, L. (1963) Relaxation methods. *Tech. Org. Chem.* 8(2):895–1054.

Fredericq, E., and Houssier, C. (1973) *Electric Dichroism and Electric Birefringence.* Clarendon Press, Oxford.

Neumann, E. (1986) Chemical electric field effects in biological macromolecules. *Prog. Biophys. Mol. Biol.* 47:197–231.

———, Sowers, A. E., and Jordan, C. A., Eds. (1989) *Electroporation and Electrofusion in Cell Biology.* Plenum Press, New York.

Polk, C., and Postow, E., Eds. (1986) *Handbook of Biological Effects of Electromagnetic Fields.* CRC Press, Boca Raton, FL.

ELECTRON MICROSCOPY OF BIOMOLECULES

Claus-Thomas Bock, Hanswalter Zentgraf, and John Sommerville

1 Principles

2 Techniques
  2.1   General
  2.2   Nucleic Acids
  2.3   Chromatin
  2.4   Proteins
  2.5   Macromolecular Assemblies

3 Applications
  3.1   Use of Antibodies
  3.2   Detection of Radiolabeled Molecules
  3.3   Electron Spectroscopic Imaging (ESI)
  3.4   Image Reconstruction

Key Words

Cryoelectron Microscopy   Imaging of unfixed, unstained biomolecules in a hydrated state after rapid freezing to low temperature ($-160°C$).

Heavy Metal Shadowing   Evaporation of a film of heavy metal (e.g., platinum–palladium or tungsten–tantalum) onto a dehydrated preparation. The deposit of metal particles around the biomolecules improves contrast and gives a shadowed, three-dimensional appearance.

High Resolution Autoradiography (EM ARG)   Detection of radiolabeled molecules by coating a stained or shadowed preparation with a film of photographic emulsion. Radioactive emissions hit silver halide crystals in the emulsion, which can be developed into silver grains.

Immunoelectron Microscopy   The application of antibodies to map specific sites on biomolecules. The antibody molecules may be visible after shadowing or negative staining, although detection can be improved by binding of a secondary antibody linked to an electron-dense tag (e.g., ferritin or colloidal gold particles).

Miller Spread   Deposition of dispersed chromatin on to a carbon-coated grid by centrifugation through a denser phase containing sucrose and formaldehyde.

Negative Staining   Instead of staining the biomolecules themselves (positive staining), a solution of heavy metal salt (e.g., uranyl acetate or sodium phosphotungstate) is deposited in the hydrated spaces around and within the molecules.

Replica Casting   Adsorption of biomolecules to a mica surface followed by heavy metal shadowing and coating of the preparation with a film of carbon. The carbon–metal replica (minus the biomolecules) is then removed from the mica and mounted on an electron microscope grid for viewing.

Support Film   A thin film of plastic or carbon that is attached to an EM grid and provides a substrate onto which biomolecules can be adsorbed.

Electron microscopy (EM) is a method appropriate for examining details of the sizes and shapes of biological macromolecules and is particularly useful in studying isolated or reconstituted macromolecular assemblies. Small amounts of material (often < 1 μg) can be used and prepared in a state suitable for viewing in as short a time as several minutes. The basic procedure involves adsorption of the biomolecules on to a support film, followed by staining or shadowing of the preparation and viewing of the dehydrated molecules in vacuo in an electron beam. A resolution of 1–2 nm is routine with conventional microscopes. The basic procedure can be adapted to give information about sites of specific epitopes, location of newly synthesized components, internal structures, and atomic composition.

Structures most suitable for EM analysis are nucleoprotein complexes, including ribosomes, nucleosomes, and virus particles. DNA and RNA molecules are also suitable, and their lengths can be directly related to the number of base pairs or nucleotide residues determined biochemically. In general, small proteins (< 50,000 Da) are difficult to visualize, but good detail can be obtained if they are isolated as multimeric complexes or can be induced to form filaments or crystalline arrays.

A wide range of applications is available using EM techniques, including virus identification, mapping of hybridized regions in heteroduplexes, detailing of macromolecular interaction, and analysis of the organization of molecular components in replication, transcription, splicing, and translation complexes.

1   PRINCIPLES

The transmission electron microscope (TEM) consists of a metal column from which air is evacuated and through which a linear beam of electrons is accelerated and focused by electromagnetic lenses. The biomolecules are adsorbed onto a support film, stained and dehydrated (or frozen in an aqueous film on the grid), and in-

troduced into the electron beam through an air lock. Whereas some of the electrons collide with atoms in the specimen, lose energy, and are scattered, the remaining electrons pass through the preparation and are focused to form an image on a phosphorescent screen (for direct viewing) or on a photographic plate (for later examination). Under ideal conditions, a resolution of 0.1–0.2 nm can be obtained, however limitations are imposed by the naturally low masses of atoms (primarily hydrogen, carbon, and oxygen) contained in biomolecules, by distortions and artifacts created during sample preparation, and by radiation damage. Techniques are designed with the following objectives:

1. To minimize distortion by immobilization of the molecules, in their native state, onto an appropriate support.
2. To stabilize molecular complexes by suitable chemical fixation.
3. To improve contrast by staining the preparation with heavy metal salt or by shadowing with heavy metal.

Irrespective of the method adopted, data derived from the electron microscopic examination of biomolecules should be entirely consistent with the known biochemical and biophysical properties of the particles. Whenever possible, apparent sizes should be checked by independent measurement of sedimentation rate, electrophoretic mobility, or gel filtration elution. Also, features revealed by EM of isolated molecules should be compared with observations made on them in situ, by EM of cell or tissue sections.

2   TECHNIQUES

2.1   GENERAL

Similar principles of sample preparation (Figure 1) apply for nucleic acids, proteins, and nucleoprotein complexes, since most of these are less than 20 nm thick and can be adsorbed directly from solution onto the support matrix. The concentration of molecules must be high enough to permit several examples to be viewed together in the one field. The efficiency of uptake onto the support is not always predictable, and a range of initial concentrations should be tried. It is important that the biomolecules be held in a solution known to maintain the proper structural features directly prior to applying to the support.

The grid consists of a fine meshwork, usually of copper, which must be coated with a thin support film of plastic (e.g., collodion, parlodion) or carbon. Carbon films are preferable because they are thin (down to 2 nm) and contribute little to the image. However, they are frequently hydrophobic and should be subjected to ionizing gases (by "glow discharging") or treated chemically (e.g., with Alcian blue 8GX) to render them hydrophilic before use.

2.2   NUCLEIC ACIDS

Nucleic acids are used at a concentration of 1–5 μg/mL and as little as 0.1 μg is required; detailed procedures for working with these molecules may be found in the literature. The molecules can be native DNA or RNA, cloned DNA, single-stranded or double-stranded forms, partially denatured duplexes, or denatured and hybridized structures (Figures 2–6). A key step in their preparation for

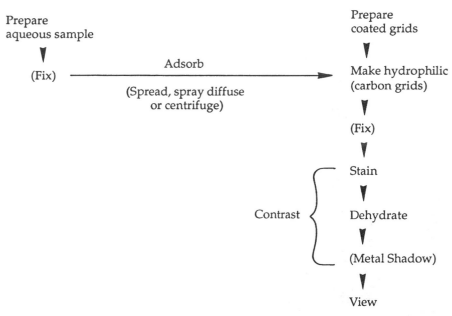

Figure 1.   Flow diagram of the basic procedure. Steps in parentheses are not always used.

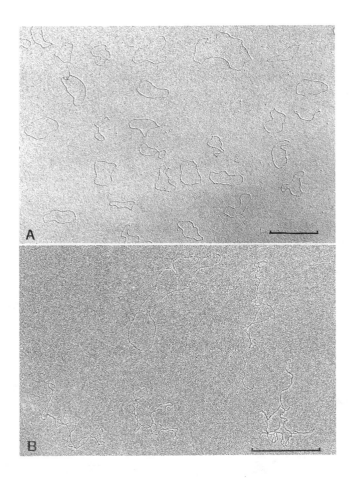

Figure 2.   (A) Simian virus 40 (SV40) DNA spread in the presence of cytochrome *c* down a glass slide and across a water hypophase. (B) *Xenopus laevis* ribosomal DNA derived plasmid DNA spread by the cytochrome *c* microdiffusion droplet technique. The DNA–protein films were picked up on parlodion-coated (A) and carbon-coated (B) copper grids. The use of carbon-coated grids yields a significantly higher proportion of supercoiled molecules and results in a smoother background than is the case for parlodion-coated grids. However, the attachment of molecules to the grid is significantly higher with parlodion. For contrast enhancement, the specimens were rotary-shadowed with platinum/palladium (80:20) at an angle of 8°. Note that an additional, very brief, unidirectional shadowing in (A) leads to further contrast enhancement. The bars represent 1 μm.

Figure 3. Herpes simplex virus (HSV) DNA as revealed by the cytochrome *c* microdiffusion droplet technique, picked up on a carbon-coated grid and shadowed as described in Figure 2A. Because shearing and stretching forces are minimal when the droplet technique is used, it is possible to spread this extremely large viral genome (150–160 kilobases) to its full length. Bacteriophage PM2 circular DNA serves as size markers. The ends of the HSV genome are denoted by arrow-heads. Bar represents 1 μm.

Figure 4. The "unraveling" of the HSV genome out of the viral core as shown by cytochrome *c* spreading. Complete intact virus particles (see Figure 17D) were briefly lysed in water and immediately spread onto a water hypophase. The white dotlike structure anchoring the spread DNA represents the unraveled part of the genome, surrounded by fuzzy material, most probably remnants of the viral core. Bar represents 1 μm.

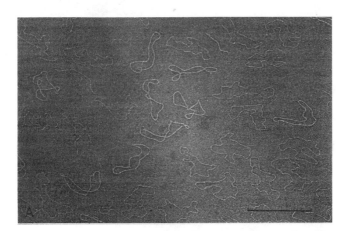
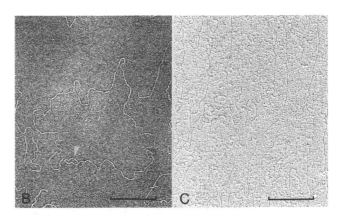

Figure 5. (A) Double-stranded and single-stranded DNA of bacteriophage fd spread in 40% formamide, 1 mM EDTA, 10 mM Tris-HCl (pH 8.5) onto a water hypophase. The cytochrome c–DNA film was picked up on a parlodion-coated grid and was shadowed as described in Figure 2B. The double-stranded DNA molecules are wider with smoother contours than the single-stranded molecules. (B) Replicative intermediate of plasmid pA2Y1, derived from a parvovirus, AAV (adenoassociated virus), spread by the microdiffusion technique in the presence of 10% formamide (parlodion-coated grids, rotary shadowing). The replication eye is denoted by an arrowhead. For contrast enhancement, (A) and (B) are printed in reverse contrast.. (C) Tobacco mosaic virus (TMV) RNA was diluted in a spreading solution made from 4 M urea dissolved in "pure" formamide, 1 mM EDTA, 10 mM Tris-HCl, (pH 8.5), heated to 80°C (5 min) and spread on a 60°C water hypophase, picked up on parlodion-coated grids, and shadowed. The bars represent 1 μm in (A), 0.5 μm in (B) and (C).

EM is the spreading out of the molecules so that they can be adsorbed in an extended, nonaggregated form on the support film. The most successful method of spreading is that devised originally by Kleinschmidt and Zahn (Table 1). The negatively charged nucleic acid is mixed with a basic protein (usually cytochrome c) in a solution appropriate for maintaining the required conformation. This spreading mix, or hyperphase, is spread, via a glass ramp, onto the surface of a second solution, the hypophase, and a molecular monolayer of nucleic acid and protein is formed at the liquid–air interface. Alternatively, the molecular monolayer can be formed by diffusion to the surface of the spreading mix contained in a small vessel or even in a droplet sitting on a hydrophobic surface (Table 2). The nucleic acid–protein complex is adsorbed through brief contact with the support film, stained with ethanolic uranyl acetate, rinsed in ethanol, and air-dried. To improve contrast, the preparation is rotary-shadowed, usually with platinum–palladium, at an angle of 6–9° from the plane of the metal vapor. It should be noted that although metal shadowing greatly exaggerates the thickness of the molecules, double-stranded and single-stranded regions can still be differentiated (Figures 5 and 6).

Cytochrome c should not be used in spreading solutions when experimentally bound proteins are being studied because it obscures fine detail of nucleic acid structure. Thus to permit detection of proteins bound to specific regions of nucleic acid molecules, protein-free spreading methods have been devised (Figure 7E). These methods employ low molecular weight substances (e.g., benzyl-dimethylalkylammonium chloride or ethidium bromide) in place of cytochrome c. In working with protein molecules bound to nucleic acids, care must be taken to stabilize the complexes before spreading by cross-linking the molecules in solution, with glutaraldehyde (Figure 7) or formaldehyde (for chromatin, see Section 2.3). Good detail of the association can be obtained by making a carbon–metal replica of the molecules bound to the surface of freshly cleaved mica. The use of platinum–carbon or tungsten for shadowing reveals finer detail.

2.3 CHROMATIN

Chromatin spreading techniques are mostly adapted from the procedure devised by Miller and co-workers (see Table 3). Large chromatin units are obtained by lysing cells or nuclei in a low salt concentration, high pH buffer (often referred to as "pH 9 water"). The chromatin is left to disperse and is then centrifuged at low speed, through a denser solution containing fixative (formaldehyde), onto the surface of a carbon-coated grid. The grid is dipped into a solution of wetting agent (Kodak Photoflo 200) and air-dried. The preparations can be positively stained (with ethanolic phosphotungstic acid), negatively stained (with aqueous uranyl acetate), or metal-shadowed. Chromatin spread from avian erythrocytes shows the typical nucleosomal configuration of dispersed chromatin (Figure 8). The features of actively transcribing chromatin can be seen in nucleolar (pre-rRNA) units (Figure 9). In contrast, nonnucleolar transcription units normally contain more widely spaced nascent transcripts (Figure 10B).

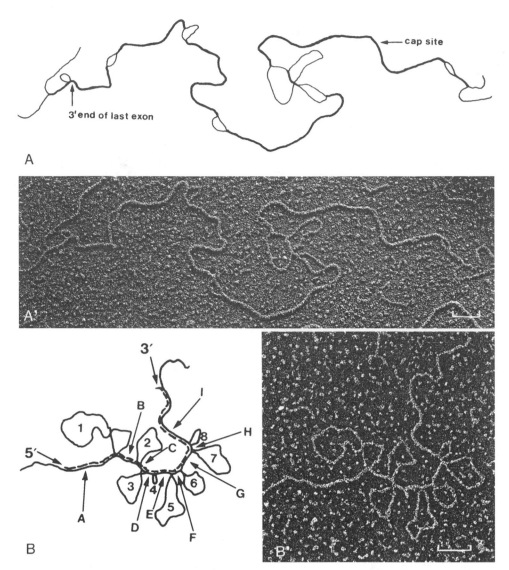

Figure 6. (A′) DNA:DNA heteroduplex between cloned mouse tyrosine aminotransferase (TAT) and rat TAT. Heteroduplex was formed after denaturation in 0.1 M NaOH, 20 mM EDTA at 20°C for 10 minutes and renaturation in the presence of 50% formamide, 20 mM Tris-HCl (pH 7.2) at 20°C for 90 minutes. Heteroduplex was prepared from a spreading solution containing 4 M urea dissolved in "pure" formamide, 1 mM EDTA, 10 mM Tris-HCl, (pH 8.5) on water hypophase followed by rotary shadowing. (A) The interpretive drawing of A′; the positions of the cap and of the 3′ end of the last exon are indicated. (B′) Heteroduplex formed between poly(A+) RNA from bovine muscle epidermis and a genomic clone that contains the gene coding for epidermal bovine keratin (λKBla). Heteroduplex was formed in 70% formamide, 0.3 M NaCl, 1 mM EDTA, 20 mM Tris-HCl (pH 8.0). The sample was kept for 17 hours at 62°C, followed by incubation at 64°C for 2–4 hours. The hybrid was prepared for electron microscopy as described in (A′). (B) The interpretative drawing of B′: DNA is represented by a continuous line, RNA by an interrupted line; the 5′ end of mRNA is identified by a change in molecular diameter, the 3′ end by a projecting [i.e., nonhybridized, poly(A) tail]. Exons are denoted by capital letters, introns by arabic numerals. For contrast enhancement, (A′) and (B′) have been printed as negatives. Bars represent 0.1 μm.

Table 1   Sequence of Steps for Spreading Nucleic Acids

1. Mix nucleic acid,[a] cytochrome *c*, and buffer.[b]
2. Pour hypophase solution[c] into a spreading trough.
3. Insert a glass slide in the trough with one end resting on the rim.
4. Apply spreading solution on the glass slide.
5. Wait until solution has run down the ramp and spread out.
6. Touch the film side of a grid[d] to the surface of the hypophase.
7. Stain the grid.[e]
8. Dehydrate the specimen.
9. Rotary shadow for contrast enhancement.[f]

[a]This can be dsDNA (Figures 2A and 4), ssDNA (Figure 5A), RNA (Figure 5C), or heteroduplex molecules (Figure 6).
[b]Spreading of ssDNA, RNA, or heteroduplex molecules needs denaturing agents, such asd formamide and/or urea (Figures 5 and 6).
[c]Usually, water, or ammonium acetate is used; for highly denaturing conditions, formamide can be included.
[d]Normally, carbon- or parlodion-coated grids are used (Figure 2).
[e]Widely used is ethanolic uranyl acetate.
[f]Usually, platinum/palladium (80:20) is used for rotary shadowing at an angle of 6–9°; specimens can also be viewed without rotary shadowing (Figure 7C).
Note: These preparation steps are also used with slight modifications for protein-free spreading procedures (e.g., in the presence of benzyldimethylalkylammonium chloride: Figure 7E. For details, see procedures given in Sommerville and Scheer (1987).

For small chromatin units, such as chromatin fragments and viral chromatin (minichromosomes), high speed centrifugation is required to deposit the chromatin on the coated grid. Alternatively, small chromatin units can be adsorbed directly by floating the grid, carbon film down, on the surface of a droplet containing the sample (Table 4). The nucleosomal configuration of simian virus 40 chromatin (the SV40 minichromosome) is shown in Figure 11A–C. Supranucleosomal particles—for instance, those obtained by brief nuclease digestion of avian erythrocyte chromatin (Figure 11D)—are also prepared by direct absorption.

2.4   Proteins

The most commonly used method for contrast enhancement of proteins is negative staining (Table 5). This method is extremely rapid and involves drying of an aqueous solution of heavy metal salt (e.g., uranyl acetate) with the protein molecules on the coated grid. The metal salt occupies hydrated regions in and around the biomolecules and, upon drying, forms a dense cast. Since the biomolecules them-

Table 2   Sequence of Steps for Spreading Nucleic Acids: Microdiffusion Droplet Technique

1. Mix nucleic acid, cytochrome *c*, and buffer.[a]
2. Put a small drop of the solution on a sheet of parafilm.
3. Leave for 10–30 minutes for diffusion.
4. Pick up nucleic acid–protein film by touching grid[b] to the surface of the drop.
5. Stain, dehydrate, and shadow.

[a]The procedure yields best results with double-stranded nucleic acids (Figures 2B, 3, and 7A–D). For spreading single-stranded nucleic acids, denaturing agents, such as formamide and/or urea, have to be added to the solution.
[b]Carbon- or parlodion-coated grids may be used (Figure 2).
Note: For details, see procedures given in Sommerville and Scheer (1987).

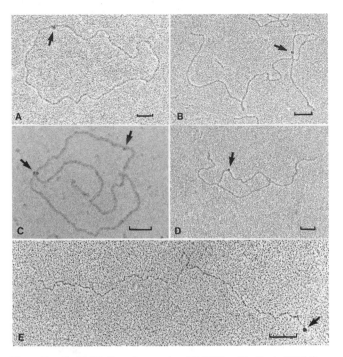

Figure 7.   (A, B) Binding of monoclonal Z-DNA antibodies to DNA from bovine papillomavirus (BPV-1). Antibodies against left-handed Z-DNA were incubated and cross-linked with 0.1% glutaraldehyde (2 h) to the pML2d-BPV-1 plasmids (A). To determine the position of the binding site, the DNA was digested with the single-cutting restriction enzyme Xba I (B). Arrows indicate the position of Z-DNA antibodies on plasmids. (C, D), Localization of large-tumor antigen (T antigen) on replicative intermediates of SV40 DNA. The molecules were immunostained with either ferritin-labeled protein A (C) or ferritin-labeled goat-anti-mouse antibodies (D) after cross-linking with 0.1% glutaraldehyde (20 min, 37°C). To localize T antigen on the replicated section, the complex was cleaved with the single-cutting enzyme Bgl I. Arrows indicate the ferritin-labeled T antigen on DNA. These complexes (A–D) were prepared for electron microscopy according to the cytochrome *c* droplet diffusion technique. The specimens were rotary-shadowed except the one shown in (C), which is only stained by ethanolic uranyl acetate. (E) DNA-relaxing enzyme molecules linked to single-stranded SV40 DNA (arrow). After enzyme reaction, the complexes were spread from a solution containing 0.25 M NaOH, indicating that the protein dot is stable in alkali and therefore linked via a covalent bond to the end of the single-stranded DNA. Enzyme–DNA complexes were prepared with a protein-free spreading procedure using benzyldimethylalkylammonium chloride (BAC) and shadowed. Bars represents 0.1 μm.

Table 3   Sequence of Steps for Spreading Large Chromatin Units: Miller Spreading

1. Incubate cells, isolated nuclei, chromosomes, or chromatin under low ionic strength conditions for chromatin dispersal.
2. Keep on ice for 15–30 minutes.
3. Glow-discharge carbon-coated grids.
4. Fill the well of the centrifugation chamber with a sucrose solution containing formaldehyde.
5. Insert the carbon-coated grid, which is now hydrophilic.
6. Layer a small volume of chromatin solution on top of the sucrose cushion.
7. Centrifuge the chamber (speed and time depend on the equipment used).
8. Remove grid and dry in Photoflo detergent.
9. Stain with phosphotungstic acid and dehydrate.
10. Rotary shadow.

Note: For details see Figures 8, 9, and 11E, and procedures given in Sommerville and Scheer (1987).

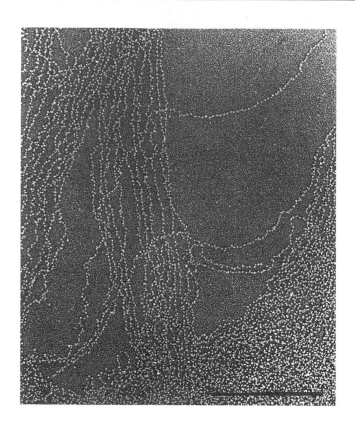

Figure 8. Appearance of the nucleosomal chromatin configuration of chicken erythrocytes lysed and swollen in 0.5 mM sodium borate buffer (pH 8.8) for 5 minutes, spread by the Miller technique, positively stained, and metal-shadowed. The dispersed chromatin shows the characteristic nucleosomal "beads-on-a-string" organization. Bar represents 1 μm.

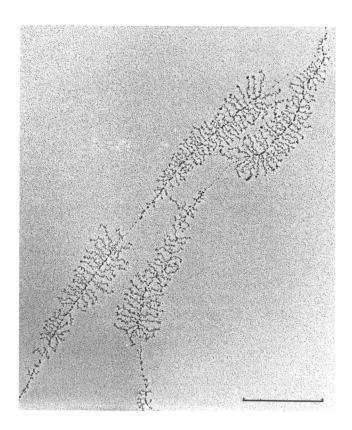

Figure 9. Organization of nucleolar chromatin from oocytes of the newt *Pleurodeles waltlii,* after spreading using the Miller technique, positive staining, and rotary shadowing. The transcribed chromatin segments are densely packed with RNA polymerases and nascent ribonucleoprotein (RNP) fibrils. The tandemly arranged active pre-rRNA genes are interspersed by the spacer regions. The axis of the spacer is relatively thin and does not show nucleosome-sized particles. Bar represents 1 μm.

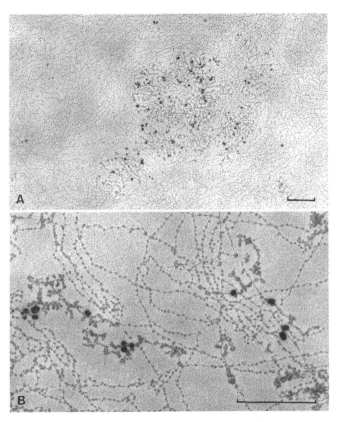

Figure 10. High resolution autoradiography of chromatin spread from mouse P815 cells after labeling for 5 minutes with [³H]uridine. (A) Radioactivity (recorded as dense silver grains) associated with densely transcribing regions showing the gradient-like features of pre-rRNA transcription units (c.f. Figure 9). The background consists of many strands of nucleosomal chromatin. (B) Radioactivity associated with individual RNP fibrils characteristic of those found in nonnucleolar transcription units. In sparsely transcribing units, nucleosomes can be seen on the chromatin strand between the nascent transcripts. Both preparations were stained with phosphotungstic acid, rotary-shadowed with platinum, and exposed with Ilford L4 emulsion. Bars represent 0.5 μm. (Courtesy of S. Fakan, with permission from Springer-Verlag.)

Table 4   Sequence of Steps for Spreading Small Chromatin Units

Part A

1. Dilute chromatin for dispersal under conditions of low ionic strength.
2. Place a drop of chromatin solution on a sheet of parafilm.
3. Place a glow-discharged carbon grid on top of the drop.
4. After 15 minutes, remove the grid and place it on top of a drop of double distilled water.
5. After 10 minutes transfer it to a second drop of water.
6. After 10 minutes remove the grid; negative-stain and air-dry the preparation.
7. Rotary-shadow for contrast enhancement (see Figures 11A, B, and D)

Part B

1. Hold a glow-discharged, carbon-coated grid with forceps.
2. Apply a drop of chromatin solution.
3. Allow the chromatin to absorb to the film for 1–2 minutes.
4. Stain and air-dry (alternatively shadow) (see Figure 11C).

Note: For details, see procedures given in Sommerville and Scheer (1987).

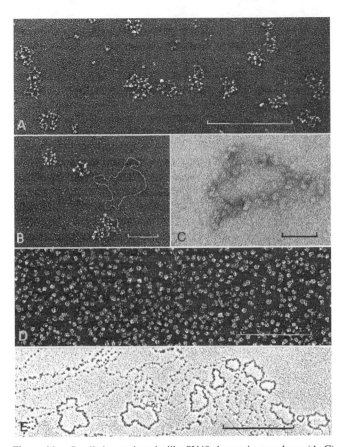

Figure 11.   Small chromatin units like SV40 chromatin complexes (A–C) or supranucleosomal particles from chicken erythrocyte chromatin (D) are prepared by direct absorption to freshly glow-discharged, carbon-coated grids. The chromatin is allowed to absorb onto the film for 1–2 minutes, stained with 2% uranyl acetate (C), and rotary-shadowed (A, B, D). Under spreading conditions involving buffers of low ionic strength (2 mM EDTA, 1 mM Tris-HCl, pH 8.4) the SV40 nucleoprotein complexes ("minichromosomes") show the typical "beads-on-a-string" morphology of the circular nucleosomal chain (A, B). The foreshortening of DNA in the SV40 "minichromosomes" due to nucleosomal organization is demonstrated by comparison with the viral DNA, which is included in the preparation (B). This packaging of the SV40 genome, at first-order level into nucleosomes, results in a foreshortening ratio of about 5.5:1. (D) At the second level of packaging, chromatin is organized into supranucleosomal structures. Fractions of supranucleosomal granular subunits, obtained by brief digestion of chromatin with micrococcal nuclease at physiological salt concentration, can be spread after fixation for 15 minutes at 4°C with glutaraldehyde (final concentration 0.2%). (E) Nonnucleosomal forms of chromatin beside nucleosomal chains are observed in Miller spread preparations of African green monkey kidney cells (RC37) after infection with herpes simplex virus. The "bubblelike" configurations, thickly coated with a single-strand DNA-binding protein, alternate with unbranched thin intercepts (staining and shadowing as in Figures 8 and 9) Bars represent 0.5 μm in (A), (D), and (E), 0.2 μm in (B), and 0.05 μm in (C).

Table 5   Sequence of Steps for Negative Staining

1. Hold a coated grid[a] with forceps.
2. Apply a drop containing biological material to the grid.
3. After 10–60 seconds drain off excess liquid.
4. Apply a drop of negative stain[b] for 10–60 seconds.[c]
5. Remove liquid and air-dry the preparation.

[a]Carbon, parlodion, or snadwich coats (made from Formvar and carbon) may be used. Glow discharge increases the amount of biological material absorbed to the film; plastic films, however, are destroyed by glow discharge.
[b]Uranyl acetate and phosphotungstic acid are widely used for negative staining [for a detailed description of different stains, see Harris (1991)].
[c]Alternatively, the grid can be transferred through several drops of stain.
Note: For details on the various procedures, see Figures 12–16, and Harris (1991).

selves are not stained, but the background is stained, a negative image is created. This technique has been used extensively for examining protein filaments. Negatively stained intermediate filaments, in this example, glial fibrillary acidic protein (GFAP), are shown in Figure 12A. Details of the arrangement of individual collagen molecules within the collagen fibril can be obtained from preparations like that seen in Figure 13.

An alternative to the adsorption of proteins from droplets is to spray them, in aerosols, onto the surface of carbon-coated grids. The spraying is most easily achieved by touching the tip of a glass capillary tube, containing both sample and stain, into a stream of nitrogen. Elongate, or rod-shaped, proteins can be mixed with glycerol (and no stain) and sprayed onto the surface of a piece of fresh-

ly cleaved mica. The deposit is then shadowed finely and carbon-coated to produce a replica cast of the protein molecules. (Details are provided in the literature.) Unidirectionally shadowed molecules, prepared by the spray/replica technique and representing early stages in the assembly of GFAP filaments, are shown in Figure 12B.

2.5   Macromolecular Assemblies

Multimeric enzyme complexes, ribosomes, viruses, and other particles can be treated in ways similar to those described for protein filaments or for small chromatin units. Particles in the size range of 10–200 nm diameter are excellent targets for negative staining (Figures 12–15), which often reveals considerable detail of surface structure (Figure 16). Nevertheless, the size of the particles of dried stain (or of the metal particles after shadowing) limits the resolution of structural detail. A superior technique, cryoelectron microscopy, involves the rapid freezing of a thin (100 nm) aqueous film containing the specimen particles. Below −143°C, water can be held in a vitrified state, which resembles the liquid state and avoids formation of ice crystals. In this condition macromolecules and macromolecular assemblies can be viewed free of artifacts from fixation, dehydration, and staining. Underfocus phase contrast is used to produce a high resolution image.

3   APPLICATIONS

3.1   Use of Antibodies

Antibody molecules (IgG and IgM) are large enough to be resolved in metal-shadowed or negatively stained preparations. Monoclonal

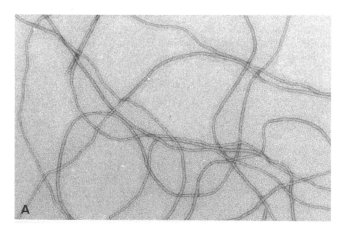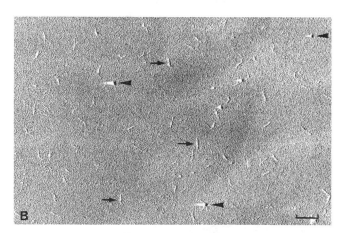

Figure 12.   Intermediate filaments as visualized by negative staining and partly assembled monomers as visualized by metal shadowing. (A) Glial fibrillary acidic protein (GFAP) intermediate-sized filaments, spread onto carbon film and negatively stained with 1% w/v uranyl acetate. GFAP was prepared from spinal cord and assembled in vitro by dialysis against the following buffer: 10 mM Tris-HCl (pH 7.0), 1 mM MgCl₂, 50 mM NaCl, and 25 mM 2-mercaptoethanol. Note that the intermediate filaments are 10 nm in width and are long with smooth edges. (B) The assembly of intermediate filaments can be arrested at various points along the assembly pathway by choosing the appropriate buffer conditions. In this preparation GFAP molecules containing chains of four proteins were formed in 10 mM Tris-HCl (pH 8.5). A sample of this was made 50% v/v in glycerol, sprayed onto freshly cleaved mica, and unidirectionally shadowed at 10° with platinum. The replica was then floated on water and picked up onto grids. The molecules of GFAP appear as short rodlets, between 45 and 65 nm long and 2–3 nm wide (arrows). Latex beads demonstrate the direction of shadow and locate the position of the droplets on the replica (arrowheads). Both micrographs are shown at the same magnification; bar represents 0.1 μm. (Courtesy of R. Quinlan and A. M. Hutcheson.)

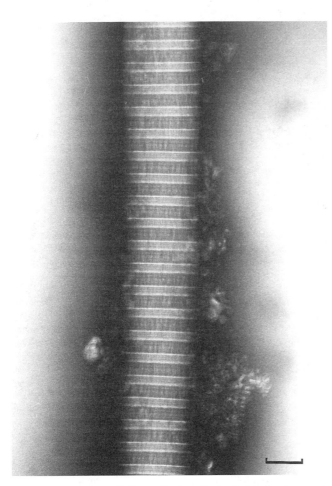

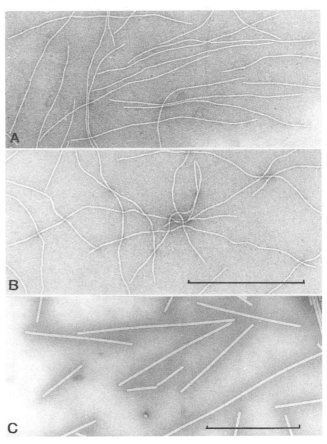

Figure 13. A collagen fibril, negatively stained with 4% phosphotungstic acid, on a glow-discharged, carbon-coated grid. The staggered arrangement of the collagen molecules in the fibril results in the striated appearance after negative staining. The bar represent 0.1 μm.

Figure 14. Examples of filamenteous bacteriophages and rodlike viruses. The bacteriophages fd were negatively stained with either 4% phosphotungstic acid (A) or 2% uranyl acetate (B). The tobacco mosaic viruses in (C) are stained with 4% phosphotungstic acid. All preparations were made on carbon-coated, glow-discharged grids. The bars represent 0.5 μm.

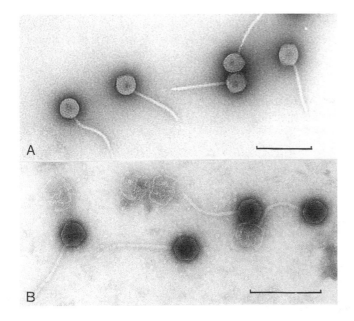

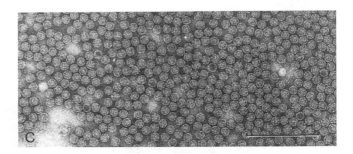

Figure 15. Examples of tailed (A, B) and globular (C) bacteriophages. Bacteriophage λ (A) and the globular bacteriophage fr (C) were stained with 4% uranyl acetate. The bacteriophage T5 (B) was stained with 2% uranyl acetate. Note the empty phage heads in (B). All preparations were performed on freshly glow-discharged, carbon-coated grids. The bars represents 0.2 μm.

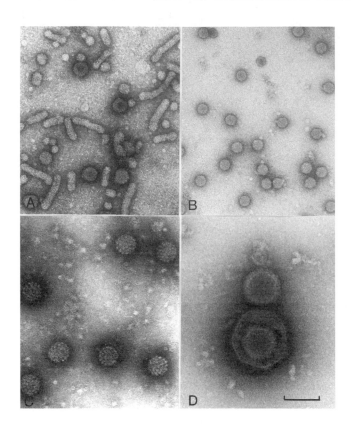

Figure 16. Details of virus structure revealed in negatively stained preparations on carbon-coated, glow-discharged grids. (A) Hepatitis B viruses (HBV) were stained with 2% uranyl acetate. (B) Polioviruses and (C) bovine papilloma viruses (BPV) were stained with 4% phosphotungstic acid. (D) Herpes simplex virus (HSV) was stained with 2% uranyl formate. HBV was obtained from a patient's serum and reveals, in addition to the mature "Dane" particles, globular and filamenteous forms of the viral S protein (A). Polioviruses show a fine-textured surface (B), while BPVs reveal a moruloid structure, which is due to the capsomere architecture (C). This organization is also visible in the HSV capsid, even when surrounded by the envelope (D). The micrographs are magnified to the same scale to give an impression of the size difference between the virus families. Bar represents 0.1 µm.

antibodies can be used to map sites on nucleic acid or protein molecules and to identify individual components within macromolecular assemblies. Antibodies can also be used to build up denser structures around small proteins, which are themselves difficult to resolve. The use of secondary antibodies tagged with electron-dense markers such as ferritin (C and D in Figure 7B) or colloidal gold particles are routinely used to improve detection. A specific example is the use of antibiotin to detect biotinylated nucleic acid probes that have been hybridized to complementary sequences in chromatin preparations or in situ in isolated chromosomes. In the example shown in Figure 17, the antibody-binding site is enhanced for EM detection by addition of a secondary antibody tagged with colloidal gold particles. Thus specific genes can be detected within chromatin masses.

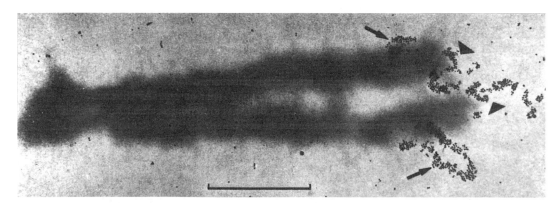

Figure 17. Whole-chromosome mount from *Xenopus* culture cell after in situ hybridization with a mixture of biotinylated DNA encoding tRNA and oocyte-specific 5S RNA. The chromosomes are partially unfolded and the sites of hybridization are detected using secondary antibodies tagged with colloidal gold. The gold particles are seen to decorate DNA loops containing tRNA genes (arrows) and 5S RNA genes (arrowheads). The bar represents 2 µm. (Courtesy of S. Narayanswami and B. A. Hamkalo, with permission from Oxford University Press.)

Table 6 Sequence of Steps for High Resolution Autoradiography of Biomolecules

1. Radioactively label[a] living cells, or subcellular fractions *in vitro*.
2. Isolate molecules or molecular complexes and prepare by direct absorption on the grid or by spreading techniques.
3. Stain,[b] dehydrate, and rotary shadow.
4. Evaporate a thin layer of carbon (4–8 nm) on the specimen.[c]
5. Coat the specimen with a layer of photographic emulsion.[d]
6. Follow the gold latensification–Elon–ascorbic acid (GEA) procedure.[e]
7. Transfer to Elon–ascorbic acid developer.
8. Transfer into the fixing bath.
9. Wash and air-dry.

[a]Generally, radioactive tritium is used.
[b]Negative or positive staining; shadowing with platinum/palladium.
[c]The fine coat prevents chemical interactions between specimen and emulsion.
[d]A detailed description of the procedures and equipment used to obtain a film of nuclear emulsion consisting of a homogeneous monolayer of silver halide crystals is given in Sommerville and Scheer (1987).
[e]The GEA development procedure increases the sensitivity and gives rise to silver grains of rather small size.
Note: For details see procedures given in Figure 10 and Sommerville and Scheer (1987).

3.2   Detection of Radiolabeled Molecules

Although high resolution autoradiography (EM ARG) is applied mostly to the detection of newly synthesized (radiolabeled) components *in situ* at the cellular level, application is possible with spread molecules and molecular complexes. However, EM ARG is limited to the use of radioisotopes emitting soft β-particles, usually ^{3}H. Molecules which have been radiolabeled either in vivo or in vitro are prepared for electron microscopy by the method most appropriate for that type of sample (Table 6). After staining, the preparation is covered, first with a thin protective coat of carbon and then with a layer of photographic emulsion. Ideally, the emulsion should consist of a homogeneous monolayer of the silver halide crystals. After being kept in the dark, to allow a sufficient amount of radioactive decay, the emulsion is developed, leaving silver grains above the sites of radiolabeling in the molecules. EM ARG is subject to two severe limitations:

1. The accuracy with which the developed silver grain can be located to the source of radiation in the specimen molecule. This depends on the thickness of the preparation and the diameter of the silver halide crystal in the emulsion and results in scattering of grains with a half distance of about 100 nm.
2. The efficiency in detecting radioactive disintegrations. Even molecules labeled to high specific activities require exposure times of generally more than a few weeks.

In spite of these limitations, EM ARG has been used successfully in studying sites of replication in isolated DNA molecules and the location of active transcription complexes in spread chromatin (Figure 10).

3.3   Electron Spectroscopic Imaging (ESI)

By adapting the electron microscope to include an imaging electron energy filter, positional information can be recovered from electrons that have lost a specific and characteristic amount of energy in colliding with atoms in the specimen molecules. This approach has been used successfully in the location of phosphorus atoms, particularly those in the phosphodiester bonds of nucleic acids. The resulting ESI has been used to delineate the path of the sugar–phosphate backbone in nucleosomes and the configuration of 7SL RNA within the signal recognition (ribonucleoprotein) particle. Along with related techniques in element mapping within biomolecules and their complexes, ESI is becoming increasingly sophisticated and promises many diverse applications.

3.4   Image Reconstruction

Individual molecules, under optimum conditions of spreading and staining, give weak and ill-defined images. To improve on the amount of structural detail, information from many molecules can be combined to smooth out random variations between images. To do this it is necessary to use molecules (e.g., protein filaments), which contain regular, repeating arrays of subunits. Alternatively, the biomolecules can be induced to form crystalline arrays of regular, tightly packed, oriented units. Electron micrographs of arrays of these types can then be used for image processing to produce enhanced structural detail.

See also DNA Structure; Immunology; Nucleic Acid Hybrids, Formation and Structure of; RNA Three-Dimensional Structures, Computer Modeling of.

Bibliography

Dubochet, J., Adrian, M., Chang, J., Homo, J.-C., Lepault, J., McDowall, A. W., and Schultz, P. (1988); Cryoelectron microscopy of vitrified specimens. *Q. Rev. Biophys.* 21:129–228.
Harris, J. R., Ed. (1991) *Electron Microscopy in Biology: A Practical Approach.* Oxford University Press, London and New York.
Ottensmeyer, F. P. (1986); Elemental mapping by energy filtration: Advantages, limitations and compromises. *Ann. N.Y. Acad. Sci.* 483:339–353.
Sommerville, J., and Scheer, U., Eds. (1987) *Electron Microscopy in Molecular Biology: A Practical Approach.* Oxford University Press, London and New York.

Electron Transfer, Biological

Christopher C. Moser, Ramy S. Farid, and P. Leslie Dutton

1   **Electron Transfer Theory**
   1.1   Tunneling
   1.2   Franck–Condon Factors

2   **Empirical Results**
   2.1   Free Energy Dependence of Intraprotein Electron Transfer
   2.2   Distance Dependence of Optimal Electron Transfer Rate

3   **Electron Transfer Protein Design**

Key Words

Franck–Condon Factors   The dependence of the electron transfer rate on the overlap of the reactant and product nuclear wave
functions.

Reorganization Energy   The energy required to distort the geometry of the electron donor-acceptor pair to resemble the equilibrium geometry after electron transfer.

Respiratory electron transfer involves the guided stepwise transfer
of electrons from a source of reducing equivalents to a sink of oxidizing equivalents. Photosythetic electron transfer adds to this
framework light-pumped cyclic electron transfer by virtue of
chlorophyll's ability to act both as an electron donor (in the excited
state) and as an electron acceptor (in the oxidized ground state). It
is because bioenergetic electron transfer proteins are intimately associated with a closed membrane that this redox energy can be converted and stored in other forms. The chemiosmotic model describes the principles by which electron transfer reactions are
oriented across the membrane and coupled to the absorption and release of protons, whereupon they generate transmembrane electric
and proton gradients.

This entry explores the underlying principles by which proteins
guide electron transfer to assure reactions that are productive within the chemiosmotic framework and also maintain engineering
efficiency. We will show that the fundamental role of protein is to
provide a scaffolding that separate redox centers by controlled distances in specific directions. Nearly as important is the influence of
the protein environment on the free energy of reactions and types of
relaxation that occur concurrent with the shift of charge.

1   ELECTRON TRANSFER THEORY

1.1   TUNNELING

Fundamentally, all biological electron transfers are tunneling reactions. Although redox reactions involving mobile redox carriers
such as water-soluble cytochrome *c* or membrane-bound ubiquinone may be rate-limited by diffusive reactions, and critical reactions that couple electron transfer to proton uptake or release may
be dependent on the dynamics of proton donors or acceptors, every
electron transfer involves the quantum mechanical tunneling of an
electron from donor to acceptor through the barrier of the intervening protein medium. The quantum mechanical nature of this reaction is most obvious in intraprotein electron transfer, especially in
the photosynthetic reaction center in which light-initiated electron
transfer is nearly independent of temperature down to liquid helium
temperatures. This relative temperature independence contrasts
with a classical transition state description in which a reactant absorbs thermal energy to surmount an activation energy barrier and
decays into products.

A simple description of the rate of intraprotein electron transfer
is provided by Fermi's Golden Rule, derived from quantum mechanical perturbation theory applied to well-separated (weakly coupled) redox centers. This equation is written to separate the influence of the electronic and nuclear wavefunctions of the reactant and
product. The electron transfer rate *(k$_{et}$)* is proportional to the electronic coupling, which is in turn proportional to the square of the
overlap of the electronic wave functions of the reactant and product

(V^2_R). The rate is also proportional to the Franck–Condon factors
(FC), which describe the overlap of the reactant and product nuclear
wave functions:

$$k_{et} = \frac{2\pi}{\hbar} V_R^2 FC \tag{1}$$

The electronic wave functions of the transferred electron have
large values only immediately around the atoms of the redox centers themselves. In the case of such small centers as FeS clusters, almost all the transferred electron density is within a sphere of a few
angstroms, while for a porphyrin-derived redox center, such as
hemes or chlorophylls, electron density will be delocalized over a
10 A disk. Electron density in the intervening medium where donor
and acceptor wavefunctions overlap will be quite small. Perhaps the
simplest way to approximate the wavefunctions in this critical area
is to model the redox centers as relatively narrow electron potential
wells surrounded by barrier medium of constant potential. In such
a case the electronic wave functions will decay exponentially with
the edge-to-edge distance between redox centers, and their decay
rate will depend on the height of the potential barrier of the intervening medium. These assumptions lead to the relation:

$$V_R^2 = V_0^2 e^{-\beta R} \tag{2}$$

where R is the edge-to-edge distance between the redox centers, β
the effect of the barrier height on the wave function decay, and V_0^2
the electronic coupling that is extrapolated to a condition where the
redox center edges overlap. This relation is a simple one, which ig-
nores the atomic heterogeneity of the protein medium between redox centers and the possibility that electron transfer may occur
along specific low potential paths following the chemical bonds of
amino acids.

1.2   FRANCK–CONDON FACTORS

Marcus has provided a simple description of the essential elements
of the nuclei-dependent Franck–Condon factors. As in the classical
transition state description, reactant and product states, representing
the electron localized on the donor and acceptor, lie at the bottom
of different potential energy wells (see Figure 1). In the Marcus description these potential energy surfaces are represented by simple
parabolas because the nuclei of the reactants and products are treated as a generalized harmonic oscillator. For simplicity, the frequency of the reactant and product harmonic oscillators are considered
unchanged. Only the geometry of the nuclei and the overall free energy change upon electron transfer. As the geometry of the reactant
is distorted from the lowest energy equilibrium configuration, the
energy of the reactant state will rise. The energy required to distort
the geometry of the reactant to the equilibrium geometry of the
product defines the reorganization energy (λ). The greatest value of
the reactant nuclear wave function will tend to be near the equilibrium nuclear geometry. However, as the temperature is raised, higher vibrational energy levels will be populated in a Boltzmann distribution, with the effect of spreading the wave functions and
generally increasing the overlap with the product wave functions.
The calculated overlap of the nuclear wave functions is given by the
Gaussian expression:

$$FC = \sqrt{4\pi\lambda k_B T} \exp{-\frac{(-\Delta G^0 - \lambda)^2}{4\lambda k_B T}} \tag{3}$$

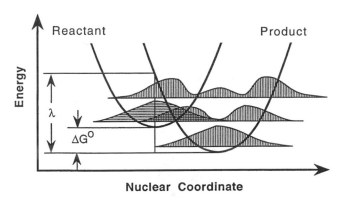

Figure 1. Parabolic harmonic oscillator potential surfaces representing the energy of the reactant and product states before and after electron transfer. The minimum of each potential well represents the equilibrium position of the nuclei when the electron is being transferred in on the donor or acceptor. The energy required to distort the nuclei of the reactant to the equilibrium geometry of the product before electron transfer defines the reorganization energy (λ). Superimposed on the classical potential surfaces are vibrational quantum energy levels and representations of the density of harmonic oscillator wave functions. The Franck–Condon factors will be proportional to the overlap of reactant and product nuclear wave functions, shown schematically as shaded areas.

This expression in combination with Fermi's Golden Rule predicts a Gaussian dependence of the electron transfer rate on the free energy of the reaction at a given fixed distance. The maximum rate is found when the free energy of the reaction matches the reorganization energy, a situation that corresponds to the intersection of the reactant and product potential surfaces at the minimum of the reactant parabola. The Marcus relation predicts the surprising behavior that increasing the free energy of a reaction beyond the reorganization energy (into what is called the inverted region) actually decreases the expected rate of the reaction.

The intersection of the reactant and product parabolas in the Marcus picture occurs at a free energy $(-\Delta G^\circ - \lambda)^2/4\lambda$ above the bottom of the reactant well. This energy formally corresponds to a classical activation energy and leads to an Arrhenius-type temperature dependence. Yet a number of biological electron transfer reactions are known to become essentially temperature independent at low temperatures, suggesting that a quantum mechanical modification of the Franck–Condon factors must be considered. A semiclassical relation presented by Hopfield maintains the overall Gaussian dependence of the FC factors on free energy, but replaces the Marcus variance of $4\lambda kT$ with a new variance

$$\sigma^2 = \lambda \hbar \omega \coth\left(\frac{\hbar\omega}{2k_BT}\right) \qquad (4)$$

introducing the quantum vibrational energy of the oscillator as $\hbar\omega$. At high temperatures (thermal energy significantly greater than half the quantum vibrational energy) the coth term reduces to $2kT/\hbar\omega$ *and the variance to $2\lambda kT$*. This is the same variance found in the Marcus expression, predicting the same Arrhenius temperature dependence. At low temperature, the coth term reduces to 1, the variance to $\lambda\hbar\omega$, and the FC factors become nearly independent of temperature.

A completely quantum mechanical expression for the FC factors, using quantum harmonic oscillator wave functions, has been de-

scribed by DeVault. The form of the expression is more difficult to follow intuitively because a term rising exponentially in free energy competes with a simultaneously falling modified Bessel function I_p.

$$FC = \frac{1}{\hbar\omega} \exp\left\{-S(2n+1)\left(\frac{n+1}{n}\right)^{P/2} I_p[2S\sqrt{n(n+1)}]\right\} \qquad (5a)$$

$$S = \frac{\lambda}{\hbar\omega} \; ; P = \frac{-\Delta G^\circ}{\hbar\omega} \; ; n = \frac{1}{e^{\hbar\omega}-1} \qquad (5b)$$

The net effect of the opposing functions is a rising and falling of the electron transfer rate with free energy that is nearly Gaussian for relatively small $\hbar\omega$ but becomes asymmetric as the reorganization energy approaches $\hbar\omega$.

Equations (5) strictly define an FC factor only for free energies corresponding to exact multiples of the quantum vibrational energy. However, all these expressions can be modified to accommodate the coupling of electron transfer to more than one nuclear vibration by convoluting together independent expressions for each vibrational frequency. This tends to smooth out the discrete behavior of equations (5). The convolution of two Gaussians is another Gaussian with a summed variance. For expression (4), the variance becomes a reorganization energy weighted sum of all vibrational frequencies, with vibrations smaller than $2kT$ contributing an effective vibrational energy of $2kT$. In other words, in the case of a spectrum of vibrations coupled to electron transfer, it becomes meaningful to consider a total reorganization energy and an overall characteristic frequency.

2   EMPIRICAL RESULTS

2.1   FREE ENERGY DEPENDENCE OF INTRAPROTEIN ELECTRON TRANSFER

Systems in which the parameters of the foregoing theoretical expressions are accurately measured and systematically varied are still relatively rare in both chemistry and biology. However, several systems of covalently linked redox centers display a conspicuous Gaussian-like rate versus free energy behavior predicted by the Marcus relation. In bacterial photosynthetic reaction centers, X-ray diffraction provides well-defined structures, while quinone substitution reveals a similarly Gaussian-like behavior within an approximately order of magnitude experimental scatter (see Figure 2). The rate maxima suggests a reorganization energy of about 700 meV and the breadth suggests a characteristic frequency of about 70 meV. Six other electron transfer reactions between reaction center redox components have been described, although with a less extensive free energy dependence. Approximately Gaussian relationships with a rate maximum at about 1100 meV have been found in various heme proteins in which surface histidine residues have been ruthenium-modified to create light-activatable redox centers. A similar reorganization energy is anticipated for pulse radiolysis induced electron transfer from a disulfide bridge to the Cu center in azurin.

The vast majority of experimental intraprotein electron transfer systems have limited free energy variation and uncertain reorganization energies and distances. However, we note that reorganization energy for various biological reactions generally tends toward moderate values of 500 to 700 meV for reactions that take place within the relatively nonpolar protein interior, increasing toward 1000 to 1200 meV for reactions that involve redox centers at the relatively

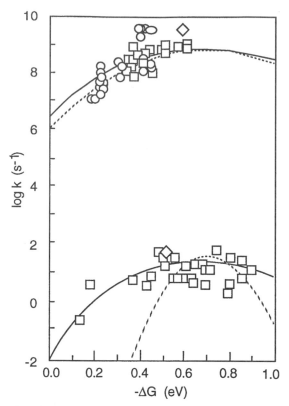

Figure 2. The free energy dependence of the rate of a number of intraprotein electron transfers appears to be approximately Gaussian (parabolic on a log plot). Examples of the free energy dependence of the rate of physiologically productive electron transfer over 10 Å (edge to edge) from photoreduced bacteriopheophytin (BPh-) to primary quinone Q_A, and physiologically unproductive charge recombination over 22.5 Å from Q_A to the photooxidized bacteriochlorophyll dimer (BChl$_2$). The data for the latter reaction are shown at a temperature of 35 K. The maximum of the parabola defines the reorganization energy and the free energy optimal rate. Superimposed on these data are the theoretical relationships of equation (4) (solid lines) and (3) (dashed lines). At low temperature the classical relation of equation (3) narrow considerably.

polar exteriors. Unusually low reorganization energy values seem to be associated with ultrafast picosecond or sub picosecond electron transfer between chlorins in the initial charge separation, perhaps because of the delocalization of the electrons over the chlorin ring structure.

We also note that a characteristic frequency of about 70 meV is adequate to model the width of the Gaussian rate-free energy relationships within the scatter of the data available. This frequency generates curves that are very similar to the classical Marcus relation at room temperature and do not narrow as dramatically at low temperatures (see Figure 2). Since lowering characteristic frequencies below 50 meV should have no effect on electron transfer rates at physiological temperatures, natural selection is not expected to favor low values. On the other hand, on the basis of the data sets we have examined, it is clear that characteristic frequencies higher than about 100 meV have not been selected. It appears that the typical vibrational frequency spectrum of a condensed organic medium such as protein provides the basis for a characteristic frequency in

this range and that natural selection cannot effectively modify this parameter.

2.2   DISTANCE DEPENDENCE OF OPTIMAL ELECTRON TRANSFER RATE

Comparing the free energy optimized rate of various intraprotein electron transfer reactions both minimizes the variations in electron transfer rate due to FC factors and reveals the dependence of the rate on the electronic wavefunction overlap found in equation (1). If the logarithms of the optimal rates just described are plotted versus edge-to-edge distance, a surprisingly linear relationship is found over 12 orders of magnitude (see Figure 3). This result suggests that a model of simple exponential decay of electronic wave functions through a medium barrier of constant average height ($\beta = 1.4\,A^{-1}$) is usefully accurate in intraprotein electron transfer.

The same β value does not seem to apply to chemically synthesized covalently linked systems. It appears that most synthetic systems provide a more direct through- (covalent) bond link between redox centers that has the effect of more efficiently propagating the electronic wave functions of the donor and acceptor ($\beta \sim 0.7\,A^{-1}$). It appears that the relatively large size of typical biological redox centers (e.g., porphyrins) and the typical geometry of amino acid packing in a protein has led to multiple paths for wave function propagation and average β values similar to those found in an organic solvent glass. Furthermore, it is clear from the reaction center examples presented that natural selection has not modified β

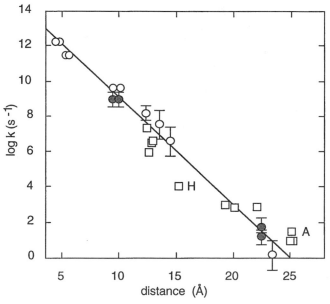

Figure 3. A plot of the log of the free energy optimal rate as a function of the edge-to-edge distance between redox centers in proteins generates a surprisingly linear relationship over 12 orders of magnitude of rate. Rates of charge separating and recombination reactions in two species of photosynthetic reaction centers are shown as circles; solid circles are not shown in Figure 2. Rates of nonphysiological reactions in modified electron transfer proteins are shown as squares. Two rates with obvious deviations above and below the average β are labeled H (His-62 ruthenated cytochrome c) and A (disulfide bride to Cu center in azurin).

of protein medium to systematically favor physiologically productive electron transfer over energy-wasting charge recombination reactions.

In some cases it appears to be possible to construct an electron transfer system that operates at a rate significantly different from that predicted by the average protein β. Figure 3 shows that a cytochrome c ruthenated at histidine 62 gives an electron transfer rate two orders of magnitude less than that anticipated for the distance given the expected free energy and reorganization energy, while the azurin reactions are two orders of magnitude faster. (Generally, edge-to-edge distance includes within redox centers all atoms that make up the conjugated π-ring system of porphyrin and quinone-based redox centers, and ligands of the smaller metal atom centers.) Indeed, a detailed bond-by-bond tunneling pathway analysis can be successfully applied to these examples. The medium between redox centers in cyt c-Ru-(His 62) is found to be unusually free of covalent links, while the beta barrel structure of azurin leads to an unusually large number of covalent links.

There may be several factors in biologically relevant reactions that disfavor the appearance of highly linked or unlinked paths and favor an average β. Molecular dynamics and the large size of common biological redox centers can increase the averaging effect of multiple competing pathways between redox centers. Other concerns such as protein assembly and stability may dominate long-range structural choices that affect connectivity, while changes in free energy and reorganization energy can effect rate changes without significant structural change. While rate modulation greater than two orders of magnitude by connectivity changes may be physiologically possible, it remains to be seen whether an expanded study of natural electron transfer systems will uncover examples of β modification driven by natural selection.

3   ELECTRON TRANSFER PROTEIN DESIGN

So far we have considered optimal electron transfer rates in what is essentially "solid state," nonadiabatic intraprotein electron transfer. In these systems the only first-order parameters of electron transfer theory that appear to be modified to set electron transfer rates are distance, free energy, and reorganization energy. A first-order numerical estimate for the upper limit of the electron transfer rate is

$$\log k_{et} = 15 - 0.6\,R - 3.1\,\frac{(-\Delta G^0 - \lambda)^2}{\lambda} \qquad (6)$$

where the electron transfer rate is expressed in reciprocal seconds, the edge-to-edge distance in angstrom units, and the free energy and reorganization energy in electron volts.

Nevertheless, many electron transfer rates are slowed by rate-limiting diffusion (e.g., with the redox carriers cytochrome c or ubiquinone) or coupling to other events such as proton binding or release. Biological electron transfer systems seem generally tolerant of coupling reactions in the micro- and millisecond range as long as the overall rate exceeds 10^3 s^{-1}. For a typical small free energy respiratory reaction, this rate can be met, provided the redox centers are separated by an edge-to-edge distance of less than 17 Å. Because the low dielectric membrane thickness is 35 Å, this means that redox proteins in a chemiosmotic scheme must have at least two electron transfer reactions or three redox centers to complete trans-

membrane electron transfer. This 17 Å distance also represents the amount of protein insulation that is required to prevent electron transfer to adventitious redox centers that may encounter the redox protein.

Nanosecond and faster reactions are essential in the face of rapid decay mechanisms, such as are found in light reactions of photosynthesis. Efficient light trapping by photosynthetic systems requires electron transfer to be significantly faster than the roughly nanosecond fluorescence decay of chlorophyll. In the bacterial reaction centers this speed of electron transfer appears to be assured by placing the chlorins of the first two electron transfer reactions at approximately 5 Å edge-to-edge spacing. In addition, the greater than 1 eV energies associated with the creation of an excited state pose a special problem for photosynthetic systems in that energy dumping charge recombination reactions to ground state will have a very large free energy that is potentially competitive with productive charge separation. The Marcus "inverted region" discussed in Section 2.2 provides a means to assure that such large free energy reactions will be slow, provided the reorganization energy for the reaction is less than half the total available energy. Indeed, the reorganization energies of the initial charge separation reactions appear to be among the lowest observed. The failure to satisfy these criteria often results in relatively poor quantum efficiency and charge separation stability in synthetic redox systems.

The general first-order description of intraprotein electron transfer described here should prove useful for systems less well characterized than the bacterial photosynthetic reaction centers. In the many bioenergetic proteins for which an X-ray crystal structure does not exist, equation (6) in combination with rough estimates of the free energy and reorganization energy of the reactions will predict edge-to-edge distances within a few angstroms. This relation will also prove useful in the design of de novo protein electron transfer systems. Indeed, if one is armed with an awareness of the importance of a low reorganization energy for initial charge separation, equation (6) can provide a description of the distances and free energies to construct charge-separating proteins that match the reaction center in quantum efficiency and exceed it in overall efficiency of energy conversion.

See also CELLULAR ENERGETICS, BIOCHEMICAL BASIS OF.

Bibliography

Beratan, D. N., Betts, J. N., and Onuchic, J. N. (1991) Protein electron transfer rates set by the bridging secondary and tertiary structure. *Science,* 252:1285–1288.

Bolton, J., McLendon, G. L., and Mataga, N. Eds. (1991) *Electron Transfer in Inorganic, Organic, and Biological Systems.* American Chemical Society, Washington, DC.

Kirmaier, C., and Holten, D. (1987) Primary photochemistry of reaction centers from the photosynthetic purple bacteria. *Photosynth. Res.* 13: 225–260.

Marcus, R. A., and Sutin, N. (1985) Electron transfers in chemistry and biology. *Biochim. Biophys. Acta,* 811:265–322.

Moser, C. C., and Dutton, P. L. (1992) Engineering protein structure for electron transfer function in photosynthetic reaction centers. *Biochim. Biophys. Acta,* 1101:171–176.

——, Keske, J. M., Warncke, K., Farid, R. S., and Dutton, P. L. (1992) The nature of biological electron transfer. *Nature,* 355:796–802.

ELECTROPORATION AND ELECTROFUSION

Donald C. Chang

Key Words

Dielectrophoresis   Attraction between neighboring cells in an electric field due to an induced dipole–dipole interaction.

Electropermeabilization   The permeabilization of a cell membrane by applying a pulse of intense electric field.

Electropore   An aqueous pathway in the cell membrane that is permeabilized by an applied electric field.

Membrane Breakdown   The destruction of the normal bilayer structure of a membrane as a result of an excessive membrane potential.

In electroporation, a cell membrane is temporarily permeabilized by exposure to an intense electric field. This phenomenon can be used to introduce a variety of exogenous molecules, particularly DNA, into living cells. Electrofusion is a related phenomenon by which neighboring cells can be induced to fuse by applying a pulsed electric field. Both electroporation and electrofusion are results of the electrical breakdown of the cell membrane. Electroporation is now a principal method of gene transfer, in both prokaryotic and eukaryotic cells. Electrofusion is found to be the most efficient method of cell fusion and has important uses in agriculture and in the production of hybridomas.

1   PRINCIPLES

1.1   ELECTRIC FIELD INDUCED MEMBRANE BREAKDOWN

The phenomenon of electroporation has been known for some time; studies can be traced back to the early 1960s. The application of electroporation in molecular biology, however, is relatively recent.

The first report of using electroporation for gene transfer appeared only in 1982.

Electroporation is the result of membrane breakdown induced by an applied electric field. Suppose a spherical cell with diameter r is exposed to an external electric field for a time period t. The cell membrane will develop an induced membrane potential V_m according to the following equation:

$$V_m = 1.5Er \cos \theta \{1 - \exp(-t/\tau_c)\} \quad (1)$$

where E is the field strength, θ is the angle between the field and the normal of the membrane, and τ_c is the relaxation time of the cell. For a typical biological cell, τ_c is approximately

$$\tau_c = r\,C_m(\rho_i + 0.5\,\rho_o) \quad (2)$$

where C_m is the unit capacitance of the cell membrane and ρ_i and ρ_o are the resistivity of the cytoplasm and the external medium, respectively. For small cells, such as red blood cells, τ_c is on the order of a microsecond. Then, when the external electric field is applied for a sufficiently long time period, say, 20 μs, the induced membrane potential will reach the steady value:

$$V_m = 1.5Er \cos \theta \quad (3)$$

From this equation, it is apparent that the induced membrane potential is not uniform across the entire cell membrane. The induced membrane potential is highest at places where the membrane faces the poles of the electric field (Figure 1).

The preceding analysis holds true as long as the cell membrane remains intact and its electrical properties do not change. When the external field strength is so high that the induced membrane potential exceeds a threshold potential V_{th}, the bilayer membrane will experience an electrical breakdown; at this point the cell membrane can no longer maintain its structural integrity and becomes permeable. This phenomenon is called "electroporation" or "electropermeabilization." For most cells, the threshold potential is typically on the order of 0.5–1 V, which is about 10 times larger than the normal resting potential of an animal cell. The occurrence of electrical breakdown is related to the thinness of the cell membrane. The insulating property of the cell membrane depends on its lipid bilayer, which is only 5 nm thick. When the potential across the cell mem-

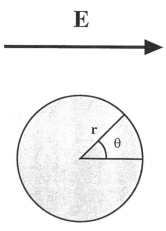

Figure 1.   Schematic diagram showing a spherical cell exposed to an external electric field, E.

brane is increased to 1 V, the electric field within the bilayer reaches 2×10^6 V/cm! Unable to withstand an electric field of such magnitude, the membrane undergoes a dielectric breakdown.

After a cell membrane has been permeabilized by an applied field, molecules that normally could not penetrate the membrane can now pass through easily. These molecules include ions, metabolites, carbohydrates, proteins, and nucleic acids. Some of the molecules that enter the cell during the electropermeabilized state are very large. It has been reported that DNA molecules of sizes up to 150 kb can be taken up by electropermeabilized cells. The uptake of these very large molecules, however, may involve some complex processes. As for small molecules, such as small carbohydrates or proteins of low molecular weight, it is evident that the particles can enter the cell by simple diffusion through the permeabilized cell membrane.

If the external electric field is not excessively high, the electropermeabilized cell can recover by resealing its membrane after the external field has been removed. A model membrane, such as a lipid bilayer, can be used to study the processes of electroporation and recovery by monitoring the changes of electrical properties of the membrane in response to the application of the external field. For the lipid bilayer, the electroporation process involves three distinct stages.

1. *Nonpermeabilized state.* This is the state of the membrane before application of the external field. The conductance of the membrane is extremely low.
2. *Permeabilized state.* When the external field is applied, the induced membrane potential (V_m) of the membrane builds up exponentially. When V_m approaches the threshold potential (V_{th}), the conductance of the membrane increases drastically, indicating that the membrane has become permeabilized to certain ions.
3. *Recovery state.* If the external field is applied only for a brief period and the field strength is not too high, the bilayer will recover by gradually resealing itself after the external electric field has been turned off. In this case the bilayer membrane is said to experience a "reversible breakdown." The recovery process can be monitored by observing an exponential decrease of the membrane conductance following the termination of the applied field. This recovery process appears to be very rapid; the time constant is generally on the order of submilliseconds. If the external field is applied for too long or the field strength is too high, the bilayer will not be able to recover and it will rupture even after the removal of the applied field. In this situation, the membrane is said to undergo "irreversible breakdown."

In electroporation of real cell membranes, the situation is more complicated. There is evidence suggesting that the electroporation process may involve several substages. Furthermore, the membrane may undergo secondary structural changes following the removal of the external field. Thus, when an electric field is applied to electroporate cells, the cell membrane may undergo at least four different stages of change:

1. *Nonpermeabilized state.* Before the external field is turned on, the cell membrane is in its normal intact state.
2. *Permeabilized state due to primary structural changes.* After the external field has been applied, the induced membrane potential quickly reaches the threshold potential. As a result, the membrane experiences an electrical breakdown and undergoes a structural change to become permeabilized. This electric field mediated membrane breakdown is commonly referred to as "primary structural changes."
3. *Permeabilized state due to secondary structural changes.* Once the cell membrane has become permeabilized, molecules can pass through it to enter or leave the cell. Such flow of material can induce further structural changes in the membrane, making it even more permeable to other molecules. Furthermore, breakdown of the membrane structure in a local region may affect the stability of the membrane structures in neighboring regions, and thus could further enlarge the membrane pores. These secondary structural changes can continue even after the removal of the external electric field.
4. *Recovery state.* After the external field has been turned off, and the forces causing the pores to expand have subsided, the membrane will repair itself and start to reseal. The mechanism of membrane healing is not yet well understood. But, it is well known that a cell can repair its own membrane following puncture or scratch. This recovery process takes considerably more time than is needed in the case of a lipid bilayer model membrane. Moreover, the resealing process itself is believed to involve several substages. The larger membrane pores may reseal first, with a time constant on the order of seconds. They, however, do not reseal completely, but reduce into residue pores that may linger for considerable longer times. Depending on the cell type, the resealing process may take several minutes to half an hour at room temperature. At lower temperatures, such as at 4°C, it will take even longer.

The structure and properties of the electric field induced pores have been investigated actively in the last two decades. The results are still far from complete. There have been different suggestions about the structures of these membrane pores. Some thought that these pores may appear in the form of cracks in the membrane. Others proposed that they are craterlike round pores. There are also conflicting estimates of the size of the pores. One study concluded, on the basis of an examination of transport properties of the electropermeabilized cell membrane to carbohydrate molecules of different sizes, that the membrane pores are very small, with diameters on the order of 1 nm. Another group examined the size of labeled molecules that can be loaded into red cell ghosts and concluded that the size of the membrane pores could reach 10 nm. Since, however, large macromolecules such as DNA can pass through the cell membrane during the electropermeabilized state, it is suspected that the electric field induced pores may be significantly large. One recent study used a rapid-freezing electron microscopy (EM) technique to examine the dynamic changes of membrane structure during the electropermeabilized state. Large volcano-shaped, porelike structures with diameters as large as 100 nm were observed in the freeze-fractured membrane (Figure 2). Such porelike structures appeared to undergo resealing within a few seconds after the application of the electrical pulse.

The different estimates of pore size may be partially related to differences in measurement techniques and partially to the dynamic nature of the pores, which can appear differently at different stages. The large porelike structures observed in the rapid-freezing EM study may represent transient membrane pores, through which DNA molecules can enter the cell. The small pores found in the carbohydrate study, on the other hand, may represent the long-lasting

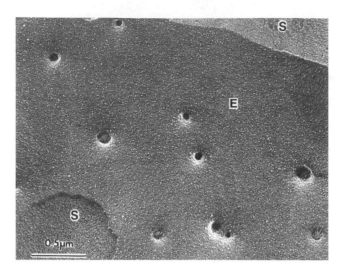

Figure 2. External membrane face (E) of an electropermeabilized human red blood cell frozen at 40 ms after the application of an electrical pulse. The frozen extracellular medium is labeled S. [For details, see D. C. Chang and T. S. Reese, *Biophys. J.* 58:1–12 (1990).]

residue pores at the recovery stage. Since measurements of carbohydrate transport require many minutes, most of the large membrane pores would have undergone partial resealing during the measurement period.

1.2 Uptake of Exogenous Molecules During Electroporation

Molecules with a variety of shapes and molecular weights can be taken up by cells during the electropermeabilized state. These molecules include DNA, RNA, proteins, drugs, metabolites, ions, and molecular probes. The process of molecular uptake can differ depending on the size and charge of the transported molecules. Small molecules could enter cells by simple diffusion during the electropermeabilized state. In studies using either lipid vesicles or red cell ghosts, it has been demonstrated that loading of small molecules (e.g., ions, carbohydrates of low molecular weight, dyes) during electroporation can be explained simply based on a diffusion model. The uptake of larger molecules such as DNA is more complicated; the process is believed to involve a number of sequential steps. At this point, the exact mechanism of DNA uptake during electroporation is not clearly understood, but there is evidence suggesting that the following processes are involved:

1. Before application of the external electric field, a certain fraction of DNA molecules may have already been bound to the cell surface. Such binding could be a required intermediate step for DNA to enter the cell when the cell membrane is electropermeabilized.
2. Application of an external field creates pores on the cell membrane. Some of these pores may have transient openings that are large enough to allow passage of DNA molecules.
3. Through an electrophoretic process, the negatively charged DNA molecules may be driven toward the newly created pores and enter the cell by the applied electric field.
4. Alternatively, the entire DNA molecule may not pass through the cell membrane during the electropermeabilized state. In-

stead, the applied electric field may disrupt the membrane structure to create a phospholipid–DNA complex, which would then stimulate the internalization of DNA by endocytosis.

The foregoing processes have been examined in a number of studies and are supported by some of the experimental results. For example, in some systems, concentration of magnesium ions was found to have a significant effect on gene transfer efficiency. Such a finding may be explained on the basis of DNA binding to the cell surface. Several laboratories, including our own, found that in order to have an effective gene transfer, DNA must be added to the cell mixture before the electric field is applied. This finding suggests that either the large transient pores have a very short life or DNA is driven into the cell by the applied field. Using cultured cells grown on a filter membrane, it was possible to examine the effect of the polarity of the electric field on gene uptake. It was found that when DNA was applied on the cathode side, the gene transfection efficiency was much higher. This finding supports the notion that DNA enters cells by an electrophoretic process.

1.3 Electric Field Induced Cell Fusion

A pulsed electric field can be applied not only to electropermeabilize cells but also to induce cells to fuse. When two neighboring cells are placed very close to each other, applying a pulsed electric field similar to those used in electroporation will induce these cells to fuse together. This process is called "electrofusion." The development of electrofusion has a history slightly shorter than that of electroporation. Induction of fusion of plant cells by electrical stimulation was first reported in 1979. Electrofusion of several other cell types was announced in the following years. Because of its many important applications, including cell hybridization, animal cloning, and generation of antibody-producing hybridomas, electrofusion became a popular subject of study in the early 1980s.

To induce cell fusion, the cells must first be brought in contact with each other. There are three main methods to achieve this:

1. *By mechanical means.* A specially designed container can be used to force cells to aggregate in close proximity. Or, attached cells can be cultured side by side.
2. *By chemical means.* Cells can be coated with certain chemicals that bring the cells together. For example, one cell type can be coated with biotin, the other with avidin. Or, one cell type may express an antibody against a surface protein of its fusion partner. The immunoreaction would bring the proper pair of cells together.
3. *By dielectrophoresis.* This is probably the most elegant way to bring cells together for electrofusion. When a suspended cell is exposed to an external electric field, it develops an induced dipole moment in parallel to the field. Thus, under a weak electric field, suspended cells would tend to attract each other through the induced dipole–dipole interaction. They often form a long chain of cells (commonly referred to as a "pearl chain"). This process is called "dielectrophoresis" (Figure 3). The external field used to produce dielectrophoresis is not limited to a dc field; it can also be an oscillating field. For technical reasons, the dielectrophoresis field used in electrofusion is a continuous oscillating field with a frequency on the order of 10^4–10^6 Hz; the field strength is usually quite low, in the

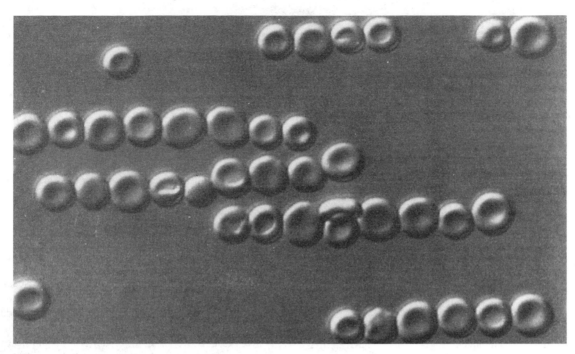

Figure 3.   Formation of "pearl chains" of human red blood cells induced by applying a low intensity ac field (200 V/cm, 60 kHz). (Reproduced from Chang, et al., 1992, p. 319.)

range of 100 V/cm. Such a low field is sufficient to bring about dielectrophoresis while not causing electropermeabilization.

It has been demonstrated in studies using red blood cells that in electrofusion, two kinds of fusion can occur: "membrane fusion" and "lumen fusion." In membrane fusion, only the membranes of the neighboring cells fuse, while the cell cytoplasms remain separate. In lumen fusion, the fusing cells truly fuse their cytoplasms together to form a syncytium. Membrane fusion can be observed using a hydrophobic dye such as DiI, which is selectively distributed in the membrane. If cells labeled with DiI are mixed with unlabeled cells, the dye will be seen to transfer from a labeled cell to a nonlabeled cell when the membranes of the two cells fuse. Lumen fusion can be monitored by preloading one of the fusing partner with fluorescent dye, such as dextran conjugated with fluorescein isothiocyanate. It was discovered that even after cell membranes have fused, the lumina of the two cells may or may not fuse together. In cell types other than red blood cells, most of the fusion products of electrofusion are of the lumen fusion type.

The time required to fuse two cells using electrofusion varies with the cell type. The electric field is usually applied for a very brief period, about a millisecond. In a few minutes after the application of the electrical pulse, fusion between neighboring cells can be seen. In red blood cells, the fusion process may be completed in about 5 minutes. For attached mammalian cultured cells, however, the fusion process may take many hours.

The mechanism of electrofusion is not fully understood at present. It is generally believed that the phenomenon is closely related to electroporation. To induce cells to fuse, the membrane structure of the two neighboring cells must be disrupted, so that the units can merge together. This electric field induced disruption of the membrane structure could be very similar to that of the electroperme-

abilization process. In the literature, several models have been proposed to explain how the electric field may bring about cell fusion:

1. *Joining of electropores in the apposing membranes.* In the early model of electrofusion, it was suggested that when two cells are in close contact during the electroporation state, electropores on the apposing cell membranes may line up with each other. During the membrane resealing process, the pores may merge to become a continuous intercellular channel, with the apposing membranes now fusing together at the rim of the merged pore. Such a membrane structure is unstable owing to its sharp curvature. As a result, the merged pore expands quickly. This expansion will gradually lead to the fusion of the two cells.

2. *Electric field induced compaction.* When cells are lined up to form a pearl chain, application of an intense electrical pulse produces two major effects on the cell membranes. Besides breaking down the membranes, the applied electric field induces a strong dipolar interaction between the adjacent cells. Such interaction generates a strong attractive force between the apposing cells and forces the adjacent membranes to come together to make physical contact, which eventually leads to fusion.

3. *Weakening of the hydration force.* In this model, it is argued that there exists a strong repulsive force when two cells are brought into close proximity. This force originates from the ordering of several water layers at the cell surface. When cells are brought into contact, their respective water layers repel each other. This repulsive force, called the "hydration force," is the major force for preventing spontaneous cell fusion. When an intense electric field is applied, it will disrupt part of the membrane structure and partially destroy the orderly or-

ganization of the cell membrane. As a result, the ordering of the surrounding hydration layer will also be destroyed. The strength of the hydration force is greatly reduced, and the adjacent membranes of the two cells can now come into contact and subsequently fuse.

In the early studies of electrofusion, it was thought that cell fusion could not occur unless cells were brought into contact before application of the intense electric field. Later experiments demonstrated that application of an intense electric field can induce fusogenic properties even in isolated cells. When cells are first exposed to an intense electric field and then brought together in the absence of the field, some of them can still fuse with each other. Such electric field induced fusogenic property was cited as supporting evidence for model 3, just discussed (weakening of the hydration force).

2   TECHNIQUES

2.1   WAVEFORMS USED FOR ELECTROPORATION AND ELECTROFUSION

The external electric field used in electroporation or electrofusion is always applied in the form of a pulse. Such a pulse can have different shapes, or waveforms, which include rectangular pulse, exponential decay pulse, and oscillating pulse (Figure 4). In the early stage of development of electroporation and electrofusion, most of the studies used only rectangular pulses. The pulse width, or τ, was usually very short. The major requirement for the pulse width is that it be longer than the relaxation time of the cell, which is typically on the order of microseconds. Another requirement is that the field be applied long enough to allow the membrane to undergo an electrical breakdown. There is also a limitation on the length of τ. In a study using electron microscopy to examine structural changes of electropermeabilized cells, it was shown that when the pulse width was shorter than 20 μs, cells were more or less intact over a very wide range of applied field strength. If τ is much longer than 20 μs, however, the same field strength causes excessive cell damage. Thus, in most of the early works, the pulse width is usually very short, and the intensity of the applied electric field is relatively high, typically several kilovolts per centimeter for mammalian cells. To electroporate smaller cells, such as bacteria, the required field strength may be as high as 20 kV/cm.

Recently, the exponential decay pulse has become a more popular waveform for use in electroporation or electrofusion. This pulse is usually generated by discharging a large capacitor, which is precharged to a high voltage and is thus frequently referred to as a "capacitor discharge (CD) pulse." The CD pulse requires only sim-

ple equipment to generate such a pulse. It is difficult to deliver multiple pulses, however, because it takes time to repeat the process of discharging and recharging the capacitor.

Besides being simple in equipment requirements, the CD pulse has another advantage over the rectangular pulse. In 1987 a team of scientists from Stanford University conducted a comparative study in which mammalian cells were transfected with reporter genes using either short CD pulses with high field strength or a long CD pulses with low field strength. It was found that the long CD pulse with low field strength gives a much higher transfection efficiency. Results of this work suggest that a long CD pulse may be more efficient than a narrow rectangular pulse in transfecting mammalian cells. Since the late 1980s, then, the CD pulse has become a more popular choice for electroporation use.

More recently, a new type of waveform that utilizes an oscillating field has been developed. It is called a "dc-shifted radio frequency (RF) pulse." This RF pulse is found to have several advantages. First, the waveform is more effective in electroporating (or electrofusing) cells of heterogeneous size. As indicated in equation (3), the induced membrane potential is proportional to the radius of the cell. Thus, with a given applied electric field, the induced membrane potential will vary depending on the cell size. This variability may cause large cells to have too high a V_m and suffer excessive damage, while small cells have too low a V_m, with the result that their membranes do not break down. This problem can be solved by using the RF pulses. Under an oscillating field, V_m is given by the following equation:

$$V_m = \frac{3Er\cos\theta}{2[1 + (\omega\tau_c)^2]^{1/2}} \qquad (4)$$

where ω is the angular frequency of the applied electric field; τ_c and other symbols are as defined in connection with equations (1) and (2). At a reasonably high frequency, V_m is almost inversely proportional to τ_c, which is in turn roughly proportional to the cell radius r (see equation 2). Therefore, under an oscillating field, V_m is not very sensitive to r. As a result, both large and small cells will have a similar V_m under the same field and will both be efficiently electroporated without excessive cell damage.

Second, the RF pulse is more effective in inducing membrane breakdown. Not only can the oscillating field cause a compression in the cell membrane, it can drive the charged membrane lipids and proteins to undergo an oscillating motion. This motion in turn will create structural fatigue in the membrane and make it more susceptible to electrical breakdown.

In mammalian cells, such as CV-1 cells, COS-M6 cells, and human red blood cells, it has been demonstrated in experiments that the RF waveform can provide a higher efficiency in electroporation and electrofusion than the rectangular pulse or the CD pulse.

2.2   BASIC PROCEDURES OF ELECTROPORATION AND ELECTROFUSION

The equipment required for electroporating cells is relatively simple, consisting mainly of an electrical pulse generator and a cell chamber (Figure 5). The major function of the pulse generator is to provide a high voltage pulse with large current output. This is necessary because the cell sample may have a very low resistance. The sample chamber not only contains the cell sample but also provides a pair of electrodes for applying the electric field. The electrodes are

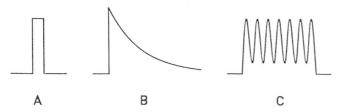

Figure 4.   Major types of waveform used in electroporation and electrofusion: (A) a rectangular pulse, (B) a CD pulse generated by discharging a capacitor, and (C) a dc-shifted radiofrequency pulse.

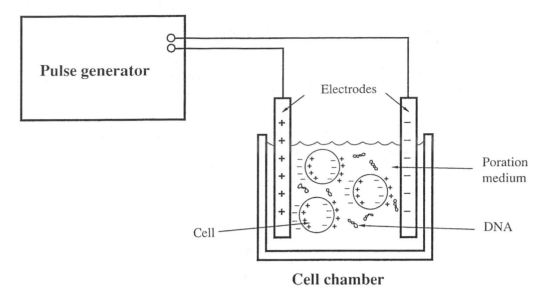

Cell chamber

Figure 5.   Schematic diagram showing the basic device required for using electroporation to introduce exogenous genes into cells.

usually two parallel metal plates made of platinum, stainless steel, or aluminum, typically separated by 0.1–0.4 cm.

The procedure of electroporation is also very simple. Suppose one wants to use electroporation to introduce certain recombinant DNA into a cultured cell line. The basic steps are as follows:

1. When cells are growing in the mid-log phase, harvest them using the standard cell detach treatment.
2. Wash cells at least once with a poration medium (PM) and resuspend them in PM. Although PM composition varies greatly between studies, it is basically an isotonic buffer with the majority of salt replaced by nonelectrolytes, such as sucrose or mannitol.
3. Load the cell sample into the chamber.
4. Apply the electrical pulse (or pulses) to the sample.
5. Transfer cells from the sample chamber into the culture plate (or flask) and culture them in their normal conditions.
6. After a sufficient length of time, the cells can be assayed for expression of the tranfected gene or put under a selection process.

The procedures of electrofusion are similar to those of electroporation, except that cells must be placed in contact with their fusion partners before applying the intense electric field. Depending on the objective of the fusion, cells can be brought together using various methods, including mechanical, chemical, or electrical means (see Section 1.3). The buffer used in electrofusion, called "fusion medium" (FM), is also similar to the PM used in electroporation, except that FM must contain very little salt. This is particularly important when dielectrophoresis is used to bring cells together. The conductivity of the FM must be kept low to avoid generating excessive current in the cell sample.

2.3   FACTORS AFFECTING THE EFFICIENCY OF ELECTROPORATION AND ELECTROFUSION

There are many factors that can affect the efficiency of electroporation. The major ones are discussed in Sections 2.3.1 to 2.3.5.

2.3.1 Field Strength

Probably the most important factor that can affect the efficiency of electroporation and cell viability is field strength. To electroporate cells, the induced membrane potential V_m must be large enough to approach the threshold potential V_{th}. As indicated in equation (3), V_m is directly proportional to the applied field. If the applied electric field is too low, V_m will be much smaller than the threshold potential, and electrical breakdown of the cell membrane will not occur. Under these conditions, cells will not be properly electroporated or will fail to fuse. On the other hand, if the applied field strength is too high, V_m will be much greater than the threshold potential and will lead to irreversible membrane break-down. The membrane will fail to reseal, and the cells can no longer be viable. Thus, selecting a proper field strength is a critical stepin both electroporation and electrofusion experiments.

As can be seen in equation (3), to generate a fixed V_m, the field strength must be inversely proportional to the radius of the cell. Thus, the field strength use to porate small cells, such as *E. coli* bacteria, is very high, typically on the order of 10 kV/cm. For larger animal cells, such as mammalian cultured cells, the applied field strength is typically about 1 kV/cm.

2.3.2 Pulse Width

The second most important parameter is the pulse width, τ. When τ is too short, there will not be enough time for the membrane to develop an electrical breakdown. Too long a τ, however, will cause excessive damage to the membrane, and the cell will not survive. When the rectangular pulse is used, the typical pulse width is less than 0.1 ms. For CD pulse or RF pulse, the pulse width may be significantly longer, on the order of milliseconds.

The parameters of field strength and pulse width are found to complement each other. For a given cell type, the optimal value of E will shift upward when the pulse width is reduced.

2.3.3 Number of Pulses

For some cells, a single pulse is sufficient for effective induction of electroporation or electrofusion. In most cell types, however, application of multiple pulses can greatly improve the efficiency of gene transfer.

2.3.4 Growth Condition of the Cells

To achieve high efficiency of gene transfer, cells must be harvested during the active growth phase prior to electroporation. If the cultured cells are grown for too long, the transfection efficiency will usually be low. Replating the cells a day or two prior to electroporation is recommended.

2.3.5 Other Factors

Other relevant and influential factors include temperature and the presence of divalent ions in the cell culture medium. For example, the concentration of Mg^{2+} may be critical to the efficiency of electroporation in some cell types. Some studies recommend the addition of carrier DNA to enhance the efficiency of gene uptake during the electroporation process.

Most of the factors affecting the efficiency of electroporation can also affect the efficiency of electrofusion. There are also other factors that are uniquely important to electrofusion. For example, the alignment of cells can greatly influence fusion efficiency. To fuse cells, one must bring the cells together before applying the high intensity pulse. As discussed in Section 1.3, there are different methods of bringing cells together, and the most elegant is to use a low amplitude oscillating field, which can induce cells to form a "pearl chain" by the process of dielectrophoresis. Optimal frequency and amplitude of the applied dielectrophoretic field for pearl chain formation depend on the diameter and the surface charge properties of each individual cell. Thus, to maximize the efficiency of electrofusion, one needs to optimize the pearl chain formation process.

It has been suggested that certain chemical treatments of the cell surface can considerably increase electrofusion efficiency. For example, it was reported that pretreating red blood cells with pronase can greatly enhance the efficiency of electrofusion. Such an enhancement may result from the removal of some of the glycoproteins at the cell surface, thus allowing the neighboring cells to be closer to each other.

Another factor that can affect fusion efficiency is the osmolarity of the fusion medium. The cell membrane can be expanded using a partial osmotic shock to increase cell volume. Recently, this technique has been applied in the generation of human hybridoma. In this work, cells were treated with a hypo-osmotic medium before the intense electric field was applied. It was reported that with this treatment, the efficiency of electrofusion was greatly enhanced.

In our experience, the waveform of the electric pulse can also significantly affect fusion efficiency. We found that by using a train of RF pulses, the electrofusion efficiency can be greatly increased compared to protocols using rectangular pulses.

3   APPLICATIONS

3.1   APPLICATIONS OF ELECTROPORATION IN GENE TRANSFER

The most important application of electroporation is in the introduction of recombinant DNA into biological cells. Electroporation has been used for introducing genes into many different cell types, both eukaryotic and prokaryotic, including animal cells, plant cells, unicellular organisms such as algae or yeast, parasites, and bacteria. In fact, electroporation is now the method of choice for introducing genes into *E. coli;* the resulting transfection efficiency can be as high as 10^9 colonies per microgram of DNA, which is almost a thousandfold higher than that normally obtained by conventional methods.

Electroporation can be used for both transient and stable gene transfection in mammalian cells. In 1982 T. K. Wong and E. Neumann first demonstrated, using electrical pulses, that a plasmid DNA containing the simplex virus thymidine kinase (TK) gene could be successfully introduced into mouse L cells deficient in the TK gene. Stable transformants were obtained in this experiment. Later, electroporation was applied to transfect myeloma cells and a neuronal cell line. Within a few years, electroporation was used to introduce exogenous DNA into a variety of cells, including fibroblasts, lymphocytes, endocrine cells, primary animal cultures, hepatoma cells, hematopoietic stem cells, and mammary carcinoma cells.

Today, electroporation is widely used in many studies of molecular biology for introducing genes into cell types of all sorts and for a wide variety of purposes. One common application of electroporation-mediated gene transfer is the study of gene regulation. Usually a reporter gene is inserted into a plasmid vector with specific transcription promoter and enhancer sequences. By comparing the efficiency of gene expression under various conditions, such as the availability of certain transcriptional factors, one can gain knowledge about how the gene of interest is regulated. Alternatively, one may mutate the upstream sequences of a certain gene, fuse it with the reporter gene, and introduce it into cells by electroporation. By comparing the resulting level of gene expression under various mutation conditions, one can identify the major promoter or enhancer sequences.

In addition, functional genes can be introduced into mammalian cells by stable transfection using electroporation. This can be done by coinserting a drug-resistant gene with the gene of interest into the cell genome and then subjecting the cell to a selection medium containing the specific drug (such as G418). Using this method, a large number of different types of DNA have been transfected into a variety of mammalian cells. For example, the complete human immunoglobulin κ (Igκ) gene was stably transfected into mouse L cells and mouse Pre-B cells by electroporation. Northern blot analysis of RNA indicated that the expression of the Igκ gene is highly tissue specific.

Not only can electroporation be used for introducing genes into cells to produce stably transfected cell lines, but a similar method can be used to introduce genes directly into eggs or stem cells for the generation of transgenic animals. For example, it has been reported that transgenic zebrafish can be produced using electroporation. Electroporation has been used to introduce genes into plant cells, as well: protoplasts of a number of plants, including carrot, maize, and tobacco leaf mesophile cells have been transfected. In addition, this technique can be applied to make transgenic plants. It has been reported that electroporation can be used for the generation of transgenic cereal plants such as rice and sorghum.

In comparison to other methods of gene transfer, such as those using calcium phosphate or retroviruses, electroporation has several distinctive advantages. First, it is a simple method. The procedure takes only minutes. And, there is no need for prior treatment of the DNA to be inserted. Second, electroporation is efficient. In comparison with the chemical methods using calcium phosphate or

DEAE dextran, electroporation usually offers much higher transfection efficiency. Third, since electroporation is a physical method, it is widely applicable to many cell types. Unlike the chemical method or the virus vector, electroporation is less dependent on the surface properties of the cell membrane; hence, electroporation can be used to introduce genes into a very wide spectrum of different cell types. Finally, electroporation is a relatively "clean" method because it involves little chemical or biological hazard in comparison to the chemical and retroviral methods.

3.2   ELECTROINJECTION OF OTHER MOLECULES

Besides genes, electroporation can be applied to introduce many other types of molecule into living cells. In fact, the early usage of electroporation was to introduce ions, such as calcium, and small molecules, such as EGTA and ATP, into cells. Lately, electroporation has also been used to introduce fluorescently labeled dextrans and molecular probes such as Calcium Green-1. Electroporation can also be used for the introduction of large molecules such as antibodies. For example, using electroporation, monoclonal antibodies have been introduced into HeLa cells and murine lymphoma cells. Furthermore, electroporation has been used for the introduction of a number of active biological compounds, including restriction enzymes, Epstein–Barr virus episomal replicons, substrates of enzymes, and drugs that block certain signal transduction pathways, such as heparin.

3.3   STUDY OF THE MECHANISM OF MEMBRANE AND CELL FUSION

Electrofusion has been used for studying the mechanism of membrane fusion. Most of these works were done using either model membranes such as lipid vesicles, or cells with simplified membrane structure, such as red blood cells or red cell ghosts. More recently, electrofusion has been used in the investigation of cytoplasmic fusion. Cytoplasmic fusion is more difficult to study than membrane fusion because of a number of technical requirements associated with the former. The fusing cells must be maintained in their normal physiological state for a long period for observation. Such a requirement excludes some of the common fusogens, like polyethylene glycol or virus, that may have toxic effects. Furthermore, to study the dynamic changes of cytoplasmic structures, cell fusion must be induced by a treatment that requires very little time, and the fusion process must be relatively synchronized.

These requirements can be satisfied using the electrofusion technique. The fusion event can be induced by applying a short pulse of electric field; the fusion efficiency is very high—about 70%. In addition, all cells exposed to the same electrical pulse would fuse almost in synchrony. Because of these advantages, the electrofusion method has been used to study the reorganization of various cytoplasmic components, including cytoskeletons and organelles, during the fusion of mammalian cells.

3.4   PRODUCTION OF HYBRIDOMAS AND ANTIBODIES

One of the major applications of electrofusion is to make hybridomas for the production of monoclonal antibodies. For example, mouse myeloma cells can be fused with mouse lymphocytes and then selected for clones producing a specific antibody. Traditional methods using a chemical fusogen, such as polyethylene glycol, are not as efficient as electrofusion. Several comparative studies have shown that electric field induced cell fusion can be a thousandfold more efficient than fusion events initiated by chemical means. Lately, there has been a strong interest in making human hybridomas for producing human antibodies, which would obviously be much more useful than animal antibodies for medical purposes. The production of human hybridomas is particularly difficult because, with the conventional fusion methods, the fusion between antibody-producing cells and human plasma cells usually has very low yield. Recently, by using the electrofusion technique, several groups were able to develop new protocols to produce human hybridomas successfully.

3.5   USES OF ELECTROFUSION IN AGRICULTURE

One of the most important applications of electrofusion in agriculture is plant hybridization. It has been demonstrated that the plant protoplasts from a number of species can be fused using an electric field to make hybrids. For example, *Nicotiana tabacum* has been fused with *Nicotiana plumbaginitolia*. Such a fusion technique, which can complement the traditional methods of hybridization by genetic crossing, will be highly useful for agricultural research in the future.

Another interesting use of electrofusion is animal cloning and the production of transgenic animals. In the livestock industry, it is desirable to produce a large stock of animals with identical genetic strains that encode preferred characteristics, such as rapid growth and disease resistance. Using the electrofusion method, animals of identical genome can be cloned. The technique is relatively simple in principle. Basically it involves dissociating a 32-cell embryo and then using electrofusion to join each of the dissociated embryonic cells with an anuclear egg. Even with a 50% fusion efficiency, one can produce 16 fused eggs, each of which contains the same genomic information of the original embryo. The same process can then be repeated to further multiply the number of cloned embryos, which can later be transferred into the wombs of female animals for development. By this means, the genetic information for a specific trait or property can be amplified to clone a large number of genetically identical animals. This powerful method can be used to rapidly create a large stock of animals with a preferred genotype. The technique is currently used by the livestock industry.

4   PERSPECTIVES

4.1   FUTURE APPLICATIONS OF ELECTROPORATION IN MOLECULAR BIOLOGY

One of the most important applications of electroporation as a gene transfer technique in molecular biology is gene targeting. At this time, it is difficult to use other methods of gene transfer to conduct homologous DNA recombination. For example, in the retroviral method, the inserted gene is flanked by long terminal repeat sequences, which have a tendency to insert the recombinant gene into random sites on the genome, defeating the purpose of homologous recombination. Chemical methods, such as those using calcium phosphate, have the disadvantage of inserting multiple copies of the recombinant gene into the genome; the number of copies is uncontrollable. These problems can be avoided by using electroporation as the gene transfer method. This method allows the investigator to control the copy number of the recombinant DNA to be inserted into the genome. Also, by linkage with proper flanking sequences, the gene introduced by electroporation can be directed to integrate into the proper site of the genome. Hence, electroporation will become a very important tool for gene transfer in the future, when gene targeting work is more advanced.

A related use of electroporation is for gene therapy. It is widely recognized that gene therapy will become an important branch of medicine in the future, providing the ultimate way to cure certain inherited diseases. But the gene therapy technology, at present, still has many limitations. One of the most critical concerns is safety, since the cells inserted with the new gene must remain inside the patient's body. Currently, the most common method of introducing therapeutic genes into patient cells is the retroviral method. Although the retrovirus vectors employed are always reengineered so that under normal conditions they will not be active, there is still a concern that some of these retrovirus vectors could pose a risk for patients under certain unusual conditions, such as becoming infected with certain wild-type viruses. With the electroporation method, such concerns do not exist. Being a physical method, electroporation has practically no biological side effects and thus would be a much safer method for use in gene therapy.

In addition, the electroporation method is less cell-type-dependent than the chemical methods and the retrovirus method, which depend on the surface properties of the target cells, rendering certain cell types resistant to transfection mediated by chemicals or virus. It happens that some of the cell types that are highly important for gene therapy, such as lymphocytes and stem cells, are generally resistant to the common retrovirus vectors. Since electroporation is a physical method, it is not particularly sensitive to the composition of the membrane proteins, and thus, is less dependent on cell type. So, in principle, once the electroporation method has been properly optimized, it can be applied for introducing genes into most target cells.

4.2 ELECTROPORATION AS AN IMPORTANT TOOL FOR STUDYING CELL SIGNALING

At present, electroporation has been mainly used in molecular biology. With time, this technique will have the potential to become an important tool in the study of cell biology. Electroporation can be used as a massive microinjection technique. Microinjection by means of a glass pipette, the current method, is tedious, requiring expensive equipment and extensive training. It is also difficult to inject a large number of cells at one time. With electroporation, thousands or even millions of cells can be injected within a few minutes! Thus, electroporation can become a very efficient injection method.

One of the major uses of electroporation is to introduce molecular probes, such as ion indicators (e.g., for Ca^{2+} or H^+) or cell-labeling dyes. At present, most of the calcium indicators, such as fura-2 or fluo-3, are introduced into the cell either by direct injection or by modifying the indicators into triacetoxymethyl ester (AM) forms, which can permeate the cell membrane. The amount of molecules introduced by the latter methods is difficult to control, however. Certain portions of the AM form indicator could also be taken up by the intracellular organelles, thus making it more difficult to interpret the experimental results. These problems can be avoided by using electroporation to introduce non-AM-form indicators. Electroporation has been demonstrated to be efficient in injecting several different molecular probes into a wide variety of cells, including animal, plant, and unicellular cells. The amount of molecules injected usually can be controlled by varying the electroporation conditions.

In the future, the most potentially useful application of electroporation is probably the introduction of various substances into cells to study the processes of signal transduction. To study a hypothetical signal transduction pathway, for example, one can introduce kinases or phosphatases, and their substrates, activators, or inhibitors into cells. It is well known that many of the cell functions are regulated by phosphorylation and dephosphorylation of certain proteins. The signal pathway usually involves multiple components and a cascade of kinase/phosphatase activities. It is very difficult to study such a pathway. With the electroporation technique, reagents can be introduced into the living cell to specifically inhibit or activate a selected step of the signaling process. In this way, one can examine the functional role of a specific signaling molecule and eventually gain a better overall understanding of the entire signal transduction pathway under study.

Similarly, electroporation can be used to introduce many other factors that are important for biological regulations, such as transcription factors, gene products of oncogene or tumor suppressor genes, small G proteins, and intracellular receptors of hormones or other signaling molecules. The ability to introduce these factors into a large number of living cells will open up new opportunities for examining the molecular mechanisms of many important cellular functions.

See also GENETIC IMMUNIZATION; GENETIC VACCINATION, GENE TRANSFER TECHNIQUES FOR USE IN; LIPID-BASED GENE VECTORS; MOLECULAR GENETIC MEDICINE; PLANT CELL TRANSFORMATION, PHYSICAL METHODS FOR; VACCINE BIOTECHNOLOGY.

Bibliography

Chang, D. C., Chassy, B. M., Saunders, J. A., and Sowers, A. E., Eds. (1992) *Guide to Electroporation and Electrofusion.* Academic Press, San Diego, CA.
Cole, K. S. (1968) *Membranes, Ions, and Impulses.* University of California Press, Berkeley.
Neumann, E., and Sowers, A. E., Eds. (1989) *Electroporation and Electrofusion in Cell Biology,* pp. 215–27. Plenum Press, New York.
Potter, H. (1988) Electroporation in biology: Methods, applications and instrumentation. *Anal. Biochem.* 174:361–373.
Sowers, A. E., Eds. (1987) *Cell Fusion,* pp. 479–496. Plenum Press, New York.
Zimmermann, U., and Vienken, J. (1982) Electric field-induced cell-to-cell fusion. *J. Membrane Biol.* 67:165–182.

ENDOCRINOLOGY, MOLECULAR
Franklyn F. Bolander, Jr.

1 **Introduction**

2 **Hormones**

3 **Receptors**
  3.1 Nuclear Receptors
  3.2 Membrane Receptors

4 **Second Messengers**
  4.1 G Proteins and Cyclic Nucleotides
  4.2 Calcium, Calmodulin, and Phospholipids
  4.3 Miscellaneous Mediators

Key Words

Adaptor   A small molecule consisting only of phosphotyrosine (SH2) and polyproline helix (SH3) binding domains and used to associate proteins.

G Protein   A GTP-binding protein with intrinsic GTPase activity; it acts as a molecular switch (which is "on" when bound to GTP: guanosine 5'-triphosphate) with a built-in timer.

Hormone   A chemical, nonnutrient, intercellular messenger that is effective at very low concentrations.

Hormone Response Element (HRE)   A DNA sequence recognized by a transcription factor that is primarily regulated by hormones.

Oncogene   A gene that usually encodes a component of the growth factor pathway and has undergone a mutation resulting in constitutive activity and tumor formation.

Endocrinology is the study of hormones, chemicals that cells use to communicate with each other. Traditionally, endocrinology was studied at the organismal level: the effects of hormones on tissue or animal growth, reproduction, or metabolism. Molecular endocrinology seeks to determine the molecular mechanisms for these gross effects: the actions of hormones, direct or through mediators, on enzymes, transport processes, the cytoskeleton, and transcription factors. Such information is vital to understanding how complex metazoans coordinate cellular functions. This knowledge can be useful, as well, in the evaluation and treatment of various endocrine diseases that have genetic bases. In addition, many tumors arise when the genes for various growth-promoting hormones or for components of their signaling pathway undergo mutations that render them constitutively active.

1   INTRODUCTION

Multicellular organisms have two major coordinating systems: the nervous and the endocrine systems. Both utilize chemical messengers: in the endocrine system, these molecules are called hormones and are usually secreted into a circulatory system for general distribution. In the nervous system, the molecules are called neurotransmitters and are released into synapses for more precise effects. The distinction is not always this clear; many of the chemicals used by each system and their mechanisms of action are identical. In addition, some hormones are made and act locally in a very defined area, while some neurotransmitters can leak out into the general circulation. Thus, there is a general tendency to consider all chemical messengers as a single functional group.

2   HORMONES

Structurally, hormones are extremely diverse; nearly every organic group is represented. Proteins and peptides are the largest group. In addition, there are hormones that are derivatives of amino acids, sterols, fatty acids, phospholipids, nucleotides, and carbohydrates. There are even several gaseous hormones: ethylene and nitric oxide.

However, one property divides all hormones into two major groups and will determine their mechanisms of action: water solubility. Hydrophobic ligands have no problem crossing the plasma membrane; as such, these hormones have a direct mechanism of action. In particular, they migrate to the nucleus, where they interact with transcription factors. Since their mechanism is primarily genomic, they tend to have delayed, long-term effects; they are often involved in developmental processes. Hydrophilic hormones cannot cross the plasma membrane and must interact with binding proteins, the receptors. Since these receptors are integral membrane proteins, they must generate another signal on the cytosolic side; if the hormone is the primary messenger, then this subsequent factor becomes a second messenger. These latter messengers often activate kinases, which can have acute effects on metabolism and cell structure or phosphorylate transcription factors for more long-term effects.

3   RECEPTORS

3.1   Nuclear Receptors

As noted in Section 2, the receptors for lipophilic hormones are ligand-regulated transcription factors. These receptors are all homologous: the amino terminus possesses a transcription activation domain (TAD), the center has a DNA-binding region, and the carboxy terminus binds the hormone and heat shock proteins (hsps), and contains a dimerization domain and a second TAD. The hsps are necessary to maintain the receptors in a conformation required for ligand binding; once the hormone has bound the hsps dissociate (Figure 1). Ligand binding also induces receptor dimerization, phosphorylation, nuclear translocation (although some receptors

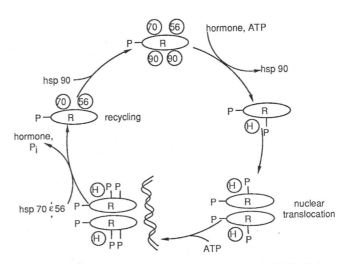

Figure 1.   Activation and recycling of a nuclear receptor: 56, 70, 90, heat shock proteins; H, hormone; P, phosphorylation site; R, receptor.

are constitutively located in the nucleus), DNA binding, and transcription activation. After the hormone has dissociated, the receptor is dephosphorylated and recycled.

The nuclear receptors can be divided into three families based on their structures and the DNA sequences to which they bind. The glucocorticoid family is the most recently evolved group and contains the cortisol, aldosterone, androgen, and progesterone receptors. The members of this family are basically homodimers, require hsp 90, and bind inverted repeats of the hormone response element (HRE) TGTTCT. The thyroid hormone family is the oldest and most diverse group and includes receptors for the thyroid hormone, vitamins A and D, ecdysone, and arachidonic acid. They are most active as heterodimers, do not require hsp 90, and can bind either direct or inverted repeats of TGACC. The estrogen family contains only the estrogen receptor and a few related receptors whose ligands are still unidentified. Its properties lie between the two other groups: it binds the thyroid hormone HRE but only as inverted repeats; in addition, it forms homodimers and requires hsp 90, like the glucocorticoid family.

3.2 MEMBRANE RECEPTORS

Membrane receptors are considerably more diverse. Four basic superfamilies are recognized: the enzyme-linked, fibronectin-like, serpentine, and ion channel receptors (Figure 2). The enzyme-linked receptors are the simplest: the basic structure consists of a single protein that traverses the plasmalemma once via an α-helix. The aminoterminal extracellular domain binds the hormone, while the carboxyterminal cytosolic domain possesses a catalytic site. Ligand binding induces dimerization and enzyme activation. Three types of enzymatic activity have been identified: the tyrosine ki-

nases are represented by four families (the epidermal growth factor, insulin, platelet-derived growth factor, and nerve growth factor groups); the serine-threonine kinases are found only in the transforming growth factor β family; and the guanylate cyclases generate cyclic GMP (cGMP) in response to atrial natriuretic factors.

In the case of the receptor tyrosine kinases (RTKs), the major substrates are themselves. Phosphorylated tyrosines are recognized by SH2 domains that are found in many enzymes and adaptors; these latter proteins will bind the autophosphorylated receptors and then mediate many of the biological activities of RTKs (see Section 4).

The fibronectin-like receptors have the same structure as the enzyme-linked binding proteins except that there is no recognizable catalytic site in the cytosolic domain. The extracellular region is composed of two modified units first identified in fibronectin, a matrix protein. In the cytokine (class 1) receptors, these units form two seven-stranded β-sheets that join at right angles to create a ligand-binding pocket (Figure 2B). This core may be further embellished by immunoglobulin loops and/or additional fibronectin domains. The class 2 receptors form repeats of five-stranded β-sheets that extend over the hormone like fingers.

In the cytosolic juxtamembrane region of the fibronectin-like receptors is a conserved proline-rich region that constitutively binds soluble tyrosine kinases. Like RTKs, these kinases are activated by receptor aggregation induced by ligand binding. As such, the fibronectin-like receptors can be considered to be RTKs whose catalytic site is located on a separate subunit.

The next superfamily has many names: G protein coupled, seven transmembrane segment, and serpentine receptors. Here we use the latter term because of its simplicity. The serpentine receptors are probably the oldest and most diverse of the membrane receptors and mediate both sensory and endocrine transduction. The protein tra-

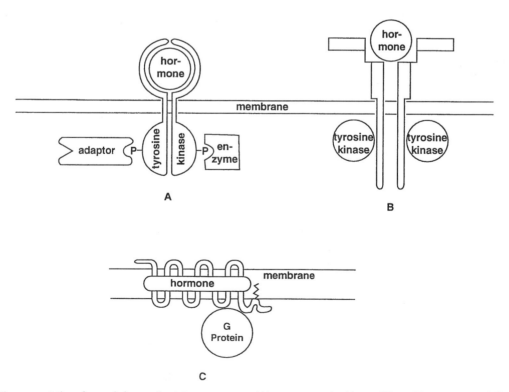

Figure 2. Schematic representation of several classes of membrane receptors: (A) receptor tyrosine kinase, (B) cytokine receptor, and (C) serpentine receptor.

verses the plasma membrane seven times; several of these transmembrane α-helices have conserved prolines that kink the helices, resulting in the formation of a ligand-binding pocket (Figure 2C). As such, most of the hormones using these receptors are small (e.g., the catecholamines, histamine, prostaglandins, etc.). Receptors with larger ligands often have extended amino termini to aid in hormone binding. Several of the intracellular loops bind and activate G proteins (see Section 4.1), especially the third loop and the juxtamembrane portion of the carboxy terminus. The latter also forms a loop because its midsection is held to the plasma membrane by a palmitic acid that is covalently bound to a conserved cysteine in the carboxy terminus.

The last superfamily consists of the ion channel receptors. The ligand-gated channels are pentamers of homologous subunits; each subunit contributes an α-helix toward forming the wall of the channel. The hormone appears to bind between subunits and to open the channel either by pushing the subunits farther apart or by twisting them. Acetylcholine activates sodium channels; glycine and γ-aminoisobutyric acid, chloride channels; and serotonin and glutamate, calcium channels.

The voltage-gated channels are homotetramers; each subunit has six transmembrane helices. The pore is formed by a β-loop between the last two helices; together, the four subunits create an eight-stranded β-barrel channel. Two families from this group have become regulated by second messengers: first, the cyclic nucleotide gated channels are cation pores opened by cAMP or cGMP. The second family are channels that release internal stores of calcium; they include the cyclic ADP–ribose receptors and the inositol 1,4,5-trisphosphate (IP_3: see Section 4.2).

4   SECOND MESSENGERS

4.1   G Proteins and Cyclic Nucleotides

G proteins are molecular switches: they are active when GTP is bound to them and inactive when GDP is bound. In addition, they have intrinsic, although weak, GTPase activity; as such, they eventually turn themselves off when they hydrolyze the bound GTP to GDP. In the active state, they can stimulate enzymes and ion channels, and affect the cytoskeleton and vesicular trafficking. There are several G-protein families, but only two that have been closely associated with hormone action are discussed here: a small G protein (Ras) and a large G protein (the αβγ trimer).

For a hormone to "flip" this switch on, it must facilitate the exchange of GDP for GTP. For example, the conversion of Ras·GDP to Ras·GTP is accomplished by a Ras GDP dissociation stimulator (RasGDS) (Figure 3), which is under hormone regulation. In fact, there are several different types of RasGDS, each with its own mechanism of control. One RasGDS (called mammalian SOS) is stimulated indirectly: receptor or soluble tyrosine kinases phosphorylate either themselves or some substrate. An adaptor (Grb2) binds to the phosphotyrosine via its SH2 domain; Grb2 also has an SH3 domain that binds polyproline helices. Such a helix is present in RasGDS, and as a result, a complex is formed among a tyrosine

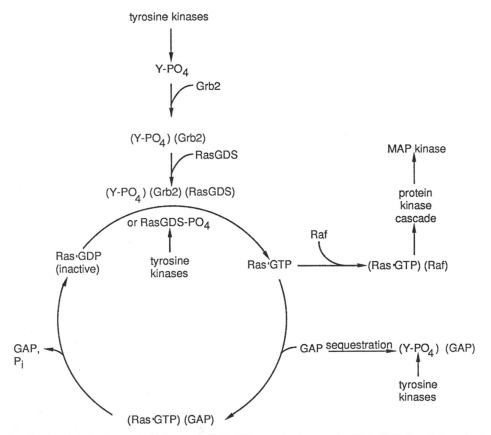

Figure 3.   Activation–inactivation cycle for Ras, a small G protein: GAP, GTPase-activating protein; GDS, GDP dissociation stimulator; Grb2, an adaptor; Raf, a protein kinase; Ras, a small G protein; Y-PO$_4$, tyrosine phosphorylation site.

phosphorylated protein, Grb2, and RasGDS. This complex activates RasGDS, which then stimulates nucleotide exchange on Ras. Once formed, Ras·GTP binds to Raf, a serine/threonine kinase and brings it to the plasmalemma, where it is stimulated by a still unknown process. Finally, Raf initiates a protein kinase cascade that eventually activates mitogen-activated protein kinase (MAPK), a critical kinase in transcription regulation (see Section 5.2).

The GTPase activity of small G proteins is so weak that accessory proteins, the GTPase activating proteins (GAPs), are required. This is another potential site for hormone input: RasGAP has an SH2 domain that binds phosphorylated tyrosines. Upon autophosphorylation, RTKs can bind and sequester RasGAP, thereby prolonging the activated state of Ras·GTP.

The large G proteins occur as heterotrimers: the α subunit is the GTPase, while the $\beta\gamma$ dimer is involved with membrane localization and protein association. Although the GTPase activity of α is slow, the large G proteins do not need GAPs. There are four subfamilies: G_s stimulates adenylate cyclase, G_i inhibits adenylate cyclase, G_q stimulates phospholipase Cβ (PLCβ), and $G_{12/13}$ activates the (Na^+, H^+) exchanger, NHE-1. G_s and G_i also directly stimulate several ion channels. The hormone–receptor complex directly accelerates nucleotide exchange, resulting in the activation of α_s and the dissociation of $\beta\gamma$ (Figure 4). As long as GTP remains bound to α_s, adenylate cyclase will be stimulated. However, α_s will eventually hydrolyze GTP and reassociate with $\beta\gamma$. G_i represent a counterregulatory mechanism that operates in a parallel manner except that both α_i and $\beta\gamma$ participate in the inhibition of adenylate cyclase.

The ultimate output to this pathway is cAMP, which can activate several enzymes and ion channels; however, the major effector for cAMP is a serine/threonine kinase (protein kinase A, PKA). This tetramer has two catalytic subunits and two inhibitory subunits (Table 1). cAMP binds to the regulatory subunits and causes them to dissociate, thereby removing the inhibition.

cGMP is another cyclic nucleotide used as a mediator of hormone action; it may be generated by two different pathways. The first way, discussed in Section 3.2, involves a hormone directly binding and activating a membrane-bound guanylate cyclase. cGMP may also be synthesized by a soluble cyclase; this enzyme is activated indirectly by hormones that elevate calcium (see Section 4.2). Like cAMP, cGMP stimulates both ion channels and a homologous kinase, protein kinase G (PKG). In this kinase, the regulatory and catalytic subunits are fused into a single protein.

4.2   Calcium, Calmodulin, and Phospholipids

Calcium is an abundant cation in extracellular fluids; in addition, it is concentrated within several cellular organelles, such as mitochondria and elements of the smooth endoplasmic reticulum. However, cytosolic levels are kept extremely low, because many cellular processes are dramatically affected by calcium. Thus hormones can regulate these cellular functions by controlling the cytoplasmic calcium concentrations. The most direct mechanism for elevating calcium would be for hormones to activate ligand-gated calcium channels, such as the glutamate receptor; by merely opening these channels, hormones would cause external calcium to flood the cytoplasm.

However, a major source of hormonally released calcium is internal (Figure 5). Briefly, hormones stimulate a PLC to hydrolyze a phospholipid (polyphosphoinositide) into diacylglycerol (DG) and the former head group, IP_3. There are several PLC groups distinguished by both their structure and their hormone regulation. PLCγ possesses two SH2 domains through which it binds to autophos-

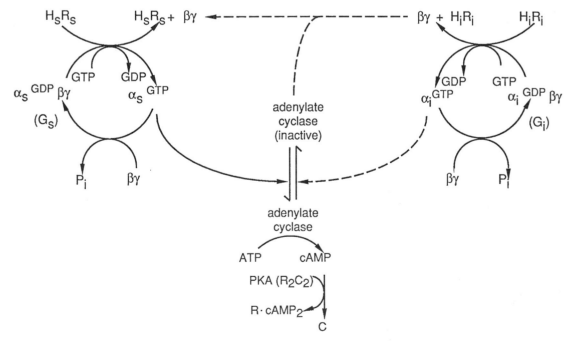

Figure 4.   Regulation of adenylate cyclase by several G-protein trimers: H_iR_i, adenylate cyclase inhibiting hormone and its receptor; H_sR_s, adenylate cyclase stimulating hormone and its receptor; R_2C_2, PKA tetramer containing two regulatory (R) and two catalytic subunits (C); inhibitory effects are depicted by dashed lines.

Table 1 Some Multifunctional Serine Threonine Kinases Involved with Hormone Action

Kinase	Abbreviation	Structure	Activators
Protein kinase A	PKA	Tetramer (R_2C_2)	cAMP binds and dissociates inhibitory R subunits
Protein kinase C	PKC	Monomer	
Conventional	cPKC		Calcium and diacylglycerol
Novel	nPKC		Diacylglycerol
Atypical	aPKC		Ceramide; arachidonic acid; polyphosphoinositides
Protein kinase G	PKG	Homodimer	cGMP binding
Calmodulin-dependent protein kinase II	CaMKII	Homodecamer or homooctamer	Calmodulin masks autoinhibitory domain
Casein kinase II	CKII	$\alpha\beta$ dimer	Polyamines and phosphorylation
Mitogen-activated protein kinase	MAPK	Monomer	Ras and PKC via a kinase cascade

phorylated RTKs; this association brings the enzyme into close proximity to its substrate and allows RTKs to phosphorylate and stimulate the PLCγ. On the other hand, PLCβ, and probably PLCδ, are activated by G proteins.

Once released, IP_3 diffuses through the cytosol to its receptor, a calcium channel on the endoplasmic reticulum. IP_3 binding opens the channel and allows the internally stored calcium to enter the cytoplasm. The DG remains in the plasma membrane, where it stimulates a calcium-activated, phospholipid-dependent protein kinase, protein kinase C (PKC). Actually, there are three PKC groups that differ in their calcium and DG requirements (Table 1).

Phospholipids may give rise to many other second messengers; for example, phospholipase A_2 can liberate arachidonic acid, which may stimulate enzymes or activate a nuclear receptor. (Table 2 lists some nuclear receptors and other transcription factors that mediate hormone action.) Arachidonic acid can also be converted to prostaglandins or other eicosanoids. These fatty acid derivatives are hormones in their own right: they are secreted, bind to specific membrane receptors, and stimulate various second-messenger pathways. However, they usually act locally; such molecules are called parahormones, and they may adjust a cell's response to other hormones or to local conditions. A lysophospholipid is what remains after the arachidonic acid has been removed, and it is also biologically active. Finally, another membrane lipid, sphingomyelin, can be the source of several mediators of hormone action. Sphingomyelin is structurally similar to phosphatidylcholine except that it has a serine rather than a glycerol backbone. Its hydrolysis can generate ceramide and sphingosine 1-phosphate, which can then affect protein phosphorylation and transcription.

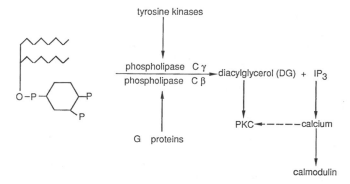

Figure 5. Brief summary of the polyphosphoinositide pathway: IP_3, inositol 1,4,5-trisphosphate.

Once elevated, calcium alone can directly affect many enzymes, the cytoskeleton, and other biological processes. However, it frequently acts through a calcium-binding protein, calmodulin (CaM). CaM is a small dumbbell-shaped peptide: two globular ends separated by an α-helix. Calcium binding to the globular ends allows the groove found in each end to wrap around an α-helix in a target protein. The central α-helix of CaM helps to determine binding selectively: it has an acidic and a hydrophobic end and is attracted to target proteins having an exposed α-helix with a basic and a lipophilic end. Such amphipathic helices often inhibit the enzymes that possess them, and their masking by CaM prevents this autoinhibition. For example, the CaM-dependent protein kinase II, a general-purpose serine/threonine kinase, has a CaM-binding site that blocks the ATP-binding site and inhibits kinase activity. In the presence of calcium, CaM binds this site, prevents it from interfering with ATP binding, and activates the kinase. There are many other enzymes that operate on this same principle.

Another enzyme activated by CaM is nitric oxide synthetase; this enzyme generates nitric oxide, which then activates a soluble guanylate cyclase. This series of events represents an alternate pathway for the production of cGMP.

4.3 MISCELLANEOUS MEDIATORS

Several other second-messenger systems have been postulated, but they have not been sufficiently explored to permit a determination of their physiological relevance to hormone action. Polyamines are small basic molecules that are required for transcription and translation. They can also activate a protein kinase, the casein kinase II. Hormonal regulation is achieved by the induction of the enzymes for polyamine synthesis; hormones can also stimulate polyamine transport into cells from the surrounding medium.

The hydrogen ion concentration represents another potential mediator of hormone action. All mitogenic stimulation is accompanied by a slight rise in cellular pH. This alkalinization has been shown to be necessary, but not sufficient, for subsequent cell division. It is not clear what the output for this signal is; although there are many pH-sensitive cellular reactions, pH optima are rarely so sharp that a few tenths of a pH unit would make a significant difference in overall activity. Growth factors raise the pH by activating NHE-1 by either PKC phosphorylation, G proteins, or calcium/calmodulin binding to an autoinhibitory site.

Finally, some species of phosphoinositides have additional sugar residues attached to the head group. A special hormone-sensitive PLC can separate the oligosaccharide component from the diacyl-

Table 2 Some Transcription Factors Mediating Hormone Action

Transcription Factor	Abbreviation	Primary Activator(s)
cAMP response element binding protein	CREB	PKA phosphorylation
CCAAT/enhancer-binding protein β	C/EBPβ	CaMKII and MAPK phosphorylation
Jun-Fos	AP-1	PKC phosphorylation (indirect)
Myc-Max		CKII and MAPK phosphorylation
Nuclear factor of activated T lymphocytes	NFAT	Dephosphorylation by PP-2B, a CaM-regulated phosphatase
Nuclear factor for κ genes of B lymphocytes	NF-κB	PKC phosphorylation of inhibitory subunit
Nuclear receptors	GR, ER, Tr, etc.	Ligand binding by steroids, retinoids, thyroid hormones, etc.
Peroxisome proliferator activated receptor	PPAR	Fatty acid binding
Serum response factor	SRF	CKII and MAPK phosphorylation
Signal transducers and activators of transcription	STAT	Direct phosphorylation by receptor or soluble tyrosine kinases

glyceride backbone. The liberated oligosaccharide can then affect enzyme activity and even gene expression, presumably as an allosteric regulator. Although this oligomer has been best studied as a possible mediator of insulin action, it has also been associated with the actions of several other hormones, such as some growth factors and cytokines.

5   BIOLOGICAL EFFECTS

5.1   NONGENOMIC

There are two major mechanisms by which hormones can affect cellular processes: allosterism and phosphorylation. For example, G proteins can directly bind and alter the activity of both enzymes (e.g., adenylate cyclase, phospholipases, several kinases) and transporters (e.g., ion and glucose channels, the NHE-1). Similarly, cyclic nucleotides can stimulate nucleotide-gated channels and protein kinases. Lipophilic hormones, which have direct access to the cellular interior, have also been shown to bind directly several enzymes, but the physiological relevance of these observations is still controversial.

However, the broadest effects are generally achieved by phosphorylation. Indeed, virtually every known second-messenger pathway activates at least one protein kinase, and several also affect pro-

Table 3 The Hormone Regulation of Several Protein Phosphatases

Protein Phosphatase (PP)	Regulation
Serine/threonine Phosphatases	
PP-1	Subunit composition affects localization
	Phosphorylation (± depending on the kinase)
PP-2A	Polyamines, ceramide, phosphorylation, and carboxymethylation (+)
	ATP and several oncogene products can shift PP-2A from a serine/threonine to a tyrosine phosphatase
PP-2B (calcineurin)	Calcium/calmodulin (+) phosphorylation and several immunosuppressants (−)
PP-2C	Calcium (in plants only)
Protein Tyrosine Phosphatases (PTPs)	
SH-PTP	Allosteric activation when its SH2 domain binds a phosphotyrosine; effects of phosphorylation are controversial
PEST-PTP	PKA and PKC phosphorylation (−)

tein phosphatases (Tables 1 and 3): the extent of this modification is controlled by a balance between these two enzyme groups. Phosphorylation is a major regulatory mechanism in metabolism; glycogen metabolism is one of the best-known examples. Hormones that stimulate glycogen breakdown activate PKA, which initiates a protein kinase cascade leading to the phosphorylation and stimulation of glycogen phosphorylase. In addition, PKA and other kinases modify and inhibit glycogen synthetase; that is, glycogen breakdown is activated, while its synthesis is blocked. Insulin, on the other hand, favors glycogen synthesis by activating the phosphatases that reverse these phosphorylations. Many other metabolic cycles are also regulated by this modification.

In addition, phosphorylation can affect other cellular processes: it can alter the function of various channels and transporters and trigger the breakdown of the cytoskeleton.

5.2   GENOMIC

Hormones and their mediators can also affect gene expression by the same mechanisms described earlier (Table 2). For example, nuclear receptors are actually transcription factors that are allosterically regulated by their ligands, usually hormones but also second messengers like arachidonic acid. Many other transcription factors are controlled by phosphorylation. The simplest mechanism would be for an RTK or a cytokine receptor with associated soluble tyrosine kinase to phosphorylate directly a transcription factor that would then migrate to the nucleus and activate gene expression. The STAT family is activated by such a mechanism (Figure 6A).

However, most transcription factors are phosphorylated via second-messengers pathways. For example, hormones whose activities are mediated by cAMP first activate G_s; the α subunit then stimulates adenylate cyclase, and the resulting accumulation of cAMP activates PKA. Finally, PKA modifies a cAMP response element binding protein (CREB); in particular, the phosphorylation occurs in a TAD and renders CREB a more efficient transcription activator (Figure 6B).

Phosphorylation can also occur on accessory proteins: NF-κB is a transcription factor frequently associated with defense genes. It is held in the cytoplasm by an inhibitory subunit, IκB. After being modified by PKC, IκB dissociates, allowing NF-κB to translocate to the nucleus, bind its HRE, and activate transcription (Figure 6C, top). Phosphorylation is not always stimulatory: NFAT, another transcription factor associated with defense genes, is tonically inhibited by phosphorylation. Activation occurs when hormones ele-

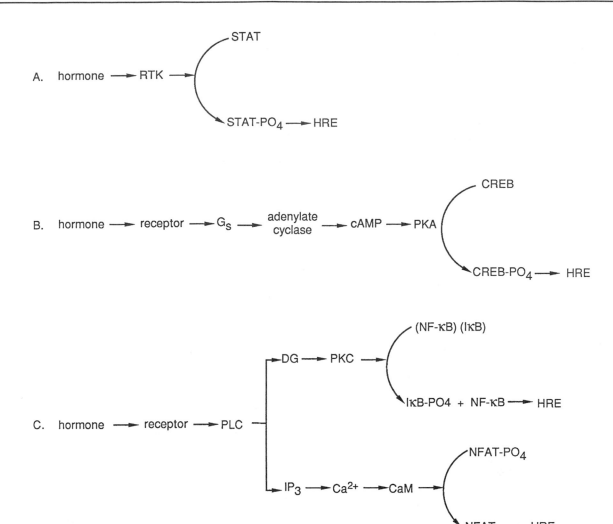

Figure 6. Activation of several transcription factors by various second-messenger pathways: CaM, calmodulin; DG, diacylglycerol; HRE, hormone response element; PLC, phospholipase C; RTK, receptor tyrosine kinase (see Tables 1 and 2 for other abbreviations).

vate calcium concentrations that stimulate PP-2B (Table 3), leading to the dephosphorylation of NFAT (Figure 6C, bottom).

The reader should be warned that the preceding discussion is an oversimplification. In fact, many transcription factors may be modified by several different kinases. This cross talk allows for the integration of multiple signals, although it may at times appear confusing. Therefore, this general overview has concentrated on the dominant regulators of several major pathways.

Finally, hormones can also affect gene expression at posttranscriptional sites. They can induce enzymes that process RNA, RNA binding proteins that stabilize or destabilize mRNA, and activate kinases that can phosphorylate components of the translational machinery.

6   CLINICAL APPLICATIONS

Nowhere have the benefits of molecular biology been more obvious than in the diagnosis and management of endocrine diseases. Hormones are so active that their serum concentrations are in or below the nanomolar range; their receptors rarely exceed more than a few thousand per cell. With such small numbers, conventional protein

purification and characterization techniques are inadequate. Most defects in hormones, receptors, and transducers are identified by nucleotide sequencing after amplification by the polymerase chain reaction. Such techniques have been so successful that a review of this size cannot begin to cover all the endocrinopathies whose molecular bases have been determined; however, all the general categories are mentioned, and a single example of each is given.

6.1   Endocrine Deficiencies

6.1.1   Hormones

A peptide hormone deficiency can most simply arise from a deletion or mutation in the gene for that hormone. The former occurs with the growth hormone (GH) gene in *isolated growth hormone deficiency, type IA,* and the latter has been reported in the insulin gene in some forms of *diabetes mellitus.* Other mutations may affect the processing of hormones. For example, mutations in the cleavage sites for the signal sequence of parathormone (PTH) or between the different chains of insulin are responsible for *familial isolated hypoparathyroidism* and *hyperproinsulinemia,* respectively. Antidiuretic hormone (ADH) concentrates the urine and is synthesized as a polyprotein: the amino terminus forms this hormone, while the carboxy terminus is a transport protein for ADH. Mutations in the transport protein region impair ADH packaging and transport, leading to ADH deficiency (called *familial neurohypophyseal diabetes insipidus*) even though ADH itself is normal. A final processing defect is represented by mutations in the protease responsible for cleaving polyproteins. Corticotropin, formerly known as adrenal corticotropic hormone (ACTH), is a pituitary hormone that stimulates the adrenal glands to produce steroids; it is synthesized as a polyprotein containing corticotropin and several other hormones. The enzyme that cuts out these hormones is postulated to be defective in the condition still called *isolated congenital ACTH deficiency.*

Steroids are not coded for by genes; they are synthesized by enzymes. Therefore, steroid deficiencies are actually deficiencies in steroid synthetic enzymes. Defects in virtually every enzyme have been reported and are grouped under the umbrella of *congenital adrenal hyperplasia.* A similar situation exists for the thyroid hormones; defects have been noted in both thyroid peroxidase and its substrate, thyroglobulin, in several forms of *goiter.*

Finally hormones may be deficient because of regulatory defects. Pit-1 is a transcription factor found exclusively in the anterior pituitary gland and stimulates transcription of the GH gene; mutations in Pit-1 lead to *familial human dwarfism.* Glucokinase represents another regulatory site; this metabolic enzyme phosphorylates glucose as soon as it is transported into the cell. In the insulin-secreting β cells of the pancreas, it doubles as a glucose sensor; elevated glucose levels stimulate the secretion of insulin, which then activates metabolic pathways that store the glucose. Mutations in glucokinase prevent the β cell from detecting glucose, and insulin fails to be secreted; this form of diabetes is called *maturity onset diabetes of the young.*

6.1.2 Receptors

In many endocrinopathies, the hormone is normal in structure and concentration, but the target tissue is unresponsive. The first cellular component the hormone encounters is its receptor, and many defects in this element have been reported. Mutations for almost every nuclear receptor have been described, but those for the androgen receptor are the most numerous, probably because androgen function is not essential for survival. The default body plan for mammals is female; males are created when the testes produce androgens that masculinize the embryo. Partially active androgen receptors leave this process incomplete, producing a *male pseudohermaphrodite.* Totally inactive mutants result in a phenotypic female; this condition is called *testicular feminization.*

Membrane receptors can also be defective. Mutants in the insulin receptor, an RTK, lead to severe forms of diabetes mellitus; those in the GH receptor, a cytokine receptor, result in *Laron dwarfism;* and defects in the ADH receptor, a serpentine receptor, produce *hereditary nephrogenic diabetes insipidus.* Even mutations in ligand-gated ion channels have been identified: the glycine receptor is a chloride channel whose activation hyperpolarizes, and therefore inhibits, neurons. Defects in this receptor leave stimulatory signals unchecked and are responsible for *hereditary hyperekplexia,* a disease characterized by overstimulation of the muscles and an exaggerated startle response.

6.1.3 Transduction

Examples of defective transduction pathways are much rarer. Many different hormones activate the same mediators; therefore, mutations in these second-messenger pathways would have much greater effects than the absence of a single hormone or its receptor. Since most reported defects have been in special isoforms restricted to certain tissues or organs, their effects were limited. For example, the eye possesses several unique isoenzymes of the phosphoinositide pathway; mutations in any of these components would result only in blindness. Another example is found in the kidneys, where ADH stimulates the translocation of a water channel, aquaporin-2, from the cellular interior to the plasmalemma. This change allows the kidney tubule to resorb water and concentrate urine. In one form of nephrogenic diabetes insipidus, this water channel is mutationally inactivated. ADH and its receptor are normal, but the kidney cannot respond to ADH because the water channel is no longer operational.

One of the few examples of a generalized defect is *pseudohypoparathyroidism, type IA,* where G_s is inactive or absent. Although this defect would affect any hormone utilizing cAMP as a second messenger, the actions of PTH appear to be most severely impacted. PTH maintains serum calcium levels by resorbing this mineral from bone and recovering it from the urine. The loss of responsiveness to PTH leads to skeletal abnormalities.

6.2 ENDOCRINE EXCESSES

Perhaps more fascinating than loss-of-function mutations are those that actually lead to overactivity. No mutations that increase the intrinsic activity of a hormone have been reported, but one mutation in angiotensinogen appears to increase its half-life in serum, allowing it to accumulate. Angiotensin is a very potent vasoconstrictor, and this mutation has been associated with some types of *essential hypertension.* Altered catabolism also forms the basis for one form of hyperaldosteronism. Aldosterone is steroid that regulates sodium metabolism; excess aldosterone activity causes sodium to be retained and blood pressure to rise. The aldosterone receptor cannot distinguish between aldosterone and cortisol; since cortisol is present in serum at concentrations much higher than aldosterone, cortisol should flood the aldosterone receptor and keep it continuously stimulated. This does not happen because in target tissues 11β-hydroxysteroid dehydrogenase rapidly metabolizes cortisol before it can reach the receptor. However, a deficiency of this enzyme leads to the overstimulation of the aldosterone receptor by cortisol.

Another form of hyperaldosteronism is caused by a regulatory mutation. The enzymes that synthesize cortisol and aldosterone are very homologous, and their genes are contiguous on the chromosome; ACTH stimulates the former via elements in the 5′ end of the gene. Because the genes are similar, they can accidentally pair up during meiosis, resulting in a nonreciprocal crossing over. In this way, the gene for aldosterone synthetase acquires the regulatory region from the other gene and the latter is lost. Because there is no cortisol, corticotropin is elevated; but instead of stimulating cortisol synthesis it stimulates the aldosterone synthetase. The patient is treated with cortisol, which feeds back to inhibit secretion of corticotropin; the aldosterone synthetase gene, in turn, is shut off. For this reason, the disease is called *glucocorticoid remediable aldosteronism.*

Finally, hormone secretion can be elevated by faulty negative feedback. As noted earlier, PTH restores serum calcium concentrations by resorbing it from bones and recovering it from urine. As calcium levels rise, PTH secretion shuts off. A mutant calcium sensor, however, blocks this negative feedback and results in the persistent secretion of PTH, which in turn elevates calcium. The heterozygous state, called *familial hypocalciuric hypercalcemia,* manifests only mild symptoms; but the homozygous condition, called *neonatal severe hyperparathyroidism,* produces a more grave disease.

Mutations can also render receptors constitutively active in the absence of their natural ligand. Those in RTKs and cytokine receptors often produce tumors, since these receptors usually serve growth factors; they are discussed in this context in Section 6.3. Several activating mutations have been reported in serpentine receptors; they usually occur in either the transmembrane helices or the cytosolic loops. The former are thought to mimic the ligand-occupied state, while the latter improve coupling to G proteins. Luteinizing hormone (LH) is a pituitary hormone that stimulates the gonads to produce sex steroids. Activating mutations in its receptor can stimulate steroidogenesis prematurely, giving rise to *familial male precocious puberty.* Because women require a second hormone for estrogen production, mutations in the LH receptor alone do not lead to any symptoms.

Finally, activating mutations have been identified in a transducer. G_s inactivates itself when it hydrolyzes GTP to GDP (see Section 4.1); mutations that adversely affect its GTPase activity without altering its ability to interact with its effectors would result in the constitutive activation of G_s. Because G_s is integral to so many pathways, this type of mutation would be lethal if it were genomic; however, somatic mutations would produce a genetic mosaic in which only some cells would possess the defective gene. Individuals with this type of G_s mosaicism have *McCune–Albright syndrome,* and their symptoms reflect overstimulation from all the hormones that utilize cAMP as a second messenger. These hormones include those that stimulate pigment cells (café au lait spots), the gonads (precocious puberty), the thyroid gland (hyperthyroidism), the adrenal glands (adrenal hyperplasia), and bone (dysplasias).

6.3 Oncogenesis

Molecular endocrinology has made substantial contributions in the area of cancer diagnosis and treatment.

6.3.1 Oncogenes

Before defining an oncogene, it would be advantageous to define a proto-oncogene. A proto-oncogene is a normal cellular gene that is usually involved in cell growth. If this gene is mutated so that its product is constitutively active, cell growth may proceed unchecked and tumor formation may occur. Such a modified gene is now called an oncogene. Oncogenes may arise spontaneously within the organism's genome, or they may be introduced by oncogenic viruses. In the latter case, it is believed that these viruses originally acquired proto-oncogenes from their hosts and later converted them to oncogenes because the latter conferred some reproductive advantage on the virus. For example, a rapidly dividing cell may have an abundance of replicative machinery that the virus can then commandeer for itself.

Most viruses have small genomes, and it is more efficient for a virus to plug into a preexisting proliferative pathway than to control

all the processes of cell division. Thus most oncogenes are components of the growth factor pathway. Some are growth hormones, like the platelet-derived growth factor and the fibroblast growth factor. Others are receptors for growth factors; the most abundant groups are the RTKs and the cytokine receptors with their associated tyrosine kinases. Many mutations involve deletions of the extracellular domain and other changes that favor aggregation. Remember, the only function of the ligand in these receptors is to induce oligomerization, which then activates the intrinsic or associated tyrosine kinases. Other mutations may incapacitate autoinhibitory sites within the kinases.

Second messengers are also well represented, especially the small G proteins like Ras and various adaptors. In the case of G proteins, mutations often occur in the GAP-binding region; without the ability to interact with GAP, GTP cannot be hydrolyzed and the G proteins remain permanently active. Finally, several oncogenes are transcription factors. Although most of these factors are constitutively active like the receptors and G proteins already noted, a few are dominant inhibitors. The active factors are involved with cell proliferation, while the inhibitory ones are involved with either differentiation or defense responses. Because a cell has limited resources, growth and differentiation often compete with each other; therefore, tumor formation requires both the induction of cell proliferation and the inhibition of differentiation. A mutated thyroid hormone receptor and a mosaic retinoic acid receptor are examples of factors in the latter group.

6.3.2 Tumor Resistance to Therapy

When hormone target tissues become malignant, they often retain their endocrine sensitivity. Such tumors can be initially treated by hormone therapy: breast cancer may response to antiestrogens; prostate cancer, to antiandrogens; and leukemias and lymphomas, to glucocorticoids. Hormone therapy not only is effective, it also is among the most easily tolerated treatments, because of its relative lack of undesirable side effects. Unfortunately, the effectiveness of this therapy is too often short-lived as the tumor eventually develops resistance. Quantitative and structural studies of nuclear receptors can help to determine which tumors will initially respond to hormone therapy and can also explain why many of them later become resistant.

Many resistant tumors have been found to have inactive receptors. The simplest case is seen in leukemias and lymphomas that undergo apoptosis in response to glucocorticoids. Resistant tumors either lack glucocorticoid receptors or possess defective receptors and, therefore, can no longer respond to these steroids. It is felt that these deletions and mutations arise spontaneously; those that inactivate the receptor offer a selective advantage by allowing the cells that have them to evade the cytotoxic effects of glucocorticoids. Such cells survive and reproduce to create a cancer resistant to these steroids.

On the other hand, breast and prostate cancers are dependent on sex steroids for survival; as such, they are treated by hormone ablation or antagonists. Therefore, simple deletion or inactivation of the receptor would prove detrimental to the survival of the cancer. Two types of mutation are observed in these cases. First, some mutations render the receptor constitutively active regardless of the presence or absence of hormone agonists or antagonists. Second, mutations in the ligand-binding domain may reduce the binding specificity of the receptor, to allow other steroids, or even steroid antagonists, to activate the receptor. In either instance, the tumors have either re-duced or eliminated their hormone dependence.

7   PERSPECTIVES

Single cells require careful regulation of metabolism, growth, and reproduction to survive in a competitive and dangerous environment. The development of multicellularity introduced the additional problem of intercellular communications to coordinate these processes. Nature's frugal solution was simply to couple external chemical signals to these preexisting intracellular regulators, which then became second messengers. Therefore, knowledge about molecular endocrinology is knowledge about the very essence of how cells control all their internal functions. In addition, this information provides valuable insights into ourselves via the various disease states that can now be explained by molecular biology. Molecular endocrinology thus affords a critical understanding of many basic physiological processes of life.

See also HETEROTRIMERIC G PROTEINS, SIGNAL TRANSDUCTION BY; HUMAN GENETIC PREDISPOSITION TO DISEASE; ONCOGENES; RECEPTOR BIOCHEMISTRY; STEROID HORMONES AND RECEPTORS.

Bibliography

Blumer, K. J., and Johnson, G. L. (1994) Diversity in function and regulation of MAP kinase pathways. *Trends Biochem. Sci.* 19:236–240.

Bolander, F. F. (1994) *Molecular Endocrinology,* 2nd ed. Academic Press, San Diego, CA.

Cohen, P., and Foulkes, J. G., Eds. (1991) *The Hormonal Control of Gene Transcription.* Elsevier, Amsterdam.

Exton, J. H. (1994) Phosphoinositide phospholipases and G proteins in hormone action. *Annu. Rev. Physiol.* 56:349–369.

McPhaul, M. J., Marcelli, M., Zoppi, S., Griffin, J. E., and Wilson, J. D. (1993) The spectrum of mutations in the androgen receptor gene that causes androgen resistance. *J. Clin. Endocrinol. Metab.* 76:17–23.

Moudgil, V. K., Ed. (1994) *Steroid Hormone Receptors: Basic and Clinical Aspects.* Birkhäuser, Boston.

Petersen, O. H., Petersen, C.C.H., and Kasai, H. (1994) Calcium and hormone action. *Annu. Rev. Physiol.* 56:297–319.

Raymond, J. R. (1994) Hereditary and acquired defects in signaling through the hormone–receptor–G protein complex. *Am. J. Physiol.* 266: F163–F174.

Weintraub, B. D., Ed. (1994) *Molecular Endocrinology: Basic Concepts and Clinical Correlations.* Raven Press, New York.

Engineering, Bioprocess: *see* Bioprocess Engineering.

ENVIRONMENTAL STRESS, GENOMIC RESPONSES TO

John G. Scandalios

Key Words

Circadian Rhythm   A biological rhythm with a period of about 24 hours.

Gene   A segment of the genome (DNA) that codes for a functional product.

Genome   The totality of a cell's genetic information, including genes and other DNA sequences.

Genomic Fluidity   The capacity of the genome to reorganize rapidly in response to a given stimulus or signal.

Oxidative Stress (Oxystress)   An elevation in the steady state concentration of reactive oxygen species. Occurs when the balance between the mechanisms triggering oxidative conditions and cellular antioxidant defenses is impaired.

Promoter   A sequence of nucleotides on DNA that is required for the initiation of transcription by RNA polymerase.

Reactive Oxygen Species (including "Free Radicals")   Toxic by-products of reduced oxygen. These include hydrogen peroxide, superoxide anion, hydroxyl radical, and singlet oxygen, which can cause damage by initiating oxidation of various macromolecules.

Signal Transduction   The mechanism by which cells or organisms perceive a signal and transmit it via the proper pathway(s) to elicit a response.

Transposable Elements (also, Mobile Elements; "Jumping Genes")   DNA segments that can move from one place to another in a genome.

All living organisms are affected by the environment in which they exist. Differences among individual organisms in response to environmental stresses are common, whether the stressing factors be infectious agents, natural environmental variations, environmental chemicals, or any other natural or anthropogenic environmental variables. During their evolution, organisms have evolved a variety of ways of adapting to environmental changes. However, the underlying mechanisms by which cells or organisms perceive environmental adversity and mobilize their defenses to it are far from understood. Such an understanding is essential in any future attempts to engineer organisms for greater tolerance or resistance to more frequent and rapid environmental changes, be they due to natural or anthropogenic causes. As a consequence of recent developments in molecular biosciences and dissection of the human and other genomes, a deeper appreciation of the mechanisms by which genes may perceive environmental signals and start a cascade of biochemical events to effect a response to a given signal is now emerging.

1   INTRODUCTION

Every living organism is affected by its environment. The environment, whether internal or external to the organism, is continually changing, and the organism must adapt if it is to survive. However, an organism apparently well adapted to the environment at any one time may be poorly adapted only a short time later if it cannot modify its physiology or behavior. Organisms that can adjust to changes in the environment are likely to exhibit a greater degree of adaptiveness than those that cannot. Thus, a dominant theme in modern biology is gene expression: it is desired to determine how these units of heredity are expressed in a selective manner in response to an external or internal signal or stimulus.

Environmental changes, irrespective of source, cause a variety of "stresses" or "shocks" that a cell must face repeatedly, and to which its genome must respond in a programmed manner for the organism to survive. Examples are response to light, oxidative stress, pathogenicity, wounding, anaerobiosis, and thermal shock, and the "SOS" response in microorganisms. For cells of any organism to respond to external cues, they must be able to perceive these cues or signals and transduce such perceptions into the appropriate response. Some sensing mechanism(s) must be present to alert the cell to imminent danger, and to trigger the orderly sequence of events that will mitigate this danger. In addition, there are genomic responses to unanticipated, unprogrammed challenges for which the genome is unprepared, but to which it responds in discernible though initially unforeseen and unpredictable ways.

Many, though not all, signals are perceived at the cell surface by plasma membrane receptors. Activation of such receptors by mechanisms such as ligand binding may lead to alterations in other cellular components, ultimately resulting in alterations in cell shape, ion conductivity, gene activity, and other cellular functions. Identification and isolation of mutants that are unable to respond, or that respond abnormally to a particular signal, may provide ways to decipher the mechanisms by which a particular signal is transduced into a given response.

Long before humans began manipulating and altering their environment, organisms from the simplest to the most complex began evolving methods to cope with stressful stimuli. Consequently, most living cells possess an amazing capacity to cope with a wide diversity of environmental challenges, including natural and synthetic toxins, extreme temperatures, high metal levels, and radiation. Many studies in the past have demonstrated clear "cause-and-effect" relationships upon exposure of a given organism or cell to a particular environmental factor or stressor. But only recently have certain environmental insults been shown to elicit specific genomic responses. At present, little is known of the underlying molecular mechanisms by which the genome perceives environmental signals and mobilizes the organism to respond. Such information is not only interesting in and of itself but is also essential in any future attempts to engineer organisms for increased tolerance to environmental adversity.

The recent dramatic advances in molecular biology have made it possible to investigate the underlying mechanisms utilized by organisms to cope with environmental stresses. Investigations of genomic responses to challenge are beginning to shed some light on unique DNA sequences capable of perceiving stress signals, thus allowing the cell or organism to mobilize its defenses. The general picture emerging from recent studies involves the sensing of a signal and the transduction of the signal to the transcription apparatus to catalyze transcription initiation. The steps involved in such a process may be summarized as follows:

- A signal, normally the intracellular or extracellular concentration of a small molecule, is perceived by a sensor.
- The signal is then transmitted to the regulatory "activator" protein.
- The signal transduction changes the conformation of the activator protein.
- The altered activator protein binds to a specific DNA site.
- The DNA–protein interaction catalyzes the binding or activity of RNA polymerase to facilitate transcription initiation.

2   GENE RESPONSES

Terminally differentiated cells express an array of genes required for their stable functioning and precise metabolic roles. A genome can respond in a rapid and specific manner by selectively decreasing or increasing the expression of specific genes. Genes whose expression is increased during times of stress presumably are critical to the organism's survival under adverse conditions. Examination and study of such "stress-responsive" genes has implications for human health and well-being, for agricultural productivity, and for furthering basic biological knowledge. In addition to aiding the organism under stress, genomes that are modified by stress can be utilized to study the molecular events that occur during periods of increased or decreased gene expression. The mechanisms by which an organism recognizes a signal to alter gene expression and responds to fill that need are important physiologically and render possible the examination of gene regulation under various environmental regimens.

The mechanisms of induction of stress response genes are similar among various organisms examined. Similarities in stress-induced changes in gene expression have been observed for a variety of stressors. Some that have been studied in some detail are thermal shock, pathogenic infections, anaerobiosis, photostress, oxystress, water stress, and heavy metals. In all cases, specific changes in transcript and/or protein expression have been observed in various organisms subjected to such challenges. Some of these are discussed in the sections that follow.

3   THERMAL STRESS

3.1   HEAT SHOCK

Organisms have evolved a variety of ways to adapt to fluctuations in temperature. The most readily discernible response to thermal stress in most organisms is the "heat shock" response. Cells from virtually every kind of organism react to hyperthermic shock by activating a small number of genes, thus inducing the synthesis of a set of proteins, the "heat shock proteins (hsps)," which protect the cell from thermal damage. These genes were initially recognized in the fruit fly, *Drosophila melanogaster,* as "puffs" in polytene chromosomes arising shortly after the embryos had been subjected to a heat shock—shifting them from their normal growth temperature (25°C) to an increased temperature (37°C). Subsequently, the protein products of these genes were identified and characterized. This gene activation is rapid—HSPs appear within a few minutes after heat shock initiation—and it is reversible. A transition from hardly detectable levels of transcription at the normal temperature to extremely high transcription rates occurs during heat shock, leading to rapid accumulation of high levels of HSPs.

3.2   HEAT SHOCK GENES AND PROTEINS

Three major types of HSP are found in most organisms: (1) the large HSPs ranging in subunit molecular weights from 80,000 to 100,000, (2) the intermediate size group (65,000–75,000); and (3) the small HSP group (15,000–30,000). The structures of these major HSPs are strongly conserved among animals, plants, yeast, and bacteria. The activation mechanism, with its components, is virtually identical among higher eukaryotes and is similar to the mechanisms found in lower eukaryotes; prokaryotes are less similar in this respect. For example, the maize (corn) *hsp70* gene has a 75% sequence homology to the *Drosophila hsp70* gene. Detailed promoter analysis of the *hsp70* gene from different species also shows commonality in the presence of a short DNA sequence upstream of the TATA box that is required for heat inducibility. A palindromic consensus regulatory sequence (CT-GAA—TTC-AG) was shown to be sufficient for conferring heat inducibility on a heterologous gene. This sequence is referred to as the "heat shock regulatory element" (HSE) and can be found within the first 400 base pairs upstream of every *hsp* gene from every higher eukaryote that has been sequenced. In addition, there are several secondary heat shock consensus elements located further upstream.

3.3   ALTERNATIVE INDUCERS OF HSPs

Many of the molecular events accompanying heat shock are also apparent in cells that have been subjected to other kinds of stress. Many of the stressors seem to interfere with oxidative phosphorylation, or energy production in general. Such agents as metal poisons, sulfhydryl oxidants, and amino acid analogues induce proteins that are identical to the HSPs. In plants, the synthesis of certain HSPs can readily be induced by factors in addition to heat shock, including osmotic stress, arsenite, anaerobiosis, high concentrations of growth factors such as abscisic acid, auxin, and ethylene, and high salt. In addition, some of the *hsp* genes are activated and expressed during normal development. For example, low molecular mass HSPs, particularly HSP70, are transiently expressed during the normal cell cycle. Such findings implicate these proteins in basic metabolic activities of the cell.

3.4   FUNCTIONS OF HSPs

Many studies have clearly demonstrated that a preshock treatment can render a biological system more resistant to a subsequent thermal stress (thermotolerance) and that this protective effect may be transient. Thermotolerance appears to be important for survival under stress conditions. For example, thermotolerant maize (corn) plants are able to survive the lethal temperature, and to outgrow the

control plants when returned to normal temperatures. It has also been demonstrated that HSPs accumulate in field grown, heat-stressed plants, suggesting that these proteins may be critical for plants that must cope with natural stress adversity. Other functions attributed to HSPs include a role (HSP70) in translocation of proteins into various eukaryotic organelles, and their participation in the folding and assembly of polypeptides. It is apparent that the *hsp* genes have pleiotropic effects and that other roles, in addition to conferring thermotolerance, will be uncovered as further studies are executed.

3.5   GENERALIZED "STRESS RESPONSE"

The heat shock response is an ideal paradigm toward comprehending how cells recognize and respond to acute and chronic exposures to environmental and physiological stresses. The heat shock genomic response has contributed significantly toward understanding the molecular and cellular mechanisms of adaptation, ranging from the regulation of heat shock gene expression to the function of stress proteins.

Activation and expression of stress genes resulting in increased synthesis of a family of stress-induced (heat shock) proteins ensure survival under stressful conditions, which, if left unchecked, would lead to irreversible cell damage and ultimately death. In addition to a protective role, some of the heat shock proteins serve as molecular chaperones, with essential roles in protein biosynthesis, and in the transport, translocation, and folding of proteins in the cell.

The genes encoding HSPs are highly conserved, with representatives from distant prokaryotic and eukaryotic species having at least 50% identity. Most studies have concentrated on the eukaryotic 70 kDa protein heat shock genes, which are ancestrally related to *Escherichia coli dnaK* and encode a large multigene family of proteins. Expression of the heat shock genes is inducible by a diversity of chemical and physiological signals. As a consequence, the more restrictive term "heat shock" has given way to the now generally accepted "*stress response*" genes and proteins. For example, conditions known to induce *hsp70* gene expression include (1) *environmental* stresses such as heat shock, heavy metals, and amino acid analogues, (2) *pathological* and/or *varied physiological states* such as aging, injury, chemotherapy, and ischemia, and (3) *nonstress conditions* such as normal cell growth, development, and differentiation. Given the diversity of inducers of the "stress response," the remaining underlying question is: How is the information transduced from inducer to the transcriptional apparatus? With the recent isolation and characterization of the promoter region of the human *hsp70* gene, which has been shown to contain multiple heat shock elements that confer stress inducibility, this question can now be tackled.

3.6   COLD ACCLIMATION

Much less is currently known about cold acclimation than is the case with heat adaptation. What is presently known is based primarily on recent work with plants. Perhaps the most dramatic manifestation of cold acclimation (cold hardiness) is the increased freezing tolerance that occurs in many plant species. Extreme examples of cold-adapted perennial species include the dogwood and birch. Species not acclimated to cold are killed by temperatures of about −10°C, whereas some cold-hardy species can survive experimental temperatures of −196°C.

Some biochemical alterations found to accompany cold acclimation include changes in lipids, increases in soluble protein and sugars, expression of new isozymes, and changes in mRNA populations. It was demonstrated as early as 1912, by wheat breeding experiments, that frost hardiness has a complex quantitative genetic basis. More direct evidence has recently been obtained from molecular investigations leading to the isolation and characterization of specific "cold-regulated" *(cor)* genes from various plant species. Although correlations between *cor* gene expression and freezing tolerance have been found, details of the exact regulation, expression, and role of these genes await further resolution.

The more thoroughly investigated heat shock response is quite different from the cold acclimation response. Unlike heat shock, the changes in *cor* gene expression that accompany cold acclimation are relatively mild and not transient, and the genes expressed at normal temperatures continue to be expressed at the low temperatures; expression of *hsp* genes, on the other hand, is generally accompanied by suppression of preexisting mRNAs. Thus, cold acclimation and heat shock appear to be distinct responses.

4   OXIDATIVE STRESS

Oxygen is essential for life on earth. In its ground state (its normal configuration, O_2) oxygen is relatively unreactive. However, during normal metabolism, and as a consequence of various environmental perturbations and pollutants (e.g., radiations, drought, air pollutants, cigarette smoke, temperature stress, herbicides), oxygen (O_2) gives rise to various highly toxic and lethal intermediates. These intermediates, referred to as active oxygen species and free radicals, include the superoxide radical $O_2^{\cdot-}$), hydrogen peroxide (H_2O_2),

and the hydroxyl radical ($^{\cdot}OH$). These and the physiologically energized form of dioxygen, singlet oxygen ($^1O_2$), are the biologically most important active oxygen species. All theses are extremely reactive and cytotoxic to all organisms. For example, $^{\cdot\cdot}OH$, one of the most potent oxidizing agents known, reacts with most macromolecules (DNA, proteins, lipids, etc.) to cause severe cellular damage leading to physiological dysfunction and cell death (Figure 1). Some of the biological consequences of oxidative damage include peroxidation of membrane lipids, loss of organelle function, mutations, enzyme inactivation, reduced metabolic efficiency, and reduced carbon fixation, leading to impaired photosynthetic capacity in plants. Free radicals and derivatives have been implicated as causative agents in the aging process, and in many human diseases, including cancer, emphysema, and immunologic impairments. Thus, the effective and rapid elimination of active oxygen species is essential to the proper functioning and survival of all living organisms.

4.1   PROTECTIVE ANTIOXIDANT DEFENSES

Nature has thus presented us with the "oxygen paradox." For life to be sustained, oxygen is required; yet in its reduced state this sustainer of life becomes a deterrent to life. As a consequence of this paradox, organisms evolved antioxidant defense mechanisms to protect themselves. Such defenses include nonenzymatic as well as enzymatic mechanisms. Among the former are such substances as β-carotene, vitamins C and E, flavonoids, and hydroxyquinones. Enzymatic defenses include enzymes capable of removing, neutralizing, or scavenging free radicals and oxyintermediates. Without such defenses, plants could not efficiently convert solar into

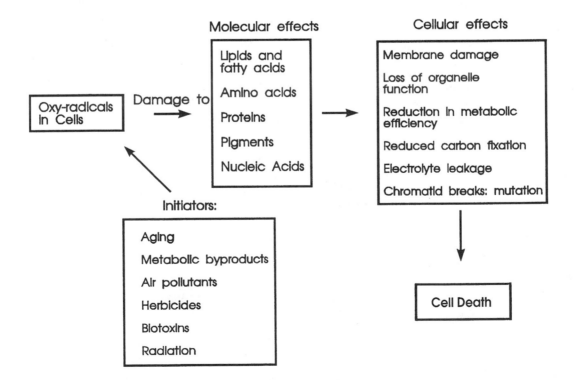

Figure 1.   Scheme showing some initiators (inducers) of oxyradicals and the biological consequences leading to a variety of physiological dysfunctions and cell death.

chemical energy, and life on earth, as we know it, would not be possible.

Ascorbate peroxidase and glutathione reductase, believed to scavenge hydrogen peroxide in chloroplast and mitochondria, respectively, are examples of enzymatic antioxidant defenses; others include catalase (CAT) and peroxidases, which efficiently remove hydrogen peroxide from cells, and superoxide dismutase (SOD), which scavenges the superoxide anion. The CAT and SOD enzymes are the most efficient antioxidant enzymes: their combined action converts the superoxide radical and hydrogen peroxide to water and molecular oxygen, thus abating the formation of the most toxic and highly reactive hydroxyl radical, averting cellular damage.

Although many defenses to cope with oxidative stress exist in all aerobic organisms, it is not yet clearly understood how these organisms mobilize their defenses to respond at the appropriate time. Increases in oxystress often lead to correlative increases in some antioxidant defenses. However, little is currently understood of how the genome perceives oxidative insult and mobilizes a response to it. Such information is essential in any future efforts to engineer organisms for increased tolerance to environmental oxidative stress. To understand the underlying mechanisms, a great deal of effort has been expended in recent years to elucidate the responsive antioxidant defense genes. State-of-the-art molecular techniques are being used to resolve gene structure, regulation, and expression, giving some insight as to how such genes act to mobilize cellular defenses. Some recent findings are discussed in Section 4.2 and 4.3.

4.2   ENZYMATIC ANTIOXIDANT DEFENSES

The genes encoding various catalases and superoxide dismutases have been isolated and characterized, and found to be evolutionarily conserved, from numerous organisms. These genes and their products have been effectively utilized in some novel experiments that have clearly demonstrated that these genes indeed play major protective roles against oxidative damage. For example, there is a yeast mutant that lacks the gene responsible for producing the manganese superoxide dismutase (MnSOD), which translocates to its mitochondria where it normally functions. The mutation is lethal, since this yeast is unable to grow in an oxygen environment. In recent experiments, the *MnSod (Sod3)* gene of maize was isolated and successfully transferred to the MnSOD-deficient yeast cells. The maize gene was properly transcribed and translated, and its protein product (the MnSOD maize protein) was properly imported and processed into the yeast's mitochondria. Most significantly, the corn MnSOD protein functioned correctly, resulting in the rescue of the transgenic yeast cells that now grew normally in an oxygen environment and were able to cope with imposed oxidative stress. Whether such approaches will prove successful in engineering various other organisms for greater tolerance to oxystress remains to be seen. However, the prospects appear fruitful. Isolation of such genes provides the opportunity to determine intron–exon boundaries, start and stop sites of transcription, and *cis*-regulatory regions, and to identify sequences responsive to oxidative stress signals that might indicate how these genes are regulated to respond and protect cells against oxystress.

4.3   ANTIOXIDANT GENE RESPONSES

Oxidative stress in bacteria has been shown to activate the transcriptional regulator *oxyR,* which in turn induces genes whose prod-

ucts protect cells from oxidative damage. The OxyR protein is directly activated by the metabolic stimulus, an oxidant, to become the transcriptional activator. Thus, OxyR protein is both the sensor of oxidative stress and the mediator of enhanced transcription of genes whose products are components of the protective response. For example, H_2O_2 added to bacterial growth medium interacts with OxyR to change its oxidation state, causing a conformational change that affects the way the protein interacts with its target promoters. The OxyR protein is encoded in a single dominant regulatory gene *(oxyR).* The several antioxidant defense structural genes scattered around the genomes of *E. coli* and *Salmonella typhimurium* and controlled by *oxyR* constitute a "regulon." Although the exact molecular mechanism for the activation of OxyR by H_2O_2 is not clear, it is known that oxidized OxyR, but not reduced OxyR, activates transcription, and that both forms can bind to DNA.

Regulons parallel to the bacterial "OxyR regulon" have not been identified in eukaryotic cells. However, as the structure of eukaryotic antioxidant defense genes is being unraveled, other signal recognition factors are being encountered. For example, an 11 base pair motif (5′-puGTGACNNNGC-3′), called the antioxidant response element (ARE), has been found in the promoter region of several eukaryotic antioxidant defense genes, including the maize *Cat1* gene. There exists convincing evidence from rat studies that the ARE might represent a *cis*-acting element that activates genes that protect eukaryotic cells against oxidative stress.

It is becoming clear that as more genes responsive to environmental stimuli are isolated and characterized, motifs or elements are identified which may be responsible for perceiving the stimuli and initiating the appropriate defensive response. Thus, a fruitful research direction involves the identification of both *cis*-acting elements and *trans*-acting substances (like the OxyR protein) relative to gene systems that respond to environmental stimuli. It is also interesting that genes responsive to environmental stimuli are often found to be highly polymorphic and to have additional functions such as roles in development. It is likely that polymorphism in genes and systems that participate in stress defense strategies is essential for the organism to survive varied and continually changing environmental stresses.

5   RAPID GENOMIC RESPONSES

There are programmed responses to threats that are initiated within the genome itself, which can lead to new and irreversible genomic modifications. Thus, in many organisms, the genome facing an adversity for which it is unprepared may reorganize itself to ensure the organism's survival. For example, cells are able to sense the presence in their nuclei of ruptured ends of chromosomes and then to activate mechanisms that will bring together and unite these ends, one with another. This is a most revealing example of the sensitivity of cells to what is occurring within them.

Although the capacity of the genome to reorganize and respond rapidly ("genomic fluidity") has been considered for a long period, it was not fully appreciated until the pioneering work of Barbara McClintock, demonstrating that discrete genetic loci could transpose in the genome of maize, was fully accepted and shown to be universal among all organisms studied.

McClintock demonstrated that transposable elements regulate gene expression following their insertion into a given locus (she used the term "controlling elements"). Transposable elements provide a regulated disruption of the chromosomes in cells in which they are active. Activation of transposable elements is a recognizable consequence of the cell's response to trauma. The mobility of these elements allows them to enter different loci and to take control of the action of the gene wherever one may enter. In addition to modifying gene action, transposable elements can restructure the genome at various levels from changes involving a few nucleotides to gross modifications of large chromosome segments.

6 ADDITIONAL GENOMIC RESPONSES TO STRESS

6.1 PLANT GALLS

Many cases are known in nature of one organism being forced to produce a wholly new structure for the benefit of another. Plant galls provide many such examples. The exact structural organization of a given gall imparts a uniqueness that begins with an initial stimulus, characteristic of each invasive species. The galls on legume roots associated with nitrogen-fixing bacteria provide a clear example of reprogramming of a plant (legume) genome by a stimulus received from foreign organisms (bacteria). In other examples, a single plant may have on its leaves several distinctly different galls, each catering to a different insect species. The stimulus for placing the insect egg into the leaf initiates reprogramming of the plant's genome, forcing it to make a unique structure adapted to the needs of the developing insect. Such genomic reprogrammings by a variety of organisms, including bacteria, fungi, and insects, are not a requisite response of the plant host genome during its normal developmental cycle.

6.2 TISSUE CULTURE INDUCED CHANGES

An interesting example of how a genome may modify itself when confronted with unusual conditions is provided by culturing cells, particularly plant cells, in artificial conditions. When cells are removed from their normal locations and placed in tissue culture, a variety of changes are observed. It is common to observe novel phenotypes emerge in many species when plants have been regenerated from protoplast or tissue cultures. Much of the variation observed in cultures is a result of genomic change ("somaclonal variation"). The passage from cell isolation, callus formation, and the ensuing production of whole plants inflicts on the cells a succession of traumatic stresses. These stresses may result in the abnormal reprogramming of the genome in a way that it does not follow the same orderly sequence that occurs in natural conditions, causing a wide range of unexpected altered phenotypes, which may or may not be heritable. Somaclonal variation provides fertile ground for selection, by breeders, of useful traits. It thus appears that tissue culture represents a form of stress on cultured cells, which in turn respond by decidedly restructuring and resetting the genome.

6.3 PHOTORESPONSES

Light, the most important environmental stimulus to which plants react, plays an indispensable role throughout the life of higher plants. Light provides the energy for photosynthesis, so essential for the existence of life on earth, and it has profound effects on gene expression by serving as a trigger and modulator of complex developmental and regulatory mechanisms. Fluctuations in light quality and quantity lead to alterations in the activity of specific genes, which culminate in a variety of developmental responses, ranging from seed germination and differentiation to flowering. Light responses are mediated by at least four photoreceptors: protochlorophylide, phytochrome, blue light receptors, and an ultraviolet B receptor. These photoreceptors, probably in conjunction with different proteins, give signals to regulatory sequences associated with light response genes. In fact, many genes are now known whose amount of mRNA transcribed changes directly in response to light. Several such genes have been shown to have a consensus "light response element" (LRE) within their promoter regions. The transacting factors that interact with such elements to effect the appropriate gene response to light are yet to be identified.

Although essential, light imposes considerable stress (see Section 4). Consequently, plants and other organisms have incorporated significant light response signals in their developmental pathways to cope with a photodependent lifestyle.

The exquisite precision by which light governs growth, development, and aging in plants is just now beginning to be understood and appreciated. Unraveling the underlying mechanisms of the light-induced responses, and ascertaining how they are modulated or modified by other factors, will be a challenging task that will provide a deeper understanding of the regulatory networks within cells. It will also provide a better understanding of how cells are able to selectively derive the beneficial aspects of the photoenvironment and to minimize its detrimental features.

6.4 CIRCADIAN RHYTHMS

Eukaryotic, and some prokaryotic, organisms exhibit daily rhythms in a wide variety of biochemical, physiological, and behavioral characteristics. Such rhythms are controlled by endogenous timing mechanisms called "circadian clocks" that serve to couple the biological rhythms with environmental day/night cycles. For the biological clock, the light/dark cycle serves as the primary source of information concerning environmental time. Thus, circadian rhythms are endogenous 24-hour oscillations in various cellular or organismal processes. The spectrum of biological processes controlled by these clocks ranges from the daily sleep/wake cycle in humans to rhythmicity in bioluminescence and cell cycle control in marine organisms, and photoresponses in plants.

All circadian rhythms have three elements in common: (1) input pathways that convey environmental information to a circadian "pacemaker" for entrainment, (2) a circadian pacemaker that generates the oscillation, and (3) input/output pathways by which the pacemaker regulates the various output rhythms. Recent analyses of the genetic basis of circadian rhythms are beginning to provide information about how clocks might function at the molecular level. The first two "clock genes" to be cloned, the *period (per)* gene from the fruit fly *Drosophila melanogaster* and the *frequency (frq)* gene from the fungus *Neurospora crassa,* share some common characteristics genetically and in the respective proteins each encodes.

In several species, single gene mutations have been found that alter various clock properties. In plants, the expression of a number of genes encoding components of the photosynthetic apparatus (e.g., *Cab* encoding part of the light-harvesting chlorophyll *a/b* binding protein complex of chloroplasts) has been shown to vary under circadian control. The same is true of some genes unrelated to photosynthesis (e.g., the maize *Cat3* catalase gene: Figure 2). The mechanism of the circadian clock in a variety of systems examined appears to involve both transcriptional and translational processes. In a number of systems, clock control seems to be mediated by *cis-*

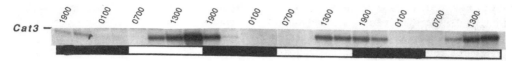

Figure 2. Circadian regulation of the *Cat3* catalase gene in maize: S_1 nuclease protection analysis of maize RNA using a *Cat3* gene-specific probe. The steady state *Cat3* transcript level varies within a 24-hour period. The diurnal pattern of *Cat3* gene expression observed was determined to be the result of a circadian-regulated transcriptional repressor. The circadian rhythm of *Cat3* is independent of phytochrome. Numbers at the top indicate the hour on a 24-hour clock. Seedlings were grown under a 24-hour photoperiod (12 h light/12 h dark), and RNA was isolated from leaves at 3-hour intervals. Bars indicate dark/light cycle periods.

acting, clock-responsive regulatory elements ("clock boxes") adjacent to or within the clock genes. Such "circadian clock responsive elements" (CCREs) have been mapped to small 5′ upstream sequences in two specific cases: the wheat *Cab1* gene and the *Cab1* and *Cab2* genes of *Arabidopsis*. The CCREs have been shown to be distinct from the light responsive elements described earlier. Thus, the study of such genes, in a variety of organisms, might provide an entry point to the mechanisms that underlie circadian rhythmicity.

7   CONCLUSIONS AND SPECULATIONS

Genomic flexibility is an extraordinary adaptability of organisms to their environment. Although the capacity of the genome to reorganize in the face of adversity had been considered since Mendel's laws of heredity were formulated, it was the seminal contributions of Barbara McClintock that set the stage for our current understanding of genomic fluidity.

In the course of evolution, organisms that had the capacity to adapt to new environments survived. This capacity was due to inclusion in the biological repertoire of each organism or species that survived of the means to cope with environmental adversity. Where multiple environments existed, organisms possessing motility (e.g., animals) could exchange one environment for another; those with a stationary lifestyle (e.g., plants) had to rely on alternate defenses. However, in adverse environments (natural or anthropogenic) from which organisms cannot escape, they must rely on diverse and unique sets of responses encoded within, or resulting from the flexibility of their genome. Recent work, as discussed here, has demonstrated that environmental stress itself can instigate genome modifications by mobilizing cell mechanisms that can restructure genomes in various ways. Such genomic reorganization allows the organism to cope with stress and, in the long term, may provide the bases for evolution of new species.

As genomes of various organisms are being investigated with state-of-the-art molecular techniques, more knowledge is being gained about structural components involved in the response of specific genes to environmental insult. However, we know virtually nothing about how the cell perceives danger and initiates the truly remarkable observed responses to it. When such information is available, and a better understanding of the defense response is attained, the engineering of organisms for resistance or sustained tolerance to adverse environmental stresses will be possible.

It is now assumed, based on substantive data, that a signal transduction pathway contains elements that enable a signal to be transmitted within and between cells and to be translated into an appropriate response. As discussed, cells can respond to a variety of environmental, physical, and chemical stimuli using a diverse range of transduction and response mechanisms. The essential features of a signal transduction pathway comprise a *receptor* (recognition element) capable of detecting a stimulus, *second messengers* (transmission elements) such as calcium or phosphorylation cascades, and *response elements* (e.g., gene transcription). Such signaling networks are now amenable to study and dissection by biochemical and genetic approaches that may elucidate the underlying mechanisms and lead to the identification of the molecules responsible. A thorough understanding of how organisms perceive, respond, and adapt to changing environmental and developmental stimuli may not be too distant.

See also FREE RADICALS IN BIOCHEMISTRY AND MEDICINE; GENETIC DIVERSITY IN MICROORGANISMS.

Bibliography

Bienz, M. (1985) Transient and developmental activation of heat-shock genes. *Trends Biochem. Sci.* 10:157–161.

Dunlap, J. C. (1990) Closely watched clocks: Molecular analysis of circadian rhythms in *Neurospora* and *Drosophila*. *Trends Genet.* 6:159–165.

McClintock, B. (1984) The significance of responses of the genome to challenge. *Science*, 226:792–801.

Scandalios, J. G., Ed. (1987) *Molecular Genetics of Development.* Academic Press, San Diego, CA.

———, Ed. (1990) *Genomic Responses to Environmental Stress.* Academic Press, San Diego, CA.

———, Ed. (1992) *Molecular Biology of Free Radical Scavenging Systems.* Cold Spring Harbor Laboratory Press, Cold Spring Harbor, NY.

ENZYMES, ENERGETICS AND REGULATION OF BIOLOGICAL CATALYSIS BY

Stephen W. Raso, Jon A. Christopher, Miriam M. Ziegler, and Thomas O. Baldwin

Key Words

Activation Energy ($\Delta G^{\ddagger}$)   The energy barrier that must be overcome for a chemical reaction to occur; formally, the free energy needed to promote reactants from the ground state to the transition state.

Active Site   The region of an enzyme where the substrate binds and chemistry occurs.

Allostery   Means of enzyme regulation in which a regulatory molecule acts on the enzyme by binding to it at a site distant from the active site; usually an induced conformational change is responsible for the change in enzymatic activity.

Coenzyme   Any of a number of organic or organometallic molecules required by certain enzymes for their catalytic function.

Cofactor   An inorganic ion or a coenzyme required by certain enzymes for their catalytic function.

Equilibrium   The state of a chemical reaction in which the forward and backward reactions are equal in rate, so that there is no net change in product or reactant concentration.

Equilibrium Constant (K_{eq})   The equilibrium ratio of product concentrations to reactant concentrations for a given reaction under a given set of conditions (such as temperature); K_{eq} is independent of *initial* concentrations of products and reactants.

Gibbs Free Energy (ΔG)   The measure of energy available to do work in a chemical system.

Substrate   The specific compound on which an enzyme acts; the reactant.

Subunit   A single polypeptide chain in a protein.

Transition State   The short-lived, unstable (high energy) species of a molecule (or molecules) that forms transiently during a chemical reaction, in which chemical bonds are partially formed and/or broken. Transition states cannot be isolated.

V_{max}   The theoretical maximum velocity (or rate) of an enzymatic reaction, which is asymptotically approached as the initial substrate concentration is increased.

Enzymes are biological catalysts (usually proteins) that help carry out nearly all chemical reactions in living systems; indeed, life as we know it would not be possible without them. Enzymes are true catalysts, which means that they increase the rate of a reaction but do not participate in it. Like all catalysts, enzymes are returned to their original form at the end of the reaction cycle. Enzymes are special as catalysts because their specificity and rate enhancements are unparalleled by any other catalysts, whether man-made or naturally occurring. Enzymatic rate enhancements can range from 10^3 to 10^{16}

times the rate of uncatalyzed reactions. Enzymes thus allow otherwise exceedingly slow chemical reactions to occur on a time scale that is biologically meaningful. Like other catalysts, enzymes lower the activation energy of a reaction, making the product kinetically accessible.

The science of enzymology has made remarkable strides in the last 50 years, largely because of advances in protein purification methods, recombinant DNA techniques, X-ray crystallography, and various types of spectroscopy, as well as the ready availability of isotopes for kinetic and structural studies. Our understanding of enzymatic mechanisms of catalysis is increasing every day. Continued study of enzymes may lead to the development of supercatalysts for use in industry, in the bioremediation of toxic waste, and as therapeutic agents for the treatment of disease. Many diseases are the result of a missing or defective enzyme; research in this area is bringing us closer to the ability to repair or create substitutes for these essential molecules.

1   INTRODUCTION

Enzymes are responsible for carrying out the highly complex chemical reactions necessary to sustain life. Like chemical catalysts of nonbiological origin, enzymes only increase the rate of a reaction that would occur naturally; they cannot catalyze a reaction that would not ordinarily happen. For example, even in the absence of enzymes, the starch in last night's baked potato would break down into simpler sugars, although it would take hundreds if not thousands of years. In the presence of some of your digestive enzymes the job is done by the time of this morning's drive to work—an enzyme has increased the rate of a spontaneous reaction. No enzyme, however, could catalyze the conversion of simple sugars to starch (in a potato, for example) without the input of additional energy; this process simply would not occur naturally, and therefore no enzyme can catalyze it. In chemical terms, enzymes cannot alter the equilibrium ratio of products and reactants. Another feature that enzymes have in common with nonbiological catalysts is the regeneration of the catalyst at the end of the reaction. Enzymes are not consumed in the course of a reaction; they are returned to their initial form by the end of the catalytic cycle.

Enzymes may be *highly* specific for the reactions they catalyze (see Figure 1), selectively binding their *substrates* (the chemicals on which they act, also called *reactants*), and efficiently converting these substrates to different chemical forms, the *products*. Often an enzyme will selectively bind one substrate out of the many thousands of chemicals in a cell, and increase the rate of one specific re-

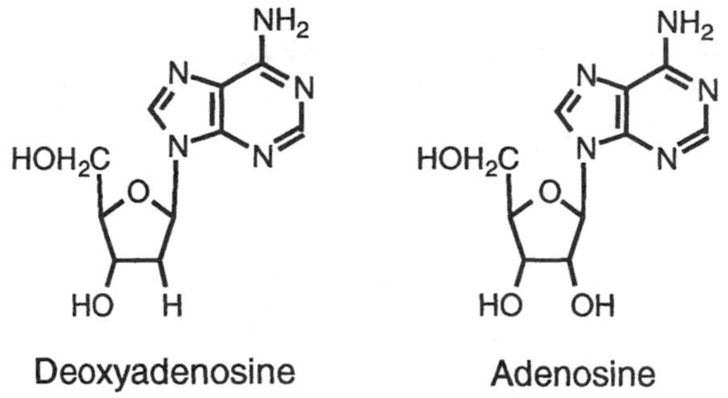

Figure 1.   Enzymes can easily distinguish between these two very similar molecules.

action by millions of times. Other enzymes are much less specific in their substrate *binding* but the reaction type they catalyze is always the same. For example, mammalian cytochrome P_{450} binds a variety of substrates, but in each case the chemical reaction is the addition of a hydroxyl group (—OH) to the substrate.

Specificity is a major feature that distinguishes enzymes from nonbiological catalysts. Another feature that distinguishes enzymes from chemical catalysts is regulation. Enzymes may be regulated in a variety of ways which may raise or lower their catalytic efficiency. Finally, enzymes may be distinguished from most chemical catalysts by the phenomenon of *saturation*. Enzymes have a maximum rate with respect to substrate concentration, whereas the majority of chemical catalysts do not.

All enzymes were once believed to be proteins, polymers of amino acids—the so-called building blocks of life. However, it has recently been shown that ribonucleic acids which are nonprotein biopolymers, can catalyze certain reactions as well, and as biological catalysts, these RNA molecules fit the old definition of an enzyme. Most biochemists now have narrowed the definition of the term "enzyme" to apply only to protein catalysts and have accepted the term "ribozyme" to describe catalytic RNA molecules. Another recent development has been the discovery of the means to produce antibodies that can catalyze specific reactions. These catalytic antibodies can also be regarded as enzymes, although not all the reactions they catalyze are of biological significance.

2   THE HISTORY OF ENZYMOLOGY

Probably the first recognition of an enzyme occurred in 1833 when Payen and Persoz found a thermolabile malt extract that would convert starch to sugar. They named this substance diastase, from the Greek for "separation," and soon "diastase" became a general term for enzymes. The parallel between the action of enzymes and the action of yeast in fermentation was noted early in the study of enzymology, and thus the term "ferment" was also used to describe enzymes.

In the latter half of the nineteenth century, a controversy arose between two schools of thought, one headed by Liebig, who held that fermentation was due to chemical substances, and the other led by Pasteur, who maintained that fermentation was inseparable from living cells. Each group had its own (equally unsatisfying) nomenclature. To avoid this controversy, in 1878 W. Kühne introduced the name *enzyme* (Greek, "in yeast") to denote that *in yeast* "something occurs which exerts this or that activity which is considered . . . fermentative," stating that the activity was due to something *in* the yeast rather than due to the yeast itself. The Pasteur–Liebig controversy was ended when Buchner accidentally obtained fermentation from a cell-free yeast extract. In 1898 Ducaux proposed that enzyme names be formed from a root indicating the nature of the reactant in the reaction, with -*ase*, the last three letters of the name diastase, added as

a suffix. Thus diastase became amylase, from amylose (the substrate) and -ase (indicating enzymatic activity). This system has been modified but still forms the basis of modern enzyme nomenclature.

At the end of the nineteenth century, Emil Fischer proposed the "lock and key" hypothesis, which states that there is a complementary relationship between the active site of an enzyme and the substrate; the substrate fits into the active site of an enzyme the same way a key fits into a lock. This concept, with some alterations, survives today.

In 1913 Michaelis and Menten, building on earlier work of Brown and Henri, provided a mathematical model of enzyme kinetics that provides information regarding the chemical mechanism of an enzyme-catalyzed reaction based on the study of its rate.

Enzymes were first obtained in nearly pure form in the 1920s, largely as a result of the work of Willstätter. The first enzyme to be crystallized was urease; the task was accomplished by Sumner in 1926. Crystallization of enzymes is a necessary first step in determining the three-dimensional chemical structure of a complete enzyme by X-ray diffraction from single crystals.

In 1948 Linus Pauling proposed what has become one of the central tenets of enzymology: enzymes perform catalysis by stabilizing the transition state of a reaction, thereby lowering the activation energy required for a reaction and thus increasing its rate. In 1951 Pauling and Richard Corey discovered two of the fundamental structural elements of proteins, α-helices and β-pleated sheets.

Another landmark of biochemistry was reached in 1953 when Frederick Sanger completed the determination of the amino acid sequence of an entire protein, insulin (see Figure 2). This work paved the way for other sequence determinations, contributing much to our knowledge of protein structure. Complete knowledge of the three-dimensional structure of a protein was not available, however, until 1961, when Perutz and Kendrew determined the structure of the oxygen-binding proteins myoglobin and hemoglobin using X-ray crystallography. In 1986 Kurt Wüthrich introduced the use of nuclear magnetic resonance (NMR) spectroscopy as a second method for structural determination of small proteins (< 200 amino acids).

In 1986 two different developments changed the way enzymologists think about enzymes. One was Thomas Cech's demonstration that RNA molecules can catalyze certain reactions, the first example of nonprotein biological catalysis. Next, Lerner and Schultz, working independently, produced antibodies that catalyzed chemical reactions, producing the first artificial enzymes.

3   CLASSIFICATION AND NOMENCLATURE

Since the first observations of enzymatic activity, attempts have been made to name the phenomenon descriptively. Until comparatively recently this has not, however, been the general practice. In many cases enzymes were given names that said nothing about their

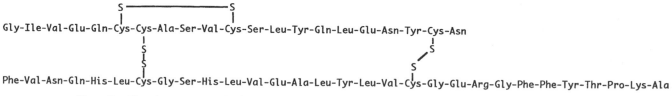

Figure 2.   The primary structure (amino acid sequence) of the bovine insulin hormone. Note the three disulfide bonds.

catalytic function. As more and more enzymes were discovered, it became necessary to devise a systematic method of nomenclature that could help classify these substances and provide some information about an enzyme's function.

As stated earlier, the basis for modern enzyme nomenclature was proposed by Ducaux in 1898. In this system enzymes were named for their substrate with the suffix -ase. Even though this method was generally accepted, it was not mandated, and there was still much confusion in the literature. Many enzymes maintained their uninformative trivial names. In some cases, separate research groups gave the same enzyme different names. Sometimes different enzymes had the same name. Although other nomenclature schemes were suggested over the years, none had official status until 1961, when the Commission on Enzymes of the International Union of Biochemistry (IUB) released a comprehensive scheme for enzyme classification and nomenclature.

Currently, enzymes are given names that help to classify them. The agency responsible for naming enzymes is the Commission on Biochemical Nomenclature of the IUB, which periodically publishes updates to a work entitled *Enzyme Nomenclature*. Under these guidelines, there are three acceptable ways to refer to an enzyme: (1) its Enzyme Commission (EC) code number, (2) its systematic name, and (3) its recommended name.

3.1   EC CODE

The EC code number provides a rigid, systematic method for classifying enzymes. An EC code number takes on the general form of EC *N.N.N.N*, where *N* represents an integer. The four integers indicate the enzyme's class, subclass, sub-sub-class, and serial number, respectively. Table 1 specifies the EC codes for some common enzymes, including many of those mentioned in this chapter. For example, alcohol dehydrogenase is denoted by EC 1.1.1.1, because it belongs to the class of oxidoreductases, the subclass using alcohols as electron donors, and the sub-sub-class using NAD^+ or $NADP^+$ as electron acceptors, and it has a serial number of 1. A 99 can be used in the subclass or sub-sub-class position to indicate that an enzyme does not fit any of those subclasses. The serial number is assigned sequentially in each sub-sub-class, giving each enzyme a unique code.

3.2   SYSTEMATIC NAME

Strict guidelines dictate the assignment of systematic names. A systematic name must include the EC class, the complete names of all substrates, and the names of any coenzymes involved. For example, EC 1.1.1.1 and EC 1.1.1.2 have systematic names "alcohol : NAD^+ oxidoreductase" and "alcohol : $NADP^+$ oxidoreductase," respectively. Therefore one can determine that EC 1.1.1.1 is dependent on the coenzyme NAD^+, and EC 1.1.1.2 is dependent on the coenzyme $NADP^+$.

3.3   RECOMMENDED NAME

Recommended names are not as detailed as systematic or EC code names, although here, too, there are guidelines to follow. Recommended names should make clear distinctions between similar enzymes. If a trivial name causes no confusion, it may be kept as the recommended name. It is preferred that the recommended name indicate the function of the enzyme. There are exceptions such as cata-

lase (hydrogen peroxide : hydrogen peroxide oxidoreductase), which has an uninformative trivial name, yet it has been accepted as the recommended name. Digestive enzymes like chymotrypsin and pepsin also have uninformative recommended names but they do not have systematic counterparts. The recommended names of EC 1.1.1.1 and EC 1.1.1.2 are "alcohol dehydrogenase" and "alcohol dehydrogenase ($NADP^+$)," respectively.

4   CHEMICAL ENERGETICS

To understand how enzymes work, we must first examine the energetics of chemical reactions. There are two major sets of properties common to all chemical reactions: *kinetic* and *thermodynamic*. In a simple sense, the kinetic parameters of a reaction describe how fast the reaction (or chemical steps within the reaction) will take place. Thermodynamics, on the other hand, gives an indication of the extent to which reactants will be converted to products in the long run. These thermodynamic predictions rely on the relative stabilities of products and reactants. Thermodynamically favored reactions may be kinetically inaccessible; that is, they would not achieve equilibrium on a biologically meaningful time scale. This is the contribution of enzymes to biochemical processes: accelerating the rates of thermodynamically favored reactions. Since enzymes, like all other chemical catalysts, affect *only* the *kinetics* of a reaction, they cannot alter the equilibrium ratio of products to reactants.

4.1   THERMODYNAMICS

The sole thermodynamic criterion for determining whether a reaction will occur spontaneously is the change in Gibbs free energy (ΔG), named for the American chemist J. W. Gibbs (1839–1903). The general relationship between ΔG and other thermodynamic parameters for any chemical reaction is as follows:

$$\Delta G = \Delta H - T\Delta S \qquad (1)$$

where ΔG is the change in free energy, ΔH is the change in enthalpy, T is the temperature in degrees Kelvin (Kelvin = °C + 273), and ΔS is the change in entropy. Enthalpy is a measure of stored energy, and entropy, in a broad sense, is the degree of disorder of a system.

All chemical reactions are theoretically reversible, such that reactants are constantly being re-formed from products. The rates of the forward and backward reactions depend on the concentrations of the reactants and products as well as on "rate constants" for the forward and back-reactions. The higher the concentration of reactants, the faster the forward reaction will go; the higher the concentration of products, the faster the reverse reaction will go. During the course of a chemical reaction, the reactants are depleted as they become transformed into products, so the rate of the forward reaction is decreasing, and the rate of the reverse reaction is increasing. Eventually, the reaction will reach a state of equilibrium; at equilibrium there is no *net* formation of products or reactants—products are converted to reactants at the same rate that reactants are converted to products.

To determine whether a reaction will occur, it is useful to look at the reaction quotient (Q) and the equilibrium constant (K_{eq}) for the reaction. For the following chemical reaction:

$$a\text{A} + b\text{B} \rightleftharpoons c\text{C} + d\text{D} \qquad (2)$$

Table 1 Nomenclature of Some Common Enzymes[a]

EC Number	Common Name	Recommended Name	Systematic Name
Class 1, Oxidoreductases			
1.1.1.1	Alcohol dehydrogenase	Alcohol dehydrogenase	Alcohol : NAD$^+$ oxidoreductase
1.11.1.6	Catalase	Catalase	Hydrogen-peroxide : hydrogen-peroxide oxidoreductase
1.14.15.1	Cytochrome P450	Camphor 5-monooxygenase	Camphor, reduced putida-ferredoxin : oxygen oxidoreductase
Class 2, Transferases			
2.7.5.1	Phosphoglucomutase	Phosphoglucomutase	α-D-Glucose-1,6-bisphosphate : α-D-glucose-1-phosphate phosphotransferase
Class 3, Hydrolases			
3.4.21.1	Chymotrypsin	Chymotrypsin	None
3.4.21.4	Trypsin	Trypsin	None
3.4.23.1	Pepsin A	Pepsin A	None
Class 4, Lyases			
4.1.2.13	Aldolase	Fructose–bisphosphate aldolase	D-Fructose-1,6-bisphosphate D-glyceraldehyde-3-phosphate-lyase
Class 5, Isomerases			
5.3.1.1	Triosephosphate isomerase	Triosephosphate isomerase	D-Glyceraldehyde-3-phosphate ketol-isomerase
Class 6, Ligases			
6.3.4.14	Biotin carboxylase	Biotin carboxylase	Biotin-carboxyl-carrier-protein : carbon-dioxide ligase (ADP forming)

[a]The table shows the Enzyme Commission (EC) numbers, common names, systematic names, and recommended names. The common names are *not* sanctioned by the Enzyme Commission but are often used in informal settings. For formal presentations the recommended or systematic name should be used.

where a, b, c, and d are coefficients representing how many molecules of the given type participate in the reaction, the reaction quotient is defined:

$$Q = \frac{[C]^c\,[D]^d}{[A]^a\,[B]^b} \qquad (3)$$

where $[X]$ indicates the initial molar concentration of X, and the equilibrium constant is

$$K_{eq} = \frac{[C]_{eq}^c\,[D]_{eq}^d}{[A]_{eq}^a\,[B]_{eq}^b} \qquad (4)$$

where $[X]_{eq}$ indicates molar concentration of X *at equilibrium.* Now, if Q is less than K_{eq} the forward reaction is spontaneous, if Q equals K_{eq} the reaction is at equilibrium, and if Q exceeds K_{eq} the backward reaction is spontaneous. It is important to realize that a change in any of the initial concentrations will alter the concentrations of reactants and products at equilibrium required to satisfy the ratio specified by K_{eq}.

Now that we have defined the reaction quotient, its relationship to the Gibbs free energy can be shown:

$$\Delta G = \Delta G^\circ + RT \ln(Q) \qquad (5)$$

where ΔG° is the standard free energy change, which is a constant for a given chemical reaction, under a given set of conditions; R is the ideal gas constant, and T is the temperature in degrees Kelvin. By definition, at equilibrium $\Delta G = 0$, it can easily be shown that:

$$\Delta G^\circ = -RT \ln(K_{eq}) \qquad (6)$$

The importance of ΔG and ΔG° is now evident. Any reaction that has a negative ΔG will proceed in the forward direction. Armed with the initial concentrations of reactants and products and with the value of the constant ΔG° (or with the value of K_{eq}, which can be used to calculate ΔG°), it becomes an easy task to calculate ΔG and to predict the direction of the reaction (i.e., whether more C and D will be formed, or more A and B). Therefore the value of ΔG° indicates the extent to which the reaction will favor products over reactants at equilibrium.

There are two points that should be noted:

1. The ΔG for a reaction changes with the concentrations of reactants and products. If a reaction has a positive ΔG, it simply means that the *forward* reaction will not go spontaneously under that set of conditions, not that the reaction will not go at all. In fact, a positive ΔG value means that under those conditions, there will be a net *back*-reaction. A change in the initial concentrations or temperature might make the forward reaction favorable.

2. The value of ΔG indicates nothing about how fast a reaction will proceed. The ΔG can only provide information about the direction and extent of the reaction in order to achieve equilibrium under a certain set of conditions. Even though the products of a reaction may be greatly favored at equilibrium (large K_{eq}), it could take many, many years for that state to be reached. Thus, some reactions although thermodynamically favored, will not occur on a biologically meaningful time scale (unless, of course, they are catalyzed).

Having considered how enzymes influence chemical reactivity, we may take a closer look at an enzyme's effect on kinetics.

4.2   Kinetics

We now turn our attention to the rates of chemical reactions. The rate of most chemical reactions is dependent on the concentrations of the substrates (reactants). Consider reaction (7), from which the coefficients have been eliminated for simplicity:

$$A + B \rightleftarrows C + D \qquad (7)$$

Here the rate *(v)* in the forward direction can be expressed as the product of the concentrations of A and B times a rate constant k:

$$v = k[A][B] \qquad (8)$$

As stated in Section 4.1, thermodynamics does not provide information about the rate of a reaction. However, by using transition state theory, we can obtain the rate law for reaction (7). The most convenient way to apply the properties of chemical reactions just described to the explanation of rates and catalysis is to construct a *reaction coordinate diagram.* Figure 3 shows reaction coordinate diagrams for an uncatalyzed and an enzyme-catalyzed reaction. Since the final products of the reaction possess less free energy (lower free energy is a more stable, thermodynamically favored state), thermodynamics tells us that the forward reaction is favored. However the reactants must first assume a very unstable, high energy state called the *transition state.* In the transition state, the chemical bonds involved in the reaction are partially formed and partially broken. Since this is a very high energy state, it exists for a very short time, on the order of a single vibration of a chemical bond, about 10^{-12} s. The free energy needed to promote a molecule (or molecules) from the ground state to the transition state is called the *activation energy* (denoted $\Delta G^\ddagger$).

Transition state theory assumes that the reactants (A and B) in a chemical process are in equilibrium with the transition state (AB‡). Because of this equilibrium, we may apply the formula for Gibbs free energy to the activation energy, giving:

$$\Delta G^\ddagger = -RT \ln\left(\frac{[AB]^\ddagger}{[A]\,[B]}\right) \qquad (9)$$

Statistical thermodynamics allows us to calculate the probability that a molecule will have a particular energy. If we take $\Delta G^\ddagger$ as that energy, we obtain the probability p of a molecule having enough energy to reach the transition state:

$$p = \frac{\kappa T}{h} \exp\left(\frac{-\Delta G^\ddagger}{RT}\right) \qquad (10)$$

where the new parameters are κ, the Boltzmann constant, and h, Planck's constant.

It is obvious that the rate of a reaction is governed by the number of molecules that have enough energy to reach the transition state (of the *slowest* step in the reaction, the rate-determining step). Assuming that once the transition state has been reached, half the molecules will go forward to products and half will go back to reactants, we can express the rate law of the reaction as follows:

$$v = \frac{\kappa T}{2h} \exp\left(\frac{-\Delta G^\ddagger}{RT}\right) [A]\,[B] \qquad (11)$$

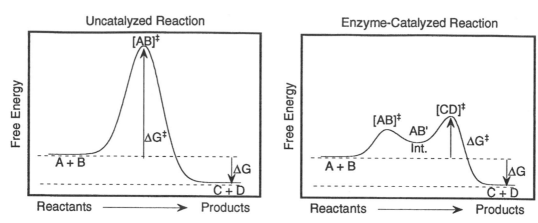

Figure 3. Reaction coordinate diagrams for the uncatalyzed and catalyzed reaction A + B ⇌ C + D. Transition states are indicated by peaks (they are high energy, short-lived species) and are indicated by a double-dagger superscript. The intermediates, which are in "valleys" (local minima), are lower energy, stable species and are labeled "Int." In the uncatalyzed reaction shown there is a single transition state. (Some reactions may have more than one.) The same reaction catalyzed by an enzyme has two transition states separated by an intermediate (AB'), indicating that enzyme-catalyzed reactions and their uncatalyzed counterparts may proceed by different mechanisms. The activation energy for the reaction ($\Delta G^{\ddagger}$) is the difference between the substrates and the transition state with the highest energy. Note that although the overall free energy change ΔG for the reaction is the same in both cases, the enzyme has lowered the activation energy for the reaction and therefore increased its rate.

and therefore the rate constant is given by

$$k = \frac{\kappa T}{2h} \exp\left(\frac{-\Delta G^{\ddagger}}{RT} \right) \qquad (12)$$

which depends only on $\Delta G^{\ddagger}$ and the temperature.

We can now see how catalysts increase the rate of a reaction. Since the rate constant of a reaction is inversely proportional to the activation energy, reactions with high activation energies have slow rates. For uncatalyzed reactions, such as the one shown on the left in Figure 3, very few molecules possess the energy needed to attain the transition state. The probability that many molecules in the population would have sufficient energy to overcome the activation barrier is small, so the reaction proceeds slowly. On the right in Figure 3, however, the reaction coordinate diagram of an enzyme-catalyzed reaction shows an alternate chemical pathway with a more accessible, lower free energy transition state (and thus a larger rate constant). Two points should be noted: first, the relative stability of reactants and products remains unchanged, with the result that their equilibrium ratio (K_{eq}) is the same; second, catalysis also increases the rate of the reverse reaction.

We may discuss the hydrolysis reaction of a peptide bond (Figure 4) as a real example of this processes. In living systems this reaction is catalyzed by enzymes called *proteases*, whose function is to cleave the peptide bonds that link amino acids in proteins. There are many different proteases which carry out a variety of tasks, ranging from specialized activation of cellular machinery to steps that make it possible for amino acids to be incorporated into new proteins, such as digestion of the protein in your food and simple degradation of old, useless proteins. Not all proteases have the same specificity, nor do they all use the same mechanism of action.

Figure 5 presents the reaction coordinate diagrams for the uncatalyzed hydrolysis of a peptide bond and the same reaction catalyzed by the digestive enzyme, chymotrypsin. Chymotrypsin is a *serine protease*, a type of protease in which the amino acid serine is in the active site and plays an active role in the catalytic process. The proposed mechanism of action of chymotrypsin is shown in Figure 6. Note that a reaction coordinate diagram depicts a relative free energy level for each chemical species in a reaction. Comparison of Figures 4, 5, and 6 reveals that for each chemical species shown in the catalyzed or uncatalyzed mechanisms, there is a corresponding free energy peak or valley in the reaction coordinate diagram. Peaks represent transition states; valleys represent intermediates, which are stable, isolatable species formed during a reaction.

4.3   Coupling

We have established that the equilibrium ratio of products and reactants is controlled by their relative stabilities (ΔG°) and that the rate of a reaction is determined by the activation energy necessary to achieve the highest energy transition state ($\Delta G^{\ddagger}$). However, one might be led to wonder how an organism can function metabolically

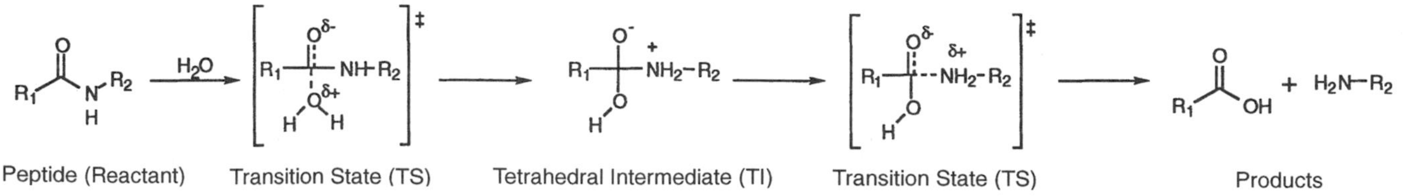

Peptide (Reactant)   Transition State (TS)   Tetrahedral Intermediate (TI)   Transition State (TS)   Products

Figure 4. Some steps in the hydrolysis of a peptide. For each species shown there is a corresponding label on the left-hand reaction coordinate diagram in Figure 5.

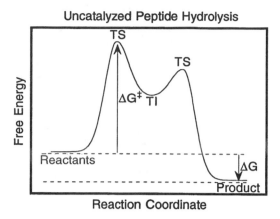

Figure 5. Reaction coordinate diagrams for the uncatalyzed (left) and chymotrypsin-catalyzed (right) cleavage reactions of a peptide: S, substrate; P, product; E, enzyme; TS, transition state; TI, tetrahedral intermediate; ES, the enzyme–substrate complex. Note that the enzyme-catalyzed reaction uses a different mechanism, which results in a lower activation energy ($\Delta G^{\ddagger}$). The first transition state in the enzyme-catalyzed case represents the binding of substrate to enzyme, and the following trough represent the enzyme–substrate (ES) complex described in Section 6.1. A water-binding step that takes place after the acyl–enzyme intermediate is formed is not shown. Compare the left panel to Figure 4 and the right panel to Figure 6.

when it needs the *product* of a reaction whose ΔG is positive. Fortunately, chemical reactions can be *coupled* so that a thermodynamically unfavored reaction may be paired with a favored one, driving both reactions forward. In the living cell, several enzymes may be involved in a catalytic series in which the product from one enzyme is the substrate for the next. This is called a metabolic pathway. A major metabolic pathway for releasing energy is the glycolytic pathway (Table 2), in which the sugar glucose is broken down to form pyruvic acid and "energy." The energy from these chemical processes is stored in the chemical bonds of adenosine triphosphate (ATP), which is used as an energy source in other reactions. At pH 7 (biological pH), the synthesis of ATP from adenosine diphosphate (ADP) and inorganic phosphate has a $\Delta G°$ of 7.3 kcal/mol, a very *unfavorable* reaction. However, the hydrolysis of phosphoenolpyruvate (PEP) to pyruvate and phosphate has a $\Delta G°$ of -14.8 kcal/mol—a very *favorable* reaction. In reaction (10) of glycolysis (Table 2), these two reactions are coupled so that the $\Delta G°$ for the overall reaction is -7.5 kcal/mol. The metabolic energy in sugar has thus been partly converted to ATP. The energy stored in ATP can then be used to do work for the organism: building proteins, making DNA, contracting muscles, and a variety of other life processes.

5   HOW ENZYMES WORK

Enzymes enhance reaction rates by factors of 10^3–10^{16} relative to the rates of the uncatalyzed reactions. Some enzymes, such as carbonic anhydrase, triosephosphate isomerase, and catalase, are considered to be "catalytically perfect" because the rate of the reaction is limited only by the rate at which the substrates diffuse through water to encounter the catalyst. Figure 6 showed the mechanism of peptide hydrolysis as catalyzed by chymotrypsin. Chymotrypsin enhances the rate of hydrolysis by 10^{10} times. Enzymes achieve these phenomenal rate enhancements by lowering the reaction's activation energy. To accomplish this, enzymes use a variety of strategies, including stabilization of the transition state, catalysis by approximation, covalent catalysis, acid–base catalysis, and redox catalysis. Occasionally enzymes employ cofactors to carry out some types of catalysis. These strategies are now discussed in more detail.

5.1   STRATEGIES ENZYMES USE TO ACHIEVE RATE ENHANCEMENTS

5.1.1   Transition State Stabilization

A very important aspect of enzymatic catalysis is the ability of the enzyme to bind to (and therefore stabilize) the transition state of a reaction, thereby lowering the activation energy of the overall reaction. Since the structure of the substrate changes during the process of its conversion into the product, the enzyme cannot be precisely complementary to all forms. The transition state is the highest energy form that exists between substrate and product, and it is to this form that enzymes are designed to bind most favorably. Because of the favorable binding interaction between the enzyme and the transition state, the free energy (ΔG) of the transition state is lowered relative to the uncatalyzed reaction, lowering the activation energy ($\Delta G^{\ddagger}$), which serves to accelerate the reaction. (Compare, e.g., the two panels of Figure 5.)

To achieve favorable binding, the active site of the enzyme may be complementary to the transition state in size, in shape, and in charge distribution, so that the *transition state* fits into the enzyme as a key fits into a lock. This is a modification of the hypothesis of Emil Fischer, who suggested that the *substrate* fits in the enzyme as a key fits in a lock. It is now felt that when the enzyme binds the substrate, it actually "bends" (distorts) the substrate toward the shape of the transition state. In this way, some of the "binding energy" (energy available due to the favorable binding interactions) is used to stabilize the transition state. For efficient catalysis, however, the enzyme should not bind the *product* very tightly to ensure that the enzyme may quickly release the product to "turn over" (complete one reaction cycle) and be ready to begin the process again. Failure to release product results in "product inhibition."

In the active site of serine proteases like chymotrypsin, the substrate interacts with complementary chemical groups that stabilize the interaction. For example, the oxygen involved in the peptide linkage to be cleaved is electron rich, and it becomes increasingly negative as the transition state is formed (see Figure 6). This negative character interacts with amide hydrogen atoms of Ser 195 and Gly 193 on the *enzyme's* peptide backbone to form hydrogen bonds.

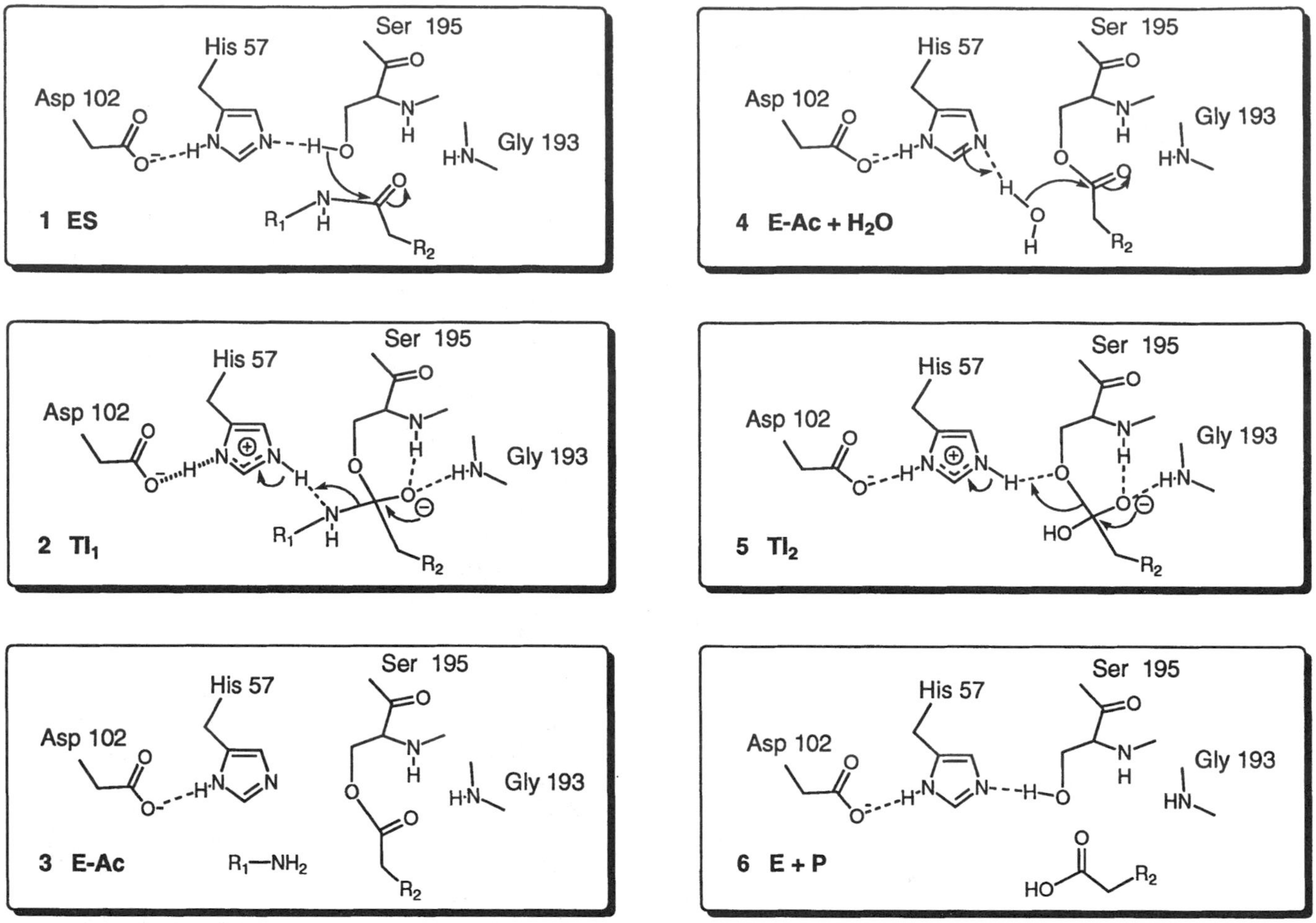

Figure 6. The chymotrypsin-catalyzed hydrolysis of a peptide bond. The dashed lines represent hydrogen bonds, the dotted lines in panel 2 represent a low barrier hydrogen bond, and the arrows show the movement of electrons during the reaction. The three residues of the catalytic triad (Asp 102, His 57, Ser 195) are identified, as well as Gly 193, which forms part of the oxyanion hole. The oxyanion hole stabilizes the transition state by donating hydrogen bonds to the substrate oxygen, which carries a negative charge in transition states and intermediates. Each panel corresponds to a species on the reaction coordinate diagram depicted in Figure 5. (1) The enzyme–substrate complex. The hydrogen bonding network extends from Asp 102 through His 57 to polarize the hydroxyl bond of Ser 195. In the transition state (not shown), there is a low barrier hydrogen bond between Asp 102 and His 57, as well as a partial bond between His 57 and the hydroxyl hydrogen of Ser 195. Attack of the oxygen from Ser 195 has led to a partial bond between Ser 195 and the carbonyl of the substrate, resulting in partial breakage of the carbonyl double bond and a partial negative charge on the carbonyl, which is stabilized by hydrogen bonds from the oxyanion hole. This is the completely concerted transition state that is currently believed to exist. (2) The first tetrahedral intermediate, showing the hydrogen bonding (especially the oxyanion hole) that stabilizes this intermediate. Collapse of this intermediate breaks the peptide bond. (3) The acyl–enzyme covalent intermediate. The first product diffuses away at this point and a water molecule diffuses into the active site. These steps are not shown in the reaction coordinate diagram (Figure 5). (4) The acyl–enzyme intermediate being attacked by a water molecule. His 57 is believed to act as a general base to deprotonate the water molecule, facilitating attack on the carbonyl. (5) The second tetrahedral intermediate, again stabilized by hydrogen bonding. Collapse of this intermediate breaks the covalent enzyme–substrate bond, releasing the product. (6) Enzyme and product. The second product may now diffuse away and return the enzyme to its initial state.

Residues 193 and 195 are arranged as a pocket, called the *oxyanion hole,* that specifically binds the negatively charged oxygen. These interactions are just some of many in the enzyme–substrate complex that make binding of the transition state a favorable reaction. This idea of transition state stabilization suggests that if a molecule could be synthesized, which is a good structural analogue of the transition state, it should bind very tightly indeed, since no binding energy would be expended in distorting the substrate into the transition state. In fact, transition state analogues are found to be potent inhibitors of enzymes. Figure 7 presents the reaction catalyzed by pro-

line racemase, the proposed transition state, and a transition state analogue that effectively inhibits the enzyme: The structural similarities between the inhibitor and the transition state should be noted.

5.1.2 Catalysis by Approximation

In the uncatalyzed reaction, molecules must collide in a very precise orientation if they are to react. In contrast, when a molecule is bound to an enzyme, it is held in an optimal orientation and prox-

Table 2   The Reactions of the Glycolytic Pathway[a]

Step	Reaction	Enzyme	$\Delta G°$	ΔG
1	Glucose + ATP $\rightleftharpoons$ glucose 6-phosphate + ADP + H⁺	Hexokinase	−4.0	−8.0
2	Glucose 6-phosphate $\rightleftharpoons$ fructose 6-phosphate	Phosphoglucose isomerase	+0.4	−0.6
3	Fructose 6-phosphate + ATP $\rightarrow$ fructose 1,6-bisphosphate + ADP + H⁺	Phosphofructokinase	−3.4	−5.3
4	Fructose 1,6-bisphosphate $\rightleftharpoons$ dihydroxyacetone phosphate + glyceraldehyde 3-phosphate	Aldolase	+5.7	−0.3
5	Dihydroxyacetone phosphate $\rightleftharpoons$ glyceraldehyde 3-phosphate	Triosephosphate isomerase	+1.8	+0.6
6	Glyceraldehyde 3-phosphate + P_i + NAD⁺ $\rightleftharpoons$ 1,3-bisphosphoglycerate + NADH + H⁺	Glyceraldehyde 3-phosphate dehydrogenase	+1.5	−0.4
7	1,3-Bisphosphoglycerate + ADP $\rightleftharpoons$ 3-phosphoglycerate + ATP	Phosphoglycerate kinase	−4.5	+0.3
8	3-Phosphoglycerate $\rightleftharpoons$ 2-phosphoglycerate	Phosphoglyceromutase	+1.1	+0.2
9	2-Phosphoglycerate $\rightleftharpoons$ phosphoenolpyruvate + H_2O	Enolase	+0.4	−0.8
10	Phosphoenolpyruvate + ADP + H⁺ $\rightarrow$ pyruvate + ATP	Pyruvate kinase	−7.5	−4.0

[a]The reactions of the glycolytic pathway. ΔG's were calculated from the known $\Delta G°$'s and estimated physiological concentrations of reactants and products

imity for reaction with another chemical group. The entropy (ΔS) involved in obtaining the correct reactive conformation is greatly reduced when the reactants are bound to the enzyme, resulting in a smaller $\Delta G^{\ddagger}$. In step 4 of the mechanism of chymotrypsin (Figure 6), a water molecule must react with the covalently bound product. The enzyme active site is situated in such a way that both the water molecule and the intermediate product are oriented in the most reactive geometry relative to each other: when they interact, the reaction will most certainly take place. In solution, without the enzyme, these molecules would have to collide randomly a great many times before finding the proper orientation for the reaction to take place.

5.1.3 Covalent Catalysis

The chemical side chains of an enzyme's constituent amino acids present in the active site may actually take part in the catalyzed re-

action, being regenerated in their original state by the end of the reaction cycle. A commonly observed mechanism is one in which an amino acid side chain displaces a chemical group of a substrate molecule to form a covalently bound enzyme–substrate intermediate, which is more reactive than the original substrate. A second substrate may then react with the intermediate complex to yield product(s) and the enzyme in its original, catalytically active form. By use of this strategy, the enzyme provides an alternative reaction mechanism with a smaller $\Delta G^{\ddagger}$. This strategy is frequently used in hydrolysis and other transfer reactions.

The chymotrypsin-catalyzed reaction involves the side chain of a serine residue in this manner. The serine oxygen displaces the amino end of the peptide bond, breaking that bond (Figure 6, step 2). In the process, however, a new bond is formed (an ester bond) between the enzyme and the other product of the reaction. This form of the enzyme is called an acyl–enzyme intermediate (Figure 6, step 3). This bond is more easily hydrolyzed than the peptide bond, es-

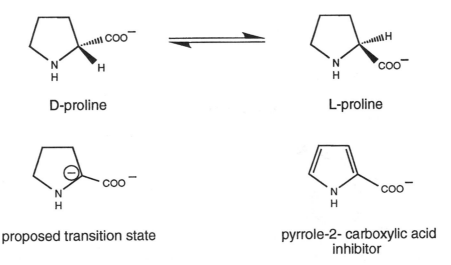

D-proline   L-proline

proposed transition state   pyrrole-2- carboxylic acid
                                        inhibitor

Figure 7.   The reaction catalyzed by proline racemase and a transition state analogue, which has been shown to be an inhibitor. Note the similarities between the transition state analogue and the proposed transition state.

pecially in the environment of the chymotrypsin active site. A water molecule may now cleave the ester to release the second product and return the enzyme to its original, catalytically active state. This reaction pathway has a lower $\Delta G^{\ddagger}$ than direct cleavage of the amide by water.

5.1.4 Acid–Base Catalysis

Amino acid side chains may also cause a chemical reaction to occur more readily by acting as acids (donors of protons) or bases (acceptors of protons). It is important to note that other (nonprotein) acids or bases can also act as catalysts, but in enzymes, the reactive groups are oriented in the active site to permit the necessary proton transfer to occur most efficiently.

Once again, chymotrypsin provides good examples of this strategy; during the course of the catalytic cycle several groups must be protonated or deprotonated for the reaction to proceed efficiently. When the enzyme is not catalyzing a reaction, the serine oxygen is protonated and involved in a hydrogen bond with His 57 (Figure 6, step 1). Upon binding of the substrate, His 57 helps deprotonate the serine so that it can react with the peptide bond. To facilitate the peptide bond breakage, the amide nitrogen is protonated by His 57 (Figure 6, step 2). A protonated amine is a much better "leaving group" in hydrolysis reactions. In a later step of the reaction mechanism (Figure 6, step 4), His 57 abstracts a proton from a water molecule, making it more reactive toward ester hydrolysis, and finally reprotonates the serine after the hydrolysis reaction has occurred. When the hystidyl residue donates a proton, it is acting as a general acid; when abstracting a proton, it is acting as a general base.

5.1.5 Redox Catalysis

Electrons may be transferred during the course of a reaction; such reactions are called *redox* (reduction–oxidation) reactions. Reduction occurs when a molecule gains electrons, and oxidation occurs when electrons are lost. Of course, there are nonenzymatic redox catalysts, but the active site of an enzyme is arranged so that everything is in the correct orientation to optimize catalytic efficiency. Enzymes most commonly catalyze redox reactions by means of either tightly bound transition metals or *coenzymes*. Coenzymes are specialized organic molecules associated with many enzymes. They provide chemical functionalities not available in the 20 common amino acids, and lead to more efficient catalysis. Redox enzymes often take advantage of transition metals because they can easily exist in a variety of oxidation states, facilitating electron transfer. During the reduction of hydrogen peroxide to water and molecular oxygen by catalase, a bound iron is able to transfer electrons (one at a time) from substrate (H_2O_2) to oxygen by changing back and forth between the 2^+ and 3^+ oxidation states.

5.1.6 Coenzymes and Cofactors

Enzyme-bound metals and coenzymes, generally referred to as *cofactors,* may take part in acid–base chemistry and covalent catalysis as well as redox reactions. Like transition metals, some coenzymes are also stable in more than one oxidation state. The most common of these are the nicotinamides and flavins (see Figure 8). These coenzymes are often associated with enzymes involved in electron transfer reactions. In many cases two or more of these cofactors can work together. The electron transfer may be between co-

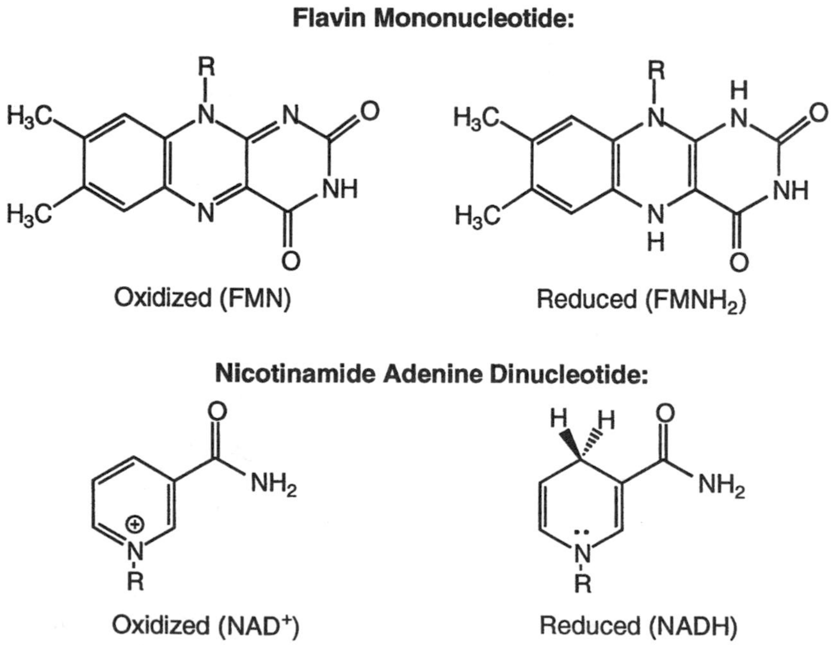

Figure 8. The functional groups of flavin mononucleotide (FMN) and nicotinamide adenine dinucleotide (NAD), two commonly occurring coenzymes, in their oxidized and reduced forms. The R in each structure represents organic components of the molecule that do not take part in the chemistry of the reaction but quite often play an important role in specific interactions with the enzyme or substrates.

factors on the same enzyme or on different enzymes. Flavin-mediated reactions of this type are common. For example, the pyruvate produced from glycolysis is fed into a series of reactions known as the Krebs cycle (also known as the citric acid cycle or the tricarboxylic acid cycle), in which isocitrate dehydrogenase uses the energy originally in the pyruvate to transfer electrons, reducing the cofactor NAD^+ to NADH. The NADH so generated may then transfer these electrons to a series of enzyme complexes that make up the *electron transport system*. Electrons are transported from enzyme to enzyme in the chain, and finally to molecular oxygen, the ultimate electron acceptor. The electron transport system is the final series of reactions in aerobic metabolism and provides the energy needed to produce most of the ATP used by aerobic organisms.

As stated before, many of these electron transfer enzymes contain metals; some use coenzymes to hold the metal in place. Examples of metal-containing coenzymes are vitamin B_{12} and heme, which contain cobalt and iron, respectively. Both catalase and hemoglobin are heme-containing proteins. The heme iron in hemoglobin is used for binding oxygen, not for redox chemistry. (Hemoglobin is the protein that carries oxygen in the blood, but it is not an enzyme.) Another common motif for iron binding in enzymes is the iron–sulfur cluster; such clusters usually contain two to four irons and inorganic sulfur bound to the enzyme via cysteinyl residues. Examples of iron–sulfur clusters can be found in the enzyme nitrogenase and in several of the enzymes found in the electron transport chain.

5.2 Serine Proteases: A Case Study in Catalysis

In recent years, determining the relative contribution of each strategy used by enzymes to the overall rate enhancement has been the focus of much study. The serine proteases such as trypsin and chymotrypsin have been well studied in this regard. The determination of the mechanism of action of the serine proteases also provides an interesting insight into the way scientific theories evolve as greater understanding is obtained; the mechanism as it is currently understood is depicted in Figure 6. It was observed that an aspartate, a histidine, and a serine residue were aligned, and probably hydrogen-bonded together. These three residues, which are present in all serine proteases, became known as the *catalytic triad*. The aspartate in this triad was believed to deprotonate the histidine, which would in turn deprotonate the serine. The deprotonated serine (an alkoxide) would then be responsible for attacking the peptide bond at the carbonyl carbon. This stepwise mechanism was dubbed the charge-transfer or charge relay mechanism, as the negative charge originally present on the aspartate was transferred through the histidine to the serine. This is an example of acid–base catalysis (the polarizing effects), catalysis by approximation (the peptide substrate is in the optimal position for reaction), and covalent catalysis (the serine participating directly in the reaction). However, this model was not without its problems. First, for the reaction to proceed by the mechanism suggested, the serine alkoxide anion should be able to react with other electrophilic chemical modification reagents such as iodoacetic acid, which was never observed. Furthermore, the pK_a of the hydroxyl group of serine is above 15, while aspartate's pK_a is around 3.9. It is difficult to imagine how an amino acid with a pK_a of 3.9 could efficiently drive the deprotonation of one with a pK_a of 15.

The charge-relay mechanism was modified based on additional X-ray crystal structure studies of other serine proteases. There are backbone amides in serine proteases that point toward and hydrogen bond to the carbonyl oxygen of the peptide substrate. These hydrogen bonds form what is called the "oxyanion hole," which serves to polarize the carbonyl group and help to stabilize the negative charge that is formed on the oxygen as the reaction proceeds. In the oxyanion hole, the hydrogens of the amides draw electron density (the stuff of which bonds are made) away from the oxygen of the carbonyl. As a result, electron density is drawn away from the carbonyl carbon as well, leaving it with a slight positive charge, which renders it more electrophilic. At the same time, the catalytic triad is drawing electron density away from the hydrogen–oxygen bond of the serine, making it more nucleophilic. The enzyme is able to function, then, because the nucleophilicity of the serine is increased, making it more likely to attack, and the electrophilicity of the carbonyl carbon is increased, making it more likely to *be* attacked. While still utilizing acid–base and covalent catalysis, and catalysis by approximation, the enzyme is now also employing transition state stabilization, in that the oxyanion hole is stabilizing the transition state at the carbonyl. This model, which continues to emphasize the involvement of the charge-relay system, however, was called into doubt when it was discovered that a mutant form of trypsin in which the aspartate was replaced with an asparagine (which lacks a negative charge) retained marginal activity. Furthermore, NMR and neutron diffraction data indicated that both the aspartate and the histidine remain charged during the catalytic cycle, which is also inconsistent with the charge transfer mechanism.

Recently, a new version of the mechanism has been proposed, based primarily on NMR data from the laboratory of Perry Frey. In this model there is a low barrier hydrogen bond between the aspartate and the histidine, in which the hydrogen is shared almost equally between the two amino acids. The reaction's transition state itself is now believed to include the entire catalytic triad, as well as the substrate and the oxyanion hole. Our understanding has progressed from the stepwise description in the charge-relay mechanism to the currently accepted completely concerted description. Future experiments may call even this mechanism into doubt, as chemical mechanisms can only be disproved, never proved. Currently, however, transition state stabilization is thought to play a major role in catalysis by the serine proteases.

It is the continued detailed study of enzymes and the specific contributions of different strategies to catalysis that may one day give scientists the ability to create supercatalysts with the efficiency and specificity of enzymes.

6 ENZYME KINETICS

One of the characteristics distinguishing enzymes from most nonenzymatic catalysts is the phenomenon of *saturation*. As shown in Figure 9, a plot of the initial velocity of an enzyme-catalyzed reaction versus the initial concentration of the substrate indicates that the rate of the reaction varies linearly with low substrate concentrations: the reaction is *first-order* with respect to substrate concentration (the rate is directly proportional to the substrate concentration). However, at high substrate concentrations, the rate soon asymptotically approaches a limiting value, called V_{max}, becoming *zero-order* with respect to substrate concentration (not varying with concentration). By comparison, a similar plot typical for a reaction catalyzed by a nonenzymatic catalyst shows no such limit; it is first-order with resect to substrate concentration over the entire range of substrate concentrations that can be obtained.

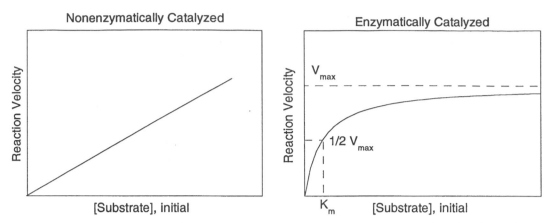

Figure 9. Effect of substrate concentration on the initial velocity of two reactions. The nonenzymatically catalyzed reaction—it's catalyzed, but not by an enzyme—shows a first-order rate increase with concentration, and no saturation. The enzymatically catalyzed reaction, on the other hand, shows saturation kinetics in which the rate approaches a maximum value.

6.1   MICHAELIS–MENTEN EQUATION

Early in the twentieth century, Henri, Michaelis, and Menten developed a mathematical treatment to explain substrate saturation, which was later extended by Briggs and Haldane. Consider a simple enzyme-catalyzed reaction as depicted in reaction (13), in which substrate (S) binds to enzyme (E) in a reversible equilibrium and is then converted into a product (P) with the regeneration of the enzyme.

$$E + S \underset{k_{-1}}{\overset{k_1}{\rightleftharpoons}} ES \overset{k_2}{\to} E + P \tag{13}$$

Here the concentration of the enzyme–substrate complex is [ES], the concentration of the free enzyme is [E], and thus the total enzyme concentration, $[E_0]$, is [E] + [ES]. The concentrations of substrate and product are represented by [S] and [P], respectively. The rate constants for the two forward reactions are indicated by k_1 and k_2, while the reverse reaction is indicated by k_{-1}.

Mathematically, a first-order rate (or velocity) is expressed as a rate constant times a concentration. For example, the rate expression for the conversion of ES to E and P would be written as k_2[ES], the rate constant times the concentration of ES. The rate of *change* of the concentration of the intermediate (d[ES]/dt) is equal to the rate of its formation from E and S (k_1[E][S]), minus its rate of breakdown. There are two modes of breakdown, one to E and S (k_{-1}[ES]), the other to E and P (k_2[ES]). The overall rate of change of the concentration of ES is therefore given by:

$$\frac{d[ES]}{dt} = k_1 [E][S] - k_{-1}[ES] - k_2[ES] \tag{14}$$

Briggs and Haldane introduced the concept of the *steady state*. The steady state assumption postulates that although initially the rate of production of ES would be positive (because initially no ES is in solution), very shortly after mixing E and S, the rate of formation of ES equals the rate of its breakdown. Therefore d[ES]/dt = 0, implying a steady state, so that

$$k_1 [E][S] - k_{-1}[ES] - k_2[ES] = 0$$
$$k_1 \{[E_0] - [ES]\} [S] - k_{-1}[ES] - k_2[ES] = 0 \tag{15}$$

Solving for [ES], one obtains

$$[ES] = \frac{k_1 [E_0][S]}{k_{-1} + k_2 + k_1 [S]} \tag{16}$$

Since the rate of the overall reaction is the rate of the formation of P from ES, the rate expression is $v = k_2$[ES]. The rate expression then becomes:

$$v = \frac{k_1 k_2 [E_0][S]}{k_{-1} + k_2 + k_1 [S]} \tag{17}$$

Finally, substituting k_0 for k_2, where k_0 may be a combination of several rate constants, not simply a single step as indicated in our example, and defining an aggregate constant $K_m = (k_{-1} + k_2)/k_1$, one arrives at the Michaelis–Menten equation:

$$v = \frac{k_0 [E_0][S]}{K_m + [S]} \tag{18}$$

This equation, which describes a rectangular hyperbola, explains the substrate saturation behavior characteristic of enzyme-catalyzed reactions. One notices that as the substrate concentration [S] becomes very large ($>> K_m$), the value of K_m + [S] may be approximated by [S]. The rate equation then reduces to:

$$v = k_0[E_0] \tag{19}$$

explaining the zero-order behavior at high concentrations of substrate. The maximal velocity V_{max} is thus mathematically equal to $k_0[E_0]$.

At [S] equal to K_m, the rate equation becomes:

$$v = \frac{V_{max}}{2} \tag{20}$$

Therefore the K_m value is not only a combination of rate constants as defined earlier; it is also the substrate concentration necessary to obtain half of V_{max}.

6.2   REGULATION

Another distinguishing feature of enzymatic catalysis is that it can be regulated. For the organism to continue to function, enzymes must respond to changing conditions in the environment. There are many levels of regulation of biological activity, starting with gene expression. The simplest way to keep an enzyme from doing its catalytic work is simply not to make the enzyme in the first place! Once an enzyme has been synthesized, there are often other methods of regulating its activity. However, not all enzymes are regulated. Often enzymes at key branch points in metabolism are extensively regulated, while other enzymes in the same metabolic pathway are not regulated. Reactions with large free energy changes (not necessarily at branch points) are also good targets for regulation. Often, reactions with large free energy changes are catalyzed by different enzymes in the two directions. Regulation of the two enzymes allows control of metabolic flow in both directions. For instance, in the glycolytic pathway shown in Table 2, the catalytic efficiency of the enzyme phosphofructokinase (PFK) is reduced by high levels of ATP, the body's energy currency. This makes perfect sense, because if the body has a lot of energy (plenty of ATP), it should not be burning glucose to release more energy; it should be storing energy for later. This reaction, then, is regulated. Note that this reaction has a high $\Delta G°$ of -3.4 kcal/mol and that there is another enzyme (fructose 1,6-bisphosphatase), which catalyzes the reverse reaction. In contrast, the enzyme phosphoglucose isomerase, immediately before PFK in the glycolytic pathway, is unregulated; this is not a metabolic branch point, the $\Delta G°$ for the reaction is low (only $+0.4$ kcal/mol), and the same enzyme catalyzes the reverse reaction.

Regulation of enzymes is often accomplished by the binding of small molecules to the enzyme. These molecules, either activators or inhibitors, change the conformation of the enzyme in such a way as to make it either more or less efficient, respectively, depending on the needs of the organism. Enzymes that are regulated in this manner still obey Michaelis–Menten kinetics, although more terms are needed to account for the added complexity. The two processes of activation and inhibition are accomplished in a number of ways.

6.2.1 Inhibition

Molecules binding to an enzyme can function to slow the enzyme down, and thus act as inhibitors. Inhibition may be either reversible or irreversible. Irreversible inhibition is characterized by a slow dissociation (or no dissociation) of the inhibitor from the enzyme, which essentially renders it permanently incapable of performing its function, effectively "killing" the enzyme. Reversible inhibition, on the other hand, is characterized by a rapid dissociation of the enzyme–inhibitor complex. There are different types of reversible inhibition, distinguished on the basis of the ability to form a ternary complex of enzyme, inhibitor, and substrate, designated EIS.

Perhaps the conceptually simplest form of inhibition is competitive inhibition. In this form of inhibition, the inhibitor competes for the same form of the enzyme as the substrate. Once the enzyme has bound inhibitor, the substrate cannot bind. Thus competitive inhibitors function by reducing the amount of active enzyme in solution, via partitioning the available enzyme between two forms: the fully active ES form, and the inactive EI form. Since, however, the binding of substrate and inhibitor are reversible reactions, and the partitioning is concentration dependent, higher substrate concentrations will partition more of the total enzyme in the active ES form, with the result that sufficient substrate can outcompete the inhibitor. In kinetic terms (see Figure 10) the presence of the inhibitor raises the Michaelis constant, K_m (recall that K_m is the substrate concentration required to reach $\frac{1}{2} V_{max}$). Since sufficient substrate can outcompete the inhibitor, V_{max} is unchanged.

Slightly more complicated are the concepts of noncompetitive, uncompetitive, and mixed inhibition, whose effects on kinetics are also depicted in Figure 10. In these forms of inhibition, the enzyme can bind both substrate and inhibitor simultaneously, but the ternary complex, EIS, cannot react to yield product; therefore, no amount of substrate can outcompete the inhibitor. The inhibitor is envisioned to cause this effect by altering the conformation of the active site in a way that renders the EIS complex inactive. In the very rare

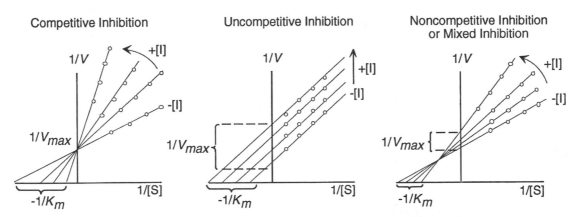

Figure 10. The effects of competitive, uncompetitive, and noncompetitive or mixed inhibitors on V_{max} and K_m. The rate of the enzyme in the presence ($+[I]$) and absence ($-[I]$) of inhibitor is shown. Inhibitor concentration increases in the direction of the arrows. For each line, the y intercept gives the apparent value of $1/V_{max}$; the x-intercept, the apparent value of $1/K_m$. As shown, competitive inhibitors decrease K_m but do not affect V_{max}; sufficient substrate is able to outcompete the inhibitor. Uncompetitive inhibitors affect both V_{max} and K_m to the same degree, yielding parallel plots. Mixed inhibitors, on the other hand, alter K_m and V_{max} to different degrees. In pure noncompetitive inhibition (not shown), all the lines intersect on the x axis, resulting in an identical K_m, but with values of V_{max} that decrease with increasing inhibitor concentration. Notice that for all types of inhibition except competitive inhibition, no amount of substrate can outcompete the inhibitor, and the intrinsic maximum rate of the enzyme has been lowered.

case of "pure" noncompetitive inhibition, the binding of inhibitor to the enzyme has no effect on *substrate binding,* and thus the affinity of E for S is the same as the affinity of EI for S. In this case, the effect of the inhibitor is to reduce the amount of enzyme (by the amount [EI] + [EIS]) able to yield product, thereby reducing V_{max}, but without affecting K_m. In the other limit, called uncompetitive inhibition, the inhibitor does not bind to the free enzyme; instead, it binds only to the ES complex, preventing the reaction from proceeding to product. In this case, the inhibitor functions by altering both substrate binding and catalytic efficiency *to the same degree* which means that if one defines a constant, α (related to the degree of inhibition), the apparent V_{max} and K_m will be equal to a αV_{max} and αK_m, respectively. This form of inhibition is also rare. It is more common that inhibitor binding alters both substrate binding *and* catalytic efficiency but to different degrees, yielding mixed inhibition, in which K_m and V_{max} are altered to different degrees.

Simple diagrams can provide a physical explanation for these behaviors (Figure 11). Competitive inhibition occurs when inhibitor binding blocks substrate binding. Of course, if the inhibitor binds in the substrate-binding site, the binding of one will prevent binding of the other, but one cannot conclude from competitive binding alone that the binding sites overlap. In noncompetitive, uncompetitive, and mixed inhibition, however, the inhibitor binds to a different site. Since the binding sites for the inhibitor and the substrate are different, increasing the substrate concentration will not displace the inhibitor and therefore V_{max} can never be attained.

The digestive enzyme trypsin can be both reversibly and irreversibly inhibited. Trypsin is synthesized in the pancreas in an inactive form called a zymogen. One would not want active trypsin in the pancreas, because it would start digesting the pancreas itself! Normally this zymogen is not activated until it is transported from the pancreas to the intestine. Occasionally the zymogen is prematurely activated in the pancreas, but in the pancreas there is a small protein, pancreatic trypsin inhibitor (PTI), which is responsible for inhibiting any prematurely activated trypsin present. PTI is a 6 kDa protein that binds extremely tightly to the active site of trypsin, a case of irreversible competitive inhibition. (Recall that irreversible inhibition need not involve covalent bonding, but a slow dissociation of the enzyme–inhibitor complex, or no dissociation at all.) PTI binds so tightly to trypsin because its structure closely resembles the transition state of the hydrolysis reaction catalyzed by trypsin and is therefore almost perfectly complementary to the active site of the enzyme. In fact, the complementarity is so good that the bound form is preferred over the unbound form by a factor of 10^{13}. Trypsin can also be irreversibly inactivated by the action of diisopropylfluorophosphate (DIFP), which forms a covalent bond with the active site serine, and the enzyme is thus incapable of carrying out its catalytic function. (DIPF is one of a family of inhibitors that are used in nerve gases because they also inhibit acetylcholinesterase, a serine *esterase* with a mechanism similar to that of the serine proteases.) There are several reversible competitive inhibitors of trypsin as well, such as benzamidine and leupeptin. These compounds bind in the active site of trypsin and reduce the concentration of enzyme that can bind substrate. However, the binding of these inhibitors is not too tight (the bound forms are preferred by "only" 10^5 or 10^7, respectively), and it is rapidly reversible. Therefore, at high concentrations of substrate, the binding of the substrate effectively competes with the binding of inhibitor and the catalytic reaction can occur.

The detection of natural products or the design of synthetic compounds that inhibit specific enzymes is the basis of much modern pharmaceutical chemistry. For example, the drugs AZT (azidothymidine) and ddC (dideoxycytosine), commonly used for AIDS treatment, bind in the active site of HIV reverse transcriptase and form complexes that are competitive inhibitors. Reverse transcriptase can also be noncompetitively inhibited: the drug Nevirapine binds in a cleft on HIV reverse transcriptase away from the active site but interferes with the catalytically necessary movement of one region of the enzyme relative to another, a natural example of "throwing a wrench into the works." The enzyme consequently has a lower maximal velocity. When this reaction is run in a test

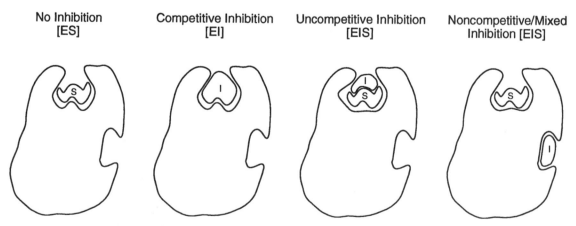

Figure 11.   A simplistic visualization of various modes of inhibition. For the purposes of this illustration the ridge in the substrate binding site is important for catalysis, but not for binding. In the case of the competitive inhibitor, binding of inhibitor interferes with *binding* of substrate (although they need *not* bind to the same site, as shown). In the case of uncompetitive binding, the inhibitor being bound has changed the active site: the "catalytic ridge" is missing and the shape of the binding site is changed, with the result that catalysis and binding are affected equally. Since in this form of inhibition, the inhibitor binds only to the ES complex, the inhibitor-binding site would not exist on the free enzyme. In the case of pure noncompetitive inhibition, only catalysis is affected, since the altered catalytic ridge is not important for binding. If one assumes that the catalytic ridge is also important for binding, the rightmost cartoon becomes an example of mixed inhibition, inasmuch as catalytic and binding efficiency have been altered to different degrees.

tube, raising the concentrations of RNA and nucleoside triphosphates (the substrates) has no effect, because the Nevirapine is still present.

6.2.2 Activation

A common form of activation involves covalent modification, a change in the chemical structure of the enzyme. Some enzymes can be phosphorylated at one of several amino acid residues to activate the enzyme. (See Figure 12 for examples.) This extra phosphate group changes the conformation of the enzyme in such a way that the enzyme is more active. Similarly, other small molecules (including methyl, adenyl, and myristyl groups) may be covalently attached to the enzyme to activate it. Enzymes may also be activated by cleavage of the peptide backbone. The inactive zymogens of trypsin and chymotrypsin are activated by just such a mechanism. Once the enzymes have moved from the pancreas and into the intestine, where they are needed, a sequence of highly specific cleavages of the backbone renders them active. The hormone insulin (Figure 2) is also created by proteolytic processing of the inactive prohormone.

Skeletal muscle glycogen phosphorylase is an enzyme that helps muscle tissue break down glycogen to utilize its stored energy. The phosphorylation of a single serine residue converts glycogen phosphorylase from its inactive (b) form to its active (a) form. One of the enzymes responsible for the *activation* of glycogen phosphorylase is also responsible for the *in*activation of glycogen synthase, the enzyme that makes glycogen (see Figure 13). This is a typical example of the coordination of regulation of enzymes: when an enzyme that *makes* a metabolite is activated, the enzymes that *break down* that metabolite are simultaneously inactivated. Interestingly enough, the inactivation of glycogen synthetase is also accomplished by a phosphorylation reaction. Covalent modification is therefore seen as a common method to both activate and inactivate enzymes.

Another form of activation involves noncovalent binding of a small molecule to the enzyme. This molecule, called an activator or positive effector, changes the conformation of the enzyme in a way that makes it *more* efficient, in contrast to the binding of inhibitors, which change the conformation of the active site to make the enzyme *less* efficient.

6.2.3 Allostery

A second form of regulation of enzyme activity is the phenomenon of allostery, meaning "other site." Allosteric enzymes are often composed of multiple subunits. The essence of allostery is that ligand binding to one subunit of an enzyme induces conformational changes in that subunit *and also in the other subunits,* altering the enzymatic properties of those other subunits. A "ligand" in this case may be a substrate, or an allosteric effector (either an activator or inhibitor). The kinetic behavior of allosteric enzymes is frequently more complex than can be described by the Michaelis–Menten equation. Regulation of some allosteric enzymes is the result of binding of the allosteric ligand(s) altering the apparent affinity of the enzyme for substrate, a property that is reflected in the K_m. Allosteric regulation may also occur at the level of catalytic efficiency, resulting in a change in maximum velocity, reflected in V_{max}. Binding of allosteric ligands to some enzymes will alter both V_{max} and K_m.

The classic example of allostery in a protein is hemoglobin. Although hemoglobin is not technically an enzyme, the allosteric phenomenon is the same in enzymes as in other proteins. A plot of the fraction of filled binding sites on hemoglobin versus O_2 concentration (Figure 14, analogous to the velocity–concentration plots in Figure 9), shows that hemoglobin does not have a hyperbolic binding curve, whereas myoglobin, the oxygen-binding protein in muscle, does. At low O_2 concentrations, the O_2 affinity of hemoglobin is low. However, as the concentration of oxygen increases, the oxygen affinity of hemoglobin increases. Each hemoglubin is a tetramer (contains four subunits) with four binding sites for O_2, one per monomer. Filling one of the sites increases the O_2 affinity of the other three.

To explain the physiological basis for this behavior, one must keep in mind the flexible nature of proteins, as well as the possibility of having proteins made up of more than one subunit. Hemoglobin, for instance, is composed of four subunits, each of which may be in a low affinity or high affinity state. In the absence of oxygen, all four subunits are in the low affinity or tense state, designated by T. As the oxygen concentration increases, a point will be reached at which a single subunit will bind an oxygen molecule. The binding of this one small molecule induces a conformational shift in the oxygen-binding site, which is then communicated throughout the subunit, resulting in a shift from the low affinity T state to the high affinity R, or relaxed, state. This one R state subunit then transmits its conformational change through the subunit interface and makes it more likely that the other subunits will assume the high affinity R state, and therefore the oxygen affinity of all the binding sites in the tetramer increases.

Another aspect of allosteric regulation of hemoglobin with an important physiological impact is called the Bohr effect, a decrease in the oxygen affinity of hemoglobin at lower pH (higher hydrogen ion concentration). In body tissues the carbon dioxide produced by metabolism of sugars and fats combines with water to make an acid, carbonic acid. This acid raises the hydrogen ion concentration (lowers the pH) in the tissues. These hydrogen ions then bind to hemoglobin, shifting the conformational equilibrium to the T state, thus lowering its binding affinity for oxygen, and so oxygen is released. The net effect of this process is to ensure that the tissues that most need oxygen are most likely to get it—a striking example of protein regulation by allosteric effects. The reverse process occurs in the lungs. As you exhale CO_2, the pH increases, and therefore the O_2 affinity of hemoglobin increases. As a result, the hemoglobin "reloads" with oxygen for another trip to the oxygen-requiring peripheral tissues.

Skeletal muscle glycogen phosphorylase, mentioned above as being regulated by phosphorylation, is also regulated allosterically by adenosine monophosphate (AMP). [High levels of AMP are indicative of a "low energy" state, since AMP is produced by the cleavage of adenosine triphosphate (ATP), a reaction that releases energy.] AMP binds to the inactive ("b") form of glycogen phosphorylase and makes it more active; as a result, when energy is needed, the tissue can begin the process of releasing the sugar (and therefore, eventually, energy) stored in glycogen (see Figure 15).

7   NONTRADITIONAL ENZYMES

For many years it was thought that all biological catalysts were enzymes, the highly evolved proteins described thus far. However, in

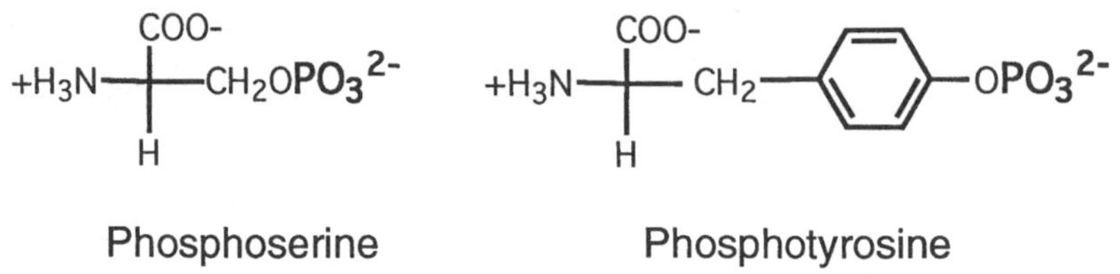

$$+H_3N - \underset{\underset{H}{|}}{\overset{\overset{COO-}{|}}{C}} - CH_2OPO_3{}^{2-}$$

Phosphoserine

$$+H_3N - \underset{\underset{H}{|}}{\overset{\overset{COO-}{|}}{C}} - CH_2 - \langle \bigcirc \rangle - OPO_3{}^{2-}$$

Phosphotyrosine

Figure 12.   Phosphorylated forms of serine and tyrosine, two commonly phosphorylated amino acids.

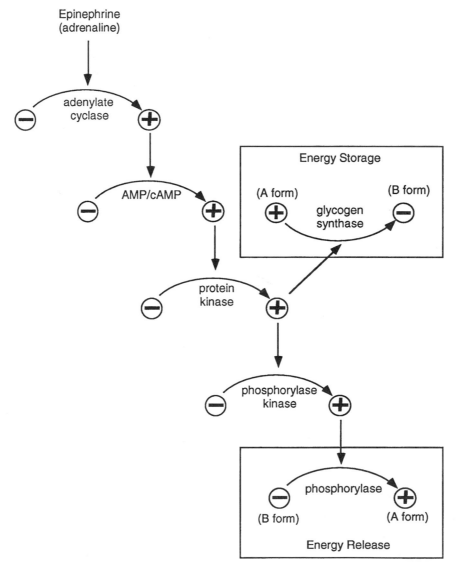

Figure 13.   A phosphorylation cascade. Plus signs indicate active forms, while minus signs refer to inactive forms. The binding of epinephrine to its receptor triggers the activation of adenylate cyclase. Once activated, an enzyme is free to turn over many times, which leads to amplification of the signal. If each step in the cascade multiplied the signal by only 10 times—surely a conservative estimate—the overall amplification that would result from the binding of *one* hormone molecule to the cell is 10,000 times. Note the coordinated regulation of the energy storage and energy release systems. See Figure 15 for more detail on the regulation of phosphorylase.

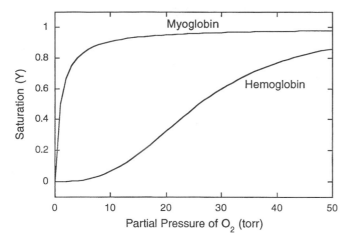

Figure 14. Oxygen-binding curves of hemoglobin and myoglobin. Myoglobin shows a hyperbolic binding curve, similar to enzymes that obey Michaelis–Menten kinetics. Hemoglobin, on the other hand, shows a sigmoidal binding curve indicative of cooperativity and allosteric regulation. Note that the oxygen affinity of hemoglobin is always lower than that of myoglobin, leading to hemoglobin's ability to release oxygen to myoglobin in peripheral tissues.

1986 two new types of biological catalyst were reported: catalytic RNA molecules and catalytic antibodies.

Catalytic RNA molecules, or *ribozymes,* discovered by Thomas Cech, are the first known example of nonprotein biological catalysts. Ribonucleic acid is another kind of biological polymer which is involved in the synthesis of proteins using the genetic information stored in a cell's DNA. RNA and DNA have different characteristics; the most important one for catalysis is the ability of RNA to assume complex three-dimensional structures. Biological catalysts must possess the ability to bind a specific substrate; the three-dimensional folds in an enzyme allow selective substrate binding. Catalytic RNA molecules are able to cut and splice themselves into a form that can then act catalytically on other RNA molecules. To date, the only ribozyme-catalyzed reactions known are cleavages of RNA and DNA. Ribozymes are true catalysts because they enhance

reaction rates without any net change to themselves. They are capable of turnover (recycling), and they show kinetics typical of enzymes.

A very interesting aspect of the discovery of catalytic RNA lies in the implications for evolution. It is conceivable that RNA molecules were in existence before DNA or proteins. It has been proposed by Walter Gilbert that RNA molecules first catalyzed their own replication and then developed other catalytic abilities. It is not difficult to envision that RNA molecules eventually developed the ability to synthesize proteins. Proteins, which were able to catalyze many more reactions with greater efficiency than ribozymes, then became the primary biological catalysts. It is speculated that DNA was first synthesized by reverse transcription of the genetic information on RNA and that DNA became the carrier of genetic information because it is chemically more stable than RNA.

Recent developments in the field of immunology have led to the development of another type of nontraditional enzyme, catalytic antibodies. Antibodies are protein molecules which bind foreign substances as part of the immune response. Although the technology to produce catalytic antibodies is relatively new, the ideas behind them date back almost half a century. As previously mentioned, in 1948 Linus Pauling proposed that enzymes bind to and stabilize the transition state of a reaction, leading to catalysis (see Section 5.1.1). The catalysis of reactions by antibodies is not a natural occurrence, because in nature antibodies, unlike enzymes, are designed to bind the *ground state* of molecules. In 1968 William Jencks suggested that antibodies could be selected for their ability to bind an analogue to the *transition state* of a reaction. Hopefully, these antibodies could be able to bind *substrates* and induce the true transition state of the reaction, just like an enzyme. It was not until the mid-1980's that techniques for obtaining pure (monoclonal) antibodies became available, and serious work on producing catalytic antibodies was possible.

In 1986, Richard Lerner and Peter Schultz separately showed enzymatic catalysis performed by an antibody. Each was trying to produce a catalyst for the hydrolysis of an ester. The general reaction scheme, shown in Figure 16, depicts the structural similarities between the transition state and its analog. Mice were injected with the transition state analogue, attached to a larger carrier molecule. The mouse's immune response then produced antibodies against

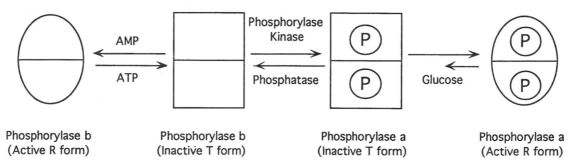

Figure 15. An abbreviated depiction of the regulation of glycogen phosphorylase. The function of this enzyme in the muscle is to release glucose stored as glycogen in response to some external stimulus that may, for example, cause fright. In response to extensive muscular activity, such as running, AMP is liberated. Binding of AMP to the inactive phosphorylase b will convert it to an active form so that it can begin the break down of glycogen. A more prolonged activation of phosphorylase is accomplished by the action of phosphorylase kinase which phosphorylates phosphorylase b converting it into the active form, phosphorylase a. Phosphorylase kinase is activated in response to the hormonal signal that is a central component of the fright response (see Figure 13). If the liberated glucose is not used by the muscle, its concentration will increase, resulting in binding to the active phosphorylase a thereby reducing its activity by stabilizing an inactive form of phosphorylase a.

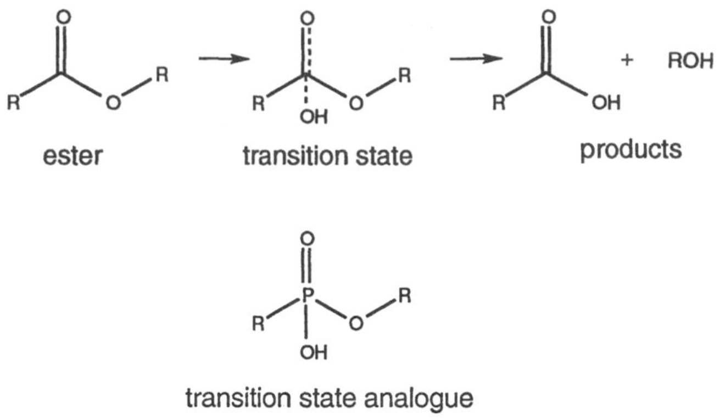

Figure 16. The partial mechanism for the hydrolysis of an ester and a transition state analogue.

the foreign molecule. As expected, these antibodies were able to catalyze the ester hydrolysis by transition state stabilization. The fact that catalytic antibodies can be produced by this method is a major validation of the hypothesis that transition state stabilization contributes significantly to reaction rate enhancement by enzymes.

Since this landmark event, many other kinds of reaction catalyzed by antibodies have been reported. All show the Michaelis–Menten saturation kinetics expected for enzymatic reactions. Many catalytic antibodies have been engineered to be more efficient by making use of catalytic side chains, transition metals, and coenzymes. To date there has not been a catalytic antibody with the impressive selectivity or rate enhancement of an enzyme, but work in this field is progressing rapidly.

8   PERSPECTIVES

Enzymology has a long and exciting history. In large part because of the development of recombinant DNA technology, we are now able to isolate in pure form large amounts of numerous enzymes that otherwise could not be isolated from the native biological source. By recombinant DNA methodology, scientists are now able to alter the catalytic activity of natural enzymes. Recent improvements in physical methods, including NMR spectroscopy and X-ray diffraction from crystals of large molecules, are allowing dramatic advances to be made in our understanding of the mechanisms and modes of communication between enzymes and other biological macromolecules. It would appear that the glory days of enzymology are not behind us, but before us.

See also POST- AND CO-TRANSLATIONAL MODIFICATIONS OF PROTEINS, ENZYME CATALYZED; PROTEIN FOLDING.

Bibliography

Baldwin, T. O., Raushel, F. M., and Scott, A. I., Eds. (1991) *Chemical Aspects of Enzyme Biotechnology—Fundamentals.* Plenum Press, New York.
Cornish-Bowden, A., and Wharton, C. W. (1988) *Enzyme Kinetics.* IRL Press, Washington, DC.
Denbigh, K. (1992) *The Principles of Chemical Equilibrium,* 4th ed. Cambridge University Press, Cambridge.
Dixon, M. D., and Webb, E. C. (1979) *Enzymes.* Academic Press, New York.
Fersht, A. (1985) *Enzyme Structure and Mechanism,* 2nd ed. Freeman, New York.
Jencks, W. P. (1969) *Catalysis in Chemistry and Enzymology.* McGraw-Hill, New York.
Stryer, L. (1988) *Biochemistry,* 3rd ed. Freeman, New York.
Walsh, C. (1979) *Enzymatic Reaction Mechanisms.* Freeman, New York.

ENZYMES, HIGH TEMPERATURE

Michael W. W. Adams

1   **Introduction**

2   **Sources of High Temperature Enzymes**

3   **Properties of High Temperature Enzymes**

4   **Molecular Characterization of High Temperature Enzymes**

5   **Protein Hyperthermostability**

Key Words

Amylase   An enzyme that degrades starch.

Archaea   Formerly Archaebacteria, one of the three domains into which all life-forms have been classified, in addition to Bacteria, which includes the majority of microorganisms, and Eukarya, which consists of all higher life-forms.

Dehydrogenase   An enzyme that catalyzes a reaction involving electron transfer to or from an external electron carrier, which is typically NAD or NADP.

Ferredoxin   A small iron–sulfur cluster containing protein that transfers electrons between enzymes.

Hydrogenase   An enzyme that catalyzes the reversible activation of hydrogen gas.

Hydrothermal Vent   A deep-sea volcanic site from which superheated, mineral-rich water emanates at temperatures approaching 400°C.

Hyperthermophile   A microorganism with an optimum growth temperature of at least 80°C and a maximum growth temperature of 90°C or above.

Mesophile   An organism that grows optimally in the range of 20–40°C, which includes virtually all known life-forms.

Oxidoreductase   An enzyme that catalyzes a reaction involving electron transfer to or from an external electron carrier, which is typically a redox protein.

Protease   An enzyme that catalyzes the hydrolysis of (specific) peptide bonds within proteins.

Universal Ancestor   The original life-form(s) from which all present-day organisms evolved.

Microorganisms have been discovered in the last few years that grow optimally at temperatures near and above 100°C, the normal

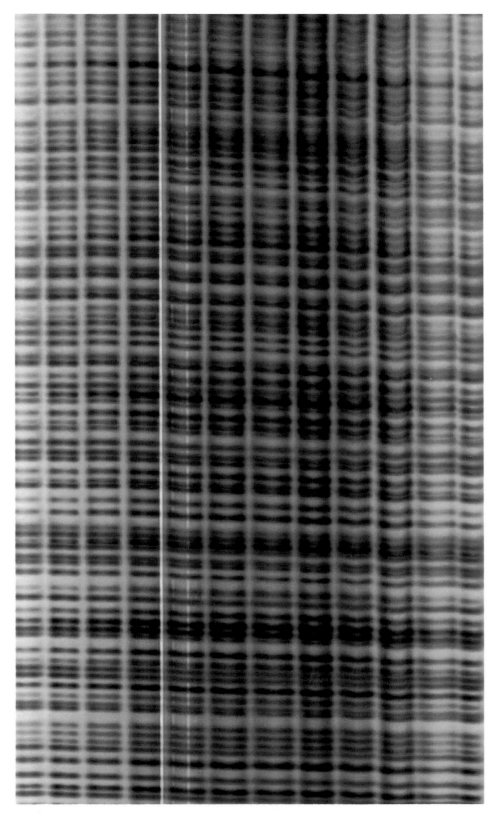

Plate 1. Portion of an "image file" obtained from a HUGE sequencing gel of 12 identical samples. The different colored bands correspond to the four different bases in the DNA sequence. (See DNA Sequencing in Ultrathin Gels, High Speed, Figure 5.)

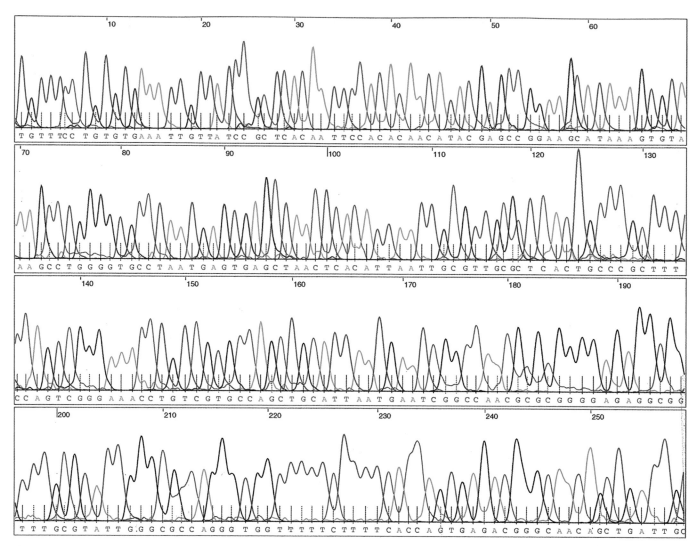

Plate 2. Portion of a set of base-called sequence data generated by fluorescence-based sequencing on the HUGE system. The region shown was obtained in about 40 minutes. (See DNA Sequencing in Ultrathin Gels, High Speed, Figure 6.)

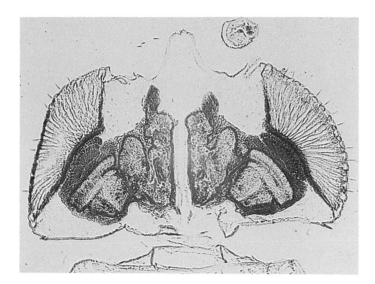

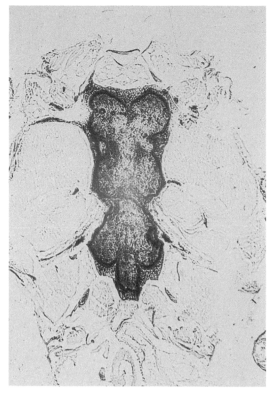

Plate 3. Acetylcholinesterase (AChe) in the central nervous system of the adult. *D. melanogaster;* the major features are shown by staining for the activity of this neurotransmitter-metabolizing enzyme. The histochemical (non-antibody-related) procedure was performed on horizontal sections through the head (top panel) and the thorax (bottom). Within each panel, anterior is to the top. This particular brain section was taken at the level of the esophagus (the empty-appearing channel running through the head's CNS); flanking this part of the alimentary system are ganglia of the CNS per se; in more lateral locations are four pair of optic ganglia, situated between the brain ganglia and the compound eyes. The latter tissue appears red because of its inherent pigment (i.e., the photoreceptor and other cells within the eye are not AChE-positive). The scale of this organism's anterior CNS is such that the distance between the outermost optic ganglia (those just underneath the eyes proper) is about 450 μm. The bottom panel shows (at approximately the same scale) the ventral, thoracic CNS (and a few AChE-positive neuronal processes, which were in part caught within this plane of section). This portion of the fly's nervous system consists of four pair of fused ganglia, the three thoracic ones per se, and (at the bottom of the image), the "abdominal" ganglion (located in the posterior thorax, but projecting axons into the abdomen). (See Drosophila Mating, Genetic Basis of, Figure 3.)

Plate 4. Fluorescence in situ hybridization of YAC 209G4 to chromosome preparations of a fragile X patient (B). A centromere probe specific for the X chromosome was also hybridized to confirm the identity of the chromosome examined. Staining by 4-,6-diamino-2-phenylindole (DAPI) clearly shows the fragile site (A). (In situ hybridizations performed by B. H. J. Eussen.) (See Fragile X Linked Mental Retardation, Figure 4.)

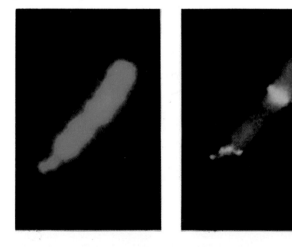

A   **B**

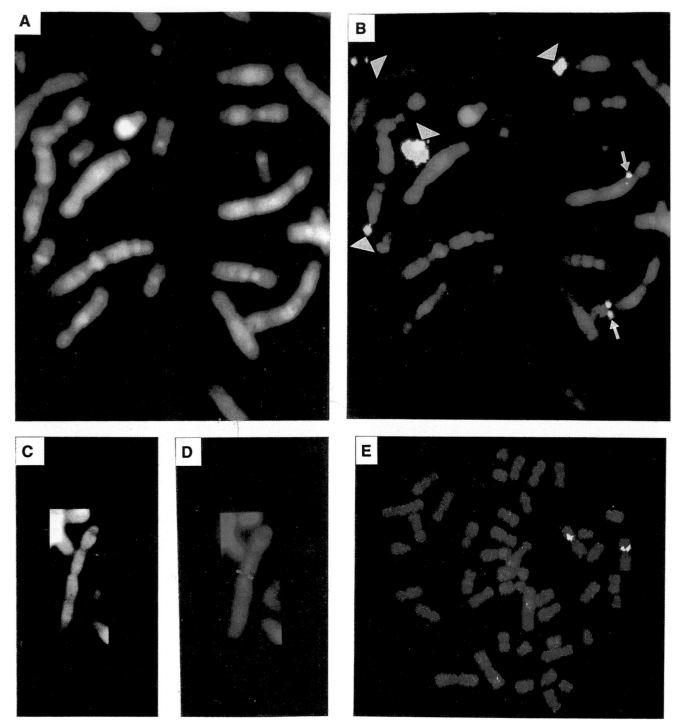

Plate 5. (A) Partial metaphase of Hoechst 33258 stained human chromosomes. This fluorescent dye preferentially binds to AT-rich stretches of DNA creating a G-(Q-)like banding pattern. (B) The same image after computer alignment, merging, and assignment of false colors. The hybridization signal for a small single-copy probe is seen on both homologues (small arrows). In comparison, the large signal obtained with a repeat sequence probe localized to the acrocentric chromosomes is shown (arrowheads). Images were obtained using an epifluorescence microscope, RCA-ISIT camera, MRC 500 workstation, and a Sony video printer. (C) DAPI stained Q-banded metaphase chromosome. (D) The same image as (C), showing two single-copy probes labeled with Fitc (green), and Texas red (red), respectively, localized relative to each other on the DAPI-stained (blue) chromosome. Images were obtained using an epifluorescence microscope, CCD camera, and MRC 500 workstation. The photographs were generated by sequential photography from the computer color monitor screen. (E) Fluorescent dye DAPI used in conjunction with propidium iodide gives an enhanced R-banding pattern (red). The image is merged and assigned false colors using appropriate computer software. The yellow Fitc hybridization signal is that of a YAC probe. All four sister chromatids are consistently labeled with such large probes. The image was obtained using a confocal microscope, BioRad MRC 600 workstation, and screen photography. (See Gene Mapping by Fluorescence In Situ Hybridization, Figure 3.)

Plate 6. Ordering three closely linked cosmid clones using hybridization to the extended chromatin fiber released by alkali treatment of fixed nuclei. The cosmids, each approximately 40 kb long, are differentially labeled by Texas Red (red), FITC (green), and a 50:50 mixture of Texas Red and FITC (giving an orange-yellow color). There is a 5 kb overlap between the red and green labeled cosmids indicated by the yellow signal. The yellow cosmid is separated from the red cosmid by a 50 kb gap. (See Gene Order by FISH and FACS, Figure 2.)

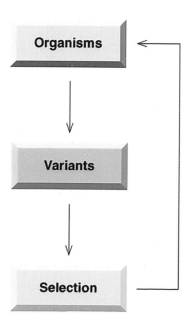

Plate 7. A conventional view of evolution. The environment functions only at the selection step. (See Genetic Intelligence, Evolution of, Figure 1.)

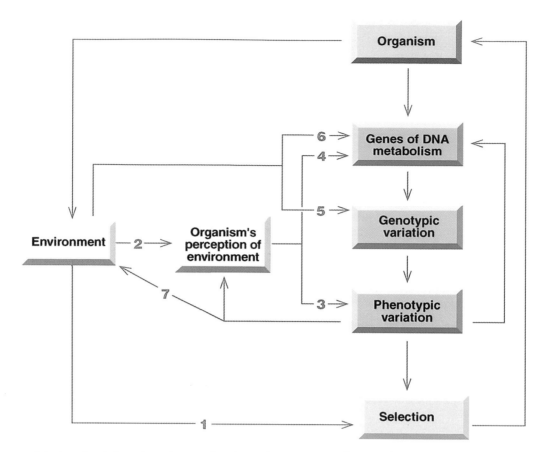

Plate 8. A more complete view of evolution—the environment functions at three stages according to the following protocol: 1, the environment is the proximate agent of selection; 2, the environment is perceived by the organism; 3, organisms use their perception of the environment to modify their physiology, as in operon induction; 4, organisms use their perception of the environment to modify their genetic metabolism, as in the SOS and p53 pathways; 5, the environment directly impinges on the DNA directly via such agents as radiation and chemical mutagens; 6, the environment can interact indirectly on DNA via the genes of DNA metabolism, as in topoisomerase inhibitors in chemotherapy; 7, the organism modifies environmental interaction with the genome as in metabolic activation, or, detoxification. (See Genetic Intelligence, Evolution of, Figure 2.)

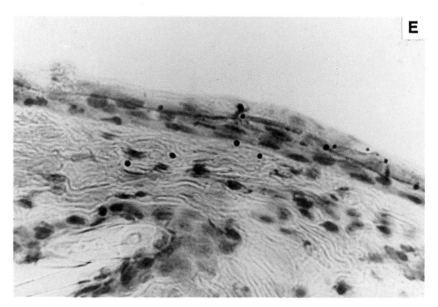

Plate 9. Design of the electric discharge mediated particle bombardment device for gene transfer. (E) Histochemical localization to superficial layers of epidermis of β-galactosidase activity encoded by naked DNA following particle bombardment. (See Genetic Vaccination, Gene Transfer Techniques for Use in, Figure 2E.)

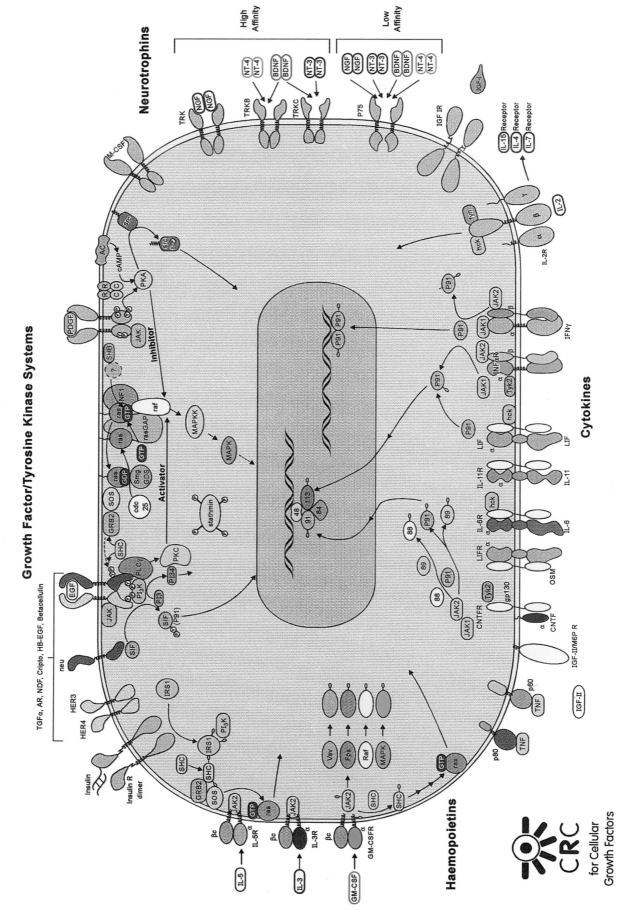

Plate 10. Growth factors/cytokines and their receptor signaling systems. (See Growth Factors, Figure 2.)

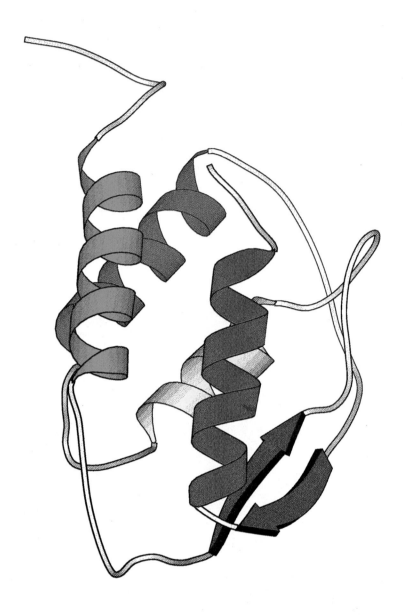

Plate 11. The three-dimensional structure of GM-CSF illustrates the typical four-helical bundle found in many of the hemopoietic regulators. (See Growth Factors, Figure 3.)

boiling point of water. Most of these so-called hyperthermophiles have been found in marine volcanic environments, which include deep-sea hydrothermal vents. They are the most slowly evolving organisms known at present. The majority are strict anaerobes and depend on the reduction of elemental sulfur (S^0) to hydrogen sulfide (H_2S) for optimal growth. Most of them utilize proteins, peptides, and sugars as growth substrates, and these are converted to simple organic acids and gases such as hydrogen and carbon dioxide. They are potentially rich sources of a variety of "hyperthermostable" enzymes, which are able to catalyze reactions at extreme temperatures. So far about 25 different enzymes and proteins have been purified and characterized from the sulfur-reducing species. Virtually all have an optimal temperature for catalysis above 100°C, and some have half-lives at this temperature of several days. However, characterization of these enzymes at the molecular level using recombinant DNA techniques is at an early stage. Moreover, detailed structural information is available on only two hyperthermophilic proteins, and insight into the molecular mechanisms by which enzymes are stabilized at temperatures near and above 100°C is just beginning to emerge.

1   INTRODUCTION

A revelation of great significance in the field of microbiology with profound implications in microbial metabolism, biochemistry, and biotechnology occurred in 1982 when microorganisms that grew above 100°C were isolated from shallow marine volcanic vents. Although we know little about the novel biochemistry that must be required to sustain these organisms under such conditions, they are potential sources of a range of "hyperthermostable" or high temperature enzymes. What is currently known about hyperthermophilic organisms and the enzymes that have been characterized from them is discussed, with an emphasis on the enzymes from the "sulfur-dependent" species, the predominant type of hyperthermophile known at present. The majority of these enzymes are intrinsically

thermostable and catalytically active at temperatures above the normal boiling point of water. However, the molecular mechanisms that impart such properties are far from clear.

2   SOURCES OF HIGH TEMPERATURE ENZYMES

"Hyperthermophiles" are defined here as organisms able to grow at 90°C and above with an optimal growth temperature of at least 80°C. Although this is a somewhat arbitrary definition, inspection of Figure 1 shows that such organisms were discovered only in the early 1980s, and since that time there has been an explosion in the number of species isolated. They now form a fairly cohesive group, which is in contrast to the large number of moderately thermophilic organisms growing in the range of 50–70°C that have been isolated over the past several decades. Figure 1 also shows that only two of the 20 or so hyperthermophilic genera currently known are conventional bacteria. The majority of them are classified as *Archaea* (formerly archaebacteria). Based on molecular (rRNA) sequence analyses, *Archaea* were recognized as a third domain of life in the late 1970s, thus distinguishing them from *Bacteria* and higher life-forms or *Eukarya.* Interestingly, these molecular studies also indicated that the archaea and eukarya have a common ancestor that is not shared by the bacteria. Moreover, both the hyperthermophilic archaea and the two hyperthermophilic bacterial genera are the most slowly evolving organisms within their respective domains, the first to have diverged from some universal ancestral life-form. Thus, the rest of present-day biology may well be the result of evolutionary pressures to adapt to low (< 100°C) temperatures. Accordingly, present-day hyperthermophiles may be the key to understanding the biochemical evolution of the enzymes and metabolic pathways found in more conventional organisms, and particularly those in higher life-forms.

All of hyperthermophilic genera currently known have been isolated from geothermally heated environments. As indicated in Table

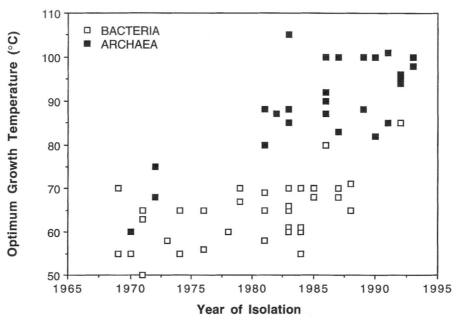

Figure 1.   Isolation of thermophilic microorganisms over the past three decades.

Table 1 Hyperthermophilic Genera: Organisms That Grow at 90°C and Above

Genus	T_{max} (C)[a]	%GC	Habitat[b]
Sulfur-Dependent Archaea			
Thermoproteus	102	56	c
Staphylothermus	98	35	d
Desulfurococcus	90	51	d,c
Thermofilum	100	57	c
Pyrobaculum	102	46	c,m
Acidianus	96	31	m/c
Pyrodictium	110	62	d/m
Thermodiscus	98	49	m
Pyrococcus	105	38	d/m
Thermococcus	97	57	d/m
Hyperthermus	110	56	m
Sulfate-Reducing Archaea			
Archaeoglobus	95	46	d/m
Methanogenic Archaea			
Methanococcus	91	31	d/m
Methanothermus	97	33	c
Methanopyrus	110	60	d
Bacteria			
Thermotoga	90	46	m
Aquifex	95	40	m

[a]Maximum growth temperature.
[b]Species have been isolated from shallow marine vents (m), deep-sea hydrothermal vents (d), and/or continental hot springs (c).

Table 2 Physiology of Sulfur-Reducing Hyperthermophiles

Species	T_{opt} (°C)	Obligately S[0] Dependent[a]
Obligate Anaerobes		
Heterotrophic, Proteolytic		
Desulfurococcus mobilis, D. mucosus	87	Yes
Desulfurococcus strain "SY"	90	Yes
"ES-1"	82	Yes
Hyperthermus butylicus	100	Yes
Pyrobaculum organotrophum	100	Yes
Pyrococcus abyssi	98	No
Pyrococcus "GB-D"	95	Yes
Staphylothermus marinus	92	Yes
Thermococcus celer	88	Yes
Thermodiscus maritimus	90	Yes
Thermofilum librum	88	Yes
Thermofilum pendens	88	Yes
Thermoproteus uzoniensis	90	No
Heterotrophic, Proteolytic, and Saccharolytic		
Desulfurococcus amylolyticus	90	Yes
"ES-4"	100	Yes
Pyrococcus furiosus	100	No
Pyrococcus woesei	100	No
Pyrodictium abyssi	105	No
Thermococcus litoralis	88	No
Thermococcus stetteri	80	Yes
Thermotoga maritima	80	No
Thermotoga neapolitana	80	No
Autotrophic		
Pyrobaculum islandicum	100	Yes
Pyrodictium brockii	105	Yes
Pyrodictium occultum	105	Yes
Thermoproteus neutrophilus	88	Yes
Thermoproteus tenax	88	Yes
Facultative Anaerobes		
Autotrophic		
Acidianus infernus	90	No
Aquifex pyrophilus	85	No
Heterotrophic		
Pyrobaculum aerophilum	100	No[b]

[a]Species grow poorly, if at all, if S[0] is not added to the growth medium.
[b]The growth of *Pyrobaculum aerophilum* is uniquely inhibited by S[0].

1, some are found in continental hot springs, but the majority are of marine origin, which includes hot sediments found in coastal waters at depths down to 100 m or so, and deep-sea hydrothermal vents, located up to 4000 m below sea level. Table 1 also shows that the majority of the hyperthermophiles are referred to as sulfur-dependent organisms, since their growth depends to a greater or lesser extent on the reduction of elemental sulfur (S^0) to hydrogen sulfide (H_2S). This group also includes the two bacterial genera, *Thermotoga* and *Aquifex*. The physiological properties of these S^0-reducing species are summarized in Table 2. Almost all of them grow only under strictly anaerobic conditions. In fact, of the three species able to utilize oxygen, two of them (the bacterium *Aquifex pyrophilus* and the archaeon *Acidianus infernus*) are microaerophilic and tolerate oxygen only up to 5% (v/v).

In addition, Table 2 shows that most of the S^0-reducing hyperthermophiles are obligately dependent on S^0 reduction for growth. Therefore, those that can grow well in the absence of S^0, which include species of the archaeal genera *Pyrococcus, Thermococcus, Thermoproteus,* and *Pyrodictium,* and species of the bacterial genus *Thermotoga,* have fermentative- rather than respiratory-type metabolisms. Table 2 also shows that most hyperthermophiles are strict organotrophs, which use as their carbon and nitrogen sources, such complex protein-containing mixtures as yeast and meat extracts, tryptone, peptone, and casein.

Several of the hyperthermophiles are also able to use carbohydrates as additional or primary C sources. In this regard the archaea have a limited substrate range, which typically includes both complex carbohydrates such as starch and glycogen, and disaccharides like maltose and cellobiose. In contrast, the saccharolytic bacterium *Thermotoga maritima* is able to utilize a wide range of mono-, di-, and also more complex carbohydrates. In any event, the heterotrophic hyperthermophiles are potential sources of a variety of hydrolytic-type enzymes, such as proteases, isomerases, and amylases, enzymes that are of some industrial importance.

3   PROPERTIES OF HIGH TEMPERATURE ENZYMES

The enzymes that have been purified so far from the S^0-dependent hyperthermophilic genera are listed in Table 3. Because of the prob-

Table 3   Enzymes Purified from S^0-Reducing Hyperthermophiles

Enzyme	T_{opt} [a] (°C)[a]	$t_{50\%}$ (h/°C)[b]	Catalytic Activity	Organism[c]
Hydrolases				
Protease	115	33/98	Peptide hydrolysis	Pf
Protease	100	1.5/95	Peptide hydrolysis	Dm
Amylase	100	2/120	Starch hydrolysis	Pf
Amylase	100	6/100	Starch hydrolysis	Pw
Amylopullulanase	118	20/98	Starch degradation	ES
α-Glucosidase	110	48/98	Maltose hydrolysis	Pf
β-Glucosidase	105	13/110	Cellobiose hydrolysis	Pf
Sucrose α-glucohydrolase	105	48/95	Sucrose hydrolysis	Pf
Xylanase	105	1.5/95	Xylan hydrolysis	TsF
Cellobiohydrolase	105	1.1/108	Cellulose hydrolysis	TsF
Oxidoreductases (ferredoxin-dependent)				
Pyruvate oxidoreductase	>95	0.3/90	Pyruvate to acetyl CoA	Pf
Pyruvate oxidoreductase	>95	15/80	Pyruvate to acetyl CoA	Tm
Aldehyde oxidoreductase	>95	6/80	Aldehydes to acids	Pf
Indolepyruvate oxidoreductase	>95	6/80	Indolepyruvate to aryl CoA	Pf
Formaldehyde oxidoreductase	>95	2/80	Aldehydes to acids	Tl
Hydrogenases/Redox Proteins				
Hydrogenase	>95	1/90	H_2 production	Tm
Hydrogenase	>90	1/98	H_2 oxidation	Pb
Hydrogenase	>95	2/100	H_2 production	Pf
Ferredoxin	>95	>24/95	Electron transfer	Pf
Ferredoxin	>95	>24/95	Electron transfer	Tl
Ferredoxin	>95	>24/95	Electron transfer	Tm
Rubredoxin	>95	>24/95	Electron transfer	Pf
Dehydrogenases (NAD/P-dependent)				
Glutamate dehydrogenase	>95	10/100	Glutamate to 2-OG[d]	Pf
Glutamate dehydrogenase	>95	1.5/105	Glutamate to 2-OG	Pw
Glutamate dehydrogenase	>95	10/100	Glutamate to 2-OG	ES
Glutamate dehydrogenase	>95	2/100	Glutamate to 2-OG	Tl
GAPDH	Nd[e]	0.7/100	GAP to BPG	Pw
GAPDH	Nd	0.3/100	GAP to BPG	Tt
GAPDH	>90	2/95	GAP to BPG	Tm
Sulfide dehydrogenase	>95	12/95	Sulfur to hydrogen sulfide	Pf
Alcohol dehydrogenase	80	2/85	Aldehydes to alcohols	Tl
Lactate dehydrogenase	>90	1.5/90	Pyruvate to lactate	Tm
Nucleic Acid Modifying Enzymes				
DNA polymerase	>75	20/95	DNA replication	Pf
DNA polymerase	>75	7/95	DNA replication	Tl
RNA polymerase	Nr[f]	Nr	RNA synthesis	Tm
Other Enzymes				
Aromatic transaminases (I, II)	>95	Nr	Amino acids to 2-oxoacids	Tl
Enolase	>95	0.7/100	2-PG to PEP	Pf
Glucose isomerase	110	0.2/115	Glucose to fructose	Tm

[a] Optimum temperature for catalytic activity in vitro.

[b] Time required to lose 50% of catalytic activity after incubation at the indicated temperature.

[c] The organisms and the abbreviations used are: Pf, *Pyrococcus furiosus;* Dm, *Desulfurococcus mucosus;* Pw, *Pyrococcus woesei;* ES, ES-4; Tl, *Thermococcus litoralis;* Pb, *Pyrodictium brockii;* Tt, *Thermoproteus tenax;* Tm, *Thermotoga maritima;* Tsf, *Thermotoga* sp. strain FjSS3-B.1; Tc, *Thermococcus celer.*

[d] The abbreviations used are: 2-OG, 2-oxoglutarate; GAPDH, glyceraldehyde 3-phosphate dehydrogenase; GAP, glyceraldehyde 3-phosphate; BPG, 1,3-bisphosphoglycerate; 2PG, 2-phosphoglycerate; PEP, phosphoenolpyruvate.

[e] Nd, not determined because of the instability of the substrates at high temperature.

[f] Nr, not reported.

lems of culturing organisms that produce toxic and corrosive H_2S, especially at high temperatures, most of these enzymes have been obtained from species of *Pyrococcus, Thermococcus,* and *Thermotoga,* organisms that can be grown to high cell densities in the absence of S^0. In general, the types of enzyme purified so far reflect the ability of these organisms to grow by fermentative-type metabolisms using both proteins and various carbohydrates as substrates. In most cases, although not all, analogous enzymes have been characterized from more conventional mesophilic organisms.

The hydrolytic-type enzymes that have been purified are all involved in metabolizing the primary growth substrates of the hyperthermophiles. High protease activity has been measured in extracts of many of the hyperthermophiles, and some appear to contain perhaps as many as a dozen different enzymes with proteolytic activity. So far, however, proteases have been purified from only two species. An extracellular protease was obtained from a species of *Desulfurococcus.* Substrate specificity studies indicated that the enzyme had a preference for hydrophobic residues. The pure enzyme was active up to 125°C.

In addition, a serine-type protease has been purified from the cytoplasm of *Pyrococcus furiosus.* It was isolated by boiling cell-free extracts with the detergent, sodium dodecyl sulfate (SDS, 1%), for 24 hours. This procedure destroyed all but four of the cellular proteins, and two of these, termed S102 and S66 (to reflect their relative molecular weights) had proteolytic activity. S66, which appeared to be a proteolytic product of S102, was purified further. This has a half-life of activity of about 1.5 days at 98°C.

In addition to proteases, hyperthermophiles that metabolize carbohydrates produce a variety of enzymes that are able to utilize starch, maltose, cellulose, and/or xylan as substrates, and several of these enzymes have been purified and characterized. α-Amylase, which hydrolyzes the α-1,4-glucosidic linkages of starch, is produced both intra- and extracellularly by *Pyrococcus furiosus,* and the purified enzyme is active up to 140°C. This organism also contains an intracellular α-glucosidase, which further degrades the products of starch hydrolysis to glucose. The activity of the pure enzyme has been measured up to 130°C.

On the other hand, the novel hyperthermophile ES-4 produces an unusual extracellular amylolytic enzyme, amylopullulanase, which hydrolyzes both α-1,4- and α-1,6-glucosidic linkages. This enzyme is stabilized by calcium ions, which increased the upper temperature limit for activity from 130°C to 140°C. Other saccharolytic enzymes purified from *Pyrococcus furiosus* include a β-glucosidase, which hydrolyzes the growth substrate cellobiose (a disaccharide), and sucrose α-glucohydrolase (or invertase), which hydrolyzes sucrose. The latter is not a growth substrate, and the function of this enzyme in *Pyrococcus furiosus* is not known.

Additional types of saccharolytic enzyme have been purified from species of the bacterial genus *Thermotoga,* reflecting their ability to utilize a wider range of carbohydrates than the saccharolytic archaea. These include xylanase and glucose isomerase, which are primary enzymes in the utilization of the polysaccharide xylan and the monosaccharide xylose, respectively.

One *Thermotoga* strain also produces an extracellular cellulose-degrading cellobiohydrolase, although the organism does not utilize cellulose as a growth substrate. Nevertheless, this is the most thermostable cellulase yet reported, with a temperature optimum of 105°C. Indeed, some of the saccharolytic enzymes from the hyperthermophiles represent the most stable enzymes yet characterized from any organism, with half-lives of 1–2 days near 100°C (Table 3). The activities of *Pyrococcus woesei* amylase and ES-4 amylo-

pullulanase have been measured at 130 and 135°C, respectively, and the immobilized xylanase from the *Thermotoga* strain retained 25% of its activity after an hour at 130°C in molten sorbitol.

Many of the other enzymes that have been purified from hyperthermophiles are involved in the further metabolism of the products of carbohydrate and protein hydrolysis. In the archaea these include a range of unusual oxidoreductases that couple the oxidation of various oxo acids and aldehydes to the reduction of a low molecular weight protein termed ferredoxin. Reduced ferredoxin then functions to provide reducing equivalents for both H_2 production and the reduction of S^0. The ferredoxin-linked oxidoreductases shown in Table 3 that oxidize formaldehyde (FOR) to produce formate and indolepyruvate (IOR) to produce indole acetyl CoA are thought to be involved in the fermentation of the amino acids produced by proteolytic activity. The precise role of FOR is not known, whereas the substrates of IOR, which include indolepyruvate, *p*-hydroxyphenylpyruvate, and phenylpyruvate, are the transaminated forms of the aromatic amino acids. These derivatives are generated by two aromatic amino transferases, which have been purified from both *Pyrococcus furiosus* and *Thermococcus litoralis.* The various aryl CoAs produced by IOR are believed to be used both for energy conservation and as carbon sources for biosynthesis.

The two other ferredoxin-linked oxidoreductases listed in Table 3 utilize either aldehydes to produce the corresponding acid (AOR) or pyruvate (POR) to generate acetyl CoA. Both enzymes appear to be involved in the fermentation of glucose, which is generated by the saccharolytic enzymes. Although POR and IOR catalyze virtually identical reactions, POR from *Pyrococcus furiosus* is a heterotetrameric enzyme, whereas IOR from the same organism has only two types of subunit. POR has also been purified from *Thermotoga maritima,* and this is also a heterotetrameric enzyme. On the other hand, the aldehyde-metabolizing enzymes AOR and FOR are closely related structurally despite their apparently different physiological roles. Both contain an iron–sulfur cluster, and they are unusual in that their catalytic sites contain tungsten, an element seldom used in biological systems. Enzymes analogous to AOR, FOR, and IOR have not been found in conventional mesophilic organisms, nor in species of *Thermotoga;* hence these enzymes appear to be unique to the hyperthermophilic and heterotrophic archaea. Indeed, the pathways utilized to metabolize carbohydrates and peptides by *Thermotoga* appear to be quite different from those in the archaea, with those of *Thermotoga* more resembling the metabolic routes found in more conventional bacteria.

Ferredoxin has been purified from several of the hyperthermophilic archaea and also from *Thermotoga maritima.* Analogous proteins have also been purified from a wide range of mesophilic bacteria. Hyperthermophilic ferredoxins are typical of this class of redox protein. They are comprised of about 65 amino acids, and they contain a single iron–sulfur cluster as the entity that serves to donate and accept electrons. The cluster comprises four iron and four inorganic sulfur atoms in a cube-type arrangement, which is covalently attached to the protein via the sulfur atoms of four cysteine residues. Such [4Fe–4S] clusters are ubiquitous electron carriers in biological systems in general and are found in many oxidoreductase-type enzyme. These also include those from the hyperthermophiles just described (AOR, FOR, IOR, and POR) and those in hydrogenase and sulfide dehydrogenase, which are discussed shortly. In all these enzymes, the 4Fe cluster serves an electron transfer role. In addition, several mesophilic enzymes employ a [4Fe–4S] cluster at their catalytic site to bind substrates. Of course, the hyperthermophilic ferredoxins are distinguished from their mesophilic cousins, the most stable of which denature at 80°C, by their remarkable thermal stability, being unaffected by a 24-hour incubation at 95°C. That of *Pyrococcus furiosus* is also unusual in that its [4Fe–4S] cluster is liganded by only three rather than four cysteine residues. This imparts unusual spectroscopic properties to the cluster, and also chemical reactivity such that in vitro it will bind exogenous ligands such as CO and cyanide. Hence, because the *Pyrococcus furiosus* protein is so stable, it has been extensively analyzed by an array of spectroscopic techniques to gain fundamental electronic and magnetic information on the properties of its cluster. In addition, this ferredoxin is being used as a stable model system to mimic the ligand-binding properties of more complex mesophilic enzymes.

In the fermentative hyperthermophiles, reduced ferredoxin is utilized to provide reducing equivalents ultimately for the production of hydrogen gas (H_2) or for the reduction of S^0 to H_2S. Hydrogenase is the enzyme responsible for catalyzing H_2 production, and this enzyme has been purified from the archaeon *Pyrococcus furiosus* and from the bacterium *Thermotoga maritima*, both of which produce H_2 during growth. An H_2-oxidizing hydrogenase has been also purified and characterized from *Pyrodictium brockii*, and obligately autotrophic organism that obtains energy for growth by using H_2 to reduce S^0 to H_2S via a membrane-associated electron transport system. In fact, *Pyrodictium brockii* hydrogenase is the only enzyme listed in Table 3 that is membrane-bound: all the others are cytoplasmic enzymes. The hydrogenases of *Pyrodictium brockii* and *Pyrococcus furiosus* both contain iron–sulfur clusters of the type found in the ferredoxins, but H_2 catalysis by these enzymes takes place at a monomeric nickel site. Nickel-containing hydrogenases have also been purified from a variety of H_2-metabolizing mesophilic organisms. On the other hand, *Thermotoga maritima* hydrogenase lacks nickel, and H_2 production occurs at a novel type of iron-sulfur termed the "H" cluster, the structure of which is not known. This cluster was first identified in hydrogenases that were purified from conventional anaerobic bacteria. Hence, there is a fundamental difference between the hydrogenases of the hyperthermophilic archaea and the hyperthermophilic bacteria.

In addition to ferredoxin, a low molecular weight redox protein known as rubredoxin has been purified from *Pyrococcus furiosus*. Rubredoxin contains a mononuclear iron atom coordinated to the sulfur atoms of four cysteine residues. Such proteins have been purified from anaerobic bacteria of several different types, although their physiological function is not known. As with *Pyrococcus furiosus* ferredoxin, the high stability of the rubredoxin from this organism makes it particularly useful for detailed spectroscopic analyses to investigate the nature of its metal site as well as metal–protein interactions, and several such studies have been carried out. In addition, *Pyrococcus furiosus* rubredoxin, together with the aldehyde oxidoreductase of this organism, are of some importance because they are the only hyperthermophilic proteins for which three-dimensional structures are available. The results from these studies are discussed later.

Several dehydrogenase-type enzymes have also been purified from the hyperthermophiles. These utilize the nicotinamide nucleotides NAD and/or NADP as electron carriers rather than the redox protein ferredoxin. However, these cofactors have half-lives near 100°C of only a few minutes, and it is not known how they are stabilized inside the cell. Interestingly, when NAD was added to lactate dehydrogenase from *Thermotoga maritima*, the half-life of catalytic activity at 90°C increased from 2 minutes to 150 minutes, suggesting that the cofactor and the enzyme may serve to stabilize each other at the growth temperature of the organism. Similarly,

substrate stability may also limit the characterization of hyperthermophilic enzymes. For example, the substrates for glyceraldehyde 3-phosphate dehydrogenase (GAPDH), which has been purified from *Pyrococcus woesei*, *Thermoproteus tenax*, and *Thermotoga maritima*, are very unstable and reliable enzymatic assays are not possible much above 70°C. Potential means of stabilizing such compounds within the cytoplasm of hyperthermophiles are discussed in Section 5.

The heterotrophic hyperthermophilic archaea contain surprisingly high intracellular concentrations of glutamate dehydrogenase, which can represent up to 20% of the cytoplasmic protein. This enzyme has been found in all the heterotrophic archaea so far examined, but its activity is not detectable in cell-free extracts of *Thermotoga maritima*. In the archaea, glutamate dehydrogenase is thought to catalyze glutamate oxidation when peptides are used as the growth substrate. This reaction generates 2-oxoglutarate, which can be used for biosynthesis. However, why this enzyme is present at such high intracellular concentrations is not known. So far it has been purified and characterized from four archaeal species (Table 3). The properties of all these enzymes resemble those of their analogues from mesophilic organisms, except of course, in their stability at high temperature. Although glutamate dehydrogenase is not present (or is present only at low levels) in species of *Thermotoga*, the latter contain significant amount of GAPDH, an enzyme that is present only in very low concentrations in the hyperthermophilic archaea. Similarly, lactate dehydrogenase has been purified from *Thermotoga*, but this enzyme is not present in species of *Pyrococcus* or *Thermococcus*, again, reflecting the different metabolic pathways used by these heterotrophic organisms for the metabolism of peptides and carbohydrates. Paralleling the case of glutamate dehydrogenase from the archaea, the molecular and catalytic properties of lactate dehydrogenase from *Thermotoga maritima* are very similar to those of the enzyme from mesophilic organisms.

Alcohol dehydrogenase, which interconverts alcohols and aldehydes, has been purified from *Thermococcus litoralis* and is present in the novel deep-sea isolate ES-1 (although not yet purified). This activity is not found in *Thermotoga* and is also absent (or present at very low levels) in *Pyrococcus furiosus*, indicating that metabolic differences also exist between the hyperthermophilic archaea. In addition, the archaeal alcohol dehydrogenases contains iron and is only the second example of such an enzyme. The function of hyperthermophilic alcohol dehydrogenases is not known, as *Thermococcus litoralis* and ES-1 do not use alcohols as growth substrates, nor are alcohols end products of their fermentative pathways.

Two S^0-reducing enzymes have been purified from the cytoplasm of *Pyrococcus furiosus*. The first is a new enzyme, previously unknown in mesophilic organisms, termed sulfide dehydrogenase. This enzyme utilized NADPH as an electron donor (it has no S^0-reducing activity with NADH) for the production of H_2S from both S^0 and polysulfide. Polysulfide is a soluble form of S^0 that is generated from S^0 in sulfide-containing media at high temperatures. Polysulfide, rather than insoluble S^0, is thought to be the actual substrate for H_2S production by the hyperthermophiles. For example, sulfide dehydrogenase is a cytoplasmic enzyme and would therefore be unable to utilize insoluble S^0. Like many of the other enzymes already discussed, sulfide dehydrogenase contains iron–sulfur, and it also contains flavin. In addition to polysulfide reduction, the enzyme catalyzes in vitro the reduction of *Pyrococcus furiosus* rubredoxin using NADPH as the electron donor, and the reduction of NADP using the reduced form of *Pyrococcus furiosus* ferredoxin as the electron donor. Hence, in addition to H_2S production, this multifunctional enzyme may have other functions.

The second S^0-reducing enzyme obtained from *Pyrococcus furiosus* is the hydrogenase discussed earlier. This enzyme not only reduces protons to H_2, it also reduces S^0 and polysulfide to H_2S. It is therefore referred to as "sulfhydrogenase", to indicate its dual catalytic activities. Interestingly, its S^0-reducing activity, but not its H_2 production activity, is stimulated in vitro by the rubredoxin from *Pyrococcus furiosus*. Since this protein is also reduced by sulfide dehydrogenase, rubredoxin may play a role in S^0 reduction. However, the pathways of electron flow between the oxidative carbohydrate- and peptide-metabolizing enzymes of *Pyrococcus furiosus* and the two S^0-reducing enzymes is complicated by the ability of sulfide dehydrogenase to function as a ferredoxin: NADP oxidoreductase. A third type of S^0-reducing enzyme—a membrane-bound polysulfide reductase—is present in hyperthermophiles such as *Pyrodictium brockii* whose growth is obligately dependent on S^0 reduction. However, such an enzyme has yet to be purified.

Other enzymes that have been characterized from the S^0-dependent hyperthermophiles include those that modify nucleic acids. Such enzymes that are able to function at extreme temperatures have recently become of particular interest because of their utilization in the polymerase chain reaction (PCR) and related technologies. PCR is now used as a routine tool in molecular biology and requires a thermostable DNA polymerase. Three hyperthermophilic DNA polymerase are currently commercially available. These enzymes, termed Vent, Deep Vent, and Pfu DNA polymerases, are purified from *Thermococcus litoralis,* strain GB-D, and *Pyrococcus furiosus,* respectively. The purification and properties of two of them, from *Thermococcus litoralis* and *Pyrococcus furiosus,* have been described. The only other polymerase-type enzyme obtained so far from the hyperthermophiles is the RNA polymerase from *Thermotoga maritima.* This organism is also a source of two other proteins: an elongation factor (EF-Tu), which is involved in protein synthesis, and an outer membrane protein termed Ompα.

4   MOLECULAR CHARACTERIZATION OF HIGH TEMPERATURE ENZYMES

Complete amino acid sequences are known for 13 of the enzymes that have been purified from the hyperthermophilic S^0-reducing organisms. As indicated in Table 4, most of these are derived from

Table 4   Molecular Characterization of Enzymes from S^0-Reducing Hyperthermophiles

Organism/Enzyme	Source of Amino Acid Sequence[a]	Expression in *E. coli*[b]
Pyrococcus furiosus		
Ferredoxin	Gene and protein	Yes (purified)
DNA polymerase	Gene	Yes (purified)
Amylase	Gene	Yes (not purified)
Glutamate dehydrogenase	Gene	No
Aldehyde Fd oxidoreductase[c]	Gene	No
Rubredoxin	Protein	No
Pyrococcus woesei		
GAPDH	Gene	Yes (purified)
Glutamine synthetase	Gene (protein not purified)	No
Thermococcus litoralis		
DNA polymerase	Gene	Yes (purified)
Ferredoxin	Protein	No
ES-4		
Glutamate dehydrogenase	Gene	Yes (not purified)
Thermotoga maritima		
Ferredoxin	Gene and protein	Yes (purified)
GAPDH	Gene and protein	Yes (not purified)
Lactate dehydrogenase	Gene	Yes (not purified)
RNA polymerase $(\alpha\beta,\beta'\sigma)$[d]	Gene (β, β' subunits)	No
Glutamine synthetase	Gene (protein not purified)	Yes (not purified)
Trp EGD[e]	Gene (protein not purified)	No
4-α-Glucanotransferase	(Not sequenced)	Yes (purified)
β-Galactosidase	(Not sequenced)	Yes (purified)
β-Glucosidase	(Not sequenced)	Yes (purified)

[a] Indicates whether amino acid sequence was derived from the native protein or was inferred from the gene sequence. In the latter case, it is also indicated if the native protein has yet to be purified.

[b] Indicates whether the cloned gene has been expressed in *E. coli* and if the recombinant protein was purified.

[c] Fd, ferredoxin.

[d] The sequences are known for two of the four types of subunit in RNA polymerase.

[e] Trp EGD are the genes for anthranilate synthetase I (E), anthranilate synthetase II (G), and anthranilate phosphoribosyltransferase (D).

gene sequences, and most are for proteins purified from species of *Pyrococcus* and *Thermotoga*. In some cases, such with as the glutamine synthetases from *Pyrococcus woesei* and *Thermotoga maritima*, and some of the tryptophan biosynthetic enzymes from *Thermotoga maritima*, the native protein has yet to be purified. Similarly, the gene for *Thermotoga maritima* glutamine synthetase has been expressed in an active form in *Escherichia coli*, but neither the recombinant protein nor the native enzyme has been purified. In fact, as discussed in this section, only in a few cases has a hyperthermophilic protein been purified from both natural and recombinant sources.

One of the most extensively characterized hyperthermophilic proteins is the ferredoxin from *Pyrococcus furiosus*. Its amino acid sequence has been obtained from the protein, from the corresponding cloned gene, and from the recombinant protein. Interestingly, the gene for this small protein (molecular weight 7500) codes for an N-terminal methionine residue, but this is not present in the purified protein. Moreover, when the gene for the ferredoxin was expressed in *E. coli*, the recombinant protein also lacked methionine at the N-terminus. Hence, both *Pyrococcus furiosus*, which is a hyperthermophilic archaeon, and *E. coli*, a mesophilic bacterium, appear to have analogous posttranslational mechanisms involving a methionyl aminopeptidase. In addition, the recombinant ferredoxin contained a single [4Fe–4S] cluster and the spectroscopic, redox, thermostability, and physiological electron transfer properties of the protein were identical to those of the native ferredoxin. Similarly, although less extensively characterized, the gene for the ferredoxin from *Thermotoga maritima* has been expressed in *E. coli* to give a protein that was also indistinguishable from the native ferredoxin.

In addition to the hyperthermophilic redox proteins, the genes for several much larger hyperthermophilic enzymes have been expressed in *E. coli* (Table 4). However, few of the recombinant enzymes have been characterized. For example, in addition to the hyperthermophilic ferredoxins, only the products of the genes encoding GAPDH from *Pyrococcus woesei* and the DNA polymerases of *Pyrococcus furiosus* and *Thermococcus litoralis* have been purified from *E. coli*. Comparisons between the purified forms of the native and recombinant enzymes have shown that the catalytic and thermal stability properties of the recombinant versions are more or less identical to the native forms. The two DNA polymerases are both monomeric proteins of molecular weight 93,000, whereas *Pyrococcus woesei* GAPDH is a homotetramer of molecular weight 150,000. The results with GAPDH are of some importance because they show that multimeric hyperthermophilic proteins can be obtained in an active recombinant form. However, other studies have shown that this may not be a general phenomenon. For example, the α-amylase of *Pyrococcus furiosus* is a homodimer of molecular weight 132,000 but the recombinant form in crude extracts of *E. coli*, although active, was of much higher molecular weight suggesting it might represent some aggregated form of the enzyme. Similarly, the gene from the novel strain ES-4 for glutamate dehydrogenase, which is a hexahomomeric protein of molecular weight 270,000, has been expressed in *E. coli* but no activity was detected in cell-free extracts. Moreover, only a small fraction of the recombinant subunits, which are presumably inactive, appeared to assemble into hexamers.

On the other hand, the genes for three carbohydrate-metabolizing enzymes, 4-α-glucanotransferase, β-galactosidase, and β-glucosidase, from *Thermotoga maritima*, have been cloned and expressed in *E. coli*; all three recombinant enzymes were catalytically active and were purified. The problem is that none of these enzymes have been purified from *Thermotoga maritima*, so it is not known how the properties of the recombinant enzymes compared to those of the native forms. Moreover, no heteromeric hyperthermophilic protein has yet been expressed in a catalytically active form because each of these enzymes contains a single type of subunit. Clearly, much is to be understood about the expression of genes from hyperthermophiles and the production of active, recombinant enzymes.

5   PROTEIN HYPERTHERMOSTABILITY

Understanding the mechanisms that stabilize globular proteins, and particularly at high temperatures, has been one of the most challenging and as yet unresolved problems in biochemistry and biotechnology. Proteins from both mesophilic and "moderately" thermophilic organisms are typically destroyed very rapidly at temperatures in the range of 80–100°C. This occurs by the hydrolysis of certain peptide bonds, deamination of side chains, and cleavage of disulfide bonds. However, as shown in Table 3, the enzymes that have been purified from hyperthermophilic organisms reveal a new aspect of protein stability: virtually all are catalytically active above 95°C and most require incubation for many hours, and even days, at temperatures near 100°C before significant losses in catalytic activity are observed. So, what are the mechanisms by which proteins are stabilized at extreme temperatures, and what do we know about the upper temperature limits at which enzymes may still be able to function?

The question of how hyperthermophilic organisms stabilize their proteins at temperatures near and above 100°C begs the question of how stability is conferred on other biological molecules—in particular, various cofactors and organic intermediates, many of which are rapidly denatured at such temperature in vitro. For example, NADPH and NADH have half-lives of minutes near 100°C. Do the cytoplasms of these organisms contain "thermoprotectants" to stabilize such molecules, thereby allowing cellular function at extreme temperatures? For some of the hyperthermophiles, the answer is yes. One species of *Pyrococcus* has been shown to contain high intracellular concentrations (near 0.8 M) of both potassium ions and a novel sugar phosphate. Similarly, hyperthermophilic methanogens contain an unusual form of diphosphoglycerate at concentrations near 1.0 M, with correspondingly high potassium levels. Such compounds have been shown to increase the stability of certain enzymes purified from these organisms at extreme temperatures in vitro. However, "thermoprotectants" are not a universal mechanism for stabilizing biological molecules, as several of the S^0-reducing hyperthermophiles have been shown to contain unexceptional concentrations ($\approx$ 100 mM) of both potassium and sodium ions.

A further argument against the specific intracellular stabilization of proteins comes from studies with several of the proteins listed in Table 3. These have established that their enhanced thermostability is an intrinsic property of the pure protein. That is, these proteins do not contain organic or inorganic factors that are not present in the homologous protein from mesophilic organisms. Moreover, the stability of some of the hyperthermophilic proteins appears to be close to a proposed theoretical limit. Recent studies on the thermodynamic behavior of various mesophilic proteins and model systems have shown that there is a convergence temperature for enthalpy and entropy changes with an upper limit near 113°C. At this point, the apolar contributions to protein structure are predicted to be zero. In other words, hydrophobic interactions are thought to cease to contribute to the stability of any protein at this temperature, and globular proteins will unfold to a denatured and inactive state. So far, only limited calorimetric data are available on hyperthermophilic proteins, but the results appear to support this theory. Of the enzymes, only the glutamate dehydrogenases from *Pyrococcus furiosus* and ES-4, and *Pyrococcus furiosus* α-amylase have been examined, and all three show transition temperatures (T_m values) near 113°C. On the other hand, *Pyrococcus furiosus* ferredoxin, which contains a [4Fe–4S] cluster, exhibits a T_m value of 117°C, and the zinc-substituted form of *Pyrococcus furiosus* rubredoxin undergoes thermal denaturation at 124°C. Hence, it remains to be seen whether proteins that lack covalently coordinated metal ions are stable above the proposed theoretical limit of 113°C.

That the high thermal resistance of hyperthermophilic proteins is an intrinsic property is also readily demonstrated by recombinant techniques. For example, the recombinant form of *Pyrococcus furiosus* ferredoxin produced in *E. coli* is as stable as the native form purified from *Pyrococcus furiosus,* and the same results have been obtained using the gene for GAPDH from *Pyrococcus woesei*. Similarly, except for those unique to the hyperthermophiles, the enzymes listed in Table 3 are comparable in size and complexity to their relatives obtained from mesophilic organisms. Thus, enhanced thermostability is predominantly a result of specific amino acid changes in the sequence of a given protein. Accordingly, one would expect that significant insight into the stabilizing mechanisms involved should be revealed by direct comparisons of the amino acid sequences of homologous proteins from hyperthermophilic and mesophilic organisms. Of course, an understanding of such mechanisms might lead to a generic protocol which, through site-specific changes, might significantly improve the stability at extreme temperatures of thermolabile mesophilic proteins.

However, in spite of the attractiveness and simplicity of comparative sequence approach, it appears to be quite unrealistic. Several studies have been carried out in which the amino acid sequences of the same enzyme type obtained from mesophilic moderately thermophilic and hyperthermophilic organisms were extensively analyzed. These have shown that there are no uniform differences between them, and that there are no general "traffic rules" of amino acid exchanges in shifting from thermolabile to thermostable proteins. Indeed, the results have frequently contradicted earlier studies comparing mesophilic and moderately thermophilic proteins. For example, the amino acid sequence of the rubredoxin from *Pyrococcus furiosus* has been systematically compared with those of the dozen or so mesophilic rubredoxins that have been purified. These proteins contain only about 50 residues, and there are several residues unique to the *Pyrococcus furiosus* protein, which shows in the range of 40–60% sequence identity with the mesophilic proteins. However, such analyses offer no clue to why the hyperthermophilic protein is stable at 95°C for at least 24 hours, while the most stable of the others denatures at 80°C. Clearly, trying to assess stabilizing mechanisms by amino acid sequence comparisons with much larger enzymes is of limited, if any, value. Such mechanisms can be apparent only from detailed comparisons at the level of three-dimensional structures.

Detailed structural information is available for three hyperthermophilic proteins, and all have been obtained from *Pyrococcus furiosus*. The three-dimensional structure of its rubredoxin was determined by NMR spectroscopy, by X-ray crystallography (to 1.8 Å), and by molecular modeling. The latter investigation was based on the known crystal structures of four mesophilic rubredoxins. All three techniques gave virtually the same structure for the hyperthermophilic protein. Moreover, it was virtually identical to that of mesophilic rubredoxins, with a highly conserved hydrophobic core and the same folding topology. Hence, in this case hyperthermostability is conferred by rather minor changes, and these involve mainly surface residues. The most striking difference was at the N-terminal region, which in the *Pyrococcus furiosus* protein was incorporated into a hydrogen bonding network. In contrast, the N-termini of the mesophilic rubredoxins is highly disordered and is thought to be the first part of the protein to denature at high temperatures.

That hyperthermostability is a result of relatively small structural changes, at least in redox-type proteins, has been confirmed using the ferredoxin of *Pyrococcus furiosus*. The secondary structure of this small protein (66 amino acids) has been determined by two-dimensional NMR spectroscopy. Compared to the known crystal structure of a mesophilic ferredoxin, the *Pyrococcus furiosus* protein exhibits high structural homology, and its increased thermal stability appears to be the result of an extension of secondary structural elements. For example, in place of a loop–helix–double-stranded β-sheet in the mesophilic protein, *Pyrococcus furiosus* ferredoxin contains a slightly longer helix, a tighter loop, and a triple-stranded sheet. Moreover, the latter sheet incorporates the N- and C-termini, a configuration that is not seen in the mesophilic protein. Hence, the structural data from the hyperthermophilic redox proteins suggest that highly flexible terminal regions may well be the initiation site for thermal denaturation in mesophilic proteins.

The only hyperthermophilic enzyme (rather than redox protein) whose crystal structure has been determined is the aldehyde ferredoxin oxidoreductase (AOR) from *Pyrococcus furiosus* (at 2.3 Å resolution). AOR consists of two identical subunits of molecular weight 66,000 (605 amino acids), each containing a catalytic site comprising a tungsten atom with a ferredoxin-type [4Fe–4S] center in close proximity. The tungsten is coordinated to an organic moiety known as molybdopterin, which is a derivative of folate. Molybdopterin was first identified in molybdenum-containing enzymes, in which it coordinates a molybdenum atom. Molybdenum is closely related chemically to tungsten, and although tungsten is seldom used in biological systems, molybdoenzymes are ubiquitous in nature and are found in both microorganisms and higher life-forms. However, structural information is not yet available for any molybdoenzyme. Thus, the crystal structure of AOR is the first for a molybdopterin-containing enzyme, as well as the first for a tungsten-containing enzyme. Surprisingly, the AOR structure revealed that the W atom in each subunit is coordinated not to one, but to two molybdopterins. It is not known whether this is also the case in molybdoenzymes.

The absence of structural information on any other tungsten- or molybdenum-containing enzyme from mesophilic organisms makes it difficult to rationalize the extreme thermal stability of *Pyrococcus furiosus* AOR in structural terms—there is no mesophilic counterpart for comparison. Moreover, the secondary and tertiary structural elements of AOR were typical of those seen for other globular proteins. However, in comparison with more than 30 mesophilic proteins (containing 300–2000 residues), it was shown that AOR exhibits at the same time both the minimum solvent-exposed surface area and the maximum fraction of buried atoms. In general, this combination would be expected to increase stability by reducing the unfavorable surface energy while at the same time increasing interior packing. Thus, hyperthermostability may reflect a number of subtle interactions that minimize a protein's ratio of surface area to volume. How this is achieved in AOR is completely unknown, and whether this is a phenomenon unique to hyperthermophilic enzymes must obviously await structural information on enzymes of other types, ideally those having a mesophilic counterpart that has been well characterized structurally. If changes in ratios of surface area to volume are a primary factor in determining thermal stability, it does not bode well for understanding the molecular basis for hyperthermostability, or for the chances of finding means to stabilize labile mesophilic enzymes, even if high resolution structures of them are available. It suggests that simple approaches such as just one or two site-specific amino acid changes in a complex enzyme are very unlikely to lead to a global change that would result, for example, in a significantly decreased surface area. Clearly, we have only begun to get the barest of glimpses of what stabilizes enzymes from hyperthermophilic organisms, and detailed structural data on a range of both hyperthermostable enzymes and on their mesophilic relatives are sorely needed.

See also BACTERIAL GROWTH AND DIVISION; ENZYMES, ENERGETICS AND REGULATION OF BIOLOGICAL CATALYSIS BY.

Bibliography

Adams, M.W.W. (1990) The metabolism of hydrogen by extremely thermophilic, sulfur-dependent bacteria. *FEMS Microbiol. Rev.* 75:219–238.
————. (1993) Enzymes and proteins from microorganisms that grow near and above 100°C. *Annu. Rev. Microbiol.* 47:627–658.
————, and Kelly, R. M., Eds. (1992) *Biocatalysis at Extreme Temperatures: Enzyme Systems Near and Above 100°C,* ACS Symposium Series 498. American Chemical Society, Washington, DC.
Stetter, K. O., Fiala, G., Huber, G., Huber, R., and Segerer, G. (1990) Hyperthermophilic microorganisms. *FEMS Microbiol. Rev.* 75:117–124.

ENZYMOLOGY, NONAQUEOUS

Sudipta Chatterjee, Shabbir Bambot, Richard Bormett, Sanford Asher, and Alan J. Russell

1 **Introduction**
2 **Areas of Nonaqueous Enzymology**
  2.1 Enzymes in Reversed Micelles and Biphasic Mixtures
  2.2 Enzymes Chemically Modified for Solubilization in Organic Media
  2.3 Protein Engineering of Enzymes for Organic Biocatalysis
  2.4 Freeze-Dried and Immobilized Enzymes in Organic Media
  2.5 Freeze-Dried Enzymes in Supercritical Fluids
3 **Effects of Solvents on Organic Biocatalysis**
  3.1 Effects of Solvents on Enzyme Morphology
  3.2 Interaction of Solvent with the "Essential" Water of Suspended Enzymes
  3.4 Interactions of Solvents with Substrates and Products
  3.5 Structural Integrity of Enzymes Suspended in Organic Solvents
  3.6 Thermostability of Enzymes in Organic Media
  3.7 Effects of Organic Solvents on Enzyme Kinetics
4 **Solvent Engineering**
  4.1 Control of Enzyme Activity by Solvents
  4.2 Control of Selectivity of Enzymes by Solvents
  4.3 Control of Enzyme Stability by Organic Solvents
5 **Mechanistic Integrity of Enzymes in Organic Media**
6 **The Future of Nonaqueous Enzymology**

Key Words

Enzyme A thermolabile proteinaceous biological catalyst.

Lyophilized Enzyme Powder A freeze-dried preparation of biological catalysts, which, if desired, can be suspended directly into a water-free organic solvent to promote activity.

Supercritical Fluid A material above its critical temperature and pressure, the physical properties of which lie in between those of liquids and gases

Modern day nonaqueous enzymology evolved in a number of distinct steps. Initially, water-miscible organic solvents, such as acetone and ethanol, were added to aqueous enzyme solutions to determine the maximal concentration of solvents that enzymes can tolerate ($\approx$ 50% for most conventional enzymes). Next biphasic mixtures in which aqueous solutions of enzyme were emulsified in water immiscible solvents, such as chloroform and ethyl acetate, were evaluated. This approach was further developed by the use of reversed micelles to stabilize enzymes in aqueous organic mixtures, and the application of freeze-dried enzyme powders suspended in anhydrous organic solvents and supercritical fluids. This article lists the advantages of nonaqueous enzymology over the aqueous form, as well as the disadvantages of the newer technology. It also reviews hydrolases, explores the effects of solvents on organic biocatalysts, and discusses commercial processes that use enzymes, as well as control aspects of solvent engineering.

1 INTRODUCTION

Enzymes are thermolabile, biological, proteinaceous catalysts capable of enhancing the rates of chemical reactions by up to 10^{12}-fold. They are distinguished from other catalysts by properties such as remarkable specificity and the ability to work at mild temperatures and pH. Conventional biocatalysis is carried out in aqueous solutions. Recently, however, interest has shifted from aqueous to nonaqueous enzymology, since most industrial chemistry takes place in organic solvents. Some of the advantages of nonaqueous enzymology over aqueous enzymology are as follows:

- Greater solubility of hydrophobic substrates in organic media
- Ease of recovery of enzymes from organic media due to their insolubility in organic solvents eliminating the need for enzyme immobilization
- Reduced microbial contamination of bioreactors
- Shifting of thermodynamic equilibria to favor synthesis over hydrolysis (e.g., in peptide and ester synthesis)

- Ease of product recovery from low boiling, high vapor pressure solvents
- Effect of solvent on substrate specificity

Hence, it is not difficult to envisage enzymes as ideal catalysts in industrial processes in the not too distant future. There are, of course, certain disadvantages such as:

- Inactivation of enzymes by many organic solvents
- Limited enzyme activity in most organic solvents (compared to their activity in water)
- Absence of any predictive model to describe the effect of solvents on enzyme activity, specificity, and structural stability

Hydrolases, such as lipases, proteases, and esterases, are the most widely used enzymes in organic solvents. In aqueous media, these enzymes catalyze reactions in which water acts as a nucleophile. In organic media, the hydrolases are able to accept other nucleophiles such as alcohols, amines, thio esters, and oximes. Thus, hydrolysis reactions in water can be substituted by synthetic reactions such as peptide synthesis, transesterification, esterification, and aminolysis. At present there are four commercial processes that utilize enzymes in organic media:

- Lipase-catalyzed interesterification of fats
- Thermolysin-catalyzed synthesis of the sweetener aspartame
- Lipase-catalyzed production of optically active 2-halopropionic acid
- Peroxidase-catalyzed production of phenolic resin

2   AREAS OF NONAQUEOUS ENZYMOLOGY

The various strategies that enable the utilization of proteins in nonaqueous media are shown in Figure 1. The areas covered range from homogeneous systems as represented by reversed micelles, soluble polyethylene glycol (PEG) modified enzymes and other soluble chemically modified enzymes, to heterogeneous systems as represented by enzyme powders dispersed in anhydrous organic solvents and supercritical fluids, enzymes in biphasic mixtures of immiscible organic solvents and water, and immobilized enzyme systems in organic solvents. The choice of any system is, of course, dictated by the relative advantages and disadvantages of each.

2.1   ENZYMES IN REVERSED MICELLES AND BIPHASIC MIXTURES

Reversed micelles (microemulsions) and biphasic mixtures are water–solvent mixtures that rely on protein interactions at the water–organic solvent interface. A reversed micelle is an aggregate of amphiphilic molecules in organic solvents which can encapsulate a water pool. The polar head groups of the surfactant molecules orient themselves toward the interior aqueous phase, whereas the hydrophobic chains arrange themselves in the continuous organic phase.

Reversed micellar systems provide a means for protein solubilization in organic media without direct interaction between enzyme and solvent. The presence of the surfactant layer between the aqueous and organic media effectively shields the enzyme from the dele-

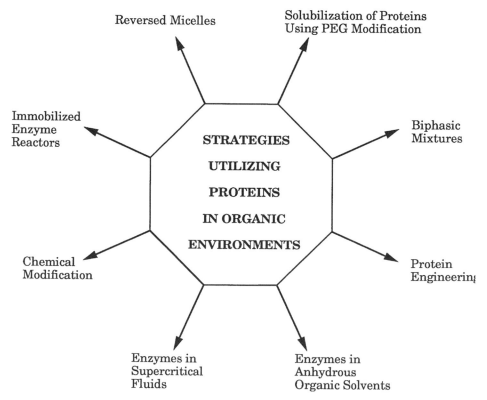

Figure 1.   Methods enabling the use of proteins in organic environments.

terious effects of organic solvents, thereby enabling the protein to exist and function in its native aqueous conformation. Water-soluble proteins such as subtilisin and chymotrypsin, as well as integral membrane proteins such as cytochrome oxidase, rhodopsin, and ATPase, have been successfully solubilized in reversed micelles.

Biphasic mixtures are composed of two immiscible phases. Choice of the organic solvent phase is determined by a number of factors such as the differential solubility of substrates in the aqueous and organic phase, the partitioning of products between the two phases, and the extent to which the organic phase is able to influence enzyme stability and activity. Synthesis and hydrolytic reactions have been studied in biphasic systems using a wide variety of enzymes such as lipases, acid phosphatase, β-glucosidase, and β-fructofuranosidase. Such studies have demonstrated the potential of biphasic systems for increasing the thermostability, and in some cases the β-glucosidase and acid phosphatase activity, of hydrolytic reactions.

2.2 Enzymes Chemically Modified for Solubilization in Organic Media

A recent development in organic biocatalysis is the use of PEG-modified enzymes, which are soluble in a variety of organic solvents. Modification by means of polyethylene glycol involves the covalent attachment of a hydrophobic polymer to the surface of an enzyme, thus facilitating proteins solubilization. Enzymes such as PEG–lipase, PEG–chymotrypsin, PEG–catalase, and PEG–peroxidase have been used to catalyze a wide spectrum of reactions in organic media. Inada and co-workers have analyzed the structures of PEG–catalase and PEG–horseradish peroxidase in organic media. The absorption spectra for PEG–catalase and PEG–horseradish peroxidase in benzene were found to be similar to those of the unmodified enzymes in water, indicating that the association of prosthetic groups and protein was not affected by PEG modification. More recently Sakurai and colleagues have demonstrated the possibility of using PEG–chymotrypsin, PEG–papain, PEG–thermolysin, and PEG–pepsin for solid state peptide synthesis in organic media. In another study, Gaertner and Puigserver successfully used PEG–chymotrypsin for peptide synthesis in organic media.

Other examples of enzymes chemically modified for solubilization in organic media include lipases from *Pseudomonas fluorescens* modified with a copolymer of poly(oxyethylene allylmethyl diether) maleic anhydride and lipoxygenase modified using *N*-acyloxysuccinimide. The modification of lipases rendered the enzyme soluble and active in organic media and also improved its thermostability. The modification of lipoxygenase indicated an increase of hydrophobicity at positions distant from the active site and showed promise for future use of this enzyme in organic media.

An interesting variation on the concept of solubilization by chemical modification is the dissolution of hydrophilic chymotrypsin in nonpolar organic solvents (*n*-octane, cyclohexane, and toluene) up to micromolar concentration levels by prolonged shaking at room temperature.

2.3 Protein Engineering of Enzymes for Organic Biocatalysis

A recent trend in nonaqueous enzymology is the use of protein engineering techniques to redesign enzymes for improved activity and stability in organic media. Initial interest in this type of work was generated by the work done with crambin, a small plant protein, which is soluble in organic solvents. Arnold and co-workers recently modified subtilisin E and α-lytic protease by site-directed mutagenesis for enhanced compatibility with polar organic media. Both enzymes exhibited enhanced stability and activity in mixtures of water and dimethyl formamide. In a separate study, Wong et al. successfully engineered subtilisin BPN′ for enhanced stability in dimethylformamide. The results of these two studies augur well for adoption of protein engineering tactics for optimization of organic biocatalysis.

2.4 Freeze-Dried and Immobilized Enzymes in Organic Media

When an enzyme powder is suspended in an organic medium, the total concentration of water in the system can be as low as 0.01% ($V_{water}/V_{solvent}$). The total amount of water necessary to support activity is actually less than a monolayer per molecule of enzyme in many cases. Under these conditions, the enzyme is protected from solvent and therefore functions in an essentially anhydrous media. Suspensions of freeze-dried (lyophilized) enzyme powders and immobilized enzyme dispersed in anhydrous organic media are examples of heterogeneous biphasic systems of low water concentration (as opposed to immiscible organic solvent/water mixtures). As mentioned in Section 1, such systems offer the advantage of ease of enzyme recovery, improved enzyme thermostability, and the reduction of undesirable side reactions that require water as a substrate. However, a significant problem faced in heterogeneous systems is mass transfer limitation. Mixing and sonication are usually employed in such systems to overcome diffusional limitations.

Applications of enzymes in anhydrous organic media include the use of chymotrypsin, lipases, and subtilisin for peptide synthesis, regioselective acylation of carbohydrates, synthesis of biosurfactants, and lipolysis of fats and horseradish peroxidase for the manufacture of sensors used to detect unacceptable extremes in temperature.

Immobilized enzymes systems for use in organic solvents may be prepared either by physical adsorption on a matrix or by covalent attachment to a stationary phase. Stark and Holmberg have successfully immobilized lipase from species of *Rhizopus* on tresyl silica in hexane buffer and microemulsions based on hexane. The immobilized enzyme was found to catalyze both hydrolysis and transesterification reactions. In another study with immobilized lipases Rizzi and colleagues looked at the kinetics of synthesis of isoamyl alcohol via transesterification reactions in *n*-hexane. Proteases have been immobilized on porous chitosan beads. Compared to the free enzyme, the immobilized enzymes were found to exhibit enhanced stability and activity in ester and peptide synthesis in organic solvents such as methanol, dimethylformamide, and acetonitrile. Blanco et al. have evaluated the effect of immobilization of chymotrypsin on its activity and stability in anhydrous organic solvents. They, too, observed better activity with immobilized enzyme than with free enzyme.

Continuous reactors utilizing immobilized enzymes have been designed for hydrolysis of olive oil by lipases, peptide synthesis with thermolysin, and the resolution of racemic amines by subtilisin in organic media. The racemate resolution process was developed for the production of compounds such as *(R)*-1-aminoindan and *(R)*-1-(1-napthyl)ethylamine, which are considered to be important intermediates in the pharmaceutical industry.

2.5 Freeze-Dried Enzymes in Supercritical Fluids

The use of enzymes in supercritical fluids is a relatively new concept in nonaqueous enzymology. A supercritical fluid is described as a material above its critical temperature and pressure, exhibiting physical properties between those of a gas and a liquid. For biocatalytic processes these substances offer advantages such as high diffusivity, ease of downstream processing, recyclability, and low toxicity. Studies in supercritical fluids have been carried out in a number of solvents such as carbon dioxide, ethane, sulfur hexafluoride, and fluoroform.

3 EFFECTS OF SOLVENTS ON ORGANIC BIOCATALYSIS

Organic solvents may influence nonaqueous biocatalysis on a macroscopic level by interacting with the enzyme, with concomitant effects on its structure and morphology. In addition, interactions of the solvent medium with substrates and products of reaction may influence enzyme stability, kinetics, and reaction mechanism.

3.1 Effects of Solvents on Enzyme Morphology

Complete characterization of enzyme function in organic media requires a detailed study of the effects of environment on enzyme structure, activity, and specificity. Morphological studies on protein powders suspended in organic solvents have shown considerable changes in the powder as a function of the water associated with the protein. An environmental scanning electron micrograph (ESEM) showed the powder to be flaky (≈ 0.2 μm thick) with dimensions exceeding 100 μm at a vapor pressure of 9.2 torr. When the humidity of the analysis chamber was increased above 50%, the flakes were observed to swell, and at 85% humidity they coalesced to form branched structures. The effect of an organic solvent, toluene, on particle morphology was also investigated in these studies. Specifically, toluene was evaporated by slowly drying a suspension of lyophilized subtilisin. Comparison of the images of the dried enzyme and native enzyme showed that the exposure to toluene had no marked effect on enzyme morphology. Particle morphology becomes particularly important when one is attempting to understand the impact of organic solvents on enzyme-catalyzed reactions that proceed relatively fast but may be diffusionally limited.

3.2 Interaction of Solvent with the "Essential" Water of Suspended Enzymes

Noncovalent interactions such as hydrogen bonding, hydrophobic forces, and electrostatic and van der Waals interactions help to maintain an enzyme in its catalytically active form. Water plays an important role in all these interactions. Replacement of water by organic solvents should, therefore, directly affect the native structure and activity of an enzyme.

Organic solvents can interact with the water molecules bound to lyophilized enzyme powders as well as with the enzyme molecule itself. These interactions may take place in two steps. In the first step, water may be stripped from the enzyme. In the second step, the solvent may penetrate and interact directly with the protein, thereby affecting its native configuration and activity. In this regard, solvent hydrophilicity will determine the ability of the solvent to "strip" away the essential water. The more hydrophilic the solvent, the greater will be its tendency to strip water. Clearly, the direct in-teraction of a nonaqueous solvent with an enzyme would be expected to be deleterious to overall enzyme activity, and thus it is expected, a priori, that hydrophobic solvents are favorable when the goal is to minimize solvent–enzyme interactions.

Halling has analyzed the interaction between protein and water on a molecular level. His studies show that organic solvents have little effect on tightly bound water, since there is little penetration of the primary hydration layer in most solvents. Gorman and Dordick have looked at solvent-induced desorption of tritiated water from lyophilized chymotrypsin, subtilisin, and horseradish peroxidase suspended in anhydrous organic solvents. They report the highest degree of desorption by polar solvents such as n-propanol, n-butanol, methanol, and dimethylformamide. Wasacz et al., via Fourier transform infrared (FTIR) spectroscopy studies, have provided direct evidence of the stripping of water from albumin by methanol.

3.4 Interactions of Solvents with Substrates and Products

Organic solvents may reduce an enzyme's activity by adversely affecting either stability or partitioning of substrate or product. Chloroform, by acting as a phenoxy radical quencher, has been shown to reduce the activity of horseradish peroxidase catalyzed polymerization of phenols (a reaction initiated by the generation of phenoxy radicals). The partitioning of products between the enzyme and the reaction medium also depends on the relative hydrophobicities of the enzyme and all organic media. For instance, if the products formed are hydrophilic, they will tend to partition more onto the hydrated enzyme particle rather than into the bulk organic solvent. This may subsequently lead to product inhibition.

3.5 Structural Integrity of Enzymes Suspended in Organic Solvents

The first direct demonstration of structurally intact enzymes in anhydrous organic dispersants was made with electron paramagnetic resonance (EPR) spectroscopy. This study on an immobilized alcohol dehydrogenase indicated that the enzyme had not unfolded. While EPR can investigate only the environment and mobility of a spin label, the data nevertheless suggest that a protein can maintain its native conformation in an organic dispersant. Klibanov and colleagues have shown that the microenvironment of the active site histidine of α-lytic protease suspended in anhydrous acetone is identical in water and in organic dispersants. Solid state [¹⁵N] nuclear magnetic resonance (NMR) spectroscopy was used to analyze the catalytic triad at the active site of this serine protease, and it was demonstrated that the hydrogen bond network at the active site of this serine protease does not change when the enzyme is suspended in organic solvents.

A complete understanding of structure–environment relationships requires the application of all possible techniques that provide information about protein structure. EPR and NMR are excellent techniques for investigating local regions of the protein. EPR, as stated earlier, is limited to the environment of a spin label, whereas NMR is limited by protein size and the type of investigation being performed. Many classical techniques for protein structure determination, such as circular dichroism and X-ray crystallography, are, of course, not possible for enzyme powders suspended in anhydrous organic solvents.

FTIR is a widely accepted method for the study of global protein structure in a variety of environments. The use of infrared spectroscopy in the study of proteins was pioneered by Elliott and Ambrose in 1950 and has been extensively applied since. Of particular relevance to this study is the use of FTIR techniques to detect small changes in global and local structure of proteins. The IR spectra of polymers such as proteins can be interpreted in terms of vibrations of structural repeat units. The vibration of a single repeat unit such as an α-helix or β-sheet can be separated from an otherwise complicated spectrum, enabling quantification of secondary structure. Nine groups of vibrational frequencies, manifested as characteristic bands in FTIR protein spectra, have been identified. Of these, amide I and amide II are the most useful infrared probes of protein structure. FTIR spectroscopy also is readily applicable to the study of enzymes in anhydrous environments, since the interference from the very strong absorption of water in the spectral region of interest is dramatically decreased. On the basis of a detailed study of conformationally sensitive infrared absorption frequencies, it is possible to detect changes in protein secondary structure that arise from alterations in the protein environments.

Earlier investigators have used FTIR methods to study the effect of organic solvents on global protein structure. In a series of papers, Wasacz and colleagues studied the effect of exchanging the solvent of a protein from water to alcohols. The expected findings were reported: namely, when water is gradually replaced by methanol (and other alcohols) as bulk solvent, structural changes are induced in the protein, and FTIR analysis indicates an increased helix content. This work does not contradict the long-recognized loss of activity of proteins that have been diluted with high concentrations of alcohols. In the 1970s solid protein samples were also dispersed in mineral oil, and their FTIR spectra were compared to protein dissolved in water. The authors expressed surprise that the spectra (for lysozyme) were similar, although their data agree with the earlier work of Dastoli and Price, who were the first investigators to measure kinetic constants for active lyophilized enzymes dispersed in a variety of solvents.

More recently Mantsch and colleagues have investigated the effect of halogenated alcohols on protein secondary structure. They report that partial unfolding, aggregation, and helix structure alterations result from increasing the mole fraction of solvent in the system. They also indicate that the helical structure of myoglobin in pure chloroethanol is indistinguishable from that in water, supporting the claim that protein secondary structure is not affected by neat organic dispersants.

Subtilisin is the most intensely studied enzyme that functions in organic media, although there is little information on its structural integrity in organic media. FTIR spectroscopy has also been used to analyze the global structure of subtilisin suspended in organic media and to determine the effect of organic dispersants on the global and local structure of myoglobin.

Figure 2 shows typical FTIR spectra for highly concentrated samples of subtilisin and myoglobin solubilized in aqueous solution and suspended in organic dispersants. The spectra were generated by deconvolution of spectra followed by subtraction of the water spectrum. The choice of organic solvent was guided by the need to match the density of dispersant and enzyme particle to prevent settling of the sample during data collection. As expected, subtilisin and myoglobin solubilized in aqueous media have different spectra, confirming that the FTIR technique is sufficiently sensitive to detect the structural differences between individual proteins. Table 1 presents

a quantitative summary of the data. To test the validity of this method for the detection of protein unfolding, denaturation of subtilisin was induced by solvation with dimethyl sulfoxide (DMSO) followed by FTIR analysis. In a separate experiment, phenylmethylsulfonyl fluoride inhibited subtilisin dissolved in water was denatured by boiling. The use of the inhibitor was merely to prevent autolysis during the experiment. There is a clear difference in the spectra between the native and denatured proteins. In each case the amide I band splits into two discrete bands. It is interesting that the intensity of the bands is reversed for protein dissolved in aqueous and organic solvents. It has been recognized that solvents vary considerably in their helicogenicity. Since both DMSO and water can solubilize proteins, but have distinct helicogenicity, one would not expect the FTIR spectra to be similar for the same protein.

Figure 2 also presents the FTIR spectra for subtilisin (native and denatured) and myoglobin in carbon tetrachloride and mineral oil. As just described, differences in the shape, magnitude, and position of FTIR peaks indicate an altered structure of the protein. Clearly the global structures of subtilisin and myoglobin do not change radically when the solvent is changed from water to mineral oil or carbon tetrachloride. If subtilisin was unfolded in organic dispersants, or even if a fraction of the enzyme present was denatured, the global change in structure would be detectable by FTIR spectroscopy. In the organic dispersants tested one sees no such protein denaturation. Indeed, the spectra for both myoglobin and subtilisin in organic media are almost identical to those in aqueous solution.

There is a close correlation between the area of the amide I and II bands and protein secondary structure. Table 1 provides estimates for the total areas of each band for myoglobin and subtilisin in aqueous solution, and organic dispersant. In both environments there is a striking similarity not only between band positions but also with respect to the intensity of the amide peaks. Size and shape of the amide I and II bands are related to the α-helix, random coil, and β-pleated sheet content of proteins. An approximation (using the data shown) of the ratio of α-helix to β-sheet content supports the notion that the global structure of subtilisin does not change drastically upon suspension in an anhydrous solvent.

A separate experiment investigated the interaction between myoglobin and its prosthetic group in aqueous and organic environments. When the azide ion binds to the heme iron of hemoproteins the antisymmetric stretching frequency shifts from 2049 cm^{-1} to 2045 and 2023 cm^{-1}, which corresponds to azide bound to high and low spin irons, respectively. The half-width of the antisymmetric stretch also decreases from 22 cm^{-1} to about 10 cm^{-1}. The active site of lyophilized horse azidometmyoglobin suspended in mineral oil or carbon tetrachloride has been examined and compared to that of the acid denatured suspension of azidometmyoglobin in water.

In the organic dispersant studied, the positions of the azide antisymmetric stretches do not appear to shift relative to the aqueous protein, although the half-widths do increase to about 20 cm^{-1} (Figure 3A–C). It is also clear from our measurements that the iron spin state equilibrium, which favors the low spin species, is not significantly perturbed, since the shoulder corresponding to the high spin species appears to be only half as intense as the low spin peak. The acid-denatured azidometmyoglobin shows only a single peak at 2055 cm^{-1}, with a half-width of about 21 cm^{-1} (Figure 3D). This is higher in energy than the free azide (2049 cm^{-1}) but still lower than the aqueous hydrogen azide (2148 cm^{-1}).

From Figures 2 and 3 one can conclude that there is no change in the secondary structure of subtilisin and myoglobin upon exposure

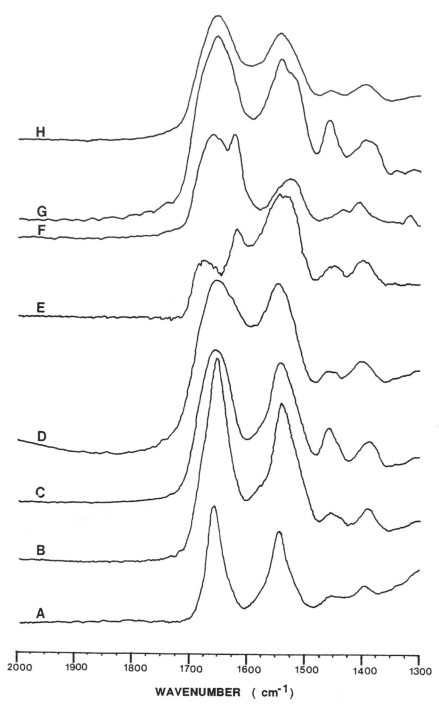

Figure 2. FTIR spectra of myoglobin and subtilisin in aqueous and organic media: (A) myoglobin in aqueous solution, (B) lyophilized myoglobin dispersed in carbon tetrachloride, (C) lyophilized myoglobin dispersed in mineral oil, (D) subtilisin in aqueous solution, (E) boiled phenazine methosulfate–subtilisin in aqueous solution, (F) lyophilized subtilisin in dimethyl sulfoxide, (G) lyophilized subtilisin dispersed in mineral oil, (H) lyophilized subtilisin dispersed in carbon tetrachloride.

of powdered preparations of the enzyme to carbon tetrachloride and mineral oil. In addition, the data for azidometmyoglobin suggest that there is no significant change in the protein structure around the heme-binding site, although protein structural changes that affect the bandwidths of the azide antisymmetric stretch may occur.

Subtilisin, a serine protease, is catalytically active in carbon tetrachloride, although the dispersant is more hydrophilic than those rec-

ommended for use with biocatalysts. It is interesting to note that when subtilisin is suspended in carbon tetrachloride, the water associated with the enzyme is partitioned into the dispersant, which hypothetically results in decreased activity of the enzyme. Clearly, the removal of water from the enzyme surface by the dispersant does not result in a significant structural alteration, and the lowered activity must be explained in other ways.

Table 1   Amide I and II frequencies for myoglobin and subtilisin in aqueous and nonaqueous environments.

Protein	Solvent/Dispersant	Amide I Frequency (cm^{-1}, (%))[a]	Amide II Frequency (cm^{-1}, (%))
PMS-Subtilisin	Water	1653.2, (62±1)	1547.1, (38±1)
Subtilisin (Denatured)[b]	DMSO	1662, 1626, (28±5)	1526, (72±10)
PMS-Subtilisin (Denatured)[c]	Water	1676, 1617, (80±10)	1539, (20±4)
Subtilisin	Mineral oil	1651.3, (62±1)	1537.4, (38±1)
Subtilisin	CCl$_4$	1649.3, (63±1)	1541.3, (37±1)
Myoglobin	Water	1655.1, (52±1)	1547.1, (48±1)
Myoglobin	Mineral oil	1655.1, (55±1)	1541.3, (45±1)
Myoglobin	CCl$_4$	1655.0, (57±1)	1541.0, (43±1)

[a]The individual areas for the spectra in Figure 1 were approximated using the trapezoidal rule.
[b]Subtilisin was resuspended in dimethylsulfoxide to facilitate denaturation.
[c]Subtilisin was first inactivated with phenylmethylsulfonyl fluoride, then denatured by boiling in buffer for 5 minutes prior to performing FT-IR.

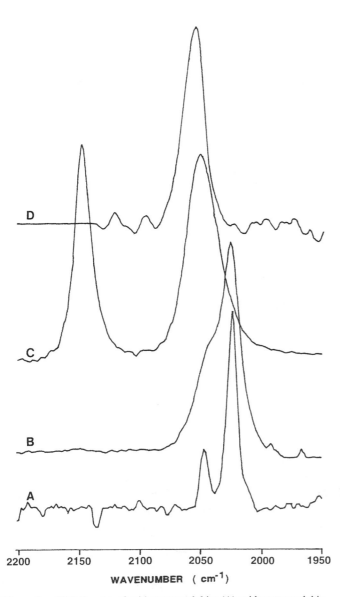

Figure 3. FTIR Spectra of azidometmyoglobin: (A) azidometmyoglobin in aqueous solution, (B) azidometmyoglobin in mineral oil, (C) free azide and hydrogen azide in aqueous solution, (D) acid-denatured azidometmyoglobin in aqueous solution.

3.6   THERMOSTABILITY OF ENZYMES IN ORGANIC MEDIA

At high temperatures, enzyme inactivation in aqueous media is caused by protein unfolding, covalent alterations in the primary structure of the protein, or both. Water plays a primary role in both types of inactivation by facilitating protein unfolding through increasing protein mobility; water also participates as a substrate in the processes of disulfide bond interchange, deamidation of glutamine and asparagine residues, and hydrolysis of peptide bonds. Hence, it is not surprising that the absence of water in nonaqueous environments positively affects enzyme thermostability.

Studies with lipases, mitochondrial F$_1$-ATPase, and cytochrome oxidase have shown these enzymes to be more active and stable in organic solvents at higher temperatures than when dissolved in water. Often, increased thermostability is accompanied by increased stability toward other denaturing effects such as proteolysis.

Volkin et al. have analyzed the thermoinactivation of three unrelated enzymes (ribonuclease, chymotrypsin, and lysozyme) between 110 and 145°C in anhydrous organic media. They found these enzymes to be more thermostable in hydrophobic solvents, such as butanol and nonane, than in hydrophilic solvents such as dimethylformamide. The thermostability of ribonuclease in nonane was attributed to increasing water content of the lyophilized enzyme powder. At high temperature the aggregation of ribonuclease molecules was found to be due to both physical association and chemical cross-linking. In another study, Ahern and Klibanov analyzed the mechanism of irreversible thermoinactivation of hen egg lysozyme at 100°C. Their results showed that deamidation of asparagine residue, hydrolysis of peptide bonds at aspartic acid residues, and the formation of incorrect (scrambled) structures all contribute to enzyme inactivation.

3.7   EFFECTS OF ORGANIC SOLVENTS ON ENZYME KINETICS

As mentioned earlier, organic solvents may affect rates of enzymatic reactions either directly or indirectly. One can consider the binding of solvent molecules to be a direct effect; indirect effects include the partitioning of substrates and products between the solvent and enzyme, the shift of chemical equilibria, and certain mass transfer limitations that are due to the organic solvent. The overall effects are manifested in changes of kinetic parameters for biocatalyzed reactions upon manipulation of the solvent environment. In a study with methanol–water mixtures, methanol was shown to compete with water for the binding sites in the active site of α-chymotrypsin, resulting in a decrease of enzyme activity during the hydrolysis of an ester of hydrocinnamic acid.

Serine proteases, such as α-chymotrypsin and subtilisin, are excellent model systems for studying the effect of solvent on biocatalyst kinetic properties. These enzymes catalyze the hydrolysis of ester and amide substrates via the acyl–enzyme mechanism. In the first step the enzyme and substrate associate noncovalently. Next the hydroxyl group of the serine residue attacks the carbonyl group on the substrate, leading to the formation of a tetrahedral intermediate, which collapses to release the first product, an alcohol or an amine, and the covalent acyl–enzyme complex. This is followed by the attack on the acyl-enzyme complex by the second nucleophile, water, to give rise to an enzyme–product complex, which subsequently breaks down to form the second product, an acid, and releases the free enzyme. It has been reported that in the α-chymotrypsin-catalyzed hydrolysis of an amide substrate (where acylation is rate determining), the addition of organic solvents to water has no effect on k_2, the acylation rate constant. This is because water is not involved in the rate-determining acylation step. However, for an ester substrate (where deacylation is rate controlling), the deacylation rate constant, k_3, was found to increase when the concentration of water in an organic solvent reaction system was increased, implicating the direct involvement of water in the rate-determining deacylation step.

4   SOLVENT ENGINEERING

Conventional biocatalysis is carried out in aqueous media; therefore, it is not surprising that most methods developed to optimize enzyme performance are water based. Nonaqueous enzymology offers a new route for control of enzyme properties by "solvent engineering," which affords a means for the control of enzyme function by changing reaction medium, rather than by changing the enzyme itself.

4.1   CONTROL OF ENZYME ACTIVITY BY SOLVENTS

Solvents can affect enzyme activity by affecting substrate partitioning between the enzyme's active site and the solvent. For example, in water, hydrophobic substrates like esters of phenylalanine and tyrosine easily partition into the active sites of the enzymes, which are lined with hydrophobic amino acid residues. In organic media, however, hydrophobic solvents reduce the substrate partitioning into the active site thereby causing a decrease in rate. Zaks and Klibanov have measured rates of transesterification of N-acetyl-L-phenylalanine ethyl ester with propanol in a series of solvents using both subtilisin and chymotrypsin. Their study showed enzyme activity to be a function of solvent hydrophilicity. It was hypothesized that the more hydrophilic the solvent, the greater will be its tendency to interact with the enzyme's essential water, and thus adversely influence enzyme structure and activity.

4.2   CONTROL OF SELECTIVITY OF ENZYMES BY SOLVENTS

In general, enzymes retain their stereoselectivity in organic media. The release of water from a hydrophobic binding site upon correct binding of a substrate will be more energetically favored in water than in organic solvents. One would not expect incorrect binding of inappropriate enantiomers to be affected to such a significant extent.

Hence, in some cases nonaqueous media relax an enzyme's enantioselectivity by weakening the interaction of an enzyme with its preferred substrate.

Changing of the reaction medium can also bring about an alteration in substrate specificity. In water, α-chymotrypsin catalyzes the hydrolysis of N-acetyl-L-phenylalanine ethyl ester, a hydrophobic substrate, 10^5 times faster than that of N-acetyl-L-serine methyl ester, a hydrophilic substrate. However, in organic solvents the enzyme is five times more reactive toward the hydrophilic substrate. It has been suggested that the nature of the solvent can modulate enzyme specificity. In water, for instance, the binding of hydrophobic substrates to hydrophobic enzyme active sites is driven by the entropy increase resulting from the expulsion of the ordered water molecules surrounding both the enzyme pocket and substrate. In organic solvents this process will be unfavorable relative to organic media. The process will, however, be favored in hydrophilic rather than in hydrophobic solvents. Thus changing the solvent will have an effect on the binding of substrates to enzymes (in terms of binding, hydrophilic will be better than hydrophobic). In addition, by changing solvents it may be possible to fine-tune enantioselectivity to a predetermined degree.

Substrate specificity of α-chymotrypsin-catalyzed esterification in organic media was investigated by Clapes and Adlercreutz. The study shows that the specificity of chymotrypsin toward the side chains of the amino acid substrates is the same as that in water. However, the enzyme specificity toward the N-protecting group is the reverse of that in water. In water, the specificity constant of esterification was found to increase proportionally with the hydrophobicity of the N-protecting group, whereas in organic solvents, the reverse was found to be true. These studies demonstrate the potential of using solvent engineering to control enzyme selectivity and specificity.

4.3   CONTROL OF ENZYME STABILITY BY ORGANIC SOLVENTS

Engineering of proteins to improve conformational stability and stability toward denaturants, oxidants, proteolysis, and high temperatures is an elusive goal. Protein engineering uses techniques such as site-specific mutagenesis to engineer disulfide bonds into globular proteins for the purpose of enhancing conformational and proteolytic stability. Phenotype screening of random mutants is often used to screen for mutants with enhanced thermostability. Enzymes in organic media have shown improvements in both catalytic and storage stability. α-Chymotrypsin incubated in anhydrous octane exhibited a half-life of 6 months, whereas in water the half-life is reduced to 1 week. At the same time, the half-life was observed to be a function of the nature of the medium. The enzyme had half-lives of 130, 80, 35, and 1.5 minutes in butyl ether, tert-amyl alcohol, dioxane, and pyridine, respectively. Obviously, in respect of stability, the best solvents will be those able to keep an enzyme in its native conformation via solvophobic interactions.

Khmelnitsky has evaluated the denaturation capacity of a series of organic solvents while working with α-chymotrypsin, trypsin, lactase, chymotrypsinogen, cytochrome c, and myoglobin. Denaturation capacities were reported in terms of a threshold concentration, which is defined as the concentration of added organic solvent at which the enzyme loses half its initial activity. For the enzymes studied, abrupt changes in spectra of the dissolved enzyme were noted at this threshold concentration. This implies that the drop in enzyme activity was specifically due to protein denaturation. They also noted that there is a critical water content that is necessary for enzyme activity. Once the enzyme has dehydrated beyond this critical value, denaturation sets in. In a separate study, Zaks and Klibanov, while working with alcohol dehydrogenase, polyphenol oxidase, and alcohol oxidase, also noticed that the catalytic activity of the enzyme was a function of the water content of the enzyme.

Schulze and Klibanov have investigated the effects of solvents, substrates, water content of medium, and pH prior to lyophilization on the stability of subtilisin in acetonitrile and in *tert*-amyl alcohol. They observed that enzyme lyophilization at the pH optima for activity or incubation in the presence of substrates/ligands had a positive effect on enzyme stability. On the other hand, the addition to the reaction medium of exogenous water adversely affected enzyme stability. Other efforts at solvent engineering include the investigation of the effects of additives such as glycerols and polyols in aiding a protein to maintain its native structure in organic media via solvophobic interactions. Efforts have also been made to optimize esterification reactions by removal of water produced during reactions using loop reactors or reduced pressure. Finally, another example of solvent engineering is the alteration of enzyme activity by the addition of water. It has been observed that the addition of water to α-chymotrypsin and lipases elevates enzyme activity and influences enzyme stereoselectivity, respectively.

5   MECHANISTIC INTEGRITY OF ENZYMES IN ORGANIC MEDIA

Before comparing an enzyme's activities in different media, it is important to determine its structural and mechanistic integrity of enzyme in these media. For if the mechanism and structure were to change from solvent to solvent, it would be impossible to correlate activity to the properties of the solvent alone. Initial studies on mechanisms in organic solvents with porcine pancreatic lipase and PEG–chymotrypsin have indicated that the enzymes obey Michaelis–Menten kinetics in anhydrous organic solvents. Analyses of reaction mechanisms via the conventional parallel lines method have indicated that the acyl–enzyme mechanism holds true even in nonaqueous media.

Linear free energy relations have been used to prove the transition state structures in organic solvents by Hammett analyses. Hammett analysis, which investigates the charge distribution of the transition state, is very sensitive to reaction mechanism. For esters, it has been reported that the Hammett values are almost the same for hydrolysis in water, and for transesterification in tetrahydrofuran, acetone, butyl ether, acetonitrile and *tert*-amyl alcohol using subtilisin. This result not only provides evidence of the acyl–enzyme mechanism but also implies that the active site of subtilisin and the transition state structure are equivalent in water and organic media.

Investigation of kinetic isotope effects on labeled substrates of subtilisin in various organic solvents and in water has shown that the magnitude of the primary deuterium kinetic isotope effect is the same in different solvents. This study, too, indicates that the transition state structure formed during transesterification and hydrolysis are independent of the nature of the reaction medium. Although the above-mentioned mechanistic studies point toward an acyl–enzyme mechanism for subtilisin-catalyzed transesterification reactions in organic media, there is no direct proof of the mechanism, since the acyl–enzyme complex has never been isolated or detected in organic solvents. Hence, studies on the mechanism of subtilisin-catalyzed reactions in organic media are by no means complete.

The log P model, which relates enzyme activity to solvent hydrophobicity, is most frequently used to define the ability of a solvent to support catalytic activity. However, this simple model provides neither a kinetic nor a mechanistic basis for the explanation of biocatalysis, nor does it consider the diminished binding of substrate to enzyme in organic media. It has value only because it can be used as the first step to choose solvents that support catalytic activity and also to determine suitable reaction conditions.

To successfully control enzyme activity and specificity in organic media, the effects of other solvent properties (e.g., solvent viscosity, density, hydrophobicity, dielectric constant, dipole moment) also need to be considered. To date, all the kinetic studies in organic solvents have related enzyme function and structure to the apparent kinetic constant k_{cat}, K_m, and k_{cat}/K_m. The macroscopic kinetic constants are a function of the microscopic rate constants for the individual steps of a reaction mechanism. Determination of the individual rate constants is therefore necessary for an understanding of the effect of solvent on the individual steps of a multistep reaction mechanism.

6   THE FUTURE OF NONAQUEOUS ENZYMOLOGY

The opportunities presented by nonaqueous biocatalysis are limited only by our imaginations. In just one decade, this area of research has become one of the most active research areas in science and engineering. While the technology has already been commercialized, full-scale adoption of nonaqueous enzymology is dependent on the generation of a predictive model to correlate the effects of solvent properties to the binding and catalytic steps of an enzyme-catalyzed reaction. The design of bioreactors, and the development of upstream and downstream processing units for nonaqueous enzymology, are also central to further utilization of this core technology. The clear need for biocatalysts in industry should not obscure what the use of enzymes in extreme environments can teach us about the molecular basis of enzyme catalysis. Current research is already demonstrating how the delicate interaction between environment and biological molecules serves to modulate function.

See also Chirality in Biology; Enzymes, Energetics and Regulation of Biological Catalysis by.

Bibliography

Arnold, F. H. (1988) *Protein Eng* 2:21–25.
Burke, P. A., Smith, S. O., Bachovchin, W. W., and Klibanov, A. M. (1989) *J. Am. Chem. Soc.* 111:8290–8291.
Clark, D. S., Creagh, L., Skerker, P., Guinn, R. M., Prausnitz, J., and Blanch, H. (1989) *ACS Symp. Ser.* 392:104–110.
Dastoli, F. R., Musto, N. A., and Price, S. (1966) *Biochem. Biophys.* 115:44–47.
Dordick, J. S. (1989) *Enzyme Microb. Technol.* 11:194–201.
Elliott, A., and Ambrose, E. J. (1950) *Nature,* 165:921–922.
Fersht, A. R. (1985) *Enzyme Structure and Mechanism.* Freeman, New York.
Kamat, S., Barrera, J., Beckman, E. J., and Russell, A. J. (1992) *Biotechnol. Bioeng.* 40:158–166.
Klibanov, A. M. (1989) *Trends Biochem. Sci.* (*Pers. Ed.*) 14:141–144.
Klibanov, A. M. (1986) *Chem. Technol.* 16:354–359.
Laane, C., Boeren, S., Vos, K., and Veeger, C. (1987) *Biotechnol. Bioeng.* 30:81–87.

Russell, A. J., Chatterjee, S., Rapanovich, I., and Goodwin, J. (1992) In *CRC Critical Reviews,* A. Gomez-Puyou, Ed.; pp. 92–109. CRC Press, Boca Raton, FL.

Sakurai, T., Margolin, A., Russell, A. J., and Klibanov, A. M. (1988) *J. Am. Chem. Soc.* 110:7236–7237.

Tanford, C. (1968) *Adv. Protein Chem.* 23:121–282.

Wong, C. H., Chen, S. T., Hennen, W. J., Bibbs, W. J., Wang, Y. F., Liu, J.L.C., Pantobiano, M. W., Whitlow, M., and Bryan, P. N. (1990) *J. Am. Chem. Soc.* 112:945–953.

Zaks, A., and Russell, A. J. (1988) *J. Biotechnol.* 8:259–269.

EPIDEMIOLOGY, MOLECULAR

Paul A. Schulte, Frederica P. Perera, and Nathaniel Rothman

1 **Principles**
  1.1 Biologic Markers of Exposure
  1.2 Biologic Markers of Effect
  1.3 Biologic Markers of Susceptibility
  1.4 Utility of Molecular Epidemiology

2 **Techniques**
  2.1 Representational Validity of Molecular Biological Markers
  2.2 Validation of the Behavior of Molecular Biological Markers
  2.3 Etiologic Studies
  2.4 Public Health Applications

Key Words

Biological Markers (Biomarkers) Biochemical, molecular, genetic, immunologic, or physiologic signals of events in biological systems.

Biologic Marker of Effect A measurable cellular, biochemical, or molecular alteration within an organism that, depending on magnitude, can be recognized as an established or potential health impairment or disease.

Biologic Marker of Exposure A xenobiotic chemical or its metabolite, or the product of an interaction between a chemical, physical, or biologic agent and some target cell or biomolecule.

Biologic Marker of Susceptibility An inherited or acquired indicator of the response of an individual or a population to a specific xenobiotic agent.

Molecular epidemiology is the use of molecular biological techniques to identify exposures, effects, or susceptibility factors in studies of human populations. Molecular epidemiology and traditional epidemiology utilize the same paradigm. However, the former presents the opportunity to use the enhanced resolving power of molecular biology in the assessment of exposure–disease relationships. The resolving power, to elucidate a continuum of events between xenobiotic exposure and disease, can provide stronger approaches to research, prevention, and intervention.

1 PRINCIPLES

The use of molecular biological techniques in epidemiology provides a potentially powerful tool for medical and public health researchers. These techniques allow for the identification of biological markers (Figure 1) that can indicate exposure to a xenobiotic agent, reveal a biological effect early in the natural history of disease, or represent unique disease subtypes or susceptibility to the development of disease. Although the use of biological markers in epidemiology is not new, the current generation of molecular biological markers enhances past approaches. Epidemiology is an observational science: one makes inferences about disease and health based on comparing groups of people in terms of disease incidence and mortality. Ideally, the groups being compared should be similar in all respects except for the risk factor in question. A benefit of molecular epidemiology is that instead of comparing two groups on the basis of environmental exposure, investigators can compare populations with respect to dose of an environmental agent as measured in critical macromolecules, such as DNA or surrogate proteins for DNA. That is presumably a more accurate means of classifying the subjects' true exposure. At the other end of the exposure–disease

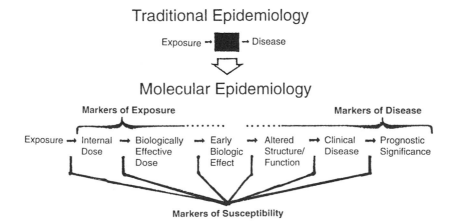

Figure 1. Enhancement of the traditional epidemiologic paradigm by the use of biological markers resulting in a molecular epidemiologic approach. In traditional epidemiology the mechanism of action is often a "black box," and associations between an exposure and disease are made by inference. In molecular epidemiology a continuum between an exposure and a disease is defined, and various markers are identified.

continuum, instead of using frank disease as an outcome variable in an epidemiologic study, it is possible to use a validated biologic marker of effect.

1.1 BIOLOGICAL MARKERS OF EXPOSURE

The utility of a biological marker of exposure depends, in part, on its half-life, the pattern of the exposure it is measuring (e.g., regular daily exposure vs. infrequent episodic exposure), and whether secular trends have occurred in that exposure (e.g., smoking cessation). In addition, the information it provides must be compared to the availability and quality of other sources of data (e.g., questionnaires, environmental monitoring, medical records). Essentially all exposure measures misclassify some subjects—it is the relative ability of different sources of data to correctly place individuals into exposure categories that must be considered. For example, subjects can generally report average smoking habits and smoking duration in an accurate manner, permitting the calculation of cumulative exposure. Since internal dose markers associated with tobacco smoke have relatively short half-lives, and thus reflect only recent exposure (e.g., cotinine, a metabolite of nicotine), they have limited utility by themselves to directly assess a smoking–cancer relationship.

In contrast, it is difficult to obtain accurate information about dietary exposure to aflatoxin by questionnaire because exposure is sporadic and is present in a spectrum of different food types. In this instance, even a short-term internal dose marker might classify the long-term exposure status of subjects more accurately than questionnaire data. For example, a nested case–control study conducted in Shanghai, China, demonstrated that aflatoxin exposure, assessed by measuring several aflatoxin metabolites (and an N^7-guanine aflatoxin adduct) in banked urine samples was associated with an increased risk of hepatocellular carcinoma, while aflatoxin exposure assessed by questionnaire was not associated with elevated risk.

1.2 BIOLOGIC MARKERS OF EFFECT

In terms of disease, techniques are available for identifying biological changes earlier in the continuum between homeostatic response to pathological agents or conditions and development of frank disease. This advance has implications for identifying opportunities for prevention. Biologic markers intermediate between exposure and disease, if validated for disease or risk of disease, can be used to screen people or to allow for early disease detection. These markers can also be used in intervention trials as outcome indicators rather than waiting for subjects to develop disease.

There is a popular notion that molecular epidemiology has the potential to contribute to assessment of risk of an individual. However, that point needs clarification. Historically, it has been possible to use an epidemiologic data set, that is, data on a group consisting of sick and healthy people, with and without certain risk factors, to develop risk functions that provide for an estimate of individual risk. This was accomplished in the 1960s using data from the Framingham Longitudinal Study of cardiovascular disease. Based on knowing the risk for people with a certain aggregation of characteristics, it was possible to predict the risk of an individual. Molecular epidemiologic approaches provide a means for more confident estimation because more mechanistic information can be utilized. Still the resultant assessment is only probabilistic determination, not deterministic. While the prediction is that a certain individual is likely to

develop disease in the future, there is no guarantee that he or she will develop it.

A biomarker may have utility to screen populations at high risk of disease as part of a primary or secondary prevention effort. However, a substantial amount of information is required before a biomarker can be used for this purpose. In particular, the cumulative probability that an individual will develop disease over a defined period, given a constellation of biologic and nonbiologic risk factors, must be estimated along with a calculation of its uncertainty.

1.3 BIOLOGIC MARKERS OF SUSCEPTIBILITY

One of the greatest contributions of molecular epidemiology is the ability to discern the role of host factors, particularly genetic factors, in accounting for variation in response. Why similarly exposed people do not get the same diseases is a target question for molecular epidemiology. In most disease systems, susceptibility markers are being identified and evaluated. These markers can be incorporated into epidemiologic models as modifiers of the relationship between an exposure and an effect (see Section 2.1.3).

Traditional epidemiologic approaches can also be enhanced by using molecular genetic techniques to identify host factors that could account for differences in disease risk. Thus, for example, a metabolic polymorphism can be detected from peripheral blood lymphocyte DNA, and groups at potentially greater and lesser risk of disease can be distinguished.

The category of markers of susceptibility includes polymorphisms in genes responsible for chemical activation or detoxification, DNA repair, and genomic stability. Susceptibility genes of some types may interact with chemical exposures of very specific types (e.g., cytochrome P450 enzyme subtypes and phase II conjugating enzymes), while others may confer more general susceptibility (e.g., p53 mutations in Li–Fraumeni syndrome). Markers can be measured at the DNA level, if the genetic basis of a polymorphic phenotype has been identified, or at the phenotypic level (e.g., drug probes of hepatic enzyme activity, DNA repair measured in peripheral lymphocytes).

1.4 UTILITY OF MOLECULAR EPIDEMIOLOGY

Molecular epidemiology is a useful temporary blanket term that reflects the reality that increasingly disease, causal exposures, and risk factors are being defined at the molecular level. The utility of this term is that it serves as a signpost for epidemiologists to consider using independent (risk factor and dependent (outcome) variables that are derived from molecular biological techniques and assays. So, for example, DNA adducts may be used in addition to breathing-zone measurement of a carcinogen; gene mutational patterns may be used as an indicator of disease rather than a nosological death certificate; and genotyping based on the polymerase chain reaction can be used in addition to race and sex to stratify populations for comparisons.

In summary, molecular epidemiology has the potential to contribute the following opportunities and capabilities to public health:

- Delineation of a continuum of events between an exposure and a resultant disease
- Identification of exposures to smaller amounts of xenobiotics and enhanced dose reconstruction
- Identification of events earlier in the natural history of clinical diseases and on a smaller scale

- Reduction of misclassification of dependent and independent variables
- Indication of mechanisms by which an exposure and a disease are related
- Better accounting for variability and effect modification
- Enhanced individual and group risk assessments

2   TECHNIQUES

If molecular biological markers in epidemiology and public health can be used in the ways described in Section 1, they must be demonstrated to be valid in terms of both the assay and the marker. An assay will be valid if it measures what it is expected to measure. A marker will be valid for epidemiologic purposes in two ways. First, it will be valid to the extent that it represents exposure, disease, or susceptibility. Second, it will be valid insofar as the extent of variation in groups with different demographic, behavioral, or medical characteristics is known. A key question is the prevalence of a particular biomarker (e.g., a mutation) in different ethnic or racial groups, in smokers or drinkers, or in people with various hereditary and acquired diseases. Molecular epidemiologic approaches can be useful in the validation of molecular biological markers in these two ways.

2.1   Representational Validity of Molecular Biological Markers

2.1.1 Validation of the Relationship Between Exposure and Dose

$$\text{Exposure} \rightarrow \underset{\text{dose}}{\text{Internal}} \rightarrow \underset{\substack{\text{effective}\\\text{dose}}}{\text{Biologically}}$$

Molecular markers of exposure may be validated by assessing the relation of an exogenous exposure to an internal dose or biologically effective dose. Critical in validation studies is the need to have an effective exposure assessment. It may be necessary to use a combination of personal and environmental monitoring and questionnaires, record review, and modeling to reconstruct exposure history. The approach also requires understanding of the pharmacokinetics associated with the particular xenobiotics. Related to this are the needs to understand the natural history of the marker and to use it in the validation study. For example, in a study of hydroxyethyl hemoglobin adducts in workers exposed to ethylene oxide, the life span of the erythrocyte, hence the constituent hemoglobin molecule ($\approx$ 4 months), was used as a dosimeter of cumulative exposure. There is also a need to account for factors that might influence the appearance of a molecular biological marker. In the aforementioned studies, when mean values were adjusted for important covariants such as age, cigarette smoking, and education, an exposure response relationship was found, at levels below the permissible exposure level.

2.1.2 Validation of the Relationship Between Biological Effects and Disease

$$\underset{\text{effects}}{\underset{\text{biologic}}{\text{Early}}} \rightarrow \underset{\text{function}}{\underset{\text{structure/}}{\text{Altered}}} \rightarrow \text{Disease}$$

Validation information in the biological effects–disease category is limited. The often repeated question is, What do the data mean concerning health and disease? Validation studies of these types are difficult to accomplish because of the temporal factor. To identify an early effect—that is, an effect in pathogenesis or an effect predictive of disease—generally requires a prospective study design, although cross-sectional clinical studies of diseased and heavily exposed individuals can be used to great advantage. When a prospective design is not used, however, care must be taken to avoid biased associations. This is often difficult; hence prospective studies are the best approach for validation.

Prospective studies are expensive and time-consuming, and few are conducted. For example, despite the large number of studies on most of the cytogenetic markers, there is little consensus on their predictive value because most of the studies have been cross-sectional and markers were not linked to disease. Specifically, in epidemiologic terms, predictive value is evaluated in terms of the percent of those who test positive for a marker who actually develop the disease. To perform the appropriate prospective studies of sister chromatid exchanges would take a large population and a relatively long time.

The best example of such a study is the Nordic prospective study on the relationship between peripheral lymphocyte chromosome damage and cancer morbidity in occupational groups. Ten laboratories in four Nordic countries participated in a study of a combined cohort of persons (mostly from occupational groups) who had been cytogenetically tested. The cohort is being followed prospectively for cancer morbidity. The cohort is comprised of 3190 subjects, of whom 1986 subjects (62%) have been scored for chromosome aberrations and 2024 subjects (63%) scored for sister chromatid exchanges. Preliminary analysis indicates that chromosomal aberrations are associated with cancer. These biologic markers in peripheral lymphocytes represent carcinogenic changes elsewhere in the body. This is the critical criterion of a useful biologic marker.

To serve as a valid outcome measure, an intermediate marker must be correlated with disease risk. The criteria for validating intermediate biomarkers have been extensively discussed and include the sensitivity of the marker (i.e., the proportion of subjects who develop cancer who are positive for the biomarker), the relative risk (a measure of the strength) of the association between the marker and disease, and a judgment about the extent to which the exposure–disease relationship is mediated through a process reflected directly or indirectly by the marker.

2.1.3 Validation of Markers of Susceptibility

$$\text{Exposure} \rightarrow \text{Susceptibility} \rightarrow \text{Disease}$$

The tools of molecular biology and analytical chemistry have allowed researchers to identify a degree of interindividual variability not previously imagined. Validated biological markers of susceptibility can serve as effect modifiers in epidemiologic studies. Effect modification is a term with statistical and biological aspects. Statistically, the examination of joint effects of two or more reactors is often discussed in the context of effect modification. It depends on the statistical method (e.g., multiplicative or additive) used to model interaction. From the biological perspective, effect modification contributes to answering the question of why all similarly exposed individuals do not develop a disease. The answer, in part, lies in individual variability in metabolic and detoxification capabilities, their ability to repair genetic damage, or other host factors.

To validate a susceptibility marker, it is important to minimize misclassification, which can occur as a result of laboratory or epidemiologic factors that affect phenotyping or genotyping. Next, it is necessary to demonstrate that the marker increases the risk of disease.

The issue of the correlation of acetylation phenotype and bladder cancer from aromatic amines illustrates the concept of susceptibility. Some aromatic amines are detoxified by the enzyme N-acetyltransferase, and the slow phenotype of this enzyme has been associated with bladder cancer in exposed individuals. Despite a plethora of studies, the scientific literature is not conclusive on the extent to which being a slow acetylator modifies the risk for bladder cancer in people exposed to aromatic amines. Generally, most studies have been too small, have had weak exposure characterization, and were not designed to allow proper determination of whether exposure or susceptibility was the key factor. An example of how partial validation of a susceptibility marker might occur without using disease as the outcome involved the formation of hemoglobin adducts (which are documented surrogates for DNA adducts believed to be involved in carcinogenesis) in slow and fast acetylators who had been exposed to 4-aminobiphenyl. Slow acetylators had an average of 1.5-fold greater frequency of adducts than the fast acetylators. Despite these encouraging efforts at validation, few markers of susceptibility have been validated, and none are ready for use in population screening.

2.2 Validation of the Behavior of Molecular Biological Markers

Before biomarkers can be used for etiological and prevention research, they need to be validated in populations. This calls for the development of analytical methods for use in large-scale populations. Currently, there is inadequate research support for scaling-up efforts needed for population studies. The validation and scaling efforts discussed here require close collaboration between laboratory scientists and population scientists (clinicians, epidemiologists, industrial hygienists, and exposure assessors). Transitional studies bridge the gap between the development of molecular markers in the laboratory and their application in population-based studies. These studies generally involve the initial evaluation and application of biomarkers in healthy human populations. Their objective is to address issues in sample processing, evaluate assay accuracy and precision, collect information about potential confounders and effect modifiers, and study early biologic effects of selected exposure. Transitional studies can be divided into three broad categories to clarify their distinctive research goals: developmental, characterization, and applied studies. In practice, however, elements of all three types of study are often incorporated into a single field investigation.

2.2.1 Developmental/Characterization Studies

Identification of a promising new molecular biomarker in the laboratory does not mean that the biomarker is ready for use in an epidemiologic study. Other basic issues need to be resolved before its application in human studies can be considered. First, the reliability (i.e., the repeatability of the assay) of a marker must be determined. As long as an assay is reliable, the ordering of subjects by the measure is preserved. Since this is all that is required for studying a marker–disease relationship, reliability, and not accuracy, is of initial importance. Reliability of laboratory assays may be assessed by the analysis of blind replicate samples representative of the range of values likely to be found in human populations. After the assay reliability and accuracy has been determined, it is important to define the optimal conditions for collecting, processing, and storing biological specimens for eventual assay, since not uncommonly, small variations in these conditions determine the subsequent analyzability of samples.

These studies are generally designed to address questions about the presence or levels of a newly developed marker in the general population. In addition, they serve to identify factors that are confounders or effect modifiers of a marker (e.g., age, gender, medications) that need to be measured and taken into account when applying the marker in subsequent studies.

2.2.2 Applied Studies

Applied studies are investigations performed on subjects with particular patterns of exposure to xenobiotics (e.g., occupational exposures, smokers) or on patients receiving chemotherapy. In these studies, the biomarker is treated as the outcome variable. At this stage of research, the biomarker has not been shown to predict an increase in risk of disease. The marker, however, can often be used to provide insight into the association between external exposure and biologic processes early in the exposure–disease relationship.

Applied studies can help establish the biologic plausibility of associations detected in etiologic studies. Applied studies generally cannot establish a causal relationship between a given exposure, or a given level of exposure, and risk for developing disease. The results of applied studies using the biomarker as outcome are suggestive only until a marker is shown to predict disease risk, which can be established only by comparing risk of disease in individuals with and without the marker. In these studies, the biomarker may be overly sensitive (i.e., it may respond to low levels of chemical exposures without biological relevance), it may be insensitive, or it may reflect phenomena that are irrelevant to the disease process. Until these relationships have been sorted out, the findings are merely suggestive.

2.3 Etiologic Studies

The major objective of molecular epidemiology is to conduct etiologic and applied research. Etiologic studies [i.e., ecologic, case–control, case–case (also referred to as a case series), prospective cohort, family, twin, and intervention studies] can be distinguished from transitional studies in that the former involve either clinically ill subjects, asymptomatic subjects with early disease, or subjects positive for an intermediate process known to be associated with increase risk of disease (e.g., colon adenomas and risk of colon cancer). Case–control, case–case, and prospective etiologic studies can effectively utilize molecular epidemiologic approaches.

2.3.1 Case–Control Studies

A case–control study involves the comparison of cases (people with a particular disease) with controls (people without that disease) for various risk factors. The risk factors could be an exposure, trait, or biomarker. Traditionally, case–control studies have involved patients with clinically confirmed disease, identified either through the presence of symptoms or as a result of incidental findings on routine clinical examinations. Increasingly, cases are being defined as asymptomatic subjects whose early preclinical disease has been ascertained by screening (e.g., early breast disease, colon polyps, cervical dysplasia).

The case–control study design is used far more frequently than the prospective cohort study design because of its relatively greater efficiency and lower costs. Therefore, maximizing opportunities to creatively integrate biomarkers into case–control studies is important. Because some markers are affected by disease itself, which raises complications of reverse causality, it is important to define which biomarker categories can most effectively be used in this study design.

2.3.2 Case–Case Studies

A case–case study involves a series of cases of the same types that are compared on the basis of a particular exposure and a particular biologic characteristic. The accumulation of a spectrum of *ras* oncogene mutations in leukemia cases with benzene exposure compared to those in leukemia cases without benzene exposure is an example of a case–case study. Such studies have the potential of identifying an exposure-specific effect at the molecular level. Case–case studies, however, cannot be used to estimate the relative risk of disease from a specific exposure. A nondisease control group is required for this purpose.

2.3.3 Prospective Cohort Studies

Prospective studies involve healthy people, characterized by presence or absence of a risk factor, who are followed forward to determine the risk of disease. In prospective studies, the biological samples may be collected from subjects at various times. These samples are either analyzed at the time of collection or banked for later analysis. One way to utilize biologic markers in a prospective study is to follow the groups of subjects forward in time: subjects who develop disease are identified, and premorbid biomarker levels in the group with disease are compared with those without disease. Often, a nested case–control approach is used. Samples from cases and only a sample of the controls (noncases) are analyzed, which considerably reduces the laboratory requirements and costs. Although the prospective study design is by far the most time-consuming and expensive type of observational epidemiologic study, it is the only method available to test the association of biomarkers with disease risk when the markers are transient or may be directly or indirectly affected by disease.

Prospective studies may yield banks of biological specimens that are useful for future studies. Large cohort studies initiated in the 1980s and 1990s are, in general, banking most or all fractions of the peripheral blood sample to allow a far wider range of biologic assays to be performed, particularly those that require DNA.

2.4   PUBLIC HEALTH APPLICATIONS

Public health practice incorporates the end use of validated biomarkers for risk assessment by government agencies, population screening (both active and passive), and clinical and preventive medical practice. The use of molecular biomarkers in public health practice is still in its infancy, although several very promising markers may soon find their way into the public health arena. The standard principles of biological monitoring and medical screening are applicable to the use of any biomarker, however. These include assay reliability and cost, strength of the association between a mark-

er and disease risk, prevalence of the marker in the population, availability of effective, preventive strategies that can be employed in subjects who are positive for the marker, and a host of ethical considerations, such as informing subjects of test results.

See also CANCER; HUMAN GENETIC PREDISPOSITION TO DISEASE; PATHOLOGY, MOLECULAR; XENOBIOTIC METABOLISM.

Bibliography

Gledhill, B. L., and Mauro, F., Eds. (1990) *New Horizons in Biological Dosimetry.* Wiley-Liss, New York.
Groopman, J. D., and Skipper, P. C., Eds. (1991) *Molecular Dosimetry and Human Cancer: Analytical, Epidemiological and Social Considerations.* CRC Press, Boca Raton, FL.
Hulka, B. S. (1991) Epidemiological studies using biological markers: issues for epidemiologists. *Cancer Epidemiol. Biomarkers Prev.* 1:13–19.
———, Wilcosky, T. C., and Griffith, J. C., Eds. (1990) *Biological Markers in Epidemiology.* Oxford University Press, New York.
National Research Council (1992) *Biologic Markers in Immunotoxicology.* National Academy Press, Washington, DC.
Perera, F. P., and Weinstein, I. B. (1982) Molecular epidemiology and carcinogen–DNA adduct detection: New approaches to studies of human cancer causation. *Chronic Dis.* 35:581–600.
Rothman, N., Stewart, W. F., and Schulte, P. A. (1995) Incorporating biomarkers into cancer epidemiology: A matrix of biomarker and study design categories. *Cancer Epidemiol. Biomarkers Prev.* 4:301–312.
Schatzkin, A., Freedman, L. S., Schiffman, M. H., Dawsey, S. M. (1990) Validation of intermediate endpoints in cancer research. *J. Natl. Cancer Inst.* 82:1746–1752.
Schulte, P. A. (1993) The use of biological markers in occupational health research and practice. *J. Toxicol Environ Health,* 40:359–366.
———, and Perera, F. P., Eds. (1993) *Molecular Epidemiology Principles and Practices.* Academic Press, San Diego, CA.

EPIDERMAL KERATINOCYTES, MOLECULAR BIOLOGY OF

Miroslav Blumenberg

Key Words

Activated Keratinocytes   Epidermal cells under hyperproliferative or inflammatory conditions; migratory, highly responsive to extracellular stimuli, and producing growth factors and cytokines. Distinct from basal keratinocytes, which differentiate normally.

Desmosomes   Very strong mechanical protein plaques found on the lateral surfaces of keratinocytes, connecting adjacent cells and giving epidermis its mechanical resilience. Basal surfaces connect to the basement membrane via hemidesmosomes.

Differentiation   The orderly process by which more primitive progenitor cells give rise to specialized cells, elaborating new proteins and performing specific functions. In epidermis, the basal layer differentiates into the layers above.

Keratinopathy   Inherited disease that is due to a mutation in one of the keratin genes or genes whose product interacts with keratins.

Keratins   The predominant proteins of epidermis; cytoskeletal proteins of all epithelial cells; largely α-helical members of two related families.

Stem Cells   Primitive, undifferentiated cells with unlimited proliferative potential and capability to differentiate along multiple pathways.

The main function of the epidermis is protection: protection from desiccation, from UV damage, and from mechanical injury as well as from immunological insult (e.g., from microbial invasion). The predominant cell type of the epidermis, the keratinocyte, can provide mechanical protection, prevent water loss, and form the first line of immunological defense as well. Keratinocyte differentiation is a dynamic process in which the cells at the bottom layer constantly and rapidly divide to give rise to the differentiating cells above. During normal epidermal differentiation, within 4–6 weeks, keratinocyte stem cells pass through several stages of development, from basal to cornified. Basal cells are in contact with the basement membrane, are mitotically active, and express keratins K#5 and K#14. Basal keratinocytes detach from the basement membrane, stop dividing, become spinous, and initiate the process of terminal differentiation. They start to express copious amounts of a large number of markers of differentiation, such as keratins K#1 and K#10, which are fully expressed at high levels in the spinous and granular layers. Filaggrin and precursors of the cornified envelopes, such as loricrin and involucrin, as well as epidermal transglutaminase are expressed in the granular layers. The final stages of differentiation include complete dissolution of all cellular organelles and nuclei, structural reorganization, and cross-linking of the protein content, resulting in the formation of cornified envelopes that are assembled into a metabolically inert stratum corneum.

1   THE BASAL LAYER

The keratinocytes in the basal layer are attached to the basement membrane. They are strikingly polarized with clearly distinct basal, lateral, and apical surfaces. These cells are progenitors of the differentiating cells superficial to them, and they provide the mechanical attachment of the epidermis to the substratum. Basal keratinocytes produce characteristic markers to accomplish these functions.

1.1   STEM CELLS

Among the most interesting and important questions in keratinocyte differentiation (Figure 1) is the one concerned with the existence and position of the stem cells. All continuously dividing, self-renewing tissues contain a population of cells that have a very large proliferative capacity. Usually these cells proliferate rather slowly. They give rise to one daughter stem cell and another daughter cell whose progeny cells are destined to differentiate, usually after dividing several additional times.

In stratified epithelia, stem cells are found in characteristic physical locations (Figure 2), which provide mechanical as well as chemical protection to the stem cells. The chemical protection is particularly important because the stem cells are sensitive to oncogenic damage that can lead to carcinomas. Furthermore the location of stem cells must expose them to the inductive signaling to divide, signaling that comes from both the overlaying epithelium and the underlying stroma.

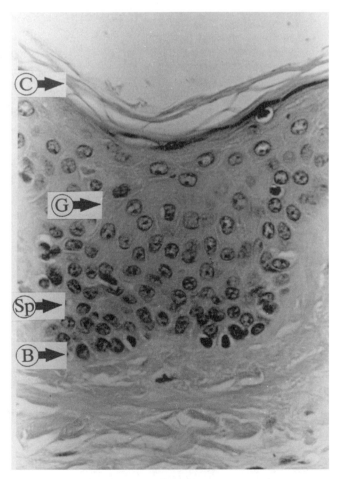

Figure 1.   A cross section of human epidermis. Four differentiation stages can be identified: B, Basal; Sp, spinous; G, granular; C, cornified c. (Photomicrograph courtesy of Dr. H. Kamino.)

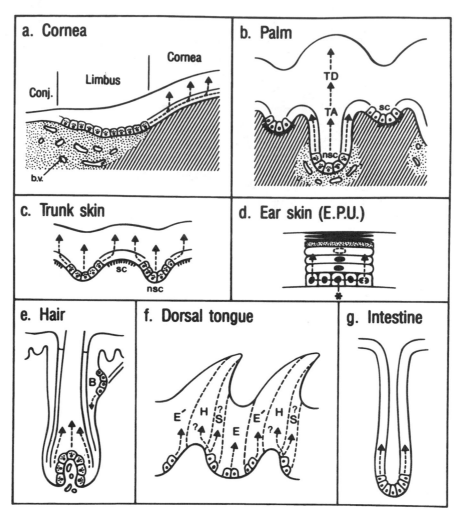

Figure 2. Position of stem cells in stratified epithelia. (a) In cornea, the stem cells are found in the limbus and move centripetally. (b) In palms and soles, the stem cells are the nonserrated basal cells in the deep ridges. (c) In trunk skin, stem cells are probably in the shallow ridges and can give rise to transiently amplifying cells shown in (d), thus forming an epidermal proliferative unit or (E.P.U.). (e) In hair, the bulge area B contains the stem cells, which migrate downward into the papilla before differentiating. (f) In the dorsal tongue, several different epithelia can be recognized; it is not known whether each has a characteristic stem cell population or whether the same stem cell can give rise to various phenotypes. (g) In the intestine, the stem cells are at the bottoms of the crypts. (Drawing courtesy of Dr. T.-T. Sun.)

For example, in cornea the stem cells are found in the limbus at the basement membrane. In the palmoplantar epithelium, which is characterized by deep rete ridges, basal cells of two types have been characterized: a population of rapidly cycling serrated cells that give rise to transiently amplifying suprabasal cells, and a population of very slowly cycling, undifferentiated, nonserrated cells believed to represent the palmoplantar stem cells. Similarly, in the intestine the bottom of the crypts harbors a population of stem cells that are pluripotent and give rise to all the epithelial cells above them.

In the trunk skin, slowly cycling cells, possibly stem cells, have been found not only in the interfollicular epidermis but in the bulge area of the hair follicle as well. These cells, which are relatively undifferentiated, can give rise both to the sebaceous gland and to the epithelial cells of the hair follicle. The cells from the bulge area of the hair follicle can regenerate a fully differentiated multilayered epidermis in culture. A subpopulation of basal cells has a large proliferation potential and increased amounts of specific integrins, basement membrane attachment proteins. These cells may be the interfollicular stem cells.

1.2   ATTACHMENT TO THE BASEMENT MEMBRANE

Basal keratinocytes contain several integrins, the proteins that recognize extracellular matrix components. Integrins are transmembrane proteins composed of one α and one β subunit. The particular $\alpha\beta$ combination determines the specificity of binding of integrin to its extracellular matrix protein ligand. Healthy human keratinocytes express integrins $\alpha_2\beta_1$, $\alpha_3\beta_1$, $\alpha_5\beta_1$, $\alpha_6\beta_4$, and $\alpha_v\beta_5$, receptors for collagen, laminin, fibronectin, kalinin, and vitronectin, respectively. Some of these integrins (e.g., $\alpha_6\beta_4$) are localized at the basal aspect, while others, the β_1 family, are predominantly localized on the lateral surfaces of the basal keratinocytes.

The binding of the integrins to their ligands provides one of the components that anchor the basal keratinocytes to the basement membrane. However, recently it has been proposed that integrins can also serve to transduce signals from the extracellular milieu into the cell. In other words, the binding of the integrins to the basement membrane can inform the cell about the composition and nature of the basement membrane it sits on. This property may be particularly important dur-

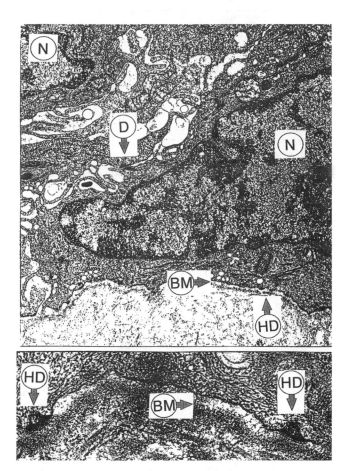

Figure 3. Electron micrograph showing hemidesmosomes (HD) and desmosomes (D). Note the nuclei (N) of two adjacent basal cells and the basement membrane (BM). The lower micrograph, at higher magnification, reveals the intracellular association of the hemidesmosomes with keratins, and extracellular with basement membrane.

ing wound healing, when the cells deposit a provisional, fibronectin-rich substrate, which is later replaced by a permanent, laminin-rich basement membrane. Indeed, during wound healing additional integrins appear at the keratinocyte surfaces; their distribution is dramatically changed, however, and they persist into the suprabasal layers.

The most prominent structures on the basal surface of keratinocytes are the hemidesmosomes (Figure 3). Hemidesmosomes physically resemble halves of desmosomes, the most prominent structures of the lateral cell-to-cell contacts, and both desmosomes and hemidesmosomes are linked to the keratin cytoskeletal network in the cytoplasm. Biochemically, hemidesmosomes and desmosomes are quite different. Among the components of hemidesmosomes, not found in desmosomes, are the bullous pemphigoid antigens. Hemidesmosomes also contain a specific integrin comprising $\alpha_6\beta_4$ subunits. The β_4 integrin subunit contains an extraordinarily long cytoplasmic tail, thought to provide the interaction between the hemidesmosomes and keratin intermediate filaments of the cytoskeleton.

2   KERATINS AS MARKERS OF EPITHELIAL DIFFERENTIATION

Keratins are phenotypic markers of epithelial development and differentiation. The intermediate filament network in all epithelial cells consist of keratin proteins, a large family of approximately 30 proteins. Keratins K#5 and K#14 are specific to basal cells. Differentiating keratinocytes are recognized by the presence of keratins K#1 and K#10, and K#6 and K#16 are expressed in activated keratinocytes.

2.1   STRUCTURE OF KERATIN PROTEINS

All intermediate filament proteins have similarities with the amino acid sequences of the keratin proteins. They have a central α-helical rod domain region that contains approximately 310 amino acids, disrupted at conserved sites by nonhelical linkers and bracketed by end domains that are highly variable in sequence and structure. The α-helical segments are highly conserved in sequence and in length, while the linkers vary considerably among different keratins both in sequence and in length (Figure 4).

The precise and conserved arrangement of α-helical and linker subdomains was thought to be essential for the assembly, structure, and function of keratin filament conformation. Recent experiments show that the structure of the central domain tolerates significant alterations, because either disrupting the α-helices or making the linkers α-helical is compatible with filament formation and structure. Disruption of the ends of the central domain, however, destroyed the filaments. The sequences that have been most preserved in all intermediate filament proteins are the ends. The sequence $LNDR^L_FAX^{YIE}_{FLD}K$ is found in the amino-proximal end, whereas $TYRXLLEGE^E_D$ is in the carboxyproximal end of keratin proteins. Disruption of the α-helical conformation at the ends would shorten the length of the α-helical rod and prevent its proper alignment with its neighbors.

The conserved central domains are responsible for the common features characteristic of all keratins (interactions with other proteins, assembly, etc.). The terminal end domains vary greatly among keratins and give each keratin its individual character. Glycine-rich termini are characteristic of epidermal keratins, cysteine-rich of hair keratins, and so on.

After several keratin genes had been cloned and sequenced, it was found that they are clustered in linkage groups. The acidic-type human keratin genes are located on chromosome 17q12-q21, while basic-type human keratin genes are on chromosome 12q11-q13. The murine genome, interestingly, has several additional loci that house skin and hair genes: basic-type keratin genes are linked on chromosome 15 to *Ca* (caracul, curled hair), *Sha* (shaven), and *Ve* (velvet),

end domain	alpha-helical domain	end domain

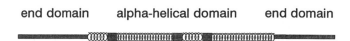

alpha-helical domains
Dimerization & cytoskeleton formation

end domains

Hair	Cysteine-rich	S-S crosslinking
Epidermis	Glycine-rich	Tight packing

Figure 4.   Keratin protein structure.

whereas *Re* (with kinky hair), *Den* (denuded), and *Bsk* (bare skin) are linked to the acidic-type keratin genes on murine chromosome 11, thus potentially including the keratin genes into several "epidermis-specific" chromosomal loci.

2.2   BASAL-CELL-SPECIFIC MARKERS: KERATINS K#5 AND K#14

Within their respective families, keratins K#5 and K#14 are relatively small. They are very rich in serine, which may make them relatively less insoluble than other keratins. Because they are serine-rich, they are highly susceptible to phosphorylation. Phosphorylation has been associated with dissolution of the keratin protein network during cell mitosis. In this respect it may be important that keratins K#5 and K#14 are synthesized in the basal, proliferative compartment of all multilayered epithelia, not only of the epidermis.

2.3   EARLY MARKERS OF DIFFERENTIATION: KERATINS K#1 AND K#10

Keratins K#1 and K#10 are two of the earliest markers of keratinocyte differentiation. They have very long, glycine-rich terminal domains with subdomains consisting of short (4–10 amino acid) segments repeated 3–15 times. Glycine is the ultimate hydrophobic amino acid, because it has no side chain. Polyglycine can pack extremely tightly into paracrystaline, hydrophobic aggregates, making it resistant to proteolysis. These are the ideal materials for building the protective stratum corneum.

2.4   MARKERS OF ACTIVATED KERATINOCYTES: KERATINS K#6 AND K#16

In all activated keratinocytes one finds the expression of a specific keratin protein pair, K#6 and K#16. K#6 and K#16 are serine-rich keratin proteins more similar in sequence to the basal-cell-specific keratins K#5 and K#14, respectively, than to the differentiation-specific keratins K#1 and K#10. Keratins K#6 and K#16 are not expressed in healthy interfollicular epidermis. However, they are found in conditions associated with hyperproliferation—for example, in psoriasis, wound healing, and carcinomas. A reciprocal relationship exists between the production of keratins K#6 and K#16 and differentiation-specific keratins, such as K#1 and K#10 in epidermis or K#3 and K#12 in cornea. K#6 and K#16 can be expressed in healthy tissue—for example, in the developing mammary epithelium, in the outer root sheet of the hair follicle, and in fungiform and filiform papillae of the tongue.

Epidermal growth factor (EGF) and transforming growth factor α (TGFα) specifically induce transcription of both K#6 and K#16 keratin genes through specific DNA sites and nuclear transcription factors. Proto-oncoproteins Fos and Jun also induce synthesis of K#6 and K#16, which may be relevant in carcinomas.

2.5   REGULATION OF KERATIN GENE EXPRESSION

Regulation of keratin gene expression occurs predominantly if not exclusively at the level of transcription initiation. The factors involved in keratin gene transcription may be divided into three broad categories: (1) general transcription factors that bind to many different genes and function in many different cell types, (2) cell-type-specific and keratin-gene-specific factors, which are functional only in the cell types in which a given keratin is normally expressed, and (3) modulators of the level of transcription (i.e., the transcription factors and recognition elements that are affected by hormones, vitamins, growth factors, and other environmental signals).

As expected, most of the important sites for regulation are found in the upstream region, although in several instances introns and even the downstream sequences contain regulatory elements. For example, keratin genes K#8 and K#18 have AP-1 and Ets sites in the intronic and downstream sequences, and downstream from the K#1 gene there is a vitamin-D3-responsive site, although it is not clear whether this site is actually responsible for regulation of K#1 expression or, possibly, regulates another nearby keratin gene. Interestingly, the human K#18 keratin gene is insulated from the influence of nearby sites by a set of *Alu* repetitive sequences that bracket the gene.

In the promoters of keratin genes one finds sites for common transcription factors (Figure 5). Most notable are Sp1 and AP2 sites. These are essential for the expression of the genes and were at times thought to be responsible for the epithelia-specific expression. NF-1 sites are also common, and so are the retinoic acid responsive elements (RAREs). These promoters also contain binding sites for less well known or even hitherto uncharacterized transcription factors. An example of the former is E2BP, a transcription factor that is a member of heterogeneous RNA-binding proteins; an unnamed protein that binds to keratin gene promoters at sites adjacent to Sp1 and AP2 sites was characterized only recently. Indeed, transcription factors often bind to keratin genes at adjacent sites in tight clusters. It is possible that such clusters confer cell type specificity to the expression of keratin genes.

Several transcriptional factors were found to be specific for epidermal keratinocytes. Among these is basonuclin, an unusual transcription factor that contains many pairs of Zn fingers and a serine stripe running down its α-helical segment. Importantly, this protein is found exclusively in the nuclei of the basal and the first suprabasal layer of epidermal keratinocytes, and it may play a role in specifying basal-layer-specific transcription.

Transcription factors Skn1 and Skn2 are also specifically found in skin. They are functionally distinct, belong to the Oct 2 family, and contain Pou domains. Therefore they belong to a family of transcription factors that specify the differentiation pathway of a given cell. It is not yet known whether Skn1 and Skn2 play a role in epidermal differentiation.

3   THE DIFFERENTIATION PROCESS

3.1   THE DIFFERENTIATION SIGNAL

At some point the basal cell leaves the basal layer and commences differentiation. The signal that causes initiation of differentiation is unknown, but the most likely candidate is the $\alpha_5\beta_1$ integrin, the fibronectin receptor. When occupied by its ligand, $\alpha_5\beta_1$ integrin holds keratinocytes attached to the basement membrane. After release of the ligand it may convey to the cell the message that the attachment to the substrate has been weakened, which could be the signal to commence differentiating. It is important to note that $\alpha_5\beta_1$ integrin can convey this signal without being removed from the surface. Free, unliganded $\alpha_5\beta_1$ integrin, while physically still present on the cell surface, may be in a different conformation from the one it holds when attached to its ligand. Certain antibodies that bind in-

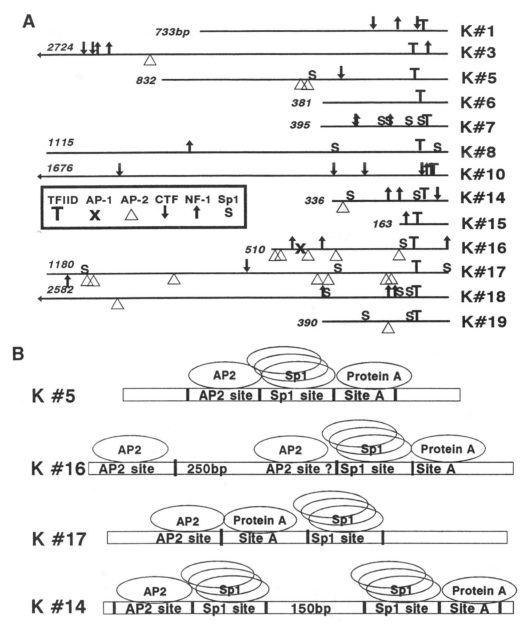

Figure 5. (A) Computer analysis of transcription binding sites in the sequences of the promoters of keratin genes. (B) Gel-shift analysis showing the existence of clusters of transcription factor binding sites in keratin gene promoters.

tegrin $\alpha_5\beta_1$ can mimic the effect of fibronectin: presumably they keep the receptor in the "attached" conformation and thus inhibit the signal for differentiating. Signaling by $\alpha_5\beta_1$ in keratinocytes may be analogous in this respect to the signaling by the fibrin receptor, the GPIIb-GPIIIa integrin, in platelets. Although this is a very attractive mechanism, at present it is only hypothetical and has to be confirmed experimentally.

3.2   MARKERS OF DIFFERENTIATION

Certain keratinocyte protein markers are destined to become incorporated into the cornified envelope. They are therefore localized just below the cell membrane. Perhaps the most prominent of these is involucrin, a polymorphic, 68 kDa, largely α-helical, flexible, rod-shaped protein. It has a highly repetitive structure that apparently

arose through a directional expansion mechanism. The consensus sequence of the 10 amino acid repeats contains three glutamines that are the substrate for cross-linking to other proteins by the epidermal transglutaminase. The rod-shaped structure and the large number of glutamines allow involucrin to become extensively cross-linked to many other proteins. The cross-linking converts the soluble protein in the cytoplasm into an insoluble form.

Among the proteins that appear relatively early during epidermal differentiation is loricrin. Loricrin is extremely rich in glycines and serines, which comprise, respectively, 47 and 23% of the protein. Cornified envelopes are also extremely rich in these two amino acids (34 and 20%, respectively), and it appears that loricrin is a substantial contributor to the cornified envelope. The glycine and serine-rich domains contain tandem repeats of short motifs, similar to the end domains of keratins. The four glycine/serine-rich domains are separated by lysine- and glutamine-rich linkers that give the molecule its alkaline isoelectric point (pI). Lysine and glutamine are substrates for the transglutaminases that cross-link loricrin during cornified envelope formation. Because of its unusual amino acid composition, loricrin is highly insoluble, even prior to its cross-linking and incorporation into cornified envelopes.

Filaggrin is a relatively small (37 kDa) basic, intermediate filament-associated protein found in the stratum corneum. Its eponymic function is to aggregate the filaments consisting of keratin into laterally associated thick, insoluble bundles. In stratum corneum, filaggrin is degraded to free amino acids, which may play a role in regulating the osmolarity of the stratum corneum milieu.

Filaggrin is synthesized in the granular layer as a large precursor, profilaggrin. Profilaggrin contains approximately 10–12 virtually identical copies of the mature filaggrin protein, arranged in tandem, separated by linkers and bracketed by the amino-terminal and carboxyterminal polypeptides. Nascent profilaggrin is extensively phosphorylated by several kinases and packaged into keratohyalin granules. Phosphorylation protects profilaggrin from proteolysis until filaggrin is needed in the upper layers for the aggregation of keratin filaments. Dephosphorylation is also accomplished by multiple enzymes. The dephosphorylation of the linker regions and their vicinity exposes them to proteolysis, which first yields intermediates that contain several units of filaggrin, but ultimately results in filaggrin monomers (Figure 6).

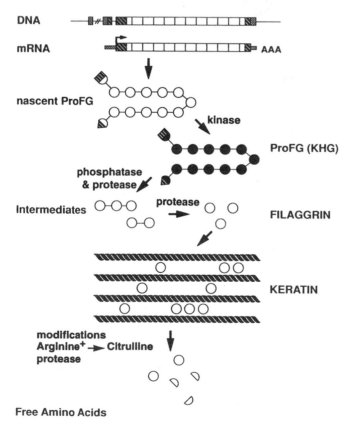

Figure 6. Processing of profilaggrin. The DNA sequence of the profilaggrin gene consists of many tandem units that are transcribed into mRNA, which is translated into profilaggrin (ProFG). Phosphorylation and dephosphorylation regulate the processing of profilaggrin to filaggrin monomers, which are competent to aggregate keratin filaments. In the stratum corneum, filaggrin is degraded to free amino acids.

Several steps of profilaggrin processing require calcium. Interestingly, the amino-terminal domain contains two helix–loop–helix segments homologous to the calcium-binding domains of other proteins. The function of the calcium-binding domain is at present unknown, but it is tempting to speculate that considering the importance of calcium in epidermal differentiation, this domain is the calcium-dependent trigger for profilaggrin processing.

Specific proteins synthesized in the granular layer assemble under the cell membrane and are cross-linked into an insoluble cornified cell envelope by epidermal transglutaminases. The cross-linked bond is resistant to proteolytic and peptidolytic enzymes, which makes cross-linked proteins less susceptible to proteolysis in general. Therefore, in combination with the cysteine disulfide bonds, the cross-linking by transglutaminase makes cornified envelopes very stable.

The transglutaminase important in epidermal cornification is the keratinocyte-specific, membrane-bound transglutaminase K, the gene for which has been mapped to chromosome 14q11.2. It is not linked to the epidermis-specific loci described earlier. Recent cloning of the gene for this enzyme will allow analysis of its regulation and biochemistry, as well as elucidation of its role in pathological defects of cornification.

Transglutaminase K is specifically expressed in the granular layer. Agents that promote keratinocyte differentiation induce expression of this enzyme, whereas inhibitors of differentiation inhibit it. Posttranslational acylation of the protein with palmitic and myristic acids anchors transglutaminase K to the plasma membrane, placing the enzyme in perfect position to cross-link and assemble proteins at the membrane. Thus, transglutaminase K is distinct from the tissue-type transglutaminase. Interestingly, TGFβ, an agent that promotes the basal-cell-specific phenotype, increases the level of tissue transglutaminase but not of transglutaminase K. This would suggest that the tissue transglutaminase plays a role in creating basal-cell-specific, possibly basement-membrane-associated, supramolecular structures, whereas transglutaminase K specifically creates cornified envelopes.

3.3   THE EFFECTS OF HORMONES AND VITAMINS

Various extracellular signals, such as growth factors, hormones, and vitamins, are important regulators of development and differentiation processes in general, and of keratinization in particular. Here we describe the effects of vitamins and hormones, such as retinoic acid (RA), vitamin D3 (D3), and thyroid (T3) and steroid hormones, which mediate their signals through nuclear receptors.

Retinoids, which include dietary vitamin A, as well as retinol and retinoic acid, have profound effects on the development, growth, and differentiation of various tissues and organs. They are especially important for vision, reproduction, and maintenance of epithelia. The first effects of vitamin A on skin were observed by Mori in 1922. Since that time, skin and epidermis have been model tissues to study RA action. It has been shown that hypovitaminosis A causes hyperkeratinization of the epidermis, while nonkeratinizing tissues, such as conjunctiva and cornea, become keratinized. Conversely, hypervitaminosis A causes inhibition of keratinization, hyperplasia, and a block of terminal differentiation. Increased levels of RA have been shown to suppress keratinization in vitro by suppressing the synthesis of late differentiation markers on the transcriptional level.

RA and its receptors regulate expression of many genes during the process of keratinization. A number of keratinocyte differentiation markers such as loricrin, filaggrin, and transglutaminase are inhibited by RA. One exception is involucrin, which is not regulated by RA. RA also regulates expression of number of keratin genes, markers of epidermal differentiation, and physiology. It suppresses the expression of the K#5, K#14, K#6, K#16, K#1, and K#10 genes. RA-responsive elements in keratin genes have been identified, and they consist of a cluster of several receptor-binding half-sites with different orientations and spacings.

In contrast to vitamin A, which in excess inhibits keratinization and expression of differentiation markers, vitamin D3 promotes the keratinization process. The active derivative of vitamin D3, 1, α25-dihydroxycholecalciferol (D3), effects a decrease in proliferation, enhanced formation of cornified envelopes, and increased levels of transglutaminase. The differentiating effects may involve calcium homeostasis, polyphosphoinositide metabolism, or suppression of the EGF receptor and *c-myc* oncogene synthesis.

Because of the similarity of their DNA-binding domains, the receptors for RA and D3 can recognize the same RE sequences. In contrast to RA, which directly regulates keratin gene expression through RARE, D3 and its receptor D3R do not have a direct effect on transcription of keratin genes through the common recognition elements with RARE. The effect of D3 on keratin synthesis appears to be only indirect—that is, by modulation of the keratinocyte phenotype (Figure 7).

Although not extensively studied in skin, thyroid hormone (T3) has effects on cornification and lipogenesis, as well as on levels of transglutaminase and plasminogen activator. Hypothyroidism causes numerous skin changes such as eczema, palmoplantar hyperkeratosis, ichthyosis, and scaly and hyperkeratotic skin. Interestingly, one of the early signs of vitamin A deficiency in chickens is hypothyroidism.

Glucocorticoids affect all components of the skin. Among the processes affected are epidermal cell differentiation and replication, as well as the proliferation of dermal fibroblasts and the synthesis of matrix proteins. Glucocorticoids are widely used both in topical and systemic therapy of a large number of dermatological diseases such as acute contact dermatitis, atopic eczema, pemphigus vulgaris, systemic lupus erythematosus, and cystic acne. The mechanism of action of glucocorticoids in the skin is not yet clear, although it is known that they cause immunosuppression and inhibition of the inflammatory response. It is known that the skin contains glucocorticoid, estrogen, androgen, and progesterone receptors and that it is the site of active metabolism of androgens and progesterone. The role of estrogen and progesterone receptors in skin malignancy is not clear, however, and the exact role of these hormones in keratinocyte differentiation remains to be elucidated.

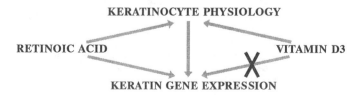

Figure 7.   Both retinoic acid and vitamin D3 regulate keratinocyte physiology, which determines qualitatively the type of keratin protein expressed in the cell. However, retinoic acid, but not vitamin D3, can directly regulate the quantity of expressed keratin protein.

4   KERATINOCYTES AS IMMUNE CELLS

The role of epidermal keratinocytes in defense against mechanical injury and desiccation has been appreciated for a long time, but their role in immunological defense became apparent only recently, when it was realized that keratinocytes can produce a cornucopia of growth factors, chemoattractants, and cytokines. Furthermore, keratinocytes express receptors for many polypeptide factors, respond to autocrine stimulation, and also respond to the signals produced by the immune system. The importance of signaling between keratinocytes and lymphocytes is apparent in the cutaneous disorders that involve both these cell types, including delayed-type hypersensitivity, cutaneous T-cell lymphoma, psoriasis, and atopic dermatitis. The immunological function of the keratinocyte appears in pathological conditions (e.g., during wound healing) or in allergic and inflammatory reactions. In response to epidermal injury, keratinocytes become "activated": they produce and respond to growth factors and cytokines, become migratory, and can produce components of the basement membrane. Activated keratinocytes express a specific pair of keratin proteins, K#6 and K#16, distinct from the keratins in the healthy epidermis. The extracellular signals that induce keratinocytes to start differentiating or to become activated are at present not known.

The initial signal for activation of keratinocytes may be the release of activated interleukin 1 (IL-1). Once activated, keratinocytes synthesize additional signaling growth factors and cytokines. These include TGFα, IL-3, IL-6, IL-8, G-CSF, GM-CSF, and MCSF. The effects on these signaling molecules produced by keratinocytes are not only paracrine and chemotactic for white blood cells, but also autocrine for keratinocytes themselves. They may lead to secondary effects of keratinocyte activation. Several extracellular markers are specifically expressed by the activated keratinocytes. These include cell surface proteins CD13, CD14. CD68, ICAM-1, and HLA-DR, as well as components of the extracellular matrix, integrins, and receptors for both the autocrine factors and those produced by infiltrating immune cells.

In a feedback loop, the increase in the expression of cell surface receptors may augment the initial activation signal. The various signaling molecules may be synergistic or antagonistic with respect to one another. This allows the activated phenotype to be specifically modified, which can lead to different activated phenotypes. In other words, keratinocytes activated in psoriasis are different from keratinocytes activated during wound healing.

Among the most extensively studied cellular receptor signaling pathways are those involving epidermal growth factor and its receptor. Binding of the appropriate ligands to the epidermal growth factor receptor (EGFR) can activate keratinocytes. Upon ligand binding, the EGFR dimerizes, activating its intracellular protein tyrosine kinase. EGFR activation signals are conveyed to the nucleus by a system of protein phosphorylation and dephosphorylation. These phosphorylation–dephosphorylation signals are eventually conveyed to nuclear proteins that regulate both gene expression and cell division. Among the regulated genes are those encoding additional regulators that when activated, cause major morphological, developmental, and differentiation changes. In many cell types, activation of EGFR results in major pleiotropic changes including increased motility, degradation of extracellular matrix, and proliferation. In response to the activation of the EGFR, keratinocytes proliferate, degrade components of the extracellular matrix, and become migratory. In adult epidermis, EGFR is primarily expressed in the basal and, to a lesser degree, the deepest suprabasal layers.

5   INHERITED KERATINOPATHIES

In the past few years inherited skin diseases have been a successful area of research that correlates epithelial cell structure with molecular defects. In general, two different approaches are used in these studies: first, linkage studies and gene mapping in affected families, and second, transgenic animal models that include incorporation of the mutated gene(s) of interest into the genome of the host animal and analysis of its effects on development and differentiation of epithelium. Combined, these two approaches have revealed the cause for such epidermal disorders as epidermolysis bullosa simplex (EBS), caused by mutation(s) in keratin genes K#5 and K#14; epidermolytic hyperkeratosis (EHK), linked to mutation(s) in keratin genes K#1, K#2, and K#10; and palmoplantar hyperkeratosis (PPK), caused by mutations in keratin K#9 gene (see Table 1). Recent genetic studies have determined that a gene for Darier's disease is located in chromosome 12q, and a gene for familial psoriasis susceptibility is mapping to the distal end of chromosome 17q. It has recently also been shown that overexpression of the human K#16 keratin gene in transgenic mice causes aberrant keratinization of the outer root sheath of the hair follicle and adjacent epidermis. Taken together, all these studies of inherited skin diseases are expanding our knowledge of cellular structures and their function and opening new possibilities for more effective treatments in the future.

6   CONCLUSIONS AND PROSPECTS

Although many details are known about the differentiation of keratinocytes, much more work needs to be done. Of course additional markers, enzymes, and pathways will be discovered. But the fundamental, overall picture of the mechanism of keratinocyte differentiation is not clear. Figure 8 suggests several hypotheses:

1. *Orderly pathway:* step A triggers step B triggers step C. While elegant, this model is hard to reconcile with, for instance, inversion of steps in certain diseases, notably psoriasis.
2. *Parallel pathways:* initial signal(s) trigger(s) steps A, B, and C; then A triggers DEF while B triggers G and H; then D triggers J and K while F and G trigger L, and so on. This complicated model could account for all observations—for example, the retinoic acid suppressible markers could be on one pathway and other markers, notably involucrin, on another. However there is absolutely no mechanistic evidence that the early steps are causing the later steps. This leaves us with a third possibility.

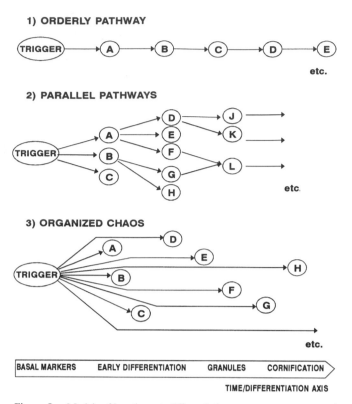

Figure 8.   Models of keratinocyte differentiation; arrows represent causal relations.

3. *Organized chaos:* once the initial signal has been received, all subsequent steps are programmed to fall independently into place at the right stage. At least some of the details of the regulatory circuits will become clarified in the next several years through the study of keratinocyte transcription factors and their activation and inhibition.

The second fundamental problem lies in the need to determine how the differentiation program is disturbed in disorders of keratinization. Although some of these, notably the inherited ones, are amenable to reverse genetics, we are far from understanding what happens in most pathological conditions. Basic research in cutaneous biology will eventually provide the answers.

See also DNA REPLICATION AND TRANSCRIPTION; GENE EXPRESSION, REGULATION OF; PROTEIN ARCHITECTURE AND ANALYSIS.

Bibliography

Blumenberg, M. (1993) Molecular biology of human keratin genes. In *Molecular Biology of the Skin: The Keratinocyte,* M. Darmon and M. Blumenberg, Eds., pp. 1–24. Academic Press, New York.

Dale, B. A., Presland, R. B., Fleckman, P., Kam, E., and Resing, K. A. (1993) Phenotypic expression and processing of filaggrin in epidermal differentiation. In *Molecular Biology of the Skin: The Keratinocyte,* M. Darmon and M. Blumenberg, Eds., pp. 79–106. Academic Press, New York.

Darmon, M., and Blumenberg, M. (1993) Retinoic acid in epithelial and epidermal differentiation. In *Molecular Biology of the Skin: The Keratinocyte,* M. Darmon and M. Blumenberg, Eds., pp. 181–198. Academic Press, New York.

Table 1   Loci of Genes for Some Inherited Keratinopathies

Disease	Locus	Gene
Epidermolysis	12q	K5
bullosa simplex	17q	K14
Epidermolytic	12q	K1
hyperkeratosis	17q	K10
Palmoplantar	17q	K9
keratoderma		
Darier's disease	12q	?
Ichthyosis bullosa	12q	K2
of Siemens		

Fuchs, E. (1990) Epidermal differentiation: The bare essentials. *J. Cell Biol.* 111:2807–2814.

———. (1994) Mini-review on the cellular mechanisms of disease. Intermediate filaments and disease: Mutations that cripple cell strength. *J. Cell Biol.* 125:511–516.

Kupper, T. S. (1990) Role of epidermal cytokines. In *Immunophysiology. The Role of Cells and Cytokines in Immunity and Inflammation,* J. J. Oppenheim and E. M. Shevach, Eds., pp. 285–305. Oxford University Press, London and New York.

Miller, S. J., Lavker, R. M., and Sun, T. T. (1993) Keratinocyte stem cells of cornea, skin and hair follicle: Common and distinguishing features. *Semin. Dev. Biol.* 4:217–240.

Schweitzer, J. (1993) Murine epidermal keratins. In *Molecular Biology of the Skin: The Keratinocyte,* M. Darmon and M. Blumenberg, Eds., pp. 33–72. Academic Press, New York.

Steinert, P. M., and Roop, D. R. (1988) Molecular and cellular biology of intermediate filaments. [Review]. *Annu. Rev. Biochem.* 57:593–625.

Evolution: *see* **Genetic Analysis of Populations; Genetic Diversity in Microorganisms; Mitochondrial DNA, Evolution of Human; Paleontology, Molecular.**

EXPRESSION SYSTEMS FOR DNA PROCESSES

Ka-Yiu San and George N. Bennett

Key Words

Antibiotic Resistance   The ability of the cell to grow in the presence of a chemical (antibiotic) that normally inhibits an essential cell function.

Bacteriophage   A virus that infects bacteria.

Coding Region   The nucleic acid segment that contains the linear arrangement of codons specifying the order of amino acids in the protein.

Codon   A three-nucleotide unit in molecule of messenger RNA that specifies the particular amino acid or stop signal in the mRNA as it undergoes translation.

Fusion Protein   A genetic construct that attaches a gene encoding a protein or a portion of the gene to another coding sequence such that the protein formed during translation is a linear combination of the two proteins.

Partition   The usual segregation of the genetic material of the cell into the two daughter cells as the cell divides.

Plasmid Stability   Two kinds of stability are considered in production situations. Structural stability refers to the structural integrity of the plasmid and lack of mutations within the plasmid DNA molecule, while segregational stability refers to the ability of the plasmid to be maintained through proper partitioning at cell division and not generate plasmid free cells.

Promoter   The DNA site at which the RNA polymerase specifically binds to initiate transcription.

Protease   An enzyme that cleaves the peptide bonds between amino acids comprising the protein.

Ribosome-Binding Site   The region of the mRNA near the site of translation initiation that is important for ribosome recognition.

RNA Secondary Structure   A structure formed within an RNA molecule by bending the molecule such that a series of specific base pairs can form between nearby complementary nucleotides, resulting in a small region that has double-stranded character.

Translation   The process of protein synthesis in which the order of codons on the individual mRNA is used to specify the order of linkage of the amino acids of that particular protein as it is made on the ribosome.

Virus   An agent naturally able to infect a host cell and bearing genes that allow its complete reproduction within a specific host cell.

The production of a specific protein by recombinant DNA technology entails (1) cloning the gene that encodes the protein from the desired organism, (2) forming a suitable genetic construct that permits the gene to be expressed in a host organism, and (3) introducing and maintaining the construct in the host, to allow an adequate yield of functional protein to be produced. The ability to take a gene from any organism and express it in adequate quantity in a system in which the structure and function of the protein encoded by the gene can be analyzed has great importance in the research laboratory. Physical and enzymatic features of the protein can be studied, and its functional location in cells or tissues can be determined, in an effort to better define the biological function of a newly discovered gene. Industrial applications include the production of therapeutic and diagnostic proteins with great impact in the field of medicine, and the production of stable bulk enzymes for the food and specialty chemical industries.

1   INTRODUCTION TO HOST–VECTOR SYSTEMS

To produce a specific RNA or protein product in a particular host cell, a suitable DNA construct must be prepared. This article discusses the essential features of expression systems that are widely used for overproduction of proteins for a variety of purposes. The expression system is composed of an expression vector and a specific host cell. Not only are selection and construction of the ideal expression vector important, but the choice of the appropriate host can affect production efficiency considerably. We consider features of the expression vectors first, and then discuss the important attributes of various hosts. A number of organisms have been used as hosts for expression of foreign proteins. Since, however, the most complete picture is available for *Escherichia coli,* that prokaryotic organism serves as an example and is discussed in more detail.

Expression vectors usually consist of small, circular plasmids specifically designed with several key features that allow a foreign gene inserted into the plasmid to be expressed in the host cell. Four important elements of the plasmid are (1) an origin of replication, which allows the plasmid to be replicated in the host, (2) a selectable genetic marker, which allows cells bearing the plasmid to preferentially grow on a specifically composed medium, (3) transcription and translation signals recognized by the host cell, which ensure that the gene introduced into the vector can be expressed into a protein product, and (4) a suitable unique restriction site, located appropriately with regard to the transcription and translation signals, where the plasmid can be cleaved and the foreign DNA can be inserted. The generalized structure typical of many expression vectors is presented in Figure 1. The sections that follow explain the variation and importance of each of these features in the overall expression system.

2   COMPONENTS OF *E. Coli* PLASMID VECTORS

2.1   ORIGIN OF REPLICATION

The origin of replication of the plasmid is a specific sequence that defines where replication of the plasmid DNA begins. Because of its essential function in this event, the structure also defines the frequency with which replication initiates. This in turn defines the copy number of the plasmid, that is, the number of individual plasmid molecules present in each cell. The mechanism of copy number control is complicated and involves the synthesis and interaction of regulatory proteins and RNAs. Thus the copy number can be altered by mutations in the plasmid that affect the binding sites of these regulators or their rate of synthesis. The plasmid copy number is important because of the gene dosage effect. If each plasmid carries a copy of the gene, and each copy can give rise to a certain maximal amount of product per unit time, the more copies of the plasmid present in the cell, the higher the level of production attained per cell. Of course, there is a limit to this effect, and at a certain level the protein and nucleic acid synthesis machinery of the cell becomes saturated and the increased metabolic burden of maintaining the plasmid overcomes the advantage of the gene dosage in the production process.

The most widely used expression vectors for *E. coli* are derived from the origin of replication of the ColE1 plasmid and have copy numbers of between 10 and 100. Other origins of replication used on some vectors include R1, RK2, RSF1010, and R6K. The beneficial properties of high gene dosage have been used advantageously (e.g., in derivatives of the R1 origin) through the construction of temperature-sensitive copy number control systems in which the copy number can be raised dramatically by increasing the temperature.

Another feature specified by the origin of replication is the compatibility of the plasmid. Since it is sometimes desirable to maintain two distinct plasmids within the same cell, two different origin of replication types would be employed on the two different plasmids. Otherwise, one of the two plasmids bearing the same type of origin would dominate without special selection and the other would be lost from the cell as the cells grew and divided. This phenomenon is called incompatibility, and plasmids with the same origin type are said to be in the same incompatibility group.

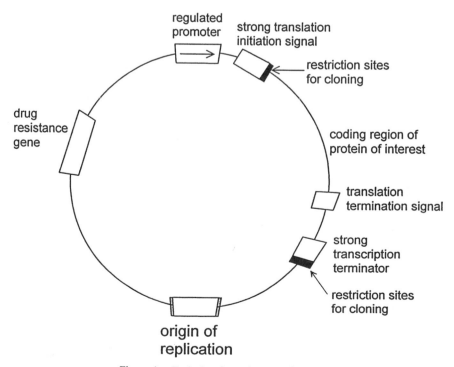

Figure 1.   Typical prokaryotic expression vector.

Other functions defined by the origin of replication are the host range of the plasmid—that is, which bacterial species are able to allow replication of the plasmid—and the ability of the plasmid to be transferred to another bacterial cell by natural genetic processes. To reduce the possibility of unintentional movement of the plasmid to another bacterial species, most vectors have been constructed so as not to contain the elements needed for transfer.

2.2   ELEMENTS FOR SELECTION AND MAINTENANCE OF THE VECTOR

The ability to select the vector and maintain the plasmid in the host cell is an important aspect of the expression system, especially in production regimens that involve a longer time course or continuous culture fermentations. Obviously, cells that have lost the expression plasmid will no longer be capable of production of the desired protein; therefore, if a significant proportion of plasmid-free cells arises in the culture, the overall yield will decrease. This factor is made more troublesome because plasmid-free cells generally grow faster in the absence of selection, since they do not have the extra burden of plasmid-associated DNA and protein synthesis. To counter these tendencies, the system for the selection and maintenance of plasmids must be effective and inexpensive, and it must not interfere with other aspects of the expression system. The selection should be strong and dominant, and the frequency of spontaneous resistance arising in the host strain should be very low. It is also useful for practical purposes if the selectable gene is expressed well from a plasmid and is small enough to be placed in the vector while keeping the overall plasmid size to a minimum.

The most common type of selectable marker incorporated on the plasmid is an antibiotic resistance element allowing selection on ampicillin, kanamycin, or tetracycline. These selections are fine in ordinary laboratory manipulations; in large-scale cultures, however, the cost of the antibiotic must be considered.

Other approaches place an essential gene on the vector and use a special host strain in which that gene is deleted from the chromosome (e.g., the genes for the valine tRNA synthetase *valS* and the single-stranded DNA binding protein *ssb* have been used in this way). In such a system a drug additive is not needed, and if the plasmid is lost from the cell, the host cell alone is unable to grow.

The exploitation of naturally occurring partition systems (par), which maintain copy number and select against plasmid loss, has also been successful. An example of this approach has been the incorporation of the *parB* locus on vectors. The *parB* locus selects

strongly against plasmid loss by killing the cells that no longer carry the *parB* locus. An advantage of the *parB* system is that it does not require any specially constructed host strain.

2.3   TRANSCRIPTION AND TRANSLATION SIGNALS

The transcription and translation signals allow the coding region of the gene to be efficiently recognized by the host cell machinery, that is, RNA polymerase and ribosomal complexes, which decode the nucleotide sequence of the DNA of the gene into the designated sequence of amino acids in the protein. Figure 2 is a schematic diagram of typical prokaryotic transcription and translation signals. The signal specifying where transcription starts is called the promoter. The sequence determines the affinity of the DNA segment for RNA polymerase and is important in determining maximal transcription levels. The more efficient the promoter, the greater the synthesis of messenger RNA of the gene under control of that promoter. Certain expression vectors employ a relatively strong promoter that is called unregulated or constitutive. This approach is feasible if the protein to be made is very stable in the cell and does not cause any toxic effects or retard growth. As a rule, however, investigators use a regulated promoter, which can allow induction of production of the desired protein at a suitable time. This approach has the advantages of a short period of protein synthesis and induction at the most opportune time in the growth of the batch culture, where the optimal combination of cell density and specific protein production can be achieved.

The most widely used control systems for transcription initiation are those regulated by temperature, or addition of a chemical agent or metabolite. A strong promoter from the bacteriophage λ is normally prevented from binding RNA polymerase by a repressor. If an altered form of the repressor is used that is unstable at high temperature, an increase in temperature will release this negative control of the promoter, allowing the synthesis of large quantities of the desired messenger RNA. This promoter system can also be released from repressor control through addition of DNA-damaging agents. This method is sometimes used instead of an increase in temperature.

Another widely used regulatory system is based on the repressor control of the lactose promoter. The DNA sequence to which the lactose repressor binds (called the operator) can be combined with a variety of other stronger promoters (e.g., the tryptophan or lipoprotein promoters), which then become controlled by the lactose repressor. The addition of isothiopropylgalactoside

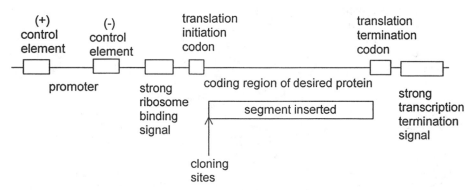

Figure 2.   Typical prokaryotic transcription and translation features.

(IPTG), an analogue of lactose, is most commonly used to induce the system.

A number of other regulation systems have been advocated for process control of expression systems. Among these are phosphate, oxygen, and acid-induced promoters. Some promoters are regulated by specific activators that stimulate transcription rather than by a negatively acting repressor system.

An extension of the transcription initiation regulatory system is the use of a special RNA polymerase not normally found in the cell, in combination with the unique promoter sequence recognized by that RNA polymerase to produce the desired message. A system based on the bacteriophage T7 RNA polymerase and the T7 promoter it recognizes can yield a very high specific production of the desired product. In this case the T7 promoter is located on the expression vector and the T7 RNA polymerase gene is placed under the control of a regulated system typically of the type mentioned earlier. Thus, induction in this system first produces the T7 RNA polymerase, which in turn synthesizes the messenger corresponding to the gene of interest.

Once a suitable means of producing the messenger RNA that codes for the desired protein has been selected, the next consideration is to translate the message with optimal efficiency. Several factors affect this efficiency. One is the stability of the message: the longer the half-life of the mRNA in the cell, the greater the number of times it can be translated before it is degraded, thus the greater the number of molecules of protein that can be generated from each messenger. The presence of hairpin or stem–loop structures, where the nucleotide sequence allows intramolecular base pairing particularly near the end of the message, seems to offer stability from degradation. These hairpin sequences are normally included in the vector construct at the 3′ end of the message beyond the protein-coding sequence. The hairpin sequences act in transcription termination and thereby isolate the transcribed gene from the origin of replication and other genes. The stability of the messenger RNA can also be affected by growth rate and the genetic background of the host.

The major factor usually limiting translation is the ability of the ribosome to initiate translation on the messenger RNA. Optimal interaction of the ribosome with the mRNA can be impeded by secondary structures in the message which block access of the ribosome to the initiation codon. The binding of the ribosome to the messenger requires interaction between the ribosome, specifically a segment of the ribosomal RNA, and the region of the mRNA about 10 nucleotides 5′ from the translation initiation codon. Optimal sequences for this region have been analyzed for a number of genes, and a sequence specifying a highly efficient ribosome-binding site is often included in the vector just preceding the position reserved for the protein-coding sequence. Sequences frequently placed in vectors for their efficient ribosome sites include chemically synthesized optimal consensus sequences, the T7 g10-L region, or those of other highly translated genes. While these standard translation initiation systems produce a reasonable yield with most proteins, the sequence in this area often is modified to gain optimal production of an individual protein (e.g., to avoid the formation of competing secondary structure in the mRNA, which would hamper interaction with the ribosome).

Translational limitations can also occur as a result of the specific nature of the protein-coding region (e.g., unusual composition, unique codon usage, special secondary structures). These problems cannot be addressed at the expression vector level. At the C-termi-

nal end of the protein, translation is terminated and the protein is released from the ribosome by release factors. The most widely used translation termination codon is UAA; thus to ensure efficient termination and release of the protein, it is often decided to place this codon or tandem termination codons at the end of the gene.

2.4 Unique Cleavage Sites for Cloning

The location of the restriction enzyme site at which the protein-coding sequence is introduced into the plasmid can allow the protein to be expressed in its most useful form. In many cases it is desired to produce the complete intact natural protein beginning with its N-terminal amino acid and ending with the native C-terminal amino acid. This would frequently be the situation when the exact structure of a mature, natural, pharmaceutically active protein is needed. To position the coding sequence correctly with respect to the ribosome-binding site, a restriction site incorporating an ATG sequence is used to place the translation initiation site appropriately in the vector. Two commonly used sites that contain an ATG and rarely exist in any given DNA sequence are *Nco* I and *Nde* I. In a second unique restriction site that is often used, the C-terminal end of the gene is joined to the vector to orient the protein-coding segment during cloning. Such a construct would then enable translation of the complete protein-coding region.

Not all mature proteins have an N-terminal methionine; however, this is the primary terminus formed in bacterial systems. If the extra amino acid is not removed from the protein subsequently by in vitro methods, the desired protein can initially be made as part of a longer peptide chain, which is cleaved by a specific protease to generate the N-terminus of the mature protein product. A variety of N-terminal protein carrier elements have been used in making these fusion protein products. In the formation of fusion proteins, the vector carries the appropriate signals for translation initiation and a coding region of the N-terminal component of the fusion protein, with a restriction site at which the DNA fragment bearing the foreign gene can be placed in the proper reading frame.

The properties of the N-terminal component of the protein fusion can be used to assist in purification of the fusion protein. For example, if this segment encodes an easily purified protein such as the maltose-binding protein or glutathione *S*-transferase, the combined protein fusion product can be readily isolated by taking advantage of the interaction between the carrier protein and a suitable chromatography matrix. The addition of tracts of basic, acidic, or metal-binding amino acids to the protein can also be used to aid purification.

This protein fusion concept can be utilized to allow for protein processing in vivo. For example, if the N-terminal segment encodes a localization signal, the protein can be directed to the periplasmic space of *E. coli* or in some cases into the extracellular medium. This type of protein export can aid in the purification and stabilization of the product. A vector employing protein A as an exportable carrier is one of this type.

In some applications the expression of the entire protein is not necessary, but only a portion of it is required. Since only a recognizable epitope is needed for raising antibodies or for the detection of an antigen by antibodies, expression of only a short protein segment is sufficient. The cloned segment comprising the antigenic site can be attached to a protein carrier segment that will allow effective exposure of the foreign peptide. Placement at the C-terminal end of β-galactosidase has been used effectively for this purpose. In some cases the foreign protein segment can be localized to the surface of a cell or phage particle by using a suitable surface protein as the carrier portion, thus presenting the immunologically active antigen or antibody species in a way that allows detection, isolation, or use of the living cell or phage particle.

3 HOST FEATURES

3.1 SPECIFIC FEATURES OF *E. coli*

The use of the correct host can have a significant impact on the final yield of the desired protein product. Factors generally useful to the host include the ability to grow rapidly and to a high density, the ability to be transformed (so the DNA construct can be readily introduced), and possession of a low recombination and mutagenic rate (so the plasmid is not frequently lost, damaged, or otherwise inactivated). Desirable growth properties are maintained by using a strain that is as close to wild type as possible and bears neither mutations affecting the major carbon utilization pathways nor other mutations that are of themselves stressful to the cell. Recombination-deficient strains (e.g., *recA*), are often used to try to avoid problems with genetic instability.

The level of production of certain proteins is limited by their tendency to be degraded. Therefore, host cells with reduced protease levels have been used to enhance the in vivo stability of the foreign protein of interest. Commonly used mutations in *E. coli* that have reduced protein degradation rates are *lon* (which encodes a major ATP-dependent protease), *rpoH*, encoding a transcriptional regulator of heat shock stress induced genes, and *clp*, which encodes a second ATP-dependent protease. An alteration in capsule biosynthesis exists in certain of these strains, making the cells sticky or slimy. This aspect can be avoided by including a mutation abolishing *galE* activity, which will block biosynthesis of the polysaccharide capsule material. In some cases the protein stability can be addressed at the expression vector level by using a fusion protein construct. The presence of the longer carrier protein often will effectively stabilize the added foreign protein segment. Translation of very large proteins is also a limitation in many strains; however, certain other nonlaboratory *E. coli* strains exhibit an increased ability to translate particular long protein molecules.

In addition, special host cell requirements are necessary to carry out induction of expression). For example, to induce the expression of genes under the control of the λ promoter, a mutant λ repressor gene encoding a temperature-sensitive repressor protein is placed on the chromosome of the host rather than locating this gene on the expression plasmid. In the case of the T7 RNA polymerase expression system, the gene for the T7 RNA polymerase is placed under the control of the *lac* promoter, and this construct is incorporated in the chromosome of the host. To control the *lac* promoter more completely, mutant strains that maintain an increased level of the regulatory protein are frequently used. Other mutations to increase the sensitivity of the induction or to lower the basal expression level can also be incorporated in the genome of the host to facilitate the coordination of the host with the specific expression system and with the special requirements of the protein being produced.

3.2 OVERVIEW OF OTHER HOSTS

While *E. coli* continues to be the most widespread expression system, other host–vector systems have certain advantages. The strengths of the *E. coli* system are the variety of vectors and the number of genetically altered hosts available, and the well-studied methods for manipulating this organism. Although *E. coli* is capable of high level production, protease problems, the formation of inactive inclusion bodies containing the product, and the lack of a eukaryotic glycosylation system limit the usefulness of this organism for production of a number of proteins from mammalian sources. Other bacteria have received some attention because of their ability to grow on particular compounds, their potential for secretion of the protein, or their industrial potential. General aspects of several of these systems are presented here, and the individual systems are covered in more detail in Sections 3.2.1 to 3.2.6.

Yeasts are a suitable production system for a number of processes. Not only are they well studied like *E. coli*, and amenable to scale-up, but they are able to carry out some posttranslational processing of eukaryotic proteins such as specific cleavage or glycosylation. Fungi can produce high yields of commercial proteins and are reasonably capable of glycosylation and secretion of proteins into the medium. Methods have not been as completely developed, however, and the use of organisms of this class has not been widespread.

Viruses that infect insect cells (baculoviruses) have gained attention as a system for production glycosylated protein at reasonably high levels, while being less expensive than using mammalian cells. The vaccinia virus has been used to express foreign antigens in whole animals, demonstrating its potential for use as a vaccine vector system.

Although mammalian cell lines have the advantage of producing truly identical processed mammalian protein, the expense and difficulty in scale-up has limited commercial production with such systems. The extensive study of mammalian cells has established methods for introducing DNA, and a variety of vectors useful in manipulation of animal cells have been developed.

3.2.1 *Bacillus subtilis*

Bacillus subtilis has been considered to be a suitable host based on several potentially advantageous properties. Considerable information is available concerning its molecular biology, physiology, and genetics. The details of translation, transcription, and plasmid systems are well studied. The employment of *B. subtilis* and related *Bacillus* species in other industrial production situations also has stimulated development of *B. subtilis* for production of recombinant proteins. Perhaps the most commercially interesting reason for using *B. subtilis* is the ability of this microorganism to secrete proteins into the medium. A simple system that would allow isolation of the protein from the supernatant broth directly, without the requirement for cell disruption, would simplify the downstream processing portion of the production process.

The expectation that *B. subtilis* would play a major role in the commercial production of recombinant proteins has been unrealized largely as a result of technical factors. Although regulated expression systems are available for *B. subtilis* based on those used in *E. coli* (e.g., *lac*), and secretion systems based on α-amylase, alkaline protease, or levansucrase have been reported, their use has not been widespread. The secretion system can be made to release some heterologous proteins into the medium, but not all are processed well. Another problem has been the presence of proteases that act to degrade the desired protein. This difficulty has been approached through the development of strains in which genes for several of the proteases are inactivated. Early observation of plasmid instability led to the development of integrated expression systems, in which the promoter and the gene to be expressed were recombined onto the bacterial chromosome. This strategy takes advantage of the active recombination system of *B. subtilis* to integrate the construct and thereby stabilize it. This approach to the formation of a stable overproducing strain may become more generally used in other organisms.

3.2.2 Yeast

Yeasts have many of the advantages of bacteria—for example, they are fast growing; high cell density is achieved; industrial-scale processes are established; and molecular biology and genetics are well developed—yet as eukaryotes they are capable of more sophisticated posttranslational protein processing than is found in bacteria. Yeasts have the transcription regulatory features and processing system of a eukaryotic cell. A general diagram of eukaryotic signals found in many expression vectors is presented in Figure 3. The most commonly used yeast in laboratory research is *Saccharomyces cerevisiae,* an organism that can be cultured in the haploid state. Most industrial strains, however, have more than one copy of each chromosome per cell, which makes them more difficult to manipulate genetically.

Other yeasts, such as *Pichia pastoris* and *Hansenula polymorpha,* have also been investigated and may offer an advantage in production of glycosylated proteins. *S. cerevisiae* vectors of either the plasmid or integrating type are available. The most common plasmid-based system derives from the 2 μm plasmid found in the nuclei at about 60–100 copies per cell. There are a large number of vectors derived from this plasmid, with the copy number varying depending on the size of the insert, the selective marker, and the growth conditions. Vectors employing promoters affected by carbon source have been used to give high level expression of heterologous proteins; examples include various GAL4-regulated promoters, the alcohol dehydrogenase ADH2 promoter, the phosphoglycerate kinase promoter, the glyceraldehyde 3-phosphate dehydrogenase promoter, and temperature-regulated or copper-regulated systems.

The secretion and glycosylation abilities of yeast have received attention. The α-mating-factor secretion signal has been the most used, with the desired protein-coding sequence placed just after the α-factor segment specifying the signal for secretion and cleavage by protease. Yeast is capable of glycosylation, but the resulting addition of sugar modifications to human proteins is somewhat different from that found in the natural protein. For example, the protein produced in yeast often bears additional sugar residues at

asparagine glycosylation sites. To reduce overglycosylation, *mnn* (mannan defective) or *alg* (asparagine-linked glycosylation) mutants of yeast have been studied and used as hosts for protein production. In efforts to improve protein stability from degradation by the vacuole proteinases, mutations that disrupt these protease genes have been introduced. The use of these special yeast strains has enhanced the ability to isolate a number of proteins. Studies of the ubiquitin-mediated protein degradation process and the N-terminus removal or acetylation systems have led to the use of special strains or growth conditions to allow the desired N-terminus on the mature protein to be obtained.

3.2.3 Mammalian Cells

Expression of proteins in mammalian systems affords production of the identical animal protein and thus eliminates any question regarding the effectiveness or natural character of the protein. Several factors—the expense of mammalian cell culture, low cell density, and low productivity—have limited the use of this system to investigations in which the synthesis, processing, or role of the protein is being explored, or the protein cannot be satisfactorily expressed in a more amenable system.

While COS monkey cells are often used in combination with viral vectors for transient expression purposes for research experiments, the development of a cell line capable of sustained production in serum-free suspension culture is desired for long-term production. Suitable lines have been established in COS, CHO, mouse L cells, or other lines by linking the expression construct for the desired gene to a readily selected marker and integrating the entire DNA into a random chromosomal location.

The expression construct usually contains an enhancer element, which allows the production of high level expression (e.g., SV40, polyoma, or adenovirus enhancers; the heavy metal responsive metallothionein enhancer; glucocorticoid responsive enhancers). An active promoter (e.g., the SV40 early promoter, the RSV promoter, adenovirus major late promoter) is commonly used, and appropriate efficient splicing signals derived from SV40, adenovirus, or immunoglobulin genes are also positioned on the expression vector along with polyadenylation signals.

To select for the presence of the integrated construct in the recipient cell line, a readily selectable gene is linked to the expression construct. A selectable marker that is frequently used is that specified by dihydrofolate reductase, which allows the cells to grow in the presence of methotrexate. Other selectable genes that can be used in a similar way are those for adenosine deaminase, ornithine

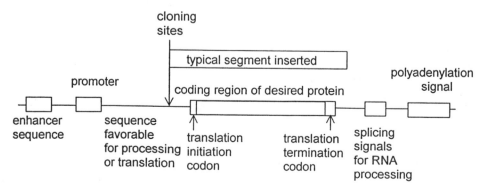

Figure 3. Typical eukaryotic transcription and translation features.

decarboxylase, or asparagine synthetase, although in some cases a specific mutant cell line is required for selection. After the construct has been introduced into the appropriate cell line by calcium phosphate, DEAE–dextran, polybrene, or liposome-mediated transfection or electroporation, the cells are grown under selection and with increasing levels of the selecting agent (e.g., methotrexate for dihydrofolate reductase gene amplification), and the genomic region at which the construct is integrated becomes amplified. This amplification provides a high gene dose in the genome of the host cell and increases the level of synthesis of the protein ten- to several hundred–fold. Once formed, such cell lines can be stably maintained if care is taken to monitor for deletions. Improvements in the production level attained by single-copy integrated vectors and their adaptation to growth in serum-free suspension culture have simplified the use and prospects for mammalian cell production.

3.2.4 Insect Virus

Insect virus expression systems for the production of proteins, particularly glycosylated mammalian proteins, have received attention because of the high level of expression (up to 50% of total cell protein) available with this technology. The baculovirus system employs the strong transcription signals for expressing the polyhedron protein of the virus. These viruses replicate in the nucleus and form inclusion bodies, which are surrounded by a protective matrix containing a large amount of the polyhedron protein. The virus commonly used is *Autographa californica* nuclear polyhedrosis virus, and it is cultured on the cells of the fall army worm, *Spodoperta frugiperda*. Silkworms can also be used as the host, providing a simple, low cost route for production.

This virus is so large (130 kb) that direct manipulation of the viral DNA is difficult, and the desired gene usually is introduced through in vivo recombination between the foreign gene construct and the viral genome. The difficulty in detecting the recombinant baculovirus has retarded work with the system. Newer methods for making the recombinants at higher efficiency by incorporating a positive selection for the recombinants coupled with the introduction of reporter or detection systems have simplified the task of generating and identifying the desired viral expression constructs.

3.2.5 Vaccinia Virus

Vaccinia virus vectors have been used in the preparation of vaccines for animal diseases. The virus is a large pox virus, which replicates in the cytoplasm of the host cell. While the virus is not a health hazard to humans, it is quite immunogenic and thus serves as a suitable carrier for the antigenic surface proteins of other viruses. The recombinant vaccinia virus is prepared by recombination in vivo, as is the case for the baculovirus system. A variety of markers are available for identification or selection of the recombinant viruses. These include β-galactosidase, thymidine kinase, neomycin resistance, and mycophenolic acid resistance. The introduced gene is placed under the control of a vaccinia virus promoter with appropriate signals, as in the case for expression in mammalian cells, and the host cells (e.g., CV-1 cells) are coinfected with the vaccinia virus and the above-described construct bearing the desired gene, the expression signals, and the selective marker. After characterization, the recombinant vaccinia virus can be propagated on any suitable host. The antigenic components of several important disease-causing agents have been placed on the virus in this way (e.g., hepatitis B; rabies and herpes viruses). The recombinant vaccinia viruses seem to be able to produce an effective immune response to the foreign protein in whole animals.

3.2.6 Transgenic Plants and Animals

Whole plants or animals have been used as production systems. This strategy entails the preparation of the expression construct as for expression in mammalian cells and then introducing it not into cultured cells but into an egg. Sometimes an integration event will occur, and as the animal develops, the altered chromosomal segment can be found in germ line cells. In this situation, the construct can then be transmitted to all cells of the progeny of the original transgenic animal and on through succeeding generations.

The most widely used system for the production of a specific protein from transgenic animals takes advantage of the large amount of concentrated protein secreted into milk by the mammary gland. The desired gene is attached to the synthesis and secretion signals used for natural production of the abundant milk protein, casein. This approach has been taken with sheep and cows, which are then able to secrete human β-interferon, α_1-antitrypsin, or human factor IX in their milk. High levels of the protein can be produced (e.g., > 10 g per liter of milk in some animals). The strategy enables long-term production and facilitates isolation of the protein in quantity without harming the animal. Once prepared and characterized, a herd of such animals would have a very low continuing production cost.

Transgenic plants can be created by the introduction of genes via infection by *Agrobacterium tumefaciens*, a bacterium that can pass DNA of its Ti plasmid into plant cells, where this segment integrates into the plant genome. In nature, this transformation of plants by the bacterium generates a gall or tumorous callus on the plant at the site of the wound at which the infection occurred. By genetically engineering the Ti plasmid to remove the tumor-causing genes while retaining the DNA transfer capability, useful vectors have been prepared. To effectively express a gene in the plant, suitable transcription and translation signals need to be incorporated surrounding the gene. Signals that have been used for this purpose are the nopaline synthetase promoter, the promoter from the gene for the small subunit of the carbon dioxide fixing enzyme, ribulose bisphosphate carboxylase, and those derived from the cauliflower mosaic virus.

The generation of a transgenic plant calls for the use of cells that are susceptible to infection and competent in regenerating a plant through culturing techniques. These two criteria have limited the system. *Agrobacterium tumefaciens* infects dicotyledonous plants, a group that does not include most cereal plants. For the cereals and many other plants, DNA is generally introduced into the cells by coating it onto an inert projectile and firing it into plant leaves or embryogenic calluses. Recombination followed by propagation from single cells under appropriate selection (usually kanamycin or phosphinothricin) can generate the transgenic plant. Fertile transgenic rice, corn, and wheat plants have now been formed. So far, the genes introduced into plants have been mainly intended to increase resistance to herbicides, insects, or viruses; such properties as ripening and taste can be altered, however.

Viral vectors can also be used with certain plant species (e.g., cowpea mosaic virus) and have been used to express animal virus surface antigens in whole plants. This approach is attractive in that the virus is easily isolated and stored and is very immunogenic but cannot harmfully infect animals.

4   PROCESS CONSIDERATIONS

The synthesis of protein from cloned genes relies solely on the metabolic machinery of the host cell. Consequently, any process factors that affect the host cell physiology will have a significant impact on the ultimate expression level. Factors such as temperature, impeller agitation speed, and medium formulation or feeding strategy in the case of a fed-batch fermentation are important variables to be considered in arriving at a production process. In industrial processes, where high volumetric productivity is the desired goal, extensive optimization studies are performed with these variables to achieve maximal overproduction of the specific protein.

Several general themes from the vast experimental observations of process conditions on *E. coli* protein production are mentioned here. The composition of the growth medium can affect both cell growth and recombinant protein yield. Complex medium generally yields a higher productivity. This is due to the presence of biosynthetic intermediates and cofactors, which the cell does not then need to synthesize, thus decreasing the metabolic burden. A balanced carbon-to-nitrogen ratio in the medium is also critical in achieving a high yield of recombinant protein. The temperature has a significant effect on cell growth and waste products accumulated. The formation of acetate, a harmful waste product, is increased at high temperature. Lower temperature favors the formation of soluble protein, while higher temperature encourages the accumulation of the protein in inclusion bodies.

Another challenging task is providing an adequate supply of oxygen to the culture, especially for dense cultures. The transfer of oxygen from the gas stream can be increased to a certain extent, by increasing the agitation speed. Oxygen supplementation is required in some cases, and genetic manipulation of the host has also been considered.

Waste product accumulation has a detrimental effect on recombinant protein production. Strategies to limit the effects of waste products have included lowering the temperature of the culture, carefully controlling the nutrient supply in a fed batch system, and the use of genetically altered hosts that exhibit reduced levels of harmful metabolic products.

5   PERSPECTIVES

The ability to clone and express high levels of proteins in other organisms has already led to great advances in the speed and detail with which biological systems can be analyzed. With the further expansion of this technology to other organisms and the construction of more complicated and sophisticated derivatives of currently studied systems, the impact of recombinant protein production seems certain to increase. Several areas likely to emerge are applications to industrial microorganisms, the formation and use of transgenic plants and animals, and the analysis of unknown proteins associated with genetic disorders.

See also DNA MARKERS, CLONED; *E. coli* GENOME; FUNGAL BIOTECHNOLOGY; GENE EXPRESSION, REGULATION OF; PLASMIDS; YEAST GENETICS.

Bibliography

Alitalo, K. K., Huhtala, M.-L., Knowles, J., and Vaheri, A., Eds. (1990) *Recombinant Systems in Protein Expression.* Elsevier, Amsterdam.

Barr, P. J., Brake, A. J., and Valenzuela, P., Eds. (1989) *Yeast Genetic Engineering.* Butterworths, Boston.

Buckholz, R. G., and Gleeson, M.A.G. (1991) Yeast systems for the commercial production of heterologous proteins. *Bio/Technology,* 9:1067–1072.

Goeddel, D. V., Ed. (1990) *Methods in Enzymology,* Vol. 185, *Gene Expression Technology.* Academic Press, San Diego, CA.

Henninghausen, L., Ruiz, L., and Wall, R. (1990) Transgenic animals—Production of foreign proteins in milk. *Curr. Opinion Biotechnol.* 1:74–78.

Hruby, D. E. (1990) Vaccinia virus vectors: New strategies for producing recombinant vaccines. *Clin. Microbiol. Rev.* 3:153–170.

Kriegler, M. (1991) *Gene Transfer and Expression: A Laboratory Manual.* Freeman, New York.

Old, R. W., and Primrose, S. B. (1990) *Principles of Gene Manipulation: An Introduction to Genetic Engineering,* 4th ed. Blackwell Scientific Publications, Oxford.

O'Reilly, D. R., Luckow, V. A., and Miller, L. K. (1992) *Baculovirus Expression Vectors—A Laboratory Manual.* Freeman, New York.

Reznikoff, W., and Gold, L., Eds. (1986) *Maximizing Gene Expression.* Butterworths, Boston.

Ridgeway, A. A. (1988) Mammalian expression vectors. *Biotechniques,* 10:467–492.

Vasil, I. L. (1990) The realities and challenges of plant biotechnology. *Bio/Technology,* 8:296–301.

Wang, L.-F., and Doi, R. H. (1992) Heterologous gene expression in *Bacillus subtilis.* In *Biology of Bacilli: Applications to Industry,* R. H. Doi and M. McGloughlin, Eds., pp. 63–104. Butterworth-Heinemann, Boston.

EXTRACELLULAR MATRIX

Linda J. Sandell

Key Words

Exon   DNA of the gene that is represented in the mature RNA product.

Extracellular Matrix   Material lying adjacent to and between cells.

Gene Structure   Organization of the gene, including the promoter, exons, and introns.

Intron   DNA of the gene that is removed from the mature RNA product.

mRNA Splicing The removal of introns from the pre-mRNA and joining of exons in mature RNA; thus introns are spliced out, while exons are spliced together.

The extracellular matrix is necessary for development and normal functioning of all cell types in an organism: it is made up of secreted proteins and glycoproteins whose complex interactions determine the matrix properties. The composition of the extracellular matrix is controlled at the level of gene expression by providing the quantity of mRNA sufficient to produce adequate proteins and by alternative splicing of pre-mRNA to generate proteins with the proper functional domains.

1 INTRODUCTION

The extracellular matrix (ECM) of all tissues is a complex mixture of secreted proteins that collectively play a critical role in determining and maintaining tissue function. The ECM is located primarily around connective tissue cells and under epithelia. ECM proteins range from the multifunctional fibronectins and thrombospondins to the large families of collagen types, proteoglycans, and laminins, among others. Fibrillar collagens types I, II, and III are the principal ECM proteins that confer the structural characteristics typical of tissues such as bone, skin, blood vessels, and cartilage. Other ECM proteins such as the fibronectins, laminins, and tenascin play critical roles in cell–cell interactions, cell migration, and cytoskeletal organization. The structure and function of extracellular matrix proteins have been reviewed recently in two very informative books by Hay (1993) and Kreis and Vale (1993). While not inclusive, this contribution discusses the molecular biology of the major categories of ECM components.

1.1 Regulation of ECM Gene Expression

It is becoming increasingly clear that the regulation of ECM gene expression, both transcriptionally (to regulate quantity) and posttranscriptionally (using alternative splicing of mRNA to include or remove functional domains), is of crucial importance during morphogenesis and cell differentiation, cell migration and proliferation, wound healing, and disease processes such as fibroses and arthroses. ECM expression is regulated by a wide variety of growth factors and cytokines, being generally stimulated by, for example, transforming growth factor beta, and insulin-like growth factors I and II, and inhibited by interleukin 1 and interferon gamma. Interestingly, as more information accumulates regarding regulation of ECM expression, it is apparent that ECM molecules are quite independently controlled in their expression and that regulation is dependent on

cell type. Table I shows some examples of regulation of ECM molecules by cytokines. This is a rapidly emerging field and more information is published every day. It is now known that certain ECM molecules can effect the expression of themselves or other ECM molecules: for example, the amino-propeptide of type I collagen is thought to down-regulate collagen synthesis, while the small proteoglycan decorin may be a component of feedback regulation of cell growth.

1.2 Gene Structure

Matrix proteins tend to be quite large, and often undergo extensive posttranslation modification. The genes are large and generally are composed of exons averaging 100–300 base pairs. Over the last 10 years, a considerable amount of information has accumulated, describing the structure, synthesis, and turnover of extracellular matrix proteins and, in particular, the genes coding for these proteins. Without exception, all these vertebrate ECM genes are single-copy, multiexon structures which, in many examples, represent among the most complex genomic structures yet described. The human $\alpha 1$ (XIII) collagen, for example, is approximately 140 kb in size and contains 39 exons; yet only 1.8% of this large gene is used to code for the 2300 bp of the $\alpha 1$ (XIII) mRNA.

1.3 Alternative Splicing of Pre-mRNA

Recently, several examples of alternate usage of particular exons were observed in a number of different ECM genes. This process of alternate exon usage, also referred to as alternative splicing, has been shown to account for isoforms of fibronectin and several proteoglycans. In addition, several investigators have demonstrated alternate exon usage to be a feature within the genes coding for several collagen types. In some instances, alternate exon usage is clearly responsible for the production of functionally distinct isoforms of an ECM protein. In other examples, the functional significance of alternative splicing of a pre-mRNA coding for a particular ECM protein is not yet clear. It does seem, however, that the mechanism of alternative splicing is useful for generating subtle differences within these large proteins while keeping the common parts of the protein identical.

2 COLLAGENS

2.1 Function

The collagens are a large family of genetically and structurally distinct proteins that collectively make up a major component of the ECM of most multicellular organisms. This family of triple-helical

Table 1 Some ECM Molecules Regulated by Cytokines

Cytokine	Molecules						
	FN	COLI	COLII	Decorin	Biglycan	Versican	Aggrecan
TGF-β	Up	Up	Up	No change	Up	Up	Up
IL-I	Down	Up	Down	Up	n.d.	Down	Down
Retinoic acid	Up	Up	Down	Up	Down	Up	Down
IFN-γ	Down	Down	Down	n.d.	n.d.	n.d.	n.d.

Abbreviations: FN, fibronectin; COLI, II, collagens I and II; TGF-β, transforming growth factor beta; Il-I, interleukin I; IF-N-γ, interferon gamma; n.d., not determined.

connective tissue proteins plays a critical role in the development and maintenance of a variety of functions including tissue architecture, tissue strength, and cell–cell interactions. Work over the last few years has clearly shown the important role that collagens play in diseases ranging from inherited skeletal dysplasias to tumor metastasis. This work has also led to the suggestion that collagen defects may be involved in such common diseases as osteoarthritis, osteoporosis, and disorders of the cardiovascular system. To date, 18 members of the vertebrate collagen family of proteins have been identified, coded for by at least 30 different genes. The collagens are the largest gene family and have been studied in most detail; consequently, the molecular biology of the collagens serves as our focus.

2.2 GENE STRUCTURE

The collagens are divided into two major groups: fibrillar and non-fibrillar. The fibrillar collagens, sometimes also called interstitial, include types I, II, III, V, and XI. The nonfibrillar include types VI, IX, X, XI, XII, XIV, and XIII. A subgroup of the nonfibril-forming collagens are classified as FACIT (fibril-associated collagens with interrupted triple helices), as they are found in close association with fibrillar collagens. The interstitial collagen chains are composed of three distinct or identical chains arranged in a "collagen" triple helix having the amino acid sequence Gly-X-Y, where X is often proline and Y, hydroxyproline. The collagens are initially synthesized as procollagens with globular propeptides at each end. The non-fiber-forming collagens are similar in overall structure, but they have noncollagenous, globular domains interspersed between Gly-X-Y domains. The protein structure is uniquely reflected in the gene structure. In general, the Gly-X-Y domains are encoded by very regularly sized exons, while the globular propeptides reflect gene structure observed in other genes. For example, the COL2A1 gene is made up of 54 exons and 53 introns. One exon (exon 2) is known to be alternatively spliced. This exon–intron structure is striking and represents a consistent pattern of all the fibrillar collagen genes (including types I, II, III, V, and XI). With very few exceptions, the conservation of exon structure is independent of species or type of fibrillar collagen. All the Gly-X-Y coding exons are 54 bp, 45 bp, or multiples of 54 or 45 (90, 108, and 162 bp). The translation of this gene structure into protein is shown diagrammatically in Figure 1. At each end of the triple-helical encoding exons are junction exons, which contain some Gly-X-Y coding domain and part of the nonhelical domain. The junction exons vary to a small extent and have even evolved to add Gly-X-Y coding triplets to type III collagen. In the propeptide-encoding domains, there are features of structural conservation within the fibrillar collagen genes, although not to the same extent as observed in the collagenous coding domain. For example, the penultimate exon, which contains a highly conserved coding sequence around a carbohydrate attachment site, is 243 bp in all the fibrillar genes where DNA sequence in this region is available. Equally, the coding region of the last exon, which contains the C-terminus of the C-propeptide in all phylogenetic species and all fibrillar collagen genes, is 144 bp in length. The other two exons that make up the N-terminus of the C-propeptide and the C-telopeptide vary in size to a small extent. The high degree of conservation of the C-propeptide between collagen genes argues for a similar and critical function of the propeptide in all collagens—perhaps the alignment of the three chains and initiation of triple-helix formation. The C-propeptide has been isolated from the extracellu-

lar matrix of cartilage (chondrocalcin) and may have an additional function, possibly in mineralization. On the other hand, the 5′ region of the fibrillar collagen genes, coding for the N-propeptide domain, clearly shows the highest degree of structural and sequence divergence, indicating that the function of the amino propeptide differs somewhat between collagen types.

2.3 ALTERNATIVE SPLICING

In the family of collagen genes, there are a number of examples of alternative splicing of pre-mRNA, thus providing further diversity to an already large gene family. The simplest alternative splicing occurs in the COL2A1 gene shown in Figure 1A. Here, exon 2, which comprises the large (69 amino acid), cysteine-rich domain of the N-propeptide, is either included (type IIA) or excluded (type IIB) from the mature mRNA. These two collagen mRNAs are tissue specific, with the type IIB mRNA found in chondrocytes resident in cartilage and the type IIA mRNA found in chondroprogenitor cells, bone precursors, skin precursor cells in avians, and in a variety of epithelial–mesenchymal junctions of the developing embryo. It is known that the type IIB procollagen provides structural integrity to the cartilage ECM. The type IIA procollagen may be involved in induction of skeletal structures or have some other function during differentiation.

Expression of specific domains of type VI and type XIII collagens also is regulated by alternative splicing of exons. While some of the alternative expression is tissue specific and may be developmentally regulated, there is no clear pattern, nor any proposed function for alternative splicing in these genes, at the present time.

2.4 ALTERNATIVE USE OF GENE PROMOTERS

The collagens also provide examples of alternative promoter use. The α1(IX) and α2(I) collagen genes also generate diversity by differential usage of exons. However, the mechanism is not by alternative splicing of the pre-mRNA, but by alternative start sites of transcription. These events eliminate most of the 5′ exons. In both the chicken and human α1(IX), the choice of transcription start site is tissue specific: in cartilage, transcription begins at exon 1 and contains a large N-terminal non-triple-helical domain of 266 amino acids (NC4); in primary corneal stroma, transcription begins in the intron between exons 6 and 7, eliminating the expression of the NC4 domain. In the chicken gene, these promoters are separated by 20 kb of DNA. The function of the alternatively spliced region is unknown; however, it has an estimated isoelectric point of pI 9.7 and 10.55, in the chicken and human, respectively, and is believed to bind to proteoglycans.

Transcription of the α2(I) collagen gene is also tissue specific; however, the reading frame of the protein is changed by the alternative use of promoter, producing a different and noncollagenous protein. In most type I collagen synthesizing tissues, all the previously described exons are used whole; in chondrocytes, transcription begins at a new site within intron 5 and includes an additional exon. The use of the cartilage transcription start site replaces exons 1 and 2 with a 96 bp exon contained within intron 2. The resulting transcripts contain a reading frame that is out of frame with the collagen coding sequence. The putative protein product of this mRNA is consequently not collagenous and may encode a protein of regulatory function. Interestingly, this anomaly provides a mechanism for inhibition of type I procollagen synthesis by chondrocytes. In

A

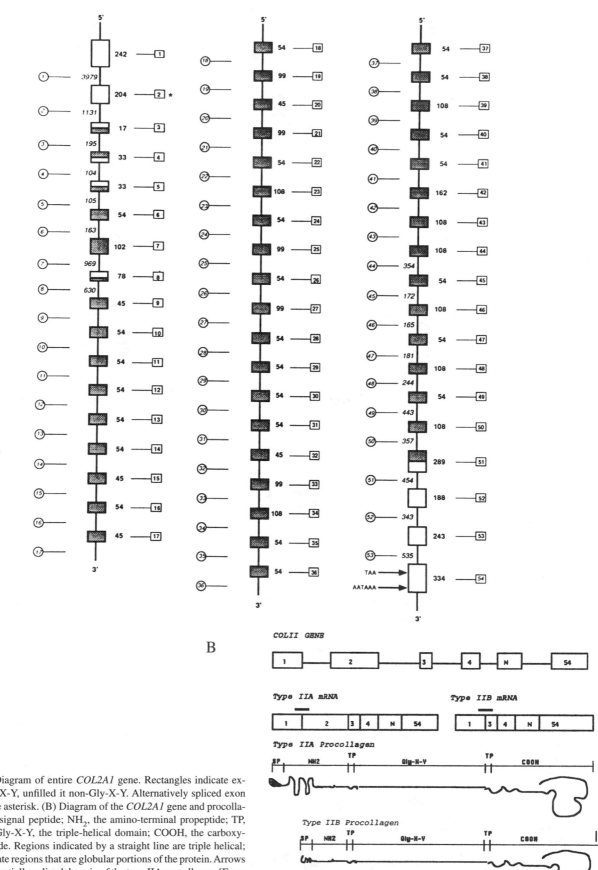

Figure 1. (A) Diagram of entire *COL2A1* gene. Rectangles indicate exons; filled it Gly-X-Y, unfilled it non-Gly-X-Y. Alternatively spliced exon is indicated by the asterisk. (B) Diagram of the *COL2A1* gene and procollagen α-chain: SP, signal peptide; NH₂, the amino-terminal propeptide; TP, the telopeptide; Gly-X-Y, the triple-helical domain; COOH, the carboxy-terminal propeptide. Regions indicated by a straight line are triple helical; curves lines indicate regions that are globular portions of the protein. Arrows indicate the differentially spliced domain of the type IIA procollagen. [From Sandell et al., 1991, *JCB* 114:1307–1319.]

prechondrogenic limb mesenchyme, transcription of the α2(I) collagen initiates at the previously described bone/tendon promoter, while during differentiation into chondrocytes, the promoter utilization is switched to the cartilage promoter.

3   ELASTIN

Tropoelastin is the soluble precursor to elastin, the major protein component of elastic fibers. Several isoforms of tropoelastin have been identified in a variety of elastic tissues; however, unlike the other connective tissue genes, there is no elastin family of genes, only a single gene. Isoforms are generated solely through alternative splicing of pre-mRNA. Tropoelastins from all vertebrates studies are approximately 70 kDa in size. The complete, derived amino acid sequence from overlapping tropoelastin cDNA and genomic DNA clones from several vertebrate species has confirmed earlier peptide sequence data suggesting the existence of several functional domains in tropoelastin—including alanine-rich, lysine-containing regions necessary for the formation of desmosine cross-links. The resilient properties of tropoelastin are conferred by multiple hydrophobic domains rich in valine, proline, and leucine. In addition to a signal sequence at the amino-terminal end of tropoelastin, there are several other domains, including the only cysteine-containing region located at the carboxy terminus of the protein. Most of these domains are encoded by separate exons within the tropoelastin gene.

The complete genomic sequence coding for tropoelastin has been analyzed in rats, humans, and cows. In all species, the tropoelastin gene is approximately 40 kb in length and is composed of about 36 exons. The majority of these exons code for either hydrophobic or cross-link domains. By DNA sequencing and S1 nuclease protection analysis, it is clear that a total of approximately 40% of these exons are subject to alternate usage in elastic tissues such as nuchal ligament, aorta, lung, and skin from varying developmental ages of rats, cows, or humans. Figure 2 summarizes all known exons or domains within exons in the vertebrate tropoelastin gene that are sub-

ject to alternative splicing. Any clear correlation with development or tissue-specific expression of the tropocollagen isoforms cannot be observed, and the function of these multiple forms is unknown.

4   FIBRONECTIN

Fibronectins (FN) are large, extracellular matrix glycoproteins that function as both plasma and ECM proteins. FN contains binding domains for fibrin, cells, heparin and proteoglycans, and collagen. The gene structure is shown in Figure 3. The plasma and cellular forms of FN have been isolated and extensively studied. The difference between these two forms and other forms with more subtle differences can be accounted for by alternative splicing.

Alternative splicing of fibronectin pre-mRNA was the first example of alternative exon usage as a means of generating ECM protein diversity. As a major cell adhesion molecule, FN has multiple sites for interaction with cell surface receptors and is an essential structural component of the extracellular network. The binding domains are composed of sets of repeating modules called types I, II, and III repeats. Single exons encode each type I and type II repeat and pairs of exons combine to code for each type III repeat (see Figure 3). The FN gene consists of 48 exons and 47 introns that are spliced to give an mRNA about 8 kb long. Within this framework, the FN mRNA varies at three sites via two different patterns of alternative splicing. The first of these regions to be identified, the variable (V) region, exemplifies exon subdivision in which a single 467-base exon containing alternate and constitutive coding sequences is partially or completely included. In rodents and cows, the 360-base V region is either entirely included or entirely excluded, or the first 75 bases are omitted during splicing. The number of splice variants at this site differs among species, with five in humans and two in chickens and frogs. The exon is never skipped because 107 bases encoding half the flanking type III repeat are constitutively included. The resulting polypeptides contain additional 120 or 95 amino acid segments or no extra residues, thus generating three types of FN subunits V120, V95, and V0, respectively. This segment has a

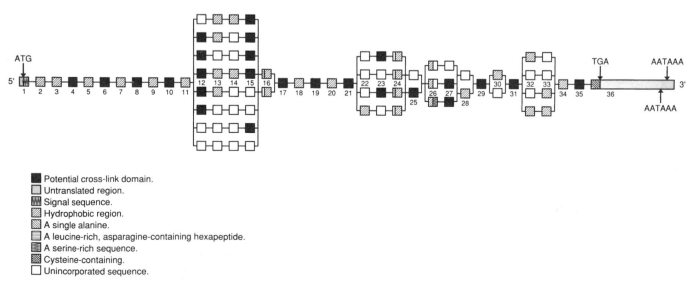

■ Potential cross-link domain.
□ Untranslated region.
▥ Signal sequence.
▨ Hydrophobic region.
▧ A single alanine.
▤ A leucine-rich, asparagine-containing hexapeptide.
▦ A serine-rich sequence.
▩ Cysteine-containing.
□ Unincorporated sequence.

Figure 2.   A summary of exons alternatively spliced in the vertebrate tropoelastin gene. Exons are numbered from the 5′ end of the gene and the various domains encoded by exons are identified by variable shading explained in the legend. Known permutations of alternatively spliced exons are illustrated. The initiation and termination codons and polyadenylation signals are also indicated. The relative size of exons and introns are not indicated [From Boyd et al. (1993).]

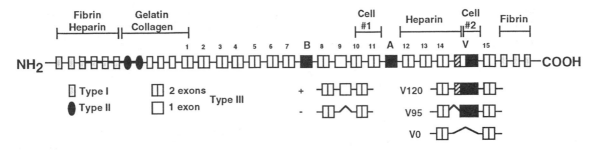

Figure 3. FN subunit variants. The repeat structure of FN is shown and consists of type I, type II, and type III repeats as indicated. Each type I and II repeat is contained within a single exon in the gene. All but three of the 17 type III repeats are encoded by two exons; the exceptions are the alternatively spliced repeats EIIIA (A) and EIIIB(B) and one constitutive repeat, exon 9. Inclusion (+) or omission (−) of EIIIA and/or EIIIB results in the structures shown. Exon subdivision of the V region generates three variants differing in the number of extra amino acids (V120, V95, V0). The function-binding sites are indicated above the gene structure. The two cell-binding sites are recognized by different integrin receptors: 1, an RGD site, is recognized by $\alpha5\beta1$; 2, a EILDV site, is recognized by $\alpha4\beta1$ integrin. [Modification of a figure from Boyd et al. (1993).]

unique sequence; unlike the remainder of FN, it is not part of the repeating structure.

The second pattern of alternative splicing is exon skipping. At two sites, the single exons EIIIA and EIIIB, encoding single type III repeats, are included or skipped independently, generating mRNAs with all combinations of these two exons. Mechanistically, the EIIIB splicing event appears to involve formation of a specific protein RNA complex on the intervening sequence 5′ of this exon.

All FN alternate exon usage is regulated in a tissue-specific manner. Liver splicing of FN is the most simple and distinctive. Liver synthesizes circulating plasma FN, a heterodimer of one V+ subunit (containing either the V120 or V95 segment) and one V0 subunit and lacking both EIIIA and EIIIB. Most, if not all, other tissues express the cellular form of FN, which is composed of V95 or V120 (V+) FNs with varying amounts of EIIIA and/or EIIIB.

The most striking shifts in the splicing repertoire occur during embryogenesis, wound healing, and oncogenesis. All three situations show an increase in the amount of EIIIA+ and/or EIIIB+ mRNA relative to normal adult tissues. This is most clearly illustrated by in situ hybridization of cutaneous wound sections. Adjacent normal tissue cells express V+ FN, while fibroblasts within the granulation tissue of the wound bed are also very positive for EIIIA+ and EIIIB+ mRNAs. In addition, quantitation of EIIIA and EIIIB mRNAs identified by RNase protection analysis demonstrates a dramatic drop in their levels at birth and increased inclusion of some oncogenically transformed cells. Thus, changes in the combination and proportions of EIIIA+ and EIIIB+ RNAs at different stages of tissue development, remodeling and tumor formation could have significant effects on the interactions and functions of FN.

5   PROTEOGLYCANS

Structurally, proteoglycans are distinguished from other glycoproteins in possessing glycosaminoglycan chains on a protein core. The glycosaminoglycan chains themselves differ in number (from 1 to 100) and glycan composition: chondroitin sulfate, dermatan sulfate, heparan sulfate, and keratan sulfate. Proteoglycans are present in a variety of basement membranes, ECMs of most connective tissue cells, and cell membranes. While functionally disperse, the proteoglycan core proteins can be classified into a number of families: the large aggregating proteoglycans (aggrecan, versican, and CD44) and the small leucine-rich proteoglycans (decorin, biglycan, fibromodulin, and lumican). The large proteoglycans of this group bind

to hyaluronic acid and have between 15 (versicon) and 100 (aggrecan) glycosaminoglycan chains. Functions for the small proteoglycans include both collagen and fibronectin binding.

Other proteoglycans include serglycin (in intracellular storage granules), perlecan (the membrane heparan–sulfate proteoglycan), syndecan and fibroglycan (transmembrane proteoglycans), and betaglycan (a TGF-β binding protein). The syndecan family (four members) is important in embryogenesis and organ development, while perlecan is the largest and most common proteoglycan of basement membranes.

The large proteoglycan mRNAs are up to 10–12 kb, and the genes can be quite large. Aggrecan, the related glycoprotein link protein CD44, and the large proteoglycan versican are all alternatively spliced. One of the domains that is alternatively spliced is an epidermal growth factor-like module. A member of the small proteoglycan family, decorin, exhibits usage of alternative promoters.

6   OTHER INTERACTIVE GLYCOPROTEINS: LAMININS, TENASCIN, THROMBOSPONDIN

Laminin is a member of a family of basement membrane proteins including merosin, S-laminin, and S-merosin. This family is expanding rapidly and is now known to include epiligrin. All the laminins are made up of three distinct gene products; A, B1, and B2 subunits (recently renamed α, β and γ, respectively), with a total molecular weight of 800 kDa. The laminins function in adhesion, spreading, migration, differentiation, and growth of cells. Many of the genes have been isolated, and there may be alternative splicing of some members of the family. The amino acid sequence of the laminin B1 and B2 chains indicates that they are the products of closely related but distinct genes, which are likely to be derived from a common ancestor gene.

Tenascin (also called hexabrachion, J1 glycoproteins, and cytotactin) is a large disulfide-linked hexmeric ECM glycoprotein with many repeated structural units such as heptad, EGF-like, and fibronectin type III repeats, as well as a homology to the globular domain of β- and γ-fibrinogen. cDNA sequences are available for chicken, mouse, and human. Interestingly, a tenascinlike transcript is encoded on the opposite strand of the human steroid 21-hydroxylase/complement C4 locus on chromosome 6p21. Tenascin is expressed during embryogenesis, during regeneration, and in tumors; it can be alternatively spliced in a tissue-specific manner.

Thrombospondin is a 420 kDa adhesive glycoprotein composed of three subunits of equal molecular weight. It is expressed in developing heart, muscle, bone, and brain, and in response to injury and inflammation in adult tissue. Thus far there are four members of the thrombospondin gene family including a protein found primarily in cartilage, cartilage oligomeric protein (COMP).

See also CELL-CELL INTERACTIONS; CYTOSKELETON-PLASMA MEMBRANE INTERACTIONS; GLYCOPROTEINS, SECRETORY.

Bibliography

Boyd, C. D., Pierce, R. A., Schwarzbauer, J. E., Doege, K., and Sandell, L. J. (1993) Alternative exon usage in matrix genes. *Matrix* 13:457–469.

Burgeson, R. E., Chiquet, M., Duetzmann, R., Ekbloom, P., Engel, J., Kleinman, H., Martin, G. R., Meneguzzi, G., Paulsson, M., Saues, J., Timpl, R., Trygguason, K., Yamada, Y., and Yukchenco, P. D. (1994) A new nomenclature for the laminins. *Matris Biol.* 14:209–211.

Hay, E. D. (1993) *Cell Biology of Extracellular Matrix,* 2nd ed. Plenum Press, New York.

Kreis, T., and Vale, R. (1993) *Guidebook to the Extracellular Matrix and Adhesion Proteins.* Sambrook and Tooze, Oxford University Press, Oxford.

Ramirez, F., and Di Liberto, M. (1990) Complex and diversified regulatory programs control the expression of vertebrate collagen genes. *FABSEB J.* 4:1616–1623.

Sandell, L. J., and Boyd, C. D. (1990) Conserved and divergent sequence and functional elements within collagen genes. In *Extracellular Matrix Genes,* pp. 1–56. Academic Press, Orlando, FL.

FEMALE REPRODUCTIVE SYSTEM, MOLECULAR BIOLOGY OF

J. K. Findlay and J. E. Mercer

Key Words

Endocrine Regulation   Regulation by a hormone made by an organ and secreted into the bloodstream, where it circulates and influences specific target organs.

Endothelins   Peptides of 21 amino acids with potent vasoconstrictor as well as mitogenic and hormone-releasing properties; made by a variety of epithelial and vascular smooth muscle cells.

FSH Isoforms   Forms of follicle-stimulating hormone that differ in the degree of glycosylation of the mature protein leading to differences in their biochemical properties and circulating half-lives in the blood.

G Proteins   Coupling proteins, so called because they bind GTP, which link receptors to effector molecules. They regulate a variety of enzymes and ion channels. The G proteins form a heterotrimeric complex of α, β, and δ subunits. There are multiple members of each family of subunits, giving rise to a complexity of responses.

Gonadotropin-Releasing Hormone   A decapeptide synthesized in hypothalamic neurones; it is released in a pulsatile fashion into the hypothalamic–pituitary portal blood system and stimulates synthesis and release of gonadotropins by the pituitary gland.

Gonadotropins   Glycoprotein hormones, called follicle-stimulating hormone, luteinizing hormone, and chorionic gonadotropin, consisting of a common α subunit linked to a hormone-specific β subunit, which stimulate ovarian and placental cells.

Leukemia Inhibitory Factor   A pluripotent cytokine that inhibits differentiation of embryonic stem cells, is expressed in endometrial epithelial cells, and has been shown to be essential for implantation in mice.

Local Regulation   Autocrine or paracrine regulation of cells by factors made locally within a tissue.

Oxytocin   A nonapeptide with uterotonic, prostaglandin-releasing and milk-letdown activities, made by hypothalamic neurones and ovarian cells.

Receptors   Membrane or intracellular polypeptides that bind hormones or other factors with a high degree of specificity and transduce the signal to the cell either as transcription factors regulating genes or via a second-messenger system.

Second-Messenger Systems   Small cytoplasmic molecules which, when activated, amplify the signal they receive, (e.g., cyclic AMP and inositol phosphate).

Signal Transduction   The process by which a signal received at the cell surface is passed from one carrier to another to achieve a cellular response.

Steroidogenic Enzymes   A family of enzymes, many of which belong to the P450 gene family, that catalyze the synthesis and metabolism of steroid hormones such as estrogens, progestogens, and androgens.

Tissue Remodeling Enzymes   Enzymes belonging particularly to the matrix metalloproteinase family, such as MMPs 1, 2, 3, 7, and 9, and their associated tissue inhibitors, the TIMPS, which are responsible for remodeling the constituents of basement membranes and the extracellular matrix during growth, angiogenesis, tissue breakdown, and wound healing.

Transforming Growth Factor β Superfamily   A family of pleiotropic factors that have high homology at the level of the gene and in their tertiary protein structure. Examples include TGF β, inhibins, activins, and Müllerian inhibiting substance.

Transmembrane Helices   G-protein-linked receptors are a family of receptors characterized by a structural motif of seven hydrophobic regions that span the cell membrane, forming a helical structure within the membrane.

Zona Pellucida Proteins   A family of highly glycosylated proteins unique to the surface coat of the oocyte, several of which serve as primary and secondary sperm-binding proteins on the egg at the time of fertilization.

The female reproductive system consists of the hypothalamus, pituitary, ovaries, breasts, and uterus. These organs make up a finely tuned system that controls puberty, the menstrual cycle, pregnancy and birth, lactation, and the menopause. With the tools available through molecular biological approaches, we are able to come to a greater understanding of the complex interactions that operate in this system. The hormones, growth factors, and enzymes that are able to exert unique and specific effects in this system do so either locally or via endocrine mechanisms. The elucidation of the molecules involved in this regulation and their mechanisms of action leads to a better understanding of the physiology and pathophysiology of reproduction, and provides the potential means for developing better methods for controlling fertility and infertility.

1 PRINCIPLES: REGULATION OF REPRODUCTIVE PROCESSES IN THE FEMALE

1.1 THE HYPOTHALAMIC–PITUITARY–OVARIAN–UTERINE AXIS

The regulation and coordination of all facets of reproduction in the female that lead to fertility are based in the hypothalamic–pituitary–ovarian–uterine axis. It is the activities of numerous hormones, growth factors, and cytokines, and their receptors and binding proteins, as well as enzymes, structural proteins, and other factors within this axis, that control the development of the reproductive system in fetal life and during puberty, as well as the cyclic production of hormones and eggs during reproductive life, the establishment and maintenance of pregnancy, the birth process, lactation, the menopause, and the secondary sex characteristics.

1.2 LEVELS OF REGULATION

The hypothalamic uterine axis was characterized originally by the positive and negative feedback relationships of the endocrine hormones that had been shown to regulate the reproductive system. For example, follicle-stimulating hormone (FSH) made by the pituitary gland stimulates production of estradiol-17β (E2) by the ovarian follicle; E2 in turn has a negative feedback effect to limit the production of FSH, and so an equilibrium is established that ultimately determines the extent of stimulation of egg and hormone production by the ovary.

In addition to the endocrine regulation of this axis, there is regulation at the organ and cellular level by locally produced factors and enzymes that have been recognized for many years, but have received more attention recently with the identification of growth factors and cytokines operating within the axis. Some of the endocrine hormones also act locally. For example, E2 is produced by the ovary and acts directly on the follicles; prostaglandins produced by the endometrium act within the uterus. The growth factors, cytokines, enzymes, and other factors identified more recently within the reproductive axis are structurally identical to those produced in other organ systems. Nevertheless they can be differentially regulated and can perform functions specific to the reproductive axis. Examples are the cyclic tissue remodeling by matrix metalloproteinases (MMPs) and angiogenic factors in the ovary and uterus, the role of activin in the functions of the pituitary, ovary, and placenta, and the actions of the insulin-like growth factors (IGFs) and their binding proteins in the ovary and endometrium.

1.3 SPECIFICITIES OF ACTION

It has already been said that many of the hormones and other regulatory factors are not unique to the female reproductive system even though they perform specific functions. Regulation of hormones and cytokines occurs at their sites of production and at their sites of action by cell-specific expression of their receptors; the presence of binding proteins and metabolizing enzymes determines the extent and specificity of their actions.

The few substances that are unique to the reproductive system can generally be found in both the female and the male; they include gonadotropin-releasing hormone (GnRH) and its receptor, FSH and its receptor, luteinizing hormone (LH) and its receptor, Müllerian inhibiting substance (MIS) and its receptor, oxytocin, and relaxin. Substances unique to the female are the placental proteins such as the chorionic gonadotropins and lactogens (and a special class of interferons, φ, in ruminants), and the zona pellucida proteins (ZP1–3 or 4), which coat the oocyte.

It is not possible in the space available to give a full account of the expression and actions of all the factors involved in regulating the female reproductive system. The reader is referred to the bibliography and to other references therein. Instead, we have briefly summarized selected examples of regulation as a means of giving some insights into the subject. Tables 1–3 summarize reproductive processes under local control in the pituitary, ovary, and uterus.

2 EXAMPLES OF REGULATION OF FEMALE REPRODUCTION

2.1 GONADOTROPIN-RELEASING HORMONE AND ITS RECEPTOR

2.1.1 Gonadotropin-Releasing Hormone

GnRH is the key hormone in the development and maintenance of the reproductive system. It is produced in the hypothalamus and provides the drive for production of LH and FSH in the pituitary gland. GnRH is secreted from the hypothalamus in a pulsatile fashion, which is critical for the regulation of synthesis and secretion of the pituitary gonadotropin hormones. GnRH is a decapeptide; it is processed from a longer precursor protein that has a signal peptide at the N-terminus and a 56 amino acid GnRH-associated peptide (GAP) at the C-terminus. The GnRH gene consists of four exons. The opposite strand of DNA is also translated in the hypothalamus and heart, but it is not known whether this transcript, known as SH, has any functional role. GnRH is expressed in specific neurones in the preoptic area of the hypothalamus, and its expression is regulated by several factors. Many neurones impinge on GnRH neurones, including dopaminergic, noradrenergic, and GABAergic neurons. The GnRH promoter has both an activating and an inhibiting region. Direct steroid regulation of GnRH is suggested by the presence of regions in the promoter with high homology to known steroid hormone response elements, including a putative estrogen response element (ERE) that is capable of binding estrogen receptor and negatively regulating expression of the gene in vitro.

2.1.2 GnRH Receptor

The pituitary GnRH receptor (GnRH-R) is a member of the seven membrane spanning domain, G-protein linked receptor family. The nucleotide sequence encodes a protein of 327 residues; however, this receptor lacks an intracellular C-terminal tail, which is also characteristic of these receptors. Among other mechanisms, the signaling of GnRH is regulated through desensitization of the receptor, which in the related thyrotropin-releasing hormone (TRH) receptor is believed to occur via the C-terminal tail. Since the GnRH-R lacks a C-terminal tail, it is possible that desensitization of this receptor may operate via a mechanism completely different from that of the TRH receptor.

The mRNA for the GnRH-R has been identified in pituitary, ovary, and testis, with three different sizes of mRNA present in rat pituitary and ovary; furthermore, the expression of GnRH-R is hormonally regulated in the pituitary gland. The structural relationships of the transmembrane helices 2 and 7 and their function in ligand binding have been studied using a reciprocal mutation in those helices, demonstrating that they are probably adjacent in the cell membrane and that they are involved in GnRH binding.

GnRH signal transduction is achieved through interaction of the receptor with G proteins that are coupled to further second-messenger systems. It has been shown that the GnRH-R is coupled to the G proteins $G\alpha_q$ and $G\alpha_{11}$. These G-protein regulatory subunits modify the activity of phospholipase C and subsequently protein kinase C during intracellular signaling.

2.2 GONADOTROPINS AND THEIR RECEPTORS

2.2.1 Gonadotropins

LH and FSH are members of the glycoprotein hormone family, which also includes thyroid-stimulating hormone (TSH) and, in humans and horses, chorionic gonadotropin (CG). Each of these hormones consists of two different subunits, α and β, which are glycosylated, contain multiple disulfide linkages, and are noncovalently linked to each other to form the active molecule. The α-subunit apoproteins are the same for all the glycoprotein hormones within a species and are encoded by a single gene, which in the human is located on chromosome 6. The β subunits are different for each hormone and determine the biological specificity of each hormone. The LHβ and FSHβ genes show similar genomic arrangements, suggesting a common ancestral origin. The human LHβ gene is located on chromosome 19 and the FSHβ gene on chromosome 11. The addition of the carbohydrate moieties to the apoproteins is important for correct synthesis of bioactive gonadotropin dimers, and for achieving appropriate biological activity of the molecule. Regulatory mechanisms may operate through modifying the sugars on these hormones (see Section 3.2).

The anterior pituitary gland in a mature animal has several different types of endocrine cell, each of which secretes particular hormones. LH and FSH are synthesized in the gonadotroph cells of the anterior pituitary, with populations of cells synthesizing either LH or FSH, or both. This ability of some gonadotrophs to produce more than one hormone makes them unusual in relation to the other endocrine cell types in the pituitary such as the corticotrophs, thyrotrophs, lactotrophs, and somatotrophs, each of which synthesizes and secretes largely a single trophic hormone in the adult. During development, however, it is likely that the different pituitary cell types arise from a common cell type under the control of cell-specific transcription factors that determine the temporal expression patterns of the hormonal genes in the pituitary (see Section 2.5.2).

LH and FSH are secreted in different patterns under different physiological circumstances, and the expression of the genes that

Table 1 Examples of Endocrine and Local Regulation of Gene Expression in the Pituitary Gland

Gene	Endocrine Regulator	Local Regulator
α Subunit of gonadotropins	GnRH, E_2, progesterone	?
LHβ	GnRH, E_2, progesterone	?
FSHβ	GnRH, E_2, inhibin	Activin, follistatin
GnRH-R	E_2, ?	?

encode them is also regulated differently under different conditions. As discussed in the introduction, GnRH provides the major stimulatory drive for expression of the β-subunit genes, and the steroid hormones in general have negative regulatory effects. The gonadal peptides inhibin and follistatin negatively regulate the expression of FSHβ, and activin stimulates expression of the FSHβ gene (see Section 2.3.2). These data are summarized in Table 1.

GnRH regulation of the expression of the α-subunit gene has been shown to occur via interactions of transcription factors with the α-subunit promoter. Transgenic mice with either 1.6 kb of the human, or 315 base pairs of the bovine α-subunit gene 5'-flanking region fused to a chloramphenicol acetyltransferase reporter gene (CAT), and treated with pulses of GnRH every 2 hours, showed increased expression of the CAT reporter gene. These studies are the first step toward identifying the exact region of the promoter involved in GnRH regulation and of subsequently identifying the trans-acting factors that might bind to this region of the gene.

2.2.2 Gonadotropin Receptors

The LH/CG and FSH receptors are also members of the seven membrane spanning domain, G-protein-linked receptor family. Many of the receptors in this family have very short, extracellular amino termini, but the LH/CG and FSH receptors are unusual in that they have very long amino termini, which are encoded by 10 and 9 exons, respectively. The seven membrane spanning domains and the carboxyterminal intracellular region are encoded by exon 11 and exon 10, respectively, the final exon in both these receptors. The extracellular domains of these receptors have 14 copies of an imperfectly repeated sequence of approximately 25 amino acids known as leucine-rich repeats. The functional significance of this structure is not known, but it may be involved in cell–cell or protein–protein interactions.

The gonadotropin receptors are expressed in the ovary, and both can be detected as multiple transcripts that may arise as a result of the use of multiple transcription start sites or alternate polyadenylation signals. Both receptors have been shown to exhibit alternate splicing of the first 10 or 9 exons, and some of these transcripts would give rise to truncated peptides. It is not clear that these truncated forms of the receptor are translated, or that they have any functional significance.

2.3 THE TRANSFORMING GROWTH FACTOR β SUPERFAMILY AND ITS RECEPTORS

2.3.1 The TGFβ Family

Transforming growth factor β was described originally as a factor that induced normal rat kidney fibroblasts to form colonies in soft agar in the presence of epidermal growth factor (EGF). Subsequently it was shown to have important actions in embryonic de-

Table 2 Examples of Ovarian Processes Under Local Control

Process	Factor(s) Involved
Folliculogenesis	TGF-β, activin, inhibin, IGF-I and II, IGF-BPs, Follistatin, TGF-α, FGF, steroids
Angiogenesis	VEGF, FGF
Oocyte maturation	c-kit ligand
Ovulation and corpus luteum formation	MMPs, TIMPs, PA, PAI, prostaglandins

velopment, tissue remodeling, and wound healing, and the peptide and its receptors can be found in almost all cells. TGFβ became the name given to a superfamily of peptides by virtue of the 30–40% amino acid homologies to TGFβs of a number of apparently disparate peptides. They include the inhibins and activins, MIS, bone morphogenic proteins, osteogenic proteins, Vg1 and Vgr-1, growth differentiation factors, the decapentaplegic complex of *Drosophila*, *nodal* and *dorsalin-1* in mice, and *ceg-1* in *Caenorhabditis elegans* and *Drosophila*. The superfamily has been divided into four subfamilies on the basis of in vitro biological activities and the intron–exon structure of their genes. These subfamilies are the TGFβ family, the DVR family (the largest), the inhibin family, and the MIS family; members of each family have been identified as having a regulatory role in the female reproductive system (Table 2).

Five forms of TGFβ have been cloned. Each is a disulfide-linked homodimer with a molecular weight of 25 kDa in the unreduced form. Each monomer is 112–114 amino acids long and occupies the carboxy-terminal sequence of its precursor. There is a 70–80% sequence homology between the monomers, and there is complete conservation of the nine cysteine residues in each monomer. TGFβ generally is secreted in a biologically inactive, latent form noncovalently associated with a dimer of the pro region of the precursor. The mechanism of activation is not well understood but may involve a plasmin.

The TGFβ2 dimer has been crystallized, and its molecular structure appears to be unique in that the dimer exists in an extended rather than a globular form. Eight of the nine cysteines in each chain are involved in intrachain disulfide bridges with only a single interchain bridge. There are two hydrophilic cavities at the interface of the monomers, but their functional significance is not clear.

2.3.2 The Inhibin Family

Inhibin was identified originally as a water-soluble testicular extract that prevented the hypertrophy of pituitary cells in response to castration. This definition was subsequently refined to designate an activity capable of selective suppression of FSH secretion, and on this basis, inhibin was isolated from ovarian follicular fluid and identified as a glycosylated heterodimer consisting of an α and a β chain joined by disulfide bonds. Like TGFβ, the mature dimer consists of the C-terminal sequences of a larger precursor of each chain, having a molecular weight of 31–32 kDa, although larger (e.g., 51, 58, 64, 105 kDa) biologically active forms exist. The α-chain precursor of 366 amino acids in the human is encoded by a single gene with only one intron, and homology of the mature peptide (133 amino acids) is between 85 and 96% for human, bovine, ovine, and porcine specimens. The seven cysteine residues are totally conserved between species, but the human α chain contains two potential N-linked glycosylation sites compared to one in the animal species.

There are also other molecules such as αN, derived from the N-terminal end of the precursor of the α chain, which do not have properly defined biological roles. There are two distinct forms of the β chain, called βA and βB, each coded by a single gene containing one intron; the human precursor peptides are 426 and 407 amino acids, respectively. The homologies of the mature β chains (116 and 115 amino acids, respectively) exceed 95%, and the nine cysteines are conserved. Indeed the distribution of the cysteine residues in the α, βA, and βB chains are very similar between the three peptides and other family members. Inhibin exists as either inhibin A (αβA) or inhibin B (αβB), and both forms have FSH-suppressing activity.

At the time that inhibin was being isolated from ovarian follicular fluid, an activity was found that stimulated FSH secretion. This activity, called activin, was identified as a nonglycosylated dimer of the β chains with a molecular weight of 25 kDa. Activin exists as three isoforms, activin A (βAβA), AB (βAβB), and B (βBβB), all of which appear to have similar FSH-stimulating activity. Subsequently, activin has been identified as a pleiotropic regulator in many tissues including the developing embryo, where it acts as an inducer of the mesoderm, and in bone marrow, where it is an erythroid-differentiating factor. Transgenic mice deficient for the inhibin βB subunit produced offspring, but the animals had developmental and reproductive disorders. There was a failure of eyelid fusion, and whereas the male offspring bred normally, the offspring of the females had high perinatal mortality. Furthermore, expression of the βA subunit was up-regulated in the ovaries of the females. These data suggest that activin βB plays a role in fetal development and is critical for female fecundity, but is not essential for mesoderm formation.

Follistatin or FSH-suppressing activity was also isolated from ovarian follicular fluid but was found to be unrelated structurally to inhibin. It exists as up to six isoforms (31–42 kDa), which arise as a result of a combination of alternate splicing of its mRNA and variable glycosylation of the mature peptide. It is encoded by a single gene, consisting of six exons, with the alternate splice site located in exon 5. All the follistatin genes and cDNAs from different mammalian species sequenced to date are more than 95% homologous at the amino acid level, with a homologue in *Xenopus* having 85% homology. The mature follistatin proteins consist of four contiguous domains, each encoded by a different exon; the last three domains are 52% homologous with each other and are also structurally related to EGF and to a pancreatic secretory inhibitor. Follistatin derives its FSH-suppressing activity from its ability to bind activin; it is expressed in the germ cells, embryo, pituitary, ovary, and placenta, as well as the brain, adrenal, bone marrow, and spleen, although not always coincidently with activin. It is also highly expressed in the kidney and pancreas, where the inhibin β subunit mRNA has not been detected. Neither follistatin nor inhibin β subunit is expressed in the liver.

2.3.3 The MIS Family

MIS, also called anti-Müllerian hormone or Müllerian-inhibiting factor, is the testicular factor responsible for regression of the Müllerian primordia, the first step of male somatic differentiation. It has subsequently been demonstrated as a product of granulosa cells of antral follicles of the ovary, although the level of production is low compared with the Sertoli cell of the testis. Native MIS is a high molecular weight (140–144 or 280–290 kDa) glycoprotein, formed by the assembly of dimers or tetramers of a 70 or 72 kDa monomer.

Like other members of the TGFβ family, the MIS monomer is derived from a larger 546 amino acid precusor in which the mature form is at the C-terminal end of the molecule. The 140 kDa homodimer is further processed to a noncovalent complex of 110 and 25 kDa dimers, although it is not known whether this step is required for activation of the biological activity of MIS. The MIS gene, located on human chromosome 19, p13.3, is only 2.7 kb and is encoded by 5 exons. There are multiple transcription initiation sites, perhaps because the human MIS promoter lacks consensus TATA and CAAT boxes, similar to many so-called housekeeping genes.

The physiological significance of expression of MIS in the female gonad, especially the adult, is unclear. There are data to suggest that MIS might affect development of the fetal ovary. For example, transgenic female mice chronically overexpressing MIS undergo total Müllerian regression and most lack ovaries; where there is ovarian tissue, it is severely depleted of germ cells and contains structures resembling seminiferous tubules. MIS has also been shown to inhibit aromatase activity in fetal ovaries.

2.3.4 TGF Superfamily Receptors

Several receptors have now been identified and have been shown to be members of the TGFβ receptor superfamily. These receptors need to interact with one another to produce a cellular response; that is, a type I and a type II receptor must associate in the cellular membrane to bind ligand and transduce an intracellular signal. The cDNA encoding a type II activin receptor (ActRII) was first identified from a mouse pituitary corticotroph cell line, AtT20, expression cDNA library. Subsequently, another member of the family (ActRIIB), which bears close homology to the original receptor, was cloned from Balb/c 3T3 cells, a mouse fibroblast cell line. More recently, several different TGFβ receptor superfamily members have been identified, including the TGFβ type I and II receptors, the MIS type I receptor, the bone morphogenic protein (BMP) type I receptor and TSR I, a type 1 receptor that can bind TGFβ or activin.

Both the type I and type II activin receptors consist of an extracellular domain, a short, hydrophobic, membrane-spanning domain, and a longer intracellular region that contains a serine/threonine–kinase domain. The activin receptor type I (ActRI) is widely expressed; the ActRII is also expressed in a wide range of tissues, including the gonads. ActRIIB, which is not encoded by the gene that encodes ActRII, has two regions close to the membrane-spanning domain which may be alternately spliced, possibly giving rise to four different mRNA transcripts (ActRIIB1–4). These four transcripts were isolated by means of the polymerase chain reaction from mouse testis, but it is not known whether they are expressed in other tissues. The intracellular, serine–kinase domains of the various type II receptors share a high degree (90%) of amino acid identity, but there is only 40% identity between type I and type II receptors in that domain. The extracellular domains of the type I and type II receptors have very little sequence similarity, apart from a cysteine cluster motif found in all superfamily members so far identified.

In addition to ActRI, a second type I receptor, TSR-I, which binds either activin or TGFβ, has been identified. This receptor is able to interact with either ActRII or the TGFβ type II receptors to bind the appropriate ligand. When it binds activin in combination with ActRII, however, it does not stimulate transcription of the ActRI/ActRII responsive reporter gene used in these studies, suggesting that it does not work via the same intracellular signalling pathway used by the ActRI and ActRII combination of receptors. At least three other TGFβ superfamily type I receptors have now been cloned; it is probable that there are further members of this receptor superfamily, and one or more of these may prove to be the elusive inhibin receptor(s).

At present, it is unclear which of these receptors is responsible for mediating the effect of activin on FSH in the pituitary gland. Oocyte injections of mRNA have been used to show that the type II receptor mediates mesoderm induction in *Xenopus*. Both type II receptors (ActRII and ActRIIB) have been demonstrated to bind activin and, with a 10-fold lower affinity, inhibin. Identification of the appropriate gonadotroph activin receptor, and the putative inhibin receptor, will help in the understanding of the interaction of these peptides in the regulation of the pituitary gonadotropins. It is probable that the expression of activin receptors is hormonally regulated, as are many other hormone receptors. Neither the second-messenger pathways through which these receptors act nor the possible transcription factors involved in gene regulation by these hormones have been identified. The mechanism of action of TGFβ has been much more extensively studied, with at least two different intracellular pathways proposed to mediate the effects of TGFβ: one pathway is believed to involve the phosphorylation of the retinoblastoma protein. However, the precise molecules involved in postreceptor events remain unidentified.

2.4 STEROIDOGENIC AND OTHER ENZYMES

2.4.1 Expression of Steroidogenic Enzymes

Steroids regulating the female reproductive axis originate primarily from the ovary and the placenta, with important contributions by the uterus, adrenal gland, and other peripheral organs such as fat cells and skin. During the ovarian cycle and pregnancy, steroid secretion by ovary and placenta occurs in a highly regulated and episodic fashion that is determined in part by the pituitary and placental gonadotropins. Changes in the levels and activities of the steroidogenic enzymes are major mechanisms determining these changes.

The synthesis of the steroids involves successive elimination of carbon atoms from the precursor, cholesterol, catalyzed by members of the cytochrome P450 gene family, namely, cholesterol side chain cleavage cytochrome (P450 (P450scc), 17α-hydroxylase (P45017α), and aromatase (P450arom). There are also 3β-, 17β-, and 20α-hydroxysteroid oxidoreductases, which catalyze the interconversion of some of the steroids as well as isomerization reactions, and 5α-reductase activity, which reduces testosterone to biologically active dihydrotestosterone (DHT). Not only the cellular expression of these enzymes but their electron donors have been extensively investigated in the female reproductive system; studies have focused on gene expression, protein synthesis, and activity assays.

A good example of the cellular compartmentalization of steroid synthesis is the "two cell, two gonadotropin" model of estrogen synthesis by the ovarian follicle. The follicle consists of an inner avascular granulosa cell layer within a basement membrane, surrounded by an outer vascular thecal cell layer. LH stimulates production of the precursor androgen by stimulating P45017α in the theca interna. The granulosa cells, which lack the P45017α, do have the P450arom which, under the influence of FSH, converts the thecal androgen to estrogen. Thus both cell types and both gonadotropins are crucial for estrogen production.

More recent work in this area has centered on the upstream regulation of the steroidogenic enzymes, and on the regulatory mechanisms of uptake and distribution of the precursor cholesterol such as the functions of sterol carrier protein 2.

2.4.2 Tissue Remodeling Enzymes

The ovary and the uterus are unique among organs in having regular cyclic remodeling of the tissues. In the ovary, this involves the processes of ovulation and formation and the regression of the corpus luteum (Table 2). In the uterus, there is the sloughing and regeneration of the endometrium at menstruation, tissue remodeling at implantation, preparation of the uterus and cervix for birth, and their reconstruction after birth (Table 3). A major part of this process involves degradation of the extracellular matrix (ECM) by MMPs. The ECM consists of either an interstitial, collagen-rich, ground substance housing the stromal–interstitial cells, or the basement membrane underlying epithelial and stromal cells. Each has a characteristic composition that determines the type of MMP responsible for its degradation. For example, MMP-1, previously known as tissue collagenase, breaks down interstitial ECM, whereas MMP-2 and -9, which are gelatinases, are specific for the basement membrane ECM, which contains collagen IV, laminin, and fibronectin. The MMPs are neutral zinc proteinases that are secreted as latent zymogens requiring proteolysis for activation. The in vivo mechanisms of activation are not known but are thought to involve specific proteases, including plasmin and some of the MMPs themselves, and an interaction with growth factors and cytokines such as tumor necrosis factor α (TNF-α), EGF, basic fibroblast growth factor (bFGF), TGF-β and interleukin-1 (IL-1) which are often found in association with the ECM. In addition to their regulation at the level of expression and activation, the activities of MMPs are inhibited by the formation of 1:1 complexes with tissue inhibitors of MMPs (TIMPs), and by α2-macroglobulin (α2-M).

The MMPs required to degrade interstitial and basement membrane ECM, and their regulatory elements, can be found in the uterus and the ovary. Potential functions of MMPs in the uterus include angiogenesis, implantation, endometrial remodeling and menstruation, activation of growth factors such as FGF and the colony-stimulating factors (CSF) by releasing them from the ECM, preparation of the cervix for birth, and collagenolysis in the myometrium after birth. In the ovary, the process of ovulation requires breakdown by MMPs of both interstitial (by MMP-1) and basement membrane ECM (by MMP-2 and MMP-9) to allow release of the egg from the follicle. MMPs are also likely to be involved in the extensive angiogenesis and tissue remodeling that accompanies the formation of the corpus luteum from the recently ruptured follicle.

2.5 TRANSCRIPTION FACTORS

2.5.1 Steroid Receptors

Steroid receptors are ligand-inducible transcription factors that interact with specific cis-acting enhancer sequences in target genes called hormone-responsive elements (HREs). They are one of a superfamily of DNA-binding transcription factors divided according to their DNA-binding domains into steroid receptors, thyroid and retinoic acid receptors, and orphan receptors (lacking known ligands). The intracellular steroid receptors are generally associated with heat shock proteins, and in the presence of ligand, they dissociate from the heat shock proteins, translocate to the nucleus, and undergo dimerization.

The estrogen (ER), progesterone (PR), and androgen (AR) receptors are of major interest in the female reproductive system. Each of these receptors has five or six important functional domains. The N-terminal A/B domains contain the cell-specific and constitutive transactivation (TAF-1) functions and the isoform-specific sequences. The C domain or DNA-binding domain is the most conserved region, containing invariant cysteine repeats that form two "zinc fingers," which are crucial for binding of the receptor to the DNA. The constitutive nuclear localization signal is found in the C to D region, which together with the hormone-binding E domain, is the site of interaction with the heat shock proteins. The E domain is also involved in hormone binding, nuclear translocalization, hormone-dependent transactivation, and dimerization. The F domain is found in ER but not PR.

As would be expected from the central role and multiple actions of estrogen in the female reproductive system, ERs have been localized in cells in the hypothalamus, pituitary, ovary, mammary gland, and uterus. The type of cell containing ER has been found to change according to the stage of differentiation or function, and in some cases such cells can be proximal to the cells eliciting the final response. For example, in the endometrium, epithelial–mesenchymal interactions involve an action of ER on the stromal cells, which then have a paracrine influence on the epithelial cells. Expression of ER is constitutive in some cases and can be up- and down-regulated by estrogen itself, and by FSH in the ovary.

PRs undergo both ligand-dependent and -independent phosphorylation, with the former essential for receptor activation. The PR protein exists in two forms, A and B, which arise by either alternative transcription or alternative translation initiation from an internal AUG; and it has been suggested that the A and B forms have different transactivating functions. Regulation of the PR gene is not well understood. The PR gene is transcriptionally activated by estrogen and inhibited by progesterone, but the mechanism(s) of these actions is not known. In target cells such as the uterine epithelium, PRs are under multifactorial control. Estrogen, progesterone, IGF, EGF, and agents that increase intracellular levels of cAMP can all modulate PR mRNA and protein. There is evidence for expression of PR mRNA and protein in all the reproductive tissues, although the cell expressing PR may vary, and it is not clear whether there are tissue-specific transcripts. The functional significance of PR in some of these cells is also unclear.

Table 3 Examples of Uterine Processes Possibly Under Local Control

Process	Factor(s) Involved
Endometrial cell proliferation	EGF, TGF-α, IGF-1, IGF-BP, IL-1, endothelin
Endometrial breakdown and repair	IL-1, TNF, MMPs, TIMPs
Endometrial bleeding and hemostasis	Prostaglandins, endothelin, MMPs, TIMPs, EDRF, VEGF
Angiogenesis	FGF, VEGF
Decidualization	Prostaglandins, RLX
Immune suppression	IFN, TNF, IL
Conceptus and placental growth	CSF-1, EGF
Implantation	LIF, GMCSF

2.5.2 CREB and Other Pituitary Transcription Factors

Choriocarcinoma cells have been used extensively to study the transcriptional regulation of the α-subunit gene of the gonadotropin hormones by cAMP in transient transfection studies. A region in the promoter of the human α-subunit gene has been identified which is capable of conferring cAMP responsiveness. This cAMP-responsive element (CRE) is an 18 base pair repeated palindromic sequence homologous to CREs found in other cAMP-responsive genes. Further studies confirmed the presence of a CRE between -132 and -99 bp, deletion of which resulted in a marked loss of basal transcriptional activity. The CRE identified in the α-subunit gene has been shown to bind the cyclic AMP response element–binding protein (CREB).

The pituitary gland represents an interesting developmental question, in that five distinct endocrine cell types arise from one cell lineage. The corticotrophs (which produce POMC) arise first, followed by the thyrotrophs (which produce TSH), gonadotrophs (LH and FSH), somatotrophs, which produce growth hormone (GH), and lactotropes, which produce prolactin. As discussed earlier, the α subunit of the glycoproteins is produced in both the gonadotrophs and the thyrotrophs. During development of the pituitary gland, these five cell types arise in a definite temporal pattern that must be regulated by tissue-specific transcription factors. Pit-1 is a pituitary-specific POU-domain transcription factor that is first detected at the time in fetal development coinciding with the start of GH gene transcription. In Snell and Jackson dwarf mice that have mutations in the gene for Pit-1, there is a complete absence of somatotrophs, lactotrophs, and thyrotrophs, indicating that Pit-1 is essential for the development of these cell lineages.

An equivalent transcription factor that is essential for the development of gonadotroph cells has not yet been characterized, but a putative gonadotroph-specific factor has been identified which binds to the α-subunit promoter and regulates its transcription. Expression of the α-subunit gene in thyrotrophs is determined by a different, as yet unidentified factor, which interacts with a separate region of the promoter. Placenta-specific expression of this gene in humans is primarily determined by the trophoblast specific element binding protein (TSEB) and a further trophoblast-specific protein, αACT. Elucidation of the transcription factors necessary for expression of LHβ and FSHβ and for development of the gonadotroph cell lineage remains an intensive area of research.

2.6   OTHER PEPTIDES

2.6.1 Endothelins

Endothelin was described originally as a product of endothelial cells that had a potent vasoconstrictor effect on vessels by means of an action on the smooth muscle cells. Subsequently its sites of production have expanded considerably to include many epithelial cells in the pituitary, ovary, endometrium, myometrium, and placenta. The endothelins are a family of three peptides (ET-1, ET-2, ET-3), encoded by three separate mammalian genes that are related to the gene for sarafotoxin in snakes. Each isoform is a 21 amino acid peptide, derived by posttranslational processing from a larger precursor peptide of around 200 amino acids. The endothelin-converting enzymes (designated ECE-1 and -2) that perform these cleavages have been cloned, but their specificities are still unclear. They may, together with neutral endopeptidase, be responsible for degradation of the ETs. The ETs act through one of three receptors (ETA, ETB, and ETC), each of which has been cloned. Each receptor has a different affinity profile for the ETs, with ETA binding preferentially to ET-1 and predominating on smooth muscle cells of blood vessels, whereas ETB binds both ET-1 and ET-2 and has a wider cellular distribution. It has been suggested that the different receptor subtypes are coupled to different transduction systems, since such a relationship could explain the pluripotent actions of ETs, which now include mitogenic and hormone-releasing activities as well as vasoconstriction.

2.6.2 Oxytocin

The uterotonic and milk-ejecting activities of the posterior pituitary and corpus luteum of the ovary were first described at the beginning of this century, but it was not until the early 1950s that the activity was identified as oxytocin, a cyclic nonapeptide. It was discovered that oxytocin is released by exocytosis of secretory granules, which also contain a carrier protein called neurophysin 1, which with oxytocin forms part of a larger precursor molecule. The single-copy genes in all species are similar in structure and contain three exons, with exon A encoding the signal peptide, oxytocin, and the amino-terminal region of neurophysin (amino acids 1–9) and exons B (amino acids 10–76) and C (amino acids 77–93) encoding the remainder of neurophysin. Complex posttranslational processing of the precursor molecule by exo- and endopeptidases, a carboxypeptidase B-like enzyme, and an amidation enzyme gives rise to neurophysin and several pre forms of oxytocin, which associate with the neurophysin in the secretory granule before release from the cell.

The oxytocin gene is expressed in the cell bodies of the supraoptic and paraventricular nuclei of the hypothalamus of all species, and the precursor is processed en route along the axons of the median eminence and pituitary stalk to the storage granules of the posterior pituitary. Oxytocin released from this source is involved in the milk ejection reflex and the stimulation of uterine contractions. Expression of the oxytocin gene in the ovary, particularly the corpus luteum, is well documented in ruminants, where it is believed to play a role in inducing the release of the uterine luteolysin prostaglandin F2α, which is responsible for regression of the corpus luteum of an infertile cycle. Clear understanding is lacking, however, of the level of expression and production of oxytocin in the ovary of other species, particularly humans, and its physiological role.

A single class of receptor sites for oxytocin has been characterized on the principal target cells, smooth muscle of the reproductive tract and myoepithelial cells of the mammary gland, as well as on the endometrium of the sheep and human. The human oxytocin receptor gene cDNA has been cloned. It encodes a 388 amino acid polypeptide with seven transmembrane domains typical of G-protein-coupled receptors and is a member of the arginine–vasopressin/oxytocin family of receptors. The mRNA was found in two sizes, 3.6 kb in breast and 4.4 kb in ovary, uterine endometrium, and myometrium. The levels of oxytocin receptor mRNA are high in the myometrium, the maternally derived decidual cells and the chorionic trophoblast cells at term, consistent with its role in human labor. The oxytocin receptor mRNA is also expressed in the sheep uterus, particularly in the endometrium, where oxytocin is responsible for releasing prostaglandin F2α. In all these cases, expression of the oxytocin receptor mRNA is coincident with detection of the receptor protein and binding activity.

2.6.3 Zona Pellucida Proteins

The zona pellucida or egg surface coat of the mouse consists almost entirely of three glycoproteins called ZP1, -2, and -3. Mouse ZP2 (120 kDa) and ZP3 (83 kDa) are present as heterodimers along the filaments that make up the extracellular coat of the egg, and the filaments are interconnected by ZP1 (200 kDa), a dimer of identical polypeptides held together by disulfide bonds. ZP3 has been identified as the primary sperm receptor on the mouse egg. There is a 22 amino acid signal sequence, and the mature protein contains 402 amino acids rich in proline, serine, and threonine residues, as well as N- and O-linked glycosylation sites that make up the final molecular weight. ZP3 is encoded by a single-copy 8.6 kb gene containing eight exons; it yields a 1.5 kb mRNA, which is expressed only in the oocyte. Maximum expression occurs during oocyte growth, with levels falling to undetectable after fertilization.

Sperm binding to mouse eggs is an example of carbohydrate-mediated cellular adhesion. Once bound, the sperm undergo the acrosome reaction, which removes their egg-binding proteins; to remain bound, the inner acrosomal membrane of the sperm binds to ZP2, which acts as a secondary receptor and allows the sperm to penetrate the zona pellucida and fuse with the egg plasma membrane.

Homologues of the mouse zona proteins are found in other species including human, pig, hamster, rabbit, and fox, and it is likely that they play the same role.

2.6.4 Leukemia Inhibitory Factor

Leukemia inhibitory factor (LIF) is a 45–56 kDa glycoprotein that has multiple activities as a cytokine, including the acute phase response in hepatocytes, regulation of differentiation and proliferation of hematopoietic cell lines, remodeling bone, and inhibiting differentiation of embryonic stem cells. LIF mRNA and protein are also detectable in the endometrial epithelial cells of mice, sheep, and women. In mice, its expression is independent of the presence of a blastocyst and occurs transiently in response to a rise in E2 on day 4 of pregnancy, just prior to implantation. This transient expression of LIF was shown to be essential for implantation using transgenic mice lacking LIF expression. These animals conceive, but the blastocysts fail to implant and develop, although the blastocysts remain viable and will develop to term if transferred to a wild-type female. It remains to be determined whether LIF has a similar role in other species.

3   PATHOPHYSIOLOGICAL CONSIDERATIONS

3.1   GENETIC DEFECTS

A number of genetic defects have been identified which have helped to define the role of several hormones and pathways in the female reproductive system. Some examples are given in this section.

Two known genetic diseases have the phenotype of loss of GnRH gene expression. The *hpg* mouse is infertile as a result of a lack of GnRH secretion from the hypothalamus due to the deletion of the GnRH gene. Transgenic insertion of a normal GnRH gene can restore reproductive function in these mice. In humans, however, individuals exhibiting a similar phenotype, hypogonadotropic hypogonadism, do not secrete GnRH despite having a normal GnRH gene. These individuals have Kallman's syndrome, in which the infertility is associated with anosmia. The defect in Kallman's patients has been shown to be an X-linked defect that arrests the migration of GnRH neurons from the olfactory placode to the hypothalamus during fetal development, such that in the adult there are no GnRH neurons in the hypothalamus. The KAL gene is located at Xp22.3 and encodes a neural adhesion molecule with structural similarities to neural cell adhesion molecule (NCAM).

In familial male precocious puberty, an autosomal dominant, male-limited, gonadotropin-independent disorder, affected males begin puberty as young as age 4. The genetic defect leading to this syndrome has been shown to be due to a single base change in the LH receptor that results in an amino acid substitution in the sixth transmembrane domain, which in turn leads to constitutive activation of the receptor. Thus cyclic AMP production is increased in the absence of LH, triggering Leydig cell activity and testosterone production. A similar syndrome is testotoxicosis, which also results in precocious puberty. The defect in this case is in the LH receptor linked G protein Gsα, which activates adenylyl cyclase. A single base mutation again results in an amino acid substitution, causing constitutively activated Gsα, which elevates adenylyl cyclase and thus cAMP in the testis. This mutation, however also results in resistance to hormones that activate cAMP (parathyroid hormone and TSH), and these two disparate effects are attributed to the mutated Gsα protein, which is unstable at 37°C but retains function at the lower temperature of the testis. Both these syndromes show a male-limited pattern of inheritance, probably because LH is sufficient to trigger production of testosterone in males, but both LH and FSH are necessary to trigger steroidogenesis in the ovary. Thus, constitutive activation of the LH receptor alone would not be sufficient to trigger precocious puberty in females.

McCune–Albright syndrome is characterized by polyostotic fibrous dysplasia, café au lait pigmentation of the skin, and multiple endocrinopathies, including precocious sexual development. The genetic defect in this syndrome is an activating mutation in the Gsα protein. This syndrome exhibits sporadic occurrence and is believed to be due to a dominant somatic mutation that occurs early in development, resulting in precocious puberty in both males and females, with females having luteinized follicular cells in the ovary at a very young age.

The importance of estrogen for normal development has been demonstrated in two patients with genetic defects. In one case, in a karyotypic female with pseudohermaphroditism, a null mutation in the aromatase P-450 cytochrome gene resulted in an effective lack of estrogen production, demonstrating that estrogen is essential in the female for normal development of secondary sexual characteristics, including a pubertal growth spurt. In the second case, a male was homozygous for a point mutation giving rise to a premature stop codon and consequently had no estrogen receptor. This patient had osteoporosis, unfused epiphyses, and continuing linear growth into adulthood. These cases demonstrate that estrogen is important in both sexes for normal skeletal growth and development and are supported by studies of transgenic mice lacking functional estrogen receptor. Both sexes of mice were infertile and showed a variety of phenotypic abnormalities of the gonads, mammary glands, reproductive tract, and skeletal system.

3.2   FSH ISOFORMS AND INFERTILITY

There is now substantial data showing that pituitary and serum LH and FSH exist as structurally heterogeneous isoforms, not as uniform structures. Variation in carbohydrate structure, particularly sialic acid content, accounts for most of this heterogeneity. Other factors (e.g., sulfation, phosphorylation, other modifications to the carbohydrate structure) are possible.

The population of isoforms can change according to the sex and endocrine status of the animal, involving mechanisms that are not understood. For example, human pituitary FSH isoforms are either larger and/or more acidic, with a longer half-life in the male than in the female. The physiological significance of changes in the relative proportions of FSH isoforms is more readily understood in terms of their differences in clearance rates than any difference in their intrinsic (in vitro) biological activity. It is not clear whether differences in the isoform profiles of gonadotropins in individuals are reflected in the subjects' fertility or infertility.

3.3   Tumors and Inhibin

After the isolation and characterization of inhibin and development of radioimmunoassays, there have been several reports of elevated levels of inhibin immunoactivity in serum of patients with granulosa cell tumors of the ovary. Because inhibin is normally undetectable in serum of postmenopausal women, the potential value of the peptide as a tumor marker is particularly significant in women without endogenous ovarian function. Reports of the utility of inhibin as a tumor marker indicate that serum inhibin is elevated in more than 80% of postmenopausal patients with mucinous carcinomas of the ovary, in 17% with serous carcinomas, and in 17% with clear cell carcinomas of the ovary. The nature of the inhibin product appears to vary with the type of tumor. Serum FSH levels vary inversely with inhibin levels in granulosa tumors, consistent with production of biologically active, dimeric inhibin. In mucinous tumors, on the other hand, FSH levels were generally high despite elevated inhibin, raising the possibility that in these tumors the biologically inactive inhibin α-subunit precursor is the major product. Alternatively, the mucinous tumors may overproduce activin, a dimer of the β subunits, which has FSH-stimulating activity. These data also suggest that expression of the three inhibin genes, α, βA, and βB, and the subsequent posttranslational processing and assembly of the subunits, are regulated differently in the different tumors.

It is of interest in this regard that 70 of 71 transgenic mice with a deletion of the α-inhibin gene developed gonadal tumors, leading the authors to conclude that inhibin α may be a tumor suppressor gene. An alternative hypothesis is that in the absence of inhibin α, activin is over-expressed and could be the agent responsible for tumor development. Whatever the explanation, the data are difficult to reconcile with the reports of elevated inhibin levels in patients with ovarian cancer, necessitating new concepts regarding the pathogenesis of ovarian tumors and their treatment.

Inhibin is also a product of the placenta, leading to the suggestion that it may be a useful marker of trophoblastic diseases, particularly hydatidiform mole, as an adjunct to the measurement of intact chorionic gonadotropin. Preliminary data support the potential of inhibin measurements for detecting molar pregnancies, where levels are elevated in comparison to those at the appropriate stage of normal pregnancy, but more data are needed.

4   PERSPECTIVES

Significant advances have been made in our understanding of the regulation of the female reproductive system in recent years, primarily in the areas of autocrine and paracrine regulation and in the receptor and second-messenger systems, which subserve the actions of hormones and local regulators. Nevertheless, caution should be exercised in interpreting the available information. The existence of some of the putative local regulators is based on detection of expression of the mRNA and protein in situ or in vitro. Their production and physiological role in vivo is less certain in many cases. Attempts to rectify this deficiency (e.g., by using mice with gene deletions) have not always provided a clear indication of physiological roles, partly because of the possible substitution or compensation of one regulator for another in these circumstances.

The second-messenger systems that serve the pleiotropic actions of many of the hormones and regulators are not well understood, particularly where one hormone can have multiple effects on the same cell. There is still much to be learned about the transcription factors that link the hormone action to the gene, and the mechanisms by which these factors regulate the genes. We now know from the example of the TGFβ receptors that association of different receptor subtypes can lead to different affinities for the ligands, which is another point at which regulation could occur. Until recently, there has been a tendency to overlook the importance of binding proteins and metabolizing enzymes in determining the local concentrations of *biologically active* hormones or factors; the molecular mechanisms controlling these substances are now under active investigation.

In addition to the reservations just noted, there are a number of key processes in the female reproductive system that we still do not understand. They include the following:

- The nature and regulation of the GnRH pulse generator
- The balance between endocrine and paracrine regulation of cells
- Control of reinitiation of growth and of atresia of ovarian follicles
- Mechanisms controlling menstruation and implantation
- Immune responses to pregnancy

Molecular approaches will play a major part of solving these inadequacies.

See also Endocrinology, Molecular; Male Reproductive System, Molecular Biology of; Receptor Biochemistry; Sex Determination.

Bibliography

Adashi, E. Y., and Leung, P.C.K., Eds. (1993) *The Ovary*. Raven Press, New York.

Findlay, J. K., Ed. (1994) *Molecular Biology of the Female Reproductive System*. Academic Press, San Diego, CA.

Knobil, E., and Neill, J. D., Eds. (1994) *The Physiology of Reproduction*, Vols. 1 and 2, 2nd ed. Raven Press, New York.

Mercer, J. E., and Chin, W. W. (1995) Regulation of pituitary gonadotropin gene expression. In *Oxford Reviews of Reproductive Biology*, Vol. 17, pp. 159–203. Oxford University Press, Oxford.

Seppala, M., Ed. (1991) *Factors of Importance for Implantation*. Baillieres Clinical Obstetrics & Gynaecology, Vol. 5(1). Baillere & Tindall, London.

Fish, Transgenic: *see* Transgenic Fish.

Fluorescence In Situ Hybridization, Gene Mapping by: *see* Gene Mapping by Fluorescence In Situ Hybridization.

Fluorescence Spectroscopy of Biomolecules

Joseph R. Lakowicz

Key Words

Anisotropy or Polarization   A measure of the rate of fluorophore or macromolecule rotational diffusion.

Fluorescence   The emission of light from molecules in excited electronic states.

Fluorescence Lifetime   The average amount of time between light absorption and emission, typically nanoseconds (10^{-9} s).

Fluorescence Resonance Energy Transfer (FRET)   The transfer of the excited state energy from the initially excited (donor) to an acceptor molecule.

Fluorophores   Fluorescent molecules, which can be intrinsic, or added to biological molecules; also called *probes*.

Fluorescence methods are widely used in the biological sciences, including biochemistry, biophysics, and cell biology, to name a few. Fluorescence is the emission of light from molecules in excited electronic states. This phenomenon is often observed in everyday life as the green or orange glow from new antifreeze or from wall posters under ultraviolet or "black" light. Fluorescence can be detected with high sensitivity for the same reason that stars are more easily observed at night—because the light is seen against a dark background.

Many biological molecules display intrinsic fluorescence, or can be labeled with extrinsic fluorophores. The emission of these probes can be used to reveal the structure and function of biological macromolecules. Additionally, fluorescence is used in DNA sequencing, for cell identification and sorting in flow cytometry, in clinical chemistry, and to reveal the localization and movement of intracellular substances by means of fluorescence microscopy. Because of

the sensitivity of fluorescence detection, and the expense and difficulties of handling radiometric substances, there is a continuing development of medical tests based on the phenomenon of fluorescence. These tests include the enzyme-linked immunoassays (ELISA) and fluorescence polarization immunoassays.

The phenomenon of fluorescence emission occurs in the nanosecond time scale. Time-resolved measurements are performed because these data provide additional information about the structure, dynamics, and functional properties of proteins and membranes. This information is available because the nanosecond lifetimes of fluorescence are comparable to the time scale of biochemical phenomena, such as the rotational diffusion of proteins or the motions of probe molecules within cell membranes. The nanosecond time scale of fluorescence requires pulsed lasers, high speed detection, and sophisticated electronics. Advances in the technology of laser light sources and detectors is providing increased resolution of biological structures. At present, time-resolved fluorescence is primarily a research tool in biochemistry and biophysics. However, medical testing based on time-resolved measurements appears to be rapidly migrating to the clinical laboratory, to the doctor's office, and even to home health care.

1   PHENOMENON OF FLUORESCENCE

1.1   TYPICAL FLUORESCENT SUBSTANCES

Fluorescence molecules are typically aromatic species that display light absorption in the ultraviolet (250–400 nm) to visible (400–700 nm) regions of the spectrum. A wide variety of fluorescent substances are known. Some of these occur naturally, such as tryptophan in proteins (Figure 1, top). The indole groups of tryptophan residues are the dominant source of UV absorbance and emission in proteins. However, tyrosine also contributes to the absorption and emission of proteins. Emission from phenylalanine is observed only when the sample protein lacks both tyrosine and tryptophan residues, which is a rare occurrence.

In the case of biological membranes there are few natural (intrinsic) fluorophores. Hence it is common to add extrinsic fluorescence labels, usually called probes, to obtain a useful signal. The probes can simply partition into the membranes, as occurs for diphenylhydantoin (DPH), or they can resemble the lipids, like a rhodamine (Rh) fatty acid (Figure 1, middle). While the nitrogenous bases of DNA absorb UV light, they do not display useful fluorescence. However, a wide variety of dyes bind spontaneously to DNA, such as acridines, ethidium bromide, and other planar cationic species. For this reason, staining of cells with dyes that bind to DNA has been used to visualize and sort the chromosomes. There are a few naturally fluorescent bases, such as the Y_t base, which occurs in the anticodon region of a particular tRNA (Figure 1, bottom).

1.2   INTRINSIC AND EXTRINSIC FLUOROPHORES

A wide variety of other substances display significant fluorescence (Figure 2). Among biological molecules, one can observe fluorescence from reduced nicotinamide adenine dinucleotide (NADH) and from oxidized flavins (FAD, the adenine dinucleotide, and FMN, the mononucleotide), as well as from chlorophyll in some instances. In addition to natural fluorophores, there exist numerous other fluorescent species, including the polynuclear aromatic hydrocarbons (PAH) such as perylene (Figure 2), which are responsi-

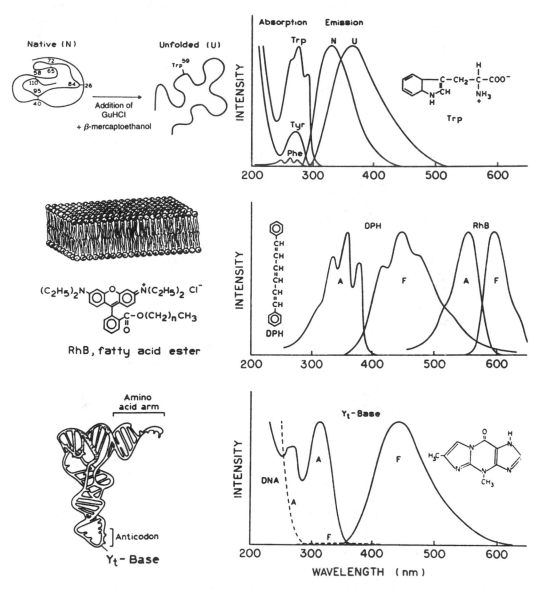

Figure 1.   Absorption and emission spectra of biomolecules: top, tryptophan emission from proteins; middle, spectra of extrinsic membrane probes; bottom, spectra of naturally occurring fluorescent base, Yt base. DNA itself (dashed curve) displays very weak emission.

ble for the blue glow occasionally seen from gasoline. One of the oldest known fluorophores is quinine (Figure 2), which was studied in 1852 by G. G. Stokes, who appears to have been the first to recognize that the "color" of the emission was at longer wavelengths compared to the incident violet–ultraviolet light. The blue glow from the surface of a glass of tonic water is due to this natural quinine fluorophore (Figure 2).

Occasionally, a species of interest is not fluorescent, or is not fluorescent in a convenient region of the UV–visible spectrum. A wide variety of extrinsic probes have been developed for labeling the macromolecules in such cases. Two of the most widely used probes, dansyl chloride (DNS-Cl, where DNS stands for dimethyamino-naphthylsulfonyl) and fluorescein isothiocyanate (FITC), are shown in Figure 2. These probes react with the free amino groups of proteins, resulting in proteins which fluorescence at blue (DNS) or green (FITC) wavelengths.

Another approach to obtaining the desired fluorescent signal from the molecule of interest is to synthesize a chemical analogue that displays both the chemical properties of the parent molecules and useful fluorescence. For example, ATP is essentially nonfluorescent. However, it is possible to create analogues that are fluorescent. The etheno-ATP (ε-ATP) analogue is fluorescent and might be expected to bind to kinases, but its base-pairing properties are obviously compromised. The benzo-AMP derivative is also fluorescent, and one can expect it to display the same base pairing as AMP (Figure 2), but it may be too large to fit into some binding sites or in a DNA helix. A useful fluorescent probe is one that displays a high intensity, is stable during continued illumination, and does not substantially perturb the biomolecule being studied.

1.3   FLUORESCENT INDICATORS

Another class of fluorophores can be regarded as fluorescent indicators. For instance, measurements and imaging of intracellular cal-

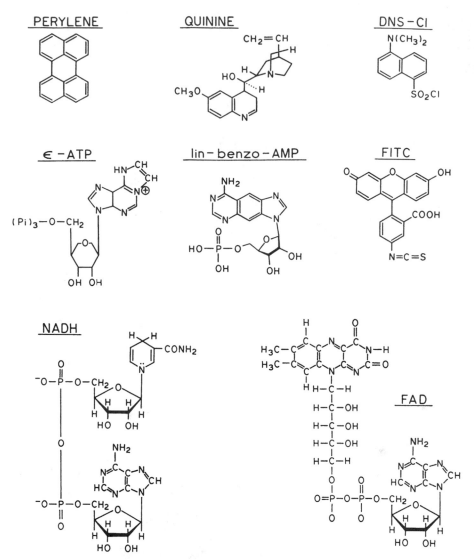

Figure 2. Typical fluorophores.

cium have elicited widespread interest because of the role of Ca^{2+} as a second messenger in cell signaling and activation. However, for all practical purposes, Ca^{2+} is spectroscopically silent. The concentration of intracellular Ca^{2+}, or its local concentrations, cannot be determined by cell disruption and subsequent analysis because the local concentrations are altered and the total concentration is distorted by the release of intracellular Ca^{2+} stores. Hence, it is desirable to determine the Ca^{2+} levels by fluorescence detection or fluorescence microscopy.

Fortunately, it has been possible to develop fluorescent indicators that change their intensity, absorption, or emission spectra in response to Ca^{2+} or other analytes (Figure 3). The most widely used Ca^{2+} probe is Fura-2, which displays a change in its absorption spectrum in response to binding Ca^{2+}. Such spectral shifts are valuable because the Ca^{2+} concentration in a sample, or in a microscopic image, can be determined from the ratio of fluorescence intensities observed at two excitation wavelengths. For example, in the case of Fura-2 the two excitation wavelengths would be near 340 and 380 nm, where the absorption coefficient is different from the Ca^{2+}-free and Ca^{2+}-bound forms of the dye. Such wavelength–ratiometric

measurements are essential in fluorescence microscopy when it is not possible to know the probe concentration at each point in the cell. The indicators can be microinjected into the cells, or placed more easily by incubation of the cells with the acetoxymethyl ester (AM) form of the dyes. These neutral forms of the dye diffuse across the membranes and upon hydrolysis, by become trapped in the cell by intracellular esterases.

A disadvantage of Fura-2 as a Ca^{2+} indicator is that it requires UV excitation, which results in background fluorescence from biological samples due to NADH, FAD, and UBF, the ubiquitous blue fluorescence, observed upon illumination of most biological samples with ultraviolet light. Consequently, there is interest in longer wavelength Ca^{2+} probes, such as CaGreen (Figure 3), which is a conjugate of a fluorescein derivative and the Ca^{2+}-specific chelating group BAPTA. Since, however, this particular probe does not display a spectral shift, wavelength–ratiometric measurements are not possible. The absence of a visible wavelength Ca^{2+} probe reveals two important areas of research: the development of improved fluorophores and the use of time-resolved methods in chemical sensing and imaging (Section 4.2).

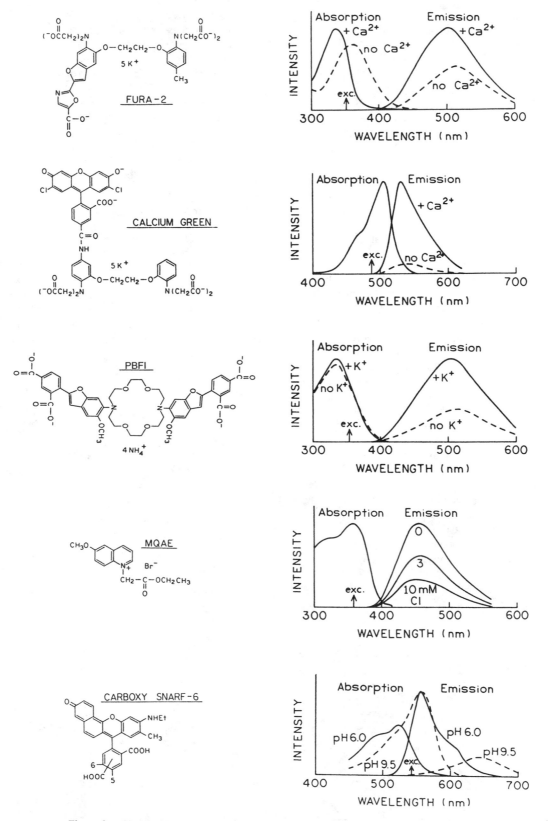

Figure 3. Chemical structures and absorption and emission spectra of fluorescent indicators.

Fluorescence indicators have also been developed for Mg^{2+}, Na^+, K^+, H^+ and Cl^-, some of which are shown in Figure 3. In the case of the K^+ sensor PBFI, the aza-crown group serves to specifically bind the K^+ ion. In the case of SNARF-6, the probe displays different absorption and emission spectra in the protonated and deprotonated states. The sensing mechanism of the chloride probe MQAE is distinct from that of the Ca^{2+}, K^+, and H^+ probes that bind to the analyte. In the case of MQAE, the probe is quenched by diffusive encounters with Cl^-, which results in a decrease of MQAE intensity in proportion to the Cl^- concentration. The Cl^- is acting as a collisional quencher of fluorescence (Section 2.2).

2   MOLECULAR INFORMATION FROM FLUORESCENCE

2.1   EMISSION SPECTRA AND THE STOKES SHIFT

The most dramatic aspect of fluorescence is its occurrence at wavelengths longer than absorption, as seen for all the fluorophores in Figures 1–3. These Stokes shifts, which are most dramatic for polar fluorophores in polar solvents, are due to interactions between the fluorophore and its immediate environment. The indole group of tryptophan residues in proteins is one such solvent-sensitive fluorophore, and the emission spectra of indole can reveal the location of tryptophan residues in proteins. The emission from an exposed surface residue will occur at longer wavelengths than for a tryptophan residue in the protein's interior. This phenomena is illustrated in Figure 1 (top), which shows a shift in the spectrum of tryptophan residue upon unfolding of a protein and the subsequent exposure of the tryptophan residue to the aqueous phase. Prior to unfolding, the residue is shielded from the solvent by the folded protein.

Because emission spectra are sensitive to the fluorophore's environment, these spectra of extrinsic probes are often used to determine a probe's location on a macromolecule. For example, one of the widely used probes for such studies is toluidinylnaphthalenesulfonate (TNS: Figure 4), which displays the additional favorable property of being very weakly fluorescent in water. Weak fluorescence in water and strong fluorescence when bound to a biomolecule is a combination of properties shared by other widely used probes, including the DNA strain ethidium bromide. Upon the addition of apomyoglobin to a solution of TNS, there is a large

increase in fluorescence intensity, as well as a shift of the emission spectrum to shorter wavelengths. This probe also binds to membranes (Figure 4). The emission spectrum of TNS bound to model membranes of dimyristoyl-L-α-phosphatidyl-choline (DMPC). This result reflects the nonpolar character of the heme-binding site of apomyoglobin to which TNS binds. The TNS-binding sites on the surface of the membrane are more polar, as seen from the longer wavelength TNS emission. From the emission spectrum, one can judge that TNS binds to the polar head group region of the membranes, rather than to the nonpolar acyl side chain region. Hence, the emission spectra of solvent-sensitive fluorophores provide information on the location of the binding sites on the macromolecules.

2.2   QUENCHING OF FLUORESCENCE

A wide variety of small molecules or ions act as quenchers of fluorescence; that is, they decrease the intensity of the emission. These substances include iodide (I^-), oxygen, chlorinated hydrocarbons, amines, and disulfide groups, to name a few. The accessibility of fluorophores to such quenchers can be used to determine the location of probes on macromolecules, or the porosity of proteins and membranes to the quenchers. This concept is illustrated in Figure 5, which shows the intensity of a protein- or membrane-bound fluo-

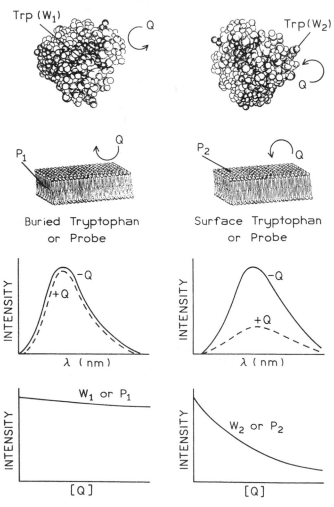

Figure 5.   Accessibility of fluorophores to quenchers.

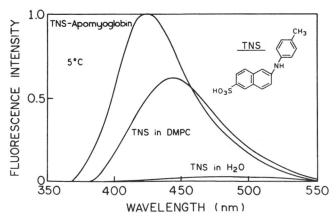

Figure 4.   Emission spectra of TNS in water, bound to apomyoglobin, and bound to lipid vesicles.

rophore in the presence of a dissolved water-soluble quencher [Q]. The emission intensity of a tryptophan on the protein's surface (W$_2$), or on surface of a cell membrane (P$_2$, right), will be decreased in the presence of a water-soluble quencher. The intensity of a buried tryptophan residue (W$_1$) or of a probe in the membrane interior (P$_1$) will be less affected by the dissolved quencher (left). Alternatively, one can add lipid-soluble quenchers, such as brominated fatty acids, to study the interior acyl side chain region of membranes.

2.3   FLUORESCENCE POLARIZATION OR ANISOTROPY

Fluorescence probes usually remain in the excited state for 1–100 ns, a duration called the fluorescence lifetime. Because rotational diffusion of proteins also occurs in 1–100 ns, the time scale of these fluorescence lifetimes is favorable for studies of the associative and/or rotational behavior of macromolecules. Measurements of fluorescence polarization (or anisotropy) allow one to measure, as well, the rotational motions of proteins in dilute (physiologically relevant) solutions. This possibility exists because of the polarization properties of light, and the dependence of light absorption on the alignment of the fluorophores with the electric vector of the incident light. If a sample is illuminated with vertically polarized light (Figure 6), the emission can also be polarized. However, rotational motions of the protein during the lifetime of the excited state can randomize the excited state population. Faster rotational diffusion results in lower polarization of the emitted light.

One reason for the widespread use of polarization measurements is the ease with which these absolute values can be measured and

compared between laboratories. The polarization P and anisotropy r are obtained from the intensity values measured through a polarized oriented parallel $I_\parallel$ or perpendicular $I_\perp$ to the vertically polarized excitation (Figure 6):

$$P = \frac{I_\parallel - I_\perp}{I_\parallel + I_\perp} \tag{1}$$

$$r = \frac{I_\parallel - I_\perp}{I_\parallel + 2I_\perp} \tag{2}$$

Both values are in widespread use, but P was used first historically, whereas r is a more "theoretically" correct value.

The phenomena of fluorescence polarization can be used to measure the apparent volume (or molecular weight) of proteins. This measurement is possible because larger proteins rotate more slowly. Hence, if a protein binds to another protein, the rotational rate decreases, and the anisotropy(s) increases (Figure 6). The rotational rate of a molecule is often described by its rotational correlation time θ. The P and r values are determined by the values of θ and the fluorescence lifetime τ according to

$$\left(\frac{1}{p} - \frac{1}{3}\right) = \left(\frac{1}{p_0} - \frac{1}{3}\right) [1 + (\tau/0)] \tag{3}$$

$$\frac{r_0}{r} = [1 + (\tau/0)] = 1 + \frac{RT}{\eta V} \tau \tag{4}$$

In equation (4), we see that the anisotropy depends on the volume of the proteins V; R is the gas constant, η the viscosity, and T the temperature. The mysterious values of P_0 and r_0 are simply a characteristic of the fluorophore that represents the polarization or anisotropy values without any rotational motion.

Fluorescence polarization measurements have also been used to determine the apparent viscosity of the side chain region (center) of membranes. Such measurements of microviscosity are typically performed using a hydrophobic probe like DPH (Figure 1), which partitions into the membrane. Fortunately, DPH does not fluoresce in the aqueous phase, simplifying the measurements inasmuch as all the detected light is from the membrane-bound probe. The apparent microviscosity of the membrane is determined by comparing the polarization of the probe measured in the membrane with that observed in solutions of known viscosity.

2.4   FLUORESCENCE RESONANCE ENERGY TRANSFER

Fluorescence resonance energy transfer (FRET) provides an opportunity to measure the distances between sites on macromolecules. The phenomenon of FRET is a predictable through-space interaction in which the excited state energy is transferred from the donor to an acceptor. This transfer does not involve an intermediate photon, but rather is a coupling of dipoles in which the energy is transferred over a characteristic distance, called the Förster distance R_0. These distances range from 15 to 60 Å, which is comparable to the diameter of many proteins and to the thicknesses of membranes. The efficiency of energy transfer depends strongly in the distance R between donor (D) and acceptor (A):

$$E = \frac{R_0^6}{R^6 + R_0^6} \tag{5}$$

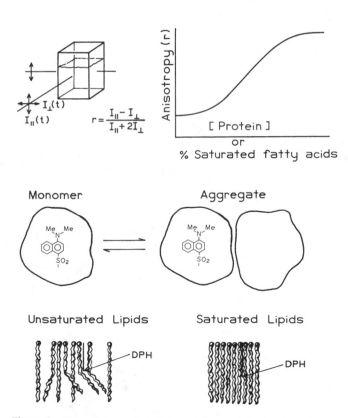

Figure 6. Fluorescence polarization, protein association, and membrane microviscosity.

Hence, if the efficiency of transfer is measured, and the Förster distance is known, FRET can be used as a molecular ruler to calculate the donor-to-acceptor distance R.

FRET allows determination of molecular distances because the Förster distances can be reliably calculated from the absorption spectrum of the acceptor and the emission spectrum of the donor. This spectral overlap (Figure 7) is referred to as the overlap integral. The use of FRET to study protein conformation is illustrated in Figure 7. In this case the protein can exist in a random coil state, or as an α-helix. Transfer occurs from the single tryptophan residue, the donor D, to a dansyl residue, the acceptor A. The dansyl group (Figure 2) is a frequently used extrinsic label for proteins. The use of modern time-resolved instrumentation allows not only distance measurements, but resolution of the range of donor-to-acceptor distances. The data can be used to show that the protein exists alternately in a single conformation α-helical state or, when in the random coil state, with a range of donor-to-acceptor distances. Other examples of the use of energy transfer include measurements of distances between sites on multidomain proteins, such as the distance between the antigen-binding sites on immunoglobulins, or the distance between a site on a membrane-bound protein and the surface of the membrane.

3   TIME-RESOLVED FLUORESCENCE

The examples described in Section 2 were steady state measurements, meaning that the data were collected while the samples were continuously illuminated. Such measurements in fact represent an

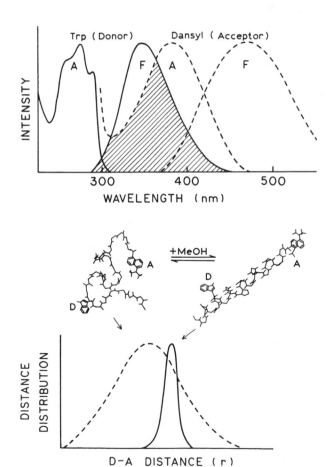

Figure 7.   Distance distribution in a random coil and α-helical protein.

average of the time-dependent precesses that occur during the lifetime of the excited state. There is considerably more information available if one measures the time-dependent decays following illumination with brief pulses of light. Such data are often referred to as the impulse response functions. The time-resolved measurements are technologically challenging because both the light pulses and the measurements of the emission must be more rapid than the nanosecond lifetimes.

3.1   Jablonski Diagram for Fluorescence

The information content of time-resolved fluorescence measurement can be understood by examining the Jablonski diagram (Figure 8), which illustrates the phenomena that can occur between absorption and emission, and thus can affect the time-dependent emission. Absorption of a photon is nearly instantaneous, occurring in 10^{-15} s, or the time it takes a photon to travel the length of a photon. The fluorophore is now in the first singlet state, typically with vibrational energy from the excess energy of the absorbed photon. The vibrational energy is lost very quickly (10^{-12} s), which is not significant in comparison to the typical nanosecond lifetimes. The fluorophore is now in a metastable state, which persists for nanoseconds. A wide variety of phenomena can occur on the nanosecond time scale, altering the emission.

As described in Section 2.1, the emission is typically shifted to longer wavelengths (compared to the absorption). These Stokes shifts occur because the excited fluorophores typically have increased charge separation in the excited state. In polar solvents, the solvent molecules can reorient around the excited state, decreasing the energy level. In fluid solvents, this process occurs in less than 0.1 ns and is often complete prior to emission.

The relaxed fluorophores now remain in the excited state for several nanoseconds. The intrinsic or radiative lifetime of the fluorophore τ_R is given by the inverse of the emission rate Γ

$$\tau_R = \frac{1}{\Gamma} \qquad (6)$$

The values of τ_R or Γ can be estimated from the absorption spectrum of the fluorophore. Typically, molecules with higher absorption coefficients display shorter lifetimes because the same transition (S_0 to S_1 or S_1 to S_0) is involved in absorption and emission. Hence, a higher probability of absorption indicates a larger value of Γ and thus a shorter radiation lifetime.

In practice, the measured lifetimes are typically shorter than the τ_R values because other less understood rate processes (Σk_i in Figure 8) also allow return to the ground state. In the absence of additional quenching processes, the measured lifetime τ_0 is

$$\tau_0 = \frac{1}{\Gamma + \Sigma k_i} \qquad (7)$$

These additional return paths to the ground state shorten the measured lifetime.

An often used quantity is the quantum yield of a fluorophore, which is the ratio of photons emitted to Q, the number of absorbed photons. It is easy to see from equations (6) and (7) that the quantum yield is given by

$$Q = \frac{\Gamma}{\Gamma + \Sigma k_i} \qquad (8)$$

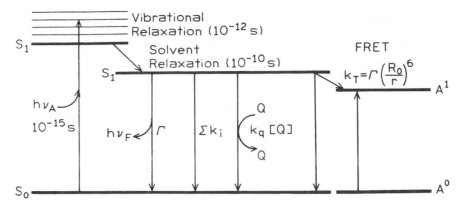

Figure 8. Jablonski diagram for the processes of absorption, solvent relaxation, emission, quenching, and/or energy transfer.

This ratio expresses the rate of emission divided by the total rate of return to the ground state. The presence of additional pathways results in a decreased quantum yield.

The concept of quantum yield is perhaps best illustrated by consideration of a tryptophan residue in a folded protein that is immediately adjacent to a group that quenches its fluorescence, that is, a larger value for k_i. The quantum yield of this residue will be low. Upon unfolding of the protein, and movement of the tryptophan residue away from the quencher, the value of k_i will decrease and the quantum yield will increase.

A number of other phenomena can alter the excited state population. For instance, collisional quenchers cause the fluorophores to return to the ground state upon contact. The rate of quenching depends on the diffusion coefficient and the local concentration of the quencher. Hence, the lifetime in the presence of a collisional quencher is

$$\tau = \frac{1}{\Gamma + \Sigma k_i + k_Q[Q]} \tag{9}$$

where k_Q depends on the diffusion coefficient. Equations (7) and (9) can be used to obtain the well-known Stern–Volmer equation, which describes collisional quenching of fluorescence,

$$\frac{I_0}{I} = \frac{\tau_0}{\tau} = 1 + k_Q\tau_0[Q] = 1 + K[Q] \tag{10}$$

For collisional quenching, the intensities I_0 and I are proportional to the lifetimes τ_0 and τ. A plot of I_0/I or τ_0/τ versus the quencher concentration can be used to determine the Stern–Volmer quenching constant $K = k_Q\tau_0$, and the accessibility of the fluorophore to contact with the quencher.

Stern–Volmer plots are widely used to investigate fluorophore accessibility. For example, Figure 9 shows the quenching of two proteins by acrylamide. Each protein contains a single tryptophan residue. This residue in ribonuclease T$_1$ is buried in the interior of the protein; it cannot be contacted and quenched by acrylamide. Hence, there is little or no quenching ($I_0/I = 1$). In contrast, the tryptophan residue in S. (staph) nuclease is quenched by acrylamide, and is then thought to be on the protein's surface.

3.2 METHODS TO MEASURE TIME-RESOLVED FLUORESCENCE

The rapid time scale of fluorescence makes the measurement of time-dependent decays a significant technological challenge. At present, two methods are widely used, the time domain (TD) and frequency domain (FD) or phase modulation methods (Figure 10). In time domain fluorometry, the sample is excited with a brief pulse of light from a flash lamp or a pulsed laser source. The width of the pulse must be shorter than the decay time. After the pulse, emission decays roughly as an exponential

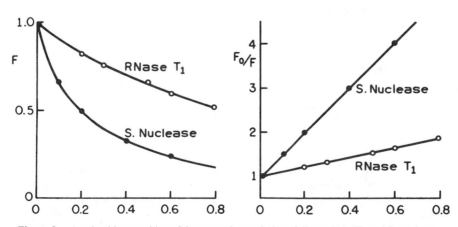

Figure 9. Acrylamide quenching of the tryptophan emission of ribonuclease T$_1$ and S. nuclease.

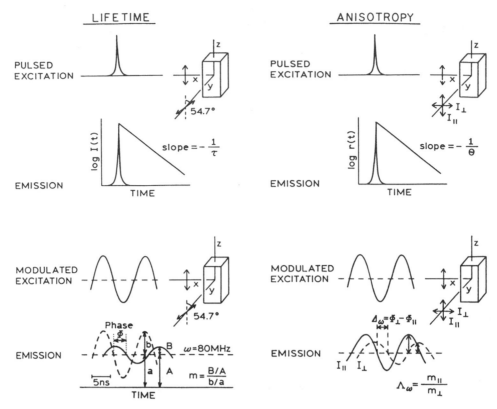

Figure 10.   Measurement of the fluorescence intensity decay (left) and anisotropy decay (right) by time domain or phase modulation fluorometry.

$$I(t) - I_0 \exp\left(-\frac{t}{\tau}\right) \qquad (11)$$

where I_0 is the initial intensity and τ the decay time. The decay time can be obtained from the inverse slope of the $\log(t)$ versus time (t) plot (Figure 10).

The alternative method of measuring the fluorescence lifetime relies on the use of an amplitude-modulated light source. When such a light source excites a fluorescent solution, the emission is delayed in time by a phase angle ϕ, and is demodulated by an amount m, relative to the incident light. The lifetime is relative to these measured quantities by

$$\tan \phi = \omega\tau \qquad (12)$$

$$m = (1 + \omega^2\tau^2)^{-1/2} \qquad (13)$$

where ω is 2π times the light modulation frequency in hertz. Both TD and FD measurements are in widespread use. Just a few years ago, measurements of time-resolved fluorescence were performed only in a few specialized laboratories. Today, advances in pulsed lasers and high speed detectors have made such measurements increasingly reliable and simple.

3.3   TIME-RESOLVED DECAYS OF BIOLOGICAL MOLECULES

Biological molecules typically display complex time-dependent decays. This occurs for a variety of reasons, such as the presence of a fluorophore in two or more distinct environments, or the presence of two or more fluorophores. For instance, consider a protein with two tryptophan residues, each having a decay time of 5 ns (Figure 11). The intensity decay is then given by

$$I(t) = I_0 \exp\left(-\frac{t}{5 \; ns}\right) \qquad (14)$$

Now consider that the protein is mixed with a quencher, and the exposed tryptophan residue is quenched to 1.0 ns, but the buried residue is not quenched and its lifetime is unchanged. The intensity decay would be

$$I(t) = 0.5 \exp\left(-\frac{t}{5 \; ns}\right) + 0.5 \exp\left(-\frac{t}{1 \; ns}\right) \qquad (15)$$

that is, a multiexponential decay. Such complex decays are often observed for biopolymers even without the addition of quenchers, and the phenomenon appears to be dependent on the conformation of the macromolecule. Many time-resolved measurements aim to quantify the decay law, and to interpret the decay law in terms of molecular features of the macromolecules.

The expected form of the TD and FD data for a multiexponential decay is shown in Figure 11. In the absence of quenching the intensity, following the excitation pulse, decays as a single exponential (Figure 11), and the plot of $\log I(t)$ versus time is linear. The 5 ns lifetime can be obtained from the slope. However, in the presence of quencher, the $\log I(t)$ plot is no longer linear, but displays curvature. The degree of curvature is modest (Figure 11, middle, dashed curve), and it is moderately difficult to recover the precise form of the decay law from the data.

Frequency domain data for the same sample are shown in the lower panel of Figure 11. These data consist of phase and modulation values measures over a range of modulation frequencies. The shape of the frequency response can be used to determine the intensity decay law (equation 15). In essence, the FD method measures the

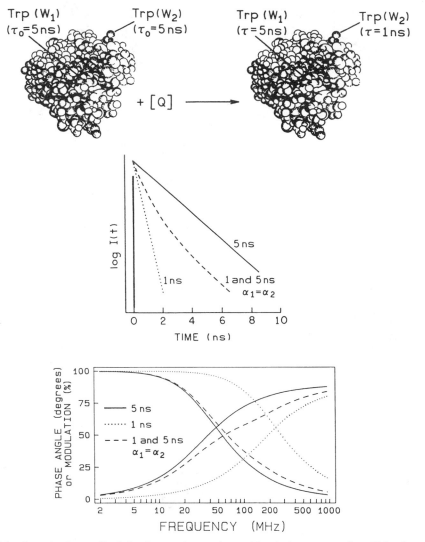

Figure 11.   Intensity decay of buried and exposed tryptophan residues in the presence of a collisional quencher.

Fourier transform of the intensity decay. Because it is difficult to resolve closely spaced exponential time constants, much effort has been made to increase the signal-to-noise ratio of both TD and FD methods.

3.4   TIME-RESOLVED ANISOTROPY DECAYS

Time-resolved anisotropy decays can provide detailed information on the rotational motion of proteins. Suppose we have a protein, with a monomer molecule weight of 20,000 Dal, which associates into tetramers (Figure 12). The rotational correlation time of the monomer in water is expected to be near 5 ns. If we measure the polarized time-dependent decays (Figure 10, right), the time-dependent anisotropy can be calculated from

$$r(t) = \frac{I_{\parallel}(t) - I_{\perp}(t)}{I_{\parallel}(t) + 2 I_{\perp}(t)} = r_0 \exp\left(-\frac{t}{\theta}\right) \qquad (16)$$

In this expression r_0 represents the anisotropy prior to any rotational motions and θ the rotational correlation time. For the monomer, the anisotropy will decay with a 5 ns correlation time.

Suppose now that the protein associates into a tetramer. The correlation time is now expected to be fourfold larger, which will be seen as a 20 ns decay time in the log $r(t)$ plot (Figure 12). We note that anisotropy decays can be used to study a variety of other phenomena, such as the shapes of proteins, protein association with membranes, rotations of fluorophores in membranes, and flexing of DNA. As for the intensity decays, the anisotropy decays can be measured by either TD or FD methods. In the FD method, the data consist of phase and modulation measurements of the polarized components of the emission (Figure 10, lower right).

4   EMERGING BIOMEDICAL APPLICATIONS OF TIME-RESOLVED FLUORESCENCE

Time-resolved fluorescence spectroscopy is presently regarded as a research tool in biochemistry, biophysics, and chemical physics. Advances in laser technology, the development of long wavelength probes, and the use of lifetime-based methods are resulting in the rapid migration of time-resolved fluorescence to the clinical chem-

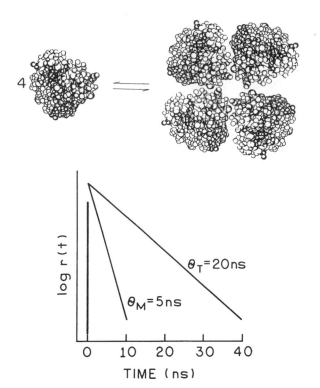

Figure 12. Anisotropy decay of a protein that self-associates into tetramers.

istry lab, to the hospital patient's bedside, and even to the doctor's office and home health care. Additionally, time-resolved imaging is now a reality in fluorescence microscopy and promises to provide chemical imaging of a variety of intracellular analytes and/or cellular phenomena.

4.1   LIFETIME-BASED FLUORESCENCE SENSING

Some schemes for fluorescence sensing are summarized in Figure 13. At present, most fluorescence assays are based on the standard intensity-based methods, in which the intensity of the probe molecules changes in response to the analytes of interest (Figure 13, left). However, it has been realized that lifetime-based methods possess intrinsic advantages for chemical sensing. If the intensity of a probe varies in response to an analyte, or if the amount of signal is proportional to the analyte, then it appears simple and straightforward to relate this intensity to the analyte concentration (left). Intensity-based methods are initially the easiest to implement because many probe fluorophores change intensity and/or quantum yield in response to analytes. Additionally, collisional quenching processes (quenching by oxygen, iodide, chloride, etc.) result in changes in intensity without significant shifts to the emission spectrum. While intensity measurements are simple and accurate in the laboratory, they are often inadequate in real-world situations. This is because the sample may be turbid, the optical surfaces may be imprecise or dirty, and the optical alignment may vary from sample to sample. In the case of fluorescence microscopy, it is often impossible to know the probe concentration at each point in the image. In any event, photobleaching, phototransformation, and/or diffusive processes cause the effective probe concentration to change continually.

In principle, the problems of intensity-based sensing can be avoided using wavelength-ratiometric probes, that is, fluorophores that display spectral changes in absorption or emission spectrum upon binding or interaction with the analyte (Figure 13). Wavelength–ratiometric probes provide a straightforward means of avoiding the difficulties of intensity-based sensing. However, few such probes are available, and it is clear that they are difficult to create.

To circumvent the difficulties of intensity-based measurements and the scarcity of probes, time-resolved or lifetime-based sensing may be used (Figure 14). Real-world sensing applications often oc-

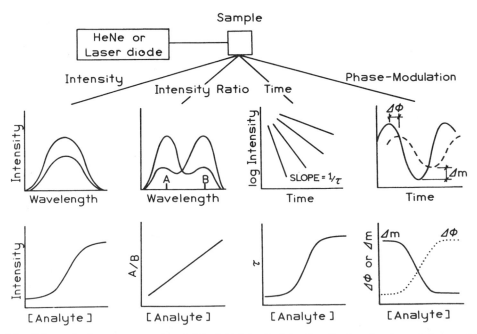

Figure 13. Schemes for fluorescence sensing. *Left to right:* intensity, intensity ratio, time domain, and phase modulation.

FLUORESCENCE SENSING METHODS

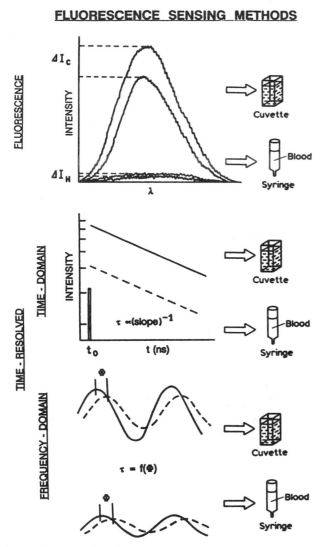

Figure 14.   Comparison of intensity and lifetime sensing: λ, wavelength.

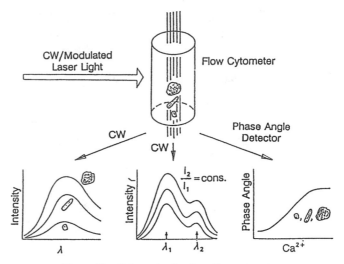

Figure 15.   Schematic of a phase flow cytometer.

cur in environments that are not equivalent to optically clear and clean cuvettes. Instead, numerous factors can affect the intensity values: imperfections or misalignment of surfaces and light losses in optical fibers, to name just a few. Additionally, many desired applications, such as homogeneous immunoassays or transdermal sensing measurements, require quantitative measurements in highly turbid or absorbing media (Figure 14). Such factors preclude quantitative measurements of intensities, or even intensity ratios.

Lifetime-based sensing can be mostly insensitive to real-world effects, which are not expected to alter the rate at which the intensity decays (Figure 14, middle). In our opinion, phase modulation/phase modulation sensing provides additional advantages (Figure 14, bottom). The FD instrumentation uses of radio frequency methods to reject noise and filter signals, resulting in reliable data in electrically noisy environments.

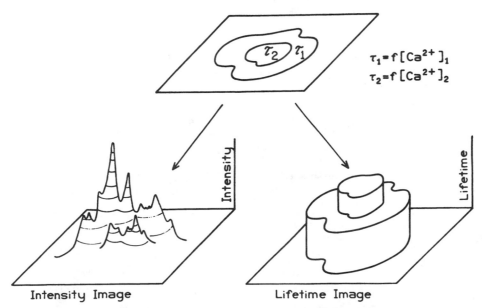

Figure 16.   Intuitive description of fluorescence lifetime imaging (FLIM).

4.2 Lifetime Measurements in Flow Cytometry and Imaging

An advantage of lifetime measurements is that the lifetime is not dependent on the total intensity. Consequently, lifetime measurements can be useful even when the intensities are not informative or cannot be reliably measured. Such situations arise with flow cytometry and/or fluorescence-activated cell sorting. In these measurements the cells flow one by one through a laser beam, which excites the fluorescence. The total signal is somewhat variable as a result of the differing volumes of the cells and the unequal amounts of probe (Figure 15). The use of lifetime measurements, on a cell-by-cell basis, can circumvent these difficulties of cell volume and/or probe concentrations, as well as the limitations of a small number of wavelength–ratiometric probes.

Lifetime-based sensing also provides significant advantages in fluorescence microscopy. Suppose the cell contains two regions with different Ca^{2+} concentrations and that the probe lifetime depends on Ca^{2+} (Figure 16). An intensity image of the cell will reveal the localization of the probe (lower left). In the emerging technology of lifetime imaging, the image contrast will be derived from the lifetime in each region of the cell (Figure 16, lower right), rather than from the local probe concentration. Such lifetime imaging will be analogous to magnetic resonance imaging (MRI), where the contrast is obtained from the proton relaxation time (analogous to a fluorescence lifetime), not from the signal intensity (analogous to the fluorescence intensity).

While not yet conveniently available, lifetime-based flow cytometry and imaging will allow cell sorting and imaging independent of total intensity. Such measurements, which seemed technologically impossible just a few years ago, are now achievable because of advances in lasers, opto-electronics, and computer technology.

5   PERSPECTIVES

There are numerous additional applications of fluorescence in the biomedical sciences. These applications are being developed because of the high sensitivity of fluorescence detection and the desire to eliminate the use of ionizing radiation (X-rays and radioactivity) in the laboratory and in clinical practice. Additionally, there is growing recognition of the value of using longer wavelength probes (red or near-IR), because tissues are nonabsorbing at these wavelengths, and there is less autofluorescence to interfere with the measurements. These properties suggest the possibility of noninvasive diagnostics based on red/near-IR probes, which can be excited with simple laser diode sources of the type used in everyday compact disk players. We have all observed the possibility of noninvasive diagnostics when, as children we saw the red light of a flashlight transmitted through our fingers. However, only recently has technology enabled the use of this observation for research and clinical purposes.

See also GENE MAPPING BY FLUORESCENCE IN SITU HYBRIDIZATION; GENE ORDER BY FISH AND FACS; LABELING, BIOPHYSICAL; WHOLE CHROMOSOME COMPLEMENTARY PROBE FLUORESCENCE STAINING.

Bibliography

Demas, J. N. (1983) *Excited State Lifetime Measurements.* Academic Press, New York.

Dewey, T. G., Ed. (1991) *Biophysical and Biochemical Aspects of Fluorescence Spectroscopy.* Plenum Press, New York.

Lakowicz, J. R. (1983) *Principles of Fluorescence Spectroscopy.* Plenum Press, New York.

———, Ed. (1991) *Topics in Fluorescence Spectroscopy:* Vol. 1, *Techniques;* Vol. 2, *Principles;* Vol. 3, *Biochemical Application.* Plenum Press, New York.

Middlehoek, S., and Cammann, K., Eds. (1993) *Sensors and Actuators B, Chem.* B11:1–565.

O'Connor, D. V., and Phillips, D. (1984) *Time-Correlated Single Photon Counting.* Academic Press, London.

Schulman, S. G. (1985) *Molecular Luminescence Spectroscopy Methods and Applications,* Part 1. Wiley-Interscience, New York.

Steiner, R. F., Ed. (1983) *Excited States of Biopolymers.* Plenum Press, New York.

Stryer, L. (1978) *Annu. Rev. Biochem.* 47:819–846.

Taylor, D. L., Waggoner, A. S., Lanni, F., Murphy, R. F., and Birge, R. R. (1986) *Applications of Fluorescence in the Biomedical Sciences.* Liss, New York.

Van Dyke, K., and Van Dyke, R., Eds. (1990) *Luminescence Immunoassay and Molecular Applications.* CRC Press, Boca Raton, FL.

Wolfbeis, O. S., Ed. (1993) *Fluorescence Spectroscopy: New Methods and Applications.* Springer-Verlag, Berlin.

Folding of Proteins: *see* Protein Folding; Protein Modeling.

FOOD PROTEINS, INTERACTIONS OF

Nicholas Parris and Charles Onwulata

1   **Protein Structure and Conformation**
   1.1   Food Proteins
   1.2   Enzymes
   1.3   Energy and Nutrition Values

2   **Protein Interactions**
   2.1   Protein–Protein
   2.2   Carbohydrates
   2.3   Lipids
   2.4   Hydration
   2.5   Soluble Ions

3   **Functional Protein Interactions**
   3.1   Gelation
   3.2   Emulsification
   3.3   Foaming
   3.4   Thermal Processing
   3.5   Thermally Induced Mutagens

Key Words

Emulsion   A thermodynamically unstable dispersion of micellar particles or globules in a liquid medium.

Foaming   A coarse dispersion of a gas in a liquid: the bulk of the phase volume is the gas, and the liquid is distributed between the gas bubbles in thin sheets called lamellae, which are formed when the liquid and gas are agitated together in the presence of a stabilizing agent.

Food Protein Interaction   The interaction of protein molecules and other compounds within their domain, which affects the protein's behavior in food products.

Gelation   The formation of an ordered, continuous matrix, entrapping another component, in which attractive and repulsive forces are balanced.

Hydrogen Bond   A highly directional noncovalent bond in which the hydrogen atom is shared with two other atoms.

Hydrophobic Interactions   The association of nonpolar molecules or groups with each other.

Phospholipids   Membrane lipids that possess polar groups linked to a diglyceride by a phosphodiester bridge.

Protein Micelle   Protein aggregates that are formed by the reversible interaction of protein monomers.

Syneresis   Shrinking of the gel with expulsion of trapped liquid.

Van der Waals Forces   Nonspecific attractive forces between two atoms.

Proteins, the most abundant macromolecules found in living cells, constitute approximate half a cell's dry weight. They are required in the food of humans, fish, and most higher animals. Historically, food proteins have been selected for their nutritional value, and they can be obtained from many naturally occurring sources. Proteins undergo a wide range of structural and conformational changes through a variety of complex interactions during processing and storage. Such changes can affect the principal purpose of dietary protein, which is to supply nitrogen and amino acids for the synthesis of proteins in the body. It is through an understanding of these interactions and their effects on functionality that food proteins have played a major role in the food supply.

The macrostructure of a protein is determined by its amino acid sequence. Amino acids are essential in basic nutrition, growth, and maintenance. Nine of the 20 identified amino acids are considered to be essential. Histidine, isoleucine, leucine, lysine, methionine, phenylalanine, threonine, tryptophan, and valine are so classified because they cannot be synthesized by humans and must therefore be supplied in the diet. When essential amino acids are not present in sufficient quantity, they are termed "limiting amino acids" because the utilization of other amino acids is restricted by the inadequacy of the supply of the former.

Processing of food proteins is designed to reduce microbial and enzymatic spoilage, inactivate antinutritional substances, improve the availability of perishable foods, and enhance the sensory quality of the food. Chemical changes that occur during food processing can influence nutritive value, sensory properties, and functional properties of the food. Heat treatment of proteins under acidic conditions can result in unfolding or denaturation of proteins and inactivation of enzymes and antinutritional substances. More severe heating can result in cross-linking of proteins. Racemization can occur in food proteins heated under alkaline conditions, with the subsequent formation of lysinoalanine due to the reaction of dehydroalanine and lysine. The presence of lysinoalanine can significantly reduce the nutritive value of proteins. Process-induced chemical interaction, which can occur between proteins and carbohydrates, results in nonoxidative, nonenzymatic browning (the Maillard reaction). Lipids, especially when unsaturated, are susceptible to oxidation. They can form lipid peroxides, which interact with proteins to form lipid–protein complexes, resulting in a decrease in the nutritive value of the food.

In addition to the aforementioned purposes of food protein processing, proteins play an important role in improving the functionality of food ingredients. Whey proteins commonly are subjected to heat denaturation before spray-drying to improve water-holding properties for bread making. Soy proteins are treated with alkali to improve solubility and textural properties and to obtain desirable rheological properties. Viscosity and solubility measurements are commonly used to obtain information about the functional behavior and physicochemical nature of proteins.

1   PROTEIN STRUCTURE AND CONFORMATION

Proteins are natural compounds composed of amino acids organized at four different levels of structure: primary, secondary, tertiary, and quaternary. The primary structure consists of amino acids, which are sequenced in a linear polypeptide chain and constitute the basic building blocks. Amino acids are joined together through an amide linkage or peptide bond, which is formed through the removal of a hydroxyl group from the carboxyl group of one amino acid and a hydrogen atom from the α-amino group of the adjoining amino acid. The C—N bond of the peptide linkage exhibits partial double-bond character and is not free to rotate, which imposes some constraints on the number of conformations the polypeptide can assume (Figure 1). There is, however, a large degree of rotational freedom around the single C_α—C and C_α—N bonds of the polypeptide. The secondary structure of a protein describes the regular folding of a polypeptide chain, which is due primarily to hydrogen bonding of the polypeptide chain, resulting in stable conformations.

The secondary structure of the polypeptide of the protein is composed of α-helices, β-pleated sheets, and random coils (Figure 2). The α-helix structure exists in polypeptides that are tightly coiled in a rodlike structure, with side chains extending outward from the helix. In contrast, the polypeptide in the β-sheet structure formed of aligned loops in a plane with adjacent strands are stabilized by hydrogen bonds between the NH and CO groups. The arrangement of member strands can be parallel, antiparallel, or a mixture of the two. Many proteins have well defined combinations of β-strands and α-helices or β-strands alone and represent a form of supersecondary structure. Some amino acids cannot form α-helices because of electrostatic repulsion or bulky side chains. As a result, the polypeptide assumes a random coil structure having minimal electrostatic free energy. In general, re-

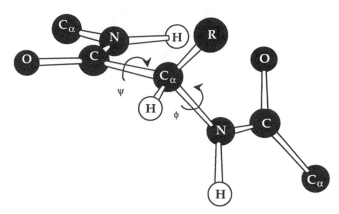

Figure 1. Polypeptide chain linking two peptide units; ϕ and ψ indicate rotation about the α-carbon. The two peptide bonds adjacent to the α-carbon are located in different planes. [From Stryer (1975); reproduced with permission of L. Stryer.]

gions of α-helices and β-sheets in the protein are well defined; random coils are not.

The tertiary structure refers to the steric relationships of amino acid residues and overall architecture in three dimensions of the polypeptide chain whose folding brings into proximity parts of the molecule otherwise widely separated along the backbone. Proteins containing more than one polypeptide chain exhibit another level of organization. This quaternary structure describes the packing of polypeptide chains, which are stabilized by hydrogen bonds and van der Waals forces. Structural changes at this level have unique biological applications, which are often mediated by enzymes. It is through systematic denaturation of the organized structures and forces of food proteins that changes in functional behavior can be achieved.

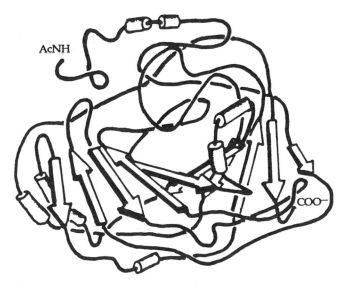

Figure 2. Schematic tertiary structure of the protein carbonic anhydrase. Helices are drawn as cylinders, β-sheet strands as arrows with N at the tail and C at the point of each arrow. [From Fennema (1985), reproduced by permission of O. R. Fennema.]

1.1   Food Proteins

Food proteins are derived from a number of sources: plants, meat and fish, milk, eggs, and microbial proteins from unicellular or multicellular organisms (e.g., bacteria, yeast, molds, algae). Plant proteins are broadly classified according to their solubility, shape, prosthetic group, and regulatory properties, as well as biological activity. They were first classified on the basis of their solubility as albumins, globulins, glutenins, and prolamines. Albumins are the most water-soluble globular proteins. Next most soluble are globulins such as conglycinin (7S) and glycinin (11S) from soybean, which are soluble in dilute salt solutions at neutral pH. Glutelins, such as wheat glutenins, are soluble in dilute acid or alkali. Prolamines are corn or wheat storage proteins and are soluble in 70% ethanol.

Muscle or contractile proteins are derived directly from animal tissue and are the most conspicuous food protein in the human diet. Muscle proteins are generally classified into three groups based on their solubility in water: sarcoplasmic proteins (soluble in water or dilute salt), myofibrillar proteins (soluble in salt solution > 0.6 M), and stromal proteins, which are the least soluble class of muscle proteins. Fibrous proteins—for example, myofibrillar proteins such as actin and myosin, or stromal proteins, such as collagen and elastin—have polypeptide chains arranged in long strands and serve a structural or protective role. Collagen, the major protein of connective tissue, occurs in several polymorphic forms consisting of three intertwined polypeptide chains.

Egg proteins are primarily globular proteins found in the albumen (egg white) and include ovalbumin, conalbumin, ovotransferrin, ovomucoid, and lysozyme. The egg yolk contains a variety of lipoproteins, which act as emulsifying agents, as well as livetin (possibly derived from hen's blood) and phosvitin, an iron carrier.

Milk proteins consist of a colloidal dispersion of casein micelles and soluble whey proteins; their stability is of tremendous practical importance in the dairy field. Casein micelles are extremely sensitive to changes in ionic environment and readily aggregate in response to increased concentration of calcium and magnesium ion or below the isoelectric point (pI).

Other proteins classified according to their prosthetic groups (tightly associated non–amino acid moieties) are lipoproteins, glycoproteins, and myoglobins (including hemoglobin) with attached fat or oil, sugar, or iron groups.

Proteins derived from unicellular organisms are grown on food-processing by-products from which the protein is harvested and subsequently purified.

1.2   Enzymes

Enzymes are proteins that function as specific biological catalysts for chemical reactions in living systems. Enzymes are much more efficient and specific than other catalysts, and they operate in restricted conditions of temperature and pH. They are frequently used in the food industry to modify the functional behavior of food proteins. For instance, proteases, which hydrolyze peptide bonds, can be recovered from plant sources such as papain and are used as commercial meat tenderizers. It is important, however, that the nutritive value of the product not be significantly reduced by the hydrolysis of peptide bond of the protein catalyzed by the enzyme.

Enzymatic action on free amino acids is of significance in food spoilage and production of flavors in fermented foods. For example, enzymatic degradation of free amino acids by microorganisms of the genera *Staphylococcus, Pseudomonas,* and *Micrococcus* produce specific enzymes that are responsible for the degradation reactions. Enzymes are also used in the manufacture of beer, bread, cheese, coffee, vinegar, vitamins, and many other products.

1.3 ENERGY AND NUTRITION VALUES

The nutritional quality of proteins is determined by their amino acid composition. Nine of the most common amino acids are classified as essential because they cannot be synthesized in the human body. The nutritional value of a protein for human consumption is based on the content of essential amino acids compared with human requirements, and by protein efficiency ratio (PER), which is determined by dividing the weight gain of a rat by the protein intake. Nutrient availability and quality of food proteins depend on the processing treatments and interaction with other food components. Protein requirements for maintaining good health and growth are listed as the recommended dietary allowance (RDA) set by the U.S. National Academy of Sciences. The RDA for a certain population varies depending on such factors as age, sex, level of activity, and individual metabolic factors. Protein requirements to meet growth and maintenance range from 2.0 g per kilogram of body weight for infants to 0.8 g/kg for adults. Energy needs increases during pregnancy and lactation as well as for work, stress, sickness, and aging. Proteins differ in nutrient value because of differences in amino acid composition. Protein quality is measured by chemical scores and biological assays that assess such qualities as content of essential amino acids, nitrogen value, limiting amino acids, and efficiency ratio. Cheese, eggs, fish, meat, and milks are sources of animal proteins. Vegetable proteins can be found in beans, grains, nuts, and a variety of vegetables.

2 PROTEIN INTERACTIONS

Molecular forces governing protein interactions determine the relationship of the structure of individual proteins to their functional properties as well as the association of a protein with other compounds in the cell. These forces include covalent bonds (intermolecular disulfide linkages), ionic interactions (salt bridges), hydrogen bonding, hydrophobic interactions, hydration, and steric repulsion.

Bonds or interactions that determine secondary and tertiary structure of proteins are shown in Figure 3. Covalent bonds include all the bonds of the primary structure of the protein as well as disulfide bonds, which are formed between cysteine residues and are dependent on the conformation and structure of the peptide chain. Of the covalent forces, disulfide bonding is the most important in protein interactions. Protein denaturation results in changes in the secondary and tertiary structures. Proteolysis by appropriate enzymes or hydrolysis with strong acid or base is necessary to break peptide bonds. Noncovalent molecular forces (e.g., hydrogen bonds, hydrophobic interaction, repulsion forces) are one to three orders of magnitude smaller than covalent bonding energy. The integrity of food proteins is maintained by association and disassociation of the protein in secondary and tertiary structure.

2.1 PROTEIN–PROTEIN

Protein–protein interactions often occur as a result of food processing that is designed to improve the functional properties of proteins for new product application. These interactions occur in two-stage processes consisting of unfolding of the native protein and exposing active sites, followed by association of the polypeptide chain by covalent and noncovalent forces. Stable protein interactions occur through cross-linking and ionic, polar, and hydrophobic interactions. Gelation is an association of proteins existing in a three-dimensional network with trapped water molecules. Protein cross-links are established through the formation of disulfide links by sulfhydryl groups or through hydrophobic interactions. To maintain a stable gel, a balance must exist between forces that promote formation of the network and opposing forces.

2.2 CARBOHYDRATES

Proteins and carbohydrates form irreversible complexes in nonoxidation reactions. Food protein heated in the presence of a reducing sugar results in the reaction of the carbonyl groups of the carbohydrate with the free amino group of the protein, followed by a cascade of reactions leading to polymers. These Maillard reactions are the major cause of protein damage in the drying of milk at moderate temperatures. However, more severe heating, required in the preparation of toasted breakfast cereals, bread, and biscuits, results

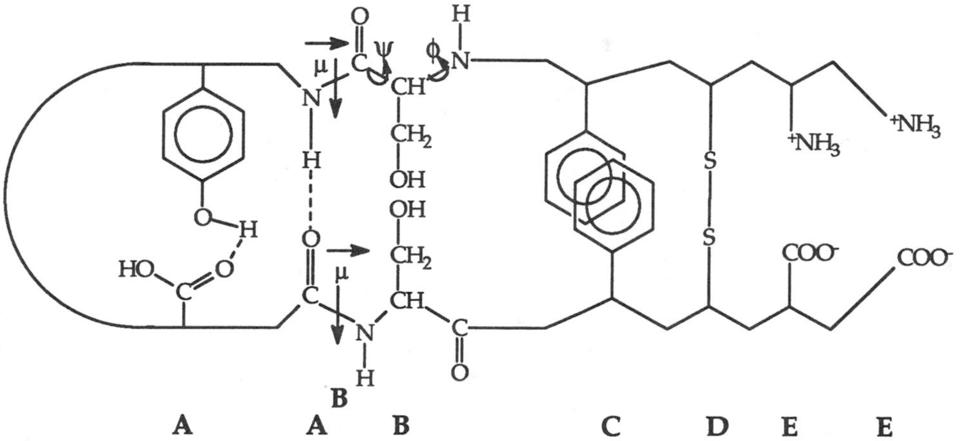

Figure 3. Bonds or interactions that determine protein secondary and tertiary structure: (A) hydrogen bond, (B) dipolar interaction (μ, dipole moment), (C) hydrophobic interaction, (D) disulfide linkage, (E) ionic interaction. [From Fennema (1985); reproduced by permission of O. R. Fennema.]

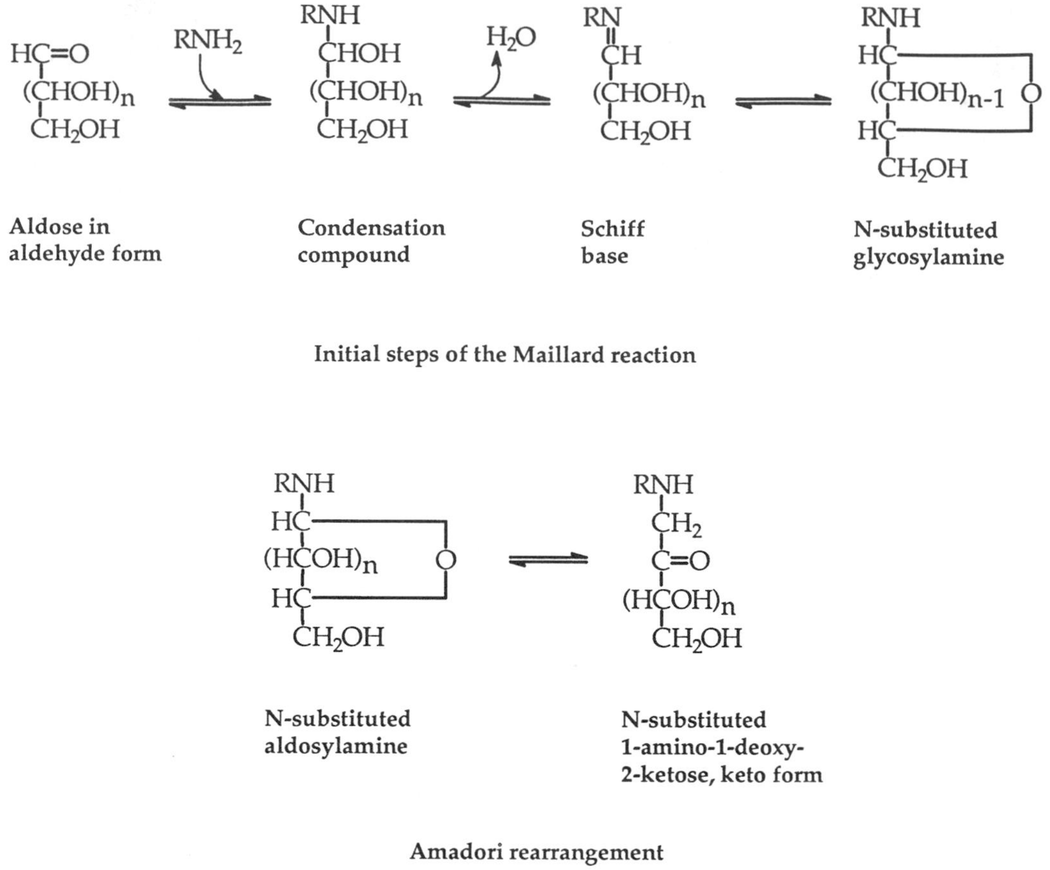

$$HC=O \quad \xrightarrow{RNH_2} \quad \begin{matrix} RNH \\ | \\ CHOH \end{matrix} \quad \xrightarrow{H_2O} \quad \begin{matrix} RN \\ || \\ CH \end{matrix} \quad \rightleftharpoons \quad \begin{matrix} RNH \\ | \\ HC \end{matrix}$$

Aldose in aldehyde form   Condensation compound   Schiff base   N-substituted glycosylamine

Initial steps of the Maillard reaction

N-substituted aldosylamine   ⇌   N-substituted 1-amino-1-deoxy-2-ketose, keto form

Amadori rearrangement

Figure 4.   Schematic representations of the Maillard reaction and the Amadori rearrangement.

in late Maillard-type damage and other undesirable reactions. In the early stages of the Maillard reaction (Figure 4), the carbonyl of the reducing sugar undergoes nucleophilic attack by the amine, followed by loss of water and ring closure to form the glycosylamine. The Amadori rearrangement follows with the formation of 1-amino-2-keto sugar, which has been isolated and identified in browned dried fruit.

2.3   Lipids

Protein–lipid interactions in nature occur in food systems such as milk and eggs. These interactions occur most notably at the cellular and intracellular membrane levels. Proteins that stabilize emulsions can adopt conformations that can interact with both lipid and aqueous phases by assuming the form of lowest free energy at the interface. Schematically the process can be represented in two steps: diffusion-controlled sorption at the interface, followed by protein unfolding (Figure 5). Emulsions are stabilized by hydrophobic interactions between the apolar region of the protein and the apolar aliphatic chain of the lipid. Confirmation of the importance of hydrophobic interactions has been demonstrated in model systems in which the energy of protein–lipid interaction reaches a maximum around the pI of the protein. Protein–lipid interaction can be increased by high pressure homogenization, which increases the number of lipid droplets and the interfacial surface area. Lipids protect proteins against thermal denaturation because of their high heat ca-

pacity and the absence of water. Strong interactions occur in flour–water mixtures, where lipids can bind with gluten proteins to form highly stable lipoglutenin complexes.

2.4   Hydration

Interactions of protein side chains with water determine the intrinsic properties of a protein (e.g., solubility, swelling, dispersibility, wettability). Water absorption is considered to be the most important step in imparting desired functional properties to proteins. Proteins interact with water through their side chains or backbone, and their solubility depends on factors such as hydrogen bonding, dipole–dipole and ionic interactions, pH, and temperature. The ability of proteins to absorb and retain water plays a major role in the textural stability of food systems. Texturized proteins may be used to form meat or seafood analogues by taking advantage of hydration properties.

2.5   Soluble Ions

The most important ionic components associated with proteins that affect solubility are sodium, potassium, and calcium. The reactive nature of proteins allows for the manipulation of the ions to accomplish food preservation by, for example, controlled dehydration. Proteins are stable within a defined range of pH; they may have net positive or net negative charges or be neutral. At pH values

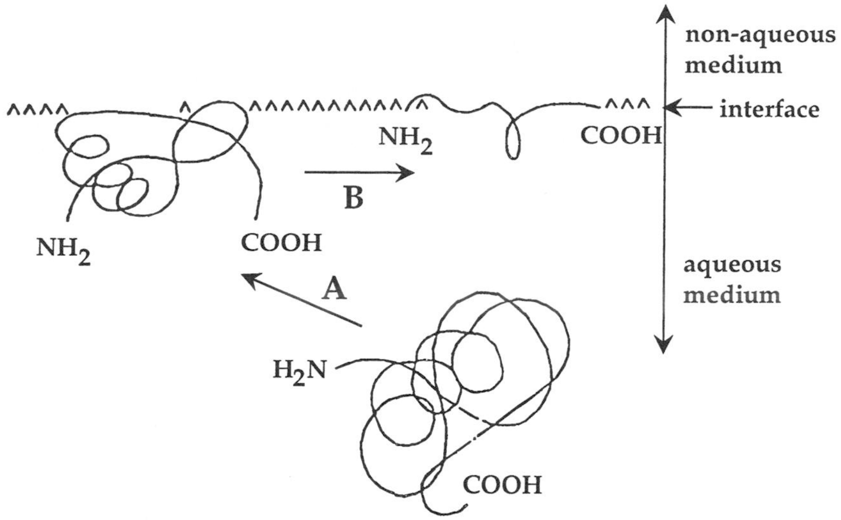

Figure 5. Schematic representation of the two-step process of protein denaturation at an interface: A, diffusion-controlled sorption at an interface; B, protein unfolding.

above their pI, proteins interact more readily with water molecules to improve solubility. Divalent cations also act as the cofactors that are essential for some enzymes to be catalytically active. Salts used in food processing can denature proteins by charge neutralization. Severe denaturation results in irreversible realignment of the native structure of the protein.

3   FUNCTIONAL PROTEIN INTERACTIONS

Physicochemical properties that enable proteins to affect the characteristics of foods during processing, storage, preparation, and consumption define the functional properties of proteins. Protein functions that affect food utilization are water absorption (viscosity and gelation), surface activity (gelation and foaming), and chemical reactivity (textural properties). Functionality of a food product is determined experimentally because, while the study of protein structure provides information on physicochemical properties, the exact relationship between structure and function is not fully understood. Changes in functionality are the result of protein interaction with water and other proteins, or the result of changes in the surface characteristics of the protein molecule.

3.1   GELATION

Gelation is the formation of an extended network of denatured protein aggregates held together by intermolecular forces. Gels can result from controlled protein denaturation and unfolding (egg white); controlled folding and realignment (collagen); controlled unfolding, disulfide interchange, and enhanced hydrophobic interactions (gluten); and enzyme hydrolysis and hydrophobic interactions (rennet milk gels). Gel stability depends on the balance of hydrophobic interactions, hydrogen bonding, and electrostatic forces, which, in turn, are affected by pH and electrolyte content. Cross-linking of proteins in the matrix via disulfide bonds results in heat-irreversible gels stabilized by hydrogen bonds. Higher molecular weight proteins, with a larger percentage of hydrophobic amino acid residues, tend to form stronger gels.

3.2   EMULSIFICATION

A stable protein emulsion is produced when proteins in a two-phase medium, accompanied by energy input (usually a combination of shear and heat), unfold slightly at the interface and align their non-polar regions toward an oil phase with their hydrophilic regions toward the aqueous phase. The balance of the hydrophilic and hydrophobic forces in the protein maintains emulsion stability. Globular proteins with a highly ordered and stable tertiary structure, such as β-lactoglobulin, bovine serum albumin, and lysozyme, are more likely to unfold and are considered to be good emulsifiers. Emulsification activity continues to increase with increasing protein denaturation, provided solubility is not compromised. Surface hydrophobicity of proteins has been correlated with emulsifying activity. Emulsions are mechanically unstable and will separate over time. The breakdown in stability results in flocculation, followed by the coalescence of particles by forces depicted in Figure 6. Although emulsifiers can minimize interfacial surface energy and reduce coalescence, emulsions gradually separate into two phases as the repulsive electrical charges decreases. Gradual increase in free energy leads to more frequent collisions of similarly charged particles. These collisions produce flocculated globules or clusters, which will continue to grow unless conditions change. Complete breakdown of the emulsion is the result of coalescence of these particles.

3.3   FOAMING

Protein foams are emulsions of gas in a continuous aqueous protein phase containing various surfactants to prevent coalescence. Foods such as ice cream and whipped toppings are examples of stable foams formed by incorporation of a large volume of air during processing. Air is forced into the medium through mechanical mixing that increases the volume more than 1000%. Foam stability is maintained by surfactants that resist the influence of gravity, pressure, and temperature, which tend to rupture the stable emulsion. Controlled denaturing of the native protein by shear, temperature, or ma-

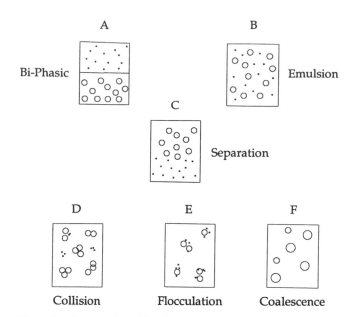

Figure 6.   Process of emulsion formation and breakdown: (A) biphasic medium of liquid and semisolid emulsion, (B) thermodynamically unstable emulsion, (C) gradual migration of emulsion based on charge separation, (D) particle collision with increasing electrostatic charges, (E) clear aggregation and flocculation of collided particles, (F) coalescence and syneresis of emulsion.

nipulation of electrostatic balance increases hydrophobicity and improves foaming. Foam-stabilizing conditions include high surface hydrophobicity, low net energy charge, minimal electrostatic repulsion, good solubility, and protein concentration.

Foam expansion (overrun) is controlled by metering the mix and air that go into the equipment in proper proportions to form air cells of the desired size; overrun is computed as the percentage increase in volume or the percentage decrease in the density of the mix. Foam stability may be controlled by addition of combinations of stabilizers and emulsifiers: salts, sugars, lipids, cations, proteins, and energy input (heat and shear) during processing. As foams age during storage, they shrink, leading to loss of air and demixing.

3.4   THERMAL PROCESSING

Food proteins are thermally processed to enhance functionality, improve textural properties, or retard deterioration. In the preservation of fruits and vegetables (blanching), heat is used to denature enzymes. Processing may lead to a slight loss in nutritional quality; however, most processes improve quality by destroying antinutritional factors through inactivation of enzymes such as trypsin inhibitor or peroxidase. Complete denaturation of peroxidase (EC.1.11.1.7) is used as an indicator of adequate processing. Protein–protein interactions may be enhanced, leading to formation of new complexes. Factors such as temperature and the presence of salts and oxidizing/reducing agents may be used to produce desirable high quality foods. Chilling or heating can denature protein. Protein structure can be altered as a result of property changes (in the electric potential, pH, etc.), and such changes can be used to enhance functionality in food. Thermal denaturation of food proteins generally occurs between 45 and 85°C, accompanied by exposure of hydrophobic groups that can lead to protein aggregation. More

severe treatments can result in the splitting of the disulfide crosslinks with release of hydrogen sulfide, as well as alteration of amino acid residues with the formation of new intra- or intermolecular covalent cross-links. Low temperature denaturation is mediated by a reduction of hydrophobic interactions in conjunction with enhanced hydrogen bonding, leading to aggregation and precipitation of proteins.

3.5   THERMALLY INDUCED MUTAGENS

Mutagens can be formed in muscle foods as a result of industrial processing or home cooking, including frying, broiling, boiling, and baking. Cooking methods that involve direct contact with high temperature sources generate the most mutagens. Three general types of mutagen are found, arising from different sources. The IQ (imidazo-quinoline) types arise from endogenous nitrogen-containing compounds such as creatine and amino acids in combination with sugars; the N-nitrosoamine types arise from endogenous secondary amines and added nitrite; and the polycyclic aromatic hydrocarbon types arise from pyrolysis of fat due to smoke treatment or upon cooking (charring) on charcoal or direct flame. These mutagens are present in meat or seafood products at a low parts-per-billion level. Extensive research aimed at developing methods to reduce or eliminate such compounds from the food supply has seen moderate success.

See also LIPID AND LIPOPROTEIN METABOLISM; NUTRITION.

Bibliography

Bodwell, C. E., and Erdman, J. W., Jr., Eds. (1988) *Nutrient Interactions.* Dekker, New York.
Creighton, T. E., Ed. (1983) *Proteins: Structures and Molecular Properties.* Freeman, New York.
El-Nokaly, M., and Cornell, D., Eds. (1991) *Microemulsions and Emulsions in Foods.* American Chemical Society, Washington, DC.
Fennema, O. R. (1985) *Food Chemistry,* 2nd ed. Dekker, New York.
Fieden, C., and Nichol, L. W., Eds. (1981) *Protein–Protein Interactions.* Wiley, New York.
Hudson, B. J. F., Ed. (1987) *Developments in Food Proteins,* Vol. 5. Elsevier Applied Sciences Publishers, Barking, U.K.
Hui, Y. H. (1991) *Encyclopedia of Food Science and Technology,* Vol. 3. Wiley-Interscience, New York.
Lehninger, A. L. (1982) *Principles of Biochemistry.* Worth Publishers, New York.
Mitchell, J. R., and Ledward, D. A., Eds. (1986) *Functional Properties of Food Macromolecules.* Elsevier Applied Science Publishers, London.
Osborne, T. B. (1924) *The Vegetable Proteins.* Longmans, Green, London.
Parris, N., and Barford, R. A., Eds. (1991) *Interactions of Food Proteins.* American Chemical Society, Washington, DC.
Phillips, G. O., Wedlock, D. J., and Williams, P. A., Eds. (1984) *Gums and Stabilisers for the Food Industry,* Vol. 2, *Applications of Hydrocolloids.* Pergamon Press, Oxford.
Stryer, L. (1975) *Biochemistry.* Freeman, San Francisco.
Waller, G. R., and Feather, M. S., Eds. (1983) *The Maillard Reaction in Foods and Nutrition.* American Chemical Society, Washington, DC.
Whitaker, J. R., and Tannenbaum, S. R., Eds. (1977) *Food Proteins.* AVI Publishing, Westport, CT.

FRAGILE SITES

Grant R. Sutherland, Elizabeth Baker, and Robert I. Richards

1 **Nomenclature**

2 **Classification**

3 **Molecular Genetics**

4 **Cytogenetics**

5 **Induction of Fragile Sites**

6 **Inheritance**

7 **Clinical Significance**
    7.1 The Fragile X Syndromes
    7.2 Other Fragile Sites
    7.3 Cancer

Key Words

Autosome Any chromosome other than the sex (X and Y) chromosomes.

Dynamic Mutation A multi-step mutational process in which the chance of each step occurring is modified by the immediately preceding step.

Fragile X Syndrome The most common form of familial mental retardation. Affected patients also have subtle dysmorphic features.

Jacobsen Syndrome A mental retardation–congenital malformation syndrome caused by loss of part of the long arm of one number 11 chromosome.

Trinucleotide Repeat A DNA sequence composed of three nucleotides repeated in tandem a number of times.

Fragile sites are specific points on chromosomes that show nonrandom gaps or breaks when the cells from which the chromosomes were prepared are exposed to a specific chemical agent or condition of tissue culture. Hence a fragile site is an area of chromatin that is not compacted when viewed during mitosis.

1 NOMENCLATURE

There are two forms of nomenclature used for fragile sites. A cytogenetic nomenclature uses "fra" followed in parentheses by the chromosome band in which the fragile site is present. The rare fragile site in band 23.3 on the long (q) arm of chromosome 11 is thus fra(11)(q23.3). Each fragile site also has a gene symbol and this is given by "*FRA*," the chromosome number, and an assigned letter of the alphabet. Thus Fra(11)(q23.3) is *FRA11B*. The gene symbol is a locus name. Alleles at the locus can be specified following an asterisk, the wild-type allele as *N*, and the mutant allele as *R* (for rare) or *C* (for common), followed by the chromosome band. Fra(11)(q23.3) in gene nomenclature would be *FRA11B*N* for the wild-type allele and *FRA11B*RQ233* for the mutant allele. In practice, the gene symbol is frequently used to specify the mutant allele and the wild-type allele is not mentioned.

2 CLASSIFICATION

Fragile sites are classified on the basis of their frequency in the population and the condition of tissue culture under which they are expressed. Rare fragile sites are seen at a specific location on 1 in 40 chromosomes at most. Common fragile sites are probably present on all chromosomes and as such are normal components of chromosome structure.

Most fragile sites are not expressed spontaneously but must be induced. When they are expressed spontaneously, the proportion of metaphases in which they are seen can usually be greatly increased by appropriate conditions of induction. The classification, based on population frequency and induction conditions, is shown in Table 1. All known human fragile sites are listed in Table 2.

3 MOLECULAR GENETICS

Only fragile sites from group 1 have been characterized at the molecular level. Nothing is known about the molecular basis of any of the other groups of fragile sites. Five of the group 1 fragile sites have been cloned, and details of these are shown in Table 3.

The common features of these fragile sites, which are likely to be common to all the group 1 fragile sites, include the following:

1. The fragile site is a dynamic mutation of a naturally occurring CCG trinucleotide repeat.
2. The CCG repeat is polymorphic in the normal population.
3. Increases above the normal range in the number of copies of the repeating unit lead to a premutation. Individuals with the premutation do not express a fragile site and, for the fragile sites on the X chromosome associated with mental handicap, there is no disease present. As the number of copies of the repeating unit increases further, the DNA in the whole area is subjected to CpG methylation. Once this has happened, the fragile site can be induced to appear, and any gene in which the repeat is located ceases to be transcribed.

For *FRAXA* the upper limit of copy number in normal chromosomes is about 55; from 55 to about 230 copies constitutes a premutation, and beyond 230 is a full mutation. The number of copies changes from generation to generation, usually increasing when transmitted by women but not changing much when transmitted by males.

Table 1 Classification of Fragile Sites

Group	Description
	Rare
Group 1	Folate-sensitive fragile sites ($n = 23$)
Group 2	Distamycin A inducible
	(a) Also inducible by BrdU ($n = 2$, *FRA16B, FRA17A*)
	(b) Not inducible by BrdU, and recorded only in Japanese subjects
	($n = 3$, *FRA8E, FRA11I, FRA16E*)
Group 3	BrdU-requiring ($n = 2$, *FRA10B, FRA12C*)
	Common
Group 4	Aphidicolin-inducible
Group 5	5-Azacytidine-inducible
Group 6	BrdU-inducible
Unclassified	*FRA4E*

Table 2   The 117 Fragile Sites in the Human Genome[a]

Gene Symbol	Chromosome Band Location	Type	Group
FRA1A	1p36	Common, aphidicolin	4
FRA1B	1p32	Common, aphidicolin	4
FRA1C	1p31.2	Common, aphidicolin	4
FRA1L	1p31	Common, aphidicolin	4
FRA1D	1p22	Common, aphidicolin	4
FRA1M	**1p21.3**	**Rare, folic acid**	**1**
FRA1E	1p21.2	Common, aphidicolin	4
FRA1J	1q12	Common, 5-azacytidine	5
FRA1F	1q21	Common, aphidicolin	4
FRA1G	1q25.1	Common, aphidicolin	4
FRA1K	1q31	Common, aphidicolin	4
FRA1H	1q42	Common, 5-azacytidine	5
FRA1I	1q44	Common, aphidicolin	4
FRA2C	2p24.2	Common, aphidicolin	4
FRA2D	2p16.2	Common, aphidicolin	4
FRA2E	2p13	Common, aphidicolin	4
FRA2L	**2p11.2**	**Rare, folic acid**	**1**
FRA2A	**2q11.2**	**Rare, folic acid**	**1**
FRA2B	**2q13**	**Rare, folic acid**	**1**
FRA2F	2q21.3	Common, aphidicolin	4
FRA2K	**2q22.3**	**Rare, folic acid**	**1**
FRA2G	2q31	Common, aphidicolin	4
FRA2H	2q32.1	Common, aphidicolin	4
FRA2I	2q33	Common, aphidicolin	4
FRA2J	2q37.3	Common, aphidicolin	4
FRA3A	3p24.2	Common, aphidicolin	4
FRA3B	3p14.2	Common, aphidicolin	4
FRA3D	3q25	Common, aphidicolin	4
FRA3C	3q27	Common, aphidicolin	4
FRA4A	4p16.1	Common, aphidicolin	4
FRA4D	4p15	Common, aphidicolin	4
FRA4B	4q12	Common, BrdU	6
FRA4E	4q27	Common, unclassified	—
FRA4C	4q31.1	Common, aphidicolin	4
FRA5E	5p14	Common, aphidicolin	4
FRA5A	5p13	Common, BrdU	6
FRA5B	5q15	Common, BrdU	6
FRA5D	5q15	Common, aphidicolin	4
FRA5F	5q21	Common, aphidicolin	4
FRA5C	5q31.1	Common, aphidicolin	4
FRA5G	**5q35**	**Rare, folic acid**	**1**
FRA6B	6p25.1	Common, aphidicolin	4
FRA6A	**6p23**	**Rare, folic acid**	**1**
FRA6C	6p22.2	Common, aphidicolin	4
FRA6D	6q13	Common, BrdU	6
FRA6G	6q15	Common, aphidicolin	4
FRA6F	6q21	Common, aphidicolin	4
FRA6E	6q26	Common, aphidicolin	4
FRA7B	7p22	Common, aphidicolin	4
FRA7C	7p14.2	Common, aphidicolin	4
FRA7D	7p13	Common, aphidicolin	4
FRA7A	**7p11.2**	**Rare, folic acid**	**1**
FRA7J	7q11	Common, aphidicolin	4
FRA7E	7q21.2	Common, aphidicolin	4
FRA7F	7q22	Common, aphidicolin	4
FRA7G	7q31.2	Common, aphidicolin	4
FRA7H	7q32.3	Common, aphidicolin	4
FRA7I	7q36	Common, aphidicolin	4
FRA8B	8q22.1	Common, aphidicolin	4
FRA8A	**8q22.3**	**Rare, folic acid**	**1**
FRA8C	8q24.1	Common, aphidicolin	4

(continued)

Table 2   *(Continued)*

Gene Symbol	Chromosome Band Location	Type	Group
FRA8E	**8q24.1**	**Rare, distamycin A**	**2b**
FRA8D	8q24.3	Common, aphidicolin	4
FRA9A	**9p21**	**Rare, folic acid**	**1**
FRA9C	9p21	Common, BrdU	6
FRA9F	9q12	Common, 5-azacytidine	5
FRA9D	9q22.1	Common, aphidicolin	4
FRA9B	**9q32**	**Rare, folic acid**	**1**
FRA9E	9q32	Common, aphidicolin	4
FRA10G	10q11.2	Common, aphidicolin	4
FRA10C	10q21	Common, BrdU	6
FRA10D	10q22.1	Common, aphidicolin	4
FRA10A	**10q23.3**	**Rare, folic acid**	**1**
FRA10B	**10q25.2**	**Rare, BrdU**	**3**
FRA10E	10q25.2	Common, aphidicolin	4
FRA10F	10q26.1	Common, aphidicolin	4
FRA11C	11p15.1	Common, aphidicolin	4
FRA11I	**11p15.1**	**Rare, distamycin A**	**2b**
FRA11D	11p14.2	Common, aphidicolin	4
FRA11E	11p13	Common, aphidicolin	4
FRA11H	11q13	Common, aphidicolin	4
FRA11A	**11q13.3**	**Rare, folic acid**	**1**
FRA11F	11q14.2	Common, aphidicolin	4
FRA11B	**11q23.3**	**Rare, folic acid**	**1**
FRA11G	11q23.3	Common, aphidicolin	4
FRA12A	**12q13.1**	**Rare, folic acid**	**1**
FRA12B	12q21.3	Common, aphidicolin	4
FRA12E	12q24	Common, aphidicolin	4
FRA12D	**12q24.13**	**Rare, folic acid**	**1**
FRA12C	**12q24.2**	**Rare, BrdU**	**3**
FRA13A	13q13.2	Common, aphidicolin	4
FRA13B	13q21	Common, BrdU	6
FRA13C	13q21.2	Common, aphidicolin	4
FRA13D	13q32	Common, aphidicolin	4
FRA14B	14q23	Common, aphidicolin	4
FRA14C	14q24.1	Common, aphidicolin	4
FRA15A	15q22	Common, aphidicolin	4
FRA16A	**16p13.11**	**Rare, folic acid**	**1**
FRA16E	**16p12.1**	**Rare, distamycin A**	**2b**
FRA16B	**16q22.1**	**Rare, distamycin A**	**2a**
FRA16C	16q22.1	Common, aphidicolin	4
FRA16D	16q23.2	Common, aphidicolin	4
FRA17A	**17p12**	**Rare, distamycin A**	**2a**
FRA17B	17q23.1	Common, aphidicolin	4
FRA18A	18q12.2	Common, aphidicolin	4
FRA18B	18q21.3	Common, aphidicolin	4
FRA19B	**19p13**	**Rare, folic acid**	**1**
FRA19A	19q13	Common, 5-azacytidine	5
FRA20B	20p12.2	Common, aphidicolin	4
FRA20A	**20p11.23**	**Rare, folic acid**	**1**
FRA22B	22q12.2	Common, aphidicolin	4
FRA22A	**22q13.1**	**Rare, folic acid**	**1**
FRAXB	Xp22.31	Common, aphidicolin	4
FRAXC	Xq22.1	Common, aphidicolin	4
FRAXD	Xq27.2	Common, aphidicolin	4
FRAXA	**Xq27.3**	**Rare, folic acid**	**1**
FRAXE	**Xq28**	**Rare, folic acid**	**1**
FRAXF	**Xq28**	**Rare, folic acid**	**1**

[a]Boldtype type indicates rare fragile sites.

Source: Modified from Sutherland (1993), where detailed references to each fragile site can be found, except for *FRA2L*, which is Schuffenhauer et al. (1996).

Table 3   Cloned Fragile Sites

	Repeat Copy number			
Site	Normal Alleles	Premutation	Full Mutation	Associated Disease
FRAXA	6–55	55–230	230–2000	Fragile X syndrome
FRAXE	6–25	? 25–200	? > 200	Mild mental retardation
FRAXF	6–29	?	? > 200	None
FRA16A	17–50	100	? > 200	None
FRA11B	11–17	85	? > 200	Possible risk of offspring with Jacobsen syndrome

The polymorphic trinucleotide repeat increases from within the normal range to give a fragile site by means of a multistep process known as dynamic mutation. The probability that a trinucleotide (or other simple sequence) repeat will change in length is a function of the number of copies of the repeat that are present without interruption or imperfection. In the normal population, most alleles of the trinucleotide repeat have some imperfections. It is the alleles with the longest stretches of perfect repeat that form a reservoir of alleles in the normal population, which can proceed via dynamic mutation to fragile sites.

Since the length of repeat determines the rate at which further increases occur, the chance that a premutation will convert to a full mutation increases with the size of the premutation alleles. This correlation has been studied in most detail for *FRAXA*. Only when this fragile site is found in the mother do offspring have full mutations. The upper limit of normal copy number is about 55. For women with premutations of fewer than 60 copies, the risk of having a child with a full mutation is close to zero. The chance that 60–80 copies will convert to a full mutation is about 30%, for 80–90 copies it is about 80%, and for greater than 90 copies, virtually 100%.

The limited numbers of normal alleles with long stretches of perfect repeat that form the reservoir of mutable alleles give rise to the phenomenon of linkage disequilibrium. Haplotypes are distributed very differently around the fragile site locus on normal and fragile site chromosomes.

The place and time of expansion of a full mutation from a premutation is uncertain. If, as some have argued, this occurs (for *FRAXA*) purely postzygotically in a small time period early in development, some imprinting mechanism would be necessary, since both the sperm and the ovum would contribute premutations but only those from the ovum would expand to full mutations. Alternatively, since the process of meiosis is so different in males and females, it is possible that the conversion of premutation to full mutation occurs during oogenesis.

The CCG repeats are adjacent to CpG islands and have been shown to be in the 5′-untranslated regions of the genes *FMR1* (at *FRAXA*) and *CBL2* (at *FRA11B*). For *FRAXA* a full mutation results in cessation of transcription of the *FMR1*, and this is the molecular basis of fragile X syndrome. *FRA11B* is in the 5′-untranslated region of the *CBL2* oncogene; genes possibly associated with other cloned fragile sites have not yet been identified.

4   CYTOGENETICS

Fragile sites are seen by examining metaphase chromosomes under the light microscope. They appear as gaps, breaks, or unusual structures (Figures 1 and 2) in a variable proportion of metaphases. The unusual chromosomal structures have arisen from breakage at the fragile site in an earlier mitosis, with the chromosomal material from either side of the fragile site ending up in different daughter cells.

Fragile sites are easier to see on plain stained chromosomes than on banded chromosomes, especially when they occur in pale G bands, as most do. It is often possible to see a fine strand of chromatin across the gap(s) at the fragile site.

The proportion of metaphases in which any particular fragile site is seen is highly variable. The common fragile sites, especially the aphidicolin-inducible ones are not often seen in more than 10% of metaphases, but occasional individuals are encountered in whom any of these may be present in up to 30% of metaphases. The com-

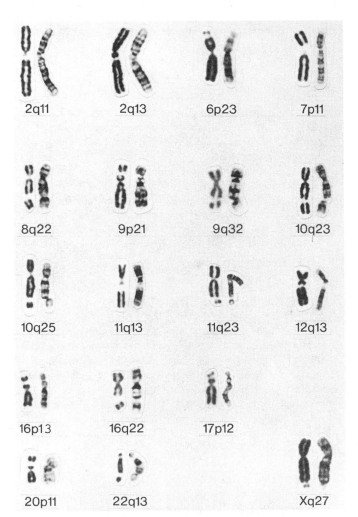

2q11	2q13	6p23	7p11
8q22	9p21	9q32	10q23
10q25	11q13	11q23	12q13
16p13	16q22	17p12	
20p11	22q13		Xq27

Figure 1.   A selection of rare fragile sites. The chromosome on the left of each pair is plain-stained and the one on the right is G-banded.

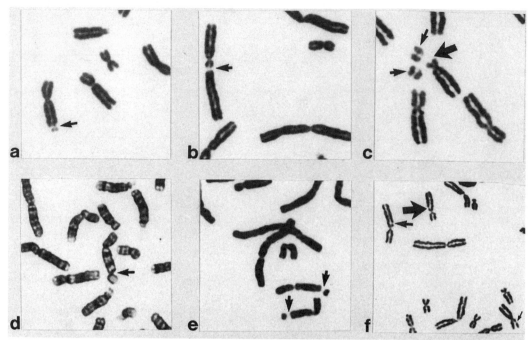

Figure 2. Partial metaphases showing the appearance of various fragile sites (arrowed): (a) the fragile X *(FRAXA)*; (b) expression of *FRA2B* as a chromosome gap; (c) *FRA9A* showing duplication of chromosome material (small arrows) distal to the fragile site (large arrow); (d) *FRA10A;* (e) homozygous expression of the common fragile site *FRA16D;* (f) BrdU-inducible fragile sites at *FRA6D* (medium arrow), *FRA9C* (small arrow) and *FRA10C* (large arrow).

mon fragile sites can be seen on both members of a chromosome pair in a single cell (homozygous expression). The rare autosomal fragile sites have never been seen in homozygotes, except for *FRA16B, FRA17A,* and *FRA10B,* where this is without apparent phenotypic effect.

The rare fragile sites are not often seen in more than 50% of metaphases and are more typically present in 20–30%, although some obligate carriers of a fragile site (from family studies) will not express it at all (presumed premutation carriers, see Section 3). The exception to these levels of cytogenetic expression is *FRA16B;* when induced with berenil, it can often be seen in more than 90% of metaphases.

5   INDUCTION OF FRAGILE SITES

Fragile sites have been studied primarily in chromosomes prepared from phytohemagglutinin-stimulated human lymphocyte cultures. While some fragile sites can be induced in chromosomes from fibroblast or lymphoblastoid cultures and somatic cell hybrids, such induction is more difficult, and the fragile sites are usually seen in lower proportions of metaphases than in lymphocyte cultures. Since the conditions required for induction of fragile sites differ for each of the groups shown in Table 1, we consider the groups separately.

Group 1. These fragile sites require a relative deficiency of deoxycytidine 5'-triphosphate (dCTP) or deoxythymidine 5'-triphosphate (dTTP) at the time of DNA synthesis. A deficiency of dTTP can be achieved by using tissue culture medium free of folic acid and thymidine, by adding an inhibitor of folate metabolism such as methotrexate, aminopterin or trimethoprim, or

specific inhibitors of thymidylate synthetase such as fluorodeoxyuridine. A high concentration of thymidine in the culture medium will lead to a deficiency of dCTP, resulting in fragile site expression. Induction conditions need to be in place for at least 24 hours before chromosome preparations are made.

Group 2. All these fragile sites are induced by distamycin A and a number of related compounds that bind to the minor groove of the DNA molecule in AT-rich regions. The most effective of these compounds is berenil. The two fragile sites in subgroup a are distinguished from those in subgroup b by also being inducible with bromodeoxyuridine (BrdU), although BrdU is not as effective an inducing agent as berenil.

Group 3. The two fragile sites are induced only by BrdU and the related compound BrdC.

Group 4. This large group of common fragile sites is seen in a small proportion of metaphases under the conditions that induce the group 1 fragile sites. They can also be specifically induced by aphidicolin. This compound is dissolved in dimethyl sulfoxide (DMSO) or ethanol for addition to cultures. Ethanol (but not DMSO) is synergistic in fragile site induction. Aphidicolin is added 24 hours before chromosome harvest, and the addition of caffeine for the last 6 hours of this period boosts the proportion of metaphases in which the fragile sites are seen.

Group 5. These fragile sites are induced by 5-azacytidine, which is added 8 hours before chromosome harvest. If the addition is much before or after this timing, the fragile sites are not seen.

Group 6. BrdU induces these fragile sites, and again the timing of addition of the compound is important. Maximum induction occurs when the addition is made 4–6 hours before chromosome preparation. If the BrdU present for 24 hours, none of the group 6 fragile sites will be seen.

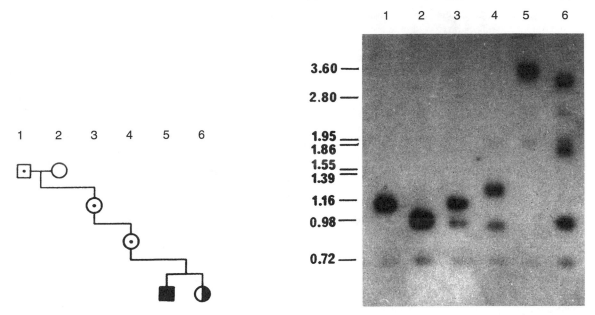

Figure 3. Inheritance of the fragile X unstable element in a four-generation lineage from a large affected pedigree. Chromosomal DNA was digested with PstI and probed with pfxa3. Pedigree symbols: normal carrier male (1); normal carrier female not expressing the fragile X (3, 4); normal carrier female expressing the fragile X (6); affected fragile X syndrome male expressing the fragile X (5); normal female (2). The carrier male in the first lane has no normal 1 kb band but a 1.15 kb band; his wife in the second lane has a 1 kb band. The daughter in lane 3 has a 1.15 kb band on her fragile X from her father and a 1 kb band from her mother. When this daughter has transmitted this band to her daughter, the fragile X band has increased to about 1.3 kb in size. This band has increased dramatically in size to about 3.5 kb in the fragile X syndrome boy, and it appears as a somatically unstable smear in the carrier girl who expresses the fragile X (last lane). She has major bands of about 3.2 and 1.9 kb and a 1.0 kb band from her normal X chromosome.

6   INHERITANCE

The patterns of inheritance of some fragile sites are very unusual. The common fragile sites have not been studied, and since they are on all chromosomes, no abnormalities in their inheritance would be expected. Of the rare fragile sites, the fragile X *(FRAXA)* associated with fragile X syndrome has been most studied. Before this fragile site had undergone molecular characterization, the inheritance patterns were particularly puzzling. Normal Mendelian segregation ratios were not seen: as Figure 3 shows, the penetrance of the fragile site and its associated syndrome depended on position in the family (this anomaly was known as the Sherman paradox: see Figure 4), and the general

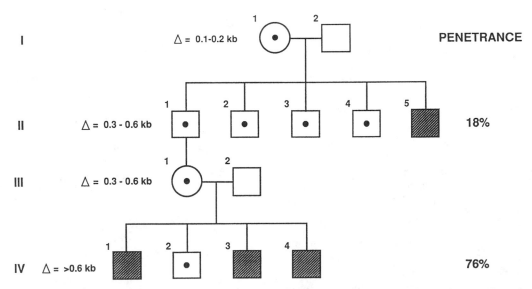

Figure 4. The Sherman paradox. A hypothetical pedigree is used to illustrate the different levels of penetrance of the fragile X mutation depending on the position of the carrier in the pedigree. The fragile X carrier I-1 has an 18% chance—approximately a one in five risk—that her fragile X sons are affected, while the fragile X carrier III-1 has a 76% chance that her fragile X sons are affected. The size of the amplification of the fragile X p(CCG)n repeat is given as Δ (i.e., for a normal individual $\Delta = 0.0$ kb). Values for Δ indicated are those expected hypothetically for individual 1 in each generation. Hatched squares indicate individuals with fragile X syndrome; dots indicate fragile X transmitters. [From Richards and Sutherland (1992).]

intellectual level of the mother. There were normal males who had transmitted the fragile X, and fragile X syndrome was apparent only when the abnormal chromosome was inherited from the mother.

These anomalies in fragile site inheritance were explained by the molecular characterization of the fragile site. Patients with premutations are asymptomatic; the chance that a premutation will expand to a full mutation in a child is a function of the size of the mother's premutation. Full mutations regress only rarely to premutations when transmitted by women, but they inevitably regress when transmitted by men.

Apart from the fragile sites on the X chromosome, there have been limited studies on the other group 1 fragile sites. For reasons that are presently unclear, these are fully penetrant when transmitted by women but only 50% penetrant when transmitted by men. While the two autosomal group 1 sites cloned so far (FRA16A and FRA11B) have molecular mechanisms similar to that of the fragile X, very few family studies at the molecular level have been performed. In at least the case of FRA16A and FRAXE a male can transmit a full mutation, but nothing is known about the likelihood of this happening.

The more frequently seen of the group 2 and 3 fragile sites (FRA16B and FRA10B) appear to have normal inheritance patterns.

7   CLINICAL SIGNIFICANCE

7.1   The Fragile X Syndromes

One fragile site on the X chromosome (FRAXA) is of major medical significance because a visible manifestation of the full mutation at this locus inactivates the FMR1 gene. This gives rise to fragile X syndrome, the most common familial form of intellectual handicap and one of the world's most common genetic diseases. The disease has been found in all population groups exposed to Western medicine, but it is not known whether it has the same frequency in all populations. Approximately 1 in every 4000 children in Caucasian populations has fragile X syndrome.

Another fragile site on the X chromosome (FRAXE) is apparently associated with a mild form of mental retardation. Only a small number of families have been studied, and not all males who carry this fragile site are mentally handicapped. The full effects of this fragile site will not be clear until more families have been characterized clinically, psychometrically, and molecularly.

7.2   Other Fragile Sites

Some mothers of children exhibiting a mental retardation–malformation condition known as Jacobsen syndrome have FRA11B. Jacobsen syndrome is due to a deletion of part of the long arm of chromosome 11, and the deletion breakpoints in some cases are at or very close to the fragile site on the mother's chromosome. This is the only autosomal fragile site with established clinical significance. Even in this situation, very few women with FRA11B will have a child with Jacobsen syndrome.

The only rare autosomal fragile sites to have been recorded in homozygous form are FRA10B, FRA16B, and FRA17A, where they are without apparent phenotypic effect. Group 1 fragile sites have never been seen in homozygotes. Any clinical effect in homozygotes would presumably depend on the function of the gene in which the fragile site was located. It is likely that the presence of a fragile site detected cytogenetically means that a full mutation is present, and any gene in which the fragile site is located is inactivated.

7.3   Cancer

There has been controversy for many years about a possible role for fragile sites in oncogenesis. The hypothesis is that breakage at the fragile sites could give rise to chromosome rearrangements of the types seen in tumor tissue. Most studies had shown that when breaks in such rearrangements were carefully mapped in relation to fragile sites, the two did not coincide. This topic has been rekindled by the recent finding that FRA11B is within the oncogene CLB2. Nevertheless, there is no evidence that individuals with rare fragile sites (everyone has all the common ones) are at any discernibly increased risk of developing malignant disease.

See also Fragile X Linked Mental Retardation; Genetic Testing; Human Chromosome 11, Genetic and Physical Map Status of.

Bibliography

Fisch, G. S., Snow, K., Thibodeau, S. N., Chalifaux, M., Holden, J. J. A., Nelson, D. L., Howard-Peebles, P. N., and Maddalena, A. (1995) The fragile X premutation in carriers and its effect on mutation size in offspring. *Am. J. Hum. Genet.* 56:1147–1165.

Hagerman, R. J., and Silverman, A. C. (1991) *Fragile X Syndrome.* Johns Hopkins University Press, Baltimore.

Jones, C., Penny, L., Mattina, T., Yu, S., Baker, E., Voullaire, L., Langdon, W. Y., Sutherland, G. R., Richards, R. I., and Tunnacliffe, A. (1995) Association of a chromosome deletion syndrome with a fragile site within the proto-oncogene CBL2. *Nature* 376:145–149.

Kunst, C., and Warren, S. T. (1994) Cryptic and polar variation of the fragile X repeat could result in predisposing normal alleles. *Cell,* 77:853–861.

Mandel, J. L., and Heitz, D. (1992) Molecular genetics of the fragile X syndrome: A novel type of unstable mutation. *Curr. Opinion Genet. Dev.* 2:422–430.

Richards, R. I., and Sutherland, G. R. (1992) Fragile X syndrome: The molecular picture comes into focus. *Trends Genet.* 8:249–255.

Schuffenhauer, S., Lederer, G., and Murken, J. (1996) A heritable folate-sensitive fragile site on chromosome 2p11.2 (FRA2L). *Chromosome Res.* (in press).

Sherman, S. C., Jacobs, P. A., Morton, N. E., Froster-Iskenius, U., Howard-Peebles, P. N., Nielsen, K. B., Partington, M. W., Sutherland, G. R., Turner, G., and Watson, M. (1985) Further segregation analysis of the fragile X syndrome with special reference to transmitting males. *Hum. Genet.* 69:289–299.

Sutherland, G. R. (1990) Human fragile sites. In *Genetic Maps,* Vol. V, S. J. O'Brien, Ed., pp. 5.193–5.196. Cold Spring Harbor Laboratory Press, Plain View, NY.

———. (1991) The detection of fragile sites on human chromosomes. In *Advanced Techniques in Chromosome Research,* K. W. Adolph, Ed., pp. 203–222. Dekker, New York.

———. (1993) Human fragile sites. In *Genetic Maps,* Vol. VI, S. J. O'Brien, Ed., pp. 6.264–6.267. Cold Spring Harbor Laboratory Press, Plainview, NY.

———, and Hecht, F. (1985) *Fragile Sites on Human Chromosomes.* Oxford University Press, New York.

———, and Richards, R. I. (1994) Dynamic mutations. *Am. Sci.* 82: 157–163.

———, and ———. (1995) Fragile X syndrome. In *Diseases of the Fetus and Newborn,* Vol. 2, 2nd ed., G. B. Reed, A. Claireaux, and F. Cockburn, Eds., pp. 1115–1119. Chapman & Hall, London.

———, and ———. (1995) Molecular basis of fragile sites in human chromosomes. *Curr. Opinion Genet. Dev.* 5:323–327.

FRAGILE X LINKED MENTAL RETARDATION

A. J. M. H. Verkerk and B. A. Oostra

Key Words

CpG Island   A DNA region rich in cytosine and guanine base pairs usually found in promoter regions of genes.

Cytogenetics   The study of (stained) metaphase chromosome preparations by means of microscopy.

Fragile Site   Can be induced in chromosomes by culturing cells in folate-deficient medium.

Normal Transmitting Male   A phenotypically normal male who is a carrier of a premutation in the *FMR-1* gene.

Premutation   A mutation in the *FMR-1* gene without phenotypic symptoms.

The fragile X or Martin Bell syndrome is an X-linked heritable disease. After trisomy 21 or Down syndrome it is the most frequent cause of mental retardation. The syndrome is associated with a (rare) fragile site (referred to as FRAXA) at the end of the long arm of the X chromosome in band Xq27.3. For an X-linked disease, this syndrome shows an unusual pattern of inheritance because part of the carrier females are affected and because of the occurrence of normal male carriers. The prevalence of the fragile X syndrome in the Caucasian population is estimated to be 1 in 1250 males and 1 in 2500 females.

1 DESCRIPTION OF THE FRAGILE X SYNDROME

1.1 SOME HISTORIAL ASPECTS

In 1943 Martin and Bell reported the first pedigree with clear familial X-linked mental retardation. In 1969 Lubs noted a secondary constriction or fragile site at the end of the long arm of the X chromosome (Xq27.3) in cultured cells from patients with X-linked mental retardation. In 1978 Turner found macroorchidism (enlarged testes) in most of these male patients. The term "Martin Bell syndrome" has been used synonymously with "fragile X syndrome."

1.2 CLINICAL CHARACTERISTICS

Developmental delay is usually noted in the second year of life. In male fragile X patients moderate to severe mental retardation is found. Other clinical symptoms are a long face, prominent forehead with relative macrocephaly, large everted ears, loose joints, and macroorchidism. Figure 1 shows typical facial characteristics of a male patient. Macroorchidism is found in 65–70% of all fragile X positive adult males and usually develops after puberty. Behavioral abnormalities include hyperactivity, hand-flapping, and poor eye contact. Approximately 30% of female carriers are mildly to moderately mentally retarded with typical facial characteristics. Fragile X patients have a normal life expectancy.

Figure 1.   Fragile X patient with typical facial characteristics.

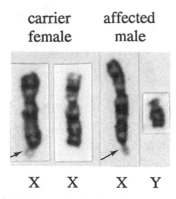

carrier   affected
female   male

X   X   X   Y

Figure 2.   Sex chromosomes of a female carrier and an affected male. The fragile site at Xq27.3 is clearly visible and indicated with an arrow. (Chromosome preparations provided by Dr. J. O. Van Hemel.)

1.3   CYTOGENETIC EXPRESSION OF THE FRAGILE X SITE

For a number of years confirmation of the clinical diagnosis was dependent on detecting the fragile site cytogenetically at Xq27.3 in cultured lymphocytes, fibroblasts, or amniocytes. The appearance of the fragile site, which is microscopically visible as a gap or break (Figure 2), is induced by culturing cells under specific conditions before chromosome spreads are made. In mentally retarded males the fragile site is usually seen in 2–60% of the cells. The site can be detected in only 50% of obligate carrier females.

1.4   UNUSUAL PATTERN OF INHERITANCE

In X-linked recessive diseases, females who carry a defective gene on one of their X chromosomes are usually unaffected. Their daughters have a risk of 50% of being unaffected carriers and their sons have a risk of 50% of being affected. Fragile X syndrome deviates from this pattern in the following points:

1. Thirty percent of the carrier females are mentally retarded, although they are often less severely affected than males. For this reason the syndrome is also referred to as an X-linked dominant disorder with incomplete penetrance. Sons of affected females have a risk of 50% of being affected.
2. Twenty percent of males who carry the fragile X mutation are phenotypically normal and are called normal transmitting males (NTMs). They will pass the mutation on to their daughters, who are also normal carriers. These daughters are again at risk of having affected children, but in this case the risk for a son to be affected is 40% instead of 50% (see Section 3). In NTMs and their daughters, no fragile sites at Xq27.3 are detected.
3. The disease phenotype is found in offspring only after transmission of the mutation by a female. Daughters of male carriers are always normal; the daughters of the few male patients that have reproduced are also normal.

2   POSITIONAL CLONING

Without knowledge about basic biochemical defects, positional cloning is the tool to identify genes. In fragile X syndrome the observation of a fragile X site was a first step in mapping the responsible gene to a location on a chromosome. Linkage analysis with polymorphic markers placed the fragile X locus between factor IX at Xq27.1 and factor VIII at Xq28, thus confirming the locus for the

gene at or near the fragile site. The next step was the isolation of markers that are closely linked to the fragile site and segregate with the disease in fragile X families. The isolation of new markers in this region eventually restricted the area of interest to 1000 kilobases (kb), and different molecular approaches were used that led to the identification of the fragile X gene. To study a chromosomal region, a physical map of such a region must be established, with markers located at fixed positions relative to the gene locus of interest. These markers can then be used to isolate larger chromosomal regions to subsequently "walk" along the chromosome to the gene locus.

2.1   USE OF HYBRID CELL PANELS

Somatic cell hybrids are hamster or mouse cells that contain a single human chromosome or part of one. Such cell hybrids are very useful for ordering markers along a human chromosome. To order markers in the fragile X region, many cell hybrids were developed containing different parts of the human X chromosome. A panel of somatic cell hybrids is shown in Figure 3. Two cell lines, termed Micro21D and Q1X, have been essential in the cloning of the fragile X gene. These cell lines, each containing a reciprocal part of the human X chromosome, were developed by Warren to specifically break in the middle of the fragile X site, the location presumed to contain the fragile X gene. Micro21D contains the Xpter-Xq27.3 part of the human X chromosome, Q1X contains the Xq27.3-Xqter part of the human X chromosome.

The rough location of a marker on the human X chromosome is determined by the presence or absence of this marker in a certain hybrid cell line. Figure 3 indicates the location of a number of markers from the fragile X region; their order is established with the use of genetic linkage studies and the different hybrid cell lines.

2.2   YACS SPANNING THE FRAGILE SITE

Cloning of human DNA in yeast artificial chromosome vectors (YACs), a technique developed by Burke and co-workers in 1987, made it possible to isolate DNA fragments from 100 kb to more than 1000 kb. Using markers tightly linked to the fragile site, including St677, p46-1.1, M759, and Do33 (see Figure 3), it was possible to isolate within the fragile X region several YACs ranging from 200 to 950 kb. Fluorescent in situ hybridization of one such YAC, 209G4, was performed on metaphase chromosomes from a fragile X patient after the fragile site had been cytogenetically induced. This 475 kb human DNA containing YAC was shown to cross the fragile site (Figure 4; see color plate 4). Hybridization signals are visible on both sides of the fragile site. This YAC was found to contain the fragile X gene. Fluorescent in situ hybridization of a 40 kb subclone of the YAC, cosmid 22.3, also disclosed signals on both sides of the fragile site (not shown). This cosmid contains a large part of the fragile X gene.

2.3   PHYSICAL MAP OF THE FRAGILE X REGION

A physical map of the fragile X region has been constructed using pulsed field gel electrophoresis and the markers shown in Figure 3, YACs and subclones of these YACs. The map shows part of the q27.3 region of the X chromosome, with markers flanking the fragile site FRAXA (Figure 5A). YAC 209G4 crosses the FRAXA site and contains the fragile X gene, *FMR-1* (Figure 5B and Section 3). The parts of the human X chromosome in the cell lines Micro21D

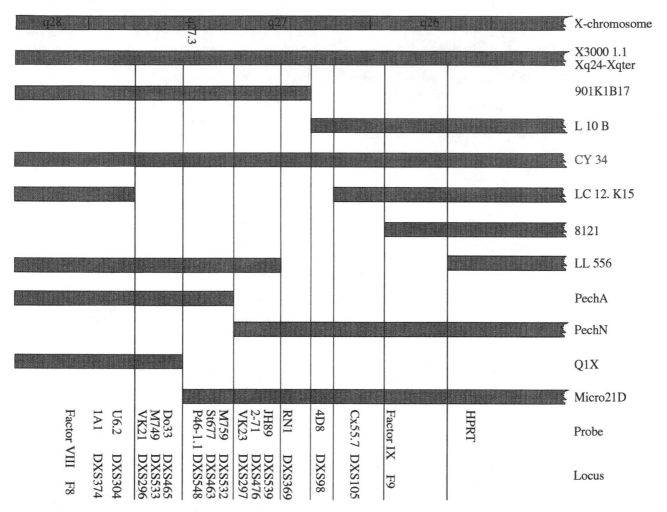

Figure 3. Hybrid cell panel consisting of several somatic hybrid cell lines containing different parts of the human X chromosome. Only the distal part of the Xq content of each hybrid is shown. The location of different markers in the fragile X region is indicated. Distances are not drawn to scale.

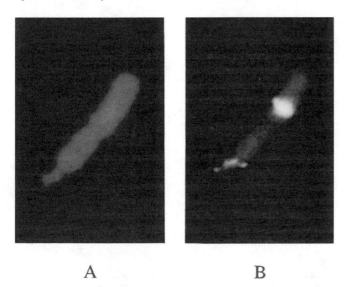

A   B

Figure 4. Fluorescence in situ hybridization of YAC 209G4 to chromosome preparations of a fragile X patient (B). A centromere probe specific for the X chromosome was also hybridized to confirm the identity of the chromosome examined. Staining by 4-,6-diamino-2-phenylindole (DAPI) clearly shows the fragile site (A). (In situ hybridizations performed by B. H. J. Eussen.) (See color plate 4.)

and Q1X break within the *FMR-1* gene. [Figure 5C, D is discussed in Section 5.]

3   THE FRAGILE X GENE, *FMR-1*

3.1   CHARACTERIZATION OF *FMR-1*

The fragile X gene, *FMR-1* (fragile X mental retardation-1), has a genomic size of approximately 40 kb and consists of 17 exons and 16 introns. The length of the mRNA is about 4.4 kb, with an open reading frame of about 1.9 kb. In the 5′ untranslated region of the *FMR-1* gene, an unusual CGG repeat is present. The ATG start codon for translation is found after the CGG repeat, indicating that the repeat is not translated into protein. A CpG island is located 250 bp proximal of the repeat. This CpG island as well as the CGG repeat itself are hypermethylated in DNA of fragile X patients. No methylation of the island or of the repeat is found in DNA of normal males and NTMs. CpG islands are found in promoter regions of genes and are C/G-rich areas that can be recognized by specific rare cutting restriction enzymes. Methylation of a CpG island is usually correlated with lack of gene expression. Methylation of the CpG island in front of the *FMR-1* gene coincides with lack of mRNA transcription from the *FMR-1* gene.

3.2   Nature of the Fragile X Mutation

The number of CGG repeats is found to be polymorphic in the human, mouse and pig *FMR-1* genes. In normal individuals the number of repeats varies from 6 to 53, with an average of 29 (Figure 6). In fragile X syndrome there is an increase in size of this CGG repeat. In general two classes of mutations are seen in fragile X families: premutations and full mutations. In NTMs and their daughters, the repeat number ranges from 43 to 200. This number of repeats is not associated with the disease phenotype and is called a premutation. In male and female patients and in many carrier females, more than 200 repeats are found, with some expansions exceeding 2000 repeats. This number of repeats is associated with the disease phenotype and is called a full mutation. In case of a full mutation, the CpG island and the amplified repeat itself are methylated, resulting in loss of transcription of the gene. Figure 7 represents schematically the molecular basis of the fragile X syndrome.

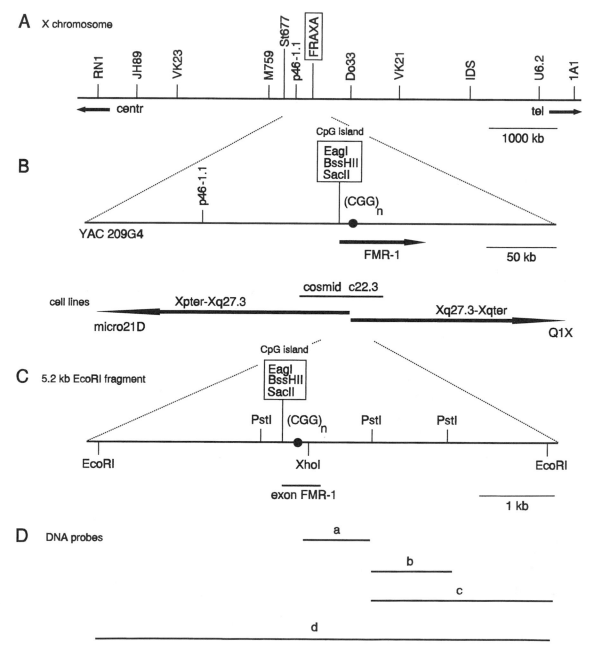

Figure 5.   Map of the fragile X region at Xq27.3. (A) Genetic map of DNA markers flanking the fragile site FRAXA. (B) Partial physical map of YAC 209G4 with the localization of the *FMR-1* gene. The CpG island of the *FMR-1* gene and the CGG repeat are indicated; restriction enzymes that recognize restriction sites in CpG islands are given in the box. The parts of the human X chromosome contained within the cell lines Micro21D and Q1X break within the CGG repeat in the *FMR-1* gene. (C) A 5.2 kb *Eco*RI fragment within the *FMR-1* gene that contains the CpG island, the CGG repeat, and the first exon of the gene. (D) Representation of DNA probes used in fragile X diagnosis: a, pfxa3 = Ox0.55 = pPX6; b, StB12.3 = pfxa7 = pP2; c, Ox1.9; d, pE5.1 = pfxa1.

3.3 INSTABILITY AND MOSAICISM OF THE CGG REPEAT

There is a small overlap between CGG repeat numbers in normal and premutation alleles, which can be distinguished by studying (in)stability of the allele in families.

Normal alleles containing 6–53 repeats are stable. This means that there is no change in repeat size when the gene is passed on to the next generation. Premutation alleles are unstable and in general increase in size by passage to the next generation. Occasionally a decrease in size is observed. Full mutations are also unstable and are found in offspring only after transmission of the premutation or full mutation by a female. Two kinds of mosaicism are observed in fragile X syndrome patients:

1. Extensive somatic instability is observed in individuals with a full mutation. During mitotic divisions the repeat increases (or sometimes decreases) in length, resulting in a variable number of repeats in different cells of the same individual. This is made visible by Southern blot analysis as a "smear" (see Sections 5.1 and 5.2). The mechanism of repeat amplification is still unknown (see Section 9.3).
2. Twenty-five percent of male patients have a premutation is found in part of their cells in addition to a full mutation. Despite transcription of the (unmethylated) premutation alleles, these males are mentally retarded. Overall, there is an apparent insufficiency of protein production in the appropriate tissues, resulting in the disease phenotype. This kind of mosaicism is also observed in fragile X females.

This somatic mosaicism probably is already established during early embryogenesis; indeed, identical patterns of mosaicism were found in different tissues of a 13-week-old fetus and in adult males. If mitotic instability occurred in every mitotic division during life, proliferation of a single cell would result in variable random patterns in different tissues. In vitro studies of proliferation of a single cell with one particular repeat size have demonstrated that the repeat length is maintained after a considerable number of cell divisions.

3.4 ALTERNATIVE SPLICING

In general splicing involves removing introns from a transcript and forming a bond between the ends of exons, resulting in one species of mRNA molecules and one species of protein. The *FMR-1* gene, however, is subject to alternative splicing. Alternative splicing of the precursor mRNA can lead the *FMR-1* gene to give rise to as many as 12 different mRNA molecules and thus to 12 possible proteins. Figure 8 gives the intron–exon distribution of *FMR-1*. The lengths of introns and exons are indicated. Alternative splicing occurs at three different locations in the gene: at exons XII, XV, and XVII. Exon XII (63 bases) can be spliced in or out of different transcripts. Of exon XV, only the first 36 bases or the first 75 bases can be spliced in or out. Of exon XVII, the first 51 bases can be spliced in or out. In all cases the open reading frame is maintained. Exon XII is a complete exon that can be spliced out in a conventional way. At exons XV and XVII, splicing occurs in the exon sequences, which is very unusual. The sequences of all (alternative) splice junction sites conform to the sequence of junction splice sites in primates. The longest possible transcript codes for a protein of 631 amino acids, the smallest for 568 amino acids. In addition, at the 3′ end of the gene, alternative use of different polyadenylation signals is found, resulting in longer or shorter 3′ untranslated regions. Not much is known about the protein function, since no significant homology is found to any other known protein. However, investigators have identified two 30 amino acid domains that resemble motifs in proteins thought to be involved in RNA binding. Antibodies directed against the FMR-1 protein recognize more than one protein product in white blood cells of healthy individuals, confirming the alternative splicing. The intracellular localization of the FMR-1 protein was found to be predominantly cytoplasmic. No different localization was found for the different splice variants of the proteins. In patients not expressing FMR-1 mRNA, no FMR-1 proteins could be detected.

3.5 mRNA EXPRESSION

Various levels of mRNA expression have been found in different organs, with high expression in human adult testis and brain, especially in the cerebellum and hippocampus. Studies of embryonal tissues of 8 and 9 weeks showed high expression in the brain, spinal cord, ganglia, neural retina, and cartilaginous structures, but not in testis. Further study of brain tissue of 25-week-old fetuses showed a high level of *FMR-1* mRNA in differentiating neurons of numerous cerebral structures. The *FMR-1* gene has been well conserved during evolution. A gene homologous to *FMR-1* is found in many species, including nematodes, chicken, mice, pigs, and different primates. In the mouse *Fmr-1* gene, 97% homology to the human gene is found at the protein level.

3.6 mRNA EXPRESSION AND METHYLATION

Methylation of the *FMR-1* CpG island has been examined in 10–11 weeks old tissues and chorionic villi of male fetuses with a full mutation. Almost complete methylation of the CpG island was found in the different fetal tissues; but no *FMR-1* mRNA, or a very low level of it, was present in these tissues. In the chorionic villi no (or only partial) methylation was found of the CpG island, with *FMR-1* mRNA present. This indicates that in the absence of methylation, transcription of the gene with a full mutation is possible. Villi and fetal tissues are derived from the same embryonal tissue and are separated late in the second week of gestation. In older fetuses complete methylation of the *FMR-1* CpG island was found. This means (1) that methylation of the expanded repeat probably occurs after the second week and is completed around or after the eleventh week of gestation and (2) that extension of the repeat is preceded by methylation. The methylation caused by the full mutation is present earlier than the methylation caused by the process of lyonization, which is not yet complete in the tenth week of gestation.

4   ELUCIDATION OF PAST OBSERVATIONS

Since the nature of the mutation is known, several previously puzzling observations can be explained.

1. In premutation alleles no methylation of the CpG island of the *FMR-1* gene is found, resulting in normal transcription of the gene. Therefore NTMs and females who carry a premutation are never mentally retarded.

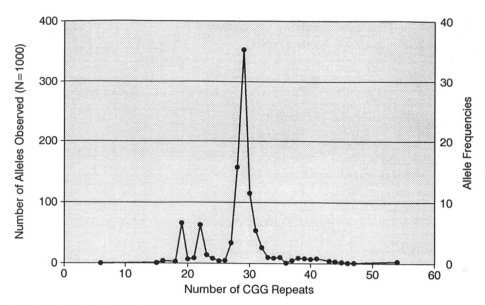

Figure 6. Distribution of number of CGG repeats in the *FMR-1* gene in the normal population. *x* axis, allele size; *y* axis, observed number of alleles.

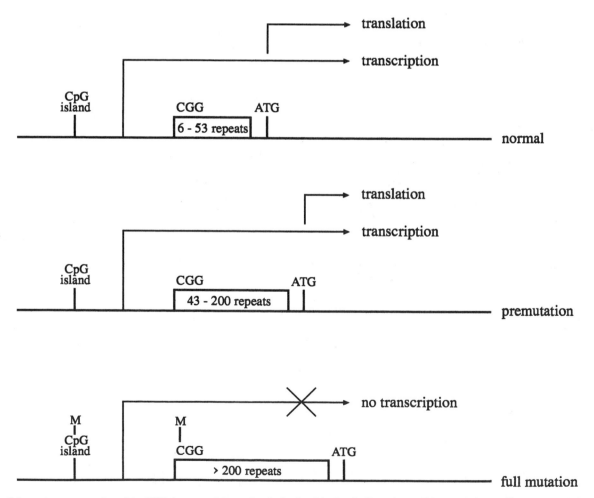

Figure 7. Schematic representation of the *FMR-1* gene and the molecular basis of the fragile X syndrome (M = methylated). Premutation alleles are found in normal transmitting males and in normal female carriers. Full mutations are found in 100% of affected males. Of the females carrying a full mutation, about 70% are affected.

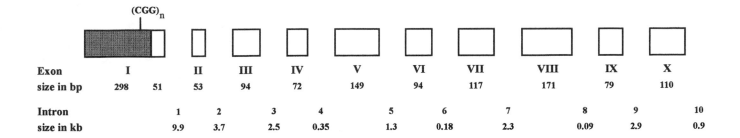

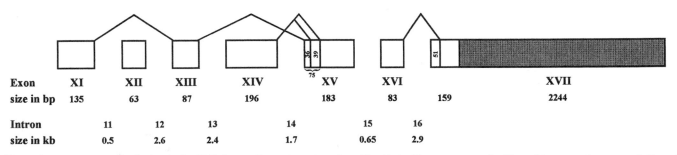

Figure 8. Intron–exon distribution in the *FMR-1* gene. Exons (open boxes) are identified with roman numerals. Sizes of introns and exons are indicated. Shaded portions at the 3' and 5' ends signify untranslated regions of the gene. The alternative splicing is depicted.

2. The length of the CGG repeat is correlated with the percentage of cells in which cytogenetic expression is found. Females who carry a premutation and NTMs never show cytogenetic expression. Practically all male patients with full mutations show cytogenetic expression.

3. It was noted by Sherman in 1985 that daughters of NTMs have a risk of 40% instead of 50% for their sons being affected. The finding that the risk for mental impairment in fragile X pedigrees is dependent on the position of individuals in the pedigree has become known as the Sherman paradox. Premutation alleles increase (and sometimes decrease) in size with almost each transmission to individuals in the next generation. Premutation alleles transmitted by NTM always stay in the premutation range in their daughters. Exclusively premutation alleles transmitted by females have a chance of increasing to a full mutation in the offspring (see Section 1.4). The risk of expansion of a pre-

mutation to a full mutation in the next generation is correlated with the size of the premutation allele of the mother. A low repeat number has a low risk, and a high repeat number has a high risk, of expanding to a full mutation. When a premutation contains more than about 90 repeats, the chance of its becoming a full mutation in the next generation is almost 100%. These risks correspond to the risks calculated by Sherman and explain the non-Mendelian pattern of inheritance in fragile X syndrome.

A premutation can be transmitted through many generations before expanding to a full mutation, resulting in the disease phenotype. Fragile X mutations are predominantly found on a limited number of haplotypes, suggesting founder chromosomes for this mutation.

Table 1 gives an overview of the different mutations in fragile X patients and carriers.

Table 1   Overview of the Different Mutations in Fragile X Patients and Carriers

Individual	Mutation	Mental Retardation	Number of Repeats	Methylation of *FMR-1*-Associated CpG Island	mRNA Transfer	Cytogenetic Expression of FRAXA	Risk of Affected Offspring
Normal		−	6–53	−	+	−	−
NTM	Pre-	−	46–200	−	+	−	−
Daughter of NTM and female carrier of a premutation	Pre-	−	46–200	−	+	−	+
Female carrier of a full mutation	Full[a]	70%[b]	>200	+[c]	+[d]	±50%[e]	+
Male patient	Full[a]	+	>200	+	−	+	−

[a]Somatic mosaicism is seen in about 25% of male patients. It is also observed in female carriers.
[b]Mild to moderate mental retardation is found in about 70% of females carrying a full mutation.
[c]Methylation is found on the inactive X and on the X chromosome carrying the full mutation.
[d]mRNA is transcribed only from the normal active X.
[e]Only in about 50% of females carrying a full mutation can the fragile site be detected cytogenetically.

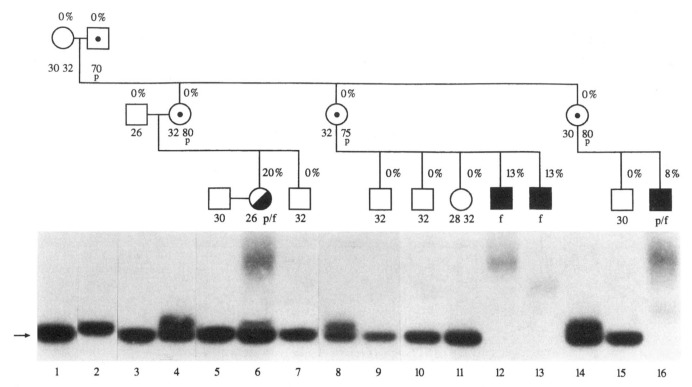

Figure 9. Detection of the *FMR-1* gene in a fragile X family by means of Southern blot analysis: ⊡, normal transmitting male; ⊙, normal female carrier; ◕, female carrier with cytogenetic expression; ■, mentally retarded male; *, percentage of cytogenetic expression; #, number of CGG repeats for the different X chromosomes (determined by means of PCR) is indicated; p, premutation; f, full mutation. (Example of fragile X DNA diagnosis provided by Dr. D. J. J. Halley and Dr. A. M. W. van der Ouweland.)

5   DIAGNOSTICS TODAY

5.1   TECHNICAL PROCEDURE

DNA fragments from individuals are visualized by Southern blot analysis and hybridization with a radioactively labeled specific probe. For DNA diagnosis of the fragile X syndrome, a specific probe is used to demonstrate the presence of a genomic DNA fragment containing the CGG repeat (Figure 5C, D). In normal individuals a 5.2 kb *Eco*RI restriction fragment is found. Premutations are visible as fragments somewhat larger than 5.2 kb (≤ 5.7 kb). Full mutations are visible as diffuse broad bands or "smears," which are due to the presence of different repeat lengths (> 200 repeats) in different cells. By means of the polymerase chain reaction (PCR) technique, it is possible to determine the CGG repeat number of normal and premutation alleles. Because the C/G content of CGG repeat containing alleles is high, it is still difficult to determine repeat numbers above 200 by PCR.

5.2   EXAMPLE OF DNA DIAGNOSIS IN A FRAGILE X FAMILY

Postnatal and prenatal diagnosis of fragile X syndrome can now be exercised with a high degree of accuracy by detecting the CGG repeat amplification at the DNA level. Figure 9 shows the pedigree of a fragile X family in which premutations and full mutations are found. The DNA of the grandmother (1) has one normal 5.2 kb band (arrow), representing her two normal X chromosomes. The grandfather (2) is a normal transmitting male with a premutation of 70 re-

peats in the *FMR-1* gene; his DNA shows a band slightly larger than 5.2 kb. He passed his premutation on to his three daughters (4, 8, and 14), who are phenotypically normal carriers. Besides the 5.2 kb band of their normal chromosome, they have a premutation band of increased size (75 and 80 repeats). Daughter 4 has one affected daughter (6), who is mosaic for a premutation and a full mutation. Daughter 8 has two affected sons (12 and 13) with full mutations. Daughter 14 has one affected son (16), who is mosaic for a premutation and a full mutation.

6   PATIENTS WITHOUT CGG REPEAT EXPANSION

Since publication of the description of the gene defect, patients with the fragile X phenotype but without CGG repeat expansion or cytogenetic expression have been reported. In one such a patient the syndrome is caused by a de novo deletion of less than 250 kb that includes the CpG island and at least five exons of the *FMR-1* gene. In another patient a deletion of more than 2 megabases, including the entire *FMR-1* gene, was found. In these patients flanking regions are also missing, and thus the involvement of other genes contributing to the disease phenotype could not be excluded. In one patient, however, a de novo point mutation was found in the *FMR-1* gene in the putative RNA-binding domain (see Section 3.4), resulting in an aberrant protein causing the fragile X phenotype. This indicates that the fragile X syndrome is a single-gene disorder and that loss of function or mutation in the *FMR-1* gene leads to the clinical phenotype of the fragile X syndrome.

7   OTHER FRAGILE SITES AT Xq27

Several families have been described in which individuals exhibiting a fragile site at Xq27 do not have a repeat amplification in the *FMR-1* gene, even though they show expression of the fragile site at levels from 5 to 75%. Mentally retarded individuals in these families were in the first instance examined for a possible fragile X syndrome diagnosis. In most cases, after clinical reexamination at a later stage, the affected individuals were assessed to have a nonspecific phenotype. In these families, another neighboring fragile site was found. Two other different fragile sites, called FRAXE and FRAXF, are found that are cytogenetically indistinguishable from FRAXA. These fragile sites can be distinguished by in situ hybridization of different DNA probes or cosmids from the Xq27-q28 region that flank the fragile sites on fragile site induced metaphase chromosomes (see Figure 10). The FRAXA site lies between cosmid 7.1 (a distal subclone from YAC 209G4) and Do33. The FRAXE fragile site lies between Do33 and VK21 and is located 600 kb distal to FRAXA. FRAXF lies between VK21 and 1A1 and is located more than 1000 kb distal from FRAXA.

The FRAXE site is associated with mild mental retardation and an extended GCC repeat (> 200 repeats) in patients. In normal individuals 6–25 copies of the GCC sequence are found, with an average of 15. The associated CpG island lies 600 kb distal to the *FMR-1*-associated CpG island and is methylated in FRAXE patients. This suggests that methylation also plays a role in the inacti-

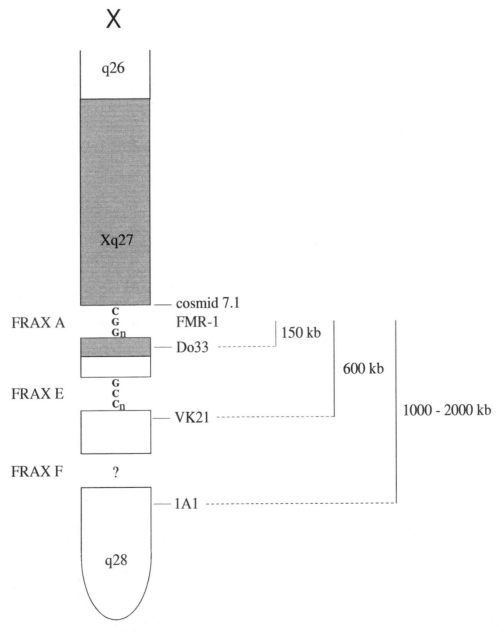

Figure 10.   Different fragile sites in the Xq27-28 region: FRAXA containing a CGG repeat, FRAXE containing a GCC repeat, and FRAXF, of which the molecular composition is unknown. These sites can be distinguished by in situ hybridization with different probes.

vation of the gene associated with the FRAXE CpG island, which in turn results in mental impairment. FRAXE patients have a normal methylation pattern across the FRAXA region.

The molecular nature of FRAXF is still unknown. Cloning this region of fragility will determine whether FRAXF is also due to a repeat amplification.

8   OTHER DISEASES CONTAINING TRINUCLEOTIDE REPEATS

Fragile X syndrome is the first reported disorder of a series of disorders in which expansion of a trinucleotide repeat is responsible for the disease phenotype. Other diseases that present the same kind of mutation mechanism are Kennedy's disease, or spinal and bulbar muscular atrophy (SBMA), myotonic dystrophy (DM), Huntington's disease (HD), spinocerebellar ataxia type 1 (SCA1), and the mental retardation associated with the FRAXE site (see Section 7). The genes associated with these disorders all contain a polymorphic trinucleotide repeat that is expanded in patients. In the cases of DM, SBMA, HD, and SCA1, the repeat is contained in the mature transcript. The location of the repeat in FRAXE is not yet known. Table 2 gives an overview of data considering the different genes involved.

1. Kennedy's disease is an X-linked recessive disease that has been described in about 50 families. Affected males show an adult onset of progressive muscle weakness and atrophy; they may have gynecomastia and reduced fertility. A CAG repeat in the androgene receptor gene on chromosome Xq11-12 is expanded in patients.
2. Myotonic dystrophy is an autosomal dominant neuromuscular disease with an estimated incidence of 1 in 8000. Patients exhibit progressive muscle weakness and wasting, myotonia, and cataracts. A CTG repeat in the 5′ untranslated region of the DM gene on chromosome 19q13.3 is expanded in patients.
3. Huntington's disease is an autosomal dominant progressive neurodegenerative disorder and affects about 1 in 10,000 individuals. A CAG repeat in the HD gene on chromosome 4p16.3 is expanded in patients.
4. Spinocerebellar ataxia type 1 is an autosomal dominant progressive neurodegenerative disease. The SCA1 gene is locat-

ed on chromosome 6p22-23 and contains a CAG repeat that is extended in patients.

These four disorders exhibit the phenomenon of "anticipation," that is, the apparent occurrence of increasing severity of symptoms of an inherited disorder (through successive generations), with progressively earlier age of onset of the disease in successive generations. The effect can be explained by the molecular findings: the number of repeats increases with the transmission of the mutation from generation to generation. In more severely affected individuals, higher copy numbers of repeats are found. The "anticipation" as seen in fragile X syndrome is an all-or-nothing phenomenon and does not fit the definition. In carriers of a premutation, normal transcription and translation take place, resulting in a normal phenotype. When the amplification passes the border of 200 CGG repeats, there is methylation of the promoter region of the FMR-1 gene and the repeat itself occurs, preventing transcription and resulting in the disease phenotype in all males. In these males the severity of the disease is not dependent on the number of repeats found in their FMR-1 gene. In DM, HD, and SCA1 the anticipation is evident; in SBMA it is present but not as evident as in DM, HD, and SCA1. In the few known FRAXE families, the degree of amplification is found to correlate with the clinical manifestation of mental retardation.

9   CONSIDERATIONS

9.1   TIMING OF EXPANSION

The risk of expansion from a premutation in the parent to a full mutation in the offspring is dependent on two factors: the size of the premutation and parental origin.

1. *Transmission by a female:* when a premutation is transmitted by a female, the chance of a full mutation in the offspring correlates with the size of the premutation in the mother (see Section 4).
2. *Transmission by a male:* when a premutation is transmitted by a (normal transmitting) male, expansion to the full mutation never occurs, irrespective of the size of the repeat. This ex-

Table 2   Data Concerning the Trinucleotide Repeat Genes Involved in Fragile X Syndrome, Kennedy's Disease, Myotonic Dystrophy, Huntington's Disease, and Spinocerebellar Ataxia Type 1

	Fragile X Syndrome	Kennedy's Disease	Myotonic Dystrophy	Huntington's Disease	Spinocerebellar Ataxia Type 1
Chromosomal location	Xq27.3	Xq11–12	19q13.3	4p16.3	6p22–23
Gene involved	FMR-1	AR	DM	IT15	SCA1
Repeat	CGG	CAG	CTG	CAG	CAG
Location in the gene	5′ UTR	5′ coding	3′ UTR	5′ coding	Nk
Number of					
Repeats in normals	6–53	12–30	5–35	11–34	25–36
Repeats in patients	200–>2000	40–62	50–>2000	42–>66	43–81
Gene characteristics					
Genomic size, kb	40	>90	14	210	Nk
mRNA transcript, kb	4.4	10.6	3.4	10–11	10
Open reading frame, kb	1.9	2.7	1.9	9.4	Nk
Protein length (amino acids)	631[a]	910	629[a]	3144	Nk

[a]Alternative splicing means that several differently sized proteins are possible. The longest possible protein is indicated.
Abbreviations: FMR-1, fragile X mental retardation 1; *AR,* androgene receptor; *MD,* myotonic dystrophy; *IT15,* Interesting transcript 15; *SCA1, Spinocerebellar ataxia type 1;* Nk, not known yet.

plains why daughters of NTMs always have a premutation and are phenotypically normal.

The few patients to have reproduced who somatically have a full mutation also have daughters carrying a premutation who are phenotypically normal.

Fragile X males having a full methylated mutation in their lymphocytes showed only unmethylated premutations in their sperm cells. One could speculate that also in oocytes of fragile X females only premutations are present. Subsequently (i.e., after fertilization in the early embryo), the expansion of a premutation to a full mutation might occur. However, female gametes have not yet been studied. The occurrence of expansion to a full mutation only after female transmission is not yet explained, but it could be due to a parental imprinting mechanism in the *FMR-1* region: a distinction is made between paternally and maternally derived premutations, and for some reason only maternally derived alleles can expend to a full mutation in the offspring.

9.2 Absence of the Full Mutation in Male Gametocytes

Several explanations are possible for the absence of the full mutation in male gametocytes:

1. Gametes with a premutation might have a selective advantage over gametes with a full mutation.

   (a) Both premutations and full mutations are present in spermatogonia of primordial germ cells, but only cells containing a premutation can develop to mature sperm cells. In normal mice a high expression of *FMR-1* mRNA is found in proliferating stages of spermatogonia and not in mature sperm. This could imply that *FMR-1* expression is necessary for germ cell proliferation and that only spermatogonia with a premutation allele, which allows for mRNA synthesis, may enter spermatogenesis.

   (b) Under selective pressure, a regression from a full mutation to a premutation might occur, but this kind of a regression has not been documented in fragile X families. If regression exists, it probably occurs much less frequently than expansion.

2. The male embryo receives a premutation allele from the mother, so a premutation is present in the early embryo. If expansion occurs in an early embryonic phase, it might occur in the somatic cells only, after separation of the germ line cells, which occurs between days 5 and 20 after fertilization.

9.3 Mechanism of Expansion

The mechanism through which the repeat expansion takes place is still unknown. Premutations can be passed on through many generations before the occurrence in a pedigree of a full mutation, and subsequently a fragile X phenotype. Different mechanisms have been proposed, but none can explain why only repeat expansions occur in fragile X syndrome (or in the other disorders subject to expansions of repeat containing genes), because according to these models repeat regressions would have to occur as often as expansions.

1. Model 1 assumes expansion during the meiotic phase in germ cells. Unequal crossover between homologous chromosomes during meiosis cannot explain the mechanism of repeat amplification for four reasons: (a) no recombinations have been found to occur at the site of the mutation; (b) in males, amplification occurs without meiotic crossover because premutation alleles passed on by an NTM (where only one X chromosome is present) also change in size (but stay in the premutation range); (c) all expansions from a premutation to a full mutation are much longer than might be expected from recombination; and (d) although deletions or decreases should occur as often as increases, far more increases than decreases have been observed.

2. Model 2 assumes mitotic phase expansion in the germ cells but also can explain expansion during mitosis in somatic cells. Before germ cells enter meiosis, they go through a number of mitotic divisions during which unequal sister chromatid exchange (SCE) can occur, resulting eventually in gametes carrying alleles with different numbers of repeats. This model, in contrast to model 1, at least explains that premutation alleles from NTMs can change in size, because SCE can occur in a single X chromosome. Why premutation alleles transmitted by females can expand further (to a full mutation) than premutation alleles transmitted by NTMs still remains a question.

3. Another possibility would be that DNA replication of C/G-rich areas is difficult and that inequality of the rate of DNA synthesis might lead to multiple incomplete single strands. Out-of-register pairing of the repeats and reinitiation might then lead to longer or shorter alleles. This might occur during replication in germ cells preceding meiosis, or in somatic cells. But again decreases in repeat lengths are rare events.

4. GAP repair may occur. That is, a loop might form in the repeat region during replication when the repeat has reached a critical length. If, to remove the loop, a nick occurred in the strand opposite the loop, stretching the DNA, there would be a gap in the other strand, which would have to be filled in, extending the DNA.

9.4 Repeats and Cytogenetic Expression

Fragile X syndrome and FRAXE mental retardation are both associated with a cytogenetically visible fragile site. The other diseases caused by repeat extensions are not associated with fragile sites. The repeats in fragile X syndrome (CGG) and in FRAXE mental retardation (GCC) both contain CpG sequences, which can be methylated on the cytosine, which occurs when these repeats expand beyond 200 repeats. Therefore it is reasonable to assume that the existence of fragility is correlated with the nature of the repeat (composed of C and G nucleotides) and/or with the methylation of this kind of repeat.

9.5 Concluding Remarks

The mechanism of repeat expansion, the time of repeat expansion, and the function of the FMR-1 protein are still unknown. In the mouse *fmr-1* gene, which has a high homology to the human *FMR-1* gene, a polymorphic CGG repeat is also present. The study of *fmr-1* in a mouse model might give an indication of the function of *FMR-1* in the human. By studying transgenic mice in which the *fmr-1* gene has been knocked out or by introducing an enlarged CGG repeat in the *fmr-1* gene, we may find some answers related to the function and timing and mechanism of repeat expansion.

Bibliography

Caskey, C. T., Pizzuti, A., Fu, Y. H., Fenwick, R. G., and Nelson, D. J. (1992) Triplet repeat mutations in human disease. *Science,* 256: 784–789.

Hagerman, R. J., and Silverman, A. C., Eds. (1991) *Fragile X Syndrome: Diagnosis, Treatment and Research.* John Hopkins Press, Baltimore.

Hirst, M. C., Knight, S. J. L., Bell, M. V., Super, M., and Davies, K. E. (1992) The fragile X syndrome. *Clin. Sci.* 83: 255–264.

McKusick, V. A. (1992) 309550 Mental retardation, X-linked, associated with marXq28. In *Mendelian Inheritance in Man. Catalogs of Autosomal Dominant, Autosomal Recessive, and X-Linked Phenotypes.* Johns Hopkins University Press, Baltimore and London.

Oostra, B. A., Willems, P. J., and Verkerk, A. J. M. H. (1993) Fragile X syndrome: A growing gene. In *Genome Analysis,* Vol. 6, *Genome Maps and Neurological Disorders,* K. E. Davies and S. M. Tilghman, Eds. Cold Spring Harbor Laboratory Press, Plainview, NY.

Richards, R. I., and Sutherland, G. R. (1992) Dynamic mutations: A new class of mutations causing human disease. *Cell,* 70: 709–712.

X-Linked Mental Retardation 5, (1992) Special issue. *Am. J. Med. Gen.* 43: 1–372.

FREE RADICALS IN BIOCHEMISTRY AND MEDICINE

Barry Halliwell

Key Words

Antioxidant   A molecule that protects a biological target against oxidative damage.

Free Radical   Any species containing one or more unpaired electrons.

Oxidative Stress   An imbalance between the generation of reactive oxygen species and antioxidant protection, in favor of the former.

Polyunsaturated Fatty Acid   A fatty acid with two or more carbon–carbon double bonds in the side chain.

Reactive Oxygen Species   A collective name given to oxygen-containing radicals ($O_2^{\cdot-}$, $OH^{\cdot}$, peroxyl, alkoxyl) and some other nonradical derivatives of oxygen, such as H_2O_2.

Free radicals and other oxygen-derived species are constantly generated in vivo, both by "accidents of chemistry" and for specific metabolic purposes. The reactivity of different free radicals varies, but some can cause severe damage to biological molecules, especially to DNA, lipids, and proteins. Antioxidant defense systems scavenge oxygen-derived species and minimize their formation, but are not 100% effective. Hence, repair systems exist to deal with molecules that have been oxidatively damaged. Damage to DNA by hydroxyl radicals appears to occur in all aerobic cells and might be a significant contributor to the age-dependent development of cancer.

1   WHAT IS A RADICAL?

In the structure of atoms and molecules, electrons usually associate in pairs, each pair moving within a defined region of space around the nucleus. This space is referred to as an atomic or molecular *orbital.* One electron in each pair has a spin quantum number of $+\frac{1}{2}$, the other $-\frac{1}{2}$. A *free radical* is any species capable of independent existence (hence the term "free") that contains one or more *unpaired electrons,* an unpaired electron being one that is alone in an orbital. The simplest free radical is an atom of the element hydrogen, with one proton and a single electron. Table 1 gives examples of other free radicals. The spectroscopic technique of *electron spin resonance* (ESR) is used to measure free radicals; it records the energy changes that occur as unpaired electrons align in response to a magnetic field. A superscript dot is used to denote free radical species (Table 1).

2   RADICALS IN VIVO

The chemical reactivity of free radicals varies. One of the most reactive is *hydroxyl radical* ($OH^{\cdot}$). Exposure of living organisms to ionizing radiation causes homolytic fission of O—H bonds in water (remember that living organisms are greater than 70% water)

$$H_2O \rightarrow H^{\cdot} + OH^{\cdot} \qquad (1)$$

to give $H^{\cdot}$ and $OH^{\cdot}$. Hydroxyl radical reacts at a diffusion-controlled rate with almost all molecules in living cells. Hence, when $OH^{\cdot}$ is formed in vivo, it damages whatever it is generated next to—it cannot migrate any significant distance within the cell. The harmful effects on living organisms of excess exposure to ionizing radiation

Table 1 Examples of Free Radicals

Name	Formula	Comments
Hydrogen atom	$H^{\bullet}$	The simplest free radical
Trichloromethyl	$CCl_3^{\bullet}$	A carbon-centered radical (i.e., the unpaired electron resides on carbon). $CCl_3^{\bullet}$ is formed during metabolism of CCl_4 in the liver and contributes to the toxic effects of this solvent.
Superoxide	$O_2^{\bullet -}$	An oxygen-centered radical.
Hydroxyl	$OH^{\bullet}$	An oxygen-centered radical. The most highly reactive oxygen radical known.
Thiyl	$RS^{\bullet}$	A group of radicals with an unpaired electron residing on sulfur.
Peroxyl, alkoxyl	$RO_2^{\bullet}, RO^{\bullet}$	Oxygen-centered radicals formed during the breakdown of organic peroxides.
Oxides of nitrogen	$NO^{\bullet}, NO_2^{\bullet}$	Both are free radicals. $NO^{\bullet}$ is formed in vivo from the amino acid L-arginine. $NO_2^{\bullet}$, made when $NO^{\bullet}$ reacts with O_2, is found in polluted air and smoke from burning organic materials (e.g., cigarette smoke).

are thought to be initiated by attack of $OH^{\bullet}$ on proteins, carbohydrates, DNA, and lipids. For example, $OH^{\bullet}$ can abstract hydrogen atoms from fatty acid side chains in membrane lipids and initiate the process of lipid peroxidation.

2.1 LIPID PEROXIDATION

Initiation of peroxidation occurs by the attack of any species ($R^{\bullet}$) capable of abstracting hydrogen from a polyunsaturated fatty acid side chain in a membrane (such fatty acid side chains are more susceptible to free radical attack than are saturated or monounsaturated side chains):

$$—CH + R^{\bullet} \rightarrow —C^{\bullet} + RH \qquad (2)$$

Species able to abstract hydrogen include $OH^{\bullet}$ and peroxyl radicals (Table 1).

The carbon-centered radical that resulted from reaction (2) reacts fast with O_2:

$$—C^{\bullet} + O_2 \rightarrow —CO_2^{\bullet} \qquad (3)$$

and a fatty acid side chain *peroxyl radical* is formed. This can attack adjacent fatty acid side chains and *propagate* lipid peroxidation:

$$—CO_2^{\bullet} + —CH \rightarrow —CO_2H + —C^{\bullet} \qquad (4)$$

The chain reaction thus continues, and *lipid peroxides* (—CO_2H) accumulate in the membrane. Lipid peroxides destabilize membranes and make them "leaky" to ions. Peroxyl radicals can attack not only lipids but also membrane proteins (e.g., damaging enzymes, receptors, and signal transduction systems), and they can oxidize cholesterol.

2.2 THE HYDROXYL AND SUPEROXIDE RADICALS

When $OH^{\bullet}$ is generated adjacent to DNA, it attacks both the deoxyribose and the purine and pyrimidine bases. Figure 1 shows the

structures of some of the products generated by attack of $OH^{\bullet}$ on the DNA bases: this wide range of products is characteristic of attack by $OH^{\bullet}$ and may be used to show that such attack has occurred in vivo. For example, if most or all of the compounds given in Figure 1 are present in DNA that has been extracted from a tissue, we know that the DNA has suffered attack by $OH^{\bullet}$. Such "$OH^{\bullet}$ fingerprint" experiments have been used to study the role of free radicals in DNA damage by radiation and toxic agents. It has also been found that there are larger amounts of these $OH^{\bullet}$-derived products in DNA from human cancerous tumors than in normal tissue. Whether this is due to increased $OH^{\bullet}$ formation or to decreased repair of the damage (see Section 7) remains to be evaluated.

Whereas $OH^{\bullet}$ is probably always harmful, other (less reactive) free radicals may often be useful in vivo. Free radicals are known to be produced metabolically in living organisms. For example, the free radical nitric oxide ($NO^{\bullet}$) is synthesized from the amino acid L-arginine by vascular endothelial cells, phagocytes, certain cells in the brain, and many other cell types. Nitric oxide is a vasodilator agent and, possibly, an important neurotransmitter. It may also be involved in the killing of parasites by macrophages in some mammalian species.

Superoxide radical ($O_2^{\bullet -}$) is the one-electron reduction product of oxygen. It is produced by phagocytic cells (neutrophils, monocytes, macrophages, eosinophils) and helps them to kill bacteria. Evidence is accumulating to suggest that smaller amounts of extracellular $O_2^{\bullet -}$ may be generated, perhaps as an intercellular signal molecule, by several other cell types, including endothelial cells, lymphocytes, and fibroblasts. In addition to this deliberate metabolic production of $O_2^{\bullet -}$, some $O_2^{\bullet -}$ is produced within cells by mitochondria and endoplasmic reticulum. This is usually thought to be an unavoidable consequence of "leakage" of electrons onto O_2 from their correct paths in electron transfer chains (i.e., an "accident of chemistry"). Aerobic organisms also contain many "autoxidizable" compounds, compounds that react directly with oxygen to generate free radicals. For example, adrenaline (epinephrine) slowly reacts with O_2 to form $O_2^{\bullet -}$, which then oxidizes more epinephrine, setting up a chain reaction. These autoxidizable compounds presumably react with O_2 in vivo, providing another source of $O_2^{\bullet -}$

2.3 HYDROGEN PEROXIDE—A NONRADICAL

Most of the $O_2^{\bullet -}$ generated in vivo probably undergoes a dismutation reaction, represented by the overall equation

$$2O_2^{\bullet -} + 2H^+ \rightarrow H_2O_2 + O_2 \qquad (5)$$

Hydrogen peroxide (H_2O_2), a nonradical, resembles water in its molecular structure and is very diffusible within and between cells. As well as arising from $O_2^{\bullet -}$, H_2O_2 is produced by the action of certain oxidase enzymes in cells, including amino acid oxidases and the enzyme xanthine oxidase. Xanthine oxidase catalyzes the oxidation of hypoxanthine to xanthine, and of xanthine to uric acid. Oxygen is simultaneously reduced both to $O_2^{\bullet -}$ and to H_2O_2. Low levels of xanthine oxidase are present in many mammalian tissues, especially in the gastrointestinal tract. Levels of xanthine oxidase often increase when tissues are subjected to insult, such as trauma or deprivation of oxygen.

Like $O_2^{\bullet -}$, H_2O_2 has certain useful metabolic functions. For example, H_2O_2 is used by the enzyme *thyroid peroxidase* to help make thyroid hormones. H_2O_2 (or products derived from it) can displace the inhibitory subunit from the cytoplasmic transcription factor NF-

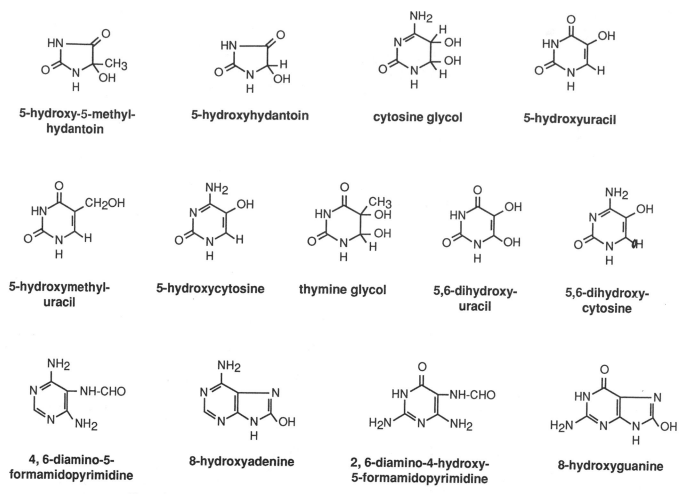

Figure 1.   Some products of attack of hydroxyl radicals on purine and pyrimidine bases of DNA.

κB. The active factor migrates to the nucleus and activates many different genes by binding to specific DNA sequences in enhancer and promoter elements. Thus, H_2O_2 can induce expression of genes controlled by NF-κB. This property is of particular interest at the moment because NF-κB can induce the expression of genes of the provirus HIV-1, the most common cause of acquired immunodeficiency syndrome.

3   RADICAL REACTIONS

If two free radicals meet, they can join their unpaired electrons to form a covalent bond. Thus, for two hydrogen atoms

$$H^. + H^. \rightarrow H_2 \qquad (6)$$

and for $NO^.$ and $O_2^{.-}$

$$NO^. + O_2^{.-} \rightarrow ONOO^- \qquad (7)$$
$$\text{peroxynitrite}$$

However, when a free radical reacts with a nonradical, a new radical results; that is, a *chain reaction* is set up. Since most biological molecules are nonradicals, the generation in vivo of reactive radicals such as $OH^.$ usually sets off chain reactions. For example, attack of $OH^.$ on fatty acid side chains in membranes and other biological lipoproteins can abstract hydrogen, leaving a carbon-centered radical and initiating the process of lipid peroxidation

(Section 2.1). Attack of $OH^.$ on the DNA bases produces radicals, which then undergo complex reactions to generate the end products shown in Figure 1.

4   THE ROLE OF TRANSITION METAL IONS

Neither $O_2^{.-}$ nor H_2O_2 is very reactive. At low levels, they are often metabolically useful. If they come into contact with iron or copper ions, however, the noxious $OH^.$ can be formed:

$$O_2^{.-} + H_2O_2 \xrightarrow{Cu, Fe} OH^. + OH^- + O_2 \qquad (8)$$

The chemistry involved is much more complicated than this simple summary equation suggests, however.

Hence, the presence in a biological system of transition metal ions able to catalyze reaction (8) can cause $O_2^{.-}$ and H_2O_2 to be damaging. Because $OH^.$ cannot migrate any significant distance, the damage will occur at the sites at which the "catalytic" metal ions are present. That is, it is *site-specific* free radical damage. For example, if such metal ions are bound to DNA and $O_2^{.-}$ plus H_2O_2 reach the DNA, $OH^.$ will be generated and damage the DNA (Figure 1). However, if the metal ions were bound to membranes, then lipid peroxidation could result (Section 2.1).

It has also been suggested that peroxynitrite (reaction 7) can generate $OH^.$:

$$ONOO^- + H^+ \rightarrow NO_2^. + OH^. \qquad (9)$$

If reaction (9) occurs in vivo, it represents a non-transition-metal-dependent mechanism of converting $O_2^{.-}$ into $OH^.$.

5 ANTIOXIDANT DEFENSES

Living organisms have evolved antioxidant defenses to remove excess $O_2^{.-}$ and H_2O_2. *Superoxide dismutase* enzymes (SODs) remove $O_2^{.-}$ by accelerating its conversion to H_2O_2 (equation 1) by about four orders of magnitude at pH 7.4. Mammalian cells have a SOD enzyme containing manganese at its active site (MnSOD) in the mitochondria. A SOD with copper and zinc at the active site (CuZnSOD) is also present, but largely in the cytosol.

Because SOD enzymes generate H_2O_2, they work in collaboration with H_2O_2-removing enzymes. *Catalases* convert H_2O_2 to water and O_2:

$$2H_2O_2 \rightarrow 2H_2O + O_2 \tag{10}$$

Catalases are present in the peroxisomes of mammalian cells and probably serve to destroy H_2O_2 generated by oxidase enzymes located within these subcellular organelles. However, the most important H_2O_2-removing enzymes in mammalian cells are the *glutathione peroxidases* (GSHPX), enzymes that require selenium (a selenocysteine residue, essential for enzyme activity, is present at the active site). GSHPX enzymes remove H_2O_2 by using it to oxidize reduced glutathione (GSH) to oxidized glutathione (GSSG). Glutathione reductase, a flavoprotein (FAD-containing) enzyme, regenerates GSH from GSSG, with NADPH as a source of reducing power (Figure 2).

An additional important antioxidant defense system consists of metal ion storage and transport proteins, which organisms have evolved to keep iron and copper safely protein-bound whenever possible, thus minimizing the occurrence of reaction (8).

Antioxidant defense enzymes are essential for healthy aerobic life. Because of impaired biosynthesis of certain amino acids, for example, SOD-negative mutants of *E. coli* (obtained by transposon insertions into the cloned structural genes, followed by an exchange between the mutated SOD allele carried by a plasmid, and the chromosomal wild-type allele) will not grow aerobically unless given a rich growth medium. Even when so supplemented, SOD$^-$ *E. coli* cells grow slowly, suffer membrane damage, are abnormally sensitive to damage by H_2O_2 (perhaps because of reaction 8) and show a high mutation rate.

In addition to antioxidant defense enzymes, some small molecule free radical scavengers exist. GSH can scavenge various free radi-

cals directly, as well as being a substrate for GSHPX enzymes. α-Tocopherol is the most important (but by no means the only) free radical scavenger within membranes and lipoproteins. α-Tocopherol inhibits lipid peroxidation by scavenging peroxyl radicals (Table 1), which are intermediates in the chain reaction described in Section 2.1:

$$\alpha\text{-TH} + LOO^. \rightarrow \alpha T^. + LOOH \tag{11}$$

where $\alpha T^.$ is the α-tocopherol radical and $LOO^.$ is the peroxyl radical. Since $\alpha T^.$ is less efficient at abstracting hydrogen than $LOO^.$, the chain reaction of peroxidation is slowed. Several biological mechanisms may exist for recycling $\alpha T^.$ back to α-tocopherol, although none of them has yet been proved rigorously to operate in vivo. Likely mechanisms include the reaction of $\alpha T^.$ with ascorbic acid at the surface of membranes and lipoproteins:

$$\alpha T^. + \text{ascorbate} \rightarrow \alpha TH + \text{ascorbate}^. \tag{12}$$

and/or with ubiquinol (reduced coenzyme Q) within membranes or lipoproteins:

$$CoQH_2 + \alpha T^. \rightarrow \alpha TH + CoQH^. + H^+ \tag{13}$$

Antioxidant defense enzymes exist as a balanced coordinated system in mammals. Thus, although SOD is an important defense, an excess of SOD in relation to the activity of peroxide-metabolizing enzymes can be deleterious. This has been shown by transfecting cells with human cDNAs encoding SOD. The consequences of excess SOD activity may be relevant to the clinical condition known as *Down syndrome* (DS), a point being explored by the use of transgenic animals, as indicated in Section 6.

6 TRANSGENIC ANIMAL TECHNOLOGY IN THE STUDY OF ELEVATED SOD

Down syndrome is the most common human genetic disorder, occurring once in every 600–800 live births. Defects that may be suffered by DS individuals include short stature, malformations of the skin around the eyes, and mental retardation. Patients who survive beyond their thirties show an increased risk of developing Alzheimer's disease (this dementing condition normally affects >10–15% of individuals 65 years of age and perhaps 20% over the age of 80: it is marked by gradually developing loss of memory combined with growing confusion and disorientation). DS is usually caused by having three copies of chromosome 21 (trisomy 21), one of whose genes encodes Cu/Zn-SOD.

Does the elevated level of Cu/Zn-SOD cause, or contribute to, DS? An approach to an answer could entail the following steps:

- Construct transgenic mice that overexpress human Cu/Zn-SOD.
- Isolate embryos from the reproductive tracts of female mice.
- Microinject DNA carrying the gene for human Cu/Zn-SOD and the DNA sequences needed for its expression into each embryo pronucleus.
- Transfer embryos into reproductive tracts of "pseudopregnant" foster mothers.
- Allow fetuses to develop to term.
- Select and breed progeny expressing human Cu/Zn-SOD in tissues, thereby raising total SOD activity.

$2GSH + H_2O_2 \rightarrow GSSG + 2H_2O$
$2GSH + \text{fatty acid—OOH} \rightarrow GSSG + \text{fatty acid—OH} + H_2O$
Glutathione peroxidase

$GSSG + NADPH + H^+ \rightarrow 2GSH + NADP^+$
Glutathione reductase

Figure 2. The glutathione system. Reduced glutathione is a tripeptide, glutamic acid–cysteine–glycine. It is present at millimolar concentrations in most mammalian cells. In oxidized glutathione (GSSG), two tripeptides are linked by a disulfide bridge. Glutathione has several additional metabolic functions. Glutathione peroxidase can also destroy fatty acid (lipid–OOH) peroxides by converting them to hydroxy alcohols (lipid–OH). Mammalian cells additionally contain a phospholipid hydroperoxide glutathione peroxidase that can apparently perform the same reaction on lipid peroxides *within membranes;* how exactly it works is uncertain as yet.

The resulting mice will have elevated levels of Cu/Zn-SOD. They are found to

- be more resistant than controls to O_2 toxicity,
- be more resistant than controls to certain toxins, and
- show abnormal neuromuscular junctions in the tongue.

In addition, such transgenic mice may show some of the other neurological defects characteristic of Down syndrome.

7   OXIDATIVE STRESS AND REPAIR SYSTEMS

Normally, the production of $O_2^{\cdot-}$ and H_2O_2 is approximately balanced by the antioxidant defense systems; that is, the antioxidants are not present in great excess. One reason for this may be that production of some $O_2^{\cdot-}$ and H_2O_2 is useful in vivo, to prevent cells from scavenging them with 100% efficiency. Indeed, it has been argued that the aging process involves cumulative free radical damage over the lifetime of a species. However, if antioxidant levels fall or production of $O_2^{\cdot-}$ and H_2O_2 increases, an imbalance occurs and *oxidative stress* is said to result. Oxidative stress can result from the following phenomena:

- Depletions of antioxidants—for example, through inadequate dietary intakes of α-tocopherol, ascorbic acid (vitamin C), sulfur-containing amino acids (needed for GSH manufacture), or riboflavin (needed to make the FAD cofactor in glutathione reductase).
- Excess production of oxygen-derived species—for example, by exposure to elevated O_2 concentrations, by the presence of toxins that are metabolized to produce free radicals, or by excessive activation of "natural" radical-producing systems (e.g., inappropriate activation of phagocytic cells in chronic inflammatory diseases, such as rheumatoid arthritis and ulcerative colitis).

Cells can tolerate mild oxidative stress, which often results in upregulation of the synthesis of antioxidant defence enzymes in an attempt to restore the balance. For example, exposure of *E. coli* to toxins that increase $O_2^{\cdot-}$ production accelerates the biosynthesis of at least 40 different proteins. Nine of these (including MnSOD) belong to the same regulon, controlled by the *sox* locus, which contains two adjacent genes, *soxR* and *soxS*. Expression of *soxS* is increased by $O_2^{\cdot-}$, and it then leads to expression of *soxR*. The excess $O_2^{\cdot-}$ forms an excess of H_2O_2, which oxidizes the protein oxy R, leading to activation of the transcription of another panel of genes, including genes encoding catalase and glutathione reductase enzymes (Figure 3).

Because antioxidant defenses do not scavenge 100% of the radicals generated in vivo, some damage to biological molecules occurs. This damage is continuously repaired, however, with the result that baseline levels of damage are kept low. Table 2 summarizes some of the repair systems known to exist.

8   CONSEQUENCES OF OXIDATIVE STRESS

Although most tissues can adapt to mild oxidative stress, severe oxidative stress can cause cell damage and death, by a series of interacting mechanisms. Thus, oxidative stress appears to cause increases in the levels of "free" Ca^{2+} within cells, as well as "free" iron, which can lead to $OH^{\cdot}$ generation. Some of this $OH^{\cdot}$ genera-

Treat with H_2O_2: Upon treatment with low doses, *E. coli* adapts and becomes resistant to high levels of H_2O_2 that normally would kill it. One of the adaptation mechanisms is as follows:

Treatment with H_2O_2

↓

Oxidation of oxy R protein

↓

Activation of transcription
of genes to increase synthesis of at least eight proteins,
including catalase and glutathione reductase

H_2O_2 also activates expression of other protective genes in *E. coli*, by mechanisms not involving *oxyR*.

Treat with $O_2^{\cdot-}$: The H_2O_2 response system above may be activated as follows:

Treatment with $O_2^{\cdot-}$

Formation of H_2O_2 and activation of its response system	Activation of *soxR* and *soxS* genes, leading to increased synthesis of at least nine proteins, including MnSOD and a DNA repair enzyme

Figure 3.   Regulation of antioxidant defenses in *E. coli*.

tion seems to occur within the nucleus, so that DNA is attacked. Hydroxyl radical attacks DNA in a multiplicity of ways (Figure 1); one of the main products of $OH^{\cdot}$ attack on the purine bases is 8-hydroxyguanine, a miscoding lesion often leading to $G \rightarrow T$ transversions. An excessive rise in intracellular free Ca^{2+} can activate endonucleases and cause DNA fragmentation.

Several toxins impose oxidative stress during their metabolism. Carbon tetrachloride is one example (Table 1): it is metabolized in vivo by endoplasmic reticulum to produce a free radical

Table 2   Repair of Oxidative Damage

Substrate of Damage	Repair System
DNA. All components of DNA can be attacked by $OH^{\cdot}$. Several other reactive oxygen species (ROS) attack guanine preferentially. H_2O_2 and $O_2^{\cdot-}$ do not attack DNA.	A wide range of enzymes exist that recognize abnormalities in DNA and remove them by excision, resynthesis, and rejoining of the DNA strand.
Proteins. Many ROS can oxidize—SH groups. $OH^{\cdot}$ attacks many amino acid residues. Proteins often bind transition metal ions, making them a target of attack by site-specific $OH^{\cdot}$ generation.	Oxidized methionine residues may be repaired by a methionine sulfoxide reductase enzyme. Other damaged proteins may be recognized and preferentially destroyed by cellular proteases.
Lipids. Some ROS (not including $O_2^{\cdot-}$ or H_2O_2) can initiate lipid peroxidation.	Chain-breaking antioxidants (especially α-tocopherol) remove chain-propagating peroxyl radicals. Phospholipid hydroperoxide glutathione peroxidase (Figure 2) can remove peroxides from membranes. Normal membrane turnover can replace damaged lipids.

(trichloromethyl peroxyl radical, $CCl_3O_2^{\cdot}$) that is very efficient at initiating lipid peroxidation by abstracting hydrogen from polyunsaturated fatty acid side chains (Section 2.1). Another is paraquat, a herbicide that causes lung damage in humans. Its metabolism within the lung leads to production of large amounts of $O_2^{\cdot-}$ and H_2O_2. Paraquat is reduced to a free radical by lung enzymes

$$PQ^{2+} + e^- \rightarrow PQ^{\cdot+} \qquad (14)$$

This radical then reacts very fast with O_2 in a nonenzymic reaction

$$PQ^{\cdot+} + O_2 \rightarrow PQ + O_2^{\cdot-} \qquad (15)$$

The paraquat is regenerated to repeat the cycle. Other examples of such "redox cycling" toxins are alloxan, diquat, and 6-hydroxydopamine.

Cigarette smoking has many deleterious health effects, and oxidate stress is believed to be involved in several of these. Table 3 summarizes the mechanisms by which cigarette smoke can impose oxidative stress.

9   OXIDATIVE STRESS AND HUMAN DISEASE

Oxidative stress is an inevitable accompaniment of tissue injury during human disease, for the reasons summarized in Figure 4. For example, when phagocytes produce excess $O_2^{\cdot-}$, H_2O_2, and other species at sites of chronic inflammation, severe damage can result. This seems to happen in the inflamed joints of patients with rheumatoid arthritis and in the gut of patients with inflammatory bowel diseases, such as Crohn's disease and ulcerative colitis. Tissue injury can release metal ions from their storage sites within cells, leading to $OH^{\cdot}$ generation. This may be a particularly important mechanism in the brain, in that iron-dependent free radical reactions can occur after injury (e.g., by trauma or ischemia), spreading the injury to adjacent cells.

Since oxidative stress occurs to some degree after every tissue injury, the main question to be asked in evaluating its role in human disease is not "Can we demonstrate oxidative stress?" but rather

Table 3   Why Cigarette Smoking May Impose Oxidative Stress

Smoke contains many free radicals (both in the gas and tar phases), especially peroxyl radicals, that might attack biological molecules and deplete antioxidants, such as ascorbic acid and α-tocopherol.

Smoke contains oxides of nitrogen, including the unpleasant nitrogen dioxide ($NO_2^{\cdot}$), a free radical that can initiate lipid peroxidation.

The tar phase of smoke contains lipid-soluble hydroquinones, which may enter the tissues and redox cycle. Some may also release iron from the iron storage protein ferritin, providing "free" iron to catalyze $OH^{\cdot}$ formation.

- Smoking may irritate lung macrophages, encouraging them to make $O_2^{\cdot-}$.
- Smokers' lungs contain more neutrophils than the lungs of nonsmokers, and smoke might activate these cells to make $O_2^{\cdot-}$.

Smokers who eat poorly and drink more alcohol than nonsmokers may have an insufficient dietary intake of nutrient antioxidants.

- The effects of cigarette smoke on phagocytes are dose-related. Low levels may stimulate them, but high levels may poison phagocytes and so depress their activity.

"Does the oxidative stress that occurs make a significant contribution to disease activity?" The answer to the latter question appears to be "yes" in at least some cases, including rheumatoid arthritis, atherosclerosis, and ulcerative colitis. However, it may well be "no" in many others. Elucidating the precise role played by free radicals has not been easy because such species are difficult to measure, but the development of modern assay techniques is helping to solve this problem.

10   ASSAY METHODOLOGY

The only technique that can detect free radicals directly is electron spin resonance (ESR) spectroscopy, but only very unreactive radi-

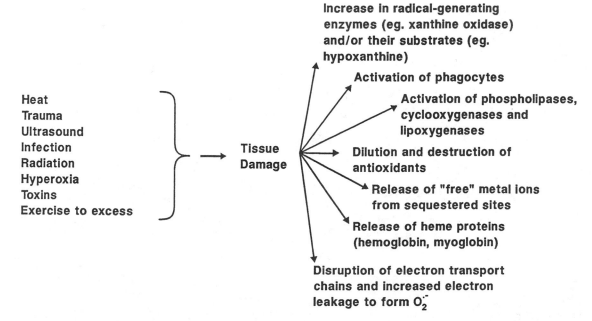

Heat
Trauma
Ultrasound
Infection
Radiation
Hyperoxia
Toxins
Exercise to excess

→ Tissue Damage →

Increase in radical-generating enzymes (eg. xanthine oxidase) and/or their substrates (eg. hypoxanthine)

Activation of phagocytes

Activation of phospholipases, cyclooxygenases and lipoxygenases

Dilution and destruction of antioxidants

Release of "free" metal ions from sequestered sites

Release of heme proteins (hemoglobin, myoglobin)

Disruption of electron transport chains and increased electron leakage to form $O_2^{\cdot-}$

Figure 4.   How tissue damage can cause oxidative stress. [From Halliwell et al. (1992), with permission.]

cals can accumulate in vivo to levels that ESR techniques can detect. Hence, ESR studies of biological material detect fairly unreactive radicals, such as ascorbate radical. Highly reactive radicals formed in biological systems can be identified by two general approaches. The first is trapping, and the second involves the measurement of the end products of free radical attack (Sections 10.2 and 10.3).

In trapping, the radical is allowed to react with a trap molecule to give one or more stable products, which are then measured. The most popular trapping method is spin trapping, in which the radical reacts with a "trap molecule" to form a more stable radical, which does accumulate to the level detectable by ESR. Spin-traps such as α-phenyl-*tert*-butyl nitrone and 5,5-dimethylpyrroline-*N*-oxide have been useful in detecting certain free radicals in vitro and in whole animals, but currently available spin traps have not succeeded in detecting $O_2^{\cdot-}$ or OH· in vivo, nor can they be administered to humans.

10.1 Aromatic Hydroxylation

There are many trapping methods other than spin trapping. The technique of aromatic hydroxylation, for example, is based on the reaction of OH· generated under physiological conditions with aromatic compounds at a diffusion-controlled rate, giving rise predominantly to hydroxylated end products. An aromatic hydroxylation assay was first applied to humans using salicylate (2-hydroxybenzoate) as a "trap" for OH·. Attack of OH· on salicylate produces two major hydroxylated products: 2,3-dihydroxybenzoate and 2,5-dihydroxybenzoate. The latter product can be produced metabolically, whereas 2,3-dihydroxybenzoate apparently cannot. Hence, the formation of the latter product may be used to detect OH· production in vivo.

Another aromatic trap for OH· is the amino acid, phenylalanine. In vivo, the L-isomer of this amino acid is hydroxylated by phenylalanine hydroxylase at position 4 on the ring to give L-*p*-tyrosine.

D-Phenylalanine is not recognized by this enzyme. By contrast, OH cannot distinguish between the two isomers: it acts on both L- and D-phenylalanine to produce a mixture of *o*-, *m*-, and *p*-tyrosines (Figure 5). Formation of these tyrosines has been used as an index of OH· production by cells, by injured organs, and in food that has been irradiated.

10.2 Uric Acid Degradation

In humans and other primates, which lack a functional urate oxidase enzyme, uric acid is an end product of purine metabolism. (A gene resembling the urate oxidase gene of rat is present in the human genome, but there is a stop codon in one of the exons.) It is widely believed that uric acid acts as an antioxidant in vivo. Hence, measuring the products of attack of oxygen-derived species on uric acid is a potential marker of oxidative damage uniquely applicable to humans and to other primates. Uric acid is oxidized by several reactive oxygen species (including OH·, but not $O_2^{\cdot-}$ or H_2O_2). The major product of uric acid oxidation in all cases is allantoin, but other products formed are oxonic acid, oxaluric acid, cyanuric acid, and parabanic acid. The concentrations of all these products are increased in synovial fluid and serum from patients suffering from rheumatoid arthritis and in serum from patients with iron overload arising as a consequence of idiopathic hemochromatosis. The blood plasma of iron-overloaded hemochromatosis patients contains "free" iron, which can catalyze free radical reactions. By contrast, healthy subjects never contain "free" plasma iron or copper—it is safely bound to transport proteins.

10.3 Assays for Oxygen-Derived Species; "Fingerprint" Assays

Instead of attempting to "trap" oxygen-derived species, one can sometimes implicate them as agents of tissue injury by examining

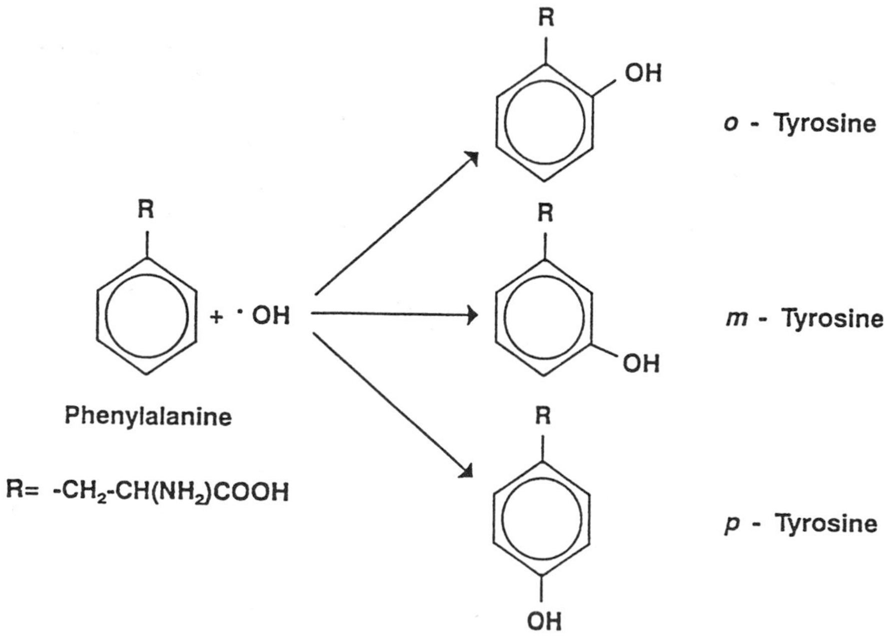

Figure 5. Products resulting from the attack of hydroxyl radicals on the amino acid phenylalanine. Hydroxyl radicals add on to the aromatic ring to produce intermediate radicals, which are converted to hydroxylated products under physiological conditions.

the pattern of chemical change they produce upon reaction with certain biological molecules. Thus end products of lipid peroxidation (e.g., lipid hydroperoxides, Section 2.1) or of free radical damage to DNA (Figure 1) and proteins can be measured. Attack on proteins of OH˙ generated in site-specific metal ion dependent reactions can generate carbonyl compounds on the proteins, and these "protein carbonyls" are often measured as an index of such protein damage in vivo.

11   CONCLUSION

Free radicals are a normal part of human metabolism: we cannot escape them. They can be both favorable and unfavorable, depending on amount, location, chemical nature, and levels of antioxidant defense.

See also DNA DAMAGE AND REPAIR; ENVIRONMENTAL STRESS, GENOMIC RESPONSES TO.

Bibliography

Amstad, P., Peskin, A., Shah, G., Mirault, M. E., Moret, R., Zbinden, I., and Cerutti, P. (1991) The balance between Cu,Zn-superoxide dismutase and catalase affects the sensitivity of mouse epidermal cells to oxidative stress. *Biochemistry,* 30:9305–9313.
Halliwell, B. (1992) Reactive oxygen species and the central nervous system. *J. Neurochem.* 59:1609–1623.
———, and Gutteridge, J.M.C. (1989) *Free Radicals in Biology and Medicine,* 2nd ed. Clarendon Press, Oxford.
———, and Gutteridge, J.M.C. (1994) *Antioxidants in Nutrition, Health, and Disease.* Clarendon Press, Oxford.
———, ———, and Cross, C. E. (1992) Free radicals, antioxidants and human disease: Where are we now? *J. Lab. Clin. Med.* 119:598–620.
McBride, T. J., Preston, B. D., and Loeb, L. A. (1991) Mutagenic spectrum resulting from DNA damage by oxygen radicals. *Biochemistry,* 30:207–213.
Packer, L., and Fuchs, J., Eds. (1993) *Vitamin E in Health and Disease.* Dekker, New York.
Rosen, D. R. (1993) Mutations in Cu/Zn superoxide dismutase gene are associated with familial amyotrophic lateral sclerosis. *Nature,* 362:59–62.
Scandalios, J. G., Ed. (1992) *Molecular Biology of Free Radical Scavenging Systems.* Cold Spring Harbor Laboratory Press, Cold Spring Harbor, NY.
Schreck, R., Albermann, K.A.J., and Baeuerle, P. A. (1992) Nuclear factor κB: An oxidative stress responsive transcription factor of eukaryotic cells (a review). *Free Radical Res. Commun.* 17:221–237.
Sies, H., Ed. (1991) *Oxidative Stress, Oxidants and Antioxidants.* Academic Press, London.
Storz, G., and Tartaglia, L. A. (1992) Oxy R: A regulator of antioxidant genes. *J. Nutr.* 122:627–630.
Tkeshelashvili, L. K., McBride, T., Spence, K., and Loeb, L. A. (1991) Mutation spectrum of copper-induced DNA damage. *J. Biol. Chem.* 266:6401–6406.

FUNGAL BIOTECHNOLOGY

Brian McNeil

Key Words

Batch Culture   A closed culture system, to which no major additions of nutrients are made after inoculation of the nutrient medium with the chosen microorganisms.

Continuous Culture (Chemostat)   A culture system in which the rate of addition of fresh medium balances the rate of removal of spent medium and cells, such that at steady state, the concentration of all reactants is constant.

Fed Batch Culture   A culture system to which nutrient is added at discrete time intervals or on a continuous basis, with the aim of avoiding either limitation or inhibition of O_2 due to the nutrient.

Filamentous Fungi (Molds)   Eukaryotic microorganisms, characterized by growth at the apical tip, which form a branched network of vegetative filaments, a mycelium.

Solid Substrate Fermentation (SSF)   A low water fermentation process in which the fungus grows on or through the solid substrate, often a cereal grain or similar material.

Submerged Liquid Fermentation (SLF)   Describing liquid fermentation systems, within which the fungus is aerated and, often, mechanically agitated.

Molds (filamentous fungi) are used in the service of man to produce a wide range of valuable products, to improve feedstuffs, to carry out biotransformations, and to effect bioremediation. The technology used to cultivate and control mold activity on either solid or liquid media allows the production of high levels (kilos per cubic meter) of the desired products from selected microbial strains. Nevertheless, many problems in cultivating these microorganisms, due directly to their morphology, remain to be fully resolved. The biotechnological role of the filamentous fungi seems set to expand, with the development of effective transformation systems and increasing use of organisms such as *Aspergillus niger* as efficient systems for the secretion of heterologous proteins. However, an understanding of the technological means of successfully cultivating these microorganisms will continue to be essential to the exploitation of their biotechnological potential.

1   INTRODUCTION

Biotechnology has been defined as "the use of whole cells or parts derived therefrom to catalyze the formation of useful products, or, to treat pollutants."

Within the broad field of biotechnology the filamentous fungi (molds) have long occupied an important position. This diverse and metabolically versatile group of eukaryotic microorganisms is characterized by the formation of vegetative branching filaments (hyphae), which together form a mycelium. This growth pattern is a result of polarization of growth at the hyphal tip (apex).

Although most groups of molds have economically significant members, biotechnologically, the most important fungi are found within the Fungi Imperfecti, Ascomycetes, and Basidiomycetes.

1.1   Physical Characteristics of Filamentous Fungi (Molds)

The hyphal mode of growth confers a number of advantages on the filamentous fungi. These include the ability to translocate nutrients within the mycelial network, the ability to grow away from the initial growth point, such that hyphae at the edge of a colony constantly encounter fresh nutrients, and the ability to grow into solid materials (penetrative growth), thus allowing the uptake of nutrients whose availability would otherwise be restricted by diffusion.

In the growth process fungi form branches, and branch frequency is closely related to nutrient availability.

1.2   Metabolic Activities of Filamentous Fungi

As a consequence of their mode of growth, the filamentous fungi possess a number of metabolic activities that are attractive to the biotechnologist. First among these is the synthesis and excretion of lytic enzymes (proteases, lipases, carbohydrases). Production of such hydrolytic enzymes aids fungal invasiveness. Under appropri-

ate conditions, some fungi can secrete large quantities of proteins (50 gL^{-1} in some cases) into their environment.

The filamentous fungi are, as a group, prodigious producers of secondary metabolites. This structurally diverse group of products, formed in significant amounts during the stationary growth phase, includes compounds such as the penicillins and the gibberellins.

2   DEVELOPMENT OF FUNGAL BIOTECHNOLOGY

2.1   Solid Substrate Fermentation (SSF) Systems

Traditionally molds have been used largely in the production and improvement of foods. Natural colonization of moist grains or cereals by molds led to the discovery that certain molds could improve the flavor, texture, and nutritional value of largely carbohydrate materials.

The next steps involved control of the physicochemical conditions of the process, and, later isolation, selection, and use of particular strains. These processes, the classic examples of solid substrate fermentation (SSF) processes, are characterized by the overgrowth of the substrate, often a grain, by the fungus, which is then used as a source of enzymes to break down substrate components or for addition to other substrates. Such processes are often referred to by the Japanese term "koji." Koji processes are still widely used in the Orient for formation of a wide range of products (Table 1). Most of these processes are now operated at industrial scale, and many have been taken up by countries outside Asia. Tempeh, for example, in which the action of the fungi produces a material high in protein, flavorsome, and useful as a meat supplement, is produced in quantity in both Europe and the United States.

Europeans traditionally have used fungi such as *Penicillium roquefortii* and *P. camembertii* to grow on or in cheeses and to enhance flavor by production of lipases and proteases. Further use of SSF came in the industrial production of mushrooms, the most common being *Agaricus bisporus*.

All SSF systems are low water systems and generally have relatively low power inputs. Conversely, they are generally recognized as being difficult to monitor, control, and scale up. For example, O$_2$ transfer and the removal of CO$_2$ and heat can be difficult. Several SSF fermentor types are illustrated in Figure 1.

2.2   Submerged Liquid Fermentation (SLF) Systems

Citric acid illustrates particularly well the development of the fermentation technology required for convenient formation of mold products in large quantities, at reasonable *rates* in liquid systems. Initially derived from lemon juice, citric acid was first produced industrially by a surface tray process using *Aspergillus niger*. The U.S.

Table 1 Processes Involving a Koji Stage

Product	Substrates	Mold Involved	Location
Shoyu (soy sauce)	Soybeans/wheat	*Aspergillus soyae* or *A. oryzae*	Japan (but similar products elsewhere)
Miso	Rice/barley soybeans	*A. soyae*	Japan
Sake (rice wine)	Rice	*A. oryzae*	Japan
Tempeh	Soybeans	*Rhizopus* sp. *R. oligosporus*	Indonesia

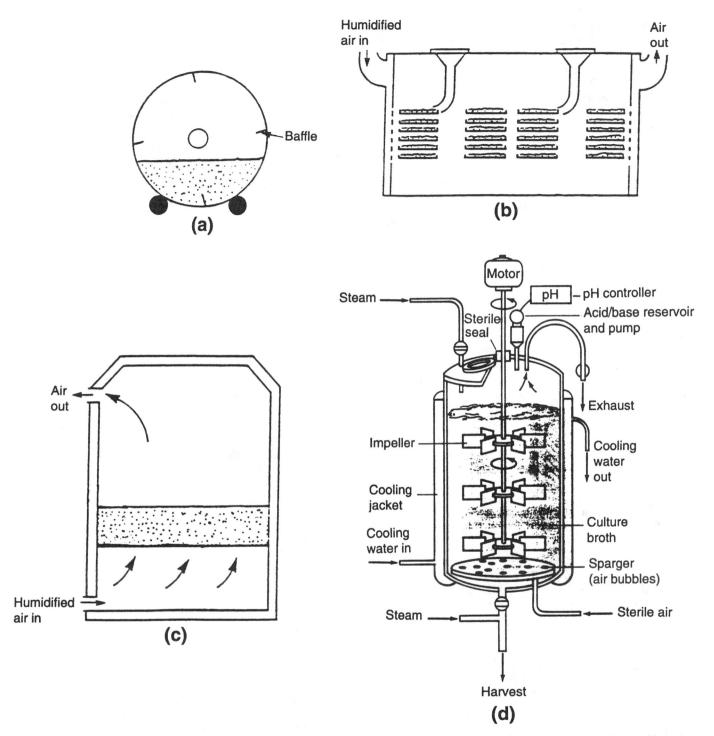

Figure 1.   Fermentor configurations, showing three SSF types: (a) rotating drum (b) tray system, and (c) forced air flow system [from Whipps and Lumsden (1989)] and one SLF type: (d) stirred tank bioreactor [from Kinghorn and Turner (1992)].

firm of Pfizer used this relatively lengthy (6–12 days), labor-intensive process until 1991.

From the late 1940s, the use of the stirred tank reactor (STR) for submerged liquid cultivation of *A. niger* resulted in citrate production processes lasting 3–5 days at 25–30°C. A typical STR is shown in Figure 1. Use of such fermentors allowed high oxygen transfer rates to be maintained, leading to high productivity with low manual labor needs. These fermentors were easier to monitor and control than SSF systems. The mold's physical environment could thus be more readily optimized. The STR is now the industry workhorse, a flexible reactor type used for the production of a wide range of products.

Some of the features developed in citrate fermentations are held in common with other STR mold fermentations, including:

- Use of highly bred (often relatively *genetically unstable*) pure cultures of production strains (Production strains are *usually* stable these days!)
- Use of relatively crude raw materials such as molasses and corn steep liquor at the industrial scale
- Continuous aeration with sterile (filtered) air, and continuous stirring
- Maintenance of asepsis during fermentation (molds grow more slowly than bacteria and yeasts; thus any contamination by these microorganisms could lead to displacement of the fungus)

Current demand for citric acid is around 500,000 tons, though this amount is growing at 3–5% per year. Increasingly, the citric acid market is being concentrated in the hands of fewer, larger products. Currently, only seven companies produce the vast bulk of citrate used.

2.2.1 Penicillins

At around the same time that submerged citrate processes developed, fermentations in STRs involving pure cultures of *Penicillium chrysogenum* for penicillin G production were also being developed. These processes are conceptually very similar to those used in citrate production. Initially, the carbon source was lactose, which permitted slow fungal growth and good excretion of the product, the secondary metabolite penicillin G. Later, fed batch processes using glucose feeds were developed. These are the current methods of production for penicillin G and V, which differ chemically only in the attached side chain (phenylacetic acid in the former, phenoxy-acetic acid in the latter). Bulk penicillins G and V are increasingly used as raw materials for conversion into a wide range of semisynthetic penicillins.

Penicillins, like citrate, are mature biotechnological products produced in bulk and characterized by "survival of the fittest" with respect to producers.

2.2.2 Enzyme Production by Fermentation

Fungi are excellent sources of a wide range of lytic enzymes. Most of these are produced using the well-understood STR system described briefly in Section 2.2. In this area the aspergilli are preeminent. The current world market for industrial enzymes, several of which are produced from aspergilli by SLF, is around $600 million per annum.

2.2.3 Other Liquid Fermentation Products

Table 2 lists some products of the liquid fermentation of fungi. In general, it is much easier to screen for antimicrobial activity of mold-derived compounds than for other activities. Indeed, some products such as cyclosporins were originally isolated for their antimicrobial characteristics (antifungal for cyclosporins), with their value as immunosuppressants being realized only later. It is not unreasonable to state that the use of cyclosporins have revolutionized transplant surgery.

Mucor species have been shown to accumulate large quantities of unsaturated fatty acids (e.g., linoleic). Although technically successful, SLF-based processes for production are currently not economic.

The production of Mycoprotein (Quorn) by ICI/RHM [Imperial Chemical Industries (now Zeneca) and Rank Hovis MacDougall] has been a considerable success. This is one of the few continuous culture SLF processes, if not the only one, operated at the industrial scale. *Fusarium graminearum* is grown in STRs on a defined sugar–mineral salts medium, and the fungal biomass produced is harvested. By virtue of its filamentous nature, the biomass is textured, and this can be controlled or modified to simulate various meat products. The bland-tasting product readily accepts other flavors. Quorn appears to be an almost perfect foodstuff, having high protein quality, low fat content, zero cholesterol, and a reasonably high fiber content. Interest is also growing in the role of some fungal polysaccharides to modulate the immune systems of test animals. Such compounds are classed as biological response modifiers (BRMs) and include the β-glucan polysaccharides lentinan and scleroglucan. These compounds may stimulate the immune system by enhancement of macrophage activity.

3   PRODUCTION MODES

3.1   BATCH CULTURE

Although batch culture is the traditional mode of cultivation in fermentors, it is not suited to the formation of products where inhibi-

Table 2   Some of the Products of Fungal Biotechnology

Organism	Product(s)	Use	Mode of Production
Aspergillus niger	Citric acid	Acidulant	SLF (usually)
A. oryzae	Amyloglucosidase and other enzymes	Starch hydrolysis, lipolysis, proteolysisis, etc.	SLF (fed batch)
Penicillium chrysogenum	Penicillins G and V	Antibacterial agents	SLF (fed batch)
Fusarium graminearum	Mycoprotein (Quorn)	Human foodstuff	SLF (continuous culture)
Gibberella fujikuroi	Gibberrellins	Plant growth regulators	SLF
Tolypocladium (Beauveria)	Cyclosporin	Immunosuppressant in transplant surgery	SLF
Sclerotium glucanicum	Scleroglucan	Tertiary oil recovery; immuno-stimulant; antitumor agent	SLF (batch)
Agaricus campestris *A. bisporus*	Fruiting bodies (mushrooms)	Human foodstuff	SSF
Lentinus edodes	Fruiting body or cell extracts	Human foodstuff; immuno-stimulant	SLF

tion by substrate, product, oxygen limitation, or product precursors is a problem. Fungal citrate may be produced in batch culture.

3.2 FED BATCH CULTURE

In fed batch culture, the fermentation, often after an initial batch phase, is fed regularly or continuously with required nutrients or precursors. The feed is controlled to ensure that the level of the added compounds is never inhibitory to the culture; or, to avoid oxygen limitation, a rate is maintained that just balances the consumption of nutrients with oxygen consumption. Penicillins and amyloglucosidase are produced by this method, as are many other mold secondary metabolites.

3.3 CONTINUOUS CULTURE

The *chemostat* is the most common type of continuous culture system encountered. Here the culture is continuously fed fresh nutrient medium, and spent medium, cells, and product are continuously drawn off. The system is not ideally suited to secondary metabolite production, but it is suitable for biomass production, where productivities can be much higher than in batch fermentations. This technique is also widely used on a small (lab) scale to study aspects of cellular biochemistry and physiology.

4 PROBLEMS IN CULTIVATION OF FUNGI IN FERMENTORS

Despite a long history of cultivation in both SSF and SLF systems, it is recognized that there are still some problems to be overcome. In relation to SSF systems, some of the problems have already been mentioned. In liquid systems, especially the STR, particular difficulties arise as a result of the particular morphology of the molds.

In these fermentor vessels, formation of a branched mycelium results in a rapid increase in the viscosity of the culture. The rheological character of the fermentation fluid may also change, as well: a low viscosity Newtonian system may become highly viscous and non-Newtonian.

The general effect of this trend is to reduce mixing efficiency, as well as oxygen and heat transfer in the fermentor. The result is heterogeneity within the fermentor, and in many cases fermentor productivity is limited.

Some producers of fungal metabolites accept these restrictions as an inevitable consequence of cultivation of filamentous fungi. Others, however, are actively seeking forms of the producing microorganisms that are less "filamentous" and have much shorter hyphae, to reduce the problems somewhat. It is only reasonable to state, however, that these problems are still far from overcome.

5 FILAMENTOUS FUNGI AND HETEROLOGOUS PROTEIN PRODUCTION

The filamentous fungi are showing particular promise in the area of producing heterologous proteins. Various fungi have been used to produce a wide range of proteins (Table 3). Initially, many of the systems were clearly aimed at examining the feasibility of using fungi as expression systems for "foreign" genes, and these early studies often involved fungi such as *Neurospora crassa* and *Aspergillus nidulans*. More recently, a number of companies have achieved worthwhile yields of recombinant products.

It is generally acknowledged that the filamentous fungi have a number of actual or potential advantages in this role, including the following.

1. Filamentous fungi are permissive in relation to the expression of foreign genes.
2. Some filamentous fungi can secrete into the extracellular environment copious quantities of homologous proteins. Under the correct physiological conditions, for example, *A. niger* can produce tens of grams of glucoamylase per liter of culture broth. One should not underestimate the attractions of this feature; here we have a group of organisms whose particular mode of existence requires them to possess efficient protein exporting machinery, thus, given point 1, they would appear to be ideal secretion systems for heterologous proteins. This initial promise of copious secretion of heterologous proteins has yet to be fully realized, having proved to be rather more difficult than was first thought. Some of the reasons for this are discussed later.
3. Posttranslational patterns of processing in the filamentous fungi are much closer to those of higher eukaryotes. Since many of the proteins of commercial interest are of mammalian origin [e.g., tissue plasminogen activator (tPA), interferons], this suggests that filamentous fungi may, at first sight, be more promising hosts than well-known prokaryotic expression systems such as *E. coli*. Specifically, patterns of glycosylation in fungal proteins and higher eukaryotic proteins have a degree of similarity.
4. Inclusion bodies (accumulations of denatured foreign proteins in the cytoplasm of bacteria) are not seen in fungi. Their absence is potentially advantageous, since to recover such proteins involves cell disruption, separation, and renaturation, all of which imply increased production costs.
5. From a regulatory viewpoint, many fungi make attractive hosts for the production of proteins for food (e.g., chymosin) or therapeutic use (e.g., tPA) because the U.S. Food and Drug Administration has given them GRAS (generally recognized as safe) status.

Table 3 Heterologous Protein Production in Molds: Some Hosts, Promoters, and Products

Host	Promoter Source	Origin of Promoter	Product
Asperillus oryzae	Amylase	*A. oryzae*	Aspartyl protease
A. niger var. *awamori*	Glucanylase	*A. niger*	Calf chymosin
Trichoderma reesei	Cellobiohydrolase 1	*T. reesei*	Calf chymosin
A. nidulans	Alcohol dehydrogenase	*A. niger*	Human tissue plasminogen activator
		A. niger	Human interferon, human growth hormone, human interleukin 6
Achlya bisexualis	SV40 promoter	Simian virus	Interferon

6. Relative to mammalian and plant cell cultures, fungal cultures are fast growing (doubling times of 3–4 h are not atypical), and the technology for cultivating molds is well characterized and readily available. Downstream processing methods are also well understood.
7. Mitotic stability of transformants is high.

Conversely, a number of potential drawbacks exist, including the following.

1. Transformation frequency is generally very low compared to transformation in bacteria and yeasts, and methods usually involve protoplast generation.
2. Filamentous fungi are a level of complexity above unicellular organisms such as bacteria and yeast, and their development of a complex three-dimensional network, with protein secretion localized to specific areas of the mycelium only, may add to the problems of operation. Despite a long industrial history, there is still a great deal we do not understand about the physiology, and the genetics, of industrially important fungi.
3. Many fungi produce other by-products, such as organic acids, which represent a diversion of substrate away from protein synthesis and may alter the culture pH. In addition, many fungal strains produce proteases (such as aspergillopepsin A), which could well degrade the potential product. In some cases production of proteases can be circumvented. For example, Genencor deleted the gene for aspergillopepsin A from the production strain of A. niger, thus increasing the secretion of calf chymosin.
4. Because fungal cultures are highly viscous and pseudoplastic, mixing in fermentor vessels is poor. Thus, even if the conditions required to optimize heterologous protein production are known, it may not be possible to ensure that these conditions prevail throughout the fermentor vessel. If uniformly good conditions are not achieved, the overall result will be reduced productivity.
5. The relationship between copy number of the transforming gene and production of the gene product is not a simple one in some species. Such complexity may be the result of the rather random nature of the integration process.

Despite the difficulties just enumerated, there have been successes also. Notable among these is the production of calf chymosin (Genencor) and phytase (Novo Nordisk). As can be seen in Table 3, there is also considerable interest in the production of therapeutic proteins using molds. The requirements here are more rigorous than for chymosin and phytase, which one might regard as industrial-scale products. In the case of chymosin, the aim is to maximize yield of product using essentially the same technology used for production of native enzymes (see Section 2.2). The companies involved, and more importantly, the regulatory authorities, are familiar with these, and provided evidence of the purity and authenticity of the product is available, progress toward industrial production is relatively swift. It is to be expected that the companies involved in homologous protein production will increasingly use the fungal expression technology developed over the past decade or so to improve production of many industrial enzyme processes. The GRAS status of many of the chosen host strains will be helpful in this endeavor.

Different criteria may apply with regard to the therapeutic proteins produced using fungal expression systems. Yields at the mil-ligram- or even the microgram-per-liter level may be acceptable initially here, when the "retail" price is so high and society's need so great. Conversely one might reasonably expect the path to regulatory approval to be somewhat longer, since the requirements for the demonstration of consistent purity and authenticity of products will be much more rigorous, and molds have no track record as sources of therapeutic proteins.

One aspect that has become apparent is that secretion of heterologous proteins from fungi rarely reaches the levels achievable with homologous protein. For example, calf chymosin from A. niger reaches about 1.2 g/L, while A. niger glucoamylase can reach up to 20–30 g/L. This discrepancy has been the subject of great interest, and its resolution may go far to ensuring the success of the fungi as industrially important expression systems.

5.1   Transcription

In general, it would appear that the process of heterologous protein secretion is not restricted at the transcriptional level. (One interesting exception is the transcriptional limitation on chymosin expression in Trichoderma reesei). In view of the highly effective promoters in current use, this is not surprising. However, site of integration (where the foreign DNA is introduced into the genome) may affect expression level. This area certainly requires further study.

5.2   Translation

Despite some evidence that specific sequences upstream of the initiation code may influence translation, it is not yet clear whether protein production is markedly affected at this stage.

5.3   Posttranslational Modifications

Many eukaryotic proteins are glycosylated, and the specific pattern of glycosylation may influence protein folding (and thus activity) and aid in passage through the secretory system. If incorrectly glycosylated, they may be very rapidly degraded in mammalian systems. Unlike yeasts, fungi generally do not appear to hyperglycosylate proteins (a notable exception is expression of T. reesei cellulases in A. nidulans). Fungal patterns of protein glycosylation are closer to those of higher eukaryotes than other microorganisms.

At present the role of glycosylation in the secretion process is unclear and requires attention.

5.4   Gene Fusion

This strategy has improved secretion levels of a number of heterologous proteins, including chymosin and human interleukin 6. In gene fusion, the foreign gene is fused with a native gene, whose product is efficiently exported. For efficient secretion, in eukaryotes, proteins usually require an N-terminal hydrophobic secretion signal peptide; likewise, other sequences lacking the signal peptide may influence the protein's passage through the secretory system. Fusion of the calf chymosin gene with the promoter and signal peptide sequence of the glucoamylase gene led to improved production of calf chymosin in A. nidulans. Where product assay and screening procedures are relatively simple, a program of mutagenesis can lead to improvements in secretion levels, presumably through alterations in the sequences lacking the signal region.

5.5   PROTEOLYSIS

The problem of proteolysis induced by native proteases may also be of significance. As one might expect, most homologous proteins secreted by fungi are not degraded by native proteases, but heterologous proteins can be rapidly degraded. The deletion of the aspergillopepsin A gene has already been mentioned, but intracellular proteolysis may restrict secretion levels in many systems.

5.6   CONCLUSION

We selected the filamentous fungi as expression systems for foreign genes because these microorganisms are excellent protein secretors and we believed we understood them well. What is now apparent is that we do not yet understand why certain fungi are effective protein exporters. In other words, we did not understand them as well as we thought. Fundamental research into posttranslational aspects of protein trafficking and the secretory system of fungi is required if we are to fully capitalize on the abilities of this group of microorganisms as expression systems.

6   BIOTECHNOLOGY OF MOLDS IN AGRICULTURE

The filamentous fungi have a very close relationship with many plants, which can take the form of a mutually beneficial link between plant and mold (a mycorrhiza). Alternatively, many fungi are plant pathogens and can cause diseases in commercially important plant species. Other aspects of fungal activity that may impact on agricultural practice are the abilities of some fungi to attack insects, mites, nematodes, and other fungi. These latter activities indicate that fungi may have potential as biocontrol agents.

6.1   BIOCONTROL USING FUNGI

The use of fungi to control growth of undesirable plant species (weeds), pests (e.g., insects, mites, nematodes), and other plant pathogenic fungi has a number of attractions. The first of these is the potential for specific control of the problem species by selection of a very narrow range pathogenic fungus. Other potential advantages include a possible reduction in chemical residues in or on foods or feeds, through reduced use of chemical insecticides, herbicides, pesticides, and fungicides, and the possibility of minimizing damage to the ecosystem (a consequence of the first two points listed). Thus, destruction of the problem species, should not be accompanied by elimination of beneficial species, or unacceptable amounts of damage to them. One final potential advantage of using living agents such as fungi in a biocontrol program is the possibility of sustained control of the pathogen or pest by permanent introduction of the biocontrol agent.

Conversely, many potential drawbacks to the use of fungi may exist, including the lack of complete control achieved by many biocontrol agents alone; the slow rate of kill of the pathogen, which may allow significant crop damage; difficulties in mass production, storage, and delivery of the infective agent itself; and reduction in efficacy due to nonideal environmental conditions. From a producer's commercial viewpoint, the possibility of persistence of the biocontrol fungus, which would reduce future sales, must also be considered. Biocontrol agents usually cost more to produce than chemical agents; and a greater degree of sophistication is required in their use.

6.2   MYCOINSECTICIDES

Broadly speaking, two classes of fungi can be distinguished, biotrophic (i.e., those that can replicate only in a living host) and those capable of growth in fermentation processes. Although more than 100 fungal genera have been shown to attack insect pests, few species have actually been widely used in the field. Prominent among those that have been so used are the deuteromycete fungi *Beauveria bassiana, Verticillium lecanii,* and *Metarhizium anisopliae* All these fungi have been shown to produce conidia, the infective form of the organism in vitro. *B. bassiana* is used in the former Soviet Union and in China to control pests such as Colorado potato beetle, pine caterpillars, and European corn borer.

Mycoinsecticides usually attack by direct penetration of the insect cuticle by means of enzyme action and, to a lesser extent, mechanical pressure. Since the first step in infection is the contact between the fungal spore and the insect cuticle, activity of the mycoinsecticide is heavily influenced by relative humidity and temperature. The ability of some molds to effectively secrete extracellular depolymerizing enzymes such as chitinase, proteases, and lipases is a vital feature in their ability to parasitize living insects. Thus, the potential for improvement by the use of modern molecular biological techniques clearly exists. The battery of enzymes produced corresponds to the major constituents of insect cuticle.

In addition to enzymes, entomopathogenic fungi produce a number of toxic metabolites that may have a role in the lethal effects observed in vivo. *B. bassiana* produces a cyclic peptide, beauvericin, and *M. anisopliae* produces a range of five related cyclic peptides, the destruxins. The synthetase gene for a very similar fungal cyclic peptide (enhiantin) has been isolated and characterized. The potential now exists for increasing the efficacy of mycoinsecticide antibiotic synthetases as probes for the isolation of synthetase genes, and for the development of genetically improved strains. A better understanding of the mechanism of action and the genetics of pathogenicity should allow us to make significant improvements in these biocontrol agents by means of a molecular biological approach.

6.3   PRODUCTION TECHNOLOGY

Ideally the aim is a production process to turn out large numbers of conidia rapidly, cheaply, and reproducibly. Processes involved are submerged (liquid) fermentation, solid substrate, and combined.

In the West, most fermentation capacity is in the form of STRs, and these have been used to produce *B. bassiana* and *V. lecanii.* Many fungi, however, do not conidiate in these systems; instead, they form blastospores. These cells can be used in formulation of the mycoinsecticide, but they are considerably less stable than conidia. Surface (SSF) culture can produce large numbers of conidia, but tends to be labor intensive. It has been successfully used in Brazil, Russia, China, Canada, and the United States for *V. lecanii* and *B. bassiana.* In many countries it is an especially attractive, low technology, low cost local solution to local problems. Indeed, the desired fungus can be grown on nonsterile agricultural wastes. Care must be taken, however, for the danger of allergic reaction to inhalation of spores may be significant.

Combined processes usually involve STR production of mycelial cultures followed by a step that achieves induction of conidiation.

6.4   Other Fungal Biocontrol Agents

Many of the preceding comments regarding difficulties of production, formulation, application, and stability apply equally to use of fungi in other biocontrol areas. However, fungi have been and are being used to control weeds in many countries. Mycoherbicidal species such as *Colleotrichum gleosporoides* (active component of Collego), used to control jointvetch, and *Phytophthora palmivora* (active component of Devine) have achieved some success, and many other potential mycoherbicides are in development worldwide. Difficulties with potential mycoherbicides, such as having too broad a host range, and loss of virulence, complicate development, but extensive research into the use of these agents continues. Both classical mutation and selection techniques, and recombinant DNA technology, could be used to produce stable, highly virulent strains, which defy heretofore resistant pests; these would be ideal candidates for use in integrated pest management schemes. So far, only natural isolates are in use, but as our ability to modify organisms genetically improves, progress will be made in development of these promising nonchemical agents.

Fungi, of course, are significant plant pathogens. In the late 1980s fungal crop diseases were estimated to cause crop damage costing several billion dollars annually in the United States alone. Legislation aimed at reducing fungicide usage inevitably means that alternatives must be found. Fortunately, many fungal species can attack and kill other species, including plant pathogenic types. Species such as *Peniophora gigantea* and *Trichoderma viride* have demonstrated antifungal activity. Again, despite considerable research, the use of such agents is still at an early stage. Before rapid progress using gene technology can be made, there must be fundamental studies into the biochemistry, genetics, and ecology of these microorganisms, to improve on the "natural" strains in current use.

Fungal biocontrol agents have great promise as elements of integrated pest management systems, where they may be used in conjunction with chemical agents at much reduced levels. Development of such agents, which are currently of questionable economic viability, will be driven forward by public demand, since biological control is perceived as ecologically more acceptable than chemical control.

6.5   Ectomycorrhizal Fungi

Ectomycorrhizas, mutually beneficial close associations between the root systems of plants and a fungus (or fungi), occur on about 10% of world flora. The ectomycorrhizal fungi generally improve plant growth, increase the available area for nutrient uptake from soil, reduce the chance of infection by root pathogenic fungi, and increase tolerance to drought, toxins, and extremes of pH.

Ectomycorrhizal associations are particularly valuable in many economically important forest species from families such as Pinaceae (pine, larch, fir) Betulaceae (alder, birch), and Fagaceae (oak, chestnut, beech).

Deliberate introduction of mycorrhizal fungi to forest or nursery soils to stimulate growth has many attractions. Three methods have been used: addition of soil containing the fungi, incorporation of sporophore material, and, most reliably, inoculation of seedlings with pure cultures of vegetative mycelia. The main problem with the last approach is the difficulty of large-scale axenic cultivation of ectomycorrhizal fungi in STRs.

In addition to their widespread use in commercial forestry plantations (about 7–8 million trees annually are being "tailored" with specific ectomycorrhizal fungal symbionts in the United States alone), mycorrhizal fungi may also be of value in the reclamation of landfill sites. Revegetation of such sites may be hindered by presence of toxic leachates and gases, and also by absence of suitable fungal symbionts. It has been shown, however, that when fungal symbionts are present, host plant tolerance to physical and chemical stresses improves. Ectomycorrhizal fungi may also be of help in afforestation of sites containing industrial wastes, such as anthracite waste, metallic mine tailings, and coal spoils.

6.6   Prospects

The future for ectomycorrhizal fungi in reforestation and afforestation programs looks very bright. Biotechnology is used to enhance natural processes in this instance. The remaining difficulties concerning consistent large-scale cultivation are being resolved. Further studies into the maintenance of the infectivity of ectomycorrhizal fungi after cultivation in vitro are still required.

7   FUNGAL BIOTRANSFORMATIONS

Certain fungi can carry out highly specific transformations of complex organic molecules to produce molecules of high pharmaceutical potency and high optical purity. Because of their oxidative character, fungi have great advantages in oxygen-mediated biotransformation, especially when stereospecificity is a requirement.

7.1   Steroids from Fungi

The advantages of an oxidative nature were demonstrated in the transformation of steroids by fungi such as *Curvularia, Rhizopus,* and *Penicillium raistrickii*. These fungi can carry out highly specific modifications to steroids—for example, hydroxylations, dehydrogenations, and side chain degradations—which would be extremely difficult to carry out chemically. Such modifications can alter the pharmacological activity considerably. Examples include conversion of progesterone to 11-β-OH-cortexolone by *Curvularia*.

7.2   Other Biopharmaceuticals from Fungi

The search for pharmaceutically active nonantibiotic drugs continues unceasingly, but screening for activities such as immunostimulation or immunosuppression and antihypercholesteremia is difficult compared to antibiosis screens. It is perhaps unsurprising then that two very successful fungal products were discovered recently almost by accident. Cyclosporin A, for example, was found to be a potent immunosuppressant. However, initially, the producing organism *(Tolypocladium)* was being examined for antifungal activity and toxicity to mosquito larvae; its true potential discovered later. The identification of the antihypercholesteremic agent mevinolin, originally isolated as an antifungal agent, was similarly serendipitous.

The discovery that cyclosporin A inhibits a cytosolic peptidyl–prolyl isomerase activity may point the way toward a more specific screen for immunosuppressive drugs based on this enzyme.

Edible fungi have long been reported to have many beneficial effects on consumption. Recent research has shown that compounds such as lentinan (from *Lentinus edodes*) and scleroglucan (from *Sclerotium glucanicum*) may be potent antitumor agents. Both compounds are high molecular weight branched glucans. Leninan has also been shown to be capable of boosting a depressed immune system under trial conditions. Molds represent a tremendous (largely untapped) reservoir of biopharmaceuticals; the major hindrance to the biotechnological development of this resource is the current lack of directed and effective screens for desirable activities.

8 FUNGI AND TOXIC MATERIALS

White rot fungi, which have the ability to degrade a vast range of hazardous organic compounds, may be of use in systems designed to reduce the environmental impact of waste streams containing such materials or to reclaim land containing such compounds (bioremediation).

A great deal of work has been carried out on the lignin-degrading fungus *Phanerochaete chrysosporium,* and evidence suggests that lignin peroxidases are responsible for degradation of some xenobiotics.

The list of xenobiotics that can be degraded by fungi lengthens constantly. Presently included are insecticides such as DDT, lindane, and wood preservatives (e.g., pentachlorophenol), through to chlorinated biphenyls and dioxins. Despite the demonstrated potential of fungal systems in xenobiotic degradation and in the treatment of kraft pulp wastes (the toxicity of which is largely due to the presence of chlorinated phenols, catechols, and guaiacols), much work relating to how these organisms mineralize the organic materials remains to be done.

By virtue of their large surface area and the presence of charged molecules on their surfaces (polysaccharides, proteins), filamentous fungi have the ability to bind metal ions in significant amounts. Several potential uses can be made of this capacity, including the cleanup of metal-containing waste streams, the concentration of desirable metal ions such as Ni^{2+}, and the concentration of ions of uranium and thorium from process streams generated by the nuclear industry. The advantages of removing up to 90% of the metal ions from such process streams and concentrating them in the fungal biomass, which can be further processed to reduce its volume, are obvious.

9 GENETIC IMPROVEMENT OF FUNGAL STRAINS

Once a fungus expressing a desirable trait, such as excretion of valuable metabolites or a potentiality for biocontrol, has been recognized and isolated, there begins almost immediately a search to improve that characteristic. While much can be achieved in terms of metabolic or physiological control, the primary focus for strain improvement is at the genetic level.

In many fungi of commercial significance, one or more of the following methods have been adopted for strain improvement.

1. Selective screening and selection.
2. Mutagenesis, screening, and selection.
3. Use of selective "breeding," sometimes by the use of the parasexual cycle, to combine desirable characteristics of two strains to give a novel genotype.
4. In the past decade or so, the ability to directly transform fungi by the introduction of a desired gene or genes has been developed.

Methods 1 and 2 have served biotechnology well. In the case of penicillin G, for example, they have contributed to increases in titer from the early isolates to current production strains of several thou-

sandfold. They do however, have limitations. In particular, in some systems mutagenesis to increase metabolite synthesis may adversely affect other desired characteristics (e.g., strain vigor and fitness) that are especially important for strains used for biocontrol or as bioinoculants.

More importantly, a point of diminishing returns will be reached with respect to the conventional methods; indeed, it has been reached for organisms such as *Penicillium chrysogenum.*

The parasexual cycle, first recognized in *A. nidulans,* has been widely exploited to improve a range of industrial fungi by combining characteristics from related strains. The usefulness of this technique is limited by the existence of fungal incompatibility systems that restrict the occurrence of anastamosis to closely related strains. This obvious restriction on selective breeding (it narrows the genetic pool from which one can select) can sometimes be overcome by the generation of protoplasts. Such systems have been successfully applied to a range of fungal types including *T. harzianum, B. bassiana,* and *M. anisopliae.* Interspecies transfer is also a possibility.

At best, such methods are somewhat random even in selective breeding programs. Ideally, there is a need to specifically transfer the gene or genes responsible for the desired activity from one organism to another. The potentialities of such direct transformation have been illustrated, not only in heterologous protein production, but also in improvement of β-lactam producing fungi by Lilly Research Laboratories.

Since the early studies on transformation of *A. nidulans* and *N. crassa,* the number of selective markers has steadily increased, and now a range of nutritional and dominant selective markers is available. This second group has been applied in transformation studies of many economically important metabolite- or protein-producing fungi and some plant pathogens. They include bleomycin, hygromycin, and mutant β-tubulin genes conferring benonyl resistance.

Fungal transformation commences with cell wall digestion in the presence of osmotic stabilizers, using preparations such as Novozym 234, sometimes in conjunction with other enzymes (e.g., driselase and glucuronidase). This stage may critically affect transformation frequency.

Transforming DNA has been shown to integrate into the chromosomal DNA. In *A. nidulans* it has been possible to study direct/indirect gene replacement, and gene disruption effects.

In the work by Lilly on β-lactam producers, dominant selectable markers such as hygromycin, phleomycin, and *amid S* (acetamidase gene of *A. nidulans*) have been used in the transformation systems of *Cephalosporium acremonium* and *Penicillium chrysogenum.* Using these systems, the investigators made progress in three distinct areas. In the first application targeted gene disruption was used to locate and clone a β-lactam synthetic gene *(pcbAB)* in *C. acremonium.* The second application involved use of gene dosage to improve cephalosporin C titer in *C. acremonium.* Addition of one extra copy of *cefEF* gene (which encodes for the enzyme catalyzing the rate-limiting step in cephalosporin C synthesis) led to a 15% increase in titer at the pilot scale. By comparison, mutagenesis and screening of very large numbers of derivatives of the parent strain failed to lead to the identification of any that were capable of increased productivity. This work demonstrates the great potency of the new techniques in molecular biology in overcoming identified metabolic limitations in fungi.

The final application involved biosynthetic pathway engineering, that is, the alteration of the β-lactam synthesis pathway to give a new end product, which is either an antibiotic with altered capabilities or a useful intermediate for further synthesis. This project involved the expression in *P. chrysogenum* of the *cefE* gene of *Streptomyces clavuligerus,* which led to significant synthesis of a desired enzyme (deacetoxycephalosporin C synthetase).

Taken together, these studies illustrate the tremendous capabilities of new gene technology to alter, in a planned fashion, the capabilities of existing fungal strains. Currently, such alterations require a major directed effort, with highly specific aims, and cannot realistically be justified unless they result in the improvement of an organism or the introduction of a process of great economic significance. As the technology matures, its applicability will broaden, and undoubtedly the bioengineering approaches discussed here will be used in many fungi of current interest.

10   PERSPECTIVE

The use of molds in man's service has a long history. Conversely, their industrial exploitation can be measured in decades. These microorganisms occupy a central role in the fermentation industry and, in the near future, will continue to do so.

However, the future of fungal biotechnology will be heavily dependent on new product development and the development of new areas in which molds can usefully be applied. There are particularly promising developments in the areas of biotransformations, bioremediation using fungi, and use of filamentous fungi as systems for the expression and production of heterologous proteins. The application of the new techniques of gene transfer to filamentous fungi will leave no current application area untouched and may revolutionize our use of them.

See also Bioprocess Engineering; Drug Synthesis.

Bibliography

Kinghorn, J. R., and Turner, G., Eds. (1992) *Applied Molecular Genetics of Filamentous Fungi.* Blackie, Glasgow.

Leatham, E. F., Ed. (1992) *Frontiers in Industrial Mycology.* Chapman & Hall, New York.

Leong, S., and Berka, R. M. (1992) *Molecular Industrial Mycology.* Dekker, New York.

McNeil, B., and Harvey, L. M. (1990) *Fermentation: A Practical Approach.* Oxford University Press, Oxford.

Schmauder, M. P., *et al.* (1991) Application of immobilised cells for biotransformations of steroids. *J. Biol. Microbiol.* 31:453–477.

Ward, I. P. (1989) *Fermentation Biotechnology.* Open University Press, Milton Keynes, U.K.

Whipps, J. M., and Lumsden, R. D., Eds. (1989) *Biotechnology of Fungi for Improving Plant Growth*. Cambridge University Press, Cambridge.

GAUCHER DISEASE

Ernest Beutler

Key Words

Frameshift   A change in a gene that causes it to be read so that the sets of three nucleotides that compromise a codon are out of synchrony. The wrong triplets are, therefore, read.

Gaucher Disease   A hereditary disorder characterized by the accumulation of a glycolipid in macrophages throughout the body. Disease manifestations include enlargement of the spleen and liver and bone fractures.

Glycolipids   Complexes of sugars and fats.

Hydrophilic   Literally "water loving"; mixes freely with water.

Hydrophobic   Literally "water hating"; does not mix freely with water.

Phenotype   The observed effect caused by a mutation in an organism.

Pseudogene   A copy of a functional gene that has undergone mutational change that render it no longer functional.

Gaucher disease is the most common of the glycolipid storage disorders. It was first described in 1882 by Phillipe Charles Ernest Gaucher (1854–1918) in a 32-year-old woman with massive enlargement of the spleen. Gaucher regarded the disorder an "epithelioma of the spleen." Now it is recognized to be the consequence of the deposition of glucosylceramide (glucocerebrosidase) in the macrophages of the body. This glycolipid is a catabolic product of globosides and gangliosides, complex glycolipids that are constituents of cell membranes. Normally the enzyme glucocerebrosidase cleaves glucose from glucocerebrosidase, leaving ceramide, but in Gaucher disease a deficiency of glucocere-brosidase leads to accumulation of glucocerebrosidase. Rarely, glucocerebrosidase activity of patients with glucocerebrosidase storage has been normal, and the origin of the disease in such patients is a deficiency in a cofactor, saposin, required for normal activity of glucocerebrosidase.

In 1927 Oberling and Woringer recognized similarities between the visceral pathologies of "Gaucher disease" and a rapidly progressive, fatal infantile disease involving the central nervous system. This condition has been designated acute neuronopathic, or type 2, disease. In 1959 Hillborg described the "Norrbottnian" form of slowly progressive neuronopathic Gaucher disease in a Swedish isolate above the Arctic Circle, type 3 Gaucher disease.

Thus three major types of Gaucher disease have been delineated based on the absence (type 1) or presence and severity (types 2 and 3) of primary central nervous system involvement. Type 1 disease is sometimes designated as the "adult" form of the disease; this practice is misleading, however, since type 1 disease often becomes apparent in early childhood. Within each type, even within the same ethnic group, the phenotypes and genotypes can be markedly heterogeneous.

1   THE GLUCOCEREBROSIDASE GENE

Located on chromosome 1 at q21, the gene for glucocerebrosidase is approximately 7 kilobases in length and contains 11 exons. A 5 kb pseudogene is located about 16 kb downstream. The pseudogene has maintained a high degree of homology with the functional gene. Although it is transcribed, an unusual property for a pseudogene to possess, the pseudogene does not contain a long open reading frame and has a 55 base pair deletion from what was once the coding region. *Alu* sequences that have been inserted into introns give the functional gene greater length than the pseudogene. The putative TATA- and CAAT-like boxes of the promoter have been identified about 260 bp upstream from the upstream ATG start codon.

The cDNA is about 2 kb in length. There are two in-frame ATGs at the 5' end, and both are utilized as start codons. The relative importance or function of these two start sites is unknown. Transfer into murine cells of human acid β-glucosidase cDNAs mutated to ablate one or the other of the start sites has shown that either site can produce functional enzyme in vivo. The sequence between the upstream and downstream ATG is hydrophilic, while that between the downstream ATG and the codon that represents the amino end of the mature protein is the typical hydrophobic sequence expected in a leader sequence. Messenger RNA of several different lengths has been detected, probably because of the existence of alternate polyadenylation sites, alternative splicing, or the presence of pseudogene mRNA.

Table 1   Nucleotides at 12 Positions in the Four Glucocerebrosidase Haplotypes

Designation	Frequency	Glucocerebrosidase Haplotypes at Nucleotides:											
		−802	−725	−614	2128	2834	3297	3747	3854	3931	4644	5135	6144
1 +	Common	a	c	c	a	c	g	g	t	g	del	c	g
2 −	Common	g	t	t	g	g	a	g	c	a	a	a	a
3 African	Common African	g	t	t	g	a	a	g	c	a	a	a	a
4 Uncommon	Rare	a	c	c	a	c	g	a	t	g	del	c	g

Table 2   Gaucher Disease Mutations

		Point Mutations That Cause Gaucher Disease					
cDNA Number[a]	Amino Acid Number	Genomic Number	Nucleotide Substitution	Amino Acid Substitution		Effect	Population Frequency[b]
IVS2+1		1067	G→A[c]			? Severe	Uncommon
476	120	3060	G→A	Arg→Gln		? Mild	Rare
481	122	3065	C→T	Pro→Ser		Mild	Rare
535[d]]	140	3119	G→C	Asp→His		?	Rare
1093]	326	5309	G→A	Glu→Lys		?	
586	157	3170	A→C	Lys→Gln		Severe	Rare
751	212	3545	T→C	Tyr→His		? Mild	Rare
754	213	3548	T→A[c]	Phe→Ile		Severe	Uncommon
764	216	4113	T→A	Phe→Tyr		Mild	Rare
1043	309	5259	C→T	Ala→Val		Mild	Rare
1053	312	5269	G→T	Trp→Cys		Mild	Rare
1090	325	5306	G→A[c]	Gly→Arg		Severe	Rare
1141	342	5357	T→G	Cys→Gly		Severe	Rare
1208	364	5424	G→C	Ser→Thr		Mild	Rare
1226	370	5841	A→G	Asn→Ser		Mild	Common
1297	394	5912	G→T	Val→Leu		Severe	Uncommon
1342	409	5957	G→C[c]	Asp→His		Severe	Uncommon
1343	409	5958	A→T	Asp→Val		Severe	Rare
1361	415	5976	C→G	Pro→Arg		Severe	Rare
1448	444	6433	T→C[c]	Leu→Pro		Severe	Common
1504	463	6489	C→T	Arg→Cys		Mild	Uncommon
1604	496	6683	G→A	Arg→His		Mild	Uncommon

	Insertions and Deletions That Cause Gaucher Disease				
cDNA Number[a]	Genomic Number	Nucleotide Substitution	Amino Acid Substitution	Effect	Population Frequency[b]
84	1035	G→GG		Severe	Common
1263–1317 del	5879–5933[c] del			? Severe	Rare

	Recombination Events That Cause Gaucher Disease		
Location of	Crossover Event(s)[e]	Effect	Population Frequency[b]
>1343 <1388	>5957 <6272[f]	Severe	Uncommon
>455 <475][d,g]	>3039 <3059][d,g]	? Severe	Rare
>754	>3548		
>1317 <1343	>5932 <5957	Severe	Uncommon
>1343 <1388	>5957 <6272	Severe	Uncommon
>1225 <1263	>5588 <5878	Severe	Rare

[a] Nucleotide or amino acid position.

[b] Common, high frequency in at least one population; Uncommon, found in a number of unrelated patients; Rare, found in only one or two individuals.

[c] Pseudogene sequence.

[d] Both found in one gene.

[e] Only approximate ranges can be given, since the pseudogene and functional gene contain long identical sequences.

[f] Physical fusion with loss of intergenic segment.

[g] The first range represents crossover from gene to pseudogene; the second from pseudogene to gene. This region contains seven mutations, only six of which are identical with pseudogene sequence. At genomic nt 3474, within the region, the nucleotide conforms to the active gene, not the pseudogene. Thus "conversion" seems imperfect.

2 POLYMORPHISMS AND DEFICIENCY MUTATIONS

Twelve polymorphic sites are known to exist in the introns and flanking regions of the acid β-glucosidase gene, but surprisingly only four major haplotypes have been found. Table 1 summarizes these sites and the haplotypes that have been identified. Many point mutations that cause glucocerebrosidase deficiency, hence Gaucher disease, have been identified. These are summarized in Table 2.

Some genes with the 1448C mutation contain other mutations, each corresponding to the sequence of the pseudogene. Sometimes this type of mutation, designated XOVR, represents a crossover between the functional gene and pseudogene with loss of the genetic material between the two. In other cases the mechanism by which the abnormal allele has been formed is not known. Such multiple mutations have been referred to as complex alleles, pseudopattern, or "rec" (for recombinant). There are probably several different mechanism by which these anomalies arise, including the formation of fusion genes, as in the XOVR mutation and the occurrence of gene conversion events.

Some mutations are relatively mild in their phenotypic effect and have been associated only with type 1 disease. Others are more severe and are observed also in type 2 or type 3 (neuronopathic) disease. In the case of some mutations the number of cases that have been observed is so limited that the severity of the mutation is not known. A severity score has been devised that permits quantitation of clinical severity of patients with Gaucher disease. As shown in Figure 1, patients who are homozygous for the 1226G mutation tend to have relatively mild disease with a late onset, while patients who also carry one of the more drastic mutations have more severe disease with early onset. However, there is clearly overlap in the severity of disease manifestations between groups with a severe genotype, such as 1226G/84GG and a mild genotype such as 1226G/1226G.

3 POPULATION GENETICS

Of the many mutations that have now been documented, only three appear to approach polymorphic frequencies. The most common is the A→G transition, at nucleotide (nt) 1226, which produces a protein with an Asn→Ser substitution at amino acid 370 of the mature protein. It is present in slightly more than 6% of the Ashkenazi Jewish population. This mutation is the principal cause of the high incidence of Gaucher disease in this ethnic group. An insertion of a guanine at nt 84 of the cDNA is the second most common Jewish mutation. It is found in approximately 0.6% of the Jewish population. Because it produces a frameshift even before the amino terminus of the mature protein, this drastic mutation produces no enzyme protein. A T→C transition at nt 1448, producing a Leu→Pro substitution at amino acid 444 of the mature protein, is present at a relatively high frequency in the Norbottnian population of northern Sweden. The same mutation occurs in other populations at low frequencies. It is noteworthy that the homologous position in the pseudogene is occupied by a C, just as in the mutation that produces Gaucher disease.

Each of the two most common Jewish mutations is found only in a single haplotype. The 1226G mutation is always in the context of the − haplotype, as defined by the *Pvu* II polymorphism (Pv1.1⁻), while the 84GG mutation is always in the context of the + haplotype. Even the less common IVS2(+1) mutation appears to be limited to a single haplotype, in this case +. Moreover, if the haplotype

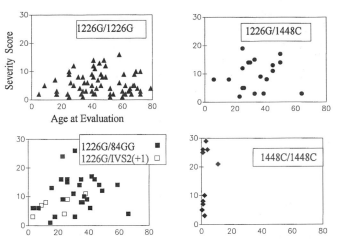

Figure 1. The relation between age at the time of evaluation and severity score of patients with four different Gaucher disease genotypes. The most severe disease is seen in the 1448C homozygotes, while many of the patients who were homozygous for the 1226G mutation are elderly and have very mild disease. The severity score is obtained by assigning points for various disease manifestations (e.g., for enlargement of the spleen, abnormally low blood counts, abnormal liver function tests, fractures of the bones, or involvement of the lungs). Thus, patients with a high score have severe multiorgan disease, while those with a score of zero have no disease manifestations at all.

is extended to include a common polymorphism in the nearby gene for type L pyruvate kinase, the mutations are still found in the context of a single haplotype. The finding of each of these mutations in its own haplotype is consistent with, and argues for, a single and evolutionarily relatively recent origin of each mutation. While it might be argued that the existence of a single mutation causing Gaucher disease in the Jewish population was due to genetic drift— an accident, as it were—this explanation becomes quite untenable when three different mutations, each of unique origin, are found in the same population. An additional circumstance argues for some sort of heterozygote advantage to account for the high frequency of Gaucher disease in the Ashkenazi Jewish population: that at least two other mutations that affect closely related metabolic reactions are also most common in the Jewish population. These are hexosaminidase A deficiency, leading to Tay–Sachs disease, and sphingomyelinase deficiency, leading to Niemann–Pick disease.

While the circumstances just mentioned clearly indicate that heterozygotes for these mutations have enjoyed a selective advantage, probably one that was relatively unique to the circumstances of Jews within the past few thousand years, the nature of this advantage is unknown. Suggestions have included resistance to infection with tuberculosis and some sort of advantageous neuropsychological trait. The population data that have been collected thus far do not support either of these possibilities. It is entirely possible that the selective pressure was exerted by an infectious agent that is no longer prevalent.

4 BIOCHEMISTRY

The storage of massive amounts of glucocerebroside is the defining feature of Gaucher disease. Small quantities of this glycolipid normally are widely distributed in many tissues as a metabolic intermediate in the synthesis and degradation of complex glycosphingolipids, such as the gangliosides or globoside. Glucocerebrosidase is lysosomal in location, for glucocerebroside and its precursors are normally degraded in this organelle, which functions to cleave glucocerebroside into ceramide and glucose. When a deficiency of the enzyme is inherited, glucocerebrosidase accumulates in the tissues, giving rise to the typical Gaucher cell, a macrophage engorged with the glycolipid.

Glucocerebrosidase requires the presence of a heat-stable protein cofactor, either saposin A or C, to function normally in the hydrolysis of glucocerebroside. The result of proteolytic cleavage from a larger precursor molecule that contains saposins A, B, and C, a deficiency of saposins is an extremely rare cause of a severe neuronopathic from of Gaucher disease.

Plasma levels of glucosylceramide are elevated 2- to 20-fold in patients with Gaucher disease, but such levels do not correlate with the type of Gaucher disease nor the degree of involvement within a type. The predominant site of storage is, of course, the liver and spleen. Here the glycolipid fatty acid chains are predominantly 20–24 carbons in length. Glucosylsphingosine (i.e., glucocerebrosidase that has lost its fatty acid chain) is also present in increased amounts.

5   CLINICAL MANIFESTATIONS

5.1   TYPE 1 GAUCHER DISEASE

The clinical manifestations of type 1 Gaucher disease are a consequence of the accumulation of glucocerebroside in macrophages. This results in dysfunction of the liver and spleen, displacement of normal bone marrow by the storage cells, and damage to bone leading to infarctions and fractures. Occasionally primary or secondary involvement of other organs may contribute to the total clinical picture. Children with Gaucher disease generally show growth retardation, but their mental development is quite normal. Indeed, many patients with Gaucher disease are extraordinarily gifted.

The broad spectrum of severity of Gaucher disease is a striking characteristic of type 1 disease. At one extreme patients have been diagnosed as late as the eighth and even the ninth decade of life, and some "patients" with the disease have been ascertained during family studies or population surveys. Almost invariably the latter are found to have inherited the 1226G/1226G genotype. At the opposite extreme are children with massive hepatosplenomegaly associated with severe abnormalities of liver function, pancytopenias, and extensive skeletal abnormalities. Such children may die of the complications of Gaucher disease in the first or second decade of life. Usually their genotypes include at least one very severely deficient allele such as the IVS2(+1) or 84GG mutation.

Some patients with severe visceral (liver and spleen) enlargement are largely spared skeletal involvement; others with severe bone disease have minimal visceral disease. In some, visceral and skeletal involvement are approximately equal in severity. In contrast to its effect on overall disease severity, the type of mutation seems to have little influence on the site of disease involvement.

As pointed out in connection with Figure 1, there is a tendency for some mutations to be associated with mild and others with more severe disease. However, a puzzling and important aspect of Gaucher disease is that considerable variability in disease manifestations is observed among patients with the same genotype. This variability is not due to additional mutations within the glucocerebrosidase gene, because it is observed even within a sibship, in which sibs must be carrying the same two mutant genes. There may be modifying genes that play a role in the severity of expression of Gaucher disease, but

it is notable that among the few sets of identical twins that have been observed, some discordance in expression is also sometimes evident. This suggests that environmental factors, such as infectious agents, may have a profound influence on disease expression.

Of the abnormalities that may exist in the peripheral blood, thrombocytopenia is most frequent. As a result, bleeding is the most common presenting finding in patients with Gaucher disease. Factor IX deficiency is also common in patients with Gaucher disease, and this appears to be a coincidental association, because both genetic disorders are common among Ashkenazi Jews. Early in the course of the disease, thrombocytopenia is usually due to splenic sequestration of platelets and invariably responds to splenectomy. Later, in patients who have already undergone splenectomy, replacement of the marrow by Gaucher cells may be more important.

Anemia is usually mild, but occasionally hemoglobin levels fall to levels as low as 5 g/dL and transfusions may be required. Leukopenia also occurs in some patients. These changes are probably due to a combination of increased splenic sequestration, when the spleen is present, and decreased production because of replacement of the marrow by Gaucher cells.

Splenic enlargement is usually present and is commonly the presenting sign. In severely affected patients the spleen may be huge, filling the abdomen. The bulk of the spleen may interfere with normal food intake, resulting in considerable discomfort. The liver is usually enlarged, and when the disorder is severe the liver may fill the entire abdomen. Frank hepatic failure and/or cirrhosis with portal hypertension and ascites occurs in a small percentage of patients, and bleeding from varices may cause death.

The skeletal manifestations of Gaucher disease can be totally incapacitating. Generalized demineralization of the skeleton is common, and fractures of vertebrae and aseptic necrosis of the femoral head are serious consequences of the disease. Many patients with radiographically significant to severe Gaucher disease have few or no bony symptoms. Episodic, sometimes excruciatingly painful "bone crises" occur in about 20–40% of patients.

Albeit uncommon, pulmonary failure is one of the most serious consequences of Gaucher disease. It may occur as the result of frank infiltration of lung by Gaucher cells, particularly in children, but also occurs in patients without demonstrable Gaucher cells in the lungs. In such patients, right-to-left intrapulmonary shunting, probably secondary to liver disease, seems to be the cause.

By definition, primary central nervous system involvement does not occur in type 1 Gaucher disease. However, occasionally central nervous systems can be observed as a secondary manifestation of true type 1 disease. For example, bleeding, fat emboli, and compression of the spinal cord may all be causes of central nervous system damage in type 1 disease.

5.2   TYPE 2 GAUCHER DISEASE

Oculomotor abnormalities, represented by the appearance of bilateral fixed strabismus or of oculomotor apraxia, are often the first manifestations of type 2 disease. Additional findings include hypertonia of the neck muscles with extreme arching of the neck (opistotonus), bulbar signs, limb rigidity, and seizures. Infants with type 2 disease usually die within the first two years of life.

5.3   TYPE 3 GAUCHER DISEASE

The severity of type 3 disease is intermediate between that of types 1 and 2. Massive visceral involvement is usually present. Neurologic symptoms similar to those in type 2 disease are present, but with a later onset and relatively lower severity. The prototype of type 3 disease is the Norrbottnian form, which is due to a point mutation at nt 1448. The first symptoms are usually the results of visceral involvement, with neurologic findings developing in about half the children during the first decade of life. As in type 2 disease, disorders of eye movement are the usual first symptoms, with other neurologic manifestations, such as ataxia, developing later.

6   THERAPY

Much can be done to improve the quality of life of patients with Gaucher disease without actually affecting the basic disease process. Joint replacement, splenectomy, and treatment with antibiotics of infections that may occur are all helpful. However, understanding of the basic pathophysiology, biochemical genetics, and molecular genetics has created the possibility of interventions that remove storage lipid or correct the genetic defect.

6.1   BONE MARROW TRANSPLANTATION

The manifestations of type 1 Gaucher disease are entirely a result of changes in macrophages, progeny of the hematopoietic stem cell. It was therefore to be expected that allogeneic marrow transplantation should be curative. A few patients with type 1 disease have undergone transplantation. Although the response has been favorable in surviving patients, transplantation is still a high risk treatment modality, and there have been deaths secondary to the procedure. Transplantation has also been carried out in patients with type 3 disease, but the effect on the neurologic disease is uncertain.

6.2   GENE TRANSFER

Correction of the defect in hematopoietic stem cells of the patient should be an effective way to treat Gaucher disease, since replacing the patient's defective hematopoietic stem cell with a genetically normal stem cell from another individual will effect a cure. However, stem cells that produce glucocerebrosidase would have no proliferative advantage over those that do not. Hence, cure would be expected only if the patient's untransformed cells were ablated by chemotherapy or irradiation. Thus, a rational strategy for the treatment of Gaucher disease would be marrow ablation followed by autologous transplantation with transformed hematopoietic stem cells. Considerable effort has been expended to develop the required efficient gene transfer technology.

The transfer of glucocerebrosidase cDNA into a variety of cells using retroviral vectors has been accomplished. Included are cultured fibroblasts, transformed lymphoblasts, and, most importantly, hematopoietic stem cells. Recently, high efficiency transfer of the human acid β-glucosidase cDNA into murine and human hematopoietic stem cells or progenitors has been accomplished with some evidence of sustained gene expression. The adenoassociated virus vector has also been used for gene transfer.

6.3   ENZYME REPLACEMENT

Macrophages are by their very nature cells that ingest material from their environment. The idea of restoring the missing enzyme to these cells by exogenous administration has been pursued by several investigators since it was first suggested by DeDuve. With the commercial production of large amounts of glucocerebrosidase from human placenta and subsequently by recombinant technology (alglucerase, Genzyme, Boston), considerable success has attended the administration of the enzyme in the treatment of Gaucher dis-

ease. Alglucerase is mannose-terminated glucocerebrosidase, an enzyme from which terminal sugars have been removed based on the hypothesis that the enzyme will be targeted to the mannose receptor of macrophages. There is serious doubt about the mechanisms of its action because alglucerase binds to another mannose-dependent receptor that is ubiquitous in distribution and presumably does not deliver its ligand to the lysome. Nonetheless, alglucerase is effective in the treatment of Gaucher disease, particularly if small amounts are administered at frequent intervals.

7   CLINICAL GENETICS AND PREVENTION

7.1   HETEROZYGOTE DETECTION

While the assay of leukocyte or fibroblast acid β-glucosidase activity detects many heterozygotes for Gaucher disease, there is considerable overlap between enzyme levels in normal subjects and in carriers of a Gaucher disease gene. Detection of heterozygotes by DNA analysis is much more satisfactory. When the specific mutation(s) occurring in the family being examined are known, heterozygote detection is highly accurate. Barring technical errors or the existence of additional unsuspected Gaucher disease genes (as may happen when one of the parents of the propositus actual has unsuspected Gaucher disease), every family member can be correctly classified with respect to genotype. When one or more of the mutations in a family are unknown, DNA analysis will be accurate only with respect to the known mutation; in population studies, those who have mutations for which the DNA is not examined will, of course, elude detection. In the Ashkenazi Jewish population, the occurrence of this problem is relatively infrequent, since examination of DNA of Gaucher patients for just four mutations [84GG, 1226G, IVS2(+1), and 1448C] will account for some 96.5% of the mutations. Since the homozygous state for the 1226G mutation frequently causes mild late onset disease, this mutation is underrepresented in the patient population as contrasted with the general population, and is about twice as frequent in the general population in relation to the other mutations. Examination for additional mutations that have been found in more than just a single family—those at nt 1297, 1504, and 1604—brings the percentage of mutants detected in the patient population to 98% and that in the Ashkenazi Jewish general population to 99%. In the non-Jewish population the situation is somewhat less favorable, since only about 75% of the mutations are encompassed by examining the DNA for the 1226G and 1448C mutations, and adding additional mutations such as 84GG to the mutations detected does not increase the number of mutant genes ascertained greatly, since each of these mutations individually is quite rare in the non-Jewish population. The frequency of the mutations causing Gaucher disease in Jewish and non-Jewish patients, as found in 186 patients in which examination for 29 mutations was performed, is summarized in Table 3.

7.2   PRENATAL DIAGNOSIS

β-Glucosidase activity can be estimated in amniotic fluid cells and chorionic villus samples using 4-methyl umbelliferyl-β-glucoside as a substrate. Thus, prenatal diagnosis can be achieved for all types of Gaucher disease. Recently, the availability of DNA-based technology provides another facile means for identification of the genotype of the fetus. If both mutations are known in a sib, prenatal diagnosis can be achieved. If not, linkage with the *Pvu* II polymorphic site and with the pyruvate kinase polymorphism may be helpful.

Table 3 Occurrence of Gaucher Disease Mutations in Jewish and Non-Jewish Patients

Mutation[a]	Occurrence in Jewish Patients	Occurrence in Non-Jewish Patients
1226G	187	37
1448C	8	47
84GG	30	1
IVS2(+1)	6	1
1297	5	0
XOVR	1	5
1604A	3	1
1504T	0	6
751C	1	0
1549A	0	1
476A	0	1
1085T	0	1
481T	0	2
72del	0	1
764A	0	2
1342C	0	1
del	0	1
?	3	20

[a]XOVR, a fusion gene resulting from recombination between the glucocerebrosidase gene and pseudogene; ?, none of the 29 mutations tested were present.

7.3 Genetic Counseling

When a couple already has a child with Gaucher disease and both parents are heterozygotes, the recurrence risk, as in other autosomal recessive disorders, is 1:4. Because homozygotes for the 1226G mutation often have very mild disease, there are many instances in which one of the parents is a homozygote. Indeed, this happens sufficiently frequently that Gaucher disease was once regarded as being transmitted as an autosomal dominant disorder. When one of the parents is a homozygote and the other heterozygous, the recurrence risk is 1:2.

If the couple being counseled already has children with Gaucher disease, prediction of disease severity is easier than if there are no children with the disease. Although, as noted, there are disturbing exceptions, there is a considerable degree of concordance between sibs. However, substantial differences between sibs, particularly in disease manifestations if not severity, do exist. That is, it is not uncommon for one sib to have primarily visceral disease while another sib has mostly skeletal disease. It is less common for one sib to be severely affected while another is virtually free of disease manifestations, although this occasionally occurs.

If there are no affected children in the family, the counselor must depend on the genotype of the parents, if it is known. In the most common circumstance, when both are 1226G heterozygotes, they can be told that (1) neurologic disease will not occur, (2) it is possible but not likely that a child homozygous for this mutation will have severe disease with early onset, and (3) it is much more likely that the child will have mild disease or be essentially asymptomatic.

When the fetus is at risk for the 1226G/84GG or the 1226G/1448C genotype, the information to be given to the parents is quite different. In these genotypes mild disease is possible, but much less probable. The median age of onset of disease in children with these genotypes is about 5 and 8 years, respectively, and generally moderately severe disease develops. However, there are occasional exceptions, comprising about 10% of the patients, in which either of these genotypes is associated with quite mild disease. Again, neurological disease is not expected in these genotypes.

When the child is at risk for the 1448C/1448C genotype or one of its even more serious variants in which additional mutations are present (variously designated as XOVR, *rec*, pseudopattern, and complex allele; see Section 2), the prognosis is quite grave. The likelihood of eventual, if not early, neurologic involvement is high, and in any case severe visceral and skeletal disease is to be expected. Counseling may be more difficult when one of the less common alleles is present. However, as a general rule, if such an allele has been found in a patient with neurologic disease, it may be considered to be in the same class with the 1448C and related alleles; if it has not resulted in neurologic disease in combination with the 1448C allele, it may be considered to be milder in its phenotypic effect, more closely resembling the 1226G allele.

See also Human Genetic Predisposition to Disease.

Bibliography

Beutler, E. (1991) Gaucher's disease. *New Engl. J. Med.* 325:1354–1360.
———. (1992) Gaucher disease: New molecular approaches to diagnosis and treatment. *Science,* 256:794–799.
———, and Grabowski, G. (1993) Gaucher disease. In *The Metabolic Basis of Inherited Disease*, C. R. Scriver, A. L. Beaudet, W. S. Sly, and D. Valle, Eds. McGraw-Hill, New York.
Karlsson, S., Correll, P. H., and Xu, L. (1993) Gene transfer and bone marrow transplantation with special reference to Gaucher's disease. *Bone Marrow Transplant.* 11 suppl. 1:124–127.
Kohn, D. B., Nolta, J. A., Weinthal, J., Bahner, I., Yu, X. J., Lilley, J., and Crooks, G. M. (1991) Toward gene therapy for Gaucher disease. *Hum. Gene Ther.* 2:101–105.
Martin, B. M., Sidransky, E., and Ginns, E. I. (1989) Gaucher's disease: Advances and challenges. *Adv. Pediatr.* 36:277–306.

Gel Electrophoresis of Proteins, Two-Dimensional Polyacrylamide

Peter J. Wirth

1 Principles

2 Techniques
   2.1 First Dimension (IEF)
   2.2 Second Dimension (SDS-PAGE)
   2.3 Metabolic Labeling with [^{35}S]Methionine and [^{32}P]Orthophosphate

3 Polypeptide Detection Methods
   3.1 Coomassie Brilliant Blue Staining
   3.2 Silver Staining
   3.3 Radioactive Labeling
   3.4 Double- and Triple-Label Analysis

4 Polypeptide Identification and Characterization
   4.1 Western Immunoblot Analysis
   4.2 Protein Microsequencing

5 Concluding Remarks

Key Words

Amphoteric Describing zwitterionic compounds containing both positively and negatively charge groups within the same molecule, hence able to function as either an acid or base.

Carrier Ampholytes Low molecular weight amphoteric poly-aminopolycarboxylic acids with isoelectric points from pH 2.5 to 11 which generate a continuous pH gradient in an electric field.

Isoelectric Point (pI) The pH at which a protein exhibits no net charge (i.e., the molecule has an equal number of positive and negative charges), hence fails to migrate in an electric field.

Western Electroblotting Electrotransfer of separated proteins from gels to a thin support matrix, such as nitrocellulose, to which they bind and on which they are immobilized.

The electrophoretic separation of proteins in polyacrylamide gels has become the method of choice for the fractionation and characterization of proteins at both the analytical and preparative levels. Separation is dependent on both the relative molecular mass, or "molecular weight" (M_r), and shape, as well as the net charge of the individual proteins, hence is directly related to the primary amino acid composition of the individual proteins. In practice, however, proteins are typically separated either as a function of their M_r (molecular sieving) independent of their net charge or, alternatively, on the basis of their net charge independent of their M_r. Although the resolving power of each technique is quite good (on the order of 100 distinct protein bands), single-dimensional protein separations are insufficient for certain analytical problems. For example, it has been estimated that the total number of proteins synthesized by a typical mammalian cell (e.g., liver cell) is in the range of 5000–20,000, with as few as 3000–6000 playing any significant role in cellular maintenance. Therefore single-dimensional electrophoretic separation protocols do not provide the resolution necessary for the analysis of whole-cell lysate mixtures. To circumvent this problem Patrick O'Farrel and Joachim Klose in 1975 introduced the technique of high resolution two-dimensional polyacrylamide gel electrophoresis (2D-PAGE).

2D-PAGE is the combination of two relatively simple and independent electrophoretic techniques, namely, isoelectric focusing (IEF), in which proteins are separated first on the basis of their intrinsic charge or isoelectric point (pI), followed by electrophoresis in the presence of the detergent sodium dodecyl sulfate, in which proteins are then fractionated on the basis of their M_r. This relatively simple combination of two widely used electrophoretic procedures provides an extremely powerful analytical technique for the simultaneous analysis of 2000–3000 individual proteins on a single electrophoretogram. As a result, 2D-PAGE has found widespread application in a broad range of biological systems. This technique has been used in the analysis of phenotypically dependent alterations of protein expression during both normal and abnormal growth and differentiation, including cancer development and other disease states, as well as in the identification of specific protein changes induced by mutagens and carcinogens, hormone treatment, mitogen stimulation, and nutrient changes, and for the characterization of human and animal tissues and cells. In addition to these analytical applications, 2D-PAGE has recently been utilized preparatively for the isolation of extremely pure proteins for antibody production and N-terminal Edman amino acid microsequence analysis for subsequent molecular biological applications.

1   PRINCIPLES

Figure 1 illustrates the procedures by which two-dimensional polyacrylamide gel electrophoresis (2D-PAGE) is carried out. First, protein mixtures (isolates from intact cells, whole tissues, or protein

Two-Dimensional Polyacrylamide Gel Electrophoresis (2D-PAGE)

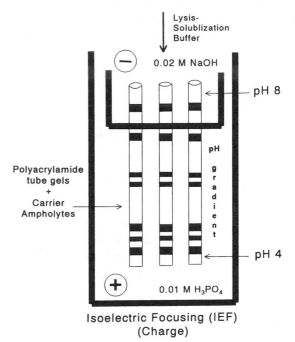

Isoelectric Focusing (IEF)
(Charge)

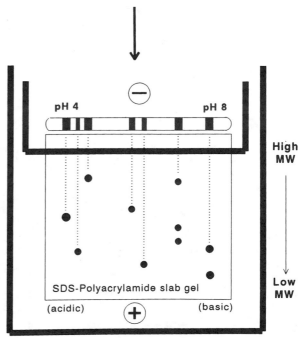

SDS-Polyacrylamide Gel Electrophoresis (SDS-PAGE)
(Molecular Weight)

Figure 1. Schematic representation of first-dimension (IEF) and second dimension (SDS-PAGE) procedures of 2D-PAGE.

fractions) are solubilized in a denaturing solution containing non-ionic detergents (e.g., CHAPS, NP-40, Triton-X), urea, a reducing agent (β-mercaptoethanol, dithiothreitol, etc.), and carrier ampholytes and applied to first-dimension cylindrical polyacrylamide tube gels containing an appropriate mixture of carrier ampholytes. Sufficient voltage is applied to the tube gels to induce the individual proteins to migrate to their respective isoelectric points (pIs) as determined by each protein's content of specific acidic and basic amino acid residues.

The polyacrylamide gels are then extruded from the glass tubes using a transfer buffer containing sodium dodecyl sulfate (SDS), which results in the stoichiometric binding of SDS to the individual proteins and the gels affixed directly onto the top of a second-dimension SDS polyacrylamide sieving gel. When voltage is applied, the individual proteins, which are negatively charged as a result of their association with SDS, migrate out from the cylindrical tube gels and into the slab gel, where they are separated by molecular sieving according to their relative molecular mass, or "molecular weight," M_r. The resultant electrophoretogram is a two-dimensional map in which individual proteins (polypeptides) are displayed as discrete spots with the slower migrating, higher M_r proteins located toward the top of the gel and the more acidic proteins to the left.

Figure 2 is the autoradiogram of a typical 2D-PAGE separation of approximately 1000 [^{35}S]methionine labeled whole-cell lysate proteins from cultured rat liver epithelial (RLE) cells over the pH range of 5–7 and M_r of 14,000–205,000. Actin (pI 5.7 and MW 43,000) a ubiquitous structural protein found in most cell types, is identified for reference.

Recently, an alternative procedure to the use of carrier ampholytes in the first-dimension isoelectric focusing (IEF) mode of 2D-PAGE has been developed and is rapidly gaining widespread acceptance in more and more laboratories. This procedure, common-

ly known as immobilized pH gradient (IPG) electrophoresis, utilizes commercially available ultrathin precast polyacrylamide gel strips containing covalently bound, low molecular weight, mono-substituted acrylamido acids and bases to set up well-established, highly stable, preformed pH gradients. The use of IPG gel strips offers many advantages not afforded by the use of carrier ampholyte IEF, including higher resolution, better reproducibility, greater protein loading capacities, and wider pI range of separations. Whereas IEF permits the accurate separation of proteins having pIs between pH 4.5 and 7.5, protein separation utilizing IPG is easily extended to pI 9.0 with no protein streaking. IPG gel strips are routinely run on a flatbed horizontal gel electrophoresis apparatus and are equilibrated with SDS-containing transfer buffer prior to affixation on the second-dimension SDS slab gels. From a technical view, IPG strips are considerably easier to manipulate than tube gels, since the IPG gels are precast on Mylar support sheets and are not subject to breakage like tube gels.

Although 2D-PAGE maps appear highly complex and complicated, with practice, the technique is highly reproducible, and routine analysis of the 2D-PAGE patterns can be performed by superimposing one photographic image over another on a light box. For more complex work, however, sophisticated computer-assisted analysis programs have been developed and are available.

2   TECHNIQUES

There is no single method for the preparation, running, and analysis of 2D-PAGE gels. Detailed descriptions regarding the exact protocols for the running of 2D gels is beyond the scope of this article, and readers are referred to the Bibliography for additional details and citations. In the following sections I briefly describe generalized experimental procedures and techniques common to most 2D-PAGE separation protocols, referring to ongoing research applications in my laboratory.

2.1   FIRST DIMENSION (IEF)

First-dimension IEF is typically performed in 4.5% polyacrylamide tube gels (1 mm × 160 mm) containing 2% carrier ampholytes (1.6%, pH 4–8; 0.4%, pH 3–10). Protein samples are dissolved in denaturing lysis buffer (9.5 M urea, 1% NP-40, 1% CHAPS, 10 mM dithiothreitol, and 2% carrier ampholytes) to a final protein concentration of 10 μg/μL and 10–100 μg of protein loaded at the basic end of the IEF tube. Samples are electrophoresed at room temperature at constant power (0.02 W/tube) for total of 13.5 kV·hs.

If IPG gel strips are utilized, first-dimensional flatbed horizontal electrophoresis is performed at 20°C using constant voltage (3000 V) for a total of 100–200 kV·hs. Sample preparation for IPG is the same as that utilized for carrier ampholyte IEF.

2.2   SECOND DIMENSION (SDS-PAGE)

Second-dimension SDS-PAGE electrophoresis is typically performed using vertical slab gels containing 0.1% SDS and constant percentage of polyacrylamide (10%), although gradient gels containing variable concentrations of polyacrylamide (4–20%) are also used for increased resolution. Following separation in the first dimension, the polyacrylamide gel rods are carefully extruded from the glass tubes directly onto the surface of the second-dimension SDS slab gel using an SDS-containing transfer buffer and held in place with 1% agarose dissolved in running buffer.

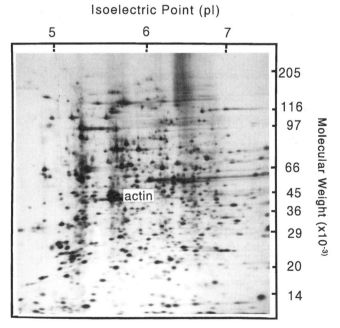

Figure 2.   2D-PAGE autoradiograph of [^{35}S]methionine-labeled polypeptides from cultured rat liver epithelial cells. Polypeptides were separated in the first dimension by IEF in the pH range of 4.5 (left) to 7.5 (right); molecular weight range, 14–205 kDa.

If IPG electrophoresis is performed in the first dimension, gel strips are equilibrated for 10 minutes in the SDS transfer buffer before placement on the SDS slab gels. IPG strips are similarly held in place with 1% agarose dissolved in running buffer. Electrophoresis is performed at constant current (20 mA/gel) at 10°C using a running buffer of 0.192 M glycine, 25 mM Tris base, and 0.1% SDS. Following electrophoresis (4–5 h) individual polypeptides are routinely visualized using a variety of techniques including organic dye or inorganic metal salt based staining protocols as well as radioactivity (Sections 3.3 and 3.4) detection procedures.

2.3   Metabolic Labeling with [^{35}S]Methionine and [^{32}P]Orthophosphate

Monolayer cell cultures are routinely grown to approximately 75–80% confluence in 20–100 mm plastic tissue culture dishes and prelabeled for 4 hours at 37°C in 5 mL of serum-free cell culture growth medium (e.g., Dulbecco's, Hams F-12, RPMI 1640) lacking methionine and cysteine and containing 50 μCi/mL of [^{35}S]methionine (specific activity > 1000 Ci/mmol).

Alternatively, cells can be prelabeled with [^{32}P]orthophosphate (specific activity > 8000 Ci/mmol) in serum-free, phosphate-free growth medium for 4 hours. After labeling, cell surfaces are washed twice at 4°C with 10 mM Tris HCl buffer, pH 7.0 containing protease inhibitors (0.2 mM phenylmethylsulfonyl fluoride, 1 μg/mL each of leupeptin and aprotinin), and phosphatase inhibitors (50 mM sodium fluoride, 1 mM o-vanadate, and 5 mM EDTA). Cells are scraped from dishes and pelleted by centrifugation. Cell pellets are solubilized in the lysis buffer given in Section 2.1 and clarified by high speed (> 250,000 g) centrifugation; radioactivity incorporation is then determined by trichloroacetic acid precipitation and scintillation counting.

3   POLYPEPTIDE DETECTION TECHNIQUES

Practically all analytical techniques that can be applied following 1D-PAGE can similarly be utilized following 2D-PAGE, including the commonly used protein staining and autoradiographic and fluorographic procedures. Detection of protein species is usually performed postelectrophoretically using a variety of organic dyes and stains; among these are Amido Black 10B, Ponceau S, Fast Green FCF, Procion Blue, Nigrosin, India ink, and Coomassie Brilliant Blue (CBB) R-250 and CBB G-250, as well as a plethora of "silver" staining methods. If cell lysates or protein samples are isotopically labeled either metabolically or chemically with ^{35}S, ^{14}C, ^{3}H, ^{32}P, or ^{125}I, then either autoradiography or fluorography (^{3}H) of the dried gels is performed, thereby imparting to 2D-PAGE its extremely high degree of sensitivity.

3.1   Coomassie Brilliant Blue

Coomassie Brilliant Blue is probably the most popular organic protein stain for 2D-PAGE gels because of its ease of use and relatively low cost. Staining is usually performed for 1–2 hours with 0.2% (w/v) CBB R-250 in methanol/acetic acid (45:10, v/v) followed by removal of excess dye (3–18 h) with the same methanol/acetic acid solution. In an alternative CBB staining procedure, which eliminates the need for background destaining, a colloidal suspension of CBB G-250 is dispersed in 12.5% trichloroacetic acid. In this one-step procedure, the colloidally dispersed dye does not penetrate the

polyacrylamide gel matrix and only the protein bands or spots themselves are stained.

Although the standard CBB staining procedures are relatively insensitive, they are, capable of detecting protein bands (spots) of 30–40 ng, and this degree of sensitivity is usually sufficient for most applications. If, however, a higher degree of sensitivity is required, an alternative to CBB dye staining, namely, silver staining, is also available.

3.2   Silver Staining

Silver staining was introduced in 1979 and in less than two decades, over 100 variations of the original procedure have appeared, and numerous commercial silver staining kits have been developed. Silver staining is on the order of 20–200 times more sensitive than the commonly used CBB stains and is capable of detecting as little as 1–5 ng of protein. Figure 3 exemplifies the enhanced sensitivity of silver staining over CBB staining. Duplicate 2D-PAGE separations

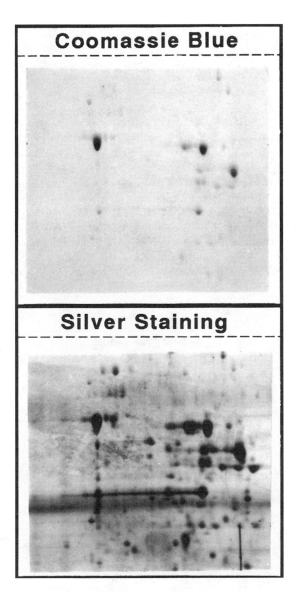

Figure 3. Comparison of CBB and silver staining. 2D-PAGE separations of duplicate samples were performed and individual gels were stained with CBB and silver stain, respectively. A selected area of each gel is shown.

of 100 μg of RLE cellular protein were performed, and each gel was stained with either CBB or silver. A selected area of each stained gel is shown in Figure 3.

Although silver staining is the most sensitive nonradioactive protein detection method available, a number of drawbacks limit its usefulness. Among these are the following: (1) background staining is high; (2) staining procedures are multistepped, more time-consuming, and considerably more costly than CBB staining, (3) silver staining seriously quenches detection of ^{3}H-labeled samples by fluorographic methods, (4) and the most serious, certain proteins (e.g., calmodulin) fail to stain or stain very poorly when subjected to single-step silver staining procedures. Despite these limitations, silver staining has found wide applications in the detection of trace polypeptides not amenable to labeling with radioactive isotopes, the most sensitive of all detection procedures. Table 1 briefly summarizes one of the most commonly used silver staining procedures (Morrisey).

3.3 Radioactive Labeling

Proteins can be radiolabeled either metabolically or chemically in any number of ways. Metabolic labeling of cells or tissues during in vitro growth or maintenance in culture, by incorporation of either single amino acids ([^{35}S] methionine or [^{3}H] leucine) or amino acid mixtures labeled with ^{14}C or ^{3}H, is probably the most common method (Section 2.3) and usually results in high radioactivity incorporation. Techniques also exist for the detection of posttranslationally modified proteins using either [^{32}P]orthophosphate or [^{3}H]glucosamine for the detection of phosphorylated or glycosylated proteins, respectively. Whole-animal labeling with [^{35}S]methionine, [^{3}H]leucine, and [^{32}P]orthophosphate has also been performed, but because large quantities of radioactivity are required to ensure sufficient incorporation of isotope, in vivo labeling is not routinely performed. Proteins not amenable to metabolic labeling can be labeled by chemical modification using either iodination with the isotope ^{125}I or reductive methylation ([^{14}C]formaldehyde and sodium borohydride) or detected postelectrophretically utilizing radioactive stains (e.g., [^{59}FeCl$_3$- or [^{59}Fe]ferrous bathophenanthroline), although these applications are somewhat limited in their use.

Whatever their method of preparation, radiolabeled proteins are usually detected using either autoradiography or fluorography and protein visualization on appropriate (Kodak SB-5, Kodak XAR, or Kodak X-OMAT) X-ray films.

3.4 Double- and Triple-Label Analysis

In certain instances (e.g., biological studies of protein synthesis/degradation/phosphorylation), the comparison of protein patterns following 2D-PAGE may be simplified using double or even triple labeling techniques. This "triple labeling" procedure involves the independent labeling of two aliquots of cells from the same cell population with two different isotopes (e.g., [^{35}S] methionine and [^{32}P] orthophosphate), mixing the samples prior to 2D-PAGE separation, dye staining (CBB or silver) of the resulting 2D-gels, and autoradiography of the dried, stained gels using two Kodak SB-5 (or equivalent) X-ray films placed back to back. The first film, which is positioned in direct contact with the dried stained gel, visualizes exposure to both ^{35}S and ^{32}P, while the second film, because of the differential energy levels of the two isotopes (β-particles from ^{35}S are unable to pass through the film emulsion and Mylar backing of the first film), records only exposure to ^{32}P. Whereas silver staining presents serious quenching problems for both autoradiographic and fluorographic detection of ^{3}H-labeled proteins (Section 3.3), only minor (< 10%) quenching is observed using higher β-emitters such as ^{35}S, and ^{14}C. The juxtapositioning of the stained images with the autoradiographic film images permits the unambiguous mapping of the phosphorylated proteins back to their corresponding stained and [^{35}S]methionine-labeled counterparts. This technique, which has wide application to a variety of biological studies, including phosphoprotein-mediated signal transduction, allows one to analyze constitutive levels of protein expression (dye staining) with rates of protein synthesis and/or degradation ([^{35}S]methionine labeling) and protein phosphorylation and/or dephosphorylation ([^{32}P]orthophosphate labeling).

In the samples shown in Figure 4, polypeptides 1–3 are readily detected both with silver staining and metabolic labeling (^{35}S and ^{32}P) and are major phosphorylated proteins. The autoradiographic detection (film darkening) of polypeptides 1–3 in Figure 4B represents the sum of exposure to both ^{35}S and ^{32}P while Figure 4C results from exposure to ^{32}P alone. Polypeptides 4 and 5, on the other hand, stain heavily with silver, and readily incorporate [^{35}S]methionine but are not visualized with ^{32}P. Polypeptides 6 and 7 fail to stain or stain very weakly with silver, yet are heavily labeled with both [^{35}S]methionine and [^{32}P]orthophosphate, suggesting that 6 and 7 may represent rapidly turning over, short half-life phosphoproteins. Polypeptide 8 is weakly detected with [^{35}S]methionine and does not incorporate [^{32}P]orthophosphate but stains heavily with silver. These results suggest that polypeptide 8 is a relatively stable, constitutively expressed, nonphosphorylated protein with a slow turnover rate.

4 POLYPEPTIDE IDENTIFICATION AND CHARACTERIZATION

Whereas 2D-PAGE is an excellent method for the separation of complex protein mixtures and provides useful pI and M_r information, it affords little or no insight concerning either the physico-

---

Table 1   Silver Staining Procedure

Day 1: Preparation

1. Remove gels from glass plates and place gels in glass trays (2 gels/tray). Soak, with gentle shaking, in distilled water for 5 minutes.
2. Fix proteins in a solution of ethanol, acetic acid, and distilled water (EtOH/HOAc/dH$_2$O) (25:10:65 v/v/v) for 1 hour.
3. Wash and store gels in EtOH/HOAc/dH$_2$O (5:5:90) for 3 hours to overnight.

Day 2: Staining

1. Soak gels in freshly prepared 0.001% dithiothreitol (DTT) (10 mg/L) for 30 minute.
2. Stain gels 15 minutes in 0.2% freshly prepared silver nitrate.
3. Wash gels twice in distilled H$_2$O, for 3 minutes each time.
4. Develop for 5 minutes in solution containing 3% sodium carbonate and 38% formaldehyde (0.5 mL/L).
5. Stop development in 2% HOAc.
6. Store gels until drying in glycerol/ethanol/dH$_2$O (7:10:83).
7. Soak gels in 2% glycerol for 5 minutes prior to drying.

Note: Do not touch gels with bare hands, and use only high quality, double distilled water for all solutions.

$(M_r \times 10^{-3})$

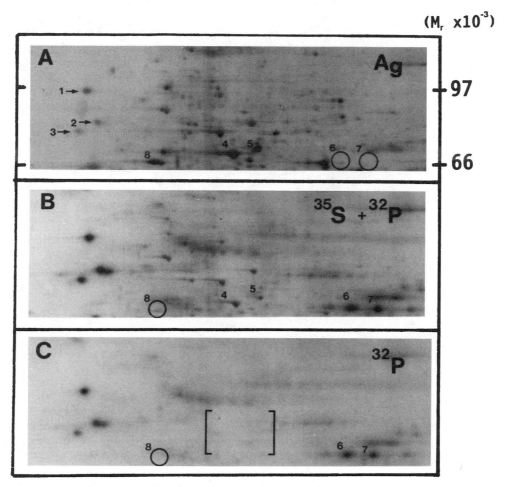

Figure 4. Simultaneous comparison of silver-stained, [^{35}S]methionine-, and [^{32}P]orthophosphate-labeled rat liver epithelial cellular protein: (A) silver-stained proteins, (B) ^{35}S- plus ^{32}P-labeled proteins, and (C) ^{32}P only. Numbered proteins represent polypeptides, which exhibit varying degrees of expression depending on mode of detection (silver vs. metabolic labeling) [Reproduced from Luo and Wirth (1993).]

chemical or biochemical function of the separated proteins. Therefore ancillary procedures are required for further characterization of individual proteins.

4.1 Western Immunoblot Analysis

Western electroblotting and immunoblot analysis using monoclonal and polyclonal antibodies, originally based on DNA Southern blotting principles, is one of the most widely used procedures to identify specific proteins on the 2D maps. Immediately following 2D-PAGE, proteins are transferred (electroblotted) from gels to thin matrix supports, usually nitrocellulose, nylon, or polyvinylidene difluoride (PVDF) membranes, as illustrated schematically in the Figure 5. Membranes are incubated initially with dilute protein solutions (e.g., bovine serum albumin, gelatin, or instant nonfat dry milk) to block nonspecific binding sites, after which the membrane-bound proteins are incubated with either specific monoclonal or polyclonal antibodies or group-specific ligands (e.g., concanavalin A for the detection of glycoproteins, or [^{59}Fe]FeCl$_3$ to detect iron-binding proteins). If antibodies are used, blots are incubated with a second radiolabeled (^{125}I) or enzyme conjugated (e.g., horseradish peroxidase, alkaline phosphatase, β-galactosidase) antibody that has been directed toward the primary antibody. The resulting pro-

tein (antigen)–antibody–antibody "sandwich" is detected either autoradiographically or colorimetrically (Figures 5 and 6).

Figure 6 depicts the 2D-PAGE separation of tropomyosin-associated proteins from an untransformed normal RLE cell line, a chemically transformed malignant RLE cell line (AFB-B6), and a spontaneously transformed malignant cell line (ST-C3T). Cellular proteins were separated by 2D-PAGE and transferred to nitrocellulose membranes by Western electroblotting (Figure 5), and immunoblot analysis was performed using a rabbit antitropomyosin polyclonal antibody. Visualization was accomplished using a horseradish peroxidase conjugated goat anti-rabbit second antibody. In normal RLE cells, all six tropomyosin isoforms (1–6) are expressed. However, during transformation there appears to be a selective loss of expression of specific tropomyosin isoforms (4, 5, and 6), dependent on the mode of transformation (chemically induced or spontaneous).

4.2 Protein Microsequencing

Although the vast majority of 2D-PAGE applications to date have been for analytical purposes, two new and exciting applications utilizing 2D-PAGE are rapidly emerging. 2D-PAGE is beginning to be used more frequently for the "micropreparative" isolation of suffi-

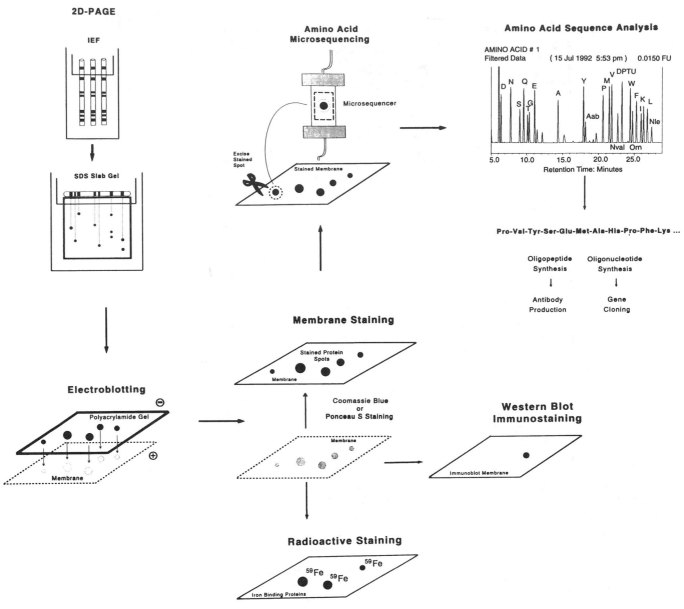

Figure 5.   Schematic illustration of 2D-PAGE and ancillary procedures.

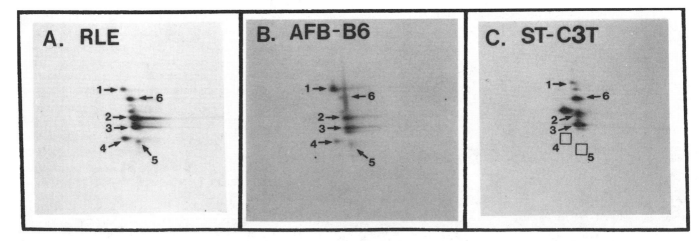

Figure 6.   Western transfer and immunoblot analysis of tropomyosin associated proteins from normal and transformed RLE cells. Small numbers represent individual tropomyosin isoforms and have been numbered according to Matsumura and colleagues; for abbreviations, see text.

Unknown (0.9 pmol)

```
       10      16
 --EDKKeEALLR---M
```

Blocked

1-Phosphatidylinositol-4,5-biphosphate phosphodiesterase - Rat (13.7 pmol)

```
 -DVLELTDENFE-RV
 ::::::::::::: ::
 SALLASASDVLELTDENFESRVSDTGSAGLMLVEFF
           20        30        40
```

Aldehyde dehydrogenase (2.3 pmol)

```
 aTSAvPApNQqpQvfen--
 ::::::::::::::: :
 ATSAVPAPNQQPEVFCNQ
```

Blocked

Protein disulfide isomerase (9.2 pmol)

```
 lYsssddVIeltpdnf
 :::::::::::: ::
 LYSSSDDVIELTPSNF
```

Unknown (1.6 pmol)

```
 1   5    10   15   20
 LHTKGALPLDTVcNYmViPtk
 g gr fr f    L     g
```

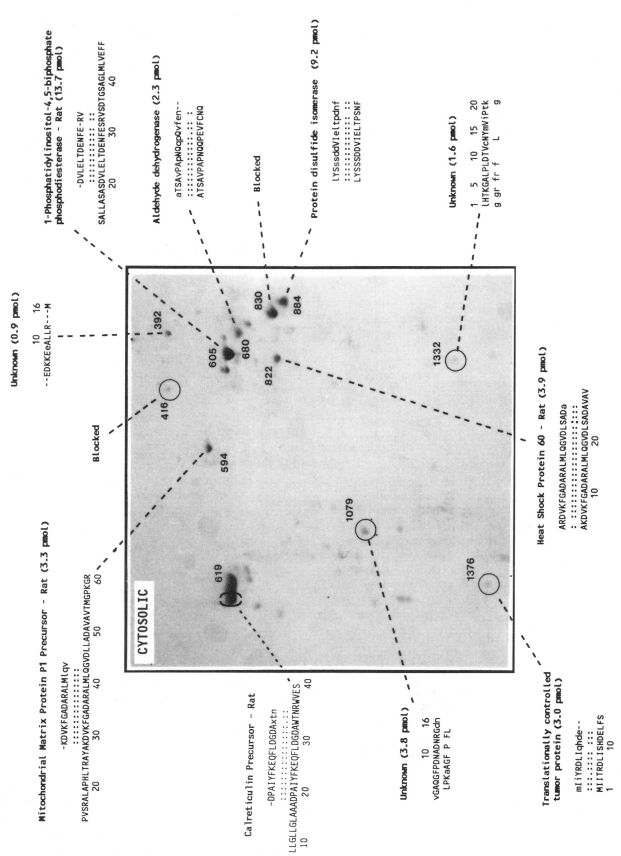

CYTOSOLIC

Mitochondrial Matrix Protein P1 Precursor - Rat (3.3 pmol)

```
 -KDVKFGADARALMLqv
 ::::::::::::::::
 PVSRALAPHLTRAYAKDVKFGADARALMLQGVDLLADAVAVTMGPKGR
          20        30        40        50        60
```

Calreticulin Precursor - Rat

```
             10      16
 -DPAIYFKEQFLDGDAxtn
 ::::::::::::::::
 LLGLLGLAAADPAIYFKEQFLDGDAWTNRWVES
 10        20        30        40
```

Unknown (3.8 pmol)

```
     10      16
 mIiYRDLIqhde--
 :: :::: :::
 MIITRDLISHDELFS
 1        10
```

Translationally controlled tumor protein (3.0 pmol)

```
 vGAQGFPDNADNRGdn
 LPKaGF P FL
```

Heat Shock Protein 60 - Rat (3.9 pmol)

```
 ARDVKFGADARALMLQGVDLSADa
 : :::::::::::::::::::::::
 AKDVKFGADARALMLQGVDLSADAVAV
                10        20
```

Figure 7. N-Terminal microsequencing of RLE nuclear cytosolic "nuclear sap" polypeptides. Amino acid residues are given by one-letter notation; x, assignment of a phenylthiohydantoin amino acid was not possible. Amino acids in the lower line represent the most probable assignment. Upper amino acid sequences are obtained from excised protein spots and the lower sequences are derived from the respective protein sequence databases. [Reproduced from Wirth et al. (1993).]

359

cient quantities of extremely pure polypeptides to be used directly for the generation of both monoclonal and polyclonal antibodies. Under certain circumstances, individual CBB-stained protein spots have been excised from whole gels and used for the direct immunization of rabbits for polyclonal antibody formation.

In addition to antibody production, recent analytical advances in the area of Edman N-terminal amino acid microsequencing have made it possible to obtain N-terminal and internal amino acid microsequence information for polypeptide spots isolated directly from 2D-PAGE gels, as schematically illustrated in Figure 5.

Following 2D-PAGE, proteins are electrophoretically transferred to PVDF membrane supports and visualized using either CBB or Ponceau S staining. CBB staining is two to three times more sensitive that Ponceau S staining, but Ponceau S is more desirable if N-terminal blockage necessitates subsequent enzymatic hydrolysis of individual proteins. Regardless of the mode of visualization, individual protein spots of interest are then cut from the stained membranes and inserted directly into a gas phase microsequencer and N-terminal amino acid sequencing is performed.

Figure 7 is a 2D-PAGE map of RLE "nuclear sap" proteins electroblotted to PVDF membrane and stained with Ponceau S. If a protein is not N-terminally blocked (e.g., acetylated, glycosylated, etc.), proteins expressed as little as 5–10 pmol (250–500 ng of 50 kDa protein) can be easily sequenced. N-Terminally blocked proteins require additional chemical and/or enzymatic cleavage and purification by high performance liquid chromatography prior to sequence analysis. Under optimal conditions 10–15 amino acid residues can usually be obtained from as little as 1–10 pmol of unblocked protein (Figure 7: proteins 392 and 1332), with some proteins such as heat shock protein 60 yielding up to 25 residues. The sequence data can then be used to search various protein sequence databases (e.g., PIR, PATCHX, SWISSPOT) for protein identification or sequence homology comparisons. If a protein has not been characterized or its identity is unknown (Figure 7: proteins 392, 1079, 1332), the partial sequences can be used for additional characterization studies. These include the large-scale chemical synthesis of corresponding poly(oligo)-peptides for both polyclonal and monoclonal antibody preparation as well as for the design and synthesis of specific oligonucleotide probes for cDNA isolation, gene cloning, and subsequent genetic analysis.

5   CONCLUDING REMARKS

The technique of 2D-PAGE provides an extremely powerful analytical method for the separation of complex protein mixtures. With recent refinements in protein microsequencing techniques, it provides a valuable link between protein biochemistry and molecular biology.

See also HPLC OF BIOLOGICAL MACROMOLECULES; PROTEIN ARCHITECTURE AND ANALYSIS; PROTEIN PURIFICATION; PROTEINS AND PEPTIDES, ISOLATION FOR SEQUENCE ANALYSIS OF.

Bibliography

Celis, J. E., and Bravo, R. (1984) *Two-Dimensional Electrophoresis of Proteins: Methods and Applications.* Academic Press, New York.

Dunbar, B. S. (1987) *Two-Dimensional Electrophoresis and Immunological Techniques.* Plenum Press, New York.

Hames, B. D., and Rickwood, D. (1990) *Gel Electrophoresis of Proteins: A Practical Approach.* IRL Press, New York.

Luo, L-di and Wirth, P. J. (1993) Consecutive silver staining and autoradiography of ^{35}S and ^{32}P-labeled cellular proteins: Applications for the analysis of signal transducing pathways. *Electrophoresis* 14, 127–136.

Pharmacia LKB Biotechnology AB. (1993) *2-D Electrophoresis Protocol Using IPG Immoboline DryStrips: Instruction Manual.* Pharmacia, Uppsala, Sweden.

Wirth, P. J., Luo, L-di, Benjamin, T., Hoang, T. N., Olson, A. D. and Parmalee, D. C. (1993) The rat liver epithelial (RLE) cell nuclear protein database. *Electrophoresis* 14, 1199–1215.

GENE DISTRIBUTION IN THE HUMAN GENOME

Giorgio Bernardi

1   **Introduction**

2   **Genome Organization and Gene Distribution**

3   **Gene Distribution in Vertebrate Evolution**

Key Words

Chromosomal Bands   During mitosis, chromosomes condense and, at metaphase, they are characterized by specific staining properties. Under standard conditions, Giemsa staining produces a total of about 400 bands that comprise, on average, 7.5 million base pairs (Mb) of DNA.

Genome   Every living organism contains, in its genome, all the genetic information that is required to produce its proteins (as well as some ribonucleic acids, which are not translated into protein) and is transmitted to its progeny. The genome consists of DNA, which is made up of two complementary strands wound around each other to form a double helix. The building blocks of each DNA strand are deoxyribonucleotides. These are formed by a phosphate ester of deoxyribose (a sugar), linked to one of four bases: two purines [adenine (A) and guanine (G)] and two pyrimidines [thymine (T) and cytosine (C)]. In the DNA double helix, purines pair with pyrimidines (A with T, G with C).

Isochores   Long regions of DNA characterized by homogeneity of base composition. Isochore size is 0.2–1.5 Mb or more. Isochores belong to a small number of families having distinct compositions.

This article deals with the organization of nucleotide sequences in the human genome and with the evolutionary history of such organization. Far from being an ensemble of genes scattered over vast expanses or intergenic sequences, the genome is highly ordered from the nucleotide level to the chromosomal level. Nucleotide sequences, whether in the 3% of the genome formed by coding sequences or in the 97% formed by noncoding sequences, obey precise rules that amount to a genomic code.

1   INTRODUCTION

The term *genome* was coined three-quarters of a century ago by the German botanist Winkler, to designate the haploid chromosome set. While current textbooks of molecular biology do not yet go beyond the purely operational definition (i.e., the eukaryotic genome is the sum total of genes and intergenic sequences), a number of molecular biologists have been thinking for some time that the genome is more than the sum of its parts. This implies the existence of structural and functional interactions between the coding sequences (the minority) and the noncoding sequences (the great majority). This general rather vague concept has been changed into a precise one by the discovery, in our laboratory, of compositional genome properties. These properties, which have mainly been defined by investigations on the nuclear genome of vertebrates, are briefly outlined here. They comprise the isochore organization, the compositional patterns of DNA fragments (or molecules) and of coding sequences, the compositional correlations between coding and noncoding sequences and, above all, the gene distribution and its associated functional properties.

2   GENOME ORGANIZATION AND GENE DISTRIBUTION

The mammalian genomes are mosaics of *isochores* (Figure 1)—long (> 300 kb) DNA segments that are homogeneous in base composition and cover a 30–60% GC range. This is an extremely wide

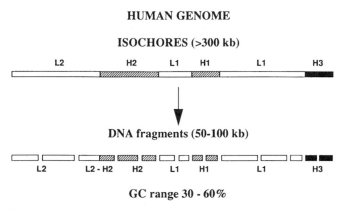

Figure 1. Scheme of the isochore organization of the human genome. This genome, which is a typical mammalian genome, is a mosaic of large (> 300 kb) DNA segments, the isochores. These are compositionally homogeneous (above a size of 3 kb) and can be subdivided into a small number of families: GC-poor (L1 and L2), GC-rich (H1) and (H2), and very GC-rich (H3). The GC range of the isochores from the human genome is 30–60%. [From Bernardi (1993).]

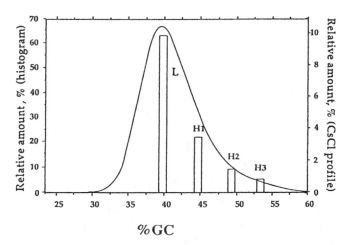

Figure 2. Histogram of the isochore families from the human genome. The relative amounts of major DNA components derived from isochore families L (i.e., L1 + L2), H1, H2, and H3 are superimposed on the cesium chloride profile of human DNA. [From Mouchiroud et al. (1991).]

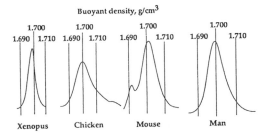

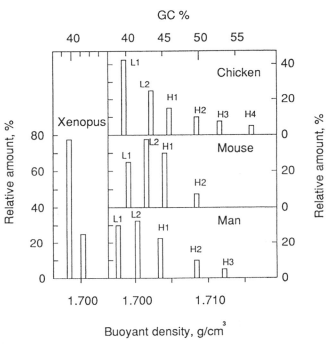

Figure 3. Compositional patterns of vertebrate genomes. *Top:* Cesium chloride (CsCl) profiles of DNAs from *Xenopus,* chicken, mouse, and man. [From Thiery et al. (1976)]. *Bottom:* Histograms showing the relative amounts, modal buoyant densities, and GC levels of the major DNA components from *Xenopus,* chicken, mouse, and man, as estimated after fractionation of DNA by preparative density gradient centrifugation in the presence of a sequence-specific DNA ligand [Ag⁺ or bis(acetatomercurimethyl) dioxane) (BAMD)]. The major DNA components are the families of large DNA fragments (see Figure 1) derived from different isochore families. Satellite and minor DNA components (such as ribosomal DNA) are not shown in these histograms. (From Bernardi (1993).]

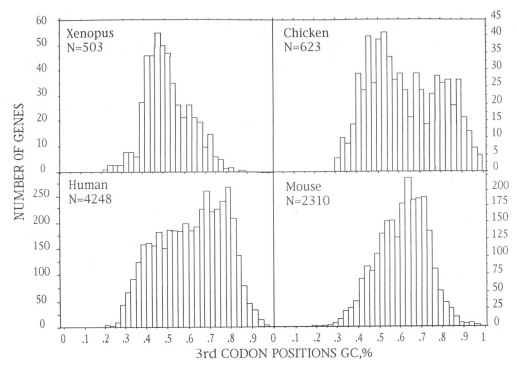

Figure 4. Compositional distribution of third-codon positions from vertebrate genes. The number of genes taken into account is indicated. A 2.5% GC window was used. [From Bernardi (1993).]

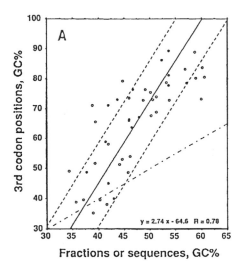

Figure 5. (A) GC levels of third-codon positions from human genes are plotted against the GC levels of DNA fractions (dots) or extended sequences (circles) in which the genes are located. The correlation coefficient and slope are indicated. The dot–dash line is the diagonal line (slope = 1). GC levels of third-codon positions would fall on this line if they were identical to GC levels of surrounding DNA. The broken lines indicate a ±5% GC range around the slope. [From Mouchiroud et al. (1991).]

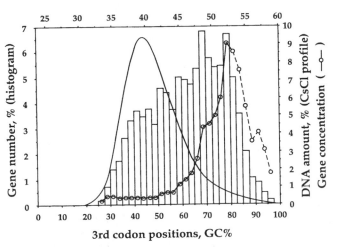

Figure 6. Profile of gene concentration in the human genome as obtained by dividing the relative amounts of genes in each 2.5% GC interval of the histogram by the corresponding relative amounts of DNA deduced from the CsCl profile. The apparent decrease in gene concentration for very high GC values (broken line) is due to the presence of rDNA in that region. The last concentration values are uncertain because they correspond to very low amounts of DNA. [From G. Bernardi (1993).]

range, almost as wide as that covered by all bacterial genomes (25–72% GC). In the human genome, isochores can be assigned to two GC-poor families (L1 and L2), representing two-thirds of the genome, and to three GC-rich families (H1, H2, and H3), forming the remaining third (Figure 2).

The *compositional distributions* of large (> 100 kb) genome fragments, such as those forming routine DNA preparations, of exons (and particularly of their third-codon positions) and of introns, represent *compositional patterns*. These correspond to *genome phenotypes* in that they differ characteristically not only between cold- and warm-blooded vertebrates, but also between mammals and birds and even between murids and most other mammals (Figures 3 and 4).

Compositional correlations exist between exons (and their codon positions) and isochores (Figure 5), as well as between exons and introns. These correlations concern, therefore, coding and noncoding sequences and are not trivial, since coding sequences make up only about 3% of the genome, whereas noncoding sequences correspond to 97% of the genome. The compositional correlations represent a *genomic code*. It should be noted that a *universal correlation* holds among GC levels of codon positions (third positions against first and/or second positions). Both the genomic code and the universal correlation are apparently due to compositional constraints working in the same direction (toward GC or AT), although to different extents on coding and noncoding sequences as well as on different codon positions.

The compositional correlations between GC_3 (the GC level of third-codon positions) and isochore GC have a practical interest in that they allowed us to position the coding sequence histograms of Figure 4 relative to the cesium chloride profile of Figure 3 and to assess the *gene distribution* in the human genome. In fact, if one divides the relative number of genes per histogram bar by the corresponding relative amount of DNA, one can see that gene concentration is low and constant in GC-poor isochores L1 and L2, increases with increasing GC in isochore families H1 and H2, and reaches a maximum in isochore family H3, which exhibits at least a 20-fold higher gene concentration compared to GC-poor isochores (Figure 6).

The H3 isochore family has been called the *human genome core* because it corresponds to the functionally most significant part of the human genome. Indeed, the H3 isochore family is endowed not only with the highest gene (and CpG island) concentration, but also with an open chromatin structure (as witnessed by the accessibility to DNases, as well as by the scarcity of histone H1, the acetylation of histones H3 and H4, and the wider nucleosome spacing), with the highest transcription and recombination levels and with an early replication timing. The genes of the genome core have the highest GC_3 levels relative to their flanking sequences, have the shortest exons and introns, exhibit an extreme codon usage, and encode proteins characterized by amino acid frequencies differing from those of proteins encoded by GC-poor isochores.

The human genome core is located in T (telomeric) bands which are essentially formed by GC-rich isochores (mainly of the H2 and H3 families). In contrast, R′ bands—that is, the subset of R bands comprising reverse bands exclusive of T bands, consist of both GC-rich isochores (of the H1 family) and GC-poor isochores. Finally, G(iemsa) bands are formed almost exclusively by GC-poor isochores. The difference in GC level between G bands and T bands is about 15%. About 20% of genes are present in G bands and about 80% in R bands (60% of them in T bands). The location of a majority of genes in T bands is of interest in view of the association of telomeres with the nuclear matrix and envelope.

3 GENE DISTRIBUTION IN VERTEBRATE EVOLUTION

It should be stressed that the gene distribution reported for the human genome seems to have been conserved in evolution, since genes show their highest concentration in the GC-richest isochores of all vertebrates.

In the case of homologous mammalian genes, it has been possible to show that third-codon-position synonymous substitutions exhibit frequencies and compositions that strongly suggest natural selection. Under these circumstances, the compositional changes in noncoding sequences, which are correlated with those occurring in third-codon positions, suggest that noncoding sequences are not junk DNA, but must fulfill some functional role.

As already mentioned, the compositional pattern of the human genome, which is typical of the genomes of most mammals and similar to the genomes of birds, is strikingly different from the compositional patterns of cold-blooded vertebrates, which exhibit a much lower degree of heterogeneity and are characterized by metaphase chromosomes, which do not show R-banding. These different genome phenotypes of warm- versus cold-blooded vertebrates are due to compositional changes. While the gene-poor, GC-poor isochores have undergone little or no compositional change in vertebrates genomes, the gene-rich, GC-rich isochores are those that underwent compositional changes in evolution.

See also MAMMALIAN GENOME; MOUSE GENOME; TANDEMLY REPEATED NON-CODING DNA SEQUENCES.

Bibliography

Aïssani, B., D'Onofrio, G., Mouchiroud, D., Gardiner, K., Gautier, C. and Bernardi, G. (1991) The compositional properties of human genes. *J. Mol. Evol.* 32:497–503.

Bernardi, G. (1995) The human genome: Organization and evolution. *Ann. Rev. Gen.* 29:445–476.

———, and Bernardi, G. (1986) Compositional constraints and genome evolution. *J. Mol. Evol.* 24:1–11.

———, Olofsson, B., Filipski, J., Zerial, M., Salinas, J., Cuny, G., Meunier-Rotival, M., and Rodier, F. (1985) The mosaic genome of warm-blooded vertebrates. *Science,* 228:953–958.

Cacciò, S., Perani, P., Saccone, S., Kadi, F., and Bernardi, G. (1995) Single-copy sequence homology among the GC-richest isochores of the genomes from warm-blooded vertebrates. *J. Mol. Evol.* 39:331–339.

de Lange, T. (1992) Human telomeres are attached to the nuclear matrix. *EMBO J.* 11:717–724.

D'Onofrio, G., Mouchiroud, D., Aïssani, B., Gautier, C., and Bernardi, G. (1991) Correlations between the compositional properties of human genes, codon usage and amino acid composition of proteins. *J. Mol. Evol.* 32:504–510.

Duret, L., Mouchiroud, D., and Gautier, C. (1995) Statistical analysis of vertebrate sequences reveals that long genes are scarce in GC-rich isochores. *J. Mol. Evol.* 40:308–317.

Macaya, G., Thiery, J. P., and Bernardi, G. (1976) An approach to the organization of eukaryotic genomes at a macromolecular level. *J. Mol. Biol.* 108:237–254.

Mouchiroud, D., Fichant, G., and Bernardi, G. (1987) Compositional compartmentalization and gene composition in the genome of vertebrates. *J. Mol. Evol.* 26:198–204.

———, D'Onofrio, G., Aïssani, B., Macaya, G., Gautier, C., and Bernardi, G. (1991) The distribution of genes in the human genome. *Gene,* 100:181–187.

Saccone, S., De Sario, A., Della Valle, G., and Bernardi, G. (1992) The highest gene concentrations in the human genome are in T-bands of metaphase chromosomes. *Proc. Natl. Acad. Sci. U.S.A.* 89:4913–4917.

———, ———, Wiegant, J., Rap, A. K., Della Valle, G., and Bernardi, G. (1993) Correlations between isochores and chromosomal bands in the human genome. *Proc. Natl. Acad. Sci. U.S.A.* 90:11929–11933.

Tazi, J., and Bird, A. (1991) Alternative chromatin structure at CpG islands. *Cell,* 60:909–920.

Thiery, J. P., Macaya, G., and Bernardi, G. (1976) An analysis of eukaryotic genomes by density gradient centrifugation. *J. Mol. Biol.* 108: 219–235.

Winkler, H. (1920) *Verbreitung und Ursache der Parthenogenesis im Pflanzen- und Tierreich.* Fischer, Jena, 1920.

Zoubak, S., D'Onofrio, G., Cacciò, S., Bernardi, G., and Bernardi, G. (1995) Specific compositional patterns of synonymous positions in homologous mammalian genes. *J. Mol. Evol.* 40:293–307.

GENE EXPRESSION, REGULATION OF

Göran Akusjärvi

Key Words

Epigenetic Modification   Changes in the phenotype that are not due to alterations in the genotype (i.e., mutations in the DNA).

Exons   Eukaryotic genes are encoded in discontinuous segments, where the coding portions are interrupted by noncoding sequences of unknown function (see **introns**). Both exonic and intronic sequences are transcribed into a nuclear precursor RNA. The segments of a eukaryotic gene that is preserved in the mature mRNA are the exon sequences. Prokaryotic genes usually are not split and, thus, are encoded by a contiguous DNA sequence.

Introns   Represent a segment of DNA that is transcribed into RNA in the nucleus of the eukaryotic cell, but is excised by RNA splicing before the exportation of mature mRNA to the cytoplasm.

Nucleosome   The basic structural subunit used to condense DNA in a cell. The nucleosome consists of approximately 200 base pairs of DNA wrapped around a protein core made up of histone proteins.

Promoter   The nucleotide sequence in the DNA to which the RNA polymerase binds when it begins transcription.

RNA Splicing   The process by which introns are removed from a nuclear precursor RNA during formation of a functional mRNA.

TATA Box   A conserved TATAAA sequence found about 25–30 base pairs upstream of the transcription initiation site in eukaryotic RNA polymerase II promoters. (A similar sequence element is also found in prokaryotic promoters, the −10 element.) The TATA box binds the general transcription factor TFIID and helps position the RNA polymerase for correct initiation.

TFIID   Transcription factor D for RNA polymerase II: a general transcription factor that interacts with the TATA box. It consists of the TATA-binding protein (TBP), which makes contact with the TATA box and several TBP-associated factors (TAFs), which are required for regulation of transcription.

Translation   The process by which the ribosome reads the nucleotide sequence of the mRNA and directs the incorporation amino acids into protein.

U snRNP   The uridine-rich, small nuclear ribonucleoprotein particles contained in large quantity in the nuclei of eukaryotic cells. The U1, U2, U4, U5, and U6 snRNPs have all been shown to be involved in RNA splicing. Other U snRNPs serve other functions in the cell.

Genetic information is transmitted between generations of a species in the form of a stable DNA molecule. This molecule is replicated before cell division to ensure that all offspring receive the same genetic constitution. Expression of the genetic information has been summarized in the so-called central dogma, according to which the flow of genetic information in a cell is transmitted from the DNA to an RNA intermediate to a protein.

Organisms are divided into two major groups, depending on whether their cells possess a nucleus: the prokaryotes, which include the bacteria and the blue-green algae, do not have a nucleus, and the eukaryotes, comprising animals, plants, and fungi. Cells of eukaryotes have a nucleus that encapsulates the DNA. The mechanisms to regulate gene expression in eu- and prokaryotes are similar, although eukaryotes generally use more complex regulatory pathways. In prokaryotes on–off switches of transcription appear to be the key mechanisms in the control of gene activity, although other mechanisms such as transcriptional attenuation, transcriptional termination, translational control, and mRNA and protein turnover play significant roles in the regulation of specific genes. In eukaryotes similar mechanisms are in operation. In addition, an increasing number of eukaryotic genes is now being shown to be regulated at the level of RNA processing.

1   DEFINING A TRANSCRIPTION UNIT

A transcription unit represents the combination of DNA segments that together constitute an expressible unit, whose expression leads to synthesis of a functional gene product(s) that often is a protein but also may be an RNA molecule. In prokaryotes, proteins in a specific metabolic pathway are often encoded by genes that are clustered at the DNA level and transcribed into one polycistronic mRNA that is used to translate the different proteins. Since prokaryotes lack a nucleus, the ribosomes are able to assemble on the nascent RNA chain and, thus, can initiate translation before transcription of the mature mRNA has been completed (Figure 1; see also Section 3.2). In eukaryotes, in contrast, the primary transcription product is a precursor RNA that is processed by RNA splicing before it is transported to the cytoplasm and presented to the ribosomes. This gives

eukaryotes a unique, very important level at which gene expression may be regulated (see Section 4). An additional difference between eukaryotes and prokaryotes is that eukaryotic mRNA usually is functionally monocistronic. This means that the mRNA may encode two open translational reading frames. As a rule, however, only the open reading frame closest to the 5′ end of the mRNA is translated into protein. This asymmetry results from the mechanism by which ribosomes binds to a mRNA (see Section 6.2). Transcription involves synthesis of an RNA chain that is identical in sequence to one of the two complementary DNA strands.

Transcription can be subdivided into three stages: I, initiation which begins by RNA polymerase binding to the double-stranded DNA molecule and incorporating the first nucleotide(s); II, elongation, during which the RNA polymerase moves along the DNA template, in a 5′-to-3′ direction, and extends the growing RNA chain by adding one nucleotide at the time; and III, termination, the stage at which RNA synthesis ends and the RNA polymerase complex disassembles from the transcription unit.

For reasons of space, this article concentrates on RNA synthesis and maturation of protein-encoding messenger RNAs (mRNAs). However, similar mechanisms are used to regulate synthesis of RNA molecules of other types.

2   REGULATION OF TRANSCRIPTION IN EUKARYOTES

Eukaryotic cells contain three RNA polymerases, designated I, II, and III, which are responsible for synthesis of specific RNA molecules within the cell. RNA polymerase I controls synthesis of ribo-

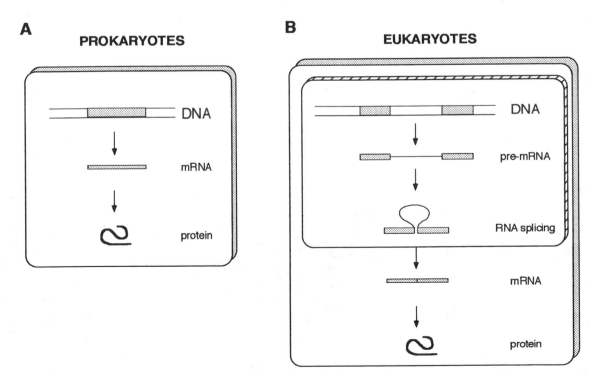

Figure 1. The transfer of genetic information in eukaryotic and prokaryotic cells. In eukaryotes the coding regions in the DNA (called exons) are separated by noncoding sequences (called introns) that are removed by RNA splicing before the mature mRNA is transported to the cytoplasm. In prokaryotic cells, which lack a cell nucleus, the nascent RNA chain can be used for translation (see Figure 9). These differences in organization are used by prokaryotes and eukaryotes to create unique levels for regulation of gene expression (see text).

somal RNA (rRNA), RNA polymerase II is responsible for synthesis of protein-coding mRNA, whereas RNA polymerase III takes care of transfer RNA (tRNA) and 5S RNA synthesis. The three polymerases are large enzymes consisting of approximately 10 subunits each. Some subunits are shared among the different polymerases, whereas others are unique and probably determine the specificity of the transcription process. Recent data have also shown that the different classes of polymerases use some common transcription factors. For example, the TATA-binding protein (TBP, see Section 2.2) which originally was thought to be specifically used in RNA polymerase II transcription, has been shown to be a universal transcription factor required for transcription by all three classes of polymerases, although only RNA polymerase II genes contain a binding site for TBP.

2.1 STRUCTURE OF A EUKARYOTIC PROMOTER

The transcriptional activity of a prototypical RNA polymerase II gene is regulated by a series of DNA elements that can be subdivided into a core promoter element, consisting of the TATA element and the transcription initiation site, and upstream activating sequences (UAS), which usually are positioned upstream of the core promoter element (Figure 2) but can in some cases also be found downstream of the transcription start site. In many genes several UAS motifs are clustered to entities called enhancers. Enhancer sequences are DNA segments, containing binding sites for several transcription factors, that activate transcription in a position- and orientation-independent manner. It is likely that enhancer sequences

are able to active transcription in a position-independent manner because they become spatially positioned close to the core promoter through bending of the DNA molecule. Each UAS motif in an enhancer is the binding site for a specific protein, a transcription factor, which may have a positive or negative effect on core promoter activity. The basal transcription factor TFIID has been shown to play a central role in this process by binding to the TATA element in the core promoter and facilitating the recruitment of the RNA polymerase to the promoter (Figure 2). In some promoters the TATA element is functionally substituted by an initiator region (INR), which is positioned at the site of transcription initiation (Figure 2). The assembly of an initiation-competent RNA polymerase at a promoter is a complex process that requires the participation of many basal transcription factors. Figure 2 represents only TFIID and TFIIB, since stable binding of these factors to a core promoter appears to be the decisive event committing the promoter for transcription. TFIID is a multiprotein complex consisting of the TATA box binding protein (TBP) and approximately seven tightly associated proteins (TAFs).

2.2 REGULATION OF PROMOTER ACTIVITY

From a regulatory standpoint it is important to note that recombinant TBP has been shown to be sufficient to direct basal transcription from a core promoter element. However, the TBP-associated factors (TAFs) have been shown to be essential for regulated transcription by the UAS-binding factors (Figure 2). Thus, TBP is sufficient for constitutive transcription but TAFs are necessary for reg-

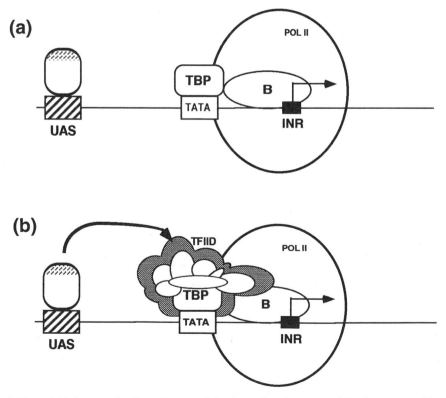

Figure 2.   A simplistic model for preinitiation complex formation on a minimal core element promoter (a) and a promoter (b) regulated by an upstream activating sequence (UAS). [the contributions of only a few of the general transcription factors have been indicated.] The principal difference between these two promoters is that the TATA-binding protein (TBP) is sufficient for basal promoter activity (a), whereas regulation of transcription by upstream binding transcription factors (b) requires TFIID and consists of TBP plus TBP-associated factors (TAFs): INR, initiator region; A, TFIIA; B, TFIIB; pol II, RNA polymerase II.

ulated transcription. Many UAS-binding transcription factors have been shown to transmit a signal to the core promoter element by directly making contact with the general transcription factors TBP or TFIIB (Table 1). UAS-binding transcription factors have been shown to have activation domains with different properties: for example, acidic blobs; proline-rich, glutamine-rich, or serine/threonine-rich sequences. Current evidence suggests that the different classes of UAS-binding transcription factors may transmit a signal to the basal promoter complex by making specific contacts with different TAFs. For example, an interaction between transcription factor SP1 and TAF-110 has been shown to mediate SP1 transactivation of transcription. Similarly, other UAS-binding transcription factors, with activation domains of other types, have been shown to interact with other TAFs, or directly with TBP or TFIIB (Table 1). Collectively these protein–protein interactions are likely to facilitate stable RNA polymerase binding to the core promoter element, hence to increase promoter activity. It is not yet clear whether UAS-binding factors cause an isomerization from a closed to an open complex as has been shown for some prokaryotic enhancer factors (see Section 3.1). When the RNA polymerase leaves the promoter, TFIID remains bound at the TATA element and is ready to help a second RNA polymerase to bind and initiate transcription at the promoter. The activity of TFIID appears also to be regulated by inhibitory proteins that interact with TBP. Such TBP-inhibitory protein complexes may serve an important regulatory role by keeping in a repressed but rapidly inducible state genes that have been removed from inactive chromatin.

2.3   REGULATION OF TRANSCRIPTION FACTOR ACTIVITY

Regulated transcription is largely determined by the properties of the UAS-binding transcription factors. There are in principle three ways of tuning the activity of a UAS-binding transcription factor (Figure 3): covalent or noncovalent modification of the UAS-binding factor, and variation of the subunit composition. These mechanisms may be used individually or in combination to regulate transcription.

To illustrate the flexibility of transcriptional control in eukaryotic cells, two examples are presented. The first concerns the activation of steroid hormone dependent gene transcription (Figure 4). Steroid hormones, a group of substances derived from cholesterol, exert a wide range of effects on processes such as growth, metabolism, and sexual differentiation. A prototypical member of a steroid

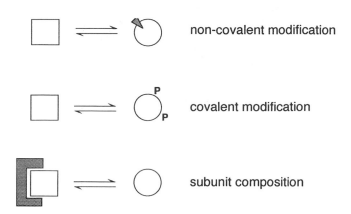

Figure 3. Three principal mechanisms for regulating transcription factor activity.

hormone inducible transcription factor is the glucocorticoid receptor (GR). In the absence of hormone, this receptor is found in the cytoplasm in an inactive form complexed to the heat-inducible protein hsp90. Treatment with steroid hormones results in the release of GR, which becomes free to dimerize and move to the nucleus and

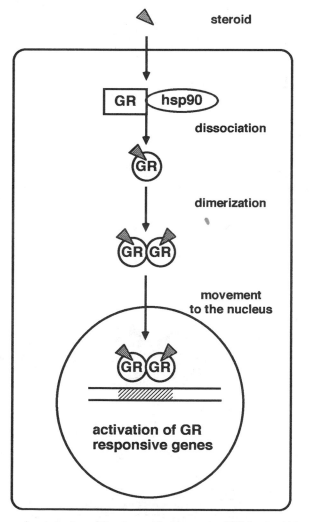

Figure 4. Activation of the glucocorticoid receptor (GR) by steroid hormone binding, which results in activation of GR by dissociation of the inactive GR–hsp90 complex.

Table 1   Examples of UAS Transcription Factors That Interact with the General Transcription Factors TBP or TFIIB[a]

Transcription Factor	TBP	TFIIB
ATF	++	
VP16	+	++
E1A	++	+
EBV zta	++	
CMV IE2	++	
p53	++	
SV40 LT	++	++
AP1	++	
E2F	++	++
NFκB	++	
Estrogen receptor		++
Progesteron receptor		++

[a]Number of plus signs indicates the preferred binding.

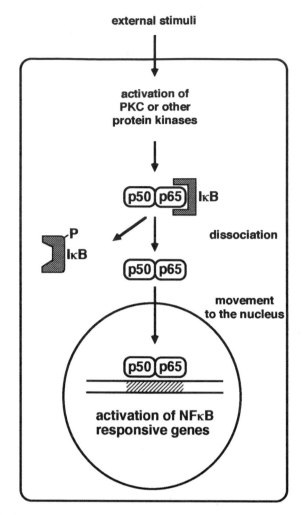

Figure 5. Activation of NFκB by dissociation of the inhibitory IκB. A number of external stimuli (including viruses, lipopolysacharides, and TNFα) result in activation of intracellular kinases, such as protein kinase C (PKC), which phosphorylates IκB. NFκB consists of two polypeptides with molecular weights of 50,000 Da (p50) and 65,000 Da (p65).

activate transcription. In addition to inducing a dissociation of receptor for hsp90, ligand binding may have a second effect for specific receptor activation of target promoters.

The second example concerns NFκB activation of transcription (Figure 5). NFκB was originally found as a transcription factor required for immunoglobulin κ light chain gene expression. It is now clear that NFκB is present in all tissues and regulates the activity of a large number of cellular and viral promoters. NFκB is present in the cytoplasm in an inactive form complexed to the inhibitory protein IκB in a wide variety of cells. Treatment of cells with various agents such as lipopolysaccharides or phorbol esters results in a dissociation of the inactive NFκB/IκB complex by inducing a phosphorylation of the IκB protein. Now, in the absence of the inhibitory protein, NFκB can dimerize and migrate to the cell nucleus, where it functions as a transcriptional activator of NFκB responsive promoters.

The activity of most characterized transcription factors that are regulated falls into one of the three general pathways schematized in Figure 3; or, a combination of these pathways may be used to regulate activity. However, there are additional parameters that may control the activity of UAS-binding transcription factors.

Heterodimerization among different UAS-activating transcription factors has been shown to increase the repertoire by which transcription factors with different binding specificities can be generated. For example, the prototypical AP1 transcription factor consists of a heterodimer consisting of c-jun and c-fos. It binds to its cognate DNA motif with a higher affinity than, for example, a c-jun–c-jun homodimer or a Jun B–c-fos heterodimer (Figure 6). The combinatorial complexity is further increased by the ability of c-jun to form heterodimers with members of the ATF family of transcription factors. Thus, heterodimerization among members of a transcription factor family or members from different transcription factor families is an important mechanism of generating factors with alternative DNA-binding specificity.

2.4   REGULATION OF TRANSCRIPTION DURING DEVELOPMENT AND DIFFERENTIATION

We are far from understanding how transcription is regulated during development and differentiation. However, some important parameters have been defined. For example, genes that are expressed in specific organs often contain binding sites for cell-specific tran-

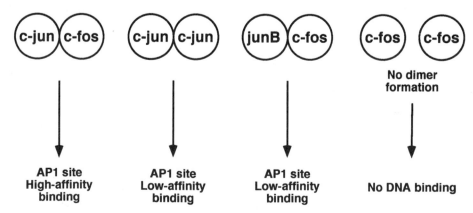

Figure 6. Heterodimer formation between different members of the fos and jun families of transcription factors results in AP1 complexes with different affinities to the consensus AP1 DNA-binding site.

scription factors that may vary from cell type to cell type. Thus, tissue-specific transcription is often regulated by the precise arrangement of regulatory UAS motifs in the promoter, the availability of the cognate transcription factors, and the way these transcription factors influence the activity of the promoter. Since transcription factors most often are dimeric proteins, the exact composition of the two partners may vary across cell types and may have different transcription regulatory properties. Furthermore, nucleosomes appears to be general repressors of transcription. Thus, there is evidence that transcription factors and histones are mutually exclusive residents at a promoter. In activating transcription, UAS-binding transcription factors help TFIID to bind to the core promoter element, hence to prevent repression of the promoter by nucleosome formation. Also, epigenetic modification of a promoter (e.g., methylation of cytosines in the promoter) may change the activity of a gene without altering the basic nucleotide sequence. Increased methylation usually correlates with a reduced expression of a neighboring gene, whereas a reduction of methylation correlates with higher levels of expression.

Although only the efficiency of transcription initiation has been considered, it seems likely that RNA polymerase elongation also is an important parameter for regulation of gene expression in eukaryotes. Thus, there are several examples of RNA polymerase halting at specific pause sites during elongation. To be able to complete a transcript, the polymerase must be able to override this premature transcription stop signal.

3   REGULATION OF TRANSCRIPTION IN PROKARYOTES

Prokaryotic cells contain only one type of RNA polymerase, which is responsible for synthesis of all types of RNA: mRNA, rRNA, and tRNA. The core enzyme is a four-subunit enzyme, consisting of two α, one β, and one β' subunit. However, the holoenzyme, which is the complete enzyme, also contains the sigma factor (the σ polypeptide). The sigma factor is required for proper RNA polymerase binding to a prokaryotic promoter. After the initiation reaction, the sigma factor leaves the polymerase complex, and the core polymerase takes care of elongation.

A prototypical prokaryotic core promoter contains two conserved sequence motifs, at position −10 and −35 relative to the transcription start site (Figure 2). The −10 consensus sequence is TATAAT, and it resembles the eukaryotic TATA element. The −35 consensus sequence is TTGACA. The spacing between the two elements is of critical importance for the efficiency with which the RNA polymerase binds to the promoter. The exact sequence at the −10 and −35 positions varies slightly among transcription units. Usually promoters that have a better homology to the consensus sequences also are stronger promoters. An important mechanism for regulating the transcriptional activity of a prokaryotic promoter is found in the ability to provide the core enzyme with different sigma factors. The multitude of sigma factors may be regarded as the prokaryotic version of upstream binding transcription factors found in eukaryotes. It has been shown that sigma factors determine promoter specificity by recognizing −10 and −35 elements with different base sequences. This strategy mediates the heat shock response and the regulated expression of genes during developmental processes. An example is sporulation in *Bacillus subtilis,* which uses a cascade of different sigma factors to cause the differentiation from a vegetative bacterium to a spore (Figure 7).

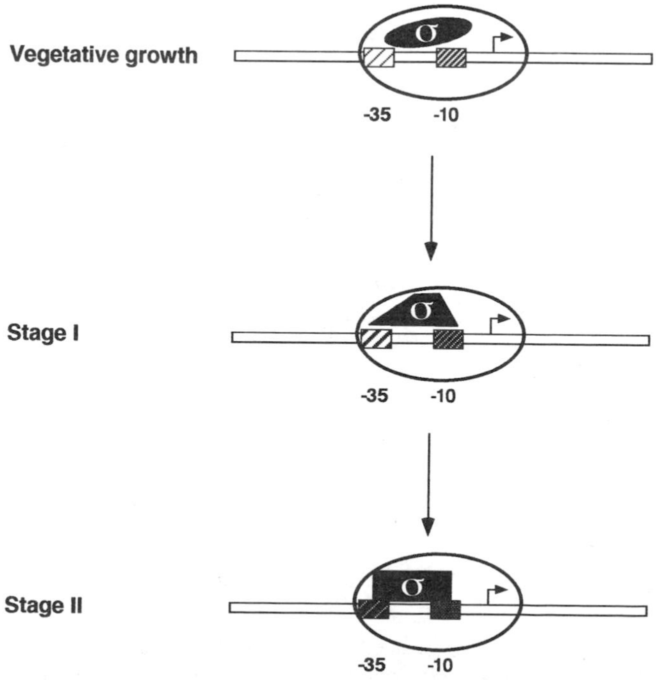

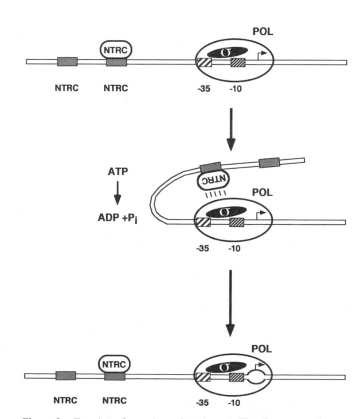

Figure 7. Sporulation in *Bacillus subtilis* involves successive changes in the sigma factor, which controls the specificity of RNA polymerase transcription initiation.

Figure 8. Function of a prokaryotic enhancer. The nitrogen regulatory protein C (NTRC) activates the *glnA* promoter in *Salmonella typhimurium* by inducing a melting of the two DNA strands at the site of transcription initiation (open complex formation). This reaction requires ATP hydrolysis.

3.1   TRANSCRIPTIONAL ENHANCERS IN PROKARYOTES

Expression of prokaryotic genes appears to be subjected to a more direct regulation by fewer trans factors acting at fewer regulatory sites than is the case with eukaryotic promoters. However, a few transcriptional enhancers, similar to those found in eukaryotic cells, have also been described in prokaryotes. For example, the enhancer-binding protein NTRC (nitrogen regulatory protein C) activates the *glnA* promoter from *Salmonella typhimurium* from a distance by means of DNA looping (Figure 8). In contrast to the postulated role for eukaryotic enhancer-binding proteins, NTRC stimulates transcription by a transient contact between the activator and the polymerase. These interactions catalyze an isomerization of a closed complex to an open complex, which then is competent for initiation of transcription. A second difference from eukaryotic transcriptional activator proteins is that NTRC-induced formation of an open complex requires an enzymatic activity that hydrolyzes ATP.

3.2   TRANSCRIPTIONAL REPRESSORS

Transcriptional repressors are important proteins used to control the activity of prokaryotic promoters (Figure 9). Enzymes used in a spe-

cific metabolic pathway are often organized into an operon that is transcribed into a polycistronic mRNA. Then the synthesis of such proteins can be controlled by transcriptional induction or repression. In many prokaryotic operons, specific repressor proteins control the transcriptional activity of the promoter. Repressors are DNA-binding proteins that predominantly block either RNA polymerase access to the −10 and/or −35 regions in the promoter, or transcription elongation, by associating with an operator sequence that is positioned downstream of the start site and interferes with polymerase elongation (Figure 9). Usually these regulatory proteins undergo allosteric changes in response to binding of a specific ligand.

The paradigm is the *lac* operon in *E. coli*. In this system synthesis of proteins necessary for the use of lactose as a carbon source is repressed by the *lac* repressor if cells can possibly use glucose for growth. Specifically, the *lac* repressor binds to an operator sequence immediately downstream of the transcription start site in the *lac* operon (Figure 9) and precludes synthesis of the polycistronic mRNA encoding the proteins necessary for metabolism of lactose. If cells are grown on lactose as the carbon source, lactose functions as an inducer of *lac* operon transcription by binding to the *lac* repressor and converting it to an inactive form, which does not bind

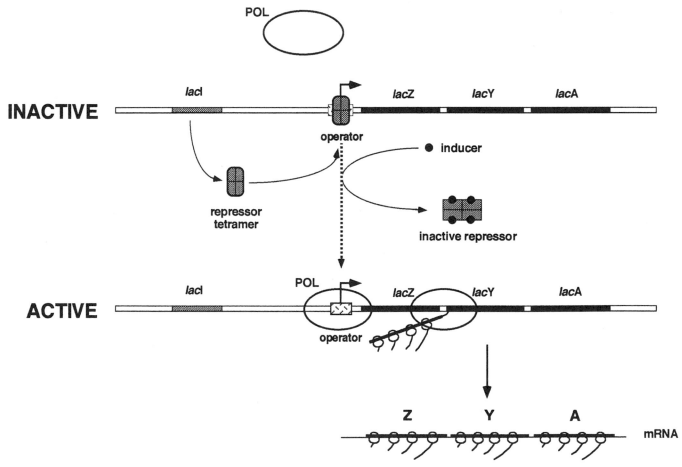

Figure 9.   Regulation of the *lac* operon in *E. coli*. The enzymes required for *E. coli* to grow on lactose are induced only when needed. The *lacI* gene encodes a repressor protein whose tetramer binds to the operator in the *lac* gene and prevents transcription of the structural genes. If cells are grown on lactose as the sole source of carbon, the β-galactoside inducer converts the repressor into an inactive form that cannot bind to the operator. As a consequence, transcription starts at the promoter, and the β-galactosidase (*lacZ*), the permease (*lacY*), and the β-galactosidase transacetylzase (*lacA*) enzymes are synthesized.

DNA (Figure 9) and therefore is unable to inhibit transcription of the *lac* operon.

The *lac* operon represents an example of an inducible system in which an inducer activates transcription. However, inducers can also be used to inhibit transcription of an operon. A good example is the *trp* operon in *E. coli*. In this system, the repressor is normally inactive and RNA polymerase transcribes the operon encoding the enzymes necessary for biosynthesis of tryptophan. When the cell has acquired sufficient quantities of tryptophan, the tryptophan binds to the repressor, which then becomes activated and binds to the operator and inhibits further synthesis of the enzyme responsible for tryptophan synthesis. Thus, a balanced level of the amino acid is autoregulated through a feedback loop. The *lac* and *trp* operons are examples of the versatile way specific DNA-binding proteins, namely, repressors, may be used for the negative or positive control of gene expression.

3.3   REGULATION OF TRANSCRIPTION ELONGATION AND TERMINATION

Regulation of gene expression can also occur at the level of transcription termination. The 3′ end of a prokaryotic mRNA is usually generated by transcription termination rather than posttranscriptional cleavage. There are two modes of termination of bacterial RNA polymerase transcription; rho-independent and rho-dependent transcription termination. Rho-independent transcription termination is characterized by a hairpin in the secondary structure of the RNA followed by a run of U residues. RNA synthesis probably slows down or pauses at the hairpin structure. The string of U residues immediately after the hairpin probably then provides the signal that allows the RNA polymerase to dissociate from the DNA template. Rho-dependent transcription termination requires a specific protein, the rho factor.

Proper termination does not occur in all termination signals, however. In some transcription units the polymerase pauses at a hairpin structure but then resumes transcription if the rho factor is not present. There is no U stretch in rho-dependent transcription terminator signals. However, polymerase pausing at secondary RNA structures is an important determinant for rho-dependent transcription termination, as well. The rho factor binds RNA and is believed to move along the nascent RNA chain after the RNA polymerase. When the rho factor catches up to a stalled RNA polymerase that has paused at a termination signal, it unwinds the DNA–RNA hybrids and causes termination.

Synthesis of enzymes required for tryptophan production is also regulated at the level of attenuation of transcription (Figure 10). The attenuator region is located between the promoter and the first structural gene in the *trp* operon. It causes RNA polymerase termination in response to high concentrations of tryptophan in the cell. This occurs because the leader region in the *trp* operon is translated into a very short peptide that encodes two tryptophans. The leader mRNA can adopt two alternative conformations depending on whether the amino acid tryptophan is available for incorporation into the leader peptide. If tryptophan is available, the *trp* leader peptide is synthesized and the *trp* leader mRNA adopts a stem—loop structure in which complementary segments are paired such that a rho-independent transcription termination signal is created. If cells contain low concentrations of tryptophan, translation of the *trp* leader peptide is prematurely terminated and the leader region adopts an alternative conformation that permits transcription to continue to the end of the operon.

As a consequence, the *trp* operon is expressed and the amino acid tryptophan is synthesized in the cell. This type of regulation is not unusual in the biosynthesis of enzymes required for amino acid synthesis in prokaryotes. It is unique to the prokaryotes, however, since it requires that transcription and translation be coupled. In eukaryotes this type of regulation cannot take place because the two processes are physically separated by the nuclear membrane (Figure 1).

4   REGULATION OF GENE EXPRESSION AT THE LEVEL OF RNA SPLICING

Virtually all prokaryotic genes are contained as a contiguous DNA segment. In contrast, most eukaryotic genes are discontinuous, with the coding sequences (exons) interrupted by stretches of noncoding sequences (introns) (Figure 1). The introns, which are present at the DNA level and in the primary transcription product of the gene, are removed by RNA splicing before the mature mRNA is transported to the cytoplasm. Introns have been found in all types of eukaryotic RNA: mRNA, rRNA, and tRNA. The number of introns in mRNA-encoding genes varies considerably among genes. For example, the α-*interferon* gene has no introns, whereas the gene for *dystrophin* has as many as 70 introns. Also the size of introns can vary from fewer than 100 nucleotides to more than 200,000 nucleotides.

4.1   MECHANISM OF RNA SPLICE SITE CHOICE DURING SPLICEOSOME ASSEMBLY

Short conserved sequence motifs at the beginning (5′ end) and the end (3′ end) of the intron are used as recognition sequences to guide the assembly of a large RNA protein particle, the spliceosome (Figure 11), which catalyzes the cleavage and ligation reactions necessary to mature the final cytoplasmic mRNA. The ends of the introns are, in part, identified by RNA–RNA base pairing between the precursor RNA and a number of uridine-rich; small abundant nuclear ribonucleoprotein particles, the so-called U snRNPs. For example, the 5′ splice site is recognized through a short base pairing between U1 snRNA, and the 3′ splice site is similarly defined by a base pairing between U2 snRNA and the branch point. The conserved sequences at the 5′ and 3′ ends of the intron are surprisingly short, considering the precision with which very large introns are excised during splicing.

Assembly of the spliceosome (Figure 11) has been shown to proceed over two stable intermediate stages, the commitment complex and the prespliceosome. Formation of the commitment complex, which is the earliest detectable stable precursor to the spliceosome, requires interactions between U1 snRNP and the 5′ splice site and between the U2 snRNP auxiliary factor (U2AF) and the polypyrimidine tract at the 3′ splice site of the pre-mRNA. The recognition and functional interaction of the 5′ and the 3′ splice sites in the commitment complex appear also to require one or several members of the newly discovered SR family of proteins—for example, SC35 and ASF/SF2. All these proteins have a conspicuous amino acid domain consisting of repeated serine and arginine residues, hence the name. The commitment complex is converted to the prespliceosome by incorporation of U2 snRNP, and the prespliceo-

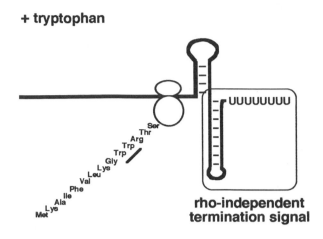

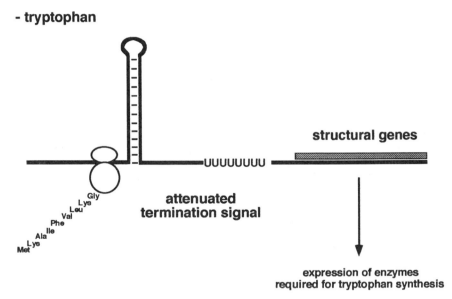

Figure 10. Attenuation of tryptophan operon expression. The *trp* mRNA encodes a short leader polypeptide that has two neighboring tryptophan residues. The position of the ribosome on the RNA determines which of the two alternative stem–loop structures is formed, and thus whether transcription can continue or the transcriptional terminator structure is generated.

some is converted to the mature spliceosome by incorporation of the triple snRNP (U4/U6-U5), in two reactions (Figure 11), which require a number of ill-characterized protein factors.

It is widely believed that once the pre-mRNA has been committed to the splicing pathway, its fate is irreversibly determined. Since the exon–intron boundaries are fixed during this process, formation of the commitment complex is likely to be the key regulatory step at which splicing is regulated. The apparent essential role of the SR proteins in commitment complex formation makes them likely candidates as targets for regulation during splicing.

4.2 ALTERNATIVE SPLICING AS A MECHANISM OF GENERATING PROTEIN DIVERSITY

Since the 5′ and 3′ splice site sequences are short, and not precisely conserved between introns, they occur frequently in the primary sequences of many natural precursor RNAs. This creates conditions that often provide the spliceosome with an opportunity to combine different 5′ and 3′ splice sites in a precursor RNA, to produce several alternatively spliced cytoplasmic mRNAs from a single nuclear gene. This, of course, means that multiple proteins, differing in pri-

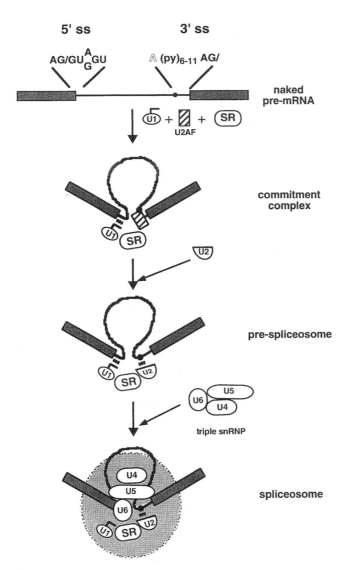

Figure 11. Spliceosome assembly pathway. Some of the U snRNP–pre-mRNA contacts are drawn arbitrarily. For simplicity, most non-snRNP factors required for spliceosome formation have been omitted. At the top, the conserved nucleotide sequences at the 5′ and 3′ splice sites are indicated. The run of pyrimidines necessary for the binding of U2AF is denoted by (py)6–11. See text for details.

mary amino acid sequence and in biological activity, can be produced from a single eukaryotic gene. Of specific interest is the demonstration that the production of alternatively spliced mRNAs, in many cases, is regulated in either a temporal, developmental, or tissue-specific manner.

One of the most spectacular examples of the use of alternative RNA splice site choice to regulate gene expression is the somatic sex determination pathway in *Drosophila melanogaster* (Figure 12). In this system sex determination has been shown to involve a cascade of regulatory events taking place at the level of alternative RNA splice site choice. In females, which contain two X chromosomes, an early promoter of the *Sex-lethal* (*Sxl*) gene is activated. Once made, the female specific Sxl protein autoregulates its expression and controls the splicing of the precursor RNA from two

other *Drosophila* genes, *transformer* (*Tra*) and *double-sex* (*Dsx*), to produce female flies (Figure 12). In males, where a biologically inactive Sxl protein is expressed, the *Sxl, Tra,* and *Dsx* precursor RNAs are processed by a default splicing pathway, resulting in the development of male flies. The Dsx protein, the final protein in the cascade, is believed to be a transcriptional repressor that depending on the outcome of the *Dsx* splicing reaction, creates proteins regulating the development of each fly on the male or female pathway.

A second, extreme, example of translation of a very complex set of proteins created by alternative splicing is illustrated by the *troponin* T gene (Figure 13). The production of the α or β group of *troponin* T mRNAs is developmentally regulated. Thus, exon 16 is specific to adult muscle *troponin* T mRNAs (α group), and exon 17 is used in both adult and embryonic mRNAs (β group). Each of the 31 possible alternative splices involving exons 4–8 is spliced in with exon 16 and with exon 17, to make up a total of 62 alternatively spliced α- and β-*troponin* T mRNAs.

Today it is widely accepted that alternative splicing is an important mechanism in the regulation of gene expression during growth and development in eukaryotic cells. A large number of eukaryotic genes have been shown to contain mature, alternatively spliced mRNAs. Examples include growth factors, growth factor receptors, intracellular messengers, transcription factors, oncogenes, and muscle proteins; the number is increasing every month.

In some lower eukaryotes, mRNAs have been shown to be produced by a mechanism termed trans-splicing—the splicing together of two initially unconnected RNA molecules to form a functional mRNA. This process has so far not been documented in higher eukaryotes. However, it should be noted that human cells are capable of trans-splicing suitable precursor RNAs. Thus, it may just be a question of time before the first example is described. If such an anomaly were to occur, it would add an extra level at which RNA splicing could be used to regulate gene expression.

5   REGULATION OF EUKARYOTIC GENE EXPRESSION AT THE LEVEL OF 3′ END FORMATION

Transcription termination in eukaryotes is an ill-defined process. AT-rich sequences, in combination with secondary structures, probably determine the position at which RNA synthesis ends. Usually the 3′ end of a eukaryotic mRNA is generated by an endonucleolytic cleavage of the primary transcript. Thus, the RNA polymerase transcribes past the site that specifies the 3′ end of the mature mRNA and sequences in the precursor RNA are then recognized as targets for an endonucleolytic cleavage, followed by a nontemplated addition of 100–200 A residues to the 3′ end, thus forming the polyadenylated tail. A highly conserved AAUAAA sequence, 11–30 nucleotides upstream of the cleavage site, serves as a key signal for specifying the position of the cleavage/polyadenyl-ation reaction. Since eukaryotic genes often encode multiple potential poly(A) sites, the precise usage of one or another poly(A) site can regulate gene expression. For example, if a poly(A) signal further downstream in a transcription unit is used, a novel exon may be spliced into the mRNA, as in the production of secreted or membrane-bound forms of immunoglobulin M. Alternatively, a new translational reading frame, encoding another protein, may be spliced from such a precursor RNA—for example, production of calcitonin (thyroid cells) and the calcitonin-related protein (brain) (Figure 14). How-

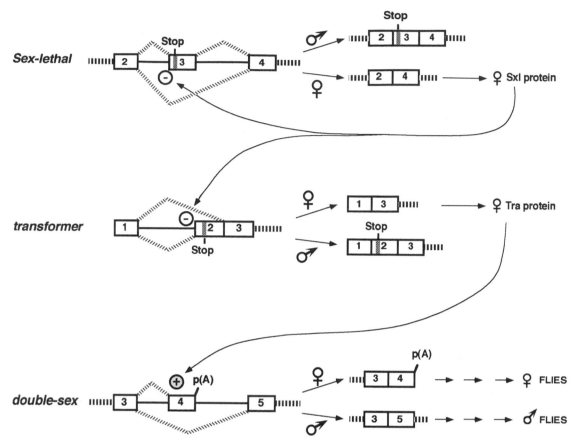

Figure 12. The cascade of regulated alternative splicing events in the sex determination pathway of *Drosophila*. Numbered boxes represents exons and connecting lines introns. For clarity, only the exons and introns involved in the regulated splicing events are shown. Positions of translational termination codons that will cause premature termination of protein synthesis if that exon is spliced into the mRNA are indicated by Stop. See text for further details.

ever, it is not clear whether the use of alternative splice site dictates use of different polyadenylation sites or whether selection of a poly(A) site results in alternative splicing of the precursor RNA.

In contrast to eukaryotes, the primary transcript of prokaryotic protein-coding genes usually serves as mRNA without any modification or processing. However, there are exceptions. For example, prokaryotic ribosomal RNAs and transfer RNAs are often matured by RNA cleavage from larger precursor RNAs.

6   CYTOPLASMIC REGULATION OF GENE EXPRESSION

Genes are most frequently regulated at the level of RNA production: synthesis or processing. However, gene expression can also be regulated at other levels, such as translational efficiency of mRNA, RNA and protein stability, or protein modification. As is the case for transcriptional regulation, control of gene expression at the level of translation often occurs at the initiation step of the decoding process. Thus, not all mRNAs that reach the cytoplasm are used directly to synthesize protein. It has been calculated that as much as 10% of genes in a eukaryotic cell are regulated at the level of translation. Translational control enables a cell to change the concentration of a protein rapidly and reversibly without rapid turnover of its mRNA.

6.1   REGULATION OF GENE EXPRESSION AT THE LEVEL OF TRANSLATION

Initiation of translation in eukaryotes is catalyzed by a number of proteins, called initiation factors. Among these, enzyme initiation factor 2 (eIF2) plays a central role by feeding ribosomes onto the mRNA (Figure 15). In a reaction requiring GTP hydrolysis, eIF2 helps to assemble the mature ribosome on the mRNA (80S initiation complex). To be able to catalyze a second round of translation, eIF2-GDP has to be converted to an active eIF2-GTP complex. The charging of eIF2 is a key step at which the translational efficiency in a cell is regulated. For example, many virus infections result in the synthesis of the interferon-induced eIF2 α-protein kinase, which phosphorylates one subunit of eIF2 and withdraws it from its essential role in 80S ribosome initiation complex formation. As a consequence, protein synthesis stops. However, since viruses require protein synthesis for formation of new viral particles, many viruses have evolved mechanisms to circumvent the negative effect of the interferon-induced eIF2 α-protein kinase on cellular translational efficiency. For example, adenovirus encodes for a low molecular weight RNA (the VA RNA) that inhibits the activity of the eIF2 α-protein kinase, thus allowing for efficient viral protein synthesis also in interferon-stimulated cells.

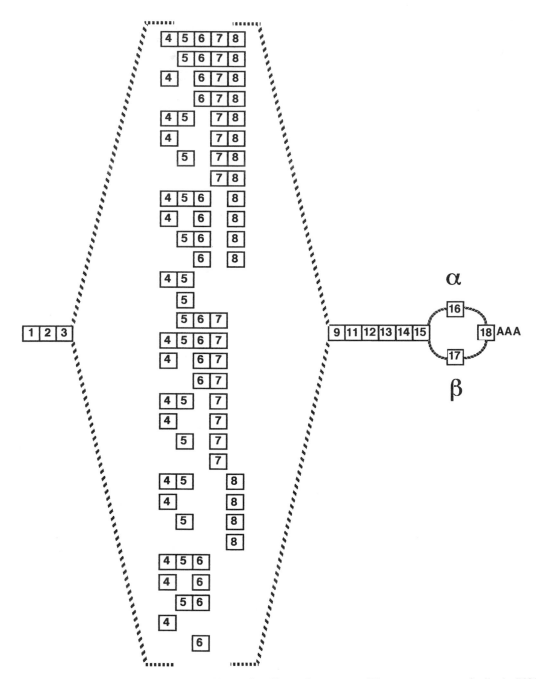

Figure 13.   Alternative splicing of the *troponin* T precursor RNA. Numbered boxes denotes exons. There are two groups of spliced mRNAs, resulting from alternative inclusion of exon 16 (α type) or 17 (β type). There are 31 variants of each type, depending on which of exons 4–8 are retained during the splicing process, and in what order.

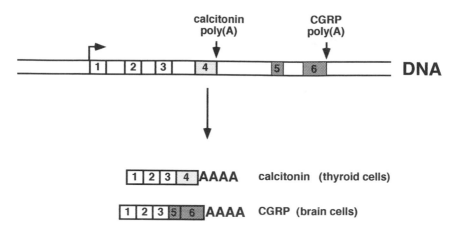

Figure 14. An example of alternative polyadenylation. Two different spliced mRNAs, one encoding calcitonin and the other encoding calcitonin-related protein (CGRP), are generated, depending on which of the two polyadenylation signals is used. Numbered boxes represent exons; AAAA denotes the poly(A) tail at the 3′ end of eukaryotic mRNAs.

6.2 REGULATION OF PROTEIN SYNTHESIS BY TRANSLATIONAL REPRESSORS OR ENHANCERS

Translational repressors have been described that bind to the 5′ end of mRNAs and thereby control protein production. This type of mechanism was originally described in bacteria, where it was shown that ribosomal proteins are able to autoregulate their synthesis by repressing excess ribosomal protein translation. Similar mechanisms have also been shown to occur in eukaryotes. For example, synthesis of the intracellular iron storage protein ferritin is rapidly adjusted in response to the level of soluble iron atoms (Figure 16). In the absence of iron, an approximately 30 nucleotide long

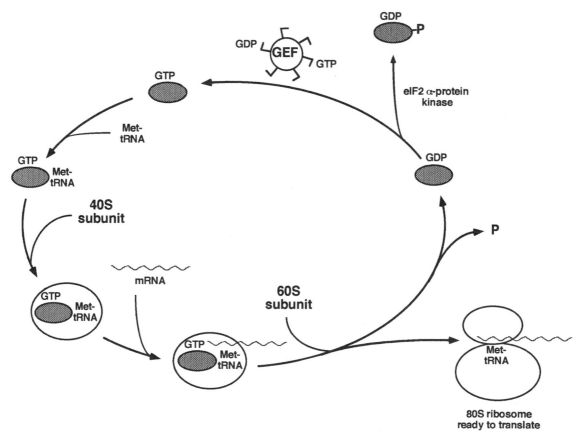

Figure 15. Eukaryotic initiation factor 2 (eIF2) has a central role in initiation of protein synthesis in eukaryotic cells. Under normal conditions, eIF2 helps to assemble the 80S ribosomal initiation complex. This reaction is associated with GTP hydrolysis and the release of an eIF2–GDP complex. In this form eIF2 is inactive for further rounds of initiation until the GDP has been exchanged for GTP, a reaction catalyzed by the guanine nucleotide exchange factor (GEF). In virus-infected cells, the interferon-induced eIF2 α-protein kinase phosphorylates the eIF2–GDP complex. As a consequence, translation ceases in virus-infected cells because GEF is unable to exchange the GDP for GTP.

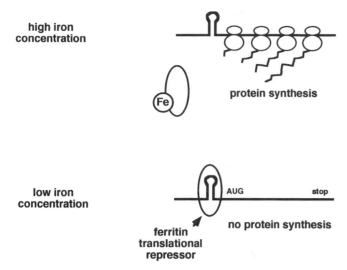

Figure 16. Regulation of ferritin synthesis by a translational repressor. When cells contain low concentrations of iron, and thus do not need ferritin, the translational repressor binds to a short stem–loop structure in the mRNA and prevents synthesis of ferritin. In response to an increase in the intracellular concentration of iron, the repressor is dissociated from the mRNA and ferritin is synthesized.

RNA duplex region in the 5' leader region of the mRNA binds a regulatory protein that prevents translation of the downstream RNA sequences. The addition of iron results in the dissociation of the repressor protein from the mRNA and, as a consequence, ferritin protein synthesis. In this system, synthesis of the iron-binding ferritin protein is closely regulated by the availability of iron.

The opposite type of regulation—that is, positive control of translation—is also possible. Certain RNA viruses (picornaviruses) have been shown to contain RNA sequences that serve as translational enhancers and can direct ribosomes to initiate translation at AUG sequences positioned internally in a mRNA. Such translational enhancers override the rule that the first AUG, counted from the 5' cap structure of a eukaryotic mRNA, is used for initiation of protein synthesis. The internal ribosome-binding mechanism is vital for picornavirus growth because the virus inhibits cellular translation in infected cells by inactivating the normal cap-dependent mechanism used for translation initiation on cellular mRNAs. This enables picornavirus translation to proceed also in infected cells, since the virus uses the translational enhancer to ensure efficient viral protein synthesis.

Because most mRNAs in bacteria have a very short half-life ($\approx$ 3 min), a given bacterium can rapidly adjust its gene expression in response to environmental changes. Eukaryotic mRNAs are usually much more stable, ranging from less than 30 minutes to more than 20 hours. Many short-lived mRNAs, such as those encoding for certain growth factors, contain long A- and U-rich sequences in the 3' noncoding region, which seems to control the mRNA instability. Furthermore, extracellular signals may be used to control the stability of the mRNA within a cell.

In addition, gene products that require posttranslational modification or transport to specific cellular compartments may be regulated at each level. For further readings, see the Bibliography.

See also GENOMIC IMPRINTING, MOLECULAR GENETICS OF; HOMEODOMAIN PROTEINS: PROTEIN DESIGNS FOR THE SPECIFIC RECOGNITION OF DNA; REPRESSOR-OPERATOR RECOGNITION; TRANSLATION OF RNA TO PROTEIN.

Bibliography

Alberts, B., Bray, D., Lewis, J., Raff, M., Roberts, K., and Watson, J. (1994) *Molecular Biology of the Cell.* Garland Publishing, New York.
Gesteland, R. F., and Atkins, J. F. (1993) *The RNA World.* Cold Spring Harbor Laboratory Press, Plainview, NY.
Latchman, D. S. (1993) *Eukaryotic Transcription Factors.* Academic Press, Harcourt Brace Jovanovich, New York.
Lewin, B. (1990) *Genes IV.* Cell Press, Cambridge.
McKnight, S. L., and Yamamoto, K. R. (1992) *Transcriptional Regulation.* Cold Spring Harbor Laboratory Press, Plainview, NY.
Singer, M., and Berg, P. (1991) *Genes and Genomes.* University Science Books, Mill Valley, CA.

Gene Expression Regulation in Plants: *see* Plant Gene Expression Regulation.

Gene Gun Method for Introduction of Cloned DNA: *see* Plant Cell Transformation, Physical Methods for.

GENE MAPPING BY FLUORESCENCE IN SITU HYBRIDIZATION

Amanda C. Heppell-Parton

1 **Introduction**

2 **Hybridization Targets**
  2.1 Categorization
  2.2 Preparation
  2.3 Identification

3 **Hybridization Probes**
  3.1 Categorization
  3.2 Probe Labeling
  3.3 Optimization

4 **Hybridization Conditions**
  4.1 Target Preparation
  4.2 Hybridization
  4.3 Posthybridization

5 **Signal Detection**
  5.1 Direct Versus Indirect
  5.2 Amplification
  5.3 Fluorochromes
  5.4 Multicolor, Multiprobe FISH

6 **Signal Visualization**
  6.1 Fluorescence Microscopy
  6.2 CCD Cameras
  6.3 Confocal Microscopy
  6.4 Photography
  6.5 Electronic Image Recording

7 **Conclusion**

Key Words

Chromosome Band   A part of a chromosome that can be clearly distinguished from its adjacent segments by applying available banding techniques that cause it to appear darker or lighter.

Haptens   Reporter molecules conjugated to nucleotides, enabling their incorporation into nucleic acid probes and subsequent detection of labeled sequence.

Image Analysis   A system for analyzing images recorded with computer software programs.

Network   Random cross-hybridization between vector DNA fragments, providing increased signal amplification at the target site.

Probe   A DNA sequence used to detect its homologous location on a chromosome.

Probe Detection   A system for visualizing the site of a probe nucleic acid after in situ binding of complementary sequences.

Signal   The fluorescent region indicating the site of bound probe DNA.

Target (sequence)   A stretch of DNA "in situ" with complementary sequence to the intact DNA probe insert.

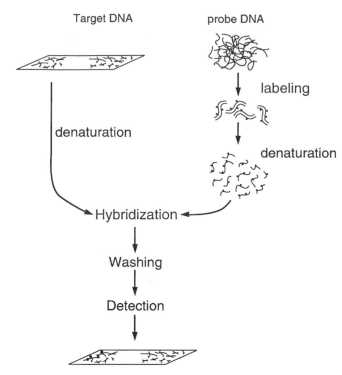

Figure 1.   Fluorescence in situ hybridization.

To map a gene is to identify the position it occupies on a chromosome. Such gene localization can be achieved using a number of different approaches. However, in situ hybridization (ISH), in which a DNA fragment (probe) representing the gene is seen directly on the chromosome, is one of the most precise. A compilation of these individual localizations enables a gene map to be constructed. In practice, relatively few genes have been isolated. The majority of available probes are anonymous DNA fragments. However, once assigned to a specific position on a chromosome, these fragments play an important part in connecting up isolated data to create a complete map. FISH, the powerful fluorescence-based form of in situ hybridization, is the only technique to combine gene localization with relative gene order, thus permitting the simultaneous analysis of two sets of information required for a complete gene map.

1   INTRODUCTION

In situ hybridization (Figure 1) is a molecular genetics technique, the basis of which is the complementary base pairing of the DNA molecule. Increased temperature separates (denatures) the two DNA strands, and as the temperature decreases, rejoining (rean-

nealing) occurs. DNA probes recognize and bind (hybridize) to their complementary sequences in the intact chromosomes fixed on a microscope slide. The probe is tagged (labeled) with a reporter molecule (hapten). Both probe and chromosomal DNA are denatured, put together on the slide, and allowed to reanneal. The probe DNA recognizes its complementary sequence within the intact chromosome and reanneals, forming a hybrid molecule at the target site. Unbound probe is then washed off. In the case of fluorescence in situ hybridization (FISH), bound probe is detected either directly, by means of the fluorescent nature of the hapten (Figure 2A), or by using binding proteins that recognize the hapten within the probe. These binding proteins either are themselves fluorescently tagged (Figure 2B) or are subsequently recognized by a second fluorescently tagged binding protein (Figure 2C). Visualization using a fluorescence microscope reveals fluorescently stained chromosomes

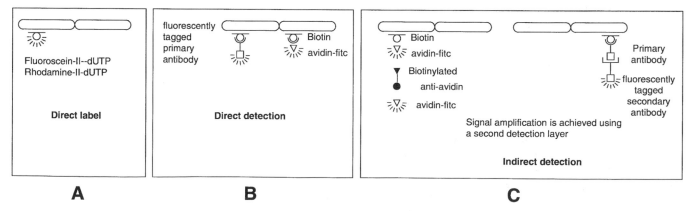

Figure 2.   Signal detection and amplification.

displaying a chromosome-specific signal in a second fluorescent color at the target site (Figure 1).

Because several fluorochromes are available, one of the major advantages of FISH over other mapping techniques is the ability to map probes relative to each other on the same chromosome in multiple colors. This system allows the determination of an unambiguous order, which is essential in attempts to connect isolated pieces of mapping data.

2   HYBRIDIZATION TARGETS

2.1   CATEGORIZATION

A number of possible hybridization targets are available. As mentioned, the DNA target is visualized in the form of chromosomes. The DNA at this stage is at its most condensed; the probe target sites along the length of the DNA are tightly packed close together. For the purposes of ordering probes located close together, the target DNA should be as loosely packed as possible, to give the largest possible separation distance between target sites. Such less condensed material enables the resolution limits of FISH to be increased. Although the vast majority of FISH gene mapping is still performed on metaphase chromosomes, the increasing requirement for more refined localizations for probes and indeed distinction of two probes located within the same chromosomal subband has resulted in the development of FISH on DNA targets that are increasingly less condensed.

During the cell cycle, DNA is for the majority of the time present in interphase as a loosely tangled mass. As the cell enters mitosis, this tangled mass of DNA becomes more and more condensed, progressing through prophase, then prometaphase, to metaphase. The preparation of chromosomes at both prophase and prometaphase, and in fact interphase chromatin, has resulted in improved resolution of the FISH technique. Probe sequences recognizing target sites separated by 2 megabases (Mb) can be distinguished on metaphase chromosomes. Longer prometaphase chromosomes increase the resolution to a 1 Mb distinction. Probe hybridization sites separated by as little as 50 kilobases (kb) can be distinguished when hybridized to interphase chromatin. In fact, interphase chromatin is an ideal target for probe target sequences separated by 50 kb up to 1 Mb, although discrepancies have been found in telomeric regions of the chromosomes.

Recently this approach for achieving increased resolution has been pushed even further. Hamster egg–human sperm fusions provide highly decondensed pronuclei, for high resolution gene mapping in the 20–800 kb range. Pronuclei may be helpful for mapping centromeric sequences, because heterochromatin, prevalent in centromeric regions, is less condensed relative to euchromatin in pronuclei compared to interphase. Finally, nuclear extraction providing extended DNA loops (halos) around the nuclear matrix increases resolution to a few kilobases. Careful selection of the hybridization target thus enables the analyst to distinguish probe target site separation from several megabases to as little as a kilobase. In addition to a requirement to distinguish two independent signals, it is necessary to identify these two independent signals as representative of two distinct probes. FISH has the ability to achieve this because numerous fluorochromes are available with which to label or subsequently detect numerous probes simultaneously, using different colors (see Section 5).

2.2   PREPARATION

Standard culture techniques are routinely used to produce chromosome preparations. To accumulate cells at mitosis, the growing cells are treated with a mitotic spindle inhibitor, colcemid. Cells enter mitosis but are unable to progress further than metaphase. At this stage the chromosomes are easily visualized and identified. The longer the arrest time, the greater the number of metaphase cells accumulated. However, the metaphase chromosomes continue to condense, becoming shorter with increasing time. The shorter the chromosomes, the poorer the quality of bands obtained and the harder it is to identify chromosomes and subbands. To overcome this problem, and still obtain increased numbers of mitotic cells, it is possible to synchronize the cell growth using methotrexate, bromodeoxyuridine (BrdU), or thymidine (ultimately blocking DNA synthesis S phase). When such a block is released, all cells proceed in synchrony. The advantages of this approach are multifold: for example, the number of prophase, prometaphase, or metaphase figures obtained can be easily controlled by altering the duration of the colcemid treatment. Since all the cells are at the same stage in the cell cycle at any one time, a colcemid treatment stopped after 2 minutes will contain a large number of prophase chromosomes. After 5 minutes prometaphases will form the majority, and by 20 minutes, abundant metaphases are present. In addition, synchronization treatments subsequently produce chromosomes that tend to be less condensed, allowing a higher resolution mapping. A further advantage of BrdU is that its incorporation contributes toward an enhanced banding pattern when the chromosomes are fluorescently stained, leading to easier chromosome identification (see Section 2.3).

2.3   IDENTIFICATION

When FISH is performed on metaphase chromosomes, chromosome bands are required for chromosome identification and precise localization of probe signal. Chromosome banding patterns are reproducible, and they fall into two main categories: G- or (Q-) bands and R-bands, as defined by the International System for [Human] Cytogenetic Nomenclature (ISCN). Numerous probes can be relatively mapped using coordinates defined as the distance between the signal and a reference point. These coordinates are given as a fraction of the length of the total chromosome or subchromosomal region. However, such mapping coordinates, defined by fractional length values, cannot be extrapolated from the ISCN ideograms to elucidate their band locus character. To assign a regional map localization for a probe sequence, chromosome bands are critical.

Fluorochromes (fluorescent dyes) used to stain chromosomes can be categorized into two different types on the basis of their base pair affinity. Depending on their affinity, they produce basic patterns similar to Q- or R-bands. The most commonly used fluorescent dyes for chromosomes are 4',6-diamidino-2-phenylindole (DAPI) (blue), and Hoechst 33258 (green), both of which have an AT base binding affinity and produce a fluorescent Q- (G-) banding pattern (Figure 3C and 3A, respectively; see color plate 2). Propidium iodide (PI) (red) has no real base pair affinity; however, an enhanced R-banding pattern is obtained by using a combined DAPI/PI staining approach (Figure 3E). This double staining method is referred to as counterstaining. In addition, it is possible to produce distinct chromosomal banding patterns using FISH with DNA probes for in-

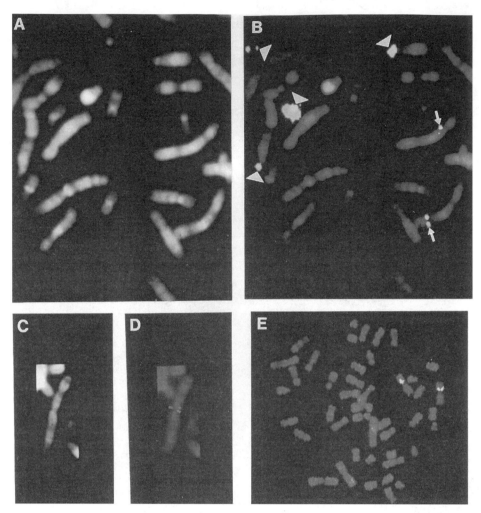

Figure 3. (A) Partial metaphase of Hoechst 33258 stained human chromosomes. This fluorescent dye preferentially binds to AT-rich stretches of DNA creating a G-(Q-)like banding pattern. (B) The same image after computer alignment, merging, and assignment of false colors. The hybridization signal for a small single-copy probe is seen on both homologues (small arrows). In comparison, the large signal obtained with a repeat sequence probe localized to the acrocentric chromosomes is shown (arrowheads). Images were obtained using an epifluorescence microscope, RCA-ISIT camera, MRC 500 workstation, and a Sony video printer. (C) DAPI stained Q-banded metaphase chromosome. (D) The same image as (C), showing two single-copy probes labeled with Fitc (green), and Texas red (red), respectively, localized relative to each other on the DAPI-stained (blue) chromosome. Images were obtained using an epifluorescence microscope, CCD camera, and MRC 500 workstation. The photographs were generated by sequential photography from the computer color monitor screen. (E) Fluorescent dye DAPI used in conjunction with propidium iodide gives an enhanced R-banding pattern (red). The image is merged and assigned false colors using appropriate computer software. The yellow Fitc hybridization signal is that of a YAC probe. All four sister chromatids are consistently labeled with such large probes. The image was obtained using a confocal microscope, BioRad MRC 600 workstation, and screen photography. (See color plate 5.)

terspersed repetitive elements (IRS) (see Section 3). For example, *Alu* repeats, when highlighted on the chromosomes using *Alu*-rich DNA fragments as probes, or primers for oligonucleotide-primed in situ DNA synthesis (see Section 5), generate an R-banding pattern referred to as *Alu* banding.

Well-established light microscope banding techniques may be preferred. These techniques can be performed with FISH prior to (prebanding) or after ISH (postbanding). Prebanding has the advantage of allowing the analyst to avoid any interference of chromosomal heat denaturation with banding quality. The main disadvantage is that metaphase spreads must be relocated and rephotographed.

3   HYBRIDIZATION PROBES

3.1   CATEGORIZATION

Similar sequences of DNA, constituting approximately 30%, are repeated throughout the genome. These repeat sequences, or interspersed repetitive sequences, fall into two categories: short interspersed repetitive elements (SINEs), such as *Alu* elements, and long interspersed, repetitive elements (LINEs), such as *Ll* elements. These sequences are common throughout the genome, so large probes carrying unique sequences complementary only to their target sequences on a specific chromosome also carry IRSs that are

Probe DNA

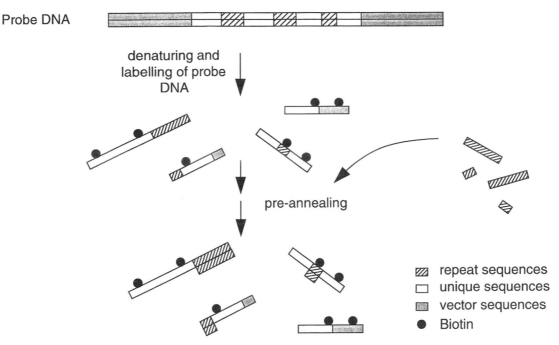

denaturing and labelling of probe DNA

pre-annealing

▨ repeat sequences
▢ unique sequences
▨ vector sequences
● Biotin

Figure 4. Probe DNA is hapten labeled and denatured. Labeled, single-stranded DNA probe fragments are then allowed to preanneal with an excess of repeat sequence single-stranded fragments. Reassociation of repeat sequences renders such sequences double-stranded and thus not available to hybridize to complementary sequences in the target DNA (chromosome). Unique sequences remain single-stranded and are subsequently able to bind to their complementary sequences on the chromosome.

shared throughout the genome. It was once necessary to avoid non-specific hybridization, by ensuring that any probe to be used for FISH was free of such repeat sequences. This requirement limited probe size, as larger sequences, by virtue of their size, carried repeat sequences and called for subcloning of smaller unique inserts (500 bp–5 kb) in *plasmid vectors*.

Smaller unique probe inserts recognize smaller target sites on the chromosome, with corresponding reductions in signal size and intensity. As a result, they are detected with decreased efficiency, with only 20–70% of the relevant chromosome carrying the signal. Such probes rely on amplification and network formation (Section 5) for visualization (Figure 3B, 3D).

The advantages of larger targets, hence larger probe inserts, are demonstrated by *repeat sequence probes* (randomly repeated stretches of unique sequence). These probes, with their large target sites, demonstrated high detection efficiency (90–99%), and the larger signal decreased their reliance on amplification and networking (Figure 3B).

Recently vectors able to carry larger DNA inserts—for example,

cosmids, bacteriophage, and yeast artificial chromosomes (YACs), with inserts from 15–20 kb up to 400–600 kb—have been increasingly available for FISH mapping. The DNA inserts carried by these vectors include repetitive stretches of DNA. FISH would result in hybridization signal detection on all chromosomes, masking the chromosome-specific signal attributable to the unique sequences. However, competitive in situ suppression (CISS) hybridization (Figure 4), in which repetitive sequences are "soaked up" by repeat fragments prior to hybridization, has enabled such larger target detection using these cosmids, bacteriophage, and YACs (Figure 3E). The repetitive elements within the probe DNA are thus rendered double stranded and, as such, are not available to hybridize to complementary sequences within chromosomes on the slide. These probes offer increased target site detection efficiency and larger signal. The disadvantage of repeats has recently been overcome by using CISS hybridization, which makes possible the exploitation of the advantages of larger probes. Probe categories are summarized in Table 1.

The probes discussed so far have all been double-stranded ge-

Table 1   The Characteristics of Probes Available for Gene Mapping

Probe	Inserts/Repeats	Target Site (kb)	Detection Efficiency (%)	Repeat Sequences	Signal
Plasmid	Small, unique insert	0.5–5	50	Absent	Small, discrete
Cosmid bacteriophage	Larger insert; repeats present	40	80–95	Present	Large, discrete
Yeast artificial chromosomes	Large insert; repeats present	400	99	Present	Large
Repeat sequence probes	Unique sequences, tandemly repeated	100	90–99	Absent	Large, diffuse

nomic DNA fragments, cloned into vectors able to carry inserts of increasing sizes. Increasing size of insert has the advantage of increased target site detection and signal size, but the disadvantage of repeat sequences. Double-stranded cDNA probes—that is, DNA formed using an RNA template, therefore corresponding to expressed sequences only—pose less serious problems with repeats. However such probes have small inserts recognizing sequences that in the target are often interrupted with introns. As a result, the signals are often quite weak.

A number of competing reactions occur during in situ hybridization with double-stranded probes. These include probe renaturation, in situ renaturation, and in situ hybridization. The first two, obviously undesirable, can be overcome by using single-stranded probes. The main sources of such probes are RNA (see Section 3.2.4) and synthetic oligonucleotides (see Section 3.2.5).

3.2   Probe Labeling

DNA probes are tagged or labeled with haptens, which either have fluorescent properties (direct labels) or can be recognized by binding proteins that are either fluorescently tagged (direct detection) or are subsequently detected by a secondary fluorescently tagged binding protein (indirect detection) (see Figure 2). Probe DNA can either be double-stranded (ds) or single-stranded (ss). The majority of available probes for FISH fall into the former category. For these probes, labeling can be achieved using nick translation, random priming, or the polymerase chain reaction (PCR). Single-stranded RNA probes are synthesized and labeled using in vitro transcription; end labeling is used for oligonucleotides. The most commonly used haptens are biotin and digoxigenin. The directly labeled haptens fluorescein and rhodamine have been used successfully in FISH, but only for large targets.

3.2.1 Nick Translation

Nick translation is a specific procedure for incorporating labeled nucleotides into dsDNA. The enzyme DNase1 randomly introduces "nicks" in the dsDNA molecule. Labeled nucleotides are then sequentially added to the DNA, and preexisting nucleotides are removed in a reaction catalyzed by the enzyme DNA polymerase 1. The final product of the nick translation reaction is a pool of dsDNA fragments uniformly labeled with the hapten of choice. The frequency with which the "nicks" are created in the DNA template molecule can be controlled by varying the DNase 1 concentration, which in turn determines the size of the final labeled DNA fragments. The DNase 1 concentration is thus optimized to produce the optimal probe length (see Section 4.2).

3.2.2 Random Priming

Random hexanucleotide priming sequences hybridize to the single-stranded DNA template. DNA polymerase "Klenow" fragment then catalyzes the addition of nucleotides in the $5' \rightarrow 3'$ direction. The length of the labeled strands is controlled by varying the ratio of primer to template or the length of the primer. A high primer concentration or a shorter primer will result in short labeled fragments, and vice versa. The ratio of primer to template is altered to produce the optimal probe length (see Section 4.2).

3.2.3 Polymerase Chain Reaction

The polymerase chain reaction can be used for probe labeling. This approach is rarely used for smaller unique sequence probes because it is necessary to know the insert DNA sequence for primer construction. PCR is, however, used to achieve labeling of larger probe DNA inserts. The sequences used as primers are the repeat sequences that occur throughout human genomic DNA. As mentioned earlier, most repeat sequences are nonrandomly distributed throughout the human genome, an arrangement leading to disadvantages in the use of these repeat sequences as priming sites to achieve uniform labeling of a large DNA probe insert. If the DNA inserts in the probe of interest originate from an *Alu*-poor region of the human genome, the fragments of DNA that have been amplified and labeled will not be representative of the whole probe insert.

One way to overcome the problem just described is to use a technique called degenerate oligonucleotide primed PCR or DOP-PCR. Such degenerate primers allow random yet uniform amplification of the DNA insert sequence of the probe. The labeled fragments are thus representative of the larger site. The basis of the technique is very similar to that of the random priming method. After strand separation (denaturation) has been induced, the primers bind to complementary sites along the length of the probe DNA. DNA polymerase then catalyzes nucleotide incorporation from the site of primer hybridization. The PCR technique differs from random priming in that when strand synthesis is complete, a further round of denaturation, primer reannealing, and strand synthesis takes place. This time twice the number of template DNA molecules are present. Each round doubles the DNA quantity. Thus labeled probes can be produced from very small quantities of starting DNA material.

3.2.4 In Vitro Transcription

RNA probes are generated by in vitro transcription using a linearized template. A promoter for RNA polymerase must be available on the vector DNA. The size of the probe is defined by the distance of the restriction site used for linearization from the start of transcription this guarantees the same length for all probes. RNA probes are inherently single-stranded. Promoters are located at either end, and thus it is possible to generate both sense and antisense probes. In hybridization experiments, the strength of the bond between probe and target plays an important role. The strength decreases in the order RNA:RNA, DNA:RNA, DNA:DNA.

3.2.5 End Labeling

Synthetic oligonucleotides have several advantages. They are readily available through automated synthesis, and they are small and single-stranded, excluding the possibility of renaturation. Their small size gives them good penetration properties, but it also means that they tend to cover less target. Hapten incorporation is usually achieved by enzymatic end labeling.

3.3   Optimization

For all labeling techniques, labeling efficiency can be determined prior to the ISH experiment. A known amount of nonradioactively labeled DNA or RNA probe is spotted on a nylon or nitrocellulose filter in parallel with the unknown labeled probe, after which each residue is detected with antibodies conjugated to enzymes that subsequently participate in an enzymatic color reaction. A visual comparison of the color intensities of the different probes gives an indication of the success of the labeling reaction.

DNA purity and reagent quality are critical for the achievement of optimally labeled probes. Failure to label is almost invariably a function of the quality of the DNA. The problem can sometimes be alleviated by phenol-extracting the DNA, or by purification on a Sephadex G50 column. Commercial kits that enable DNA to be rapidly prepared or cleaned include Prep-a-Gene, Magic DNA Clean Up, Qiagen, Plasmid-Quick, and Gene Clean. Labeling using nick translation involves the use of two enzymes, which may be selectively inhibited by different contaminants in the DNA samples.

Labeling by means of the random priming approach uses "Klenow" DNA polymerase, which is more resistant to inhibition by DNA contaminants. In theory, random priming is the more reliable and reproducible labeling choice.

Unincorporated labeled nucleotides may be removed by applying the labeling reaction to a Sephadex G50 column. Alternatively, ethanol precipitation is performed. The labeled probe coprecipitates with the carrier DNA, while incorporated mononucleotides remain in the supernatant.

4   HYBRIDIZATION CONDITIONS

4.1   TARGET PREPARATION

Once the target has been fixed on the slide, RNase treatment serves to remove endogenous RNA and may subsequently aid probe DNA accessibility. Similarly, protein-digesting enzyme treatments increase accessibility by digesting protein surrounding the target DNAs. Treatment with HCl is thought to extract proteins and achieve partial hydrolysis of target sequences, and in doing so, improve the signal-to-noise ratio.

Double-stranded DNA target must be denatured, and this is usually achieved with extremes of pH or heat. Generally, such treatments can lead to loss of morphology. Therefore, a compromise must be found between hybridization signal strength (detection) and target morphology (identification).

4.2   HYBRIDIZATION

The degree of specificity of the hybridization reaction can be controlled accurately by varying reaction conditions, such as temperature, pH, and formamide and salt concentration. The degree of mismatch that can be tolerated in a hybridization reaction is referred to as the stringency. A low stringency is achieved by using low temperature and formamide concentration and high salt concentration. Stringency increases with increasing temperature and formamide concentration and decreasing salt concentration. Thus low salt concentration and high temperature favor accurate base pairing. To understand the mechanism behind these factors, one must first consider the stability of the probe DNA and target DNA hybrid. Hybrid stability can be assessed from the melting temperature T_m, which is the temperature at which 50% of the nucleic acid duplexes become associated. Hybrid stability depends on salt and formamide concentration. Sodium chloride in the hybridization solutions exists as monovalent cations Na^+ and anions Cl^-. An increase in the salt concentration leads to an increase in the monovalent cations Na^+, which electrostatically interact with the nucleic acid phosphate groups, resulting in a decrease in electrostatic repulsion between the duplex DNA strands. Thus an increase in salt concentration leads to

an increase in hybrid stability. The base pair composition of the DNA sequence to be mapped can influence hybrid stability. Increasing molar ratio of GC pairs in the DNA leads to a more stable hybrid. Although there is little variation in hybrid stability in the pH range of 5–9, at more basic pH values (>9), hybrid stability decreases. Thus higher pH can be used to produce more stringent hybridization conditions.

Organic solvents reduce hybrid thermal stability, enabling hybridizations to be performed at lower temperatures, which in turn helps to preserve chromosome morphology. Formamide, the organic solvent most frequently used, reduces the melting temperature 0.72°C for each percent of solvent. Thus hybridization can be performed at 30–45°C if 50% formamide is present in the hybridization mixture.

All these factors are put together to create the following formula allowing a determination of the T_m, hence T_r, for optimal hybridization of a specific probe and target DNA:

$$T_m = 81.5 + 16.6 \, (\log M) + 0.41 \, (\%G + C) - \frac{650}{L} - 0.72 \, F$$
$$- 1.4 \, (\% \text{ mismatch})$$

where M is ionic strength (mol/L), L is probe length, and F is the percentage of formamide.

Another factor that influences the kinetics of hybridization and concentration is the probe length. The rate of renaturation of DNA in solution is proportional to the square root of the single-stranded fragment length. Consequently, maximal hybridization rates are obtained with long probes. Short probes are preferred for in situ hybridization, however, for reasons of increased accessibility. Fragment length also influences thermal stability.

The optimum fragment size for FISH is also dependent on the target sequence size and the probe mass. For larger targets, less probe can be included and the fragments can be larger. With increasing fragment length however, ability to access the target may become a problem. In general, for a target of 1 kb, fragment sizes of 200–500 bp are optimal. For whole-chromosome targets, 600–1000 bp fragments are used.

The probe concentration affects the rate at which the first few base pairs are formed (nucleation reaction), which is the rate-limiting step. Adjacent base pairs are formed subsequently, provided the bases are in register. The higher the probe concentration, the faster the reannealing step. The addition of dextran sulfate to the hybridization solution artificially increases probe concentration. In aqueous solution, dextran sulfate is strongly hydrated. Consequently, macromolecules have no access to the hydrating water, which results in an apparent increase in probe concentration, hence higher hybridization rates. Thus, depending on the desired stringency and the hybrid melting temperature, hybridization is undertaken at 37–60°C, in a hybridization solution containing sodium ions, 50% formamide, and dextran sulfate (pH 7). Finally, human or salmon sperm DNA fragments are included in the hybridization mix to reduce nonspecific binding of labeled probes (see CISS hybridization, Section 3.1).

4.3   POSTHYBRIDIZATION

Labeled probe may hybridize to sequences that bear homology to the probe sequence but are not identical to it. Such hybrids are less stable than the perfectly matched hybrids. They may be dissociated by performing washings of increasing stringency. In the case of RNA–DNA hybrids, nonspecifically bound RNA may be removed by RNase A treatment, ensuring a very low background signal.

5   SIGNAL DETECTION

5.1   DIRECT VERSUS INDIRECT

In the case of direct labels, signal detection is instantaneous. The hapten itself has fluorescent properties; in other words, it is the fluorochrome that is directly conjugated to the nucleotide and so incorporated into the probe. After hybridization, any unbound probe is washed away. Then the DNA target (e.g., chromosomes) is stained using a fluorescent dye, and the slide is mounted in antifade and applied directly to the fluorescent microscope for visualization (Figure 2A). A number of direct labels are commercially available. In most cases, however, signal detection is achieved using fluorescently labeled binding proteins in one or many layers, depending on the level of signal amplification required. The binding proteins are most often antibodies. The primary antibody may be conjugated to a fluorochrome, allowing signal visualization after a single detection layer. This is described as direct detection (Figure 2B). It is also possible to detect the hapten biotin using the binding protein avidin, or streptavidin conjugated to a fluorochrome. In the case of indirect detection (Figure 2C), the primary layer is then detected by a secondary layer, which again may be conjugated to a fluorochrome and immediately visualized, or may itself be subsequently detected by a tertiary layer. At present it is generally the rule that the more direct detection techniques are possible only with larger probes, hence larger detection sites with bigger signals. The small target sites are still dependent on amplification steps using multilayer detection to aid visualization.

5.2   AMPLIFICATION

Additional layers of fluorochrome-conjugated binding proteins are used to achieve amplification of the signal at the target probe hybridization site, as discussed in Section 5.1. One alternative method, in addition to the multilayer detection approach to amplify the signal, is to use PRINS: oligonucleotide-primed in situ DNA synthesis. The bound probe fragment at the target site in the intact chromosome is used as a primer from which DNA amplification incorporating additional hapten molecules is achieved. The resultant larger site for detection displays a larger signal.

For small probes detecting small target sequences, network (hyperpolymer) formation is thought to be particularly important with respect to signal intensity increase. The ability of probe fragments to hybridize to each other in overlapping complementary regions results in the creation of an enlarged region for subsequent detection at the hybridization site, with a corresponding increase in signal size. The extent and reproducibility of networking is difficult to control, probably because of the critical dependence on probe size.

5.3   FLUOROCHROMES

The most commonly used fluorescent labels are fluorescein isothiocyanate (Fitc), tetramethyl rhodamine (Tritc), Texas red, and aminomethyl/coumarin (AMCA). Table 2 shows the wavelengths required to optimally excite each of these fluorochromes and the wavelength emitted as a result in each case. Thus, blue light having a wavelength of 490 nm will optimally excite the fluorochrome flu-

Table 2   Fluorochromes Available for DNA Staining and Signal Detection in FISH

Fluorochromes	Excitation Maximum Wavelengths (nm)	Emission Maximum Wavelengths (nm)
DNA Counterstains		
Propidium iodide	330 and 520 (green)	620 (red)
DAPI	350 (UV)	460 (blue)
Hoechst 33258	360 (UV)	470 (blue/green)
Hybridization Site Labels		
Fluorescein isothiocyanate (Fitc)	490 (blue)	520 (green)
Texas red	596 (yellow)	620 (red)
Tetramethyl rhodamine (Tritc)	554 (green)	573 (orange)
Aminomethyl/coumarin (AMCA)	350 (UV)	450 (blue)

orescein isothiocyanate and cause it to emit green light having a wavelength of 520 nm. It is this green light that is seen by the eye and recorded as a positive signal. For a more detailed explanation of the microscope filters required to achieve such excitation of fluorochromes, and visualization of emitted light, see Section 6.1.

5.4   MULTICOLOR, MULTIPROBE FISH

The availability of more than one hapten for DNA labeling and numerous fluorochromes for DNA detection allows multicolor, multiprobe FISH to be performed (Figure 3D). Using such a multicolor approach, three probes can be relatively ordered when two are detected in green and one in red. This order can be confirmed when, subsequently, one of the previously "green"-tagged probes is detected in "red." Probes whose signals are apparently indistinguishable using a prometaphase or metaphase chromosome target can thus be distinguished and ordered using interphase chromatin target DNA. In this way the resolution of the FISH technique can be increased. However, there is potential to increase resolution still further by *combinatorial probe labeling* (multiple hapten labeling), and still further by *color ratio labeling.*

In combinatorial probe labeling, a probe labeled with a green fluorochrome will be seen as green, a probe labeled with a red fluorochrome will be seen as red, a third probe labeled with both green and red will be seen as yellow. Thus three-color analysis can be performed with only two fluorochromes. The target DNA in the form of stained chromosomes can be seen in blue, adding a fourth color to the composite image. In the same way, seven-color analyses can be performed with three fluorochromes.

Combinatorial probe labeling creates additional colors for probe identification by combining two single colors in equal proportions. Color ratio labeling takes this approach one step further. Instead of adding just one more color in an equal ratio combination of two single colors, one uses a variety of different ratio combinations to allow the creation of many more colors from the two originals. For example, two single colors, red and green, are combined in equal proportions to create a third color, yellow. If red and green were also combined with more red (75%) than green (25%), a fourth color is created, allowing simultaneous identification of a fourth probe. Similarly, an increase in green (75%) and a reduction in the red (25%) produces a fifth color, hence a fifth probe. An increasing number of "ratios" can be employed. Since, however, the human eye is unable to distinguish them, analysis relies on image analysis software that allows distinction of these mixed color ratio labels and assignment of clearly distinct primary colors (false colors) for identification.

6 SIGNAL VISUALIZATION

6.1 FLUORESCENCE MICROSCOPY

A standard fluorescence microscope is diagramatically represented in Figure 5. A light source supplying light of many wavelengths passes first through an excitation filter, which selects for the specific wavelength required to optimally excite the fluorochrome in the specimen. A 45° dichroic mirror reflects wavelengths below and transmits those above a set value, to suit the fluorochrome to be detected. The fluorochrome absorbs this light and emits light of a higher wavelength, which is transmitted by the dichroic mirror and thus passes through to the emission filter, which selects only for the wavelength of light emitted by the fluorochrome under observation. Filter sets (excitation and emission filters plus dichroic mirror) may be very selective for a given fluorochrome. Changing the filter sets allows a number of different fluorochromes from one image to be optimally excited and their emission spectra collected. These images require subsequent superimposition to create a final composite image. Although multicolor fluorescence images can be readily accumulated in this way, filter changes result in a shift in image registration. The separate images then require alignment, which is achieved by using an alignment signal, or avoided, by using a multi–band pass filter. The latter represents a compromise between the optimal wavelengths, producing weaker signals.

6.2 CCD CAMERAS

Smaller probes require greater sensitivity, which is achieved by using either a high sensitivity camera or confocal microscopy. The most sensitive camera system is the cooled, charge-coupled device (CCD) integrating camera. Low illumination photon counting with longer integration times is possible, allowing the detection of fluorescent signals that would not otherwise be visible. A mercury arc lamp light source supplies high peaks on or near excitation wavelengths of the popular fluorochromes. When this apparatus is used in conjunction with a wide range of excitation filters, multicolor images can be produced. Numerous software packages are available for alignment and synthetic color imaging. The small unique sequence probes shown in Figure 3 were visualized with the aid of a CCD camera (see figure legend).

6.3 CONFOCAL MICROSCOPY

In confocal microscopy, light from a point source enables focal illumination and imaging of a single point in the sample, eliminating out-of-focus elements. Coordinated sample scanning makes it possible to compose a complete image. Laser illumination provides the required bright light source, and high resolution. The registration problem is overcome by parallel excitation/emission of different fluorochromes, which can be excited by the lasers available. The YAC probe shown in Figure 3E was visualized using a confocal microscope (see figure legend).

6.4 PHOTOGRAPHY

There is a direct relation between sensitivity (ASA rating) and quality of the final print, reflected by its graininess. Low ASA films tend to produce prints of much finer grain. In fluorescence photomicroscopy, light levels emitted by the specimen are low. Therefore, long exposure times are not permissible in cytogenetic investigation because the light emitted by fluorochromes fades so rapidly. A compromise between exposure time and final photographic quality must be made.

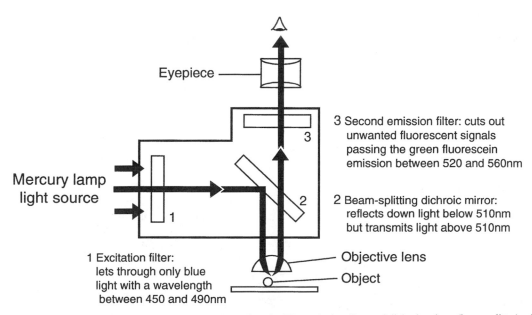

Figure 5. The mercury lamp emits light with peaks at a number of wavelengths. The excitation filter and dichroic mirror (beam splitter) select for the wavelength that will optimally excite the fluorophore to be visualised. The dichroic mirror and emission filter select for the wavelength emitted by the excited fluorophore. In this example, fluoroscein isothiocyanite (Fitc) is excited at 490 nm (blue), and the green fluorescence emitted at 520 nm is visualised.

6.5 ELECTRONIC IMAGE RECORDING

To exploit the full potential of multicolor FISH, quantitative digital fluorescence microscopy is needed. Digitized images are easier to handle and can be subject to powerful image processing. Recent methods of storing images electronically and reproducing them as photographic quality hard copy are developing rapidly. There are several advantages: no dark room is required; the prints are made instantaneously; and once the microscope image has been captured, it can be subjected to computer enhancement and manipulation techniques.

7   CONCLUSION

The temptations of speed, accuracy, and increased spatial resolution both prompted and fueled the development of FISH. This powerful technique is particularly important in gene mapping because it is the only technique to combine accurate gene localization with relative gene order, and both sets of information are required for a complete gene map.

Continuing improvements in probe preparation, detection, and visualization will allow the accumulation of even more information from a single hybridization experiment.

See also HUMAN CHROMOSOMES, PHYSICAL MAPS OF; HUMAN DISEASE GENE MAPPING; WHOLE CHROMOSOME COMPLEMENTARY PROBE FLUORESCENCE STAINING.

Bibliography

Gray, J. W., and Pinkel, D. (1992) Molecular cytogenetics in human cancer diagnosis. *Cancer,* 69 (6):1536–1542.
Lichter, P., and Cremer, T. (1992) Chromosome analysis by non-isotopic in situ hybridisation. In *Human Cytogenetics—A Practical Approach,* Vol. I, *Constitutional Analysis,* D. E. Rooney and B. H. Czepulkowski, Eds., pp. 157–190. Oxford University Press, Oxford.
Polak, J. M., and McGee, J.O'D., Eds. (1990) *In Situ Hybridisation—Principles and Practice.* Oxford University Press, Oxford.
Trask, B. J. (1991) Gene mapping by situ hybridisation. *Curr. Opinion Genet. Dev.* 1:82087.
———. (1991) Fluorescence in situ hybridisation: Applications in cytogenetics and gene mapping. *Trends Genet* 7(5):149–154.
Wilkinson, D. G., Ed. (1992) *In Situ Hybridisation: A Practical Approach.* Oxford University Press, Oxford.

GENE ORDER BY FISH AND FACS

Malcolm A. Ferguson-Smith

1   **Introduction**

2   **Techniques**
   2.1   Fluorescence In Situ Hybridization (FISH)
   2.2   Fluorescence-Activated Chromosome Sorting (FACS)

3   **Applications**
   3.1   Gene Order by FISH
   3.2   Gene Order by FACS

4   **Perspective**

Key Words

Chromosome-Specific Paint   The product of PCR amplification of sorted chromosomes, using random PCR primers. When labeled and hybridized to metaphases by fluorescence in situ hybridization (FISH) techniques, the chromosome paint will give an even distribution of FISH signals along the length of the chromosomes from which the paint was derived.

Contig   A contiguous series of overlapping cloned DNA sequences.

Cosmid   A vector capable of cloning inserts of bacterial DNA measuring 20 to 40 kilobases.

Flow Karyotype   Graphical representation of mitotic chromosomes produced by a dual laser flow cytometer in which the chromosomes are arranged according to size and base pair ratio.

Gene Locus   The position of a gene on its chromosome.

Gene Probe   Cloned DNA sequence of part of a gene used to locate the gene on its chromosome and to identify restriction fragments.

Lymphoblastoid Cell Line   Cell line derived from peripheral lymphocytes transformed and immortalized by Epstein-Barr virus.

Microsatellite Marker   Polymorphic DNA sequence composed of a variable number of tandemly arranged simple di-, tri-, or tetranucleotide repeats.

Polymerase Chain Reaction (PCR)   A method for the primer-directed amplification of specific DNA sequences.

Reciprocal Translocation   The result of chromosome breakage and the exchange of chromosome material between two non-homologous chromosomes.

Somatic Cell Hybrid   Cell line derived from the fusion of cells from two different species, most often human and rodent; in the latter case, human chromosomes tend to segregate out of the hybrid.

Yeast Artificial Chromosome (YAC)   A vector capable of cloning large inserts of DNA measuring 200 to 1000 kb in yeast cells.

Two cytological techniques used in mapping and ordering genes in the Human Genome Project are fluorescence in situ hybridization (FISH) and fluorescence-activated chromosome sorting (FACS).

FISH anneals nonisotopically labeled DNA gene probes to their complementary sequences on chromosomes immobilized on microscope slides. The sites of hybridization, and thus the chromosomal location of the gene, are detected by immunological and other systems in which the antibody to the DNA label is coupled to a fluorochrome. The fluorescent signal is detected by fluorescence microscopy, is collected onto optical disks with the aid of a sensitive digital camera, and is available for detailed examination using image analysis systems. The use of different DNA labels for different probes allows the determination by multicolor FISH of the relative order of several genes along the metaphase chromosome. The order of closely linked DNA sequences may be determined from interphase nuclei or from extended chromatin fibers released from fixed nuclei.

In FACS, mitotic chromosomes in fluid suspension are passed sequentially through a dual laser flow cytometer and sorted into individual groups according to size and base pair ratio. Pure samples of most individual chromosomes can be obtained and used for the chromosome assignment of DNA sequences by the PCR technique. Regional assignment and gene order are achieved by sorting the products of reciprocal translocations.

1   INTRODUCTION

The Human Genome Project is an international venture to construct detailed genetic and physical maps of each of the 23 different human chromosomes. The genetic map provides the location of genes and sufficient reference markers to enable new genes to be located rapidly by family (genetic linkage) studies. The physical map consists of a contiguous series of overlapping DNA sequences (contigs) extending throughout the length of each chromosome. Already approximately 80% of the entire human genome is represented in the form of physical maps by such contigs. The final phase of the Human Genome Project aims to sequence the entire DNA of the genome within 10 years. The most interesting parts of the genome, namely, the genes that are transcribed into messenger RNA and translated into proteins, are being sequenced first. The map of these genes is sometimes referred to as the transcriptional or expression map.

The DNA sequence of a particular gene may be determined either from knowledge of the amino acid sequence of its protein product or by a more complex strategy known as positional cloning. Most of the common single-gene disorders that affect human populations (e.g., thalassemia, cystic fibrosis) have now been mapped and sequenced by one or other of these two methods. In positional cloning, families in which members are affected with a recognizable disorder or trait (phenotype) are tested with a series of chromosome-specific DNA reference markers spread evenly across the genome to determine which marker tends to be transmitted through the family in association with the particular phenotype. As members of a chromosome pair assort randomly during the formation of eggs and sperm, linkage of a chromosome marker and the gene locus is the first indication that the gene can be assigned to that chromosome. Genetic recombination due to meiotic crossing over may also separate marker and gene locus if they lie apart from each other on the same chromosome. The linkage analysis is therefore repeated with additional markers from the same chromosome, to assign the gene to the smallest possible chromosomal region. Once this has been achieved, the relevant contig in the physical map can be identified and used to isolate candidates for the gene in question. In the final stage in positional cloning, the various candidate genes are sequenced, whereupon it is possible to look for gene mutations in affected patients and confirm that such anomalies are absent in unaffected controls.

During the construction of genetic maps, new DNA clones are isolated and characterized for their value as genetic markers. Nowadays, the first approach to mapping such markers uses the technique of in situ hybridization, whereby the DNA probe, suitably labeled, is annealed to preparations of chromosomes fixed onto microscope slides. The DNA hybridizes to its complementary sequence on the chromosome, and the site of hybridization is recognized from the label used in making the DNA probe. Radioactive labels were used

initially, but these have been replaced by agents such as biotin and digoxigenin, which can be incorporated into the probe DNA and detected by fluorescence microscopy using fluorochromes coupled to appropriate antibodies. Fluorescence in situ hybridization (FISH) has been developed into a very sensitive technique, capable of readily detecting DNA probes containing at least 1 kilobase of DNA sequence. With multiple fluorochromes and appropriate filter systems on the fluorescence microscope, several colors can be used with different probes to map more than one sequence at a time. This capability is particularly valuable when the aim is to determine the order of different DNA markers along the same chromosome.

The most useful markers for genetic mapping are those that show extensive variation (polymorphism) between individuals, but are transmitted without change through families. Microsatellite markers, in which the variation is due to differences in the number of copies of simple di- or trinucleotide repeats, are among the most frequent of these markers and are readily distinguished by use of the polymerase chain reaction (PCR). Using specific DNA primers flanking the microsatellite, PCR amplifies the fragment containing the dinucleotide repeat, which can then be sized by ethidium bromide stained gel electrophoresis.

FISH is not applicable to mapping microsatellite markers, and so alternative strategies are required. One approach uses a series of interspecific somatic cell hybrids formed by fusing human and rodent cells. It is possible to construct hybrid cell lines so that each contains a single human chromosome within a background of rodent chromosomes. PCR amplification of a human microsatellite sequence will occur only if the appropriate chromosome is present. Panels of monospecific hybrids can therefore be assembled for assigning such markers to their respective chromosomes. Occasionally problems arise; a hybrid may be contaminated by part of a second human chromosome that has become incorporated into one of the rodent chromosomes; sometimes the PCR reaction amplifies a rodent sequence that happens to be homologous to the human sequence.

An alternative strategy for assigning microsatellite markers to their respective chromosomes depends on the ability to sort human chromosomes into their respective categories using flow cytometry. In brief, a fluid suspension of chromosomes is prepared from an actively growing cell culture by using colchicine to arrest cells in mitosis. The mitotic cells are disrupted and the released chromosomes are stained with a mixture of two fluorescent dyes, which detect AT-rich and GC-rich DNA, respectively. The stained chromosomes are passed sequentially at a speed of up to 2000 per second through two lasers tuned to excite the fluorescent emission from each dye in a fluorescence-activated cell sorter (FACS). The amount of fluorescence in each chromosome is collected and stored, and this information is used to produce a bivariate flow karyotype (Figure 1) in which large numbers of individual chromosomes form 20 discrete clusters in characteristic relationships. The technique effectively separates all chromosomes except chromosomes 9 through 12, and the distinct pattern of fluorescence in each chromosome enables the FACS to sort each category into separate tubes. Thus a panel of sorted chromosomes can be produced, each tube with 300 to 500 chromosomes, and this is sufficient material to allow a microsatellite probe to be assigned by PCR to its respective chromosome. Chromosomes 9 through 12 can be separated for mapping purposes by exploiting heteromorphisms (for chromosome 9) and translocation derivatives. Similarly, regional assignments may also be made by sorting translocation chromosomes with appropriate breakpoints.

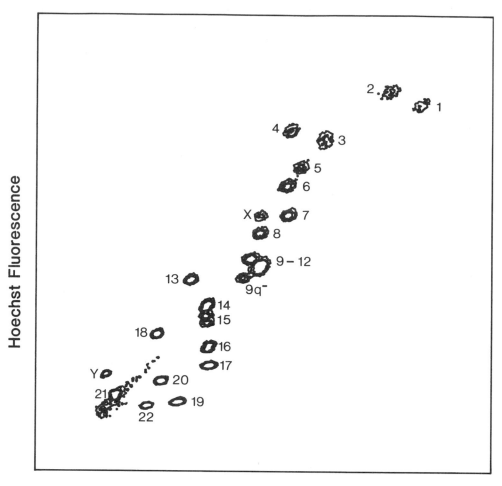

Chromomycin A3 Fluorescence

Figure 1.   Bivariate flow karyotype showing separation of individual chromosomes according to size and base pair ratio. In this preparation, one homologue of chromosome 9 is separated from the remainder of the 9–12 cluster because it contains substantially less than the average complement of centric heterochromatin.

The DNA of chromosomes sorted by FACS can be amplified by PCR and labeled to produce chromosome paints. The amplified DNA is useful in hybridization experiments to assist in the analysis of chromosomes rearrangements, for the chromosome paint will anneal only to the chromosomes from which it is derived. Once again the labeled DNA is detected by FISH. A modification of the method, termed reverse painting, depends on the ability to sort and make paints from rearranged chromosomes. Hybridization onto normal metaphase spreads reveals the origins of the various chromosomes involved in the rearrangement. This is one of the most sensitive techniques for the analysis of complex chromosome arrangements.

2   TECHNIQUES

2.1   FLUORESCENCE IN SITU HYBRIDIZATION (FISH)

FISH depends on hybridizing labeled nucleic acid probes to cytological preparations of chromosomes and chromatin and detecting the presence of the annealed sequences by fluorochrome-conjugated reagents using a fluorescence microscope.

2.2   FLUORESCENCE-ACTIVATED CHROMOSOME SORTING (FACS)

High resolution chromosome sorting can be achieved with a commercial fluorescence-activated cell sorter equipped with two argon ion lasers. Preparation of the chromosome suspension, FACS analysis, sorting, and collection of sorted chromosomes are the main steps in the process.

2.2.1 Chromosome Preparation

Chromosome preparations are made by standard methods from short-term peripheral blood cultures, lymphoblastoid cell lines, and a variety of tissue culture cells including skin fibroblasts. Cells are arrested in metaphase by the addition of colcemid (0.1 μg/mL) to the culture for 6 to 12 hours. The cells are then resuspended in low ionic strength buffer (45 mM KCl, 10 mM MgSO$_4$, 3 mM dithiothretol, and 5 mM HEPES at pH 8.0) and incubated for 10 minutes at room temperature. Triton X-100 is added to a final concentration of 0.25% and the sample left on ice for 10 minutes.

An alternative technique is to treat the cells in 75 mM KCl and resuspend the cell pellet in a buffer containing polyamines and Triton X-100. Chromosomes are released into suspension by 10 seconds of rapid vortexing and stained immediately in Chromomycin A3 (final concentration 40 μg/mL) and Hoechst 33258 (final concentration 2 μg/mL), followed by 2 hours of incubation at 4°C. Fifteen minutes prior to flow analysis, sodium citrate (final concentration 10 mM) and sodium sulfate (final concentration 25 mM) are added to the sample.

2.2.2 Analysis, Sorting, and Collection

Chromosomes are sorted using a dual laser flow cytometer (FACStar Plus or equivalent) equipped with two 5 W argon ion lasers. One laser is tuned to emit 300 mW of light in the UV (351–364 nm) to excite Hoechst fluorescence, and the second laser is tuned to emit 300 mW at 458 nm to excite Chromomycin fluorescence. The chromosome suspension is passed through the two laser beams sequentially, to permit the fluorescence emitted from each chromosome to be collected separately and used to create a flow karyotype (Figure 1) in which large numbers of individual chromosome are represented by discrete clusters in characteristic relationships. Each chromosome cluster has a position in the flow karyotype determined by its relative size and base pair ratio. AT-rich chromosomes tend to sort above a diagonal line drawn through the middle of the chromosome clusters, while GC-rich chromosomes sort below this line. As the chromosomes pass through the two lasers, the fluid system breaks into droplets, some of which will contain one chromosome. Sorting is achieved by giving an electrical charge to the droplets containing the chromosome of interest so that they can be deflected into a container as they pass between two high voltage plates. Highly pure chromosome samples can be collected in this way.

3   APPLICATIONS

3.1   Gene Order by FISH

Early methods for distinguishing between different human chromosomes depended on their size and centromere position at the metaphase stage of mitosis. Unequivocal identification of every chromosome became possible only when banding techniques of staining were introduced in 1970. At that stage each chromosome could be numbered. Assignment of gene loci to individual chromosome numbers has thus depended on cytological techniques complemented by genetic linkage studies, which are based on the knowledge that if one locus maps to a particular chromosome, all other loci that are linked must also be located on the same chromosome. The relative position of a gene on the chromosome is now most easily determined by FISH, and this requires that at least part of the gene be included in a cloned DNA sequence that is more than one kilobase long. Probes made from cosmid (≤40 kb) or yeast artificial chromosome (YAC) clones (≤ one megabase) are most frequently used for FISH mapping because they give strong hybridization signals—characteristically, twin signals in the same position on both chromatids of the chromosome. If two or more probes that map to the same chromosome are hybridized at the same time, it is often possible to determine their order on the chromosome. However, it has been found that closely linked probes cannot be ordered on a metaphase chromosome unless they are at least one megabase apart. This is because the DNA molecule within the chromosome is condensed by several orders of coiling and supercoiling and because the supercoiled chromatin fiber is attached to the chromosome scaffold in a tightly packed series of loops that radiate out from the center in every direction. DNA sequences that are less than one megabase apart on the chromatin fiber may thus appear to be superimposed on one another, or to be in the wrong order relative to the ends of the chromosome.

One strategy for resolving the correct order of two probes that map to the same location at metaphase is to hybridize them to preparations made from a panel of cell lines containing reciprocal translocations that span the region of interest. Suitable translocations can often be obtained from the commercial and private cell banks, which specialize in making such collections available to gene mappers. Clearly, when a translocation breakpoint separates two probes, a single signal will appear on each of the translocation derivatives; if the translocation breakpoint is on either side of the probes, only one translocation derivative will show the signal. The two signals can be distinguished by two-color FISH.

The order of sequences less than one megabase apart may also be determined by FISH, using interphase nuclei in which the chromosomes are extended to 10 times their metaphase length and are not visible as discrete entities. At least three probes are required: one for the sequence whose location is to be determined in relation to the second sequence, and a third whose position in relation to the second is already known. Cytological preparations are fixed and air-dried as for chromosome preparations and cosmid clones are labeled for detection using at least two colors. Nuclei in the Gl phase (i.e., before DNA replication) are chosen for scoring. A typical nucleus will show three signals on each of the two copies of the chromosome in question, and their consensus order along the chromosome usually can be determined from the analysis of 20 to 50 nuclei. Two signals can be resolved if they are more than 50 kb apart, but greater resolution is not possible. As at metaphase; the limitation of resolution is mostly due to the organization of the chromatin fiber into loops radiating from a central scaffold. Measurements made in cytological preparations of the distance between hybridization signals can therefore provide only crude estimates of the relative distance between DNA segments.

The construction of genetic maps has been greatly facilitated by interphase FISH, which has helped to resolve the order of DNA genes and markers too close together to be determined by genetic linkage studies alone. In the construction of physical maps, FISH has played a similar role in the correct ordering of YAC contigs along the chromosome. FISH can readily resolve inconsistencies due to such problems as colligation of two separate DNA sequences within the same YAC. More recently, FISH has been applied to the ordering of cosmid subclones within a single YAC. Interphase FISH does not have sufficient resolution for this purpose, and so alkali treatment and other methods have been developed to release the chromatin fiber from its protein scaffold and fix the extended fiber in the microscope slide in a form suitable for FISH. Cosmid probes hybridized to such extended chromatin fibers typically show a series of interrupted signals in the form of a string of particles along the fiber. By using different combinations of fluorochromes, the order of different cosmids along the fiber can be determined and overlaps and gaps of 5 to 10 kb between cosmids readily detected (Figure 2; see color plate 6). In this particular application, FISH demonstrates its power to bridge the resolution gap between genetic and physical maps.

Figure 2. Ordering three closely linked cosmid clones using hybridization to the extended chromatin fiber released by alkali treatment of fixed nuclei. The cosmids, each approximately 40 kb long, are differentially labeled by Texas Red (red), FITC (green), and a 50:50 mixture of Texas Red and FITC (giving an orange-yellow color). There is a 5 kb overlap between the red and green labeled cosmids indicated by the yellow signal. The yellow cosmid is separated from the red cosmid by a 50 kb gap. (See color plate 6.)

3.2 GENE ORDER BY FACS

Ordering DNA sequences by FISH depends on the availability of a genomic DNA clone that is both large enough and contains at least part of the DNA sequence of interest. For most practical purposes this means isolating a cosmid clone from an appropriate genomic library. However, many of the most polymorphic DNA markers used in genetic mapping are microsatellite markers tested by PCR. As discussed in Section 1, the markers may be mapped directly, without recourse to FISH, by amplification, using either interspecific so-

Table 1 Flow Dot-Blot Analysis of Chromosome 9 Translocations[a]

Probe	9T05 9q12	9T10 9q22	9T02 9q22.1	9T06 9q32	9T08 9q33	9T14 9q34.1	9T03 9q34.1	9T04 9q34.1	9T12 9q34.2	9T01 9q34.3
ALAD	D	D	D	*P*	*P*	*P*	*P*	*P*	*P*	*P*
MCOA12	D	D	D	D	*P*		*P*	*P*	*P*	*P*
ORM	D	D	D	D	*P*	*p*	*P*	*P*	*P*	*P*
GSN	D	D	D	D	*P*	*P*	*P*	*P*	*P*	*P*
HXB	D	D	D	D	*P*	*P*	*P*	*P*	*P*	*P*
CRIP111	D	D	D	D	D	*P*	*P*	*P*	*P*	*P*
AK1	D	D	D	D	D	D	*P*	*P*	*P*	*P*
SPTAN1	D	D	D	D	D	D	*P*	*P*	*P*	*P*
ASSg3	D	D	D	D	D	D	D	*P*	*P*	*P*
T39-2-2	D	D	D	D	D	D	D	*P*	*P*	*P*
ABL3	D	D	D	D	D	D	D	D	*P*	*P*
DBH	D	D	D	D	D	D	D	D	*P*	*P*
MCT136	D	D	D	D	D	D	D	D	*P*	*P*
MCT96.1	D	D	D	D	D	D	D	D	D	*P*

[a]Ten cell lines from balanced translocation heterozygotes with different translocation break points in chromosome 9q were used to sort the two translocation derivatives and prepare dot blots of approximately 10,000 chromosomes onto nitrocellulose filter disks. DNA probes for each of the loci listed on the left were hybridized to the filter disks: D and P indicate the site of hybridization relative to the translocation break point (i.e. on the derivative chromosome carrying either the *distal* or the *proximal* segment of the long arm of chromosome 9. The results give the precise order of many of the loci between 9q12 and 9q34.3.

matic cell hybrids or chromosomes sorted by FACS. If the FACS technique is used, gene order is then achieved by sorting translocation derivatives to determine on which side of the translocation breakpoint the microsatellite is located.

Chromosome sorting may also be used for mapping complementary DNA sequences. For this purpose, 10,000 chromosomes of each type are sorted onto nitrocellulose filter disks using mild aspiration from beneath the disk. The chromosomal DNA is denatured and baked onto the disk, which can then be used for filter hybridization with radiolabeled ^{32}p DNA probes in the usual way. Autoradiography of the filter panel reveals the chromosomal location of the probe. The order of various chromosome-specific probes may be determined using filters prepared from translocation chromosomes (Table 1).

4 PERSPECTIVE

For the foreseeable future, FISH will play a very important role in mapping cloned DNA sequences to their positions on the genetic map and in the construction of physical maps. In terms of simplicity and economy, it is the technique of first choice for gene localization. Because quick results are possible, the method is used for confirmation and verification of the identity of DNA clones generated in the molecular biology laboratory. It is the routine test used to exclude coligation during the characterization of clones isolated from YAC libraries. Apart from gene mapping, FISH has an increasing role in the diagnosis of chromosome aberrations, whether constitutional or associated with the pathogenesis of cancer.

The role of FISH in gene mapping is likely to diminish in the more distant future when a fully validated physical map of each chromosome has been achieved based on a contiguous series of overlapping YAC and cosmid clones extending from one end of the chromosome to the other. With such a physical map it will be possible to assign any unknown DNA sequence to a single YAC or cosmid clone in the series, hence to its position on the chromosome. Similarly, the need for chromosome sorting by FACS will diminish in terms of human gene mapping. However, it is likely that gene mapping in many other species will be facilitated by the ease of sorting individual chromosomes. In nonhuman species it is more difficult and usually impossible to construct interspecific cell hybrids containing only one chromosome from one of the hybrid parents. Mapping panels of sorted chromosomes may prove to be an important resource for mapping these genomes.

See also DNA MARKERS, CLONED; GENE MAPPING BY FLUORESCENCE IN SITU HYBRIDIZATION; HUMAN CHROMOSOMES, PHYSICAL MAPS OF; HYBRIDIZATION FOR SEQUENCING OF DNA.

Bibliography

Carter, N. P., Ferguson-Smith, M. A., Perryman, M. T., Telenius, H., Pelmear, A. H., Leversha, M. A., Glancy, M. T., Wood, S. L., Cook, K., Dyson, H. M., Ferguson-Smith, M. E., and Willatt, L. R. (1992) Reverse chromosome painting: A method for the rapid analysis of aberrant chromosomes in clinical cytogenetics. *J. Med. Genet.* 29:299–307.

Fidlerova, H., Senger, G., Kost, M., Sanseau, P., and Sheer, D. (1994) Two simple procedures for releasing chromatin from routinely fixed cells for fluorescence in situ hybridisation. *Cytogenet. Cell Gen.* 65:203–205.

Reed, T., Baldini, A., Rand, T. C., and Ward, D. C. (1992) Simultaneous visualisation of seven different DNA probes by in-situ hybridization using combinational fluorescence and digital imaging microscopy. *Proc. Natl. Acad. Sci. U.S.A.* 89:1388–1392.

Rooney, D. E., and Czepulkowski, B. H. (1992) *Human Cytogenetics: A Practical Approach.* IRL Press, Oxford.

Trask, B. J., and Pinkel, D. (1990) Fluorescence in-situ hybridization with DNA probes. In *Methods of Cell Biology*, Vol. 33. Academic Press, New York.

———, Massa, H., Kenwrick, S., and Gitschier, J. (1991) Mapping of human chromosome Xq28 by two color fluorescence in-situ hybridization of DNA sequences to interphase cell nuclei. *Am. J. Hum. Genet.* 48:1–15.

van den Engh, G., Trask, B., Lansdorp, P., and Gray, J. (1988) Improved resolution of flow cytometric measurements of Hoechst and Chromomycin-A3 stained human chromosomes after addition of citrate and sulphate. *Cytometry* 9:266–270.

Wiegant, J., Kalle, W., Mullenders, L., Brookes, S., Hoovers, J.M.N., Damverse, J. G., van Ommen, G.J.B., and Raap, A. K. (1992) High resolution in-situ hybridisation using DNA halo preparations. *Hum. Mol. Genet.* 1:587–591.

GENETIC ANALYSIS OF POPULATIONS

A. Rus Hoelzel

1 Introduction

2 Nature of Mutations and Mutation Rates

3 Dissemination of Genetic Change in Natural Populations
  3.1 Natural Selection
  3.2 Genetic Drift
  3.3 Molecular Drive

4 Methods of Analysis
  4.1 Sample Collection
  4.2 Enzyme Electrophoresis
  4.3 DNA Extraction
  4.4 RFLP Analysis
  4.5 DNA Sequencing
  4.6 PCR Analysis
  4.7 SSCP and Gradient Gel Analysis
  4.8 RAPD Analysis

5 Choice of Genetic Marker and Interpretation
  5.1 Enzymes
  5.2 Nuclear Genes
  5.3 Mitochondrial DNA
  5.4 Chloroplast DNA
  5.5 Minisatellite DNA
  5.6 Microsatellite DNA
  5.7 Analysis of Variation

6 Conclusions

Key Words

Demographics The study of populations, especially growth rate and age structure.

DNA Turnover Continual gain and loss of DNA regions due to a variety of mechanisms including DNA slippage, gene conversion, transposition, and unequal crossing over.

Electrophoresis The movement of molecules on a solid medium through an electric field. Various support mediums are used, including filter paper, starch gel, agarose, and polyacrylamide.

Molecular Clock The rate of molecular genetic change at a given gene, often including the assumption of a constant rate.

Polymorphism Two or more genetically distinct types in the same interbreeding population.

The molecular genetic analysis of populations involves three main stages: sample collection, analysis, and interpretation. The sample must be representative of the population, and therefore collections need to be random and sufficiently large. The method of analysis and choice of genetic marker will depend on the specific question being addressed. Interpretation of results will depend in part on the method of analysis. DNA analytical methods offer great potential through the analysis of specific regions of the nuclear, mitochondrial, and chloroplast genomes, but numerous mechanisms can effect change, and therefore all interpretation based on time-dependent change must be done cautiously. An understanding of the genetic structure of populations facilitates the understanding of dispersal, reproductive, and social behavior, and provides essential data for the conservation of natural variation.

1   INTRODUCTION

It has long been recognized that inherited characters vary and that the quantification of this variation can be used to distinguish local races. However, it was not until the advent of protein gel electrophoresis in the 1960s that the extent of genetic variation in natural populations began to be appreciated. Well over 1000 species have been investigated for protein polymorphisms, and most displayed extensive variation. Gel electrophoresis became the standard method for the comparison of genetic variation within and between populations, and it is still appropriate for many applications. However, most of the variation present in the genome cannot be detected by this method. The DNA sequence determines, through an RNA intermediary, the amino acid sequence in the polypeptides that combine to form proteins. Only a small proportion of the genome is translated into proteins, and the three-letter codon that determines the sequence of amino acids is degenerate, especially in the second and third positions. Therefore, even within the DNA sequence that encodes the protein, there can be considerable variation that is not expressed. Furthermore, only a small proportion of the changes that do affect the amino acid sequence will also affect the charge properties of the protein, which determine its migration through an electric current (and consequently only some changes can be detected by the electrophoretic method).

Recent innovations have enabled the analysis of DNA directly, providing greater resolution and facilitating the interpretation of variation. This article concentrates on the use of these DNA analytical methods to interpret population level genetic variation.

2   NATURE OF MUTATIONS AND MUTATION RATES

Mutations can occur that affect the sequence of nucleic acids through the conversion of a single base (e.g., a point mutation changing A to T), through the loss by deletion of one or more bases, and through the addition by insertion of one or more bases. Change

can also occur through the operation of a variety of DNA turnover mechanisms. For the purposes of this discussion I will emphasize two mechanisms that generate variation in repetitive regions of DNA by changing the number of repeated elements. These are DNA slippage (Figure 1a) and unequal crossing over (Figure 1b). In each case, a misalignment of DNA strands causes the molecule to be repaired in such a way that there is a gain or loss in an array of repeated elements. Repetitive DNA regions are very common in most vertebrate and plant genomes, and change in the copy number of re-

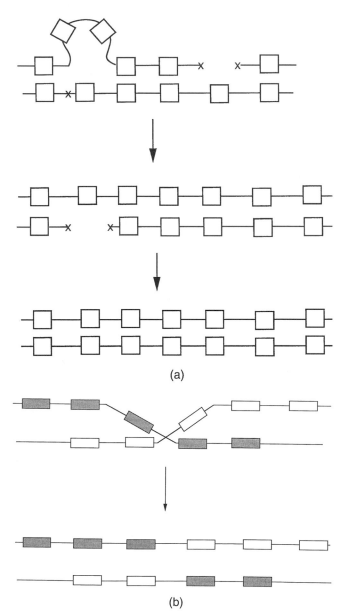

Figure 1. (a) One of several DNA slippage mechanisms: the end result is an extension of the array within a single chromatid. Similar mechanisms lead to contraction of the array and slippage during replication. The process illustrated involves endonuclease-mediated breakage, slippage of the upper strand, and polymerase-mediated repair events. The boxes represent repeated elements. (b) Unequal crossing over. Two strands, each with five repetitive elements, undergo an unequal crossing over event, resulting in the gain of one element for the upper strand and a corresponding loss from the lower strand.

peats generally occurs at a much higher rate than change due to point mutations.

Eukaryotic genes are divided into coding (exons) and noncoding regions (introns). Introns are transcribed into RNA, but not translated into proteins (Figure 2), which means that these sequences are not phenotypically expressed. Point mutations accumulate at a higher rate in noncoding regions, in part because of a diminished level of selection. Within coding regions, the second or third position of the codon is often "synonymous"; that is, the nucleotide in those positions can change without changing the encoded amino acid. For the nuclear genome of eukaryotes, the nucleotide substitution rate (per base pair, per year) in introns and synonymous sites ranges from about 1×10^{-9} to 10×10^{-9} with an average of about 5×10^{-9}. In exons the rate varies from about 0.004×10^{-9} in histone genes to 2.8×10^{-9} in interferon A. Natural selection (see Section 3.1) usually prevents major disruption of coding sequences with repetitive DNA regions, but these are a common feature of noncoding regions. The rate of DNA turnover in these repetitive sequences depends on which mechanisms are affecting change and various other factors. The rate of change due to these processes can be very fast; rates up to 10^{-2} have been documented.

3   DISSEMINATION OF GENETIC CHANGE IN NATURAL POPULATIONS

There are three principal mechanisms by which genetic change is spread through a population. These are natural selection, genetic drift, and molecular drive. The correct interpretation of genetic variation depends on how the DNA region under analysis is affected by each of these mechanisms. Although it is beyond the scope of this article to provide a detailed discussion, the basic tenets should be made clear.

3.1   NATURAL SELECTION

The biochemical, morphological, physiological, or behavioral consequences of genetic makeup (genotype), are referred to as the phenotype. Natural selection is the differential reproductive success of different phenotypes. Generally, if a region of DNA is not expressed

phenotypically (nor affected by a phenotypically expressed region), it is not exposed to natural selection. On the other hand, some genes are critical to survival or reproductive success, and therefore are under strong selective control.

Genetic variation at a locus can be maintained in a population by balancing selection or overdominance (when the heterozygote condition is selected for), or limited by directional selection (when one allele is favored at the expense of the other). Adaptation to specific habitats and the general character of the local environment can also affect genetic diversity. For example, habitat specialists might be expected to show lower levels of genetic diversity than habitat generalists, and species inhabiting relatively stable environments might be expected to show less variability. In each case the prediction is based on the idea that a more complex or variable environment will support a greater degree of genetic variation within the population as a whole.

3.2   GENETIC DRIFT

Genetic drift is a change in gene frequencies that is a consequence of the random gain and loss of individuals (and genotypes) in a population. According to the neutral theory of evolution, most evolution occurs through the random fixation of selectively neutral mutations. This theory predicts that change accumulating in this way will do so at an approximately constant rate (a molecular clock). When this is true, genetic change can be used to interpret the demographic histories of populations. Figure 3 plots the expected relationship between effective population size, genetic diversity, and the time (in generations) required for lost variability to be recovered.

Two stochastic mechanisms can reduce variation in a population: the so-called founder effect and a population bottleneck. In either case the result is a small population in which the effects of genetic drift are expected to reduce levels of variation. This happens both as a consequence of the sampling of genotypes and by inbreeding. Once variation has been lost, the process of accumulating new mutations is very slow. An equilibrium level of variation will take on the order of the inverse of the mutation rate to recover, which is in some cases longer than the expected life span of the species.

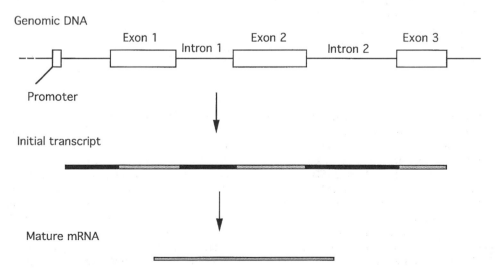

Figure 2. Division of genes into coding (exons) and noncoding (introns) regions. The initial transcript is enzymatically modified to represent only the exon regions of the gene in the mature messenger RNA.

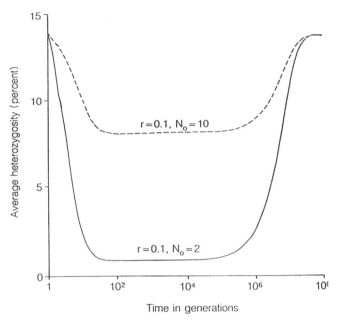

Figure 3. The theoretical changes in average heterozygosity when a population goes through a bottleneck. Logistic growth is assumed, and the intrinsic rate of growth *r* is set at 0.1. In one case (solid line), the bottleneck effective population size is $N_0 = 2$; in the other (dashed line) $N_0 = 10$. [After Nei (1987).]

3.3 MOLECULAR DRIVE

"Molecular drive" is a term used to describe the way genetic change can be spread through an array of a repeated DNA motif, and through individuals in a population, as a consequence of DNA turnover mechanisms. For example, if a mutation occurs in one of the repeats in an array, it can then be spread through the array by unequal crossing over (see Figure 1b). The continual gain and loss of repeats in the lifetime of an individual, when combined with spreading by sexual recombination, ensures that a new variant is either homogenized through the array and population, or lost. This process is directly analogous to the diffusion of alleles through a population by neutral drift (except when the turnover mechanism is biased in one direction or another). A number of molecular techniques currently in use for population studies utilize repetitive DNA regions. The correct interpretation of variation in these regions will depend to some extent on an understanding of the mechanisms involved in the generation of variation.

4 METHODS OF ANALYSIS

There are numerous methods by which genetic variation can be directly or indirectly measured. I will concentrate on those that have been fundamental to population genetic studies or show considerable promise for the near future. In general, and for a number of reasons, it is advisable to use several methods. First, different genetic markers may change at different rates and be affected by different evolutionary forces. A marker that evolves relatively slowly but is thought to be selectively neutral would allow a relatively unambiguous assessment of the phylogenetic relationship between distantly related taxa. On the other hand, genes that evolve quickly and are under strong selection (such as the genes in major histocompatibility class II) allow the typing of individuals, the relevance of

which can sometimes be interpreted. Hypervariable markers, such as minisatellite arrays (see Section 5.5) can be used to assess paternity and close kinship relationships. Markers that show an intermediate level of variation, such as some microsatellite loci, mitochondrial DNA, and allozymes, are the most suited to the comparison of populations. This is because too much variation will obscure genetic differences (saturation will lead variation within a population to be as great as that between populations), and too little variation will not permit sufficient resolution.

Another factor involves the character of the genetic marker. For example, repetitive DNA regions may be affected by DNA turnover in such a way that variants are homogenized through an array and through the population (molecular drive). This can bias the frequency of variants in a population, with the result that breeding populations are more distinct than is expected by random drift. This is most likely to be relevant for investigations of variation in a gene family (a tandem array of coding genes, such as the rDNA gene family, which codes for 18*S* and 28*S* ribisomal RNAs). Single base pair changes and repeat number changes (within the introns of the genes) can be spread through the array and through the population, depending on the relative rates of mutation and turnover.

4.1 SAMPLE COLLECTION

The first step in any project is to define the objectives. For example, it may be important to determine both the phylogeographic structure of populations and the level of variation within each population. Therefore the choice of genetic markers and sampling protocol will have to satisfy the requirements of each test. The choice of markers for these two applications should overlap; but an emphasis on uniparentally inherited markers (such as mitochondrial or chloroplast DNA; see Sections 5.3 and 5.4) can facilitate the comparison of variation between populations, and hypervariable markers (such as minisatellites and microsatellites; see Sections 5.5 and 5.6) can increase resolution to determine the degree of inbreeding or drift in small local populations. In general it is best to use a large number of markers. The reliability of summary statistics used in the comparison of populations (such as genetic distance and G_{ST}) or levels of variation within populations (such as heterozygosity) depends more on the number of different markers used than on the number of individuals sampled. The analysis of a large number of markers is also useful to minimize the impact of locus-specific effects, such as selection.

A pilot study should usually be designed to test the applicability of the sampling regime and the suitability of various markers. A small sample from nearby and distantly separated populations will provide a sense of the range of variation for measures of genetic distance and diversity. In this case markers that provide sufficient resolution are sought. Too little variation will leave real differences undetected, and too much variation can obscure differences by saturation (when there is as much variation within as between populations) and complicate interpretation (if mutation occurs frequently between a finite set of genotypes). An inappropriate level of variation is most likely to be a problem with repetitive DNA markers. When measuring variation within a population, markers should be chosen to include a range of expected variation, and the same set should be used on each population being compared for internal variation. A pilot study is not necessary because all markers will be informative, provided something is known about variation in the marker for a comparable case, or between populations of the same species.

The sample size needed to test genetic distance and differentiation will depend on the statistical resolution required, which depends on the genetic difference between populations, the variability of the markers, and the number of markers used. There are two types of error in the relevant statistical comparisons. A type I error occurs when a null hypothesis is rejected that should be accepted. In biological studies the allowance for this type of error is often set at 5%, though a more stringent criterion should be adopted if a large number of comparisons are done. A type II error occurs when a null hypothesis is accepted that should be rejected. This is usually expressed as the power of the test, which is 1 minus the type II error. For a comparison between two populations using just one marker, with a difference between allele frequencies of 5%, thousands of individuals would need be screened to statistically distinguish between them, given a type I error of 5% and a power of 90%. The solution is to use more markers with an emphasis on those with a high level of variability.

The collection of samples will depend to some extent on the biology of the study species. In mobile, sexually reproducing species, the dispersal behavior of males and females can affect the distribution of genotypes. For example, if females are phylopatric (reproducing near where they were born), genetic variation between adult males and females in local samples may vary. Furthermore, in this case the comparison of populations using a matrilineally inherited marker (such as mtDNA in mammals) will exaggerate the difference between populations. It is also important to take into account temporal and spatial patterns of movement in mobile species. A project that begins sampling in one location and finishes in another sometime later might simply follow the migration of a single reproductive population!

4.2   ENZYME ELECTROPHORESIS

The migration of proteins through an electric field in a gel matrix can be measured by enzyme electrophoresis. The charge properties of the protein determine the extent and direction of migration. Tissue homogenized in buffer can be used as the source material, and specific enzymes can then be visualized by immersing the gel into a medium containing the reaction conditions for that enzyme such that a colored dye is incorporated into the end product. The dye produces bands on the gel indicating the position of the enzyme. The number of bands will depend on the number of subunits in the enzyme, and on whether the locus is homozygous or heterozygous. Gels can be prepared from a variety of media (starch, polyacrylamide, agar, etc.) and run by a variety of methods (horizontal, vertical, disc, etc.). For screening large numbers of individuals in a population survey, the most common method (also the simplest and least expensive) is horizontal starch gel electrophoresis. The apparatus for this method is schematized in Figure 4. Gels running 10–30 samples can be cut horizontally two times (giving three slices) and each slice stained for a different enzyme. In most cases the resolution of bands is sufficient, and poor resolution can often be improved by using a different gel medium or running method. In general, the different methods are designed to improve resolution and to permit the separation of more subtle differences between alleles.

4.3   DNA EXTRACTION

DNA is extracted from tissue by removing proteins, fats, and carbohydrates and then precipitating the DNA out of an aqueous salt

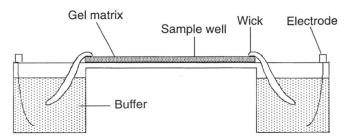

Figure 4.   Nondenaturing, horizontal gel electrophoresis of proteins.

solution. This procedure is usually straightforward for animal tissues, but it can be quite difficult with plant material because naturally occurring chemicals degrade the DNA or inhibit further analysis. The first step is to homogenize the tissue under an aqueous-based buffer that promotes cell lysis. By the most common method (Figure 5), the homogenate is mixed with phenol and chloroform, which are immiscible with water. The proteins and so on go into the organic (phenol) phase, while the DNA remains in the aqueous phase. After several repetitions, the DNA is precipitated with salt and ethanol. A careful preparation of DNA will include sheared fragments 20–100 kb long. Excessive agitation further shears (degrades) the DNA. Since nucleases (which can be present on fingers and particulate matter in the air) will also degrade DNA, the procedure must be done under sterile conditions.

4.4   RFLP ANALYSIS

Restriction fragment length polymorphisms (RFLPs) can be analyzed when DNA is digested with a special enzyme called a restriction endonuclease (Figure 6). These enzymes recognize specific short sequences of DNA (e.g., the enzyme EcoRI recognizes the sequence GAATTC) and cleave the DNA at every instance of that sequence. Mutations can occur that create new sites or eliminate old ones. The next step is to separate the DNA fragments by size in a gel matrix. This is done by electrophoresing the samples through an agarose or polyacrylamide gel under buffer. DNA naturally assorts by size (on a roughly logarithmic scale) by this method. The double-stranded DNA in the gel is then made single-stranded (usually with an alkaline solution) and transferred onto a solid membrane that binds DNA (Figure 7). Then a segment of DNA to be compared between individuals (e.g., the Adh gene or the mitochondrial genome) is labeled with a radioisotope (or some nonradioactive label) and hybridized (whereby complementary strands of DNA combine and re-form a double-stranded molecule) to the DNA bound to the membrane. Only the fragments of bound DNA that match this DNA probe will be made radioactive. After exposure to X-ray film, these relevant fragments can be visualized as black (exposed) bands on the film (Figure 8).

4.5   DNA SEQUENCING

The procedure that determines the sequence of nucleic acid bases comprising a segment of DNA offers the greatest resolution for comparisons of genetic differentiation and permits the greatest scope for interpretation. There are several methods of DNA sequencing, but the one used most frequently is chain termination. By the chain termination method, a single-stranded DNA template is di-

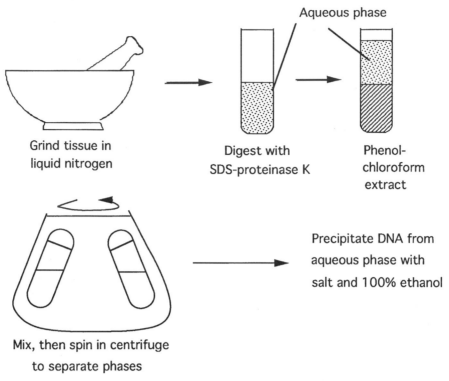

Figure 5. Phenol–chloroform extraction of DNA from solid tissue.

vided into four reaction mixes, one for each nucleotide base (A, T, C, G). A DNA polymerase (an enzyme that can synthesize DNA along a single-stranded template, starting at a double-stranded/single-stranded interphase), a short (15–25 bases) primer segment of DNA, and all four deoxynucleotide bases are included in each reaction (one of which is labeled either radioactively or with a fluorescent dye). In addition, each reaction will include a smaller quantity of one of the four dideoxynucleotides (A, T, C, or G). These bases lack the 3′-OH group (Figure 6) necessary for DNA chain elongation and

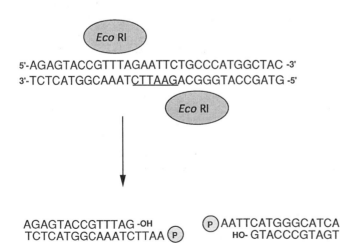

Figure 6. Restriction digestion with the enzyme *Eco*RI. Digestion leaves a 3′ hydroxyl group and 5′ phosphate group exposed, and either a 5′ overhang (shown), a 3′ overhang, or blunt ends, depending on the enzyme. *Eco*RI has a 6 bp recognition site. Many other common enzymes have 4 bp recognition sites.

therefore terminate the reaction at every occurrence of that base. A comparison of the length of strands synthesized for each of the four reactions gives the sequence of bases (Figure 9).

4.6   PCR ANALYSIS

The polymerase chain reaction (PCR) permits the enzymatic amplification of a specific segment of DNA. This technique has greatly facilitated population level screening of genetic variation by saving time and reducing expense. Two short primers of single-stranded DNA (usually 15–25 bases long) are designed flanking a region of interest. These are often based on published DNA sequences of particular genes or developed from cloned DNA. A programmable heating block is then used to vary the reaction conditions. First the reactions are heated to denature the template DNA (to separate the double helix into its two complementary strands); then the temperature is lowered to allow the primers to anneal to the template, the reaction is heated to an optimal temperature for the polymerase, and the cycle is repeated 20–40 times (Figure 10). The key to this technique is the use of a thermostable polymerase (most polymerases would be denatured and rendered inactive by the high temperatures necessary for the DNA denaturation step).

4.7   SSCP AND GRADIENT GEL ANALYSIS

Single-strand conformation polymorphisms (SSCPs), are based on the sequence-dependent mobilities of denatured (single-stranded) DNA when run through a nondenaturing gel. PCR-amplified DNA, usually 100–200 bp long, is denatured and loaded onto a gel in nondenaturing conditions. As the molecule migrates through the gel, it reanneals with itself, folding into a hairpin structure. The conformation and mobility of this structure depend on the sequence

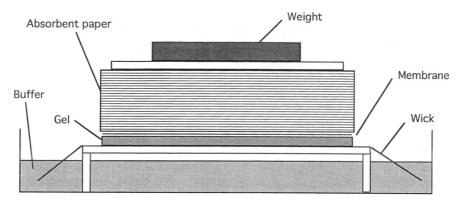

Figure 7. Southern blotting to transfer DNA from gel to membrane.

of DNA residues. As little as a single base change in 200 can alter the distance a fragment will migrate in the gel, although there is no simple correspondence between sequence composition and mobility. Therefore, differences can be detected quickly (permitting the screening of a large number of samples simultaneously), but they must be quantified by sequencing all products with distinct mobilities.

Gradient gel electrophoresis, an alternative method for detecting single base changes, is based on the melting (denaturing) properties of DNA in solution. DNA melts in domains of two dozen to several hundred base pairs. Sequence changes within a domain are characterized by mobility changes. As the double-stranded molecule passes through the denaturing gradient, it becomes increasingly denatured, and its mobility reflects the sequence within each melting domain. Changes within the last domain would not normally be detected because sequence-dependent migration is lost upon completion of strand dissociation. To correct for this degradation, a 30–40 bp GC-rich sequence (a "GC-clamp") is added to the 5′ end of the PCR primers used to amplify the sequence being analyzed.

4.8 RAPD ANALYSIS

Random amplified polymorphic DNA (RAPD) analysis uses just one short oligonucleotide (usually 10–12 bp) in a PCR reaction to amplify random segments of DNA. It works because a small num-

ber of fragments (5–10) will be amplified when the oligonucleotide anneals by chance on each template strand over a size range that can be amplified by PCR (usually <3 kb). A 10-mer, for example, will anneal by chance once every million base pairs (about a thousand times in the human genome). There are two ways that the amplified fragments can show polymorphisms. First, the amplified region may be repetitive and show length variation in the population (i.e., there will be a variable number of tandem repeats: VNTR). In this case the relevant band will show a size variation, although very large alleles may not amplify (PCR preferentially amplifies smaller fragments). Second, a single base pair polymorphism may interfere with the annealing of the primer in some individuals. In this case, the band will be present or absent, but if present, it is not expected to vary in size.

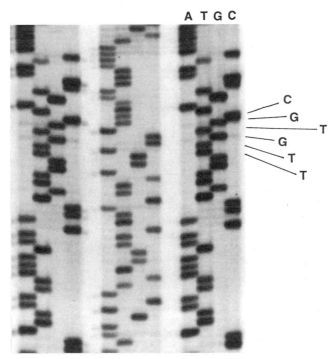

Figure 9. A small portion of the nucleic acid sequence that can be "read" from a gel subjected to the chain termination method. The four reactions (one for each nucleic acid residue) are run side by side on a 6% polyacrylamide gel.

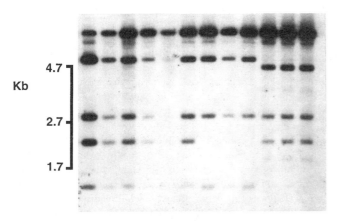

Figure 8. Panel of 12 lions digested with *Ava* II and hybridized with a heterologous mtDNA probe. (Courtesy of Robyn Hottman and Steve O'Brien.)

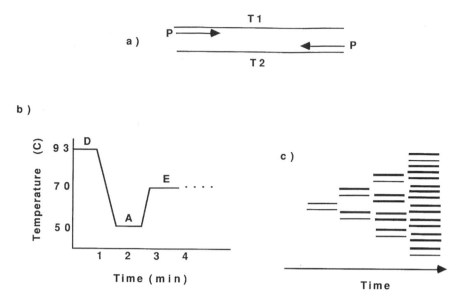

Figure 10.   The polymerase chain reaction (PCR). (a) Short oligonucleotide "primers" anneal to the denatured template DNA. (b) The reaction is cycled between denaturing, annealing, and extension temperatures. (c) Repeated cycling results in an exponential amplification of the target sequence.

5   CHOICE OF GENETIC MARKER AND INTERPRETATION

The nuclear genome is a mosaic of DNA regions under the influence of different mechanisms affecting change, and conserved by selection to different degrees. In addition, there are the mitochondrial genomes and, in plants, the chloroplast genomes, with their own unique properties. A population geneticist can use this variation in the mode and rate of evolution to advantage, but it also means that results must be interpreted cautiously. I will describe six types of genetic marker, each with different properties and in some cases, best suited to particular levels of analysis.

5.1   ENZYMES

Enzymes are a functional part of cell physiology, and therefore they are exposed to natural selection. The apparent rate of evolution in these markers varies greatly, from about 10^{-9} to 10^{-6} change per gene (given an average length of 1000 bp) per year. In some cases—for example, in the histone genes—there is clearly strong selection, which complicates interpretation. However, as long as the evolutionary rate is relatively constant over time, these markers are useful for phylogenetic comparisons. As described in Section 4.2, the analytical procedure (enzyme electrophoresis) is relatively fast and inexpensive. The level of resolution possible is relatively low, but this can be an advantage when the time of divergence between taxa under comparison is great.

The data generated by this method are for the most part Mendelian, and the patterns resulting from segregation at a single locus can usually be distinguished. This means that data can be interpreted with respect to theoretical predictions about the frequency of genotypes in a population, based on the frequency of alleles. Deviations from the expected frequency (as predicted by Hardy-Weinberg) can be quantified statistically and interpreted. Furthermore, knowing the function of the encoded protein can help with the interpretation of patterns of variation and deviations from expected frequencies. For example, an excess of heterozygotes could indicate selection for heterozygotic individuals (heterosis), while an excess of homozygotes could indicate inbreeding, drift, or directional selection. Knowledge of the gene product could help determine the likely factor. When the data are consistent with Hardy-Weinberg expectations, formulations to interpret genetic distance and relative levels of variation can be applied. This can also be accomplished using some of the other markers described here. By far the largest existing database, however, describes allozyme variation in natural populations.

5.2   NUCLEAR GENES

Sequence data of nuclear genes, including those encoding enzymes, offer an immediate advantage over enzyme electrophoresis. This is because coding and noncoding regions can be differentiated, and within the coding region of the gene, synonymous and nonsynonymous sites can be distinguished. Thus regions that are likely to be under the influence of natural selection can be separated from those that are more likely to be evolving at a neutral rate. It also becomes possible to compare sequences and to determine the mode of mutation (point mutation, deletion, insertion, rearrangement, etc.), and in some cases the mechanism. All these factors contribute to the accurate interpretation of the observed variation.

For comparisons between conspecific populations, it is necessary to identify both alleles for each individual, and this is typically more difficult in DNA sequencing than for enzyme electrophoresis because of the need to sequence a number of clones from each individual to ensure that both alleles have been sampled. One relatively easy way around such detailed sequencing is to screen the population using either the SSCP or gradient gel methods. These methods identify sequences that are distinct by as little as one base pair. In most cases a single locus will show two bands when heterozygous and one when homozygous (although sometimes there are repeatable artifact bands related to additional stable conformations of the single strands). Combined with DNA sequencing to distinguish the different phenotypes indicated by SSCP or the gradient gel, this approach provides an efficient and very powerful protocol for the screening of variation at the population level. It shares most of the advantages of allozyme electrophoresis (apart from considerations of difficulty and expense) and provides considerably greater resolution.

Variation in nuclear genes can also be quantified using the RFLP method, which permits the rapid screening of a large number of samples. RFLP is limited, however, in that the precise character of the variable restriction sites is not known (unless the region has been sequenced and the sites can be mapped), and the allelic relationship at a locus cannot always be readily determined. Another consideration is the effect of methylation, to which some restriction enzymes are sensitive. Methylated sites are not cut by sensitive enzymes. This property originally evolved in bacteria as a defense against restriction enzymes produced by bacteriophage. Especially in the nuclear genome, some sites may be methylated, potentially creating within or between populations an RFLP that does not reflect a real sequence difference.

5.3   MITOCHONDRIAL DNA

In multicellular animals, the mitochondrial genome is a double-stranded, closed-circular molecule ranging in length from 15.7 to 19.5 kb. It evolves at a higher rate than comparable nuclear genomic regions (5–10 times faster: Figure 11) and in most cases shows strict maternal transmission (no paternal contribution to the F1 generation). These two factors have made mitochondrial DNA (mtDNA) an attractive marker for the genetic comparison of animal populations. Higher levels of variation permit greater resolution, and the lack of sexual recombination facilitates interpretation. In plants the transmission is matrilineal, but the level of variation is generally very low.

Variation across the mitochondrial genome is heterogeneous, as a result of varying levels of sequence conservation, and to some extent, DNA turnover. A common and very useful method of analysis is to sequence mtDNA regions that have been amplified by PCR. Small regions (300–400 bp) can be amplified and sequenced directly. Sequencing each strand is usually sufficient to read through a sequence of this length. This method is useful in that it permits a focused analysis of variation in a specific region. A phylogenetic comparison of different species can give a sense of the pattern of variation for that region. The most common method, however, has been RFLP analysis. This is done either by using an mtDNA probe to screen whole-cell DNA digests or by isolating mtDNA and visualizing restriction fragments by radioactive labeling or silver staining. The former method is relatively fast and easy (because of the high copy number of mitochondrial genomes). The latter method can offer considerably more resolution, however, because smaller restriction fragments can be resolved when they are visualized in the gel, compared to probing DNA that has been transferred to a membrane (see Section 4.3).

Demographic considerations are important to the interpretation of mtDNA variation for two main reasons. First, mtDNA is more labile to extreme fluctuations in population size than nuclear DNA, since a single mating pair will pass on four copies of the nuclear genome, but only one copy of the mitochondrial genome. Also, the local extinction of matrilines can limit the diversity of mtDNA, especially in small populations. Second, the geographic distribution of mitochondrial haplotypes will depend on the dispersal behavior of females. If females are less likely to disperse (as is commonly the case for mammals), then mtDNA variation may show greater geographic structuring than nuclear DNA variation. Interpretation can also be complicated by the existence of more than one form of the

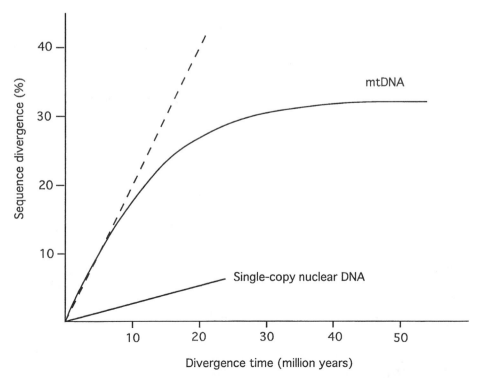

Figure 11.   Rates of nucleotide substitution in single-copy nuclear DNA and mtDNA, plotted with the theoretical rate without saturation effects in mtDNA (dashed line). [After Brown et al. (1979) in Hoezel and Dover (1991).]

mitochondrial genome in an individual, but this condition (hetero-plasmy) is relatively rare.

5.4 Chloroplast DNA

Sequence variation in the chloroplast genome (cpDNA) is highly conserved, as is variation in the size of the genome for most taxa, though the range for all plant species is large (120–217 kb). Inheritance can be biparental, strictly maternal or strictly paternal, depending on the species. Most population studies have found very low levels of cpDNA variation, though there are exceptions. The utility of this genome as a population genetic marker will depend on the mode of transmission and level of variation for a given taxa. Strict paternal or maternal transmission may provide useful information even if the relative level of variation is low.

5.5 Minisatellite DNA

There is an abundance of repetitive "satellite" DNA in the nuclear genome of most complex organisms. These regions are classified according to their composition and structure. Minisatellite arrays are nested repeats of a highly conserved core sequence, 15–40 bp long, interspersed on most chromosomes. A minisatellite array can be up to 20–25 kb long. A high rate of unequal crossing over generates a high level of length variation in these arrays. The "DNA fingerprinting" method takes advantage of the conserved minisatellite core sequence and the large number of loci sharing that core repeat. A probe for the core sequence is used in a multilocus RFLP analysis, which reveals a ladder of hypervariable bands that can identify an individual or determine paternity with a very small chance of error (Figure 12). While DNA fingerprinting is a powerful tool for the assessment of close kinship, the interpretation of variation in this marker at the population level is not straightforward. One problem is that variation saturates, such that variation within a population is often as great as variation between populations. This marker is typically more useful to a population geneticist as a tool to reveal reproductive strategy, which in turn has an important effect on the genetic structure of a population.

Exceptions to this general rule exist, especially when genetic variation for a population or species has been reduced. This can occur by genetic drift in small populations or by inbreeding. In this case variation for the multilocus pattern must be statistically treated as a phenotype, since the allelic relationship between bands in the fingerprint cannot in most cases be determined. Typically a relative measure of "band-sharing" is used as an index of genetic relatedness. An alternative is to clone an individual locus and use that DNA as a probe. Single-locus minisatellite length variation can be treated as a Mendelian character, although interpretation of levels and patterns of variation should be tempered by the understanding that various mechanisms (e.g., unequal crossing over and slippage) may be promoting change at different rates. It should not be assumed that these rates are constant over the relevant time frames for population level comparisons.

5.6 Microsatellite DNA

Microsatellites are small arrays (typically <100 bp) of simple di- and trinucleotide repeats (longer repeat elements are less common). These arrays vary in length over time because of DNA slip-

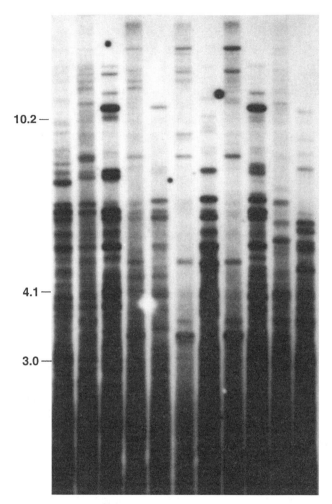

Figure 12.   Minisatellite DNA variation (DNA fingerprints). DNA from 11 southern elephant seals was digested with *Hae* III and probed with (CAC)$_n$. Fragment sizes shown in kilobases.

page. Some loci are hypervariable, but most are less variable than minisatellite loci. They are visualized by designing oligonucleotide primers on either side of the array and amplifying by PCR (see Section 4.6). A radioactive nucleotide will be incorporated during the amplification. Alternatively, one of the primers is radioactively labeled. The amplification products are then run on a long polyacrylamide gel (to allow single base pair resolution), and each individual should be represented by one (homozygote) or two (heterozygote) bands (Figure 13). Polymerase slippage during amplification sometimes creates artifact bands.

In principle, variation at these loci can be interpreted in the same way as for multiple alleles at an enzyme locus. However, there are several problems. First, little is known about the constancy of DNA slippage rates over time or between loci, and what is known suggests that the rate can vary considerably. This lack of uniformity will compromise interpretation if the rate is changing over a shorter period than that separating the populations under comparison. Second, the high level of variation at these loci means that variation in individual reproductive success can have a large effect on overall genotype frequencies, meaning that frequencies could change significantly

(In the figure, the fragment size labels shown are: 10.2 —, 4.1 —, 3.0 —)

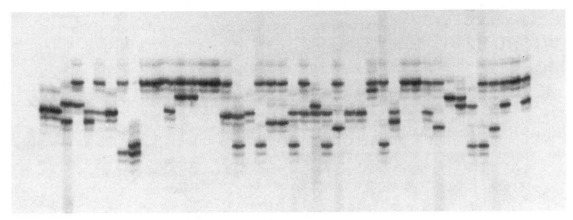

Figure 13. Microsatellite DNA variation for 45 southern elephant seals.

from one generation to the next in the same population, especially in highly polygenous species. Third, this type of variation violates assumptions about the novelty of new mutations. In the "infinite allele" model, all new mutations are considered to be unique. Since, however, DNA slippage can increase or decrease the number of repeats in an array, mutations could easily switch back and forth between a finite number of states. Therefore, these markers are best applied at the population level only in conjunction with other genetic markers and data on behavior and demographics for the subject species.

5.7 ANALYSIS OF VARIATION

A mathematical description of the statistical methods for assessing and comparing levels of variation within and between populations is beyond the scope of this article. However, I will briefly describe a few of the general assumptions on which most of these analyses are based. For genetic changes that are selectively neutral, the accumulation of change can be interpreted to estimate time-dependent differentiation within and between populations. For the analysis of DNA, a statistical interpretation is generally based on the following assumptions: (1) that nucleotides are distributed randomly in the genome, (2) that variation arises solely by base substitution, (3) that substitution rates are the same for all nucleotides, and (4) that all relevant bands on a gel can be detected, and similar bands are not scored as identical.

The fourth condition can be controlled by careful methodology. The first three conditions do not strictly hold, but Nei (1987) has argued that small deviations will not significantly alter the results. Also, since there are only four bases in the genetic code, there is a certain probability that mutations will revert (e.g., from A to T then back to A). Formulations have been derived to compensate for this effect.

Allozyme variation is measured phenotypically (as the electrophoretic mobility of proteins). Therefore it is especially important to establish that these characters follow the Mendelian rules of segregation during sexual recombination. This is usually done by conducting pedigrees or, when this is not practical, by testing for correspondence to the Hardy-Weinberg rule (which states that the gene and genotype frequencies will remain constant between generations in an infinitely large, random mating population). Deviations from the expected Hardy-Weinberg proportions of alleles could also indicate the action of selection.

6   CONCLUSIONS

In general, being able to choose between genetic markers that differ in rates and modes of change has greatly facilitated the interpretation of variation at the population level. However, data still need to be interpreted cautiously, and the best approach will be to use several complementary genetic markers. Different markers offer both different levels of resolution and different characteristics. For example, mtDNA traces matrilineal lineages while cpDNA in some plant species and Y-chromosome-specific markers trace paternal lineages. The comparison of relatives within a social group requires a high level of resolution. Hypervariable minisatellite and microsatellite loci are well suited to this application. In the comparison of populations of a species occupying geographically separate ranges, mtDNA variation investigated together with a nuclear genetic marker can indicate both the degree of separation and patterns of dispersal.

See also DNA DAMAGE AND REPAIR; GENETIC DIVERSITY IN MICROORGANISMS; MITOCHONDRIAL DNA, EVOLUTION OF HUMAN; POPULATION-SPECIFIC GENETIC MARKERS AND DISEASE; TANDEMLY REPEATED NON-CODING DNA SEQUENCES.

Bibliography

Hames, B. D., and Rickwood, D. (1990) *Gel Electrophoresis of Proteins, A Practical Approach.* Oxford University Press, Oxford.

Hillis, D. M., and Moritz, C., Eds. (1990) *Molecular Systematics.* Sinauer Associates, Sunderland, MA.

Hoelzel, A. R., Ed. (1992) *Molecular Genetic Analysis of Populations, A Practical Approach.* Oxford University Press, Oxford.

———, and Dover, G. A. (1991) *Molecular Genetic Ecology.* Oxford University Press, Oxford.

Maynard Smith, J. (1989) *Evolutionary Genetics.* Oxford University Press, Oxford.

Nei, M. (1987) *Molecular Evolutionary Genetics.* Columbia University Press, New York.

Genetic Code Expansion and Expression: *see* Protein Expression by Expansion of the Genetic Code.

Genetic Diversity in Microorganisms

Werner Arber

Key Words

Biological Evolution   A nondirected, dynamic process of diversification resulting from the steady interplay between spontaneous mutagenesis and natural selection.

DNA Rearrangement   Results from mostly enzyme-mediated recombination processes, which can be intra- or intermolecular.

Gene Acquisition   Results from horizontal transfer of genetic information from a donor cell to a receptor cell in transformation, conjugation, or phage-mediated transduction.

Spontaneous Mutation   Defined here as any alteration of nucleotide sequences occurring to DNA without the intended intervention of an investigator.

Transposition   DNA rearrangement mediated by a mobile genetic element such as an insertion sequence element or a transposon.

Variation Generator   An enzyme or enzyme system whose involvement in the generation of genetic variation has been documented.

Bacterial genetics took its start about 50 years ago and strongly influenced the development of the molecular genetic strategies and techniques now available to study gene structure and functions of living organisms. Bacterial genetics also revealed natural processes of horizontal gene transfer (transformation, conjugation, phage-mediated transduction) as well as systems (e.g., restriction–modification systems) for holding such gene transfer in tolerably low frequencies to ensure a certain degree of genetic stability. Work with

bacterial and bacteriophage systems has also helped to unravel both homologous and nonhomologous enzyme-mediated recombination processes at the molecular level. The acquired knowledge now promotes an understanding of the molecular processes that contribute to the generation of genetic variation. Among these processes, special attention is given to DNA rearrangements resulting from transposition and from site-specific recombination, which sometimes occurs at secondary crossover sites.

Present knowledge on the genetic plasticity of haploid microorganisms offers insights into the molecular basis for the natural interplay between mutagenesis and selection. This approach greatly profits from the short generation times and relatively small genome sizes of haploid microorganisms, which allow one to investigate population genetic questions and to draw conclusions on the mechanisms of evolutionary processes. A general conclusion relates to the existence in the microbial genome of genes with specific evolutionary functions involved in the generation and the limitation of genetic variation.

1 INTRODUCTION

Microbial genetics is at the root of molecular genetics and gene technology, whose strategies and methods have become applicable to genetic studies with any kind of living organism. This new advantage has opened interesting possibilities for the investigation and better comprehension of the mechanisms of interactions between biological macromolecules supporting life processes. Microbial genetics also offers deeper insights into the process of biological evolution. This contribution concentrates on molecular mechanisms involved in microbial evolution, mainly at the level of spontaneous mutagenesis. The insight obtained improves our knowledge on the generation of genetic variation. Instances of genetic diversity can be seen as static pictures of the dynamic process known as biological evolution. At least some of the principles identified in studies with microbial populations are likely to be of general relevance for biological evolution of all forms of life.

2 IMPORTANT ROOTS OF MOLECULAR GENETICS

Five sets of discoveries, largely based on work carried out with microorganisms between 1943 and 1953, the decade preceding the publication of the double-helix model for DNA, were essential for the later development of molecular genetics. These are listed briefly as follows.

1. It was realized that bacteria and bacteriophages have genes that can mutate, and that spontaneous mutations normally arise independently of the presence of selective agents. It was also learned that the genetic information of bacteria and of some bacteriophages is carried in DNA molecules rather than in other biological macromolecules such as proteins.
2. The newly discovered phenomena of DNA transformation, bacterial conjugation, and bacteriophage-mediated transduction demonstrated natural means of horizontal gene transfer between different bacterial cells.
3. It was seen that horizontal gene transfer has natural limits, including systems of host-controlled modification, today known as DNA restriction–modification systems.
4. Mobile genetic elements were identified as sources of genetic instability and were seen to represent mediators of genetic re-

arrangements. While such rearrangements are often caused by transposition, they can also result from the integration of a bacteriophage genome into the genome of its bacterial host strain, which is thereby rendered lysogenic.

5. Structural analysis of DNA molecules led to the double-helix model, offering an understanding of semiconservative DNA replication at the molecular level and thus of information transfer into progeny.

3   EASE OF MOLECULAR AND POPULATION GENETIC INVESTIGATIONS WITH BACTERIA

Many classical microbial genetic investigations were carried out with *Escherichia coli* K-12. Its genome is a single circular DNA molecule (chromosome) of about 4.7×10^6 base pairs (bp). This microorganism now has an extremely well-established genetics with about 1500 identified genes, which represent, according to estimates, about half of all genes present. The haploid nature of *E. coli* brings about a rapid phenotypic manifestation of mutations. In periods of growth the rate of spontaneous mutagenesis is about 10^{-9} per bp and per generation. This represents one new mutation in every few hundred cells in each generation. *E. coli* has several well-studied bacteriophages and plasmids. This material facilitates investigations on life processes in these bacteria.

Under good growth conditions, the generation time of *E. coli,* measured between one cell division and the next, is very short, on the order of 30 minutes. Upon exponential growth, this leads to a multiplication factor of 1000 every 5 hours. Thus, a population of 10^9 cells representing 30 generations is reached from an inoculum of a single cell in only 15 hours. This rapid growth rate greatly facilitates population genetic studies and thus investigations on the evolutionary process.

4   MECHANISMS AND EFFECTS OF SPONTANEOUS ALTERATIONS OF GENOMIC DNA SEQUENCES

On the filiform DNA molecules of *E. coli* and its bacteriophages and plasmids, the genetic information is stored as linear sequences of nucleotides or base pairs. Genes depend on the presence of continuous sequences of base pairs (reading frames) that encode specific gene products, usually proteins, and of expression control signals that ensure the occurrence of gene expression at the relevant time with the needed efficiency. Mutations can affect reading frames as well as control signals, both of which represent specific DNA sequences. In addition, some specific DNA sequences relate to the control of the metabolism of the DNA molecules, in particular their replication.

For simplicity, we will use the term "spontaneous mutation" to label any type of alternation of a DNA sequence unintended by the investigator. This definition says nothing about whether a mutation relates to a phenotypic change. It is well known that only a relatively small fraction of mutations can provide an advantage to the organism and thus can be considered to be useful. Those useful mutations become "selected" because they allow the organisms that carry them to succeed by eventually outgrowing other members of a given microbial population. More often, mutations either provide a selective disadvantage or are lethal. (In the longer term, such mutations are eliminated from populations by the selection pressure.)

Finally, some mutations are neutral or silent. They do not immediately affect the life of the organism and are inert to selection, at least as long as living conditions are constant. Because of the relatively frequent occurrence of lethal mutations and mutations providing selective disadvantage, a tolerable mutation rate of any haploid organism should be lower than one mutation per genome and per generation. As we have already seen, this criterion is fulfilled by *E. coli.*

Molecular genetic studies have revealed that overall spontaneous mutagenesis is brought about by a large number of specific molecular mechanisms. Studies on spontaneous mutagenesis are difficult because not all these mechanisms act with comparable efficiencies and these efficiencies may depend on environmental conditions such as temperature. Considerable knowledge has been accumulated, however, and on the basis of available data, we group the mechanisms of spontaneous mutagenesis (i.e., alterations of DNA sequences occurring without intervention of an investigator) into four categories. These are defined in Sections 4.1 through 4.4.

4.1   INFIDELITY OF DNA REPLICATION

The process of DNA replication is one of the important sources of genetic variation, which may depend both on structural features of the substrate DNA and on functional characteristics of the replication fork. Some of the infidelities of DNA replication are likely to depend on tautomeric forms of nucleotides: that is, a structural flexibility inherent to these organic compounds. Base pairing depends on specific structural forms and can result in a long-term mispairing if short-living, unstable tautomeric forms are "correctly" used in the synthesis upon DNA replication. For this reason, we consider mutations resulting in this process not as errors but rather as infidelities. This process is a classical source of nucleotide substitution, and it has an important role in the long-term development of new biological functions.

Other replication infidelities that may relate to slippage in the replication fork can result in either the deletion or the insertion of one or a few nucleotides in the newly synthesized DNA strand. If such mutations occur within reading frames for protein synthesis, the phenotypic effect may be drastic, since in the protein synthesized from a gene affected by such a frameshift mutation, the amino acid sequence downstream of the site of mutation can strongly differ from that of the nonmutated product. In addition, the size of such a mutated protein is usually altered depending on the chance occurrence of an appropriate stop codon in the new reading frame.

Proofreading devices and other enzymatic repair systems prevent replication infidelities from producing mutations at intolerably high rates. Generally, they act rapidly after replication by screening for imperfect base pairing in the double helix. Successful repair thereby requires specific means to distinguish the newly replicated DNA strand from its template, the complementary parental strand. Because of these correction activities, a vast majority of primary mispairings are removed before they have the opportunity to become fixed as mutations.

Genetic information of some viruses, but sometimes also segments of genetic information of chromosomal origin, may pass through RNA molecules, which may later become retrotranscribed into DNA. No efficient repair systems are known to act at the level of RNA. Therefore RNA replication shows a higher degree of infidelity than DNA replication. In consequence, genetic information that becomes replicated as RNA molecules generally suffers increased mutation rates.

4.2   Internal and Environmental Mutagens

A relatively large number of internal and environmental agents exert mutagenic effects by means of molecular mechanisms that in many cases are well understood. Internal mutagenesis can originate within the building blocks of DNA themselves, the nucleotides, which are not fully stable. They may, for example, suffer desamination, in which case specific base pairing may be affected. On the other hand, some intermediate products of the normal metabolic activities of a cell may be mutagenic and may thus contribute to spontaneous mutagenesis. The mutagens of the environment include a multitude of chemical compounds, but also ultraviolet and ionizing radiations as well as some physicochemical constraints such as elevated temperature. Each of these mutagens and mutagenic conditions contributes in a specific way to the generation of genetic variation.

Again, some of the sequence alterations brought about by internal and environmental mutagens are efficiently repaired by enzymatic systems. Since, however, the efficiency of such repair is rarely 100%, evolutionarily relevant mutations persist. Some of the repair processes depend on genetic recombination systems, which by themselves can also contribute to spontaneous mutagenesis.

4.3   DNA Rearrangements

Indeed, various recombination processes are well known to mediate DNA rearrangements, which often result in new nucleotide sequences. While in haploid organisms general recombination is not essential for propagation, it influences genetic stability at the population level in various ways as a generator of new sequence varieties. For example, it can bring about sequence duplications and deletions by acting at segments of homology that are carried at different locations in a genome.

Two other, widely spread types of recombination systems are dealt with separately in Sections 5.1 and 5.2: site-specific recombination and transposition. Both are known to contribute to genetic variation. Still other recombination processes, such as the one mediated by DNA gyrase, can perhaps best be grouped as illegitimate recombinations. This group may contain several different molecular mechanisms that act with low efficiency and have remained at least in part unexplained.

4.4   DNA Acquisition

While the mutagenesis mechanisms belonging to the three categories explained in Sections 4.1 through 4.3 are exerted within the microbial genome and can affect any part of the genome, an additional category of spontaneous sequence alterations depends on an external source of genetic information. In DNA acquisition, genetic information indeed originates from an organism other than the one undergoing mutagenesis. DNA acquisition can occur by means of transformation, conjugation, or virus-mediated transduction. In the latter two strategies a plasmid or a viral genome, respectively, acts as natural gene vector.

DNA acquisition represents a particularly interesting source of new genetic information for the receptor bacterium because the chance that the acquired DNA exerts useful biological functions is

quite high—most likely, it has already assumed the same functions in the donor bacterium. Besides the specific DNA transfer mechanisms already mentioned, DNA acquisition largely depends on DNA rearrangements belonging to the processes described in Section 4.3. Indeed, some of the mechanisms of horizontal gene transfer depend on a recombinational interaction between the donor genome and the gene vector. Furthermore, the acquired genetic information must find in the receptor cell the possibility of becoming stably inherited. This can be ensured either by recombination into the receptor genome or by its independent maintenance as a plasmid. The recombination mechanism entailed may be general recombination in gene conversion; other possibilities include transposition and site-specific integration.

5   ENZYME-MEDIATED DNA REARRANGEMENTS CAN GENERATE GENETIC VARIATIONS

Sections 5.1 and 5.2 present selected cases of enzyme-mediated generation of genetic variation by DNA rearrangement and analyze them with regard to their evolutionary role.

5.1   Site-Specific DNA Inversion at Secondary Crossover Sites

Genetic fusions represent the results of joining together segments of two genes (gene fusions) or of two operons (operon fusions) that are not normally together. An operon is a set of often functionally related genes that are copied into messenger RNA (i.e., transcribed) as a single unit. As a result of this organization, those genes are coordinately regulated; that is, they are turned on or off at the same time. Therefore, in an operon fusion, one or more genes are put under a different transcription control, but the genes per se remain unchanged. In contrast, gene fusion results in a hybrid gene composed of sequence motifs and often functional domains originating in different genes.

In site-specific DNA inversion, a DNA segment bordered by specific DNA sequences acting as sites of crossing over becomes periodically inverted by the action of the enzyme DNA invertase. Depending on the location of the crossover sites, DNA inversion can give rise to gene fusion or to operon fusion. The underlying flip-flop system can result in microbial populations composed of organisms with different phenotypic appearances: if, for example, the DNA inversion affects the specificity of phage tail fibers, as is the case with phages P1 and μ of *E. coli,* phage populations with two different host ranges will result.

Occasionally, a DNA sequence that deviates considerably from the efficiently used crossover site, a so-called secondary crossover site, can serve in DNA inversion, which thus involves a normal crossover site and a secondary crossover site. This process results in novel DNA arrangements, many of which may not be maintained because of lethal consequences or reduced fitness; but if a few new sequences are beneficial for the life of the organism, these may be selectively favored. This DNA rearrangement activity can thus be looked at as evolutionarily important. Since many different DNA sequences can serve in this process as secondary crossover sites, although at quite low frequencies, site-specific DNA inversion systems act as variation generators in large populations of microorganisms. I have thus postulated that this evolutionary role of DNA inversion systems may be more important than their much more efficient flip-flop mechanism, which can at most help a microbial population to more readily adapt to two different, frequently encountered environmental conditions. As a matter of fact, other strategies could be used as well for the latter purpose.

5.2   TRANSPOSITION OF MOBILE GENETIC ELEMENTS

Already nine different mobile genetic elements have been found to reside, often in several copies, in the chromosome of E. coli K-12 derivatives. This adds up to occupation of about 1% of the chromosomal length by such insertion sequences, also called IS elements. At rates on the order of 10^{-6} per individual IS element and per cell generation, these mobile genetic elements undergo transpositional DNA rearrangements. These include simple transposition of an element and more complex DNA rearrangements such as DNA inversion, deletion formation, and the cointegration of two DNA molecules. Because of different degrees of specificity in the target selection upon transposition, the IS-mediated DNA rearrangements are neither strictly reproducible nor fully random. Transposition activities thus also act as variation generators. In addition to DNA rearrangements mediated by the enzyme transposase, which is usually encoded by the mobile DNA element itself, other DNA rearrangements just take advantage of extended segments of DNA homologies at the sites of residence of identical IS elements, at which general recombination can act. Altogether, IS elements represent a major source of genetic plasticity of microorganisms.

Transposition occurs not only in growing populations of bacteria, but also in prolonged phases of rest. This is readily seen with bacterial cultures stored at room temperature in stabs (little vials containing a small volume of growth medium in agar). Stabs are inoculated with a drop of a bacterial culture taken up with a platinum loop, which is inserted ("stabbed") from the top to the bottom of the agar. After overnight incubation, the stab is tightly sealed and stored at room temperature. Most strains of E. coli are viable in stabs during several decades of storage. That IS elements exert transpositional activities under these storage conditions is easily seen as follows.

A stab can be opened at any time, a small portion of the bacterial culture removed, and the bacteria well suspended in liquid medium. After appropriate dilution, bacteria are spread on solid medium. Individual colonies grown upon overnight incubation are then isolated. DNA from such subclones is extracted and fragmented with a restriction enzyme. The DNA fragments are separated by gel electrophoresis. Southern hybridization with appropriate hybridization probes can then show whether different subclones reveal restriction fragment length polymorphisms (RFLPs), which are indicative of the occurrence of mutations during storage.

If this method is applied to subclones isolated from old stab cultures, and if DNA sequences from IS elements serve as hybridization probes, an extensive polymorphism is revealed. No or only little polymorphism is seen with hybridization probes from unique chromosomal genes. Good evidence is available that transposition represents a major source of this genomic plasticity observed in stabs, which at most allow for a very residual growth at the expense of dead cells. One can conclude that the enzymes promoting transposition are steadily present in the stored stabs. Indeed, the IS-related polymorphism increases linearly with time of storage for periods as long as 30 years. For a culture of E. coli strain W3110, it turned out to be impossible to determine precisely which genome

structure was at the origin of the bacterial population studied after 30 years of storage. On the average, each subclone had suffered about a dozen RFLP changes as identified with hybridization probes from eight different residential IS elements, of which IS5 was the most active. Nevertheless a very interesting unrooted pedigree of these individual subclones could be drawn, and this offered an amazing insight into the genetic plasticity of E. coli.

6   PROMOTION AND LIMITATION OF GENE ACQUISITION

The association and dissociation of chromosomal genes with natural gene vectors arises from transpositional activities and general recombination acting at IS elements that are at different chromosomal locations. These mechanisms have been well studied with conjugative plasmids and with bacteriophage genomes serving in specialized transduction. For example, composite transposons, which are defined as two identical IS elements flanking a segment of genomic DNA (often with more than one gene unrelated to the transposition process), are known to occasionally transpose into a natural gene vector and, after their transfer into a receptor cell, to transpose again into the receptor chromosome. Hence, together with other mechanisms, such as site-specific and illegitimate types of recombination, transposition also represents an important promoter of horizontal gene transfer.

Several natural factors seriously limit gene acquisition. Transformation, conjugation, and transduction depend on surface compatibilities of the bacteria involved. Furthermore, upon penetration of donor DNA into receptor cells, the DNA is very often confronted with restriction endonucleases. These enzymes cause a fragmentation of the invading foreign DNA, which is subsequently completely degraded. Before fragments become degraded, however, they are recombinogenic and may find a chance to incorporate all or part of their genetic information into the host genome. Therefore, we interpret the role of restriction systems as follows: they keep the rate of DNA acquisition low, and at the same time they stimulate the fixation of relatively small segments of acquired DNA to the receptor genome. This strategy of acquisition in small steps can best offer microbial populations the chance to occasionally extend their biological capacities without extensive risk of disturbing the functional harmony of the receptor cell by acquiring too many different functions at once. These considerations have their relevance at the level of selection of hybrids resulting from horizontal DNA transfer. This selection is exerted as one of the last steps in the acquisition process.

7   GENETIC DIVERSITY REFLECTS THE PRESENT STATE OF BIOLOGICAL EVOLUTION

Biological evolution is a steady, dynamic process. The four categories of molecular mechanisms of mutagenesis described in Section 4 continuously exert their interactions on microbial populations, so that genetic diversity should steadily increase. However, while some of the newly arising DNA arrangements disappear rapidly by appropriate repair processes, others may at first remain but then submit to selection pressure. Normally, genetic diversity may be kept at a more or less constant level by such selection pressure, together with sampling due to the limited size of the biosphere of our planet, which offers space to only about 10^{30} living cells. This balance may hold as long as the overall environmental conditions of life do not undergo drastic alterations, which would of course seriously affect selection. Hence the importance of a relatively large pool of gene functions and life-forms, together with some genetic plasticity, to ensure good chances of genetic adaptation in times of changing living conditions.

Gene acquisition can be seen as a very efficient means by which a microbial strain extends its genetic capacities. This function may be of particular relevance if selective pressure undergoes drastic changes—as has been the case, for example, for enterobacteria since antibiotics became widely used in human and veterinary medicine. Much of our present knowledge of gene acquisition stems indeed from studies on the spreading and selection of drug resistance determinants.

In drawing the evolutionary tree of bacteria, DNA acquisition, which we can consider as a strategy for sharing in the success of others, should be accounted for by adding horizontal shunts between individual branches. Although usually only small DNA segments flow through such shunts at one time or another in horizontal gene transfer, in the vertical flux of genes from one generation to the next in the growing branches of the tree, the entire genome is of course steadily a target of any one of the mechanisms belonging to the first three categories of spontaneous mutagenesis.

In comparison with DNA acquisition, and in considering efficiencies per single event, DNA rearrangements internal to a genome may be evolutionarily less efficient and nucleotide substitution still less. However, these processes make important contributions to the evolutionary process. Nucleotide substitution is a major source of new biological functions, and it also contributes to the amelioration of existing functions. DNA rearrangements can bring about important improvements of existing capacities by, for example, the fusion of functional domains and of DNA sequence motifs.

In considering the evolutionary process, we should keep in mind that it largely depends both on the diversity and on the kind of genetic information already available, since time spans needed to develop completely new functions without making use of existing sequences are very long, not in the least because of the extremely large number of specific sequences that are possible in the linear arrangement of nucleotides of a gene and of a genome.

8   CONCLUSIONS

Rather than being the result of an accumulation of errors, biological evolution appears to depend both on a certain degree of structural plasticity of organic compounds (nucleic acids and proteins interacting with nucleic acids) and on the availability of many different specific biological functions. Enzymes acting as variation generators make fundamental contributions to spontaneous mutagenesis, including the rearrangement of existing sequences. In addition, more or less complex enzyme systems and organelles promote horizontal gene transfer and regulate its efficiency. With only a few exceptions, enzymes involved in these processes are not required for life of individuals in populations, and the biological significance of these agents becomes manifested only at the population level. As a matter of fact, spontaneous mutagenesis often does more to hamper the life of an affected individual than it does to convey benefit. Any benefit becomes obvious only at the population level insofar as mutations that occur make their contributions to a steady evolution.

It is a very interesting concept for our worldview to know that among the genes encoded in the genomes of living organisms there are, besides a large number of genes essential for each individual life, other genes, the products of which can ensure long-term development of life in a wide variety of phenotypic forms. Molecular genetic studies carried out with haploid microorganisms have brought about good evidence for these conclusions. It is quite likely that they might also apply to higher organisms, but these lifeforms are much less accessible than microorganisms to experimental approaches in search of evidence.

See also BACTERIAL GROWTH AND DIVISION; *E. COLI* GENOME; GENETIC ANALYSIS OF POPULATIONS; GENETICS; MUTAGENESIS, MOLECULAR BASIS OF.

Bibliography

Arber, W. (1991) Elements in microbial evolution. *J. Mol. Evol.* 33:4–12.
———. (1993) Evolution of prokaryotic genomes. *Gene*, 135:49–56.
———. (1994) Bacteriophage transduction. In *Encyclopedia of Virology*, pp. 107–113. Academic Press, London.
———. (1995) The generation of variation in bacterial genomes. *J. Mol. Evol.* 40:7–12.
———, Naas, T., and Blot, M. (1994) Generation of genetic diversity by DNA rearrangements in resting bacteria. *FEMS Microbiol. Ecol.* 15:5–14.
Drake, J. W. (1991) Spontaneous mutation. *Annu. Rev. Genet.* 25:125–146.
Echols, H., and Goodman, M. F. (1991) Fidelity mechanisms in DNA replication. *Annu. Rev. Biochem.* 60:477–511.
Galas, D. J., and Chandler, M. (1989) Bacterial insertion sequences. In *Mobile DNA*, D. E. Berg and M. M. Howe, Eds., pp. 109–162. American Society for Microbiology, Washington, DC.
Glasgow, A. C., Hughes, K. T., and Simon, M. I. (1989) Bacterial DNA inversion systems. In *Mobile DNA*, D. E. Berg and M. M. Howe, Eds., pp. 637–659. American Society for Microbiology, Washington, DC.
Kucherlapati, R., and Smith, G. R., Eds. (1988) *Genetic Recombination.* American Society for Microbiology, Washington, DC.
Lorenz, M. G., and Wackernagel, W. (1994) Bacterial gene transfer by natural genetic transformation in the environment. *Microbiol. Rev.* 58:563–602.
Mizuuchi, K. (1992) Transpositional recombination: Mechanistic insights from studies of Mu and other elements. *Annu. Rev. Biochem.* 61:1011–1051.
Moses, R. E., and Summers, W. C., Eds. (1988) *DNA Replication and Mutagenesis.* American Society for Microbiology, Washington, DC.
Sandmeier, H. (1994) Acquisition and rearrangement of sequence motifs in the evolution of bacteriophage tail fibers. *Mol. Microbiol.* 12:343–350.
Shapiro, J. A., Ed. (1983) *Mobile Genetic Elements.* Academic Press, New York.
West, S. C. (1992) Enzymes and molecular mechanisms of genetic recombination. *Annu. Rev. Biochem.* 61:603–640.
Woese, C. R. (1987) Bacterial evolution. *Microbiol. Rev.* 51:221–271.

Genetic Engineering of Antibody Molecules: *see* Antibody Molecules, Genetic Engineering of.

Genetic Intelligence, Evolution of

David S. Thaler and Bradley T. Messmer

Key Words

Adaptive, Directed, or Cairnsian Mutation   Mutation that is more likely to occur in circumstances in which the mutation is advantageous (i.e., nonrandom mutation). "Strong" adaptive mutations are those that do not appear to be generalized stress responses and are highly specific to the stress invoked.

Environment   Systems are often defined as having an "inside" and an "outside." The environment of an system consists of anything "outside." The outside of an organism is defined by its skin; the outside of a cell is defined by its membrane. The outside of a chromosome is the cellular milieu, and the outside of a DNA base is all of the above as well as its sequence context.

Evolution   Change with time or in the series (generations) of a lineage in the same direction as the flow of time.

Fixation of Genetic Change in an Individual Organism   Irrevocable commitment that a new allele will be inherited by progeny.

Genetic Algorithms (GAs)   A class of computer algorithms with certain characteristics. GAs consist of bit strings that give rise to progeny bit strings. Some progeny are identical to the parent and some are different. Units in the population of bit strings compete with one another for survival and reproduction according to selection criteria mediated by the program environment.

Genetic Change   Alterations of hereditary information. Typically—but not always!—genetic change occurs concomitantly with replication of the organism.

Genetic Evolution   Change in the hereditary information with time, especially if transferred to subsequent generations.

Genetic Intelligence   Hereditary information responsible for the mechanisms that generate genetic change.

Genetics   Mechanisms by which information is transferred between generations. The term "genetics" has its origins in biology, and "generations" refers to progeny cells, viruses, or multicellular organisms. The genetic properties of descent with modification are present in other contexts, such as computer programs known as genetic algorithms.

Intelligence   Tools and mechanisms for receiving information from the environment and adjusting response.

Lethal Environmental Selection   The process by which certain environments kill unfit organisms.

Mutation   Genetic change in which preexisting hereditary information is not transferred to a new context.

Nonlethal Environmental Selection   Describing the prevention, stopping, or slowing of the growth of less fit organisms without killing them, as these processes are mediated by the environment.

Random Mutation   Mutation that is no more likely to occur in environments that select for it than in environments that do not. In the context of genetics and mutation, "random" does not mean that every possible change is equally likely to occur. Even in the narrow context of point mutations, there are inherent biases in all examined systems.

Recombination   Genetic change in which preexisting hereditary information is transferred to a new context.

Recombination, Homologous, Legitimate, Generalized   Recombination in which the similarity of two sequences is a requisite for their ability to recombine. To a first approximation, homologous recombination depends on similarity, not on the specific sequence.

This article frames the intellectual context of "adaptive," "directed," or "Cairnsian" mutation, an area of research that is—and will likely remain—controversial. We review points to consider when critically reading in the field and also examine the question of how mechanisms for an "intelligent" generation of diversity could evolve. In the latter context, we suggest new connections between the fields of genetic algorithms in computer science and those of biological evolution proper (carbon- rather than silicon-based life-forms).

1   HISTORICAL AND INTELLECTUAL CONTEXT

Lamarckian interpretations persisted longer in microbiology than in other domains of biology. In part, this was due to the plethora of traits and organisms being observed and interpreted with a limited set of conceptual tools. Different traits and mechanisms turned out to be involved which are now realized to include at least five classes: operon induction, mutation, transposition, chromosome rearrangement (especially duplication), and phase variations such as those mediated by the site-specific recombination systems, lysogenic transformations, and other extragenic elements. The difficulty in inferring genotype from phenotype for these quite different mechanisms contributed to the confusion. The polemic reached high levels, and resolution of the controversy—albeit temporary— had a major effect on the development of modern molecular biology.

Luria and Delbruck made an enormous dent in the debate with their classic experiments that were designed to address one aspect of the generation of diversity (G.O.D.) question: Do mutations arise in response to the conditions that select for them, or are mutagenic variants preexisting in the population but unseen in the absence of selection? The Luria–Delbruck experiments had the following form. A number of replica liquid cultures were each inoculated with a small number of bacteria. "Small" means so few cells that no T1 phage-resistant cells are present at the start. The cultures were then allowed to grow to saturation, and the number of mutant cells in a sample was determined. Mutant cells were enumerated by spreading samples of the cultures onto petri plates that had been seeded with several million T1 phage. Only cells resistant to T1 phage could give rise to colonies. The question is, Did the T1-resistant phage exist in the liquid culture before there existed any selection for them, or was the resistance to T1 phage induced by the selective condition itself? In the former case, the role of the environment is restricted to selection for preexisting mutants; in the later case, the environment also plays a role in generating the mutants it subsequently selects.

Luria and Delbruck made the quantitative argument that resistance to T1 phage arose in the absence of selection. The argument was based on two elements:

1. If each cell has a given probability (e.g., one in a million) of mutating in response to the challenge, when a large number of bacteria (typically 10^8 cells present in 0.1 mL of a saturated overnight culture of the bacteria) were spread onto a plate, the number of resistant mutants found on each plate should represent a Poisson distribution around the arithmetic average (e.g., 100 in the numerical example given).

2. On the other hand, suppose that mutants arise during the nonselective growth of the overnight, with the same rate of one in a million. Mutant cells arising before the culture is saturated will give rise to mutant progeny, and cultures will therefore contain clones of mutants rather than single mutants.

Luria and Delbruck's model, in which mutants arise independently of their subsequent selection predicted a wide (i.e., wider than Poisson distribution) variation in the number of mutants in a series of cultures. The quantitative analysis of Luria and Delbruck assumes that mutation occurs at a constant probability per cell division.

The respreading experiments of Newcomb, as well as the replica plating of Lederberg and Lederberg, were both advanced as confirmation of the conclusion reached by Luria and Delbruck. Replica plating used sib selection to achieve a direct demonstration that clones of resistant (to the drug streptomycin) mutants existed prior to and independent of selection.

The formulation of Luria and Delbruck states that mutation is a function of the number of cell generations with a fixed probability of mutation at each generation. This way of thinking about mutation resonated well with a molecular mechanism that involved misinsertion of bases during normal replication. It also fit with a model of gradualistic evolution that occurs via the misinsertion of bases during normal replication. Molecular clocks based on a constant rate of random base substitution made tidy icing on the Hostess Twinkie of the neo-Darwinian synthesis.

Many interpretations of these three classic experiments erroneously conflate the elements of clonality with the time at which a mutant arises. The incorrect assumption being that if mutants arise independently of the selection, then mutants will give rise to clones before the assay. On the other hand, if mutants arise after selection is imposed, each mutation event will result in a separate enumerable colony. In the former case enumerating the number of mutants is an indirect measure of the mutation rate; in the latter case the rate of mutation is given simply by counting the number of mutant cells and dividing by the total number of cells tested.

A shrill stridency associated with the view that all mutations arise without regard to their subsequent selection is well illustrated by Monod in his 1970 *Chance and Necessity*:

In modern biological research some of the most outstanding work, both as to methodology and to significance, bears upon the area known as molecular genetics (Benzer, Yanofsky, Brenner and Crick). In particular this work has made it possible to analyze the different types of discrete accidental alterations a DNA sequence may suffer. Various mutations have been identified as due to

1. The substitution of a single pair of nucleotides for another pair;
2. The deletion or addition of one or several pairs of nucleotides; and
3. Various types of "scrambling" of the genetic text by inversion, duplication, displacement, or fusion of more or less extended segments.

We call these events accidental; we say that they are random occurrences. And since they constitute the *only* possible source of modifications in the genetic text, itself the *sole* repository of the organism's hereditary structures, it necessarily follows that chance *alone* is at the source of every innovation, of all creation in the biosphere. Pure chance, absolutely free but blind, at the very root of the stupendous edifice of evolution: this central concept of modern biology is no longer one among other possible or even conceivable hypotheses. It is today the *sole* conceivable hypothesis, the only one that squares with observed and tested fact. And nothing warrants the supposition—or the hope—that on this score our position is likely ever to be revised. (pp. 112–113; Italics in original)

Monod goes on to say:

There is no scientific concept, in any of the sciences, more destructive of anthropocentrism than this one, and no other so rouses an instinctive protest from the intensely telonomic creatures that we are.

Monod is correct in the consequences of his last statement, but he is wrong in his dogmatic emphasis on the impossibility of contradiction. The authors of the present article believe that "modern biological research," including molecular genetics, does not preclude the opportunity for theory and experiment in which mind and nature are not *a priori* at odds.

2   CAIRNS AND PRECURSORS

In 1988 John Cairns and colleagues published a paper that continues to stimulate intense work and controversy among molecular biologists, geneticists, evolutionists, and even philosophers. The results suggest that some mutations arise more frequently when they are being selected for, in apparent contradiction to the idea that mutations arise without regard for their utility. The role of the environment in selecting among preexisting variants is not challenged by these approaches; rather, the environment is also endowed with roles in generating the variation on which selection acts.

An essential aspect of protocols that search for adaptive mutagenesis is to use nonlethal selection (e.g., lactose prototrophy). Luria and Delbruck's selection was for resistance to phage T1. A cell becomes resistant to T1 by mutating its gene for the phage receptor. *E. coli* has about 200 phage receptors on its surface. Even if a bacterium were "smart" enough to immediately mutate the gene for the receptor as soon as it "saw" itself threatened by phage, it would still be dead by virtue of already-existing protein. Phage receptors are passed along with the wall at cell division. Vulnerability to phage lysis continues for several generations even after a mutation prevents the gene from encoding new phage receptor. For this reason the experimental protocol of phage resistance is capable of documenting only mutations that had occurred several generations before selection was applied. This protocol was able to address only the point that mutants can and do arise in the absence of selection. It was never capable of addressing the question of whether selection can induce mutation. A similar limitation is inherent in the experiments of Lederberg and Lederberg, because of the nature of the streptomycin resistance assay. For the resistance phenotype to be expressed in a cell, sensitive ribosomes must be diluted out by growth after the mutation has occurred.

Cairns's 1988 paper and the questions raised by it were not without precursors. Some of these precursors will be mentioned with an eye to what was different in the formulation of Cairns.

Max Delbruck may have been the first to recognize the limitations that could be rigorously inferred from his paper with Luria. In the 1946 Cold Spring Harbor Symposium (published in 1947) Delbruck says, "In view of our ignorance of the causes and mechanisms of mutations, one should keep in mind the possible occurrence of specifically adaptive mutations . . . or should exclude their occurrence in a variety of specific cases."

The conclusion that mutation occurs as a consequence of errors arising during replication is incomplete on several counts. Frances Ryan in the late 1950s and early 1960s demonstrated that prototrophic revertants of *E. coli* histidine auxotrophs could accumulate while the cells were not growing. Meiosis is another case in which genetic change is massively increased during a single replicative generation. Interallelic recombination per cell division in yeast meiosis is increased between 10^3 and 10^4 times when compared to the mitotic recombination rate between the same alleles. Radiation and chemical stimulation of genetic change certainly made it clear that the environment can create in DNA the lesions that subsequently are processed to fix mutations. Premutagenic lesions of many sorts in many sorts of cells are induced by chemicals or radiation in nongrowing cells and require subsequent growth for enumeration. The makings for adaptive mutagenesis lie in the possibility that they may also require growth for their fixation.

Mutagenic mechanisms can alter replication fidelity even at nonlesioned sites. One of the products of the SOS DNA damage repair response regulon acts as a negative regulator of the $5' \rightarrow 3'$-exonuclease of DNA polymerase. When the $5' \rightarrow 3'$-exonuclease functions, the polymerase is between 10^3 and 10^4 times more accurate. This nuclease activity is known as "editing" or "proofreading" because it preferentially excises terminal nucleotides that are paired incorrectly (i.e., not in Watson–Crick fashion). During the SOS response, overall DNA replication is decreased in fidelity. The qualitative nature of genetic change is also altered during the SOS response; that is, some types of mutation arise relatively more frequently relative to others. Nontargeted mutagenesis during the SOS response was the subject of considerable polemic around ideas of "inducible evolution" that were the immediate precursors of controversy around specifically adaptive mutation.

Barbara McClintock considered genomic change to be an important mechanism for both differentiation and evolution. These ideas were formed before the rise of operon theory partly in the context of somatic adaptation. McClintock remained concerned with the possibilities and properties of adaptive genomic change. In her 1983 Nobel Prize lecture she said, "A goal for the future would be to determine the extent of knowledge the cell has of itself and how it utilizes this knowledge in a 'thoughtful' manner when challenged." James Shapiro, in the discussion of his 1984 paper, pointed out that the work of Luria and Delbruck, Newcomb, and Lederberg and Lederberg involved immediately lethal selection and could detect only mutants that had originated in the absence of selection. He ended this important paper with the statement:

Indeed, now that we know about mobile genetic elements, inducible mutator systems and multiple biochemical activities that reorganize DNA molecules, the most pertinent questions in studies of hereditary change must be questions of control and regulation.

The special contribution of Cairns and colleagues is that the distinction between generalized stress responses and "thoughtful" reactions of the organism was for the first time explicitly incorporated into experimental protocols. The adequacy of the methods used has been the subject of a much criticism and refinement, but the field of adaptive mutation, while not getting calmer, has—on average—been reaching higher intellectual levels ever since.

If Ryan had included nonselected genes in his experiments, adaptive mutation as presently discussed would have been discovered before 1960. Shapiro's 1984 paper does not include nonselected genes; when the experimental design was extended to include tests inherent in Cairns et al. 1988, the results of Mittler and Lensky indicated a more generalized stress response. On the other hand, more recent work by Shapiro in 1994 indicated that the sequence of mutations differs depending on whether the mutations arose under selective or nonselective conditions.

3   CRITICAL READING OF THE ADAPTIVE MUTATION LITERATURE

The field of science that entails the investigation of adaptive mutation issues is likely to remain populated by mixed herds of sheep, goats, and unicorns. To aid critical reading, some hints and questions to keep in mind while examining the literature are enumerated.

1. If mutants occur prior to selection but grow more slowly, the appearance of Luria–Delbruck "jackpots" will be suppressed. Slow growers also may come up later under selection, giving a false clue that the mutation arose after selection had been imposed.

2. Always consider the possibility of, and the amount of, growth under conditions where growth is not supposed to happen. For example, if a lactose-negative cell is asked to mutate such that it can grow on medium with lactose as the sole carbon source, might there be a smidgen of glucose in lactose medium? Even if cell numbers do not increase, there may be cell turnover via cannibalism. Is there a chance for cross-feeding between cell types that are present, or could cross-feeding arise under the selective conditions? In the absence of growth, are other processes such as phage induction, conjugation, or meiosis still possible, and how might they contribute to the results observed?

3. Is the type of change occurring under selection the same type that occurs during nonselective growth? If the spectra of mutations recovered are different, the evidence is much stronger that a different mechanism is operating under one condition than another. Be on the lookout for unexpected sources of selection and for reconstructions to address them.

4. Are the genetic dependencies of mutation the same under selective and nonselective conditions? Different genetic dependencies are bona fide evidence of different pathways, although one must always be cautious of the "kicking `em when they're down" syndrome. That is, some mutations may result in a cell that is weaker and not able to grow well under challenged conditions.

5. Is the selection completely lethal or nonlethal? If nonlethal, how much of a stress is it? Is there a general stress response

pathway—such as SOS, carbon starvation (cAMP), or amino acid starvation (stringent response)—that could be invoked under the experimental conditions? In what ways does the experimental design distinguish between known stress responses and keep in mind the possibility of new ones?

6. What is the evidence for specificity? Are there control genes that are not under selection but whose mutation or recombination rate can be measured? What is the relation of control genes and their assays to the experimental ones? The best experimental designs have symmetry (i.e., selections for one or the other gene with measurement of the rate of change in both). Are the results reported as both absolute quantities of change (e.g., mutation rate per cell) and as a ratio of different types of change? Is the evidence for "adaptive" mutation based only on the time of mutants arising, or is it also based on ratios and the spectrum of mutation?

4   MECHANISMS PROPOSED FOR ADAPTIVE MUTAGENESIS

Many mechanisms have been proposed for the apparently Lamarckian inheritance that occurs under some regimes of nonlethal selection. This section mentions several such mechanisms. Some of the following classifications are orthogonals. The ideas in this list should be treated rather like a deck of cards, which the thinker can draw on in interpreting results and controversies.

1. Mechanisms in which the mutation rate is increased over a small portion (say 1%) of the genome in each cell such that in a population, all regions will be highly mutagenized. This category was proposed by Barry Hall, who subsequently provided counterevidence to the corollary that sequence changes should be clustered.
2. Mechanisms in which the generation of mutations is increased in response to properly perceived stress (PPS). This class of events is to generalize most directly from SOS-induced mutagenesis and from the facts of meiosis (i.e., that there are physiological circumstances of PPS in which the overall mutation rate is altered). Slightly more focused and specific are models in which the mutation rate for a subset of the total genes in the organism is altered as a function of some condition. Candidate mechanisms to focus mutation onto a subset of genes include genes and processes that regulate operon and regulon expression.
3. Mechanisms that invoke error-prone DNA synthesis, either over small regions of the genome (e.g., gaps created during conditions of starvation) or over an entire round of replication.
4. Homologous recombination (part of many proposals). Rosenberg has reviewed recombination models of several sorts, including those in which error-prone synthesis is concomitant with recombination and others in which the degree of sequence similarity required for homologous recombination is varied under conditions of properly perceived stress.
5. Mechanisms in which the generation of mutation and its fixation are separated. This separation is accomplished by making fixation dependent on growth at the critical moment. This most intriguing class of mechanisms was first suggested in the seminal 1988 article by Cairns et al. These authors proposed a specific mechanism in which error-prone transcription occasionally results in the production from a mutant locus of a

wild-type transcript. This mRNA will allow the cell to grow momentarily. The model further proposes that the change in physiology that accompanies momentary growth results in a burst of reverse transcriptase activity that acts on mRNA, resulting in DNA that homologously recombines with the chromosomal locus to fix the mutation.

F. W. Stahl proposed a mechanism in which gaps in the DNA arise during starvation and are filled in by error-prone DNA synthesis. New DNA synthesis is normally undermethylated and is subject to mismatch correction in favor of the old chain. However if the cell starts to grow (e.g., because error-prone synthesis has resulted in a wild-type allele), the mismatch will be resolved by replication and the mutation will be fixed in one of the progeny cells. Shapiro and colleagues have proposed and presented supporting evidence that gene fusions mediated by phage Mμ are modified by local transcription and replication.

5   HOW MIGHT GENETIC INTELLIGENCE HAVE EVOLVED?

The generation of genetic variation is in large part a function of the genes and physiology of DNA metabolism. Genes responsible for the generation of particular alleles will tend to be coinherited with the alleles that they played a role in generating. The components exist for feedback between the generators of genetic diversity and the environment that selects among variants. The utilization of this capability to generate among the range of all possible variation that subset which is "anticipated" to be of use is admittedly a radical proposition, but we contend that it should be taken seriously. Figure 1 (see color plate 7) illustrates the conventional view of evolution. In this simple view the sole act of the environment is to select among variants; there is no room for the environment to lay any role in the generation of diversity. In Figure 2 (see color plate 8), a more complex and realistic view of evolution, the environment acts both on the generation of variation and as a selective agent among variants.

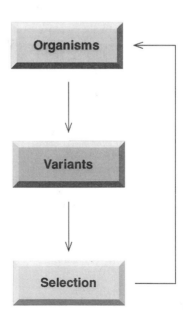

Figure 1.   A conventional view of evolution. The environment functions only at the selection step. (Reproduced with permission from Thaler, D. S. (1994) *Science* 264:224–225.) (See color plate 7.)

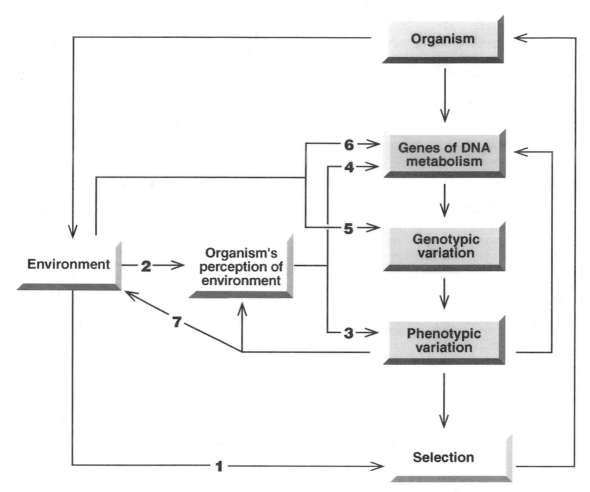

Figure 2. A more complete view of evolution—the environment functions at three stages according to the following protocol: 1, the environment is the proximate agent of selection; 2, the environment is perceived by the organism; 3, organisms use their perception of the environment to modify their physiology, as in operon induction; 4, organisms use their perception of the environment to modify their genetic metabolism, as in the SOS and p53 pathways; 5, the environment directly impinges on the DNA directly via such agents as radiation and chemical mutagens; 6, the environment can interact indirectly on DNA via the genes of DNA metabolism, as in topoisomerase inhibitors in chemotherapy; 7, the organism modifies environmental interaction with the genome as in metabolic activation, or, detoxification. (Reproduced with permission from Thaler, D. S. (1994) *Science* 264:224–225.) (See color plate 8.)

The action of the environment on the generation of variation takes several forms and goes via several routes. In direct action on DNA, radiation or chemicals can alter the chemistry, leading to a change in base sequence. These agents can also act less directly by modifying proteins involved in DNA metabolism. This is the route by which topoismerase inhibitors or agents that alter nucleotide pools can promote mutation and recombination. To become fixed as a mutation, even the most direct action of the outside environment on DNA requires processing by DNA repair and replication proteins. The "environment" of a DNA sequence means the cell's internal milieu and state as well as that which is outside an organism's skin or membrane. Even its sequence context is properly considered to be part of the "environment" of a particular base.

At no stage of evolution could the generation of diversity ever be purely "random" if by "random" it is meant that every sort of change is equally probable. For the initial discussion let us limit the case to substitution mutations in which the polymerase allows the pairing and ligation of a non-Watson–Crick base pair and uses that mispair as the primer for further extension. Until another round of replication, the mutation is not fixed; rather, it remains vulnerable to mis-

match correction. A purely random model for the generation of missense mutations would suppose that the all substitutions are equally common and that the rate of substitution is invariant. These points are not accurate statements. The rate of each possible substitution mutation is different both because the polymerase has a different propensity to make each sort of mistake and because the mismatch correction system has different affinities for each of the possible mismatched base pairs. The rate of substitution mutation is strongly affected by the nucleotide pools, and such pools are altered with cell physiology. The proofreading of polymerase is specifically inhibited by the SOS-induced gene *umuD*. Transcriptional focusing could allow regulation of genetic change to be as tightly coupled to the organism's perception of its environment as is gene expression.

Much of what is inherited is not due to a direct effect on survival. It is, instead, the product of neutral selection. The information that is mediated by perception and modulates physiology may or may not be congruent with that which alters genetic metabolism. In Figure 2, what is the relationship of information flowing through arrows 3 and 4, and what modulates that relationship via physiology and evolution? We are not claiming that evolution moves inexorably toward anticipatory mutation. We simply point out that the tools and feedback mechanisms exist such that life *could* evolve the ability to focus the generation of variation where an evolved mechanism "guesses" or "anticipates." Mutation rate is sometimes portrayed as a compromise of fidelity, economy, and the occasional need to generate variation. Differential variation with respect to genes and times, particularly if sensitive to the environment, offers a way out of compromise.

Rapid bacterial evolution to antibiotic resistance and clonal evolution in oncogenesis both involve concerted genetic changes that may include aspects of selection-promoted mutation. Study of these processes implies a further feedback loop.

6   GENETIC ALGORITHMS

Genetic algorithms are computer codes inspired by analogies to biology and the chromosome and capable of evolving. This section discusses the means of generating diversity in genetic algorithms and the possibility of using genetic algorithms to evolve a pseudo-genetic intelligence. Computer algorithms are not constrained to mimic the biological systems that inspired them. However, we believe that in terms of mechanisms to generate diversity there is a richness of carbon-based genetics that has not yet found its way into silicon.

Evolution of a computer began by constructing analogies for natural components. The binary code sequence of 0s and 1s is analogous to the nucleotide sequences of G, A, T, and C. The next analogy involves natural selection. Selection is typically taken to be a process whereby individuals of greater fitness are more likely to survive and propagate. The fitness evaluation of genetic algorithms is not directly related to the generation of diversity and is not discussed here further.

The basic outline of a genetic algorithm is as follows:

1. Create a population of chromosomes.
2. Evaluate the fitness of all chromosomes.
3. Select parents from among fitter chromosomes.
4. Produce offspring from parents while employing diversity generating operators.
5. Apply selective pressure to remove less fit chromosomes.
6. Return to step 2.

(Detailed discussions of the anatomy of genetic algorithms are provided by Holland and Kauffman; see Bibliography.) Each of these steps requires variables that may be chosen arbitrarily. In particular, the variable operators in step 4 could be developed to give genetic algorithms greater potential for generating diversity in manners similar to natural systems.

Genetic algorithms utilize bit strings analogous to chromosomes. The length of the chromosome is arbitrary for any given algorithm. The following string is an example of a binary chromosome of length 40:

```
1010010010000101111001100101010101010101
```

The reproductive aspect of genetic algorithms varies widely in method, but there are some common features that are essential to any genetic algorithm. The most important of these features is the generation of diversity among the offspring of the parent chromosomes.

It is the generation of diversity and the subsequent selection for the fittest individuals that allows genetic algorithms to display evolutionary behavior. Natural systems have evolved a wide variety of techniques for altering the arrangement of their chromosomes, and analogous techniques can be instituted in genetic algorithms. Doing so might have the double advantage of allowing genetic algorithms to adapt to more complex situations while also providing a tool for observing evolutionary behavior of these algorithms.

The simplest means of generating diversity in a genetic algorithm is random mutation concomitant with replication. A bit flip (from 0 to 1 or vice versa) occurs at an identical and fixed probability for each position at each replication. This approach is analogous to random base substitutions in the nucleotide sequence of an organism. The probability with which mutation occurs is set by the algorithm design. We doubt that random mutations alone can provide the optimal generation of diversity for genetic algorithms.

Insertion and deletion of single bases are simple extensions of base substitutions. Polymerase replication errors are most commonly cited as the cause for these mutations in natural systems. They are manifested in a genetic algorithm in the same manner as random bit flips, at some given probability in each reproduced chromosome. Figure 3 gives examples from nucleotide and binary sequences.

It is important to note that the three mutations illustrated in Figure 3 do not always occur at the same rate at every location of a natural chromosome. They frequently depend on sequence context. For example, inserted bases resulting from a looping-out error on a newly synthesized DNA strand occur more often in stretches of repeated bases (e.g., GACCTAAAAATCGAT). Therefore, the genetic algorithm that seeks to model this phenomenon must vary the probability of the mutation's occurring in specific sequence environments. Mutations of these types can provide diversity from par-

	Natural Chromosome
Parent	GATCCGTACGATCGGGGGCTACATCCTGAGA
Offspring	GATCCGTACGATCGGAGGCTACATCCTGAGA

	Binary Chromosome
Parent	01001010101001011100101010101010001
Offspring	01001010101001001100101010101010001

(a)

	Natural Chromosome
Parent	GATCCGTACGATCGCGGGCTACATCCTGAGA
Offspring	GATCCGTACGATCGCGGGGCTACATCCTGAGA

	Binary Chromosome
Parent	01001010101001011100101010101010001
Offspring	010010101010010111100101010101010001

(b)

	Natural Chromosome
Parent	GATCCGTACGATCGCGGGCTACATCCTGAGA
Offspring	GATCCGTACGATCGGGGCTACATCCTGAGA

	Binary Chromosome
Parent	01001010101001011100101010101010001
Offspring	0100101010100101100101010101010001

(c)

Figure 3. Three types of replication error: (a) random mutation, (b) random insertion, and (c) random deletion.

ent to offspring, but they do not allow any exchange of information between existing parents. For that to occur, there must be mating and some form of genetic recombination.

Genetic recombination is a common occurrence in natural systems, and it forms the basis of sexual reproduction. The ability to exchange and reorder DNA is fundamental to the evolution of sexual organisms; it is also important to genetic algorithms. Currently, genetic algorithms recombine in very limited, but powerful, ways. Figure 4 illustrates a common form of recombination used. In this technique, the two parent chromosomes are first passed through a crossover probability function, which determines the probability that the two parents will recombine. Once passed, a point on the chromosome is randomly chosen (in this case the tenth position) and the codes are swapped from parent to parent.

This procedure is powerful because it allows the exchange of large blocks of code, blocks that could contain useful groups of code. Studies have shown that blocks of code that tend to increase the fitness of the chromosome increase their frequency in the population over many generations. These blocks have been termed schemata, and they are roughly analogous to genes. Other extensions of this procedure are sometimes used, typically allowing for more than one crossover point. However, despite its power, this technique is not an accurate representation of natural systems. Biological organisms do not experience completely random recombinations. Instead, there are a number of regulatory conditions that affect the probabilities of where and when recombination occurs. In homologous (= generalized = legitimate) recombination, such as that which occurs during meiosis, recombination is dependent on similarity between two sequences. The exact degree of similarity required is itself a genetically regulated parameter. Similarity matching in genetic algorithms would require an operator that compares lengths and similarity and, if satisfied, recombines after an appropriate probability function. For example, the two parent chromosomes in Figure 5 will only recombine if there is similarity in nine bits out of ten, and then only 50% of the time.

This form of recombination has several advantages. It possesses a greater number of variables that can be altered to study their effect. The size of the region of similarity, the degree of similarity required, where and how often within the region of similarity the recombination will occur, and the probability of recombination then occurring are all arbitrary values. The analogies of these values are poorly understood in natural systems, and genetic algorithms can provide insight on their interactions. Also, unlike recombinational operators that predetermine the number of crossover points, this method has the inherent potential for multiple crossovers. In biological systems the probability of multiple crossovers is often different from the simple product of individual events. This genetic

Parent 1
1111111111111111111111111111111111111111

Parent 2
0000000000000000000000000000000000000000

Offspring 1
1111111111000000000000000000000000000000

Offspring 2
0000000000111111111111111111111111111111

Figure 4.   Recombination involving a crossover probability function.

Parent 1
10101001*00111011***010100101011010010101011**

Parent 2
00010110*00101011*01111100110101101011100

Offspring 1
10101001001110110111110011010110101111100

Offspring 2
00010110000101**0110101001010110100010101011**

Figure 5.   Recombination involving similarity matching.

property called "interference" has not yet been incorporated into genetic algorithms.

The recombinational model just described allows for the creation of recombinational hot spots and cold spots. Hot spots are regions in which generalized recombination occurs with increased frequency; they are a common phenomenon in biological organisms. Hot spots would be invoked at the point of the probability check. The probability that recombination occurs once homology has been established could then be directly influenced by the sequence of the similarity. In addition to specific sequences influencing the probability of recombination, the probability could be made a function of the ratio of 1s to 0s, or it could depend on the prevalence of repeated bits (analogous to the proposal that GC regions or perfect palindromes are hotspots in biological systems).

Homologous recombination is not the only means by which natural organisms recombine. Site-specific recombination occurs not as a result of sequence similarity but because certain sequences are recognized by specific proteins that bind to those sequences. The chemical pathways are not significant for genetic algorithms, but their results can be simulated to good effect. Typically, site-specific recombination removes, inserts, or copies regions of DNA, as is the case with many transposons. Including an analogous function in genetic algorithms would allow another means of generating diversity among binary chromosomes. For example, in the chromosome in Figure 6, the sequence **0000011111** is a transposable element that, at some probability, will move itself into the center of any sequence of 011001.

Occasionally in natural systems transposons will incorporate and relocate pieces of chromosomal DNA. Genetic algorithms can also model this behavior. For example, the preceding transposable element could be made to recognize only the repeat 0s and 1s of the sequence, regardless of code in between them (Figure 7).

Such a system allows useful groups of code (i.e., genes) to be moved around the chromosome. Alternately, they could be deleted and, once excised, copied into the new location without being removed from their original site, inverted, or a combination of the latter two (Figure 8).

As with homologous recombination, site-specific recombination-

Parent 1
11010100011**0000011111**0011010011*0110*010101011

Offspring 1
1101010011001101011011**0000011111**0010101011

Figure 6.   Recombination involving a transposable element.

```
                   Parent 1
1101010011000001010111111110011010110110010
```

```
                   Offspring 1
1101010011001101011011000001010111111110010
```

Figure 7. Recombination involving incorporation and relocation of transposons.

al techniques require many arbitrary values. The sequences recognized, the size of the transposable unit, the probability that a recombinational event will occur, and the rate at which each type of recombinational event occurs all must be given to the algorithm.

Most cells contain a great deal more DNA than is required to encode their genes. Eukaryotic chromosomes are estimated to be up to 98% irrelevant DNA. These vast quantities of unused DNA are considered to be a sort of blackboard on which evolution can scribble without affecting those regions of the chromosome encoding essential genes. Therefore some expansion of the bit sequences available must occur. One means of doing this is to elongate the chromosomes with regions of "junk" sequence that are not included in the fitness evaluation.

Biological evolution is horizontal as well as vertical. Intergeneric recombination moves DNA between widely separated lineages. Plasmids and viruses move DNA from cell to cell with great promiscuity. Many prokaryotic organisms are metabolically adapted for the uptake of DNA they encounter in the environment. Given that natural organisms have a great deal more DNA than just their chromosomes, genetic algorithms could profit from analogous forms of binary code.

Regions of code that do not initially encode useful genes must at some point become recruited into the fitness evaluation if they are to ever be more than "junk" code. Regulation of which regions of the chromosome are to be used in the fitness evaluation is under the control of the chromosome in natural systems. Aspects of gene regulation and expression can be incorporated into genetic algorithms to provide similarity to natural systems. An analogy with promoter sequences can be used to determine which region of the chromosome is to be evaluated, thus defining the genes. Promoter sequences also make it possible to hand over some controls to the chromosome itself. For example, the probability that homologous recombination will occur could be given to the chromosome by encoding the probability value on the chromosome after a specific promoter. As a result, one of the variables that

```
                   Parent 1
1101010011000001010111111110011010110110010
```

```
                   Deleted
110101001100110101100110010
```

```
                    Copied
1101010011000001010111111110011010110110000001010111111110010
```

```
                   Inverted
11010100111111111101010000000011010110110010
```

```
              Copied and inverted
```

Figure 8. Additional forms of recombination.

influences the generation of diversity is itself subject to genetic change. Any value usually specified in the structure of a genetic algorithm could be relocated to the chromosome in this way, in effect giving the chromosome the ability to alter its own means of changing. It is even possible to allow the chromosomes to evolve promoters by predefining both default values for the variables and the sequence of a promoter, initially absent from the population, that would encode the value of that variable. In this way we hope to evolve genetic intelligence as the binary chromosome takes control of its own means of changing. New insights are almost certain to arise through the simultaneous study of carbon- and silicon-based life-forms.

See also MITOCHONDRIAL DNA, EVOLUTION OF HUMAN; MOLECULAR DIVERSITY, THEORETICAL ASPECTS OF; MUTAGENESIS, MOLECULAR BIOLOGY OF; NUCLEIC ACID ENZYMES, EVOLUTIONARY ENGINEERING OF; ORIGINS OF LIFE, MOLECULAR BASIS OF.

Bibliography

Bateson, G., and Bateson, M. C. (1986) *Angels Fear. Towards an Epistemology of the Sacred.* Bantam Books, New York.
Brock, T. D. (1990) *The Emergence of Bacterial Genetics.* Cold Spring Harbor Press. Cold Spring Harbor, NY.
Cairns, J., Overbaugh, J., and Miller, S. (1988) The origin of mutants. *Nature,* 335:142–145.
Delbruck, M. (1986) *Mind from Matter? An Essay on Evolutionary Epistemology.* Blackwell Scientific Publications, Palo Alto, CA.
Foster, P. L. (1993) Adaptive mutation: The uses of adversity. *Annu. Rev. Microbiol.* 47:467–504.
Holland, J. H. (1992) *Adaptation in Natural and Artificial Systems: An Introductory Analysis with Applications to Biology, Control and AI,* Bradford Series in Complex Adaptive Systems. MIT Press, Cambridge, MA.
Kauffman, S. A. (1993) *The Origins of Order: Self-Organization and Selection in Evolution.* Oxford University Press, Oxford.
Lenski, R. E., and Mittler, J. E. (1992) The directed mutation controversy and neo-Darwinism. *Science,* 259:188–194.
McClintock, B. (1983) The significance of responses of the genome to challenge. *Science,* 226:792–801.
Rosenberg, S. M. (1994) In pursuit of a molecular mechanism for adaptive mutation. *Genome,* 137:893–899.
Shapiro, J. A. (1993) Natural genetic engineering of the bacterial genome. *Curr. Opinion Genet. Dev.* 3(6):845–848.
Thaler, D. S. (1994) The evolution of genetic intelligence. *Science,* 264:224–225. Correspondence and reply: *Science,* 265:1994–1996.
———. (1994) Sex is for sisters: Intragenomic recombination and homology-dependent mutation as sources of evolutionary variation. *Trends Ecol. Evol.* 9:108–110.

Genetic Manipulation of Plant Cells: *see* Plant Cells, Genetic Manipulation of.

Genetic Markers: *see* Population-Specific Genetic Markers and Disease.

Genetic Medicine: *see* Molecular Genetic Medicine.

Genetic Mutations and Aging: *see* Aging, Genetic Mutations in.

GENETICS

D. Peter Snustad

Key Words

Codon   The unit of three contiguous nucleotides in messenger RNA specifying the incorporation of one amino acid in the polypeptide produced by translating that mRNA on the polyribosomes.

Complementation Test   The introduction of two recessive mutations into the same cell but on different chromosomes (a trans heterozygote) to determine whether the mutations are in the same gene or in two different genes. If both mutations are in the same gene, the $m_1 + / + m_2$ heterozygote will exhibit a mutant phenotype, whereas if they are in two different genes, the trans heterozygote will exhibit the wild-type phenotype.

Gene   The basic unit of genetic information. The unit of function specifying the structure of one primary product, usually one polypeptide chain, and defined operationally by the complementation test.

Mutation   Heritable change in the structure of the genetic material of an organism. When used in the broad sense, mutations include both "point mutations," involving changes in the structure of individual genes, and gross changes in chromosome structure (chromosome aberrations). When used in the narrow sense, the term "mutation" applies only to "point mutations." Mutation is used to refer both to the *process* by which the change occurs and to the *result* of the process, the *alteration* in the gene or genetic material.

Recombination   The generation in progeny of combinations of genes that were not present together in either parent, by (1) independent assortment of nonhomologous chromosomes during meiosis or (2) crossing over (breakage and exchange of parts) of homologous chromosomes during meiosis or mitosis.

The phenotype of a living organism is controlled by its genotype, the summation of its genetic information, acting within the constraints imposed by the environment in which the organism exists. Much of the genetic material of an organism is organized into basic functional units called genes, with each gene encoding one primary gene product, most commonly one polypeptide. The genetic information of all living organisms, whether viruses, bacteria, corn plants, or humans, is stored in the sequence of bases (purines and pyrimidines) in the deoxyribonucleic acid (DNA) present in their chromosomes. In some viruses, the genetic information is stored in the sequence of bases in ribonucleic acid (RNA). The genetic information is encoded using a four-letter alphabet, namely, the four bases: adenine (A), guanine (G), cytosine (C), and thymine (T). In RNA, uracil (U) replaces the thymine present in DNA. The genetic material of an organism must carry out two essential functions: (1) the genotypic function, transmission of the genetic information from generation to generation, and (2) the phenotypic function, directing the growth and development of the offspring into mature, reproductive adults. The genetic material of an organism is not static; it changes or mutates on occasion to provide new genetic variability, which provides the raw material for evolution. Recombination of genetic material occurs by the independent assortment of nonhomologous chromosomes and by crossing over between homologous chromosomes to provide new combinations of genes and thus new phenotypes on which natural selection can act to allow evolution to progress.

1   GENETIC INFORMATION

The genetic information of living organisms is stored in large macromolecules called nucleic acids. These nucleic acids are of two types: DNA, which contains the sugar 2'-deoxyribose, and RNA, which contains the sugar ribose. In all eukaryotic organisms, the genetic information is stored in giant DNA molecules located in from one to many chromosomes, the number depending on the species. In some viruses that contain no DNA, the genetic information is stored in RNA.

1.1   FOUR-LETTER ALPHABET

The genetic information is stored in nucleic acids using a four-letter alphabet composed of the four bases adenine (A), guanine (G), cytosine (C) and thymine (T) in DNA; uracil (U) replaces thymine in RNA. Although there are only four letters in this genetic alphabet, a vast amount of information can be stored because the nucleic acids utilized for storage are very large. For example, one complete copy of the human genome (all the genetic information in one complete set of the human chromosomes) contains three billion (3×10^9) base pairs of DNA. Since the number of possible sequences of four letters used n at a time is 4^n, one can see that with $n = 3 \times 10^9$, the human genome has the capacity to store a huge amount of information.

1.2   THE GENE: THE BASIC UNIT OF FUNCTION

The basic functional unit of genetic information is the gene, defined operationally by the complementation test and most commonly specifying the amino acid sequence of one polypeptide chain. Different forms of a given gene are called alleles. The wild-type alleles of a gene are those that exist at relatively high frequencies in natural populations and yield wild-type or "normal" phenotypes; they are usually symbolized by a + or by a symbol with a + superscript (e.g., w^+ for the allele that yields wild-type red eyes in fruit flies). Alleles of a gene that result in abnormal or nonwild-type phenotypes are called mutant alleles; they are usually symbolized by one to three italic letters (e.g., w and w^{ap} for the alleles that cause white

A. Two mutations in the same gene.

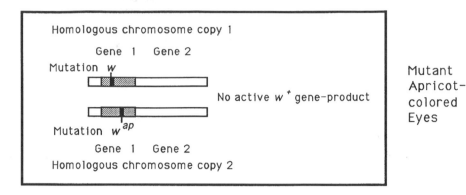

B. Two mutations in two different genes.

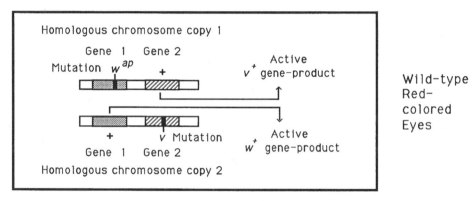

Figure 1. Principle of the complementation test used in the operational definition of genes, the basic units of function of genetic material. The operation is to place the two recessive mutations in question in the same cell or cells of a multicellular organism on two separate chromosomes (i.e., to construct a trans heterozygote) and to determine whether this cell or organism has a mutant or a wild-type phenotype. If the phenotype is mutant, the two mutations are in the same gene; this is illustrated for the w (white eyes) and w^{ap} (apricot eyes) mutations of *Drosophila* in (A). If the phenotype of the trans heterozygote is wild type, the mutations are in two different genes and are said to complement each other. Complementation between the w^{ap} and v (vermilion eye color) mutations of *Drosophila* is illustrated in (B); note that active (wild-type) products of both genes (w^+ and v^+) are present in the trans heterozygote (B)—thus, the wild-type phenotype.

and apricot eye color, respectively, in fruit flies). Many eukaryotes, such as corn plants, fruit flies, and humans, contain two copies of their genome in most cells, thus two copies of each of their chromosomes; such eukaryotes are called diploids. Diploid organisms may contain two different alleles of any given gene, in which case they are heterozygous (e.g., w/w^+; w^{ap}/w^+; w/w^{ap}) or two identical copies of a given gene, in which case they are homozygous (e.g., w^+/w^+; w/w; w^{ap}/w^{ap}). A w^{ap}/w^+ heterozygous fruit fly has wild-type red eyes. The w^+ allele is expressed in this heterozygous fly; w^+ is thus called the dominant allele. The w^{ap} allele is not expressed in this heterozygous fly; it is said to be recessive because its effect on the phenotype is masked by the w^+ allele.

The complementation test is performed by producing cells or organisms that contain two recessive mutant genes located on two different chromosomes (i.e., trans heterozygotes) and determining whether these cells or organisms have mutant or wild-type phenotypes. If the two mutant genes are allelic—that is, if the defects or mutations are in the same gene—the trans heterozygote will have a

mutant phenotype. If the two mutant genes are not allelic—that is, if the mutations are in two different genes—the trans heterozygote will have the wild-type phenotype. The rationale behind the complementation test is illustrated in Figure 1.

2 DNA REPLICATION: THE GENOTYPIC FUNCTION

The genetic information of an organism is transmitted from cell to cell during development and from generation to generation during reproduction by the accurate replication of the sequence of bases in nucleic acids based on the precise base pairing in double-stranded nucleic acids: A with T and G with C. Two properly base-paired strands of DNA are complementary and contain the same genetic information. Thus, when the two strands of a parental double helix of DNA separate, the base sequence of each parental strand will serve as a template for the synthesis of a new complementary strand, as shown in Structure 1.

Parental DNA Molecule

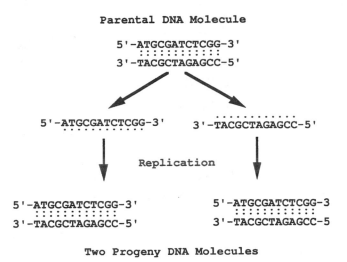

5'-ATGCGATCTCGG-3'
3'-TACGCTAGAGCC-5'

5'-ATGCGATCTCGG-3'   3'-TACGCTAGAGCC-5'

Replication

5'-ATGCGATCTCGG-3'   5'-ATGCGATCTCGG-3
3'-TACGCTAGAGCC-5'   3'-TACGCTAGAGCC-5

Two Progeny DNA Molecules

Structure 1

3   GENE EXPRESSION: THE PHENOTYPIC FUNCTION

The genetic information controls the morphogenesis of the organism, be it a virus, a bacterium, a plant, or an animal. This genetic infor-

mation must be expressed accurately both spatially and temporally to produce the appropriate three-dimensional form of the organism. In multicellular organisms, the genetic information must control the growth and differentiation of the organism from the single-celled zygote to the mature adult. To accomplish this phenotypic function, each gene of an organism must be expressed at the proper time and in the proper cells during development. The initial steps in the pathways of gene expression, transcription, and translation are quite well elucidated; Figure 2 illustrates these steps for the expression of the human β-hemoglobin gene. In contrast, we are just beginning to understand morphogenesis at the cell, tissue, and organ levels.

3.1   TRANSCRIPTION

The first step in gene expression, transcription, involves converting genetic information stored in the form of base pairs in double-stranded DNA into the sequence of bases in a single-stranded molecule of messenger RNA. This process, which is catalyzed by enzymes called RNA polymerases, occurs when one strand of the DNA is used as a template to synthesize a complementary strand of RNA using the same base-pairing rules that apply in DNA replication except that uracil is incorporated into RNA at positions where thymine would be present in DNA (see Figure 2, "Transcription" row).

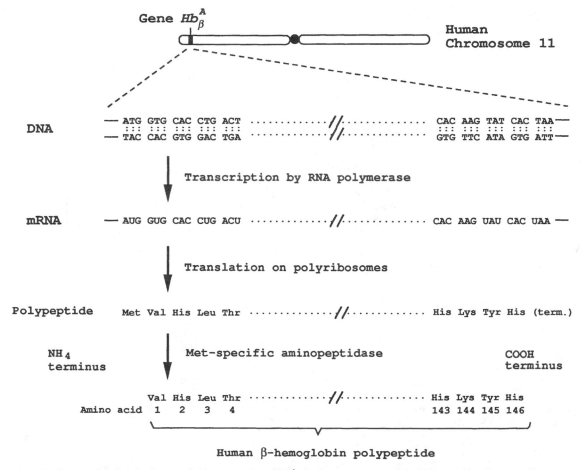

Figure 2.   Schematic diagram showing the fist steps in the expression of Hb_β^A, the human gene encoding β-hemoglobin. The steps shown are transcription, translation, and proteolytic removal of the amino-terminal methionine residue from the primary translation product. For simplicity, only the terminal portions of the coding sequence and the polypeptide product are shown.

3.2   TRANSLATION

During translation, the sequence of bases in the mRNA molecule is converted ("translated") into the specified sequence of amino acids in the polypeptide gene product according to the rules of the genetic code (Figure 2, "Translation" row). Each amino acid is specified by one codon, a triplet of three adjacent bases in the mRNA. Translation is a complex process occurring on cytoplasmic macromolecular structures called ribosomes and requiring the participation of many other macromolecules.

3.3   COMPLEX PATHWAYS OF GENE ACTION

The pathway through which a gene exerts its effect on the phenotype of the organism is often long and complex, especially in multicellular eukaryotes (Figure 3). The pathways of gene action frequently involve protein–protein and other macromolecular interactions, cell–cell interactions and intercellular communication by hormones and other signal molecules, tissue and organ interactions, and restrictions imposed by environmental factors.

3.4   REGULATION OF GENE EXPRESSION

In all organisms, gene expression is highly regulated; thus energy is used to synthesize gene products only when those products are needed for growth and differentiation of the organism. In higher eukaryotes, only a small proportion of the genes in the genome are being expressed in any one cell type. Thus, gene expression is highly programmed to ensure that genes needed to make neurons are turned on only in developing nerve cells, genes needed to make red blood cells are expressed only in progenitors of erythrocytes, and so on. The mechanisms by which gene expression is regulated are numerous and beyond the scope of this entry.

4   MUTATION: THE ULTIMATE SOURCE OF ALL NEW GENETIC VARIABILITY

Although genetic information must be transmitted from generation to generation with considerable accuracy, it is not static. Rather, it undergoes occasional change or mutation to produce new genetic variability, which provides the raw material for ongoing evolution. The new variant genes produced by mutation are called mutant alleles and often result in abnormal or mutant phenotypes. When used in the narrow sense, mutation refers only to changes in the structures of individual genes. However, in the broad sense, mutation refers to any heritable change in the genetic material and includes gross changes in chromosome structure or chromosome aberrations. There are four types of gross chromosome rearrangement: duplications, deletions, inversions, and translocations. A duplication is the occurrence of a segment of a chromosome in two or more copies per genome. A deletion results from the loss of a segment of a chromosome. An inversion occurs when an internal segment of a chromosome is turned end-for-end relative to its orientation in a normal chromosome. A translocation results when a segment of a chromosome is broken off and becomes attached to another chromosome.

Point mutations within individual genes may be either base pair substitutions or the insertion or deletion of one or a few contiguous base pairs. The insertion or deletion of one or two base pairs within the coding sequence of a gene alters the codon reading frame in the mRNA; thus, such mutations are referred to as frameshift mutations. Frameshift mutations usually result in totally nonfunctional gene products. In contrast, base pair substitutions usually result in the substitution of a single amino acid in the mutant polypeptide gene product. For example, sickle cell anemia occurs in humans that are homozygous for a β-hemoglobin gene that differs from the normal adult β-hemoglobin gene by a single base pair substitution. This one base pair change in the Hb_β^S gene changes the sixth amino acid of the β-hemoglobin polypeptide from glutamic acid in Hb_β^A homozygotes to valine in Hb_β^S homozygotes, as follows:

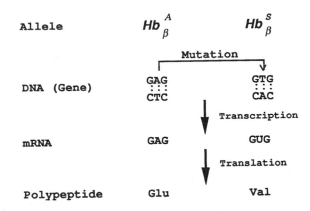

This single amino acid change in the human β-hemoglobin chain results in sickle-shaped red blood cells and in sickle cell anemia in individuals homozygous for the Hb_β^S allele. Thus, a single base pair substitution in DNA can have a very large effect on the phenotype of the organism harboring the mutation.

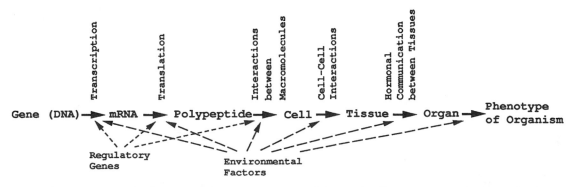

Figure 3.   Pathway of gene expression, showing some of the components that may influence the effect a given gene will have on the phenotype of an organism.

A. Independent Assortment of Genes on Nonhomologous Chromosomes during Meiosis

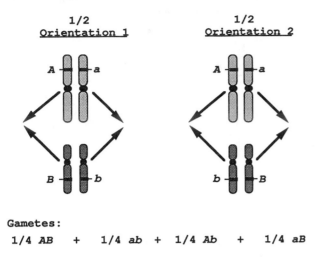

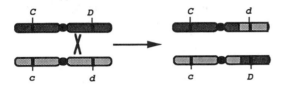

B. Crossing-over between Genes Located on Homologous Chromosomes

Figure 4. The generation of new combinations of genes by recombination: either (A) the independent assortment of genes on nonhomologous chromosomes during meiosis or (B) crossing over between genes located on homologous chromosomes during meiosis or mitosis.

5   RECOMBINATION: NEW COMBINATIONS OF GENES TO BE ACTED ON BY NATURAL OR ARTIFICIAL SELECTION

Mutation produces new genetic variability, but the resulting mutant genes must be placed in new combinations with previously existing genes so that natural selection (or, in the case of plant and animal breeding, artificial selection) can preserve the combinations that produce the organisms best adapted to specific environments (or desired by the breeder). These new combinations are produced by recombination mechanisms that are essential to the process of evolution. New combinations of genes on nonhomologous chromosomes are produced by the independent assortment of chromosomes during the first or reductional division of meiosis (Figure 4A). New combinations of genes on the same chromosome are produced by crossing over (breakage and exchange of parts) between homologous chromosomes during meiosis and mitosis (Figure 4B).

See also AGING, GENETIC MUTATIONS IN; DNA REPLICATION AND TRANSCRIPTION; GENETIC DIVERSITY IN MICROORGANISMS; HUMAN GENETIC PREDISPOSITION TO DISEASE.

Bibliography

Gardner, E. J., Simmons, M. J., and Snustad, D. P. (1991) *Principles of Genetics,* 8th ed. Wiley, New York.
Kornberg, A. (1980) *DNA Replication.* Freeman, San Francisco.
———. (1982) *Supplement to DNA Replication.* Freeman, San Francisco.
Lewin, B. (1990) *Genes IV.* Oxford University Press, New York.
Singer, M., and Berg, P. (1991) *Genes and Chromosomes.* University Science Books, Mill Valley, CA.
Watson, J. D., Hopkins, N. H., Roberts, J. W., Steitz, J. A., and Weiner, A. M. (1987) *Molecular Biology of the Gene,* 4th ed., Vols. I and II. Benjamin/Cummings, Menlo Park, CA.

GENETIC TESTING

Frank K. Fujimura

1   **Background**

2   **Direct and Indirect Genetic Analysis**
    2.1   Direct Mutation Detection
    2.2   Indirect Linkage Analysis

3   **Methodology**

4   **Potential Complications**

5   **Specific Examples**
    5.1   Autosomal Recessive Conditions
    5.2   Autosomal Dominant Conditions
    5.3   X-Linked Recessive Conditions
    5.4   Trinucleotide Repeat Conditions
    5.5   Hereditary Cancer Susceptibilities

6   **Limitations and Implications**

Key Words

Anticipation   Increase in penetrance and/or severity of a genetic disease from generation to generation.

Autosomal   Related to the nonsex chromosomes.

Carrier   An individual with one abnormal and one normal copy of a recessive gene. Carriers, which are either male or female for autosomal recessive genes, and female for X-linked recessive genes, generally do not have clinical symptoms of the disease.

Denaturing Gradient Gel Electrophoresis (DGGE)   A method for point mutation scanning by electrophoresis in a gel containing a gradient of denaturants (urea and formamide). Sequence-dependent differences in denaturation conditions can be detected.

Dominant   Describing a phenotype expressed in heterozygotes.

Epigenetic   Describing genetic properties not related to the primary nucleotide sequence of a gene (e.g., methylation).

Gene   A segment of DNA that contains all the information for the regulated biosynthesis of an RNA product.

Hemoglobinopathy   A disease of one or more of the globin genes whose protein products make up hemoglobin.

Heteroduplex   A double-stranded DNA molecule in which the strands are derived from two different alleles of a particular locus.

Heterogeneity   The condition of more than one gene (genotype) independently leading to an apparently single condition (phenotype).

Heterozygote An individual carrying two different alleles of a particular locus.

Homozygote An individual carrying two copies of the same allele at a particular locus.

Imprinting The selective expression of one parental allele of a gene due to gamete-specific modification; the imprinted allele is inactive.

Linkage Analysis Indirect analysis of a genetic disease gene by tracking polymorphic markers that are physically associated (linked) with the gene of interest.

Locus A location on a chromosome or DNA molecule corresponding to a gene or a physical or phenotypic feature.

Mosaicism The presence in one individual of cells having different (diploid) genotypes.

Penetrance The qualitative expression of a particular genotype.

Recessive Describing a phenotype expressed only in homozygotes.

Severity The quantitative expression of a phenotype.

Short Tandem Repeats (STRs) Polymorphic loci with relatively small (2–4 base pairs) tandemly duplicated sequences, useful as markers for linkage analysis.

Single-Strand Conformational Polymorphism (SSCP) A structural variation in single-stranded DNA due to sequence-dependent intrastrand secondary structure. SSCP is detectable by electrophoresis in nondenaturing gels and is useful for point mutation scanning.

Uniparental Disomy The condition of inheriting both alleles of a gene from one parent.

Variable Number of Tandem Repeats (VNTR) A polymorphic locus with relatively large (>50 base pairs) tandemly duplicated sequences. VNTRs are useful as markers for linkage analysis.

X-Linked Describing a genetic condition mapping to the X-chromosome; when recessive, the phenotype is expressed only in males.

The clinical application of DNA probe technologies for molecular genetic testing is proceeding rapidly. The recent discoveries of the human genes for numerous genetic conditions, including Duchenne muscular dystrophy, cystic fibrosis, fragile X syndrome, and Huntington's disease have quickly resulted in the transfer of these new technologies into the clinical arena. These new technologies, while very powerful, also have limitations that need to be considered for correct interpretation of test results. The pace of technology transfer challenges the ability of the clinical community and the general public to comprehend and deal with the diverse technical, medical, ethical, legal, and social implications of genetic testing.

With the impending expansion of gene discovery from relatively rare genetic conditions such as Huntington's disease to more common conditions (e.g., heritable cancers, heart disease), the potential impact of genetic testing on society is very significant. The results of genetic testing can affect not only the health of the patient but also many other aspects of the patient's life, including reproductive options, lifestyle considerations, employability, and insurability. Furthermore, genetic test results can have significant clinical and social implications for some of the relatives of the individual tested. These many issues make the delivery of clinical molecular genetic services a challenging but critical task.

1 BACKGROUND

The recent interest and activity in the cloning of human genes has resulted in the isolation and characterization of numerous genes relevant to genetic disease. In many cases, genes were isolated because their respective protein products were known. Clinical testing for these genetic diseases often was available by biochemical methods prior to gene isolation. Although direct genetic analysis may become possible with the isolation of a particular gene, biochemical or other methods, if available, are often preferred over DNA probes at present because of lower costs, shorter turnaround times, and superior or comparable sensitivity and specificity.

For many genetic diseases, however, biochemical tests are unavailable because the nature of the gene products responsible for these diseases has not been determined. The identification of these genes has been the focus of intense activity. The first triumph of "reverse genetics" or "positional cloning"—the use of molecular methods to map, isolate, and characterize genes with unknown gene products—came in 1986 with the cloning of the genes for chronic granulomatous disease, Duchenne muscular dystrophy, and retinoblastoma. Several other genes, notably those responsible for cystic fibrosis, fragile X syndrome, myotonic dystrophy, and Huntington disease, have been isolated subsequently. The importance of identifying genes with unknown gene products for diagnosis is that knowledge of the gene structure often leads to immediate capability for analyzing mutations in the gene or alterations in the expressed gene products.

Genetic diseases can be classified as autosomal or X-linked, depending on chromosomal association, and as dominant or recessive depending on the nature of phenotypic expression. The dominant or recessive nature of genetic disease is not always clearly distinguishable and depends on the phenotype that is being monitored. Molecular methods, by measuring genotype rather than phenotype, provide an analytical method that is independent of phenotypic variability. For clinical utility, correlation of genotype with phenotype is important.

The current applications of DNA testing for genetic diseases include diagnosis or confirmation of diagnosis, carrier detection, and prenatal diagnosis. Although limited in scope at present, future genetic testing very likely will include presymptomatic diagnosis and risk assessment for disease susceptibilities, particularly for such relatively common conditions as cancer, heart disease, and neurological disorders.

2 DIRECT AND INDIRECT GENETIC ANALYSIS

2.1 DIRECT MUTATION DETECTION

Ideally, DNA analysis for genetic disease involves testing for the specific mutation or mutations relevant for the specific condition. This ideal situation is rarely realized in practice. Direct detection of mutations requires knowledge about the nucleotide sequence of the gene and the specific mutations in the gene that are associated with the disease phenotype. With rare exceptions, the complete characterization of mutations has proven to be a formidable task. For example, in the case of cystic fibrosis (CF), identification of one mutation, F508, which can account for as much as 80% of the CF mutations in some populations, led to great optimism that a relatively small number of mutations could account for a majority of CF cases. However, this optimism was soon quenched upon realization that CF could result from a very large number of different mutations within the CF gene. More than 500 mutations have been identified to date, and the number is increasing.

Because of the large number of different mutations found for most genetic disease genes, identification of specific mutations is very important for accurate diagnosis. Key to development of a diagnostic test is the characterization of the spectrum, frequency, and clinical significance of mutations within a gene. These mutations can be point mutations, short deletions or duplications, or large deletions or duplications. If a mutation cannot be identified in a patient specimen, the accuracy of the result depends on the frequency of mutations that cannot be ruled out by the analysis. This frequency may not always be known, and it can vary among different ethnic populations. Therefore, while direct detection of mutations provides very accurate results when the mutations are detectable, the utility of this approach is limited if a significant proportion of mutations within a gene cannot be detected. In the case of CF, ΔF508 and six other mutations comprise approximately 80–85% of the CF mutations in Caucasians of northern European background. Testing for an additional 15 mutations may increase the detection level to approximately 90% for this ethnic group. To achieve 95% detection of mutations may require analysis for as many as 25–50 more mutations.

2.2 Indirect Linkage Analysis

Although direct detection of mutations is the ideal method for genetic analysis using DNA probes, indirect analysis using linkage can be performed when the gene of interest or the specific mutation in the disease gene has not been characterized. Linkage analysis tracks a disease allele indirectly by following the pattern of inheritance of polymorphic alleles that map very close to the disease gene. This method requires probes for polymorphisms that are linked to the gene of interest. Linkage analysis makes use of polymorphic markers of several types, including restriction fragment length polymorphisms (RFLPs), variable number of tandem repeat (VNTR) sequences, and dinucleotide or other short tandem repeat (STR) sequences.

Although very powerful, linkage analysis has certain limitations. First, the requirement for testing multiple individuals in a family can be met only when the family members are available and cooperative. Analysis generally requires testing of an affected individual who may or may not be available. Testing of the affected individual is usually critical to establish the particular polymorphic allele that is linked to the disease allele for a specific family. Because analysis of several family members is necessary, confidentiality of individual patient information may be compromised. The interpretations of a linkage analysis depend on the accuracy of the family pedigree and the clinical diagnosis of affected individuals. Nonpaternity can lead to erroneous interpretations if not detected or to social complications if it is detected. Finally, because linkage analysis is an indirect assay for the gene of interest, there is the possibility of false interpretations due to recombination between the marker being analyzed and the disease gene. The recombination frequency increases proportionally to the distance between the two genetic loci.

In spite of these limitations, the method is very useful for carrier analysis and prenatal diagnosis of numerous genetic diseases. For disease genes that have been mapped but not cloned, linkage analysis may provide the only means of genetic testing.

Table 1 lists some of the more common genetic conditions that currently can be analyzed by nucleic acid probes; also shown are the modes of inheritance and the chromosome locations of these disease genes. The current status of these disease genes with regard to cloning is indicated. The list is neither comprehensive nor static, and changes are occurring rapidly.

3 METHODOLOGY

Gross genetic changes can be visualized directly using cytogenetic methods. Classical cytogenetics has been enhanced by the use of DNA probes to visualize specific chromosomes or genes. These probes are labeled with a fluorochrome and can be seen with a fluorescence microscope. Fluorescence in situ hybridization (FISH) methodology provides a very useful adjunct to standard cytogenetic methods for rapid detection of gross genetic changes such as aneuploidies, translocations, and large deletions and duplications. Unlike standard cytogenetics, which requires observation of metaphase chromosomes, many FISH analyses can be performed on interphase cells, eliminating the need for cell culture. The applications of FISH should increase in cytogenetic laboratories in the future.

Many mutations in disease genes involve relatively subtle changes that are not detectable by cytogenetics or by FISH. These genetic alterations call for more sensitive tools of molecular biology. Southern blot analysis is the classic method for RFLP analysis. More recently, nucleic acid amplification, usually using the polymerase chain reaction (PCR) coupled with various detection methods, has been replacing Southern blotting. Virtually all RFLPs can be analyzed by PCR with great savings in time and specimen quantity. Furthermore, PCR offers greater flexibility and diversity for analysis of other polymorphisms including VNTRs and STRs, as well as specific mutations. Other amplification methods, including the ligase chain reaction (LCR), also can be used for linkage analysis, as well as for direct mutation detection.

For detection of large deletions, Southern blotting is the general method of choice. Pulse field electrophoresis is used if the DNA fragments analyzed are very large (>15–20 kilobase pairs). If specific deletions are of interest, PCR methods can be devised for their detection. Multiplex PCR reactions can be used to scan relatively large portions of the genome for cases of deletions that tend to cluster in "hot spots," effectively replacing multiple Southern blots with a single reaction tube. The clear advantage of multiplex PCR over Southern blotting is the ability to scan mutation "hot spots" using less than 1% the amount of DNA (low specimen size requirement) and taking less than 10% the time (short turnaround time). Furthermore, PCR is much more tolerant of poor DNA specimen quality than is Southern blotting.

Several methods can be used for detection of specific mutations. The most widely utilized method uses allele-specific oligonucleotides (ASOs) for the mutations of interest. ASO probes usually are about 20 nucleotides long and, upon labeling, are used as hybridization probes for target sequences that are immobilized onto a solid support such as a nylon membrane. Labeling of probes can use isotopic (^{32}P) or nonisotopic (usually biotinylated or digoxigenin-modified substrates) methods. The target sequences invariably are amplified by PCR or some other method, and in some cases are sep-

Table 1   Genetic Conditions for Molecular Diagnosis

Condition	Mode	Gene Product	Symbol	Chromosome	Cloned
Adrenoleukodystrophy	XR		ALD	X	+
Adult polycystic kidney disease	AD		PKD1	16	+
Agammaglobulinemia, X-linked	XR		AGMX2	X	+
α1-Antitrypsin deficiency	AR	α1-Antitrypsin	AAT	14	+
α-Thalassemia	AR	α-Globin	HBA	16	+
Alport syndrome	XD	Collagen IV, α5-subunit	COL4A5	X	+
Alzheimer disease, familial early onset	AD	Amyloid precursor protein	APP	21	+
Amyotrophic lateral sclerosis	AD	Superoxide dismutase 1	SOD1	21	+
Angelman syndrome	AR		ANCR	15	−
Beckwith–Wiedemann syndrome	AD		BWS	11	−
β-Thalassemia	AR	β-Globin	HBB	11	+
Breast cancer susceptibility	AD		BRCA1	17	−
Charcot–Marie–Tooth disease 1A	AD	Peripheral myelin protein 22	PMP22	17	+
Charcot–Marie–Tooth disease 1B	AD	Myelin protein zero	MPZ	1	+
Chronic granulomatous disease, X-linked	XR	Cytochrome b-245 β subunit	CYBB	X	+
Congenital adrenal hyperplasia	AR	21-Hydroxylase	CAH	6	+
Cystic fibrosis	AR	Cystic fibrosis transmembrane regulator	CFTR	7	+
Dentatorubral–pallidoluysian atrophy	AD		DRPLA	12	+
Duchenne/Becker muscular dystrophy	XR	Dystrophin	DMD	X	+
Fabry disease	XR	α-Galactosidase	GLA	X	+
Familial adenomatous polyposis	AD	APC protein	APC	5	+
Familial hypercholesterolemia	AD	Low density lipoprotein receptor	LDLR	19	+
Fragile X syndrome	XD	Fragile X metal retardation	FRAXA	X	+
Friedrich ataxia	AR		FRDA	9	−
Fructose intolerance, hereditary	AR	Aldolase B	ALDOB	9	+
Gaucher disease	AR	Glucocerebrosidase	GBA	1	+
Glycerol kinase deficiency	XR	Glycerol kinase	GK	X	+
Hemochromatosis	AR		HFE	6	−
Hemophilia A	XR	Factor VIII	F8C	X	+
Hemophilia B	XR	Factor IX	F9	X	+
Hereditary nonpolyposis colon cancer	AD	*mut L* Homologue 1	MLH1	3	+
Hereditary nonpolyposis colon cancer	AD	*mut S* homologue 2	MSH2	2	+
Huntington disease	AD	Huntingtin	HD	4	+
Kennedy disease (spinal/bulbar muscular atrophy)	XR	Androgen receptor	AR	X	+
Lesch–Nyhan syndrome	XR	Hypoxanthine phosphoribosyltransferase	HPRT	X	+
Li–Fraumeni syndrome	AD	Tumor suppressor protein p53	TP53	17	+
Malignant hyperthermia	AD	Ryanodine receptor	RYR1	19	+
Marfan syndrome	AD	Fibrillin	FBN	15	+
Medium chain acyl–CoA dehydrogenase deficiency	AR	Medium chain acyl–CoA dehydrogenase	MCAD	1	+
Melanoma susceptibility	AD		MLM	9	−
Menkes disease	XR		MNK	X	+
Multiple endocrine neoplasia 1	AD		MEN1	11	−
Multiple endocrine neoplasia 2A	AD	ret oncoprotein	MEN2A	10	+
Myotonic dystrophy	AD	Myotonin protein kinase	DM	19	+
Neurofibromatosis type 1	AD	Neurofibromin	NF1	17	+
Neurofibromatosis type 2	AD	Merlin, schwannomin	NF2	22	+
Norrie disease	XR		NDP	X	+
Ornithine transcarbamylase deficiency	XR	Ornithine transcarbamylase	OTC	X	+
Osteogenesis imperfecta	AD	Collagen I, α$_1$ subunit	COL1A1	17	+
Osteogenesis imperfecta	AD	Collagen I, α$_2$ subunit	COL1A2	7	+
Phenylketonuria	AR	Phenylalanine hydroxylase	PAH	12	+
Prader–Willi syndrome	AR		PWCR	15	−
Retinoblastoma	AD		RB1	13	+
Sandhoff disease	AR	Hexosaminidase, β subunit	HEXB	5	+
Sickle cell anemia	AR	β-Globin	HBB	11	+
Spinal cerebellar ataxia type 1	AD		SCA1	6	+
Steroid sulfatase deficiency	XR	Steroid sulfatase	STS	X	+
Tay–Sachs disease	AR	Hexosamididase A, α subunit	HEXA	15	+
Tuberous sclerosis	AD		TSC1	9	−
Von Hippel–Lindau disease	AD		VHL	3	+
Werdnig–Hoffmann disease (spinal muscular atrophy)	AR		SMA1	5	−
Williams syndrome	AD	Elastin 1	ELN	7	+
Wilms' tumor type 1	AR		WT1	11	+
Wilson disease	AR		WND	13	+

arated by gel electrophoresis prior to immobilization. If gel separation is used, the target sequences are transferred onto the support membrane by Southern blotting. Usually, separation is not necessary, and the target sequences are placed directly onto the support as dot or slot blots. An alternative to the dot/slot blot method is the reverse dot blot method in which the ASOs are immobilized onto a membrane or some other support medium, and the membranes are hybridized with target sequences that have been labeled during amplification.

Several other methods have been used for specific mutation detection. One of the first to be employed depends on coincidental restriction site alteration. Certain mutations alter restriction enzyme cleavage sites and can be distinguished from the corresponding wild-type sequence by digestion with the appropriate enzyme. For example, the S mutation in the β-globin gene, which when homozygous results in sickle cell anemia, causes a loss of a restriction site for the enzyme Mst II. This obviously is not a general method, since very few mutations affect restriction enzyme sites. A modification of this method involves the use of a slightly mismatched PCR primer containing sequences that when combined with the mutation sequence of interest, create a restriction enzyme cleavage site at the site of the mutation. This modification extends the utility of the restriction site alteration method of mutation detection, but the need at the mutation site for appropriate sequences that can be altered to create a restriction site limits it, nevertheless.

Allele-specific amplification and allele-specific ligation, utilizing primers complementary to either the wild-type or the mutant sequence, provide two alternative means for the detection of specific mutations. In addition, some methods are available to screen for the presence of mutations without identifying the specific mutation itself. These methods include single-strand conformational polymorphism (SSCP) analysis, denaturing gradient gel electrophoresis (DGGE), and mismatch cleavage analysis by enzymatic (RNase A) or chemical (piperidine) means. SSCP and DGGE detect differences in electrophoretic mobilities between wild-type and mutant sequence that are due to sequence-specific differences in intrastrand structure (SSCP) or in duplex melting properties (DGGE). Mismatch cleavage methods detect unpaired regions in heteroduplex molecules of wild-type and mutant DNA. Of these methods, SSCP is the most widely utilized, because it is technically less challenging than the other two. The efficiency of SSCP is about 80–90% for detection of point mutations in regions of 200 nucleotides. A recent modification of SSCP called dideoxy fingerprinting may increase the efficiency of mutation detection to over 90%. DGGE and chemical mismatch cleavage methods are reported to give greater than 90% detection of point mutations in regions of 500–600 nucleotides. These mutation screening methods, although very useful as research tools, have yet to be utilized on a widespread basis in clinical laboratories.

DNA sequencing provides another option for mutation detection. Sequencing, in theory, should be able to detect every possible mutation within a gene. As a practical matter, DNA sequencing had been used for screening genes for mutations, but its utilization as a clinical diagnostic tool is limited currently by its high cost and labor intensiveness. Should the cost of DNA sequencing drop in the future, genetic analysis by direct sequencing may become feasible. New sequencing strategies coupled with automation will be important factors in future utilization of this method for routine genetic disease testing. Critical to the success of direct sequencing for routine clinical use will be the ability to detect heterozygotes in patient DNA specimens.

One approach to increasing the efficiency of sequencing has been the use of automated sequencing apparatuses capable of approximately fourfold increase in throughput over manual sequencing. Current instruments use slab gels of polyacrylamide for electrophoretic separations. Use of capillaries with appropriate separation media may give another three- to five-fold increase in sequencing throughput over today's automated sequencers. Current methods for sequencing involve creation of four nested sets of single-stranded DNA fragments with one common end and the other end determined by the location of each of the four nucleotides. The four sets of nested fragments are separated by electrophoresis on the basis of size, resulting in sequencing "ladders." The process is relatively time-consuming and labor intensive. One attempt to increase throughput entails sequencing by hybridization, where a nucleotide sequence is determined by the pattern of hybridization to an ordered, solid state set of oligonucleotides (e.g., a "chip" containing all possible octamer sequences). Although there are technical challenges to be overcome, sequencing "chips" may prove to be an efficient sequencing method in the future, with some obvious applications in the area of genetic disease analysis.

The task of scanning genes for mutations can be formidable, particularly if mutations are located throughout a large gene. In special cases, functional assays may be devised as alternatives to direct genetic analysis. The development of functional assays depends on the nature of the gene, the types of mutation found in the gene, and the effect of these mutations on the function of the gene. In the case of *TP53*, a tumor suppressor gene involved in Li–Fraumeni syndrome, most mutations are missense and lead to protein product that lacks the transcriptional activation properties of the wild-type protein. These properties of the *TP53* gene mutations have been used to develop an in vitro transcriptional activation assay for mutations of this gene. Contrast this to the tumor suppressor gene *APC*, involved in familial adenomatous polyposis colon cancer. Most *APC* mutations are nonsense or frameshift, leading to truncated protein product. Protein expression assays detecting premature termination in the *APC* gene due to these mutations have been developed. Note that while functional assays may overcome the difficulty of directly detecting many possible mutations within a gene, they are specific for the gene of interest and the types of mutation within the genes. In some cases, functional assays may be as technically challenging as direct mutation detection; nevertheless, they have their place, if not as alternatives to direct mutation analysis, as important adjuncts to molecular genetic testing.

4   POTENTIAL COMPLICATIONS

The technical limitations of mutation detection and linkage analysis have been noted. In addition, failure to consider the unique features of some genetic diseases can lead to incorrect interpretations or predictions. One possible complication is mosaicism, which can exist in both germ line and somatic cells. Germ line mosaicism will affect the risk of a mosaic for having affected children. Direct detection of germ line mosaicism usually is not possible because upon analysis, somatic cells like blood may not be representative of gametes. Somatic cell mosaicism may influence the expression of a disease phenotype. As with germ line mosaicism, the genotype of the tissue tested may not reflect the genotype of the relevant tissue. The mechanism for mosaicism, although not understood, may differ for different diseases. Mosaicism can complicate correlation of genotype with phenotype. For example, somatic mosaicism may be responsible for one reported case of discordant phenotype with respect to fragile X syndrome between monozygotic twins.

Uniparental disomy and imprinting, although apparently rare, can confound genetic analysis. These phenomena were recognized relatively recently, and their effects are still in the process of being elucidated. One of the effects of uniparental disomy is the inheritance of an autosomal recessive condition from only one carrier parent, as has been observed for cystic fibrosis and some other conditions. Imprinting can result in gender-specific transmission of autosomal dominant conditions, as has been observed for some cases of Prader–Willi and Angelman syndromes. The genes for these two conditions have not been cloned, but they are distinct and map within chromosome 15q11-12. Paternally derived deletions of 15q11-12 can lead to Prader–Willi syndrome due to maternal imprinting of this gene, while maternally derived deletions of 15q11-12 lead to Angelman syndrome due to paternal imprinting. Uniparental disomy of an imprinted allele can result in a genetic disease through epigenetic mechanisms. Again for Prader–Willi and Angelman syndromes, respectively, maternal and paternal uniparental disomy can specifically result in each condition.

5 SPECIFIC EXAMPLES

5.1 AUTOSOMAL RECESSIVE CONDITIONS

5.1.1 α1-Antitrypsin Deficiency

α 1-Antitrypsin (AAT) is a serum protease inhibitor encoded on chromosome 14. AAT deficiency is an autosomal recessive trait, and two clinically relevant alleles of the gene are designated Z and S. Homozygotes of the Z allele and SZ compound heterozygotes are highly sensitive to dust and smoke, resulting in often fatal lung and liver disease. These two alleles are due to separate single-nucleotide mutations within the AAT gene. Genetic detection of AAT deficiency due to the Z or S mutations is straightforward but is usually limited to prenatal diagnosis. For more routine testing, direct analysis of AAT protein in serum using standard gel electrophoretic or isoelectric focusing methods is performed.

5.1.2 Congenital Adrenal Hyperplasia

Congenital adrenal hyperplasia (CAH) is an autosomal recessive condition characterized by virilization, which may be accompanied by complications due to sodium loss (salt wasting). CAH is due to a defect in the biosynthesis of cortisol, and most cases are due to mutations in the 21-hydroxylase gene *(CYP21B)*. Analysis of this gene, located on chromosome 6, is complicated by the presence of a adjacent pseudogene with several inactivating mutations that often alter the active gene by a process called gene conversion. Analysis generally involves PCR amplification and direct mutation detection. For routine diagnosis of CAH, the analysis of 17-hydroxyprogesterone usually is performed. DNA probe analysis is generally limited to prenatal diagnosis.

5.1.3 Cystic Fibrosis

Cystic fibrosis is the most common severe genetic disease among Caucasians. It is an autosomal recessive condition, and carriers show no symptoms of the disease. Affected individuals carry two mutant *CF* genes and can exhibit impairments of varying severity in respiratory, digestive, and reproductive functions. The frequency

of CF in Caucasian populations is about one in 2500 births, meaning that about one in 25 Caucasians is a carrier. More than 500 mutations have been described in the *CF* gene, which is approximately 250 kb in size and produces an mRNA of about 6.5 kb. Of these mutations, one called ΔF508 accounts for approximately 70% of the *CF* mutations in the U.S. Caucasian population. None of the other mutations is highly abundant, but particular mutations are very frequent in specific ethnic groups such as the Ashkenazic Jews and the Hutterites.

CF gene analysis generally involves testing for the more common mutations. There is yet no consensus on the number or the set of mutations to be analyzed. Different laboratories are analyzing anywhere from as few as four to six mutations to as many as twenty or more mutations. Detection frequencies are between 80 and 90% in Caucasians, and appreciably lower in most other ethnic groups.

5.1.4 Gaucher Disease

An autosomal recessive condition, Gaucher disease is a lysosomal storage disease resulting from a deficiency of the enzyme glucocerebrosidase, whose gene (*GBA*) is located on chromosome 1. The deficiency results in a pathological phenotype in macrophages leading to problems of the spleen, liver, and bone. Gaucher disease is relatively common in the Ashkenazic Jewish population, and although numerous mutations in the *GBA* gene can lead to the condition, in this ethnic group five mutations can account for about 98% of the disease mutations. Mutation screening for diagnosis is possible, but biochemical analysis is more routinely performed. The value of mutation detection for diagnosis of Gaucher disease is somewhat limited because of the variability of symptoms even for those with the same apparent genotype. Nevertheless, numerous mutations have been characterized with reasonable correlation with regard to severity. Diagnosis of Gaucher disease is useful because enzyme replacement therapy is available, although at very high cost.

5.1.5 Hemoglobinopathies

Among the most well characterized of the genetic diseases are the homoglobinopathies, conditions affecting the hemoglobin genes. The genes encoding the polypeptides of hemoglobin are located in two clusters, one on chromosome 11 representing the β-globin family of genes and the other on chromosome 16 representing the α-globin family. The various hemoglobin molecules are tetrameric, consisting of two α-globin chains and two β-globin chains. The major form of normal adult hemoglobin is hemoglobin A ($\alpha_2\beta_2$). Other normal forms of hemoglobin include A2 ($\alpha_2\delta_2$), fetal ($\alpha_2\gamma_2$), and embryonic ($\alpha_2\varepsilon_2$). Defects in the α- and β-globin genes result in the thallasemias. Very specific mutations within the β-globin gene result in the sickling hemoglobinopathies like sickle cell anemia.

Sickle Cell Anemia The prototype of single-mutation analysis is sickle cell anemia, which is an autosomal recessive condition due to a single mutation in the β-globin gene on chromosome 11. This mutation, resulting in the *S* allele, causes a single amino acid change in β-globin. Individuals homozygous for this mutation have red blood cells that, under certain conditions, form a characteristic crescent or "sickle" shape, resulting in hemolytic anemia and various other clinical complications. Two other mutations, *C* and *E,* of the β-globin gene can also result in sickling hemoglobinopathies, usually with less severe phenotype than the *S* allele. The detection of these mutations is very straightforward by PCR using ASOs or restriction endonuclease digestion, but DNA probe analysis for sickling hemoglobinopathies, in general, is limited to prenatal diagnosis. Routine screening and confirmation of diagnosis is best done by analysis of the protein by other methods (e.g., hemoglobin electrophoresis).

β-Thalassemia   Although biochemical tests are available for diagnosis of β-thalassemia, DNA probe analysis does have utility for prenatal diagnosis. Mutations in the β-globin (*HBB*) gene have been well characterized and encompass virtually the entire spectrum of mutations including deletions, gene fusions, frameshift, missense, nonsense, splicing, and promoter mutations. The vast majority of these mutations are point mutations resulting in missense. The disease phenotype can vary significantly from β-thalassemia major, a severe lethal form of the disease, to very mild, virtually asymptomatic forms. The variety of mutations differs greatly by ethnic group, and genetic testing generally involves the use of a panel of mutations specific for the ethnic group of the patient being tested. Alternatively, direct sequencing of the exons *HBB* gene can be performed.

α-Thalassemia   The α-thalassemias result from various defects of the α-globin genes on chromosome 16. Unlike β-thalassemia, a large number of α-thalassemia mutations involve large deletions. Because most individuals have two α-globin loci per chromosome, several types of deletion resulting in α-thalassemia can exist. Deletion of all four copies of the gene is lethal, resulting in a fatal condition called hydrops fetalis. An individual with only one copy of the α-globin gene has a relatively severe form of α-thalassemia called hemoglobin H disease. Although deletion is the major mechanism, numerous other mutations can exist in the α-globin genes, and their detection requires use of mutation scanning methods. As with the other hemoglobinopathies, hemoglobin electrophoresis is routinely used for diagnosis of α-thalassemia, except in the case of prenatal diagnosis, where direct mutation detection or linkage analysis can be used.

5.1.6 Medium Chain Acyl Coenzyme A Dehydrogenase Deficiency

Deficiency of medium chain acyl coenzyme A dehydrogenase (MCAD), an autosomal recessive condition, can result in life-threatening episodes of hypoglycemia and coma, resembling sudden infant death syndrome (SIDS). Some cases of SIDS, as well as unexplained infant death during the first four years, may be due to MCAD deficiency. About 90% of the MCAD gene mutations are due to one point mutation of A to G resulting in a change from lysine to glutamic acid at position 304 of the mature MCAD protein (K304E). Direct analysis of the K304E mutation is highly specific and relatively simple and is preferable to the more difficult biochemical analysis of urine for diagnosis of MCAD deficiency in the case of homozygotes of K304E. Even though the K304E mutation may account for 90% of MCAD mutations, 20% of those with MCAD deficiency will not be homozygous for this mutation and cannot be diagnosed definitively by testing for only this one mutation. This example illustrates one of the main difficulties that needs to be addressed for genetic testing: mutation analysis of most human disease genes is very specific (low false positive rate) but can have low sensitivity (relatively high false negative rate) because of

the number of mutations to be analyzed. In spite of this difficulty, MCAD genotyping for the K304E allele should become more widely utilized in the future because this mutation is relatively frequent and the biochemical assay is difficult.

5.1.7 Tay–Sachs Disease

Tay–Sachs disease, mapped to chromosome 15, is an autosomal recessive neural degenerative condition due to a deficiency of the enzyme hexosaminidase A. The enzyme consists of subunits of types, α and β, with the gene for the α subunit (*HEXA*) mutated in Tay–Sachs disease. Hexosaminidase B contains only the β subunit, and mutations in the gene for the β subunit on chromosome 5 result in a condition called Sandhoff disease, which has symptoms very similar to Tay–Sachs disease. Manifestations of Tay–Sachs disease can range from severe infantile onset forms to relatively mild adult forms, with the more severe forms being fatal by the third year of life. Severity and genotype are correlated certain mutations. The frequency of Tay–Sachs disease is relatively high in the Ashkenazic Jewish population (≈ 1 in 2000) and about 10,000-fold lower in frequency in the overall U.S. population. Other populations showing relatively high frequencies for Tay–Sachs disease are the French Canadians and the Cajuns.

The spectrum of mutations in the *HEXA* gene in the Ashkenazic Jewish population has been characterized. Two mutations associated with infantile onset Tay–Sachs disease comprise about 90% of the mutations in this ethnic group. Many other mutations have been described, including a large deletion of the 5′ end of the *HEXA* gene, which is relatively prevalent in the French Canadian population. Otherwise, most of these mutations are relatively rare. Although DNA probe analysis is possible for Tay–Sachs disease, biochemical analysis of hexosaminidase A is more commonly done.

5.2   AUTOSOMAL DOMINANT CONDITIONS

5.2.1 Amyotrophic Lateral Sclerosis

Amyotrophic lateral sclerosis (ALS), sometimes called Lou Gehrig disease, is a fatal degenerative neural disorder with variable age of onset. About 10% of ALS is familial and is transmitted as an autosomal dominant disorder. Recently, the superoxide dismutase 1 (*SOD1*) gene on chromosome 21 has been associated with some cases of familial ALS. The spectrum of mutations in the *SOD1* gene is being characterized, with indications that most mutations are missense and that some cases of apparently sporadic ALS are due to germ line mutations in the *SOD1* gene. Genetic heterogeneity does exist for familial ALS, and other genes in addition to *SOD1* can independently lead to this disease. It will be important to establish the relative proportion of *SOD1* mutations in the etiology of ALS before screening of this gene for susceptibility to ALS is feasible.

5.2.2 Charcot–Marie–Tooth Disease

Charcot–Marie–Tooth (CMT) disease is a genetically heterogeneous condition affecting peripheral motor and sensory neurons. There are several forms of CMT, which show autosomal dominant, autosomal recessive, and X-linked forms of inheritance. Although severity and age of onset can vary significantly, symptoms include muscle weakness and atrophy, sensory loss, and complications of the gastrointestinal, cardiac, and skeletal systems. The most common forms of CMT, designated type 1, show autosomal dominant inheritance and are characterized by slow motor nerve conduction velocity. Two of these type 1 forms of CMT have been linked to separate genes on chromosomes 17 *(CMT1A)* and 1 *(CMT1B)*.

CMT1A is the more common form and is due to mutations within the peripheral myelin protein 22 *(PMP22)* gene. A relatively common mutation is a duplication of the gene, suggesting that gene dosage of *PMP22* is one mechanism for expression of *CMT1A*. This mutation can be detected cytogenetically using FISH. Other mutations within the *PMP22* gene leading to *CMT1A* have been described. The other mutations include missense as well as an intragenic deletion, indicating that mechanisms other than gene dosage can be involved in the etiology of *CMT1A*. Other mutations within the *PMP22* gene can result in a condition known as Dejerine–Sottas syndrome, a neuropathy similar to *CMT1A* but with generally more severe symptoms.

CMT1B is due to mutations in the myelin protein zero *(MPZ)* gene on chromosome 1. The spectrum of mutations in the *MPZ* gene responsible for *CMT1B* has not been characterized fully, although initial indications are that numerous missense mutations are present. As with *CMT1A*, analysis of *CMT1B* is generally limited to families with appropriate histories. For such families, mutation scanning can be performed, followed by linkage analysis if no mutation can be detected.

5.2.3 Connective Tissue Diseases

Adult Polycystic Kidney Disease   Adult or late onset polycystic kidney disease (APKD) is linked to chromosome 16, and a candidate gene *(PKD1)* has been isolated recently. Mutations within the gene are being characterized. At present, linkage analysis is performed, but direct mutation analysis should be available in the near future. APKD is characterized by renal cysts and kidney failure. The condition is genetically heterogeneous, with the form linked to chromosome 16 being the most prevalent. Analysis of APKD is of value for screening potential kidney donors for affected relatives.

Alport Syndrome   Alport syndrome is a genetically heterogeneous condition characterized by nephritis, sometimes accompanied by impairment or loss of vision and hearing. The most common form of Alport syndrome is an X-linked dominant trait that has been associated with the gene for the α5 chain of type IV collagen *(COL4A5)*. Type IV collagen is the major component of basement membranes, which are important structural organizers of different tissues. Alport syndrome results from various types of mutation found in the *COL4A5* gene. Mutation detection is possible but difficult for this large gene. Linkage analysis of families with X-linked Alport syndrome is also possible.

Marfan Syndrome   Marfan syndrome is a disease of connective tissue affecting skeletal, optic, and cardiovascular features. Severity is highly variable, with some fatalities due to cardiovascular problems such as aortic aneurism. Age of onset varies dramatically from infancy to adulthood. In addition to cardiac problems, Marfan syndrome is characterized by long limbs and digits, loose joints, and ocular problems. The fibrillin 1 *(FBN1)* gene on chromosome 15 has been shown to be mutated in Marfan disease patients. Direct detection of mutations within the *FBN1* gene is complicated because of the large size of the gene (65 exons) and the apparently large number of mutations in the gene that lead to Marfan syndrome. Most of the mutations have yet to be characterized, and a large amount of effort will be required to identify and characterize mutations in such a large gene.

Osteogenesis Imperfecta   Osteogenesis imperfecta (OI) is a heterogeneous, autosomal dominant condition with highly variable severity. The common feature of OI is extreme bone fragility, usually due to defects in type I collagen, which is a major structural component of bone. Type I collagen consists of two subunits, α1 and α2, and OI can result from mutations in the genes encoding either subunit. The gene for the α1 subunit *(COL1A1)* is located on chromosome 17, while that for the α2 subunit *(COL1A2)* is located on chromosome 7. Direct mutation detection is possible for analysis of OI, but in practice it is severely limited by the size and nature of collagen genes, which contain many small exons encoding the repetitive tripeptide, Gly-x-y. Linkage analysis is possible for appropriate families in which association of OI with a particular collagen gene can be established.

5.2.4 Familial Hypercholesterolemia

Familial hypercholesterolemia (FH) is an autosomal dominant condition characterized by elevated serum cholesterol levels due to defects in the low density lipoprotein receptor *(LDLR)* gene located on chromosome 19. FH patients are at high risk for early onset coronary artery disease and other cardiovascular problems. Homozygotes show significantly more severe symptoms than heterozygotes. Numerous mutations of the *LDLR* gene have been described for FH, and mutation screening methods are generally used to identify mutations in specific families. The large number of mutations makes routine genetic analysis for FH problematic; serum cholesterol measurement along with pedigree analysis provides a more practical method for diagnosis of this condition.

5.3   X-Linked Recessive Conditions

5.3.1 Duchenne Muscular Dystrophy

Duchenne muscular dystrophy (DMD) is the most common of the childhood muscular dystrophies. Affected males usually present with muscle weakness at around 2–6 years of age, are unable to walk at around 12 years, and seldom survive beyond their twenties. The *DMD* gene encodes a protein designated dystrophin that is found in muscle and brain cells. A milder form of the disease, called Becker muscular dystrophy (BMD), is due to alterations in the same gene as *DMD*. Approximately 60% of cases of DMD and BMD are due to deletions or duplications that can be detected by Southern blot analysis using the complete 14 kb dystrophin cDNA probe. Up to 80% of the deletions detectable by Southern blot analysis can be detected rapidly by multiplex PCR using up to 18 pairs of primers in deletion "hot spots" in the *DMD* gene. While 60% of *DMD* mutations can be detected as deletions and duplications, the remaining 40% are presumably due to point mutations or short deletions that lie within the 2,500 kb *DMD* gene. Point mutation scanning is unfeasible by present methods because of the extremely large size of this gene.

Mutation detection of *DMD* deletions, when successful, provides a rapid and convenient method for diagnosis of affected males and identification of carrier females (by quantitative PCR or Southern blotting). When mutation detection is unsuccessful, linkage analysis may be an option if appropriate family members are available.

5.3.2 Hemophilias A and B

Hemophilias A and B are two separate X-linked clotting disorders due to mutations in the genes encoding clotting factors VIII *(F8C)* and IX (*F9*), respectively. Hemophilia A is more common than hemophilia B, probably reflecting the relative sizes of the genes. Many different mutations have been found in both genes, and mutation screening is possible but labor intensive. As with the hemoglobinopathies, molecular genetic analysis is usually limited to prenatal diagnosis.

5.3.3 Lesch–Nyhan Syndrome

Lesch–Nyhan syndrome is a relatively rare X-linked recessive condition resulting in mental retardation, cerebral palsy, and self-mutilation. The condition is due to defects in the gene encoding hypoxanthine–guanine phosphoribosyltransferase (HPRT), an enzyme involved in purine metabolism. The gene is relatively small, making mutation scanning feasible. Mutations have been identified using mismatch cleavage methods as well as by direct sequencing. Most of the mutations are point mutations, and many different mutations exist in the gene. Molecular genetic analysis is generally limited to prenatal diagnosis.

5.4   TRINUCLEOTIDE REPEAT CONDITIONS

Recently, a new mechanism for mutagenesis has been described for several genetic conditions. These conditions involve genes that contain CTG (CAG) or CGG (CCG) trinucleotide repeats that can exist in stable normal form or unstable expanded form. The number of copies of the trinucleotide determining stable from unstable forms varies from disease to disease, and the mechanism causing instability of these repeats is not known. Expansion of trinucleotide repeats can occur during gametogenesis or embryogenesis or both, depending on the disease. The trinucleotide repeat conditions described to date share some common features. They are all neurological disorders that more or less exhibit a phenomenon called "anticipation" (i.e., the severity of the disease generally increases from generation to generation). These diseases also can show gender-specific bias in transmission of the disease or in disease severity. In addition to fragile X syndrome, Huntington disease, and myotonic dystrophy, several other conditions, including Kennedy disease (spinal and bulbar muscular atrophy), spinal cerebellar ataxia type 1, and dentatorubral–pallidoluysian atrophy, are due to expansion of trinucleotide repeat sequences. Undoubtedly, other genetic diseases involving trinucleotide repeat expansion will be discovered.

5.4.1 Fragile X Syndrome

Fragile X syndrome is the most frequently reported of the heritable forms of mental retardation and exhibits unusual genetic features. Although it is an X-linked genetic disease, females can show symptoms of varying severity. Furthermore, there are clinically normal males that are carriers of fragile X syndrome and transmit the disease through their clinically normal daughters to their grandchildren. The major fragile X locus (FRAXA) contains a gene designated *FMR1* with a CGG trinucleotide repeat in the 5' untranslated region of the gene. Normal individuals have between 5 and 50 copies of the CGG repeat; affected individuals have greater than 200 copies of the repeat. Individuals with 50–200 copies of the CGG repeat are said to be carriers of a fragile X premutation. These individuals, very often, do not have symptoms of fragile X but have a high probability of transmitting the disease to their children or grandchildren. Apparently, the CGG repeat becomes destabilized as it increases in size. Thus, alleles with fewer than 50 copies of the

CGG repeat appear to be stable and do not change in future generations. Alleles with 50–200 CGG repeats are less stable and tend to expand in size when transmitted to offspring. The stability of premutations appears to be size dependent, and alleles with more than 100 repeats have very high probability of expanding to more than 200 repeats. The expansion of trinucleotide repeats for fragile X apparently occurs in somatic cells during early embryonic development, and the instability is limited to the maternally inherited X chromosome.

The full fragile X mutation, having more than 200 trinucleotide repeats, shows hypermethylation of the expanded allele in patient DNA samples. Thus, molecular analysis of fragile X syndrome involves detection of expanded DNA using Southern blotting and/or PCR, as well as determination of the methylation status of the fragile X locus using methylation-sensitive restriction endonucleases.

5.4.2 Huntington Disease

A fatal, late onset neurodegenerative disorder, Huntington disease (HD) is an autosomal dominant condition linked to chromosome 4. One of the first disease genes to be mapped by the RFLP method, isolation of the *HD* gene proved especially elusive, and isolation of the gene took almost 10 years of intense effort by an international consortium of laboratories. The putative protein product of the *HD* gene has been designated huntingtin and shows no homology to any other protein. A CAG trinucleotide repeat lies within the coding sequence of the gene and is presumably translated into multiple glutamine residues near the N-terminus of huntingtin. Studies by different laboratories indicate that non-*HD* chromosomes contain from 9 to 37 copies of the CAG repeat, *HD* chromosomes from 16 to over 100 copies of the repeat. With regard to diagnosis, having more than 40 copies of the CAG repeat appears to be highly indicative of HD, while having fewer than 30 copies suggests a low probability. There is some anticipation in HD, although not as pronounced as in fragile X syndrome or myotonic dystrophy. Also, there is a tendency for greater severity of disease arising from paternally inherited alleles. The clinical significance of 30–40 copies of the CAG repeat remains to be clarified and may require interpretation within the context of a family study.

HD, more than most other genetic diseases, has raised emotional ethical and social issues regarding genetic testing. In addition to being dominant, HD is completely penetrant; that is, anyone inheriting the mutated form of the gene is certain to become affected, with no hope for a cure at the present time. The age of onset, however, cannot be predicted precisely. The value of presymptomatic diagnosis of a condition for which no intervention exists has been argued pro and con, and extensive counseling is warranted before at-risk individuals make a decision regarding testing for HD.

5.4.3 Myotonic Dystrophy

Myotonic dystrophy (DM) is the most common adult form of muscular dystrophy, characterized by myotonia and muscle weakness. The gene for DM, located on chromosome 19, apparently encodes a protein kinase designated myotonin. A CTG trinucleotide repeat is located within the 3' untranslated region of the myotonin mRNA. Normal individuals have *DM* alleles containing from 5 to 35 copies of the trinucleotide repeat, while affected individuals have 150–2000 copies of the repeat. Individuals with 40–150 CTG repeats are usually either asymptomatic or mildly affected, perhaps analogous to carriers of fragile X premutations. In general, the age of onset decreases and the severity increases with increasing numbers of CTG repeats. Furthermore, the severity of the disease, in some instances, is greater when the expanded allele is maternally inherited. Direct analysis of trinucleotide repeats is the preferred method for prenatal diagnosis and for confirmation of diagnosis of DM. Recent results suggest that the CTG repeats do not directly affect the myotonin gene, but have a general deleterious effect on accumulation of mRNA.

5.5   Hereditary Cancer Susceptibilities

Numerous factors, some genetic and some environmental, contribute to the development of cancer. Genetic factors predisposing individuals to certain types of cancer are beginning to be elucidated. Heritable mutations in certain cancer susceptibility genes have been shown to predispose individuals carrying these mutations to certain types of cancer. The relative contribution of germ line mutations to any specific form of cancer is not known and probably varies from cancer to cancer. For example, 5–20% of breast cancers may be due to heritable mutations. The genetic heterogeneity for different familial cancers can be variable. For example, Li–Fraumeni syndrome is always due to germ line mutations in the *TP53* gene, while hereditary nonpolyposis colon cancer can result from predisposing mutations in at least two separate genes. The degree of heterogeneity for most hereditary cancers is not known.

Heritable cancer susceptibility mutations can, in theory, be analyzed using the molecular tools used for monogenic disease. The spectrum of mutations in these cancer genes appears to be broad, necessitating efficient screening methods for mutation detection. Furthermore, polygenic conditions like cancer introduce into genetic analysis additional complications that usually are not a factor for monogenic disease analysis. One complication is genetic heterogeneity. When this condition exists (i.e., when several different genes independently confer susceptibility to a particular type of cancer), the problem of mutation detection is multiplied by the number of different genes that exist. In the absence of a detectable mutation within a susceptibility gene, linkage analysis generally is not feasible because linkage with one gene cannot be inferred for a particular family without prior analysis of many family members through multiple generations. Other complications are differences in penetrance for different mutations, which may be a difficult property to assess, and variations in severity of different mutations. Furthermore, penetrance and severity can be influenced by other genetic, epigenetic, and environmental factors.

In spite of these limitations and complications, there has been progress in analyzing members of families at risk for some of the rarer forms of heritable cancers, such as retinoblastoma and Li–Fraumeni syndrome. Linkage analysis is possible for large families in which disease association with a specific locus can be determined. Thus far, genetic analysis of familial cancer has been limited. Greater utilization will depend on the isolation and characterization of additional cancer susceptibility genes and on clinical correlation studies establishing the effectiveness and the clinical significance of detecting mutations within these genes. The rapid pace of gene discovery, coupled with the relatively high frequency of certain malignancies such as breast and colon cancer, suggest that significant progress leading to the routine genetic analysis for heritable cancers is in the offing.

5.5.1   Breast Cancer

It is estimated that 2–20% of breast cancers may be due to inherited predisposition. One predisposing gene, designated *BRCA1*, has been mapped to a small region of chromosome 17q21. Mutations of *BRCA1* may be involved in a significant fraction of heritable breast cancer susceptibility. Linkage analysis of families clearly showing linkage with 17q21 is available, but ascertainment of such families requires large pedigrees with data on multiple affected members. After intense effort, the *BRCA1* gene has been cloned. The gene is large and mutations are located throughout the gene. Recently, a second breast cancer predisposing gene, designated *BRCA2*, has been identified. The *BRCA2* gene, located on chromosome 13, also appears to be a large gene with a complex spectrum of potential mutations.

5.5.2   Colon Cancer

Familial adenomatous polyposis (FAP) is a rare cancer syndrome characterized by the early development of numerous colonic polyps that inevitably become malignant if left untreated. FAP is due to germ line mutations in the *APC* gene. The *APC* gene, located on chromosome 5, is large, and mutations can be located throughout the gene. Most mutations lead to premature chain termination due to nonsense or frameshift changes. These mutations can be detected directly, a difficult task given the size of the gene. Alternatively, as described earlier, several functional assays measuring chain termination have been developed for *APC* (see Section 3).

A more common form of colon cancer is hereditary nonpolyposis colon cancer (HNPCC), a genetically heterogeneous condition with some cases linked to chromosome 2 and other cases linked to chromosome 3. An observed characteristic of HNPCC tumor cells is instability of genomic DNA, manifested by somatic variability of dinucleotide repeat sequences. A similar type of genetic instability exists in bacterial cells defective in DNA mismatch repair. Human homologues of the bacterial DNA repair genes *mutS* and *mutL*, designated *MSH2* and *MLH1*, respectively, were isolated as candidate genes for HNPCC susceptibility. *MSH2* and *MLH1* are mutated in some carriers of HNPCC, indicating a direct relationship between these genes and susceptibility to this form of colon cancer. Mutation scanning of *MSH2* and *MLH1* is possible, but clinical utility will depend on characterization of the spectrum, frequency, and penetrance of predisposing mutations in these genes. Initial indications are that the number of mutations is large, and efficient mutation detection methods may be required for their detection.

5.5.3   Li–Fraumeni Syndrome

Li–Fraumeni syndrome (LFS) is a very rare cancer syndrome characterized by the early occurrence of tumors of the breast, brain, soft tissue, bone marrow, bone, or adrenal cortex. LFS is due to germ line mutations in the *TP53* gene, which encodes a tumor suppressor protein, p53. The spectrum of mutations is varied, and mutation scanning methods can be used for detection of specific mutations. Alternatively, a functional assay for p53 protein may be performed. Although the clinical market for analysis of germ line *TP53* mutations is very limited because of the low number of LFS families, analysis of somatic mutations in the *TP53* gene may become an important tool in the diagnosis and prognosis of different cancers. Somatic mutation of the *TP53* gene is one of the most common genetic changes in cancer, occurring in about half of all tumors.

5.5.4 Multiple Endocrine Neoplasia

Multiple endocrine neoplasia (MEN) is a genetically heterogeneous group of conditions resulting in neoplasms of the pituitary, parathyroid, and pancreas. Two genetic loci implicated in MEN are located on chromosomes 10 and 11. The *RET* proto-oncogene on chromosome 10 has been associated with *MEN2A*. The spectrum of mutations found in the *RET* gene for *MEN2A* appears to be limited, with a majority altering a cysteine residue at codon 380. Direct mutation detection is feasible for this gene. The association of *RET* with a heritable cancer susceptibility is unique in that a germ line mutation exists in a proto-oncogene, rather than in a tumor suppressor gene.

The MEN locus of chromosome 11, *MEN1* has yet to be cloned. Analysis of *MEN1* families is currently being performed indirectly by linkage.

5.5.5 Neurofibromatosis

There are two forms of neurofibromatosis (NF). The more common form, NF1 or von Recklinghausen disease, shows peripheral nerve manifestation of café au lait spots and cutaneous neurofibromas. The *NF1* gene, located on chromosome 17, has been cloned. The protein product of the gene, called neurofibromin, is a member of the GTPase-activating protein (GAP) family and apparently acts as a tumor suppressor by downregulation of the p21(ras) protein. The *NF1* gene is very large, encoding a protein comprising more than 2800 amino acids and spanning over 300 kb of genomic sequences. Identification of mutations in this large gene is difficult, limiting genetic testing for NF1. For some families, linkage analysis can be performed.

The rarer, more morbid form of neurofibromatosis, NF2, differs from NF1 by its manifestation of schwannomas, meningiomas, and other central nervous system tumors. The *NF2* gene on chromosome 22 recently was cloned, and the protein product of this gene, called merlin or schwannomin, was shown to be related to a family of structural proteins that link the cytoskeleton with integral membrane proteins. Germ line mutations of the *NF2* gene have been detected in carriers of *NF2*. Furthermore, somatic mutations of this gene have been found in sporadic cases of schwannoma and meningioma, consistent with the classification of the *NF2* gene as a classic tumor suppressor gene.

5.5.6 Retinoblastoma

Retinoblastoma is a rare childhood tumor of the retina. The retinoblastoma gene, *RB1,* on chromosome 13, was one of the first genes isolated by positional cloning. The gene is large, encoding a protein of 928 amino acids. Many different mutations are found in the *RB1* gene, with some indications of hot spots. About half the retinoblastoma patients have germ line mutations, but most of these are presumably due to new mutations, since germ line mutations usually are not found in either parent. Mutation screening is difficult because of the size of the gene, and as with *NF1,* linkage analysis can be performed for families with hereditary retinoblastoma. All sporadic cases of retinoblastoma apparently involve somatic mutations in the *RB1* gene. Thus, the *RB1* gene is a classic tumor suppressor gene specific for the development of this type of tumor. Furthermore, somatic mutations in the *RB1* gene are often found in tumors of many other types, indicating the involvement of this gene in the etiology of various neoplasms.

5.5.7 Von Hippel–Lindau Disease

Von Hippel–Lindau (VHL) disease is a rare cancer syndrome; affected individuals are susceptible to a variety of tumors including those of the central nervous system, eye, and kidney. Renal cell carcinoma (RCC) is one of the more frequent tumors observed in VHL patients. The VHL gene has been mapped to chromosome 3 and has recently been cloned; it appears to behave as a classical tumor suppressor gene. Various types of mutations of this gene have been found in patients with VHL disease. Furthermore, about half of sporadic cases of RCC of the clear cell type have mutations in the *VHL* gene. *VHL* gene mutations appear to be very specific for clear-cell RCC, inasmuch as mutations of this gene are not detected in numerous other tumor cells, including RCC of the papillary type. Although VHL disease families may benefit from genetic analysis of the *VHL* gene, the clinical significance of detecting *VHL* gene mutations in sporadic cases of RCC remains to be determined.

5.5.8 Wilms' Tumor

Wilms' tumor (WT) is an embryonic cancer syndrome characterized by early childhood renal tumors. Although the condition is recognized as a familial cancer syndrome, only about 1% of WT cases are familial, with the rest apparently resulting from germ line new mutations. At least two genetic loci have been associated with WT, a genetically heterogeneous condition. One locus, at chromosome 11p13, contains the *WT1* gene, which is mutated in about 10% of tumor cells from WT patients. A second locus, at 11p15, is associated with Beckwith–Wiedemann syndrome (BWS), another embryonic cancer syndrome that can give rise to WT. The 11p15 region is often duplicated in BWS patients, and this chromosomal abnormality can be detected by standard cytogenetic methods or by FISH. The BWS gene has not been isolated, but two potential candidate genes, *H19* and *IGF2,* have been suggested. Both these candidate genes are imprinted, complicating the analysis of WT related to BWS. Further complication arises from the possibility that the two genes at 11p13 and 11p15 loci may act together in some cases of WT. These complications need to be resolved before molecular genetic analysis of WT becomes routine.

6   LIMITATIONS AND IMPLICATIONS

Although very powerful, genetic testing does have limitations and pitfalls. Linkage analysis depends on correct pedigree and diagnosis, generally requires a specimen from an affected individual, and can be confounded by recombination. Mutation detection methods utilizing amplification technologies are subject to contamination and can lead to errors if there are sequence polymorphisms at primer or probe sites. Mutation scanning methods must be able to distinguish relevant mutations from benign sequence changes. Because DNA technology is relatively new to the clinical arena, it is essential to educate patients and physicians of its limitations, as well as its benefits.

In addition to technical issues, a number of social, economic, ethical, and legal issues surround genetic testing. Nonpaternity was alluded to earlier. Other considerations include reproductive choice, discrimination with respect to insurability and employment, confidentiality of test information, and the psychological impact of test results on individuals. There are reported incidents of misuse of genetic testing information by health maintenance organizations, insurance companies, and employers.

As the range of conditions that can be tested at the DNA level increases, the scope of genetic testing will continue to expand; the potential benefits of the technology are immense. The technology needs improvement, however, to standardize the testing, and to make it less labor intensive, as well as accessible at reasonable cost. Furthermore, there is a need to provide individuals with the ability to deal with the complex information that can result from such testing. Education is critical to this process, and appropriate mechanisms are necessary to ensure the confidentiality of genetic information and to protect individuals against the misuse or abuse of the information.

See also DNA IN NEOPLASTIC DISEASE DIAGNOSIS; HUMAN GENETIC PREDISPOSITION TO DISEASE; LIGASE-MEDIATED GENE DETECTION; MOLECULAR GENETIC MEDICINE; POPULATION-SPECIFIC GENETIC MARKERS AND DISEASE.

Bibliography

Antonarakis, S. E. (1989) Diagnosis of genetic disorders at the DNA level. *New Engl. J. Med.* 320:1153–1163.

Knudson, A. G. (1993) Antioncogenes and human cancer. *Proc. Natl. Acad. Sci.* (U.S.A.) 90:10914–10921.

McKusick, V. A., Francomano, C. A., Antonarakis, S. E., and Pearson, P. L. (1994) *Mendelian Inheritance in Man,* 11th ed. Johns Hopkins University Press, Baltimore.

Molecular Advances in Genetic Disease (special issue). (1992) *Science,* 256:766–813.

Ostrer, H., and Hejtmancik, J. F. (1988) Prenatal diagnosis and carrier detection of genetic diseases by analysis of deoxyribonucleic acid. *J. Pediatr.* 112:679–687.

Scriver, C. R., Beaudet, A. L., Sly, W. S., and Valle, D., Eds. (1989) *The Metabolic Basis of Inherited Disease,* 6th ed. McGraw-Hill, New York.

Tsui, L.-C. (1992) The spectrum of cystic fibrosis mutations. *Trends Genet.* 8:392–398.

GENETIC VACCINATION, GENE TRANSFER TECHNIQUES FOR USE IN

Stephen G. Kayes and Jon A. Wolff

1 **Principles**
    1.1   Immunological Considerations for Vaccine
          Development
    1.2   Foreign Gene Transfer and Expression

2 **Techniques**
    2.1   Physicochemical Methods of Gene Transfer
    2.2   Biological Methods of Gene Transfer

3 **Applications**
    3.1   Bacterial Diseases
    3.2   Viral Diseases
    3.3   Protozoal Diseases
    3.4   Helminthic Infections
    3.5   Cancer

4 **Perspectives**

Key Words

Direct Gene Transfer   The direct administration into a living organism of a polynucleotide encoding a protein.

Genetic Vaccination   The generation of the immunized state following administration of a polynucleotide vaccine or genetically modified cells from culture.

Indirect Gene Transfer   The insertion of a polynucleotide encoding a protein into a cell in culture, which is then transferred into the whole organism. The gene transfer can be effected by either physicochemical or biological means (ex vivo).

Naked DNA Injection   Typically refers to the in situ administration of plasmid DNA directly into the tissues of an animal. Relatively high levels of foreign gene expression have been achieved following plasmid DNA injection into rodent muscle, with lesser amounts of expression from other species.

Physicochemical Method of Gene Transfer   The use of nonviral methods to transfer polynucleotides into cells in culture or *in vivo* involving naked DNA, electroporation, the gene gun, calcium phosphate, polylysine conjugates, and liposomes, including cationic lipids.

Viral Vectors   The use of engineered viruses carrying a foreign gene. These viruses usually lack the necessary genes for their own replication and therefore require a packaging or helper cell line for their propagation.

The essence of vaccination is the ability to present the immune system with one or more pathogen-specific antigens in advance of an actual infection. This allows the immunological machinery to process the antigen through a primary immune response and to generate long-lived memory cells. Then, when the antigens are encountered a second or subsequent time, the memory cells will ensure that there can be a rapid and effective neutralizing or even sterilizing protective immune response.

The immune response usually consists of a cell-mediated and an antibody-mediated reaction. The humoral immune response is mediated by B lymphocytes, which develop into plasma cells that secrete antigen-specific antibody. The cellular immune response is carried out by the T lymphocytes, which mature in the thymus prior to engaging foreign antigens. Both T and B lymphocytes are necessary for the complete immune response; and for any given infective organism, the immune response will comprise varying contributions of both the T- and B-cell arms of the immune system.

A variety of methods are being developed to express foreign genes in cells in culture and in the whole organism. Application of these methods to the development of novel immunization approaches has produced some very encouraging results. Viral envelope genes have been successfully transfected into mouse skeletal muscle and the mice shown to be resistant (i.e., vaccinated) against subsequent challenge with the virulent live virus. The use of gene transfer technology for vaccination offers great promise over conventional vaccine methods because of the ability to stimulate the immune system with only the antigens of interest, possibly without requiring the use of toxic adjuvants to boost the strength of the resulting immune response and to control or influence the type and location of the immune response. A major concern of using attenuated live vaccines is the potential risk of reversion back to the wild-type virulent organism, with the development of a fulminant infection. Using gene transfer techniques can circumvent this problem. Under ideal circumstances, a genetic vaccine should be completely protective against the specific infection after a single administration, and the protection should persist for a lifetime. At this juncture in the evolution of genetic vaccine technology there is still much to learn. Based on the first reports, however, the promise seems very exciting, indeed.

1   PRINCIPLES

If vaccines are to be targeted to select for the most effective type of host response, the presentation of the relevant antigens must accommodate the major histocompatibility complex (MHC). The MHC is a series of associated genes that code lymphocyte surface proteins, which determine the nature of the actual response to antigen. Two major classes of gene products can interact to produce immune responses of two different kinds. Class I MHC cellular determinants, when appropriately stimulated by antigen, will lead to the generation of killer T lymphocytes. These cells function most notably in transplant rejection and in the elimination of cells harboring intracellular parasites such as pathogenic viruses. Class II MHC reactions invariably lead to the production of antigen-specific antibodies. Once released, these antibodies bind to determinants on the antigen and promote their internalization by phagocytic cells, which then degrade the antigen. Most commonly, class II responses are seen in bacterial infections, although the antibodies can also serve to neutralize soluble toxins secreted by pathogenic organisms. While it is true that the antibodies are made by cells of the B-lymphocyte lineage, they frequently require assistance from helper T lymphocytes. These helper T lymphocytes express a surface protein designated CD4 (i.e., cellular determinant 4+ T helper cells), which serves as an adhesion molecule for docking with other lymphocytes or accessory cells. In contrast to the helper cells, the killer T cells generated as a consequence of MHC class I interactions usually express a different adhesion molecule designated CD8. The manner in which antigen is presented to the immune system determines to a great extent whether the immune response will be class I or II restricted.

1.1   IMMUNOLOGICAL CONSIDERATIONS FOR VACCINE DEVELOPMENT

To mount an immune response to any antigen more complex than a repeating polysaccharide requires the involvement of T helper cells. It is now known that there are at least two subsets of T helper cells, designated Th1 and Th2. Although these cells were first demonstrated in mice, there is now substantial evidence that there are Th1- and Th2-like cells in humans also. Th1 cells when stimulated by antigen produce and secrete the soluble mediators interleukin 2 and interferon gamma (IL-2 and IFN-γ, respectively) and carry out cell-mediated immune responses such as delayed hypersensitivity and macrophage activation. Th2 cells, in contrast, secrete IL-4, IL-5, and IL-10. These cytokines drive the immune reaction toward the humoral side of the immune response. IL-10 has the interesting property of being able to inhibit Th1 responses, while IFN-γ seems to be able to down-regulate Th2 responses.

The question of what determines whether an antigen elicits a Th1 response such as granuloma formulation or cell-mediated killing of antigen-expressing target cells, or a Th2 response accompanied by the production of large amounts of antibody such as occurs in a helminth infection, has led to the recognition that early in the course of an infection, macrophages and some B lymphocytes produce a newly discovered cytokine designated IL-12. IL-12 is thought to stimulate a population of immature T cells to mature into functional Th1 cells, which then go on to effect a cell-mediated immune response. In contrast, IL-4 drives the same population of naive T cells to become Th2 cells, and the result is an antibody response. While the source of this IL-4 is not known, mast cells, basophils, or even T cells are currently considered to be possible candidates. Because IL-10, which can down-regulate Th1 responses, is also produced by macrophages, it is apparent that macrophages play a crucial role in determining the kind of immune response a given antigen will provoke based on the relative amounts of IL-10 and IL-12 produced. The properties of the antigen that favor one induction pathway over the other are not currently known.

By considering the role of IL-12, it is possible to design vaccines for diseases that can be controlled by cell-mediated immunity. IL-12 might very well be given as an adjuvant concomitantly with the induction of antigen-specific responsiveness. As progress is made toward identifying the source of IL-4, it will soon be possible to select with confidence the arm of the immune response that best protects against a given pathogen, and this advantage in turn may raise the practice of vaccination from an art to a science.

When designing vaccination strategies, it must be remembered that how antigen is presented to the immune system determines to a very large extent the kind of immune response that will be generated. It is now known that the expression of transgenes results in a presentation to the immune system that is not the same as that which occurs after the foreign protein is injected into the body as during a conventional vaccination. For exogenously injected antigens that can be taken up by antigen-processing cells (i.e., macrophages, Langerhans cells, dendritic cells) or which by their nature are small biochemical moieties (e.g., toxins), the predominant response will be carried out by CD4+ T cells. This means that antibodies will be made against the antigen. In contrast, if immunization against a viral agent or tumor is desired, the immune system must process the infected and/or altered host cell. Expressed proteins from transfected genes more often than not elicit cell-mediated immunity. Because some viral vectors such as fowlpox virus can accept large DNA inserts, it may be possible to design a vaccine platform that contains not only a pathogen-specific transgene but one or more cytokine genes, to steer the immune response to the desired outcome.

1.2   FOREIGN GENE TRANSFER AND EXPRESSION

A prerequisite of using a gene-based vaccine is increasingly being met as more and more genes of pathogenic organisms are being identified, isolated, and cloned. The next step is to transfer the genetic material into cells of the host to allow expression of the inserted gene. The expressed protein must then interact with the cells of the immune system (Figure 1). There are numerous techniques by which the genetic transfer can be effected. Some methods use infectious vectors to gain entry into the cell and access its protein synthesizing apparatus, while other methods are of a physicochemical nature. Both viral and nonviral methods can be used on cells that have been removed from the vaccinee (ex vivo methods) or on the vaccinee directly (in vivo methods).

In ex vivo methods, a cell population is removed, placed into tis-

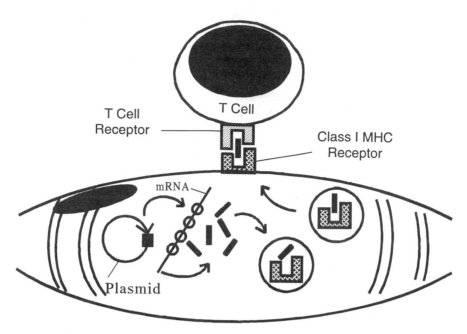

Figure 1. Schematic representation of key events occurring when naked DNA is expressed by skeletal muscle and the gene product presented to the T cell during genetic vaccination. The injected plasmid, encoding a pathogen-specific antigen, is transcribed into mRNA, which is translated to the corresponding protein (antigen). The antigen is then associated with the class I major histocompatibility complex (MHC) receptor within a Golgi transfer vesicle and translocated to the muscle cell surface. Here the antigen in the context of the MHC receptor is presented to the T cell to initiate a cell-mediated immune response. [Modified from J. Cohn, *Science,* 259:1691 (1993).]

sue culture, transformed (clonally expanded, if necessary), and returned to the host. This method ensures that only the cells of interest are transduced and, in the case of viral vectors, that no other cells of the body become infected. This may be a most attractive path for the development of cancer vaccines wherein a piece of the patient's tumor or tumor-infiltrating lymphocytes can be removed and engineered to be highly immunogenic. Genes can be transduced with viral vectors, or inserted into tissue culture cells by any of several methods including calcium phosphate precipitation, electroporation, liposome-mediated transfer, and DNA-coated particle bombardment (the gene gun). All but the latter method constitute the so-called physiochemical methods of gene transfer.

Despite the promise of these in vivo (direct) and ex vivo (indirect) gene transfer techniques for treating genetic and acquired disorders, the ability to sustain high levels of long-term expression has stymied actual clinical success. For purposes of vaccination, it is not necessary to achieve a level of expression and its longevity to the degree that it would be required for gene therapy.

2   TECHNIQUES

2.1   Physicochemical Methods of Gene Transfer

In physicochemical methods of gene transfer the desired sequence is cloned into plasmid DNA that contains essential promoter–enhancer sequences required for eukaryotic transcription. Additional sequences contained in the plasmid are the SV40 T antigen intron and a polyadenyl addition signal for processing mRNA. Because plasmid DNA is maintained as an episomal element within the nucleus, physicochemical methods are not particularly well suited for use in rapidly dividing or regenerating tissues. The development of human artificial chromosomes and plasmids capable of insertion is theoretically possible.

2.1.1 Calcium Phosphate Precipitation

The precipitation of calcium phosphate was one of the first methods used to insert genes into mammalian cells maintained in tissue culture and is considered to be a relatively simple technique. Essentially, a small quantity of DNA is precipitated with $CaCl_2$ in a phosphate buffer at neutral pH, and the resulting phosphate–DNA precipitate is added to cells in tissue culture. The cells take up the DNA precipitate over the next 4–24 hours. The mechanism by which cells take up the DNA is not well understood but most likely involves the endocytic pathway, with escape from the endosome followed by transposition to the nucleus of the cell. This method offers the advantages of economical operation. The major disadvantage is that it does not transfect all cell types efficiently.

2.1.2 Electroporation

Electroporation refers to the process whereby DNA enters the cytoplasm of cells that have been subjected to an electric pulse while the cells and DNA are suspended in the gap of closely apposed electrodes. The DNA passes through pores that have been induced in the cell membrane by the electric current. Critical aspects of the procedure include buffer composition and properties of the electric pulse. This procedure is not frequently used because of (a) the relatively large amount of labor involved in placing the target cells into the electroporator apparatus and removing them, (b) the need for a specific apparatus, (c) safety precautions related to the large voltages used, and (d) the large percentage of cells that are killed. The technique has been very useful for putting foreign genes into lymphocytes, however.

2.1.3 Liposome-Mediated Transfer

Cationic liposomes are quite effective at moving DNA into cells. The positively charged lipids can be coated with DNA, whose negatively charged phosphate residues interact with the lipids. Liposome gene delivery systems offer many advantages over other non-viral gene transfer strategies, especially those using retroviruses. Liposomes are biologically safe, can be used repeatedly without fear of an untoward immune response, and do not depend on cellular division. For example, liposomes carrying the chloramphenicol acetyltransferase (CAT) reporter gene were injected into the vasculature of mice, and several tissues showed significant expression 2 months later.

In many situations targeted delivery of specific genes will be desired and/or required for successful immunization. For example, targeting by aerosolization of liposomes containing the cystic fibrosis transmembrane regulator (CFTR) gene has successfully delivered a functional gene to lung. Other strategies featuring liposome–DNA complexes include the use of tissue-specific promoters (especially for genes destined for the liver), the incorporation into the liposome of monoclonal antibodies that can selectively target expressed cellular receptor proteins, and the use of lipids that show tissue-specific affinities. Liposomes containing the gene for HLA-B7, a costimulatory molecule required in immunoactivation, have been injected directly into melanomas, and the tumors regressed, apparently in response to the treatment. Genes in liposomes have been delivered to specific segments of blood vessel walls using catheters specially designed to occlude a vascular segment both proximally and distally to a centrally located dispensing port in the catheter.

2.1.4 DNA Particle Bombardment

Genes can also be inserted into target cells or even into specific organs by accelerating DNA-coated particles to high velocity. This "biolistic" technique most often employs the so-called gene gun. A typical particle bombardment device either uses pressurized helium or consists of a spark discharge chamber having two closely apposed electrodes (Figure 2; see color plate 9 for part E). The chamber is situated beneath a film of gold particles that have been coated with plasmid DNA coding for the gene of interest. The target is just beyond the film of particles and is most often an inverted tissue culture dish containing an adherent monolayer. The gun is fired when a high voltage capacitor discharges, vaporizing a drop of water suspended between the two electrodes. This creates a shock wave, which propels the gold particles into the target. Like liposome-mediated gene transfer, particle-accelerated gene transfer obviates most biological constraints such as the involvement of cell receptors and the need for cellular replication, or any other factor that can limit DNA uptake and/or gene expression. To date, the gene gun has been used extensively for in vitro transformation, but it has also been used to deliver genes in vivo to muscle, pancreas, kidney, spleen, and liver, with reporter genes being detected for periods up to one year after transfer. The gene gun seems to work best on skin. Interestingly, not only did mice "shot" in the pinna of the ear with gold particles coated with DNA coding for either human growth hormone or α-1 antitrypsin express these human proteins, but their immune systems recognized the proteins as foreign and made antibodies against them. The highest levels of protein expression and subsequent antibody production were associated with particle delivery to cells of the epidermis (i.e., the epithelium) rather than the

dermis and were in proportion to the number of particles delivered. The amount of DNA necessary to obtain an optimal antibody titer was 100 ng, which is considerably less than is required for the intramuscular injection technique. From the perspective of gene-based vaccine development, this finding is very encouraging.

2.1.5 Direct Muscle Injection

Injection of genetic constructs directly into striated muscle in vivo results in expression and, in many instances, secretion of the gene product into the circulation. Following intramuscular injection, plasmid DNA is found distributed throughout the muscle and is able to diffuse throughout muscle-associated connective tissues. Foreign transgenes have been expressed for more than 2 years, but the efficiency of transfer is relatively low (2% of treated muscle) (Figure 3). Localization studies using plasmid DNA conjugated to colloidal gold and compared to gold-tagged polyethylene glycol have shown that nucleic acids (but not the polyethylene glycol) are able to traverse the external lamina and enter myofibers through the T-tubule system, apparently associated with caveolae in the sarcolemma.

Since the original observations by Wolff et al. in 1990 that striated muscle can express injected transgenes or mRNAs, many of the variables of the technique have been clarified, though many questions still remain. Of particular interest is the tendency of the level of expression of transgenes to be unpredictable, although in most cases the degree of expression seems to be proportional to the amount of nucleic acid taken up. For example, in one series of experiments, a 100 µg injection into the mouse quadriceps muscle resulted in approximately 1.5% of the fibers expressing β-Gal activity as demonstrated by enzyme histochemistry 7 days after the injection. Staining could be observed up to 400 µm from the injection site. At present, the best evidence available indicates that plasmid DNA, when taken up by skeletal muscle, remains circular and neither integrates into chromosomal DNA nor replicates.

The expression of naked transgenes was found to vary substantially following injection into different mice under identical conditions of injection. Subsequent studies indicated that injection technique per se is not responsible for this observed variation in transgene expression. Nor does the volume of fluid containing the genetic construct appear to be critical in determining uptake or expression. Likewise, neither the rate of injection (< 30 s to 5 min) nor the type of needle used to administer it (27 gauge with 0.2 cm limiting collar vs. glass capillary) seems to affect the outcome (Figure 4). However, conditions that lead to increased degradation of DNA can result in less uptake and expression. Interestingly, DNA dissolved in sterile saline results in a 5- to 10-fold greater transgene activity than the same DNA dissolved in 20% sucrose, while an injection of 25% sucrose 15–30 minutes prior to DNA injection was reported to reduce the variability in the technique. This result presumably reflects a property of saline, namely, to cause less muscle damage than the sucrose vehicle in the former case, perhaps actually stimulating pinocytotic uptake in the latter instance. Animals that received multiple injections of the same construct over several weeks expressed less activity than did mice receiving a single injection, and because naked DNA taken up by skeletal muscle tends to persist for substantial periods of time, better expression can be expected with fewer administrations. The choice of promoter sequence to drive gene transcription can influence gene expression by a factor of 1000 or more.

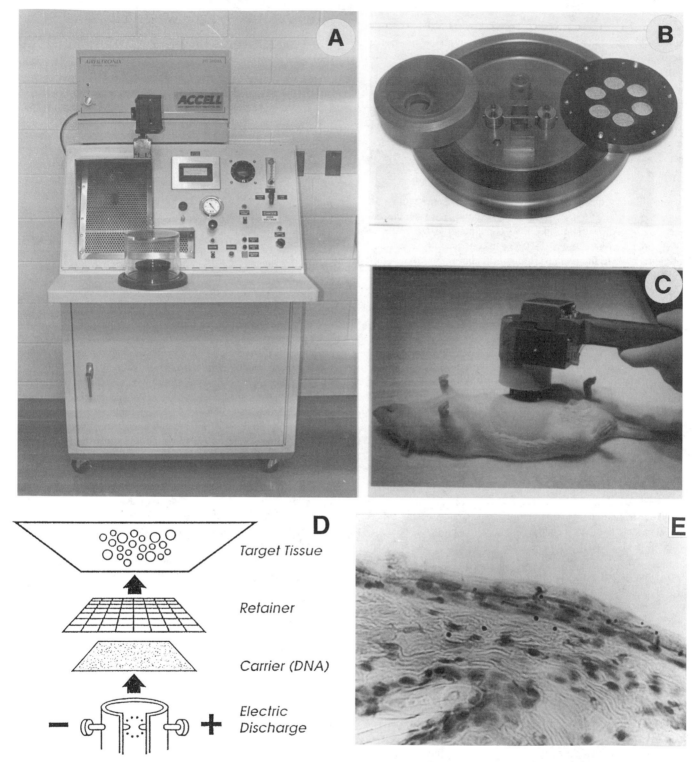

Figure 2.   Design of the electric discharge mediated particle bombardment device for gene transfer. (A) General design. (B) The electric discharge and explosion chamber. (C) The hand held device for in vivo gene transfer. (D) Diagram of the particle acceleration mechanism of the Accel instrument. (E) Histochemical localization to superficial layers of epidermis of β-galactosidase activity encoded by naked DNA following particle bombardment (see color plate 9). In (D) the motive force is generated in a spark discharge chamber containing two electrodes. A 10-μL water droplet is placed between the electrodes, and a high voltage capacitor is discharged through the water droplet, which vaporizes instantly, creating a shock wave. A polyvinyl chloride (PVC) pipe with an internal diameter of 13 mm is adequate for use as the spark discharge chamber. The electrodes are located opposite each other, project into the interior of the chamber approximately 5 mm below the top, and are protected at the tips with an arc-resistant alloy. The gap between the two electrodes can be adjusted by appropriately threading them into or out of the spark chamber. A spacer ring is placed above the spark chamber. A removable spacer ring allows the distance from the spark discharge to the carrier sheet to be varied so that the force of the shock wave can be adjusted. The motive force can also be adjusted by varying the voltage of the discharge. The carrier sheet on which the DNA-coated gold particles are precipitated is placed on top of the spacer; the function of this sheet is to

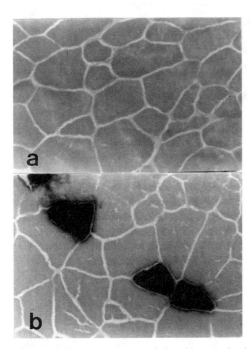

Figure 3. Histochemical demonstration of plasmid expression by murine skeletal muscle. The quadriceps muscles of mice were injected with 100 μL of either a plasmid encoding the *E. coli* β-galactosidase gene or a control. Three weeks after injection, the muscles were removed and frozen sections prepared and stained with X-gal in the presence of KFeFeCN and counterstained with eosin: (a) β-gal(−) section, and (b) β-gal(+) section. The blackened myofibers in (b) indicate expression of the bacterial gene product by the mammalian skeletal muscle. (Original micrographs 400×.)

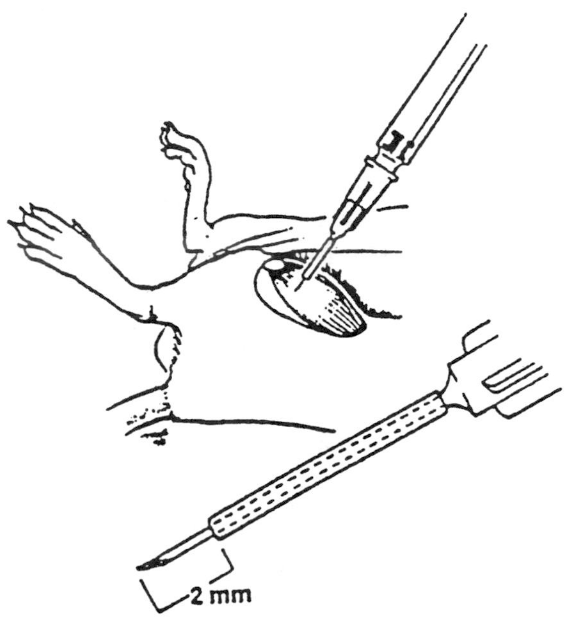

Figure 4. Injection of naked DNA into quadriceps of a mouse: schematic diagram and Williams collar, which ensures proper placement of needle tip. [Reproduced from J. A. Wolff et al., *BioTechnology,* 11:474–485 (1991), permission of Eaton Publishing Co., 154 E. Central St., Natick, MA 01760.]

More recently, the drug Marcaine has been found to dramatically increase uptake and expression when given a few days prior to the DNA injection. The drug is slightly myotoxic, and satellite cells surrounding necrotic fibers undergo mitosis, which may account for the enhanced uptake and expression.

Other variables that affect uptake and expression of transgenes by skeletal muscle also have been examined. Whether the muscle is denervated appears to have no effect on uptake or expression. However, electrical stimulation of the muscle before injection decreases expression, although electrical stimulation after injection has no effect. Developmentally immature muscle was also found to be capable of uptake and expression of transgenes.

Following exposure to naked plasmid DNA in vitro, primary muscle cells express transgenes much better than primary or immortalized cells of other types. Currently, this is thought to correlate with the presence of caveolae in the T-tubule system of the stri-

ated muscles. This is supported by the observation that unlike cardiac and skeletal muscle, smooth muscles which lack T tubules, do not take up or express DNA.

Intramuscular injection of plasmid DNA has been used by several laboratories to immunize experimental animals, and direct injection has been shown to elicit both cellular and humoral immune responses. Mice have been successfully immunized by intramuscular injection of plasmid DNA carrying an influenza gene, as determined by their ability to survive a lethal challenge infection. Likewise, chickens were given plasmid DNA injections by one of three routes (intravenous, intraperitoneal, or intramuscular) and subjected to challenge. In these studies, gene vaccination resulted in 60% protection, which was attributed to use of a gene from a heterologous virus (i.e., a virus related to the influenza virus that normally infects chickens but differing by 16% in amino acid sequence). Mice that have been vaccinated with plasmid DNA encoding bovine herpes

transfer the force of the shock wave from the spark discharge to the carrier particles. Located above the carrier sheet is a 100-mesh stainless steel screen that retains the sheet so that it does not proceed to the target tissue. The target can be inverted over the retaining screen so that the tissue is in the direct path of the gold particles. In (B) the blast chamber consists of a PVC block forming a rectangular cavity divided into two chambers by a partial wall. In one of these chambers are two arc points spaced equidistant from the sides and rear wall. The points are set 0.5 mm apart and bridged by a 10 μL drop of water prior to each discharge. The spark chamber is covered by a wafer of PVC, which reflects the primary shock wave back into the chamber, avoiding energy loss. The cavity is partially obstructed by a partition that separates the spark chamber from the reflection chamber. The partition serves to prevent the primary shock wave from interacting directly with the carrier sheet. As secondary waves are formed by reflection from the chamber walls, their combined fronts distribute their force evenly over the carrier sheet, an 18 mm² piece of 0.5 mil metalicized Mylar (DuPont, Wilmington, DE). This sheet, which forms the roof of the reflection chamber, is accelerated upward until it encounters the retaining screen (100-mesh stainless steel) as the gold particles on its surface proceed to the target tissue. The energy for the arc is provided by a 25 kV, 2 μF capacitor, charged from a 25-kV DC variable power supply. (Figure courtesy of Dr. N. S. Yang, Agracetus, Inc., Middleton, WI; Modified from Yang et al.; *Methods Neurosci.* 21:427, by permission.)

glycoprotein have gone on to make significant amounts of antigly-coprotein antibody, and the sera from these mice contained significant amounts of viral neutralizing activity compared to sera from mice receiving the plasmid platform lacking the herpes gene. In addition, cattle that have been injected intramuscularly with plasmids carrying bovine herpesvirus genes exhibited significantly elevated titers of anti–bovine herpesvirus glycoprotein, and they shed significantly less virus than the control animals following infectious challenge. Recently, nonhuman primates (cynomolgus macaques) were injected intramuscularly with a plasmid expressing the *gp160* gene of the human immunodeficiency virus. Animals were inoculated three times and serum collected 2 weeks latter. This regimen led to seroconversion of all test animals, and the sera were shown to contain neutralizing activity as determined by the reduction in number of microsyncytia forming in cultures of T cells exposed to HIV-I.

2.2 Biological Methods of Gene Transfer

The use of viruses as shuttle vectors in gene therapy strategies is well established. Viral vectors are attractive gene transfer platforms because they all depend on receptor-mediated entry into cells and therefore, in principle, only cells expressing the appropriate receptor will be transduced.

In many instances however, the viral receptor may be expressed on large numbers of different cell types. Each of the viruses discussed in Sections 2.2.1 to 2.2.4 has some special attribute or biological property that may offer an experimental advantage over other vectors or methods of gene transfer. An overriding concern when using a viral vector is that it not lead to a productive viral infection of the host when introduced during the vaccination procedure. This unwanted result usually is prevented by either inserting the gene of interest into one of the genes required for viral replication (insertional inactivation) or removing genes that are required for the packaging of new viral particles in the infected cells. In the latter case, a packaging cell line designed to supply the genes the virus lacks is required to produce vector virus. These replication-deficient viruses are capable of infecting cells; since, however, they cannot replicate themselves, they cannot infect new cells within the host. Nevertheless, the potential for subsequent viral infection following recombination with a wild-type virus leading to rescue of the vaccine vector should always be borne in mind.

2.2.1 Retrovirus-Mediated Gene Transfer

Retroviruses are positive-sense, single-stranded RNA viruses. Their RNA is converted into DNA using the viral reverse transcriptase that is carried in the nucleocapsid. This DNA then enters the host cell nucleus and integrates more or less randomly into the host genome. It is this integrated viral DNA, also known as the *provirus,* which is transcribed by the host polymerase into mRNA. A typical retrovirus expresses three structural genes. These are the *gag* gene, which encodes a series of proteins involved in viral assembly; the *pol* gene, which encodes a polyprotein containing several activities including the reverse transcriptase and integrase; and the *env* gene, which encodes the glycoprotein expressed on the outer capsid and serves as the ligand for the host cell receptor. The provirus contains a single transcriptional unit flanked by repeated sequences at both the 5' and 3' ends known as long terminal repeats (LTRs). In practice, the *gag, pol,* and *env* genes are deleted and replaced with the transgene. The LTRs and the ψ region responsible for RNA assembly into new par-

ticles are preserved. Retrovirus vectors can accommodate large inserts up to about 7 kb. Transcription is initiated by the 5' LTR. A packaging cell line that contains proviral DNA from a different retrovirus (i.e., the helper virus) is transfected with the transgene-modified retrovirus, and the helper virus then supplies the *gag, pol,* and *env* genes, which were excised from the carrier retrovirus. The helper virus is disabled by virtue of having its ψ region removed and therefore cannot package its own RNA. Vector RNA is then converted to DNA by a reverse transcriptase *(pol)* derived from the helper virus, and it is this DNA that will randomly integrate into the host cell genome. The host cell will then produce the retrovirally encoded proteins; because essential viral genes are not present, however, the host will not go on to produce infectious viral particles.

Retroviruses are frequently employed as gene insertion vectors because they stably integrate into the cellular genome. To be infected by a retrovirus, cells must express appropriate surface proteins or receptors, and the cell must be undergoing cellular division. The selective expression of viral receptor proteins allows the transduction of limited populations of cells. Retroviral particles are somewhat more labile than other vector viruses, and this may limit their use as vaccine platforms. While there is little chance of viral spread using such particles, "accidents" have occurred. Genes inserted into retroviral vectors can be regulated by the long terminal repeat promoter and spliced, or they can be expressed from an internal promoter of one's choosing. Interestingly, LTR-initiated transcription does not seem to work well in cells in vivo.

2.2.2 Adenovirus-Mediated Gene Transfer

Unlike retroviruses, adenoviruses (AVs) can transduce nondividing cells, leading to the expression of relatively large amounts of gene product. AVs reside within the nuclear envelop but remain extrachromosomally located. Replication-deficient AVs, which lack all or portions of the E1 region of the genome, can be produced in packaging cells engineered to express the missing genes. Typically, vector adenoviruses have parts of the E1A–E1B and/or E3 genes deleted. Expression of inserted genes is regulated by the E1A promoter, the E3 promoter, the major late promoter, or a spliced promoter sequence chosen to correspond to the desired circumstances. It is possible that at high multiplicity of infection, E1A-deleted AVs may become replication-competent. AVs encode a large number of viral proteins. There is some concern that one or more of these proteins may elicit a strong immune response directed against the transduced cells. AV particles are relatively stable (unlike retroviruses), and this property may be important for vaccines used in developing countries.

2.2.3 Adeno-Associated Virus-Mediated Gene Transfer
    (Targeted Insertion)

Adeno-associated virus (AAV) is a member of the *Dependoviridae* genus and is the only known animal DNA virus requiring the presence of a second, unrelated virus as a coinfectant to undergo viral replication. The role of the coinfectant is most often served by either an adenovirus or a herpesvirus. Because the AAV cannot replicate by itself, it is considered to be a defective or dependent virus. A primary AAV infection induces almost no cellular changes, and to date, no known disease has been associated with the presence of the AAV genome in otherwise healthy host cells. AAVs are considered to be good shuttles for insertion of foreign genes because they can infect a wide variety of host cells, and the virus remains stably integrated in the host cell genome for upward of 200 passages without loss of viral DNA. AAVs are not as efficient as retroviruses in transducing cells, but AAVs can integrate into the host genome in nondividing cells.

There is now considerable evidence that AAV is capable of targeted insertion to the long arm of chromosome 19 and in fact, a 100 base pair sequence has been identified to which AAV seems to home. AAV vectors lacking the *rep* gene integrate randomly. The usefulness of AAV for gene transfer is diminished by its helper cell needs and by its limited ability to accommodate inserts (≤5 kb). A more complete understanding of how AAV achieves this targeted localization into the host genome will go a long way toward permitting targeted vaccine development as well as facilitating site-specific gene therapy.

2.2.4 Poxvirus-Mediated Gene Transfer (Cytoplasmic Expression of Genes)

The poxvirus family includes vaccinia virus (VV), which replicates and expresses genes in the cytoplasm of infected cells. These viruses do not integrate into the host genome. Vaccinia vectors have a wide host range, and they tend to be stably maintained in the infected cells. The vaccinia viral genome is approximately 185,000 bp in length, and to date, inserts of up to 26 kb of foreign DNA have been produced. The large genome is noteworthy for permitting large numbers of vaccinia genes to be deleted without appreciable loss of viral replicative capacity. Three regulatory classes of VV genes (early, intermediate, and late) have been described. For vaccination purposes the intermediate genes may be the most efficacious for short-term expression of immunogenic transgenes. Cells are often infected with two different vaccinia vectors, one expressing a bacteriophage polymerase such T7 RNA polymerase (G1) and the other containing the foreign gene 3′ to its cognate promoter. Target cells are first infected with VV and subsequently with courier plasmid. Following VV infection, host protein synthesis is virtually inhibited.

Recently, an attenuated vaccinia virus has been engineered for use as a vector for animal vaccines. Designated NYVAC, this construct has a deletion of 18 open reading frames such that it replicates only in avian cells. In mammalian cells the virus is blocked at a stage preceding DNA replication. Human immunodeficiency virus *env* gene has been inserted into a similar recombinant construct and is currently undergoing human trials as a vaccine delivery vehicle. Other vaccination protocols using vaccinia virus include vaccination of monkeys against simian D type retrovirus and cattle against rinderpest, as well as the development of an oral bait vaccine against rabies (for use in feral skunks and raccoons).

Another virus related to VV that has potential for use in vaccination is the avipoxvirus, which replicates only in avian cells. This inability to replicate in mammalian cells suggests that avipoxvirus might be especially well suited to shuttle genes into mammalian systems.

3   APPLICATIONS

3.1   BACTERIAL DISEASES

The mechanisms whereby bacteria cause disease are as diverse as the bacteria themselves. Sources of bacterial antigens depend to a large extent on the lifestyle of the bacteria. Thus, surface proteins or enzymes often induce strong cell-mediated immune responses accompanied by humoral responses. Lipopolysaccharides (LPS) from the Gram-negative bacteria are well characterized as B-cell mitogens, and the ability to neutralize LPS has been shown to enable the prevention of septic shock. Many bacteria gain their invasive advantage by producing one or more toxins that herald infection. Finally, many bacteria have fimbria, or filaments, which express structure-specific proteins (which often define serotypes), and antibodies that bind these proteins often can interfere with the function of the fimbrium or filament, thereby inhibiting bacterial growth or replication. Therefore, choosing an antigen or group of antigens or their respective genes for vaccination will most likely be an empirical process, and a solid understanding of the biology of the bacterium and the disease process itself will be required to make the best choice.

3.2   VIRAL DISEASES

Viruses as a group represent one of the simplest life-forms, and all are obligate intracellular parasites. Essentially, viruses consist of little more than the instructions for their own replication. Those instructions can be in the form of either DNA or RNA (never both in a single virus), and occasionally some of the enzymes necessary to carry out viral replication. These instructions are packaged within the viral capsid, a protein structure that can serve as the ligand for viral receptors on host cells. All viruses require the use of the host cell organelles to carry out viral replication.

Viral structure and nucleic acid encoded proteins must account for all the potential antigens available for vaccine development. The simplest virus is encapsulated nucleic acid, which may contain some viral proteins (enzymes) within the capsid. More complex are the enveloped viruses, which are surrounded by host cell plasma membrane in which most of the host proteins have been replaced by viral coded proteins. Often the envelope is decorated with proteinaceous spikes (peplomers) that function as docking proteins, hemagglutinins, or neuraminidases. Neutralizing antibodies often block the function of these proteins, thereby preventing the infection of new cells.

It is now becoming apparent that viral replication and the diseases associated with it often depend on induced host immunological responsiveness. In particular, the cytokine responses that arise may be orchestrated more by the pathogen than by the host. For example, a cytokine that favors the generation of cytotoxic T cells, which would kill infected host cells, would not be in the best interest of the virus. On the other hand, induction of a nonlethal response, such as occurs when IL-8 is produced, leads to the accumulation of large numbers of neutrophils that are more or less ineffectual in limiting the virus and may even contribute to the destruction of host tissues. Thus, choosing antigens that cause the immune system to respond in the opposite direction from that elicited by the virus can often moderate or completely inhibit viral spread or replication.

3.3   PROTOZOAL DISEASES

The protozoal diseases are difficult at best to characterize based on common properties. A complete understanding of the complex life cycles is mandatory to begin choosing potential targets for immune intervention. Of the protozoan organisms that cause the greatest degree of human morbidity and mortality, malaria, trypanosomiasis, leishmaniasis, and amoebiasis are the diseases requiring vaccines.

Amoebiasis, which is transmitted as a result of poor public hygiene, is not considered further. The remaining diseases all result from bites of vector insects. The significance of the insect vector is that all the organisms have the ability to thrive in a cold-blooded host as well as in a warm-blooded host. Additionally, all three of the remaining class of organisms go through complicated life cycles marked by morphologically different forms. For a vaccine to be effective, for example, against the mosquito-injected malarial sporozoite, its design must accommodate the following schedule: the immune system has approximately 30 minutes from the time of the insect bite until the parasites enter the liver and are sequestered from immune attack.

There are likewise many forms of trypanosomes and *Leishmania* (e.g., trypomastigotes, epimastigotes, and amastigotes). The American trypanosomes and the *Leishmania* spp. are intracellular parasites, while the African trypanosomes live extracellularly. To complicate the matters for the vaccinologist, each of these forms expresses stage-specific antigens. Finally, the African trypanosomes have evolved the highly ingenious mechanism of immune subversion know as antigenic variation. Antigenic variation is manifested as a population of organisms all expressing a single major surface glycoprotein known as the variant antigenic type (VAT). As the host begins to mount an IgM response (the first response to any antigen), this population of trypanosomes dies, but not before it has been replaced by another population, all of whose members express a new VAT. A new VAT is expressed about every 7–9 days, and as a result the humoral immune response is always struggling to stay current, the serum is filled with now useless IgM molecules, and the parasites go blissfully on with the infection. To date, more than 1000 genes encoding VATs have been discovered. The sequence in which the genes are expressed tends to not be predictable. The first few VATs that are expressed following transmission by the tsetse fly, which are known as metacyclical antigens, in fact are predictable, but immunologists have not been able to successfully exploit this observation.

Resistance and susceptibility to protozoan infections may be determined in part by the way in which relevant antigens (whatever they may be) are presented to the immune system. As noted earlier, responses in which killer lymphocytes and activated macrophages predominate may correlate with resistance. These responses are associated with IFN-γ, IL-2, and IL-12. If the antigens are presented in such a way that antibody is favored over cellular immunity, a susceptible state may be established and the infection will persist and flourish. An interesting variation on recombinant viral vaccines has been utilized in a murine model of leishmaniasis. In this model a gene encoding a leishmanial surface proteinase has been cloned and expressed in a vaccine strain of *Bacillus Calmette-Guerin* (BCG, a bacterium with a long history of use in vaccinating against tuberculosis). Following injection of the recombinant BCG into mice, which were subsequently challenged with either *Leishmania major* or *L. mexicana* promastigotes (the extracellular form of the life cycle) or amastigotes (the intracellular form), there was evidence of significant protection against establishment of both states of both species.

3.4   HELMINTHIC INFECTIONS

The helminths of agricultural importance or that infect man can be divided into three classes. These are the roundworms, which are protected by an outer cuticle, the trematodes or flukes, and the ces-

todes or tapeworms. The latter two classes collectively are known as the flatworms, and these worms have outer teguments or membranous syncytia. The size of these worms is extremely variable, from macroscopic to microscopic, and some, like *Trichinella spiralis,* are both an extracellular (the adults located in the intestinal lumen) and intracellular parasite (infective larvae within host muscle cells). As was seen in the protozoans, the life cycles are extremely complex, and the host is usually exposed to many life cycle stage-specific antigens. For almost all parasites of man and animal, there is good evidence of cellular and humoral immunity induced as a result of infection. It has little consequence on the outcome of the infection, however. The only successful antihelminth vaccine developed to date uses radiation-attenuated live infective larvae of *Dictyocaulus viviparus* to prevent parasitic bronchitis in cattle. There seems to be no shortage of candidate antigens for use in antiparasitic vaccines. However, most of these molecules have failed to engender adequate protection or even protection comparable to that seen with prior infection or vaccination with attenuated whole organisms. This disappointing performance has been attributed to the possibility that the highly purified or recombinant molecules are presented to the immune system in such a way that inappropriate T-cell subsets or their cytokines are stimulated or secreted, respectively.

3.5   CANCER

Immunological control of cancer is extremely difficult because the tumor cells often divide as fast or faster than the lymphocytes, there is competition for limited calories necessary to fuel an immune response, and many of the chemotherapeutic agents used are frankly immunosuppressive. In addition, the expectation of those in the oncology community who believe in the concept of a "cancer vaccine" is different from that of the infectious disease community. The strategy in the latter group tends to be prophylactic in nature, while the oncologists see their vaccine being administered after the disease is well established. Cancer patients may have already been treated with surgery, radiation, or drugs before the "vaccine" is even considered "appropriate." Many of the ideas for cancer vaccines revolve around a paradigm of ex vivo gene therapy wherein a gene, such as tumor necrosis factor (TNF), thymidine kinase (which makes cells susceptible to the lethal effect of the antiviral drug gangcyclovir), or a cytokine (to activate cytotoxic killer T cells) is transferred into cells recovered from the patient's own tumor load. Right from the start these methods are hindered by virtue of possibility that the patient will die before these personalized, genetically engineered cells can grow up in the laboratory.

Having said all this, it is encouraging to note that some successful efforts have been reported in an attempt to develop a cancer vaccine. Studies from mouse models of cancer in which genetically altered tumor cells are injected back into tumor-bearing mice indicate that tumor rejection early on is not T-cell dependent but long-term surveillance clearly is. The higher the expression of the transgene, the better is the protection afforded by the vaccine. Insertion of the human cDNA encoding IL-2 into renal cell carcinoma lines has been accomplished using a retroviral vector. When human renal cell tumors were initiated in nude mice (congenitally athymic, hence having no T cells), subcutaneous injection of the retrovirally transfected cells inhibited tumor growth while similarly transfected cells expressing human IFN-γ were without effect.

In another promising approach to cancer vaccination, mice were vaccinated with plasmid DNA encoding human carcinoembryonic antigen (CEA). The resulting immune response of these mice was comparable to that of mice previously immunized with a vaccinia virus vector carrying the CEA gene. The vaccinia-immunized mice were subsequently shown to be protected against a CEA-expressing syngeneic mouse colon carcinoma challenge.

4 PERSPECTIVES

Many latent viral and other types of infection require good cell-mediated immunity. Genetic vaccination holds much promise against viral disorders because in a sense, the vaccine platform simulates the actual viral infection; also, there is more flexibility in design of administration, and only antigens and immunomodulatory genes required for a custom immune response need to be included. Genetic vaccines have a distinct advantage over conventional vaccines in that they do not require the use of adjuvants and they can be administered multiple times if necessary without untoward side effects. From a gene therapy perspective, a major stumbling block in actual clinical settings of gene therapy may have been eliminated, namely, the need to achieve a high level of expression for effective immunization.

The vaccine needs of the developing countries are extremely great, and so too are the logistics of administering a vaccine when it is available. Another major factor that affects the development of vaccines for use in economically disadvantaged countries is the cost factor. The cost is obviously high, and many countries simply have no budget with which to purchase and distribute the vaccine once it becomes available. Thus, for a vaccine to be effective under these constraints, it must be economical, stable for long periods of time, easily administered by minimally trained personnel, and ideally, effective after just one or two administrations. Naked DNA vaccines may be better able to meet these criteria than live organism or protein-based conventional vaccines, especially when one considers that the diseases of the developing countries have generally resisted vaccination by more conventional means.

Health care workers are ever more vigilant for emerging new infectious agents. The ability to bring molecular biological techniques to bear on these agents lends itself to the rapid production of a genetic vaccine without the need to produce large quantities of proteins, as is currently done for the yearly influenza vaccines. These are the fruits of recombinant DNA that most clearly offer the clinic the promise of great disease prevention. It is interesting that microbiology grew out of the quest of early scientists to vaccinate against smallpox using the vaccinia virus, while genetic vaccination grew out of the quest of microbiologists to exploit molecular biological techniques to transfer genes into living organisms.

> . . . and the end of all our exploring will be to arrive where
> we started and know the place for the first time.
> —T. S. Eliot

See also ELECTROPORATION AND ELECTROFUSION; MOLECULAR GENETIC MEDICINE.

Bibliography

Ellis, R. W., Ed. (1992) *Vaccines: New Approaches to Immunological Problems.* Butterworth-Heinemann, Boston.
Mulligan, R. C. (1993) The basic science of gene therapy. *Science,* 260:926–932.
Walsh, C. E., Liu, J. M., Miller, J. L., Nienhuis, A. W., and Samulski, R. J. (1993) Gene therapy for human hemoglobinopathies. *Proc. Soc. Exp. Biol. Med.* 204:289–300.
Wolff, J. A., Ed. (1994) *Gene Therapeutics: Methods and Applications of Direct Gene Transfer.* Birkhauser, Boston.
Yang, N.-S. (1992) Gene transfer into mammalian somatic cells in vivo. *Crit. Rev. Biotechnol.* 12:335–356.

Genome: *see* specific genome, e.g., *E. coli* Genome; Mouse Genome.

GENOMIC IMPRINTING
Wolf Reik, Robert Feil, and Nicholas D. Allen

1 **Parental Imprinting**
   1.1 Embryological Evidence
   1.2 Genetic Evidence

2 **Imprinted Genes in Mouse and Human**

3 **Chimeric Rescue**

4 **Molecular Mechanism**

5 **Genetic Disease**

6 **Why Imprinting?**

Key Words

Androgenetic Embryo An embryo with two paternal genomes and no maternal genome, produced by nuclear transplantation.

DNA Methylation Attachment of a methyl (CH_3) group to the bases of the DNA. Most of the methylation in the mammalian genome is in dinucleotides consisting of cytosine followed by guanine (CpG).

Epigenetic Modification A reversible but heritable alteration of DNA above the level of sequence (e.g., the methylation of C residues in CpG dinucleotides). The additional layer of information so constituted in the chromosome may indicate, for example, the parental origin of the chromosome.

Fetal Growth Factors Imprinted genes include those with a strong effect on fetal growth; paternal inheritance of some of these genes increases the size of the fetus, whereas maternal inheritance can decrease size.

Imprinting and Cancer Syndromes Because proliferation and differentiation of fetal tissues are affected by imprinting, imprinted genes can be involved in cancer syndromes in the human. This is especially true with respect to childhood tumors, where both monoparental disomy and somatic loss of imprinting have been observed.

Modifier Gene A gene that can affect or control DNA methylation or imprinting.

Nuclear Transplantation The moving of early embryonic nuclei or pronuclei between embryos, using micromanipulation equipment.

Nutrient Transfer   In mammals, there is substantial transfer of nutrients across the placenta from mother to fetus. An evolutionary theory, based on genetic conflict, predicts that genes exerting a major influence on this process should be under selective pressure to become imprinted.

Parental Imprinting   The expression of certain genes from either the maternal or paternal chromosomes; achieved by epigenetic modification of chromosomes.

Parthenogenesis   The production of offspring from eggs only (no sperm). Parthenogenesis is possible in some groups of animals but is absent in mammals because of the functional inequivalence of the parental genomes caused by imprinting.

Phenotype   The observable or measurable characteristics of an organism.

Transgenic Mice   Mice that have additional genetic material (DNA) incorporated into their genome; micromanipulation techniques are used to effect such modifications.

Uniparental Disomy   Inheritance of particular chromosomes in two copies from one parent, with corresponding absence of the chromosome from the other parent. The presence of imprinted genes on this chromosome will lead to an altered balance of gene product, hence to specific phenotypes or disease in the human.

Genomic imprinting is a genetic mechanism in mammals and other organisms by which genes are expressed or repressed depending on their inheritance from mother or father. Expression or repression is achieved by epigenetic modification of DNA and chromosomes, including DNA methylation. A number of imprinted genes have been identified, and these have important functions in mammalian development including the control of fetal growth and fetal viability, as well as aspects of animal and human behavior after birth. Malregulation of imprinted genes is involved in a large number of genetic disorders in the human.

1   PARENTAL IMPRINTING

In its widest definition, genetic imprinting is considered to be any process in which epigenetic information is introduced into chromosomes and is stably replicated together with the chromosome as cells divide. It may have a role in some aspects of cell memory in the development of multicellular organisms. Imprinting is carried over to progeny cells at cell division; it can affect gene expression (often many cell generations later), and it is reversible. Unlike heritable changes due to mutation or directed gene rearrangement (as in the immunoglobulin genes), imprints can normally be removed from chromosomes without leaving behind any permanent alteration of the genetic material. Imprinting can be seen as a remote control switch; once turned, it may have no effects for many cell generations, but eventually gene expression and thus phenotype will depend on this initial switch.

Here we consider a particular class of imprints, those that mark the parental origin of genomes, chromosomes, and genes in mammals. Parentally imprinted genes are those whose expression depends on whether they are carried on a maternally or paternally derived chromosome. Some are expressed from maternal chromo-

somes, others from paternal chromosomes. In the past 5 years or so, parental imprinting has progressed from an intriguing embryological observation to a firmly established biological principle, with far-reaching consequences for mammalian development and genetics, and human disease. Genes subject to parental imprinting have been identified and possible molecular mechanisms of imprinting suggested. This article focuses exclusively on parental imprinting of autosomes. Parental imprinting is also known in invertebrates and plants, but no such species are discussed here.

1.1   EMBRYOLOGICAL EVIDENCE

Embryological investigations provided the first indication of the differential expression of essential genes for development by maternal and paternal chromosomes. In mammals, the presence of both parental genomes in the zygote seems to be essential to the production of a viable fetus. In the mouse, in which most work on parental imprinting has been done, monoparental diploid embryos cannot develop to term, irrespective of whether they carry two maternal or two paternal genomes, or whether the ones with two maternal genomes are derived parthenogenetically or gynogenetically (by nuclear transplantation). Such embryos die soon after implantation at the latest.

The phenotype of parthenogenetic (or gynogenetic) and androgenetic embryos is different from the earliest stages. Whereas parthenogenetic embryos develop in culture at a rate comparable to that of normal embryos at least to the blastocyst stage, only a variable proportion of androgenetic embryos reach that stage, and in general they progress more slowly. Following implantation, parthenogenetic and androgenetic embryos have very different phenotypes indeed (Figure 1). Parthenogenetic conceptuses can reach the 25-somite stage on day 10 of gestation with a relatively well-developed embryo proper, but their extraembryonic tissues are severely deficient, especially the trophoblast, which is crucially important for placentation, hence nutrient transfer from the mother. Androgenetic embryos, by contrast, show very much the opposite phenotype: extraembryonic structures are well developed, but the embryo itself is severely retarded and does not progress beyond the stage of 4–6 somites.

In the mouse therefore, and probably in other mammals including humans, both parental genomes are necessary for development, and they must provide the correct gene dosage at certain gene loci, which are termed imprinted loci. The different phenotypes of the two types of experimental embryo indicate the cumulative actions of maternally or paternally imprinted genes.

1.2   GENETIC EVIDENCE

The role of small subsets of imprinted genes can be investigated genetically in embryos in which defined regions of one or the other parental genome are disomic (present in two copies), a condition known as uniparental disomy. Uniparental disomic embryos can be obtained by crossing mice that are heterozygous for particular chromosomal translocations. Because of nondisjunction at meiosis, some of the gametes will contain no copies of the translocation or the corresponding normal chromosome, whereas others will contain both. Embryos arising from fusion of two such gametes will therefore contain two copies of a chromosome or part of a chromosome derived from the same parent. Uniparental disomic embryos have

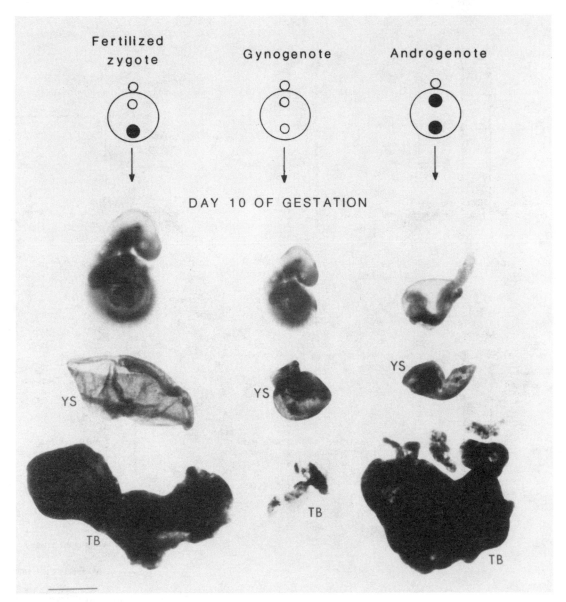

Figure 1. The phenotypes of normal, gynogenetic, and androgenetic mouse embryos and extraembryonic tissues at 10 days of gestation. Gynogenetic and androgenetic embryos are obtained by pronuclear transplantation: TB, trophoblast; YS, yolk sac.

been obtained for most chromosomes, and their development has been analyzed.

Some of these embryos show altered phenotypes, indicating the presence of imprinted genes on the disomic segment. For example, mice that are paternally disomic for the proximal portion of chromosome 11 are larger than their normal littermates, whereas mice maternally disomic for the same chromosome are smaller (Figure 2). This suggests that an imprinted gene or genes whose difference in expression causes the growth differences, is present on proximal chromosome 11.

All imprinted regions identified so far determine phenotypes associated with growth and viability of the embryo, and in some cases, perhaps, behavior of neonates. The total number of parentally imprinted genes in unknown, but a minimum estimate can be made from the number of identified imprinted chromosome segments (Figure 2).

2   IMPRINTED GENES IN MOUSE AND HUMAN

Ten genes residing in five chromosomal regions (Figure 2, Table 1) in the mouse have now been identified as undergoing parental imprinting; for many, the imprinting status is conserved in the human genome (Table 1). Since most of these genes have known functions, it is now possible to begin to interpret the monoparental disomy phenotypes in the context of the expression or repression of these genes.

Mice inheriting from the father a null allele of the insulin-like

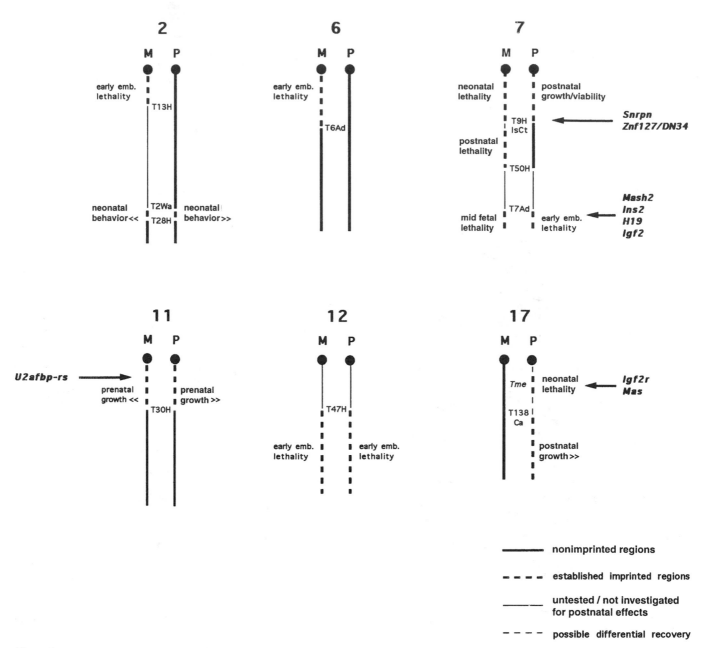

Figure 2. Imprinted regions in the mouse genome. For six mouse chromosomes, imprinted regions have been established so far. The translocation breakpoints that define these regions are shown. The phenotypes of maternal (M) or paternal (P) disomy of these regions are indicated, as well as the location of the imprinted genes, identified so far in the mouse. [Adapted from C. V. Beechey and B. M. Cattanach, Genetic imprinting map. *Mouse Genome*, 91:102–104 (1994).]

growth factor II gene *(Igf2)* on distal chromosome 7 turn out to be considerably smaller than their littermates, whereas when the mutant allele is maternally inherited, offspring show no growth deficiency. Consistent with this phenotype, *Igf2* transcripts are produced from the paternal allele, but the maternal allele is almost completely repressed. The only exception to this allele-specific expression is found in the leptomeninges and choroid plexus in brain, where both alleles are expressed and expression continues into adulthood. This indicates that imprinting may be specific for both tissue and developmental stage.

Additional evidence for maternal imprinting of *Igf2* comes from

the extremely low levels of *Igf2* RNA in mouse embryos that are maternally disomic for the distal region of chromosome 7, where the *Igf2* gene resides. These embryos are smaller than their littermates and die in utero during the last third of the gestation period. *Igf2* is expressed widely throughout the embryo, predominantly but not exclusively in extraembryonic tissues and in the mesodermal lineage, and is known to be an embryonic mitogen.

In the genomic region containing the *Igf2* gene, three other imprinted genes have been identified. At 75 kb downstream of *Igf2*, the *H19* gene, which does not appear to encode a protein, is expressed from the maternal chromosome but repressed on the pater-

Table 1   Imprinted Genes in Mouse and Human

Expression	Mouse		Human	
	Gene	Chromosome	Gene	Chromosome
Paternal	*Igf2*	Distal 7	*Igf2* chr	11p15.5
	Insulin 2	Distal 7		
	Mas	Proximal 17		
	Xist	Mid X		
	U2afbp-rs	Proximal 11		
	nrpn	Mid 7	*SNRPN*	15q11–13
	ZNF127/DN34	Mid 7	*ZNF127/DN34*	15q11–13
			PAR1	15q11–13
			PAR5	15q11–13
			IPW	15q11–13
Maternal	*H19*	Distal 7	*H19*	11p15.5
	Igf2r	Proximal 17	*IGF2R*(?)	6q
	Mash2	Distal 7		

nal one. In its temporal and spatial expression, it is very similar to *Igf2*. In contrast to *Igf2*, the product of the *H19* gene may down-regulate cellular proliferation.

The insulin 2 gene (*Ins2*), located 18 kb upstream of *Igf2*, is imprinted in a tissue-specific manner. In the yolk sac (an extraembryonic tissue), it is the paternal chromosome that expresses the *Ins2* gene, whereas the maternal chromosome is repressed. In the same chromosomal region, further upstream of *Igf2*, the *Mash2* gene is imprinted in trophoblast tissue, with the maternal allele being expressed. Both *IGF2* and *H19* are also imprinted (in the same direction) in the human, whereas for *INS* and *MASH2* no information is available at present.

The *Snrpn* gene (small nuclear ribonuclear protein N), on the central part of mouse chromosome 7, is expressed from the paternal chromosome, predominately in brain and heart. It encodes a polypeptide thought to be involved in tissue-specific RNA splicing. Its absence in the human neurobehavioral disease known as Prader–Willi syndrome (PWS) (see also Section 5) may contribute to some of the symptoms of this condition.

Four other imprinted human genes recently identified close to the *Snrpn* gene could be imprinted in the mouse as well. *IPW* (imprinted gene in the Prader–Willi syndrome region), at 150 kb distal to *Snrpn*, is expressed from the paternal allele in fetal tissues and appears, like the imprinted *H19* gene, not to encode a protein. The *ZNF127* gene, which codes for a zinc finger protein and is located in the same chromosomal region, displays a parent-specific methylation imprint. Two additional imprinted genes with paternal expression and unknown function are also located in the PWS region (*PAR1* and *PAR5*). Given the phenotypic features of PWS, maternal duplication of the central part of mouse chromosome 7 may constitute a mouse model for this syndrome.

Another imprinted locus is *Tme* (T maternal effect), on proximal mouse chromosome 17. Deletions of the *Tme* locus such as T^{hp} or t^{wlub2} are lethal at late fetal stages when the maternal chromosome is involved but not when the paternal chromosome is implicated. The gene for the IGF-II receptor (*Igf2r*) (the cation-independent mannose 6-phosphate receptor) has been mapped to this region of mouse chromosome 17 and has been shown to be imprinted, with the maternal copy being the active one. The phenotype of mice in which the *Igf2r* gene has been inactivated by homologous recombination clearly overlaps with that described for the *Tme* mutation. Recently, the *Mas* proto-oncogene, which is only 300 kb from the

Igf2r gene, was found to be subject to parental imprinting as well. This intronless gene encodes a peptide receptor that may transduce extracellular signals to G proteins and is expressed from the paternal allele exclusively in various embryonic tissues; the transcriptional repression of *Mas* is cell-type and developmental stage specific.

Another intronless imprinted gene has been identified in the proximal part of mouse chromosome 11, an imprinted chromosomal region that had been identified earlier by analysis of uniparental disomic mice (Figure 2). This gene, "U2af binding protein related sequence" (*U2afbp-rs*), is expressed only from the paternal allele, predominantly in the brain. The protein it encodes is homologous to the splicing factor U2af.

Whereas some genes have already been shown to be imprinted in mouse and human (Table 1), this may not be true in all cases: the imprinting status of the *IGF2R* gene in the human is unclear at present.

It is interesting to note that imprinted genes appear to be clustered in the genome. Whether this is for mechanistic or adaptive (evolutionary) reasons is not clear at present.

The molecular mechanism of imprinting of these genes is at present unknown (but see Section 4 for some possible mechanisms). Indeed it is not even known at precisely what level of gene regulation repression occurs. However, differences in DNA methylation and chromatin structure have recently been detected between parental alleles of some imprinted genes.

3   CHIMERIC RESCUE

Cells from parthenogenetic or androgenetic embryos can be "rescued" by incorporating them into chimeric embryos that also contain normal cells and in which their later development and their effect on the chimeric embryo can be tested. Allocation of cells to different lineages does not seem to be affected, but selection against parthenogenetic (PG) and androgenetic (AG) cells occurs once lineages have been set apart. PG cells are consistently lost from the extraembryonic tissues (in aggregation chimeras) and in the fetus from mesodermal tissues, but are usually well represented in neuroectodermal tissues. In contrast, AG cells are selected against in neuroectodermal tissues but contribute substantially, perhaps to an even greater extent than normal cells, to mesodermal tissues such as skeletal muscle, heart muscle, and skeleton. AG chimeras that survive to term (because they have a small proportion of AG cells) show severe skeletal abnormalities that are due partly to hyperplasia of cartilage.

The basis of cell selection is at present unknown, but one possibility is a greater (or reduced) propensity to cell proliferation within a lineage due to overexpression or lack of autocrine or short-range paracrine growth factors (such as IGF-II) or their receptors. It is certainly suggestive that PG cells are notoriously absent in the very tissues in which *Igf2*, *Igf2r*, and *H19* are predominantly expressed. Chimeric embryos have also been made from cells with maternal or paternal disomy of distal chromosome 7, on which *Igf2*, *H19*, and *Ins2* reside. Paternal disomic cells expressing a double dose of IGF-II cause an increase in size of the chimeric embryo. This agrees with observations that AG chimeras, which are disomic for the whole paternal genome, can be up to 50% larger than controls.

4   MOLECULAR MECHANISM

The molecular mechanism of imprinting must involve epigenetic modifications of DNA. Some transgenes in the mouse become imprinted, and such transgenic mice therefore have provided a model system for studying parental imprinting. Indeed, it has been demonstrated that DNA methylation and other features of chromatin may be part of the imprinting mechanism. Imprinted genes have therefore been searched extensively for differences in DNA methylation and chromatin structure between the parental alleles (Figure 3). The *Igf2* gene shows no differences in methylation or chromatin in the promoter region of the gene; however, sites within the 3′ part of the gene (fourth intron up to the 5′ part of sixth exon) and a region upstream of the first promoter show parent-specific methylation, with the paternal (expressed) allele being more highly methylated. By contrast, the *H19* gene shows extensive promoter methylation of the repressed (paternal) copy associated with decreased DNAse I

sensitivity. Interestingly, these allelic methylation differences are established after fertilization and therefore may not constitute primary gamete-specific imprints. The tight linkage and reciprocal imprinting of *Igf2* and *H19* has led to a hypothesis that the two genes constitute an imprinted domain with a shared regulatory region. If this is correct, then primary gamete-specific imprints may reside outside the immediate vicinity of each gene.

The *Igf2r* gene also has a promoter methylation imprint (on the repressed paternal allele), but it also has an intronic region in which the expressed (maternal) copy is more highly methylated. This intronic methylation imprint is present in the egg before fertilization occurs and thus may be a primary imprinting signal. Whether imprinting mechanisms are different for different genes remains to be seen. The somatic methylation patterns observed for imprinted genes (*Igf2*, *H19*, and *Igf2r*) are illustrated in Figure 3.

In addition to differences in methylation and chromatin, differences have been detected in the timing of replication of imprinted genes in the cell cycle. In all imprinted genes studied, the paternal copy of genes replicates earlier than the maternal one, regardless of whether the genes are maternally or paternally imprinted. In addition, such replication domains may be large (in the megabase range) and may contain genes that are not themselves imprinted.

Genetic screens are also being developed to identify genes that may play a role in controlling imprinting, so-called modifier genes. If methylation is indeed involved in imprinting, the methyltransferase gene itself is of course important. Mice have now been created in which this methylase gene has been mutated, and the effect of this on imprinted genes is as predicted from the methylation imprints: *Igf2* and *Igf2r* expression decreases, whereas *H19* expression increases. This experiment demonstrates directly that DNA methylation is involved at least in the somatic maintenance of parental imprints.

At present it is necessary to postulate other genes that direct the

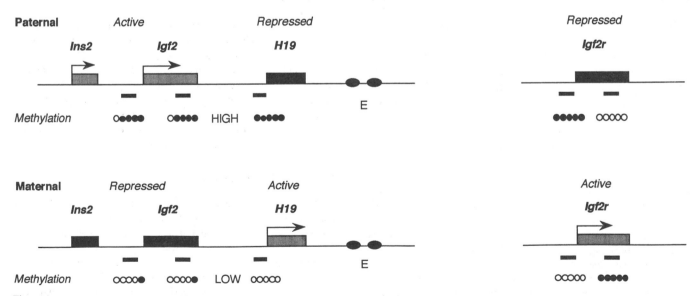

Figure 3.   Imprinted genes possess differentially methylated parental chromosomes. The tightly linked *Ins2*, *Igf2*, and *H19* genes on mouse chromosome 7 are illustrated together with the downstream *H19* enhancer region E. The *Igf2r* gene on mouse chromosome 17 is also shown. Solid circles indicate CpG methylation in the defined regions indicated by the solid bars. Gene transcription is represented by an arrow. In somatic cells, the paternal alleles of the *Igf2* and *H19* genes are more methylated than the maternal alleles. Thus, methylation of *Igf2* is associated with gene expression, while for *H19*, methylation is associated with gene repression. Differential methylation is also seen in the *Igf2r* gene. Maternal methylation in the body of the gene is already present in the unfertilized egg; paternal methylation in the promoter region arises postzygotically and is associated with gene repression. The *Ins2* gene is imprinted only in the yolk sac, not in the pancreas. Methylation differences in *Ins2* have so far not been reported.

specificity of methylation imprinting. Work toward isolation of such genes has made use of the observation that transgene methylation can depend on genotype-specific modifier genes, different alleles of which segregate in inbred strains. One such modifier gene has been chromosomally mapped, but none has been cloned so far. Whether there exist in interbreeding populations different alleles of modifier genes that lead to large effects on imprinted genes is a matter of speculation at present. Modifier genes may also be involved in controlling the specificity of DNA methylation patterns in mammalian cells, and they may affect penetrance and expressivity in genetic disease.

The nature of the modifier genes that can affect epigenetic programming in mammals is at present unknown. There are, however, overt similarities between transgene modification in the mouse and position effect variegation in *Drosophila melanogaster;* and it is possible that the modifier genes work in a similar way to the enhancers and suppressors of position effect variegation. In the fly, a number of these genes have now been cloned, and some encode proteins that may be involved in the formation of heterochromatin domains (which can form stable epigenetic switches). A family of genes with homology to the *Drosophila* heterochromatin protein gene *HP1* has recently been identified in the mouse, and it will be important to see what kind of phenotype variant alleles of these genes can elicit.

5   GENETIC DISEASE

Some of the imprinted genes discovered in the mouse, notably *Igf2, H19,* and *Snrpn,* have also been shown to be imprinted in the human (Table 1). If an imbalance arises therefore between maternal and paternal chromosomes—for example, in the form of a monoparental disomy of particular chromosomes—there will be an increased or decreased gene dosage of imprinted loci. Many different diseases have been associated with disomies of specific chromosomes. Alternatively, imprints can be lost (or gained) during development by mutational or epigenetic mechanisms, resulting in functional disomy. Altered imprinting could thus result in disease phenotypes.

The neurobehavioral syndromes Prader–Willi (PWS) and Angelman syndrome (AS) arise from paternal deletion or maternal disomy and from maternal deletion or paternal disomy of a region on chromosome 15q11-13, respectively. This pattern indicates that at least two loci are involved, and in fact recent work reveals nonoverlapping deletions in this region that cause either AS or PWS. The imprinted genes *Snrpn, ZNF127, PAR1, PAR5,* and *IPW* (all paternally expressed) in the PWS segment may contribute to at least part of the PWS phenotype. In both AS and PWS there are also a number of cases with no deletion or disomy, and these may arise from mutational derepression or repression of an imprinted gene. Indeed, cases with small deletions have now been discovered in which the imprinting pattern of the whole region is altered ("imprinting mutation"); hence a modifier gene that acts to imprint other genes in the region may have been deleted.

The Beckwith–Wiedemann syndrome is a fetal overgrowth syndrome associated with a variety of embryonal tumors (Wilms' tumor, rhabdomyosarcoma, etc.). Paternal disomy of chromosome 11p15.5 has been found in this syndrome, and this segment contains the imprinted *IGF2* and *H19* genes. Hence overexpression of *IGF2* (and lack of *H19*) may contribute to the aberrant growth phenotype. Interestingly, nondisomic cases have now been found in which the maternal *IGF2* allele is expressed, suggesting that mutational or epigenetic derepression of this imprinted gene can cause the disease.

In Wilms' tumors there is often loss of heterozygosity of chromosome 11p15, with preferential loss of the maternal allele. As in BWS, there are also tumors with no loss of heterozygosity in which both parental *IGF2* alleles are expressed. In these tumors, both copies of *H19* are methylated, lending further support to the idea of coregulation of imprinting of both genes.

There are also parental transmission effects in monogenic disorders, such as Huntington disease (HD), where early onset cases have predominantly inherited their mutant allele from the father. Fragile X syndrome and myotonic dystrophy, in which onset or severity is dependent on maternal transmission, are also in this category. In these diseases, including HD, there is however expansion of trinucleotide repeats in the mutant genes that generally correlate with expressivity; thus the influence of imprinting is more difficult to interpret. In other cases of parental transmission effects—for example, in glomus tumor—the genes have not been identified yet and the interpretation awaits molecular analysis.

6   WHY IMPRINTING?

Evolutionary biologists now believe that they have the answer to this question, and their explanation is both elegant and consistent with most of the observations. Consider in a placental mammal a gene that acts in the fetus to acquire nutritional resources from the mother. Any growth factor gene that acts during intrauterine life, such as *Igf2,* would qualify. Paternally derived alleles at this locus will tend to increase growth of the fetus, since this would maximize their chances of spreading through the population. Maternal alleles are in principle subject to the same selective pressure. A fetus bearing maternal alleles at this locus, however, would tend to impose on the mother a burden of increased resource transfer, which might compromise reproductive success for all the offspring produced by one mother. Over evolutionary time, there will therefore be a tradeoff, a parental tug-of-war over the most successful combination of expression levels of maternally and paternally derived alleles.

Such a theory predicts the behavior of the genes for IGF-II and its receptor very accurately. Increased dosage of IGF-II can increase the size of the fetus. The *Igf2* gene should therefore be imprinted to be expressed when paternally inherited. The receptor, by contrast, presumably acts to bind IGF-II and thereby to decrease its availability for binding to the type I receptor, which is thought to be the main route of IGF-II growth factor activity. *Igf2r* expression therefore acts negatively on the IGF-II axis and is consequently predicted to be imprinted in the opposite direction, as it indeed is.

See also GENE EXPRESSION, REGULATION OF; HOMEODOMAIN PROTEINS; HUMAN CHROMOSOME II, GENETIC AND PHYSICAL MAP STATUS OF; HUMAN GENETIC PREDISPOSITION TO DISEASE.

Bibliography

Beechey, C. V., and Cattanach, B. M. (1994) Genetic imprinting map. *Mouse Genome,* 91:102–104.

Brandeis, M., Ariel, M., and Cedar, H. (1993) Dynamics of DNA methylation during development. *BioEssays,* 15:709–713.

Efstratiadis, A. (1994) Parental imprinting of autosomal genes. *Curr. Opinion Genet. Dev.* 4:265–280.

Feinberg, A. P. (1993) Genomic imprinting and gene activation in cancer. *Nature Genet.* 4:110–113.

Monk, M., and Surani, A., Eds. (1990) *Genomic Imprinting. Development, 1990 Supplement.* The Company of Biologists Ltd., Cambridge.

Moore, T., and Haig, D. (1991) Genomic imprinting in mammalian development: A parental tug-of-war. *Trends Genet.* 7:45–49.

Reik, W. (1989) Genomic imprinting and genetic disorders in man. *Trends Genet.* 5:331–336.

Sapienza, C. (1990) Parental imprinting of genes. *Sci. Am.* 263:26–32.

Surani, A., and Reik, W., Eds. (1992) Genomic imprinting in mouse and man. *Semin. Dev. Biol.* 3:73–160.

Genomic Responses to Environmental Stress: *see* Environmental Stress, Genomic Responses to.

GLYCOBIOLOGY

Akira Kobata

1 **Introduction**

2 **Principles of Studying the Sugar Chains of Glycoconjugates**

   2.1   Release of Sugar Chains of Glycoconjugates as Oligosaccharides

   2.2   Fractionation of Oligosaccharides

   2.3   Structural Analysis of Oligosaccharides

3 **Rules Detected in the Sugar Chain Structures of Glycoconjugates**

4 **Functional Roles of the Sugar Chains of Glycoconjugates**

Key Words

Anomers   The two isomeric structures, formed because the aldose C-1 atom becomes asymmetric in making the sugar chains, are called the α and β anomers.

CD15   One of the surface markers of human granulocytes and acute myelogenous leukemia cells.

Fc receptor   A cell surface receptor that binds the hinge region of immunoglobulin G.

Furanose   The isomeric ring structure formed by four carbons and one oxygen atom of aldoses and ketoses.

Glycoconjugates   The generic name of carbohydrates that are linked to various biomolecules. They can be classified into glycoproteins, glycolipids, and proteoglycans.

Glycohormones   Peptide hormones containing sugar chains.

Glycon and Aglycon   Glycosidases, which cleave the sugar chains, have two kinds of specificity: one for glycon and one for aglycon. Glycon specificity is directed to the sugar moiety released from the portion of a sugar chain containing a reducing terminal, and its anomeric configuration. The aglycon specificity is directed to the structure of the portion to which the released sugars are linked. This specificity varies by enzymes. For example, diplococcal β-galactosidase cleaves the Galβ1 → 4GlcNAc linkage but not the Galβ1 → 3GlcNAc and

the Galβ1 → 6GlcNAc linkages. In contrast, the enzyme from *Streptococcus* 6646K hydrolyzes all three linkages.

Lectin   The general name for proteins other than antibodies, which bind specifically to particular sugar chain structures.

MEL-14   The peripheral lymph node lymphocyte homing receptor.

Oligosaccharide Alcohols   Oligosaccharides, the reducing termini of which are converted to alditols.

Protein C   A serine protease zymogen; becomes a potent inhibitor of blood coagulation upon activation by thrombin.

Most proteins produced by multicellular organisms contain sugar chains and are called glycoproteins. Because of the difficulties associated with the structural study of the sugar chains of glycoproteins, however, the functional aspects of the sugar moieties of glycoproteins were ignored during the long history of protein research.

In addition to the structural multiplicity, another factor makes the structural study of the sugar chains of glycoproteins difficult. Many glycoproteins contain more than one sugar chain in one molecule. Even in the case of a glycoprotein with only one sugar chain, there is widespread microheterogeneity of sugar chain structure because the absence of a template in the biosynthetic machinery of sugar chains makes possible the formation of incomplete chains. Therefore, each sugar chain must be separated before it can be subjected to structural study.

Interest in the glycoconjugate research was stimulated early in the 1960s by elucidation of the antigenic determinants of human blood types and the molecular basis of antigenic conversion of bacteria. This research area has further attracted the interest of biologists, because many studies on cell biology have suggested the possibility that the sugar chains of glycoproteins and glycolipids play an important role as signals of cell-to-cell recognitions, which are crucial in multicellular organisms.

Development of gene technology in recent years also has accelerated the functional study of glycoprotein sugar chains. This technology has opened a way to obtaining substantial amounts of bioactive proteins, which are useful but occur in small amounts in the animal body. However, many proteins produced by animal cells occur as glycoproteins. Since many nonglycosylated recombinant proteins, obtained by using bacterial hosts, do not express the expected biological activities, the importance of the study of their sugar moieties has been noticed. "Glycobiology" is a novel scientific field, recently established on the basis of the possibility of elucidating the biological information in the sugar chains of glycoconjugates and using these data to increase our knowledge of biology.

1   INTRODUCTION

In contrast to nucleic acids and proteins, sugar chains of many different structures can be formed by using a small number of monosaccharide units. Let us consider the smallest unit of chains: A—B. In the case of nucleic acid, only one structure is made by assigning adenylic acid to A and guanylic acid to B. In the case of protein also, only one structure is made when A and B are assigned to valine and serine, respectively. In the case of sugar chains, however, many isomeric structures are formed. Suppose that A and B are assigned to galactose and mannose, respectively. As shown in Figure 1, galactose can be linked at the four hydroxyl groups of mannose: C-2,

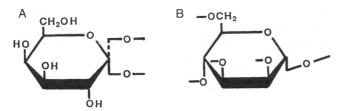

Figure 1.   Construction of sugar chains.

C-3, C-4, and C-6. Accordingly, four isomeric structures can be formed. By virtue of the ability of the galactose residue to take two anomeric configurations, α and β, the number of possible isomeric structures becomes eight. Furthermore, galactose residues can occur in the furanose form as well as in the pyranose form shown in Figure 1. Thus 16 isomeric structures are possible for the disaccharide Gal → Man. When the number of units increases (to three, four, etc.), only one structure can be formed by assigning a particular unit at each position in the case of nucleic acids and proteins, because they are linear constructs. In contrast, the number of isomeric sugar chains increases by geometrical progression, because branching can be formed in sugar chains larger than disaccharide. This means that sugar chains, but not nucleic acids and proteins, have the following characteristic feature: they can form multiple structures with a small number of units.

2   PRINCIPLES OF STUDYING THE SUGAR CHAINS OF GLYCOCONJUGATES

2.1   RELEASE OF SUGAR CHAINS OF GLYCOCONJUGATES AS OLIGOSACCHARIDES

Each sugar chain of a glycoconjugate must be isolated for structural study. The sugar chains of glycoproteins can be classified into two groups, O-linked and N-linked. An O-linked or mucin-type sugar chain contains at its reducing terminus an *N*-acetylgalactosamine residue that is linked to the hydroxyl group of either a serine or a threonine polypeptide residue. Such sugar chains can be released as oligosaccharide alcohols by heating the glycoprotein at 48°C in a solution of 1 M NaBH$_4$ and 0.05 N NaOH for 16 hours. The use of hydrazinolysis to remove acyl groups from the *N*-acetylamino sugar residues of oligosaccharide alcohols, followed by re-*N*-acetylation with ^{14}C- or ^{3}H-labeled acetic anhydride, will produce isotope-labeled oligosaccharide alcohols, which are useful for fractionation and structural analyses.

The other group of sugar chains of glycoprotein, the *N*-linked or asparagine-linked sugar chains, are linked to the amide group of an asparagine polypeptide residue; each one contains an *N*-acetylglucosamine residue at its reducing terminus. These sugar chains are quantitatively released as oligosaccharides by heating thoroughly dried glycoproteins in anhydrous hydrazine at 80–100°C for 8–10 hours. Although some of the acetyl groups of the *N*-acetylamino sugars of oligosaccharides are removed by hydrazinolysis, they can be recovered by *N*-acetylation of the oligosaccharide fraction. The oligosaccharides can be converted quantitatively to ^{3}H-labeled oligosaccharide alcohols by reduction with [^{3}H]NaBH$_4$.

N-Glycanases (glycopeptidases) from several biological sources are commercially available. These enzymes hydrolytically cleave the GlcNAc → Asn linkages of glycoproteins and leave aspartic acid residue on the polypeptide portion. The glycosylamine residues of the released oligosaccharides are spontaneously hydrolyzed into oligosaccharides and ammonia. With this catalytic action, the enzymes release all N-linked sugar chains of glycoproteins without destroying the polypeptide moieties. However, one must take care for the limits of application of these enzymes, because they can have substrate specificities including the steric hindrance of polypeptide portion.

Several endo-β-*N*-acetylglucosaminidases (hereafter referred to as endos), which act on N-linked sugar chains, were reported. These enzymes hydrolytically cleave the *N, N'*-diacetylchitobiose moieties located at the reducing termini of N-linked sugar chains. Accordingly, the enzymes release most of the sugar chains of glycoproteins intact by leaving either the *N*-acetylglucosamine residue or the Fucα1 → 6GlcNAc group on the polypeptides. Some of these enzymes are now commercially available, and each one has a unique substrate specificity (Table 1) Endo D requires Manα1→ 3Manβ1 → 4GlcNAc for its glycon specificity. The α-mannosyl

Table 1   Structural Requirements of Endoglycosidases[a]

Enzyme Name	Substrate Structure
Endo-β-*N*-acetylglucosaminidase	

Endo D

$$R \qquad\qquad\qquad R$$
$$\downarrow \qquad\qquad\qquad \downarrow$$
$$6 \qquad\qquad\qquad 6$$
$$R{\rightarrow}4Man\alpha1{\rightarrow}3Man\beta1{\rightarrow}4GlcNAc\beta1{\overset{|}{\rightarrow}}4GlcNAc{\rightarrow}Asn$$

Endo H

$$R \qquad R$$
$$\downarrow \qquad \downarrow$$
$$2 \qquad 6$$
$$Man\alpha1{\rightarrow}3Man\alpha1\searrow \qquad\qquad\qquad R$$
$$6 \qquad\qquad\qquad \downarrow$$
$$R{\rightarrow}4Man\beta1{\rightarrow}4GlcNAc\beta1{\overset{|}{\rightarrow}}4GlcNAc{\rightarrow}Asn$$
$$3$$
$$R\nearrow$$

Endo C$_{II}$

$$R \qquad R$$
$$\downarrow \qquad \downarrow$$
$$2 \qquad 6$$
$$Man\alpha1{\rightarrow}3Man\alpha1$$
$$\searrow 6$$
$$R{\rightarrow}4Man\beta1{\rightarrow}4GlcNAc\beta1{\overset{|}{\rightarrow}}4GlcNAc{\rightarrow}Asn$$
$$3$$
$$R{\rightarrow}2Man\alpha1\nearrow$$

Endo-α-*N*-acetylgalactosaminidase

$$Gal\beta1{\rightarrow}3GalNAc\alpha1{\overset{|}{\rightarrow}}Ser(Thr)$$

Endo-β-galactosidase

S. pneumoniae

$$GalNAc\alpha1{\rightarrow}3Gal\beta1{\overset{|}{\rightarrow}}4GlcNAc\beta1{\rightarrow}$$
$$(Gal) \qquad 2 \quad (Glc)$$
$$\uparrow$$
$$Fuc\alpha1$$

E. freundii
B. fragilis

$$GlcNAc\beta1{\rightarrow}3Gal\beta1{\rightarrow}4GlcNAc\beta1{\rightarrow}3Gal\beta1{\overset{|}{\rightarrow}}4GlcNAc\beta1{\rightarrow} ----$$
$$3$$
$$\uparrow$$
$$Fuc\alpha1$$

C. perfringens

$$Gal\alpha1{\rightarrow}3Gal\beta1{\overset{|}{\rightarrow}}4GlcNAc\beta1{\rightarrow} ---$$

Endosialidase

$$NeuAc\alpha2{\rightarrow}8NeuAc\alpha2{\rightarrow}8NeuAc\alpha2{\overset{|}{\rightarrow}}8NeuAc\alpha2{\overset{|}{\rightarrow}}8NeuAc\alpha2{\rightarrow}$$
$$(Gc) \qquad (Gc) \qquad (Gc) \quad (Gc) \quad (Gc)$$

[a]Dashed lines, positions of hydrolysis; R, either hydrogen or sugars.

residue of the trisaccharide group is the most important and should not be linked by other sugars except at its C-4 position. Endo H requires for its specific glycon a tetrasaccharide structure: Manα1 → 3Manα1 → 6Manβ1 → 4GlcNAc. Again, the most important sugar residue is the α-mannosyl residue at the nonreducing terminal. Addition of sugars at the hydroxyl group except for the C-2 position of this residue will make the oligosaccharide resistant to the enzyme action. Endo C$_{II}$ requires for its specific glycon a branched pentasaccharide: Manα1 → 3Manα1 → 6(Manα1 → 3)Manβ1 → 4GlcNAc. The two terminal α-mannosyl residues must be nonsubstituted or replaced by other sugars only at their C-2 positions. These enzymes have rather broad aglycon specificities, since they cleave the sugar chains linked to N-acetylglucosamine, the Fucα1 → 6GlcNAc, the GlcNAc → Asn (peptides), and the Fucα1 → 6GlcNAc → Asn (peptides groups. The enzymes can also cleave the sugar chains linked to N-acetylglucosaminitol and the Fucα1 → 6N-acetylglucosaminitol groups, but the rate of hydrolysis is much smaller than others.

Several other endoglycosidases shown in Table 1 were also found and have been used for the study of sugar chains. Endo-α-N-acetylgalactosaminidase hydrolytically cleaves the GalNAcα1 → Ser (or Thr) linkage of the Galβ1 → 3GalNAcα1 → Ser (or Thr) peptides and is commercially available as O-glycanase. Endo-β-galactosidases with different substrate specificities have been obtained from various bacteria. The endogalactosidase from *Streptococcus pneumoniae* releases the GalNAcα1 → 3(Fucα1 → 2)Gal and the Galα1 → 3(Fucα1 → 2)Gal groups from blood group A and B active sugar chains, respectively. The enzyme has been successfully used to characterize the blood group substances. Endo-β-galactosidases from *Escherichia freundii* and from *Bacteroides fragilis* hydrolytically cleave the β-galactosidic linkages in the (Galβ1 → 4GlcNAcβ1 → 3)$_n$ groups (N-acetyllactosaminoglycan) included in the sugar chains of glycoconjugates. Although the enzyme from *E. freundii* cleaves most of the β-galactosyl linkages in both linear and branched N-acetyllactosaminoglycan, the enzyme from *B. fragilis* cannot hydrolyze the β-galactosyl linkage of the → GlcNAcβ1 → 6(→ GlcNAcβ1 → 3)Galβ1 → 4GlcNAc group. Another endo-β-galactosidase, purified from the culture medium of *Clostridium perfringens*, releases Galα1 → 3Gal from the Galα1 → 3Galβ1 → 4GlcNAc group located at the nonreducing termini of the sugar chains of glycoconjugates. The glycon specificity of the enzyme is very strict, and blood group B determinant cannot be hydrolyzed.

Endosialidase is purified from a bacteriophage K1. It hydrolyzes both N-acetylneuraminic acid polymers and N-glycolylneuraminic acid polymers in which sialic acid residues are linked by an α2 → 8 linkage. The enzyme cannot hydrolyze oligomers smaller than tetrasaccharide and O-acetylated N-glycolylneuraminic acid polymers.

With the specificities shown in Table 1, endoglycosidases are quite useful for the structural study of sugar chains. For example, endo H cleaves all high mannose type sugar chains but not complex-type sugar chains, which are described later. Accordingly, this enzyme is effective in discriminating between the two subgroups of N-linked sugar chains. Endo D is effective in discriminating between two isomeric sugar chains:

$$—GlcNAc\beta1 → 2Man\alpha1 → 6(Man\alpha1 → 3)Man\beta1$$
$$→ 4GlcNAc\beta1 → 4GlcNAc$$

$$Man\alpha1 → 6(—GlcNAc\beta1$$
$$→ 2Man\alpha1 → 3)Man\beta1 → 4GlcNAc\beta1 → 4GlcNAc$$

because the former is cleaved by the enzyme but the latter is not.

Glycosphingolipids are the major glycolipids found in animal tissues. Since any glycosphingolipid contains just one sugar chain in one molecule, isolation of a single glycosphingolipid is enough for the structural study of that glycolipid's sugar chain. A glycosphingolipid with a single sugar chain is still a mixture of different molecules, however, as a result of the structural heterogeneity characterizing the ceramide portion of the molecule. To alleviate this complicated situation, glycoceramidases are used to release the sugar chains of glycosphingolipids intact as oligosaccharides. The oligosaccharides can then be labeled as described for the oligosaccharides released from glycoproteins and fractionated by using appropriate methods (see Section 2.2).

As described later, the sugar chains of proteoglycans are unique and are called glycosaminoglycans. Many glycosaminoglycan chains in proteoglycans are linked to the serine residues, constructing the Ser-Gly-X-Gly sequence of core proteins through the Galβ1 → 3Galβ1 → 4Xyl group located at their reducing termini. Accordingly, these sugar chains can be almost quantitatively released as polysaccharides by alkaline–borohydride treatment as described for O-linked sugar chains of glycoproteins. Structures of the released polysaccharides are then analyzed by using appropriate endoglycosidases, which cleave particular glycosidic linkages in sugar chains. Among them, several eliminases listed in Table 2 have been most effectively used to analyze the detailed structures of proteoglycans. As shown in Figure 2, all these enzymes cleave the amino sugar–uronic acid linkages of glycosaminoglycan and produce various disaccharides, summarized in Table 3. The enzymes in Table 2 differ in substrate specificity. For example, chondroitinase ABC cleaves both the sulfated GalNAcβ1 → 3GlcA and the sulfated GalNAcβ1 → 3IdoA linkages, while chondroitinase AC cannot cleave the sulfated GalNAcβ1 → 3IdoA linkage. On the contrary, chondroitinase B cleaves only the sulfated GalNAcβ1 → 3IdoA linkage. Chondroitinase C cleaves the SO$_4$ → 6GalNAcβ1 → 3GlcA linkage but not the SO$_4$ → 4GalNAcβ1 → 3GlcA and SO$_4$ → 6(SO$_4$ → 4)GalNAcβ1 → 3GlcA linkages. Since the disaccharides in Table 3 can be mutually separated by high performance liquid chromatography (HPLC) and quantitatively analyzed by using

Table 2   Bacterial Eliminases

Enzymes	Source	Main Substrates[a]
Hyaluronidases		
Hyaluronidase	*Streptomyces hyalurolyticus*	HA
Hyaluronidase SD	*Streptococcus dysgalactiae*	HA
Chondroitinases		
Chondroitinase ABC	*Proteus vulgaris*	CS, DS, HA
Chondroitinase AC	*Flavobacterium heparinum*	CS, HA
Chondroitinase ACII	*Arthrobacter aurescens*	HA, CS
Chondroitinase B	*Flavobacterium heparinum*	DS
Chondroitinase C	*Flavobacterium* sp. Hp 102	CS
Heparitinases		
Heparitinase I	*Flavobacterium heparinum*	HS
Heparitinase II	*Flavobacterium heparinum*	HS, Hep
Heparinase	*Flavobacterium heparinum*	Hep

[a]HA, hyaluronic acid; CS, chondroitin sulfate; DS, dermatan sulfate; HS, heparan sulfate; Hep, heparin.

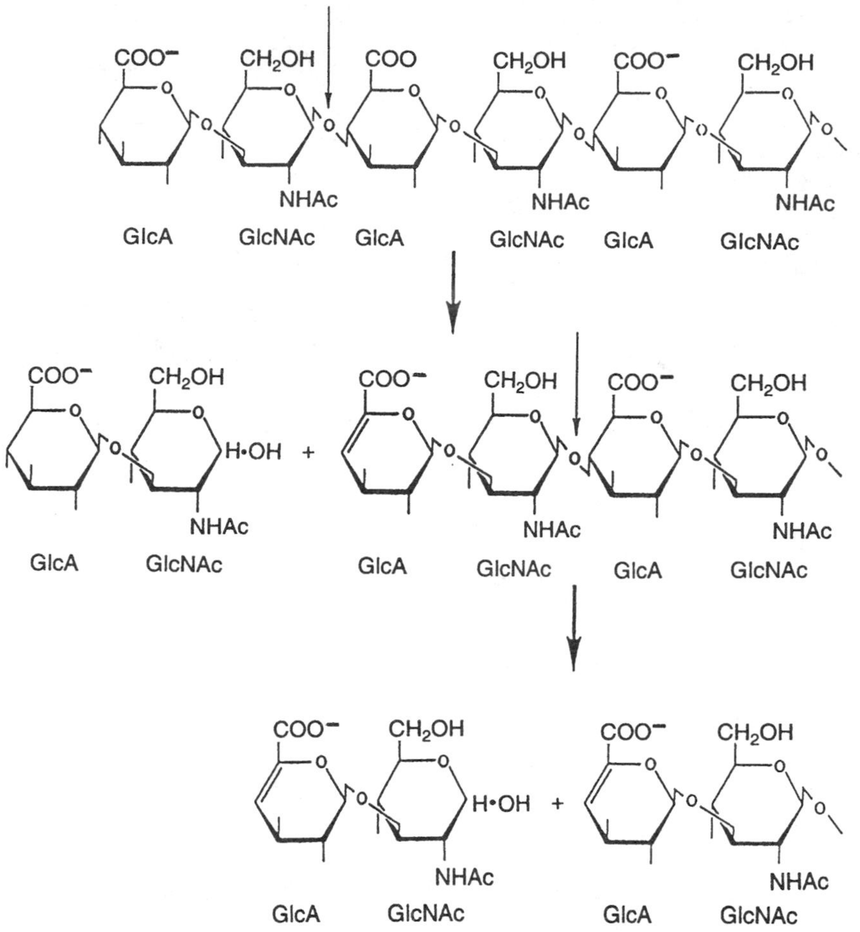

Figure 2.   Hydrolysis of hyaluronic acid by *Streptococci* hyaluronidase.

approximately 0.5 nmol of sample, molecular constructions of chondroitin sulfates, hyaluronic acids, heparins, and heparan sulfates can be analyzed in detail, including their structural heterogeneities.

Recently, endo-β-xylosidase and endo-β-galactosidase, which cleave the sugar–polypeptide linkage region of proteoglycans, were isolated from the midgut gland of a mollusk, *Patinopecten*. Since the enzymes cleave all glycosaminoglycans containing the Galβ1 → 3Galβ1 → 4Xyl group at their reducing termini, and the released polysaccharides contain reducing termini, which can be tagged by ^{3}H or fluorescent groups, they will become useful tools for the study of proteoglycans.

Table 3   Unsaturated Disaccharides Released from Glycosaminoglycans by Digestion with Eliminases

	Δ4,5GlcAβ1–3 GalNAc				Δ4,5GlcAβ1–4 GlcN		
	R_1	R_2	R_3		R_1	R_2	R_3
ΔDi-0S	H	H	H	ΔDiHS-0S	H	H	Ac
ΔDi-UA2S	SO_3^-	H	H	ΔDiHS-UA2S	SO_3^-	H	Ac
ΔDi-4S	H	SO_3^-	H	ΔDiHS-NS	H	H	SO_3^-
ΔDi-6S	H	H	SO_3^-	ΔDiHS-6S	H	SO_3^-	Ac
ΔDi-diS$_B$	SO_3^-	SO_3^-	H	ΔDiHS-diS$_1$	H	SO_3^-	SO_3^-
ΔDi-diS$_D$	SO_3^-	H	SO_3^-	ΔDiHS-diS$_2$	SO_3^-	H	SO_3^-
ΔDi-diS$_E$	SO_3^-	SO_3^-	H	ΔDiHS-diS$_3$	SO_3^-	SO_3^-	Ac
ΔDi-triS	SO_3^-	SO_3^-	SO_3^-	ΔDiHS-triS	SO_3^-	SO_3^-	SO_3^-

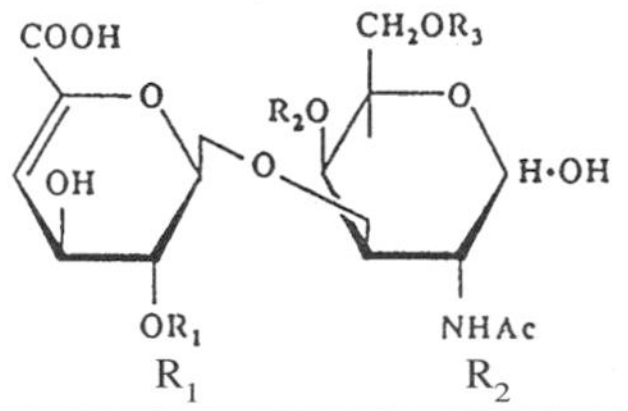

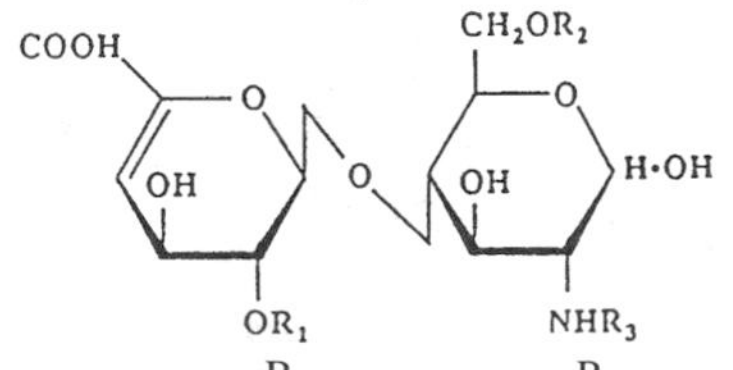

2.2   FRACTIONATION OF OLIGOSACCHARIDES

Acidic oligosaccharides containing sialic acids, phosphate, or sulfate can be fractionated by either paper electrophoresis or ion exchange column chromatography. Gel permeation chromatography supplemented with ultrafine Bio-Gel P-4 and HPLC also is widely used. In addition, affinity chromatography with use of immobilized lectin columns offers a unique and effective method of fractionating oligosaccharides. Table 4 summarizes the binding specificities of immobilized lectin columns. Since the behavior of an oligosaccharide is mainly determined by the structural requirement of each immobilized lectin column and is not affected by additional sugar chain moieties, a mixture of many oligosaccharides can be separated into its components by means of an appropriately chosen series of lectin columns.

2.3   STRUCTURAL ANALYSIS OF OLIGOSACCHARIDES

The monosaccharide sequence of an oligosaccharide can be determined by sequential exoglycosidase digestion. Since an exoglycosidase releases a particular monosaccharide from the nonreducing terminal of an oligosaccharide, the radioactivity incorporated at the reducing terminal of an oligosaccharide remains in the oligosaccharide portion until it has been completely digested to the tritium-labeled monosaccharide alcohol located at the reducing end. Because many exoglycosidases have narrow aglycon specificities, not only the sequence but also the linkage position of each monosaccharide can be determined by using such enzymes. For example, β-N-acetylhexosaminidase purified from the culture medium of *Streptococcus pneumoniae* can cleave the Galβ1 → 4GlcNAc linkage but not the Galβ1 → 3GlcNAc and the Galβ1 → 6GlcNAc linkages.

Table 4   Binding Specificities of Immobilized Lectin Columns

Lectin	Structure Necessary for Binding
Concanavalin A	R→2Manα1 ↘ 6 Manβ1→R R→2Manα1 ↗ 3 Manα1→2Manα1→R
Phytohemagglutinin E₄	Strong binding (retarded at 20°C and 2°C) ±Fucα1 R→3Galβ1→4GlcNAcβ1→2Manα1 ↘ ↓ 6 6 GlcNAcβ1→4Manβ1→4GlcNAcβ1→4GlcNAc R ↘ 3 4 ↗ Manα1 2 R→4GlcNAcβ1 ↗ Weak binding (retarded at 2°C only) ± Fucα1 R→3Galβ1→4GlcNAcβ1→2Manα1 ↘ ↓ 6 6 Manβ1→4GlcNAcβ1→4GlcNAc 3 R ↗
Datura stramonium agglutinin	Strong binding (bound) R→3Galβ1→4GlcNAcβ1 ↘ 6 Manα1→R 2 R→3Galβ1→4GlcNAcβ1 ↗ R→3Galβ1→4GlcNAcβ1→3Galβ1→4GlcNAcβ1→R Weak binding (retarded) R→3Galβ1→4GlcNAcβ1 ↘ 4 Manα1→R 2 R→3Galβ1→4GlcNAcβ1 ↗
Aleuria aurantia lectin	Fucα1 ↓ 6 R→4GlcNAc
Allomyrina dichotoma lectin	Neu5Acα2→6Galβ1→4GlcNAcβ1→R

Therefore, the enzyme is an effective reagent for the determination of the Galβ1 → 4GlcNAc structure, which is widely distributed in many glycoconjugates. Two α-mannosidases with different aglycon specificities were found in the mycellium of *Aspergillus saitoi*. α-Mannosidase I cleaves the Manα1 → 2Man linkage only. Accordingly, the enzyme is an effective reagent for identifying a series of high mannose type sugar chains (see Section 3), which are all converted to Manα1 → 6(Manα1 → 3)Manα1 → 6(Manα1 → 3)Manβ1 → 4GlcNAcβ1→4GlcNAc by incubation with the enzyme. Another enzyme of *A. saitoi*, α-mannosidase II, is also useful because it releases one mannose residue from the R→Manα1 → 6(Manα1 → 3) Manβ1 → group but not from the Manα1 → 6(R → Manα1 → 3)Manβ1 → group. Therefore, the enzyme is an effective reagent when it is desired to assign positions to the outer chains in a complex-type sugar chain (see Section 3).

In addition to the analytical techniques so far described, methylation analysis is essential for the complete structural determination of oligosaccharides. Smith degradation, which analyzes the periodate oxidation products of oligosaccharides, is also useful for the study of many oligosaccharides. Physical methods such as nuclear magnetic resonance and fast atom bombardment mass spectrometry are also becoming useful techniques for the structural analysis of sugar chains.

3   RULES DETECTED IN THE SUGAR CHAIN STRUCTURES OF GLYCOCONJUGATES

Since the N-linked sugar chains found in glycoproteins usually contain more than 10 monosaccharides, the structural multiplicity represented is theoretically enormous. Indeed, it might be impossible to elucidate the biological information in sugar chains if we had to handle such a large number of isomers. Fortunately, studies of the sugar chain structures of various glycoproteins have revealed that a series of structural rules exists in them, and variable regions are limited to parts of their structures.

The N-linked sugar chains of glycoproteins contain as a common core the pentasaccharide Manα1 → 6(Manα1 → 3)Manβ1 → 4GlcNAcβ1 → 4GlcNAc, which we shall call the "trimannosyl core." Based on the structures and locations of the sugar residues added to the trimannosyl core, N-linked sugar chains are classified into three subgroups (Figure 3). *Complex-type* sugar chains contain no mannose residues other than those in the trimannosyl core. Outer chains with an *N*-acetylglucosamine residue at their reducing termini are linked to the two α-mannosyl residues of the trimannosyl core. *High mannose type* sugar chains contain only α-mannosyl residues in addition to the trimannosyl core. Manα1 → 6(Manα1 → 3)Manα1 → 6(Manα1 → 3)Manβ1 → 4GlcNAcβ1 → 4GlcNAc, is commonly included in this type of sugar chain (dashed line in Figure 3). *Hybrid-type* sugar chains were so named because they have the characteristic features of both complex and high mannose type chains. One or two α-mannosyl residues are linked to the Manα1 → 6 arm of the trimannosyl core, as in the case of the high mannose type, and the outer chains found in complex-type sugar chains are linked to the Manα1 → 3 arm of the core. The presence or absence of the α-fucosyl residue linked to the C-6 position of the proximal *N*-acetylglucosamine residue and the β-*N*-acetylglucosamine residue linked to the C-4 position of the β-mannosyl residue of the trimannosyl core (bisecting GlcNAc) contributes to the structural variation of complex-type and hybrid-type sugar chains.

Among the three subgroups of N-linked sugar chains, the largest structural variation resides in the complex-type sugar chains. This variation is formed by two structural factors. As shown in Figure 4A, mono-, bi-, tri-, tetra-, and pentaantennary sugar chains are

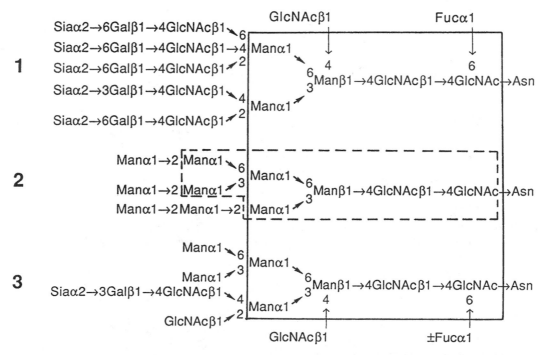

Figure 3.   Three subgroups of N-linked sugar chains: 1, complex type; 2, high mannose type; 3, hybrid type. Within solid lines are the trimannosyl core structures common to all N-linked sugar chains. Structure enclosed by dashed line is the common heptasaccharide core of high mannose type sugar chains; for structures outside, variation in sugar chains is possible.

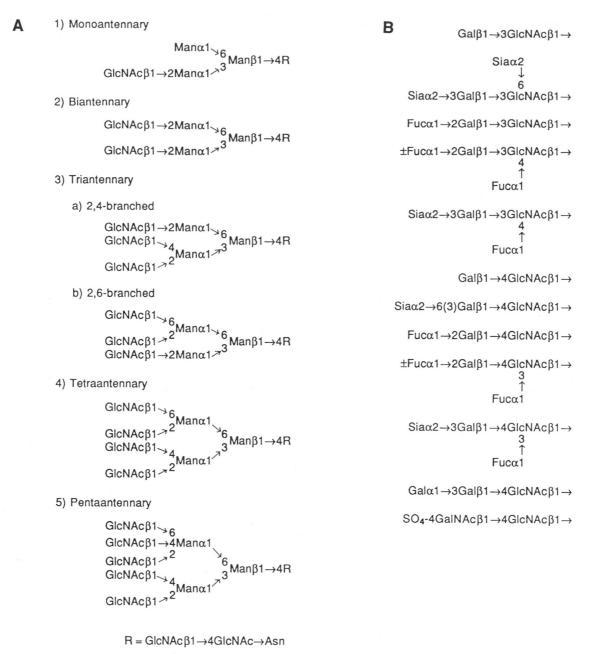

Figure 4.   Two major elements in the formation of the various structures of complex-type sugar chains: (A) branching of complex-type sugar chains and (B) various outer chain structures found in complex-type sugar chains.

formed by adding from one to five *N*-acetylglucosamine residues to the trimannosyl core. Two isomeric triantennary sugar chains, containing either the GlcNAcβ1 → 4(GlcNAcβ1 → 2)Manα1 → 3 group or the GlcNAcβ1 → 6(GlcNAcβ1 → 2)Manα1 → 6 group, are found. These isomeric sugar chains are called 2,4-branched and 2,6-branched triantennary sugar chains, respectively. On these *N*-acetylglucosamine residues, various outer chains are formed (Figure 4B) by the concerted action of many glycosyltransferases and sulfotransferase. It is important to note that bisecting GlcNAc is never elongated by the action of glycosyltransferases. Combination of the antennary and the various outer chains will form a large number of different complex-type sugar chains.

In contrast to N-linked sugar chains, O-linked chains have fewer

structural rules. At present, the latter chains are grouped into the four classes by their core structures, as shown in Figure 5. In addition, O-linked sugar chains with the GlcNAcβ1 → 6GalNAc core and the GalNAcβ1 → 3GalNAc core are found in a limited number of glycoproteins.

More than 200 glycosphingolipids, which are different in their sugar moieties, have been isolated from mammalian tissues. Most of these glycosphingolipids fall into five groups by virtue of their tetrasaccharide core structures (Figure 6). In addition to these major groups, there are minor groups containing trisaccharides, disaccharides, and galactose as common cores. These glycosphingolipids are formed by a complicated biosynthetic network. Figure 7 summarizes the biosynthetic pathway of gangliosides, the generic name

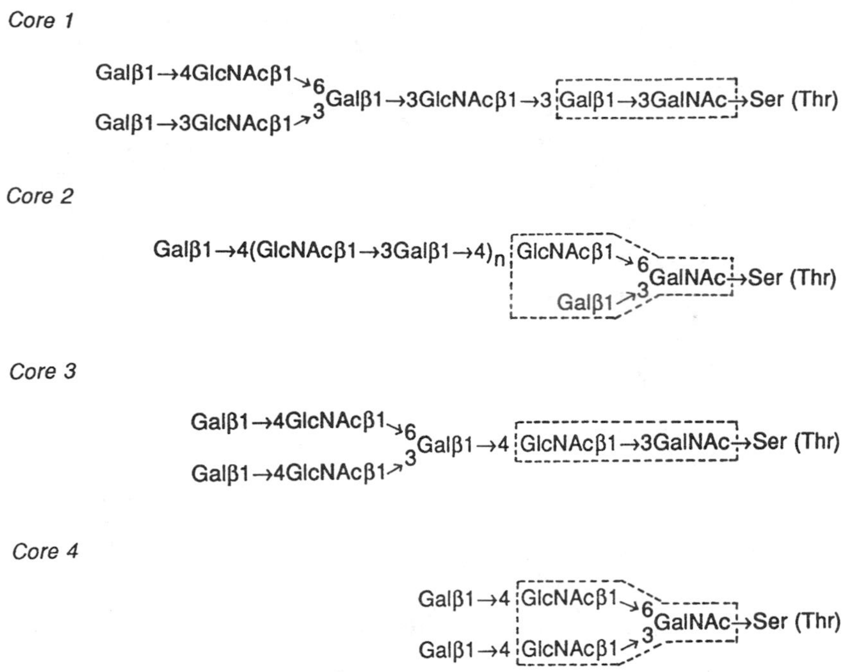

Core 1

$$\begin{array}{l}\text{Gal}\beta1{\rightarrow}4\text{GlcNAc}\beta1{\searrow}6 \\ \qquad\qquad\qquad\qquad\text{Gal}\beta1{\rightarrow}3\text{GlcNAc}\beta1{\rightarrow}3\;[\text{Gal}\beta1{\rightarrow}3\text{GalNAc}]{\rightarrow}\text{Ser (Thr)} \\ \text{Gal}\beta1{\rightarrow}3\text{GlcNAc}\beta1{\nearrow}3 \end{array}$$

Core 2

$$\begin{array}{l}\text{Gal}\beta1{\rightarrow}4(\text{GlcNAc}\beta1{\rightarrow}3\text{Gal}\beta1{\rightarrow}4)_n\;[\text{GlcNAc}\beta1{\searrow}6 \\ \qquad\qquad\qquad\qquad\qquad\qquad\qquad\qquad\text{GalNAc}]{\rightarrow}\text{Ser (Thr)} \\ \qquad\qquad\qquad\qquad\qquad\quad\text{Gal}\beta1{\nearrow}3 \end{array}$$

Core 3

$$\begin{array}{l}\text{Gal}\beta1{\rightarrow}4\text{GlcNAc}\beta1{\searrow}6 \\ \qquad\qquad\qquad\qquad\text{Gal}\beta1{\rightarrow}4\;[\text{GlcNAc}\beta1{\rightarrow}3\text{GalNAc}]{\rightarrow}\text{Ser (Thr)} \\ \text{Gal}\beta1{\rightarrow}4\text{GlcNAc}\beta1{\nearrow}3 \end{array}$$

Core 4

$$\begin{array}{l}\text{Gal}\beta1{\rightarrow}4\;[\text{GlcNAc}\beta1{\searrow}6 \\ \qquad\qquad\qquad\quad\text{GalNAc}]{\rightarrow}\text{Ser (Thr)} \\ \text{Gal}\beta1{\rightarrow}4\;[\text{GlcNAc}\beta1{\nearrow}3 \end{array}$$

Figure 5. Core structures found in O-linked sugar chains; cores are enclosed by dashed lines.

for glycosphingolipids containing sialic acid residue. Although the structures of the gangliosides listed in Figure 7 are well elucidated, the biosynthetic pathways leading to some of them have not been established.

Although proteoglycans should also be considered to be glycoproteins inasmuch as sugar chains are linked to protein cores, their sugar chains, called glycosaminoglycans, are much longer (100–200 monosaccharide residues) than regular N- and O-linked sugar chains and contain many anionic residues, such as uronic acid and sulfate. Furthermore, their structures are unique in that their basic construction is disaccharide repeats. Based on the disaccharide structures, glycosaminoglycans have long been classified into six groups: hyaluronic acids, chondroitin 4-sulfates, chondroitin 6-sul-

fates, dermatan sulfates, heparan sulfates, and keratan sulfates. However, detailed study of the structures of glycosaminoglycans has revealed that many microheterogeneities occur in the sugar chains, and it is rather hard to discriminate chondroitin 4-sulfates, chondroitin 6-sulfates, and dermatan sulfates. Accordingly, glycosaminoglycans are currently classified into four groups (Figure 8).

4   FUNCTIONAL ROLES OF THE SUGAR CHAINS OF GLYCOCONJUGATES

The accumulation of information about the structural characteristics of glycoconjugate sugar chains has enabled us to consider their functional roles in molecular terms. Glycoconjugates are distributed widely inside and outside the cells of multicellular organisms. Furthermore, the molecular construction of a single glycoconjugate group is quite variable. Figure 9 presents the molecular structures of three typical glycoproteins. Two different types of glycoprotein are found as extracellular components. Many serum glycoproteins have a limited number of sugar chains in one molecule (as in Figure 9A). Many epithelial cells lining the alimentary tracts and respiratory tracts secrete viscous glycoproteins called mucins. As shown in Figure 9C, mucins have many short sugar chains, which are distributed all over the polypeptide moiety or in clusters. Plasma membrane integrated glycoproteins are anchored in the phospholipid bilayer by their hydrophobic amino acid cluster portion, and the sugar chains are linked to the polypeptide portion extended outside the cells (Figure 9B). Although because of the structural multiplicity of glycoproteins, the functional roles of their sugar chains are not simple, it is possible to classify them roughly into two groups: one that acts to confer physicochemical properties on proteins, and another that acts as signals of cell surface recognition phenomena. O-Linked sugar chains, which occur in the form of Figure 9C, mainly work for the former function, and N-linked sugar chain for the latter.

Gangliotetraose (ganglio-series)

$$\text{Gal}\beta1{\rightarrow}3\text{GalNAc}\beta1{\rightarrow}4\text{Gal}\beta1{\rightarrow}4\text{Glc}$$

Globotetraose (globo-series)

$$\text{GalNAc}\beta1{\rightarrow}3\text{Gal}\alpha1{\rightarrow}4\text{Gal}\beta1{\rightarrow}4\text{Glc}$$

Isoglobotetraose (isoglobo-series)

$$\text{GalNAc}\beta1{\rightarrow}3\text{Gal}\alpha1{\rightarrow}3\text{Gal}\beta1{\rightarrow}4\text{Glc}$$

Lacto-*N*-tetraose (lacto-series)

$$\text{Gal}\beta1{\rightarrow}3\text{GlcNAc}\beta1{\rightarrow}3\text{Gal}\beta1{\rightarrow}4\text{Glc}$$

Lacto-*N*-neotetraose (neolacto-series)

$$\text{Gal}\beta1{\rightarrow}4\text{GlcNAc}\beta1{\rightarrow}3\text{Gal}\beta1{\rightarrow}4\text{Glc}$$

Figure 6. Structures of major tetrasaccharide cores found in animal glycosphingolipids.

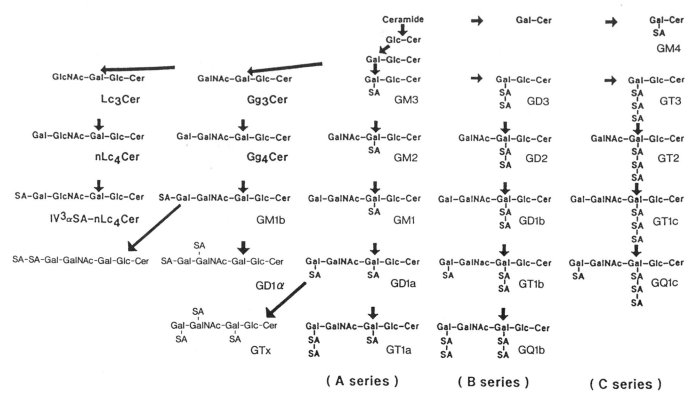

Figure 7. Biosynthetic pathway of gangliosides.

Hyaluronic acid type

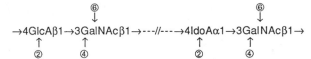

Chondroitin/dermatan sulfate type

Heparan sulfate/heparin type

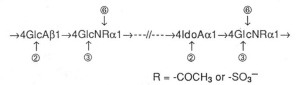

R = -COCH₃ or -SO₃⁻

Keratan sulfate type

Figure 8. Classification of glycosaminoglycans and their microheterogeneities. Except for hyaluronic acid, epimerization at the C-5 position of uronic acids and N- and O-sulfation (locations indicated by circled numbers) are the sources of microheterogeneities.

Four glycohormones have been known to occur in a variety of mammals. Three of them are produced in the same organ: luteinizing hormone and follicle-stimulating hormone are produced by gonadotrophs, while thyroid-stimulating hormone is made by thyrotrophs in the anterior pituitary. Chorionic gonadotropin, however, is produced by placental trophoblasts. All these hormones are composed of two noncovalently linked subunits of different sizes, designated α and β. Since the α subunits of all four glycohormones have identical amino acid sequences within an animal species, the specificity of each hormone to bind to its target cells had been considered to reside in its β subunit. However, elucidation of the structures of N-linked sugar chains of the four glycohormones revealed that the α subunits of these hormones are not in fact the same. Furthermore, deglycosylated human chorionic gonadotropins, obtained by enzymatic or chemical means, were found to express no hormonal activity even though they bind to the target cells more strongly than their natural counterparts. Therefore, the N-linked sugar chains of glycohormones might play crucial roles in the biological functions of these substances.

It is known from recent studies in immunology that both cellular and humoral immunological systems are controlled by a complicated network connecting the interactions of immunocompetent cells. In response to soluble immunological and inflammatory factors, leukocytes adhere to each other and to cells of other types (e.g., platelets, vascular endothelial cells). These interactions are characteristic in that they quickly form a strong contact between cells within a relatively short period compared to the permanent cell adhesion

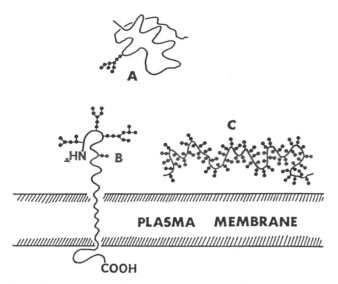

Figure 9. Structural multiplicity of glycoproteins: (A) a serum glycoprotein, (B) a plasma membrane glycoprotein, and (C) a mucin: solid circles and lines represent monosaccharides and polypeptide chains, respectively.

observed in tissue-forming cells. Therefore, such interaction should be mediated by the special adhesive molecules that appear on the surface of activated cells.

Actually, a membrane-integrated glycoprotein that mediates the binding of monocytes and neutrophils is found on the surface of vascular endothelial cells. This glycoprotein, called ELAM-1, is not constitutively expressed on the endothelial cells but is rapidly induced on the cells activated by the action of cytokines such as interleukin 1 and tumor necrosis factor. Accordingly, the glycoprotein plays an important role in accumulating monocytes and neutrophils at the site of inflammation in an animal body. Adhesion of cells to ELAM-1 correlated with presence on the cell surface of the CD15 determinant Galβ1 → 4(Fucα1 → 3)GlcNAcβ1 → — and was eliminated by sialidase treatment of the cells, suggesting that the sialylated form of the CD determinant may be a ligand. This model is supported by the recent finding that the Neu5Acα2 → 3Galβ1 → 3 or 4(Fucα1 → 4 or 3)GlcNAc group works as the ligand of ELAM-1.

Mature lymphocytes migrate from the blood circulation to the lymphatic circulation and return to the blood circulation through the thoracic duct. This phenomenon, called lymphocyte homing or lymphocyte recirculation, is specific to lymphocytes and is not found in other blood cells. When lymphocytes migrate from the bloodstream to the lymphatic organs, they initially bind to the high endothelial (HE) cells of the peripheral lymph node high endothelial venules (HEVs) and pass through the HE cells by an endocytotic mechanism. Since lymphocytes pretreated with MEL-14 monoclonal antibody failed to bind to the HE cells, this antibody was considered to recognize a homing receptor of lymphocytes. Purification of a MEL-14 antigen (gp90 MEL), followed by cloning of the gene from a cDNA library of mouse splenic T cells, revealed that the amino terminus of the protein contains the domain homologous to the C-type animal lectin. The binding of lymphocytes on a cryosection of lymph node containing HEVs was inhibited haptenically by the addition of mannose, mannose 6-phosphate, fucose 4-sulfate, fu-

cosidan, sulfatides, and related compounds, including sialic acid. Accordingly, the combination of a negatively charged group such as phosphate, sulfate, or sialic acid with mannose or fucose should be involved in the ligand of gp90 MEL.

Mutual interactions of immunocompetent cells are mediated by cell surface glycoproteins encoded by the genes within the major histocompatibility complex. Pretreatment of allogeneic stimulator cells with tunicamycin, which specifically inhibits the biosynthesis of N-linked sugar chains, prevents them from inducing the blastogenic response in thymic lymphocytes. This result indicates that the N-linked sugar chains of the surface stimulator cells play an essential role in the cell–cell interaction involved in the regulation of immune responses: not only the direct interaction of immunocompetent cells, but several soluble mediators were found to connect them. Although all these mediators are glycoproteins, no evidence is presented to indicate the functional role of their sugar chains. However, recent reports that interleukins 1 and 2 have lectin activities suggest that sugar–receptor interaction may also be included in their repertoires.

Immunoglobulin G (IgG) plays a major role in humoral immunity. It is unique among serum glycoproteins in that it contains sugar chains of extremely high microheterogeneity produced by the presence or absence of the two galactoses: the bisecting GlcNAc and the fucose residue underlined in the biantennary sugar chain in Figure 10. In spite of this complex nature, oligosaccharides of IgG samples purified from whole human sera have almost identical ratios. An interesting finding is that the IgG samples purified from sera of patients with rheumatoid arthritis are prominently lacking in galactose residue. This galactose deletion phenomenon is limited to IgG among serum glycoproteins and is induced by lower affinity to UDP-galactose of the β-galactosyltransferase in B cells. The degalactosylated human IgG sample, obtained by *Streptococcus* 6646K β-galactosidase digestion, binds less effectively to the subcomponent C1q of the first component of human complement and the Fc receptor. No decrease in binding to protein A was observed in the degalactosylated IgG. This result indicates that the function of IgG molecules can be modified by different degrees of maturation of their sugar moieties.

Thus there is evidence that sugar chains play important roles in many aspects of immunology.

Fertilization of mammals starts when a capacitized sperm meets a matured egg within an oviduct. Mammalian eggs are covered with a layer called the zona pellucida (ZP), which is synthesized and secreted from the oocytes in developing follicles. It has been proposed that the ZP plays important roles in fertilization, including facilitation of species-specific recognition by sperm, prevention of polyspermy, and protection of fertilized eggs until the blastocyst stage.

In mice, the ZP is reported to be composed of three glycoproteins: ZP1, ZP2, and ZP3. It was reported that O-linked sugar chains con-

```
                                                    Fucα1
                                                      ↓
Galβ1→4GlcNAcβ1→2Manα1                                6
                        ↘6
                          GlcNAcβ1→4Manβ1→4GlcNAcβ1→4GlcNAc
                        ↗3
Galβ1→4GlcNAcβ1→2Manα1
```

Figure 10. The largest desialylated N-linked sugar chain of human IgG. Variation occurs in the presence or absence of the underlined sugar residues.

taining α-galactosyl residue in ZP3 work as sperm receptors and that β-galactosyltransferase on the sperm plasma membrane mediates sperm–egg binding by interacting with the sugar chains of ZP glycoproteins. The rationale of the latter hypothesis is partly proven by the recent finding in porcine ZP of the N-linked sugar chains enriched in nonreducing terminal *N*-acetylglucosamine residues, which will interact with β-galactosyltransferase.

As already described, many recombinant glycoproteins have been obtained by using various animal cell lines as hosts. However, as evidenced through comparative study of the sugar chains of γ-glutamyltranspeptidases purified from the kidneys and livers of various mammals, both organ- and species-specific differences occur in the sugar chains of glycoproteins (Figure 11). In addition, an altered glycosylation phenomenon is reflected in the sugar chains of

recombinant glycoproteins because many of the cell lines used are malignant cells. Accordingly, the sugar chains of recombinant glycoproteins may display structural variations according to the type of the cells used, even though the polypeptide structures are the same. This phenomenon affords us a way to elucidate the function of the sugar chains of glycoproteins by investigating the biological activities and the sugar chain structures of recombinant glycoproteins. Comparative study of the sugar patterns and the in vivo activities of several preparations of recombinant human erythropoietin revealed that the activity was proportional to the ratio of tetraantennary to biantennary oligosaccharides.

Elucidation of the detailed structures of glycosaminoglycans is also opening a new age in proteoglycan study. Heparin was identified more than 70 years ago as an anti–blood coagulation material.

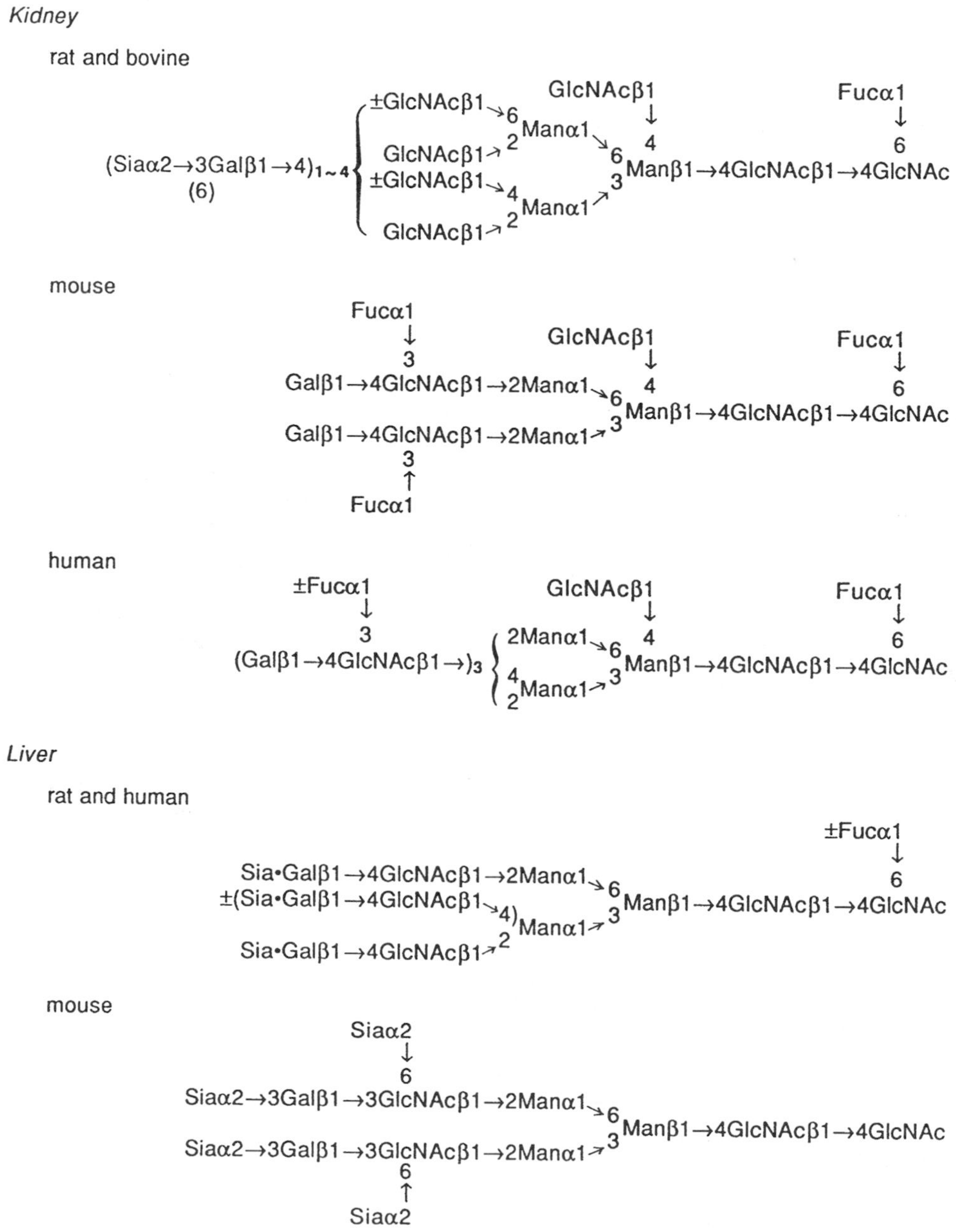

Figure 11.   Major sugar chain structures of γ-glutamyltranspeptidases purified from the kidney and the liver of various mammalian species.

The recent finding that a pentasaccharide structure in the heparin molecule binds specifically to antithrombin led to the elucidation of the molecular mechanism of the anticoagulatant activity of this proteoglycan. With this finding as a turning point, many bioactive segments of glycosaminoglycans have been determined. Since blood clots are not formed on healthy blood vessel walls, some anticoagulation material may be included on the wall surfaces. Based on the observation that the introduction into the bloodstream of heparinase solution prominently decreases anticoagulation activity, the occurrence of heparin-like sugar chains in the endoderm has been suspected. Actually, a heparan sulfate proteoglycan, which has a core protein different from that of heparin proteoglycan, is synthesized by the cultured endoderm obtained from bovine and rat artery. Another candidate anticoagulation factor is thrombomodulin, found on the surface of endoderm. This protein binds to thrombin and suppresses its fibrin clotting activity. A thrombomodulin–thrombin complex also enhances degradation of factors Va and VIIa by activating protein C. Accordingly, thrombomodulin protects the organism from thrombin overproduction. Upon chondroitinase ABC digestion, thrombomodulin was revealed to be a proteoglycan containing chondroitin sulfate. An interesting finding is that the activity of thrombin-dependent suppression of fibrin clotting is prominently decreased by treating thrombomodulin with chondroitinase ABC, while thrombin-dependent protein C activation is not affected. This information will be effective in developing new oligosaccharide drugs in the near future.

Cartilage is a tissue rich in extracellular matrix. Approximately 50% of the dry weight of cartilage is occupied by a proteoglycan called aggrecan. The structure of the core protein of aggrecan was elucidated by successful cloning of its cDNA. A lethal recessive mouse mutant, which is cartilage-matrix-deficient (cmd/cmd), is an effective model to use in the investigation of the functional role of aggrecan. The skeleton of the embryo of this mutant is strikingly different from that of age-paired normal mouse embryo, and the limbs are drastically shortened and deformed. The mutant mouse was found to be genetically deficient in the biosynthesis of the core protein of aggrecan. When the mesenchymal cells obtained from normal mouse are cultured, they form nodules rich in extracellular matrix molecules. The cells within these nodules are round. Cultures of the mesenchymal cells of a cmd mouse also form nodules. However, these nodules are devoid of extracellular matrix, being simple clusters of cells of various types and sizes. When aggrecan is added to the culture medium, the cells from a cmd mouse form nodules similar to those observed in the normal cells culture. Immunohistochemical observation of the apparently normalized nodules with use of antiaggrecan and anti–type II collagen antibodies revealed that these two molecules coexist in the extracellular matrix as in the case of nodules formed by normal cells. The level of fibronectin mRNA in the cultured cmd cells was four to eight times higher than in the cultured normal cells, and abnormal accumulation of fibronectin was observed in the nodule. This abnormal formation of fibronectin in cmd cells was also suppressed to normal level by the addition of aggrecan. These data indicated that proteoglycan plays an essential role in the histogenesis of cartilage.

It was found in 1971 that brain gangliosides bind cholera toxin and block its toxic effect on cells. Later, GM1 was found to be the major molecule participating in the expression of the inhibitory effect. Actually, GM1 in the plasma membrane of cells works as a receptor of cholera toxin, which will lead to penetration of the A subunit of the toxin within the cells. The A subunit then expresses the physiological effect of cholera toxin by triggering activation of adenylate cyclase. In the decades that followed this discovery, several other bacterial toxins were reported to interact with particular gangliosides. However, the binding specificities of these toxins have not been well established, as in the case of cholera toxin.

Glycosphingolipids can be separated by thin-layer chromatography. Upon overlaying the plate containing bacteria or viruses labeled with ^{125}I and washing the plate with buffer, one can use radioautography to detect glycosphingolipids that bind to the bacteria or viruses. This technique permitted the elucidation of the sugar chain structures required for binding to more than 12 bacteria and 14 viruses. Interestingly, some of the structures recognized by bacteria are located not in the nonreducing termini but within the sugar chains.

Functional aspects of the sugar chains of glycosphingolipids in mammalian cell differentiation have been mainly investigated on gangliosides. It was found that the ganglioside profile of HL-60 cells, a human promyelocytic leukemia cell line, changes differently according to the cells' differentiation to granulocytes and monocytes induced by dimethyl sulfoxide and a phorbol ester, respectively. Ganglioside GM3 increases remarkably by macrophage-like cell differentiation, while neolacto series gangliosides increase by granulocytic differentiation. An important set of observations suggests that GM3 is a highly potent inducer, able to lead the monocytic differentiation of HL-60 cells, while neolacto series ganglioside induces granulocytic differentiation of the cells.

The method of investigating the effect of gangliosides added to cultured cells was also successfully used to find the neurite outgrowth promoting activity of GQ1b, a tetrasialoganglioside (see Figure 7). This ganglioside increases the ectocellular phosphorylation of several surface proteins of human neuroblastoma cell lines. Inhibition of the ectocellular phosphorylation by K-252b, a reagent that does not pass through the plasma membrane, resulted in suppression of the GQ1b-dependent promotion of neuritogenesis. The evidence suggests that the two events induced by the addition of GQ1b are quite well correlated.

See also GLYCOPROTEINS, SECRETORY.

Bibliography

Allen, H. J., and Kisailus, E. C., Eds. (1992) *Glycoconjugates: Composition, Structure, and Function.* Dekker, New York.
Carbohydrate Recognition in Cellular Function: Ciba Foundation Symposium 145. (1989) Wiley, New York.
Chin, W. W., and Boime, I., Eds. (1990) *Glycoprotein Hormones.* Serono Symposia, USA, Norwell, MA.
Fukuda, M., Ed. (1992) *Cell Surface Carbohydrates and Cell Development.* CRC Press, Ann Arbor, MI.
Functions of the Proteoglycans: Ciba Foundation Symposium 124. (1986) Wiley, New York.
Ginsburg, V., and Robbins, P., Eds. (1981 and 1984) *Biology of Carbohydrates,* Vols. 1 and 2. Wiley, New York.
Horowitz, M. I., Ed. (1982) *The Glycoconjugates,* Vols III and IV. Academic Press, New York.
———, and Pigman, W., Eds. (1977 and 1978) *The Glycoconjugates,* Vols. I and II. Academic Press, New York.
Welply, J. K., and Jaworski, E., Eds. (1990) *Glycobiology.* Wiley-Liss, New York.

Glycolipids: *see* Lipopolysaccharides.

GLYCOPROTEINS, SECRETORY

Alistair G. C. Renwick

Key Words

Microheterogeneity   The occurrence of a particular carbohydrate in forms that differ in the structure of one or more of its monosaccharide constituents.

Oligosaccharide   A linear or branched carbohydrate that consists of two to twenty monosaccharides, joined by glycosidic bonds.

Secretory Glycoproteins   Proteins having covalently linked carbohydrate chains which are synthesized by specialized cells for export to extracellular destinations.

The least discerning consider carbohydrates to be dull molecules, fit to burn as metabolic fuel but devoid of the glamour and sophistication of nucleic acids and proteins. This rarely expressed minority opinion, however, is untenable. When carbohydrates are covalently attached to protein molecules, the resultant glycoproteins reveal an array of fundamental activities that include cellular adhesion, enzymic transformations, hormonal action, intercellular communication, intracellular sorting, molecular and cellular recognition and transport in the circulatory system: these phenomena and more constitute the emerging field of glycobiology. Glycoproteins are thus a major class of macromolecules, and those secreted by cells, the secretory glycoproteins, are synthesized by specialized cells for export to extracellular destinations—close by, in the case of mucus-secreting cells in the alimentary tract, or distant, as exemplified by the glycoprotein hormones of the anterior pituitary gland, which are released into the vascular system to affect peripheral endocrine tissues.

The chemistry, biochemistry, and molecular biology of glycoproteins are burgeoning areas of intense research that mirror contemporary advances in analytical chemistry and biophysics, besides reflecting the ubiquity and importance of these macromolecules in biology and medicine. Progress has been facilitated, sustained, and enhanced by concomitant advances in separation science, mass spectrometry, nuclear magnetic resonance spectroscopy, and computing. These powerful methods enable rigorous analysis of the most complex biological molecules from minute amounts of starting material, thereby establishing a firm structural foundation for the molecular biologist and the molecular geneticist.

1   OLIGOSACCHARIDE STRUCTURES

An oligosaccharide may be defined as a short chain of two to approximately twenty monosaccharide residues linked glycosidically; in this article, the term is synonymous with "glycan" and "sugar chain." Oligosaccharides obtained from secretory glycoproteins form two groups, the N- and O-linked. The N- or asparagine-linked glycans have at their reducing termini an N-acetylglucosamine unit that is bonded to the amide nitrogen of an asparagine residue in the polypeptide backbone. The enormous diversity of the N-linked oligosaccharides lends credence to the opinion that they are essential for biological functions. In contrast, the O-glycans contain at their reducing ends an N-acetylgalactosamine residue that is attached to a serine or threonine constituent in the protein component; they are often described as "mucin type" because of their abundance in mucus.

Structural determination of many glycans, which may coexist in the same glycoprotein molecule, has led to the formulation of rules that permit further classification of both groups. The asparagine-linked sugar chains fall into three categories labeled high mannose, complex, and hybrid types (Figure 1). High mannose structures consist only of mannose and N-acetylglucosamine in glycosidic linkage, and they commonly contain a heptasaccharide core, which may be modified by the addition of up to four mannose $\alpha1$-2 residues to its three terminal α-mannosyl constituents.

Glycans of the complex type have a core pentasaccharide in common with the high mannose and hybrid varieties, while structural variations in the complex type stem from outer chain moieties linked to its α-mannosyl constituents. Furthermore, complex-type sugar chains often bear a fucose residue linked to the C-6 position of the proximal N-acetylglucosaminyl unit, and a bisecting N-acetylglucosamine is frequently encountered at the C-4 position of the core β-mannosyl residue. These N-glycans are often highly branched di-, tri-, tetra-, and pentaantennary structures, and the peripheral components of many complex-type N-glycans in animal glycoproteins frequently contain repeating disaccharide units (poly-N-acetyllactosamine) in linear or branched arrangements. These permit a huge range of possible structures that include A, B, H, and Lewis blood group antigens and those associated with differentiating and tumor tissues. N-Glycans of the complex type may be highly fucosylated, phosphorylated, or sulfated; some are rich in polysialic acid constituents, and more exotic forms have been characterized from plants and primitive organisms.

The third class of N-glycans, the hybrid type, contains structural elements of the high mannose and complex sugar chains. One or two α-mannosyl residues are linked to the mannose $\alpha1$-6 component of the trimannosyl core as in high mannose oligosaccharides. The peripheral structures found in N-glycans of the complex type are attached to the mannose $\alpha1$-3 constituent. The core structure may also bear a bisecting N-acetylglucosamine residue and may be fucosylated at its proximal end.

Examination of the O-glycans reveals at least six classes of the serine/threonine—N-acetylgalactosamine type (Table 1) which may be elongated by the addition of monosaccharides, usually L-fucose, galactose, N-acetylgalactosamine, N-acetylglucosamine, and sialic acid, in β linkage. The larger O-linked sugar chains often carry terminal antigenic structures such as the human blood groups A, B, H, Le[a], and Le[b] and the X and Y determinants, all of which consist of α-linked monosaccharides. In addition, internal or terminal residues may be sulfated.

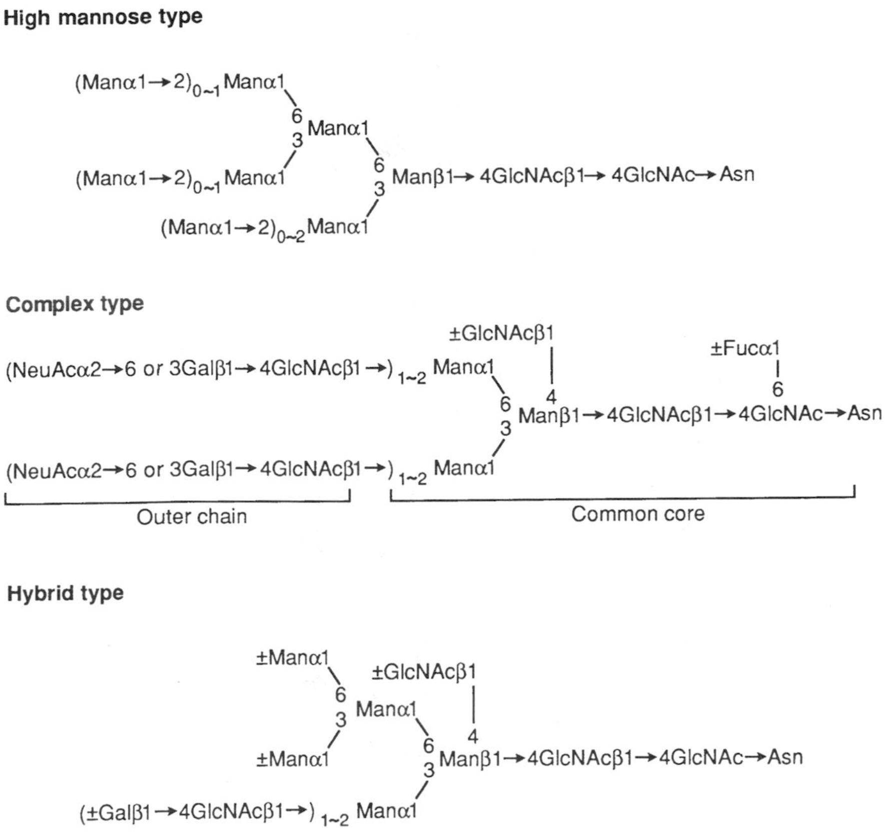

High mannose type

$(Man\alpha1\rightarrow2)_{0\sim1}Man\alpha1$

$(Man\alpha1\rightarrow2)_{0\sim1}Man\alpha1$

$(Man\alpha1\rightarrow2)_{0\sim2}Man\alpha1$

$\overset{6}{_{3}}Man\alpha1$

$Man\alpha1$

$\overset{6}{_{3}}Man\beta1\rightarrow4GlcNAc\beta1\rightarrow4GlcNAc\rightarrow Asn$

Complex type

$\pm GlcNAc\beta1$ $\pm Fuc\alpha1$

$(NeuAc\alpha2\rightarrow6\text{ or }3Gal\beta1\rightarrow4GlcNAc\beta1\rightarrow)_{1\sim2}Man\alpha1$

$(NeuAc\alpha2\rightarrow6\text{ or }3Gal\beta1\rightarrow4GlcNAc\beta1\rightarrow)_{1\sim2}Man\alpha1$

$\overset{6}{_{3}}\overset{4}{Man\beta1}\rightarrow4GlcNAc\beta1\overset{6}{\rightarrow}4GlcNAc\rightarrow Asn$

Outer chain   Common core

Hybrid type

$\pm Man\alpha1$ $\pm GlcNAc\beta1$

$\pm Man\alpha1$

$(\pm Gal\beta1\rightarrow4GlcNAc\beta1\rightarrow)_{1\sim2}Man\alpha1$

$\overset{6}{_{3}}Man\alpha1$

$\overset{6}{_{3}}\overset{4}{Man\beta1}\rightarrow4GlcNAc\beta1\rightarrow4GlcNAc\rightarrow Asn$

Figure 1. General structures of asparagine - or N-linked sugar chains. The N-linked oligosaccharides derived from glycoproteins can be classified into three subgroups. The high mannose type glycans contain only mannose and *N*-acetylglucosamine residues attached to a common heptasaccharide core. Complex-type sugar chains contain a pentasaccharide core, and structural variations occur through the number and structure of the outer chain components attached to the α-mannosyl units in that core. Removal of the terminal neuraminic (NeuAc) or sialic acid residues exposes galactose, which may interact with hepatic receptors. Hybrid-type oligosaccharides show structures characteristic of high mannose and complex-type sugar chains.

2   BIOSYNTHESIS

In contrast to the modes of biosynthesis of nucleic acids and proteins, where fidelity of replication and translation is paramount, the *N*- and *O*-glycans are not formed on a template but are fashioned in the endoplasmic reticulum (ER) and in the Golgi complex (Figure 2). Details of the transit of nascent secretory and membrane proteins from rough ER to cell surface have yet to be resolved, but many components of the secretory pathways appear to be highly conserved in nature.

Although it is expedient to consider the formation of *N*-glycans in three stages, those of early, middle, and end stage processing, the

Table 1   Core Classes of *O*-Glycans[a]

Class	Structure
1	Galβ1-3GalNAc-α-Ser(Thr)-R
2	GlcNAcβ1-6(Galβ1-3)GalNAc-α-Ser(Thr)-R
3	GlcNAcβ1-3GalNAc-α-Ser(Thr)-R
4	GlcNAcβ1-6(GlcNAcβ1-3)GalNAc-α-Ser(Thr)-R
5	GalNAcα1-3-GalNAc-α-Ser(Thr)-R
6	GlcNAcβ1-6GalNAc-α-Ser(Thr)-R

[a]The structures of *O*-glycans are both numerous and variable, which makes classification difficult. However, this system of six core classes, introduced by Schachter and Brockhausen, is advantageous in that it summarizes current knowledge and allows for the addition of new information.

phenomenon is not discontinuous. The first step occurs when an oligosaccharyl transferase recognizes an asparagine residue in the tripeptide asparagine-X-serine (or threonine), where X is any amino acid except proline. This enzyme transfers carbohydrate from a dolichol phosphate-linked oligosaccharide to the nascent polypeptide chain but not at every opportunity; fewer than half of the known consensus sequences in secreted glycoproteins are glycosylated.

Higher animals and plants synthesize $Glc_3Man_9GlcNAc_2$-P-P-dolichol, but glucose residues are not always found. Once the oligosaccharide has been transferred to recipient protein, it forms a substrate for glycosidases in the ER and Golgi complex. The early stage is initiated in the ER by at least three enzymes, α-glucosidase I, α-glucosidase II, and an α-mannosidase. Glucosidase I removes the single, terminal α1-2–linked glucose residue before the remaining α1-3–linked glucose units are cleaved by glucosidase II. The α-mannosidase then detaches a single mannosyl residue. The formation of the complex-type structures requires further processing of the mannose chains on *N*-glycans, which is achieved by α-mannosidases in the cis and medial Golgi. However, not all *N*-glycans are recognized by α-mannosidase I because many mature secretory and surface glycoproteins contain high mannose chains that are essentially unprocessed.

Midstage processing occurs in the cis and medial Golgi, where glycosyltransferases effect elongation before the endstage modifications are carried out in the trans region, which is the exit face of

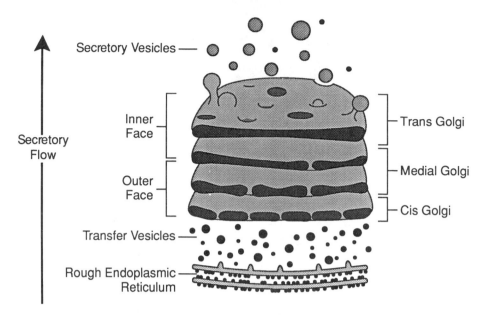

Figure 2.   Protein secretion in eukaryotes. Newly synthesized proteins and glycoproteins are carried unfolded across the membrane of the endoplasmic reticulum (ER), which contains chaperone proteins that catalyze the folding of nascent polypeptides. The chaperones carry a tetrapeptide sorting signal that allows them to be separated from secretory proteins in the Golgi complex and to be retrieved and returned to the ER. Secretory proteins are transported in small vesicles that bud from each compartment in sequence before fusion with the next.

the Golgi stack. The cis and trans faces appear to act as centers for sorting and distribution, while the pathways of secreted glycoproteins, whose ultimate destinations lie in different organelles, diverge in the trans-Golgi network.

The transit of glycoproteins and others in the membranous system appears to be unidirectional, facilitated by transport vesicles that bud from each component in sequence before fusing with the next. This is not a simple, bulk export process; transit times are variable, and the phenomena of intracellular transport are correspondingly asynchronous.

O-Glycans are made by the stepwise addition of sugars from nucleotide sugars to acceptors, and synthesis is initiated by enzymic transfer of an *N*-acetylgalactosamine residue from uridine diphosphate-*N*-acetylgalactosamine to a serine or threonine constituent in the recipient protein. No lipid intermediate is involved, nor is there a requirement for a consensus sequence of amino acids. However, O-linked sugar chains seem to predominate at β turns, where serine, threonine, and proline residues tend to cluster. While some *O*-glycans are formed early, most are synthesized after the release of nascent peptides when they have moved to the smooth ER or Golgi.

3   GLYCOSYLTRANSFERASES

The structural complexities of the *N*- and *O*-glycans and their subtle modes of biosynthesis and secretion increasingly engage the attention of investigators eager to apply contemporary methods of molecular biology and genetics to the glycosyltransferases. This is a large group of more than a hundred membrane-bound enzymes, some of which are located on many cell surfaces. However, those that demand our immediate attention are associated with the rough endoplasmic reticulum and the Golgi complex, where they catalyze the general reaction:

$$nucleotide\ sugar\ +\ acceptor \rightarrow sugar\ acceptor\ +\ nucleotide$$

This process entails specific transport of sugar nucleotides across the intracellular membranes.

A most striking feature of the glycosyltransferases is their substrate specificity, especially their ability to discriminate between different oligosaccharide structures; this property imparts direction to biosynthesis and is important in its regulation. Outside the cell, in the laboratory, the substrate specificity of these enzymes is being increasingly exploited in the preparation of oligosaccharides and neoglycoconjugates—that is, compounds such as peptides or lipids to which mono- or oligosaccharides are attached.

A number of full-length cDNA clones of a group of glycosyltransferases from the Golgi have revealed that these enzymes share a similar domain structure with little or no sequence homology, except in the case of two human α1-3 fucosyltransferases and the blood group Lewis α1-4/3-fucosyltransferase, which may have evolved from a common ancestor. Each enzyme behaves as a type II membrane-bound protein with a characteristic domain structure that consists of a short, positively charged amino-terminal cytoplasmic domain, a single transmembrane component, and a large carboxyterminal constituent that contains the active center. The last-named unit is attached to the transmembrane domain through a stem region, a potentially glycosylated segment that consists of about 40–60 amino acids with a relatively high content of glycyl and prolyl residues. The primary structures of the stem regions of different glycosyltransferases are not conserved, and this stretch of polypeptide appears to be flexible and largely free of secondary structure. Certain of these enzymes contain proteolytic sites which, after enzymic hydrolysis, yield catalytically active, soluble proteins that lack the cytoplasmic and transmembrane domains. The emerging view is that the transmembrane domain is important for retention of glycosyltransferases in the Golgi complex.

The handful of enzymes characterized so far from the endoplasmic reticulum shows considerable diversity, unlike their Golgi counterparts, and the most intensely studied are those that participate in the assembly of the dolichol phosphate-linked oligosaccharide. This structure is initiated and partially processed by enzymes whose active sites are located in the cytoplasm, whereupon the intermediate "flips" into the lumen of the endoplasmic reticulum to be modified by a group of enzymes whose catalytic centers lie in the lumen.

4 VARIATIONS IN GLYCOSYLTRANSFERASE ACTIVITY

In numerous situations, fluctuations occur in glycosyltransferase activity; these include physiological events such as cellular differentiation, lymphocytic activation, maturation and growth, and pathological conditions, particularly inflammation and malignant transformation. The expression of these enzymes also varies among species, organs, and cell types.

5 GLYCOSIDASES

Glycosidases constitute the second group of enzymes concerned in the transformation of *N*- and *O*-glycans, and they catalyze the hydrolysis of specific linkages. They may be classified into endoglycosidases, which cleave internal glycosidic bonds, and exoglycosidases, which effect the removal of sugar residues from the nonreducing termini of *N*- and *O*- glycans. Examples of both categories are used extensively in analytical procedures.

In mammalian cells the major sites of oligosaccharide degradation are the lysosomes, where hydrolysis is achieved by acid glycosidases. A large number of lysosomal storage diseases result from the defective expression of specific glycosidases and because of their clinical importance, a great deal of attention is focused on these glycoproteins in the diagnosis, treatment, genetic screening, and the elucidation of the molecular bases of these conditions.

6 INTRACELLULAR TRANSPORT OF SECRETORY GLYCOPROTEINS

Secretory glycoproteins are synthesized as precursors with an additional N-terminal signal peptide that permits recognition and facilitates the introduction of the attached polypeptide into the ER. The discovery in this organelle of chaperone proteins (i.e., proteins that enable the folding of nascent polypeptide chains) has led to the concept of the ER as a "folding factory," replete with the machinery necessary for this function and with enzymes capable of modifying the oligosaccharide or amino acid side chains of newly synthesized glycoproteins. The availability of chaperones in the ER allows eukaryotic cells to secrete a broad spectrum of proteins, some of which are destined for secretion while others will remain in that organelle. But how does sorting occur?

It appears that no specific signal is needed for secretion, but one is borne by resident proteins in the form of characteristic tetrapeptide sequences, KDEL, HDEL or a closely related structure. KDEL has been shown to be a functional sorting signal, and sequences that resemble KDEL are necessary and sufficient for the retention of protein in the ER. This signal does not prevent the exit of resident proteins from the organelle; rather, it allows their retrieval from subsequent stages in the secretory pathway. Receptors for these signals have been described in various organisms, and this method of sorting appears to be highly conserved. In the human, the KDEL receptor is normally concentrated in the Golgi, and movement of the receptor is controlled by ligand binding.

7 CONTROL OF BIOSYNTHESIS

The absence of a template mechanism in the biosynthesis of N- and O-linked oligosaccharides and the nature and complexities of the secretory process make exploration of regulatory events extremely difficult. Existing knowledge may be rudimentary, but there is sufficient to permit delineation of the main features. Most is known about the genes encoding the peptide sequences of many secretory glycoproteins and their regulation will not be considered further. On the other hand, investigation of the control of oligosaccharide synthesis is of recent origin and is fast gathering momentum.

It has been estimated that a hundred or so membrane-bound glycosyltransferases act cooperatively and in competition to synthesize specific *N*- and *O*-glycans. A number have been cloned, and while these enzymes reveal little in the way of sequence homology, they share domains characteristic of the type II membrane-bound proteins. Not only are these enzymes highly specific for substrate and acceptor molecules, but some have strict specificities for branched structures. Local controls exist in their subcellular localization, and catalysis cannot occur if there is inadequate provision of dolichol phosphate and dolichol-linked oligosaccharides, sugar nucleotides, acceptors, and cofactors such as divalent cations. The rate of export to the cell surface is also important in oligosaccharide synthesis because it may exceed that of glycosylation, resulting in the secretion of incomplete structures. The distribution of the glycosyltransferases and glycosidases is of special interest because of the species, tissue and cell specificities of these enzymes. Their intracellular concentrations fluctuate during differentiation and transformation, in physiological circumstances, and in pathological conditions.

The protein components of most glycoproteins appear to be of significance in the local control of N- and O-glycosylation. In some but not all cases, glycosylation at one or more specific sites is obligatory for correct folding of the polypeptide. In many instances improper folding results in rapid degradation and little, if any, secretion. However, a significant number of proteins are secreted without carbohydrate.

The oligosaccharides also affect glycosylation of the protein backbone by limiting or prohibiting access of processing enzymes. Many make substantial contributions to the size and apparent shapes of intact glycoproteins, and these are well illustrated in the glycoprotein hormones of the human anterior pituitary gland and placenta. The attached glycans affect the association of the two subunits in the formation of the intact, biologically active hormone and its interaction with its receptor. Other examples are found in enzyme–substrate interactions and in associations of glycoproteins with lectins. Such effects most likely result from steric hindrance and/or ionic shielding, but glycans may also affect biological activity by evoking conformational changes in the protein. In addition to local and genetic regulation, one must superimpose nervous and hormonal effects in animals that possess such systems.

8 FUNCTIONS OF SECRETORY GLYCOPROTEINS

The effects of covalently attached carbohydrate on the physicochemical properties of proteins are well known in general (Table 2) and include altered viscosity, changes in isoelectric point, altered solubility and thermal stability, and variations in degree of hydra-

Table 2 Functions of *N*- and *O*-Glycans in Glycoproteins

Altered Physicochemical Properties

Degree of hydration
Isoelectric point
Protein folding
Resistance to proteolysis
Solubility
Stabilization of tertiary structure
Thermal stability
Viscosity

Physiological Functions

Cell–cell interactions
Cell–matrix interactions
Cell secretion
Circulatory half-life
Intracellular transport
Pinocytosis
Receptor-mediated biological reactions

Table 3 Some Diseases for Which Carbohydrate-Based Drugs Have Been or Are Being Developed

Arthritis
Asthma
Atherosclerosis
Autoimmune disease
Cancer
Diabetes
Enzyme defects
Infections
Prevention of surgical adhesions
Prevention and treatment of thrombosis
Promotion of wound healing

tion and resistance to proteolysis, none of which will be considered further. Instead, the focus is directed toward aspects of function that are of present and future concern to biology and medicine.

A major impediment to the investigation of the biological activities of glycoproteins, especially the contributions of their carbohydrate constituents, is the occurrence of marked heterogeneity in the oligosaccharides attached at each glycosylation site. These glycoforms may vary in their evocation of biological responses, but in most cases this feature remains unexplored.

The pharmaceutical industry has proclaimed that the next decade is that of carbohydrate-based drugs because carbohydrates are intimately concerned in numerous physiological and pathological processes, especially as components of internal and cell surface receptors, in immune and inflammatory phenomena, and as cell surface epitopes on malignant cells. Already various enzyme inhibitors with carbohydrate structures are known to block certain digestive enzymes of the alimentary tract, but in no instance is the function of the carbohydrate component of a single glycoprotein understood.

9   PERSPECTIVES

The emergence of glycobiology and its growing impact on other fields, especially medicine, are among the most significant developments in contemporary science. The opportunities for research are immense, given the breadth, precision, and sensitivity of available analytical methods.

Chemical synthesis of the oligosaccharide components of glycoproteins and glycolipids is developing rapidly, and the use of glycosyltransferases as specific reagents has greatly facilitated progress. Solid phase synthesis of oligosaccharides is now viewed as possible, if some years away, and polymer-supported solution synthesis offers potential on a manufacturing scale.

The vitally important secretory glycoproteins that occur in minute amounts in biological systems are notoriously resistant to study, but recent developments in gene technology offer routes for exploitation. The production of human erythropoietin, a hormone necessary for the treatment of certain forms of anemia, is one signal achievement. This glycoprotein consists of 165 or 166 amino acid residues to which are attached three *N*-glycans and one O-linked oligosaccharide, which constitute some 40% of the molecular mass. The cloning

of the structural gene with transfection to transformed Chinese hamster ovary cells enabled the manufacture of recombinant erythropoietin that differs from the native hormone obtained from urine by only one disaccharide. However, not all attempts are so successful.

Advances in glycobiology have been carefully noted by the chemical and pharmaceutical industries, and already their horizons have been extended beyond the extraction, purification, and modification of natural glycoconjugates. Carbohydrate-based drugs have been or are being developed for the treatment of a broad spectrum of diseases, including those listed in Table 3.

There is also a need to modify additional therapeutic products with carbohydrate to deceive the immune system or to prolong the drugs' half-lives in recipients. But none of this is possible without detailed knowledge of the chemical structures involved and their molecular dynamics.

While these and many important problems demand solutions, the main obstacle to the elucidation of the biological role of oligosaccharides in glycoproteins is rooted in the lack of three-dimensional structures. Such molecules have resisted crystallization except in a very few instances, and information gleaned from X-ray diffraction, though enlightening, will be insufficient for the definition of structures in solution. Nuclear magnetic resonance is the method of choice. It is not yet possible to determine the three-dimensional structures of intact glycoproteins by this means, but rapid developments in this form of analysis will surely meet the need.

Thus the biochemist and the chemist face a superfluity of problems that appear to be insuperable when one considers the glycoproteins that perform vital biological functions but exist only in minute amounts. Herein are the challenges to the molecular biologist and the geneticist: new methods are required to enhance existing analytical procedures. However, there is no room for dichotomy of interests, the successful approach must employ the combined talents of chemist and biologist.

See also GLYCOBIOLOGY.

Bibliography

Allen, H. J., and Kisailus, E. C., Eds. (1992) *Glycoconjugates: Composition, Structure and Function.* Dekker, New York, Basel, and Hong Kong.

Drickamer, K., and Carver, J., Eds. (1992) Carbohydrates and glycoconjugates. *Curr. Opinion Struct. Biol.* 2:653–709.

Pelham, H. R. B. (1992) The Florey Lecture, 1992. The secretion of proteins by cells. *Proc. R. Soc. London B,* 250:1–10.

Rothman, J. E., and Orci, L. (1992) Molecular dissection of the secretory pathway. *Nature,* 355:409–415.

Stockell Hartree, A., and Renwick, A. G. C. (1992) Molecular structures of glycoprotein hormones and functions of their carbohydrate components. *Biochem. J.* 287:665–679.

GROWTH FACTORS

Antony W. Burgess

Key Words

Apoptosis   Programmed cell death resulting from genetic damage and/or absence of growth regulators.

Autocrine   Describing the actions of a growth factor/cytokine on the same cell that produces the growth factor.

Cancer   The cell mass that results from the uncontrolled production of genetically altered cells in an animal.

Cell Production   The process of hierarchical proliferation and maturation of tissue-specific stem cells, which leads to the generation of mature cells in a specific tissue.

Chemokines   Structurally related smaller cytokines (8–10 kDa) that act on inflammatory cells.

Colony-Stimulating Factors   Proteins that stimulate the proliferation and maturation of myeloid precursor cells.

Cytokines   Proteins released from one cell which modulate the proliferation, differentiation, and/or function of cells in a specific lineage.

Differentiation   The alteration in gene expression as precursor cells divide and progress toward their functional forms.

Endocrine   Describing the effects of a growth factor/cytokine/hormone produced in one organ on the cells of another organ.

Growth Factors   Proteins that influence the proliferation and maturation of tissue-specific precursor cells.

Hormones   Molecular signals released from one cell that modulate the function or production of other cells. In the broadest sense, the term "hormone" includes steroids, peptides, and lipids. In traditional physiology, the term was used to describe the molecules released by one organ that acted on the cells of another organ.

Interleukins   Proteins released from stromal or hematopoietic cells that influence the proliferation, self-renewal, commitment, and/or maturation of hematopoietic cells.

Lymphokines   Proteins that modulate the proliferation, maturation, or function of lymphoid cells.

Paracrine   Describing the actions of a growth factor/cytokine on cells in close vicinity to the cells producing the growth factor.

Receptors   Cell surface proteins that are stimulated by exogenous ligands (e.g., growth factors) to transfer a signal to the cytoplasmic compartment of a cell.

Signal Transduction   Transfer of extracellular (membrane) interactions to intracellular (cytoplasmic and/or nuclear) responses such as enzyme activation and/or modulation of gene transcription.

Stem Cells   The cells responsible for the renewal of cells in specific lineages. Stem cells have the capability of dividing to renew themselves or dividing to produce differentiated progeny that are committed to the production of mature cells.

Cell production in all tissues of multicellular organisms is under the control of a network of tissue-specific protein regulators called growth factors. Although the growth factor network may be modulated by circulating hormones, these regulators are usually produced in the tissue in which they function. Growth factors have been identified in the epithelial, neural, lymphoid, myeloid, and hepatic systems; in many cases a number of closely related growth factors are produced in each tissue. As well as controlling the normal production of mature cells, growth factors are important for regulating inflammatory, wound-healing, and antiviral responses. Many diseases appear to be associated with inappropriate growth factor responses. Overproduction of growth factors leads to accumulation of cellular deposits or infiltration of tissues by inflammatory cells. Although cancers arise as the result of numerous genetic lesions, many metastatic cancers appear to be driven by the autocrine action of growth factors.

New therapeutic approaches to cancer involve the use of specific growth factors to improve hematopoietic recovery from cytotoxic or radiation therapy. Furthermore, growth factor antagonists are being developed to suppress the proliferation of the tumor cells. An understanding of the mechanism of action of growth factors and the range of action of these proteins is important for developing effective therapies for autoimmune diseases, cancer, rheumatoid arthritis, skin diseases, and diabetes.

1   INTRODUCTION

Multicellular organisms need to control the production, maturation, and function of cells in specific tissues or organs. It is now clear that locally secreted proteins modulate the cell physiology of each tissue. These regulatory proteins have been given a variety of names, including growth factors, cytokines, and interleukins. It is well known that the endocrine systems consist of a network of circulating regulatory molecules (hormones) that includes proteins, peptides, and steroids; growth factors can be considered to be an extension of the endocrine system. While there is a direct overlap between the growth factor and endocrine systems, it is often helpful to consider that the major actions of endocrine hormones occur in tissues remote from their tissue of origin, whereas growth factors appear to have a major function within the tissues that produce them.

In the main there is no functional or structural characteristic that distinguishes different classes of growth factors—most appear to be capable of stimulating multiple biological responses that depend critically on the differentiation state of their target cells. For example, one of the hematopoietic growth factors, granulocyte colony-stimulating factor (G-CSF), stimulates the proliferation of immature bone marrow cells as well as activating bacterial killing by mature neutrophils. For most growth factors so many distinct biological effects have been observed in vitro that it is not yet possible to define their specific role in particular tissue systems. However, both pharmacological and genetic studies have demonstrated that growth factors can alter cell production, organogenesis, and disease susceptibility in animals. It should be noted that the action of a particular growth factor will usually depend on the presence of other growth factors, the availability of target cells, and the display of the appropriate cell surface receptors (Figure 1).

Growth factors stimulate cells by binding to specific cell surface proteins called receptors. Upon binding the growth factor, many receptors dimerize and activate intracellular kinase networks. Some receptors, such as the epidermal growth factor receptor, are ligand-dependent tyrosine kinases; others, such as the interleukin-2 receptor, associate with and activate particular membrane-associated tyrosine kinases. Activation of these enzyme systems can have effects on cell metabolism, membrane turnover, cytoskeletal organization, cell movement, and cell division.

Cell production occurs continuously throughout embryonic and adult life. During the earliest stages of development, cytokines such as leukemia inhibitory factor (LIF) appear to be necessary both to stimulate cell division and to prevent differentiation (i.e., to maintain pluripotentiality). At the later stages of blastogenesis and gastrulation, cells display a bewildering array of growth factor/cytokine receptors, and the availability of particular ligands can determine the motility, connectivity, and proliferative state of the cells. Often organs are remodeled through the absence of a tissue-specific growth factor. Late in gestation, the vitreous humor of the eye is full of hyalocytes—macrophage-like cells that depend on granulocyte/macrophage colony-stimulating factor (GM-CSF) for

their survival (and perhaps function). Normally, these hyalocytes persist until a few days after birth, when GM-CSF levels must decrease and hyalocytes undergo rapid, apoptotic death. In mice that overexpress GM-CSF, the hyalocytes persist in the vitreous humor.

The trafficking of mature cells between organs, and in particular from the blood into tissue spaces and even the mucosal layers, appears to be controlled by growth factors/cytokines and/or chemokines. The presence of several cytokines and membrane proteins from foreign organisms can trigger a strong proliferative response or the activation of effector functions. The action of bacterial cell wall proteins and lipids synergizes with regulators such as G-CSF or GM-CSF to induce an intense antibacterial response.

Recently our concepts of growth factor physiology have been challenged by the existence of genetically deficient animals. Often the physiological actions of specific growth factors have been inferred from the properties of these molecules in vitro. For example, G-CSF stimulates the production of neutrophils. Mice have been produced that have a genetic disruption of the G-CSF locus. Despite a profound neutrophil deficiency, these mice still produce some neutrophils. Thus, G-CSF cannot be the only growth factor controlling neutrophil production. Similarly, mice that lack the ability to produce transforming growth factor α (TGF-α) have normal skin, esophageal, stomach, and colon mucosa. Although there are several growth factors closely related to TGF-α, it is not clear whether mucosal cell production continues because of growth factor redundancy or whether the actual function of TGFα is quite different from our current hypotheses.

As well as the growth factor/cytokine families, there are families of receptors for specific growth factor families. There are at least four members of the epidermal growth factor receptor (EGFR) family: *EGFR, erb B2 (HER 2), erb B3 (HER 3),* and *erb B4.* These receptors form both homodimer and heterodimer signaling complexes, all of which are capable of stimulating specific intracellular signaling pathways.

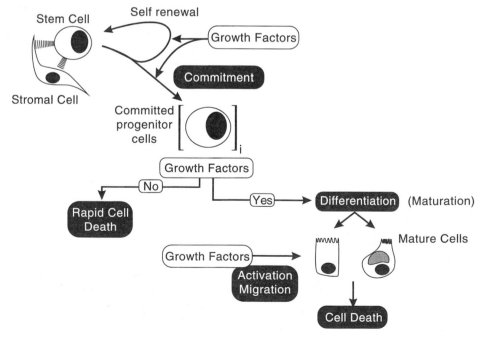

Figure 1.   There are several sites of action of growth factors during the production of cells. The action of the growth factor is likely to be quite different at the different stages of maturation.

Although there is a vast array of growth factors and cytokines, the receptor systems are built up from sets of domain structures. Thus the hematopoietin receptor families have homologous hematopoietin, fibronectin, and immunoglobulin domains, which form the extracellular binding site, attached to a short intracellular tail. The nerve growth factor/tumor necrosis factor receptor family is made up of four to eight "six-cysteine" domains in the extracellular region, and these are attached to intracellular domains associated with apoptosis or CD-40-like domains. The other major class of receptors encodes ligand-activated intracellular tyrosine kinases. Again, the extracellular regions have immunoglobulin and fibronectin-like clusters which are attached to the intracellular kinase.

Receptor activation appears to require ligand-induced aggregation, which triggers either the receptor kinase or associated intracellular kinases. Once activated, the enzymes act to stimulate several intracellular signaling cascades. While the relative importance of the different signaling cascades has not been determined in detail, it appears that many tyrosine kinase receptors activate the *ras-raf-map* kinase pathways, and many of the cytokine (hematopoietin) receptors activate the *JAK/STAT* signaling system. These signaling pathways are dependent on continuous occupancy of the receptor by ligand and even then, receptor downmodulation and phosphatase activation tend to dampen the signaling processes.

Clearly, the normal growth and differentiation activities of cells are controlled by the action of the growth factor/cytokine network. However, the complex events triggered by the cytokine network are also the major targets for tumorigenic lesions [e.g., autocrine growth factor production, constitutively activated receptors, or intracellular signaling proteins forced into the active state by mutation(s)]. If two or more of these lesions occur in cells that have damaged tumor suppressor genes, there is a considerable likelihood that the cell will grow uncontrollably—that is, become cancerous.

2   DISCOVERY

The incredible potency and diverse biological activities of each growth factor have led to the multiple, independent discoveries for many of the growth factors. A typical example is the identification of interferon β2 as a member of a family of proteins capable of inducing antiviral responses and its discovery as a growth factor in the lymphoid system, where it was called interleukin 6. Some growth factors, such as fibroblast growth factor, have been identified independently in as many as 20 different biological systems. While these studies emphasize the multiple actions of growth factors, some confusion has developed with regard to their nomenclature and their likely physiological function(s). It should not be assumed that a particular action of a growth factor in vitro is automatically related to its role in the whole animal.

The first growth factor was identified in 1906: erythropoietin (epo). This discovery arose from physiological observations on the humoral control of production of red blood cells. While nerve growth factor (NGF) and epidermal growth factor (EGF) were discovered in the late 1940s, as a result of observations on the innervation of transplanted tumors, the discovery of most other growth factors has been due to their activity in laboratory culture systems. More recently, growth factors such as stem cell factor (also called the c-*kit* ligand) have been identified as a consequence of the characterization of their respective cell surface receptors. Table 1 lists

Table 1   Chromosomal Locations of Growth Factors and Their Receptors

Growth Factor	Chromosome Location	
	Human	Mouse
Insulin	11p15	
IGF-I	12	10
IGF-II	11	7
Nerve growth factor (NGF)	1p22	3
NGF Receptor	17q–12–22	11
EGF	49–25–27	3
TGF-α	2p13	
EGF Receptor	7p14/12	11
neu	17	11
TGF-β1	19q13.1–13.3	7
TGF-β2		1
TGF-β3	14q–23–24	12
Inhibin α	2q33	1
Inhibin β	2q33	1
Müllerian inhibitory substance	19	10
TNF-α/β	6p	17
TNF-receptor (type 1)	1p36	4 distal
TNF-receptor (type 2)	12p13	6 distal
PDGF A-chain	7	
PDGF B-chain	22	15
PDGF receptor α	4q–11–12	
PDGF receptor β	5q31–32	18
a-FGF	5	
b-FGF	4	
wnt-2	11	
*hst/ks*3		17
Hepatocyte growth factor (HGF)	7q11.2–11.21	
HGF receptor (c-*met*)	7q	6
IL-1α/β: (α-chains)	2q–13–21	2
IL-2	4q–26–28	3B–C
IL-3	5q–23–31	11A5–B1
IL-4	5q31	11A5–B1
IL-5	5q31	11A5–b1
IL-6	7P15–P21	5 proximal
IL-7	8q12–13	3
IL-9	5q22–35	13
IL-11	19q13.3–13.4	
IL-12A (p35)	3p12–3q13.2	
IL-12B (p40)	531–33	
Receptors		
Interleukin 1 (type 1)	2q12	1 centro
Interleukin 2α	10p14–15	2A2–A3
Interleukin 2β	22q11.2–q12	
Interleukin 4	16p11.2–12.1	7 distal
Interleukin 5α		6 distal
Interleukin 6	1	
Interleukin 7		15 proximal
M-CSF (also called CSF-1)	1p13–p21	3F3
M-CSF receptor (c-*fms*)	5q33.2–33.3	18D
GM-CSF	5q23–31	11A5–B1
GM-CSF receptor α	αxp21-pter, Ypter-p11.2	
GM-CSF receptor β	229–12.3–13.1	
G-CSF	17q11.2–21	11D-E1
G-CSF receptor	1p34.2–35.1	4 distal
Stem cell factor	12q–22–24	10
SCF receptor (c-*kit*)	4q11–12	5
Erythropoietin	7q11–22	5G
epo receptor	19pter-q12	9
Leukemia inhibitory factor	22q12.1–12.2	11A1–A2

the chromosomal locations on mouse and human of the growth factors mentioned here.

The discovery of growth factors is, in many ways, a reflection of the serendipity and excitement of science. Time and time again they turned up in the most unexpected way. When Stanley Cohen was analyzing the nature of the molecule responsible for nerve innovation of tumors, he planned some simple experiments to test whether it was a protein, a lipid, or a nucleic acid. One of these tests required the use of a phosphodiesterase (a hydrolytic enzyme, extracted from snake venom, which breaks down nucleic acids). Instead of destroying the growth factor, the enzyme appeared to increase its activity: in fact, the enzyme preparation was contaminated with nerve growth factor. Cohen recognized the connection between snake venom and the mouse salivary gland—a unique tissue that contains almost 1 mg/g wet weight of both NGF and EGF. Eventually both NGF and EGF were isolated from the mouse salivary gland. Curiously, only the male mouse salivary glands have this enormous store of growth factors, and to this day it is not clear why they are there.

The human equivalent of EGF was first recognized by an activity (urogastrone) in urine which inhibited gastric acid secretion (and acid-induced ulceration) when injected into dogs. Harold Gregory designed the purification procedure for urogastrone and after many years obtained sufficient quantities to determine its amino acid sequence. This substance was homologous to the mouse salivary gland EGF and consequently, urogastrone was renamed as human EGF.

The growth factors/cytokines that control the production of white blood cells were discovered at the same time, but independently, by scientists in Australia and Israel. It was several years before these groups realized they were working on the same set of molecules. The two Australians, Ray Bradley and Donald Metcalf, had been trying to grow leukemic cells; the two Israelis, Dov Pluznik and Leo Sachs, thought that they had grown tissue mast cells in the laboratory. When Bradley finally "succeeded" in devising a method to use normal cells to stimulate the growth of small colonies of leukemic cells, he rushed the cultures over to Metcalf, who realized that some of the normal cells were in fact producing the colonies. The leukemic cells found to be were producing a "colony-stimulating factor" (CSF) that allowed the proliferation and differentiation of blood cell precursors to form thousands of mature neutrophils and macrophages. Pluznik and Sachs also realized that their colonies were made up of mature white blood cells.

At first only one CSF was recognized, but it gradually became clear that there were several distinct CSFs: macrophage CSF (called CSF-1 or M-CSF), granulocyte/macrophage (GM)-CSF, granulocyte (G)-CSF, and even a molecule that stimulated cell production in almost all of the hematopoietic lineages [initially named multi-CSF, now known as interleukin 3 (IL-3)]. These CSFs were initially recognized by their activities in laboratory culture systems; indeed, many scientists believed they were artifacts and were simply nutrients missing from imperfect tissue culture media. Although our knowledge of the physiology of the CSFs is still rudimentary, genetics studies and clinical trials have shown how powerful these regulators are in vivo. Many cancer patients are now treated with G-CSF or GM-CSF to improve the rate of hematopoietic recovery after cytotoxic chemotherapy.

3   GROWTH FACTOR FAMILIES (see Figure 2)

3.1   INSULIN-LIKE GROWTH FACTORS

In 1922 Banting and Best discovered insulin, the regulator of glucose metabolism, in pancreatic extracts. Insulin was the first protein

to have its amino acid sequence determined. It is now known that insulin is synthesized as a single precursor protein, which is processed and secreted as a disulfide-linked heterodimer. As well as controlling blood glucose levels, insulin stimulates cell proliferation; however, two other closely related proteins, insulin-like growth factors I and II (IGF-1 and IGF-II), are more potent stimulators of cell proliferation. IGF-I was discovered as the protein in plasma responsible for nonsuppressible insulin-like activity. IGF-I, which has been known for some time as a somatomedin, also mediates the action of growth hormone. IGF-II was the first growth factor to be associated with the aberrant growth of tumor cells in the laboratory. While the members of this family circulate in the serum, their availability is controlled by specific binding proteins. The cell surface receptors for insulin and IGF-I are composed of four chains (α_2, β_2). The β-chains are transmembrane, ligand-dependent tyrosine kinases, and the extracellular α-chains bind the ligand. The IGF-II receptor is a large (2000 amino acids) single-chain structure with no obvious intracellular catalytic domain. The IGF-II receptor also functions as the receptor for mannose-6-phosphate, which targets acid hydrolases to liposomes. It has been suggested that these two binding specificities might regulate complementary processes during tissue remodeling.

The action of the IGFs is modulated by a complex set of circulating and tissue-specific binding proteins (BPs). The IGFs circulate in a complex with IGFBP-3 (this prevents degradation and excretion). Upon release from the IGFBP-3 complex, the IGFs complex with two other IGFBPs to be transported into the tissue compartments. Again the binding proteins appear to control the release and action of the IGFs in their target tissues. This complex set of interactions ensures that the effective concentrations of free IGFs are kept under control and maintained for longer times than many other growth factors.

3.2   NERVE GROWTH FACTORS/NEUROTROPHINS

Nerve growth factor, a humoral substance produced by some tumors, is responsible for innervation of these tumors. The discovery of NGF in the 1940s initiated the search for other tissue-specific growth factors and led quickly to the identification of epidermal growth factor (see Section 3.3). Initially NGF was isolated from the salivary glands of male mice as a complex of three proteins; however, the active component is a single protein of molecular weight 30,000, and the three-dimensional structure of NGF has revealed a novel arrangement of three extended segments of twisted antiparallel β-sheet. Several other neurotrophins [brain-derived neurotrophic factor (BDNF), NT-3, and NT4/5] have been discovered. BDNF stimulates peripheral sensory ganglia as well as cholinergic and dopaminergic neurons. More recently there have been reports of a glial growth factor that is a differentially spliced form of a ligand for the *erb B3* or *erb B4* receptors.

Detailed structure–function studies have identified many of the residues required for the binding of NGF and BDNF to the low affinity (1–5 nM) neurotrophin receptor p75. However, the role of this receptor in NGF signaling is still unclear. The other class of neurotrophin receptors, *trk A, trk B,* and *trk C,* are ligand-activated tyrosine kinases. While *trk A* binds NGF, *trk B* binds BDNF or NT-4, and *trk C* recognizes NT-3.

During embryological development, the neurotrophins regulate the survival of specific classes of neurones. NGF appears to function on peripheral neurones of the sympathetic system, whereas

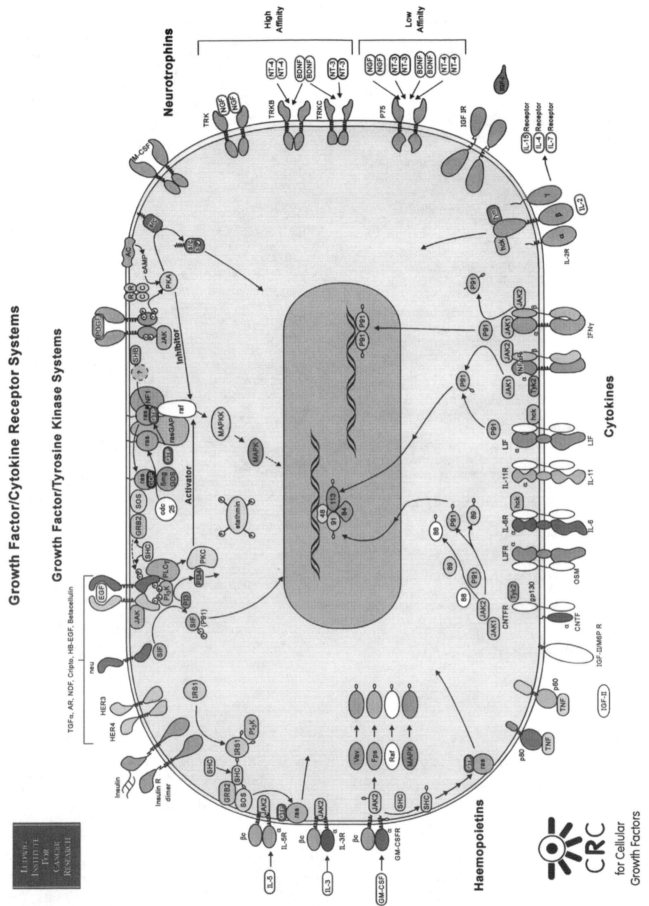

Figure 2. Growth factors/cytokines and their receptor signaling systems. (See color plate 10.)

467

BDNF has an effect on neural crest cells and NT-3 acts on oligo-dendrocyte precursors.

3.3 Epidermal Growth Factor Family

It is now 40 years since Stanley Cohen described the discovery of the factor that induces in mice both premature opening of eyelids and tooth eruption. A small protein (53 amino acids) with three disulfide bonds and two distinct folding domains, EGF stimulates lung maturation and the formation of gastrointestinal mucosa, and inhibits the secretion of acid from the gastric mucosa. EGF binds to and stimulates a single-chain tyrosine kinase receptor. There are several growth factors related to EGF—namely, transforming growth factor α, amphiregulin, cripto, heparin binding-EGF (HB-EGF), heregulin, and β-cellulin. TGF-α was discovered as an autocrine growth factor induced by some tumor viruses. EGF and TGF-α are expressed as precursor molecules. The TGF-α precursor is expressed on the surface of cells, and the mature protein is released by specific proteolysis. Cripto has been implicated in the growth of colon cancers. Interestingly, several myxoma and sarcoma viruses also encode proteins homologous to EGF and TGF-α. Amphiregulin and the c-*neu* ligand bind to distinct receptors within the EGF receptor family.

There are four closely related tyrosine kinase receptors in the EGFR family: *EGFR, HER 2, HER 3,* and *HER 4.* The sequence homology in the ligand-binding domain is between 40 and 50%, and in the kinase domain up to 80% similarity. However, it is clear that the ligand-binding specificity for each receptor is distinct, with the *EGFR* having a preference for EGF and TGF-α, whereas *HER 3* and *HER 4* bind to the *neu*-ligand and the heregulins. There is now considerable evidence that heterodimers between *EGFR* and *HER 2, HER 2* and *HER 3,* and so on, bind appropriate ligands with high affinity and activate distinct signaling pathways.

Activation of the *EGFR* kinase involves ligand-induced oligomerization. It is still not clear how many proteins are required to form the mitogenic signaling complex, but the receptor becomes autophosphorylated as well as adding phosphate to an adaptor protein called SHC. Phosphorylated SHC appears to influence the subcellular location of a complex between grb2 and SOS, the exchange protein that adds GTP to *ras*. Formation of the GTP–*ras* complex initiates a signaling cascade required for mitogenic responses to EGF. While the *ras*-activated pathway may need to be activated for mitogenesis, *EGFR* kinase must also activate other essential pathways for mitogenesis to occur.

The inappropriate expression of the *EGFR* and modified forms of *HER 2* has been associated with several tumors including gliomas, where almost 30% of high grade tumors overexpress the *EGFR. HER 3* is expressed in many breast and pancreatic tumors. It also appears that β-cellulin, *HER 3,* and activated *ras* are in part responsible for many pancreatic carcinomas. By blocking autocrine growth factor production or *EGFR,* it is possible to sensitize colonic tumors to chemotherapeutic drugs such as *cis*-platinum.

3.4 Transforming Growth Factor (TGF-β) Family

TGF-β was discovered as one of the growth factors released after tumor virus infection of fibroblasts. Indeed, TGF-β synergizes with TGF-α to induce large colony formation by normal rat kidney (NRK) cells. TGF-β was also isolated independently by its ability to inhibit the proliferation of epithelial cells: it inhibits, as well, the proliferation of both hematopoietic and lymphoid cells. TGF-β induces the differentiation of bronchial epithelial cells and prechondrocytes, but inhibits the differentiation of adipocytes and myocytes. A latent form of TGF-β is released from platelets as a complex with a specific binding protein. Active TGF-β can be released from the complex either by proteolysis or acid treatment. Several other cytokines belong to the TGFβ superfamily: two forms of TGF-β have been isolated from platelets and three other related sequences detected by screening cDNA libraries. Inhibin, the Müllerian inhibitory substances (MIS), the bone morphogenic proteins (BMPs), and activin are all related to TGF-β. The availability of the three-dimensional structure for TGF-β will allow models to be developed for the other family members.

TGF-β can induce the synthesis of extracellular matrix components in vivo and in vitro. TGF-β is also a powerful enhancer of monocyte function and a suppressor of lymphopoiesis. Mice that lack a functional TGF-β1 gene have been produced; while they are born as apparently normal animals, within 2 weeks they develop severe autoimmune and hematopoietic defects.

TGF-β appears to control the production and action of extracellular matrix (ECM) proteins by regulating secretion, increasing the level of integrin receptors, and decreasing the breakdown of ECM proteins. Taken together, these actions appear to have the capacity to control organogenesis during embryological development. In adults, TGF-β acts as a profound immunosuppressant. Indeed, some patients with brain tumors secrete so much TGF-β that their immune systems fail to function. TGF-β may enhance the granulation phase of wound healing, but the presence is presumably timed carefully, inasmuch as excess levels inhibit epithelialization. Interestingly, both TGF-α and TGF-β are produced by many tumors. It has been postulated that the tumors often down-regulate the TGF-β receptors and are not responsive to the inhibitory actions of this growth factor, while the action of the TGF-β on surrounding cells leads to extra ECM and a more favorable environment for tumor proliferation.

Three receptors have been identified for TGF-β: types I, II, and III. Types I and II bind TGF-β with high affinity (K_D = 5–25 pM). The type III TGF-β receptor is a large transmembrane proteoglycan with no obvious cytoplasmic signaling motif. The TGF-β type II receptor has ligand-dependent Ser/Thr kinase activity, which appears to signal growth inhibition. The type II receptor requires the type I receptor for ligand binding.

3.5 Platelet-Derived Growth Factors (PDGFs) and Vascular Endothelial Growth Factors (VEGFs)

For more than 70 years it has been known that serum, but not plasma, stimulates cells to proliferate in culture. Not until the early 1970s, however, was it discovered that the growth factors in serum are actually released from platelets. It took almost 10 years of protein chemistry to purify and sequence the first PDGF—a disulfide-linked heterodimer of two related polypeptides (A and B), each containing approximately 100 amino acids. At the same time that the amino acid sequence for PDGF was determined, the nucleotide sequence for the oncogene (v-*sis*) encoded by the simian sarcoma virus was characterized. The protein sequence predicted from the v-*sis* gene is closely related to the PDGF-β chain.

Different cell types are known to secrete either PDGF-AA or PDGF-BB homodimers, both of which act as growth factors for fibroblasts, smooth muscle cells, and glial cells. There appear to be two forms of the PDGF receptor monomer (α and β), and these can combine to produce functional receptor dimers: αα, αβ, and ββ. Whereas the αα receptor responds to all forms of PDGF (i.e., AA, AB, and BB) the ββ receptor responds only to PDGF-BB. The PDGF receptors are typical ligand-dependent tyrosine kinases, but the cytoplasmic domain contains an insert that modulates the intracellular signaling processes. PDGF increases the rate of wound healing in rodents, and initial reports indicate that it may be a valuable agent for treating cutaneous ulcers in diabetic patients.

Both the PDGF-A and PDGF-B chains are synthesized in a wide variety of cells—endothelium, muscle, and macrophages. The A-chain is induced by fibroblasts in response to other growth stimuli—that is, a normal autocrine loop appears to control the proliferation of some cell types. This autocrine function can be disturbed in several diseases, such as sarcomas, atherosclerosis, and rheumatoid arthritis.

Although the PDGF family is small, a related factor—vascular endothelial growth factor (VEGF)—appears to control the production of endothelial cells. Several isoforms of VEGF are produced from a single gene by differential splicing of its mRNA precursor. The smallest form, $VEGF_{121}$, is a weakly acidic protein that does not bind to heparin and is secreted from cells in a freely diffusible form. The longer forms of VEGF bind to heparin and appear to be released in response to plasmin.

VEGF stimulates the production of endothelial cells from both small and large blood vessels. As well as promoting angiogenesis, VEGF stimulates monocyte chemotaxis. VEGF is produced by many neoplastic cells. Interestingly, antibodies against VEGF can inhibit the production of blood vessels in vivo, and as a consequence, inhibit the growth of human tumor cell lines. Similarly, negative dominant VEGF receptors can inhibit the vascularization of experimental tumors.

3.6   FIBROBLAST GROWTH FACTORS (FGFs)

In the 1930s it was discovered that brain and pituitary extracts stimulate the proliferation of fibroblasts in culture. By the mid-1980s two forms of FGFs had been identified, purified, and sequenced. Both the acidic (a-FGF) and basic (b-FGF) forms are single-chain proteins of approximately 140 amino acids, and 55% of their amino acid sequences are identical. There are several other members of the FGF family: *wnt*-2, *hst/k*53, *FGF*-5, *FGF*-6, *FGF*-7 (KGF), *FGF*-8, and *FGF*-9 (glial activating factor), and there appears to be a distant homology (20–25%) to interleukins 1α and β. Both a- and b-FGFs stimulate the proliferation of cells, including colonic and breast epithelia, endothelial cells, and muscle cells. Although a-FGF and b-FGF are found associated with the extracellular matrix, neither protein has a classical signal peptide that would direct their secretion. It is still not known how these proteins are released from cells.

Although a-FGF and b-FGF bind to the same set of cell surface receptors, b-FGF has a higher affinity. While the high affinity of the FGFs for heparin-sulfate is useful for their purification, this property often interferes with binding studies to the cell surface. FGFs are potent angiogenic agents and are also capable of accelerating the healing of cutaneous wounds and damaged nerves. b-FGF is mitogenic for oligodendrocytes, astrocytes, and Schwann cells, and it acts to prolong the survival of neuronal cells in culture. In pharmacological doses, b-FGF can enhance the remyelination of neuronal

sheaths and can prevent the death of neurons in the dorsal root ganglion.

FGFs have also been purported to play a role in tumor blood vessel formation. A retrovirus encoding *FGF-4* (also called K-FGF) induces tumors in mice. Furthermore, the *FGF-4* gene is amplified in a number of breast cancers. *FGF-7* (keratinocyte growth factor, KGF) and its receptor (*FGFR-2*) are both expressed in prostate cancer cells. A variant of *FGFR-2* (K-sam) has also been detected in stomach cancer cells and appears to be associated with the tumor phenotype.

3.7   HEPATOCYTE GROWTH FACTOR (HGF)

Originally discovered as a factor that increases the motility and spreading of cells (called scatter factor), hepatocyte growth factor (HGF) stimulates the proliferation of primary hepatocytes and increases the invasiveness of endothelial and epithelial cells. HGF appears to be involved in liver regeneration, tumor progression, and several embryological processes. HGF is a 92 kDa disulfide-linked heterodimer consisting of a light chain (33 kDa) and a heavy chain (62 kDa), which are produced from a single-chain precursor.

The HGF receptor (c-*met*) is a 190kDa heterodimer in which the α-chain (50 kDa) appears to be extracellular and is linked to the β-chain (145 kDa) which is a membrane-spanning protein with a cytoplasmic tyrosine kinase domain, via a disulfide bond. It is expressed in a number of epithelial tissues, including liver and colon. Interestingly, a fragment of HGF containing two Kringle domains binds to c-*met* and stimulates the motility response and the c-*met* tyrosine kinase activity, but does not stimulate a mitogenic response in primary rat hepatocytes. Mutations in c-*met* have been associated with a number of cancer types. There are two other receptors related to c-*met*: c-*ron* and c-*sea*.

3.8   HEMATOPOIETIC GROWTH FACTORS

At least 20 cytokines or growth factors have been identified by their action on the production or function of blood cells. The major classes of hematopoietic regulators are interleukins, lymphokines, colony-stimulating factors, erythropoietin, and stem cell factor (SCF). Table 2 lists some of the biological actions of these molecules; however, it must be emphasized that as with the other growth factors, each one has multiple activities, which are not necessarily limited to the hematopoietic system. Several hematopoietic growth factors are being used clinically: in particular, interleukin 2 in conjunction with LAK cells for cancer therapy; granulocyte/macrophage CSF and G-CSF to accelerate neutrophil recovery after chemotherapy or bone marrow transplantation; epo to stimulate red blood cell production in kidney dialysis or cancer patients; and thrombopoietin (tpo) to improve the rate of platelet recovery in thrombocytopenic patients. Three-dimensional structures have been determined for a number of hematopoietic growth factors—many consist of four helical bundles packed to form a central cylinder (Figure 3; see color plate 11). The structure–function relationships of this class of growth factors are being examined in detail, and already several potent antagonists (e.g., for IL-1 and IL-4) have been produced. As might be expected from such a large number of hematopoietic growth factors, several classes of cell surface receptor have been identified: tyrosine kinase (e.g., M-CSF), multichain receptors (e.g., IL-2, IL-3, IL-6, GM-CSF), and single-chain receptors (e.g., G-CSF). Several of the hematopoietic growth factor receptors share

Table 2   Biological Actions of the Hematopoietic Growth Factors

Name	Action
Interleukin 1α/β	Stimulates thymocyte proliferation
	Increases expression of IL-2 receptor on T lymphocytes
Interleukin 2	Stimulates proliferation of T lymphocytes
Interleukin 3 (also called multi-CSF)	Stimulates production of cells in all myeloid lineages
Interleukin 4	Stimulates proliferation of B lymphocytes
	Stimulates production of IgM and potentiates mast cell proliferation in response to IL-3
Interleukin 5	Induces B-cell proliferation and differentiation in the mouse only
	Stimulates production and activation of eosinophils
Interleukin-6 (also called interferon β2)	Stimulates hematopoietic progenitor cells, B-lymphoid, and myeloma cells; induces acute phase responses
Interleukin 7	Stimulates proliferation of thymocytes, T lymphocytes (including cytotoxic T cells), and early B-cell precursors
Interleukin 8	Stimulates neutrophil chemotaxis and activates bacterial killing
	Inhibits IgE production by B lymphocytes
Interleukin 9	Stimulates proliferation of T lymphoid cell lines
	Potentiates proliferative effect of IL-2 on fetal thymocytes
Interleukin 10	Inhibits cytokine production by T lymphocytes
Interleukin 11	Stimulates production of IgG generating B cells
	Induces acute phase response
	Augments ability of IL-3 to stimulate megakaryocyte colonies
Interleukin 12	Stimulates cytokine production by NK cells and T cells
Interleukin 13	Inhibits cytokine production by monocytes and is a costimulator of B-cell lymphocyte proliferation
Interleukin 14	Induces proliferation of activated B lymphocytes
Interleukin 15	Stimulates the proliferation of cytotoxic T lymphocytes
M-CSF (also called CSF-1)	Stimulates production, activation, and proliferation of monocytes and macrophages
GM-CSF	Stimulates production and activation of neutrophils, eosinophils, and macrophages
G-CSF	Stimulates production and activation of neutrophils
Stem cell factor	Stimulates production of all hematopoietic progenitor cells
Erythropoietin	Stimulates production of red blood cells
Leukemia inhibitory factor	Inhibits differentiation of embryonal stem cells
	Induces acute phase responses
	Stimulates megakaryocyte and platelet production
	Stimulates differentiation of M1 leukemic cell line
Thrombopoietin	Stimulates production of platelets

adaptor subunits (e.g., GM-CSF, IL-3, and IL-5 share a β subunit, and these, along with the LIF oncostatin M, all appear to signal via complexes with the cell surface glycoprotein gp-130).

Cytokine receptor binding has proved to be a complex and interesting process. IL-2 binds to either of two receptor chains (α or β) with relatively low affinity: K_D = 3 and 70 nM, respectively. This low affinity binding does not stimulate cells. However, if both the α- and β-chains are present as a complex, the IL-2 binds with high affinity (K_D = 5 pM) and stimulates T-cell proliferation. The β-chain also associates with another cell surface protein called the γ-chain, which can increase the affinity of the β-chain for IL-2. The highest affinity state of the IL-2 receptor is achieved when the αβγ complex forms (K_D = 2 pM). The β- and γ-chains also complex with the IL-15 receptor to form the functional complex for IL-15 signaling, and the γ-chain complexes with the IL-4 receptor to form the high affinity IL-4 binding complex.

The IL-2 receptor binding subunits do not encode intracellular kinases; however, upon binding of IL-2, cytoplasmic tyrosine kinases in the *src* family are activated. In particular, *lck* associates with the IL-2 receptor complex, but it is still not clear whether the physical association is required for the signaling process. The γ-chain cytoplasmic region appears to be essential for activating the *src* family kinases and for stimulating the pathways leading to c-*fos* and c-*jun* expression.

Interleukin 6 binds to a specific receptor on the cell surface (IL-6Rα), but this interaction is insufficient for signaling. Once formed, the IL-6:IL-6Rα complex binds to another cell surface protein, gp130, and the trimer self-associates to form the hexameric signal-

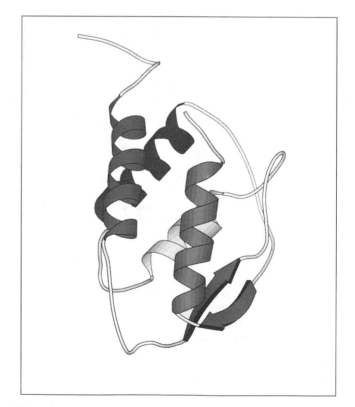

Figure 3.   The three-dimensional structure of GM-CSF illustrates the typical four-helix bundle found in many of the hematopoietic regulators. (See color plate 11.)

ing complex. gp130 also acts as the signal transduction receptor for ciliary neurotrophic factor (CNTF), LIF, and oncostatin M.

The initial signaling events stimulated by the (IL-6:IL6Rα: gp130)$_2$ complex involve the activation of the *JAK* kinase enzymes. It is not clear whether *JAK* is associated with gp130 before the signaling complex forms, but after binding of the IL-6:IL-6Rα complex and aggregation, the *JAK* kinase is activated and phosphorylates the receptor, as well as *JAK* and *STAT* (p91). Phosphorylation of the *STAT* occurs via the *JAK* kinase after the *STAT* binds to the *JAK*-phosphorylated gp130. Once phosphorylated, *STAT* is released from the receptor and associates with a smaller protein (p48) and translocates to the nucleus, where it acts as a transcriptional regulator. Similar activation systems operate in response to many of the cytokines (e.g., growth hormone, epo, interferons, IL-3, GM-CSF, G-CSF).

Interestingly, the cytokine/*JAK/STAT* signaling system is also regulated negatively. A phosphatase competes for binding to specific *JAK* phosphorylation sites on the receptor. Once bound, the phosphatase removes the phosphate from the *JAK* kinase and deactivates the enzyme. A mutant form of the epo receptor has been discovered which lacks the hematopoietic cell phosphatase-binding site. In these people, the *JAK* kinase remains activated and associated with the epo receptor, leading to an increase in red blood cell production.

4   PERSPECTIVES

The biological action of a particular growth factor will be dependent on the differentiation state of the target cell, the presence of other growth factors, and/or the existence of cell–cell contacts (see Figure 3). Cell–cell contacts and/or multiple growth factors acting early in many cell production pathways appear to be required to induce self-renewal of the tissue-specific stem cells. In a number of tissue systems the presence of particular growth factors is essential to maintain cell viability at all stages of the differentiation process. If the growth factor concentration decreases, many of the immature cells will die. Mature cells usually have a definite lifetime before disintegrating; however, the functional activity of many mature cells can be activated by growth factors; for example, GM-CSF and G-CSF will prime mature neutrophils to kill bacteria more effectively.

See also CELL DEATH AND AGING, MOLECULAR MECHANISMS OF; CYTOKINES.

Bibliography

Bradshaw, R. A., Blundell, T. L., Lapatto, R., McDonald, N. Q., and Murray-Rust, J. (1993) Nerve growth factor revisited. *Trends Biochem. Sci.* 18:48–52.

Burgess, A. W., and Tran, T. T. (1994) GROCYT, available from World Wide Web site. URL: http://www.ludwig.edu.au/

Callard, R. E., and Gearing, A. J. H., Eds. (1994) *The Cytokine Facts Book.* Academic Press, London.

de Vos, A. M., Ultsch, M., and Kossiakoff, A. A. (1992) Human growth hormone and extracellular domain of its receptor: Crystal structure of the complex. *Science,* 255:306–312.

Heath, J. K. (1993) *Growth Factors.* Oxford University Press, Oxford.

Lieschke, G. J., and Burgess, A. W. (1992) Granulocyte colony-stimulating factor and granulocyte–macrophage colony-stimulating factor. *New Engl. J. Med.* 327:28–35.

Meagher, A. (1990) *Cytokines.* Open University Press, Milton Keynes.

Miyazono, K., Ochijo, H., and Heldin, C.-H. (1993) Enhanced bFGF expression in response to transforming growth factor-β stimulation of AkR-23 cells. *Growth Factors,* 8:11–22.

Nicola, N. A., Ed. (1994) *Guidebook to Cytokines and Their Receptors.* Oxford University Press, Melbourne.

Sporn, M. B., and Roberts, A. B., Eds. (1991) *Peptide Growth Factors and Their Receptors,* Vols I and II. Springer-Verlag, Berlin.

Glossary of Basic Terms

The most basic terms in molecular biology are defined below. These, in combination with the key words listed at the head of each article, provide definitions of all essential terms in this desk reference.

Alleles   Alternative forms of a given gene, inherited separately from each parent, differing in nucleotide base sequence and located in a specific position on each homologous chromosome, affecting the functioning of a single product (RNA and/or protein).

Amplification (gene)   The process of replication of specific DNA sequences in disproportionately greater amounts than are present in the parent genetic material, e.g., PCR is an in vitro amplification technique.

cDNA (complementary DNA)   A DNA copy of an RNA molecule synthesized from an mRNA template in vitro using an enzyme called reverse transcriptase; often used as a probe.

Chromatin   The complex of nucleic acids (DNA and RNA) and proteins (histones) comprising eukaryotic chromosomes.

Chromosome   In prokaryotes, the usually circular duplex DNA molecule constituting the genome; in eukaryotes, a threadlike structure consisting of chromatin and carrying genomic information on a DNA double helix molecule. A viral chromosome may be composed of DNA or RNA.

Cloning   Asexual reproduction of cells, organisms, genes or segments of DNA identical to the original.

Cloning vector   See: *vector*.

Complementary base pairing   Nucleic acid sequences on paired polymers with opposing hydrogen bonded bases adenine (designated A) bonded to thymine (T), guanine (G) to cytosine (C) in DNA and adenine to uracil (U) replacing adenine to thymine in RNA.

DNA (deoxyribonucleic acid)   The molecular basis of the genetic code consisting of a poly-sugar phosphate backbone from which project thymine, adenine, guanine and cytosine bases. Usually found as two complementary chains (duplex) forming a double helix associated by hydrogen bonds between complementary bases.

DNA polymerase   Enzymes that catalyze the replication of DNA from the deoxyribonucleotide triphosphates using single- or double-stranded DNA as a template.

E. coli (Escherichia coli)   A colon bacillus which is the most studied of all forms of life and whose genome is presently the best sequenced and mapped.

Eukaryote   An organism (or cell) whose cells contain a true nucleus; all living matter except viruses, bacteria and blue-green algae.

Expression   The process of making the product of a gene, which is either a specific protein giving rise to a specific trait or RNA forms not translated into proteins (e.g., transfer ribosomal RNAs).

Gene   A DNA sequence, located in a particular position on a particular chromosome, which encodes a specific protein or RNA molecule.

Heterozygous   Having two different alleles for a given trait in the homologous chromosomes.

Homologies   Similarities in DNA or protein sequences between individuals of the same species or among different species.

Homologous chromosomes   Chromosome pairs, each derived from one parent, containing the same linear sequence of genes, and as a consequence, each gene is present in duplicate (e.g., humans have 23 homologous chromosome pairs but the toad has 11 pairs, the mosquito has three pairs, and so on).

Homozygous   Having two identical alleles for a given trait in the homologous chromosomes.

Hybridization   The formation of a double-stranded polynucleotide molecule when two complementary strands are brought together at moderate temperature. The strands can be DNA or RNA or one of each; a technique for assessing the extent of sequence homology between single strands of nucleic acids.

Ligation   The formation of a phosphodiester bond to join adjacent terminal nucleotides (nicks) to form a longer nucleic acid chain (DNA of RNA); catalyzed by ligase.

Marker   A gene or a restriction enzyme cutting site with a known location on a chromosome and a clear-cut phenotype (expression), or pattern of inheritance, used as a point of reference when mapping a new mutant.

mRNA (messenger RNA)   RNA used to translate information from DNA to ribosome where the information is used to make one or several proteins.

Nucleotide   The monomer which, when polymerized, forms DNA or RNA. It is composed of a nitrogenous base bonded to a sugar (ribose or deoxyribose) bonded to a phosphate.

Oligonucleotide   A polynucleotide 2 to 20 nucleotide units in length.

Operon   A series of prokaryote genes encoding enzymes of a specific biosynthesis pathway and transcribed into a single RNA molecule.

Plasmid   An extrachromosomal circular DNA molecule found in a variety of bacteria encoding "dispensable functions," such as a resistance to antibiotics. Often found in multiple copies per cell and reproduces every time the bacterial cell reproduces. May be used as a cloning vector.

Polymorphism   Difference in DNA sequence among individuals expressed as different forms of a protein in individuals of the same interbreeding population.

Polynucleotide   The polymer formed by condensation of nucleotides.

Probe   A radioactively, fluorescent or immunologically labeled oligonucleotide (RNA or DNA) used to detect complementary sequences in a hybridization experiment (e.g., identify bacterial colonies that contain cloned genes or detect specific nucleic acids following separation by gel electrophoresis.

Prokaryote   An organism that lacks a true nucleus, a bacterium, virus or blue-green algae.

Replication   The copying of a DNA molecule duplex yielding two new DNA duplex molecules, each with one strand from the original DNA duplex. Single-stranded DNA replication results in a single-stranded DNA molecule.

Repressor   A protein that binds to a specific location (operator) on DNA and prevents RNA transcription from a specific gene or operon.

Restriction mapping   Uses restriction endonuclease enzymes to produce specific cuts (cleavage) in DNA, allowing preparation of a genome map describing the order and distance between cleavage sites.

Reverse transcription   The synthesis of cDNA from an RNA template as catalyzed by reverse transcriptase.

RFLP (restriction fragment length polymorphism)   DNA fragment cut by enzymes specific to a base sequence (restriction endonuclease) generating a DNA fragment whose size varies from one individual to another. Used as markers on genome maps and for screening for mutations and genetic diseases.

Ribosomes   Small cellular components composed of proteins plus ribosomal RNA that translate the genetic code into synthesis of specific proteins.

RNA (ribonucleic acid)   A single-stranded polynucleotide with a phosphate oxyribose backbone and four bases that are identical to those in DNA, with the exception that the base uracil is substituted for thymine.

RNA polymerase   The enzyme (peptide) that binds at specific nucleotide sequences, called promoters, in front of genes in DNA, catalyzing transcription of DNA to RNA.

Transcription   Synthesis of an RNA molecule from a DNA template (gene) catalyzed by RNA polymerase.

tRNA (transfer RNA)   RNA molecules that transport specific amino acids to ribosomes into position in the correct order during protein synthesis.

Vector   DNA molecule originating from a virus, a plasmid or cell of a higher organism into which another DNA fragment can be integrated without loss of the vector's capacity for self-replication. Vectors introduce foreign DNA into host cells where it can be reproduced in large quantities.